(*Continued on page E1*)

Thinking Mathematically

Thinking Mathematically

Third Edition

ROBERT BLITZER

Miami-Dade College

PEARSON

Prentice
Hall

Upper Saddle River, New Jersey 07458

Library of Congress Cataloging-in-Publication Data

Blitzer, Robert.
    Thinking Mathematically / Robert Blitzer.—3rd ed.
      p. cm.
    Includes index.
    ISBN 0-13-143243-5
    1. Mathematics. I. Title.
   CIP data available

Executive Acquisition Editor: *Petra Recter*
Editor-in-Chief: *Sally Yagan*
Senior Managing Editor: *Linda Mihatov Behrens*
Executive Managing Editor: *Kathleen Schiaparelli*
Vice President/Director of Production and Manufacturing: *David W. Riccardi*
Production Editor: *Bayani Mendoza de Leon*
Assistant Manufacturing Manager/Buyer: *Michael Bell*
Manufacturing Manager: *Trudy Pisciotti*
Marketing Manager: *Krista Bettino*
Marketing Assistant: *Annett Uebel*
Art Director: *Heather Scott*
Assistant Managing Editor, Math Media Production: *John Matthews*
Project Manager: *Jacquelyn M. Riotto*
Creative Director: *Carole Anson*
Director of Creative Services: *Paul Belfanti*
Art Editor: *Thomas Benfatti*
Photo Researcher: *Melinda Alexander*
Photo Editor: *Beth Boyd*
Interior Designer: *ESM Design*
Cover Designer: *John Christiana*
Editorial Assistant/Supplements Editor: *Joanne Wendelken*
Art Studio: *Scientific Illustrators*
Compositor: *Preparé, Inc.*
Cover image: *Catherine Ledner / Getty Images, Inc*

© 2005, 2003, 2000 Pearson Education, Inc.
Pearson Prentice Hall
Pearson Education, Inc.
Upper Saddle River, New Jersey 07458

Pearson Prentice Hall® is a trademark of Pearson Education, Inc.

PRINTED IN THE UNITED STATES OF AMERICA

10 9 8 7 6 5

ISBN 0-13-143243-5 (College Edition)
ISBN 0-13-192011-1 (School Edition)

Pearson Education LTD., *London*
Pearson Education Australia PTY, Limited, *Sydney*
Pearson Education Singapore, Pte. Ltd
Pearson Education North Asia Ltd, *Hong Kong*
Pearson Education Canada, Ltd., *Toronto*
Pearson Educación de Mexico, S.A. de C.V.
Pearson Education—Japan, *Tokyo*
Pearson Education Malaysia, Pte. Ltd

For Harley
and to all those who have loved animals and have
been loved by them

May 28, 1995—November 20, 2003

CONTENTS

CHAPTER

1

Problem Solving and Critical Thinking   1

CHAPTER

2

Set Theory   38

CHAPTER

3

Logic   87

Statistics   630

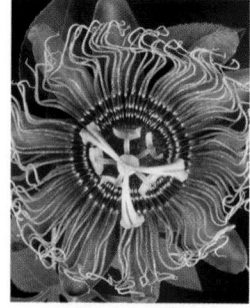

Mathematical Systems   701

Mumble.
Grumble.
Complain.
Wallow.
Hope.
Despair.
Worry.

Vote.

Voting and Apportionment   725

CHAPTER

15

Graph Theory 778

ABOUT THE
AUTHOR

Bob Blitzer is a native of Manhattan and received a Bachelor of Arts degree with dual majors in mathematics and psychology (minor: English literature) from the City College of New York. His unusual combination of academic interests led him toward a Master of Arts in mathematics from the University of Miami and a doctorate in behavioral sciences from Nova University. Bob is most energized by teaching mathematics and has taught a variety of mathematics courses at Miami-Dade College for nearly 30 years. He has received numerous teaching awards, including Innovator of the Year from the League for Innovations in the Community College, and was among the first group of recipients at Miami-Dade for an endowed chair based on excellence in the classroom. In addition to *Thinking Mathematically*, Bob has written *Introductory Algebra for College Students*, *Intermediate Algebra for College Students*, *Introductory and Intermediate Algebra for College Students*, *Algebra for College Students*, *College Algebra*, *Algebra and Trigonometry*, and *Precalculus*, all published by Prentice Hall.

PREFACE

Thinking Mathematically, Third Edition, provides a general survey of mathematical topics that are useful in our contemporary world. My primary purpose in writing the book was to show students how mathematics can be applied to their lives in interesting, enjoyable, and meaningful ways. The book's variety of topics and flexibility of sequence make it appropriate for a one- or two-term course in liberal arts mathematics, finite mathematics, mathematics for education majors, as well as for courses specifically designed to meet state-mandated requirements in mathematics.

I wrote the book to help diverse students, with different backgrounds and career plans, to succeed. *Thinking Mathematically, Third Edition*, has three major goals:

1. To help students acquire knowledge of fundamental mathematics.
2. To show students how mathematics can solve authentic problems that apply to their lives.
3. To enable students to develop problem-solving skills, while fostering critical thinking, within an interesting setting.

One major obstacle in the way of achieving these goals is the fact that very few students actually read their textbook. This has been a regular source of frustration for me and my colleagues in the classroom. Anecdotal evidence gathered over years highlights two basic reasons why students do not take advantage of their textbook:

- "I'll never use this information."
- "I can't follow the explanations."

As a result, I've written every page of this book with the intent of eliminating these two objections. See the book's Walkthrough, beginning on page xviii, or the ideas and tools I've used to do so.

What's New in the Third Edition?

I believe students and instructors will welcome the following new features:

1. **New Applications and Updated Real-World Data.** I read/skimmed/explored excerpts from approximately 40 sources to prepare the Third Edition. As a result of my research, many new, innovative applications, supported by data that extend as far up to the present as possible, appear throughout the book. For example, Section 8.6 (Investing in Stocks, Bonds, and Mutual Funds) includes discussion of corporate scandals and ramifications for the long-range investor at a time when the perception of deception is so widespread.

2. **Over 700 New Examples and Exercises.** There are new worked-out examples in many of the sections and new problems in many of the exercise sets. I replaced examples and exercises from the Second Edition because I was able to find even more interesting data and/or better illustrated concepts. The examples are very abundant in number because students learn by example. They are thoroughly annotated to the right of the mathematical steps to explain key steps and ideas, reinforcing the problem-solving approach inherent in the text.

3. **New Enrichment Essays ("Blitzer Bonuses").** The Third Edition contains a large variety of new enrichment essays, now called *Blitzer Bonuses*, ranging from *Unit Prices and Sneaky Pricejacks* (page 24) to *Creating an Inaccurate Picture by Leaving Something Out* (page 640).

4. **Fine-Tuned and Polished Explanations.** Based on user and reviewer feedback, I revisited the presentations of topics throughout the book with the goal of making them even more accessible. Some of the changes include the following:

   - Section 5.3 (The Rational Numbers) now includes a slow, example-by-example, approach to operations with rational numbers, replacing the compressed table in the Second Edition.

   - Section 5.6 (Exponents and Scientific Notation) now uses the form $a \times 10^n$ rather than a verbal description, to present scientific notation, leading to clearer, more concise, explanations.

   - Section 8.1 (Percent) now uses the basic percent formula to unify the applications.

   - Section 11.4 (Fundamentals of Probability) now uses the sample space in the definition of theoretical probability, adding a more unified approach to the illustrative examples that follow.

   - Section 11.6 (Events Involving *Not* and *Or*, Odds) contains a rewritten presentation on odds that directly gives students formulas from probability to odds and odds to probability.

   - Section 11.7 (Events Involving *And*; Conditional Probability) uses a more accessible approach to conditional probability within the framework of a restricted sample space, immediately applying this concept to real-world data in a table.

5. **Increased Study Tip Boxes.** The book's study tip boxes offer suggestions for problem solving, point out common errors to avoid, and provide informal hints and suggestions. These invaluable hints appear in greater abundance in the Third Edition.

6. **More than 30 New Technology Boxes.** These appear integrated into the text and exercises and offer guidelines and hints for using both calculators and computers. They provide students valuable assistance with the correct way to enter problems in the calculator, getting roots, doing powers, using factorials, and other technical aids.

7. **New Chapter Review Grids.** The chapter summaries, presented as outlines in the Second Edition, are now organized into two-column review charts. The left column summarizes the definitions and concepts for every section of the chapter. The right column refers students to illustrative examples (by example number and page number) that illustrate these key concepts.

8. **Expanded Supplements Package.** The Third Edition is supported by a wealth of supplements designed for added effectiveness and efficiency, many of these new to this edition.

What Content and Organizational Changes Have Been Made to the Third Edition?

- Section 2.5 (Surveys and Cardinal Numbers) concludes with a new example—a survey problem with three subsets—that gives a better sense of closure to the chapter.

- Section 3.4 (Truth Tables for the Conditional and Biconditional) contains a new discussion on determining the truth value of a compound statement for a specific case.

- Section 5.3 (The Rational Numbers) contains a new discussion of mixed numbers and improper fractions.

- Section 8.2 (Simple Interest) is now devoted entirely to simple interest, allowing a more thorough development of simple interest formulas, including applications requiring their algebraic manipulation. The section concludes with a new discussion of discounted loans, an application that uses the simple interest formula.

- Section 8.3 (Compound Interest) is now devoted entirely to compound interest, focusing the topic in more detail and allowing for a more meaningful algebraic discussion of effective annual yield.

- Section 8.5 (The Cost of Home Ownership) still contains a table for computing the monthly mortgage payments, but now also contains a formula for determining this amount.

- Section 11.2 (Permutations) contains a new discussion of permutations of duplicate items.

- Chapter 14 (Voting and Apportionment) and Chapter 15 (Graph Theory), available only in the Expanded Second Edition, are included in the Third Edition for instructors interested in exploring either of these topics.

I hope that my love for learning, as well as my respect for the diversity of students I have taught and learned from over the years, is apparent in the hundreds of applications that appear throughout the book. By connecting mathematics to the whole spectrum of learning, it is my intent to show students that their world is profoundly mathematical and , indeed, π is in the sky.

Robert Blitzer

To The Student

I have written this book to give you control over the part of your life that involves numbers and mathematical ideas. Gaining an understanding and appreciation of mathematics will help you participate fully in the twenty-first century. In some ways, you cannot get along in life without the mathematics in this book. For example, if you do not understand the basic ideas of investment, you may find it impossible to achieve your financial goals. If you do not have at least a rudimentary understanding of logic, you may not be able to analyze information objectively, possibly arriving at conclusions hastily without a method for considering all the evidence.

This book has been written so that you can learn about the power of mathematics directly from its pages. All concepts are carefully explained, important definitions and procedures are set off in boxes, and worked-out examples that present solutions in a step-by-step manner appear in every section. Each example is followed by a similar matched problem, called a Check Point, for you to try so that you can actively participate in the learning process as you read the book. (Answers to all Check Points appear in the back of the book.) Study Tip boxes offer hints and suggestions and often point out common errors to avoid. A great deal of attention has been given to applying mathematics to your life to make your learning experience both interesting and relevant.

As you begin your studies, I would like to offer some specific suggestions for using this book and for being successful in this course:

1. **Attend all lectures.** No book is intended to be a substitute for valuable insights and interactions that occur in the classroom. Arrive for lecture on time and prepared. You might also find it helpful to read a section before it is covered in lecture. This will give you a clear idea of the new material that will be discussed.

2. **Read the book.** Read each section with pen (or pencil) in hand. Move through the worked-out examples with great care. These examples provide a model for doing exercises in the exercise sets. As you proceed through the reading, do not give up if you do not understand every single word. Things will become clearer as you read on and see how various procedures are applied to specific worked-out examples.

3. **Work problems every day and check your answers.** The way to learn mathematics is by doing mathematics, which means working the Check Points and assigned exercises in the exercise sets. The more exercises you work, the better you will understand the material.

4. **Prepare for chapter exams.** After completing a chapter, study the two-column review chart, work the exercises in the Chapter Review, and work the exercises in the Chapter Test. Answers to all these exercises are given in the back of the book.

5. **Use the supplements available with this book.** A Student Study Pack is available with this book containing a wealth of materials to assist with comprehension and mastery of the course material. It includes the student solutions manual, which consists of the worked-out solutions to the book's odd-numbered exercises, all review exercises, and all Check Points. It also includes the CD Lecture Series that provides access to a one-on-one tutoring service. Ask your instructor or bookstore for information on where to obtain the Student Study Pack to accompany this text.

I wrote this book in Point Reyes National Seashore, 40 miles north of San Francisco. The park consists of 75,000 acres with miles of pristine surf-washed beaches, forested ridges, and bays bordered by white cliffs. It was my hope to convey the beauty and excitement of mathematics using nature's unspoiled beauty as a source of inspiration and creativity. Enjoy the pages that follow as you empower yourself with the mathematics needed to succeed in college, your career, and in your life.

Regards,

Bob

Robert Blitzer

ACKNOWLEDGMENTS

I would like to express my appreciation to all the reviewers for their helpful criticisms and suggestions, frequently transmitted with wit, humor and intelligence. In particular, I would like to thank:

Gerard Buskes, *University of Mississippi*
Jimmy Chang, *St. Petersburg College*
David Cochener, *Austin Peay State University*
Elizabeth T. Dameron, *Tallahassee Community College*
Tristen Denley, *University of Mississippi*
Suzanne Feldberg, *Nassau Community College*
Margareth Finster, *Erie Community College*
Lyn Geisler III, *Randolf-Macon College*
Dale Grussing, *Miami-Dade Community College*
Virginia Harder, *College at Oneonta*
Joseph Lloyd Harris, *Gulf Coast Community College*
Julia Hassett, *Oakton Community College*
Sonya Hensler, *St. Petersburg Junior College*
Larry Hoehn, *Austin Peay State University*
Alec Ingraham, *New Hampshire College*
Linda Kuroski, *Erie Community College—City Campus*
Jamie Langille, *University of Nevada Las Vegas*
Veronique Lanuqueitte, *St. Petersburg Junior College*
Linda Lohman, *Jefferson Community College*
Mike Marcozzi, *University of Nevada, Las Vegas*
Diana Martelly , *Miami-Dade Community College*
Jim Matovina, *Community College of Southern Nevada*
Erik Matsuoka, *Leeward Community College*
Marcel Maupin, *Oklahoma State University*
Diana McCammon, *Delgado Community College*
Mex McKinley, *Florida Keys Community College*
Paul Mosbos, *State University of New York-Cortland*
Lawrence S. Orilia, *Nassau Community College*
Richard F. Patterson, *University of North Florida*
Frank Pecchioni, *Jefferson Community College*
Anthony, Pettofrezzo, *University of Central Florida*
Evelyn Popplo-Cody, *Marshall University*
Virginia S. Powell, *University of Louisiana at Monroe*

Kim Query, *Lindenwood College*
Anne Quinn, *Edinboro University of Pennsylvania*
Shawn Robinson, *Valencia Community College*
Gary Russell, *Brevard Community College*
Mary Lee Seitz, *Erie Community College*
Laurie A. Stahl, *State University of New York-Fredonia*
Ann Thrower, *Kilgore College*
Mike Tomme, *Community College of Southern Nevada*
Bill Vaughters, *Valencia Community College*
Karen Villareal, *University of New Orleans*
Don Warren, *Edison Community College*
Shirley Wilson, *North Central College*
James Wooland, *Florida State University*
Marilyn Zopp, *McHenry County College*

Additional acknowledgments are extended to Dan Miller for preparing the solutions manuals, Teri Lovelace and the team at LaurelTech for preparing the answer section and serving as accuracy checker, the Prepare Inc. formatting team for their expertise paging the book, Aaron Darnell and Brian Morris at Scientific Illustrators for superbly illustrating the book, Melinda Alexander, photo researcher, for obtaining the book's photographs, and Bayani Mendoza de Leon, whose talents as production editor kept every aspect of this complex project moving through its many stages.

I would like to thank my executive editor at Prentice Hall, Petra Recter, who guided and coordinated the book from manuscript through its new design through production. My thanks also to Joanne Wendelken, print supplements editor, Jacquelyn Riotto, media project manager, Krista Bettino, marketing manager, Annett Uebel, marketing assistant, Mike Bell, assistant manufacturing manager/buyer, Heather Scott, art director, Tom Benfatti, art editor, Linda Mihatov Behrens, senior managing editor, Sally Yagan, executive editor-in-chief, for her continuing support, and to the entire Prentice Hall sales force for their confidence and enthusiasm about the book.

"When Will I Use This?"

This text integrates over 800 dynamic applications that connect mathematics to the entire spectrum of students' interests. (See Applications Index on inside front cover.)

The interesting and diverse applications...

● represent a wide range of disciplines.

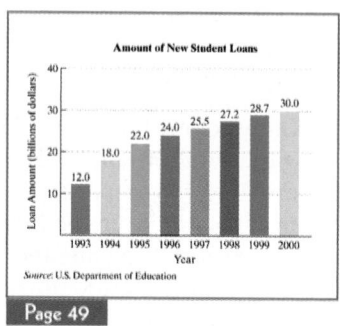

Page 49

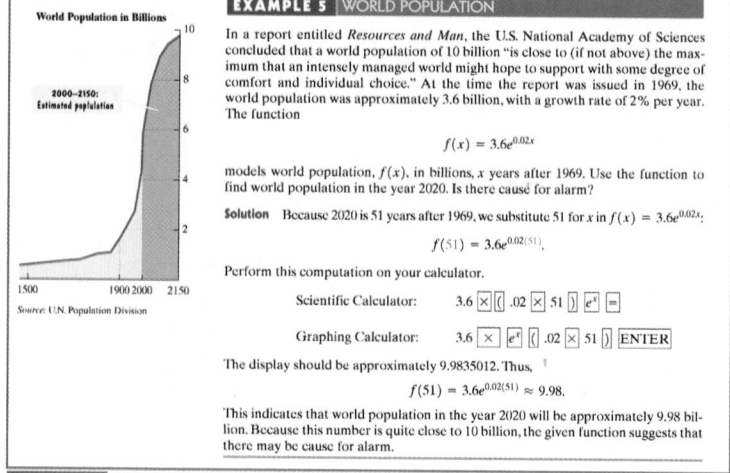

EXAMPLE 5 WORLD POPULATION

In a report entitled *Resources and Man*, the U.S. National Academy of Sciences concluded that a world population of 10 billion "is close to (if not above) the maximum that an intensely managed world might hope to support with some degree of comfort and individual choice." At the time the report was issued in 1969, the world population was approximately 3.6 billion, with a growth rate of 2% per year. The function

$$f(x) = 3.6e^{0.02x}$$

models world population, $f(x)$, in billions, x years after 1969. Use the function to find world population in the year 2020. Is there cause for alarm?

Solution Because 2020 is 51 years after 1969, we substitute 51 for x in $f(x) = 3.6e^{0.02x}$:

$$f(51) = 3.6e^{0.02(51)}.$$

Perform this computation on your calculator.

Scientific Calculator: 3.6 $\times$ $($ .02 $\times$ 51 $)$ e^x $=$

Graphing Calculator: 3.6 $\times$ e^x $($ .02 $\times$ 51 $)$ ENTER

The display should be approximately 9.9835012. Thus,

$$f(51) = 3.6e^{0.02(51)} \approx 9.98.$$

This indicates that world population in the year 2020 will be approximately 9.98 billion. Because this number is quite close to 10 billion, the given function suggests that there may be cause for alarm.

Page 369

Application Exercises

The points in the graph show costs for private and public four-year colleges projected through the year 2017. According to these projections, your daughter's college education at a private four-year school could cost about $250,000. Use the graph to estimate each of the ratios in Exercises 23–24. First express the estimated ratio as a fraction reduced to lowest terms. Then rewrite the ratio using the reduced fraction and a colon.

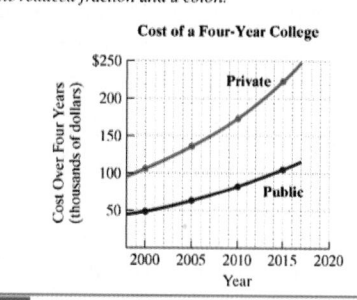

Page 308

● feature unique and interesting data that show students that mathematics can be applied in many settings.

● are driven by real and sourced data, illustrating the power of mathematics to model contemporary issues and problems.

51. In the United States, deaths from car accidents, per 100,000 persons, are decreasing at a faster rate than deaths from gunfire, shown by the blue and red lines that model the data points in the figure.

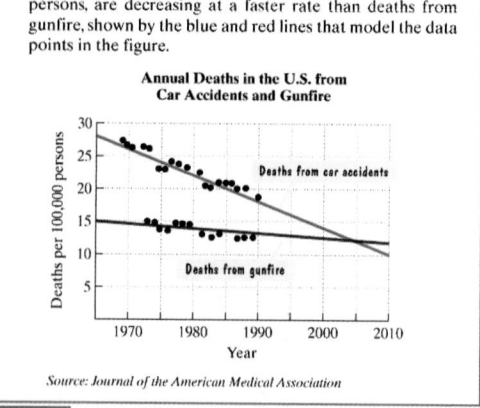

Page 383

Unique Chapter and Section Opening Vignettes

● Each chapter and section begins with a vignette highlighting an everyday scenario, posing a question about it, and exploring how the chapter section subject can be applied to answer the question. These scenarios are revisited in the course of the chapter or section in an example, discussion, or exercise.

SECTION 5.2   THE INTEGERS; ORDER OF OPERATIONS

OBJECTIVES

1. Define the integers.
2. Graph integers on a number line.
3. Use the symbols $<$ and $>$.
4. Find the absolute value of an integer.
5. Perform operations with integers.
6. Use the order of operations agreement.

People are going to live longer in the twenty-first century. This will put added pressure on the Social Security and Medicare systems. How insecure is Social Security's future? In this section, we use subtraction and a set of numbers called the *integers* to describe numerically one aspect of the insecurity.

1   Define the integers.

Defining the Integers

In Section 5.1, we applied some ideas of number theory to the set of natural, or counting, numbers:

$$\text{Natural numbers} = \{1, 2, 3, 4, 5, \dots\}.$$

Page 202

Blitzer Bonuses—Enrichment Essays

● These *Blitzer Bonuses* provide historical, interdisciplinary, and otherwise interesting connections to the math under study, showing students that math is an interesting and dynamic discipline.

BLITZER BONUS

A Poky Species

For a species that prides itself on its athletic prowess, human beings are a pretty poky group. Lions can sprint at up to 50 miles per hour; cheetahs move

Page 294

BLITZER BONUS

A Revolution at the Supermarket

0 76950 45026 4

Computerized scanning registers "read" the universal product code on packaged goods and convert it to a base two numeral that is sent to the scanner's computer. The computer calls up the appropriate price and subtracts the sale from the supermarket's inventory.

Page 178

To subtract in bases and subtract colum ing" is necessary to ample, when we bor 2s in base two, 3s in

EXAMPLE 3

Subtract:

SOLUTION We sta Because 2_{four} is gre We are working in b or 5, in base ten. No

Now we perform th

You can check the $13_{four} = 7$. Because

"I Can't Follow the Explanations."

Clear and Friendly Writing Style

- Blitzer's language is clear, direct, and simple. He breaks down concepts in a conversational style, providing analogies and drawing connections to students' experiences whenever possible.

Detailed Illustrations, Examples, and Check Points

Examples:

- are abundant because students learn by example.
- provide students with a step-by-step problem-solving approach to reach the solution. The solution contains both a numerical and verbal explanation.
- are thoroughly annotated to the right of the mathematical steps. These annotations are in a conversational style, reinforcing the voice of an instructor, explaining key steps and ideas as the problem is solved.
- offer students the opportunity to test their understanding of the example by working a similar exercise immediately following the example, called a Check Point. The answers to the Check Points are provided in the answer section.

Exercise Sets that Precisely Parallel Examples

- An extensive collection of exercises is included at the end of each section.
- Exercises are organized by level within six category types: Practice Exercises, Application Exercises, Writing in Mathematics, Critical Thinking Exercises, Technology Exercises, and Group Exercises.
- The order of the practice exercises is exactly the same as the order of the section's illustrative examples. This parallel order enables students to refer to the titled examples and their detailed explanations to achieve success working the practice exercises.

EXAMPLE 2 WILL ANYONE EVER RUN A THREE-MINUTE MILE?

One yardstick for measuring how steadily—if slowly—athletic performance has improved is the mile run. In 1923, the record for the mile was a comparatively sleepy 4 minutes, 10.4 seconds. In 1954, Roger Bannister of Britain cracked the 4-minute mark, coming in at 3 minutes, 59.4 seconds. In the half-century since, about 0.3 second per year has been shaved off Bannister's record. If this trend continues, by which year will someone run a 3-minute mile?

Solution In solving this problem, we will express time for the mile run in seconds. Our interest is in a time of 3 minutes, or 180 seconds.

STEP 1. Let x represent one of the quantities. The critical information in the problem is as follows:

- In 1954, the record was 3 minutes, 59.4 seconds, or 239.4 seconds.
- The record has decreased by 0.3 second per year since then.

We are interested in when the record will be 180 seconds. Let x = the number of years after 1954 when someone will run a 3-minute mile.

STEP 2. Represent other quantities in terms of x. There are no other unknown quantities to find, so we can skip this step.

STEP 3. Write an equation in x that describes the conditions.

The 1954 record time	decreased by	0.3 second per year for x years	equals	the 3-minute, or 180-second, mile.
239.4	−	0.3x	=	180

STEP 4. Solve the equation and answer the question.

$$239.4 - 0.3x = 180 \qquad \text{This is the equation for the problem's conditions.}$$
$$239.4 - 239.4 - 0.3x = 180 - 239.4 \qquad \text{Subtract 239.4 from both sides.}$$
$$-0.3x = -59.4 \qquad \text{Simplify.}$$
$$\frac{-0.3x}{-0.3} = \frac{-59.4}{-0.3} \qquad \text{Divide both sides by } -0.3.$$
$$x = 198 \qquad \text{Simplify.}$$

Using current trends, by 198 years (gasp!) after 1954, or in 2152, someone will run a 3-minute mile.

STEP 5. Check the proposed solution in the original wording of the problem. The problem states that the record time should be 180 seconds. Do we obtain 180 seconds if we decrease the 1954 record time, 239.4 seconds, by 0.3 second per year for 198 years, our proposed solution?

$$239.4 - 0.3(198) = 239.4 - 59.4 = 180$$

This verifies that, using current trends, the 3-minute mile will be run 198 years after 1954.

Check Point 2: Rent-a-Heap Agency charges $125 per week plus $0.20 per mile to rent a small car. How many miles can you travel for $335?

Page 294

Explanatory Voice Balloons

Voice balloons are used in a variety of ways to demystify mathematics. They:

- translate mathematical ideas into plain English.
- help clarify problem-solving procedures.
- present alternative ways of understanding concepts.
- connect complex problems to the basic concepts students have already learned.

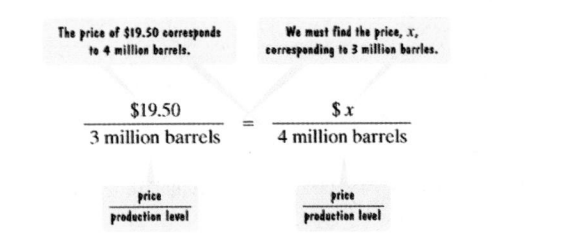

STEP 2. Set up a proportion. We will set up a proportion comparing price and production level. Because this is an inverse variation situation, we place corresponding values in opposite ratios.

The price of $19.50 corresponds to 4 million barrels.

We must find the price, x, corresponding to 3 million barrels.

$$\frac{\$19.50}{3 \text{ million barrels}} = \frac{\$ x}{4 \text{ million barrels}}$$

$\dfrac{\text{price}}{\text{production level}}$ $\dfrac{\text{price}}{\text{production level}}$

Page 307

When converting a negative improper fraction to a mixed number, copy the negative sign and then follow the previous procedure. For example,

$$-\frac{29}{8} = -3\frac{5}{8}.$$

Copy the negative sign.

Convert $\frac{29}{8}$ to a mixed number.

$$\begin{array}{r} 3 \leftarrow \text{quotient} \\ 8\overline{)29} \\ 24 \\ \hline 5 \leftarrow \text{remainder} \end{array}$$

Page 217

Study Tips

Study Tip Boxes appear in abundance throughout the book. They:

- offer suggestions for problem solving.
- point out common mistakes.
- provide informal tips and suggestions.

STUDY TIP

If your proposed solution is incorrect, you will get a false statement when you check your answer. For example, 8 is not a solution of $2x + 3 = 17$. Look what happens when we substitute 8 for x:

$$2x + 3 = 17$$
$$2 \cdot 8 + 3 \overset{?}{=} 17$$
$$16 + 3 \overset{?}{=} 17$$
$$19 = 17, \text{False}$$

Check Now we che
equation.

Because the check r
given equation is 7,

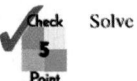

Check
5
Point

Solve and

EXAMPLE 6

Solve the equation:

Solution

$$2(x - 4)$$
$$2x - 8$$
$$7x$$
$$7x - 8$$

STUDY TIP

With increased practice in solving linear equations, try working some steps mentally. For example, consider

$$7x - 8 = -22,$$

the simplified equation on the right. Add 8 to both sides without listing all the steps:

$$7x = -14.$$

Now, divide both sides by 7:

$$x = -2.$$

Page 283

■ Technology

Optional technology boxes appear throughout the text, both in the discussion and exercises. They:

- ● suggest hints for using both graphing calculators and computers.
- ● are easy to find, or to skip, and typically follow examples.

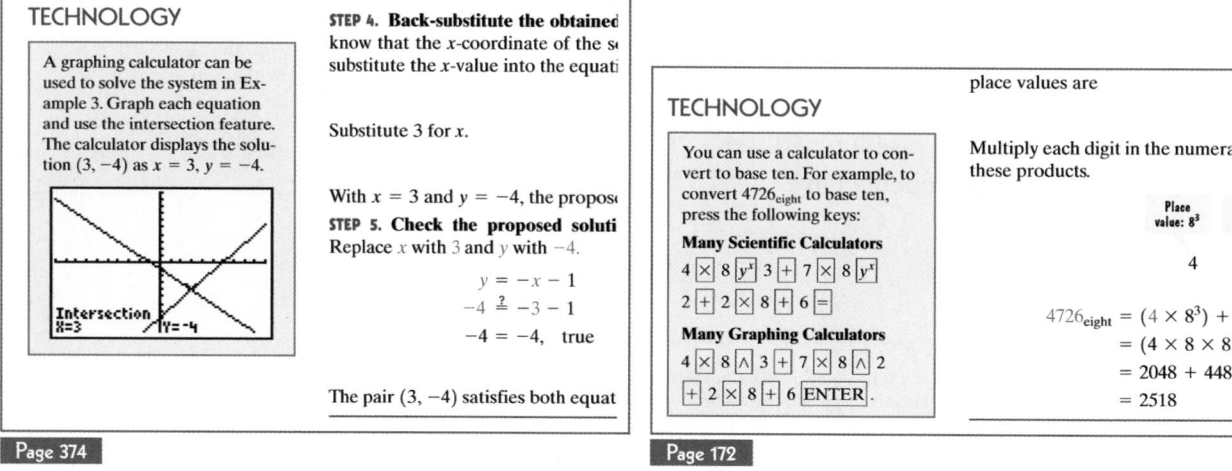

Page 374

Page 172

■ Clearly Stated Section Objectives

Learning Objectives:

- ● are clearly stated at the beginning of each section.
- ● help students recognize and focus on the most important ideas.
- ● appear in the margin at their point of use.

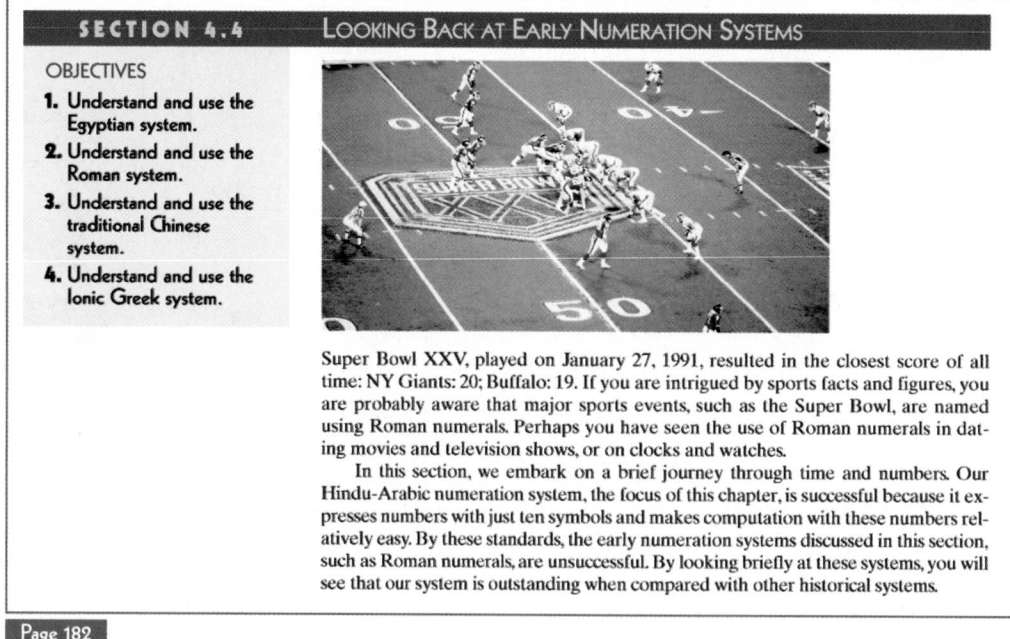

Page 182

◼ Chapter Review Grids

- ● summarize definitions and concepts for every section of the chapter.
- ● refer students to the examples that illustrate these key concepts.

Chapter Summary, Review, and Test

SUMMARY

DEFINITIONS AND CONCEPTS	EXAMPLES

5.1 Number Theory; Prime and Composite Numbers

Divisibility

   a. Natural Numbers: $N = \{1, 2, 3, 4, 5, \ldots\}$

   b. $b|a$ (b divides a: a is divisible by b) for natural numbers a and b if the operation of dividing a by b leaves a remainder of 0.

   c. Rules of divisibility are given in Table 5.1 on page 194.   Ex. 1, p. 195

Prime and Composite Numbers

   a. A prime number is a natural number greater than 1 that has only itself and 1 as factors.

   b. A composite number is a natural number greater than 1 that is divisible by a number other than itself and 1.

   c. The Fundamental Theorem of Arithmetic: Every composite number can be expressed as a product of prime numbers in one and only one way (if the order of the factors is disregarded).   Ex. 2, p. 196

Greatest Common Divisors and Least Common Multiples

   a. The greatest common divisor of two or more natural numbers is the largest number that is a divisor (or factor) of all the numbers.

   b. The procedure for finding the greatest common divisor is boxed on page 197.   Ex. 3, p. 197;

   c. The least common multiple of two or more natural numbers is the smallest natural number that is divisible by all of the numbers.   Ex. 4, p. 198

   d. The procedure for finding the least common multiple is boxed on page 199.   Ex. 5, p. 199;
                                                                              Ex. 6, p. 200

◼ Chapter Tests

Allow students to assess their understanding of the contents from the chapter.

CHAPTER 5 TEST

1. Which of the numbers 2, 3, 4, 5, 6, 8, 9, 10, and 12 divide 391,248?

2. Find the prime factorization of 252.

3. Find the greatest common divisor and the least common multiple of 48 and 72.

Perform the indicated operations in Exercises 4–6.

4. $-6 - (5 - 12)$ **5.** $(-3)(-4) \div (7 - 10)$

6. $(6 - 8)^2(5 - 7)^3$

7. Express $\frac{7}{12}$ as a decimal.

8. Express $0.\overline{64}$ as a quotient of integers in lowest terms.

In Exercises 9–11, perform the indicated operations. Where possible, reduce the answer to its lowest terms.

9. $\left(-\frac{3}{7}\right) \div \left(-2\frac{1}{7}\right)$ **10.** $\frac{19}{24} - \frac{7}{40}$ **11.** $\frac{1}{2} - 8\left(\frac{1}{4} + 1\right)$

12. Find a rational number halfway between $\frac{1}{2}$ and $\frac{2}{3}$.

13. Multiply and simplify: $\sqrt{10} \cdot \sqrt{5}$.

14. Add: $\sqrt{50} + \sqrt{32}$.

15. Rationalize the denominator: $\dfrac{6}{\sqrt{2}}$.

16. List all the rational numbers in this set:
$$\left\{-7, -\frac{4}{5}, 0, 0.25, \sqrt{3}, \sqrt{4}, \frac{22}{7}, \pi\right\}.$$

In Exercises 17–18, state the name of the property illustrated.

17. $3(2 + 5) = 3(5 + 2)$ **18.** $6(7 + 4) = 6 \cdot 7 + 6 \cdot 4$

In Exercises 19–21, evaluate each expression.

22. Multiply and express the answer in decimal notation.
$$(3 \times 10^8)(2.5 \times 10^{-5})$$

23. Divide by first expressing each number in scientific notation. Write the answer in scientific notation.
$$\frac{49,000}{0.007}$$

24. A human brain contains 3×10^{10} neurons and a gorilla brain contains 7.5×10^9 neurons. How many times as many neurons are in the brain of a human as in the brain of a gorilla?

25. Write the first six terms of the arithmetic sequence with first term, a_1, and common difference, d.
$$a_1 = 1, d = -5$$

26. Find a_9, the ninth term of the arithmetic sequence with the first term, a_1, and common difference, d.
$$a_1 = -2, d = 3$$

27. Write the first six terms of the geometric sequence with first term, a_1, and common ratio, r.
$$a_1 = 16, r = \frac{1}{2}$$

28. Find a_7, the seventh term of the geometric sequence with the first term, a_1, and common ratio, r.

STUDENT STUDY PACK

A single, easy-to-use package, available bundled with your textbook or standalone, for purchase through your bookstore. This invaluable study package contains the following:

➤ *Student Solutions Manual*

A printed manual containing fully worked out solutions to odd-numbered textbook exercises.

➤ *Pearson Tutor Center*

Tutors provide one-on-one tutoring for any problem with an answer at the back of the book. You can contact the Tutor Center via a toll-free phone number, fax, or email. Available only to the college students in the U.S. and Canada.

➤ *CD-ROM Lecture Series*

A comprehensive set of textbook-specific CD-ROMs containing short video clips of textbook objectives being reviewed and key examples being solved. These videos provide excellent support for those who require extra assistance, or for those in distance learning and/or self-paced courses.

TUTORIAL AND HOMEWORK OPTIONS

MyMathLab®

MyMathLab® is a complete online course to help you succeed in learning. MyMathLab contains an online version of your textbook with links to multimedia resource, such as video clips and practice exercises, correlated to the examples and exercises in the text. MyMathLab provides you with online homework and tests and generates a personalized study plan based on your results. The study plan links you to unlimited tutorial exercises for further study, so you can practice until you have mastered the skills. All of the online homework, tests, and tutorial work you do is tracked in your MyMathLab gradebook.

MyMathLab provides a rich and flexible set of course materials, featuring free-response exercises that are algorithmically generated for unlimited practice. Students can use online tools such as video lectures and a multimedia textbook to improve their performance. Instructors can use MyMathLab's homework and test managers to select and assign online exercises correlated to the textbook, and can import TestGen tests for added flexibility. The online gradebook—designed specifically for mathematics—automatically tracks students' homework and test results and gives the instructor control over how to calculate final grades. MyMathLab is available to qualified adopters. For more information, visit our website at www.mymathlab.com or contact your Prentice Hall sales representative for a product demonstration.

MyMathLab contains ALL the resources listed in the Student Study Pack description above in an online format (Student Solutions Manual, CD Lecture Series, Tutor Center).

Note: MyMathLab must be set up and assigned by your instructor

1 Problem Solving and Critical Thinking

The ability to solve problems using thought and reasoning is indispensable to every area of our lives. We are all problem solvers. Our work in solving problems in school, on the job, and even in our personal lives involves understanding the problem, devising a plan for solving it, and then carrying out the plan. The aim of this chapter is to help you to better develop your skills of reasoning, estimation, and problem solving. By the time you complete the chapter, you will be a better, more confident problem solver. This will help you as you study the topics throughout the book. You will be able to apply problem-solving techniques used in mathematics to real-life situations such as the one that the department manager is trying to solve. (We will return to this problem in the third section of the chapter.) As you begin your journey through this book, your growing ability to solve problems will result in growing confidence in your mathematical abilities.

You are the department manager of a company. One of your best employees who is too valuable to fire arrives to work late nearly every morning. What can you do to solve the problem of this employee's tardiness?

This problem appears as Example 7 in Section 1.3.

OBJECTIVES

1. Understand and use inductive reasoning.

2. Understand and use deductive reasoning.

Calvin and Hobbes   by Bill Watterson

Many people associate mathematics with tedious computation, meaningless algebraic procedures, and intimidating sets of equations. The truth is that mathematics is the most powerful means we have of exploring our world and describing how it works. The word *mathematics* comes from the Greek word *mathematikos*, which means "inclined to learn." To be mathematical literally means to be inquisitive, open-minded, and interested in a lifetime of pursuing knowledge!

Mathematics and Your Life

A major goal of this book is to show you how mathematics can be applied to your life in interesting, enjoyable, and meaningful ways. Here is a brief, although far from complete, preview of what you can expect.

An understanding of logic will help you to determine what is meant and what is not necessarily meant in spoken and written language. It will give you special tools to analyze information objectively, and to avoid arriving at conclusions hastily without considering all the evidence.

Literacy in numbers will provide you with an understanding of what numbers really mean. For example, the national debt of the United States is about $6.8 trillion. What exactly does this huge number mean? How much would each citizen have to pay the government to retire the debt? In Chapter 5, you will be given a technique that will provide a surprising answer to this question.

An understanding of probability theory will enable you to study outcomes of events to determine what to expect in the long run. Basic ideas of probability will give you a tool for peering into the future and seeing aspects of the universe that would otherwise be invisible.

Managing your personal finances is not an easy task. Achieving your financial goals depends on understanding basic ideas about savings, loans, and investments. Knowledge of consumer mathematics can pay, literally, by making your financial goals a reality.

You can hardly pick up a newspaper without being bombarded with numbers and graphs that describe how we live, what we want, and even how we are likely to die. These numbers influence which products are sold in stores, which movies are shown in theaters, which treatments are available for disease, and which policies are likely to be supported by politicians. An understanding of statistics will help you comprehend where these numbers come from and how they are used to make decisions.

Mathematics and Your Career

Generally speaking, the income of an occupation is related to the amount of education required. This, in turn, is usually related to the skill level required in language and mathematics. With our increasing reliance on technology, the more mathematics you know, the more career choices you will have.

"It is better to take what may seem to be too much math rather than too little. Career plans change, and one of the biggest roadblocks in undertaking new educational or training goals is poor preparation in mathematics. Furthermore, not only do people qualify for more jobs with more math, they are also better able to perform their jobs."

OCCUPATIONAL OUTLOOK
QUARTERLY

Mathematics and Your World

Mathematics is a science that helps us recognize, classify, and explore the hidden patterns of our universe. Focusing on areas as different as planetary motion, animal markings, shapes of viruses, aerodynamics of figure skaters, and the very origin of the universe, mathematics is the most powerful tool available for revealing the underlying structure of our world. Within the last 30 years, mathematicians have even found order in chaotic events such as the uncontrolled storm of noise in the nerve cells of the brain during an epileptic seizure.

1 Understand and use inductive reasoning.

Inductive Reasoning

Mathematics involves the study of patterns. In everyday life, we frequently rely on patterns and routines to draw conclusions. Here is an example:

> The last six times I went to the beach, the traffic was light on Wednesdays and heavy on Sundays. My conclusion is that weekdays have lighter traffic than weekends.

This type of reasoning process is referred to as *inductive reasoning*, or *induction*.

INDUCTIVE REASONING

Inductive reasoning is the process of arriving at a general conclusion based on observations of specific examples.

EXAMPLE 1   A VISIT TO THE FLORIDA KEYS

A tourist who visited Key West met a number of people from the Keys who were relaxed and laid back. The tourist concluded that all people living in Key West were relaxed and laid back. What reasoning process led to this conclusion?

SOLUTION   This conclusion was reached by induction. The tourist reached a general conclusion about people living in Key West based on observing a number of specific people.

Although inductive reasoning is a powerful method of drawing conclusions, we can never be absolutely certain that these conclusions are true. Even if the visitor to Key West met 100 relaxed people who lived on the island, it is possible that Key West resident number 101 is a stressed-out, nervous wreck. Resident number 101 reveals that the conclusion

> All people living in Key West are relaxed and laid back

is false.

 A study of ten people revealed that as they spent more time per week exercising, they experienced fewer headaches per month. You conclude that all people will experience fewer headaches with an increase in time spent exercising. What reasoning process led to your conclusion? Explain.

As you use inductive reasoning to find patterns, keep in mind that this form of reasoning does not prove that the pattern applies to all cases. For this reason, the conclusion is called a **conjecture**, a **hypothesis**, or an educated guess. If there is just one case for which the conjecture does not work, then the conjecture is false. Such a case is called a **counterexample**. Key West resident number 101, who is a stressed-out wreck, is a counterexample that shows "All people living in Key West are relaxed and laid back" is false.

"When I'm walking in the woods, I find it quite difficult not to look at a fern or the bark of a tree and wonder how it was formed—why is it like that?"

JAMES MURRAY,
mathematician

EXAMPLE 2   FINDING A COUNTEREXAMPLE

The ten symbols that we use to write numbers, namely 0, 1, 2, 3, 4, 5, 6, 7, 8, and 9, are called **digits**. In each example shown below, the sum of two two-digit numbers is a three-digit number.

$$\begin{array}{r} 47 \\ +73 \\ \hline 120 \end{array}$$ Two-digit numbers $$\begin{array}{r} 56 \\ +46 \\ \hline 102 \end{array}$$ Three-digit sums

Is the sum of two two-digit numbers always a three-digit number? Find a counterexample to show that the statement

The sum of two two-digit numbers is a three-digit number

is false.

SOLUTION   There are many counterexamples, but we need to find only one. Here is an example that makes the statement false:

Two-digit numbers $$\begin{array}{r} 56 \\ +43 \\ \hline 99 \end{array}$$ This is a two-digit sum, not a three-digit sum.

This example is a counterexample that shows the statement

The sum of two two-digit numbers is a three-digit number

is false.

 Check 2 Point

Find a counterexample to show that the statement

The product of two two-digit numbers is a three-digit number

is false.

Avid lottery players use a variety of inductive strategies for choosing numbers. Some will study lists of numbers that have already won lotteries and use these numbers to draw conclusions about future winners. These strategies have no more chance of success than using your birth date or other personal numbers. Inductive reasoning has its place, but not here.

Consider, for example, a lottery in which each player chooses six different numbers from 1 through 53. Two possible combinations are 2, 4, 6, 8, 10, 12 and 8, 13, 20, 29, 38, 42. In Chapter 11, you will learn that there are nearly 23 million combinations. Each six-number combination has the same probability of being randomly selected, about 1 in 23 million. Because a "memorable" combination with an obvious pattern like 2, 4, 6, 8, 10, 12 never seems to win, it is tempting to use inductive reasoning and incorrectly conclude that number sequences with obvious patterns are worse bets than nonmemorable sequences without patterns. (The reason that six-number combinations with obvious patterns never seem to win is that of the nearly 23 million possible combinations, only a tiny fraction have such patterns.)

Although patterns are irrelevant in choosing lottery numbers, they are extremely important to mathematicians. Discovery in mathematics often begins with a search for patterns about numbers.

EXAMPLE 3 | USING INDUCTIVE REASONING

Identify a pattern in each list of numbers. Then use this pattern to find the next number.

a. 4, 12, 20, 28, 36, _____   **b.** 5, 15, 45, 135, _____

SOLUTION

a. Each number in the list

4, 12, 20, 28, 36

is obtained by adding 8 to the previous number. The next number is 36 + 8, or 44.

b. Each number in the list

5, 15, 45, 135

is obtained by multiplying the previous number by 3. The next number is 135 × 3, or 405.

 Identify a pattern in each list of numbers. Then use this pattern to find the next number.

a. 3, 9, 15, 21, 27, _____   **b.** 2, 10, 50, 250, _____

Mathematics is more than recognizing number patterns. It is about the patterns that arise in the world around us. For example, by describing patterns formed by various kinds of knots, mathematicians are helping scientists investigate the knotty shapes and patterns of viruses. One of the weapons used against viruses is based on recognizing visual patterns in the possible ways that knots can be tied.

Our next example deals with recognizing visual patterns.

EXAMPLE 4 | FINDING THE NEXT FIGURE IN A VISUAL SEQUENCE

Describe two patterns in this sequence of figures. Use the patterns to draw the next figure in the sequence.

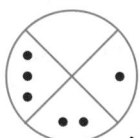 , 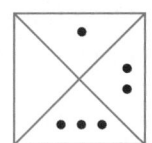 , 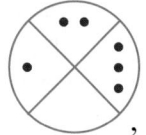 , 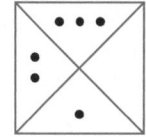 , _____

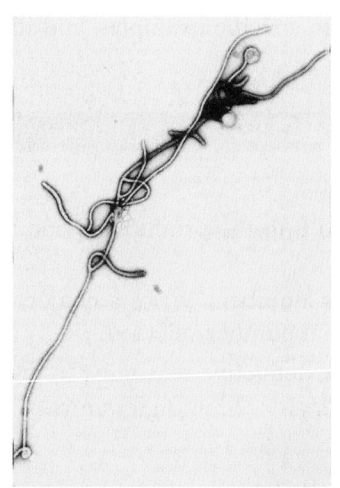

This electron microscope photograph shows the knotty shape of the Ebola virus.

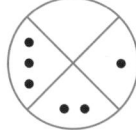

FIGURE 1.1

SOLUTION   The more obvious pattern is that the figures alternate between circles and squares. We conclude that the next figure will be a circle. We can identify the second pattern in the four regions containing no dots, one dot, two dots, and three dots. The dots are placed in order (no dots, one dot, two dots, three dots) in a clockwise direction. However, the entire pattern of the dots rotates counterclockwise as we follow the figures from left to right.

This means that the next figure should be a circle with a single dot in the right-hand region, two dots in the bottom region, three dots in the left-hand region, and no dots in the top region. This figure is drawn in Figure 1.1

 Describe two patterns in this sequence of figures. Use the patterns to draw the next figure in the sequence.

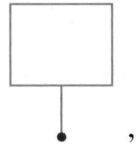 ,  ,  ,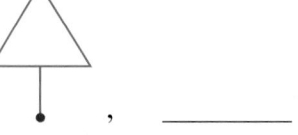

Understand and use deductive reasoning.

Deductive Reasoning

We use inductive reasoning in everyday life. Many of the conjectures that come from this kind of thinking seem highly likely, although we can never be absolutely certain that they are true. Another method of reasoning, called *deductive reasoning*, or *deduction*, can be used to prove that some conjectures are true.

> ### DEDUCTIVE REASONING
>
> **Deductive reasoning** is the process of proving a specific conclusion from one or more general statements. A conclusion that is proved true by deductive reasoning is called a **theorem**.

Deductive reasoning allows us to draw specific conclusions based on a general statement. For example, suppose that all college students are required to take mathematics. This general statement applies to every college student. Further suppose that Alberta Einstein is a college student. A specific conclusion is that Alberta Einstein is required to take mathematics. In Chapter 3, you will learn how to prove this conclusion.

Our next example illustrates the difference between inductive and deductive reasoning. The first part of the example involves reasoning that moves from specific examples to a general statement, illustrating inductive reasoning. The second part of the example begins with the general case rather than specific examples, and illustrates deductive reasoning.

EXAMPLE 5   USING INDUCTIVE AND DEDUCTIVE REASONING

Consider the following procedure:

Select a number. Multiply the number by 6. Add 8 to the product. Divide this sum by 2. Subtract 4 from the quotient.

a. Repeat this procedure for at least four different numbers. Write a conjecture that relates the result of this process to the original number selected.

b. Represent the original number as ■ or n, and use deductive reasoning to prove the conjecture in part (a).

SOLUTION

a. First, let us pick our starting numbers. We will use 4, 7, 11, and 100, but we could pick any four numbers. Next we will apply the procedure given in this example to 4, 7, 11, and 100, four individual cases, in Table 1.1.

TABLE 1.1 APPLYING A PROCEDURE TO FOUR INDIVIDUAL CASES

Select a number.	4	7	11	100
Multiply the number by 6.	$4 \times 6 = 24$	$7 \times 6 = 42$	$11 \times 6 = 66$	$100 \times 6 = 600$
Add 8 to the product.	$24 + 8 = 32$	$42 + 8 = 50$	$66 + 8 = 74$	$600 + 8 = 608$
Divide this sum by 2.	$\dfrac{32}{2} = 16$	$\dfrac{50}{2} = 25$	$\dfrac{74}{2} = 37$	$\dfrac{608}{2} = 304$
Subtract 4 from the quotient.	$16 - 4 = 12$	$25 - 4 = 21$	$37 - 4 = 33$	$304 - 4 = 300$

Because we are asked to write a conjecture that relates the result of this process to the original number selected, let us focus on the result of each case.

Original number selected	4	7	11	100
Result of the process	12	21	33	300

Do you see a pattern? Our conjecture is that the result of the process is three times the original number selected. We have used inductive reasoning.

b. Now we begin with the general case rather than specific examples. We can use  to stand for any number. We can also use the letter n to represent any number.

	Method 1: Using ■	Method 2: Using n
Select a number:	■	n
Multiply the number by 6:	■ ■ ■ ■ ■ ■	$6n$ (this means 6 times n)
Add 8 to the product:	■ ■ ■ ■ ■ ■ ⠿ ⠿ ⠿ ⠿	$6n + 8$
Divide this sum by 2:	■ ■ ■ • • • •	$\dfrac{6n + 8}{2} = \dfrac{6n}{2} + \dfrac{8}{2} = 3n + 4$
Subtract 4 from the quotient:	■ ■ ■	$3n + 4 - 4 = 3n$

If we use ■ to represent any number, the result of the process is ■ ■ ■, or three times the number ■. If we use n to represent any number, the result is $3n$, or three times the number n. Either approach proves that the result of the procedure is three times the original number selected for any number. We have used deductive reasoning.

BLITZER BONUS

The Patterns of Chaos

A magnification of the Mandelbrot set. R. F. Voss, "29-Fold M-set Seahorse" computer-generated image.

One of the new frontiers of mathematics suggests that there is an underlying order in things that appear to be random, such as the hiss and crackle of background noises as you tune a radio. Irregularities in the heartbeat, some of them severe enough to cause a heart attack, or irregularities in our sleeping patterns, such as insomnia, are examples of chaotic behavior. Chaos in the mathematical sense does not mean a complete lack of form or arrangement. In mathematics, chaos is used to describe something that appears to be random but is not actually random. The patterns of chaos appear in images like the one shown above, called the Mandelbrot set. Magnified portions of this image yield repetitions of the original structure, as well as new and unexpected patterns. The Mandelbrot set transforms the hidden structure of chaotic events into a source of wonder and inspiration.

✓**Check**
5
Point

Consider the following procedure:
Select a number. Multiply the number by 4. Add 6 to the product. Divide this sum by 2. Subtract 3 from the quotient.

a. Repeat this procedure for at least four different numbers. Write a conjecture that relates the result of this process to the original number selected.

b. Represent the original number as ▪ or n, and use deductive reasoning to prove the conjecture in part (a).

Exercise Set 1.1

Practice Exercises

1. Which reasoning process is shown in the following example? Explain your answer.

We examine the fingerprints of 1000 people. No two individuals in this group of people have identical fingerprints. We conclude that for all people, no two people have identical fingerprints.

2. Which reasoning process is shown in the following example? Explain your answer.

An HMO does a follow-up study on 200 randomly selected patients given a flu shot. None of these people became seriously ill with the flu. The study concludes that all HMO patients be urged to get a flu shot in order to prevent a serious flu.

3. Which reasoning process is shown in the following example? Explain your answer.

All mammals are warm-blooded animals. No snakes are warm-blooded. I have a pet snake. We conclude that my pet snake cannot be a mammal.

4. Which reasoning process is shown in the following example? Explain your answer.

All freshmen must live on campus. No students who live on campus can own cars. Joan is a freshman. We conclude that Joan cannot own a car.

In Exercises 5–12, find a counterexample to show that each of the statements is false.

5. No U.S. president has been younger than 65 at the time of his inauguration.

6. No singers appear in movies.

7. If a number is multiplied by itself, the result is even.

8. The sum of two three-digit numbers is a four-digit number.

9. Adding the same number to the top and bottom of a fraction does not change the fraction's value.

10. If the difference between two numbers is odd, then the two numbers are both odd.

11. If a number is added to itself, the sum is greater than the original number.

12. If 1 is divided by a number, the quotient is less than that number.

In Exercises 13–32, identify a pattern in each list of numbers. Then use this pattern to find the next number. (More than one pattern might exist, so it is possible that there is more than one correct answer.)

13. 8, 12, 16, 20, 24, _____

14. 19, 24, 29, 34, 39, _____

15. 37, 32, 27, 22, 17, _____

16. 33, 29, 25, 21, 17, _____

17. 3, 9, 27, 81, 243, _____

18. 2, 8, 32, 128, 512, _____

19. 1, 2, 4, 8, 16, _____

20. 1, 5, 25, 125, _____

21. 1, 4, 1, 8, 1, 16, 1, _____

22. 1, 4, 1, 7, 1, 10, 1, _____

23. 4, 2, 0, −2, −4, _____

24. 6, 3, 0, −3, −6, _____

25. $\frac{1}{2}, \frac{1}{6}, \frac{1}{10}, \frac{1}{14}, \frac{1}{18},$ _____

26. $1, \frac{1}{2}, \frac{1}{3}, \frac{1}{4}, \frac{1}{5},$ _____

27. $1, \frac{1}{3}, \frac{1}{9}, \frac{1}{27},$ _____

28. $1, \frac{1}{2}, \frac{1}{4}, \frac{1}{8},$ _____

29. 64, −16, 4, −1, _____

30. 125, −25, 5, −1, _____

31. $(6, 2), (0, -4), \left(7\frac{1}{2}, 3\frac{1}{2}\right), (2, -2), (3,$ _____$)$

32. $\left(\frac{2}{3}, \frac{4}{9}\right), \left(\frac{1}{5}, \frac{1}{25}\right), (7, 49), \left(-\frac{5}{6}, \frac{25}{36}\right), \left(-\frac{4}{7},$ _____$\right)$

In Exercises 33–36, identify a pattern in each sequence of figures. Then use the pattern to find the next figure in the sequence.

33.

34.

35.

36.

Exercises 37–40 describe procedures that are to be applied to numbers. In each exercise,

a. *Repeat the procedure for four numbers of your choice. Write a conjecture that relates the result of the process to the original number selected.*

b. *Represent the original number as ▪ or n, and use deductive reasoning to prove the conjecture in part (a).*

37. Select a number. Multiply the number by 4. Add 8 to the product. Divide this sum by 2. Subtract 4 from the quotient.

38. Select a number. Multiply the number by 3. Add 6 to the product. Divide this sum by 3. Subtract the original selected number from the quotient.

39. Select a number. Add 5. Double the result. Subtract 4. Divide by 2. Subtract the original selected number.

40. Select a number. Add 3. Double the result. Add 4. Divide by 2. Subtract the original selected number.

In Exercises 41–44, use inductive reasoning to predict the next line in the pattern. Then perform the arithmetic to determine whether your conjecture is correct.

41.
$$1 + 2 = \frac{2 \times 3}{2}$$
$$1 + 2 + 3 = \frac{3 \times 4}{2}$$
$$1 + 2 + 3 + 4 = \frac{4 \times 5}{2}$$
$$1 + 2 + 3 + 4 + 5 = \frac{5 \times 6}{2}$$

42.
$$3 + 6 = \frac{6 \times 3}{2}$$
$$3 + 6 + 9 = \frac{9 \times 4}{2}$$
$$3 + 6 + 9 + 12 = \frac{12 \times 5}{2}$$
$$3 + 6 + 9 + 12 + 15 = \frac{15 \times 6}{2}$$

43.
$$1 + 3 = 2 \times 2$$
$$1 + 3 + 5 = 3 \times 3$$
$$1 + 3 + 5 + 7 = 4 \times 4$$
$$1 + 3 + 5 + 7 + 9 = 5 \times 5$$

44.
$$\frac{1}{1 \times 2} + \frac{1}{2 \times 3} = \frac{2}{3}$$
$$\frac{1}{1 \times 2} + \frac{1}{2 \times 3} + \frac{1}{3 \times 4} = \frac{3}{4}$$
$$\frac{1}{1 \times 2} + \frac{1}{2 \times 3} + \frac{1}{3 \times 4} + \frac{1}{4 \times 5} = \frac{4}{5}$$

Application Exercises

*The ancient Greeks studied **figurate numbers**, so named because of their representations as geometric arrangements of points. Exercises 45–46 involve figurate numbers.*

45. The numbers 1, 3, 6, 10, 15, 21, and so on, are called **triangular numbers.**

1   3   6   10   15   21

a. Subtract the first number from the second $(3 - 1)$, the second from the third $(6 - 3)$, the third from the fourth, and so on. Use inductive reasoning to describe the pattern in the successive differences.

b. Use the pattern in part (a) to write the five triangular numbers that follow 21.

46. The numbers 1, 4, 9, 16, 25, and so on, are called **square numbers.**

1   4   9   16   25

a. Subtract the first number from the second $(4 - 1)$, the second from the third $(9 - 4)$, the third from the fourth, and so on. Use inductive reasoning to describe the pattern in the successive differences.

b. Use the pattern in part (a) to write the five square numbers that follow 25.

47. The following table relates a man's foot length, in inches, to his shoe size.

Foot length	9	10	11	12	13	14
Shoe size	5	8	11			

a. Use inductive reasoning to fill in the missing portions of the table.

b. According to the 1998 *Guinness Book of World Records*, the biggest feet currently known are those of Matthew McGrory (1973–) of Pennsylvania, who wears size 29 shoes. Use the pattern in the table to predict the length of his foot.

48. The triangular arrangement of numbers shown below is known as **Pascal's triangle**, credited to French mathematician Blaise Pascal (1623–1662). Use inductive reasoning to find the six numbers designated by question marks.

$$
\begin{array}{ccccccccccc}
 & & & & & 1 & & & & & \\
 & & & & 1 & & 1 & & & & \\
 & & & 1 & & 2 & & 1 & & & \\
 & & 1 & & 3 & & 3 & & 1 & & \\
 & 1 & & 4 & & 6 & & 4 & & 1 & \\
? & & ? & & ? & & ? & & ? & & ?
\end{array}
$$

Writing in Mathematics

Writing about mathematics will help you to learn mathematics. For all writing exercises in this book, use complete sentences to respond to the questions. Some writing exercises can be answered in a sentence; others require a paragraph or two. You can decide how much you need to write as long as your writing clearly and directly answers the question in the exercise. Standard references such as a dictionary and a thesaurus may be helpful.

49. What is a conjecture?

50. What is a counterexample? Give an example of a conclusion that can be reached using inductive reasoning. Then give a counterexample.

51. Describe what is meant by deductive reasoning. Give an example.

52. Can inductive reasoning be used to prove a theorem? Explain your answer.

53. The word *induce* comes from a Latin term meaning to lead. Explain what leading has to do with inductive reasoning.

54. Give an example of a decision that you made recently in which the reasoning you used to reach the decision was induction. Describe your reasoning process.

55. The March 29, 1999 issue of *Time* magazine is a special issue on the greatest thinkers of the twentieth century. Consult this issue of *Time* and select one of the great thinkers who used inductive or deductive reasoning in his or her work.

    a. Explain how the person you selected used one of these types of reasoning.

    b. Explain why you selected this particular person.

Critical Thinking Exercises

56. What reasoning process is shown in the following example? Explain your answer.

I see the same woman leaving the college administration building every day at 5:00 P.M. She must be an administrator at the college.

In Exercises 57–58, identify a pattern and find the next number in the list.

57. 1, 1, 2, 3, 5, 8, 13, 21, 34, _____   **58.** 15, 23, 33, 45, 59, _____

In Exercises 59–60, use inductive reasoning to predict the next line in the pattern. Then use a calculator or perform the arithmetic by hand to determine whether your conjecture is correct.

59.
$$33 \times 3367 = 111{,}111$$
$$66 \times 3367 = 222{,}222$$
$$99 \times 3367 = 333{,}333$$
$$132 \times 3367 = 444{,}444$$
_____

60.
$$1 \times 8 + 1 = 9$$
$$12 \times 8 + 2 = 98$$
$$123 \times 8 + 3 = 987$$
$$1234 \times 8 + 4 = 9876$$
$$12{,}345 \times 8 + 5 = 98{,}765$$
_____

61. Write a list of numbers that has two patterns so that the next number in the list can be 15 or 20.

Technology Exercises

62. a. Use a calculator to find 6×6, 66×66, 666×666, and 6666×6666.

    b. Describe a pattern in the numbers being multiplied and the resulting products.

    c. Use the pattern to write the next two multiplications and their products. Then use your calculator to verify these results.

    d. Is this process an example of inductive or deductive reasoning? Explain your answer.

63. a. Use a calculator to find 3367×3, 3367×6, 3367×9, and 3367×12.

    b. Describe a pattern in the numbers being multiplied and the resulting products.

    c. Use the pattern to write the next two multiplications and their products. Then use your calculator to verify these results.

    d. Is this process an example of inductive or deductive reasoning? Explain your answer.

Group Exercise

64. Reasoning in real life is often not as clean and simple as the examples of inductive and deductive reasoning presented in this section. Advertisements often use deceptive reasoning to persuade us to buy products. Each group member should find a magazine, newspaper, or television advertisement that uses deceptive logic to get viewers to purchase what they are selling. Upon returning to the group, each member should display or describe the advertisement and discuss how the advertiser is attempting to fool the viewer into buying something.

SECTION 1.2   ESTIMATION AND GRAPHS

OBJECTIVES

1. Use estimation techniques to arrive at an approximate answer to a problem.

2. Apply estimation techniques to information given by graphs.

One effect of the presidential election of 2000 was to convince Americans that their votes count. Although turnout is crucial in any democracy, since the 1960s, the percentage of Americans who actually vote has been decreasing. In this section, you will learn to apply estimation techniques to graphs so that you can analyze voting

STUDY TIP

A detailed presentation on the mathematics of voting can be found in Chapter 14.

1 Use estimation techniques to arrive at an approximate answer to a problem.

STUDY TIP

In case you have forgotten, here are the place values for a number.

3 , 7 8 3 , 9 6 2

Ones place
Tens place
Hundreds place
Thousands place
Ten thousands place
Hundred thousands place
Millions place

The place that a digit occupies in a number tells us its value in that number.

trends from elections in the 1800s through the contested election of George W. Bush in 2000.

Estimation

Estimation is the process of arriving at an approximate answer to a question. For example, companies estimate the amount of their products consumers are likely to use, and economists estimate financial trends. If you are about to cross a street, you may estimate the speed of oncoming cars so that you know whether or not to wait before crossing. Rounding numbers is also an estimation method. You might round a number without even being aware that you are doing so. You may say that you are 20 years old, rather than 20 years 5 months, or that you will be home in about a half-hour rather than 25 minutes.

You will find estimation to be equally valuable in your work for this class. Making mistakes with a calculator or computer is easy. Estimation can tell us whether the answer displayed for a computation makes sense.

In this section, we demonstrate several estimation methods. In the second part of the section, we apply these techniques to information given by graphs.

Performing computations using rounded numbers is one way to check whether an answer displayed by a calculator or a computer is reasonable.

ROUNDING NUMBERS

1. Look at the digit to the right of the digit where rounding is to occur.

2. a. If the digit to the right is 5 or greater, add 1 to the digit to be rounded. Replace all digits to the right by zeros.

   b. If the digit to the right is less than 5, do not change the digit to be rounded. Replace all digits to the right by zeros.

EXAMPLE 1   ROUNDING A NUMBER

Round 37 to the nearest ten.

SOLUTION

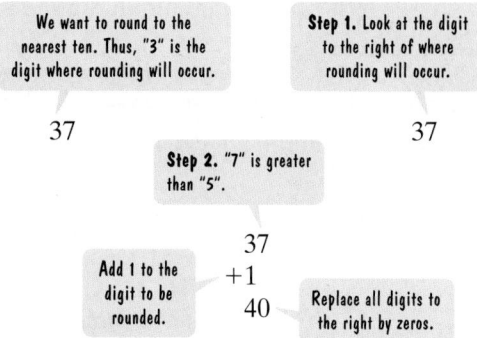

Therefore, 37 rounded to the nearest ten is 40.

The symbol ≈ means *is approximately equal to*. Based on our rounding in Example 1, we can write

$$37 \approx 40.$$

 Round 58 to the nearest ten.

EXAMPLE 2 | ESTIMATION BY ROUNDING

You purchased bread for $2.59, detergent for $2.17, a sandwich for $3.65, an apple for $0.47, and coffee for $5.79. The total bill was $18.67. Is this amount reasonable?

SOLUTION   If you are in the habit of carrying a calculator to the store, you can answer the question by finding the exact cost of the purchase. However, estimation can be used to determine if the bill is reasonable even if you do not have a calculator. We will round the cost of each item to the nearest dollar.

<table>
<tr><td></td><td>Round to the
nearest dollar.</td><td>Use digits in this column
to do the rounding.</td></tr>
</table>

Bread	$2.59 ≈ $3.00
Detergent	$2.17 ≈ $2.00
Sandwich	$3.65 ≈ $4.00
Apple	$0.47 ≈ $0.00
Coffee	$5.79 ≈ $6.00
	$15.00

The total bill, $18.67, seems a bit high compared to the $15.00 estimate. You should check the bill before paying it.

 You and a friend ate lunch at Ye Olde Cafe. The check for the meal showed soup for $2.40, tomato juice for $1.25, a roast beef sandwich for $4.60, a chicken salad sandwich for $4.40, two coffees totaling $1.40, apple pie for $1.85, and chocolate cake for $2.95.

a. Round the cost of each item to the nearest dollar and obtain an estimate for the food bill.

b. The total bill before tax was $21.85. Is this amount reasonable?

Although different rounding results in different estimates, the whole idea behind the rounding process is to make calculations simple.

EXAMPLE 3 | USING ESTIMATION IN A CALCULATION

Find a reasonable estimate of

$$\frac{0.47996 \times 88}{0.249}.$$

SOLUTION   Each decimal appears to complicate the calculation. Our goal is to round the decimals to simplify this computation. The decimal 0.47996 is approximately 0.5, which is $\frac{1}{2}$. The decimal 0.249 is approximately 0.25, which is $\frac{1}{4}$. Thus,

$$\frac{0.47996 \times 88}{0.249} \approx \frac{\frac{1}{2} \times 88}{\frac{1}{4}} = \frac{44}{\frac{1}{4}} = 44 \times \frac{4}{1} = 176.$$

Using a calculator, the answer displayed for the given computation is 169.624417671. Our estimate, 176, convinces us that the displayed answer makes sense.

It is possible to obtain an estimate in Example 3 without using fractions:

$$\frac{0.47996 \times 88}{0.249} \approx \frac{0.5 \times 88}{0.2} = \frac{44}{0.2} = 220.$$

By rounding 0.249 to the nearest tenth $(0.249 \approx 0.2)$ rather than to the nearest hundredth $\left(0.249 \approx 0.25 = \frac{1}{4}\right)$, this estimate is not as accurate as the one in Example 3.

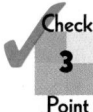 **Check Point 3** Find a reasonable estimate of

$$\frac{0.2489 \times 48}{0.5103}.$$

EXAMPLE 4   ESTIMATION BY ROUNDING

A carpenter who works full time earns $28 per hour.

a. Estimate the carpenter's weekly salary.

b. Estimate the carpenter's annual salary.

SOLUTION

a. In order to simplify the calculation, we can round the hourly rate of $28 to $30. Be sure to write out the units for each number in the calculation. The work week is 40 hours per week and the rounded salary is $30 per hour. We express this as

$$\frac{40 \text{ hours}}{\text{week}} \quad \text{and} \quad \frac{\$30}{\text{hour}}.$$

The word "per" is represented by the division bar. We multiply these two numbers to estimate the carpenter's weekly salary. We cancel out units that are identical if they are above and below the division bar.

$$\frac{40 \ \cancel{\text{hours}}}{\text{week}} \times \frac{\$30}{\cancel{\text{hours}}} = \frac{\$1200}{\text{week}}$$

Thus, the carpenter earns approximately $1200 per week, written $\approx \$1200$.

b. For the estimate of annual salary, we may round 52 weeks to 50 weeks. The annual salary is approximately the product of $1200 per week and 50 weeks per year:

$$\frac{\$1200}{\cancel{\text{week}}} \times \frac{50 \ \cancel{\text{weeks}}}{\text{year}} = \frac{\$60,000}{\text{year}}$$

Thus, the carpenter earns approximately $60,000 per year, or $60,000 annually, written $\approx \$60,000$.

 **Check Point 4** A landscape architect who works full time earns $52 per hour.

a. Estimate the landscape architect's weekly salary.

b. Estimate the landscape architect's annual salary.

Making estimates is an excellent way to put numbers into perspective. For example, what does the number 1 million really mean? Our next example may help you to comprehend this large number.

EXAMPLE 5   COMPREHENDING LARGE NUMBERS

Imagine that you counted 60 numbers per minute and continued to count non-stop until you reached 1 million. What is a reasonable estimate of the number of days it would take you to complete the counting?

STUDY TIP

In case you have forgotten, here is what you need to do in a division situation such as the one on the right.

1. Copy the number (and unit) of the number being divided into.
2. Change the division to multiplication.
3. Reverse the positions of the numbers (and units) of the number by which we divide.

SOLUTION We begin by finding how many minutes it will take to count to 1 million. You count 60 numbers per minute. To determine how many minutes it will take to reach 1 million, we divide 1 million numbers by 60 (numbers per minute).

$$\frac{1,000,000 \text{ numbers}}{\dfrac{60 \text{ numbers}}{\text{minute}}} = 1,000,000 \text{ numbers} \times \frac{\text{minute}}{60 \text{ numbers}}$$

$$= \frac{1,000,000 \text{ minutes}}{60} = 16,666\tfrac{2}{3}\text{ minutes}$$

Rounding to the nearest thousand, it will take about 17,000 minutes to count to 1 million.

Now we will convert 17,000 minutes to hours and then the hours to days. There are 60 minutes per hour. We divide 17,000 minutes by 60 (minutes per hour) to find how many hours it will take to reach 1 million.

$$\frac{17,000 \text{ minutes}}{\dfrac{60 \text{ minutes}}{\text{hour}}} = 17,000 \text{ minutes} \times \frac{\text{hour}}{60 \text{ minutes}}$$

$$= \frac{17,000 \text{ hours}}{60} = 283\tfrac{1}{3}\text{ hours}$$

Rounding to the nearest ten, it will take about 280 hours to count to 1 million.

There are 24 hours per day. Thus, we can divide 280 hours by 24 (hours per day) to find how many days it will take to reach 1 million.

$$\frac{280 \text{ hours}}{\dfrac{24 \text{ hours}}{\text{day}}} = 280 \text{ hours} \times \frac{\text{day}}{24 \text{ hours}} = \frac{280}{24}\text{ days} = 11\tfrac{2}{3}\text{ days}$$

Rounding to the nearest day, it will take about 12 days of nonstop counting to reach 1 million.

Were you surprised that 12 days of nonstop counting are required to reach 1 million? If it is difficult to comprehend 1 million, can you imagine trying to comprehend the $35 billion that Bill Gates of Microsoft is worth? It would take Mr. Gates nearly a quarter of a century to exhaust his fortune if he spent close to $4 million per day!

Check 5 Point Imagine that you counted 60 numbers per minute and continued to count nonstop until you reached 10,000. What is a reasonable estimate of the number of hours it would take you to complete the counting?

2 Apply estimation techniques to information given by graphs.

Estimation with Graphs

Magazines and newspapers often display information using circle, bar, and line graphs. The following examples illustrate how estimation techniques can be applied to each of these graphs.

Circle graphs, also called **pie charts**, show how a whole quantity is divided into parts. Circle graphs are divided into pieces, called **sectors**. Figure 1.2 on the next page is an example of a typical circle graph. The graph shows the expected makeup of the U.S. population by race and Hispanic origin for the year 2050.

EXAMPLE 6 | APPLYING ESTIMATION TECHNIQUES TO A CIRCLE GRAPH

Using Figure 1.2, answer the following:

a. What information is shown by the sector labeled "White, non-Hispanic"?

b. If the U.S. population for the year 2050 is expected to be 393,931,000, estimate the population for persons of Hispanic origin.

Projected U.S. Population in 2050

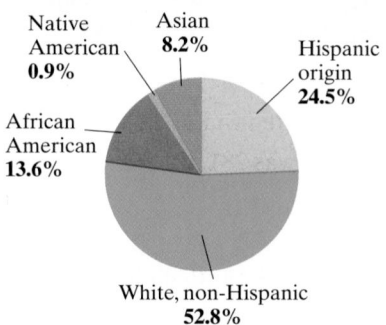

Source: U.S. Census Bureau

FIGURE 1.2

SOLUTION

a. The sector labeled "White, non-Hispanic" shows that 52.8% of the U.S. population for the year 2050 will be white, non-Hispanic. This group will be the largest racial group in 2050. (To put this number into perspective, in the year 2000, whites made up 75.1% of the U.S. population.)

b. The circle graph indicates that 24.5% of the population will be Hispanic in 2050. The actual number of Hispanics is 24.5% of the total 393,931,000.

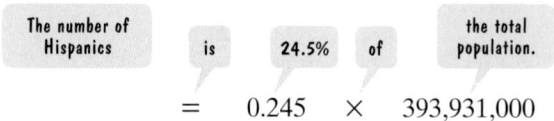

$$= \quad 0.245 \quad \times \quad 393{,}931{,}000$$

Because 24.5% can be rounded to 25%, which is $\frac{1}{4}$ of the total, we may obtain an estimate as follows:

$$\begin{array}{ccc} 393{,}931{,}000 & \rightarrow & 400{,}000{,}000 \\ \times \qquad 0.245 & \rightarrow & \times \qquad \frac{1}{4} \end{array} : \frac{400{,}000{,}000}{4} = 100{,}000{,}000.$$

An estimate of the Hispanic population for the year 2050 is 100 million. A more accurate prediction can be found by multiplying 0.245 and 393,931,000 with a calculator. The product is 96,513,095, or almost 97 million. Our estimate is close to the actual number.

Check 6 Point

Using Figure 1.2, answer the following:

a. What information is shown by the sector labeled "African American"?

b. If the U.S. population for the year 2050 is expected to be 393,931,000, estimate the population for people of Asian descent.

Bar graphs are convenient for showing comparisons among items. The bars may be either horizontal or vertical, and their heights or lengths are used to show the amounts of different items. Figure 1.3 is an example of a typical bar graph. The graph shows the number of cars per 100 people in ten selected countries.

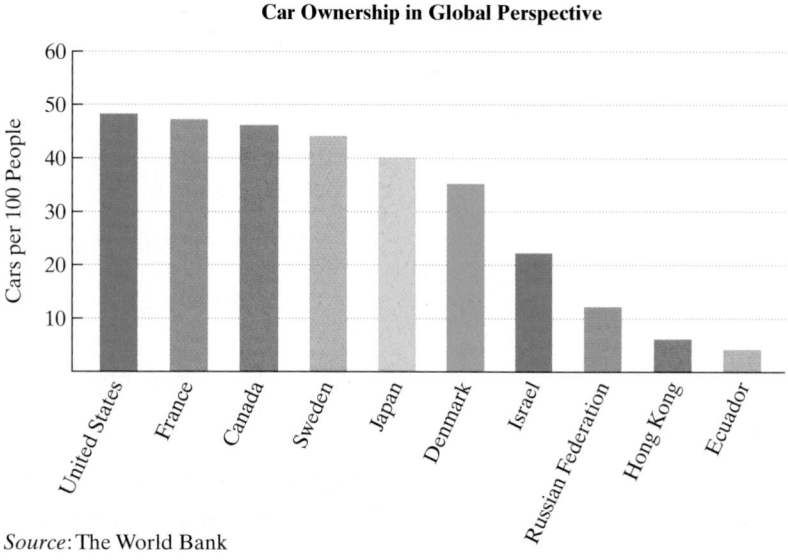

Source: The World Bank

FIGURE 1.3

EXAMPLE 7 | APPLYING ESTIMATION TECHNIQUES TO A BAR GRAPH

Using Figure 1.3, answer the following:

a. Estimate the number of cars per 100 people in the United States.

b. The population of the United States for the year 2000 was 281,421,906. Estimate the number of cars in the country at that time.

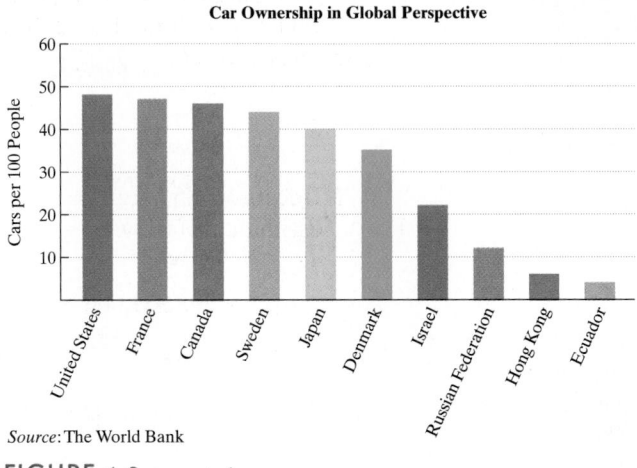

Car Ownership in Global Perspective

Source: The World Bank

FIGURE 1.3, repeated

SOLUTION

a. To estimate the number of cars per 100 people in the United States, look at the top of the bar representing the United States and then read the cars-per-100-people scale on the vertical axis. The bar extends more than midway between 40 and 50 but is less than 50 by a few units. It appears that 48 is a reasonable estimate. Thus, there are approximately 48 cars per 100 people in the United States.

b. The number of cars in the United States for the year 2000 can be determined as follows:

U.S. population

There are 48 cars per 100 people.

$$\text{Number of cars} = \frac{281,421,906}{100} \times 48$$

Divide by 100 to find the number of groups of 100 people.

To estimate the number of cars, we round the population to 280,000,000 and the cars per 100 people to 50.

$$\text{Number of cars} \approx \frac{280,000,000}{100} \times 50$$
$$= 2,800,000 \times 50$$
$$= 140,000,000$$

In 2000, there were approximately 140,000,000 (140 million) cars in the United States.

Check 7 Point

Using Figure 1.3, answer the following:

a. Estimate the number of cars per 100 people in Israel.

b. The population of Israel for the year 2000 was 6,040,000. Estimate the number of cars in the country at that time.

Line graphs are often used to illustrate trends over time. Some measure of time, such as months or years, frequently appears on the horizontal axis. Amounts are generally listed on the vertical axis. Points are drawn to represent the given information. The graph is formed by connecting the points with line segments.

Figure 1.4 is an example of a typical line graph. The graph shows the average age at which women in the United States married for the first time over a 110-year period. The years are listed on the horizontal axis and the ages are listed on the vertical axis. The symbol ⌇ on the vertical axis shows that there is a break in values between 0 and 20. Thus, the first tick mark on the vertical axis represents an average age of 20.

Figure 1.4 shows how to find the average age at which women married for the first time in 1980.

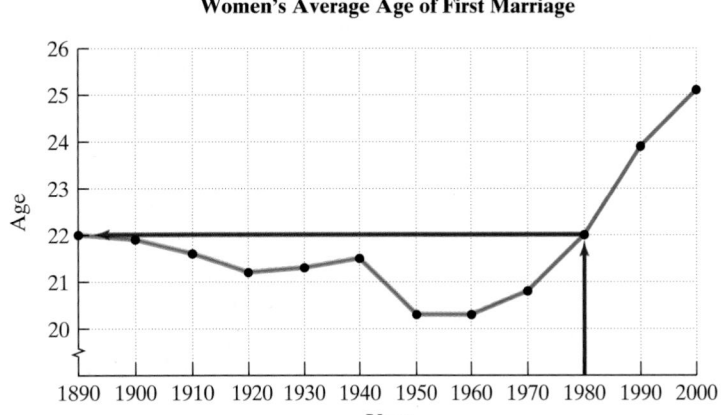

Women's Average Age of First Marriage

Source: U.S. Census Bureau

FIGURE 1.4   Average age at which U.S. women married for the first time

STEP 1. Locate 1980 on the horizontal axis.

STEP 2. Locate the point above 1980.

STEP 3. Read across to the corresponding age on the vertical axis.

The age is 22. Thus, in 1980, women in the United States married for the first time at an average age of 22.

EXAMPLE 8 | APPLYING ESTIMATION TECHNIQUES TO A LINE GRAPH

The line graph in Figure 1.5 shows the percentage of federal prisoners in the United States sentenced for drug offenses from 1970 through 2001.

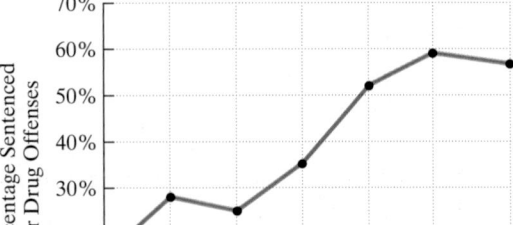

Percentage of U.S. Federal Prisoners Sentenced for Drug Offenses

Source: Frank Schmalleger, *Criminal Justice Today, Seventh Edition*, Prentice Hall, 2003

FIGURE 1.5

TABLE 1.2 NUMBER OF U.S. FEDERAL PRISONERS

Year	Federal Prisoners
1990	58,838
1995	89,538
2000	131,496
2001	140,741

Source: Bureau of Justice Statistics

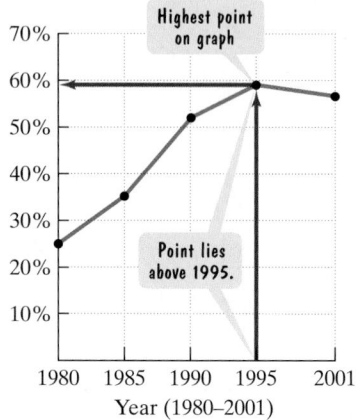

FIGURE 1.6

a. For the period shown, estimate the maximum percentage of federal prisoners sentenced for drug offenses. When did this occur?

b. Table 1.2 shows the number of federal prisoners in the United States for four selected years. Estimate the number of federal prisoners sentenced for drug offenses for the year in part (a).

SOLUTION

a. See Figure 1.6 . The maximum percentage of federal prisoners sentenced for drug offenses can be found by locating the highest point on the graph. This point lies above 1995 on the horizontal axis. Read across to the corresponding percent on the vertical axis. The number falls below 60% by approximately one unit. It appears that 59% is a reasonable estimate. Thus, the maximum percentage of federal prisoners sentenced for drug offenses is approximately 59%. This occurred in 1995.

b. Table 1.2 shows that there were 89,538 federal prisoners in 1995. The number sentenced for drug offenses can be determined as follows:

$$\text{Number sentenced for drug offenses} = \underbrace{0.59}_{59\% \text{ of}} \times \underbrace{89,538}_{\text{the total population}}$$

To estimate the number sentenced for drug offenses, we round the percent to 60% and the total federal prison population to 90,000.

$$\text{Number sentenced for drug offenses} \approx 60\% \text{ of } 90,000$$
$$= 0.6 \times 90,000$$
$$= 54,000$$

In 1995, approximately 54,000 federal prisoners were sentenced for drug offenses.

Check 8 Point

Use Figure 1.5 and Table 1.2 to estimate the number of federal prisoners sentenced for drug offenses in 2001.

Exercise Set 1.2

Practice Exercises

To help you learn estimation, we suggest that you work this exercise set without a calculator. Different rounding results in different estimates, so there is not one single correct answer to these exercises. Use rounding to make the resulting calculations simple.

1. Estimate the cost of 12 bananas at $0.19 each.

2. Estimate the total distance run by all 297 runners who each complete a 19-mile marathon.

3. Estimate the distance traveled when driving 48 miles per hour for 7 hours, 8 minutes.

4. Estimate the amount earned by a worker who works for 32.7 hours at $8.95 per hour.

5. Estimate the total cost of six grocery items if their prices are $3.47, $5.89, $19.98, $2.03, $11.85, and $0.23.

6. Estimate the total weight of four people if their individual weights are 137 pounds, 146 pounds, 172 pounds, and 197 pounds.

7. Estimate the tax on a computer that sells for $2037 with a 5% sales tax.

8. If 21% of the population suffers from depression, estimate the number of people who suffer from this disorder in a city with 585,000 residents.

9. If a person earns $19.50 per hour, estimate that person's annual income.

10. If a person earns $9.87 per hour, estimate that person's annual income.

11. Find an estimate of $\dfrac{0.57 \times 68}{0.493}$.

12. Find an estimate of $17\frac{7}{8} \div 2\frac{11}{12}$.

13. Estimate the number of hours in a year.

14. Estimate the number of hours in a decade (ten years).

15. The average life expectancy in Greece is 78 years. Estimate the country's life expectancy in hours.

16. The average life expectancy in Liberia is 39 years. Estimate the country's life expectancy in hours.

17. If a person who works 40 hours per week earns $61,500 per year, estimate that person's hourly wage.

18. If a person who works 40 hours per week earns $38,950 per year, estimate that person's hourly wage.

19. A raise of $310,000 is evenly distributed among 294 professors. Estimate the amount each professor receives.

20. A raise of $310,000 is evenly distributed among 196 professors. Estimate the amount each professor receives.

21. You lease a car at $605 per month for 3 years. Estimate the total cost of the lease.

22. You lease a car at $415 per month for 4 years. Estimate the total cost of the lease.

23. You spend $21.36 for a meal. If you want to leave a 15% tip, estimate the amount of the tip.

24. You spend $28.70 for a meal. If you want to leave a 15% tip, estimate the amount of the tip.

Use this map that shows some of the cities on the main line of a railroad to answer Exercises 25–26.

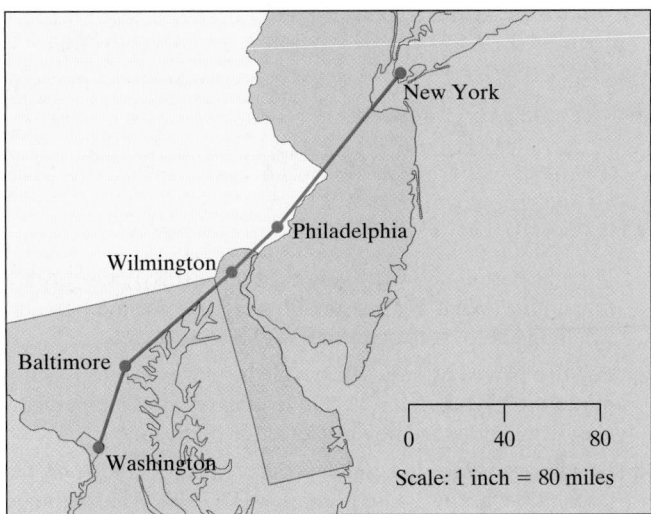

25. **a.** By rounding to the nearest inch, estimate the distance from New York to Washington.

   b. If a train travels at an average rate of 40 miles per hour, estimate the traveling time from New York to Washington.

26. **a.** Estimate the distance from Wilmington to Philadelphia.

   b. If a train travels at an average rate of 40 miles per hour, estimate the travel time from Wilmington to Philadelphia.

Application Exercises

The circle graph shows the percentage of Americans who belong to various religious groups and who are members of a congregation. The population of the United States in 2001 was 283,882,604. Use the graph to solve Exercises 27–28.

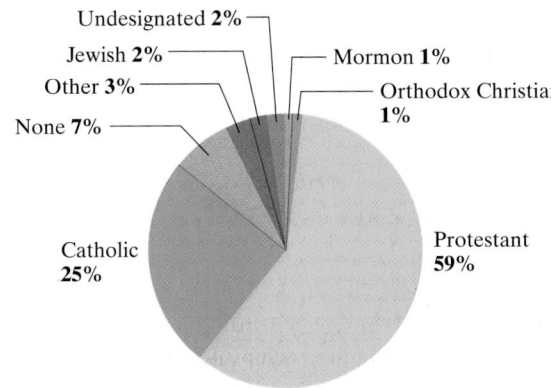

Religious Affiliation

Source: The Gallup Organization

27. Estimate the number of Protestants.

28. Estimate the number of Catholics.

In 2000, there were 9976 hate crimes reported to the FBI in the United States. The circle graph shows the motivation for these reported incidents. Use the graph to solve Exercises 29–30.

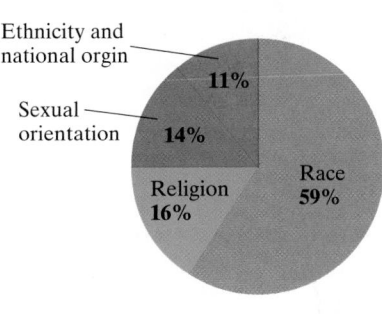

Motivation For U.S. Hate-Crime Incidents

Source: FBI

29. Estimate the number of hate-crime incidents that were motivated by race.

30. Estimate the number of hate-crime incidents that were motivated by sexual orientation.

The bar graph shows the percentage of vacations that included the activity listed on the left. Use the graph to solve Exercises 31–34.

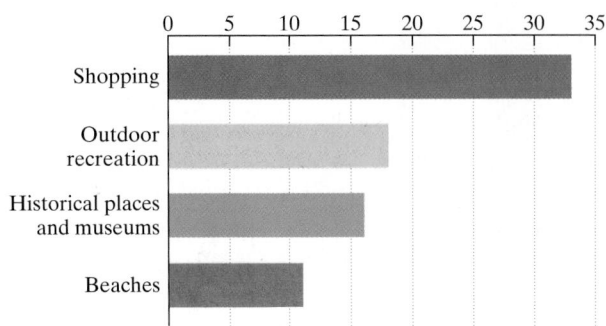

Percentage of U.S. Vacations That Included the Activity

Source: Travel Industry Association of America

31. Estimate the percentage of vacations that include shopping.

32. Estimate the percentage of vacations that include beaches.

33. Which activities are included on more than 15% of vacations?

34. Which activities are included on at least 14% of vacations and at most 30%?

The bar graph shows life expectancy in the United States by year of birth. Use the graph to solve Exercises 35–38.

Life Expectancy in the U.S. by Birth Year

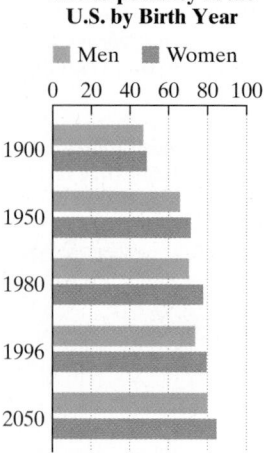

Men   Women

Source: National Center for Health Statistics, U.S. Census Bureau

35. Estimate the life expectancy for men born in 1900.

36. Estimate the life expectancy for women born in 2050.

37. By approximately how many more years can women born in 1980 expect to live as compared to men born in the same year?

38. By approximately how many more years can women born in 1996 expect to live as compared to men born in the same year?

The bar graph shows the average yearly amount spent in U.S. households buying sports gear for the sports listed on the horizontal axis. Use this graph to solve Exercises 39–42.

The Most and Least Expensive Sports in the U.S.

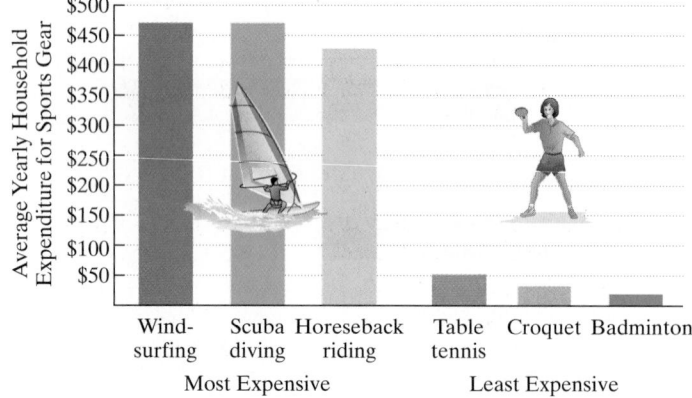

Most Expensive   Least Expensive

Source: National Sporting Goods Association

39. Estimate the amount spent buying sports gear for scuba diving.

40. Estimate the amount spent buying sports gear for horseback riding.

41. Approximately how much more is spent buying sports gear for table tennis than for croquet?

42. Approximately how much more is spent buying sports gear for table tennis than for badminton?

The poverty rate in the United States is the percentage of the population whose income falls below the government's official poverty level, which is adjusted each year for inflation. The line graph shows the poverty rate in the United States for five-year intervals from 1960 through 2000, and for 2001. Use the graph to solve Exercises 43–46.

U.S. Poverty Rate

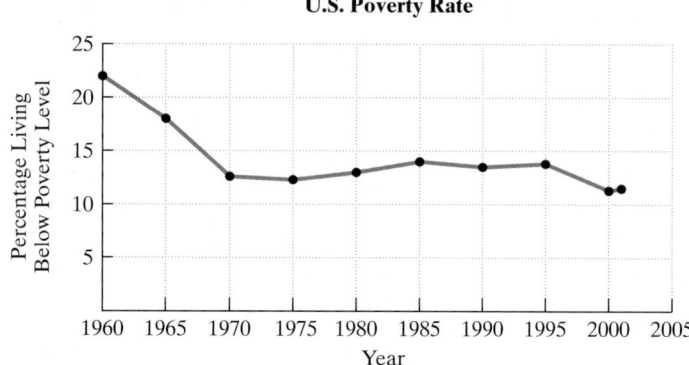

Source: U.S. Census Bureau

43. For the period shown, when did the poverty rate reach a maximum? What is a reasonable estimate for the percentage living below the poverty level for that year?

44. For the period shown, when did the poverty rate reach a minimum? What is a reasonable estimate for the percentage living below the poverty level for that year?

45. For the year of the maximum poverty rate in Exercise 43, the population of the United States was 179,323,175. Estimate the number of Americans living below the poverty level at that time.

46. For the year of the minimum poverty rate in Exercise 44, the population of the United States was 281,421,906. Estimate the number of Americans living below the poverty level at that time.

One effect of the presidential election of 2000 was to convince Americans that their votes count. Although turnout is crucial in any democracy, since the 1960s, the percentage of Americans who actually vote has been decreasing. Use the graph showing the percentage of Americans who voted in presidential elections from 1828 through 2000 to solve Exercises 47–50. Each tick mark on the horizontal axis represents 4 years.

Percentage of Americans Voting in Presidential Elections

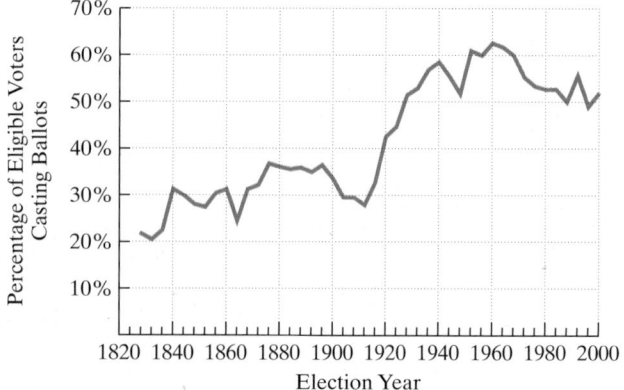

Source: National Journal

47. During which year was the percentage of Americans voting the lowest? What percentage of Americans voted in that year?

48. During which year was the percentage of Americans voting the highest? What percentage, to the nearest percent, voted in that year?

49. Approximately what percentage of Americans voted in the 2000 presidential election? Round to the nearest percent.

50. Approximately what percentage of Americans voted in the 1828 presidential election? Round to the nearest percent.

Writing in Mathematics

51. Explain how to round 163
  a. to the nearest ten.
  b. to the nearest hundred.

52. What is estimation? When is it helpful to use estimation?

53. What does the $\approx$ symbol mean?

54. In this era of calculators and computers, why is there a need to develop estimation skills?

55. Describe a circle graph.

56. What is a sector?

57. Describe a bar graph.

58. Describe a line graph.

59. Describe one way in which you use estimation in a nonacademic area of your life.

60. A forecaster at the National Hurricane Center needs to estimate the time until a hurricane with high probability of striking South Florida will hit Miami. Is it better to overestimate or underestimate? Explain your answer.

Critical Thinking Exercises

61. About the size of New Jersey, Israel has seen its population soar to more than 6 million since it was established. With the help of U.S. aid, the country now has a diversified economy rivaling those of other developed Western nations. By contrast, the Palestinians, living under Israeli occupation and a corrupt regime, endure bleak conditions. The graphs show that by 2050, Palestinians in the West Bank, Gaza Strip, and East Jerusalem will outnumber Israelis. Use the graphs to estimate the year in which the two populations will be the same. Then estimate each population at that time.

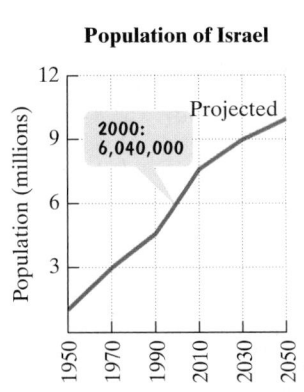

Population of Israel

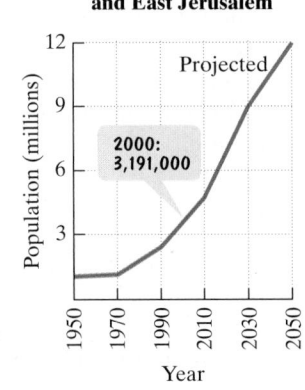

Palestinian Population in West Bank, Gaza, and East Jerusalem

Source: Newsweek

62. Find an estimate of

$$\frac{24,000 \times 5124}{14,730}.$$

63. Estimate the height, in miles, of a stack of 1 million dimes.

64. If you spend $1000 each day, estimate how long it will take to spend a billion dollars.

Technology Exercise

65. Using a spreadsheet, enter the following utility bill data into two columns. The first column should be labeled according to the month. The second column should contain the dollar amount for that month's bill. January's bill was $37.54, February's was $56.98, March's was $45.17, April's was $18.99, May's was $23.13, June's was $31.51, and July's bill was $26.80.

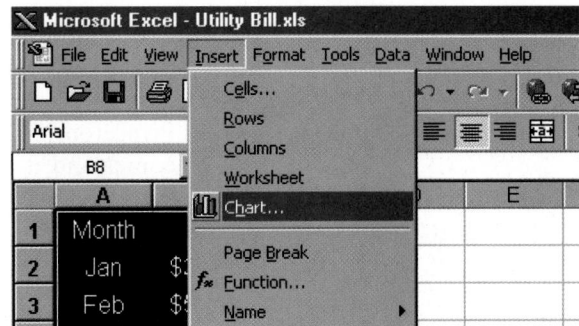

Once the data have been entered, use the computer's mouse to highlight the cells containing the data. Then activate the Chart Option of the spreadsheet by clicking on the Insert pull-down and select Chart (see image). Many types of charts are available. Of these, we will examine the Column, Line, and Pie charts that correspond to the given set of data. Upon selecting a particular type of chart, you can click the "Next" button. Work your way through the options, including the labels and cosmetic effects for each graph. Once you have decided upon all of the options (don't worry, you can go back and change them), click the "Finish" button.

  a. Create a Column Chart for the indicated data. Describe any advantages of looking at a utility bill with this type of chart.

  b. Create a Line Chart for the indicated data. Describe any advantages of looking at a utility bill with this type of chart.

  c. Create a Pie Chart for the indicated data. Describe any advantages of looking at a utility bill with this type of chart.

Group Exercise

66. Group members should devise an estimation process that can be used to answer each of the following questions. Use input from all group members to describe the best estimation process possible.

  a. Is it possible to walk from San Francisco to New York in a year?

  b. How much money is spent on ice cream in the United States each year?

SECTION 1.3   PROBLEM SOLVING

OBJECTIVE

1. Solve problems using the organization of the four-step problem-solving process.

Critical thinking and problem solving are essential skills in both school and work. A model for problem solving was established by the charismatic teacher and mathematician George Polya (1887–1985) in *How to Solve It* (Princeton University Press, Princeton, N.J., 1957). This book, first published in 1945, has sold more than 1 million copies and is available in 17 languages. Using a four-step procedure for problem solving, Polya's book demonstrates how to think clearly in any field.

 Solve problems using the organization of the four-step problem-solving process.

STUDY TIP

Problem solving can be thought of in terms of Polya's four steps. Think of these steps as guidelines that will help you organize the process of problem solving, rather than a list of rigid rules that need to be memorized. You may be able to solve certain problems without thinking about or using every step in the four-step process.

POLYA'S FOUR STEPS IN PROBLEM SOLVING

STEP 1. Understand the problem. Read the problem several times. The first reading can serve as an overview. In the second reading, write down what information is given and determine exactly what it is that the problem requires you to find.

STEP 2. Devise a plan. The plan for solving the problem might involve one or more of these suggested problem-solving strategies:

- Use inductive reasoning to look for a pattern.
- Make a systematic list or a table.
- Use estimation to make an educated guess at the solution. Check the guess against the problem's conditions and work backward to eventually determine the solution.
- Try expressing the problem more simply and solve a similar simpler problem.
- Use trial and error.
- List the given information in a chart or table.
- Try making a sketch or a diagram to illustrate the problem.
- Relate the problem to similar problems that you have seen before. Try applying the procedures used to solve the similar problem to the new one.
- Look for a "catch" if the answer seems too obvious. Perhaps the problem involves some sort of trick question deliberately intended to lead the problem solver in the wrong direction.
- Use the given information to eliminate possibilities.
- Use common sense.

STEP 3. Carry out the plan and solve the problem.

STEP 4. Look back and check the answer. The answer should satisfy the conditions of the problem. The answer should make sense and be reasonable. If this is not the case, recheck the method and any calculations. Perhaps there is an alternate way to arrive at a correct solution.

"If you don't know where you're going, you'll probably end up some place else."

YOGI BERRA

The very first step in problem solving involves evaluating the given information in a deliberate manner. Is there enough given to solve the problem? Is the information relevant to the problem's solution, or are some facts not necessary to arrive at a solution?

EXAMPLE 1   FINDING WHAT IS MISSING

Which necessary piece of information is missing and prevents you from solving the following problem?

> A man purchased five shirts, each at the same discount price. How much did he pay for them?

SOLUTION

STEP 1. Understand the problem. Here's what is given:

Number of shirts purchased: 5

We must find how much the man paid for the five shirts.

STEP 2. Devise a plan. The amount that the man paid for the five shirts is the number of shirts, 5, times the cost of each shirt. The discount price of each shirt is not given. This missing piece of information makes it impossible to solve the problem.

Which necessary piece of information is missing and prevents you from solving the following problem?
The bill for your meal totaled $20.36, including the tax. How much change should you receive from the cashier?

EXAMPLE 2   FINDING WHAT IS UNNECESSARY

In the following problem, one more piece of information is given than is necessary for solving the problem. Identify this unnecessary piece of information. Then solve the problem.

> A roll of E-Z Wipe paper towels contains 100 sheets and costs $1.38. A comparable brand, Kwik-Clean, contains five dozen sheets per roll and costs $1.23. If you need three rolls of paper towels, which brand is the better value?

SOLUTION

STEP 1. Understand the problem. Here's what is given:

E-Z Wipe: 100 sheets per roll; $1.38
Kwik-Clean: 5 dozen sheets per roll; $1.23
Needed: 3 rolls.

We must determine which brand offers the better value.

STEP 2. Devise a plan. The brand with the better value is the one that has the lower price per sheet. Thus, we can compare the two brands by finding the cost for one sheet of E-Z Wipe and one sheet of Kwik-Clean. The price per sheet, or the *unit price*, is the price of a roll divided by the number of sheets in the roll. The fact that three rolls are required is not relevant to the problem. This unnecessary piece of information is not needed to find which brand is the better value.

STEP 3. Carry out the plan and solve the problem.

E-Z Wipe:

$$\text{price per sheet} = \frac{\text{price of a roll}}{\text{number of sheets in a roll}}$$

$$= \frac{\$1.38}{100 \text{ sheets}} = \$0.0138 \approx \$0.01$$

Kwik-Clean:

$$\text{price per sheet} = \frac{\text{price of a roll}}{\text{number of sheets in a roll}}$$

$$= \frac{\$1.23}{60 \text{ sheets}} = \$0.0205 \approx \$0.02$$

> 5 dozen = 5 X 12, or 60 sheets

By comparing unit prices, we see that E-Z Wipe, at approximately $0.01 per sheet, is the better value.

STEP 4. Look back and check the answer. We can double-check the arithmetic in each of our unit-price computations. We can also see if these unit prices satisfy the problem's conditions. The product of each brand's price per sheet and the number of sheets per roll should result in the given price for a roll.

E-Z Wipe: Check $0.0138   Kwik-Clean: Check $0.0205

$0.0138 × 100 = $1.38 $0.0205 × 60 = $1.23

> These are the given prices for a roll of each respective brand.

The unit prices satisfy the problem's conditions.

A generalization of our work in Example 2 allows you to compare different brands and make a choice between various products of different sizes. When shopping at the supermarket, a useful number to keep in mind is a product's *unit price*. The **unit price** is the total price divided by the total units. Among comparable brands, the best value is the product with the lowest unit price, assuming that the units are kept uniform.

The word *per* is used to state unit prices. For example, if a 12-ounce box of cereal sells for $3.00, its unit price is determined as follows:

$$\text{Unit price} = \frac{\text{total price}}{\text{total units}} = \frac{\$3.00}{12 \text{ ounces}} = \$0.25 \text{ per ounce.}$$

 **Check 2 Point** Solve the following problem. If the problem contains information that is not relevant to its solution, identify this unnecessary piece of information.

A manufacturer packages its apple juice in bottles and boxes. A 128-ounce bottle costs $5.39 and a 9-pack of 6.75-ounce boxes costs $3.15. Which packaging option is the better value?

EXAMPLE 3 | APPLYING THE FOUR-STEP PROCEDURE

By paying $100 cash up front and the balance at $20 a week, how long will it take to pay for a bicycle costing $680?

SOLUTION

STEP 1. Understand the problem. Here's what is given:

Cost of the bicycle: $680

Amount paid in cash: $100

Weekly payments: $20.

If necessary, consult a dictionary to look up any unfamiliar words. The word "balance" means the amount still to be paid. We must find the balance to determine the number of weeks required to pay off the bicycle.

STEP 2. Devise a plan. Subtract the amount paid in cash from the cost of the bicycle. This results in the amount still to be paid. Because weekly payments are $20, divide the amount still to be paid by 20. This will give the number of weeks required to pay for the bicycle.

STEP 3. Carry out the plan and solve the problem. Begin by finding the balance, the amount still to be paid for the bicycle.

$$\begin{array}{rl} \$680 & \text{cost of the bicycle} \\ -\$100 & \text{amount paid in cash} \\ \hline \$580 & \text{amount still to be paid} \end{array}$$

Now divide the $580 balance by $20, the payment per week. The result of the division is the number of weeks needed to pay off the bicycle.

$$\frac{\dfrac{\$580}{\text{week}}}{\dfrac{\$20}{\text{week}}} = \$580 \times \frac{\text{week}}{\$20} = \frac{580 \text{ weeks}}{20} = 29 \text{ weeks}$$

It will take 29 weeks to pay for the bicycle.

STEP 4. Look back and check the answer. We can certainly double-check the arithmetic either by hand or with a calculator. We can also see if the answer satisfies the conditions of the problem.

answer ⟶
$$\begin{array}{rl} \$20 & \text{weekly payment} \\ \times\ 29 & \text{number of weeks} \\ \hline \$580 & \text{total of weekly} \\ & \text{payments} \end{array} \qquad \begin{array}{rl} \$580 & \text{total of weekly payments} \\ +\$100 & \text{amount paid in cash} \\ \hline \$680 & \text{cost of bicycle} \end{array}$$

The answer of 29 weeks satisfies the condition that the cost of the bicycle is $680.

 Check 3 Point By paying $350 cash up front and the balance at $45 per month, how long will it take to pay for a computer costing $980?

Making lists is a useful strategy in problem solving.

EXAMPLE 4 SOLVING A PROBLEM BY MAKING A LIST

Suppose you are an engineer programming the automatic gate for a 50-cent toll. The gate should accept exact change only. It should not accept pennies. How many coin combinations must you program the gate to accept?

SOLUTION

STEP 1. Understand the problem. The total change must always be 50 cents. One possible coin combination is two quarters. Another is five dimes. We need to find all such combinations.

STEP 2. Devise a plan. Make a list of all possible coin combinations. Begin with the coins of larger value and work toward the coins of smaller value.

STEP 3. Carry out the plan and solve the problem. First we must find all of the coins that are not pennies but can combine to form 50 cents. This includes half-dollars, quarters, dimes, and nickels. Now we can set up a table. We will use these coins as table headings.

STUDY TIP

An effective way to see if you understand a problem is to re-state the problem in your own words.

Half-Dollars	Quarters	Dimes	Nickels

Each row in the table will represent one possible combination for exact change. We start with the largest coin, the half-dollar. Only one half-dollar is needed to make exact change. No other coins are needed. Thus, we put a 1 in the half-dollars column and 0 in the other columns to represent the first possible combination.

Half-Dollars	Quarters	Dimes	Nickels
1	0	0	0

Likewise, two quarters are also exact change for 50 cents. We put a 0 in the half-dollars column, a 2 in the quarters column, and 0s in the columns for dimes and nickels.

Half-Dollars	Quarters	Dimes	Nickels
1	0	0	0
0	2	0	0

In this manner, we can find all possible combinations for exact change for the 50-cent toll. These combinations are shown in Table 1.3.

TABLE 1.3 EXACT CHANGE FOR 50 CENTS: NO PENNIES

Half-Dollars	Quarters	Dimes	Nickels
1	0	0	0
0	2	0	0
0	1	2	1
0	1	1	3
0	1	0	5
0	0	5	0
0	0	4	2
0	0	3	4
0	0	2	6
0	0	1	8
0	0	0	10

Count the coin combinations shown in Table 1.3. How many coin combinations must the gate accept? You must program the gate to accept 11 coin combinations.

STEP 4. Look back and check the answer. Double-check Table 1.3 to make sure that no possible combinations have been omitted and that the total in each row is 50 cents. Double-check your count of the number of combinations.

Check 4 Point

Suppose you are an engineer programming the automatic gate for a 30-cent toll. The gate should accept exact change only. It should not accept pennies. How many coin combinations must you program the gate to accept?

Sketches and diagrams are sometimes useful in problem solving.

EXAMPLE 5 SOLVING A PROBLEM BY USING A DIAGRAM

Four runners are in a one-mile race: Maria, Aretha, Thelma, and Debbie. Points are awarded only to the women finishing first or second. The first-place winner gets more points than the second-place winner. How many different arrangements of first- and second-place winners are possible?

SOLUTION

STEP 1. Understand the problem. Three possibilities for first and second position are

> Maria-Aretha
>
> Maria-Thelma
>
> Aretha-Maria

Notice that Maria finishing first and Aretha finishing second is a different outcome than Aretha finishing first and Maria finishing second. Order makes a difference because the first-place winner gets more points than the second-place winner. We must find all possibilities for first and second position.

STEP 2. Devise a plan. If Maria finishes first, then each of the other three runners could finish second:

First place	Second place	Possibilities for first and second place
Maria	Aretha	Maria-Aretha
	Thelma	Maria-Thelma
	Debbie	Maria-Debbie

Similarly, we can list each woman as the possible first-place runner. Then we will list the other three women as possible second-place runners. Next we will determine the possibilities for first and second place. This diagram will show how the runners can finish first or second.

STEP 3. Carry out the plan and solve the problem. Now we complete the diagram started in step 2. The diagram is shown in Figure 1.7.

> Because of the way Figure 1.7 branches from first to second place, it is called a **tree diagram**. We will be using tree diagrams in Chapter 11 as a problem-solving tool in the study of uncertainty and probability.

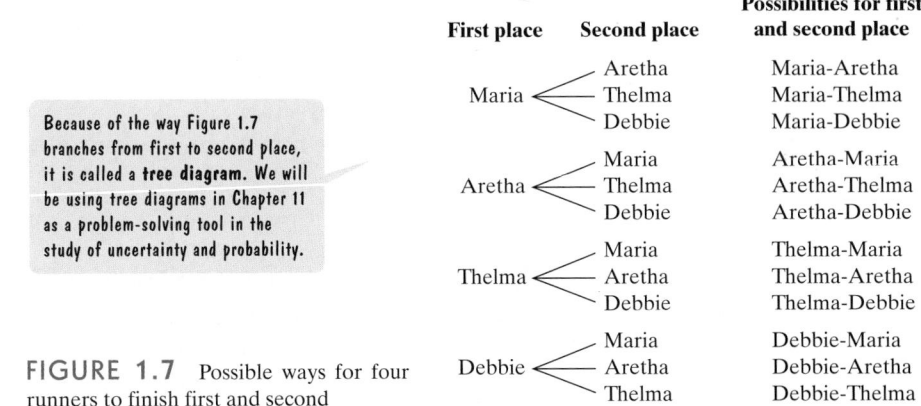

FIGURE 1.7   Possible ways for four runners to finish first and second

Count the number of possibilities shown under the third column, "Possibilities for first and second place." Can you see that there are 12 possibilities? Therefore, 12 different arrangements of first- and second-place winners are possible.

STEP 4. Look back and check the answer. Check the diagram in Figure 1.7 to make sure that no possible first- and second-place outcomes have been left out. Double-check your count for the winning pairs of runners.

Check Point 5   Your "lecture wardrobe" is rather limited—just two pairs of jeans to choose from (one blue, one black) and three T-shirts to choose from (one beige, one yellow, and one blue). How many different outfits can you form?

In Chapter 15, we will be studying diagrams, called *graphs*, that provide structures for describing relationships. In Example 6, we use such a diagram to illustrate the relationship between cities and their one-way airfares.

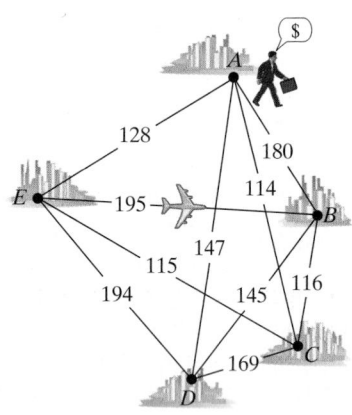

FIGURE 1.8

EXAMPLE 6 **USING A REASONABLE OPTION TO SOLVE A PROBLEM WITH MORE THAN ONE SOLUTION**

A sales director who lives in city A is required to fly to regional offices in cities $B, C,$ $D,$ and $E.$ Other than starting and ending the trip in city $A,$ there are no restrictions as to the order in which the other three cities are visited.

The one-way fares between each of the four cities are given in Table 1.4. A diagram that illustrates this information is shown in Figure 1.8.

TABLE 1.4 ONE-WAY AIRFARES

	A	B	C	D	E
A	*	$180	$114	$147	$128
B	$180	*	$116	$145	$195
C	$114	$116	*	$169	$115
D	$147	$145	$169	*	$194
E	$128	$195	$115	$194	*

Give the sales director an order for visiting cities $B, C, D,$ and E once, returning home to city $A,$ for less than $750.

SOLUTION

STEP 1. Understand the problem. There are many ways to visit cities $B, C, D,$ and E once, and return home to $A.$ One route is

$$A, E, D, C, B, A.$$ Fly from A to E to D to C to B and then back to A.

The cost of this trip involves the sum of five costs, shown in both Table 1.4 and Figure 1.8:

$$\$128 + \$194 + \$169 + \$116 + \$180 = \$787.$$

We must find a route that costs less than $750.

STEP 2. Devise a plan. The sales director starts at city $A.$ From there, fly to the city to which the airfare is cheapest. Then from there fly to the next city to which the airfare is cheapest, and so on. From the last of the cities, fly home to city $A.$ Compute the cost of this trip to see if it is less than $750. If it is not, use trial and error to find other possible routes and select an order (if there is one) whose cost is less than $750.

STEP 3. Carry out the plan and solve the problem. See Figure 1.9. The route is indicated using red lines with arrows.

- Start at $A.$
- Choose the line segment with the smallest number: 114. Fly from A to $C.$ (cost: $114)
- From $C,$ choose the line segment with the smallest number that does not lead to A: 115. Fly from C to E (cost: $115)
- From $E,$ choose the line segment with the smallest number that does not lead to a city already visited: 194. Fly from E to $D.$ (cost: $194)
- From $D,$ there is little choice but to fly to $B,$ the only city not yet visited. (cost: $145)
- From $B,$ return home to $A.$ (cost: $180)

The route that we are considering is

$$A, C, E, D, B, A.$$

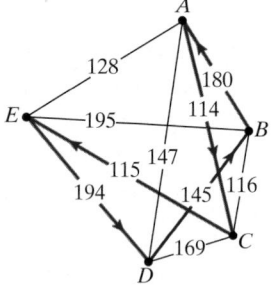

FIGURE 1.9

Let's see if the cost is less than $750. The cost is

$$\$114 + \$115 + \$194 + \$145 + \$180 = \$748$$

Because the cost is less than $750, the sales director can follow the order A, C, E, D, B, A.

STEP 4. Look back and check the answer. Use Table 1.4 or Figure 1.9 verify that the five numbers used in the sum shown above are correct. Use estimation to verify that $748 is a reasonable cost for the trip.

 **Check 6 Point** As in Example 6, a sales director who lives in city A is required to fly to regional offices in cities B, C, D, and E. The diagram in Figure 1.10 shows the one-way airfares between any two cities. Give the sales director an order for visiting cities B, C, D, and E once, returning home to city A, for less than $1460.

George Polya's *How to Solve It* demonstrates how his procedures can be used to methodically approach and solve any problem that involves thought and reasoning. In Example 7 we apply his four steps to a problem that is not mathematical.

EXAMPLE 7 | TAKING THE FIRST STEPS TO SOLVE A PROBLEM THAT IS NOT MATHEMATICAL

You are the department manager of a company. One of your employees who is too valuable to fire arrives to work late nearly every morning. Produce ideas that might help correct this employee's tardiness.

SOLUTION When you apply problem-solving techniques to real-life situations, there is often more than one solution.

STEP 1. Understand the problem. In spite of frequent lateness, this is one of your best employees. You need to correct the tardiness problem without affecting the excellent work produced by this person.

STEP 2. Devise a plan. Write down as many ideas as possible, even if they seem silly or farfetched. Quantity, not quality, is the initial goal. After making an extensive list, select the best suggestions.

STEP 3. Carry out the plan and solve the problem. Start by "brainstorming" and write down as many ideas as possible.

- Serve free coffee and donuts to people who are on time.
- Offer a cash bonus to the worker with the best punctuality record over a three-month period.
- Offer to drive the employee to work on your way in.
- Make promotions possible only with consistent on-time arrivals.
- Tell the employee that working hours will begin 30 minutes earlier. In this way, an employee who arrives 30 minutes late will be on time.
- Move the office closer to the employee's home.
- Tell the employee that arriving on time means being able to work on any chosen creative project.
- Reduce the employee's pay for tardiness.
- Hold important 9 A.M. meetings every morning for which the employee would be embarrassed to arrive late.

In the group exercise on page 36, group members will be asked to replace these suggestions with their own ideas for solving the tardiness problem. To be continued … .

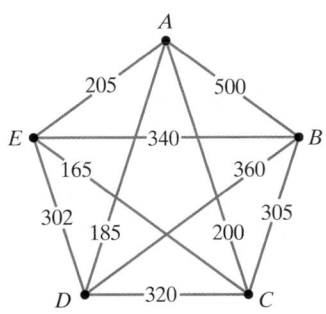

FIGURE 1.10

BLITZER BONUS

Trick Questions

Think about the following questions carefully before answering because each contains some sort of trick or catch.

> Sample: Do they have a fourth of July in England?
>
> Answer: Of course they do. However, there is no national holiday on that date!

See if you can answer the questions that follow without developing mental whiplash. The answers appear in the answer section.

1. A farmer had 17 sheep. All but 12 died. How many sheep does the farmer have left?

2. Some months have 30 days. Some have 31. How many months have 28 days?

3. A doctor had a brother, but this brother had no brothers. What was the relationship between doctor and brother?

4. If you had only one match and entered a log cabin in which there was a candle, a fireplace, and a woodburning stove, which should you light first?

 **Check 7 Point** You started college with your best friend. This semester, you and your friend are taking two classes together. However, your friend frequently misses class and is not doing the necessary homework between classes to succeed. What can you say to your friend, who values your advice and who is in danger of flunking out of college if things continue on their present course? Write four or more short solutions that might be effective in solving the problem.

Becoming more confident as a problem solver will help you as you study the topics throughout this book. Furthermore, you will be able to apply problem-solving techniques to real-life situations. Some people even use these techniques for dealing with personal problems in need of creative solutions. Here's to your continued growth as a problem solver!

Exercise Set 1.3

Everyone can become a better, more confident problem solver. As in learning any other skill, learning problem solving requires hard work and patience. Work as many problems as possible in this exercise set. You may feel confused once in a while, but do not be discouraged. Thinking about a particular problem and trying different methods can eventually lead to new insights. Be sure to check over each answer carefully!

Practice and Application Exercises

In Exercises 1–4, what necessary piece of information is missing that prevents solving the problem?

1. If a student saves $35 per week, how long will it take to save enough money to buy a computer?

2. If a steak sells for $8.15, what is the cost per pound?

3. If it takes you four minutes to read a page in a book, how many words can you read in one minute?

4. By paying $1500 cash and the balance in equal monthly payments, how many months would it take to pay for a car costing $12,495?

In Exercises 5–8, one more piece of information is given than is necessary for solving the problem. Identify this unnecessary piece of information. Then solve the problem.

5. A salesperson receives a weekly salary of $350. In addition, $15 is paid for every item sold in excess of 200 items. How much extra is received from the sale of 212 items?

6. You have $250 to spend and you need to purchase four new tires. If each tire weighs 21 pounds and costs $42 plus $2.50 tax, how much money will you have to spend after buying the tires?

7. A parking garage charges $2.50 for the first hour and $0.50 for each additional hour. If a customer gave the parking attendant $20.00 for parking from 10 A.M. to 3 P.M., how much did the garage charge?

8. An architect is designing a house. The scale on the plan is 1 inch = 6 feet. If the house is to have a length of 90 feet and a width of 30 feet, how long will the line representing the house's length be on the blueprint?

Use Polya's four-step method in problem solving to solve Exercises 9–36.

9. **a.** Which is the better value: a 15.3-ounce box of cereal for $3.37 or a 24-ounce box of cereal for $4.59?

b. The supermarket displays the unit price for the 15.3-ounce box in terms of cost per ounce, but displays the unit price for the 24-ounce box in terms of cost per pound. What are the unit prices, to the nearest cent, given by the supermarket?

c. Based on your work in parts (a) and (b), does the better value always have the lower displayed unit price? Explain your answer.

10. **a.** Which is the better value: a 12-ounce jar of honey for $2.25 or an 18-ounce jar of honey for $3.24?

b. The supermarket displays the unit price for the 12-ounce jar in terms of cost per ounce, but displays the unit price for the 18-ounce jar in terms of cost per quart. Assuming 32 ounces in a quart, what are the unit prices, to the nearest cent, given by the supermarket?

c. Based on your work in parts (a) and (b), does the better value always have the lower displayed unit price? Explain your answer.

11. One person earns $48,000 per year. Another earns $3750 per month. How much more does the first person earn in a year than the second?

12. At the beginning of a year, the odometer on a car read 25,124 miles. At the end of the year, it read 37,364 miles. If the car averaged 24 miles per gallon, how many gallons of gasoline did it use during the year?

13. A television sells for $750. Instead of paying the total amount at the time of the purchase, the same television can be bought by paying $100 down and $50 a month for 14 months. How much is saved by paying the total amount at the time of the purchase?

14. In a basketball game, the Bulldogs scored 34 field goals, each counting 2 points, and 13 foul goals, each counting 1 point. The Panthers scored 38 field goals and 8 foul goals. Which team won? By how many points did it win?

15. Calculators were purchased at $65 per dozen and sold at $20 for three calculators. Find the profit on six dozen calculators.

16. Pens are bought at 95¢ per dozen and sold in groups of four for $2.25. Find the profit on 15 dozen pens.

17. Each day a small business owner sells 200 pizza slices at $1.50 per slice and 85 sandwiches at $2.50 each. If business expenses come to $60 per day, what is the owner's profit for a ten-day period?

18. A college tutoring center pays math tutors $5.15 per hour. Tutors earn an additional $1.20 per hour for each hour over 40 hours per week. A math tutor worked 42 hours one week and 45 hours the second week. How much did the tutor earn in this two-week period?

19. A car rents for $220 per week plus $0.25 per mile. Find the rental cost for a two-week trip of 500 miles for a group of three people.

20. A college graduate receives a salary of $2750 a month for her first job. During the year she plans to spend $4800 for rent, $8200 for food, $3750 for clothing, $4250 for household expenses, and $3000 for other expenses. With the money that is left, she expects to buy as many shares of stock at $375 per share as possible. How many shares will she be able to buy?

21. Charlene decided to ride her bike from her home to visit her friend Danny. Three miles away from home, her bike got a flat tire, and she had to walk the remaining two miles to Danny's home. She could not repair the tire and had to walk all the way back home. How many more miles did Charlene walk than she rode?

22. A store received 200 containers of juice to be sold by April 1. Each container cost the store $0.75 and sold for $1.25. The store signed a contract with the manufacturer in which the manufacturer agreed to a $0.50 refund for every container not sold by April 1. If 150 containers were sold by April 1, how much profit did the store make?

23. A storeowner ordered 25 calculators that cost $30 each. The storeowner can sell each calculator for $35. The storeowner sold 22 calculators to customers. He had to return 3 calculators and pay a $2 charge for each returned calculator. Find the storeowner's profit.

24. New York City and Washington D.C. are about 240 miles apart. A car leaves New York City at noon traveling directly south toward Washington D.C. at 55 miles per hour. At the same time and along the same route, a second car leaves Washington D.C. bound for New York City traveling directly north at 45 miles per hour. How far has each car traveled when the drivers meet for lunch at 2:24 P.M.?

25. A vending machine accepts nickels, dimes, and quarters. Exact change is needed to make a purchase. How many ways can a person with five nickels, three dimes, and two quarters make a 45-cent purchase from the machine?

26. How many ways can you make change for a quarter using only pennies, nickels, and dimes?

27. If a test has four true/false questions, in how many ways can there be three answers that are false and one answer that is true?

28. There are five people in a room. Each person shakes the hand of every other person exactly once. How many handshakes are exchanged?

29. Five runners, Andy, Beth, Caleb, Darnell, and Ella, are in a one-mile race. Andy finished the race seven seconds before Caleb. Caleb finished the race two seconds before Beth. Beth finished the race six seconds after Darnell. Ella finished the race eight seconds after Darnell. In which order did the runners finish the race?

30. Eight teams are competing in a volleyball tournament. Any team that loses a game is eliminated from the tournament. How many games must be played to determine the tournament winner?

In Exercises 31–32, you have three errands to run around town, although in no particular order. You plan to start and end at home. You must go to the bank, the post office, and the dry cleaners. Distances, in miles, between any two of these locations are given in the diagram.

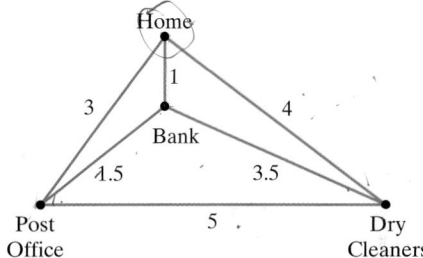

31. Determine a route whose distance is less than 12 miles for running the errands and returning home.

32. Determine a route whose distance exceeds 12 miles for running the errands and returning home.

33. The map shows five western states. Trace a route on the map that crosses each common state border exactly once.

34. The layout of a city with land masses and bridges is shown. Trace a route that shows people how to walk through the city so as to cross each bridge exactly once.

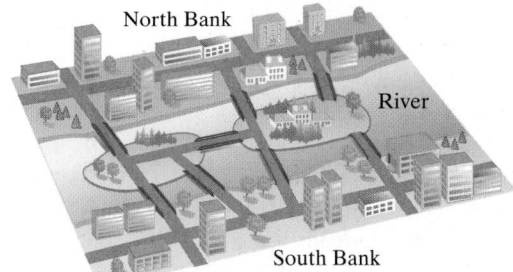

35. Jose, Bob, and Tony are college students living in adjacent dormitories. Bob lives in the middle dormitory. Their majors are business, psychology, and biology, although not necessarily in that order. The business major frequently uses the new computer in Bob's dorm room when Bob is in class. The psychology major and Jose both have 8 A.M. classes, and the psychology major knocks on Jose's wall to make sure he is awake. Determine Bob's major.

36. At the beginning of the SARS epidemic in 2003, the World Health Organization issued an advisory asking visitors to put off unnecessary travel to Toronto. Armed with the data shown in the accompanying graph, senior Canadian health official traveled to Geneva on April 28, 2003 to persuade the World Health Organization to retract their advisory. Suppose that you are a member of the World Health Organization. Based on the data shown, should the advisory be retracted? Explain your position. What other information might you find helpful in terms of whether or not to reconsider the advisory?

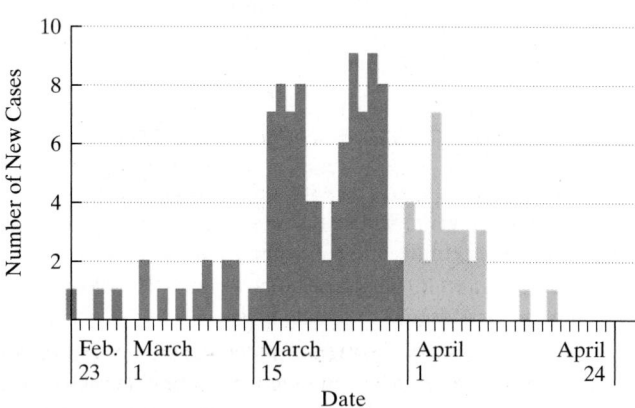

SARS Cases in Canada: February 23, 2003 Through April 24, 2003

Source: The New York Times

Solve Exercises 37–46 using the strategy or strategies of your choice.

37. Some numbers in the printing of a division problem have become illegible. They are designated below by *. Fill in the blanks.

$$
\begin{array}{r}
1** \\
\overline{)4*} \\
28 \\
\overline{*56} \\
*** \\
\overline{***} \\
*** \\
\overline{0}
\end{array}
$$

38. Three people have telephone area codes whose three digits have the same sum. One of the area codes is 252. None of the area codes contains a digit that is in one of the other area codes. None of the area codes has a first digit of 4 or 1. One of the area codes begins with 6. Another area code ends with 1. What is the area code that ends with 1?

*A **magic square** is a square array of numbers arranged so that the numbers in all rows, columns, and the two diagonals have the same sum. For the magic square shown at the top of the next column, the sum of the numbers in each row, each column, and each diagonal is 15. In Exercises 39–40, use the properties of a magic square to fill in the missing numbers.*

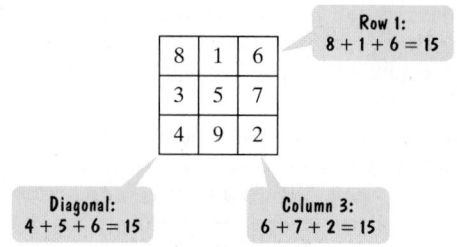

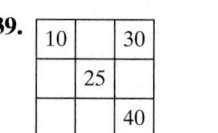
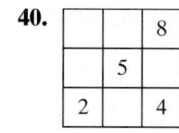

39. 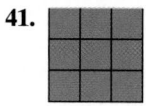   **40.**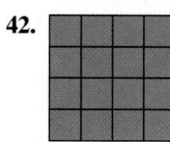

In Exercises 41–42, find the number of squares in each figure.

41. 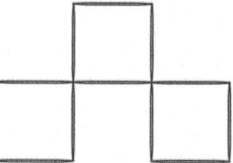   **42.**

43. Twelve toothpicks are arranged as shown in the figure. Form five squares by moving three toothpicks.

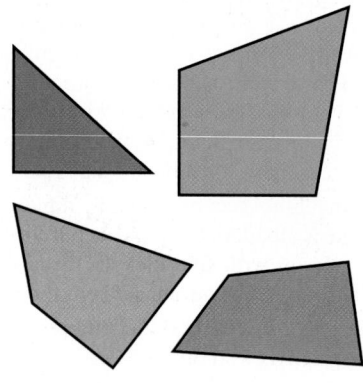

44. Make a copy of this page and then cut out the four pieces shown. First assemble them to form a triangle. Then reassemble the pieces to form a square.

45. This message is in a code used by Union prisoners during the Civil War.

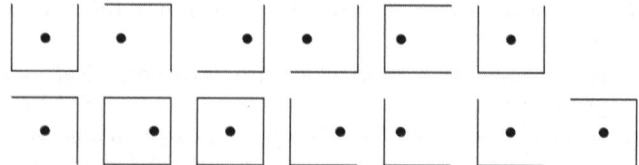

Decipher the message by determining the letter represented by each dot. Use the following array, focusing on the shape of the figure surrounding each dot and the position of the dot in the figure.

ABC	DEF	GHI
JKL	MNO	PQR
STU	VWX	YZ

46. The figure shows the number of cubes necessary to construct the first three solids in a sequence. How many cubes are needed to build the fifth solid in the sequence?

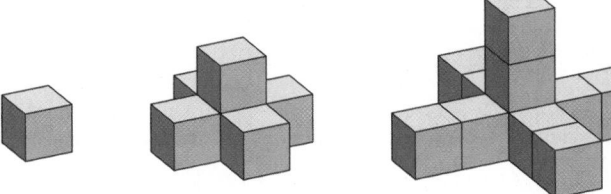

Exercises 47–49 involve problems encountered in everyday life. For each of these problematic situations, write seven or more short solutions that might be effective in solving the problem.

47. You started college with your best friend. This semester, you and your friend are taking two classes together. Your friend, who has a job and children, is failing both classes. What suggestions can you offer your friend to help pass the classes and succeed in college?

48. You have just started a job that you love and want to make a great impression with top management. You have been invited to go to an important evening party, and all of your top managers will be there. You have terrific social skills, and you know that your wit and your charm will help enhance your relationship with the supervisors. However, just before you are to go to the party, a small fire breaks out in your clothes closet. The fire causes almost no damage, although all of your clothes are destroyed. What might you do to be able to attend the party?

49. You have just been hired as the head of the automobile safety department in a large city with an enormous number of car accidents. Your first assignment is to submit a list of creative suggestions for reducing automobile accidents in the city. Your supervisor welcomes all suggestions as long as they do not cost the safety department any money. What are your suggestions?

Writing in Mathematics

In Exercises 50–52, explain the plan needed to solve the problem.

50. If you know how much was paid for several pounds of steak, find the cost of one pound.

51. If you know a person's age, find the year in which that person was born.

52. If you know how much you earn each hour, find your yearly income.

53. Write your own problem that can be solved using the four-step procedure. Then use the four steps to solve the problem.

Critical Thinking Exercises

54. Gym lockers are to be numbered from 1 through 99 using metal numbers to be nailed onto each locker. How many 7s are needed?

55. You are on vacation in an isolated town. Everyone in the town was born there and has never left. You develop a toothache and check out the two dentists in town. One dentist has gorgeous teeth and one has teeth that show the effects of poor dental work. Which dentist should you choose and why?

56. India Jones is standing on a large rock in the middle of a square pool filled with hungry, man-eating piranhas. The edge of the pool is 20 feet away from the rock. India's mom wants to rescue her son, but she is standing on the edge of the pool with only two planks, each $19\frac{1}{2}$ feet long. How can India be rescued using the two planks?

57. (This logic problem dates back to the eighth century.) A farmer needs to take his goat, wolf, and cabbage across a stream. His boat can hold him and one other passenger (the goat, wolf, or cabbage). If he takes the wolf with him, the goat will eat the cabbage. If he takes the cabbage, the wolf will eat the goat. Only when the farmer is present are the cabbage and goat safe from their respective predators. How does the farmer get everything across the stream?

58. A firefighter spraying water on a fire stood on the middle rung of a ladder. When the smoke become less thick, the firefighter moved up 4 rungs. However it got too hot, so the firefighter backed down 6 rungs. Later, the firefighter went up 7 rungs and stayed until the fire was out. Then, the firefighter climbed the remaining 4 rungs and entered the building. How many rungs does the ladder have?

59. Using each of the digits 4, 5, 6, 7, 8, and 9 exactly once, place two digits along each side of this triangle so that each side contains four digits whose sum is 17.

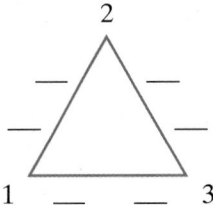

60. The Republic of Margaritaville is composed of four states, A, B, C, and D. According to the country's constitution, the congress will have 30 seats, divided among the four states according to their respective populations. The table shows each state's population.

POPULATION OF MARGARITAVILLE BY STATE

State	A	B	C	D	Total
Population (in thousands)	275	383	465	767	1890

Allocate the 30 congressional seats among the four states in a fair manner.

Group Exercises

Exercises 61–65 describe problems that have many plans for finding an answer. Group members should describe how the four steps in problem solving can be applied to find a solution. It is not necessary to actually solve each problem. Your professor will let the group know if the four steps should be described verbally by a group spokesperson or in essay form.

61. How much will it cost to install bicycle racks on campus to encourage students to use bikes, rather than cars, to get to campus?

62. How many new counselors are needed on campus to prevent students from waiting in long lines for academic advisement?

63. By how much would taxes in your state have to be increased to cut tuition at community colleges and state universities in half?

64. Is your local electric company overcharging its customers?

65. Should solar heating be required for all new construction in your community?

66. Group members should describe a problem in need of a solution. Then, as in Exercises 61–65, describe how the four steps in problem solving can be applied to find a solution.

Chapter Summary, Review, and Test

SUMMARY

DEFINITIONS AND CONCEPTS

EXAMPLES

1.1 Inductive and Deductive Reasoning

a. Inductive reasoning is the process of arriving at a general conclusion based on observations of specific examples. The conclusion is called a conjecture or a hypothesis.

Ex. 1, p. 3;
Ex. 3, p. 4;
Ex. 4, p. 5

b. A case for which a conjecture is false is called a counterexample.

Ex. 2, p. 4

c. Deductive reasoning is the process of proving a specific conclusion from one or more general statements. The statement that is proved is called a theorem.

Ex. 5, p. 6

1.2 Estimation and Graphs

a. Estimation is the process of arriving at an approximate answer to a question. The symbol ≈ means *is approximately equal to*.

Ex. 2, p. 12;
Ex. 3, p. 12;
Ex. 4, p. 13

b. Estimation is useful when interpreting information given in circle, bar, and line graphs.

Ex. 6, p. 14;
Ex. 7, p. 16;
Ex. 8, p. 17

1.3 Problem Solving

a. Understand the problem.

b. Devise a plan.

c. Carry out the plan and solve the problem.

d. Look back and check the answer.

Ex. 2, p. 23;
Ex. 3, p. 24;
Ex. 4, p. 25;
Ex. 5, p. 26;
Ex. 6, p. 28

REVIEW EXERCISES

1.1

1. Which reasoning process is shown in the following example? Explain your answer.

All books by Stephen King have made the best-seller list. *Carrie* is a novel by Stephen King. Therefore, *Carrie* was on the best-seller list.

2. Which reasoning process is shown in the following example? Explain your answer.

All books by Stephen King have made the best-seller list. Therefore, the novel that King is currently working on will make the best-seller list.

In Exercises 3–5, find a counterexample to show that each of the statements is false.

3. All U.S. states have names whose first letter comes before the letter u in the alphabet.

4. If a number is multiplied by 2 less than itself, the answer is an even number.

5. If one number is greater than a second number, the square of the first number is greater than the square of the second number.

In Exercises 6–11, identify a pattern in each list of numbers. Then use this pattern to find the next number.

6. 4, 9, 14, 19, _____

7. 7, 14, 28, 56, _____

8. 1, 3, 6, 10, 15, _____

9. $\dfrac{3}{4}, \dfrac{3}{5}, \dfrac{1}{2}, \dfrac{3}{7},$ _____

10. 40, −20, 10, −5, _____

11. 40, −20, −80, −140, _____

12. Identify a pattern in the following sequence of figures. Then use the pattern to find the next figure in the sequence.

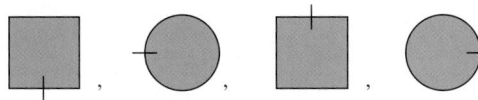

13. Consider the following procedure:

Select a number. Double the number. Add 4 to the product. Divide the sum by 2. Subtract 2 from the quotient.

a. Repeat the procedure for four numbers of your choice. Write a conjecture that relates the result of the process to the original number selected.

b. Represent the original number as ■ or n, and use deductive reasoning to prove the conjecture in part (a).

1.2

Work Exercises 14–18 without a calculator. Different rounding results in different estimates, so there is not a single correct answer to these exercises. Use rounding to make the resulting calculations simple.

14. Estimate the total cost of six grocery items if their prices are $8.47, $0.89, $2.79, $0.14, $1.19, and $4.76.

15. Estimate the salary of a worker who works for 78 hours at $6.85 per hour.

16. Estimate the number of seconds in a day.

17. Find an estimate of $27\frac{19}{20} \div 6.823$.

18. At a yard sale, Melissa bought 21 books at $0.85 each, two chairs for $11.95 each, and a ceramic plate for $14.65. Estimate the total amount that Melissa spent.

19. According to the American Water Works Association, the average U.S. household uses 356 gallons of water per day. The circle graph shows where we use that water. Estimate the number of gallons of water per month a U.S. household uses for toilets.

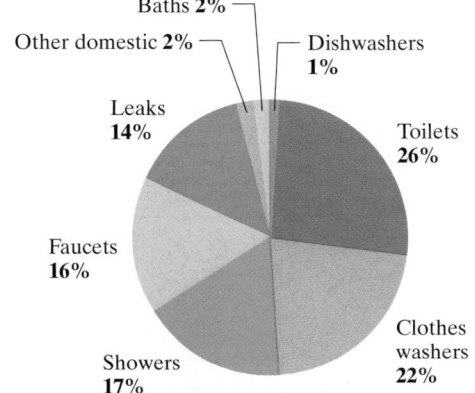

Where U.S. Households Use Water

Source: American Water Works Association

It seems that our pets' medical bills are costing us an arm and a paw. The bar graph shows veterinary costs, in billions of dollars, for dogs and cats in five selected years. Use the graph to solve Exercises 20–23.

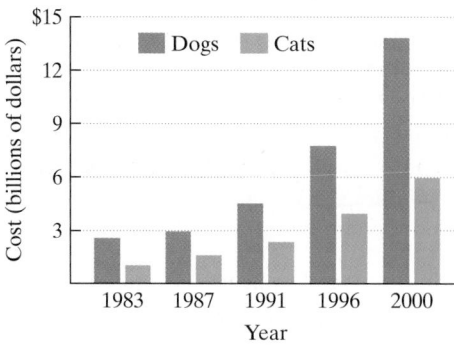

Veterinary Costs in the U.S.

Source: American Veterinary Medical Association

20. Estimate veterinary costs for cats in 1983.

21. Estimate veterinary costs for dogs in 1996.

22. Estimate the difference between veterinary costs for dogs and cats in 2000.

23. For which animals and for which years shown in the graph do veterinary costs exceed $3 billion but are at most $6 billion?

The line graph shows the number of murders per 100,000 people in the United States from 1900 through 2000. Use the graph to solve Exercises 24–26.

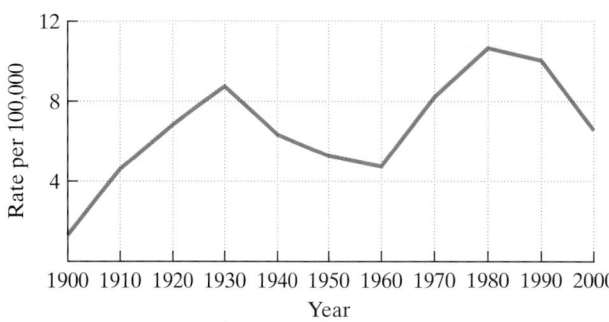

Murders Per 100,000 People in the United States

Source: National Center for Health Statistics

24. Find an estimate for the number of murders per 100,000 people in 2000.

25. For the period shown, when did the murder rate reach a maximum? What is a reasonable estimate for the number of murders per 100,000 people for that year?

26. For the period from 1970 through 2000, find an estimate for the average number of murders per 100,000 people per year.

1.3

27. What necessary piece of information is missing that prevents solving the following problem?

If 3 milligrams of a medicine is given for every 20 pounds of body weight, how many milligrams should be given to a six-year-old child?

28. In the following problem, there is one more piece of information given that is necessary for solving the problem. Identify this unnecessary piece of information. Then solve the problem.

A taxicab charges $3.00 for the first mile and 50 cents for each additional half-mile. After a six-mile trip, a customer handed the taxi driver a $20 bill. Find the cost of the trip.

Use the four-step method in problem solving to solve Exercises 29–34.

29. If there are seven frankfurters in one pound, how many pounds would you buy for a picnic to supply 28 people with two frankfurters each?

30. A car rents for $175 per week plus $0.30 per mile. Find the rental cost for a three-week trip of 1200 miles.

31. Membership in Superfit, a fitness club, is $500 per year. The club charges $1.00 for each hour the facility is used. A competing club, Healthy Bodies, charges $400 yearly plus $1.75 per hour used. If you plan to use a fitness club for 80 hours each year, which club offers the better deal? By how much?

CHAPTER 1 TEST

1. Find a counterexample to show that the following statement is false:

If a two-digit number is multiplied by a one-digit number, the answer is a two-digit number.

In Exercises 2–3, identify a pattern in each list of numbers. Then use this pattern to find the next number.

2. 0, 5, 10, 15, _____

3. $\frac{1}{6}, \frac{1}{12}, \frac{1}{24}, \frac{1}{48},$ _____

4. Identify a pattern in the following sequence of figures. Then use the pattern to find the next figure in the sequence.

, , , , , _____

5. Consider the following procedure:

Select a number. Multiply the number by 4. Add 8 to the product. Divide the sum by 2. Subtract 4 from the quotient.

a. Repeat this procedure for three numbers of your choice. Write a conjecture that relates the result of the process to the original number selected.

b. Represent the original number by ■ or *n*, and use deductive reasoning to prove the conjecture in part (a).

32. Miami is on Eastern Standard Time and San Francisco is on Pacific Standard Time, three hours earlier than Eastern Standard Time. A flight leaves Miami at 10 A.M. Eastern Standard Time, stops for 45 minutes in Houston Texas, and arrives in San Francisco at 1:30 P.M. Pacific time. What is the actual flying time from Miami to San Francisco?

33. An automobile purchased for $37,000 is worth $2600 after eight years. Assuming that the value decreased steadily each year, what was the car worth at the end of the fifth year?

34. Suppose you are an engineer programming the automatic gate for a 35-cent toll. The gate is programmed for exact change only and will not accept pennies. How many coin combinations must you program the gate to accept?

Group Exercise

35. Reread Example 7 on page 29. Group members are now the collective department manager trying to deal with the late employee. Begin by assuming that the nine suggestions for solving the tardiness problem have been tried, and none of them has worked. Group members therefore need to replace the suggestions with their own ideas for solving the problem. The group's final list should contain reasonable solutions. (For example, moving the office closer to the employee's home is not reasonable!)

Work Exercises 6–9 without a calculator. Different rounding results in different estimates, so there is not one single correct answer to these exercises. Use rounding to make the resulting calculations simple.

6. For a spring break vacation, a student needs to spend $47.00 for gas, $311.00 for food, and $405.00 for a hotel room. If the student takes $681.79 from savings, estimate how much more money is needed for the vacation.

7. The cost for opening a restaurant is $485,000. If 19 people decide to share equally in the business, estimate the amount each must contribute.

8. Find an estimate of 0.48992 × 120.

9. In 2002, world population reached 6.24 billion people. Use the circle graph to estimate the world population living in free countries.

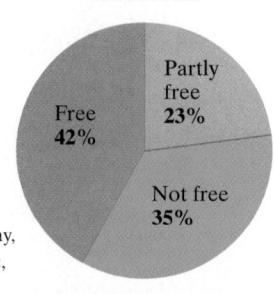

Source: Larry Berman and Bruce Murphy, *Approaching Democracy Fourth Edition*, Prentice Hall, 2003

The graph shows the number of female and minority appointments to federal judgeships for four American presidents. Use the graph to solve Exercises 10–12.

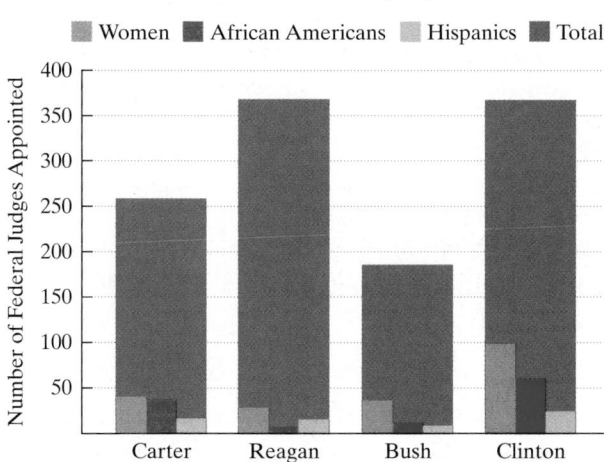

Female and Minority Appointments to Federal Judgeships

Source: James Burns et. al., *Government by the People*, Prentice Hall, 2002

10. Estimate the total federal judgeships appointed by Carter.

11. Which presidents shown appointed at least 20 and at most 50 women to federal judgeships?

12. Approximately how many more federal judgeships were appointed to African Americans by Clinton than by Bush?

13. The line graph shows the population of the United States from 1970 to 2000 for people under 16. For the period from 1970 to 2000, in which year was the population for people under 16 at a minimum? Estimate the population for that year.

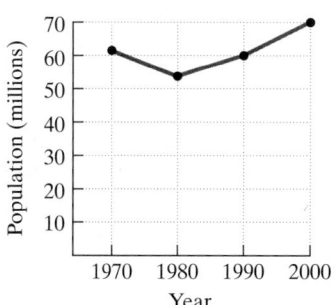

U.S. Population for People under 16

Source: U.S. Census Bureau

14. The cost of renting a boat from Estes Rental is $9 per 15 minutes. The cost from Ship and Shore Rental is $20 per half-hour. If you plan to rent the boat for three hours, which business offers the better deal and by how much?

15. A bus operates between Miami International Airport and Miami Beach, 10 miles away. It makes 20 round trips per day carrying 32 passengers per trip. If the fare each way is $11.00, how much money is taken in from one day's operation?

16. By paying $50 cash up front and the balance at $35 a week, how long will it take to pay for a computer costing $960?

Set Theory

The need to sort and organize information obtained through surveys is related to our need to find order and meaning by classifying things into collections. Mathematicians call such collections **sets**. Sets are groups of objects that usually share some common characteristic. For example, mathematicians classify numbers into sets based on whether or not their decimal representations have repeating patterns. A special set of numbers is used to provide probabilities of events. Sets are even used to describe nature's patterns, from the symmetry of a snowflake to the intricacies of a flower.

Set theory is the thread for the tapestry of mathematics. The set concept finds its way into many topics covered in this book, such as logic, algebra, and probability. Sets provide a precise way of describing and communicating mathematical ideas. Thus, learning to understand set theory will help you appreciate mathematics as a rich and living part of human culture.

You want to organize a blood drive on campus with the local Red Cross. The Red Cross asked you whether the number of potential donors warrants a commitment to provide medical staff for the blood drive. So you took a survey to obtain information. Your survey asked students two questions:

1. Would you be willing to donate blood?
2. Would you be willing to help serve a free breakfast to blood donors?

Now the survey results are in. How will you organize and present the results to the Red Cross in an efficient manner?

This problem appears as Example 1 in Section 2.5.

SECTION 2.1   BASIC SET CONCEPTS

OBJECTIVES

1. Use three methods to represent sets.

2. Use the symbols $\in$ and $\notin$.

3. Apply set notation to sets of natural numbers.

4. Determine a set's cardinal number.

5. Recognize equal sets.

6. Recognize equivalent sets.

We tend to place things in categories, allowing us to order and structure the world. For example, you can categorize yourself by your age, your ethnicity, your academic major, or your gender. Our minds cannot find order and meaning without creating collections. Mathematicians call such collections *sets*. A **set** is a collection of objects whose contents can be clearly determined. The objects in a set are called the **elements**, or **members**, of the set.

An example of a set is the set of the days of the week, whose elements are Monday, Tuesday, Wednesday, Thursday, Friday, Saturday, and Sunday.

Capital letters are generally used to name sets. Let's use W to represent the set of the days of the week.

1   Use three methods to represent sets.

Three methods are commonly used to designate a set. One method is a word description. We can describe set W as the set of the days of the week. A second method is the **roster method**. This involves listing the elements of a set inside a pair of braces, { }. The braces at the beginning and end indicate that we are representing a set. The roster form uses commas to separate the elements of the set. Thus, we can designate the set W by listing its elements:

$$W = \{\text{Monday, Tuesday, Wednesday, Thursday, Friday, Saturday, Sunday}\}.$$

BLITZER BONUS

The Loss of Sets

John Tenniel, colored by Fritz Kredel

Have you ever considered what would happen if we suddenly lost our ability to recall categories and the names that identify them? This is precisely what happened to Alice, the heroine of Lewis Carroll's *Through the Looking Glass*, as she walked with a fawn in "the woods with no names."

So they walked on together through the woods, Alice with her arms clasped lovingly round the soft neck of the Fawn, till they came out into another open field, and here the Fawn gave a sudden bound into the air, and shook itself free from Alice's arm. "I'm a Fawn!" it cried out in a voice of delight. "And, dear me! you're a human child!" A sudden look of alarm came into its beautiful brown eyes, and in another moment it had darted away at full speed.

By realizing that Alice is a member of the set of human beings, which in turn is part of the set of dangerous things, the fawn is overcome by fear. Thus, the fawn's experience is determined by the way it structures the world into sets with various characteristics.

Grouping symbols such as parentheses, ( ), and square brackets, [ ], are not used to represent sets. Only commas are used to separate the elements of a set. Separators such as colons or semicolons are not used. Finally, the order in which the elements are listed in a set is not important. Thus, another way of expressing the set of the days of the week is

$$W = \{\text{Saturday, Sunday, Monday, Tuesday, Wednesday, Thursday, Friday}\}.$$

EXAMPLE 1   REPRESENTING A SET USING A DESCRIPTION

Write a word description of the set

$$P = \{\text{Washington, Adams, Jefferson, Madison, Monroe}\}.$$

SOLUTION   Set P is the set of the first five presidents of the United States.

 **Check 1 Point**   Write a word description of the set

$$L = \{a, b, c, d, e, f\}.$$

EXAMPLE 2   REPRESENTING A SET USING THE ROSTER METHOD

Set C is the set of U.S. coins with a value of less than a dollar. Express this set using the roster method.

SOLUTION $C = \{\text{penny, nickel, dime, quarter, half-dollar}\}$

 **Check 2 Point**   Set D is the set of digits in the number 37,418. Express this set using the roster method.

The third method for representing a set is with **set-builder notation**. Using this method, the set of the days of the week can be expressed as

$$W = \{x \mid x \text{ is a day of the week}\}.$$

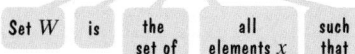

We read this notation as "Set W is the set of all elements x such that x is a day of the week." Before the vertical line is the variable x, which represents an element in general. After the vertical line is the condition x must meet in order to be an element of the set.

Table 2.1 contains two examples of sets, each represented with a word description, the roster method, and set-builder notation.

TABLE 2.1 SETS USING THREE DESIGNATIONS

Word Description	Roster Method	Set-Builder Notation
B is the set of members of the Beatles in 1963.	$B = \{$George Harrison, John Lennon, Paul McCartney, Ringo Starr$\}$	$B = \{x \mid x$ was a member of the Beatles in 1963$\}$
S is the set of states whose names begin with the letter A.	$S = \{$Alabama, Alaska, Arizona, Arkansas$\}$	$S = \{x \mid x$ is a U.S. state whose name begins with the letter A$\}$

The Beatles climbed to the top of the British music charts in 1963, conquering the United States a year later.

EXAMPLE 3   CONVERTING FROM SET-BUILDER TO ROSTER NOTATION

Express set

$$A = \{x \mid x \text{ is a month that begins with the letter M}\}$$

using the roster method.

SOLUTION   Set A is the set of all elements x such that x is a month beginning with the letter M. There are two such months, namely March and May. Thus,

$$A = \{\text{March, May}\}.$$

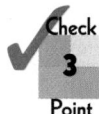

Express the set

$$L = \{x \mid x \text{ is a letter in the word } march\}$$

using the roster method.

Now let's consider a set similar to the one in Example 3. However, this time set A represents the set of all elements x such that x is a month beginning with the letter R.

$$A = \{x \mid x \text{ is a month that begins with the letter R}\}$$

There are no months that begin with the letter R, and so this set contains no elements. A set such as this that contains no elements is called the *empty set*, or the *null set*.

THE EMPTY SET

The **empty set**, also called the **null set**, is the set that contains no elements. The empty set is represented by { } or ∅.

Notice that { } and ∅ have the same meaning. However, the empty set is not represented by {∅}. This notation represents a set containing the element ∅.

The set of all lowercase letters of the English alphabet is rather long when represented by the roster method. If L is chosen as a name for this set, then we can write

$$L = \{a, b, c, d, e, f, g, h, i, j, k, l, m, n, o, p, q, r, s, t, u, v, w, x, y, z\}.$$

We can shorten the listing in set L by writing

$$L = \{a, b, c, d, \ldots, z\}.$$

The three dots after the element d, called an *ellipsis*, indicate that the elements in the set continue in the same manner up to and including the last element z.

A set must be **well defined**, meaning that its contents can be clearly determined. Using this criterion, the collection of actors who have won Academy Awards is a set. We can always determine whether or not a particular actor is an element of $\{x \mid x \text{ is an actor who won an Oscar}\}$. By contrast, consider the collection of great actors. Whether or not a person belongs to this collection is a matter of how we interpret the word *great*. In this text, we will only consider collections that form well-defined sets.

Mathematicians have developed a special notation to indicate whether or not a given element belongs to a set.

2 Use the symbols ∈ and
∉ .

THE NOTATIONS ∈ AND ∉

The symbol ∈ is used to indicate that an object is an element of a set. The symbol ∈ is used to replace the words "is an element of."
The symbol ∉ is used to indicate that an object is *not* an element of a set. The symbol ∉ is used to replace the words "is not an element of."

EXAMPLE 4 USING THE SYMBOLS ∈ AND ∉

Determine whether each statement is true or false:

a. $r \in \{a, b, c, \ldots, z\}$ **b.** $7 \notin \{1, 2, 3, 4, 5\}$ **c.** $\{a\} \in \{a, b\}$.

SOLUTION

a. Because r is an element of the set $\{a, b, c, \ldots, z\}$, the statement

$$r \in \{a, b, c, \ldots, z\}$$

is true.
Observe that an element can belong to a set in roster notation when three dots appear even though the element is not listed.

b. Because 7 is not an element of the set $\{1, 2, 3, 4, 5\}$, the statement

$$7 \notin \{1, 2, 3, 4, 5\}$$

is true.

c. Because $\{a\}$ is a set and the set $\{a\}$ is not an element of the set $\{a, b\}$, the statement

$$\{a\} \in \{a, b\}$$

is false.

 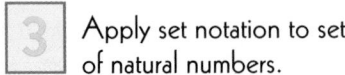

Check
4
Point

Determine whether each statement is true or false:

a. $8 \in \{1, 2, 3, \ldots, 10\}$

b. $r \notin \{a, b, c, z\}$

c. $\{\text{Monday}\} \in \{x \mid x \text{ is a day of the week}\}$.

3 Apply set notation to sets of natural numbers.

For the remainder of this section, we will focus on the set of numbers used for counting:

$$\{1, 2, 3, 4, 5, 6, 7, 8, 9, 10, 11, \ldots\}.$$

The set of counting numbers is also called the set of **natural numbers**. We represent this set by the letter **N**.

THE SET OF NATURAL NUMBERS

$$\mathbf{N} = \{1, 2, 3, 4, 5, \ldots\}$$

The three dots, or ellipsis, after the 5 indicate that there is no final element and that the listing goes on forever.

EXAMPLE 5 | REPRESENTING SETS OF NATURAL NUMBERS

Express each of the following sets using the roster method:

   a. Set A is the set of natural numbers less than 5.

   b. Set B is the set of natural numbers greater than or equal to 25.

   c. $E = \{x \mid x \in \mathbf{N} \text{ and } x \text{ is even}\}$

   d. $F = \{x \mid x \in \mathbf{N} \text{ and } x \text{ is between 6 and 10}\}$.

SOLUTION

   a. The natural numbers less than 5 are 1, 2, 3, and 4. Thus, set A can be expressed using the roster method as

$$A = \{1, 2, 3, 4\}.$$

   b. The natural numbers greater than or equal to 25 are 25, 26, 27, 28, and so forth. Set B in roster form is

$$B = \{25, 26, 27, 28, \dots\}.$$

The three dots show that the listing goes on forever.

   c. The set-builder notation

$$E = \{x \mid x \in \mathbf{N} \text{ and } x \text{ is even}\}$$

indicates that we want to list the set of all x such that x is an element of the set of natural numbers and x is even. The set of numbers that meet both conditions is the set of even natural numbers. The set in roster form is

$$E = \{2, 4, 6, 8, \dots\}.$$

   d. The set-builder notation

$$F = \{x \mid x \in \mathbf{N} \text{ and } x \text{ is between 6 and 10}\}$$

indicates the set of natural numbers between 6 and 10. The set in roster form is

$$F = \{7, 8, 9\}.$$

Notice that the word "between" excludes the endpoint values, 6 and 10, from the set.

The word "inclusive" indicates that the endpoint values between two natural numbers are included in a set. For example, the set of natural numbers between 6 and 10, inclusive, is $\{6, 7, 8, 9, 10\}$.

 **Check Point 5**   Express each of the following sets using the roster method:

   a. Set A is the set of natural numbers less than or equal to 3.

   b. Set B is the set of natural numbers greater than 14.

   c. $O = \{x \mid x \in \mathbf{N} \text{ and } x \text{ is odd}\}$

   d. $F = \{x \mid x \in \mathbf{N} \text{ and } x \text{ is between 11 and 14}\}$.

4 Determine a set's cardinal number.

The number of elements in a set is called the **cardinal number**, or **cardinality**, of the set. For example, the set {a, e, i, o, u} contains five elements and therefore has a cardinal number 5. We can also say that the set has a cardinality of 5.

DEFINITION OF A SET'S CARDINAL NUMBER

The **cardinal number** of set A, represented by $n(A)$, is the number of distinct elements in set A. The symbol $n(A)$ is read "n of A."

EXAMPLE 6   DETERMINING A SET'S CARDINAL NUMBER

Find the cardinal number of each of the following sets:

   a. $A = \{7, 9, 11, 13\}$ **b.** $B = \{0\}$
   c. $C = \{13, 14, 15, \ldots, 22, 23\}$ **d.** $\varnothing$.

SOLUTION   The cardinal number for each set is found by determining the number of elements in the set.

   a. $A = \{7, 9, 11, 13\}$ contains four elements. Thus, the cardinal number of set A is 4. We also say that set A has a cardinality of 4, or $n(A) = 4$.

   b. $B = \{0\}$ contains one element, namely 0. The cardinal number of set B is 1. Therefore, $n(B) = 1$.

   c. Set $C = \{13, 14, 15, \ldots, 22, 23\}$ lists only five elements. However, the three dots indicate that the natural numbers from 16 through 21 are also in the set. Counting the elements in the set, we find that there are 11 natural numbers in set C. The cardinality of set C is 11, and $n(C) = 11$.

   d. The empty set, $\varnothing$, contains no elements. Thus, $n(\varnothing) = 0$.

Find the cardinal number of each of the following sets:

   a. $A = \{6, 10, 14, 15, 16\}$ **b.** $B = \{872\}$
   c. $C = \{9, 10, 11, \ldots, 15, 16\}$ **d.** $D = \{\ \}$.

The sets in Example 6 and Check Point 6 are all examples of *finite sets*.

DEFINITION OF A FINITE SET

Set A is a **finite set** if $n(A) = 0$ or $n(A)$ is a natural number. A set that is not finite is called an **infinite set**.

If we have enough time, it is possible to finish counting all the elements in a finite set, thereby obtaining its cardinal number. This is not the case with an infinite set. The counting process never comes to an end. The set

$$\{3, 6, 9, 12, 15, \ldots\}$$

is an example of an infinite set. The three dots indicate that there is no last, or final, element.

Equal and Equivalent Sets

 Recognize equal sets.

We now turn to another important concept of set theory, equality of sets.

DEFINITION OF EQUALITY OF SETS

Set A is **equal** to set B means that set A and set B contain exactly the same elements, regardless of order or possible repetition of elements. We symbolize the equality of sets A and B using the statement $A = B$.

For example, if $A = \{w, x, y, z\}$ and $B = \{z, y, w, x\}$, then $A = B$ because the two sets contain exactly the same elements.

Because equal sets contain the same elements, they also have the same cardinal number. For example, the equal sets $A = \{w, x, y, z\}$ and $B = \{z, y, w, x\}$ have four elements each. Thus, both sets have the same cardinal number: 4.

EXAMPLE 7   DETERMINING WHETHER SETS ARE EQUAL

Determine whether each statement is true or false:

a. $\{4, 8, 9\} = \{8, 9, 4\}$ **b.** $\{1, 3, 5\} = \{0, 1, 3, 5\}$.

SOLUTION

a. The sets $\{4, 8, 9\}$ and $\{8, 9, 4\}$ contain exactly the same elements. Therefore, the statement

$$\{4, 8, 9\} = \{8, 9, 4\}$$

is true.

b. As we look at the given sets, $\{1, 3, 5\}$ and $\{0, 1, 3, 5\}$, we see that 0 is an element of the second set, but not the first. The sets do not contain exactly the same elements. Therefore, the sets are not equal. This means that the statement

$$\{1, 3, 5\} = \{0, 1, 3, 5\}$$

is false.

 Check **7** Point

Determine whether each statement is true or false:

a. $\{O, L, D\} = \{D, O, L\}$ **b.** $\{4, 5\} = \{5, 4, \varnothing\}$.

Repeating elements in a set does not add new elements to the set. For example, if $A = \{1, 1, 2, 2, 3\}$ and $B = \{1, 2, 3\}$, then $A = B$ since these sets contain exactly the same distinct elements. Thus, $n(A) = n(B) = 3$.

We have seen that because equal sets contain the same elements, they also have the same cardinal number. However, two sets with the same cardinal number are not necessarily equal. For example,

$$A = \{1, 2, 3, \ldots, 49, 50\}$$

and

$$B = \{x \mid x \text{ is a state in the United States}\}$$

both have a cardinality of 50. That is, $n(A) = 50$ and $n(B) = 50$. However, the sets do not contain the same elements.

Sets like these that contain the same number of elements are said to be *equivalent*.

|6| Recognize equivalent sets.

STUDY TIP

In English, the words *equal* and *equivalent* often mean the same thing. This is not the case in set theory. Equal sets contain the same elements. Equivalent sets contain the same number of elements.

DEFINITION OF EQUIVALENT SETS

Set A is **equivalent** to set B means that set A and set B contain the same number of elements. For equivalent sets, $n(A) = n(B)$.

EXAMPLE 8   RECOGNIZING EQUIVALENT SETS

Consider the following sets:

$$A = \{\text{The Graduate, The Godfather, Titanic}\}$$
$$B = \{\text{Hoffman, Brando, DiCaprio}\}.$$

a. Are these sets equal? Explain.   **b.** Are these sets equivalent? Explain.

SOLUTION

a. Sets A and B do not contain exactly the same elements. Therefore, the sets are not equal.

b. Sets A and B each contain three elements. Because they contain the same number of elements, the sets are equivalent.

Check
8
Point

Consider the following sets:

$$A = \{\textit{Frasier, The Tonight show}\}$$
$$B = \{\text{Kelsey Grammer, Jay Leno}\}.$$

a. Are these sets equal? Explain.

b. Are these sets equivalent? Explain.

BLITZER BONUS

Cardinal Numbers of Infinite Sets

P. J. Crook "*Time and Time Again*" 1981. Courtesy Barry Friedman Limited, New York.

The mirrors in the painting "Time and Time Again" have the effect of repeating the image infinitely many times, creating an endless tunnel of mirror images. There is something quite fascinating about the idea of endless infinity. Did you know that for thousands of years religious leaders warned that human beings should not examine the nature of the infinite? Religious teaching often equated infinity with the concept of a Supreme Being. One of the last victims of the Inquisition, Giordano Bruno, was burned at the stake for his explorations into the characteristics of infinity. It was not until the 1870s that the German mathematician Georg Cantor (1845–1918) began a careful analysis of the mathematics of infinity.

Although it is impossible to count the number of elements in the set of natural numbers $\{1, 2, 3, 4, 5, 6, \dots\}$, Cantor assigned this set a cardinality. He said that the set of natural numbers has the infinite cardinal number $\aleph_0$ (the first letter of the Hebrew alphabet with a subscript of zero, read "aleph-null"). There are $\aleph_0$ natural numbers, and $\aleph_0$ is called a **transfinite cardinal number**.

Sets that are equivalent to the set of natural numbers also contain $\aleph_0$ elements. Here are two examples:

Natural Numbers: $\{1, 2, 3, 4, 5, \ 6, \dots, \ n, \dots\}$

Even Natural Numbers: $\{2, 4, 6, 8, 10, 12, \dots, 2n, \dots\}$

Each natural number, n, is matched or paired with its double, $2n$, in the set of even natural numbers.

Natural Numbers: $\{1, 2, 3, 4, 5, \ 6, \dots, \qquad n, \dots\}$

Odd Natural Numbers: $\{1, 3, 5, 7, 9, \ 11, \dots, 2n - 1, \dots\}$

Each natural number, n, is matched with 1 less than its double, $2n - 1$, in the set of odd natural numbers.

These matchings mean that there are $\aleph_0$ even natural numbers and $\aleph_0$ odd natural numbers. Because there are $\aleph_0$ natural numbers, this leads to the strange observations that the "part" (the even or the odd natural numbers) has the same number of elements as the "whole" (all the natural numbers). Cantor's analysis results in unusual statements of transfinite arithmetic, such as

$$\aleph_0 + \aleph_0 = \aleph_0.$$

As Cantor continued studying infinite sets, his observations grew stranger and stranger. He even showed that some infinite sets contain more elements than others. This was too much for Cantor's colleagues, who considered his work ridiculous. His mentor, Leopold Kronecker, told him, "Look at the crazy ideas that are now surfacing with your work with infinite sets. How can one infinity be greater than another? Best to ignore such inconsistencies. By considering these monsters and infinite numbers mathematics, I will make sure that you never gain a faculty position at the University of Berlin." Although Cantor was not burned at the stake, universal condemnation of his work resulted in numerous nervous breakdowns. His final days, sadly, were spent in a psychiatric hospital. However, Cantor's work later regained the respect of mathematicians. Today, he is seen as a great mathematician who de-mystified infinity.

Exercise Set 2.1

Practice Exercises

Write a description of the sets in Exercises 1–6. There may be more than one correct description.

1. {Mercury, Venus, Earth, Mars, Jupiter, Saturn, Uranus, Neptune, Pluto}

2. {Saturday, Sunday}

3. {January, June, July}

4. {Aries, Taurus, Gemini, Cancer, . . . , Aquarius, Pisces}

5. $\{1, 3, 5, 7, \ldots, 99\}$

6. $\{5, 10, 15, 20, \ldots\}$

In Exercises 7–22, express each set using the roster method. Remember that $\mathbf{N}$ represents the set of natural numbers.

7. The set of the four seasons in a year

8. The set of months of the year that have exactly 30 days

9. $\{x \mid x$ is a month that ends with the letters b-e-r$\}$

10. $\{x \mid x$ is a letter of the alphabet that follows d and comes before j$\}$

11. The set of natural numbers less than 4

12. The set of natural numbers less than or equal to 6

13. The set of odd natural numbers less than 13

14. The set of even natural numbers less than 10

15. $\{x \mid x \in \mathbf{N}$ and x is greater than 100$\}$

16. $\{x \mid x \in \mathbf{N}$ and x is greater than 50$\}$

17. $\{x \mid x \in \mathbf{N}$ and x lies between 10 and 16$\}$

18. $\{x \mid x \in \mathbf{N}$ and x lies between 23 and 29$\}$

19. $\{x \mid x \in \mathbf{N}$ and x lies between 10 and 16, inclusive$\}$

20. $\{x \mid x \in \mathbf{N}$ and x lies between 23 and 29, inclusive$\}$

21. $\{x \mid x + 5 = 7\}$

22. $\{x \mid x + 3 = 9\}$

In Exercises 23–30, express each set using set-builder notation. (There may be more than one way to describe the condition that x must meet to be a member of the set.)

23. {Tuesday, Thursday}

24. {h, i, j, k}

25. $\{1, 2, 3, 4, 5\}$

26. $\{1, 2, 3, 4, 5, 6\}$

27. $\{3, 4, 5\}$

28. $\{4, 5, 6\}$

29. $\{3, 4, 5, \ldots\}$

30. $\{4, 5, 6, \ldots\}$

In Exercises 31–46, determine whether each statement is true or false.

31. $3 \in \{1, 3, 5, 7\}$

32. $6 \in \{2, 4, 6, 8, 10\}$

33. $12 \in \{1, 2, 3, \ldots, 14\}$

34. $10 \in \{1, 2, 3, \ldots, 16\}$

35. $5 \in \{2, 4, 6, \ldots, 20\}$

36. $8 \in \{1, 3, 5, \ldots 19\}$

37. $11 \notin \{1, 2, 3, \ldots, 9\}$

38. $17 \notin \{1, 2, 3, \ldots, 16\}$

39. $37 \notin \{1, 2, 3, \ldots, 40\}$

40. $26 \notin \{1, 2, 3, \ldots, 50\}$

41. $4 \notin \{x \mid x \in \mathbf{N}$ and x is even$\}$

42. $2 \in \{x \mid x \in \mathbf{N}$ and x is odd$\}$

43. $16 \in \{x \mid x \in \mathbf{N}$ and x is between 16 and 20$\}$

44. $18 \notin \{x \mid x \in \mathbf{N}$ and x is between 14 and 18$\}$

45. $\{3\} \in \{3, 4\}$

46. $\{7\} \in \{7, 8\}$

In Exercises 47–54, fill in the blank with either $\in$ or $\notin$ to make each statement true.

47. Mark McGwire _____ $\{x \mid x$ is a U.S. president$\}$

48. Atlanta _____ $\{x \mid x$ is a U.S. state$\}$

49. 13,791 _____ $\{x \mid x \in \mathbf{N}$ and x is odd$\}$

50. 26,794 _____ $\{x \mid x \in \mathbf{N}$ and x is even$\}$

51. $\varnothing$ ____ { }

52. 0 ____ $\varnothing$

53. 0 _____ $\mathbf{N}$

54. -1 ____ $\mathbf{N}$

Find the cardinal number for each set in Exercises 55–62.

55. $A = \{17, 19, 21, 23, 25\}$

56. $A = \{16, 18, 20, 22, 24, 26\}$

57. $B = \{2, 4, 6, \ldots, 50\}$

58. $B = \{1, 3, 5, \ldots, 47\}$

59. $C = \{x \mid x$ is a day of the week that begins with the letter A$\}$

60. $C = \{x \mid x$ is a month of the year that begins with the letter W$\}$

61. $D = \{$five$\}$

62. $D = \{$six$\}$

In Exercises 63–68, determine whether each statement is true or false.

63. $\{a, b, c, d\} = \{b, d, c, a\}$

64. $\{a, e, i, o, u\} = \{u, a, o, i, e\}$

65. $\{0, 2, 4, 6, 8\} = \{2, 4, 6, 8\}$

66. $\{0, 1, 3, 5\} = \{1, 3, 5\}$

67. $\{1, 1, 8, 8, 13\} = \{1, 8, 13\}$

68. $\{6, 9, 12\} = \{9, 12, 12, 6, 9\}$

In Exercises 69–74,

    a. *Are the sets equal? Explain.*

    b. *Are the sets equivalent? Explain.*

69. A is the set of students at your college. B is the set of students majoring in business at your college.

70. A is the set of states in the United States. B is the set of people who are now governors of the states in the United States.

71. $A = \{1, 2, 3, 4, 5\}$

    $B = \{0, 1, 2, 3, 4\}$

72. $A = \{1, 3, 5, 7, 9\}$
$B = \{2, 4, 6, 8, 10\}$

73. $A = \{1, 1, 1, 2, 2, 3, 4\}$
$B = \{4, 3, 2, 1\}$

74. $A = \{0, 1, 1, 2, 2, 2, 3, 3, 3, 3\}$
$B = \{3, 2, 1, 0\}$

Application Exercises

The bar graph shows how we spend our leisure time. Use the graph to represent each of the sets in Exercises 75–78 using the roster method.

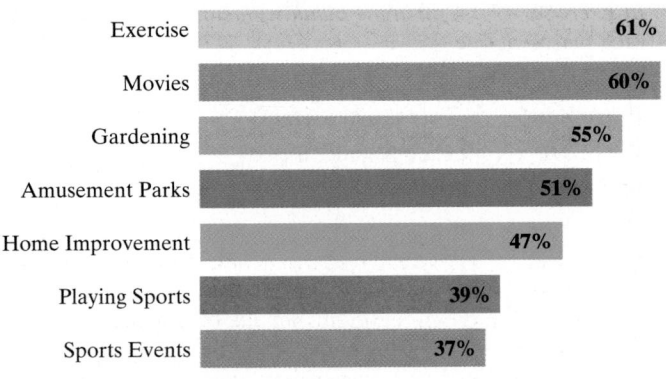

Percentage of U.S. Households Participating in Each Activity on a Regular Basis

Exercise — 61%
Movies — 60%
Gardening — 55%
Amusement Parks — 51%
Home Improvement — 47%
Playing Sports — 39%
Sports Events — 37%

Source: U.S. Census Bureau

75. The set of leisure activities in which participation exceeds 50%

76. The set of leisure activities in which participation exceeds 40%

77. $\{x \mid x$ is a leisure activity in which participation lies between 40% and 59%$\}$

78. $\{x \mid x$ is a leisure activity in which participation lies between 50% and 59%$\}$

Navigating America's airports can be tough. The bar graph shows the top ten U.S. airports with the greatest percentage of delayed arrivals and departures. Use the graph to represent each of the sets in Exercises 79–82 using the roster method.

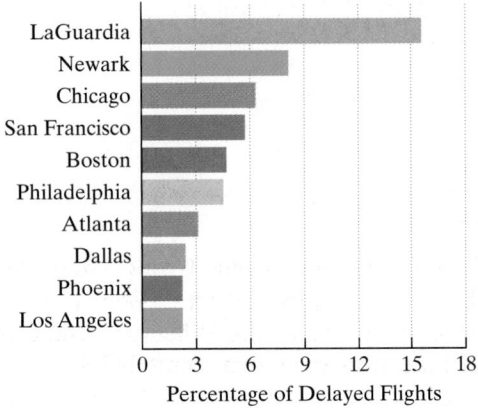

U. S. Airports with the Greatest Percentage of Delayed Flights

LaGuardia
Newark
Chicago
San Francisco
Boston
Philadelphia
Atlanta
Dallas
Phoenix
Los Angeles

Percentage of Delayed Flights: 0 3 6 9 12 15 18

Source: Newsweek

79. The set of U.S. airports where more than 6% of flights are delayed

80. The set of U.S. airports where more than 5% of flights are delayed

81. $\{x \mid x$ is a U.S. airport where between 3% and 5% of flights are delayed$\}$

82. $\{x \mid x$ is a U.S. airport where between 4% and 6% of flights are delayed$\}$

The caseload of Alzheimer's disease in the United States is expected to explode as baby boomers head into their later years. The line graph shows the percentage of Americans with the disease, by age. Based on the information in the graph, represent each of the sets in Exercises 83–84 using the roster method.

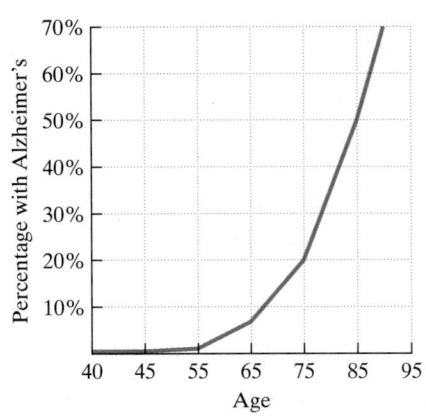

Alzheimer's Prevalence in the U.S., by Age

Percentage with Alzheimer's: 10% 20% 30% 40% 50% 60% 70%

Age: 40 45 55 65 75 85 95

Source: Centers for Disease Control

83. $\{x \mid x$ is an age at which 20% of Americans have Alzheimer's disease$\}$

84. $\{x \mid x$ is an age at which 50% of Americans have Alzheimer's disease$\}$

Writing in Mathematics

85. What is a set?

86. Describe the three methods used to represent a set. Give an example of a set represented by each method.

87. What is the empty set?

88. If a set is written in roster notation, describe how to determine if the set is a finite set or an infinite set.

89. Explain what is meant by equal sets.

90. Explain what is meant by equivalent sets.

91. The bar graph at the top of the next page shows the amount of money for new student loans from 1993 through 2000. Use the graph to write four problems similar to Exercises 75–78. Each exercise should start with a description of the amount of new student loans and require the problem solver to represent the set by listing years using the roster method. Two of the given sets should be in set-builder notation. One of the exercises should have the empty set as the answer.

Amount of New Student Loans

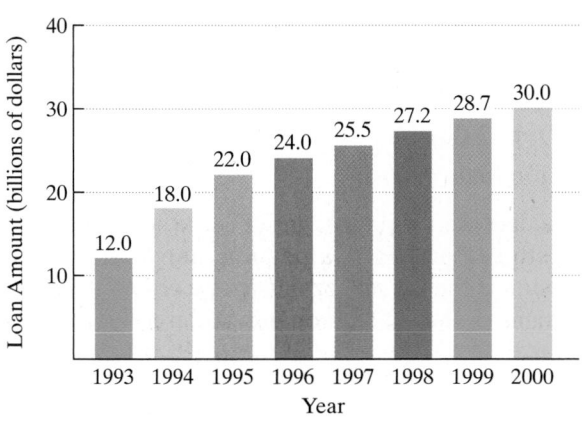

Source: U.S. Department of Education

92. Write a description of four sets, each of which contains you as an element.

93. Write a description of a set whose cardinality exceeds one million.

Critical Thinking Exercises

94. Which one of the following is true?
   a. Two sets can be equal but not equivalent.
   b. Any set in roster notation that contains three dots must be an infinite set.
   c. $n(\varnothing) = 1$
   d. Some sets that can be written in set-builder notation cannot be written in roster form.

95. Which one of the following is true?
   a. The set of fractions between 0 and 1 is an infinite set.
   b. The set of multiples of 4 between 0 and 4,000,000,000 is an infinite set.
   c. If the elements in a set cannot be counted in a trillion years, the set is an infinite set.
   d. Because 0 is not a natural number, it can be deleted from any set without changing the set's cardinality.

96. In a certain town, a barber shaves all those men and only those men who do not shave themselves. Consider each of the following sets:

$$A = \{x \,|\, x \text{ is a man of the town who shaves himself}\}$$

$$B = \{x \,|\, x \text{ is a man of the town who does not shave himself}\}.$$

The one and only barber in the town is Sweeney Todd. If s represents Sweeney Todd,

   a. is $s \in A$?
   b. is $s \in B$?

Group Exercise

97. This activity is a group research project and should result in a presentation made by group members to the entire class. Georg Cantor was certainly not the only genius in history who faced criticism during his lifetime, only to have his work acclaimed as a masterpiece after his death. Describe the life and work of three other people, including at least one mathematician, who faced similar circumstances.

SECTION 2.2   VENN DIAGRAMS AND SUBSETS

OBJECTIVES

1. Understand the meaning of a universal set.
2. Understand the basic ideas of Venn diagrams.
3. Find the complement of a set.
4. Use the symbols ⊆, ⊄, and ⊂.
5. Determine the number of subsets of a set.

One of the joys of your life is your dog, your very special buddy. Lately, however, you've noticed that your companion is not exactly the Albert Einstein of poochdom. His newest exploit is chasing everyone he sees on a bike. When you asked your vet what to do, she suggested taking the bike away from him, immediately. Despite the bad joke, you wonder about your buddy's intelligence, as well as which dog breeds are considered the smartest. In this section, you will see how sets can be used to explore the intelligence of dogs.

① Understand the meaning of a universal set.

Universal Sets

In discussing sets, it is convenient to refer to a general set that contains all elements under discussion. This general set is called the **universal set**. For example, consider the following sets:

$$A = \{\text{Louis Armstrong, Duke Ellington, Charlie Parker, Miles Davis}\}$$
$$B = \{\text{Quincy Jones, Wynton Marsalis}\}.$$

Although the people in set *A* are deceased and those in set *B* are still alive, they are all jazz musicians of the twentieth century. Thus, a possible universal set that includes all elements under discussion is the set of jazz musicians. Other choices for a universal set are the set of musicians or the set of people in the arts.

DEFINITION OF A UNIVERSAL SET

A **universal set**, symbolized by *U*, is a set that contains all the elements being considered in a given discussion or problem.

EXAMPLE 1 DESCRIBING A UNIVERSAL SET

Describe a universal set that includes all elements in sets *A* and *B*:

$$A = \{\textit{Cats, A Chorus Line, Les Misérables}\}$$
$$B = \{\textit{The Producers, Grease}\}.$$

SOLUTION The elements in sets *A* and *B* are musicals. A possible universal set is the set of Broadway musicals.

 Check 1 Point Describe a universal set that includes all elements in sets *A* and *B*:

$$A = \{\textit{Jeopardy!, Who Wants to Be a Millionaire}\}$$
$$B = \{\textit{Wheel of Fortune, Hollywood Squares}\}.$$

② Understand the basic ideas of Venn diagrams.

An Introduction to Venn Diagrams

We can obtain a more thorough understanding of sets and their relationship to a universal set by considering diagrams that allow visual analysis. **Venn diagrams**, named for the British logician John Venn (1834–1923), are used to show the visual relationship among sets.

Here are four basic ideas that are used to construct Venn diagrams:

1. In a Venn diagram, the universal set is represented by a region inside a rectangle. The figure at the right represents the universal set *U*.

STUDY TIP

The size of the circle representing set *A* in a Venn diagram has nothing to do with the number of elements in set *A*.

2. Sets within the universal set are depicted by circles, or sometimes by ovals or other shapes. In the figure at the right, set *A* is represented by the region inside the circle. All elements in set *A* are also elements of the universal set *U*.

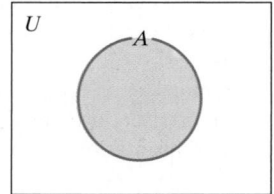

3. The region outside the circle, but within the rectangle, represents the set of elements in the universal set U that are not in set A. This region is labeled A' (read "A prime" or "A complement"). This set, called the **complement** of A, contains all elements that are contained in U but not contained in A.

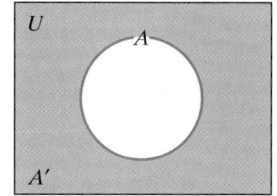

4. By combining set A, shown on the right using light blue shading, and set A', shown using dark blue shading, we obtain the universal set, U. Together, any set and its complement give the universal set.

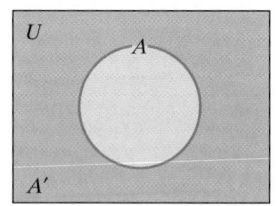

 Find the complement of a set.

DEFINITION OF THE COMPLEMENT OF A SET

The **complement** of set A, symbolized by A', is the set of all elements in the universal set that are *not* in A. This idea can be expressed in set-builder notation as follows:

$$A' = \{x | x \in U \text{ and } x \notin A\}.$$

In order to find A', a universal set U must be given. A fast way to find A' is to cross out the elements in U that are given to be in set A. A' is the set that remains.

EXAMPLE 2   FINDING A SET'S COMPLEMENT

Let $U = \{1, 2, 3, 4, 5, 6, 7, 8, 9\}$ and $A = \{1, 3, 4, 7\}$. Find A'.

SOLUTION   Set A' contains all the elements of set U that are not in set A. Because set A contains the elements $1, 3, 4$, and 7, these elements cannot be members of set A':

$$\{\cancel{1}, 2, \cancel{3}, \cancel{4}, 5, 6, \cancel{7}, 8, 9\}.$$

Thus, set A' contains $2, 5, 6, 8$, and 9:

$$A' = \{2, 5, 6, 8, 9\}.$$

✓ Check **2** Point   Let $U = \{a, b, c, d, e\}$ and $A = \{a, d\}$. Find A'.

EXAMPLE 3   ILLUSTRATING RELATIONSHIPS AMONG SETS USING A VENN DIAGRAM

Use a Venn diagram to illustrate the relationship among sets U, A, and A' from Example 2:

$$U = \{1, 2, 3, 4, 5, 6, 7, 8, 9\}$$
$$A = \{1, 3, 4, 7\}$$
$$A' = \{2, 5, 6, 8, 9\}.$$

TECHNOLOGY

Having trouble creating neat and organized Venn diagrams? Many software programs, including options in most popular word processing programs, allow you to draw neat and presentable Venn diagrams. The draw option of Corel Word-Perfect allows for the easy manipulation of individual figures within the Venn diagram. Likewise, Microsoft Paint—found in the Accessories folder of any Windows Operating System—can be used to easily create Venn diagrams. You can also add colors to the different regions. The diagram images can be inserted into any document or presentation.

SOLUTION

1. Represent the universal set, U, by the region inside a rectangle. Draw a rectangle and label it U in the upper-left-hand corner. This is shown in Figure 2.1.

2. Represent sets A and A'. Depict set A using a circle within the rectangle. Set A is the region inside the circle. A', the complement of set A, is the region outside the circle and inside the rectangle. Label A and A' as shown in Figure 2.2.

3. Because $A = \{1, 3, 4, 7\}$, place these four elements inside circle A, shown in Figure 2.3.

FIGURE 2.1

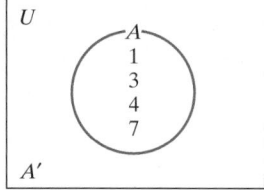

FIGURE 2.2

FIGURE 2.3 $A = \{1, 3, 4, 7\}$

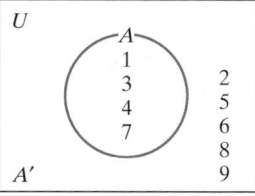

FIGURE 2.4 $A' = \{2, 5, 6, 8, 9\}$

4. Because $A' = \{2, 5, 6, 8, 9\}$, place these five elements outside circle A, in the region labeled A'. This is illustrated in Figure 2.4.

Figure 2.4 represents the required Venn diagram. By joining together the nine elements inside the rectangle, we obtain the universal set

$$U = \{1, 2, 3, 4, 5, 6, 7, 8, 9\}.$$

Check **3** Point   Use a Venn diagram to illustrate the relationship among sets U, A, and A' from Check Point 2.

Subsets of a Set

 Use the symbols $\subseteq$, $\not\subseteq$, and $\subset$.

In the twentieth century, all U.S. presidents were male. All the elements of the set of twentieth-century U.S. presidents are elements of the set of males. Situations like this, in which all the elements of one set are also elements of another set, are described by the following definition:

DEFINITION OF A SUBSET OF A SET

Set A is a **subset** of set B, expressed as

$$A \subseteq B$$

if every element in set A is also an element in set B.

Because every twentieth-century U.S. president was male, this means that the set of twentieth-century U.S. presidents is a subset of the set of males:

$$\{x \mid x \text{ is a twentieth-century U.S. president}\} \subseteq \{x \mid x \text{ is male}\}.$$

The notation $A \not\subseteq B$ means that A **is not a subset** of B. Set A is not a subset of set B if there is at least one element of set A that is not an element of set B. For example, consider the following sets:

$$A = \{1, 2, 3\} \quad \text{and} \quad B = \{1, 2\}.$$

Can you see that 3 is an element of set A that is not in set B? Thus, set A is not a subset of set B: $A \not\subseteq B$.

We can show that $A \subseteq B$ by showing that every element of set A also occurs as an element of set B. We can show that $A \nsubseteq B$ by finding one element of set A that is not in set B.

EXAMPLE 4   USING THE SYMBOLS $\subseteq$ AND $\nsubseteq$

Write $\subseteq$ or $\nsubseteq$ in each blank to make a true statement:

a. $A = \{1, 3, 5, 7\}$

$B = \{1, 3, 5, 7, 9, 11\}$
A_____B

b. $A =$ the set of music written in the twentieth century

$B =$ the set of jazz music
A_____B

c. $A = \{x \mid x$ is a planet of Earth's solar system$\}$

$B = \{$Mercury, Venus, Earth, Mars, Jupiter, Saturn, Uranus, Neptune, Pluto$\}$

A_____B.

SOLUTION

a. All the elements of $A = \{1, 3, 5, 7\}$ are also contained in the set $B = \{1, 3, 5, 7, 9, 11\}$. Therefore, set A is a subset of set B:

$$A \subseteq B.$$

b. Can you think of an example of music written in the twentieth century that was not jazz? One example is the seasonal favorite "White Christmas." This means that not every element of

$A =$ the set of music written in the twentieth century

is an element of

$B =$ the set of jazz music.

Therefore, set A is not a subset of set B:

$$A \nsubseteq B.$$

c. All the elements of

$A = \{x \mid x$ is a planet of the Earth's solar system$\}$

are contained in

$B = \{$Mercury, Venus, Earth, Mars, Jupiter, Saturn, Uranus, Neptune, Pluto$\}$.

Furthermore, the sets are equal $(A = B)$. Because all elements in set A are also in set B, set A is a subset of set B:

$$A \subseteq B.$$

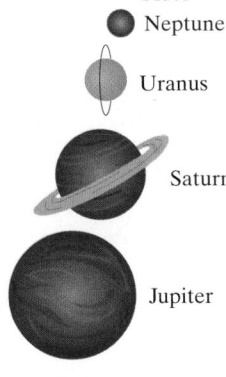

· Pluto

Neptune

Uranus

Saturn

Jupiter

● Mars

● Earth

◉ Venus

· Mercury

The nine planets in Earth's solar system

Check 4 Point

Write $\subseteq$ or $\nsubseteq$ in each blank to make a true statement:

a. $A = \{1, 3, 5, 7\}$

$B = \{1, 3, 5, 6, 9, 11\}$
A_____B

b. $A =$ the set of music written in the twentieth century

$B =$ the set of music
A_____B

c. $A = \{x \mid x$ is a day of the week$\}$

$B = \{$Monday, Tuesday, Wednesday, Thursday, Friday, Saturday, Sunday$\}$

A_____B.

In Example 4(c) and Check Point 4(c), the given equal sets suggest that every set is a subset of itself. If we know that set A is a subset of set B and we exclude the possibility of equal sets, then set A is called a *proper subset* of set B, written $A \subset B$.

DEFINITION OF A PROPER SUBSET OF A SET

Set A is a **proper subset** of set B, expressed as $A \subset B$, if set A is a subset of set B and sets A and B are not equal ($A \neq B$).

Try not to confuse the symbols for subset, $\subseteq$, and proper subset, $\subset$. Because the lower half of the subset symbol in $A \subseteq B$ suggests an equal sign, it is *possible* that sets A and B are equal, although they do not have to be. By contrast, the missing lower line for the proper subset symbol in $A \subset B$ indicates that sets A and B *cannot* be equal.

STUDY TIP

For finite sets, the subset symbols are similar to the inequality symbols from algebra.

The symbols $\subseteq$ and $\leq$

$$A \subseteq B$$

A is a subset of B.

$$a \leq b$$

The number a is less than or equal to the number b.

For example, $2 \leq 3$ and $2 \leq 2$ are true statements.

The symbols $\subset$ and $<$

$$A \subset B$$

A is a proper subset of B.

$$a < b$$

The number a is less than the number b.

For example, $2 < 3$ is true, but $2 < 2$ is false.

EXAMPLE 5   USING THE SYMBOLS $\subseteq$ AND $\subset$

Write $\subseteq$, $\subset$, or both in each blank to make a true statement:

a. $A = \{$James Joyce, Virginia Woolf$\}$

$B = \{$James Joyce, Ernest Hemingway, Ralph Ellison, Virginia Woolf$\}$

A_____B

b. $A = \{2, 4, 6, 8\}$

$B = \{2, 8, 4, 6\}$

A_____B.

SOLUTION

a. Every writer in set A is contained in set B, so A is a subset of B:

$$A \subseteq B.$$

Because the sets are not equal, set A is a proper subset of set B:

$$A \subset B.$$

The symbols $\subseteq$ and $\subset$ can both be placed in the blank to make a true statement.

b. Every number in set A is contained in set B, so A is a subset of B:

$$A \subseteq B$$

Because the sets are equal, set A is *not* a proper subset of set B. The symbol $\subset$ cannot be placed in the blank if we want to form a true statement.

Example 5(b) illustrates that **every set is a subset of itself**. If A is any set, then $A \subseteq A$ because it is obvious that each element of A is a member of A. Example 5(b) also illustrates that no set is a proper subset of itself. If A is any set, $A \not\subset A$.

Check 5 Point   Write $\subseteq$, $\subset$, or both in each blank to make a true statement:

a. $A = \{$Julia Roberts, Tom Hanks$\}$

$B = \{$Tom Hanks, Julia Roberts$\}$

A_____B

b. $A = \{2, 4, 6, 8\}$

$B = \{2, 8, 4, 6, 10\}$

A_____B.

STUDY TIP

Do not confuse the symbols $\in$ and $\subseteq$. The symbol $\in$ means "is an element of" and the symbol $\subseteq$ means "is a subset of." Notice the difference between the following true statements:

$$4 \in \{4, 8\} \qquad \{4\} \subseteq \{4, 8\}.$$

4 is an element of the set {4, 8}.

The set containing 4 is a subset of the set {4, 8}.

Subsets and the Empty Set

The meaning of $A \subseteq B$ leads to some interesting properties of the empty set.

EXAMPLE 6 THE EMPTY SET AS A SUBSET

Let $A = \{\ \}$ and $B = \{1, 2, 3, 4, 5\}$. Is $A \subseteq B$?

SOLUTION A is not a subset of B ($A \nsubseteq B$) if there is at least one element of set A that is not an element of set B. Because A represents the empty set, there are no elements in set A, period, much less elements in A that do not belong to B. Because we cannot find an element in $A = \{\ \}$ that is not contained in $B = \{1, 2, 3, 4, 5\}$, this means that $A \subseteq B$. Equivalently, $\emptyset \subseteq B$.

Check Point 6   Let $A = \{\ \}$ and $B = \{6, 7, 8\}$. Is $A \subseteq B$?

Example 6 illustrates the principle that the empty set is a subset of every set. Furthermore, the empty set is a proper subset of every set except itself.

THE EMPTY SET AS A SUBSET

1. For any set B, $\emptyset \subseteq B$.

2. For any set B other than the empty set, $\emptyset \subset B$.

---

5 Determine the number of subsets of a set.

The Number of Subsets of a Given Set

If a set contains n elements, how many subsets can be formed? Let's observe some special cases, namely sets with $0, 1, 2,$ and 3 elements. We can use inductive reasoning to arrive at a general conclusion. We begin by listing subsets and counting the number of subsets in our list. This is shown in Table 2.2.

TABLE 2.2 THE NUMBER OF SUBSETS: SOME SPECIAL CASES

Set	Number of Elements	List of All Subsets	Number of Subsets
$\{\ \}$	0	$\{\ \}$	1
$\{a\}$	1	$\{a\}, \{\ \}$	2
$\{a, b\}$	2	$\{a, b\}, \{a\}, \{b\}, \{\ \}$	4
$\{a, b, c\}$	3	$\{a, b, c\},$ $\{a, b\}, \{a, c\}, \{b, c\},$ $\{a\}, \{b\}, \{c\}, \{\ \}$	8

Table 2.2 suggests that when we increase the number of elements in the set by one, the number of subsets doubles. The number of subsets appears to be a power of 2.

Number of elements	0	1	2	3
Number of subsets	$1 = 2^0$	$2 = 2^1$	$4 = 2 \times 2 = 2^2$	$8 = 2 \times 2 \times 2 = 2^3$

The power of 2 is the same as the number of elements in the set. Using inductive reasoning, if the set contains n elements, then the number of subsets that can be formed is 2^n.

NUMBER OF SUBSETS

The number of subsets of a set with n elements is 2^n.

For a given set, we know that every subset except the set itself is a proper subset. In Table 2.2, we included the set itself when counting the number of subsets. If we want to find the number of proper subsets, we must exclude counting the given set, thereby decreasing the number by 1.

NUMBER OF PROPER SUBSETS

The number of proper subsets of a set with n elements is $2^n - 1$.

EXAMPLE 7 FINDING THE NUMBER OF SUBSETS AND PROPER SUBSETS

According to *The Intelligence of Dogs* by Stanley Coren, the four smartest breeds of dogs are

1. Border collie

2. Poodle

3. German shepherd

4. Golden retriever.

A man is trying to decide whether to buy a dog. He can afford up to four dogs, but may buy fewer or none at all. However, if the man makes a purchase, he will buy one or more of the four smartest breeds. The kennel only has one dog from each breed—that is, one border collie, one poodle, one German shepherd, and one golden retriever.

a. Consider the set that represents these breeds:

{border collie, poodle, German shepherd, golden retriever}.

Find the number of subsets of this set. What does this number represent in practical terms?

b. List all the subsets and describe what they represent for the prospective dog owner.

c. How many of the subsets are proper subsets?

Smartest Dogs
Rank/Breed

1. Border collie

2. Poodle

3. German shepherd

4. Golden retriever

Source: The Intelligence of Dogs

SOLUTION

a. A set with n elements has 2^n subsets. Because the set of breeds contains four elements, the number of subsets is $2^4 = 2 \times 2 \times 2 \times 2 = 16$. In practical terms, there are 16 different ways to make a selection.

b. Purchase no dogs: Ø

Purchase one dog: {border collie}, {poodle},

{German shepherd}, {golden retriever}

Purchase two dogs: {border collie, poodle},

{border collie, German shepherd},

{border collie, golden retriever},

{poodle, German shepherd},

{poodle, golden retriever},

{German shepherd, golden retriever}

Purchase three dogs: {border collie, poodle, German shepherd},

{border collie, poodle, golden retriever},

{border collie, German shepherd, golden retriever},

{poodle, German shepherd, golden retriever}

Purchase four dogs: {border collie, poodle, German shepherd, golden retriever}

As predicted in part(a), 16 different selections are possible.

c. Every set except

{border collie, poodle, German shepherd, golden retriever}

is a proper subset of the given set. There are 16 − 1, or 15, proper subsets. Although we obtained this number by counting, we can also use the fact that the number of proper subsets of a set with n elements is $2^n - 1$. Because the set of breeds contains four elements, there are $2^4 - 1$ proper subsets:

$$2^4 - 1 = (2 \times 2 \times 2 \times 2) - 1 = 16 - 1 = 15 \text{ proper subsets.}$$

Check 7 Point

You recently purchased three books, shown by the set

{*The Color Purple, Hannibal, The Royals*}.

Now you are deciding which books, if any, to take on vacation. You have enough room to pack up to three books, but may take fewer or none at all.

a. Find the number of subsets of the given set. What does this number represent in practical terms?

b. List all the subsets and describe what they represent in terms of vacation reading.

c. How many of the subsets are proper subsets?

BLITZER BONUS

The Smartest and the Least Intelligent Dogs

In the Venn diagram shown below, the universal set is the set of purebred dogs. Within the universal set, two subsets are shown.

U: Purebred Dogs

Set of Smartest Dogs
{Border collie, Poodle, German shepherd, Golden retriever, Doberman pinscher, Shetland sheepdog, Labrador retriever, Papillon, Rottweiler, Australian cattle dog}

Set of Least Intelligent Dogs
{Afghan hound, Basenji, Bulldog, Chow Chow, Borzoi, Bloodhound, Pekingese, Beagle, Mastiff, Basset hound}

If Afghan hounds could read, they'd probably disagree with being an element of the set of least intelligent dogs.

Source: *The Intelligence of Dogs* by Stanley Coren, 1994.

Exercise Set 2.2

Practice Exercises

In Exercises 1–4, describe a universal set U that includes all elements in the given sets. Answers may vary.

1. *A* = {Bach, Mozart, Beethoven}
    B = {Brahms, Schubert}

2. *A* = {William Shakespeare, Charles Dickens}
    B = {Mark Twain, Robert Louis Stevenson}

3. *A* = {Pepsi, Sprite}
    B = {Coca Cola, Seven-Up}

4. *A* = {Ford Taurus, Honda Accord, Toyota Camry}
    B = {Dodge Intrepid, Buick LeSabre}

In Exercises 5–8, let $U = \{a, b, c, d, e, f, g\}$, $A = \{a, b, f, g\}$, $B = \{c, d, e\}$, $C = \{a, g\}$, *and* $D = \{a, b, c, d, e, f\}$. *Use the roster method to write each of the following sets.*

5. *A'*   6. *B'*   7. *C'*   8. *D'*

In Exercises 9–12, let $U = \{1, 2, 3, 4, \ldots, 20\}$,
$A = \{1, 2, 3, 4, 5\}$, $B = \{6, 7, 8, 9\}$, $C = \{1, 3, 5, 7, \ldots, 19\}$
and $D = \{2, 4, 6, 8, \ldots, 20\}$. *Use the roster method to write each of the following sets.*

9. *A'*   10. *B'*   11. *C'*   12. *D'*

In Exercises 13–16, let $U = \{1, 2, 3, 4, \ldots\}$,
$A = \{1, 2, 3, 4, \ldots, 20\}$, $B = \{1, 2, 3, 4, \ldots, 50\}$,
$C = \{2, 4, 6, 8, \ldots\}$, *and* $D = \{1, 3, 5, 7, \ldots\}$. *Use the roster method to write each of the following sets.*

13. *A'*   14. *B'*   15. *C'*   16. *D'*

In Exercises 17–18, place the various elements in the proper regions of the following Venn diagram. Use a separate diagram for each exercise.

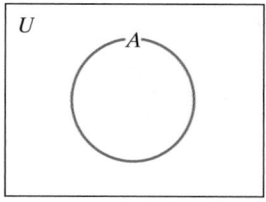

17. Let $U = \{1, 2, 3, 4, 5, 6, 7\}$ and $A = \{1, 2, 5\}$. Find *A'* and place the elements 1 through 7 in the proper region.

18. Let $U = \{1, 2, 3, 4, 5, 6, 7, 8\}$ and $A = \{2, 3, 6, 8\}$. Find *A'* and place the elements 1 through 8 in the proper region.

19. Let $U = \{a, b, c, d, e, f\}$ and $A = \{a, b, f\}$. Find *A'*. Then use a Venn diagram to illustrate the relationship among sets *U, A,* and *A'*.

20. Let $U = \{a, e, i, o, u\}$ and $A = \{a, u\}$. Find *A'*. Then use a Venn diagram to illustrate the relationship among sets *U, A,* and *A'*.

21. Let
$$U = \{0, 1, 2, \ldots, 12\}$$
and
$A = \{x \mid x \text{ is an odd natural number and } x \text{ is less than 10}\}$.
Find *A'*.

22. Let

$$U = \{0, 1, 2, \ldots, 10\}$$

and

$$A = \{x \mid x \text{ is an even natural number and } x \text{ is less than } 11\}.$$

Find A'.

In Exercises 23–34, write $\subseteq$ or $\nsubseteq$ in each blank so that the resulting statement is true.

23. $\{1, 2, 5\}$____$\{1, 2, 3, 4, 5, 6, 7\}$

24. $\{2, 3, 7\}$____$\{1, 2, 3, 4, 5, 6, 7\}$

25. $\{-3, 0, 3\}$____$\{-3, -1, 1, 3\}$

26. $\{-4, 0, 4\}$____$\{-4, -3, -1, 1, 3, 4\}$

27. $\{\text{Monday, Friday}\}$____$\{\text{Saturday, Sunday, Monday, Tuesday, Wednesday}\}$

28. $\{\text{Mercury, Venus, Earth}\}$____$\{\text{Venus, Earth, Mars, Jupiter}\}$

29. $\{x \mid x \text{ is a key on a piano}\}$____$\{x \mid x \text{ is a black key on a piano}\}$

30. $\{x \mid x \text{ is a dog}\}$____$\{x \mid x \text{ is a pure-bred dog}\}$

31. $\{c, o, n, v, e, r, s, a, t, i, o, n\}$____$\{v, o, i, c, e, s, r, a, n, t, o, n\}$

32. $\{r, e, v, o, l, u, t, i, o, n\}$____$\{t, o, l, o, v, e, r, u, i, n\}$

33. $\left\{\frac{4}{7}, \frac{9}{13}\right\}$____$\left\{\frac{7}{4}, \frac{13}{9}\right\}$

34. $\left\{\frac{1}{2}, \frac{1}{3}\right\}$____$\{2, 3, 5\}$

In Exercises 35–40, determine whether $\subseteq$, $\subset$, both, or neither can be placed in each blank to make the statement true.

35. $\{V, C, R\}$____$\{V, C, R, S\}$

36. $\{F, I, N\}$____$\{F, I, N, K\}$

37. $\{0, 2, 4, 6, 8\}$____$\{8, 0, 6, 2, 4\}$

38. $\{9, 1, 7, 3, 4\}$____$\{1, 3, 4, 7, 9\}$

39. $\{x \mid x \text{ is a person who is Christian}\}$____$\{x \mid x \text{ is a person who is Lutheran}\}$

40. $\{x \mid x \text{ is a person who is a registered voter}\}$____$\{x \mid x \text{ is a person who is a registered Republican}\}$

Determine whether each statement in Exercises 41–54 is true or false. If the statement is false, explain why.

41. Ralph $\in \{\text{Ralph, Alice, Trixie, Norton}\}$

42. Canada $\in \{\text{Mexico, United States, Canada}\}$

43. Ralph $\subseteq \{\text{Ralph, Alice, Trixie, Norton}\}$

44. Canada $\subseteq \{\text{Mexico, United States, Canada}\}$

45. $\{\text{Ralph}\} \subseteq \{\text{Ralph, Alice, Trixie, Norton}\}$

46. $\{\text{Canada}\} \subseteq \{\text{Mexico, United States, Canada}\}$

47. $\emptyset \in \{\text{Archie, Edith, Mike, Gloria}\}$

48. $\emptyset \subseteq \{\text{Charlie Chaplin, Groucho Marx, Woody Allen}\}$

49. $\{4\} \in \{\{4\}, \{8\}\}$

50. $\{1\} \in \{\{1\}, \{3\}\}$

51. $\{1, 4\} \nsubseteq \{4, 1\}$

52. $\{1, 4\} \not\subset \{4, 1\}$

53. $0 \notin \emptyset$

54. $0 \nsubseteq \emptyset$

In Exercises 55–60, list all the subsets of the given set.

55. $\{\text{border collie, poodle}\}$

56. $\{\text{Romeo, Juliet}\}$

57. $\{t, a, b\}$

58. $\{I, II, III\}$

59. $\{0\}$

60. $\emptyset$

Calculate the number of subsets and the number of proper subsets for each of the sets in Exercises 61–68.

61. $\{2, 4, 6, 8\}$

62. $\left\{\frac{1}{2}, \frac{1}{3}, \frac{1}{4}, \frac{1}{5}\right\}$

63. $\{2, 4, 6, 8, 10, 12\}$

64. $\{a, b, c, d, e, f\}$

65. $\{x \mid x \text{ is a day of the week}\}$

66. $\{x \mid x \text{ is a U.S. coin worth less than a dollar}\}$

67. $\{x \mid x \in \mathbf{N} \text{ and } x \text{ lies between 2 and 6}\}$

68. $\{x \mid x \in \mathbf{N} \text{ and } x \text{ lies between 2 and 6, inclusive}\}$

Application Exercises

Use the formula for the number of subsets of a set with n elements to solve Exercises 69–74.

69. Houses in Euclid Estates are all identical. However, a person can purchase a new house with some, all, or none of a set of options. This set includes {pool, screened-in balcony, lake view, alarm system, upgraded landscaping}. How many options are there for purchasing a house in this community?

70. A cheese pizza can be ordered with some, all, or none of the following set of toppings: {beef, ham, mushrooms, sausage, peppers, pepperoni, olives, prosciutto, onion}. How many different variations are available for ordering a pizza?

71. Based on more than 1500 ballots sent to film notables, the American Film Institute rated the top U.S. movies. The Institute selected *Citizen Kane* (1941), *Casablanca* (1942), *The Godfather* (1972), *Gone With the Wind* (1939), *Lawrence of Arabia* (1962), and *The Wizard of Oz* (1939) as the top six films. Suppose that you have all six films on video and decide to view some, all, or none of these films. How many viewing options do you have?

72. A small town has four police cars. If a radio dispatcher receives a call, depending on the nature of the situation, no cars, one car, two cars, three cars, or all four cars can be sent. How many options does the dispatcher have for sending the police cars to the scene of the caller?

73. According to the U.S. Census Bureau, the most ethnically diverse U.S. cities are New York City, Los Angeles, Miami, Chicago, Washington D.C., Houston, San Diego, and Seattle. If you decide to visit some, all, or none of these cities, how many travel options do you have?

74. Some of the movies with all-time box office grosses include *Titanic* ($601 million), *Star Wars* ($461 million), *Star Wars: Episode 1–The Phantom Menace* ($431 million), *E.T.* ($400 million), *Jurassic Park* ($357 million), *Forrest Gump* ($330 million), and *The Lion King* ($313 million). Suppose that you have all seven films on video and decide, over the course of a week, to view some, all, or none of these films. How many viewing options do you have?

Writing in Mathematics

75. Describe what is meant by a universal set. Provide an example.

76. What is a Venn diagram and how is it used?

77. Describe what is meant by the complement of a set.

78. Is it possible to find a set's complement if a universal set is not given? Explain your answer.

79. Explain what is meant by a subset.

80. What is the difference between a subset and a proper subset?

81. Explain why the empty set is a subset of every set.

82. Describe the difference between the symbols ∈ and ⊆. Explain how each symbol is used.

83. Describe the formula for finding the number of subsets for a given set. Give an example.

84. Describe how to find the number of proper subsets for a given set. Give an example.

85. In the Blitzer Bonus on page 58, what definitions might be used for "smartest" and "least intelligent" so that it becomes clear how various breeds were assigned to each of the sets?

Critical Thinking Exercises

86. Which one of the following is true?

a. The set {3} has 2^3, or eight, subsets.

b. The complement of the complement of a set is the set itself.

c. Every set has a proper subset.

d. The set {3, {1, 4}} has eight subsets.

87. Suppose that a nickel, a dime, and a quarter are on a table. You may select some, all, or none of the coins. Specify all of the different amounts of money that can be selected.

88. If a set has 127 proper subsets, how many elements are there in the set?

Technology Exercises

89. Find the draw option in your word processing program. Use it to create a Venn diagram that illustrates the relationships among the sets U, A, and A' described in Example 3 on page 51. Consult the help menu to learn how to enter text, thicken your lines, and make other modifications to your picture.

90. The symbols ∈, ∉, ⊂, ⊄, ⊆, and ⊈ are all available in any current word processing program. How can you find and use them? Use the option that allows you to insert a symbol or a character. Then, you should be presented with a group of characters to choose from. The symbols are classified in sets and subsets. To find the symbol you desire, you must choose the correct set first. Re-create the following statements using the symbols found in your word processor.

a. $2 \in \{2, 4, 6, 8\}$

b. $\{a, b, c\} \subset \{a, b, c, d, e\}$

c. ☺ ∉ {◆, ♠, ♣, ♥}

Group Exercises

91. Take some time to reread Exercises 41–54. In your group, discuss the errors that people make when using set notation, particularly the symbols ⊆ and ∈. Then present ways of helping students avoid these errors.

92. A state ballot gives citizens the option of voting for or against a particular issue. A political action group is attempting to predict the outcome of the voting. They survey a sample of the state's registered voters. As a group, you should first decide what the issue is. Then propose at least five plans for dividing the voters into subsets so that a sample of people in each subset can be interviewed. Which one of the plans contains the best subsets of the universal set of state voters in terms of predicting the election's outcome?

SECTION 2.3   VENN DIAGRAMS AND SET OPERATIONS

OBJECTIVES

1. Use Venn diagrams to visualize set relationships.

2. Perform operations with sets.

Boys and girls differ in their toy preferences. These differences are functions of our gender stereotypes—that is, our widely shared beliefs about males' and females' abilities, personality traits, and social behavior. Generally, boys have less leeway to play with "feminine" toys than girls do with "masculine" toys.

Which toys are requested by more than 40% of boys *and* more than 10% of girls? Which toys are requested by more than 40% of boys *or* more than 10% of girls? These questions are not the same. One involves sets of toys joined by the word "and"; the other involves sets of toys joined by the word "or." In this section, you will learn to use Venn diagrams to visualize these set relationships. With this skill, you will analyze childrens' toy preferences in the exercise set.

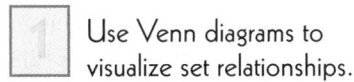

Visualizing Relationships between Two Sets

You need to determine whether there is sufficient support on campus to have a blood drive. You take a survey to obtain information, asking students

> Would you be willing to donate blood?

> Would you be willing to help serve a free breakfast to blood donors?

Set *A* represents the set of students willing to donate blood. Set *B* represents the set of students willing to help serve breakfast to donors. Possible survey results include the following:

- No students willing to donate blood are willing to serve breakfast, and vice versa.
- All students willing to donate blood are willing to serve breakfast.
- The same students who are willing to donate blood are willing to serve breakfast.
- Some of the students willing to donate blood are willing to serve breakfast.

We begin by using Venn diagrams to visualize these results. To do so, we consider four basic relationships and their visualizations.

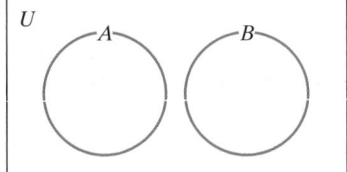

Relationship 1: Disjoint Sets   Two sets that have no elements in common are called **disjoint sets**. Two disjoint sets, *A* and *B*, are shown in the Venn diagram on the left. Disjoint sets are represented as circles that do not overlap. No elements of set *A* are elements of set *B*, and vice versa.

   If set *A* represents the set of students willing to donate blood and set *B* represents the set of students willing to serve breakfast to donors, the set diagram illustrates

> No students willing to donate blood are willing to serve breakfast, and vice versa.

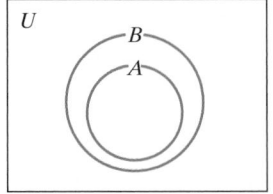

Relationship 2: Proper Subsets   If set *A* is a proper subset of set *B* ($A \subset B$), the relationship is shown in the Venn diagram on the left. All elements of set *A* are elements of set *B*. If an *x* representing an element is placed inside circle *A*, it automatically falls inside circle *B*.

   If set *A* represents the set of students willing to donate blood and set *B* represents the set of students willing to serve breakfast to donors, the set diagram illustrates

> All students willing to donate blood are willing to serve breakfast.

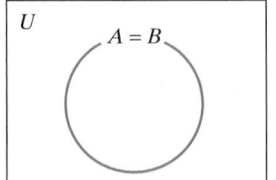

Relationship 3: Equal Sets   If $A = B$, then set *A* contains exactly the same elements as set *B*. This relationship is shown in the Venn diagram on the left. Because all elements in set *A* are in set *B*, and vice versa, this diagram illustrates that when $A = B$, then $A \subseteq B$ and $B \subseteq A$.

   If set *A* represents the set of students willing to donate blood and set *B* represents the set of students willing to serve breakfast to donors, the set diagram illustrates

> The same students who are willing to donate blood are willing to serve breakfast.

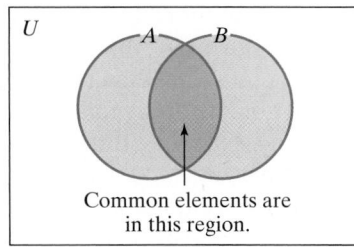

Common elements are
in this region.

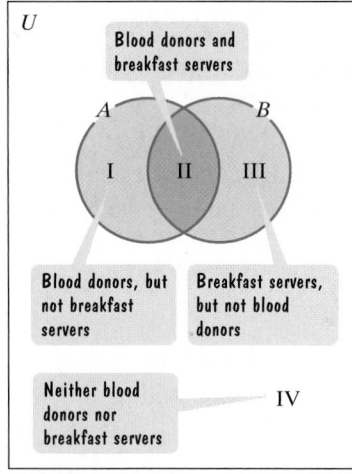

FIGURE 2.5 *A*: Set of blood donors
B: Set of breakfast servers

 Perform operations with
sets.

STUDY TIP

The word *and* is interpreted to
mean *intersection*.

Relationship 4: Sets with Some Common Elements   In mathematics, the word *some* means *there exists at least one*. If set *A* and set *B* have at least one element in common, then the circles representing the sets must overlap. This is illustrated in the Venn diagram on the left.

If set *A* represents the set of students willing to donate blood and set *B* represents the set of students willing to serve breakfast to donors, the presence of at least one student in region II in Figure 2.5 illustrates

Some students willing to donate blood are willing to serve breakfast.

Let's make sure that we understand what the other regions in Figure 2.5 represent.

Region I   This region represents the set of students willing to donate blood, but not serve breakfast. The elements that belong to set *A* but not to set *B* are in this region.

Region III   This region represents the set of students willing to serve breakfast, but not donate blood. The elements that belong to set *B* but not to set *A* are in this region.

Region IV   This region represents the set of students surveyed who are not willing to donate blood and are not willing to serve breakfast. The elements that belong to the universal set *U* that are not in sets *A* or *B* are in this region.

Operations with Sets

In arithmetic, we use operations such as addition and multiplication to combine numbers. We now turn to two set operations, called *intersection* and *union*, that will enable us to combine sets. We will also revisit the idea of the complement of a set.

Intersection   You are now sorting out the results of the blood-donor, breakfast-server survey. Recall that the Red Cross needs to decide whether the number of potential donors warrants a commitment to provide medical staff for the blood drive. Part of your report to the Red Cross involves reporting the number of students willing to both donate blood *and* serve breakfast. You must focus on the set containing all the students who are common to both set *A*, the set of donors, *and* set *B*, the set of breakfast servers. This set is called the *intersection* of sets *A* and *B*.

DEFINITION OF THE INTERSECTION OF SETS

The **intersection** of sets *A* and *B*, written $A \cap B$, is the set of elements common to both set *A* and set *B*. This definition can be expressed in set-builder notation as follows:

$$A \cap B = \{x | x \in A \quad AND \quad x \in B\}.$$

In Example 1, we are asked to find the intersection of two sets. This is done by listing the common elements of both sets. Because the intersection of two sets is also a set, we enclose these elements with braces.

EXAMPLE 1   FINDING THE INTERSECTION OF TWO SETS

Find each of the following intersections:

  a. $\{7, 8, 9, 10, 11\} \cap \{6, 8, 10, 12\}$
  b. $\{1, 3, 5, 7, 9\} \cap \{2, 4, 6, 8\}$
  c. $\{1, 3, 5, 7, 9\} \cap \varnothing.$

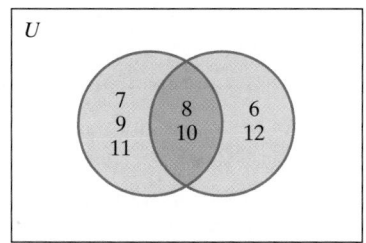

FIGURE 2.6 The numbers 8 and 10 belong to both sets.

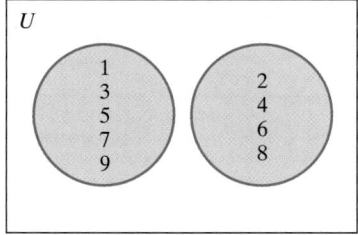

FIGURE 2.7 These disjoint sets have no common elements.

SOLUTION

a. The elements common to $\{7, 8, 9, 10, 11\}$ and $\{6, 8, 10, 12\}$ are 8 and 10. Thus,
$$\{7, 8, 9, 10, 11\} \cap \{6, 8, 10, 12\} = \{8, 10\}.$$
The Venn diagram in Figure 2.6 illustrates this situation.

b. The sets $\{1, 3, 5, 7, 9\}$ and $\{2, 4, 6, 8\}$ have no elements in common. Thus,
$$\{1, 3, 5, 7, 9\} \cap \{2, 4, 6, 8\} = \varnothing.$$
The Venn diagram in Figure 2.7 illustrates this situation. The sets are disjoint.

c. There are no elements in $\varnothing$, the empty set. This means that there can be no elements belonging to both $\{1, 3, 5, 7, 9\}$ and $\varnothing$. Therefore,
$$\{1, 3, 5, 7, 9\} \cap \varnothing = \varnothing.$$

Check 1 Point   Find each of the following intersections:

a. $\{1, 3, 5, 7, 10\} \cap \{6, 7, 10, 11\}$

b. $\{1, 2, 3\} \cap \{4, 5, 6, 7\}$

c. $\{1, 2, 3\} \cap \varnothing$.

Union   You continue to sort the results of the survey. The Red Cross also wants a report on the number of students willing to donate blood *or* serve breakfast *or* do both. You must focus on the set containing the students who are in set A, the set of donors, *or* set B, the set of breakfast servers, or both of these sets. This set is called the *union* of sets A and B.

STUDY TIP

> The word *or* is interpreted to mean *union*.

DEFINITION OF THE UNION OF SETS

The **union** of sets A and B, written $A \cup B$, is the set of elements that are members of set A or of set B or of both sets. This definition can be expressed in set-builder notation as follows:
$$A \cup B = \{x \mid x \in A \quad OR \quad x \in B\}.$$

We can find the union of set A and set B by listing the elements of set A. Then, we include any elements of set B that have not already been listed. Enclose all elements that are listed with braces. This shows that the union of two sets is also a set.

EXAMPLE 2 | FINDING THE UNION OF SETS

Find each of the following unions:

a. $\{7, 8, 9, 10, 11\} \cup \{6, 8, 10, 12\}$

b. $\{1, 3, 5, 7, 9\} \cup \{2, 4, 6, 8\}$

c. $\{1, 3, 5, 7, 9\} \cup \varnothing$.

SOLUTION   This example uses the same sets as in Example 1. However, this time we are finding the union of the sets, rather than their intersection.

a. To find $\{7, 8, 9, 10, 11\} \cup \{6, 8, 10, 12\}$, start by listing all the elements from the first set, namely 7, 8, 9, 10, and 11. Now list all the elements from the second set that are not in the first set, namely 6 and 12. The union is the set consisting of all these elements. Thus,
$$\{7, 8, 9, 10, 11\} \cup \{6, 8, 10, 12\} = \{6, 7, 8, 9, 10, 11, 12\}.$$

STUDY TIP

> When finding the union of two sets, some elements may appear in both sets. List these common elements only once, *not twice*, in the union of the sets.

b. To find $\{1, 3, 5, 7, 9\} \cup \{2, 4, 6, 8\}$, list the elements from the first set, namely 1, 3, 5, 7, and 9. Now add to the list the elements in the second set that are not in the first set. This includes every element in the second set, namely 2, 4, 6, and 8. The union is the set consisting of all these elements, so

$$\{1, 3, 5, 7, 9\} \cup \{2, 4, 6, 8\} = \{1, 2, 3, 4, 5, 6, 7, 8, 9\}.$$

c. To find $\{1, 3, 5, 7, 9\} \cup \emptyset$, list the elements from the first set, namely 1, 3, 5, 7, and 9. Because there are no elements in $\emptyset$, the empty set, there are no additional elements to add to the list. Thus,

$$\{1, 3, 5, 7, 9\} \cup \emptyset = \{1, 3, 5, 7, 9\}.$$

Examples 1 and 2 illustrate the role that the empty set plays in intersection and union.

THE EMPTY SET IN INTERSECTION AND UNION

For any set A,

1. $A \cap \emptyset = \emptyset$

2. $A \cup \emptyset = A$.

 **Check Point 2**   Find each of the following unions:

a. $\{1, 3, 5, 7, 10\} \cup \{6, 7, 10, 11\}$

b. $\{1, 2, 3\} \cup \{4, 5, 6, 7\}$

c. $\{1, 2, 3\} \cup \emptyset$.

EXAMPLE 3 | USING SET OPERATIONS

Given

$$U = \{1, 2, 3, 4, 5, 6, 7, 8, 9, 10\}$$
$$A = \{1, 3, 7, 9\}$$
$$B = \{3, 7, 8, 10\}$$

find each of the following sets:

a. $(A \cup B)'$ 　　　　　　　　　　**b.** $A' \cap B'$.

SOLUTION

a. To find $(A \cup B)'$, we will first work inside the parentheses and determine $A \cup B$. Then we'll find the complement of $A \cup B$, namely $(A \cup B)'$.

$A \cup B = \{1, 3, 7, 9\} \cup \{3, 7, 8, 10\}$ 　*These are the given sets.*
$\quad\quad = \{1, 3, 7, 8, 9, 10\}$ 　*Join (unite) the elements, listing the common elements (3 and 7) only once.*

Now find $(A \cup B)'$, the complement of $A \cup B$.

$(A \cup B)' = \{1, 3, 7, 8, 9, 10\}'$
$\quad\quad\quad = \{2, 4, 5, 6\}$ 　*List the elements in the universal set that are not listed in $\{1, 3, 7, 8, 9, 10\}$: $\{\not1, 2, \not3, 4, 5, 6, \not7, \not8, \not9, \not{10}\}$.*

b. To find $A' \cap B'$, we must first identify the elements in A' and B'. Set A' is the set of elements of U that are not in set A:

$A' = \{2, 4, 5, 6, 8, 10\}$.   　*List the elements in the universal set that are not listed in $A = \{1, 3, 7, 9\}$: $\{\not1, 2, \not3, 4, 5, 6, \not7, 8, \not9, 10\}$.*

Set B' is the set of elements of U that are not in set B:

$$B' = \{1, 2, 4, 5, 6, 9\}. \quad \text{List the elements in the universal set that are not listed in } B = \{3, 7, 8, 10\}: \{1, 2, \cancel{3}, 4, 5, 6, \cancel{7}, \cancel{8}, 9, \cancel{10}\}.$$

Now we can find $A' \cap B'$, the set of elements belonging to both A' and to B':

$$A' \cap B' = \{2, 4, 5, 6, 8, 10\} \cap \{1, 2, 4, 5, 6, 9\}$$
$$= \{2, 4, 5, 6\} \quad \text{The numbers 2, 4, 5, and 6 are common to both sets.}$$

 **Check 3 Point** Given $U = \{a, b, c, d, e\}$, $A = \{b, c\}$, and $B = \{b, c, e\}$, find each of the following sets:

a. $(A \cup B)'$ **b.** $A' \cap B'$.

It is no coincidence that we obtain the same answer in both parts of Example 3 and Check Point 3. For any two sets A and B, it is always true that $(A \cup B)' = A' \cap B'$.

We can reverse the union and intersection symbols in the above statement. For any two sets A and B, it is also true that

$$(A \cap B)' = A' \cup B'.$$

These two statements are known as *De Morgan's laws*, named for the British logician Augustus De Morgan (1806–1871).

STUDY TIP

Learn the pattern of DeMorgan's laws. They are extremely important in logic and will be discussed in more detail in Chapter 3, Logic.

DE MORGAN'S LAWS

$(A \cup B)' = A' \cap B'$:   The complement of the union of two sets is the intersection of the complements of those sets.

$(A \cap B)' = A' \cup B'$:   The complement of the intersection of two sets is the union of the complements of those sets.

Exercise Set 2.3

Practice Exercises

In Exercises 1–24, let

$$U = \{1, 2, 3, 4, 5, 6, 7\}$$
$$A = \{1, 3, 5, 7\}$$
$$B = \{1, 2, 3\}$$
$$C = \{2, 3, 4, 5, 6\}.$$

Find each of the following sets.

1. $A \cap B$	**2.** $B \cap C$	**3.** $A \cup B$	**4.** $B \cup C$
5. A'	**6.** B'	**7.** $A' \cap B'$	**8.** $B' \cap C$
9. $A \cup C'$	**10.** $B \cup C'$	**11.** $(A \cap C)'$	**12.** $(A \cap B)'$
13. $A' \cup C'$	**14.** $A' \cup B'$	**15.** $(A \cup B)'$	**16.** $(A \cup C)'$
17. $A \cup \varnothing$	**18.** $C \cup \varnothing$	**19.** $A \cap \varnothing$	**20.** $C \cap \varnothing$
21. $A \cup U$	**22.** $B \cup U$	**23.** $A \cap U$	**24.** $B \cap U$

In Exercises 25–48, let

$$U = \{a, b, c, d, e, f, g, h\}$$
$$A = \{a, g, h\}$$
$$B = \{b, g, h\}$$
$$C = \{b, c, d, e, f\}.$$

Find each of the following sets.

25. $A \cap B$	**26.** $B \cap C$	**27.** $A \cup B$	**28.** $B \cup C$
29. A'	**30.** B'	**31.** $A' \cap B'$	**32.** $B' \cap C$
33. $A \cup C'$	**34.** $B \cup C'$	**35.** $(A \cap C)'$	**36.** $(A \cap B)'$
37. $A' \cup C'$	**38.** $A' \cup B'$	**39.** $(A \cup B)'$	**40.** $(A \cup C)'$
41. $A \cup \varnothing$	**42.** $C \cup \varnothing$	**43.** $A \cap \varnothing$	**44.** $C \cap \varnothing$
45. $A \cup U$	**46.** $B \cup U$	**47.** $A \cap U$	**48.** $B \cap U$

In Exercises 49–56, let

$U = \{x | x \in \mathbf{N} \text{ and } x \text{ is less than } 9\}$

$A = \{x | x \text{ is an odd natural number less than } 9\}$

$B = \{x | x \text{ is an even natural number less than } 9\}$

$C = \{x | x \in \mathbf{N} \text{ and } x \text{ is between } 1 \text{ and } 6\}.$

Find each of the following sets.

49. $A \cup B$ **50.** $B \cup C$ **51.** $A \cap U$ **52.** $A \cup U$

53. $A \cap C'$ **54.** $A \cap B'$ **55.** $(B \cap C)'$ **56.** $(A \cap C)'$

In Exercises 57–66, use the Venn diagram to list the elements of each set in roster form.

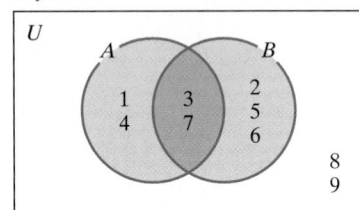

57. A **58.** B **59.** U **60.** $A \cup B$

61. $A \cap B$ **62.** A' **63.** B' **64.** $(A \cap B)'$

65. $(A \cup B)'$ **66.** $A' \cap B$

Application Exercises

As a result of cultural expectations about what is appropriate behavior for each gender, boys and girls differ substantially in their toy preferences. The graph shows the percentage of boys and girls asking for various types of toys in letters to Santa Claus. Use the information in the graph to write each set in Exercises 67–72 in roster form or express the set as ∅.

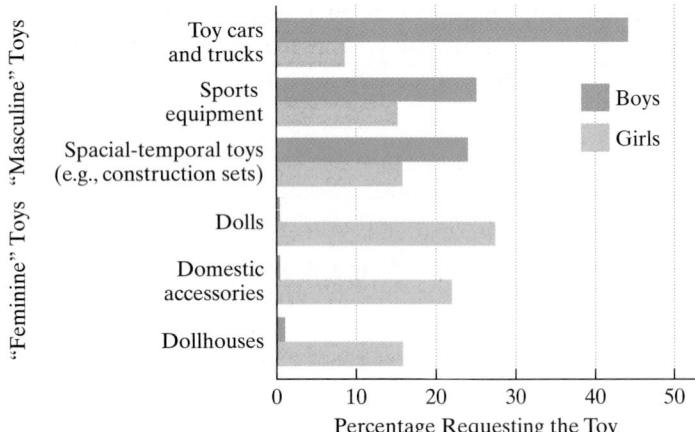

Toys Requested by Children

Source: Richard, J. G., & Simpson, C. H. (1982). Children, gender and social structure: An analysis of the contents of letters to Santa Claus. *Child Development*, 53, 429–436.

67. $\{x | x$ is a toy requested by more than 10% of the boys$\} \cap$ $\{x | x$ is a toy requested by less than 20% of the girls$\}$

68. $\{x | x$ is a toy requested by fewer than 5% of the boys$\} \cap$ $\{x | x$ is a toy requested by fewer than 20% of the girls$\}$

69. $\{x | x$ is a toy requested by more than 10% of the boys$\} \cup$ $\{x | x$ is a toy requested by less than 20% of the girls$\}$

70. $\{x | x$ is a toy requested by fewer than 5% of the boys$\} \cup$ $\{x | x$ is a toy requested by fewer than 20% of the girls$\}$

71. The set of toys requested by more than 40% of the boys and more than 10% of the girls

72. The set of toys requested by more than 40% of the boys or more than 10% of the girls

Writing in Mathematics

73. Describe the Venn diagram for two disjoint sets. How does this diagram illustrate that the sets have no common elements?

74. Describe the Venn diagram for proper subsets. How does this diagram illustrate that the elements of one set are also in the second set?

75. Describe the Venn diagram for two equal sets. How does this diagram illustrate that the sets are equal?

76. Describe the Venn diagram for two sets with common elements. How does the diagram illustrate this relationship?

77. Describe what is meant by the intersections of sets. Give an example.

78. Describe what is meant by the union of sets. Give an example.

Critical Thinking Exercises

79. Which one of the following is true?

   a. $n(A \cup B) = n(A) + n(B)$

   b. $A \cap A' = \varnothing$

   c. $(A \cup B) \subseteq A$

   d. If $A \subseteq B$, then $A \cap B = B$.

80. Which one of the following is true?

   a. $A \cap U = U$

   b. $A \cup \varnothing = \varnothing$

   c. If $A \subseteq B$, then $A \cap B = \varnothing$.

   d. If $B \subseteq A$, then $A \cap B = B$.

81. Find a universal set U and sets A and B that show that $(A \cup B)' = A' \cup B'$ is false.

Technology Exercise

82. Using the Paint Program in Windows, create four Venn diagrams, each with two sets, A and B. The diagrams must include shaded regions corresponding to the following:

   a. $A \cap B$ **b.** $A \cup B$ **c.** $A' \cup B'$ **d.** $(A \cap B)'$

Group Exercise

83. The sentence "All apes are mammals" means that the set of apes is a proper subset of the set of mammals. If set A is the set of apes and set B is the set of mammals, this relationship can be visualized with a Venn diagram in which circle A is drawn inside circle B. Group members should write four true sentences that illustrate each of the four set relationships on pages 61–62. Then, rewrite each sentence using the language of sets. Finally, use a Venn diagram to illustrate each of your group's sentences.

SECTION 2.4   SET OPERATIONS AND VENN DIAGRAMS WITH THREE SETS

OBJECTIVES

1. Perform set operations with three sets.
2. Use Venn diagrams with three sets.
3. Use Venn diagrams to illustrate equality of sets.

Our bodies are fragile and complex, vulnerable to disease, and easily damaged. The imminent mapping of the human genome—all 140,000 genes—could lead to rapid advances in treating heart disease, cancer, Alzheimer's, and AIDS. Neural stem cell research could make it possible to repair brain damage and even recreate whole parts of the brain. There appears to be no limit to the parts of our bodies that can be replaced. By contrast, at the start of the twentieth century, we lacked a basic understanding of the different types of human blood. The discovery of blood types, which can be illustrated by a Venn diagram with three sets, rescued surgery patients from random, often lethal, transfusions. In this sense, the Venn diagram that you encounter in this section reinforces our optimism that life does improve, and that we are better off today than we were one hundred years ago.

1 Perform set operations with three sets.

Set Operations with Three Sets

We now know how to find the union and intersection of two sets. We also know how to find a set's complement. In Example 1, we apply set operations to situations containing three sets. Always begin by finding the set within parentheses.

EXAMPLE 1   SET OPERATIONS WITH THREE SETS

Given

$$U = \{1, 2, 3, 4, 5, 6, 7, 8, 9\}$$
$$A = \{1, 2, 3, 4, 5\}$$
$$B = \{1, 2, 3, 6, 8\}$$
$$C = \{2, 3, 4, 6, 7\}$$

find each of the following sets:

a. $A \cup (B \cap C)$
b. $(A \cup B) \cap (A \cup C)$
c. $A \cap (B \cup C')$.

$U = \{1, 2, 3, 4, 5, 6, 7, 8, 9\}$

$A = \{1, 2, 3, 4, 5\}$

$B = \{1, 2, 3, 6, 8\}$

$C = \{2, 3, 4, 6, 7\}$

The given sets, repeated

SOLUTION

a. To find $A \cup (B \cap C)$, first find the set within the parentheses, $B \cap C$:

$$B \cap C = \{1, 2, 3, 6, 8\} \cap \{2, 3, 4, 6, 7\} = \{2, 3, 6\}.$$

> Common elements are 2, 3, and 6.

Now finish the problem by finding $A \cup (B \cap C)$:

$$A \cup (B \cap C) = \{1, 2, 3, 4, 5\} \cup \{2, 3, 6\} = \{1, 2, 3, 4, 5, 6\}.$$

> List all elements in A and then add the only unlisted element in B ∩ C, namely 6.

b. To find $(A \cup B) \cap (A \cup C)$, first find the sets within parentheses. Start with $A \cup B$:

$$A \cup B = \{1, 2, 3, 4, 5\} \cup \{1, 2, 3, 6, 8\} = \{1, 2, 3, 4, 5, 6, 8\}.$$

> List all elements in A and then add the unlisted elements in B, namely 6 and 8.

Now find $A \cup C$:

$$A \cup C = \{1, 2, 3, 4, 5\} \cup \{2, 3, 4, 6, 7\} = \{1, 2, 3, 4, 5, 6, 7\}.$$

> List all elements in A and then add the unlisted elements in C, namely 6 and 7.

Now finish the problem by finding $(A \cup B) \cap (A \cup C)$:

$$(A \cup B) \cap (A \cup C) = \{1, 2, 3, 4, 5, 6, 8\} \cap \{1, 2, 3, 4, 5, 6, 7\} = \{1, 2, 3, 4, 5, 6\}.$$

> Common elements are 1, 2, 3, 4, 5, and 6.

c. As in parts (a) and (b), to find $A \cap (B \cup C')$, begin with the set in parentheses. First we must find C', elements in U that are not in C:

$$C' = \{1, 5, 8, 9\}.$$ List the elements in U that are not in $C = \{2, 3, 4, 6, 7\}$: $\{1, \cancel{2}, \cancel{3}, \cancel{4}, 5, \cancel{6}, \cancel{7}, 8, 9\}$

Now we can identify elements of $B \cup C'$:

$$B \cup C' = \{1, 2, 3, 6, 8\} \cup \{1, 5, 8, 9\} = \{1, 2, 3, 5, 6, 8, 9\}.$$

> List all elements in B and then add the unlisted elements in C', namely 5 and 9.

Now finish the problem by finding $A \cap (B \cup C')$:

$$A \cap (B \cup C') = \{1, 2, 3, 4, 5\} \cap \{1, 2, 3, 5, 6, 8, 9\} = \{1, 2, 3, 5\}.$$

> Common elements are 1, 2, 3, and 5.

Check 1 Point

Given $U = \{a, b, c, d, e, f\}$, $A = \{a, b, c, d\}$, $B = \{a, b, d, f\}$, and $C = \{b, c, f\}$, find each of the following sets:

a. $A \cup (B \cap C)$

b. $(A \cup B) \cap (A \cup C)$

c. $A \cap (B \cup C')$.

Use Venn diagrams with three sets.

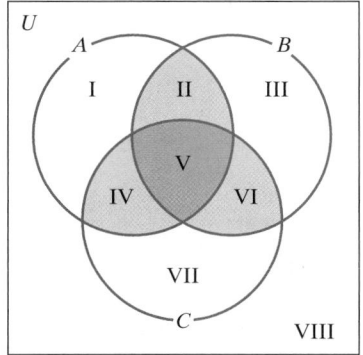

FIGURE 2.8 Three intersecting sets separate the universal set into eight regions.

Venn Diagrams with Three Sets

Venn diagrams can contain three or more sets, such as the diagram in Figure 2.8. The three sets in the figure separate the universal set, U, into eight regions. The numbering of these regions is arbitrary—that is, we can number any region as I, any region as II, and so on. Here is a description of each region:

Region I   This region represents elements in set A that are in neither sets B nor C.

Region II   This region represents elements in both sets A and B that are not in set C.

Region III   This region represents elements in set B that are in neither sets A nor C.

Region IV   This region represents elements in both sets A and C that are not in set B.

Region V   This region represents elements that are common to sets A, B, and C.

Region VI   This region represents elements in both sets B and C that are not in set A.

Region VII   This region represents elements in set C that are in neither sets A nor B.

Region VIII This region represents elements in the universal set U that are not in sets A, B, or C.

EXAMPLE 2 | UNDERSTANDING A VENN DIAGRAM WITH THREE SETS

The Venn diagram with three sets is shown again on the right.

a. Which regions represent set A?

b. Which regions represent $A \cup B$?

c. Which regions represent $B \cap C$?

d. Which regions represent A'?

e. Which region represents $A \cap B \cap C$?

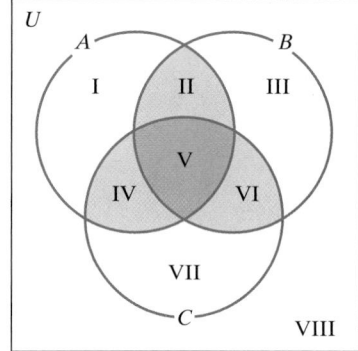

SOLUTION

a. Set A is represented by the regions inside circle A. Therefore, set A is represented by regions I, II, IV, and V.

b. Set $A \cup B$ is found in the Venn diagram by uniting, or joining, all the regions inside circles A and B. Therefore, $A \cup B$ is represented by regions I, II, III, IV, V, and VI.

c. Set $B \cap C$ is found in the Venn diagram by taking the regions common to circles B and C. Therefore, $B \cap C$ is represented by regions V and VI.

d. Set A' is represented by the regions outside circle A. Therefore, A' is represented by regions III, VI, VII, and VIII.

e. Set $A \cap B \cap C$ is found in the Venn diagram by taking the region common to circles A, B, and C. Therefore, $A \cap B \cap C$ is represented by region V.

BLITZER BONUS

Venn Diagrams and Mixing Primary Colors of Light

Venn diagrams with three intersecting circles can be used to illustrate the blending of the three primary colors of light—red, green, and blue-violet. The intersection of the three regions, white light, is the blending of all three primary colors.

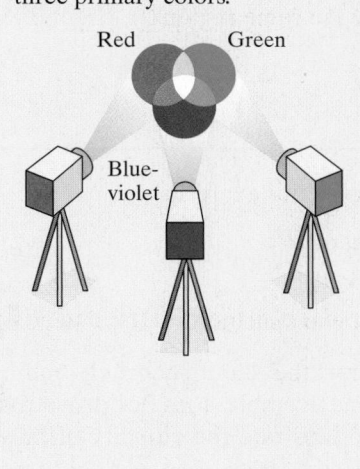

Use the Venn diagram with three sets in Example 2 to answer the following questions:

a. Which regions represent set C?

b. Which regions represent $B \cup C$?

c. Which regions represent $A \cap C$?

d. Which regions represent C'?

e. Which regions represent $A \cup B \cup C$?

Illustrating the Equality of Sets

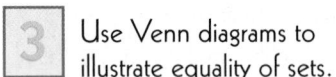

In Section 2.3, we presented one of De Morgan's laws that stated

$$(A \cap B)' = A' \cup B'.$$

To illustrate that $(A \cap B)'$ and $A' \cup B'$ are equal, we use a Venn diagram. If both sets represent the same regions of the diagram, then they are equal. Example 3 shows how this is done.

| EXAMPLE 3 | ILLUSTRATING A DE MORGAN LAW WITH A VENN DIAGRAM |

Use the Venn diagram in Figure 2.9 to show that

$$(A \cap B)' = A' \cup B'.$$

SOLUTION   Begin with the regions represented by $(A \cap B)'$.

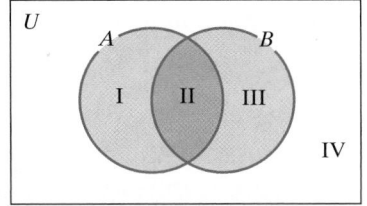

FIGURE 2.9

Set	Regions in the Venn Diagram
A	I, II
B	II, III
$A \cap B$	II (This is the region common to A and B.)
$(A \cap B)'$	I, III, IV (These are the regions in U that are not in $A \cap B$.)

Next, find the regions represented by $A' \cup B'$.

Set	Regions in the Venn Diagram
A'	III, IV (These are the regions not in A.)
B'	I, IV (These are the regions not in B.)
$A' \cup B'$	I, III, IV (These are the regions obtained by uniting the regions representing A' and B'.)

Both $(A \cap B)'$ and $A' \cup B'$ are represented by the same regions, I, III, and IV, of the Venn diagram. This result illustrates that

$$(A \cap B)' = A' \cup B'$$

for all sets A and B.

Use the Venn diagram in Figure 2.9 to solve this exercise.

a. Which region is represented by $(A \cup B)'$?

b. Which region is represented by $A' \cap B'$?

c. Based on parts (a) and (b), what can you conclude?

In Example 1, did you notice that we obtained the same set when finding $A \cup (B \cap C)$ and $(A \cup B) \cap (A \cup C)$? Although one example does not prove that these sets are equal, we can use a Venn diagram to illustrate the equality of these sets. We do this in Example 4.

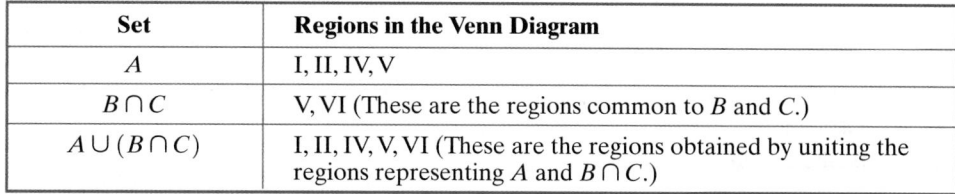

EXAMPLE 4 | ILLUSTRATING THE EQUALITY OF SETS

Use a Venn diagram to show that

$$A \cup (B \cap C) = (A \cup B) \cap (A \cup C).$$

SOLUTION   Use a Venn diagram with three sets A, B, and C, drawn in Figure 2.10. Begin with the regions represented by $A \cup (B \cap C)$.

Set	Regions in the Venn Diagram
A	I, II, IV, V
$B \cap C$	V, VI (These are the regions common to B and C.)
$A \cup (B \cap C)$	I, II, IV, V, VI (These are the regions obtained by uniting the regions representing A and $B \cap C$.)

Next, find the regions represented by $(A \cup B) \cap (A \cup C)$.

Set	Regions in the Venn Diagram
A	I, II, IV, V
B	II, III, V, VI
C	IV, V, VI, VII
$A \cup B$	I, II, III, IV, V, VI (Unite the regions representing A and B.)
$A \cup C$	I, II, IV, V, VI, VII (Unite the regions representing A and C.)
$(A \cup B) \cap (A \cup C)$	I, II, IV, V, VI (These are the regions common to $A \cup B$ and $A \cup C$.)

Both $A \cup (B \cap C)$ and $(A \cup B) \cap (A \cup C)$ are represented by the same regions, I, II, IV, V, and VI, of the Venn diagram. This result illustrates that

$$A \cup (B \cap C) = (A \cup B) \cap (A \cup C)$$

for all sets A, B, and C.

Check **4** Point

Use the Venn diagram in Figure 2.10 to solve this exercise.

a. Which regions are represented by $A \cap (B \cup C)$?

b. Which regions are represented by $(A \cap B) \cup (A \cap C)$?

c. Based on parts (a) and (b), what can you conclude?

BLITZER BONUS

Blood Types and Venn Diagrams

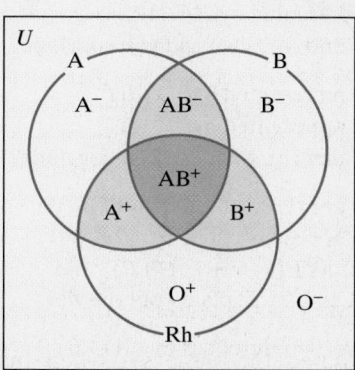

FIGURE 2.11 Human blood types

In the early 1900s, the Austrian immunologist Karl Landsteiner discovered that all blood is not the same. Blood serum drawn from one person often clumped when mixed with the blood cells of another. The clumping was caused by different antigens, proteins, and carbohydrates that trigger antibodies and fight infection. Landsteiner classified blood types based on the presence or absence of the antigens A, B, and Rh in red blood cells. The Venn diagram in Figure 2.11 contains eight regions representing the eight common blood groups.

In the Venn diagram, blood with the Rh antigen is labeled positive and blood lacking the Rh antigen is labeled negative. The region where the three circles intersect represents type AB⁺, indicating that a person with this blood type has the antigens A, B and Rh. Observe that type O blood (both positive and negative) lacks A and B antigens. Type O⁻ lacks all three antigens, A, B, and Rh.

In blood transfusions, the recipient must have all or more of the antigens present in the donor's blood. This discovery rescued surgery patients from random, often lethal, transfusions. This knowledge made the massive blood drives during World War I possible. Eventually, it made the modern blood bank possible as well.

Exercise Set 2.4

Practice Exercises

In Exercises 1–12, let

$$U = \{1, 2, 3, 4, 5, 6, 7\}$$
$$A = \{1, 3, 5, 7\}$$
$$B = \{1, 2, 3\}$$
$$C = \{2, 3, 4, 5, 6\}.$$

Find each of the following sets.

1. $A \cup (B \cap C)$ **2.** $A \cap (B \cup C)$

3. $(A \cup B) \cap (A \cup C)$ **4.** $(A \cap B) \cup (A \cap C)$

5. $A' \cap (B \cup C')$ **6.** $C' \cap (A \cup B')$

7. $(A' \cap B) \cup (A' \cap C')$ **8.** $(C' \cap A) \cup (C' \cap B')$

9. $(A \cup B \cup C)'$ **10.** $(A \cap B \cap C)'$

11. $(A \cup B)' \cap C$ **12.** $(B \cup C)' \cap A$

In Exercises 13–24, let

$$U = \{a, b, c, d, e, f, g, h\}$$
$$A = \{a, g, h\}$$
$$B = \{b, g, h\}$$
$$C = \{b, c, d, e, f\}.$$

Find each of the following sets.

13. $A \cup (B \cap C)$ **14.** $A \cap (B \cup C)$

15. $(A \cup B) \cap (A \cup C)$ **16.** $(A \cap B) \cup (A \cap C)$

17. $A' \cap (B \cup C')$ **18.** $C' \cap (A \cup B')$

19. $(A' \cap B) \cup (A' \cap C')$ **20.** $(C' \cap A) \cup (C' \cap B')$

21. $(A \cup B \cup C)'$ **22.** $(A \cap B \cap C)'$

23. $(A \cup B)' \cap C$ **24.** $(B \cup C)' \cap A$

In Exercises 25–32, use the Venn diagram shown to answer each question.

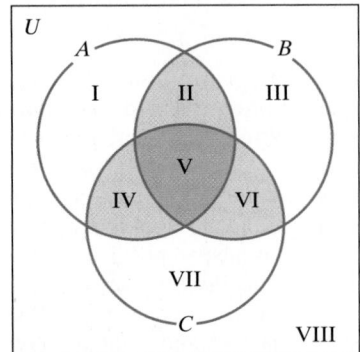

25. Which regions represent set B?

26. Which regions represent set C?

27. Which regions represent $A \cup C$?

28. Which regions represent $B \cup C$?

29. Which regions represent $A \cap B$?

30. Which regions represent $A \cap C$?

31. Which regions represent B'?

32. Which regions represent C'?

In Exercises 33–44, use the Venn diagram to list each set in roster form.

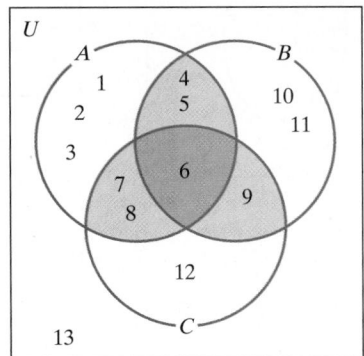

33. A **34.** B **35.** $A \cup B$

36. $B \cup C$ **37.** $(A \cup B)'$ **38.** $(B \cup C)'$

39. $A \cap B$ **40.** $A \cap C$ **41.** $A \cap B \cap C$

42. $A \cup B \cup C$ **43.** $(A \cap B \cap C)'$ **44.** $(A \cup B \cup C)'$

Use the Venn diagram shown to solve Exercises 45–54.

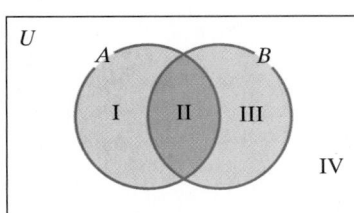

45. a. Which region is represented by $A \cap B$?
   b. Which region is represented by $B \cap A$?
   c. Based on parts (a) and (b), what can you conclude?

46. a. Which regions are represented by $A \cup B$?
   b. Which regions are represented by $B \cup A$?
   c. Based on parts (a) and (b), what can you conclude?

47. a. Which region(s) is/are represented by $(A \cap B)'$?
   b. Which region(s) is/are represented by $A' \cap B'$?
   c. Based on parts (a) and (b), are $(A \cap B)'$ and $A' \cap B'$ equal for all sets A and B? Explain your answer.

48. a. Which region(s) is/are represented by $(A \cup B)'$?
   b. Which region(s) is/are represented by $A' \cup B'$?
   c. Based on parts (a) and (b), are $(A \cup B)'$ and $A' \cup B'$ equal for all sets A and B? Explain your answer.

In Exercises 49–54, use the Venn diagram shown on page 72 to determine whether the given sets are equal for all sets A and B.

49. $A' \cup B, \qquad A \cap B'$

50. $A' \cap B, \qquad A \cup B'$

51. $(A \cup B)', \quad (A \cap B)'$

52. $(A \cup B)', \quad A' \cap B$

53. $(A' \cap B)', \quad A \cup B'$

54. $(A \cup B')', \quad A' \cap B$

Use the Venn diagram shown to solve Exercises 55–64.

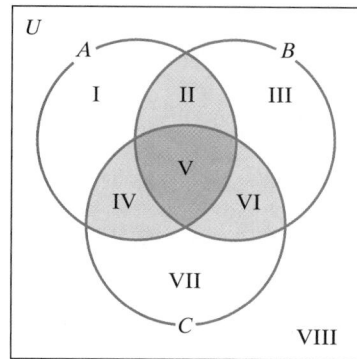

55. a. Which regions are represented by $(A \cap B) \cup C$?

   b. Which regions are represented by $(A \cup C) \cap (B \cup C)$?

   c. Based on parts (a) and (b), what can you conclude?

56. a. Which regions are represented by $(A \cup B) \cap C$?

   b. Which regions are represented by $(A \cap C) \cup (B \cap C)$?

   c. Based on parts (a) and (b), what can you conclude?

57. a. Which regions are represented by $A \cap (B \cup C)$?

   b. Which regions are represented by $A \cup (B \cap C)$?

   c. Based on parts (a) and (b), are $A \cap (B \cup C)$ and $A \cup (B \cap C)$ equal for all sets A, B, and C? Explain your answer.

58. a. Which regions are represented by $C \cup (B \cap A)$?

   b. Which regions are represented by $C \cap (B \cup A)$?

   c. Based on parts (a) and (b), are $C \cup (B \cap A)$ and $C \cap (B \cup A)$ equal for all sets A, B, and C? Explain your answer.

In Exercises 59–64, use the Venn diagram shown above to determine whether the given sets are equal for all sets A, B, and C.

59. $A \cap (B \cup C); \quad (A \cap B) \cup C$

60. $A \cup (B \cap C); \quad (A \cup B) \cap C$

61. $B \cup (A \cap C); \quad (A \cup B) \cap (B \cup C)$

62. $B \cap (A \cup C); \quad (A \cap B) \cup (B \cap C)$

63. $A \cap (B \cup C)'; \quad A \cap (B' \cap C')$

64. $A \cup (B \cap C)'; \quad A \cup (B' \cup C')$

Application Exercises

The chart shows the most popular shows on television in 2000, 2001, and 2002.

MOST POPULAR TELEVISION SHOWS

2000	2001	2002
1. Who Wants to Be a Millionaire	1. Survivor	1. Friends
2. E. R.	2. E. R.	2. CSI: Crime Scene Investigations
3. Friends	3. Who Wants to Be a Millionaire	3. E. R.
4. NFL Monday Night Football	4. Friends	4. Everybody Loves Raymond
5. Frasier	5. NFL Monday Night Football	5. Law and Order
6. 60 Minutes	6. Everybody Loves Raymond	6. Survivor

Source: Nielsen Media Research

In Exercises 65–70, use the Venn diagram to indicate in which region, I through VIII, each television show should be placed.

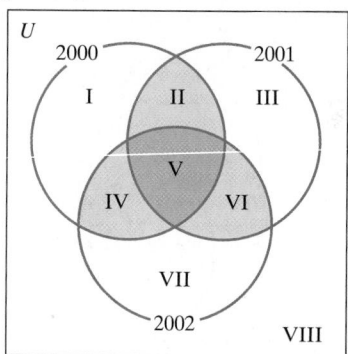

65. 60 Minutes   **66.** Law and Order

67. NFL Monday Night Football

68. Everybody Loves Raymond

69. Friends   **70.** E. R.

In Exercises 71–74, use the Venn diagram shown to indicate the region in which each company described in the chart at the top of the next page should be placed.

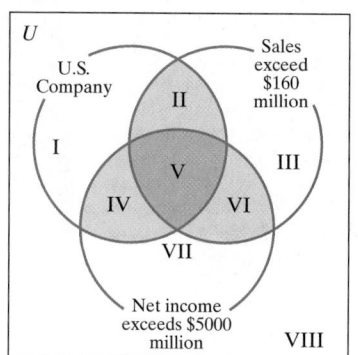

(Be sure to use the Venn diagram at the bottom of page 73 as you work these exercises.)

Name	Country	Sales in millions	Net Income in millions
71. General Motors	United States	$178,174	$6698
72. Shell Group	United Kingdom	171,657	7753
73. Ford	United States	153,627	6920
74. Mitsui	Japan	134,157	253

Source: Dow Jones Global Indexes

Writing in Mathematics

75. If you are given four sets, *A, B, C,* and *U,* describe what is involved in determining $(A \cup B)' \cap C$. Be as specific as possible in your description.

76. Describe how a Venn diagram can be used to visualize that $(A \cup B)'$ and $A' \cap B'$ are equal sets.

Critical Thinking Exercises

The eight blood types discussed in the Blitzer Bonus on page 71 are shown once again in the Venn diagram. In blood transfusions, the set of antigens in a donor's blood must be a subset of the set of antigens in a recipient's blood. Thus, the recipient must have all or more of the antigens present in the donor's blood. Use this information to solve Exercises 77–80.

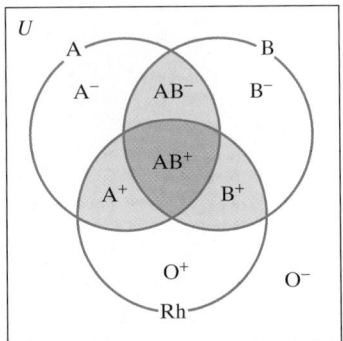

Human blood types

77. What is the blood type of a universal recipient?

78. What is the blood type of a universal donor?

79. Can an A$^+$ person donate blood to an A$^-$ person?

80. Can an A$^-$ person donate blood to an A$^+$ person?

Technology Exercise

81. Using the Paint Program in Windows, create a single Venn diagram with three sets, *A, B,* and *C,* which has the following properties:

   a. The region corresponding to $A \cap B \cap C$ is colored black.

   b. The region corresponding to $A \cap B \cap C'$ is colored orange.

   c. The region corresponding to $A' \cap B \cap C$ is colored green.

   d. The region corresponding to $A \cap B' \cap C$ is colored purple.

   e. The region corresponding to $A \cap (B \cup C)'$ is colored red.

   f. The region corresponding to $B \cap (A \cup C)'$ is colored yellow.

   g. The region corresponding to $C \cap (A \cup B)'$ is colored blue.

   h. The region corresponding to $(A \cup B \cup C)'$ is colored gray.

Group Exercise

82. Each group member should find out his or her blood type. (If you cannot obtain this information, select a blood type that you find appealing!) Read the introduction to Exercises 77–80. Referring to the Venn diagram for these exercises, each group member should determine all other group members to whom blood can be donated and from whom it can be received.

SECTION 2.5   SURVEYS AND CARDINAL NUMBERS

OBJECTIVES

1. Use Venn diagrams to visualize a survey's results.

2. Use the formula for $n(A \cup B)$.

3. Use survey results to complete Venn diagrams and answer questions about the survey.

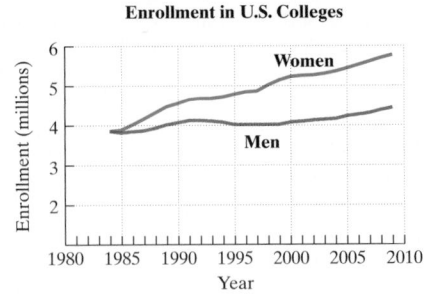

Source: Department of Education

The male minority? The graphs show enrollment in U.S. colleges, with projections through 2009. The trend indicated by the graphs is among the hottest topics of debate among college-admissions officers. Some private liberal arts colleges have quietly begun special efforts to recruit men—including admissions preferences for them.

Suppose you decide to take a survey that asks respondents the following question:

Do you agree or disagree with colleges taking special efforts to recruit men? In this section, you will see how sets and Venn diagrams are used to tabulate information collected in such a survey. In survey problems, it is helpful to remember that **and** means **intersection**, **or** means **union**, and **not** means **complement**.

Visualizing the Results of a Survey

In Section 2.1, we defined the cardinal number of set A, denoted by $n(A)$, as the number of elements in set A. Venn diagrams are helpful in determining a set's cardinality.

1 Use Venn diagrams to visualize a survey's results.

EXAMPLE 1   USING A VENN DIAGRAM TO VISUALIZE THE RESULTS OF A SURVEY

We return to the campus survey in which students were asked two questions:

Would you be willing to donate blood?

Would you be willing to help serve a free breakfast to blood donors?

Set A represents the set of students willing to donate blood. Set B represents the set of students willing to help serve breakfast to donors. The survey results are summarized in Figure 2.12. Use the diagram to answer the following questions:

a. How many students are willing to donate blood?

b. How many students are willing to help serve a free breakfast to blood donors?

c. How many students are willing to donate blood and serve breakfast?

d. How many students are willing to donate blood or serve breakfast?

e. How many students are willing to donate blood but not serve breakfast?

f. How many students are willing to serve breakfast but not donate blood?

g. How many students are neither willing to donate blood nor serve breakfast?

h. How many students were surveyed?

A: Set of students willing to
   donate blood
B: Set of students willing to
   serve breakfast to donors

FIGURE 2.12 Results of a survey

SOLUTION

a. The number of students willing to donate blood can be determined by adding the numbers in regions I and II. Thus, $n(A) = 370 + 120 = 490$. There are 490 students willing to donate blood.

b. The number of students willing to help serve a free breakfast to blood donors can be determined by adding the numbers in regions II and III. Thus, $n(B) = 120 + 220 = 340$. There are 340 students willing to help serve breakfast.

c. The number of students willing to donate blood and serve breakfast appears in region II, the region representing the intersection of the two sets. Thus, $n(A \cap B) = 120$. Therefore, 120 students are willing to donate blood and serve breakfast.

d. The number of students willing to donate blood or serve breakfast is found by adding the numbers in regions I, II, and III, representing the union of the two sets. We see that $n(A \cup B) = 370 + 120 + 220 = 710$. Therefore, 710 students in the survey are willing to donate blood or serve breakfast.

e. The region representing students who are willing to donate blood but not serve breakfast is region I. We see that 370 of the students surveyed are willing to donate blood but not serve breakfast.

f. Region III represents students willing to serve breakfast but not donate blood. We see that 220 students surveyed are willing to help serve breakfast but not donate blood.

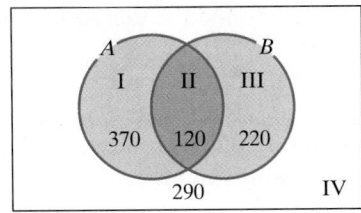

FIGURE 2.12, repeated

g. Students who are neither willing to donate blood nor serve breakfast fall within the universal set, but outside circles *A* and *B*. These students fall in region IV, where the Venn diagram indicates that there are 290 elements. There are 290 students in the survey who are neither willing to donate blood nor serve breakfast.

h. We can find the number of students surveyed by adding the numbers in regions I, II, III, and IV. Thus, $n(U) = 370 + 120 + 220 + 290 = 1000$. Therefore, 1000 students were surveyed.

 **Check Point 1** In a survey on musical tastes, respondents were asked: Do you listen to classical music? Do you listen to jazz? The survey results are summarized in Figure 2.13. Use the diagram to answer the following questions.

a. How many respondents listened to classical music?

b. How many respondents listened to jazz?

c. How many respondents listened to both classical music and jazz?

d. How many respondents listened to classical music or jazz?

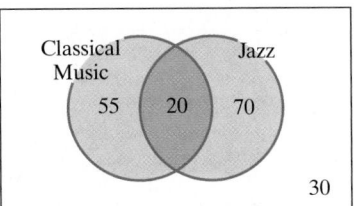

FIGURE 2.13

e. How many respondents listened to classical music but not jazz?

f. How many respondents listened to jazz but not classical music?

g. How many respondents listened to neither Classical music nor jazz?

h. How many people were surveyed?

The Cardinal Number of the Union of Two Sets

The Venn diagram for the survey in Example 1 is shown again in Figure 2.14. The diagram indicates that

$$n(A) = 490$$
$$n(B) = 340$$
$$n(A \cap B) = 120$$
$$\text{and } n(A \cup B) = 710.$$

Notice that

$$n(A \cup B) = n(A) + n(B) - n(A \cap B)$$

because

$$710 = 490 + 340 - 120.$$

This relationship is true for any two finite sets *A* and *B*.

 2 Use the formula for $n(A \cup B)$.

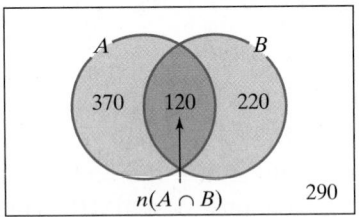

$n(A) = 370 + 120 = 490$
$n(B) = 220 + 120 = 340$
$n(A \cup B) = 370 + 120 + 220 = 710$

FIGURE 2.14

FORMULA FOR THE CARDINAL NUMBER OF THE UNION OF TWO SETS

$$n(A \cup B) = n(A) + n(B) - n(A \cap B)$$

To find the cardinal number in the union of sets *A* and *B*, add the number of elements in sets *A* and *B* and then subtract the number of elements common to both sets.

We subtract $n(A \cap B)$ in the formula for $n(A \cup B)$ so that we do not count the number in the intersection twice, once for $n(A)$, and again for $n(B)$.

> **EXAMPLE 2** **USING THE FORMULA FOR THE CARDINALITY OF THE UNION OF TWO SETS**

Set A contains 30 elements, set B contains 18 elements, and 6 elements are common to sets A and B. How many elements are in $A \cup B$?

SOLUTION

$$
\begin{aligned}
n(A \cup B) &= n(A) + n(B) - n(A \cap B) \\
&= 30 + 18 - 6 \\
&= 42
\end{aligned}
$$

There are 42 elements in $A \cup B$.

 **Check 2 Point** Set A contains 45 elements, set B contains 15 elements, and 5 elements are common to sets A and B. How many elements are in $A \cup B$?

 3 Use survey results to complete Venn diagrams and answer questions about the survey.

Solving Survey Problems

Venn diagrams are used to solve problems involving surveys. Here are the steps needed to solve survey problems:

> **SOLVING SURVEY PROBLEMS**
> 1. Use the survey's description to define sets and draw a Venn diagram.
> 2. Use the survey's results to determine the cardinality for each region in the Venn diagram. **Start with the intersection of the sets, the innermost region, and work outward.**
> 3. Use the completed Venn diagram to answer the problem's questions.

> **EXAMPLE 3** **SURVEYING PEOPLE'S ATTITUDES**

In a Gallup poll, 2000 U.S. adults were selected at random and asked to agree or disagree with the following statement:

   Job opportunities for women are not equal to those for men.

The results of the survey showed that

   1190 people agreed with the statement.

   700 women agreed with the statement.

Source: The People's Almanac

If half the people surveyed were women,

   a. How many men agreed with the statement?

   b. How many men disagreed with the statement?

SOLUTION

STEP 1. Define sets and draw a Venn diagram. The Venn diagram in Figure 2.15 shows two sets. Set W (labeled "Women") is the set of women surveyed. Set A (labeled "Agree") is the set of people surveyed who agreed with the statement. By representing the women surveyed with circle W, we do not need a separate circle for the men. The group of people outside circle W must be the set of men. Similarly, by visualizing the set of people who agreed with the statement as circle A, we do not need a separate circle for those who disagreed. The group of people outside the A (agree) circle must be the set of people disagreeing with the statement.

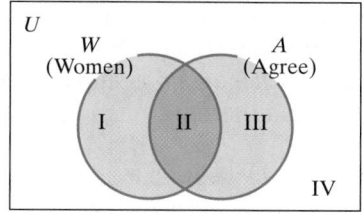

FIGURE 2.15

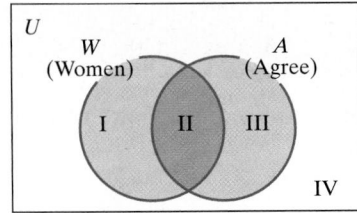

FIGURE 2.15, repeated
Our goal is to put numbers in regions I, II, III, and IV.

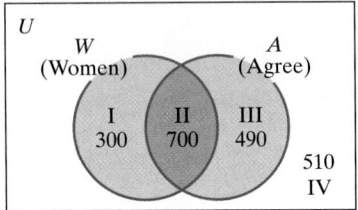

FIGURE 2.16

STEP 2. Determine the cardinality for each region in the Venn diagram, starting with the innermost region and working outward. We are given the following cardinalities:

There were 2000 people surveyed: $n(U) = 2000$.

Half the people surveyed were women: $n(W) = 1000$.

The number of people who agreed with the statement was 1190: $n(A) = 1190$.

There were 700 women who agreed with the statement: $n(W \cap A) = 700$.

We can tabulate this information on the Venn diagram in Figure 2.16. Begin with the region of intersection, region II. This region is the set of women who agreed with the statement, given to be 700. Because $n(W \cap A) = 700$, we write 700 in region II, shown in Figure 2.16. Set W contains 1000 women and consists of regions I and II. Therefore region I contains $1000 - 700$, or 300, women. We write the number 300 in region I. Set A consists of regions II and III, representing the set of people who agreed with the statement. We know that $n(A) = 1190$. With 700 people already in region II, this leaves $1190 - 700$, or 490, for region III.

The Venn diagram now shows that $n(W \cup A) = 300 + 700 + 490 = 1490$. The number of people in region IV is the difference between $n(U)$, the number of people surveyed, and $n(W \cup A)$. There are $2000 - 1490$, or 510, people in region IV.

STEP 3. Use the completed Venn diagram to answer the problem's questions.

a. The men who agreed with the statement are those members of the set of people who agreed who are not women, shown in region III. This means that 490 men surveyed agreed that job opportunities are unequal.

b. The men who disagreed with the statement can be found outside the circles of people who agreed and people who are women. This corresponds to region IV, whose cardinality is 510. Thus, 510 men surveyed disagreed that job opportunities are unequal.

 (This Check Point is based on a *Time*/CNN poll.)
In a poll, 1000 U.S. adults were were asked to agree or disagree with the following statement:

Canada is more important to the United States than Mexico.

The results of the survey showed that

550 people agreed with the statement.

50 Hispanic Americans agreed with the statement.

If 250 people surveyed were Hispanic,

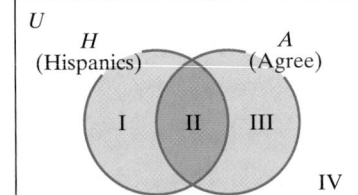

a. How many non-Hispanics agreed with the statement?

b. How many non-Hispanics disagreed with the statement?

When tabulating survey results, more than two circles within a Venn diagram are often needed. For example, consider a *Time*/CNN poll that sought to determine how Americans felt about reserving a certain number of college scholarships exclusively for minorities and women. Respondents were asked the following question:

Do you agree or disagree with the following statement: Colleges should reserve a certain number of scholarships exclusively for minorities and women?

Source: Time Almanac

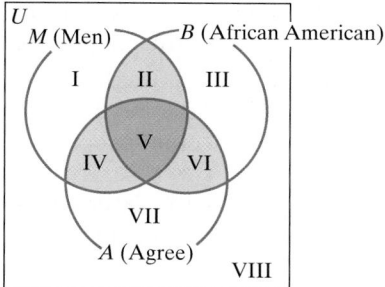

FIGURE 2.17

Suppose that we want the respondents to the poll to be identified by gender (man or woman), ethnicity (African American or other), and whether or not they agreed with the statement. A Venn diagram into which the results of the survey can be tabulated is shown in Figure 2.17.

Based on our work in Example 3, we only used one circle in the Venn diagram to indicate the gender of the respondent. We used M for men, so the set of women respondents, M', consists of the regions outside circle M. Similarly, we used B for the set of African-American respondents, so the regions outside circle B account for all other respondents. Finally, we used A for the set of respondents who agreed with the statement. Those who disagreed lie outside circle A.

In our final example, we tabulate survey results within regions of a Venn diagram containing three circles.

EXAMPLE 4   SOLVING A SURVEY PROBLEM

Sixty people were contacted and responded to a movie survey. The following information was obtained:

a. 6 people liked comedies, dramas, and science fiction.
b. 13 people liked comedies and dramas.
c. 10 people liked comedies and science fiction.
d. 11 people liked dramas and science fiction.
e. 26 people liked comedies.
f. 21 people liked dramas.
g. 25 people liked science fiction.

Use a Venn diagram to determine how many of those surveyed liked none of these categories of movies and how many liked only comedies.

SOLUTION

STEPS 1 AND 2   Draw a Venn diagram and determine the cardinality for each region.
The set of people surveyed is a universal set containing three subsets:

$$C = \text{the set of those who prefer comedies}$$
$$D = \text{the set of those who prefer dramas}$$
$$S = \text{the set of those who prefer science fiction.}$$

We draw these sets in Figure 2.18. Using condition (a), there are 6 people in $C \cap D \cap S$. Condition (b) tells us there are 13 people in $C \cap D$. Because we have counted 6 of them, there are $13 - 6$, or 7 more people in this region. This is shown in Figure 2.18(a).

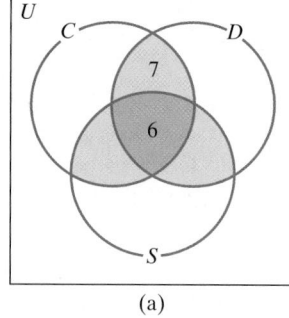
(a)

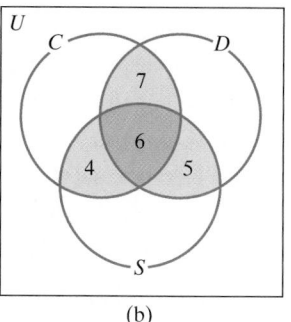
(b)

FIGURE 2.18 Determining cardinalities of $C \cap D$, $C \cap S$, and $D \cap S$

Condition (c) tells us that there are 10 people in $C \cap S$. We have counted 6 of them, so there are $10 - 6$, or 4 more people in this region. By condition (d), there are 11 people in $D \cap S$. Having counted 6 of them, there are 5 more people in the region. These cardinalities are shown in Figure 2.18(b).

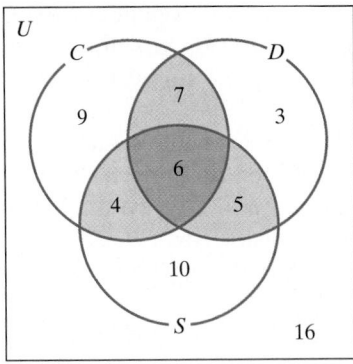

FIGURE 2.19 The cardinalities for each region of the Venn diagram

Condition (e), 26 people liked comedies, states that there are 26 people in C. We have already counted 17 of them, so there are $26 - 17$, or 9 more people in C. By condition (f), 21 people liked dramas, there are 21 people in D; with 18 already counted, there are 3 more people in D. By condition (g), 25 people liked science fiction, there are 25 people in S, and 15 are counted. This leaves 10 more people in S. These cardinalities are shown in Figure 2.19.

Figure 2.19 shows that there are $9 + 7 + 3 + 4 + 6 + 5 + 10$, or 44 people inside sets C, D, and S. Because 60 people were contacted and responded to the survey, $n(U) = 60$. This leaves $60 - 44$, or 16 people outside sets C, D, and S. This cardinality is also shown in Figure 2.19.

STEP 3 Use the completed Venn diagram to answer the problem's questions. Figure 2.19 shows that 16 people liked none of the movie categories and 9 people liked only comedies.

 **Check 4 Point**

Use the Venn diagram in Figure 2.19 to answer the following questions:

a. How many of those surveyed liked comedies or dramas?

b. How many of those surveyed liked dramas but not science fiction?

Exercise Set 2.5

Practice Exercises

Use the accompanying Venn diagram, which shows the number of elements in regions I through IV, to answer the questions in Exercises 1–8.

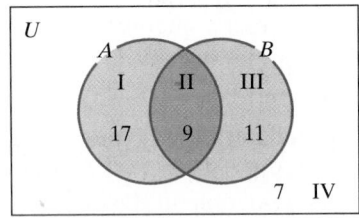

1. How many elements belong to set A?

2. How many elements belong to set B?

3. How many elements belong to set A but not set B?

4. How many elements belong to set B but not set A?

5. How many elements belong to set A or set B?

6. How many elements belong to set A and set B?

7. How many elements belong to neither set A nor set B?

8. How many elements are there in the universal set?

Use the formula for the cardinal number of the union of two sets to solve Exercise 9–12.

9. Set A contains 17 elements, set B contains 20 elements, and 6 elements are common to sets A and B. How many elements are in $A \cup B$?

10. Set A contains 30 elements, set B contains 18 elements, and 5 elements are common to sets A and B. How many elements are in $A \cup B$?

11. Set A contains 8 letters and 9 numbers. Set B contains 7 letters and 10 numbers. Four letters and 3 numbers are common to both sets A and B. Find the number of elements in set A or set B.

12. Set A contains 12 numbers and 18 letters. Set B contains 14 numbers and 10 letters. One number and 6 letters are common to both sets A and B. Find the number of elements in set A or set B.

Application Exercises

As discussed in the text on page 78, a poll asked respondents if they agreed with the statement

Colleges should reserve a certain number of scholarships exclusively for minorities and women.

Hypothetical results of the poll are tabulated in the Venn diagram. Use these cardinalities to solve Exercises 13–18.

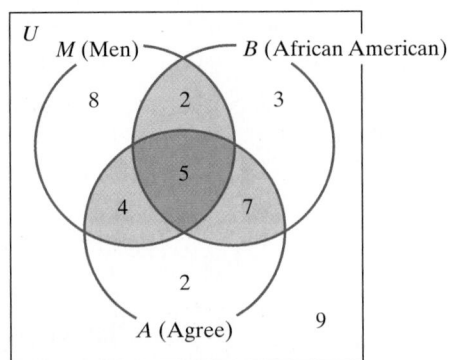

13. How many respondents agreed with the statement?

14. How many respondents disagreed with the statement?

15. How many women agreed with the statement?

16. How many people who are not African American agreed with the statement?

17. How many women who are not African American disagreed with the statement?

18. How many men who are not African American disagreed with the statement?

19. A pollster conducting a telephone poll asked two questions:

    1. Do you currently smoke cigarettes?

    2. Regardless of your answer to question 1, would you support a total ban on smoking in restaurants?

  a. Construct a Venn diagram that allows the respondents to the poll to be identified by whether or not they smoke cigarettes and whether or not they support the total ban.

  b. Write the letter b in every region of the diagram that represents smokers polled who support the total ban.

  c. Write the letter c in every region of the diagram that represents nonsmokers polled who support the ban.

  d. Write the letter d in every region of the diagram that represents nonsmokers polled who do not support the ban.

20. A pollster conducting a telephone poll at a college campus asked students two questions:

    1. Do you binge drink three or more times per month?

    2. Regardless of your answer to question 1, are you frequently behind in your school work?

  a. Construct a Venn diagram that allows the respondents to the poll to be identified by whether or not they binge drink and whether or not they frequently fall behind in school work.

  b. Write the letter b in every region of the diagram that represents binge drinkers who are frequently behind in school work.

  c. Write the letter c in every region of the diagram that represents students polled who do not binge drink but who are frequently behind in school work.

  d. Write the letter d in every region of the diagram that represents students polled who do not binge drink and who do not frequently fall behind in their school work.

21. A pollster conducting a telephone poll asked three questions:

    1. Are you religious?

    2. Have you spent time with a person during his or her last days of a terminal illness?

    3. Should assisted suicide be an option for terminally ill people?

  a. Construct a Venn diagram with three circles that can assist the pollster in tabulating the responses to the three questions.

  b. Write the letter b in every region of the diagram that represents all religious persons polled who are not in favor of assisted suicide for the terminally ill.

  c. Write the letter c in every region of the diagram that represents the people polled who do not consider themselves religious, who have not spent time with a terminally ill person during his or her last days, and who are in favor of assisted suicide for the terminally ill.

  d. Write the letter d in every region of the diagram that represents the people polled who consider themselves religious, who have not spent time with a terminally ill person during his or her last days, and who are not in favor of assisted suicide for the terminally ill.

  e. Write the letter e in a region of the Venn diagram other than those in parts (b)–(d) and then describe who in the poll is represented by this region.

22. A poll asks respondents the following question:

Do you agree or disagree with this statement: In order to address the trend in diminishing male enrollment, colleges should begin special efforts to recruit men?

  a. Construct a Venn diagram with three circles that allows the respondents to be identified by gender (man or woman), education level (college or no college), and whether or not they agreed with the statement.

  b. Write the letter b in every region of the diagram that represents men with a college education who agreed with the statement.

  c. Write the letter c in every region of the diagram that represents women who disagreed with the statement.

  d. Write the letter d in every region of the diagram that represents women without a college education who agreed with the statement.

  e. Write the letter e in a region of the Venn diagram other than those in parts (b)–(d) and then describe who in the poll is represented by this region.

In Exercises 23–26, construct a Venn diagram and determine the cardinality for each region. Use the completed Venn diagram to answer the questions.

23. A survey of 75 college students was taken to determine where they got the news about what's going on in the world. Of those surveyed, 29 students got the news from newspapers, 43 from television, and 7 from both newspapers and television.
Of those surveyed,

  a. How many got the news from only newspapers?

  b. How many got the news from only television?

  c. How many got the news from newspapers or television?

  d. How many did not get the news from either newspapers or television?

24. A survey of 120 college students was taken at registration. Of those surveyed, 75 students registered for a math course, 65 for an English course, and 40 for both math and English.
Of those surveyed,

  a. How many registered only for a math course?

  b. How many registered only for an English course?

  c. How many registered for a math course or an English course?

  d. How many did not register for either a math course or an English course?

25. A survey of 80 college students was taken to determine the musical styles they listened to. Forty-two students listened to rock, 34 to classical, and 27 to jazz. Twelve students listened to rock and jazz, 14 to rock and classical, and 10 to classical and jazz. Seven students listened to all three musical styles.
Of those surveyed,

a. How many listened to only rock music?

b. How many listened to classical and jazz, but not rock?

c. How many listened to classical or jazz, but not rock?

d. How many listened to music in exactly one of the musical styles?

e. How many listened to music in exactly two of the musical styles?

f. How many did not listen to any of the musical styles?

26. A survey of 180 college men was taken to determine participation in various campus activities. Forty-three students were in fraternities, 52 participated in campus sports, and 35 participated in various campus tutorial programs. Thirteen students participated in fraternities and sports, 14 in sports and tutorial programs, and 12 in fraternities and tutorial programs. Five students participated in all three activities.
Of those surveyed,

a. How many participated in only campus sports?

b. How many participated in fraternities and sports, but not tutorial programs?

c. How many participated in fraternities or sports, but not tutorial programs?

d. How many participated in exactly one of these activities?

e. How many participated in exactly two of these activities?

f. How many did not participate in any of the three activities?

Writing in Mathematics

27. Suppose that you are drawing a Venn diagram to sort and tabulate the results of a survey. If results are being tabulated along gender lines, explain why only a circle representing women is needed, rather than two separate circles representing the women surveyed and the men surveyed.

28. Suppose that you decide to use two sets, M and W, to sort and tabulate the responses for men and women in a survey. Describe the set of people represented by regions II and IV in the Venn diagram shown. What conclusion can you draw?

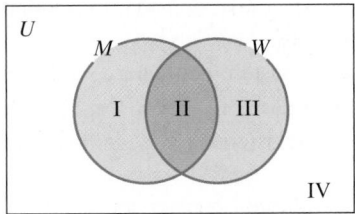

Critical Thinking Exercises

29. Which one of the following is true?

a. In a survey, 110 students were taking mathematics, 90 were taking psychology, and 20 were taking neither. Thus, 220 students were surveyed.

b. If $A \cap B = \emptyset$, then $n(A \cup B) = n(A) + n(B)$.

c. When filling in cardinalities for regions in a two-set Venn diagram, the innermost region, the intersection of the two sets, should be the last region to be filled in.

d. $n(A')$ cannot be obtained by subtracting $n(A)$ from $n(U)$.

30. In a survey of 150 students, 90 were taking mathematics and 30 were taking psychology.

a. What is the least number of students who could have been taking both courses?

b. What is the greatest number of students who could have been taking both courses?

c. What is the greatest number of students who could have been taking neither course?

31. A person applying for the position of college registrar submitted the following report to the college president on 90 students: 31 take math; 28 take chemistry; 42 take psychology; 9 take math and chemistry; 10 take chemistry and psychology; 6 take math and psychology; 4 take all three subjects; and 20 take none of these courses. The applicant was not hired. Explain why.

Group Exercise

32. This group activity is intended to provide practice in the use of Venn diagrams to sort responses to a survey. The group will determine the topic of the survey. Although you will not actually conduct the survey, it might be helpful to imagine carrying out the survey using the students on your campus.

a. In your group, decide on a topic for the survey.

b. Devise three questions that the pollster will ask to the people who are interviewed.

c. Construct a Venn diagram that will assist the pollster in sorting the answers to the three questions. The Venn diagram should contain three intersecting circles within a universal set and eight regions.

d. Describe what each of the regions in the Venn diagram represents in terms of the questions in your poll.

Chapter Summary, Review, and Test

SUMMARY

DEFINITIONS AND CONCEPTS	EXAMPLES

2.1 Basic Set Concepts

Sets and Their Designations

a. A set is a collection of objects. The objects in a set are called the elements, or members, of the set.

b. Sets can be designated by word descriptions, the roster method (a listing within braces, separating elements with commas), or set-builder notation:

$$\{ \quad x \quad | \quad \text{condition (s)} \}.$$

The set of | all elements x | such that | x meets these conditions

Ex. 1, p. 40;
Ex. 2, p. 40;
Ex. 3, p. 41

Basic Concepts and Notations

a. The empty set, or the null set, represented by $\{ \ \}$ or $\varnothing$, is a set that contains no elements.

b. The symbol $\in$ means that an object is an element of a set. The symbol $\notin$ means that an object is not an element of a set.

Ex. 4, p. 42

c. The set of natural numbers is $\mathbf{N} = \{1, 2, 3, 4, 5, \dots\}$.

Ex. 5, p. 43

d. The cardinal number of a set A, $n(A)$, is the number of distinct elements in set A.

Ex. 6, p. 44

e. Set A is a finite set if $n(A) = 0$ or if $n(A)$ is a natural number. A set that is not finite is an infinite set.

f. Equal sets have exactly the same elements. Equivalent sets have the same number of elements.

Ex. 7, p. 45;
Ex. 8, p. 45

2.2 Venn Diagrams and Subsets

The Universal Set and a Set's Complement

a. A universal set, symbolized by U, is a set that contains all the elements being considered in a given discussion or problem.

Ex. 1, p. 50

b. The complement of set A, symbolized by A', is the set of all elements in the universal set that are not in A.

Ex. 2, p. 51

Subsets

a. Set A is a subset of set B, expressed as $A \subseteq B$, if every element in set A is also in set B. The notation $A \nsubseteq B$ means that set A is not a subset of set B, so there is at least one element of set A that is not an element of set B.

Ex. 4, p. 53

b. Set A is a proper subset of set B, expressed as $A \subset B$, if A is a subset of B and $A \neq B$.

Ex. 5, p. 54

c. The empty set is a subset of every set.

Ex. 6, p. 55

d. A set with n elements has 2^n subsets and $2^n - 1$ proper subsets.

Ex. 7, p. 56

2.3 Venn Diagrams and Set Operations

Venn Diagrams

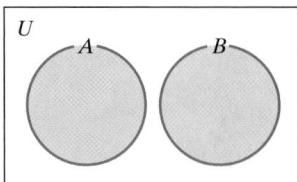

No A are B.
A and B are disjoint.

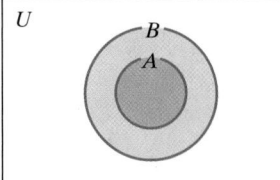

All A are B.
$A \subset B$

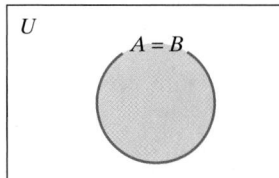

A and B are equal sets.

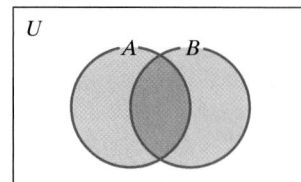

Some (at least one) A are B.

Set Operations

a. $A \cap B$ (A intersection B), which can be read set A **and** set B, is the set of elements common to both set A and set B.

Ex. 1, p. 62

b. $A \cup B$ (*A* union *B*), which can be read set *A* **or** set *B*, is the set of elements that are members of set *A* or of set *B* or of both sets. Ex. 2, p. 63

c. A' (the complement of set *A*), which can be read *A* prime or **not** *A*, is the set of all elements in the universal set that are not in *A*. Ex. 3, p. 64

De Morgan's Laws

$$(A \cap B)' = A' \cup B' \quad \text{and} \quad (A \cup B)' = A' \cap B'$$

2.4 Set Operations and Venn Diagrams with Three Sets

a. When using set operations involving three sets, begin by finding the set within parentheses. Ex. 1, p. 67

b. A Venn diagram with three sets is given in Figure 2.8 on page 69. Ex. 2, p. 69

c. If two sets represent the same regions of a Venn diagram, then they are equal. Ex. 3, p. 70; Ex. 4, p. 71

2.5 Surveys and Cardinal Numbers

a. Cardinal Number of the Union of Two Sets Ex. 2, p. 77

$$n(A \cup B) = n(A) + n(B) - n(A \cap B)$$

b. Survey problems can be solved using the three steps in the box on page 77. Ex. 3, p. 77; Ex. 4, p. 79

REVIEW EXERCISES

2.1

1. Write a description of the set $\{1, 2, 3, \ldots, 10\}$.
2. Express the set of the days of the week that begin with the letter T using the roster method.
3. Express using set-builder notation: $\{5, 6, 7\}$.

In Exercises 4–5, fill in the blank with either $\in$ or $\notin$ to make each statement true.

4. 93 ____ $\{1, 2, 3, 4, \ldots, 99, 100\}$
5. $\{d\}$ ____ $\{a, b, c, d, e\}$

Find the cardinal number for each set in Exercises 6–7.

6. $A = \{x \mid x \text{ is a month of the year}\}$
7. $B = \{18, 19, 20, \ldots, 31, 32\}$

In Exercises 8–9, fill in the blank with either $=$ or $\neq$ to make each statement true.

8. $\{0, 2, 4, 6, 8\}$ ____ $\{8, 2, 6, 4\}$
9. $\{x \mid x \text{ is a natural number greater than 7}\}$ ____ $\{8, 9, 10, \ldots, 100\}$

In Exercises 10–11, determine if the pairs of sets are equal, equivalent, both, or neither.

10. $A = \{x \mid x \text{ is a lowercase letter that comes before f in the English alphabet}\}$
    $B = \{2, 4, 6, 8, 10\}$
11. $A = \{x \mid x \in \mathbf{N} \text{ and } x \text{ is between 3 and 7}\}$
    $B = \{4, 5, 6\}$

2.2

12. Let $U = \{1, 2, 3, 4, 5, 6, 7, 8, 9\}$ and $A = \{3, 4, 5, 6, 7\}$. Find A' and place the elements 1 through 9 in the proper region of the Venn diagram shown.

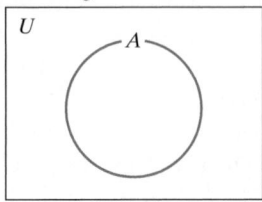

13. If U is the set of the months of the year and A is the set of months that contain exactly 30 days, use the roster method to represent A'.

In Exercises 14–16, write $\subseteq$ or $\not\subseteq$ in each blank so that the resulting statement is true.

14. $\{\text{penny, nickel, dime}\}$ ____ $\{\text{half-dollar, quarter, dime, nickel, penny}\}$
15. $\{-1, 0, 1\}$ ____ $\{-3, -2, -1, 1, 2, 3\}$
16. $\varnothing$ ____ $\{x \mid x \text{ is an odd natural number}\}$

In Exercises 17–18, determine whether $\subseteq$, $\subset$, both, or neither can be placed in each blank to make the statement true.

17. $\{1, 2\}$ ____ $\{1, 1, 2, 2\}$
18. $\{x \mid x \text{ is a person who is Baptist}\}$ ____ $\{x \mid x \text{ is a person who belongs to an organized religion}\}$

Determine whether each statement in Exercises 19–25 is true or false. If the statement is false, explain why.

19. Texas $\in \{\text{Oklahoma, Louisiana, Georgia, South Carolina}\}$
20. $4 \subseteq \{2, 4, 6, 8, 10, 12\}$
21. $\{e, f, g\} \subset \{d, e, f, g, h, i\}$
22. $\{\ominus, \varnothing\} \subset \{\varnothing, \ominus\}$
23. $\{3, 7, 9\} \subseteq \{9, 7, 3, 1\}$
24. $\{\text{six}\}$ has 2^6 subsets.
25. $\varnothing \subseteq \{\ \}$
26. List all subsets for the set $\{1, 5\}$. Which one of these subsets is not a proper subset?

Find the number of subsets and the number of proper subsets for each of the sets in Exercises 27–28.

27. $\{2, 4, 6, 8, 10\}$
28. $\{x \mid x \text{ is a month that begins with the letter J}\}$

2.3

In Exercises 29–33, let $U = \{1, 2, 3, 4, 5, 6, 7, 8\}$, $A = \{1, 2, 3, 4\}$, and $B = \{1, 2, 4, 5\}$. Find each of the following sets.

29. $A \cap B$
30. $A \cup B'$
31. $A' \cap B$
32. $(A \cup B)'$
33. $A' \cap B'$

In Exercises 34–41, use the Venn diagram to list the elements of each set in roster form.

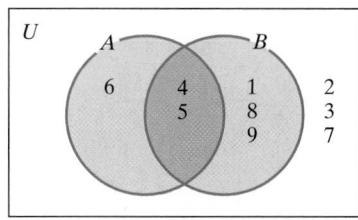

34. A **35.** B' **36.** $A \cup B$ **37.** $A \cap B$
38. $(A \cap B)'$ **39.** $(A \cup B)'$ **40.** $A \cap B'$ **41.** U

2.4

In Exercises 42–43, let

$$U = \{1, 2, 3, 4, 5, 6, 7, 8\}$$
$$A = \{1, 2, 3, 4\}$$
$$B = \{1, 2, 4, 5\}$$
$$C = \{1, 5\}.$$

Find each of the following sets.
42. $A \cup (B \cap C)$ **43.** $(A \cap C)' \cup B$

In Exercises 44–49, use the Venn diagram to list the elements of each set in roster form.

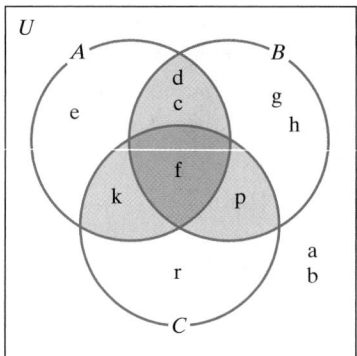

44. $A \cup C$ **45.** $B \cap C$ **46.** $(A \cap B) \cup C$
47. $A \cap C'$ **48.** $(A \cap C)'$ **49.** $A \cap B \cap C$
50. Use a Venn diagram with two intersecting circles to show that $(A \cup B)' = A' \cap B'$.
51. Use a Venn diagram with three intersecting circles to determine whether the following statement is true or false:
$$A \cap (B \cup C) = A \cup (B \cap C).$$

2.5

52. Set A contains 25 elements, set B contains 17 elements, and 9 elements are common to sets A and B. How many elements are in $A \cup B$?

53. A pollster conducting a telephone survey of college students asked two questions:
   1. Are you a registered Republican?
   2. Are you in favor of the death penalty?

a. Construct a Venn diagram with two circles that can assist the pollster in tabulating the responses to the two questions.

b. Write the letter b in every region of the diagram that represents students polled who are registered Republicans who are not in favor of the death penalty.

c. Write the letter c in every region of the diagram that represents students polled who are not registered Republicans and who are in favor of the death penalty.

In Exercises 54–55, construct a Venn diagram and determine the cardinality for each region. Use the completed Venn diagram to answer the questions.

54. A survey of 1000 American adults was taken to analyze their investments. Of those surveyed, 650 had invested in stocks, 550 in bonds, and 400 in both stocks and bonds. Of those surveyed,
  a. how many invested in only stocks?
  b. how many invested in stocks or bonds?
  c. how many did not invest in either stocks or bonds?

55. A survey of 200 students at a nonresidential college was taken to determine how they got to campus during the fall term. Of those surveyed, 118 used cars, 102 used public transportation, and 70 used bikes. Forty-eight students used cars and public transportation, 38 used cars and bikes, and 26 used public transportation and bikes. Twenty-two students used all three modes of transportation. Of those surveyed,
  a. How many used only public transportation?
  b. How many used cars and public transportation, but not bikes?
  c. How many used cars or public transportation, but not bikes?
  d. How many used exactly two of these modes of transportation?
  e. How many did not use any of the three modes of transportation to get to campus?

Group Exercise

56. We now return to the campus blood drive problem and the survey questions:

Would you be willing to donate blood?

Would you volunteer to distribute a free breakfast to blood donors?

You have learned how to sort the information obtained by these questions using a Venn diagram. However, there are still unresolved issues. Where will the free breakfast be served? Could this site be used for the blood drive itself?

  a. Group members should devise three questions that can be asked to to help settle unresolved issues related to the blood drive.

  b. Group members should construct a Venn diagram that will assist the pollster in sorting the answers to these questions.

CHAPTER 2 TEST

1. Express the following set using the roster method: $\{x \mid x$ is a natural number that is between 17 and 26$\}$.

2. Find the cardinal number for the following set:

$$\{23, 25, 27, 29, \ldots, 45\}.$$

Determine whether each statement in Exercises 3–10 is true or false. If the statement is false, explain why.

3. $\{6\} \in \{1, 2, 3, 4, 5, 6, 7\}$

4. If $A = \{x \mid x$ is a day of the week$\}$ and $B = \{2, 4, 6, \ldots, 14\}$, then sets A and B are equivalent.

5. $\{2, 4, 6, 8\} = \{8, 8, 6, 6, 4, 4, 2\}$

6. $\{d, e, f, g\} \subseteq \{a, b, c, d, e, f\}$

7. $\{3, 4, 5\} \subset \{x \mid x$ is a natural number less than 6$\}$

8. $14 \notin \{1, 2, 3, 4, \ldots, 39, 40\}$

9. $\{a, b, c, d, e\}$ has 25 subsets.

10. The empty set is a proper subset of any set, including itself.

11. List all subsets for the set $\{6, 9\}$. Which of these subsets is not a proper subset?

In Exercises 12–15, let

$$U = \{a, b, c, d, e, f, g\}$$
$$A = \{a, b, c, d\}$$
$$B = \{c, d, e, f\}$$
$$C = \{a, e, g\}.$$

Find each of the following sets.

12. $A \cup B$ 13. $(B \cap C)'$ 14. $A \cap C'$ 15. $(A \cup B) \cap C$

16. Which region or regions in the Venn diagram shown represent(s) $(A \cup B) \cap (A \cup C)$?

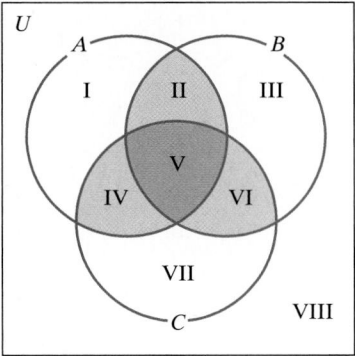

17. In a survey, 150 college students were asked if they owned a car, a computer, or both. The results showed that

   63 owned a car.

   84 owned a computer.

   27 owned both.

   a. Construct a Venn diagram with two circles, one for students owning a car, and one for students owning a computer. Fill in the number for each of the four regions in the diagram.

   b. How many students owned none of the items mentioned in the survey?

   c. How many students owned a computer only?

   d. How many students owned a car only?

   e. How many students owned a car or a computer?

3 Logic

ogic, the science of correct reasoning, enables us to analyze given information, like that in the claim about winning $7 million, and arrive at truly inescapable conclusions. In all disciplines, the ability to reason abstractly in solving problems is an important tool. We need to be able to understand the writing and speech of reasonable people to analyze information objectively and to avoid arriving at conclusions that are untrue.

In this chapter, you will analyze information through the study of symbolic logic. In symbolic logic, letters and symbols represent English sentences and their relationships to one another. This new language will enable you to think more clearly, providing you with an important tool that can extend to many areas of your life.

You are sorting through your mail and notice an envelope with this claim: If you follow the instructions inside and you return the winning number, then you will win $7 million! Are you about to become a very wealthy person or should you tear up the envelope?

This problem appears in the Blitzer Bonus on Conditional Trickery in Section 3.4.

SECTION 3.1   STATEMENTS, NEGATIONS, AND QUANTIFIED STATEMENTS

OBJECTIVES

1. Identify English sentences that are statements.
2. Express statements using symbols.
3. Form the negation of a statement.
4. Express negations using symbols.
5. Translate a negation represented by symbols into English.
6. Express quantified statements in two ways.
7. Write negations of quantified statements.

History is filled with bad predictions. Here are examples of statements that turned out to be notoriously false:

> *"The actual building of roads devoted to motor cars will not occur in the future."*
>
> Harper's Weekly, August 2, 1902

> *"As far as sinking a ship with a bomb is conceived, you just can't do it."*
>
> Rear Admiral Clark Woodward, U.S. Navy, 1939

> *"A Japanese attack on Pearl Harbor is a strategic impossibility."*
>
> George Fielding Eliot, "The Impossible War with Japan," American Mercury, September 1938

> *"Television won't be able to hold onto any market. People will soon get tired of staring at a plywood box every night."*
>
> Darryl F. Zanuck, 1949

> *"Whatever happens in Vietnam, I can conceive of nothing except military victory."*
>
> Lyndon B. Johnson, in a speech at West Point, 1967

> *"When the President does it, that means that it is not illegal."*
>
> Richard M. Nixon, TV interview with David Frost, May 20, 1977

Understanding that these statements are false enables us to mentally negate each statement and, with the assistance of historical perspective, obtain a true statement. We begin our study of logic by looking at statements and their negations.

Statements and Using Symbols to Represent Statements

1 Identify English sentences that are statements.

In everyday English, we use many different kinds of sentences. Some of these sentences are clearly true or false. Others are opinions, questions, and exclamations such as *Help*! or *Fire*! However, in logic we are concerned solely with statements, and not all English sentences are statements.

DEFINITION OF A STATEMENT
A **statement** is a sentence that is either true or false, but not both simultaneously.

Here are two examples of statements:

1. London is the capital of England.
2. William Shakespeare wrote the last episode of "Friends."

Statement 1 is true and statement 2 is false. Shakespeare had nothing to do with "Friends" (perhaps writer's block after *Macbeth*).

We do not need to know whether a given sentence is true or false to determine whether it is a statement. For example, the sentence

$$123,456,787,654,321 \text{ is } 11,111,111 \text{ multiplied by itself}$$

is a statement. Actually, the statement is true. The sentence

$$123,456,787,654,321 \text{ is } 22,222,222 \text{ multiplied by itself}$$

is also a statement, but this statement is false.

Some sentences, such as commands, questions, and opinions, are not statements because they cannot be labeled as true or false. The following sentences are not statements:

1. Read pages 23–57.   (This is an order or command.)
2. Is there order within chaos?   (This is a question.)
3. *Titanic* is the greatest movie of all time.   (This is an opinion.)

Express statements using symbols.

In symbolic logic, we use lowercase letters such as $p, q, r,$ and s to represent statements. Here are two examples:

p: London is the capital of England.

q: William Shakespeare wrote the last episode of "Friends."

The letter p represents the first statement.
The letter q represents the second statement.

Form the negation of a statement.

Negating Statements

The sentence "London is the capital of England" is a true statement. The **negation** of this statement, "London is not the capital of England," is a false statement. The negation of a statement has a meaning that is opposite that of the original meaning. Thus, the negation of a true statement is a false statement, and the negation of a false statement is a true statement.

EXAMPLE 1   FORMING NEGATIONS

Form the negation of each statement:

a. Shakespeare wrote the last episode of "Friends."
b. Today is not Monday.

SOLUTION

a. The most common way to negate this statement is to introduce *not* into the sentence. The negation is

Shakespeare did not write the last episode of "Friends."

The English language provides many ways of expressing a statement's meaning. Here is another way to express the negation:

It is not true that Shakespeare wrote the last episode of "Friends."

b. The negation of "Today is not Monday" is

It is not true that today is not Monday.

The negation is more naturally expressed in English as

Today is Monday.

 **Check Point 1**   Form the negation of each statement:

a. Paris is the capital of Spain.

b. July is not a month.

The negation of statement p is expressed by writing $\sim p$. We read this as "not p" or "It is not true that p."

4 Express negations using symbols.

EXAMPLE 2 | EXPRESSING NEGATIONS SYMBOLICALLY

Let p and q represent the following statements:

p: Shakespeare wrote the last episode of "Friends."

q: Today is not Monday.

Express each of the following statements symbolically:

a. Shakespeare did not write the last episode of "Friends."

b. Today is Monday.

SOLUTION

a. This statement is the negation of statement p.
Therefore, it is expressed symbolically as $\sim p$.

b. This statement is the negation of statement q.
Therefore, it is expressed symbolically as $\sim q$.

STUDY TIP

Letters such as p, q, or r can represent any statement, including English statements containing the word "not." When choosing letters to represent statements, you may prefer to use the symbol $\sim$ with negated English statements. However, it is not wrong to let p, q, or r represent such statements.

 **Check Point 2**   Let p and q represent the following statements:

p: Paris is the capital of Spain.

q: July is not a month.

Express each of the following statements symbolically:

a. Paris is not the capital of Spain.

b. July is a month.

In Example 2, we translated English statements into symbolic statements. In Example 3, we reverse the direction of our translation.

5 Translate a negation represented by symbols into English.

EXAMPLE 3 | TRANSLATING A SYMBOLIC STATEMENT INTO WORDS

Let p represent the following statement:

p: Harvard is a college.

Express the symbolic statement $\sim p$ in words.

SOLUTION   The symbol $\sim$ is translated as "not." Therefore, $\sim p$ represents

Harvard is not a college.

This can also be expressed as

It is not true that Harvard is a college.

Let *q* represent the following statement:

   q: Tom Hanks is a jazz singer.

Express the symbolic statement ~*q* in words.

 6 Express quantified statements in two ways.

Quantified Statements

In English, we frequently encounter statements containing the words **all, some**, and **no** (or **none**). These words are called **quantifiers**. A statement that contains one of these words is a **quantified statement**. Here are some examples:

   All poets are writers.

   Some people are bigots.

   No math books have pictures.

   Some students do not work hard.

Using our knowledge of the English language, we can express each of these quantified statements in two equivalent ways, that is, in two ways that have exactly the same meaning. These equivalent statements are shown in Table 3.1.

TABLE 3.1 EQUIVALENT WAYS OF EXPRESSING QUANTIFIED STATEMENTS

Statement	An Equivalent Way to Express the Statement	Example (Two Equivalent Quantified Statements)
All *A* are *B*.	There are no *A* that are not *B*.	All poets are writers. There are no poets that are not writers.
Some *A* are *B*.	There exists at least one *A* that is a *B*.	Some people are bigots. At least one person is a bigot.
No *A* are *B*.	All *A* are not *B*.	No math books have pictures. All math books do not have pictures.
Some *A* are not *B*.	Not all *A* are *B*.	Some students do not work hard. Not all students work hard.

7 Write negations of quantified statements.

STUDY TIP

The negation of "All writers are poets" cannot be "No writers are poets" because both statements are false. The negation of a false statement must be a true statement. In general, the negation of "All *A* are *B*" is *not* "No *A* are *B*."

Forming the negation of a quantified statement can be a bit tricky. Suppose we want to negate the statement "All writers are poets." Because this statement is false, its negation must be true. The negation is "Not all writers are poets." This means the same thing as "Some writers are not poets." Notice that the negation is a true statement.

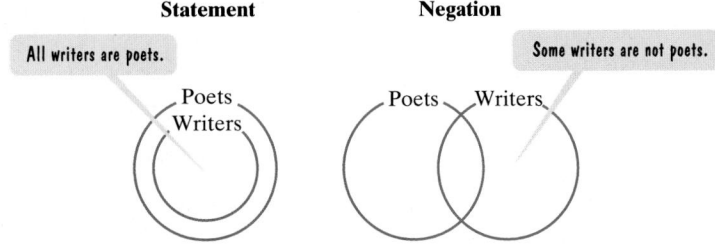

In general, the negation of "All *A* are *B*" is "Some *A* are not *B*." Likewise, the negation of "Some *A* are not *B*" is "All *A* are *B*."

Now let's investigate how to negate a statement with the word "some." Consider the statement "Some canaries weigh 50 pounds." Because "some" means "there exists at least one," the negation is "It is not true that there is at least one canary that weighs 50 pounds." Because it is not true that there is even one such critter, we can express the negation as "No canary weighs 50 pounds."

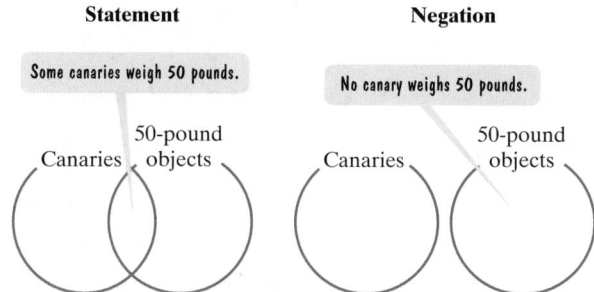

	Statement	Negation

In general, the negation of "Some *A* are *B*" is "No *A* are *B*." Likewise, the negation of "No *A* are *B*" is "Some *A* are *B*."

Negations of quantified statements are summarized in Table 3.2.

BLITZER BONUS

Logic and "Star Trek"

Four *Enterprise* officers (clockwise from lower left); Kirk, McCoy, Uhura, Spock

In the television series "Star Trek," the crew of the *Enterprise* is held captive by an evil computer. The crew escapes after one of them tells the computer, "I am lying to you." If he says that he is lying and he *is* lying, then he isn't lying. But consider this: If he says that he is lying and he *isn't* lying, then he is lying. The sentence "I am lying to you" is not a statement because it is true and false simultaneously. Thinking about this sentence destroys the computer.

TABLE 3.2 NEGATIONS OF QUANTIFIED STATEMENTS

Statement	Negation	Example (A Quantified Statement and Its Negation)
All *A* are *B*.	Some *A* are not *B*.	All people take exams honestly. The negation is Some people do not take exams honestly.
Some *A* are *B*.	No *A* are *B*.	Some roads are open. The negation is No roads are open.

(The negations of the statements in the second column are the statements in the first column.)

STUDY TIP

This diagram should help you remember the negations for quantified statements. The statements diagonally opposite each other are negations.

All *A* are *B*. ⟷ No *A* are *B*.

Some *A* are *B*. ⟷ Some *A* are not *B*.

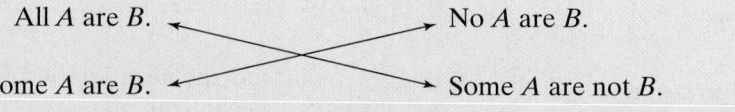

EXAMPLE 4   NEGATING A QUANTIFIED STATEMENT

The mechanic told me, "All piston rings were replaced." I later learned that the mechanic never tells the truth. What can I conclude?

SOLUTION   Let's begin with the mechanic's statement:

All piston rings were replaced.

Because the mechanic never tells the truth, I can conclude that the truth is the negation of what I was told. The negation of "All *A* are *B*" is "Some *A* are not *B*." Thus, I can conclude that

Some piston rings were not replaced.

Because "some" means "at least one," I can also correctly conclude that

At least one piston ring was not replaced.

 The board of supervisors told us, "All new tax dollars will be used to improve education." I later learned that the board of supervisors never tells the truth. What can I conclude? Express the conclusion in two equivalent ways.

Exercise Set 3.1

Practice Exercises

In Exercises 1–12, determine whether or not each sentence is a statement.

1. Manhattan Island from end to end is less than 1 million inches long.
2. There are 2,500,000 rivets in the Eiffel Tower.
3. $9 + 6 = 16$
4. $9 \times 6 = 64$
5. What's a nice person like you doing in a place like this?
6. Was Andrew Johnson impeached?
7. Give me a break.
8. Take a long walk on a short pier.
9. George W. Bush was the Republican candidate for president in 2000.
10. Al Gore was not the Democratic candidate for president in 2000.
11. No U.S. presidents were assassinated.
12. Some women have served as U.S. presidents.

In Exercises 13–16, form the negation of each statement.

13. It is raining.
14. It is snowing.
15. *Macbeth* is not a comedy show on television.
16. Albert Einstein was not offered the presidency of Israel.

In Exercises 17–20, let p, q, r, and s represent the following statements:

   p: One works hard.
   q: One succeeds.
   r: The temperature outside is not freezing.
   s: It is not true that the heater is working.

Express each of the following statements symbolically.

17. One does not work hard.
18. One does not succeed.
19. The temperature outside is freezing.
20. The heater is working.

In Exercises 21–24, let p, q, r, and s represent the following statements:

   p: "I Love Lucy" is a Broadway musical.
   q: Bruce Springsteen is a U.S. president.
   r: Benjamin Franklin did not invent the rocking chair.
   s: Bill Clinton was not impeached.

Express each of the following symbolic statements in words.

21. $\sim p$ 22. $\sim q$ 23. $\sim r$ 24. $\sim s$

In Exercises 25–28, let p, q, r, and s represent the following statements:

   p: The United States is the country with the most Internet users.
   q: New Jersey is a planet.
   r: It is not true that Michael is the most common boys' name.
   s: The final exam was not comprehensive.

Express each of the following symbolic statements in words.

25. $\sim p$ 26. $\sim q$ 27. $\sim r$ 28. $\sim s$

Exercises 29–36 contain quantified statements. For each exercise,

   a. *Express the quantified statement in an equivalent way, that is, in a way that has exactly the same meaning.*
   b. *Write the negation of the quantified statement. (The negation should begin with "all," "some," or "no.")*

29. All whales are mammals.
30. All journalists are writers.
31. Some students are business majors.
32. Some movies are comedies.
33. Some thieves are not criminals.
34. Some pianists are not keyboard players.
35. No Democratic presidents have been impeached.
36. No women have served as Supreme Court justices.

Application Exercises

In Exercises 37 and 38, choose the correct statement.

37. The City Council of a large northern metropolis promised its citizens that in the event of snow, all major roads connecting the city to its airport would remain open. The City Council did not keep its promise during the first blizzard of the season. Therefore, during the first blizzard:
   a. No major roads connecting the city to the airport were open.
   b. At least one major road connecting the city to the airport was not open.
   c. At least one major road connecting the city to the airport was open.
   d. The airport was forced to close.

38. During the Watergate scandal in 1974, President Richard Nixon assured the American people that "In all my years of public service, I have never obstructed justice." Later, events indicated that the president was not telling the truth. Therefore, in his years of public service:
   a. Nixon always obstructed justice.
   b. Nixon sometimes did not obstruct justice.
   c. Nixon sometimes obstructed justice.
   d. Nixon never obstructed justice.

Writing in Mathematics

39. What is a statement? Explain why commands, questions, and opinions are not statements.

40. Explain how to form the negation of a given English statement. Give an example.

41. Describe how the negation of statement p is expressed using symbols.

42. List the words identified as quantifiers. Give an example of a statement that uses each of these quantifiers.

43. Explain how to write the negation of a quantified statement in the form "All A are B." Give an example.

44. Explain how to write the negation of a quantified statement in the form "Some A are B." Give an example.

45. If the ancient Greek god Zeus could do anything, could he create a rock so huge that he could not move it? Explain your answer.

Critical Thinking Exercise

46. Give an example of a sentence that is not a statement because it is true and false simultaneously.

47. Give an example in which the statement "Some A are not B" is true, but the statement "Some B are not A" is false.

48. The statement

She isn't dating him because he is muscular

is confusing because it can mean two different things. Describe the two different meanings that make this statement ambiguous.

SECTION 3.2   COMPOUND STATEMENTS AND CONNECTIVES

OBJECTIVES

1. Express compound statements in symbolic form.

2. Express symbolic statements with parentheses in English.

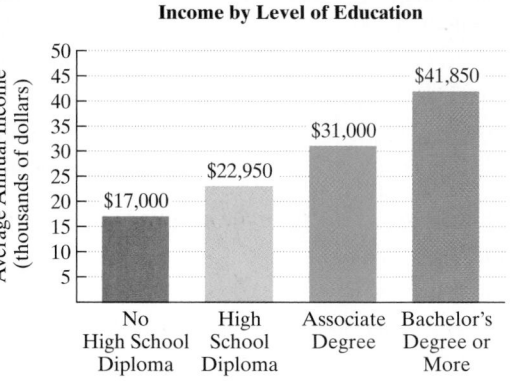

Income by Level of Education

Source: U.S. Department of Commerce

FIGURE 3.1

You are reading an article about median, or average, income by level of education. Using the graph shown in Figure 3.1, the article concludes

> Average income increases as level of education increases and average income for people with a bachelor's degree or more exceeds $41,000 per year.

We can break this conclusion down into two sentences:

> Average income increases as level of education increases.

> Average income for people with a bachelor's degree or more exceeds $41,000 per year.

These statements are called **simple statements** because each one conveys one idea with no connecting words. Statements formed by combining two or more simple statements are called **compound statements**. Words called **connectives** are used to join simple statements to form a compound statement. Connectives include words such as **and, or, if . . . then**, and **if and only if**.

Compound statements appear throughout written and spoken language. We need to be able to understand the logic of such statements to analyze information objectively. In this section, we analyze four kinds of compound statements.

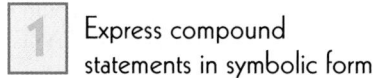
Express compound statements in symbolic form.

And **Statements**

If p and q represent two simple statements, then **the compound statement "p and q" is symbolized by $p \wedge q$.** The compound statement formed by connecting statements with the word **and** is called a **conjunction**. The symbol for *and* is $\wedge$.

EXAMPLE 1 TRANSLATING FROM ENGLISH TO SYMBOLIC FORM

Let p and q represent the following simple statements:

　　p: Harvard is a college.

　　q: Yankee Stadium is a college.

Write each of the compound statements below in its symbolic form:

　a. Harvard is a college and Yankee Stadium is a college.

　b. Harvard is a college and Yankee Stadium is not a college.

STUDY TIP

You can remember $\wedge$, the symbol for *and*, by thinking of it as an A (for And) with the horizontal bar missing.

SOLUTION

a.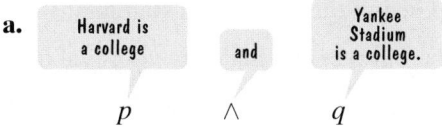

The symbolic form is $p \wedge q$.

b.

The symbolic form is $p \wedge \sim q$.

STUDY TIP

The connective *and* can also be expressed using *however* or *nevertheless*.

　In English, the connective *and* can also be expressed using *but* or *although.* Here are some examples:

　　Harvard is a college and Yankee Stadium is not.

　　Harvard is a college, but Yankee Stadium is not.

　　Harvard is a college, although Yankee Stadium is not.

Each of these statements is symbolized as $p \wedge \sim q$.

✓ Check 1 Point　Let p and q represent the following simple statements:

　　p: "Saturday Night Live" is a novel.

　　q: The Catcher in the Rye is a novel.

Write each of the compound statements below in its symbolic form:

　a. "Saturday Night Live" is a novel and *The Catcher in the Rye* is a novel.

　b. "Saturday Night Live" is not a novel and *The Catcher in the Rye* is a novel.

Or **Statements**

The connective *or* can mean two different things. For example, consider this statement:

　　I visited London or Paris.

The statement can mean

　　I visited London or Paris, but not both.

This is an example of the **exclusive or**, which means "one or the other, but not both."

By contrast, the statement can mean

I visited London or Paris or both.

This is an example of the **inclusive or**, which means "either or both."

In this chapter and in mathematics in general, when the connective *or* appears, it means the *inclusive or*. If *p* and *q* represent two simple statements, then the compound statement "*p* or *q*" means *p* or *q* or both. The compound statement formed by connecting statements with the word *or* is called a **disjunction**. **The symbol for *or* is ∨. Thus, we can symbolize the compound statement "*p* or *q* or both" by *p* ∨ *q*.**

EXAMPLE 2 | TRANSLATING FROM ENGLISH TO SYMBOLIC FORM

Let *p* and *q* represent the following simple statements:

p: The bill receives majority approval.

q: The bill becomes a law.

Write each of the compound statements below in its symbolic form:

a. The bill receives majority approval or the bill becomes a law.

b. The bill receives majority approval or the bill does not become a law.

SOLUTION

a.

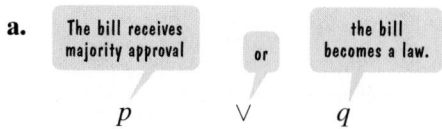

The symbolic form is *p* ∨ *q*.

b.

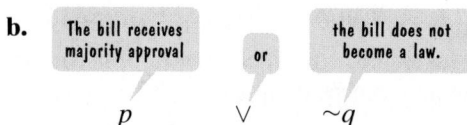

The symbolic form is *p* ∨ ~*q*.

Check 2 Point

Let *p* and *q* represent the following simple statements:

p: Taxes are increased.

q: Funding for environmental issues is increased.

Write each of the compound statements below in its symbolic form:

a. Taxes are increased or funding for environmental issues is not increased.

b. Funding for environmental issues is increased or taxes are increased.

If–Then Statements

The diagram in Figure 3.2 shows that

All poets are writers.

In Section 3.1, we saw that this can be expressed as

There are no poets that are not writers.

Another way of expressing this statement is

If a person is a poet, then that person is a writer.

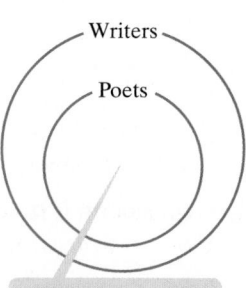

All poets are writers. If a person is a poet, then that person is a writer.

FIGURE 3.2

The form of this statement is "If p, then q." **The compound statement "If p, then q" is symbolized by $p \rightarrow q$.** The compound statement formed by connecting statements with "if–then" is called a **conditional statement**. The symbol for "if–then" is $\rightarrow$.

EXAMPLE 3   TRANSLATING FROM ENGLISH TO SYMBOLIC FORM

Let p and q represent the following simple statements:

   p: Ed is a poet.

   q: Ed is a writer.

Write each of the compound statements below in its symbolic form:

  a. If Ed is a poet, then Ed is a writer.
  b. If Ed is a writer, then Ed is a poet.
  c. If Ed is not a writer, then Ed is not a poet.

SOLUTION

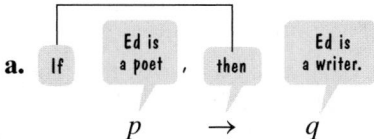

The symbolic form is $p \rightarrow q$.

The symbolic form is $q \rightarrow p$.

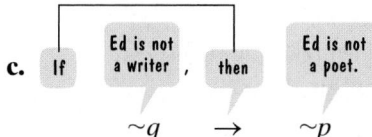

The symbolic form is $\sim q \rightarrow \sim p$.

 **Check Point 3**   Let p and q represent the following simple statements:

   p: You are a major-league baseball player.

   q: You are wealthy.

Write each of the compound statements below in its symbolic form:

  a. If you are a major-league baseball player, then you are wealthy.
  b. If you are not a major-league baseball player, then you are not wealthy.
  c. If you are not wealthy, then you are not a major-league baseball player.

Conditional statements in English often omit the word "then" and simply use a comma. When "then" is included, the comma can be included or omitted. Here are some examples:

   If Ed is a poet, then he is a writer.

   If Ed is a poet then he is a writer.

   If Ed is a poet, he is a writer.

If And Only If **Statements**

Consider the conditional statement:

> If you're in Tokyo, then you're in Japan.

This statement is true; Tokyo is the capital of Japan. Now think about this statement:

> If you're in Japan, then you're in Tokyo.

This statement is not necessarily true. For example, a person visiting another large city in Japan, such as Osaka, is certainly in Japan. However, this person is not in Tokyo. If a conditional statement is true, changing the order of its simple statements may result in a statement that is no longer true.

However, some conditional statements are true regardless of which simple statement comes first. Let's consider an example:

> Let p be: A number is greater than zero.
>
> Let q be: A number is positive.

Notice that both of the following conditionals are true:

> $p \rightarrow q$: If a number is greater than zero, then it is positive.
>
> $q \rightarrow p$: If a number is positive, then it is greater than zero.

Rather than deal with two separate conditionals, we can combine them into one **biconditional statement:**

> A number is greater than zero if and only if a number is positive.

The biconditional is symbolized by $\leftrightarrow$ **and is read "if and only if."** The phrase "if and only if" can be abbreviated as "iff."

EXAMPLE 4 | TRANSLATING FROM ENGLISH TO SYMBOLIC FORM

Table 3.3 shows that the word *set* has 464 meanings, making it the word with the most meanings in the English language. Let p and q represent the following simple statements:

> p: The word is *set*.
>
> q: The word has 464 meanings.

Write each of the compound statements below in its symbolic form:

a. The word is *set* if and only if the word has 464 meanings.

b. The word does not have 464 meanings if and only if the word is not *set*.

SOLUTION

a.

$$p \qquad \leftrightarrow \qquad q$$

The symbolic form is $p \leftrightarrow q$. Observe that each of the following statements is true:

> If the word is *set*, then it has 464 meanings.
>
> If the word has 464 meanings, then it is *set*.

b.

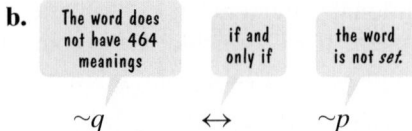

$$\sim q \qquad \leftrightarrow \qquad \sim p$$

The symbolic form is $\sim q \leftrightarrow \sim p$.

TABLE 3.3 WORDS WITH THE MOST MEANINGS IN THE *OXFORD ENGLISH DICTIONARY*

Word	Meanings
Set	464
Run	396
Go	368
Take	343
Stand	334

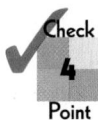

Check **4** Point

Let p and q represent the following simple statements:

> p: The word is *run*.

> q: The word has 396 meanings.

Write each of the compound statements below in its symbolic form:

a. The word has 396 meanings if and only if the word is *run*.

b. The word is not *run* if and only if the word does not have 396 meanings.

Table 3.4 summarizes the statements discussed in the first two sections of this chapter.

TABLE 3.4 STATEMENTS OF SYMBOLIC LOGIC

Name	Symbolic Form	Read
Negation	$\sim p$	Not p.
Conjunction	$p \wedge q$	p and q.
Disjunction	$p \vee q$	p or q.
Conditional	$p \rightarrow q$	If p, then q.
Biconditional	$p \leftrightarrow q$	p if and only if q.

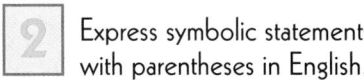
Express symbolic statements with parentheses in English.

Symbolic Statements with Parentheses

Parentheses in symbolic statements indicate which statements are to be grouped together. For example, $\sim(p \wedge q)$ means the negation of the entire statement $p \wedge q$. By contrast, $\sim p \wedge q$ means that only statement p is negated. We read $\sim(p \wedge q)$ as "it is not true that p and q." We read $\sim p \wedge q$ as "not p and q." Unless parentheses appear in a symbolic statement, the symbol $\sim$ negates only the statement that immediately follows it.

EXAMPLE 5 EXPRESSING SYMBOLIC STATEMENTS WITH AND WITHOUT PARENTHESES IN ENGLISH

Let p and q represent the following simple statements:

> p: Barry Bonds is a baseball player.

> q: Barry Bonds is a basketball player.

Write each of the following symbolic statements in words:

a. $\sim(p \wedge q)$ **b.** $\sim p \wedge q$ **c.** $\sim(p \vee q)$.

SOLUTION

a. The symbolic statement $\sim(p \wedge q)$ means the negation of the entire statement $p \wedge q$. A translation of $p \wedge q$ is

>   Barry Bonds is a baseball player and a basketball player.

A translation of $\sim(p \wedge q)$ is

>   It is not true that Barry Bonds is a baseball player and a basketball player.

We can also express this statement as

>   It is not true that Barry Bonds is both a baseball player and a basketball player.

b. A translation of $\sim p \wedge q$ is

>   Barry Bonds is not a baseball player and he is a basketball player.

In 2001, San Francisco's Barry Bonds hit 73 home runs, breaking Mark McGwire's record of 70 from 1998.

BLITZER BONUS

Lewis Carroll in Logical Wonderland

One of the world's most popular logicians was Lewis Carroll (1832–1898), author of *Alice's Adventures in Wonderland* and *Through the Looking Glass.* Carroll's passion for word games and logic is evident throughout the Alice stories.

"Speak when you're spoken to!" the Queen sharply interrupted her.

"But if everybody obeyed that rule," said Alice, who was always ready for a little argument, "and if you only spoke when you were spoken to, and the other person always waited for *you* to begin, you see nobody would ever say anything, so that—"

"Take some more tea," the March Hare said to Alice, very earnestly.

"I've had nothing yet," Alice replied in an offended tone, "so I can't take more."

"You mean you can't take *less*," said the Hatter. "It's very easy to take *more* than nothing."

Because Bonds plays baseball, not basketball, this statement is false. Can you see that this statement does not mean the same thing as the statement in part (a)? The statement is part (a) is true: Bonds is not both a baseball and a basketball player. He only plays baseball.

c. The symbolic statement $\sim(p \lor q)$ means the negation of the entire statement $p \lor q$. A translation of $p \lor q$ is

Barry Bonds is a baseball player or a basketball player.

A translation of $\sim(p \lor q)$ is

It is not true that Barry Bonds is a baseball player or a basketball player.

We can express this statement as

Barry Bonds is neither a baseball player nor a basketball player.

 **Check Point 5** Let p and q represent the following simple statements:

> p: Steven Spielberg is a director.
>
> q: Steven Spielberg is an actor.

Write each of the following symbolic statements in words:

a. $\sim(q \land p)$ **b.** $\sim q \land p$ **c.** $\sim(q \lor p)$.

Many compound statements contain more than one connective. When expressed symbolically, parentheses are used to indicate which simple statements are grouped together. When expressed in English, commas are used to indicate the groupings. Here is a table that illustrates groupings using parentheses in symbolic statements and commas in English statements:

Symbolic Statement	Statements to Group Together	English Translation
$(q \land \sim p) \rightarrow \sim r$	$q \land \sim p$	If q and not p, then not r.
$q \land (\sim p \rightarrow \sim r)$	$\sim p \rightarrow \sim r$	q, and if not p then not r.

The statement in the first row is an "if–then" conditional statement. Notice that the symbol $\rightarrow$ is outside the parentheses. By contrast, the statement in the second row is an "and" conjunction. In this case, the symbol $\land$ is outside the parentheses. Notice that when we translate the symbolic statement into English, **the simple statements in parentheses appear on the same side of the comma**.

> **EXAMPLE 6** EXPRESSING SYMBOLIC STATEMENTS WITH PARENTHESES IN ENGLISH

Let p, q, and r represent the following simple statements:

> p: A student misses lecture.
> q: A student studies.
> r: A student fails.

Write each of the symbolic statements below in words:

a. $(q \land \sim p) \rightarrow \sim r$ **b.** $q \land (\sim p \rightarrow \sim r)$.

SOLUTION

a.
$$(q \ \land \ \sim p) \ \rightarrow \ \sim r$$

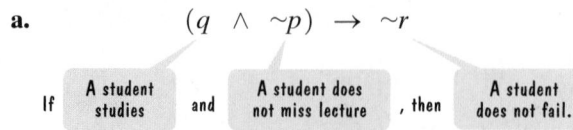
If [A student studies] and [A student does not miss lecture], then [A student does not fail.]

One possible English translation for the symbolic statement is

If a student studies and does not miss lecture, then the student does not fail.

Observe how the symbolic statements in parentheses appear on the same side of the comma in the English translation.

b. $q \;\land\; (\sim p \;\rightarrow\; \sim r)$

| A student studies | , and if | A student does not miss lecture | then | A student does not fail. |

One possible English translation for the symbolic statement is

A student studies, and if a student does not miss lecture then the student does not fail.

Once again, the symbolic statements in parentheses appear on the same side of the comma in the English statement.

Check Point 6   Let *p*, *q*, and *r* represent the following simple statements:

p: The plant is fertilized.
q: The plant is not watered.
r: The plant wilts.

Write each of the symbolic statements in words:

a. $(p \land \sim q) \rightarrow \sim r$ **b.** $p \land (\sim q \rightarrow \sim r)$.

Exercise Set 3.2

Practice Exercises

In Exercises 1–4, let p and q represent the following simple statements:

p: "I Love Lucy" *is a television show.*
q: Macbeth is a television show.

Write each compound statement in its symbolic form.

1. "I Love Lucy" is a television show and *Macbeth* is a television show.
2. *Macbeth* is a television show and "I Love Lucy" is a television show.
3. "I Love Lucy" is a television show and *Macbeth* is not a television show.
4. *Macbeth* is not a television show and "I Love Lucy" is a television show.

In Exercises 5–8, let p and q represent the following simple statements:

p: I study.
q: I pass the course.

Write each compound statement in its symbolic form.

5. I study or I pass the course.
6. I pass the course or I study.
7. I study or I do not pass the course.
8. I do not study or I do not pass the course.

In Exercises 9–12, let p and q represent the following simple statements:

p: This is an alligator.
q: This is a reptile.

Write each compound statement in its symbolic form.

9. If this is an alligator, then this is a reptile.
10. If this is a reptile, then this is an alligator.
11. If this is not an alligator, then this is not a reptile.
12. If this is not a reptile, then this is not an alligator.

In Exercises 13–16, let p and q represent the following simple statements:

p: The campus is closed.
q: It is Sunday.

Write each compound statement in its symbolic form.

13. The campus is closed if and only if it is Sunday.
14. It is Sunday if and only if the campus is closed.
15. It is not Sunday if and only if the campus is not closed.
16. The campus is not closed if and only if it is not Sunday.

In Exercises 17–24, let p and q represent the following simple statements:

p: The heater is working.
q: The house is cold.

Write each symbolic statement in words.

17. $\sim p \land q$ 18. $p \land \sim q$

19. $p \vee \sim q$
20. $\sim p \vee q$
21. $p \rightarrow \sim q$
22. $q \rightarrow \sim p$
23. $p \leftrightarrow \sim q$
24. $\sim p \leftrightarrow q$

In Exercises 25–32, let q and r represent the following simple statements:

> *q: It is July 4th.*
> *r: We are having a barbecue.*

Write each symbolic statement in words.

25. $q \wedge \sim r$
26. $\sim q \wedge r$
27. $\sim q \vee r$
28. $q \vee \sim r$
29. $r \rightarrow \sim q$
30. $q \rightarrow \sim r$
31. $\sim q \leftrightarrow r$
32. $q \leftrightarrow \sim r$

In Exercises 33–42, let p and q represent the following simple statements:

> *p: Romeo loves Juliet.*
> *q: Juliet loves Romeo.*

Write each symbolic statement in words.

33. $\sim(p \wedge q)$
34. $\sim(q \wedge p)$
35. $\sim p \wedge q$
36. $\sim q \wedge p$
37. $\sim(q \vee p)$
38. $\sim(p \vee q)$
39. $\sim q \vee p$
40. $\sim p \vee q$
41. $\sim p \wedge \sim q$
42. $\sim q \wedge \sim p$

In Exercises 43–48, let p, q, and r represent the following simple statements:

> *p: The temperature outside is freezing.*
> *q: The heater is working.*
> *r: The house is cold.*

Write each compound statement in its symbolic form.

43. The temperature outside is freezing and the heater is working, or the house is cold.

44. If the temperature outside is freezing, then the heater is working or the house is not cold.

45. If the temperature outside is freezing or the heater is not working, then the house is cold.

46. It is not the case that if the house is cold then the heater is not working.

47. The house is cold, if and only if the temperature outside is freezing and the heater isn't working.

48. If the heater is working, then the temperature outside is freezing if and only if the house is cold.

In Exercises 49–62, let p, q, and r represent the following simple statements:

> *p: The temperature is above 85°.*
> *q: We finished studying.*
> *r: We go to the beach.*

Write each symbolic statement in words.

49. $(p \wedge q) \rightarrow r$
50. $(q \wedge r) \rightarrow p$
51. $p \wedge (q \rightarrow r)$
52. $p \wedge (r \rightarrow q)$
53. $\sim r \rightarrow (\sim p \vee \sim q)$
54. $\sim p \rightarrow (q \vee r)$
55. $(\sim r \rightarrow \sim q) \vee p$
56. $(\sim p \rightarrow \sim r) \vee q$
57. $r \leftrightarrow (p \wedge q)$
58. $r \leftrightarrow (q \wedge p)$
59. $(p \leftrightarrow q) \wedge r$
60. $q \rightarrow (r \leftrightarrow p)$
61. $\sim r \rightarrow \sim(p \wedge q)$
62. $\sim(p \wedge q) \rightarrow \sim r$

Application Exercises

Exercises 63–68 contain statements made by well-known people. Use letters to represent each simple statement and rewrite the given compound statement in symbolic form.

63. "If you cannot get rid of the family skeleton, you may as well make it dance." (George Bernard Shaw)

64. "I wouldn't turn out the way I did if I didn't have all the old-fashioned values to rebel against." (Madonna)

65. "If you know what you believe then it makes it a lot easier to answer questions, and I can't answer your question." (George W. Bush)

66. "If you don't like what you're doing, you can always pick up your needle and move to another groove." (Timothy Leary)

67. "If I were an intellectual, I would be pessimistic about America, but since I'm not an intellectual, I am optimistic about America." (General Lewis B. Hershey, Director of the Selective Service during the Vietnam war) (For simplicity, regard "optimistic" as "not pessimistic.")

68. "You cannot be both a good socializer and a good writer." (Erskine Caldwell)

Writing in Mathematics

69. Describe what is meant by a compound statement.

70. What is a conjunction? Describe the symbol that forms a conjunction.

71. What is a disjunction? Describe the symbol that forms a disjunction.

72. What is a conditional statement? Describe the symbol that forms a conditional statement.

73. What is a biconditional statement? Describe the symbol that forms a biconditional statement.

74. Discuss the difference between the symbolic statements $\sim(p \wedge q)$ and $\sim p \wedge q$.

75. Suppose that a friend tells you, "This summer I plan to visit Paris or London." Under what condition can you conclude that if your friend visits Paris, London will not be visited? Under what condition can you conclude that if your friend visits Paris, London might be visited? Assuming your friend has told you the truth, what can you conclude if you know that Paris will not be visited? Explain each of your answers.

Critical Thinking Exercises

76. Use letters to represent each simple statement in the compound statement that follows. Then express the compound statement in symbolic form.
    Shooting unarmed civilians is morally justifiable if and only if bombing them is morally justifiable, and as the former is not morally justifiable, neither is the latter.

77. Using a topic on which you have strong opinions, write a compound statement that contains at least two different connectives. Then express the statement in symbolic form.

Group Exercise

78. Each group member should find a legal document that contains at least six connectives in one paragraph. The connectives should include at least three different kinds of connectives, such as "and," "or," "if–then," and "if and only if." Share your example with other members of the group, and see if the group can explain what some of the more complicated statements actually mean.

OBJECTIVES

1. Use the definitions of negation, conjunction, and disjunction.
2. Construct truth tables.
3. Use a De Morgan law.

Americans account for 4.5% of the world's population, although we consume nearly a quarter of the world's energy. Alaska has the largest area of wilderness lands in the country, as well as the nation's two largest oil fields. A trip to Alaska's Arctic refuge is a confrontation with choices: Should we drill or protect untouched wilderness?

In this section, you will work with a graph that shows the percentage of land owned by the federal government in Alaska, as well as in other Western states. By determining when statements involving negation, ~ (not), conjunction, ∧ (and), and disjunction, ∨ (or), are true and when they are false, you will be able to draw conclusions from the surprising data in this graph. Classifying a statement as true or false is called **assigning a truth value to the statement**.

1 Use the definitions of negation, conjunction, and disjunction.

Negation, ~

The negation of a true statement is a false statement. We can express this in a table in which T represents true and F represents false.

p	$\sim p$
T	F

The negation of a false statement is a true statement. This, too, can be shown in table form.

p	$\sim p$
F	T

Combining the two tables shown above results in Table 3.5, called the **truth table for negation**. This truth table expresses the idea that $\sim p$ has the opposite truth value from p.

TABLE 3.5 NEGATION

p	$\sim p$
T	F
F	T

$\sim p$ has the opposite truth value from p.

I visited London and I visited Paris.

Conjunction, ∧

A friend tells you, "I visited London and I visited Paris." In order to understand the truth values for this statement, let's break it down into its two simple statements:

p: I visited London.

q: I visited Paris.

There are four possible cases to consider.

CASE 1 Your friend actually visited both cities, so p is true and q is true. The conjunction "I visited London and I visited Paris" is true because your friend did both things. If both p and q are true, the conjunction $p \wedge q$ is true. We can show this in truth table form:

p	q	$p \wedge q$
T	T	T

CASE 2 Your friend actually visited London, but did not tell the truth about visiting Paris. In this case, p is true and q is false. Your friend didn't do what was stated, namely visit both cities, so $p \wedge q$ is false. If p is true and q is false, the conjunction $p \wedge q$ is false.

p	q	$p \wedge q$
T	F	F

CASE 3 This time, London was not visited, but Paris was. This makes p false and q true. As in case 2, your friend didn't do what was stated, namely visit both cities, so $p \wedge q$ is false. If p is false and q is true, the conjunction $p \wedge q$ is false.

p	q	$p \wedge q$
F	T	F

CASE 4 This time your friend visited neither city, so p is false and q is false. The statement that both were visited, $p \wedge q$, is false.

p	q	$p \wedge q$
F	F	F

Let's use a truth table to summarize all four cases. Only in the case that your friend visited London and visited Paris is the conjunction true. Each of the four cases appears in Table 3.6, the truth table for conjunction, ∧. The definition of conjunction is given in words to the right of the table.

THE DEFINITION OF CONJUNCTION

TABLE 3.6 CONJUNCTION

p	q	$p \wedge q$
T	T	T
T	F	F
F	T	F
F	F	F

A conjunction is true only when both simple statements are true.

Table 3.7 contains an example of each of the four cases in the conjunction truth table.

TABLE 3.7 STATEMENTS OF CONJUNCTION AND THEIR TRUTH VALUES

Statement	Truth Value	Reason
3 + 2 = 5 and London is in England.	T	Both simple statements are true.
3 + 2 = 5 and London is in France.	F	The second simple statement is false.
3 + 2 = 6 and London is in England.	F	The first simple statement is false.
3 + 2 = 6 and London is in France.	F	Both simple statements are false.

EXAMPLE 1   UNDERSTANDING THE DEFINITION OF CONJUNCTION, ∧

The federal government owns more than half the land in the West. Battles among loggers, conservationists, ranchers, artists, small-business owners, hikers, environmentalists, and oil companies are over who gets to use this land and for what purpose. The bar graph in Figure 3.3 shows the percentage of land owned by the federal government in 12 Western states. Use the information given by the graph to determine the truth value for the following conjunction:

The Federal Government owns 83% of the land in Nevada and less than 60% of the land in Utah.

**Percentage of Land Owned by
the Federal Government**

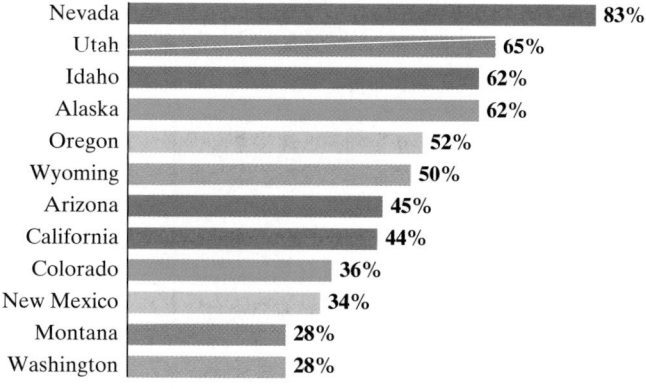

FIGURE 3.3
Source: Bureau of Land Management

SOLUTION A conjunction is true only when both simple statements are true. Consider each of the simple statements that form the conjunction. The first statement is

The federal government owns 83% of the land in Nevada.

Examine the graph in Figure 3.3 to determine the truth value of this statement. The bar on the top represents the percentage of land owned by the federal government in Nevada. Because 83% is printed to the right of the bar, this simple statement is true.
Now consider the second simple statement:

The federal government owns less than 60% of the land in Utah.

The second bar from the top represents the percentage of land owned by the federal government in Utah, given to be 65%. This is more, not less, than 60%. Thus, the second simple statement is false.

STUDY TIP

As soon as you find one simple statement in a conjunction that is false, you know that the conjunction is false.

Because a conjunction, ∧, is true only when both simple statements are true, the given conjunction is false.

 **Check 1 Point**   Use the information given by the bar graph in Figure 3.3 on page 105 to determine the truth value for the following conjunction:

> The federal government owns 62% of the land in Alaska and less than 50% of the land in California.

The statements that come before and after the main connective in a compound statement do not have to be simple statements. Consider, for example, the compound statement

$$(\sim p \lor q) \land \sim q.$$

The statements that make up this conjunction are $\sim p \lor q$ and $\sim q$. The conjunction is true only when both $\sim p \lor q$ and $\sim q$ are true. Notice that $\sim p \lor q$ is not a simple statement. We call $\sim p \lor q$ and $\sim q$ the *component statements* of the conjunction. The statements making up a compound statement are called **component statements**.

Disjunction, ∨

Now your friend states, "I will visit London or I will visit Paris." Because we assume that this is the inclusive "or," if your friend visits either or both of these cities, the truth has been told. The disjunction is false only in the event that neither city is visited. An *or* statement is true in every case, except when both component statements are false.

The truth table for disjunction, ∨, is shown in Table 3.8. The definition of disjunction is given in words to the right of the table.

THE DEFINITION OF DISJUNCTION

TABLE 3.8 DISJUNCTION

p	q	$p \lor q$
T	T	T
T	F	T
F	T	T
F	F	F

A disjunction is false only when both component statements are false.

Table 3.9 contains an example of each of the four cases in the disjunction truth table.

TABLE 3.9 STATEMENTS OF DISJUNCTION AND THEIR TRUTH VALUES

Statement	Truth Value	Reason
3 + 2 = 5 or London is in England.	T	Both component statements are true.
3 + 2 = 5 or London is in France.	T	The first component statement is true.
3 + 2 = 6 or London is in England.	T	The second component statement is true.
3 + 2 = 6 or London is in France.	F	Both component statements are false.

Percentage of U.S. Smokers, by Level of Education

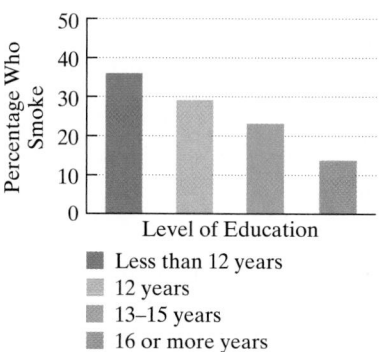

Less than 12 years
12 years
13–15 years
16 or more years

Source: Centers for Disease Control and Prevention

FIGURE 3.4

EXAMPLE 2 | UNDERSTANDING THE DEFINITION OF DISJUNCTION, ∨

The bar graph in Figure 3.4 shows the percentage of people in the United States who smoke, by level of education. Use the information given by the graph to determine the truth value for the following disjunction:

> More than 39% of people with less than a high school education smoke or the percentage of people who smoke decreases as level of education increases.

SOLUTION   A disjunction is false only when both component statements are false. Thus, we need to determine the truth value of each component statement. We can use Figure 3.4 to do so. Begin with the first component statement:

> More than 39% of people with less than a high school education smoke.

The blue bar in Figure 3.4 shows that approximately 35% of people in this group smoke. Because this is not more than 39%, the first component statement is false.
   Now consider the second component statement:

> The percentage of people who smoke decreases as level of education increases.

The decreasing bar heights in the graph illustrate that the second component statement is true.
   By definition, a disjunction, ∨, is false only when both component statements are false. In the given disjunction, only one of the two component statements is false. Thus, the given disjunction is true.

Check 2 Point   Use the information given by the graph in Figure 3.4 to determine the truth value for the following disjunction:

> More than 18% of college graduates smoke or at least 40% of people with less than a high school education smoke.

Constructing Truth Tables

2 • Construct truth tables.

Truth tables can be used to gain a better understanding of English statements. The truth tables in this section are based on the definitions of negation, ~, conjunction, ∧, and disjunction, ∨. It is helpful to remember these definitions in words.

DEFINITIONS OF NEGATION, CONJUNCTION, AND DISJUNCTION

1. Negation ~: not
   The negation of a statement has the opposite truth value from the statement.
2. Conjunction ∧: and
   The only case in which a conjunction is true is when both component statements are true.
3. Disjunction ∨: or
   The only case in which a disjunction is false is when both component statements are false.

In the examples that follow, we will be constructing truth tables for compound statements that contain the simple statements *p* and *q*. In cases like this, there are four possible truth values for *p* and *q*. We will always list the four possible combinations of

truth values for p and q in the following order. This order is not necessary, but it is a standard and makes for a consistent presentation.

p	q	
T	T	
T	F	
F	T	
F	F	

The final column heading in a truth table should be the given compound statement. Each column heading will tell you exactly how to fill in the four truth values for the statement at the top of the column. If a column heading involves negation, ~ (not), fill in the column by looking back at the column that contains the statement that must be negated. Take the opposite of the truth values in this column.

If a column heading involves the symbol for conjunction, ∧ (and), fill in the truth values in the column by looking back at two columns—the column for the statement before the ∧ connective and the column for the statement after the ∧ connective. Fill in the column by applying the definition of conjunction, writing T only when both component statements are true.

If a column heading involves the symbol for disjunction, ∨ (or), fill in the truth values in the column by looking back at two columns—the column for the statement before the ∨ connective and the column for the statement after the ∨ connective. Fill in the column by applying the definition of disjunction, writing F only when both component statements are false.

Let's unravel all this by looking at specific examples.

EXAMPLE 3 | CONSTRUCTING A TRUTH TABLE

Construct a truth table for

$$\sim(p \wedge q)$$

to determine when the statement is true and when the statement is false.

SOLUTION   The parentheses indicate that we must first determine the truth values for the conjunction $p \wedge q$. After this, we determine the truth values for the negation $\sim(p \wedge q)$ by taking the opposite of the truth values for $p \wedge q$.

STEP 1. As with all truth tables, first list the simple statements on top. Then show all the possible truth values for these statements. In this case there are two simple statements and four possible combinations, or cases.

p	q	
T	T	
T	F	
F	T	
F	F	

STEP 2. Make a column for $p \wedge q$, the statement within the parentheses in $\sim(p \wedge q)$. Use $p \wedge q$ as the heading for the column, and then fill in the truth values for the conjunction by looking back at the p and q columns. A conjunction is true only when both component statements are true.

p	q	p ∧ q
T	T	T
T	F	F
F	T	F
F	F	F

p and *q* are true, so *p* ∧ *q* is true.

STEP 3. Construct one more column for ~(p ∧ q). Fill in this column by negating the values in the p ∧ q column. Using the negation definition, take the opposite of the truth values in the third column. This completes the truth table for ~(p ∧ q).

p	q	p ∧ q	~(p ∧ q)
T	T	T	F
T	F	F	T
F	T	F	T
F	F	F	T

The final column in the truth table for ~(p ∧ q) tells us that the statement is false only when both *p* and *q* are true. For example, using

p: Bart Simpson is a cartoon character. (true)

q: Donald Duck is a cartoon character. (true)

the statement ~(p ∧ q) translates as

It is not true that Bart Simpson and Donald Duck are cartoon characters.

This compound statement is false. It *is* true that Bart Simpson and Donald Duck are cartoon characters.

Check 3 Point Construct a truth table for ~(p ∨ q) to determine when the statement is true and when the statement is false.

EXAMPLE 4   CONSTRUCTING A TRUTH TABLE

Construct a truth table for

$$\sim p \lor \sim q$$

to determine when the statement is true and when the statement is false.

SOLUTION   Without parentheses, the negation symbol, ~, negates only the statement that immediately follows it. Therefore, we first determine the truth values for ~p and for ~q. Then we determine the truth values for the *or* disjunction, ~p ∨ ~q.

STEP 1. List the simple statements on top and show the four possible cases for the truth values.

p	q	
T	T	
T	F	
F	T	
F	F	

Bart Simpson

Donald Duck
© Disney Enterprises, Inc.

STEP 2. Make columns for ~p and for ~q. Fill in the ~p column by looking back at the p column, the first column, and taking the opposite of the truth values in this column. Fill in the ~q column by taking the opposite of the truth values in the second column, the q column.

p	q	~p	~q
T	T	F	F
T	F	F	T
F	T	T	F
F	F	T	T

STEP 3. Construct one more column for ~$p \lor$ ~q. To determine the truth values of ~$p \lor$ ~q, look back at the ~p column, labeled column 3 below, and the ~q column, labeled column 4 below. Now use the disjunction definition on the entries in columns 3 and 4. Disjunction definition: An *or* statement is false only when both component statements are false. This occurs only in the first row.

p	q	~p	~q	~$p \lor$ ~q
T	T	F	F	F
T	F	F	T	T
F	T	T	F	T
F	F	T	T	T
		column 3	column 4	

~p is false and ~q is false, so ~$p \lor$ ~q is false.

 Check Point 4   Construct a truth table for ~$p \land$ ~q to determine when the statement is true and when the statement is false.

Our work in Examples 3 and 4 is summarized in Table 3.10. Notice that ~$(p \land q)$ and ~$p \lor$ ~q have the same truth value in each of the four cases.

Two statements that have the same truth value in every possible case are said to be **equivalent**. Therefore,

$$\sim(p \land q) \text{ is equivalent to } \sim p \lor \sim q.$$

This equivalency is one of the **De Morgan laws** and will be considered in more detail later in this chapter.

3 Use a De Morgan law.

TABLE 3.10

p	q	~$(p \land q)$	~$p \lor$ ~q
T	T	F	F
T	F	T	T
F	T	T	T
F	F	T	T

Truth values are the same.

EXAMPLE 5   USING A DE MORGAN LAW

Write a statement that is equivalent to

It is not true that James Joyce and Pablo Picasso are both writers.

SOLUTION   Let p and q represent the following simple statements:

p: James Joyce is a writer. (true)

q: Pablo Picasso is a writer. (false)

The given statement is of the form

$$\sim(p \land q).$$

It is not true that James Joyce and Pablo Picasso are both writers.

An equivalent statement is

$$\sim p \lor \sim q.$$

We can translate this as

James Joyce is not a writer or Pablo Picasso is not a writer.

In Example 5, the truth values of p and q correspond to the second row of Table 3.10. Thus, the given statement, $\sim(p \wedge q)$, is true. The equivalent statement that we obtained for the answer, namely $\sim p \vee \sim q$, makes sense based on our understanding of the English language. If it's not true that Joyce and Picasso are both writers, then at least one of them is not a writer.

 **Check 5 Point**   Write a statement that is equivalent to

It is not true that Will Smith and Bob Dylan are actors.

EXAMPLE 6 | CONSTRUCTING A TRUTH TABLE

Construct a truth table for

$$(\sim p \vee q) \wedge \sim q$$

to determine when the statement is true and when the statement is false.

SOLUTION   The statement is an *and* conjunction because the conjunction symbol, $\wedge$, is outside the parentheses. We cannot determine the truth values for the statement until we first determine the truth values for $\sim p \vee q$ and for $\sim q$, the component statements before and after the $\wedge$ connective:

$$\boxed{(\sim p \ \vee \ q)} \ \wedge \ \boxed{\sim q}.$$

> We'll need a column with truth values for this component statement.

> We'll need a column with truth values for this component statement.

STEP 1. The compound statement involves two simple statements and four possible cases.

p	q	
T	T	
T	F	
F	T	
F	F	

STEP 2. Because we need a column with truth values for $\sim p \vee q$, begin with $\sim p$. Use $\sim p$ as the heading. Fill in the column by looking back at the p column, labeled column 1 below, and take the opposite of the truth values in this column.

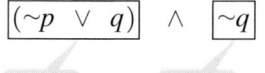

$\boxed{(\sim p \ \vee \ q)} \ \wedge \ \boxed{\sim q}$

Column needed.   Column needed.

p	q	$\sim p$
T	T	F
T	F	F
F	T	T
F	F	T
column 1		

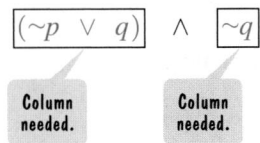

Column needed.   Column needed.

STEP 3. Now add a $\sim p \vee q$ column. To determine the truth values of $\sim p \vee q$, look back at the $\sim p$ column, labeled column 3 below, and the q column, labeled column 2 below. Now use the disjunction definition on the entries in columns 3 and 2. Disjunction definition: An *or* statement is false only when both component statements are false. This occurs only in the second row.

p	q	$\sim p$	$\sim p \vee q$
T	T	F	T
T	F	F	F
F	T	T	T
F	F	T	T
	column 2	column 3	

$\sim p$ is false and q is false, so $\sim p \vee q$ is false.

STEP 4. The statement following the $\wedge$ connective is $\sim q$, so add a $\sim q$ column. Fill in the column by looking back at the q column, column 2, and take the opposite of the truth values in this column.

p	q	$\sim p$	$\sim p \vee q$	$\sim q$
T	T	F	T	F
T	F	F	F	T
F	T	T	T	F
F	F	T	T	T
	column 2			

STEP 5. The final column heading is

$$(\sim p \vee q) \wedge \sim q$$

which is our given statement. To determine its truth values, look back at the $\sim p \vee q$ column, labeled column 4 below, and the $\sim q$ column, labeled column 5 below. Now use the conjunction definition on the entries in columns 4 and 5. Conjunction definition: An *and* statement is true only when both component statements are true. This occurs only in the last row.

p	q	$\sim p$	$\sim p \vee q$	$\sim q$	$(\sim p \vee q) \wedge \sim q$
T	T	F	T	F	F
T	F	F	F	T	F
F	T	T	T	F	F
F	F	T	T	T	T
			column 4	column 5	

$\sim p \vee q$ is true and $\sim q$ is true, so $(\sim p \vee q) \wedge \sim q$ is true.

The truth table is now complete. By looking at the truth values in the last column, we can see that the compound statement

$$(\sim p \vee q) \wedge \sim q$$

is true only in the fourth row, that is, when p is false and q is false.

Check 6 Point

Construct a truth table for $(p \wedge \sim q) \vee \sim p$ to determine when the statement is true and when the statement is false.

Exercise Set 3.3

Practice Exercises

Use the definition of conjunction, $\wedge$, to determine the truth value for each statement in Exercises 1–6.

1. The Beatles were a rock group and Ernest Hemingway was a writer.

2. John Travolta is an actor and Abraham Lincoln was a U.S. president.

3. $3 + 2 = 5$ and 3 is an even number.

4. $7 \times 5 = 35$ and 6 is an odd number.

5. Fir trees produce fur coats and some first basemen are right-handed.

6. Some dogs are not mammals and no number can exceed one million.

Use the definition of disjunction, $\vee$, to determine the truth value for each statement in Exercises 7–12.

7. $7 \times 3 = 21$ or Martin Luther King fought for racial equality.

8. $11 - 4 = 7$ or George Washington was the first U.S. president.

9. Some athletes are not college professors or Sammy Sosa was a U.S. president.

10. London is the capital of France or some students are not psychology majors.

11. All politicians have law degrees or no students are business majors.

12. All computers cost more than $1000 or no colleges have athletic programs.

In Exercises 13–20, complete the truth table for the given statement by filling in the required columns.

13. $p \vee \sim p$

p	$\sim p$	$p \vee \sim p$
T		
F		

14. $p \wedge \sim p$

p	$\sim p$	$p \wedge \sim p$
T		
F		

15. $\sim p \wedge q$

p	q	$\sim p$	$\sim p \wedge q$
T	T		
T	F		
F	T		
F	F		

16. $\sim p \vee q$

p	q	$\sim p$	$\sim p \vee q$
T	T		
T	F		
F	T		
F	F		

17. $\sim(p \vee q)$

p	q	$p \vee q$	$\sim(p \vee q)$
T	T		
T	F		
F	T		
F	F		

18. $\sim(p \vee \sim q)$

p	q	$\sim q$	$p \vee \sim q$	$\sim(p \vee \sim q)$
T	T			
T	F			
F	T			
F	F			

19. $\sim p \wedge \sim q$

p	q	$\sim p$	$\sim q$	$\sim p \wedge \sim q$
T	T			
T	F			
F	T			
F	F			

20. $p \wedge \sim q$

p	q	$\sim q$	$p \wedge \sim q$
T	T		
T	F		
F	T		
F	F		

In Exercises 21–32, construct a truth table for the given statement.

21. $p \vee \sim q$

22. $\sim q \wedge p$

23. $\sim(\sim p \vee q)$

24. $\sim(p \wedge \sim q)$

25. $(p \vee q) \wedge \sim p$

26. $(p \wedge q) \vee \sim p$

27. $\sim p \vee (p \wedge \sim q)$

28. $\sim p \wedge (p \vee \sim q)$

29. $(p \vee q) \wedge (\sim p \vee \sim q)$

30. $(p \wedge \sim q) \vee (\sim p \wedge q)$

31. $(p \wedge \sim q) \vee (p \wedge q)$

32. $(p \vee \sim q) \wedge (p \vee q)$

In Exercises 33–36, use the De Morgan law that states

$$\sim(p \wedge q) \text{ is equivalent to } \sim p \vee \sim q$$

to write an equivalent English statement for the given statement.

33. It is not true that Texas and Paris are both states.

34. It is not true that Michael Jordan and Garth Brooks are both musicians.

35. It is not the case that *Romeo and Juliet* and *Jurassic Park* were both written by Shakespeare.

36. It is not the case that "Star Wars" and "Frasier" are movies.

Application Exercises

Commuters in one-third of the largest cities in the United States spend more than 40 hours per year, equivalent to one work week, sitting in traffic. The bar graph shows the number of hours in traffic per year for the average motorist in ten cities. Use the information given by the graph to determine the truth value for each of the statements in Exercises 37–38.

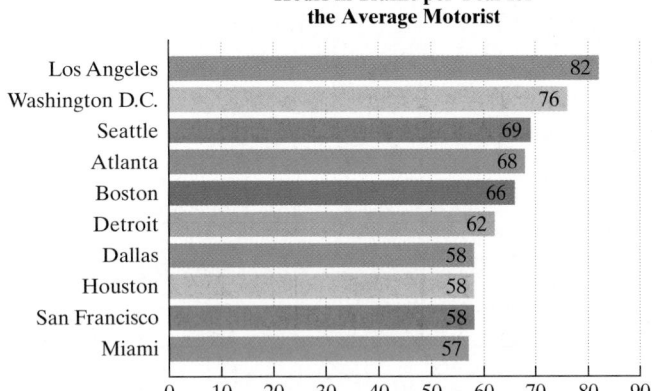

Hours in Traffic per Year for the Average Motorist

Source: Texas Transportation Institute

37. A motorist in Los Angeles averages more than 80 hours in traffic per year and motorists in Atlanta average less hours in traffic per year than in Seattle.

38. A motorist in Los Angeles averages more than 80 hours in traffic per year and motorists in Houston average less hours in traffic per year than in Dallas.

The accompanying line graphs show the life expectancy at birth for women and men in the United States born between 1900 and 2050. Use the information given by the graphs to determine the truth value for each of the statements in Exercises 39–40.

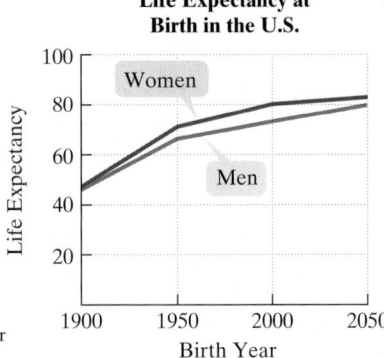

Life Expectancy at Birth in the U.S.

Source: National Center for Health Statistics

39. The average life expectancy for men born in 2000 is longer than the average life expectancy for women born in 2000 or people in the U.S. are living longer with increasing birth years.

40. Men born in 1950 have a life expectancy of 80 years or the average life expectancy for women born in any year between 1950 and 2050 is longer than the average life expectancy for men born in the same year.

The graph shows the percent distribution of divorces in the United States by number of years of marriage.

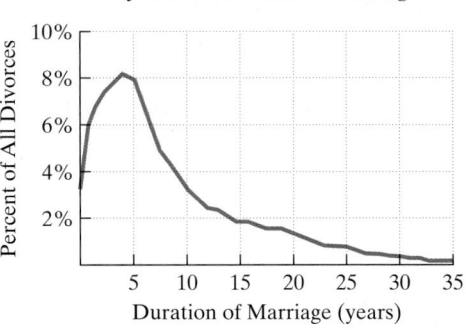

Percent Distribution of Divorces by Number of Years of Marriage

Source: Divorce Center

In Exercises 41–50, let p, q, and r represent the following simple statements:

  p: The chance of divorce peaks during the fourth year of marriage.
  q: The chance of divorce decreases after four years of marriage.
  r: More than 2% of all divorces occur during the 25th year of marriage.

Write each symbolic statement in words. Then use the information given by the graph to determine the truth value of the statement.

41. $p \wedge \sim q$ **42.** $p \wedge \sim r$ **43.** $\sim p \wedge r$ **44.** $q \wedge \sim p$

45. $p \vee \sim q$ **46.** $p \vee \sim r$ **47.** $\sim p \vee r$ **48.** $q \vee \sim p$

49. $(p \wedge q) \vee r$ **50.** $p \wedge (q \vee r)$

Writing in Mathematics

51. Under which conditions is a conjunction true?

52. Under which conditions is a conjunction false?

53. Under which conditions is a disjunction true?

54. Under which conditions is a disjunction false?

55. Describe how to construct a truth table for a compound statement.

56. Describe the information given by the truth values in the final column of a truth table.

57. Use the line graphs in Exercises 39–40 to write a true conjunction. Do not use any of the simple statements that appear in Exercises 39–40.

58. Use the line graphs in Exercises 39–40 to write a false disjunction. Do not use any of the simple statements that appear in Exercises 39–40.

59. The graph for Exercises 41–50 shows that the chance of divorce is greatest during the fourth year of marriage. What explanation can you offer for this trend?

Critical Thinking Exercises

60. Use the bar graph in Exercises 37–38 to write a true compound statement with each of the following characteristics. Do not use any of the simple statements that appear in Exercises 37–38.

   a. The statement contains three different simple statements.

   b. The statement contains two different connectives.

   c. The statement contains one simple statement with the word *not*.

61. If $\sim(p \vee q)$ is true, determine the truth values for p and q.

62. The truth table that defines $\vee$, the *inclusive or*, indicates that the compound statement is true if one or both of its component statements are true. The symbol for the *exclusive or* is $\veebar$. The *exclusive or* means *either p or q*, but *not both*. Use this meaning to write the truth table that defines $p \veebar q$.

Group Exercise

63. Each member of the group should find a graph that is of particular interest to that person. Share the graphs. The group should select the three graphs that they find most intriguing. For the graphs selected, group members should write four compound statements. Two of the statements should be true and two should be false. One of the statements should contain three different simple statements and two different connectives.

| **SECTION 3.4** | **TRUTH TABLES FOR THE CONDITIONAL AND BICONDITIONAL** |

OBJECTIVES

1. Understand the logic behind the definition of the conditional.

2. Construct truth tables for conditional statements.

3. Write a conditional statement in an equivalent form.

4. Use a truth table to determine if a compound statement is a tautology.

5. Understand the definition of the biconditional.

6. Construct truth tables for biconditional statements.

7. Construct truth tables for statements consisting of three simple statements.

8. Determine the truth value of a compound statement for a specific case.

Sisters Venus (left) and Serena Williams

The 2000 Sydney Olympics, the first games of the new millennium, featured outstanding performances from women athletes. American diver Laura Wilkinson won the 10-meter platform with a broken foot. She wore a protective boot that she tossed from the platform before diving. Venus and Serena Williams captured the doubles title in women's tennis. What's better than winning a gold medal? Winning a gold medal with your sister.

In this section's exercise set, you will work with a bar graph showing the number of women athletes in the Olympics from 1976 through 2000. By understanding when statements involving the conditional, $\rightarrow$ (if–then), and the biconditional, $\leftrightarrow$ (if and only if), are true and when they are false, you will be able to draw conclusions from the data in this graph.

1   Understand the logic behind the definition of the conditional.

Conditional Statements, $\rightarrow$

We begin by looking at the truth table for conditional statements. Suppose that your professor promises you the following:

   If you pass the final, then you pass the course.

Break the statement down into its two component statements:

   p: You pass the final.

   q: You pass the course.

**TABLE 3.11
CONDITIONAL**

	p	q	p → q
CASE 1	T	T	T
CASE 2	T	F	F
CASE 3	F	T	T
CASE 4	F	F	T

Translated into symbolic form, your professor's statement is $p \rightarrow q$. We now look at the four cases shown in Table 3.11, the truth table for the conditional.

CASE 1 (T, T) You do pass the final and you do pass the course. Your professor did what was promised, so the conditional statement is true.

CASE 2 (T, F) You pass the final, but you do not pass the course. Your professor did not do what was promised, so the conditional statement is false.

CASE 3 (F, T) You do not pass the final, but you do pass the course. Your professor's original statement talks about only what would happen if you passed the final. It says nothing about what would happen if you did not pass the final. Your professor did not break the promise of the original statement, so the conditional statements is true.

CASE 4 (F, F) You do not pass the final and you do not pass the course. As with case 3, your professor's original statement talks about only what would happen if you passed the final. The promise of the original statement has not been broken. Therefore, the conditional statement is true.

The component statement before the $\rightarrow$ connective in any conditional statement is called the **antecedent**. The component statement after the $\rightarrow$ connective is called the **consequent**:

$$\text{antecedent} \longrightarrow \text{consequent}.$$

Table 3.11 illustrates that a conditional statement is false only when the antecedent is true and the consequent is false. A conditional statement is true in all other cases.

THE DEFINITION OF THE CONDITIONAL

p	q	p → q
T	T	T
T	F	F
F	T	T
F	F	T

A conditional is false only when the antecedent is true and the consequent is false.

 Construct truth tables for conditional statements.

Constructing Truth Tables

Our first example shows how truth tables can be used to gain a better understanding of conditional statements.

EXAMPLE 1 CONSTRUCTING A TRUTH TABLE

Construct a truth table for

$$\sim q \rightarrow \sim p$$

to determine when the statement is true and when the statement is false.

SOLUTION Remember that without parentheses, the symbol $\sim$ negates only the statement that immediately follows it. Therefore, we cannot determine the truth values for this conditional statement until we first determine the truth values for $\sim q$ and for $\sim p$, the statements before and after the $\rightarrow$ connective.

STEP 1. List the simple statements on top and show the four possible cases for the truth values.

p	q	
T	T	
T	F	
F	T	
F	F	

STEP 2. Make columns for $\sim q$ and for $\sim p$. Fill in the $\sim q$ column by looking back at the q column, the second column, and taking the opposite of the truth values in this column. Fill in the $\sim p$ column by taking the opposite of the truth values in the first column, the p column.

p	q	$\sim q$	$\sim p$
T	T	F	F
T	F	T	F
F	T	F	T
F	F	T	T

STEP 3. Construct one more column for $\sim q \rightarrow \sim p$. Look back at the $\sim q$ column, labeled column 3 below, and the $\sim p$ column, labeled column 4 below. Now use the conditional definition to determine the truth values for $\sim q \rightarrow \sim p$ based on columns 3 and 4. Conditional definition: An *if–then* statement is false only when the antecedent is true and the consequent is false. This occurs only in the second row.

p	q	$\sim q$	$\sim p$	$\sim q \rightarrow \sim p$
T	T	F	F	T
T	F	T	F	F
F	T	F	T	T
F	F	T	T	T
		column 3	column 4	

$\sim q$ is true and $\sim p$ is false, so $\sim q \rightarrow \sim p$ is false.

Check Point 1

Construct a truth table for $\sim p \rightarrow \sim q$ to determine when the statement is true and when the statement is false.

3 Write a conditional statement in an equivalent form.

The truth values for $p \rightarrow q$, as well as those for $\sim q \rightarrow \sim p$ from Example 1, are shown in Table 3.12. Notice that $p \rightarrow q$ and $\sim q \rightarrow \sim p$ have the same truth value in each of the four cases. Remember that two statements that have the same truth value in every possible case are *equivalent*. Therefore,

$$p \rightarrow q \text{ is equivalent to } \sim q \rightarrow \sim p.$$

This equivalency will be considered in more detail in the next section. The box that follows expresses this equivalency in words.

TABLE 3.12

p	q	$p \rightarrow q$	$\sim q \rightarrow \sim p$
T	T	T	T
T	F	F	F
F	T	T	T
F	T	T	T

$p \rightarrow q$ and $\sim q \rightarrow \sim p$ are equivalent.

A STATEMENT THAT IS EQUIVALENT TO THE CONDITIONAL
The truth value of a conditional statement does not change if the antecedent and consequent are reversed and both are negated.

EXAMPLE 2 | WRITING A CONDITIONAL STATEMENT IN AN EQUIVALENT FORM

The following conditional statement is true:

If I live in Los Angeles, then I live in California.

Write an equivalent conditional statement.

SOLUTION First, reverse the order of the antecedent and the consequent. Then negate both statements. The equivalent conditional statement is

If I do not live in California, then I do not live in Los Angeles.

Observe that this statement is also true.

Check 2 Point The following conditional statement is true:

If the play is *Macbeth*, then Shakespeare is the author.

Write an equivalent conditional statement.

Our next example illustrates a statement that is always true. Such a statement is called a **tautology**.

> **DEFINITION OF A TAUTOLOGY**
> A **tautology** is a compound statement that is always true.

 4 Use a truth table to determine if a compound statement is a tautology.

BLITZER BONUS

Tautologies and Commercials

Think about the logic in the following statement from a commercial for a high-fiber cereal:

Some studies suggest a high-fiber, low-fat diet may reduce the risk of some kinds of cancer.

From a strictly logical point of view, "may reduce" means "It may reduce the risk of cancer or it may not." The symbolic translation of this sentence, $p \lor \sim p$, is a tautology:

p	$\sim p$	$p \lor \sim p$
T	F	T
F	T	T

The claim made in the cereal commercial can be made about any substance that is not known to actually cause cancer.

EXAMPLE 3 USING A TRUTH TABLE TO VERIFY A TAUTOLOGY

Construct a truth table for

$$[(p \lor q) \land \sim p] \to q$$

and show that the compound statement is a tautology.

SOLUTION The statement is a conditional statement because the *if–then* symbol, $\to$, is outside the grouping symbols. We cannot determine the truth values for this conditional until we first determine the truth values for the statements before and after the $\to$ connective.

$$\boxed{[(p \lor q) \land \sim p]} \quad \to \quad \boxed{q}$$

We'll need a column with truth values for this statement. Prior to this column, we'll need columns for $p \lor q$ and for $\sim p$.

We'll need a column with truth values for this statement. This will be the second column of the truth table.

The disjunction, $\lor$, is false only when both component statements are false.

The truth value of $\sim p$ is opposite that of p.

The conjunction, $\land$, is true only when both $p \lor q$ and $\sim p$ are true.

The conditional, $\to$, is false only when $(p \lor q) \land \sim p$ is true and q is false.

Show four possible cases.

p	q	$p \lor q$	$\sim p$	$(p \lor q) \land \sim p$	$[(p \lor q) \land \sim p] \to q$
T	T	T	F	F	T
T	F	T	F	F	T
F	T	T	T	T	T
F	F	F	T	F	T

Thus, the conditional statement in the last column, $[(p \lor q) \land \sim p] \to q$, is never false. Equivalently, it is true in all four cases. Because we have shown that $[(p \lor q) \land \sim p] \to q$ is always true, we have verified that this compound statement is a tautology.

Conditional statements that are tautologies are called **implications**. For the conditional statement

$$[(p \vee q) \wedge \sim p] \rightarrow q$$

we can say that

$$(p \vee q) \wedge \sim p \text{ } implies \text{ } q.$$

Using p: I am visiting London and q: I am visiting Paris, we can say that

If I am visiting London or Paris, and I am not visiting London, implies that I am visiting Paris.

 **Check Point 3** Construct a truth table for $[(p \rightarrow q) \wedge \sim q] \rightarrow \sim p$ and show that the compound statement is a tautology.

Some compound statements are false in all possible cases. Such statements are called **self-contradictions**. An example of a self-contradiction is the statement $p \wedge \sim p$:

p	$\sim p$	$p \wedge \sim p$
T	F	F
F	T	F

$p \wedge \sim p$ is always false.

If p represents "I am going," then $p \wedge \sim p$ translates as "I am going and I am not going." Such a translation sounds like a contradiction.

Biconditional Statements

5 Understand the definition of the biconditional.

In Section 3.2, we introduced the biconditional connective, $\leftrightarrow$, translated as "if and only if." The biconditional statement $p \leftrightarrow q$ means that $p \rightarrow q$ and $q \rightarrow p$. We write this symbolically as

$$(p \rightarrow q) \wedge (q \rightarrow p).$$

To create the truth table for $p \leftrightarrow q$, we will first make a truth table for the conjunction of the two conditionals $p \rightarrow q$ and $q \rightarrow p$.

EXAMPLE 4 | CONSTRUCTING A TRUTH TABLE

Construct a truth table for

$$(p \rightarrow q) \wedge (q \rightarrow p)$$

to determine when the statement is true and when the statement is false.

SOLUTION The statement is a conjunction because the conjunction symbol, $\wedge$, is outside the parentheses. We cannot determine the truth values for this conjunction until we determine the truth values for $p \rightarrow q$ and $q \rightarrow p$, the statements before and after the $\wedge$ connective.

The conditional is false only when p is true and q is false.

The conditional is false only when q is true and p is false.

The conjunction is true only when both $p \rightarrow q$ and $q \rightarrow p$ are true.

p	q	$p \rightarrow q$	$q \rightarrow p$	$(p \rightarrow q) \wedge (q \rightarrow p)$
T	T	T	T	T
T	F	F	T	F
F	T	T	F	F
F	F	T	T	T

Show four possible cases.

The truth values in the column for $(p \rightarrow q) \wedge (q \rightarrow p)$ show the truth values for the biconditional statement $p \leftrightarrow q$.

THE DEFINITION OF THE BICONDITIONAL

p	q	$p \leftrightarrow q$
T	T	T
T	F	F
F	T	F
F	F	T

A biconditional is true only when the component statements have the same truth value.

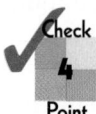

Check Point 4 Construct a truth table for $(p \rightarrow q) \wedge (\sim p \rightarrow \sim q)$ to determine when the statement is true and when the statement is false.

Before we continue our work with truth tables, let's take a moment to summarize the basic definitions of symbolic logic.

THE DEFINITIONS OF SYMBOLIC LOGIC

1. Negation $\sim$: not

The negation of a statement has the opposite truth value from the statement.

2. Conjunction $\wedge$: and

The only case in which a conjunction is true is when both component statements are true.

3. Disjunction $\vee$: or

The only case in which a disjunction is false is when both component statements are false.

4. Conditional $\rightarrow$: if–then

The only case in which a conditional is false is when the first component statement, the antecedent, is true and the second component statement, the consequent, is false.

5. Biconditional $\leftrightarrow$: if and only if

The only cases in which a biconditional is true are when the component statements have the same truth value.

6 Construct truth tables for biconditional statements.

EXAMPLE 5   CONSTRUCTING A TRUTH TABLE

Construct a truth table for

$$(p \vee q) \leftrightarrow (\sim q \rightarrow p)$$

to determine whether the statement is a tautology.

SOLUTION   The statement is a biconditional because the biconditional symbol, $\leftrightarrow$, is outside the parentheses. We cannot determine the truth values for this biconditional until we determine the truth values for the statements in parentheses.

$$\boxed{(p \vee q)} \quad \leftrightarrow \quad \boxed{\sim q \rightarrow p}$$

We need a column with truth values for this statement.   We need a column with truth values for this statement.

The completed truth table appears as follows:

p	q	$p \vee q$	$\sim q$	$\sim q \to p$	$(p \vee q) \leftrightarrow (\sim q \to p)$
T	T	T	F	T	T
T	F	T	T	T	T
F	T	T	F	T	T
F	F	F	T	F	T

We applied the definition of the biconditional to fill in the last column. In each case, the truth values of $p \vee q$ and $\sim q \to p$ are the same. Therefore, the biconditional $(p \vee q) \leftrightarrow (\sim q \to p)$ is true in each case. Because all cases are true, the biconditional is a tautology.

 **Check 5 Point**   Construct a truth table for $(p \vee q) \leftrightarrow (\sim p \to q)$ to determine whether the statement is a tautology.

Compound Statements Consisting of Three Simple Statements

Some compound statements involve three simple statements, usually represented by $p, q,$ and r. In this situation, there are eight different true-false possibilities, shown in Table 3.13. The first column has four Ts followed by four Fs. The second column has two Ts, two Fs, two Ts, and two Fs. Under the third statement, r, T alternates with F. It is not necessary to list the eight cases in this order, but this systematic method ensures that no case is repeated and that all cases are included.

 Construct truth tables for statements consisting of three simple statements.

TABLE 3.13

	p	q	r
Case 1	T	T	T
Case 2	T	T	F
Case 3	T	F	T
Case 4	T	F	F
Case 5	F	T	T
Case 6	F	T	F
Case 7	F	F	T
Case 8	F	F	F

There are eight different true-false combinations for compound statements consisting of three simple statements.

EXAMPLE 6   CONSTRUCTING A TRUTH TABLE WITH EIGHT CASES

Construct a truth table for

$$(p \wedge q) \to r$$

to determine when the statement is true and when the statement is false. Is the statement a tautology?

SOLUTION   The statement $(p \wedge q) \to r$ is a conditional statement because the *if–then* symbol, $\to$, is outside the parentheses. We cannot determine the truth values for this conditional until we have determined the truth values for $p \wedge q$ and for r, the statements before and after the $\to$ connective. The completed truth table appears as follows:

The conjunction is true only when p, the first column, is true and q, the second column, is true.

The conditional is false only when $p \wedge q$, column 4, is true and r, column 3, is false.

Show eight possible cases.

p	q	r	$p \wedge q$	$(p \wedge q) \to r$
T	T	T	T	T
T	T	F	T	F
T	F	T	F	T
T	F	F	F	T
F	T	T	F	T
F	T	F	F	T
F	F	T	F	T
F	F	F	F	T

The truth values in the final column are not all Ts. Therefore, $(p \wedge q) \to r$ is not true in all possible cases and the statement is not a tautology. We can also say that this conditional statement is not an implication.

The final column in the truth table for $(p \land q) \to r$ tells us that the statement is false only when both p and q are true and r is false. For example, an advertisement claims

>   If you use Hair Grow and apply it daily, then your bald spot will disappear.

Use the following representations:

>   *p:* You use Hair Grow.
>   *q:* You apply it daily.
>   *r:* Your bald spot will disappear.

The claim can be expressed in symbolic form as

$$(p \land q) \to r.$$

A client uses Hair Grow (p is true), applies it daily (q is true), but his bald spot does not disappear (r is false). This corresponds to the second row in the truth table for $(p \land q) \to r$. In this case the compound statement $(p \land q) \to r$, like the advertisement's claim, is false.

Check 6 Point Construct a truth table for $(p \to q) \land r$ to determine when the statement is true and when the statement is false. Is the statement a tautology?

A truth table shows the truth values of a compound statement for every possible case. In our next example, we will determine the truth value of a compound statement for a specific case in which the truth values of the simple statements are known. This does not require constructing an entire truth table. By substituting the truth values of the simple statements into the symbolic form of the compound statement and using the appropriate definitions, we can determine the truth value of the compound statement.

8 Determine the truth value of a compound statement for a specific case.

EXAMPLE 7   DETERMINING THE TRUTH VALUE OF A COMPOUND STATEMENT

The circle graph in Figure 3.5 shows the political ideology of U.S. college freshmen. Use the information in the graph to determine the truth value for the following compound statement:

>   If 17.6% of freshmen are conservative and 40% are far right, then the percentage of liberals does not exceed the percentage of conservative and far-right freshmen combined.

SOLUTION   We begin by writing the compound statement in symbolic form. Let p, q, and r represent the following simple statements:

>   *p:* 17.6% of freshmen are conservative.
>   (Figure 3.5 shows that this is true.)
>   *q:* 40% of freshmen are far right.
>   (Figure 3.5 shows that this is false:
>   0.4%, not 40%, are far right.)
>   *r:* The percentage of liberals exceeds
>   the percentage of conservative and far-right
>   freshmen combined.
>   (Figure 3.5 shows this is true: 39.6% exceeds
>   the sum of 17.6% and 0.4%.)

Using these representations, the given compound statement can be expressed in symbolic form as

$$(p \land q) \to \sim r.$$

**Political Ideology of
U.S. College Freshman**

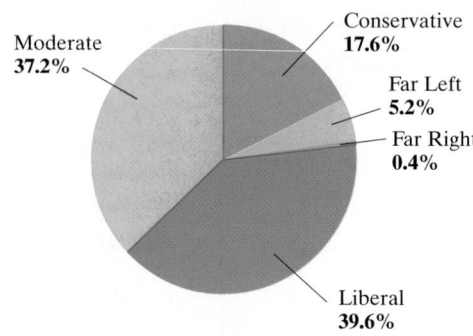

Source: *The Chronicle of Higher Education*

FIGURE 3.5

Now we substitute the truth values for *p*, *q*, and *r* that we obtained from the graph to determine the truth value for the given compound statement.

$$(p \land q) \rightarrow \sim r$$ This is the given compound statement in symbolic form.

$$(T \land F) \rightarrow \sim T$$ Substitute the truth values obtained from the graph.

$$F \rightarrow F$$ Use the definitions of conjunction and negation.

$$T$$ Use the definition of the conditional.

We conclude that the given statement is true.

Check Point 7 Use the information in the circle graph in Figure 3.5 to determine the truth value for the following compound statement:

17.6% of freshmen are conservative or 29.6% are liberal, if and only if the percentage of liberals does not exceed the percentage of conservative and far-right freshmen combined.

BLITZER BONUS

Conditional Trickery

We have all received junk mail with this claim. The conditional statement

If you follow the instructions inside and you return the winning number, then you win $7,000,000

consists of three simple statements:

p: You follow the instructions inside.   (Possibly true)

q: You return the winning number.   (False: there are only losing numbers inside, so you can't return the winning number.)

r: You win $7,000,000.   (False)

The form of the conditional statement is

$$(p \land q) \rightarrow r.$$

Because a conjunction is true only when both simple statements are true, $p \land q$ is false. Statement *r* is also false. Let's substitute these truth values into the conditional statement.

$$(p \land q) \rightarrow r$$
$$(T \land F) \rightarrow F$$
$$F \rightarrow F$$
$$T$$

A conditional is false only when the first part is true and the second part is false. This means the claim that came with the junk mail is true in spite of the fact that you're not going to win one cent!

People who do not think carefully will interpret the conditional claim to read

If you follow the instructions by returning the winning number that you will find inside this envelope, then you win $7,000,000.

This misreading is wishful thinking. There is no winning number in the envelope. This claim was not made by the conditional statement "If you follow the instructions inside and you return the winning number, then you win $7,000.000." It is this statement that entices many people to both open the mail and obediently follow the accompanying instructions. (And to buy magazines, although no purchase is necessary!)

Exercise Set 3.4

Practice Exercises

In Exercises 1–8, construct a truth table for the given statement.

1. $p \rightarrow \sim q$

2. $\sim p \rightarrow q$

3. $\sim(q \rightarrow p)$

4. $\sim(p \rightarrow q)$

5. $(p \wedge q) \rightarrow (p \vee q)$

6. $(p \vee q) \rightarrow (p \wedge q)$

7. $(p \rightarrow q) \wedge \sim q$

8. $(p \rightarrow q) \wedge \sim p$

In Exercises 9–12, use the result that states

$$p \rightarrow q \text{ is equivalent to } \sim q \rightarrow \sim p$$

to write an equivalent English statement for the given statement.

9. If I live in Seattle, then I live in Washington.

10. If I live in Salt Lake City, then I live in Utah.

11. If a plant is not watered, then it wilts.

12. If the exam is tomorrow, then I do not go out tonight.

In Exercises 13–22, construct a truth table for the given statement. Then determine whether the statement is a tautology.

13. $[(p \rightarrow q) \wedge q] \rightarrow p$

14. $[(p \rightarrow q) \wedge p] \rightarrow q$

15. $[(p \rightarrow q) \wedge \sim q] \rightarrow \sim p$

16. $[(p \rightarrow q) \wedge \sim p] \rightarrow \sim q$

17. $[(p \vee q) \wedge p] \rightarrow \sim q$

18. $[(p \vee q) \wedge \sim q] \rightarrow p$

19. $(p \rightarrow q) \rightarrow (\sim p \vee q)$

20. $(q \rightarrow p) \rightarrow (p \vee \sim q)$

21. $(p \wedge q) \wedge (\sim p \vee \sim q)$

22. $(p \vee q) \wedge (\sim p \wedge \sim q)$

In Exercises 23–32, construct a truth table for the given statement.

23. $p \leftrightarrow \sim q$

24. $\sim p \leftrightarrow q$

25. $\sim(p \leftrightarrow q)$

26. $\sim(q \leftrightarrow p)$

27. $(p \leftrightarrow q) \rightarrow p$

28. $(p \leftrightarrow q) \rightarrow q$

29. $(\sim p \leftrightarrow q) \rightarrow (\sim p \rightarrow q)$

30. $(p \leftrightarrow \sim q) \rightarrow (q \rightarrow \sim p)$

31. $[(p \wedge q) \wedge (q \rightarrow p)] \leftrightarrow (p \wedge q)$

32. $[(p \rightarrow q) \vee (p \wedge \sim p)] \leftrightarrow (\sim q \rightarrow \sim p)$

In Exercises 33–40, construct a truth table for the given statement. Then determine whether the statement is a tautology.

33. $\sim(p \wedge q) \leftrightarrow (\sim p \wedge \sim q)$

34. $\sim(p \vee q) \leftrightarrow (\sim p \wedge \sim q)$

35. $(p \rightarrow q) \leftrightarrow (q \rightarrow p)$

36. $(p \rightarrow q) \leftrightarrow (\sim p \rightarrow \sim q)$

37. $(p \rightarrow q) \leftrightarrow (\sim p \vee q)$

38. $(p \rightarrow q) \leftrightarrow (p \vee \sim q)$

39. $(p \leftrightarrow q) \leftrightarrow [(q \rightarrow p) \wedge (p \rightarrow q)]$

40. $(q \leftrightarrow p) \leftrightarrow [(p \rightarrow q) \wedge (q \rightarrow p)]$

In Exercises 41–42, complete the truth table for the given statement by filling in the required columns.

41. $(p \wedge \sim q) \vee r$

p	q	r	$\sim q$	$p \wedge \sim q$	$(p \wedge \sim q) \vee r$
T	T	T			
T	T	F			
T	F	T			
T	F	F			
F	T	T			
F	T	F			
F	F	T			
F	F	F			

42. $(p \vee \sim q) \wedge r$

p	q	r	$\sim q$	$p \vee \sim q$	$(p \vee \sim q) \wedge r$
T	T	T			
T	T	F			
T	F	T			
T	F	F			
F	T	T			
F	T	F			
F	F	T			
F	F	F			

In Exercises 43–50, construct a truth table for the given statement. Then determine whether the statement is a tautology.

43. $(p \vee q) \rightarrow r$

44. $p \rightarrow (q \vee r)$

45. $(p \wedge q) \rightarrow (p \vee r)$

46. $(p \vee r) \rightarrow (q \wedge r)$

47. $r \rightarrow (p \wedge q)$

48. $r \rightarrow (p \vee q)$

49. $[(p \rightarrow q) \wedge (q \rightarrow r)] \rightarrow (p \rightarrow r)$

50. $[(p \rightarrow q) \wedge (q \rightarrow r)] \rightarrow (\sim r \rightarrow \sim p)$

In Exercises 51–60, determine the truth value for each statement when p is false, q is true, and r is false.

51. $\sim(p \rightarrow q)$

52. $\sim(p \leftrightarrow q)$

53. $\sim p \leftrightarrow q$

54. $\sim p \rightarrow q$

55. $q \rightarrow (p \wedge r)$

56. $(p \wedge r) \rightarrow q$

57. $(\sim p \wedge q) \leftrightarrow \sim r$

58. $\sim p \leftrightarrow (\sim q \wedge r)$

59. $\sim[(p \rightarrow \sim r) \leftrightarrow (r \wedge \sim p)]$

60. $\sim[(\sim p \rightarrow r) \leftrightarrow (p \vee \sim q)]$

Application Exercises

The bar graph shows the number of women athletes in the Olympics from 1976 through 2000. Use the information given by the graph to determine the truth value for each compound statement in Exercises 61–66.

Number of Women in the Olympics

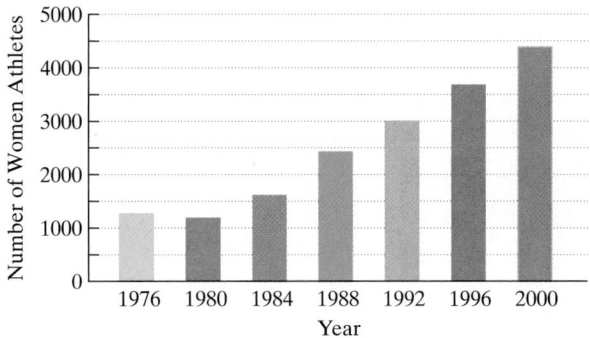

Source: International Olympic Committee

61. If there was a decrease in the number of women athletes from 1976 to 1980, then more than 4500 women participated in the 2000 Olympics.

62. If there was an increase in the number of women athletes from 1976 to 1980, then more than 4500 women participated in the 2000 Olympics.

63. If fewer than 1500 women participated in 1976 then more than 1500 women participated in 1984, or more than 2500 women participated in 1988.

64. Fewer than 1500 women participated in 1976 and more than 1500 women participated in 1984, or more than 2500 women participated in 1988.

65. Participation increased from 1984 to 1988, and if participation did not increase from 1996 to 2000 then more than 4000 women participated in the 2000 Olympics.

66. Fewer than 1500 women participated in 1976, if and only if more than 3500 women participated in 1992 or more than 3500 women participated in 1996.

The circle graph shows the results of a survey in which American adults were asked the following question:

If you got conflicting or different reports of the same news story from television, newspapers, radio, or magazines, which would you be most inclined to believe?

The circle graph shows the percentage of American adults finding each medium the most believable.

Most Believable News Source

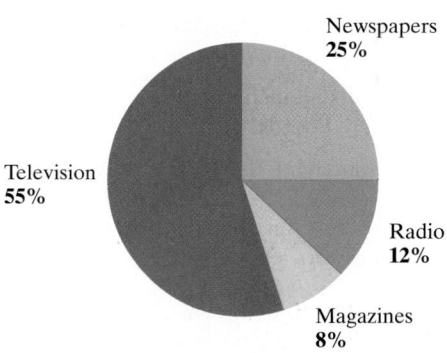

Source: Thomas R. Dye, *Who's Running America? Seventh Edition,* Prentice Hall, 2002

Use the information given by the graph to determine the truth value for each compound statement in Exercises 67–70.

67. Thirty-five percent find newspapers the most believable news source or 45% do not find television the most believable news source, if and only if more people believe a news story in a magazine than on radio.

68. Thirty-five percent find newspapers the most believable news source and 45% do not find television the most believable news source, if and only if more people believe a news story in a magazine than on radio.

69. If more people believe newspapers than radio and magazines combined, then 22% find radio the most believable news source and 8% find magazines the most believable news source.

70. If fewer people believe newspapers than radio and magazines combined, then 22% find radio the most believable news source or 8% find magazines the most believable news source.

Writing in Mathematics

71. Explain when conditional statements are true and when they are false.

72. Explain when biconditional statements are true and when they are false.

73. What is a tautology?

74. Describe how to set up the eight different true-false combinations for a compound statement consisting of three simple statements.

75. Based on the meaning of the inclusive or, explain why it is reasonable that if $p \lor q$ is true, then $\sim p \rightarrow q$ must also be true.

76. Based on the meaning of the inclusive or, explain why if $p \lor q$ is true, then $p \rightarrow \sim q$ is not necessarily true.

Critical Thinking Exercises

77. Which one of the following is true?

a. A conditional statement is false only when the consequent is true and the antecedent is false.

b. Some implications are not tautologies.

c. An equivalent form for a conditional statement is obtained by reversing the antecedent and consequent, and then negating the resulting statement.

d. A compound statement consisting of two simple statements that are both false can be true.

78. Consider the statement "If you get an A in the course, I'll take you out to eat." If you complete the course and I do take you out to eat, can you conclude that you got an A? Explain your answer.

In Exercises 79–80, the headings for the columns in the truth tables are missing. Fill in the statements to replace the missing headings. (More than one correct statement may be possible.)

79.

Do not repeat the statement from the third column.

p	q				
T	T	T	F	T	T
T	F	F	F	F	T
F	T	T	T	T	T
F	F	T	T	T	T

80.

Do not repeat the previous statement.

p	q				
T	T	T	T	F	F
T	F	T	F	T	T
F	T	T	F	T	T
F	F	F	F	T	T

81. A compound statement involving two simple statements has 4 different true-false possibilities. A compound statement involving three simple statements has 8 different true-false possibilities, and one with four simple statements has 16 different true-false possibilities. How many true-false possibilities are there for a compound statement involving n simple statements?

SECTION 3.5	EQUIVALENT STATEMENTS, VARIATIONS OF CONDITIONAL STATEMENTS, AND DE MORGAN'S LAWS

OBJECTIVES

1. Use a truth table to show that statements are equivalent.

2. Write the equivalent contrapositive for a conditional statement.

3. Write the converse and inverse of a conditional statement.

4. Write the negation of a conditional statement.

5. Write equivalent statements using De Morgan's laws.

6. Use De Morgan's laws to negate a conjunction, $\land$, and a disjunction, $\lor$.

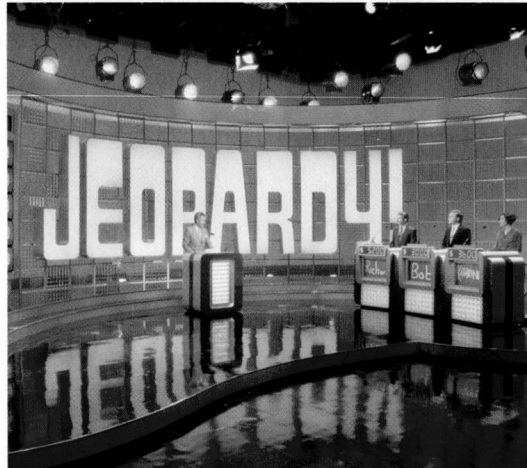

Jerry Orbach	$34,000
Charles Shaugnessy	$31,800
Andy Richter	$29,400
Norman Schwarzkopf	$28,000
Jon Stewart	$28,000
Kareem Abdul-Jabbar	$27,000
Mark McEwen	$26,700
Bob Costas	$25,000
Cheech Marin	$25,000

The answer: See the above list. The question: Who are *Celebrity Jeopardy!*'s all-time highest earners? The list indicates that if you are Jon Stewart, then you won $28,000. Does that mean if you won $28,000, then you are Jon Stewart? If you aren't Jon Stewart, then did you not win $28,000? Furthermore, if you did not win $28,000, then can you not be Jon Stewart? In this section, we will use truth tables and logic to unravel this verbal morass of conditional statements.

Equivalent Statements

We have seen that equivalent compound statements, made up of the same simple statements, have the same corresponding truth values for all true-false combinations of these simple statements. If a compound statement is true, then its equivalent statement must also be true. Likewise, if a compound statement is false, its equivalent statement must also be false.

Truth tables are used to show that two statements are equivalent. When translated into English, equivalencies can be used to gain a better understanding of English statements.

EXAMPLE 1 | SHOWING THAT STATEMENTS ARE EQUIVALENT

a. Show that $p \lor \sim q$ and $\sim p \to \sim q$ are equivalent.

b. Use the result from part (a) to write a statement that is equivalent to

The bill receives majority approval or the bill does not become law.

SOLUTION

a. Construct a truth table that shows the truth values for $p \lor \sim q$ and $\sim p \to \sim q$. The truth values for each statement are shown below.

p	q	$\sim q$	$p \lor \sim q$	$\sim p$	$\sim p \to \sim q$
T	T	F	T	F	T
T	F	T	T	F	T
F	T	F	F	T	F
F	F	T	T	T	T

Corresponding truth values are the same.

The table shows that the truth values for $p \lor \sim q$ and $\sim p \to \sim q$ are the same. Therefore, the statements are equivalent.

b. The statement

The bill receives majority approval or the bill does not become law

can be expressed in symbolic form using the following representations:

p: The bill receives majority approval.

q: The bill becomes law.

In symbolic form, the statement is $p \lor \sim q$. Based on the truth table in part (a), we know that an equivalent statement is $\sim p \to \sim q$. The equivalent statement can be expressed in words as

If the bill does not receive majority approval, then the bill does not become law.

Notice that the given statement and its equivalent are both true.

Check 1 Point

a. Show that $p \lor q$ and $\sim q \to p$ are equivalent.

b. Use the result from part (a) to write a statement that is equivalent to

I attend classes or I lose my scholarship.

1 Use a truth table to show that statements are equivalent.

BLITZER BONUS

Fuzzy Logic

The yin-yang symbol stands for a world of opposites, a world in which everything is a matter of degree. By contrast, every statement in logic is either true or false. Existing computers also perform calculations that distinguish between the logical states "true" and "false," using 1 and 0. In recent years, computer programmers have realized that in human thought distinctions are not so stark, and it is possible to work in terms of degree of truth (quite cold, very sunny, etc.). In **fuzzy logic**, every statement is assigned a truth value between 0 (absolutely false) and 1 (absolutely true), inclusive. Combined with systems that learn to solve problems in a way similar to the human brain, fuzzy logic could be used to create computers that emulate human thought and creativity.

A special symbol, $\equiv$, is used to show that two statements are equivalent. Because $p \vee \sim q$ and $\sim p \rightarrow \sim q$ are equivalent, we can express this equivalency by writing

$$p \vee \sim q \equiv \sim p \rightarrow \sim q \quad \text{and} \quad \sim p \rightarrow \sim q \equiv p \vee \sim q.$$

EXAMPLE 2 SHOWING THAT STATEMENTS ARE EQUIVALENT

Show that $\sim(\sim p) \equiv p$.

SOLUTION Determine the truth values for $\sim(\sim p)$ and p. These are shown in the truth table on the left.

The truth values for $\sim(\sim p)$ were obtained by taking the opposite of each truth value for $\sim p$. The table shows that the truth values for $\sim(\sim p)$ and p are the same. Therefore, the statements are equivalent:

$$\sim(\sim p) \equiv p.$$

p	$\sim p$	$\sim(\sim p)$
T	F	T
F	T	F

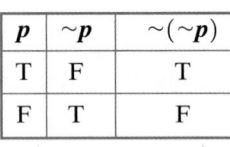
Corresponding truth values are the same.

STUDY TIP

A familiar algebraic statement should help you remember the logical equivalency

$$\sim(\sim p) \equiv p.$$

If a represents a number, then

$$-(-a) = a.$$

For example, $-(-4) = 4$.

The equivalence in Example 2 illustrates that **the double negation of a statement is equivalent to the statement**. For example, the statement "It is not true that Ernest Hemingway was not a writer" means the same thing as "Ernest Hemingway was a writer."

 **Check 2 Point**   Show that $\sim[\sim(\sim p)] \equiv \sim p$.

EXAMPLE 3 EQUIVALENCIES AND TRUTH TABLES

Select the statement that is not equivalent to

Miguel is blushing or sunburned.

a. If Miguel is blushing, then he is not sunburned.

b. Miguel is sunburned or blushing.

c. If Miguel is not blushing, then he is sunburned.

d. If Miguel is not sunburned, then he is blushing.

SOLUTION To determine which of the choices is not equivalent to the given statement, begin by writing the given statement and the choices in symbolic form. Then construct a truth table and compare each statement's truth values to those of the given statement. The nonequivalent statement is the one that does not have exactly the same truth values as the given statement.

The simple statements that make up "Miguel is blushing or sunburned" can be represented as follows:

p: Miguel is blushing.

q: Miguel is sunburned.

Here are the symbolic representations for the given statement and the four choices:

Miguel is blushing or sunburned: $p \vee q$

a. If Miguel is blushing, then he is not sunburned: $p \rightarrow \sim q$

b. Miguel is sunburned or blushing: $q \vee p$

c. If Miguel is not blushing, then he is sunburned: $\sim p \rightarrow q$

d. If Miguel is not sunburned, then he is blushing: $\sim q \rightarrow p$

Next, construct a truth table that contains the truth values for the given statement, $p \vee q$, as well as those for the four options. The truth table is shown as follows:

		Given		(a)	(b)		(c)	(d)
p	q	$p \vee q$	$\sim q$	$p \rightarrow \sim q$	$q \vee p$	$\sim p$	$\sim p \rightarrow q$	$\sim q \rightarrow p$
T	T	T	F	F	T	F	T	T
T	F	T	T	T	T	F	T	T
F	T	T	F	T	T	T	T	T
F	F	F	T	T	F	T	F	F

Equivalent (spanning Given, (b), (c), (d))

Not equivalent (a)

The statement in option (a) does not have the same corresponding truth values as those for $p \vee q$. Therefore, this statement is not equivalent to the given statement.

We can use our understanding of the inclusive *or* to see why the following statements are not equivalent:

Miguel is blushing or sunburned.

If Miguel is blushing, then he is not sunburned.

Let us assume that the first statement is true. The inclusive *or* tells us that Miguel might be both blushing and sunburned. This means that the second statement might not be true. The fact that Miguel is blushing does not indicate that he is not sunburned; he might be both.

Check 3 Point Select the statement that is not equivalent to

If it's raining, then I need a jacket.

a. It's not raining or I need a jacket.

b. I need a jacket or it's not raining.

c. If I need a jacket, then it's raining.

d. If I do not need a jacket, then it's not raining.

Variations of the Conditional Statement $p \rightarrow q$

2 Write the equivalent contrapositive for a conditional statement.

In Section 3.4, we learned that $p \rightarrow q$ is equivalent to $\sim q \rightarrow \sim p$. The truth value of a conditional statement does not change if the antecedent and the consequent are reversed and then both of them are negated. The **contrapositive** is a statement obtained by reversing and negating the antecedent and the consequent.

p	q	$p \rightarrow q$	$\sim q \rightarrow \sim p$
T	T	T	T
T	F	F	F
F	T	T	T
F	F	T	T

$p \rightarrow q$ and $\sim q \rightarrow \sim p$ are equivalent.

> **A CONDITIONAL STATEMENT AND ITS EQUIVALENT CONTRAPOSITIVE**
>
> $$p \rightarrow q \equiv \sim q \rightarrow \sim p$$
>
> The truth value of a conditional statement does not change if the antecedent and consequent are reversed and both are negated. The statement $\sim q \rightarrow \sim p$ is called the **contrapositive** of the conditional $p \rightarrow q$.

EXAMPLE 4 | WRITING EQUIVALENT CONTRAPOSITIVES

Write the equivalent contrapositive for each of the following statements:

a. If it is a Chevrolet, then it is a vehicle.

b. If the cable is out, then we do not watch TV.

c. If all people obey the law, then no prisons are needed.

d. $\sim q \rightarrow r$.

SOLUTION   In each part, first reverse the order of the antecedent and the consequent. Then negate both statements.

a. The equivalent contrapositive of

> If it's a Chevrolet, then it's a vehicle

is

> If it's not a vehicle, then it's not a Chevrolet.

Notice that both of these statements are true.

b. The equivalent contrapositive of

> If the cable is out, then we do not watch TV

is

> If we do watch TV, then the cable is not out.

Negations of
Quantified Statements

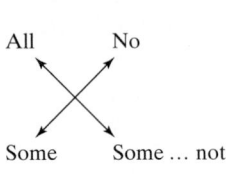

All   No

Some   Some ... not

c. The equivalent contrapositive of

> If all people obey the law, then no prisons are needed

can be expressed initially using words and symbols:

$$\sim(\text{no prisons are needed}) \rightarrow \sim(\text{all people obey the law}).$$

Recall, as shown in the margin, that the negation of "no" is "some," and the negation of "all" is "some ... not." Using the correct negations, the contrapositive is

> If some prisons are needed, then some people do not obey the law.

d. The equivalent contrapositive of $\sim q \rightarrow r$ is

$$\sim r \rightarrow \sim(\sim q).$$ *Reverse and negate the components.*

$$\sim r \rightarrow q$$ *Simplify. The double negation of a statement is equivalent to the statement.*

Thus, $\sim q \rightarrow r \equiv \sim r \rightarrow q$.

Check
4
Point

Write the equivalent contrapositive for each of the following statements:

a. If the book is overdue, then you pay a fine.

b. If it is cold, then we do not use the pool.

c. If all students take exams honestly, then no supervision is needed.

d. $p \rightarrow \sim q$

[3] Write the converse and inverse of a conditional statement.

The truth value of a conditional statement does not change if the antecedent and the consequent are reversed and then both of them are negated. But what happens to the conditional's truth values if just one, but not both, of these changes is made? If the antecedent and the consequent are reversed, the resulting statement is called the **converse** of the conditional statement. By negating both the antecedent and the consequent, we obtain the **inverse** of the conditional statement.

Converses in *Alice's Adventures in Wonderland*

Alice has a problem with logic: She believes that a conditional and its converse mean the same thing. In the passage that follows, she states that

If I say it, I mean it

is the same as

If I mean it, I say it.

She is corrected, told that

If I eat it, I see it

is not the same as

If I see it, I eat it.

The group continues its logical humiliation of poor Alice:

"Come, we shall have some fun now" thought Alice. "I'm glad they've begun asking riddles—I believe I can guess that," she added aloud.

"Do you mean that you think you can find out the answer to it?" said the March Hare.

"Exactly so." said Alice.

"Then you should say what you mean," the March Hare went on.

"I do," Alice hastily replied; "at least—at least I mean what I say—that's the same thing, you know."

"Not the same thing a bit!" said the Hatter. "Why, you might just as well say that 'I see what I eat' is the same thing as 'I eat what I see'!"

"You might just as well say," added the March Hare, "that 'I like what I get' is the same thing as 'I get what I like'!"

"You might just as well say," added the Dormouse, which seemed to be talking in its sleep, "that 'I breathe when I sleep' is the same thing as 'I sleep when I breathe'!"

VARIATIONS OF THE CONDITIONAL STATEMENT

Name	Symbolic Form	English Translation
Conditional	$p \rightarrow q$	If p, then q.
Converse	$q \rightarrow p$	If q, then p.
Inverse	$\sim p \rightarrow \sim q$	If not p, then not q.
Contrapositive	$\sim q \rightarrow \sim p$	If not q, then not p.

EXAMPLE 5 | WRITING THE CONVERSE AND INVERSE

Consider the true conditional statement

If it's a Chevrolet, then it's a vehicle.

Write the converse and inverse of this true statement.

SOLUTION   To form the converse of "If it's a Chevrolet, then it's a vehicle," we reverse the order of the antecedent and consequent:

If it's a vehicle, then it's a Chevrolet.

This statement is not necessarily true. It can be a vehicle, but it might be a make other than a Chevrolet.

To form the inverse of "If it's a Chevrolet, then it's a vehicle," negate the antecedent and consequent:

If it's not a Chevrolet, then it's not a vehicle.

This statement is also not necessarily true. Just because it's not a Chevrolet does not mean that it's not a vehicle. A Ford is not a Chevrolet, yet it certainly is a vehicle.

Example 5 illustrates that if a conditional statement is true, its converse and inverse are not necessarily true. Because the equivalent of a true statement must be true, we see that a conditional statement is not equivalent to its converse or its inverse.

The relationships among the truth values for a conditional statement, its converse, its inverse, and its contrapositive are shown in the truth table that follows:

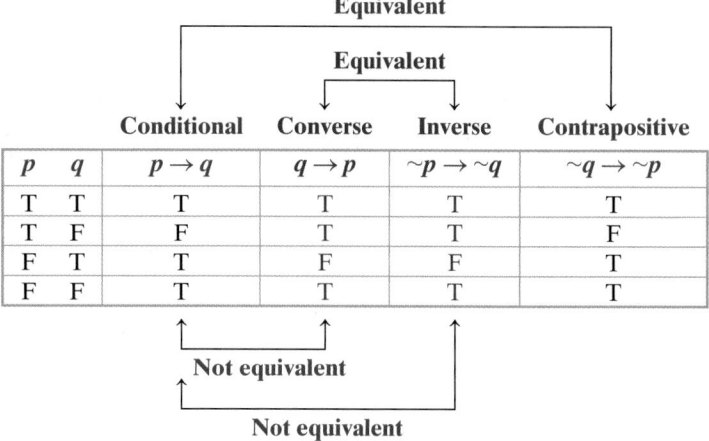

The truth table reveals that a conditional statement is equivalent to its contrapositive. The table also shows that a conditional statement is not equivalent to its converse; in some cases they have the same truth value, but in other cases they have opposite truth values. Also, a conditional statement is not equivalent to its inverse. By contrast, the converse and the inverse are equivalent to each other.

✓ Check 5 Point   Write the converse and inverse of the statement:

If it's a bird, then it can fly.

4   Write the negation of a conditional statement.

The Negation of the Conditional Statement $p \rightarrow q$

Suppose that your accountant makes the following statement:

If you itemize deductions, then you pay less in taxes.

When will your accountant have told you a lie? The only case in which you have been lied to is when you itemize deductions and you do *not* pay less in taxes. We can analyze this situation symbolically with the following representations:

p: You itemize deductions.

q: You pay less in taxes.

We represent each compound statement in symbolic form.

$p \rightarrow q$: If you itemize deductions, then you pay less in taxes.

$p \wedge \sim q$: You itemize deductions and you do not pay less in taxes.

The truth table that follows shows that the negation of $p \rightarrow q$ is $p \wedge \sim q$.

p	q	$p \rightarrow q$	$\sim q$	$p \wedge \sim q$
T	T	T	F	F
T	F	F	T	T
F	T	T	F	F
F	F	T	T	F

These columns have opposite truth values, so $p \wedge \sim q$ negates $p \rightarrow q$.

THE NEGATION OF A CONDITIONAL STATEMENT

The negation of $p \rightarrow q$ is $p \wedge \sim q$. This can be expressed as

$$\sim(p \rightarrow q) \equiv p \wedge \sim q.$$

To form the negation of a conditional statement, leave the antecedent (the first part) unchanged, change the *if–then* connective to *and*, and negate the consequent (the second part).

EXAMPLE 6 | WRITING THE NEGATION OF A CONDITIONAL STATEMENT

Write the negation of

If it rains, then I bring my umbrella.

SOLUTION

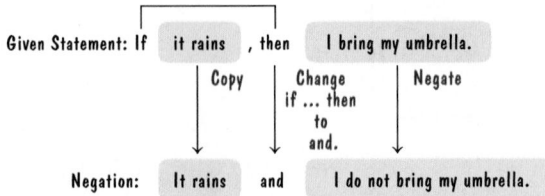

Given Statement: If   it rains   , then   I bring my umbrella.

Copy   Change if ... then to and.   Negate

Negation: It rains   and   I do not bring my umbrella.

The negation of the given statement is

It rains and I do not bring my umbrella.

Check 6 Point   Write the negation of

If the triangle is isosceles, then it has two equal sides.

The box that follows summarizes what we have learned about conditional statements:

THE CONDITIONAL STATEMENT $p \to q$

Contrapositive:

$p \to q$ is equivalent to $\sim q \to \sim p$ (the contrapositive).

Converse and Inverse:

1. $p \to q$ is not equivalent to $q \to p$ (the converse).
2. $p \to q$ is not equivalent to $\sim p \to \sim q$ (the inverse).

Negation:

The negation of $p \to q$ is $p \wedge \sim q$.

---

 5 Write equivalent statements using De Morgan's laws.

Augustus De Morgan

The Rolling Stones
—Not both rock groups

De Morgan's Laws

In Section 3.3, we learned that $\sim(p \wedge q)$ is equivalent to $\sim p \vee \sim q$. Using the symbol for equivalence, $\equiv$, we can write

$$\sim(p \wedge q) \equiv \sim p \vee \sim q.$$

p	q	$\sim(p \wedge q)$	$\sim p \vee \sim q$
T	T	F	F
T	F	T	T
F	T	T	T
F	F	T	T

Equivalent statements

This equivalency is one of two equivalences known as **De Morgan's laws**, named after Augustus De Morgan (1806–1871), an English mathematician.

DE MORGAN'S LAWS

1. $\sim(p \wedge q) \equiv \sim p \vee \sim q$
2. $\sim(p \vee q) \equiv \sim p \wedge \sim q$

EXAMPLE 7   USING A DE MORGAN LAW

Write a statement that is equivalent to

It is not true that the Rolling Stones and De Morgan's Laws are both rock groups.

SOLUTION   Let p and q represent the following simple statements:

p: The Rolling Stones is a rock group.

q: De Morgan's Laws is a rock group.

The statement is of the form $\sim(p \wedge q)$. An equivalent statement is $\sim p \vee \sim q$. We can translate this as

The Rolling Stones is not a rock group or De Morgan's Laws is not a rock group.

 **Check 7 Point**   Write a statement that is equivalent to

It is not true that Kelsey Grammer and Katie Couric are both actors.

EXAMPLE 8   USING A DE MORGAN LAW

Write a statement that is equivalent to

It is not true that *Jaws* is a musical or a comedy.

SOLUTION   Let p and q represent the following simple statements:

p: Jaws is a musical.

q: Jaws is a comedy.

The statement is of the form $\sim(p \vee q)$. An equivalent statement is $\sim p \wedge \sim q$. We can translate this as

Jaws is not a musical and *Jaws* is not a comedy.

This can also be expressed as

Jaws is neither a musical nor a comedy.

Check 8 Point   Write a statement that is equivalent to

It is not true that Oprah Winfrey is a jazz musician or a presidential candidate.

De Morgan's Laws and Negations

De Morgan's laws can be used to write the negation of a compound statement involving conjunction ($\wedge$, and) and disjunction ($\vee$, or).

> #### DE MORGAN'S LAWS AND NEGATIONS
>
> **1.** $\sim(p \wedge q) \equiv \sim p \vee \sim q$
> The negation of $p \wedge q$ is $\sim p \vee \sim q$. To negate a conjunction, negate each component statement and change *and* to *or*.
>
> **2.** $\sim(p \vee q) \equiv \sim p \wedge \sim q$
> The negation of $p \vee q$ is $\sim p \wedge \sim q$. To negate a disjunction, negate each component statement and change *or* to *and*.

6   Use De Morgan's laws to negate a conjunction, $\wedge$, and a disjunction, $\vee$.

THE TEN GREATEST AMERICANS OF ALL TIME

American	Score
1. Abraham Lincoln	36
2. George Washington	35
3. Franklin D. Roosevelt	31
4. Thomas Edison	27
5. Benjamin Franklin	24
6. Martin Luther King, Jr.	21
7. Henry Ford	19
8. Theodore Roosevelt	18
9. Clara Barton	16
10. Eleanor Roosevelt	16

EXAMPLE 9   NEGATING A DISJUNCTION, $\vee$

Who is the greatest American ever? The editors of *The Definitive Guide to the Best and Worst of Everything* (Prentice Hall, 1997) used a 0 to 40 score to arrive at the list in the margin. Write the negation of this statement that was used to form the list:

An American citizen scores more than 15 or that person is not on the greatest-of-all-time list.

SOLUTION   By using De Morgan's law for negating a disjunction, we can write the negation in English without having to first translate into symbolic form. To negate this disjunction, negate each component statement and change *or* to *and*.

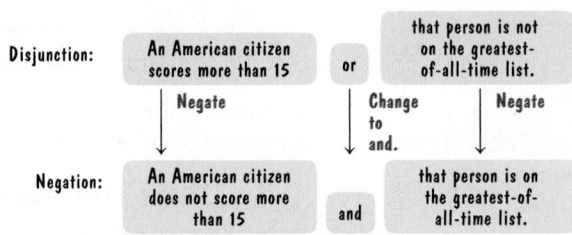

The negation of the given statement

> An American citizen scores more than 15 or that person is not on the greatest-of-all-time list

is

> An American citizen does not score more than 15 and that person is on the greatest-of-all-time list.

 **Check Point 9**   Write the negation of the statement

> You leave by 5 P.M. or you will not arrive home on time.

EXAMPLE 10   USING A DE MORGAN LAW TO FORMULATE A CONTRAPOSITIVE

Write a statement that is equivalent to

> If it rains, I do not go outdoors and I study.

SOLUTION   We begin by writing the conditional statement in symbolic form. Let p, q, and r represent the following simple statements:

p: It rains. q: I go outdoors. r: I study.

Using these representations, the given conditional statement can be expressed in symbolic form as

$$p \rightarrow (\sim q \wedge r).$$ If it rains, I do not go outdoors and I study.

An equivalent statement is the contrapositive.

$\sim(\sim q \wedge r) \rightarrow \sim p$ Form the contrapositive: Reverse and negate the components.

$[\sim(\sim q) \vee \sim r] \rightarrow \sim p$ Use a De Morgan Law to negate the conjunction: Negate each component and change $\wedge$ to $\vee$.

$(q \vee \sim r) \rightarrow \sim p$ Simplify. The double negation of a statement is equivalent to the statement.

Thus,

$$p \rightarrow (\sim q \wedge r) \equiv (q \vee \sim r) \rightarrow \sim p.$$

Using the representations for p, q, and r, a statement that is equivalent to "If it rains, I do not go outdoors and I study" is

> If I go outdoors or I do not study, it is not raining.

 **Check Point 10**   Write a statement that is equivalent to

> If it is not windy, we can swim and we cannot sail.

Exercise Set 3.5

Practice Exercises

1. a. Use a truth table to show that $p \rightarrow q$ and $\sim p \vee q$ are equivalent.

b. Use the result from part (a) to write a statement that is equivalent to

If a number is even, then it is divisible by 2.

2. a. Use a truth table to show that $\sim p \rightarrow q$ and $p \vee q$ are equivalent.

b. Use the result from part (a) to write a statement that is equivalent to

If a major dam on the upper Nile River is not in place, then the lower Nile overflows its banks each year.

3. Select the statement that is equivalent to

I saw *Rent* or *Ragtime*.

a. If I did not see *Rent*, I saw *Ragtime*.

b. I saw both *Rent* and *Ragtime*.

c. If I saw *Rent*, I did not see *Ragtime*.

d. If I saw *Ragtime*, I did not see *Rent*.

4. Select the statement that is equivalent to

Citizen Kane or *Howard the Duck* appear in a list of greatest U.S. movies.

a. If *Citizen Kane* appears in the list of greatest U.S. movies, *Howard the Duck* does not.

b. It *Howard the Duck* does not appear in the list of greatest U.S. movies, then *Citizen Kane* does.

c. Both *Citizen Kane* and *Howard the Duck* appear in a list of greatest U.S. movies.

d. If *Howard the Duck* appears in the list of greatest U.S. movies, *Citizen Kane* does not.

5. Select the statement that is *not* equivalent to

It is not true that Sondheim and Picasso are both musicians.

a. Sondheim is not a musician or Picasso is not a musician.

b. If Sondheim is a musician, then Picasso is not a musician.

c. Sondheim is not a musician and Picasso is not a musician.

d. If Picasso is a musician, then Sondheim is not a musician.

6. Select the statement that is *not* equivalent to

It is not true that England and Africa are both countries.

a. If England is a country, then Africa is not a country.

b. England is not a country and Africa is not a country.

c. England is not a country or Africa is not a country.

d. If Africa is a country, then England is not a country.

Write the equivalent contrapositive for each statement in Exercises 7–18.

7. If I am in Chicago, then I am in Illinois.

8. If I am in Birmingham, then I am in the South.

9. If the stereo is playing, then I cannot hear you.

10. If it is blue, then it is not an apple.

11. "If you don't laugh, you die." (humorist Alan King)

12. "If it doesn't fit, you must acquit." (lawyer Johnnie Cochran)

13. If the president is telling the truth, then all troops were withdrawn.

14. If the review session is successful, then no students fail the test.

15. If all institutions place profit above human need, then some people suffer.

16. If all hard workers are successful, then some people are not hard workers.

17. $\sim q \rightarrow \sim r$

18. $\sim p \rightarrow r$

Write the converse and inverse for each statement in Exercises 19–30.

19. If you touch poison oak, then you get a skin rash.

20. If I am sick, then I miss class.

21. If the play is *Macbeth*, then Shakespeare is the author.

22. If the movie is *Saving Private Ryan*, then Tom Hanks appears.

23. If you are driving the car, then you are not sleeping.

24. If you are in class, then you are not listening to the radio.

25. If I am in Charleston, then I am not in the West.

26. If I am in London, then I am not in Spain.

27. If it's St. Patrick's Day, then some people wear green.

28. If it's Thanksgiving, then some people eat turkey.

29. $\sim q \rightarrow \sim r$

30. $\sim p \rightarrow r$

Write the negation of each conditional statement in Exercises 31–40.

31. If I am in Los Angeles, then I am in California.

32. If I am in Houston, then I am in Texas.

33. If it is purple, then it is not a carrot.

34. If the TV is playing, then I cannot concentrate.

35. If he doesn't, I will.

36. If she says yes, he says no.

37. If there is a blizzard, then all schools are closed.

38. If there is a tax cut, then all people have extra spending money.

39. $\sim q \rightarrow \sim r$

40. $\sim p \rightarrow r$

Use De Morgan's laws to write a statement that is equivalent to each statement in Exercises 41–56.

41. It is not true that Australia and China are both islands.

42. It is not true that Florida and California are both peninsulas.

43. It is not the case that my high school encouraged creativity and diversity.

44. It is not the case that the course covers logic and dream analysis.

45. It is not true that Babe Ruth was a writer or a lawyer.

46. It is not true that Martin Luther King supported violent protest or the Vietnam war.

47. It is not the case that the United States has eradicated poverty or racism.

48. It is not the case that the movie is interesting or entertaining.

49. $\sim(\sim p \wedge q)$ **50.** $\sim(p \vee \sim q)$

51. If you attend lecture and study, you succeed.

52. If a number is a natural number and a whole number, the number is 0.

53. If he does not cook, his wife or child does.

54. If it is Saturday or Sunday, I do not work.

55. $p \rightarrow (q \vee \sim r)$ **56.** $p \rightarrow (\sim q \wedge \sim r)$

Use De Morgan's laws to write the negation of each statement in Exercises 57–68.

57. I'm going to Seattle or San Francisco.

58. This course covers logic or statistics.

59. I study or I do not pass.

60. I give up tobacco or I am not healthy.

61. I am not going and he is going.

62. I do not apply myself and I succeed.

63. A bill becomes law and it does not receive majority approval.

64. They see the show and they do not have tickets.

65. $p \vee \sim q$ **66.** $\sim p \vee q$

67. $p \wedge (q \vee r)$ **68.** $p \vee (q \wedge r)$

In Exercises 69–76, determine which, if any, of the three given statements are equivalent. You may use information about a conditional statement's converse, inverse, or contrapositive, De Morgan's laws, or truth tables.

69. a. If he is guilty, then he does not take a lie-detector test.
  b. He is not guilty or he takes a lie-detector test.
  c. If he is not guilty, then he takes a lie-detector test.

70. a. If the train is late, then I am not in class on time.
  b. The train is late or I am in class on time.
  c. If I am in class on time, then the train is not late.

71. a. It is not true that I have a ticket and cannot go.
  b. I do not have a ticket and can go.
  c. I have a ticket or I cannot go.

72. a. I work hard or I do not succeed.
  b. It is not true that I do not work hard and succeed.
  c. I do not work hard and I do succeed.

73. a. If the grass turns yellow, you did not use fertilizer or water.
  b. If you use fertilizer and water, the grass will not turn yellow.
  c. If the grass does not turn yellow, you used fertilizer and water.

74. a. If you do not file or provide fraudulent information, you will be prosecuted.
  b. If you file and do not provide fraudulent information, you will not be prosecuted.
  c. If you are not prosecuted, you filed or did not provide fraudulent information.

75. a. I'm leaving, and Tom is relieved or Sue is relieved.
  b. I'm leaving, and it is false that Tom and Sue are not relieved.
  c. If I'm leaving, then Tom is relieved or Sue is relieved.

76. a. You play at least three instruments, and if you have a master's degree in music then you are eligible.
  b. You are eligible, if and only if you have a master's degree in music and play at least three instruments.
  c. You play at least three instruments, and if you are not eligible then you do not have a master's degree in music.

Application Exercises

Cosmetic surgery in the United States is a multi-billion-dollar industry. Use the information given by the table to solve Exercises 77–79.

COSMETIC SURGERY IN THE U.S.

Procedure	Number of Operations in One Year	Percent of Patients Who Are Women
Chemical Peel	29,072	96%
Nose Reshaping	35,927	71%
Face Lift	32,283	92%
Collagen Injection	27,052	96%
Breast Lift	10,053	100%
Breast Enlargement	39,247	100%
Tummy Tuck	16,829	95%
Buttock Lift	314	93%
Liposuction	51,072	87%

Source: American Society of Plastic and Reconstructive Surgeons

77. Consider the statement

  If the procedure is a chemical peel, then 96% of the patients are women.

  a. Is this statement true or false?
  Write the statement's
  b. contrapositive.
  c. converse.
  d. inverse.
  e. negation.

  Use the information in the table to determine whether each of these statements is true, not necessarily true, or false.

For each statement in Exercise 78–79,
  a. *Use the information in the table shown to determine if the statement is true or false.*
  b. *Use a De Morgan's law to write the statement's negation.*
  c. *Use the table to determine if the negation in part (b) is true or false.*

78. There are 51,072 liposuctions each year and 93% of the liposuction patients are women.

79. There are 32,283 face lifts each year or 8% of the face-lift patients are men.

Writing in Mathematics

80. What are equivalent statements?

81. Describe how to determine if two statements are equivalent.

82. Describe how to obtain the contrapositive of a conditional statement.

83. Describe how to obtain the converse and the inverse of a conditional statement.

84. Give an example of a conditional statement that is true, but whose converse and inverse are not necessarily true. Try to make the statement somewhat different from the conditional statements that you have encountered throughout this section. Explain why the converse and the inverse that you wrote are not necessarily true.

85. Explain how to write the negation of a conditional statement.

86. Explain how to write the negation of a conjunction.

87. Give an example of a disjunction that is true, even though one of its component statements is false. Then write the negation of the disjunction and explain why the negation is false.

88. Read the essay on page 131. The Dormouse's last statement is the setup for a joke. The punchline, delivered by the Hatter to the Dormouse, is, "For you, it's the same thing." Explain the joke. What does this punchline have to do with the difference between a conditional and a biconditional statement?

Critical Thinking Exercises

89. Which one of the following is true?

a. $\sim q \rightarrow (p \wedge r)$ is equivalent to $\sim(p \vee r) \rightarrow q$.

b. The inverse of a statement's converse is the statement's contrapositive.

c. A conditional statement can sometimes be true if its contrapositive is false.

d. A conditional statement can never be false if its converse is true.

90. Write the contrapositive for the following statement:

I do not get a ticket if I observe the speed limit.

91. Write the negation for the following conjunction:

We will neither replace nor repair the roof, and we will sell the house.

Group Exercise

92. Can you think of an advertisement in which the person using a product is extremely attractive or famous? It is true that if you are this attractive or famous person, then you use the product. (Or at least pretend, for monetary gain, that you use the product!) In order to get you to buy the product, here is what the advertisers would *like* you to believe: If I use this product, then I will be just like this attractive or famous person. This, the converse, is not necessarily true and, for most of us, is unfortunately false. Each group member should find an example of this kind of deceptive advertising to share with the other group members.

SECTION 3.6 ARGUMENTS AND TRUTH TABLES

OBJECTIVES

1. Use truth tables to determine validity.

2. Use forms of valid arguments to draw logical conclusions.

Peanuts reprinted by permission of United Feature Syndicate, Inc.

The late Charles Schultz, creator of the "Peanuts" comic strip, transfixed 350 million readers worldwide with the joys and angst of his hapless Charlie Brown and Snoopy, a romantic self-deluded beagle. In 18,250 comic strips that spanned nearly 50 years, mathematics and logic were often featured. In the argument shown above, Sally concludes that

If you do not know how to read, then Leo Tolstoy will hate you.

Is this conclusion true, thereby making her argument valid? In this section, you will learn to use truth tables to determine whether arguments are valid or invalid.

An **argument** consists of two parts: the given statements, called the **premises**, and a **conclusion**. Here's an example of an argument:

1 Use truth tables to determine validity.

Premise 1:   If the toll ticket is lost, then the fare from the farthest point is paid.

Premise 2:   I lost the toll ticket.

Conclusion: Therefore, I pay the fare from the farthest point.

It appears that if the premises are true, then I have to pay the fare from the farthest point. The true premises force the conclusion to be true, making this an example of a **valid argument**.

DEFINITION OF A VALID ARGUMENT

An argument is **valid** if the conclusion is true whenever the premises are assumed to be true. An argument that is not valid is said to be an **invalid argument**, also called a **fallacy**.

Truth tables can be used to test validity. We begin by writing the argument in symbolic form. Let's do this for the argument involving the lost ticket. Represent each simple statement with a letter:

p: The toll ticket is lost.

q: The fare from the farthest point is paid.

Now we write the two premises and the conclusion in symbolic form:

Premise 1: $p \rightarrow q$ If the toll ticket is lost, then the fare from the farthest point is paid.

Premise 2: p I lost the toll ticket.

Conclusion: $\therefore q$ Therefore, I pay the fare from the farthest point.
(The three-dot triangle, $\therefore$, is read "therefore.")

To decide whether this argument is valid, we rewrite it as a conditional statement that has the following form:

$$[(p \rightarrow q) \wedge p] \rightarrow q.$$

If premise 1 and premise 2,   then   conclusion.

At this point, we can determine whether the conjunction of the premises implies that the conclusion is true for all possible truth values for p and q. We construct a truth table for the statement

$$[(p \rightarrow q) \wedge p] \rightarrow q.$$

If the final column in the truth table for $[(p \rightarrow q) \wedge p] \rightarrow q$ is true in every case, then the statement is a tautology and the argument is valid. If the conditional statement in the last column is false in at least one case, then the statement is not a tautology, and the argument is invalid. The truth table is shown below.

p	q	$p \rightarrow q$	$(p \rightarrow q) \wedge p$	$[(p \rightarrow q) \wedge p] \rightarrow q$
T	T	T	T	T
T	F	F	F	T
F	T	T	F	T
F	F	T	F	T

The final column in the table is true in every case. The conditional statement is a tautology. This means that the true premises imply the conclusion. The conclusion necessarily follows from the premises. Therefore, the argument is valid.

The form of the argument in the lost-toll-ticket example

$$p \rightarrow q$$
$$\underline{p}$$
$$\therefore q$$

is called **direct reasoning**. All arguments that have the direct reasoning form are valid regardless of the English statements that p and q represent.

Here's a step-by-step procedure to test the validity of an argument using truth tables:

TESTING THE VALIDITY OF AN ARGUMENT WITH A TRUTH TABLE

1. Use a letter to represent each simple statement in the argument.

2. Express the premises and the conclusion symbolically.

3. Write a symbolic conditional statement of the form

$$[(\text{premise 1}) \wedge (\text{premise 2}) \wedge \ldots \wedge (\text{premise } n)] \rightarrow \text{conclusion}$$

where n is the number of premises.

4. Construct a truth table for the conditional statement in step 3.

5. If the final column of the truth table has all trues, the conditional statement is a tautology and the argument is valid. If the final column does not have all trues, the conditional statement is not a tautology and the argument is invalid.

EXAMPLE 1 | DID THE PICKIEST LOGICIAN IN THE GALAXY FOUL UP?

In an episode of the television series "Star Trek," the starship *Enterprise* is hit by an ion storm, causing the power to go out. Captain Kirk wonders if Mr. Scott, the engineer, is aware of the problem. Mr. Spock, the paragon of extraterrestrial intelligence, replies, "If Mr. Scott is still with us, the power should be on momentarily." Moments later, the ship's power comes on and Spock arches his Vulcan brow: "Ah, Mr. Scott is still with us."

Spock's logic can be expressed in the form of an argument:

If Mr. Scott is still with us, then the power will come on.

The power comes on.

Therefore, Mr. Scott is still with us.

Determine whether this argument is valid or invalid.

SOLUTION

STEP 1. Use a letter to represent each simple statement in the argument. We introduce the following representations:

p: Mr. Scott is still with us.

q: The power will come on.

STEP 2. Express the premises and the conclusion symbolically.

$p \rightarrow q$ If Mr. Scott is still with us, then the power will come on.

$\underline{q}$ The power comes on.

$\therefore p$ Mr. Scott is still with us.

STEP 3. Write a symbolic statement of the form

$$[(\textbf{premise 1}) \wedge (\textbf{premise 2})] \rightarrow \textbf{conclusion}.$$

The symbolic statement is

$$[(p \rightarrow q) \wedge q] \rightarrow p.$$

STEP 4. Construct a truth table for the conditional statement in step 3.

p	q	$p \rightarrow q$	$(p \rightarrow q) \wedge q$	$[(p \rightarrow q) \wedge q] \rightarrow p$
T	T	T	T	T
T	F	F	F	T
F	T	T	T	F
F	F	T	F	T

STEP 5. Use the truth values in the final column to determine if the argument is valid or invalid. The entries in the final column of the truth table are not all true, so the conditional statement is not a tautology. Spock's argument is invalid or a fallacy.

The form of the argument in Spock's logical foul-up

$$p \rightarrow q$$
$$\underline{q}$$
$$\therefore p$$

is called the **fallacy of the converse**. It should remind you that a conditional is not equivalent to its converse. All arguments that have this form are invalid regardless of the English statements that p and q represent.

You may recall that a conditional statement is not equivalent to its inverse. Another invalid form of an argument is called the **fallacy of the inverse**:

$$p \rightarrow q$$
$$\underline{\sim p}$$
$$\therefore \sim q.$$

An example of such a fallacy is "If I study, I pass. I do not study. Therefore, I do not pass." For most students, the conclusion is true, but it does not have to be. If an argument is invalid, then the conclusion is not necessarily true. This, however, does not mean that the conclusion must be false.

Check 1 Point Use a truth table to determine whether the following argument is valid or invalid:

I study for 5 hours or I fail.

I did not study for 5 hours.

Therefore, I failed.

EXAMPLE 2 | A RUSH LIMBAUGH ARGUMENT

Conservative radio talk show host Rush Limbaugh directed the following argument at Al Gore:

> You would think that if Al Gore and company believe so passionately in their environmental crusading that [sic] they would first put these ideas to work in their own lives, right? ... Al Gore thinks the automobile is one of the greatest threats to the planet, but he sure as heck still travels in one of them—a gas guzzler too.
>
> (*See, I Told You So*, p. 168)

STUDY TIP

If Spock's fallacious reasoning represents television writers' best efforts at being logical, do not expect much from their commercials.

Rush Limbaugh, conservative radio talk show host, intrigues some people and infuriates others.

Limbaugh's passage can be expressed in the form of an argument:

If Gore really believed that the automobile were
a threat to the planet, he would not travel in a gas guzzler.

Gore does travel in a gas guzzler.

Therefore, Gore does not really believe that the automobile
is a threat to the planet.

Determine whether this argument is valid or invalid.

SOLUTION

STEP 1. Use a letter to represent each statement in the argument. We introduce the following representations:

p: Gore really believes that the automobile is a threat to the planet.

q: He does not travel in a gas guzzler.

> Despite the negation of the English statement, we can still label this statement as q rather than as $\sim q$.

STEP 2. Express the premises and the conclusion symbolically.

$p \rightarrow q$ — If Gore really believed that the automobile were a threat to the planet, he would not travel in a gas guzzler.

$\sim q$ — Gore does travel in a gas guzzler.

$\therefore \sim p$ — Therefore, Gore does not really believe that the automobile is a threat to the planet.

STEP 3. Write a symbolic statement of the form

$$[(\textbf{premise 1}) \wedge (\textbf{premise 2})] \rightarrow \textbf{conclusion}.$$

The symbolic statement is

$$[(p \rightarrow q) \wedge \sim q] \rightarrow \sim p.$$

STEP 4. Construct a truth table for the conditional statement in step 3.

p	q	$p \rightarrow q$	$\sim q$	$(p \rightarrow q) \wedge \sim q$	$\sim p$	$[(p \rightarrow q) \wedge \sim q] \rightarrow \sim p$
T	T	T	F	F	F	T
T	F	F	T	F	F	T
F	T	T	F	F	T	T
F	F	T	T	T	T	T

STUDY TIP

You can let q represent

Gore does travel in a gas guzzler.

The statement

$$[(p \rightarrow \sim q) \wedge q] \rightarrow \sim p$$

is a tautology, confirming that the argument is valid.

STEP 5. Use the truth values in the final column to determine if the argument is valid or invalid. The entries in the final column of the truth table are all true, so the conditional statement is a tautology. Limbaugh's argument is valid.

The form of the argument in the Limbaugh example

$$p \rightarrow q$$
$$\sim q$$
$$\therefore \sim p$$

should remind you that a conditional statement is equivalent to its contrapositive:

$$p \rightarrow q \equiv \sim q \rightarrow \sim p.$$

The form of this argument is called **contrapositive reasoning**.

The Limbaugh argument in Example 2 is valid because the conclusion is true *relative to the premises*. If the premises are true, then the conclusion is true. But are the premises indeed true? If they are not, or if they are flawed in some manner, then the relatively true conclusion of the argument might not be true in the real world. For example, the second premise, Gore does travel in a gas guzzler, is assumed to be true. But do we know whether Gore really traveled in a gas guzzler at the time of the argument?

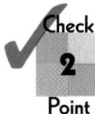

Check 2 Point Use a truth table to determine whether the following argument is valid or invalid:

> I study for 5 hours or I fail.
>
> <u>I studied for 5 hours.</u>
>
> Therefore, I did not fail.

EXAMPLE 3 | SALLY'S ARGUMENT IN A PEANUTS CARTOON

Take a moment to read the *Peanuts* cartoon shown in the section opener on page 138. We can express Sally's argument as follows:

> If you do not know how to read, then you cannot read *War and Peace*.
>
> <u>If you cannot read *War and Peace*, then Leo Tolstoy will hate you.</u>
>
> Therefore, if you do not know how to read, then Leo Tolstoy will hate you.

Determine whether this argument is valid or invalid.

SOLUTION

STEP 1. Use a letter to represent each statement in the argument. We introduce the following representations:

> *p:* You do not know how to read.
>
> *q:* You cannot read *War and Peace*.
>
> *r:* Leo Tolstoy will hate you.

STEP 2. Express the premises and the conclusion symbolically.

$p \rightarrow q$	If you do not know how to read, then you cannot read *War and Peace*.
$q \rightarrow r$	<u>If you cannot read *War and Peace*, then Leo Tolstoy will hate you.</u>
$\therefore \ p \rightarrow r$	Therefore, if you do not know how to read, then Leo Tolstoy will hate you.

STEP 3. Write a symbolic statement of the form

$$[(\textbf{premise 1}) \wedge (\textbf{premise 2}) \rightarrow \textbf{conclusion}.$$

The symbolic statement is

$$[(p \rightarrow q) \wedge (q \rightarrow r)] \rightarrow (p \rightarrow r).$$

STEP 4. Construct a truth table for the conditional statement in step 3.

p	*q*	*r*	$p \rightarrow q$	$q \rightarrow r$	$p \rightarrow r$	$(p \rightarrow q) \wedge (q \rightarrow r)$	$[(p \rightarrow q) \wedge (q \rightarrow r)] \rightarrow (p \rightarrow r)$
T	T	T	T	T	T	T	T
T	T	F	T	F	F	F	T
T	F	T	F	T	T	F	T
T	F	F	F	T	F	F	T
F	T	T	T	T	T	T	T
F	T	F	T	F	T	F	T
F	F	T	T	T	T	T	T
F	F	F	T	T	T	T	T

STUDY TIP

If you prefer using the symbol ~ with negated English statements, you can use the following representations:

> *p:* You know how to read.
>
> *q:* You can read *War and Peace*.
>
> *r:* Leo Tolstoy will hate you.

The statement

$$[(\sim p \rightarrow \sim q) \wedge (\sim q \rightarrow r)]$$
$$\rightarrow (\sim p \rightarrow r)$$

is a tautology, confirming that the argument is valid.

STEP 5. Use the truth values in the final column to determine if the argument is valid or invalid. The entry in each of the eight rows in the final column of the truth table is true, so the conditional statement is a tautology. Sally's argument is valid.

The form of Sally's argument

$$p \rightarrow q$$
$$\underline{q \rightarrow r}$$
$$\therefore p \rightarrow r$$

is called **transitive reasoning**. If p implies q and q implies r, then p must imply r. Because $p \rightarrow r$ is a valid conclusion, the contrapositive, $\sim r \rightarrow \sim p$, is also a valid conclusion. Not necessarily true are the converse, $r \rightarrow p$, and the inverse, $\sim p \rightarrow \sim r$.

 Check 3 Point

Use a truth table to determine whether the following argument is valid or invalid:

> If you do not know how to read, then you cannot read *War and Peace*.
>
> If you cannot read *War and Peace*, then Leo Tolstoy will hate you.
>
> Therefore, if you do know how to read, then Leo Tolstoy will not hate you.

Table 3.14 contains the standard forms of commonly used valid and invalid arguments. If an English argument translates into one of these forms, you can immediately determine whether or not it is valid without using a truth table.

TABLE 3.14 STANDARD FORMS OF ARGUMENTS

Valid Arguments			
Direct Reasoning	Contrapositive Reasoning	Disjunctive Reasoning	Transitive Reasoning
$p \rightarrow q$ $\underline{p}$ $\therefore q$	$p \rightarrow q$ $\underline{\sim q}$ $\therefore \sim p$	$p \vee q$ $\quad$ $p \vee q$ $\underline{\sim p}$ $\quad$ $\underline{\sim q}$ $\therefore q$ $\quad$ $\therefore p$	$p \rightarrow q$ $\underline{q \rightarrow r}$ $\therefore p \rightarrow r$ $\therefore \sim r \rightarrow \sim p$
Invalid Arguments			
Fallacy of the Converse	Fallacy of the Inverse	Misuse of Disjunctive Reasoning	Misuse of Transitive Reasoning
$p \rightarrow q$ $\underline{q}$ $\therefore p$	$p \rightarrow q$ $\underline{\sim p}$ $\therefore \sim q$	$p \vee q$ $\quad$ $p \vee q$ $\underline{p}$ $\quad$ $\underline{q}$ $\therefore \sim q$ $\quad$ $\therefore \sim p$	$p \rightarrow q$ $\underline{q \rightarrow r}$ $\therefore r \rightarrow p$ $\therefore \sim p \rightarrow \sim r$

② Use forms of valid arguments to draw logical conclusions.

Drawing Logical Conclusions

A **logical** or **valid conclusion** is one that forms a valid argument when it follows a given set of premises. Suppose that the premises of an English argument translate into any one of the symbolic forms of premises for the valid arguments in Table 3.14. The symbolic conclusion can be used to find a valid English conclusion. Example 4 shows how this is done.

EXAMPLE 4 DRAWING A LOGICAL CONCLUSION

Draw a valid conclusion from the following premises:

If all people have the same idea of what is humorous, then no jokes bomb.

Some jokes bomb.

SOLUTION

Let *p* be: All people have the same idea of what is humorous.

Let *q* be: No jokes bomb.

The form of the premises is

$p \rightarrow q$ If all people have the same idea of what is humorous, then no jokes bomb.

$\sim q$ Some jokes bomb. (Recall that the negation of "no" is "some.")

$\therefore$?

The conclusion $\sim p$ is valid because it forms the contrapositive reasoning of a valid argument when it follows the given premises. The conclusion $\sim p$ translates as

Not all people have the same idea of what is humorous.

Because the negation of "all" is "some . . . not," we can equivalently conclude that

Some people do not have the same idea of what is humorous.

Check
4
Point

Draw a valid conclusion from the following premises:

If it is midnight, then I am sleeping.

I am not sleeping.

Exercise Set 3.6

Practice Exercises

In Exercises 1–8, use a truth table to determine whether the symbolic form of the argument is valid or invalid.

1. $p \rightarrow q$
$\dfrac{\sim p}{\therefore \sim q}$

2. $p \rightarrow q$
$\dfrac{\sim p}{\therefore q}$

3. $p \rightarrow \sim q$
$\dfrac{q}{\therefore \sim p}$

4. $\sim p \rightarrow q$
$\dfrac{\sim q}{\therefore p}$

5. $p \wedge \sim q$
$\dfrac{p}{\therefore \sim q}$

6. $\sim p \vee q$
$\dfrac{p}{\therefore q}$

7. $p \rightarrow q$
$q \rightarrow p$
$\overline{\therefore p \wedge q}$

8. $(p \rightarrow q) \wedge (q \rightarrow p)$
$\dfrac{p}{\therefore p \vee q}$

In Exercises 9–14, translate each argument into symbolic form. Then use a truth table to determine whether the argument is valid or invalid. (You can ignore differences in past, present, and future tense.)

9. If it is cold, my motorcycle will not start.
My motorcycle started.
$\therefore$ It is not cold.

10. If a metrorail system is not in operation, there are traffic delays.
Over the past year there have been no traffic delays.
$\therefore$ Over the past year a metrorail system has been in operation.

11. There must be a dam or there is flooding.
This year there is flooding.
$\therefore$ This year there is no dam.

12. You must eat well or you will not be healthy.
I eat well.
$\therefore$ I am healthy.

13. If all people obey the law, then no jails are needed.
Some people do not obey the law.
∴ Some jails are needed.

14. If all people obey the law, then no jails are needed.
Some jails are needed.
∴ Some people do not obey the law.

In Exercises 15–20, use a truth table with eight cases to determine whether the symbolic form of the argument is valid or invalid.

15. $p \rightarrow q$
$q \rightarrow r$
∴ $r \rightarrow p$

16. $p \rightarrow q$
$q \rightarrow r$
∴ $\sim p \rightarrow \sim r$

17. $p \rightarrow q$
$q \wedge r$
∴ $p \vee r$

18. $\sim p \wedge q$
$p \leftrightarrow r$
∴ $p \wedge r$

19. $p \leftrightarrow q$
$q \rightarrow r$
∴ $\sim r \rightarrow \sim p$

20. $q \rightarrow \sim p$
$q \wedge r$
∴ $r \rightarrow p$

In Exercises 21–28, translate each argument into symbolic form. Then use a truth table with eight cases to determine whether each argument is valid or invalid. (Ignore differences in past, present, and future tense.)

21. If Tim and Janet play, then the team wins.
Tim played and the team lost.
∴ Janet did not play.

22. If *The Graduate* and *Midnight Cowboy* are shown, then the performance is sold out.
Midnight Cowboy was shown and the performance was not sold out.
∴ *The Graduate* was not shown.

23. If it rains or snows, then I read.
I am not reading.
∴ It is neither raining nor snowing.

24. If I am tired or hungry, I cannot concentrate.
I can concentrate.
∴ I am neither tired nor hungry.

25. If it rains or snows, then I read.
I am reading.
∴ It is raining or snowing.

26. If I am tired or hungry, I cannot concentrate.
I cannot concentrate.
∴ I am tired or hungry.

27. If it is hot and humid, I complain.
It is not hot or it is not humid.
∴ I am not complaining.

28. If I watch *Schindler's List* and *Guess Who's Coming to Dinner*, I am aware of the destructive nature of intolerance.
Today I did not watch *Schindler's List* or I did not watch *Guess Who's Coming to Dinner.*
∴ Today I am not aware of the destructive nature of intolerance.

In Exercises 29–36, translate each argument into symbolic form. Then use the forms of valid and invalid arguments in Table 3.14 on page 144 to determine whether each argument is valid or invalid.

29. If we close the door, then there is less noise.
There is less noise.
∴ We closed the door.

30. If the temperature is in the 90s, then at least two county pools are open.
It is in the 90s today.
∴ Two or more county pools are open today.

31. We criminalize drugs or we damage the future of young people.
We will not damage the future of young people.
∴ We criminalize drugs.

32. He is intelligent or an over-achiever.
He is intelligent.
∴ He is not an over-achiever.

33. If I am at the beach, then I swim in the ocean.
If I swim in the ocean, then I feel refreshed.
∴ If I am at the beach, then I feel refreshed.

34. If I'm tired, I'm edgy.
If I'm edgy, I'm nasty.
∴ If I'm not nasty, I'm not tired.

35. If I'm at the beach, then I swim in the ocean.
If I swim in the ocean, then I feel refreshed.
∴ If I'm not at the beach, then I don't feel refreshed.

36. If I'm tired, I'm edgy.
If I'm edgy, I'm nasty.
∴ If I'm nasty, I'm tired.

In Exercises 37–44, draw a valid conclusion from the given premises. (Use the standard forms of valid arguments in Table 3.14 on page 144.)

37. If a person is a chemist, then that person has a college degree.
My best friend does not have a college degree.
Therefore, . . .

38. If the Westway Expressway is not in operation, automobile traffic makes the East Side Highway look like a parking lot. On June 2, the Westway Expressway was completely shut down because of an overturned truck.
Therefore, . . .

39. The writers of "My Mother the Car" were told by the network to improve their scripts or be dropped from prime time. The writers of "My Mother the Car" did not improve their scripts.
Therefore, . . .

40. You exercise or you do not feel energized.
I do not exercise.
Therefore, . . .

41. If all electricity is off, then no lights work.
Some lights work.
Therefore, . . .

42. If all houses meet the hurricane code, then none of them are destroyed by a category 4 hurricane.
Some houses were destroyed by Andrew, a category 4 hurricane.
Therefore, . . .

43. If I vacation in Paris, I eat French pastries.
If I eat French pastries, I gain weight.
Therefore, . . .

44. If I am full-time student, I cannot work.
If I cannot work, I cannot afford a rental apartment costing more than $500 per month.
Therefore, . . .

Application Exercises

45. Read the opening paragraph to Example 1 on page 140. Rewrite the conditional statement for Mr. Spock's reply so that the pickiest logician does not foul up and his argument is valid.

46. Conservative commentator Rush Limbaugh directed the following passage at liberals and the way they think about crime.

Of course, liberals will argue that these actions [contemporary youth crime] can be laid at the foot of socioeconomic inequities, or poverty. However, the Great Depression caused a level of poverty unknown to exist in America today, and yet I have been unable to find any accounts of crime waves sweeping our large cities. Let the liberals chew on that.

(*See, I Told You So*, p. 83)

Limbaugh's passage can be expressed in the form of an argument:

If poverty causes crime, then crime waves would have swept American cities during the Great Depression.
Crime waves did not sweep American cities during the Great Depression.
∴ Poverty does not cause crime. (Liberals are wrong.)

Translate this argument into symbolic form and determine whether it is valid or invalid.

47. In Lewis Carroll's *Alice's Adventures in Wonderland*, Alice falls into a rabbit hole, eats a piece of cake, and suddenly begins to grow. Because she is no longer the same (at least not the same height), she wonders who she is. She thinks of her friend Mabel, who knows very little. After deciding that $4 \times 5 = 12$ and London is the capital of Paris, Alice realizes that she, too, knows very little. She concludes that she must be Mabel.
Express Alice's thinking in the form of an argument using the following simple statements:

p: One is Mabel.
q: One knows very little.

The argument's conclusions should be

I am Mabel.

Determine if the argument is valid or invalid.

48. In his *Autobiography*, Bertrand Russell wrote

I had had from the first a dark suspicion that the invitation [to lecture in China] might be a practical joke, and in order to test its genuineness I had got the Chinese to pay my passage money before I started. I thought that few people would spend £125 on a joke . . .

Write Russell's passage in the form of an argument using the simple statements at the top of the next column:

p: The invitation is not genuine.
q: People would not spend 125 pounds.

The argument's conclusion should be

The invitation is genuine.

Determine if the argument is valid or invalid.

Writing in Mathematics

49. Describe what is meant by a valid argument.

50. If you are given an argument in words that contains two premises and a conclusion, describe how to determine if the argument is valid or invalid.

51. Write an original argument in words for the direct reasoning form.

52. Write an original argument in words for the contrapositive reasoning form.

53. Write an original argument in words for the transitive reasoning form.

54. What is a valid conclusion?

Critical Thinking Exercises

55. Write an original argument in words that has a true conclusion, yet is invalid.

56. Draw a valid conclusion from the given premises. Then use a truth table to verify your answer.

If you only spoke when spoken to and I only spoke when spoken to, then nobody would ever say anything. Some people do say things. Therefore, . . .

57. Translate the following argument into symbolic form. Then use a truth table to determine if the argument is valid or invalid.

It's wrong to smoke in public if secondary cigarette smoke is a health threat.
If secondary cigarette smoke were not a health threat, the American Lung Association would not say that it is.
The American Lung Association says that secondary cigarette smoke is a health threat.
∴ It's wrong to smoke in public.

Group Exercise

58. In this section, we used examples from the television show "Star Trek," Rush Limbaugh's *See, I Told You So*, and the cartoon *Peanuts* to illustrate symbolic arguments.

a. From any source that is of particular interest to you (these can be the words of someone you truly admire or a person who really gets under your skin), select a paragraph or two in which the writer argues a particular point. Rewrite the reasoning in the form of an argument using words. Then translate the argument into symbolic form and use a truth table to determine if it is valid or invalid.

b. Each group member should share the selected passage with other people in the group. Explain how it was expressed in argument form. Then tell why the argument is valid or invalid.

SECTION 3.7   ARGUMENTS AND EULER DIAGRAMS

OBJECTIVE

1. Use Euler diagrams to
determine validity.

William Shakespeare

Leonard Euler

1   Use Euler diagrams to
determine validity.

He is the Shakespeare of mathematics, yet he is unknown by the general public. Most people cannot even correctly pronounce his name. The Swiss mathematician Leonhard Euler (1707–1783), whose last name rhymes with *boiler*, not *ruler*, is the most prolific mathematician in history. His collected books and papers fill some 80 volumes; Euler published an average of 800 pages of new mathematics per year over a career that spanned six decades. Euler was also an astronomer, botanist, chemist, physicist, and linguist. His productivity was not at all slowed down by the total blindness he experienced the last 17 years of his life. An equation discovered by Euler, $e^{\pi i} + 1 = 0$, connected five of the most important numbers in mathematics in a totally unexpected way.

Euler invented an elegant way to determine the validity of arguments whose premises contain the words *all, some,* and *no.* The technique for doing this uses geometric ideas and involves four basic diagrams, known as **Euler diagrams.** Figure 3.6 illustrates how Euler diagrams represent four quantified statements.

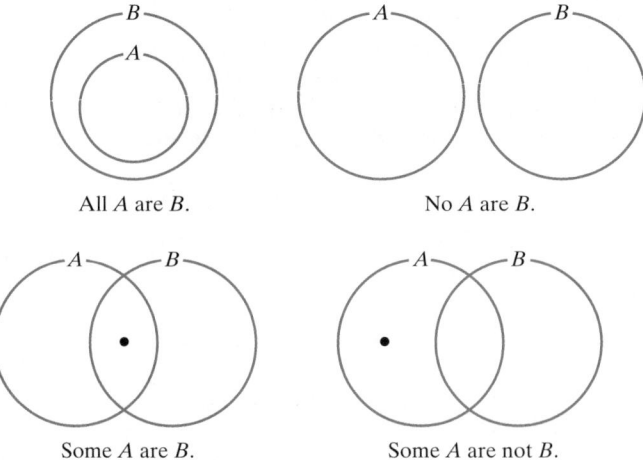

FIGURE 3.6 Euler diagrams for quantified statements

The Euler diagrams in Figure 3.6 are just like the Venn diagrams that we used in studying sets. However, there is no need to enclose the circles inside a rectangle representing a universal set. In these diagrams, circles are used to indicate relationships of premises to conclusions.

Here's a step-by-step procedure for using Euler diagrams to determine whether or not an argument is valid:

EULER DIAGRAMS AND ARGUMENTS

1. Make an Euler diagram for the first premise.
2. Make an Euler diagram for the second premise on top of the one for the first premise.
3. The argument is valid if and only if every possible diagram illustrates the conclusion of the argument. If there is even *one* possible diagram that contradicts the conclusion, this indicates that the conclusion is not true in every case, so the argument is invalid.

The goal of this procedure is to produce, if possible, *a diagram that does* **not** *illustrate the argument's conclusion*. The method of Euler diagrams boils down to determining whether such a diagram is possible. If it is, this serves as a counterexample to the argument's conclusion, and the argument is immediately invalid. By contrast, if no such counterexample can be drawn, the argument is valid.

The technique of using Euler diagrams is illustrated in Examples 1–6.

EXAMPLE 1   ARGUMENTS AND EULER DIAGRAMS

Use Euler diagrams to determine whether the following argument is valid or invalid:

> All people who arrive late cannot perform
>
> <u>All people who cannot perform are ineligible for scholarships.</u>
>
> Therefore, all people who arrive late are ineligible for scholarships.

STUDY TIP

When making Euler diagrams, remember that the size of a circle is not relevant. It is the circle's location that counts.

SOLUTION

STEP 1. Make an Euler diagram for the first premise.
We begin by diagramming the premise

> All people who arrive late cannot perform.

The region inside the smaller circle represents people who arrive late. The region inside the larger circle represents people who cannot perform.

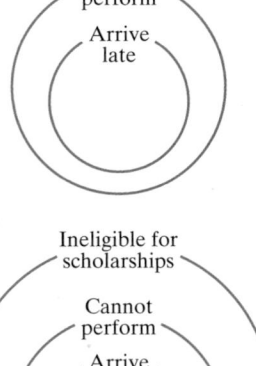

STEP 2. Make an Euler diagram for the second premise on top of the one for the first premise. We add to our previous figure the diagram for the second premise:

> All people who cannot perform are ineligible for scholarships.

A third, larger, circle representing people who are ineligible for scholarships is drawn outside the circle representing people who cannot perform.

STEP 3. The argument is valid if and only if every possible diagram illustrates the argument's conclusion. There is only one possible diagram. Let's see if this diagram illustrates the argument's conclusion, namely

> All people who arrive late are ineligible for scholarships.

This is indeed the case because the Euler diagram on page 149 shows the circle representing the people who arrive late contained within the circle of people who are ineligible for scholarships. The Euler diagram supports the conclusion and the given argument is valid.

 **Check 1 Point**   Use Euler diagrams to determine whether the following argument is valid or invalid:

All U.S. voters must register.

All people who register must be U.S. citizens.

Therefore, all U.S. voters are U.S. citizens.

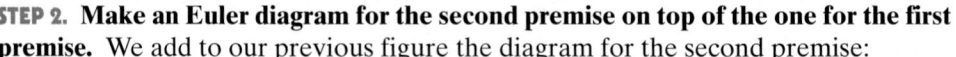
EXAMPLE 2   ARGUMENTS AND EULER DIAGRAMS

Use Euler diagrams to determine whether the following argument is valid or invalid:

All poets appreciate language.

All writers appreciate language.

Therefore, all poets are writers.

SOLUTION

STEP 1. Make an Euler diagram for the first premise. We begin by diagramming the premise

All poets appreciate language.

Up to this point, our work is similar to what we did in Example 1.

STEP 2. Make an Euler diagram for the second premise on top of the one for the first premise. We add to our previous figure the diagram for the second premise:

All writers appreciate language.

A third circle representing writers must be drawn inside the circle representing people who appreciate language. There are four ways to do this.

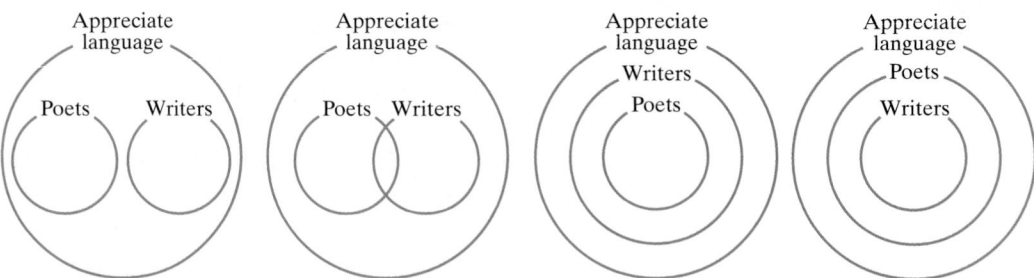

STEP 3. The argument is valid if and only if every possible diagram illustrates the argument's conclusion. The argument's conclusion is

All poets are writers.

This conclusion is not illustrated by every possible diagram shown above. One of these diagrams is repeated in the margin. This diagram shows "no poets are writers." There is no need to examine the other three diagrams.

The diagram on the left serves as a counterexample to the argument's conclusion. This means that the given argument is invalid. It would have sufficed to draw only the counterexample on the left to determine that the argument is invalid.

 **Check 2 Point** Use Euler diagrams to determine whether the following argument is valid or invalid:

> All baseball players are athletes.
>
> All ballet dancers are athletes.
>
> Therefore, no baseball players are ballet dancers.

EXAMPLE 3 | ARGUMENTS AND EULER DIAGRAMS

Use Euler diagrams to determine whether the following argument is valid or invalid:

> All freshmen live on campus.
>
> No people who live on campus can own cars.
>
> Therefore, no freshmen can own cars.

SOLUTION

STEP 1. Make an Euler diagram for the first premise. The diagram for

> All freshmen live on campus

is shown on the right. The region inside the smaller circle represents freshmen. The region inside the larger circle represents people who live on campus.

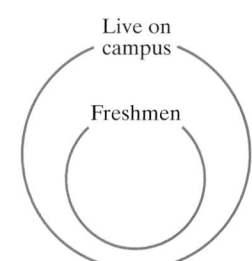

STEP 2. Make an Euler diagram for the second premise on top of the one for the first premise. We add to our previous figure the diagram for the second premise:

> No people who live on campus can own cars.

A third circle representing people who own cars is drawn outside the circle representing people who live on campus.

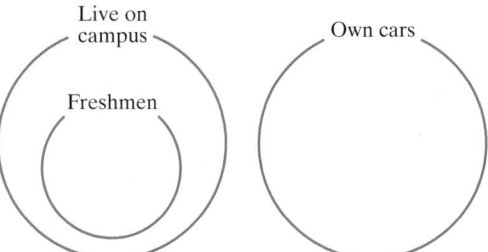

STEP 3. The argument is valid if and only if every possible diagram illustrates the argument's conclusion. There is only one possible diagram. The argument's conclusion is

> No freshmen can own cars.

This is supported by the diagram shown above because it shows the circle representing freshmen drawn outside the circle representing people who own cars. The Euler diagram supports the conclusion, and it is impossible to find a counterexample that does not. The given argument is valid.

 **Check 3 Point** Use Euler diagrams to determine whether the following argument is valid or invalid:

> All mathematicians are logical.
>
> No poets are logical.
>
> Therefore, no poets are mathematicians.

Let's see what happens to the validity if we reverse the second premise and the conclusion of the argument in Example 3.

EXAMPLE 4 | EULER DIAGRAMS AND VALIDITY

Use Euler diagrams to determine whether the following argument is valid or invalid:

All freshmen live on campus.
No freshmen can own cars.
Therefore, no people who live on campus can own cars.

SOLUTION

STEP 1. Make an Euler diagram for the first premise. We once again begin with the diagram for

All freshmen live on campus.

So far, our work is exactly the same as in the previous example.

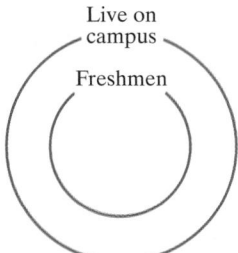

STEP 2. Make an Euler diagram for the second premise on top of the one for the first premise. We add to our previous figure the diagram for the second premise:

No freshmen can own cars.

The circle representing people who own cars is drawn outside the freshmen circle. At least two Euler diagrams are possible.

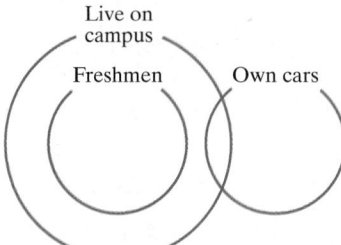   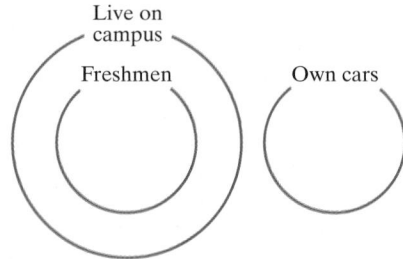

STEP 3. The argument is valid if and only if every possible diagram illustrates the argument's conclusion. The argument's conclusion is

No people who live on campus can own cars.

This conclusion is not supported by both diagrams shown above. The diagram that does not support the conclusion is repeated in the margin. Notice that the "live on campus" circle and the "own cars" circle intersect. This diagram serves as a counterexample to the argument's conclusion. This means that the argument is invalid. Once again, only the counterexample on the left is needed to conclude that the argument is invalid.

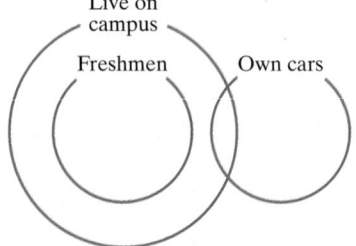

 Use Euler diagrams to determine whether the following argument is valid or invalid:

All mathematicians are logical.
No poets are mathematicians.
Therefore, no poets are logical.

So far, the arguments that we have looked at have contained "all" or "no" in the premises and conclusions. The quantifier "some" is a bit trickier to work with. Because the statement "Some A are B" means there exists at least one A that is a B, we diagram this existence by showing a dot in the region where A and B intersect, illustrated in Figure 3.7.

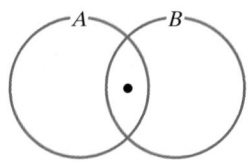

FIGURE 3.7 Some A are B.

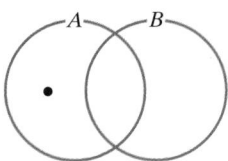

FIGURE 3.8
Illustrated by the dot is some *A* are not *B*. We cannot validly conclude that some *B* are not *A*.

Suppose that it is true that "Some *A* are not *B*," illustrated by the dot in Figure 3.8. This Euler diagram does not let us conclude that "Some *B* are not *A*" because there is not a dot in the part of the *B* circle that is not in the *A* circle. Conclusions with the word "some" must be shown by existence of at least one element represented by a dot in an Euler diagram.

Here is an example that shows the premise "Some *A* are not *B*" does not enable us to logically conclude that "Some *B* are not *A*."

    Some U.S. citizens are not U.S. senators. (true)

∴ Some U.S. senators are not U.S. citizens. (false)

EXAMPLE 5 EULER DIAGRAMS AND THE QUANTIFIER "SOME"

Use Euler diagrams to determine whether the following argument is valid or invalid:

All people are mortal.

Some mortals are students.

Therefore, some people are students.

SOLUTION

STEP 1. Make an Euler diagram for the first premise. Begin with the premise

    All people are mortal.

The Euler diagram is shown on the right.

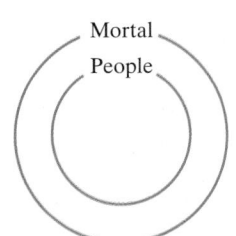

STEP 2. Make an Euler diagram for the second premise on top of the one for the first premise. We add to our previous figure the diagram for the second premise:

    Some mortals are students.

The circle representing students intersects the circle representing mortals. The dot in the region of intersection shows that at least one mortal is a student. Another diagram is possible, but if this serves as a counterexample then it is all we need. Let's check if it is a counterexample.

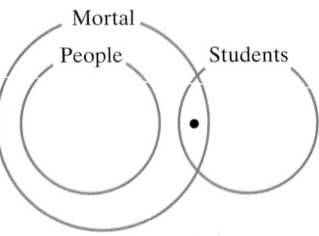

STEP 3. The argument is valid if and only if every possible diagram illustrates the conclusion of the argument. The argument's conclusion is

    Some people are students.

This conclusion is not supported by the Euler diagram. The diagram does not show the "people" circle and the "students" circle intersecting with a dot in the region of intersection. Although this conclusion is true in the real world, the Euler diagram serves as a counterexample that shows it does not follow from the premises. Therefore, the argument is invalid.

Check Point 5

Use Euler diagrams to determine whether the following argument is valid or invalid:

All mathematicians are logical.

Some poets are logical.

Therefore, some poets are mathematicians.

Some arguments show existence without using the word "some." Instead, a particular person or thing is mentioned in one of the premises. This particular person or thing is represented by a dot. Here is an example:

All men are mortal.

Aristotle is a man.

Therefore, Aristotle is mortal.

The two premises can be represented by the following Euler diagrams:

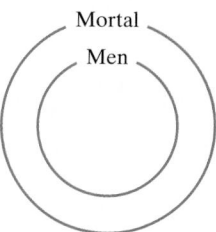
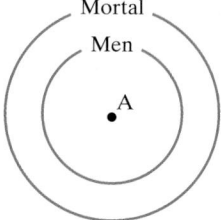

All men are mortal.   Aristotle (•) is a man.

The Euler diagram on the right uses a dot labeled A (for Aristotle). The diagram shows Aristotle (•) winding up in the "mortal" circle. The diagram supports the conclusion that Aristotle is mortal. This argument is valid.

EXAMPLE 6 | AN ARGUMENT MENTIONING ONE PERSON

Use Euler diagrams to determine whether the following argument is valid or invalid:

All children love to swim.

Jeff Rouse loves to swim.

Therefore, Jeff Rouse is a child.

SOLUTION

STEP 1. Make an Euler diagram for the first premise. Begin with the premise

All children love to swim.

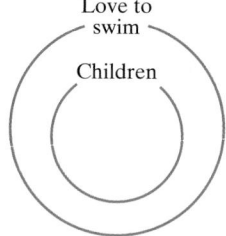

The Euler diagram is shown on the right.

STEP 2. Make an Euler diagram for the second premise on top of the one for the first premise. We add to our previous figure the diagram for the second premise:

Jeff Rouse loves to swim.

Jeff Rouse is represented by a dot labeled J. The dot must be placed in the "love to swim" circle. At least two Euler diagrams are possible.

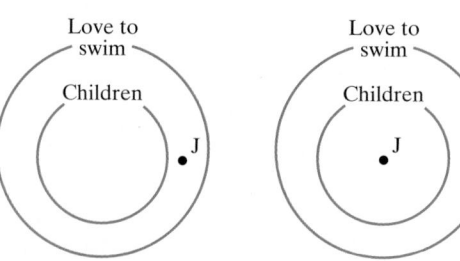

STEP 3. **The argument is valid if and only every possible diagram illustrates the conclusion of the argument.** The argument's conclusion is

Jeff Rouse is a child.

This conclusion is not supported by the Euler diagram shown on the left. The dot representing Jeff Rouse is outside the "children" circle. Jeff Rouse might not be a child. This diagram serves as a counterexample to the argument's conclusion. The argument is invalid.

Check Point 6   Use Euler diagrams to determine whether the following argument is valid or invalid:

All mathematicians are logical.
Euclid was logical.

Therefore, Euclid was a mathematician.

Exercise Set 3.7

Practice Exercises

Use Euler diagrams to determine whether each argument in Exercises 1–24 is valid or invalid.

1. All writers appreciate language.
All poets are writers.
Therefore, all poets appreciate language.

2. All physicists are scientists.
All scientists attended college.
Therefore, all physicists attended college.

3. All clocks keep time accurately.
All time-measuring devices keep time accurately.
Therefore, all clocks are time-measuring devices.

4. All cowboys live on ranches.
All cowherders live on ranches.
Therefore, all cowboys are cowherders.

5. All insects have six legs.
No spiders have six legs.
Therefore, no spiders are insects.

6. All humans are warm-blooded.
No reptiles are warm-blooded.
Therefore, no reptiles are human.

7. All insects have six legs.
No spiders are insects.
Therefore, no spiders have six legs.

8. All humans are warm-blooded.
No reptiles are human.
Therefore, no reptiles are warm-blooded.

9. All professors are wise people.
Some wise people are actors.
Therefore, some professors are actors.

10. All comedians are funny people.
Some funny people are professors.
Therefore, some comedians are professors.

11. All professors are wise people.
Some professors are actors.
Therefore, some wise people are actors.

12. All comedians are funny people.
Some comedians are professors.
Therefore, some funny people are professors.

13. All dancers are athletes.
Savion Glover is a dancer.
Therefore, Savion Glover is an athlete.

14. All actors are artists.
Kim Basinger is an actor.
Therefore, Kim Basinger is an artist.

15. All dancers are athletes.
Savion Glover is an athlete.
Therefore, Savion Glover is a dancer.

16. All actors are artists.
Kim Basinger is an artist.
Therefore, Kim Basinger is an actor.

17. Some people enjoy reading.
Some people enjoy TV.
Therefore, some people who enjoy reading enjoy TV.

18. All thefts are immoral acts.
Some thefts are justifiable.
Therefore, some immoral acts are justifiable.

19. All dogs have fleas.
Some dogs have rabies.
Therefore, all dogs with rabies have fleas.

20. All logic problems make sense.
Some jokes make sense.
Therefore, some logic problems are jokes.

21. No blank disks contain data.
Some blank disks are formatted.
Therefore, some formatted disks do not contain data.

22. Some houses have two stories.
Some houses have air conditioning.
Therefore, some houses with air conditioning have two stories.

23. All multiples of 6 are multiples of 3.
Eight is not a multiple of 3.
Therefore, 8 is not a multiple of 6.

24. All multiples of 6 are multiples of 3.
Eight is not a multiple of 6.
Therefore, 8 is not a multiple of 3.

Application Exercises

25. In the *Sixth Meditation*, Descartes writes

I first take notice here that there is a great difference between the mind and the body, in that the body, from its nature, is always divisible and the mind is completely indivisible.
Descartes's argument can be expressed as follows:

All bodies are divisible.
No minds are divisible.
Therefore, no minds are bodies.

Use an Euler diagram to determine whether the argument is valid or invalid.

26. In *Symbolic Logic*, Lewis Carroll presents the following argument:

Babies are illogical. (All babies are illogical persons.)
Illogical persons are despised. (All illogical persons are despised persons.)
Nobody is despised who can manage a crocodile. (No persons who can manage crocodiles are despised persons.)
Therefore, babies cannot manage crocodiles.

Use an Euler diagram to determine whether the argument is valid or invalid.

Writing in Mathematics

27. Explain how to use Euler diagrams to determine whether or not an argument is valid.

28. Under what circumstances should Euler diagrams rather than truth tables be used to determine whether or not an argument is valid?

Critical Thinking Exercises

29. Write an example of an argument with two quantified premises that is invalid but that has a true conclusion.

30. No animals that eat meat are vegetarians.
No cat is a vegetarian.
Felix is a cat.
Therefore, . . .

a. Felix is a vegetarian.
b. Felix is not a vegetarian.
c. Felix eats meat.
d. All animals that do not eat meat are vegetarians.

31. Supply the missing first premise that will make this argument valid.

Some opera singers are terrible actors.
Therefore, some people who take voice lessons are terrible actors.

32. Supply the missing first premise that will make this argument valid.

All amusing people are entertaining.
Therefore, some teachers are entertaining.

Chapter Summary, Review, and Test

SUMMARY

DEFINITIONS AND CONCEPTS

3.1 Statements, Negations, and Quantified Statements

3.2 Compound Statements and Connectives

EXAMPLES

Definition of a Statement
A statement is a sentence that is either true or false, but not both simultaneously.

Quantified Statements
(In the diagram, each quantified statement's equivalent is written in parentheses below the statement. The statements diagonally opposite each other are negations.)

Table 3.2, p. 92;
Ex. 4, p. 92

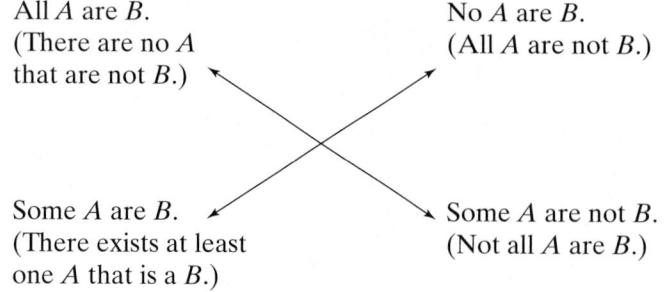

All *A* are *B*.
(There are no *A*
that are not *B*.)

No *A* are *B*.
(All *A* are not *B*.)

Some *A* are *B*.
(There exists at least
one *A* that is a *B*.)

Some *A* are not *B*.
(Not all *A* are *B*.)

Statements of Symbolic Logic

a. Negation
   ~p: Not p. It is not true that p.

b. Conjunction
   $p \wedge q$: p and q.

c. Disjunction
   $p \vee q$: p or q.

d. Conditional
   $p \rightarrow q$: If p, then q.

e. Biconditional
   $p \leftrightarrow q$: p if and only if q.

Symbolic Statements with Parentheses

a. Unless parentheses follow the negation symbol, ~, in a symbolic statement, ~ negates only the statement that immediately follows it.

b. When translating symbolic statements into English, the simple statements in parentheses appear on the same side of the comma.

3.3 Truth Tables for Negation, Conjunction, and Disjunction

3.4 Truth Tables for the Conditional and Biconditional

The Definitions of Symbolic Logic

p	q	Negation ~p	Conjunction $p \wedge q$	Disjunction $p \vee q$	Conditional $p \rightarrow q$	Biconditional $p \leftrightarrow q$
T	T	F	T	T	T	T
T	F	F	F	T	F	F
F	T	T	F	T	T	F
F	F	T	F	F	T	T

Opposite truth values from p — True only when both component statements are true — False only when both component statements are false — False only when the antecedent is true and the consequent is false — True only when the component statements have the same truth value

Constructing Truth Tables

A truth table for a compound statement shows when the statement is true and when it is false. The first few columns show the simple statements that comprise the compound statement and their possible truth values. The final column heading is the given compound statement. The truth values in each column are determined by looking back at appropriate columns and using one of the five definitions of symbolic logic. If a compound statement is always true, it is called a tautology.

3.5 Equivalent Statements, Variations of Conditional Statements, and De Morgan's Laws

Equivalent Statements

Two statements are equivalent, symbolized $\equiv$, if they have the same truth value in every possible case.

The Conditional Statement: $p \rightarrow q$

a. $p \rightarrow q$ is equivalent to ~$q \rightarrow$ ~p, the contrapositive.

b. $p \rightarrow q$ is not equivalent to $q \rightarrow p$, the converse.

c. $p \rightarrow q$ is not equivalent to ~$p \rightarrow$ ~q, the inverse.

d. The negation of $p \rightarrow q$ is $p \wedge$ ~q.

De Morgan's Laws

a. $\sim(p \wedge q) \equiv \sim p \vee \sim q$:   The negation of $p \wedge q$ is $\sim p \vee \sim q$.

b. $\sim(p \vee q) \equiv \sim p \wedge \sim q$:   The negation of $p \vee q$ is $\sim p \wedge \sim q$.

Ex. 7, p. 133;
Ex. 8, p. 134;
Ex. 9, p. 134;
Ex. 10, p. 135

3.6 Arguments and Truth Tables

Basic Definitions

An argument consists of two parts: the given statements, called the premises, and a conclusion. An argument is valid if the conclusion is true whenever the premises are assumed to be true. An argument that is not valid is called an invalid argument or a fallacy.

Testing the Validity of an Argument with a Truth Table

A procedure to test the validity of an argument using a truth table is described in the box on page 140. If the argument contains n premises, write a conditional statement of the form

Ex. 1, p. 140;
Ex. 2, p. 141;
Ex. 3, p. 143

$$[(\text{premise 1}) \wedge (\text{premise 2}) \wedge \cdots \wedge (\text{premise } n)] \rightarrow \text{conclusion}$$

and construct a truth table. If the conditional statement is a tautology, the argument is valid; if not, the argument is invalid.

Standard Forms of Arguments

Table 3.14 on page 144 contains the standard forms of commonly used valid and invalid arguments.

Ex. 4, p. 145

3.7 Arguments and Euler Diagrams

Euler Diagrams for Quantified Statements

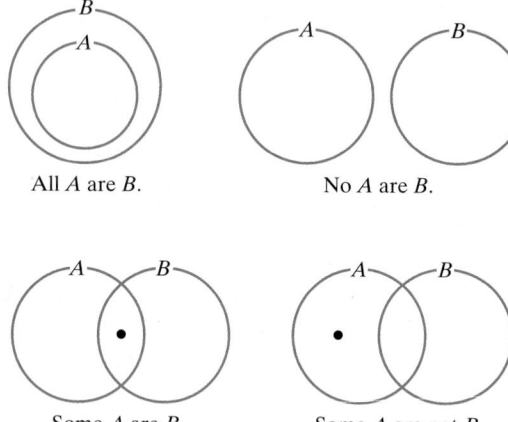

All A are B.   No A are B.

Some A are B.   Some A are not B.

Testing the Validity of an Argument with an Euler Diagram

a. Make an Euler diagram for the first premise.

b. Make an Euler diagram for the second premise on top of the one for the first premise.

c. The argument is valid if and only if every possible diagram illustrates the conclusion of the argument.

Ex. 1, p. 149;
Ex. 2, p. 150;
Ex. 3, p. 151;
Ex. 4, p. 152;
Ex. 5, p. 153;
Ex. 6, p. 154

REVIEW EXERCISES

3.1 and 3.2

In Exercises 1–6, let p, q, and r represent the following simple statements:

  p: The temperature is below 32°.
  q: We finished studying.
  r: We go to the movies.

Express each symbolic compound statement in English.

1. $(p \wedge q) \rightarrow r$

2. $\sim r \rightarrow (\sim p \vee \sim q)$

3. $p \wedge (q \rightarrow r)$

4. $r \leftrightarrow (p \wedge q)$

5. $\sim(p \wedge q)$

6. $\sim r \leftrightarrow (\sim p \vee \sim q)$

In Exercises 7–10, let p, q, and r represent the following simple statements:

  p: The outside temperature is at least 80°.
  q: The air conditioner is working.
  r: The house is hot.

Express each English statement in symbolic form.

7. The outside temperature is at least 80° and the air conditioner is working, or the house is hot.

8. If the outside temperature is at least 80° or the air conditioner is not working, then the house is hot.

9. If the air conditioner is working, then the outside temperature is at least 80° if and only if the house is hot.

10. The house is hot, if and only if the outside temperature is at least 80° and the air conditioner is not working.

In Exercises 11–14, write the negation of each statement.

11. All houses are made with wood.

12. No students major in business.

13. Some crimes are motivated by passion.

14. Some Democrats are not registered voters.

15. The speaker stated that, "All new taxes are for education." We later learned that the speaker was not telling the truth. What can we conclude about new taxes and education?

3.3 and 3.4

The bar graph shows the number of married Americans, in millions, by age. Use the information given by the graph to determine the truth value for each of the statements in Exercises 16–18.

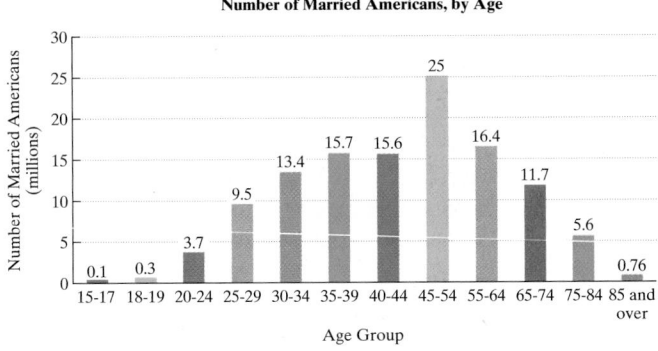

Number of Married Americans, by Age

Source: National Center for Health Statistics

16. The greatest number of married Americans are in the 45–54 age group and there are 15.7 million married Americans in the 40–44 age group.

17. The greatest number of married Americans are in the 45–54 age group or there are 15.7 million married Americans in the 40–44 age group.

18. If the number of married Americans decreases from the 20–24 to the 25–29 age group, then there are no married Americans who are 85 and over.

In Exercises 19–25, construct a truth table for the statement.

19. $p \lor (\sim p \land q)$

20. $\sim p \lor \sim q$

21. $p \rightarrow (\sim p \lor q)$

22. $p \leftrightarrow \sim q$

23. $\sim(p \lor q) \rightarrow (\sim p \land \sim q)$

24. $(p \lor q) \rightarrow \sim r$

25. $(p \land q) \leftrightarrow (p \land r)$

The bar graph shows the percentage of women in the United States military, by service branch. Use the information given by the graph to determine the truth value for each compound statement in Exercises 26–27.

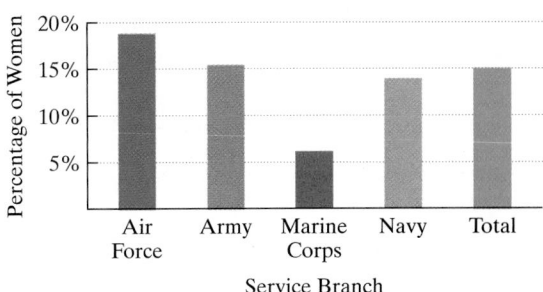

Percentage of Women in the U.S. Military, by Service Branch

Source: Newsweek, April 14, 2003

26. If the percentage of women in the American military is 15%, then the percentage exceeds 15% in the Army or is less than 10% in the Navy.

27. The percentage of women in the Air Force exceeds the percentage in the Marine Corps if and only if the percentage in the Air Force exceeds 20%, and the percentage in the Marine Corps is less than 10%.

In Exercises 28–30, determine the truth value for each statement when p is true, q is false, and r is false.

28. $\sim(q \leftrightarrow r)$

29. $(p \land q) \rightarrow (p \lor r)$

30. $(\sim q \rightarrow p) \lor (r \land \sim p)$

3.5

31. **a.** Use a truth table to show that $\sim p \lor q$ and $p \rightarrow q$ are equivalent.

   b. Use the result from part (a) to write a statement that is equivalent to

   The triangle is not isosceles or it has two equal sides.

32. Select the statement that is equivalent to

   Joe grows mangos or oranges.

   a. If Joe grows mangos, he does not grow oranges.

   b. If Joe grows oranges, he does not grow mangos.

   c. If Joe does not grow mangos, he grows oranges.

   d. Joe grows both mangos and oranges.

Write the contrapositive for each statement in Exercises 33–35.

33. If I am in Atlanta, then I am in the South.

34. If I am in class, then today is not a holiday.

35. If I work hard, then I pass all courses.

Write the converse and the inverse for each statement in Exercises 36–37.

36. If there is a storm, then classes are cancelled.

37. If the television is on, then we do not talk.

Write the negation for each conditional statement in Exercises 38–39.

38. If I am in Bogota, then I am in Colombia.

39. If I do not work hard, then I do not succeed.

Use De Morgan's laws to write a statement that is equivalent to each statement in Exercises 40–42.

40. It is not true that both Chicago and Maine are cities.

41. It is not true that Ernest Hemingway was a musician or an actor.

42. If a number is not positive and not negative, the number is 0.

Use De Morgan's laws to write the negation of each statement in Exercises 43–45.

43. I work hard or I do not succeed.

44. She is not using her car and she is taking a bus.

45. $\sim p \vee q$

In Exercises 46–48, determine which, if any, of the three given statements are equivalent.

46. a. If it is hot, then I use the air conditioner.

b. If it is not hot, then I do not use the air conditioner.

c. It is not hot or I use the air conditioner.

47. a. If she did not play, then we lost.

b. If we did not lose, then she played.

c. She did not play and we did not lose.

48. a. He is here or I'm not.

b. If I'm not here, he is.

c. It is not true that he isn't here and I am.

3.6

In Exercises 49–50, use a truth table to determine whether the symbolic form of the argument is valid or invalid.

49. $p \rightarrow q$ **50.** $p \wedge q$
       $\underline{\sim q \quad\quad}$ $\underline{q \rightarrow r \quad\quad}$
       $\therefore p$ $\therefore p \rightarrow r$

In Exercises 51–55, translate each argument into symbolic form. Then use a truth table to determine whether the argument is valid or invalid.

51. If one is a good baseball player, one must have good eye-hand coordination.
Todd does not have good eye-hand coordination.
∴ Todd is not a good baseball player.

52. If Tony plays, the team wins.
The team won.
∴ Tony played.

53. My plant is fertilized or it turns yellow.
My plant is turning yellow.
∴ My plant is not fertilized.

54. A majority of legislators vote for a bill or that bill does not become law.
A majority of legislators did not vote for bill x.
∴ Bill x did not become law.

55. If I purchase season tickets to the football games, then I do not attend all lectures.
If I do not attend all lectures, then I do not do well in school.
∴ If I do not do well in school, then I purchased season tickets to the football games.

3.7

Use Euler diagrams to determine whether each argument in Exercises 56–61 is valid or invalid.

56. All birds have feathers.
All parrots have feathers.
∴ All parrots are birds.

57. All botanists are scientists.
All scientists have college degrees.
∴ All botanists have college degrees.

58. All native desert plants can withstand severe drought.
No tree ferns can withstand severe drought.
∴ No tree ferns are native desert plants.

59. All native desert plants can withstand severe drought.
No tree ferns are native desert plants.
∴ No tree ferns can withstand severe drought.

60. All poets are writers.
Some writers are wealthy.
∴ Some poets are wealthy.

61. Some people enjoy reading.
All people who enjoy reading appreciate language.
∴ Some people appreciate language.

Group Exercise

62. Read the Blitzer Bonus on page 123. Group members are working for the advertising department of a company that sells magazine subscriptions by mail. It is time to replace the claim on the envelope with a new compound statement. The statement must be a true claim that will entice people into opening the envelope and ordering magazines. Group members should use their knowledge of logic to write as many such statements as possible.

CHAPTER 3 TEST

Use the following representations in Exercises 1–5:

   p: I'm registered.
   q: I'm a citizen.
   r: I vote.

Express each compound statement in English.

1. $(p \land q) \rightarrow r$ **2.** $\sim r \leftrightarrow (\sim p \lor \sim q)$

3. $\sim (p \lor q)$

Express each English statement in symbolic form.

4. I am registered and a citizen, or I do not vote.

5. If I am not registered or not a citizen, then I do not vote.

In Exercises 6–7, write the negation of the statement.

6. All numbers are divisible by 5.

7. Some people wear glasses.

In Exercises 8–10, construct a truth table for the statement.

8. $p \land (\sim p \lor q)$ **9.** $\sim (p \land q) \leftrightarrow (\sim p \lor \sim q)$

10. $p \leftrightarrow (q \lor r)$

In Exercises 11–12, determine the truth value for each statement when p is false, q is true, and r is false.

11. $\sim (q \rightarrow r)$ **12.** $(p \lor r) \leftrightarrow (\sim r \land p)$

13. Select the statement below that is equivalent to

Gene is an actor or a musician.

a. If Gene is an actor, then he is not a musician.

b. If Gene is not an actor, then he is a musician.

c. It is false that Gene is not an actor or not a musician.

d. If Gene is an actor, then he is a musician.

14. Write the contrapositive of

If it is August, it does not snow.

15. Write the converse and the inverse of the following statement:

If the radio is playing, then I cannot concentrate.

16. Write the negation of the following statement:

If it is cold, we do not use the pool.

17. Write a statement that is equivalent to

It is not true that the test is today or the party is tonight.

18. Write the negation of the following statement:

The banana is green and it is not ready to eat.

In Exercises 19–20, determine which, if any, of the three given statements are equivalent.

19. a. If I'm not feeling well, I'm grouchy.

   b. I'm feeling well or I'm grouchy.

   c. If I'm feeling well, I'm not grouchy.

20. a. It is not true that today is a holiday or tomorrow is a holiday.

   b. If today is not a holiday, then tomorrow is not a holiday.

   c. Today is not a holiday and tomorrow is not a holiday.

Determine whether each argument in Exercises 21–26 is valid or invalid.

21. If a parrot talks, it is intelligent.

   This parrot is intelligent._____

   ∴ This parrot talks.

22. I am sick or I am tired.

   I am not tired._____

   ∴ I am sick.

23. I am going if and only if you are not.

   You are going._____

   ∴ I'm going.

24. All mammals are warm-blooded.

   All dogs are warm-blooded._____

   ∴ All dogs are mammals.

25. All writers love language.

   No insects love language._____

   ∴ No writers are insects.

26. All rabbis are Jewish.

   Some Jews observe kosher dietary traditions._____

   ∴ Some rabbis observe kosher dietary traditions.

Number Representation and Calculation

A system for representing numbers, called a **numeration system**, gives the computer its extraordinary powers. The information fed into the computer to create the exhilarating experience of a virtual reality ski game is converted into a binary system, using only 0s and 1s. Without an understanding of how we represent numbers, the computer could not exist. In this chapter, our focus is on the system that we use to represent numbers, called the Hindu-Arabic system. The idea behind this system will eventually enable you, a "cybernaut," to see, hear, and interact with three-dimensional, digitally simulated worlds.

You don't have the money to go on the ski club trip, so you head out to the local arcade where there is a ski simulator game. Instead of the clunky old machine, there is a new virtual reality machine. You slip your money in and put on your headset. Suddenly you are at the top of a mountain, skiing down slopes at 20, 30, 40 miles per hour. Your heart races as you leap off jumps and dodge other skiers. The experience is totally realistic. As you walk home you wonder, how do they do that?

See the Blitzer Bonus on Talking to Computers in Section 4.2.

SECTION 4.1   OUR HINDU-ARABIC SYSTEM AND EARLY POSITIONAL SYSTEMS

OBJECTIVES

1. **Evaluate an exponential expression.**
2. **Write a Hindu-Arabic numeral in expanded form.**
3. **Express a number's expanded form as a Hindu-Arabic numeral.**
4. **Understand and use the Babylonian numeration system.**
5. **Understand and use the Mayan numeration system.**

FIGURE 4.1

All of us have an intuitive understanding of *more* and *less*. As humanity evolved, this sense of more and less was used to develop a system of counting. A tribe needed to know how many sheep it had, and whether the flock was increasing or decreasing in number. The earliest way of keeping count probably involved some tally method, using one vertical mark on a cave wall for each sheep. Later, a variety of vocal sounds developed as a tally for the number of things in a group. Finally, written symbols, or numerals, were used to represent numbers.

A **number** is an abstract idea that addresses the question, "How many?" A **numeral** is a symbol used to represent a number. Different symbols may be used to represent the same number. Numerals used to represent how many buffalo are shown in Figure 4.1 include

| ‖‖‖ ‖‖‖ | IX | 9. |
| Tally method | Roman numeral | Hindu-Arabic numeral |

We take numerals and the numbers that they represent for granted and use them every day. A **system of numeration** consists of a set of basic numerals and rules for combining them to represent numbers. It took humanity thousands of years to invent numeration systems that made computation a reasonable task. Today we use a system of writing numerals that was invented in India and brought to Europe by the Arabs. Our numerals are therefore called **Hindu-Arabic numerals**.

Like literature or music, a numeration system has a profound effect on the culture that created it. Computers, which affect our everyday lives, are based on an understanding of our Hindu-Arabic system of numeration. In this section, we study the characteristics of our numeration system. We also take a brief journey through history to look at two numeration systems that pointed the way toward an amazing cultural creation, our Hindu-Arabic system.

Exponential Notation

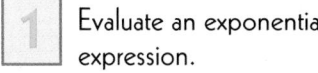 Evaluate an exponential expression.

An understanding of exponents is important in understanding the characteristics of our numeration system. Consider each of the following multiplications:

$$10 \times 10 \qquad 10 \times 10 \times 10 \qquad 10 \times 10 \times 10 \times 10.$$

$$Base \rightarrow \mathbf{10}^2 \leftarrow Exponent$$

These multiplications can be expressed more simply using what is called **exponential notation**. In this notation, 10×10 is expressed as 10^2. The 10 that is to be repeated when multiplying is called the **base**. The small ^{2} above and at the right of 10 is called the **exponent**. The exponent tells the number of times the base is to be used when multiplying. In 10^3, the base is 10 and the exponent is 3.

EXAMPLE 1   UNDERSTANDING EXPONENTIAL NOTATION

Evaluate the following:

a. 10^2 **b.** 10^3 **c.** 10^4.

SOLUTION

a. $10^2 = 10 \times 10 = 100$

b. $10^3 = 10 \times 10 \times 10 = 1000$

c. $10^4 = 10 \times 10 \times 10 \times 10 = 10,000$

 **Check Point 1**   Evaluate the following:

a. 10^5 **b.** 10^6.

Exponents are often referred to as **powers**. Because $10^2 = 100$, we also say that 100 expressed as a **power of 10** is 10^2. Similarly, because $1000 = 10^3$, 10^3 expresses 1000 as a power of 10.

Any number with an exponent of 1 is the number itself. Thus, $10^1 = 10$.

Multiplications that are expressed in exponential notation are read as follows:

10^1: "ten to the first power"

10^2: "ten to the second power" or "ten squared"

10^3: "ten to the third power" or "ten cubed"

10^4: "ten to the fourth power"

10^5: "ten to the fifth power"

Powers of Ten
$10 = 10^1$
$100 = 10^2$
$1000 = 10^3$
$10,000 = 10^4$
$100,000 = 10^5$
$1,000,000 = 10^6$

Become familiar with the powers of ten shown at the left because they play an important role in our system of numeration. Notice that **the number of zeros appearing to the right of the 1 in any numeral that expresses a power of 10 is the same as the exponent in that power of 10.** For example, in $10,000 = 10^4$, there are 4 zeros at the right of 1 in 10,000, and the exponent in 10^4 is 4. Therefore, 10^9 can be written as 1 followed by 9 zeros

$$1,000,000,000$$

or one billion.

Numbers other than 10 may also be used as the base.

TECHNOLOGY

You can use a calculator to evaluate exponential expressions. For example, to evaluate 5^3, press the following keys:

Many Scientific Calculators

5 $\boxed{y^x}$ 3 $\boxed{=}$

Many Graphing Calculators

5 $\boxed{\wedge}$ 3 $\boxed{\text{ENTER}}$.

Although calculators have special keys to evaluate powers of ten and to square bases, you can always use one of the sequences shown above.

EXAMPLE 2   USING EXPONENTIAL NOTATION

Evaluate the following:

a. 7^2 **b.** 5^3 **c.** 2^6 **d.** 9^1.

SOLUTION

a. $7^2 = 7 \times 7 = 49$

b. $5^3 = 5 \times 5 \times 5 = 125$

c. $2^6 = 2 \times 2 \times 2 \times 2 \times 2 \times 2 = 64$

d. $9^1 = 9$

 **Check Point 2**   Evaluate the following:

a. 8^2 **b.** 6^3 **c.** 2^5 **d.** 18^1.

Write a Hindu-Arabic numeral in expanded form.

Our Hindu-Arabic Numeration System

An important characteristic of our Hindu-Arabic system is that we can write the numeral for any number, large or small, using only ten symbols. The ten symbols that we use are

$$0, 1, 2, 3, 4, 5, 6, 7, 8, \text{ and } 9.$$

These symbols are called **digits**, from the Latin word for fingers.

With the use of exponents, Hindu-Arabic numerals can be written in **expanded form** in which the value of the digit in each position is made clear. In a Hindu-Arabic numeral, the place value of the first digit on the right is 1. The place value of the second digit from the right is 10. The place value of the third digit from the right is 100, or 10^2. For example, we can write 663 in expanded form by thinking of 663 as six 100s plus six 10s plus three 1s. This means that 663 in expanded form is

$$663 = (6 \times 100) + (6 \times 10) + (3 \times 1)$$
$$= (6 \times 10^2) + (6 \times 10^1) + (3 \times 1).$$

"The ingenious method of expressing all numbers by means of ten symbols appears so simple to us now that we ignore its true merits. But its very simplicity puts our arithmetic in the first rank of useful inventions."

PIERRE-SIMON LAPLACE
(French Mathematician, 1749–1827)

Because the value of a digit varies according to the position it occupies in a numeral, the Hindu-Arabic numeration system is called a **positional-value**, or **place-value**, system. The positional values in the system are based on powers of ten and are

$$\ldots, 10^5, 10^4, 10^3, 10^2, 10^1, 1.$$

EXAMPLE 3 | WRITING HINDU-ARABIC NUMERALS IN EXPANDED FORM

Write each of the following in expanded form:

a. 3407   **b.** 53,525.

SOLUTION

a. $3407 = (3 \times 10^3) + (4 \times 10^2) + (0 \times 10^1) + (7 \times 1)$
   or $= (3 \times 1000) + (4 \times 100) + (0 \times 10) + (7 \times 1)$

Because $0 \times 10^1 = 0$, this term could be left out, but the expanded form is clearer when it is included.

b. $53{,}525 = (5 \times 10^4) + (3 \times 10^3) + (5 \times 10^2) + (2 \times 10^1) + (5 \times 1)$
   or $= (5 \times 10{,}000) + (3 \times 1000) + (5 \times 100) + (2 \times 10) + (5 \times 1)$

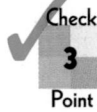 **Check Point 3**   Write each of the following in expanded form:

a. 4026   **b.** 24,232.

 Express a number's expanded form as a Hindu-Arabic numeral.

EXAMPLE 4 | EXPRESSING A NUMBER'S EXPANDED FORM AS A HINDU-ARABIC NUMERAL

Express each expanded form as a Hindu-Arabic numeral:

a. $(7 \times 10^3) + (5 \times 10^1) + (4 \times 1)$
b. $(6 \times 10^5) + (8 \times 10^1)$.

SOLUTION   For clarification, we begin by showing all powers of ten, starting with the highest exponent given. Any power of ten that is left out is expressed as 0 times that power of ten.

a. $(7 \times 10^3) + (5 \times 10^1) + (4 \times 1)$
  $= (7 \times 10^3) + (0 \times 10^2) + (5 \times 10^1) + (4 \times 1)$
  $= 7054$

b. $(6 \times 10^5) + (8 \times 10^1)$
  $= (6 \times 10^5) + (0 \times 10^4) + (0 \times 10^3) + (0 \times 10^2) + (8 \times 10^1) + (0 \times 1)$
  $= 600,080$

 Check Point 4 Express each expanded form as a Hindu-Arabic numeral:
  a. $(6 \times 10^3) + (7 \times 10^1) + (3 \times 1)$
  b. $(8 \times 10^4) + (9 \times 10^2)$.

Examples 3 and 4 show how there would be no Hindu-Arabic system without an understanding of zero and the invention of a symbol to represent nothingness. The system must have a symbol for zero to serve as a placeholder in case one or more powers of ten are not needed. The concept of zero was a new and radical invention, one that changed our ability to think about the world.

Early Positional Systems

Our Hindu-Arabic system developed over many centuries. Its digits can be found carved on ancient Hindu pillars over 2200 years old. In 1202, the Italian mathematician Leonardo Fibonacci (1170–1250) introduced the system to Europe, writing of its special characteristic: "With the nine Hindu digits and the Arab symbol 0, any number can be written." The Hindu-Arabic system came into widespread use only when printing was invented in the fifteenth century.

The Hindu-Arabic system uses powers of 10. However, positional systems can use powers of any number, not just 10. Think about our system of time, based on powers of 60:

$$1 \text{ minute} = 60 \text{ seconds}$$
$$1 \text{ hour} = 60 \text{ minutes} = 60 \times 60 \text{ seconds} = 60^2 \text{ seconds}.$$

What is significant in a positional system is position and the powers that positions convey. The first early positional system that we will discuss uses powers of 60, just like those used for units of time.

The Babylonian Numeration System

The city of Babylon, 55 miles south of present-day Baghdad, was the center of Babylonian civilization that lasted for about 1400 years between 2000 B.C. and 600 B.C. The Babylonians used wet clay as a writing surface. Their clay tablets were heated and dried to give a permanent record of their work, which we are able to decipher and read today. Table 4.1 gives the numerals of this civilization's numeration system.

The place values in the Babylonian system use powers of 60. The place values are

$$\ldots, \quad 60^3, \quad 60^2, \quad 60^1, \quad 1.$$

$60^3 = 60 \times 60 \times 60 = 216,000$ $60^2 = 60 \times 60 = 3600$

The Babylonians left a space to distinguish the various place values in a numeral from one another. For example,

means

$$(1 \times 60^2) + (10 \times 60^1) + (1 + 1) \times 1$$
$$= (1 \times 3600) + (10 \times 60) + (2 \times 1)$$
$$= 3600 + 600 + 2 = 4202.$$

Jasper Johns, "*Zero*". © Jasper Johns/ Licensed by VAGA, New York, NY.

"It took men about five thousand years, counting from the beginning of number symbols, to think of a symbol for nothing."

ISAAC ASIMOV,
Asimov on Numbers

4 Understand and use the Babylonian numeration system.

TABLE 4.1 BABYLONIAN NUMERALS

Babylonian numerals	∨	<
Hindu-Arabic numerals	1	10

EXAMPLE 5 | CONVERTING FROM A BABYLONIAN NUMERAL TO A HINDU-ARABIC NUMERAL

Write

∨ ∨   < ∨   < < ∨ ∨

as a Hindu-Arabic numeral.

SOLUTION   From left to right, the place values are 60^2, 60^1, and 1. Represent the numeral in each place as a familiar Hindu-Arabic numeral using Table 4.1. Multiply each Hindu-Arabic numeral by its respective place value. Then find the sum of these products.

means

$$(1 + 1) \times 60^2 + (10 + 1) \times 60^1 + (10 + 10 + 1 + 1) \times 1$$
$$= (2 \times 60^2) + (11 \times 60^1) + (22 \times 1)$$
$$= (2 \times 3600) + (11 \times 60) + (22 \times 1)$$
$$= 7200 + 660 + 22 = 7882$$

This sum indicates that the given Babylonian numeral is 7882 when written as a Hindu-Arabic numeral.

___

A major disadvantage of the Babylonian system is that it did not contain a symbol for zero. Some Babylonian tablets have a larger gap between the numerals or the insertion of the symbol ⩍ to indicate a missing place value, but this led to some ambiguity and confusion.

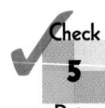 **Check 5 Point**   Write

∨ ∨ ∨   < <   < < < ∨

as a Hindu-Arabic numeral.

5 | Understand and use the Mayan numeration system.

The Mayan Numeration System

The Maya, a tribe of Central American Indians, lived on the Yucatan Peninsula. At its peak, between 300 and 1000 A.D., their civilization covered an area including parts of Mexico, all of Belize and Guatemala, and part of Honduras. They were famous for their magnificent architecture, their astronomical and mathematical knowledge, and their excellence in the arts. Their numeration system was the first to have a symbol for zero. Table 4.2 gives the Mayan numerals.

TABLE 4.2 MAYAN NUMERALS

0	1	2	3	4	5	6	7	8	9
10	11	12	13	14	15	16	17	18	19

The place values in the Mayan system are

$$..., \quad 18 \times 20^3, \quad 18 \times 20^2, \quad 18 \times 20, \quad 20, \quad 1.$$

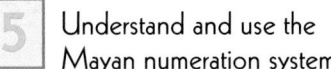

$18 \times 20 \times 20 \times 20 = 144{,}000$ $18 \times 20 \times 20 = 7200$ $18 \times 20 = 360$

Notice that instead of giving the third position a place value of 20^2, the Mayans used 18×20. This was probably done so that their calendar year of 360 days would be a basic part of the numeration system.

BLITZER BONUS

Numbers and Bird Brains

Birds have large, well-developed brains and are more intelligent than is suggested by the slur "bird brain." Parakeets can learn to count to seven. They have been taught to identify a box of food by counting the number of small objects in front of the box. Some species of birds can tell the difference between two and three. If a nest contains four eggs and one is taken, the bird will stay in the nest to protect the remaining three eggs. However, if two of the four eggs are taken, the bird recognizes that only two remain and will desert the nest, leaving the remaining two eggs unprotected.

Birds easily master complex counting problems. The sense of *more* and *less* that led to the development of numeration systems is not limited to the human species.

Numerals in the Mayan system are expressed vertically. The place value at the bottom of the column is 1.

EXAMPLE 6   USING THE MAYAN NUMERATION SYSTEM

Write

as a Hindu-Arabic numeral.

SOLUTION   The given Mayan numeral has four places. From top to bottom, the place values are 7200, 360, 20, and 1. Represent the numeral in each row as a familiar Hindu-Arabic numeral using Table 4.2. Multiply each Hindu-Arabic numeral by its respective place value. Then find the sum of these products.

Mayan numeral		Hindu-Arabic numeral		Place value		
••••	=	14	×	7200	=	100,800
⬯	=	0	×	360	=	0
••	=	7	×	20	=	140
••	=	12	×	1	=	12
						100,952

The sum on the right indicates that the given Mayan numeral is 100,952 when written as a Hindu-Arabic numeral.

 Write

as a Hindu-Arabic numeral.

Exercise Set 4.1

Practice Exercises

In Exercises 1–8, evaluate the expression.

1. 5^2 **2.** 6^2 **3.** 2^3 **4.** 4^3

5. 3^4 **6.** 2^4 **7.** 10^5 **8.** 10^6

In Exercises 9–22, write each Hindu-Arabic numeral in expanded form.

9. 36   **10.** 65   **11.** 249   **12.** 698

13. 703   **14.** 902   **15.** 4856   **16.** 5749

17. 3070   **18.** 9007   **19.** 34,569   **20.** 67,943

21. 230,007,004   **22.** 909,006,070

In Exercises 23–32, express each expanded form as a Hindu-Arabic numeral.

23. $(7 \times 10^1) + (3 \times 1)$

24. $(9 \times 10^1) + (4 \times 1)$

25. $(3 \times 10^2) + (8 \times 10^1) + (5 \times 1)$

26. $(7 \times 10^2) + (5 \times 10^1) + (3 \times 1)$

27. $(5 \times 10^5) + (2 \times 10^4) + (8 \times 10^3) + (7 \times 10^2) + (4 \times 10^1) + (3 \times 1)$

28. $(7 \times 10^6) + (4 \times 10^5) + (2 \times 10^4) + (3 \times 10^3) + (1 \times 10^2) + (9 \times 10^1) + (6 \times 1)$

29. $(7 \times 10^3) + (0 \times 10^2) + (0 \times 10^1) + (2 \times 1)$

30. $(9 \times 10^4) + (0 \times 10^3) + (0 \times 10^2) + (4 \times 10^1)$
$+ (5 \times 1)$

31. $(6 \times 10^8) + (2 \times 10^3) + (7 \times 1)$

32. $(3 \times 10^8) + (5 \times 10^4) + (4 \times 1)$

In Exercises 33–40, use Table 4.1 on page 166 to write each Babylonian numeral as a Hindu-Arabic numeral.

33. ⟨⟨ ∨∨∨   **34.** ⟨⟨⟨ ∨∨

35. ⟨⟨∨ ∨∨   **36.** ⟨⟨ ∨∨

37. ∨∨∨ ⟨∨∨ ∨∨∨   **38.** ∨∨ ⟨∨ ⟨⟨∨∨

39. ⟨∨ ⟨∨ ⟨∨ ⟨∨   **40.** ⟨⟨ ⟨⟨ ⟨∨∨ ⟨∨∨

In Exercises 41–50, use Table 4.2 on page 167 to write each Mayan numeral as a Hindu-Arabic numeral.

41. ⊷   **42.** ⊷   **43.** ⊷   **44.** ⊷

45. ⊷   **46.** ⊷   **47.** ⊷   **48.** ⊷

49. ⊷   **50.** ⊷

Application Exercises

Exercise 51 contains a Mayan numeral. Exercise 52 contains a Babylonian numeral. Write each numeral as a Hindu-Arabic numeral. The numeral that results represents a year that was not important to the Mayans or the Babylonians, but is important to us. Identify the historical event that took place in that year. If necessary, use an appropriate reference.

51. ⊷   **52.** ⟨⟨∨∨∨∨ ⟨⟨⟨⟨⟨∨∨

Writing in Mathematics

53. Describe the difference between a number and a numeral.

54. Explain how to evaluate 7^3.

55. What is the base in our Hindu-Arabic numeration system? What are the digits in the system?

56. Why is a symbol for zero needed in a positional system?

57. Explain how to write a Hindu-Arabic numeral in expanded form.

58. Describe one way that the Babylonian system is similar to the Hindu-Arabic system and one way that it is different from the Hindu-Arabic system.

59. Describe one way that the Mayan system is similar to the Hindu-Arabic system and one way that it is different from the Hindu-Arabic system.

60. *Research activity* Write a report on the history of the Hindu-Arabic system of numeration. Useful references include history of mathematics books, encyclopedias, and the World Wide Web.

Critical Thinking Exercises

61. Write ∨ ⟨∨∨ ⟨∨ as a Mayan numeral.

62. Write ⊷ as a Babylonian numeral.

63. Use Babylonian numerals to write the numeral that precedes and the numeral that follows

⟨∨ ⟨⟨⟨⟨⟨∨∨∨∨∨∨∨∨∨

Technology Exercises

In Exercises 64–67, use a calculator to evaluate each expression.

64. 4^5 **65.** 5^4 **66.** 2^9 **67.** 84^3

Group Exercises

68. Your group task is to create an original positional numeration system that is different from the three systems discussed in this section.
   a. Construct a table showing your numerals and the corresponding Hindu-Arabic numerals.
   b. Explain how to represent numbers in your system, and express a three-digit and a four-digit Hindu-Arabic numeral in your system.

69. For many individuals around the world who are blind or visually impaired, Braille numerals are used to represent numbers. Braille numerals can be detected and interpreted by touch. A group of two or three students should research how Braille numerals are formed. Many public and private organizations and schools provide materials and information for the blind and visually impaired. Using your library, resources on the World Wide Web, or local organizations, investigate the basic elements for Braille numerals, including the two-dot by three-dot cell, as well as the dot symbol used to precede a numeral. Present the results of your research to the entire class and bring examples of Braille numerals for each student to interpret by hand with no visual clues.

SECTION 4.2   NUMBER BASES IN POSITIONAL SYSTEMS

OBJECTIVES

1. Change numerals in bases other than ten to base ten.

2. Change base ten numerals to numerals in other bases.

You are being drawn deeper into cyberspace, spending more time online each week. With constantly improving high-resolution images, cyberspace is reshaping your life by nourishing shared enthusiasms. The people who built your computer talk of "bandwidth out the wazoo" that will give you the visual experience, in high-definition 3-D format, of being in the same room with a person who is actually in another city.

Because of our ten fingers and ten toes, the base ten Hindu-Arabic system seems to be an obvious choice. However, it is not base ten that computers use to process information and communicate with one another. Your experiences in cyberspace are sustained with a binary, or base two, system. In this section, we study numeration systems with bases other than ten. An understanding of such systems will help you to appreciate the nature of a positional system. You will also attain a better understanding of the computations you have used all of your life. You will even get to see how the world looks from a computer's point of view.

Changing Numerals in Bases Other Than Ten to Base Ten

 Change numerals in bases other than ten to base ten.

The base of a positional numeration system refers to the number of individual digit symbols that can be used in that system as well as to the number whose powers define the place values. For example, the digit symbols in a base five system are 0, 1, 2, 3, and 4. The place values in a base five system are powers of 5:

$$\ldots, 5^4, 5^3, 5^2, 5^1, 1$$
$$= \ldots, 5 \times 5 \times 5 \times 5, 5 \times 5 \times 5, 5 \times 5, 5, 1$$
$$= \ldots, 625, 125, 25, 5, 1.$$

When a numeral appears without a subscript, it is assumed that the base is ten. Bases other than ten are indicated with a spelled-out subscript, as in the numeral

$$122_{\text{five}}.$$

This numeral is read "one two two base five." Do not read it as "one hundred twenty two" because that terminology implies a base ten numeral, naming 122 in base ten.

We can convert 122_{five} to a base ten numeral by following the same procedure used in Section 4.1 to change the Babylonian and Mayan numerals to base ten Hindu-Arabic numerals. In the case of 122_{five}, the numeral has three places. From left to right, the place values are 5^2, 5^1, and 1. Multiply each digit in the numeral by its respective place value. Then add these products.

$$122_{\text{five}} = (1 \times 5^2) + (2 \times 5^1) + (2 \times 1)$$
$$= (1 \times 25) + (2 \times 5) + (2 \times 1)$$
$$= 25 + 10 + 2$$
$$= 37$$

Thus,

$$122_{\text{five}} = 37.$$

In base five, we do not need a digit symbol for 5 because

$$10_{\text{five}} = (1 \times 5^1) + (0 \times 1) = 5.$$

Likewise, the base ten numeral 6 is represented as 11_{five}, the base ten numeral 7 as 12_{five}, and so on. Table 4.3 shows base ten numerals from 0 through 20 and their base five equivalents.

In any base, the digit symbols begin at 0 and go up to one less than the base. In base b, the digit symbols begin at 0 and go up to $b - 1$. The place values in a base b system are powers of b:

$$\ldots, b^4, b^3, b^2, b, 1.$$

Table 4.4 shows the digit symbols and place values in various bases.

TABLE 4.3

Base Ten	Base Five
0	0
1	1
2	2
3	3
4	4
5	10
6	11
7	12
8	13
9	14
10	20
11	21
12	22
13	23
14	24
15	30
16	31
17	32
18	33
19	34
20	40

TABLE 4.4 DIGIT SYMBOLS AND PLACE VALUES IN VARIOUS BASES

Base	Digit Symbols	Place Values
two	$0, 1$	$\ldots, 2^4, 2^3, 2^2, 2^1, 1$
three	$0, 1, 2$	$\ldots, 3^4, 3^3, 3^2, 3^1, 1$
four	$0, 1, 2, 3$	$\ldots, 4^4, 4^3, 4^2, 4^1, 1$
five	$0, 1, 2, 3, 4$	$\ldots, 5^4, 5^3, 5^2, 5^1, 1$
six	$0, 1, 2, 3, 4, 5$	$\ldots, 6^4, 6^3, 6^2, 6^1, 1$
seven	$0, 1, 2, 3, 4, 5, 6$	$\ldots, 7^4, 7^3, 7^2, 7^1, 1$
eight	$0, 1, 2, 3, 4, 5, 6, 7$	$\ldots, 8^4, 8^3, 8^2, 8^1, 1$
nine	$0, 1, 2, 3, 4, 5, 6, 7, 8$	$\ldots, 9^4, 9^3, 9^2, 9^1, 1$
ten	$0, 1, 2, 3, 4, 5, 6, 7, 8, 9$	$\ldots, 10^4, 10^3, 10^2, 10^1, 1$

We have seen that in base five, 10_{five} represents one group of 5 and no groups of 1. Thus, $10_{\text{five}} = 5$. Similarly, in base six, 10_{six} represents one group of 6 and no groups of 1. Thus, $10_{\text{six}} = 6$. In general $10_{\text{base } b}$ represents one group of b and no groups of 1. This means that $10_{\text{base } b} = b$.

Here is the procedure for changing a numeral in a base other than ten to base ten:

CHANGING TO BASE TEN

To change a numeral in a base other than ten to a base ten numeral,

1. Find the place value for each digit in the numeral.

2. Multiply each digit in the numeral by its respective place value.

3. Find the sum of the products in step 2.

EXAMPLE 1   CONVERTING TO BASE TEN

Convert 4726_{eight} to base ten.

SOLUTION   The given base eight numeral has four places. From left to right, the place values are

$$8^3, 8^2, 8^1, \text{ and } 1.$$

Multiply each digit in the numeral by its respective place value. Then find the sum of these products.

Place value: 8^3	Place value: 8^2	Place value: 8^1	Place value: 1
4	7	2	6_{eight}

$$
\begin{aligned}
4726_{\text{eight}} &= (4 \times 8^3) + (7 \times 8^2) + (2 \times 8^1) + (6 \times 1) \\
&= (4 \times 8 \times 8 \times 8) + (7 \times 8 \times 8) + (2 \times 8) + (6 \times 1) \\
&= 2048 + 448 + 16 + 6 \\
&= 2518
\end{aligned}
$$

 Check 1 Point   Convert 3422_{five} to base ten.

EXAMPLE 2   CONVERTING TO BASE TEN

Convert 100101_{two} to base ten.

SOLUTION   Multiply each digit in the numeral by its respective place value. Then find the sum of these products.

Place value: 2^5	Place value: 2^4	Place value: 2^3	Place value: 2^2	Place value: 2^1	Place value: 1
1	0	0	1	0	1_{two}

$$
\begin{aligned}
100101_{\text{two}} &= (1 \times 2^5) + (0 \times 2^4) + (0 \times 2^3) + (1 \times 2^2) + (0 \times 2^1) + (1 \times 1) \\
&= (1 \times 32) + (0 \times 16) + (0 \times 8) + (1 \times 4) + (0 \times 2) + (1 \times 1) \\
&= 32 + 0 + 0 + 4 + 0 + 1 \\
&= 37
\end{aligned}
$$

 Check 2 Point   Convert 110011_{two} to base ten.

Three base systems are used in computer applications. These are base two, called a **binary system**, base eight, called an **octal system**, and base sixteen, called a **hexadecimal system**. Base sixteen presents a problem because digit symbols are needed from 0 up to one less than the base. This means that we need more digit symbols than the ten (0, 1, 2, 3, 4, 5, 6, 7, 8, and 9) used in our base ten system. Computer programmers use the letters A, B, C, D, E, and F as base sixteen digit symbols for the numbers ten through fifteen, respectively.

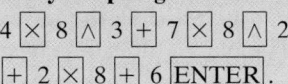
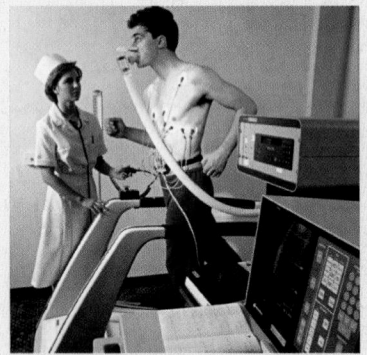

Additional digit symbols in base sixteen:

A = 10	B = 11
C = 12	D = 13
E = 14	F = 15

EXAMPLE 3 | CONVERTING TO BASE TEN

Convert EC7$_{sixteen}$ to base ten.

SOLUTION From left to right, the place values are

$$16^2, 16^1, \text{ and } 1.$$

The digit symbol E represents 14 and the digit symbol C represents 12. Although this numeral looks a bit strange, follow the usual procedure: Multiply each digit in the numeral by its respective place value. Then find the sum of these products.

Place value: 16^2	Place value: 16^1	Place value: 1
E	C	7$_{sixteen}$
E = 14	C = 12	

$$
\begin{aligned}
EC7_{sixteen} &= (14 \times 16^2) + (12 \times 16^1) + (7 \times 1) \\
&= (14 \times 16 \times 16) + (12 \times 16) + (7 \times 1) \\
&= 3584 + 192 + 7 \\
&= 3783
\end{aligned}
$$

✓ **Check Point 3**   Convert AD4$_{sixteen}$ to base ten.

Changing Base Ten Numerals to Numerals in Other Bases

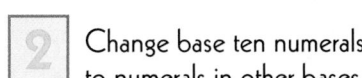

2 Change base ten numerals to numerals in other bases.

To convert a base ten numeral to a numeral in a base other than ten, we need to find how many groups of each place value are contained in the base ten numeral. When the base ten numeral consists of one or two digits, we can do this mentally. For example, suppose that we want to convert the base ten numeral, 6, to a base four numeral. The place values in base four are

$$\ldots, 4^3, 4^2, 4, 1.$$

The place values that are less than 6 are 4 and 1. We can express 6 as one group of four and two ones:

$$6_{ten} = (1 \times 4) + (2 \times 1) = 12_{four}.$$

EXAMPLE 4 | A MENTAL CONVERSION FROM BASE TEN TO BASE FIVE

Convert the base ten numeral 8 to a base five numeral.

SOLUTION The place values in base five are

$$\ldots, 5^3, 5^2, 5, 1.$$

The place values that are less than 8 are 5 and 1. We can express 8 as one group of five and three ones:

$$8_{ten} = (1 \times 5) + (3 \times 1) = 13_{five}.$$

✓ **Check Point 4**   Convert the base ten numeral 6 to a base five numeral.

If a conversion cannot be performed mentally, you can use divisions to determine how many groups of each place value are contained in a base ten numeral.

EXAMPLE 5 | USING DIVISIONS TO CONVERT FROM BASE TEN TO BASE EIGHT

Convert the base ten numeral 299 to a base eight numeral.

SOLUTION The place values in base eight are

$$\ldots, 8^3, 8^2, 8^1, 1, \quad \text{or} \quad \ldots, 512, 64, 8, 1.$$

The place values that are less than 299 are 64, 8, and 1. We can use divisions to show how many groups of each of these place values are contained in 299. Divide 299 by 64. Divide the remainder by 8.

$$
\begin{array}{r}
4 \\
64\overline{)299} \\
256 \\
\hline
43
\end{array}
\qquad
\begin{array}{r}
5 \\
8\overline{)43} \\
40 \\
\hline
3
\end{array}
$$

4 groups of 64   5 groups of 8   3 ones left over

These divisions show that 299 can be expressed as 4 groups of 64, 5 groups of 8, and 3 ones:

$$
\begin{aligned}
299 &= (4 \times 64) + (5 \times 8) + (3 \times 1) \\
&= (4 \times 8^2) + (5 \times 8^1) + (3 \times 1) \\
&= 453_{\text{eight}}.
\end{aligned}
$$

 **Check Point 5** Convert the base ten numeral 365 to a base seven numeral.

EXAMPLE 6 | USING DIVISIONS TO CONVERT FROM BASE TEN TO BASE SIX

Convert the base ten numeral 3444 to a base six numeral.

SOLUTION The place values in base six are

$$\ldots, 6^5, 6^4, 6^3, 6^2, 6^1, 1, \quad \text{or} \quad \ldots, 7776, 1296, 216, 36, 6, 1.$$

We use the powers of 6 that are less than 3444 and perform successive divisions by these powers.

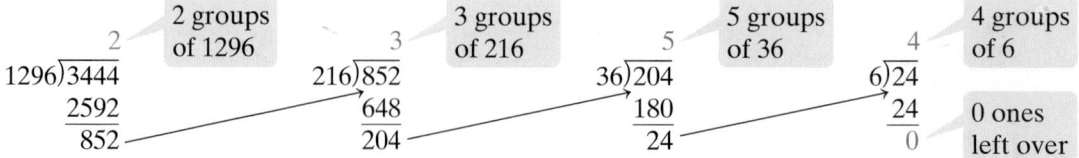

2 groups of 1296   3 groups of 216   5 groups of 36   4 groups of 6   0 ones left over

$$
\begin{array}{r}
2 \\
1296\overline{)3444} \\
2592 \\
\hline
852
\end{array}
\qquad
\begin{array}{r}
3 \\
216\overline{)852} \\
648 \\
\hline
204
\end{array}
\qquad
\begin{array}{r}
5 \\
36\overline{)204} \\
180 \\
\hline
24
\end{array}
\qquad
\begin{array}{r}
4 \\
6\overline{)24} \\
24 \\
\hline
0
\end{array}
$$

Using these four quotients and the final remainder, we can immediately write the answer.

$$3444 = 23540_{\text{six}}$$

 **Check Point 6** Convert the base ten numeral 2763 to a base five numeral.

Exercise Set 4.2

Practice Exercises

In Exercises 1–18, convert the numeral to a numeral in base ten.

1. 43_{five}

2. 34_{five}

3. 52_{eight}

4. 67_{eight}

5. 132_{four}

6. 321_{four}

7. 1011_{two}

8. 1101_{two}

9. 2035_{six}

10. 2073_{nine}

11. 70355_{eight}

12. 41502_{six}

13. 2096_{sixteen}

14. 3104_{fifteen}

15. 110101_{two}

16. 101101_{two}

17. $\text{ACE5}_{\text{sixteen}}$

18. $\text{EDF7}_{\text{sixteen}}$

In Exercises 19–28, mentally convert each base ten numeral to a numeral in the given base.

19. 7 to base five

20. 9 to base five

21. 11 to base seven

22. 12 to base seven

23. 2 to base two

24. 3 to base two

25. 13 to base four

26. 19 to base four

27. 37 to base six

28. 25 to base six

In Exercises 29–40, use divisions to convert each base ten numeral to a numeral in the given base.

29. 87 to base five

30. 85 to base seven

31. 108 to base four

32. 199 to base four

33. 19 to base two

34. 23 to base two

35. 138 to base three

36. 129 to base three

37. 386 to base six

38. 428 to base nine

39. 1599 to base seven

40. 1346 to base eight

Application Exercises

Use a procedure similar to the one used in Exercises 29–40 to solve Exercises 41–43.

41. Change 153 days to weeks and days.

42. Change 273 hours to days and hours.

43. Change $8.79 to quarters, nickels, and pennies.

Writing in Mathematics

44. Explain how to determine the place values for a four-digit numeral in base six.

45. Describe how to change a numeral in a base other than ten to a base ten numeral.

46. Describe how to change a base ten numeral to a numeral in another base.

Critical Thinking Exercises

In Exercises 47–48, write in the indicated base the counting numbers that precede and follow the number expressed by the given numeral.

47. 888_{nine}

48. $\text{EC5}_{\text{sixteen}}$

49. Arrange from smallest to largest: $11111011_{\text{two}}, 3\text{A}6_{\text{twelve}}, 673_{\text{eight}}$.

Technology Exercises

In Exercises 50–54, use the scientific calculator in Windows (or any scientific calculator that handles different base conversions) to convert each numeral to a numeral in the indicated base. Note: If you are unsure how to convert to a different base on your calculator, consult the owner's manual.

50. 45 to octal

51. 100101_{two} to decimal

52. 100101_{two} to hexadecimal

53. 567 to binary

54. 333_{eight} to hexadecimal

Group Exercises

The following topics are appropriate for either individual or group research projects. A report should be given to the class on the researched topic. Useful references include history of mathematics books, books whose purpose is to excite the reader about mathematics, encyclopedias, and the World Wide Web.

55. Societies that Use Numeration Systems with Bases Other Than Ten

56. The Use of Fingers to Represent Numbers

57. Applications of Bases Other Than Ten

58. Binary, Octal, Hexadecimal Bases and Computers

59. Negative Bases (See "Numeration Systems with Unusual Bases," by David Ballow in *The Mathematics Teacher*, May 1974, pp. 413–414.)

60. Babylonian and Mayan Civilizations and Their Contributions

SECTION 4.3   COMPUTATION IN POSITIONAL SYSTEMS

OBJECTIVES

1. Add in bases other than ten.
2. Subtract in bases other than ten.
3. Multiply in bases other than ten.
4. Divide in bases other than ten.

People have always looked for ways to make calculations faster and easier. The Hindu-Arabic system of numeration made computation simpler and less mysterious. More people were able to perform computation with ease, leading to the widespread use of the system.

All computations in bases other than ten are performed exactly like those in base ten. However, when a computation is equal to or exceeds the given base, use the mental conversions discussed in the previous section to convert from the base ten numeral to a numeral in the desired base.

Addition

1 Add in bases other than ten.

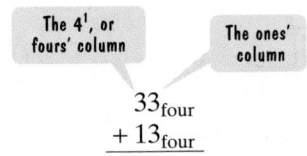

The 4^1, or fours' column

The ones' column

33_{four}
$+ 13_{four}$

EXAMPLE 1 | ADDITION IN BASE FOUR

Add:

$$33_{four}$$
$$+ 13_{four}.$$

SOLUTION   We will begin by adding the numbers in the right-hand column. In base four, the digit symbols are 0, 1, 2, and 3. If a sum in this, or any, column exceeds 3, we will have to convert this base ten number to base four. We begin by adding the numbers in the right-hand, or ones', column:

$$3_{four} + 3_{four} = 6.$$

6 is not a digit symbol in base four. However, we can express 6 as one group of four and two ones left over:

$$3_{four} + 3_{four} = 6_{ten} = (1 \times 4) + (2 \times 1) = 12_{four}.$$

Now we record the sum of the right-hand column, 12_{four}:

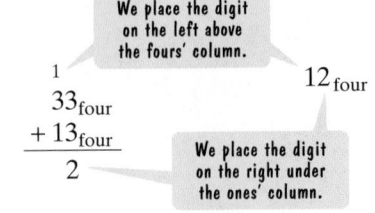

We place the digit on the left above the fours' column.

12_{four}

$\overset{1}{}$
33_{four}
$+ 13_{four}$
$\overline{2}$

We place the digit on the right under the ones' column.

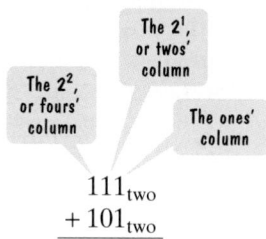

The 2^2, or fours' column

The 2^1, or twos' column

The ones' column

$$111_{\text{two}}$$
$$+ 101_{\text{two}}$$

Next, we add the three digits in the fours' column:

$$1_{\text{four}} + 3_{\text{four}} + 1_{\text{four}} = 5.$$

5 is not a digit symbol in base four. However, we can express 5 as one group of four and one left over:

$$1_{\text{four}} + 3_{\text{four}} + 1_{\text{four}} = 5_{\text{ten}} = (1 \times 4) + (1 \times 1) = 11_{\text{four}}.$$

Record the 11_{four}.

$$\begin{array}{r} 1 \\ 33_{\text{four}} \\ + 13_{\text{four}} \\ \hline 112_{\text{four}} \end{array}$$

This is the desired sum.

You can check the sum by converting $33_{\text{four}}, 13_{\text{four}},$ and 112_{four} to base ten: $33_{\text{four}} = 15, 13_{\text{four}} = 7,$ and $112_{\text{four}} = 22.$ Because $15 + 7 = 22,$ our work is correct.

✓ **Check 1 Point** Add:

$$\begin{array}{r} 32_{\text{five}} \\ + 44_{\text{five}} \end{array}.$$

EXAMPLE 2 | ADDITION IN BASE TWO

Add:

$$\begin{array}{r} 111_{\text{two}} \\ + 101_{\text{two}} \end{array}.$$

SOLUTION We begin by adding the numbers in the right-hand, or ones', column:

$$1_{\text{two}} + 1_{\text{two}} = 2.$$

2 is not a digit symbol in base two. We can express 2 as one group of 2 and zero ones left over:

$$1_{\text{two}} + 1_{\text{two}} = 2_{\text{ten}} = (1 \times 2) + (0 \times 1) = 10_{\text{two}}.$$

Now we record the sum of the right-hand column, 10_{two}:

We place the digit on the left above the twos' column.

$$\begin{array}{r} 1 \\ 111_{\text{two}} \\ + 101_{\text{two}} \\ \hline 0 \end{array} \qquad 10_{\text{two}}$$

We place the digit on the right under the ones' column.

Next, we add the three digits in the twos' column:

$$1_{\text{two}} + 1_{\text{two}} + 0_{\text{two}} = 2_{\text{ten}} = (1 \times 2) + (0 \times 1) = 10_{\text{two}}.$$

Now we record the sum of the middle column, 10_{two}:

We place the digit on the left above the fours' column.

$$\begin{array}{r} 11 \\ 111_{\text{two}} \\ + 101_{\text{two}} \\ \hline 00 \end{array} \qquad 10_{\text{two}}$$

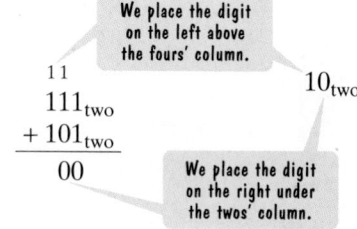

We place the digit on the right under the twos' column.

Finally, we add the three digits in the fours' column:

$$1_{two} + 1_{two} + 1_{two} = 3.$$

3 is not a digit symbol in base two. We can express 3 as one group of 2 and one 1 left over:

$$1_{two} + 1_{two} + 1_{two} = 3_{ten} = (1 \times 2) + (1 \times 1) = 11_{two}.$$

Record the 11_{two}.

$$
\begin{array}{r}
^{1\,1} \\
111_{two} \\
+\ 101_{two} \\
\hline
1100_{two}
\end{array}
$$

This is the desired sum.

You can check the sum by converting to base ten: $111_{two} = 7$, $101_{two} = 5$, and $1100_{two} = 12$. Because $7 + 5 = 12$, our work is correct.

 **Check Point 2**   Add:

$$
\begin{array}{r}
111_{two} \\
+\ 111_{two}
\end{array}.
$$

Subtraction

To subtract in bases other than ten, we line up the digits with the same place values and subtract column by column, beginning with the column on the right. If "borrowing" is necessary to perform the subtraction, borrow the amount of the base. For example, when we borrow in base ten subtraction, we borrow 10s. Likewise, we borrow 2s in base two, 3s in base three, 4s in base four, and so on.

EXAMPLE 3   SUBTRACTION IN BASE FOUR

Subtract:

$$
\begin{array}{r}
31_{four} \\
-\ 12_{four}
\end{array}.
$$

SOLUTION   We start by performing subtraction in the right column, $1_{four} - 2_{four}$. Because 2_{four} is greater than 1_{four}, we need to borrow from the preceding column. We are working in base four, so we borrow one group of 4. This gives a sum of $4 + 1$, or 5, in base ten. Now we subtract 2 from 5, obtaining a difference of 3:

We borrow one group of 4. Now there are 2 groups of 4 for this place value, not 3.

$$
\begin{array}{r}
^{2\,5} \\
3\!\!\!/1_{four} \\
-\ 12_{four} \\
\hline
3_{four}
\end{array}
$$

We add the borrowed group of 4 to 1 in base ten: $1 + 4 = 5$.

Now we perform the subtraction in the second column from the right:

We subtract 1 from 2.

$$
\begin{array}{r}
^{2\,5} \\
3\!\!\!/1_{four} \\
-\ 12_{four} \\
\hline
13_{four}
\end{array}
$$

This is the desired difference.

You can check the difference by converting to base ten: $31_{four} = 13$, $12_{four} = 6$, and $13_{four} = 7$. Because $13 - 6 = 7$, our work is correct.

 **Check Point 3**   Subtract:

$$
\begin{array}{r}
41_{five} \\
-\ 23_{five}
\end{array}.
$$

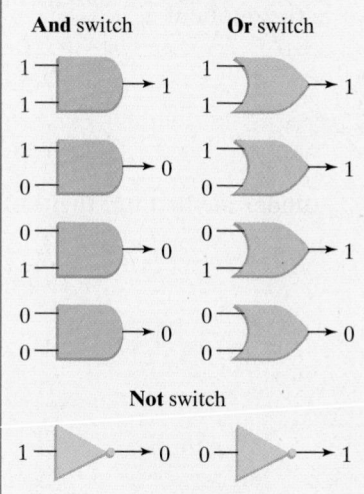
3   Multiply in bases other than ten.

EXAMPLE 4   SUBTRACTION IN BASE FIVE

Subtract:

$$3431_{\text{five}} - 1242_{\text{five}}.$$

SOLUTION

Step ① Borrow a group of 5 from the preceding column. This gives a sum of $5 + 1$, or 6, in base ten.

$$\begin{array}{r} 2\,6 \\ 343\!\!\!/1_{\text{five}} \\ -\ 1242_{\text{five}} \\ \hline 4_{\text{five}} \end{array}$$

Step ② $6 - 2 = 4$

Step ③ Borrow a group of 5 from the preceding column. This gives a sum of $5 + 2$, or 7, in base ten.

$$\begin{array}{r} 7 \\ 3\,2\,6 \\ 343\!\!\!/1_{\text{five}} \\ -\ 1242_{\text{five}} \\ \hline 34_{\text{five}} \end{array}$$

Step ④ $7 - 4 = 3$

Step ⑤ No borrowing is needed for these two columns.

$$\begin{array}{r} 7 \\ 3\,2\,6 \\ 343\!\!\!/1_{\text{five}} \\ -\ 1242_{\text{five}} \\ \hline 2134_{\text{five}} \end{array}$$

Step ⑥ $3 - 2 = 1$

Step ⑦ $3 - 1 = 2$

Thus, $3431_{\text{five}} - 1242_{\text{five}} = 2134_{\text{five}}$.

✓ Check Point 4   Subtract: $5144_{\text{seven}} - 3236_{\text{seven}}$.

Multiplication

EXAMPLE 5   MULTIPLICATION IN BASE SIX

Multiply:

$$\begin{array}{r} 34_{\text{six}} \\ \times\ 2_{\text{six}}. \end{array}$$

SOLUTION   We multiply just as we do in base ten. That is, first we will multiply the digit 2 by the digit 4 directly above it. Then we will multiply the digit 2 by the digit 3 in the left column. Keep in mind that only the digit symbols 0, 1, 2, 3, 4, and 5 are permitted in base six. We begin with

$$2_{\text{six}} \times 4_{\text{six}} = 8_{\text{ten}} = (1 \times 6) + (2 \times 1) = 12_{\text{six}}.$$

Record the 2 and carry the 1:

$$\begin{array}{r} \overset{1}{34}_{\text{six}} \\ \times\ 2_{\text{six}} \\ \hline 2_{\text{six}}. \end{array}$$

3X

Our next computation involves both multiplication and addition:

$$(2_{\text{six}} \times 3_{\text{six}}) + 1_{\text{six}} = 6 + 1 = 7_{\text{ten}} = (1 \times 6) + (1 \times 1) = 11_{\text{six}}.$$

Record the 11_{six}.

$$\begin{array}{r} 34_{\text{six}} \\ \times\ 2_{\text{six}} \\ \hline 112_{\text{six}} \end{array}$$

This is the desired product.

Let's check the product by converting to base ten: $34_{\text{six}} = 22$, $2_{\text{six}} = 2$, and $112_{\text{six}} = 44$. Because $22 \times 2 = 44$, our work is correct.

Check 5 Point   Multiply:

$$\begin{array}{r} \overset{2}{45}_{\text{seven}} \\ \times\ 3_{\text{seven}} \\ \hline 201 \end{array}$$

4   Divide in bases other than ten.

TABLE 4.5 MULTIPLICATION: BASE FOUR

×	0	1	2	3
0	0	0	0	0
1	0	1	2	3
2	0	2	10	12
3	0	3	12	21

BLITZER BONUS

Mechanical Computing

In 1944, scientists at IBM produced Mark I (shown here), the world's biggest calculator. Mark I performed basic arithmetic, multiplying and dividing huge numbers in mere seconds. Its greatest drawback was its size: 50 feet long and 35 tons!

   Calculating without thought and skill was the motivating factor in the history of mechanical computing. Today's handheld calculators provide the mathematical power of a computer, allowing the user to perform computations, graph equations, analyze data, write and use original programs, and even perform algebra's symbolic manipulations. Also, the World Wide Web now offers virtual calculators that simulate the format of actual calculators.

Division

The answer in a division problem is called a **quotient**. A multiplication table showing products in the same base as the division problem is helpful.

EXAMPLE 6 | DIVISION IN BASE FOUR

Use Table 4.5, showing products in base four, to perform the following division:

$$3_{\text{four}}\overline{)222_{\text{four}}}.$$

SOLUTION   We can use the same method to divide in base four that we use in base ten. Begin by dividing 22_{four} by 3_{four}. Use Table 4.5 to find in the vertical column headed by 3 the largest product that is less than or equal to 22_{four}. This product is 21_{four}. Because $3_{\text{four}} \times 3_{\text{four}} = 21_{\text{four}}$, the first number in the quotient is 3_{four}.

Divisor → $3_{\text{four}}\overline{)222_{\text{four}}}$ ← Dividend, with 3 ← Quotient

Now multiply $3_{\text{four}} \times 3_{\text{four}}$ and write the product, 21_{four}, under the first two digits of the dividend.

$$\begin{array}{r} 3 \\ 3_{\text{four}}\overline{)222_{\text{four}}} \\ 21 \\ \hline \end{array}$$

Subtract: $22_{\text{four}} - 21_{\text{four}} = 1_{\text{four}}$.

$$\begin{array}{r} 3 \\ 3_{\text{four}}\overline{)222_{\text{four}}} \\ 21 \\ \hline 1 \end{array}$$

Bring down the next digit in the dividend, 2_{four}.

$$\begin{array}{r} 3 \\ 3_{\text{four}}\overline{)222_{\text{four}}} \\ 21 \\ \hline 12 \end{array}$$

We now return to Table 4.5. Find in the vertical column headed by 3 the largest product that is less than or equal to 12_{four}. Because $3_{\text{four}} \times 2_{\text{four}} = 12_{\text{four}}$, the next numeral in the quotient is 2_{four}. We use this information to finish the division.

$$\begin{array}{r} 32_{\text{four}} \\ 3_{\text{four}}\overline{)222} \\ 21 \\ \hline 12 \\ 12 \\ \hline 0 \end{array}$$

This is the desired quotient.

Let's check the quotient by converting to base ten: $3_{four} = 3$, $222_{four} = 42$, and $32_{four} = 14$. Because $3\overline{)42}^{\,14}$, our work is correct.

Use Table 4.5, showing products in base four, to perform the following division:

$$2_{four}\overline{)112_{four}}.$$

Exercise Set 4.3

Practice Exercises

In Exercises 1–12, add in the indicated base.

1. $\begin{array}{r} 23_{four} \\ + 13_{four} \\ \hline \end{array}$

2. $\begin{array}{r} 31_{four} \\ + 22_{four} \\ \hline \end{array}$

3. $\begin{array}{r} 11_{two} \\ + 11_{two} \\ \hline \end{array}$

4. $\begin{array}{r} 101_{two} \\ + 11_{two} \\ \hline \end{array}$

5. $\begin{array}{r} 342_{five} \\ + 413_{five} \\ \hline \end{array}$

6. $\begin{array}{r} 323_{five} \\ + 421_{five} \\ \hline \end{array}$

7. $\begin{array}{r} 645_{seven} \\ + 324_{seven} \\ \hline \end{array}$

8. $\begin{array}{r} 632_{seven} \\ + 564_{seven} \\ \hline \end{array}$

9. $\begin{array}{r} 6784_{nine} \\ + 7865_{nine} \\ \hline \end{array}$

10. $\begin{array}{r} 1021_{three} \\ + 2011_{three} \\ \hline \end{array}$

11. $\begin{array}{r} 14632_{seven} \\ + 5604_{seven} \\ \hline \end{array}$

12. $\begin{array}{r} 53B_{sixteen} \\ + 694_{sixteen} \\ \hline \end{array}$

In Exercises 13–24, subtract in the indicated base.

13. $\begin{array}{r} 32_{four} \\ - 13_{four} \\ \hline \end{array}$

14. $\begin{array}{r} 21_{four} \\ - 12_{four} \\ \hline \end{array}$

15. $\begin{array}{r} 23_{five} \\ - 14_{five} \\ \hline \end{array}$

16. $\begin{array}{r} 32_{seven} \\ - 16_{seven} \\ \hline \end{array}$

17. $\begin{array}{r} 475_{eight} \\ - 267_{eight} \\ \hline \end{array}$

18. $\begin{array}{r} 712_{nine} \\ - 483_{nine} \\ \hline \end{array}$

19. $\begin{array}{r} 563_{seven} \\ - 164_{seven} \\ \hline \end{array}$

20. $\begin{array}{r} 462_{eight} \\ - 177_{eight} \\ \hline \end{array}$

21. $\begin{array}{r} 1001_{two} \\ - 111_{two} \\ \hline \end{array}$

22. $\begin{array}{r} 1000_{two} \\ - 101_{two} \\ \hline \end{array}$

23. $\begin{array}{r} 1200_{three} \\ - 1012_{three} \\ \hline \end{array}$

24. $\begin{array}{r} 4C6_{sixteen} \\ - 198_{sixteen} \\ \hline \end{array}$

In Exercises 25–34, multiply in the indicated base.

25. $\begin{array}{r} 25_{six} \\ \times\ 4_{six} \\ \hline \end{array}$

26. $\begin{array}{r} 34_{five} \\ \times\ 3_{five} \\ \hline \end{array}$

27. $\begin{array}{r} 11_{two} \\ \times\ 1_{two} \\ \hline \end{array}$

28. $\begin{array}{r} 21_{four} \\ \times\ 3_{four} \\ \hline \end{array}$

29. $\begin{array}{r} 543_{seven} \\ \times\ 5_{seven} \\ \hline \end{array}$

30. $\begin{array}{r} 243_{nine} \\ \times\ 6_{nine} \\ \hline \end{array}$

31. $\begin{array}{r} 623_{eight} \\ \times\ 4_{eight} \\ \hline \end{array}$

32. $\begin{array}{r} 543_{six} \\ \times\ 5_{six} \\ \hline \end{array}$

33. $\begin{array}{r} 21_{four} \\ \times\ 12_{four} \\ \hline \end{array}$

34. $\begin{array}{r} 32_{four} \\ \times\ 23_{four} \\ \hline \end{array}$

In Exercises 35–38, use the multiplication tables shown below and in the next column to divide in the indicated base.

MULTIPLICATION: BASE FOUR

×	0	1	2	3
0	0	0	0	0
1	0	1	2	3
2	0	2	10	12
3	0	3	12	21

MULTIPLICATION: BASE FIVE

×	0	1	2	3	4
0	0	0	0	0	0
1	0	1	2	3	4
2	0	2	4	11	13
3	0	3	11	14	22
4	0	4	13	22	31

35. $2_{four}\overline{)100_{four}}$

36. $2_{four}\overline{)321_{four}}$

37. $3_{five}\overline{)224_{five}}$

38. $4_{five}\overline{)134_{five}}$

Writing in Mathematics

39. Describe how to add two numbers in a base other than ten. How do you express and record the sum of numbers in a column if that sum exceeds the base?

40. Describe how to subtract two numbers in a base other than ten. How do you subtract a larger number from a smaller number in the same column?

41. Describe two difficulties that youngsters encounter when learning to add, subtract, multiply, and divide using Hindu-Arabic numerals. Base your answer on difficulties that are encountered when performing these computations in bases other than ten.

Critical Thinking Exercise

42. Divide: $31_{seven}\overline{)2426_{seven}}$.

Technology Exercises

In Exercises 43–46, use a scientific calculator that handles different base conversions to perform the calculation in the indicated base. Note: You need to put the calculator in the appropriate mode before you perform these calculations. If you are unsure how to do this, consult your owner's manual.

43. $712_{eight} + 455_{eight}$

44. $712_{eight} - 455_{eight}$

45. $65_{sixteen} \times 3_{sixteen}$

46. $516_{sixteen} \div 6_{sixteen}$

Group Exercises

47. Group members should research various methods that societies have used to perform computations. Include finger multiplication, the galley method (sometimes called the Gelosia method), Egyptian duplation, subtraction by complements, Napier's bones, and other methods of interest in your presentation to the entire class.

48. Organize a debate. One side represents people who favor performing computations by hand, using the methods and procedures discussed in this section, but applied to base ten numerals. The other side represents people who favor the use of calculators for performing all computations. Include the merits of each approach in the debate.

SECTION 4.4   LOOKING BACK AT EARLY NUMERATION SYSTEMS

OBJECTIVES

1. Understand and use the Egyptian system.

2. Understand and use the Roman system.

3. Understand and use the traditional Chinese system.

4. Understand and use the Ionic Greek system.

Super Bowl XXV, played on January 27, 1991, resulted in the closest score of all time: NY Giants: 20; Buffalo: 19. If you are intrigued by sports facts and figures, you are probably aware that major sports events, such as the Super Bowl, are named using Roman numerals. Perhaps you have seen the use of Roman numerals in dating movies and television shows, or on clocks and watches.

In this section, we embark on a brief journey through time and numbers. Our Hindu-Arabic numeration system, the focus of this chapter, is successful because it expresses numbers with just ten symbols and makes computation with these numbers relatively easy. By these standards, the early numeration systems discussed in this section, such as Roman numerals, are unsuccessful. By looking briefly at these systems, you will see that our system is outstanding when compared with other historical systems.

The Egyptian Numeration System

1 | Understand and use the Egyptian system.

Like most great civilizations, ancient Egypt had several numeration systems. The oldest is hieroglyphic notation, which developed around 3400 B.C. Table 4.6 lists the Egyptian hieroglyphic numerals with the equivalent Hindu-Arabic numerals. Notice that the numerals are powers of ten. Their numeral for 1,000,000, or 10^6, looks like someone who just won the lottery!

TABLE 4.6 EGYPTIAN HIEROGLYPHIC NUMERALS

Hindu-Arabic Numeral	Egyptian Numeral	Description
1		Staff
10		Heel bone
100		Spiral
1000		Lotus blossom
10,000		Pointing finger
100,000		Tadpole
1,000,000		Astonished person

It takes far more space to represent most numbers in the Egyptian system than in our system. This is because a number is expressed by repeating each numeral the required number of times. However, no numeral, except perhaps the astonished person, should be repeated more than nine times. If we were to use the Egyptian system to represent 764, we would need to write

100 100 100 100 100 100 100 10 10 10 10 10 10 10 1 1 1 1

and then represent each of these symbols with the appropriate hieroglyphic numeral from Table 4.6. Thus, 764 as an Egyptian numeral is

ⓞ ⓞ ⓞ ⓞ ⓞ ⓞ ⓞ ∩ ∩ ∩ ∩ ∩ ∩ ∩ | | | |.

The ancient Egyptian system is an example of an **additive system**, one in which the number represented is the sum of the values of the numerals.

EXAMPLE 1   USING THE EGYPTIAN NUMERATION SYSTEM

Write the following numeral as a Hindu-Arabic numeral:

SOLUTION   Using Table 4.6, find the value of each of the Egyptian numerals. Then add them.

1,000,000 + 10,000 + 10,000 + 10 + 10 + 10 + 1 + 1 + 1 + 1 = 1,020,034

✓ Check 1 Point   Write the following numeral as a Hindu-Arabic numeral:

𓂀 𓂀 𓂀 ⓞⓞ ∩∩ | |.

EXAMPLE 2   USING THE EGYPTIAN NUMERATION SYSTEM

Write 1752 as an Egyptian numeral.

SOLUTION   First break down the Hindu-Arabic numeral into quantities that match the Egyptian numerals:

$$1752 = 1000 + 700 + 50 + 2$$
$$= 1000 + 100 + 100 + 100 + 100 + 100 + 100 + 100$$
$$+ 10 + 10 + 10 + 10 + 10 + 1 + 1.$$

Now, use Table 4.6 to find the Egyptian symbol that matches each quantity. For example, the lotus blossom, 𓇇, matches 1000. Write each of these symbols and leave out the addition signs. Thus, the number 1752 can be expressed as

𓇇 ⓞ ⓞ ⓞ ⓞ ⓞ ⓞ ⓞ ∩ ∩ ∩ ∩ ∩ | |.

✓ Check 2 Point   Write 2563 as an Egyptian numeral.

Understand and use the Roman system.

The Roman Numeration System

The Roman numeration system was developed between 500 B.C. and 100 A.D. It evolved as a result of tax collecting and commerce in the vast Roman Empire. The Roman numerals shown in Table 4.7 were used throughout Europe until the eigh-teenth century. They are still commonly used in outlining, on clocks, for certain copyright dates, and in numbering some pages in books. Roman numerals are selected letters from the Roman alphabet.

TABLE 4.7 ROMAN NUMERALS

Roman numeral	I	V	X	L	C	D	M
Hindu-Arabic numeral	1	5	10	50	100	500	1000

If the symbols in Table 4.7 decrease in value from left to right, then add their values to obtain the value of the Roman numeral as a whole. For example, CX = 100 + 10 = 110. On the other hand, if symbols increase in value from left to right, then subtract the value of the symbol on the left from the symbol on the right. For example, IV means 5 − 1 = 4 and IX means 10 − 1 = 9.

Only the Roman numerals representing 1, 10, 100, 1000, . . . , can be subtracted. Furthermore, they can be subtracted only from their next two greater Roman numerals.

Roman numeral (values that can be subtracted are shown in red)	I	V	X	L	C	D	M
Hindu-Arabic numeral	1	5	10	50	100	500	1000

I can be subtracted only from V and X.   X can be subtracted only from L and C.   C can be subtracted only from D and M.

EXAMPLE 3   USING ROMAN NUMERALS

Write CLXVII as a Hindu-Arabic numeral.

SOLUTION   Because the numerals decrease in value from left to right, we add their values to find the value of the Roman numeral as a whole.

$$CLXVII = 100 + 50 + 10 + 5 + 1 + 1 = 167$$

 Write MCCCLXI as a Hindu-Arabic numeral.

EXAMPLE 4   USING ROMAN NUMERALS

Write MCMXCVI as a Hindu-Arabic numeral.

SOLUTION

$$
\begin{array}{ccccc}
M & CM & XC & V & I \\
\downarrow & \downarrow & \downarrow & \downarrow & \downarrow
\end{array}
$$
$$= 1000 + (1000 - 100) + (100 - 10) + 5 + 1$$
$$= 1000 + 900 + 90 + 5 + 1 = 1996$$

 **Check 4 Point** Write MCDXLVII as a Hindu-Arabic numeral.

Because Roman numerals involve subtraction as well as addition, it takes far less space to represent most numbers than in the Egyptian system. It is never necessary to repeat any symbol more than three consecutive times. For example, we write 46 as a Roman numeral using

XLVI rather than XXXXVI.

XL = 50 − 10 = 40

EXAMPLE 5 USING ROMAN NUMERALS

Write 249 as a Roman numeral.

SOLUTION

$$
\begin{aligned}
249 &= \quad\ 200 \quad+\quad 40 \quad+\quad 9 \\
&= 100 + 100 + (50 - 10) + (10 - 1) \\
&\qquad\ \downarrow\qquad\ \downarrow\qquad\quad\ \downarrow\qquad\quad\ \downarrow \\
&= \ \ \text{C}\qquad \text{C}\qquad\ \ \text{XL}\qquad\quad \text{IX}
\end{aligned}
$$

Thus, 249 = CCXLIX.

 **Check 5 Point** Write 399 as a Roman numeral.

The Roman numeration system uses bars above numerals or groups of numerals to show that the numbers are to be multiplied by 1000. For example,

$$\overline{\text{L}} = 50 \times 1000 = 50{,}000 \quad\text{and}\quad \overline{\text{CM}} = 900 \times 1000 = 900{,}000.$$

Placing bars over Roman numerals reduces the number of symbols needed to represent large numbers.

The Traditional Chinese Numeration System

The numerals used in the traditional Chinese numeration system are given in Table 4.8. At least two things are missing—a symbol for zero and a surprised lottery winner!

TABLE 4.8 TRADITIONAL CHINESE NUMERALS

Traditional Chinese numerals	一	二	三	四	五	六	七	八	九	十	百	千
Hindu-Arabic numerals	1	2	3	4	5	6	7	8	9	10	100	1000

So, how are numbers represented with this set of symbols? Chinese numerals are written vertically. Using our digits, the number 3264 is expressed as shown in the margin.

The next step is to replace each of these seven symbols with a traditional Chinese numeral from Table 4.8. Our next example illustrates this procedure.

3 Understand and use the traditional Chinese system.

$$
\left.\begin{array}{r}
3 \\
1000 \\
2 \\
100 \\
6 \\
10 \\
4
\end{array}\right\}
$$

Representing 3264 vertically is the first step in expressing it as a Chinese numeral

Writing 3264 as a Chinese numeral

EXAMPLE 6 | USING THE TRADITIONAL CHINESE NUMERATION SYSTEM

Write 3264 as a Chinese numeral.

SOLUTION First, break down the Hindu-Arabic numeral into quantities that match the Chinese numerals. Represent each quantity vertically. Then, use Table 4.8 to find the Chinese symbol that matches each quantity. This procedure, with the resulting Chinese numeral, is shown in the margin.

The Chinese system does not need a numeral for zero because it is not positional. For example, we write 8006, using zeros as placeholders, to indicate that two powers of ten, namely 10^2, or 100, and 10^1, or 10, are not needed. The Chinese leave this out, writing

Check Point 6 Write 2693 as a Chinese numeral.

④ Understand and use the Ionic Greek system.

The Ionic Greek Numeration System

The ancient Greeks, masters of art, architecture, theater, literature, philosophy, geometry, and logic, were not masters when it came to representing numbers. The Ionic Greek numeration system, which can be traced back as far as 450 B.C., used letters of their alphabet for numerals. Table 4.9 shows the many symbols (too many symbols!) used to represent numbers.

TABLE 4.9 IONIC GREEK NUMERALS

1	α	alpha	10	ι	iota	100	ρ	rho
2	β	beta	20	κ	kappa	200	σ	sigma
3	γ	gamma	30	λ	lambda	300	τ	tau
4	δ	delta	40	μ	mu	400	υ	upsilon
5	ϵ	epsilon	50	ν	nu	500	ϕ	phi
6	ι	vau	60	ξ	xi	600	χ	chi
7	ζ	zeta	70	o	omicron	700	ψ	psi
8	η	eta	80	π	pi	800	ω	omega
9	θ	theta	90	Q	koph	900	π	sampi

To represent a number from 1 to 999, the appropriate symbols are written next to one another. For example, the number $21 = 20 + 1$. When 21 is expressed as a Greek numeral, the plus sign is left out:

$$21 = \kappa\alpha.$$

Similarly, the number 823 written as a Greek numeral is $\omega\kappa\gamma$.

EXAMPLE 7 USING THE GREEK NUMERATION SYSTEM

Write $\psi\lambda\delta$ as a Hindu-Arabic numeral.

SOLUTION $\psi = 700$, $\lambda = 30$, and $\delta = 4$. Adding these numbers gives 734.

Check **7** Point   Write $\omega\pi\epsilon$ as a Hindu-Arabic numeral.

One of the many unsuccessful features of the Greek numeration system is that new symbols have to be added to represent higher numbers. It is like an alphabet that gets bigger each time a new word is used and has to be written.

Exercise Set 4.4

Practice Exercises

Use Table 4.6 on page 182 to solve Exercises 1–12.
In Exercises 1–6, write each Egyptian numeral as a Hindu-Arabic numeral.

1. ∂ ∂ ∂ ∩ ∩ | |

2. 𓀠 𓀠 𓀠 𓏢 ∂ ∂ ∩ | | | |

3. ∂ ∂ ∂ ∂ ∂ ∂ ∂ ∩ ∩ | | |

4. ∂ ∂ ∂ ∂ ∩ | | | |

5. ∂ ∩ ∩ ∩ | |

6. ∂ 𓏢 𓏺 ∂ ∂ ∂ | |

In Exercises 7–12, write each Hindu-Arabic numeral as an Egyptian numeral.

7. 423   **8.** 825   **9.** 1846

10. 1425   **11.** 23,547   **12.** 2,346,031

Use Table 4.7 on page 184 to solve Exercises 13–36.
In Exercises 13–28, write each Roman numeral as a Hindu-Arabic numeral.

13. XI   **14.** CL   **15.** XVI

16. LVII   **17.** XL   **18.** CM

19. LIX   **20.** XLIV   **21.** CXLVI

22. CLXI   **23.** MDCXXI   **24.** MMCDXLV

25. MMDCLXXVII   **26.** MDCXXVI   **27.** $\overline{\text{IX}}$CDLVI

28. $\overline{\text{V}}$MCCXI

In Exercises 29–36, write each Hindu-Arabic numeral as a Roman numeral.

29. 43   **30.** 96   **31.** 129   **32.** 469

33. 1896   **34.** 4578   **35.** 6892   **36.** 5847

Use Table 4.8 on page 185 to solve Exercises 37–48.
In Exercises 37–42, write each traditional Chinese numeral as a Hindu-Arabic numeral.

37. 八十八   **38.** 七百五   **39.** 五百二十七   **40.** 三千八十一   **41.** 二千七百七十六   **42.** 八千二百三十六

In Exercises 43–48, write each Hindu-Arabic numeral as a traditional Chinese numeral.

43. 43   **44.** 269   **45.** 583

46. 2965   **47.** 4870   **48.** 7605

Use Table 4.9 on page 186 to solve Exercises 49–56.
In Exercises 49–52, write each Ionic Greek numeral as a Hindu-Arabic numeral.

49. $\iota\beta$ **50.** $\phi\epsilon$ **51.** $\sigma\lambda\delta$ **52.** $\psi o\theta$

In Exercises 53–56, write each Hindu-Arabic numeral as an Ionic Greek numeral.

53. 43   **54.** 257   **55.** 483   **56.** 895

Application Exercises

57. Look at the back of a U.S. one dollar bill. What date is written in Roman numerals along the base of the pyramid with an eye? What is this date's significance?

58. A construction crew demolishing a very old building was surprised to find the numeral MCMLXXXIX inscribed on the cornerstone. Explain why they were surprised.

Writing in Mathematics

59. Describe how a number is represented in the Egyptian numeration system.

60. If you are interpreting a Roman numeral, when do you add values and when do you subtract them? Give an example to illustrate each case.

61. Describe how a number is represented in the traditional Chinese numeration system.

62. Describe one disadvantage of the Ionic Greek numeration system.

63. If you could use only one system of numeration described in this section, which would you prefer? Discuss the reasons for your choice.

Critical Thinking Exercises

64. Arrange these three numerals from smallest to largest.

   CCCCXLIX

65. Use Egyptian numerals to write the numeral that precedes and the numeral that follows

66. After reading this section, a student had a numeration nightmare about selling flowers in a time-warped international market. She started out with 200 flowers, selling XLVI of them to a Roman,

to an Egyptian,

$$\overset{=}{+}$$

to a traditional Chinese family, and the remainder to a Greek. How many flowers were sold to the Greek? Express the answer in the Ionic Greek numeration system.

Group Exercises

Take a moment to read the introduction to the group exercises on page 175. Exercises 67–71 list some additional topics for individual or group research projects.

67. A Time Line Showing Significant Developments in Numeration Systems

68. Animals and Number Sense

69. The Hebrew Numeration System (or any system not discussed in this chapter)

70. The Rhind Papyrus and What We Learned from It

71. Computation in an Early Numeration System

Chapter Summary, Review, and Test

SUMMARY

DEFINITIONS AND CONCEPTS   EXAMPLES

4.1 Our Hindu-Arabic System and Early Positional Systems

a. In a positional-value, or place-value, numeration system, the value of each symbol, called a digit, varies according to the position it occupies in the number.

b. The Hindu-Arabic numeration system is a base ten system with the digits 0, 1, 2, 3, 4, 5, 6, 7, 8, and 9.   Ex. 3, p. 165;
The place values in the system are   Ex. 4, p. 165

$$\ldots, 10^5, 10^4, 10^3, 10^2, 10^1, 1.$$

c. The Babylonian numeration system is a base 60 system, with place values given by   Ex. 5, p. 167

$$\ldots, \quad 60^3, \quad 60^2, \quad 60^1, \quad 1.$$
$$\text{or} \quad \text{or} \quad \text{or}$$
$$216{,}000 \quad 3600 \quad 60$$

Babylonian numerals are given in Table 4.1 on page 166.

d. The Mayan numeration system has place values given by   Ex. 6, p. 168

$$\ldots, \quad 18 \times 20^3, \quad 18 \times 20^2, \quad 18 \times 20, \quad 20, \quad 1.$$
$$\text{or} \quad \text{or} \quad \text{or}$$
$$144{,}000 \quad 7200 \quad 360$$

Mayan numerals are given in Table 4.2 on page 167.

4.2 Number Bases in Positional Systems

a. The base of a positional numeration system refers to the number of individual digit symbols used in the system as well as to the powers of the numbers used in place values. In base b, there are b digit symbols (from 0 through $b - 1$ inclusive) with place values given by

$$\ldots, b^4, b^3, b^2, b^1, 1.$$

b. To change a numeral in a base other than ten to a base ten numeral,

    1. Multiply each digit in the numeral by its respective place value.

    2. Find the sum of the products in step 1.

c. To change a base ten numeral to a base b numeral, use mental conversions or repeated divisions by powers of b to find how many groups of each place value are contained in the base ten numeral.

4.3 Computation in Positional Systems

a. Computations in bases other than ten are performed using the same procedures as in ordinary base ten arithmetic. When a computation is equal to or exceeds the given base, use mental conversions to convert from the base ten numeral to a numeral in the desired base.

b. To divide in bases other than ten, it is convenient to use a multiplication table for products in the required base.

4.4 Looking Back at Early Numeration Systems

a. A successful numeration system expresses numbers with relatively few symbols and makes computation with these numbers fairly easy.

b. By the standard in (a), the Egyptian system (Table 4.6 on page 182), the Roman system (Table 4.7 on page 184), the Chinese system (Table 4.8 on page 185), and the Greek system (Table 4.9 on page 186) are all unsuccessful. Unlike our Hindu-Arabic, system, these systems are not positional and contain no symbol for zero.

REVIEW EXERCISES

4.1

In Exercises 1–2, evaluate the expression.

1. 11^2 **2.** 7^3

In Exercises 3–5, write each Hindu-Arabic numeral in expanded form.

3. 472   **4.** 8076   **5.** 70,329

In Exercises 6–7, express each expanded form as a Hindu-Arabic numeral.

6. $(7 \times 10^5) + (0 \times 10^4) + (6 \times 10^3) + (9 \times 10^2)$
$$+ (5 \times 10^1) + (3 \times 1)$$

7. $(7 \times 10^8) + (4 \times 10^7) + (3 \times 10^2) + (6 \times 1)$

Use Table 4.1 on page 166 to write each Babylonian numeral in Exercises 8–9 as a Hindu-Arabic numeral.

8. ∨ ∨   ∨ ∨ ∨   **9.** ∨ ∨   ∨∨   ∨∨∨

Use Table 4.2 on page 167 to write each Mayan numeral in Exercises 10–11 as a Hindu-Arabic numeral.

10. ⋯   **11.** ⋯

12. Describe how a positional system is used to represent a number.

4.2

In Exercises 13–18, convert the numeral to a numeral in base ten.

13. 34_{five} **14.** 110_{two} **15.** 643_{seven}

16. 1084_{nine} **17.** $\text{FD3}_{\text{sixteen}}$ **18.** 202202_{three}

In Exercises 19–24, convert each base ten numeral to a numeral in the given base.

19. 89 to base five   **20.** 21 to base two

21. 473 to base three   **22.** 7093 to base seven

23. 9348 to base six   **24.** 554 to base twelve

4.3

In Exercises 25–28, add in the indicated base.

25. 46_{seven} **26.** 574_{eight} **27.** 11011_{two} **28.** $43C_{\text{sixteen}}$
$\quad\ \ +53_{\text{seven}}$ $+605_{\text{eight}}$ $+10101_{\text{two}}$ $+694_{\text{sixteen}}$

In Exercises 29–32, subtract in the indicated base.

29. 34_{six} **30.** 624_{seven} **31.** 1001_{two} **32.** 4121_{five}
$\quad\ \ -25_{\text{six}}$ -246_{seven} $-\ 110_{\text{two}}$ -1312_{five}

In Exercises 33–35, multiply in the indicated base.

33. 32_{four} **34.** 43_{seven} **35.** 123_{five}
$\quad\ \ \times\ 3_{\text{four}}$ $\times\ 6_{\text{seven}}$ $\times\ 4_{\text{five}}$

In Exercises 36–37, divide in the indicated base. Use the multiplication tables on page 181.

36. $2_{four}\overline{)332_{four}}$ **37.** $4_{five}\overline{)103_{five}}$

4.4

Use Table 4.6 on page 182 to solve Exercises 38–41.
In Exercises 38–39, write each Egyptian numeral as a Hindu-Arabic numeral.

38.

39.

In Exercises 40–41, write each Hindu-Arabic numeral as an Egyptian numeral.

40. 2486   **41.** 34,573

In Exercises 42–43, assume a system that represents numbers exactly like the Egyptian system, but with different symbols. In particular, A = 1, B = 10, C = 100, and D = 1000.

42. Write DDCCCBAAAA as a Hindu-Arabic numeral.

43. Write 5492 as a numeral in terms of A, B, C, and D.

44. Describe how the Egyptian system or the system in Exercises 42–43 is used to represent a number. Discuss one disadvantage of such a system when compared to our Hindu-Arabic system.

Use Table 4.7 on page 184 to solve Exercises 45–49.
In Exercises 45–47, write each Roman numeral as a Hindu-Arabic numeral.

45. CLXIII   **46.** MXXXIV   **47.** MCMXC

In Exercises 48–49, write each Hindu-Arabic numeral as a Roman numeral.

48. 49   **49.** 2965

50. Explain when to subtract the value of symbols when interpreting a Roman numeral. Give an example.

Use Table 4.8 on page 185 to solve Exercises 51–54.
In Exercises 51–52, write each traditional Chinese numeral as a Hindu-Arabic numeral.

51. 五
百
五
十
四

52. 八
千
二
百
五
十
三

In Exercises 53–54, write each Hindu-Arabic numeral as a traditional Chinese numeral.

53. 274   **54.** 3587

In Exercises 55–58, assume a system that represents numbers exactly like the traditional Chinese system, but with different symbols. The symbols are shown as follows:

Numerals in the System	A	B	C	D	E	F	G	H	I	X	Y	Z
Hindu-Arabic Numerals	1	2	3	4	5	6	7	8	9	10	100	1000

Express each numeral in Exercises 55–56 as a Hindu-Arabic numeral.

55. C   **56.** D
Y   Z
F   E
X   Y
E   B
               X

Express each Hindu-Arabic numeral in Exercises 57–58 as a numeral in the system used for Exercises 55–56.

57. 793   **58.** 6854

59. Describe how the Chinese system or the system in Exercises 55–58 is used to represent a number. Discuss one disadvantage of such a system when compared to our Hindu-Arabic system.

Use Table 4.9 on page 186 to solve Exercises 60–63.
In Exercises 60–61, write each Ionic Greek numeral as a Hindu-Arabic numeral.

60. $\chi\nu\gamma$ **61.** $\chi o\eta$

In Exercises 62–63, write each Hindu-Arabic numeral as an Ionic Greek numeral.

62. 453   **63.** 902

In Exercises 64–68, assume a system that represents numbers exactly like the Greek Ionic system, but with different symbols. The symbols are shown as follows:

Decimal	1	2	3	4	5	6	7	8	9
Ones	A	B	C	D	E	F	G	H	I
Tens	J	K	L	M	N	O	P	Q	R
Hundreds	S	T	U	V	W	X	Y	Z	a
Thousands	b	c	d	e	f	g	h	i	j
Ten thousands	k	l	m	n	o	p	q	r	s

In Exercises 64–66, express each numeral as a Hindu-Arabic numeral.

64. UNG   **65.** mhZRD   **66.** rXJH

In Exercises 67–68, express each Hindu-Arabic numeral as a numeral in the system used for Exercises 64–66.

67. 597   **68.** 25,483

69. Discuss one disadvantage of the Greek Ionic system or the system described in Exercises 64–68 when compared to our Hindu-Arabic system.

Group Exercise

70. We opened the chapter with a virtual reality ski simulator in the not-too-distant future. At the start of the twenty-first century, it is difficult to imagine the kind of future that computer technology will bring us. For this activity, group members should research cutting-edge applications of computing devices and calculators that will transform our lives in the twenty-first century. You can visit a computer store to discover what is "hot" in the industry, consult the World Wide Web, or talk with creative people in the arts who are using computers in innovative ways. Include examples in your class presentation that will truly amaze your fellow students, making this a presentation they will never forget.

CHAPTER 4 TEST

1. Evaluate 9^3.

2. Write 567 in expanded form.

3. Write 63,028 in expanded form.

4. Express as a Hindu-Arabic numeral:

   $$(7 \times 10^3) + (4 \times 10^2) + (9 \times 10^1) + (3 \times 1).$$

5. Express as a Hindu-Arabic numeral:

   $$(4 \times 10^5) + (2 \times 10^2) + (6 \times 1).$$

6. What is the difference between a number and a numeral?

7. Explain why a symbol for zero is needed in a positional system.

8. Place values in the Babylonian system are

   $$\ldots, 60^3, 60^2, 60^1, 1.$$

   Use the numerals shown to write the following Babylonian numeral as a Hindu-Arabic numeral:

   $$<< \quad < \vee\vee \quad < \vee.$$

Babylonian	$\vee$	$<$
Hindu-Arabic	1	10

9. Place values in the Mayan system are

   $$\ldots, 18 \times 20^3, 18 \times 20^2, 18 \times 20, 20, 1.$$

   Use the numerals shown to write the following Mayan numeral as a Hindu-Arabic numeral:

   .

Mayan	⬭	•	••	•••	••••	—	$\overset{\bullet}{—}$
Hindu-Arabic	0	1	2	3	4	5	6

In Exercises 10–12, convert the numeral to a numeral in base ten.

10. 423_{five}

11. 267_{nine}

12. 110101_{two}

In Exercises 13–15, convert each base ten numeral to a numeral in the given base.

13. 77 to base three

14. 56 to base two

15. 1844 to base five

In Exercises 16–18, perform the indicated operation.

16. 234_{five}
    $+ 423_{\text{five}}$

17. 562_{seven}
    $- 145_{\text{seven}}$

18. 54_{six}
    $\times 3_{\text{six}}$

19. Use the multiplication table shown to perform this division: $3_{\text{five}} \overline{)1213_{\text{five}}}$.

A MULTIPLICATION TABLE FOR BASE FIVE

×	0	1	2	3	4
0	0	0	0	0	0
1	0	1	2	3	4
2	0	2	4	11	13
3	0	3	11	14	22
4	0	4	13	22	31

Use the symbols in the tables shown below to solve Exercises 20–23.

Hindu-Arabic Numeral	Egyptian Numeral
1	|
10	∩
100	◉
1000	⚱
10,000	⌇
100,000	↝
1,000,000	𓀠

Hindu-Arabic Numeral	Roman Numeral
1	I
5	V
10	X
50	L
100	C
500	D
1000	M

20. Write the following numeral as a Hindu-Arabic numeral:

    .

21. Write 32,634 as an Egyptian numeral.

22. Write the Roman numeral MCMXCIV as a Hindu-Arabic numeral.

23. Express 459 as a Roman numeral.

24. Describe one difference between how a number is represented in the Egyptian system and the Roman system.

Number Theory and the Real Number System

Literacy with numbers, called **numeracy**, is a prerequisite for functioning in a meaningful way personally, professionally, and as a citizen. In this chapter, our focus is on understanding numbers, their properties, and their applications. After completing the chapter, you will be in a position to put numbers like 6.8 trillion in perspective, giving you a better understanding of the important policy issues that will shape the twenty-first century.

While listening to the radio, you hear politicians discussing the problem of the country's 6.8 trillion dollar deficit. They make it seem like the national debt is a real problem, but later you realize that you don't really know what that number means. How can you look at this deficit in the proper perspective? If the national debt were evenly divided among all citizens of the country, how much would each citizen have to pay? Does the deficit seem like such a significant problem now?

This problem appears as Example 9 in Section 5.6.

SECTION 5.1   NUMBER THEORY: PRIME AND COMPOSITE NUMBERS

OBJECTIVES

1. Determine divisibility.

2. Write the prime factorization of a composite number.

3. Find the greatest common divisor of two numbers.

4. Solve problems using the greatest common divisor.

5. Find the least common multiple of two numbers.

6. Solve problems using the least common multiple.

Number Theory and Divisibility

You are organizing an intramural league at your school. You need to divide 40 men and 24 women into all-male and all-female teams so that each team has the same number of people. The men's teams should have the same number of players as the women's teams. What is the largest number of people that can be placed on a team?

This problem can be solved using a branch of mathematics called **number theory**. Number theory is primarily concerned with the properties of numbers used for counting, namely 1, 2, 3, 4, 5, and so on. The set of counting numbers is also called the set of **natural numbers**. We represent this set by the letter **N**.

THE SET OF NATURAL NUMBERS

$$\mathbf{N} = \{1, 2, 3, 4, 5, 6, 7, 8, 9, 10, 11, \dots\}$$

We can solve the intramural league problem. However, to do so we must understand the concept of divisibility. For example, there are a number of different ways to divide the 24 women into teams, including

1 team with all 24 women:	$1 \times 24 = 24$
2 teams with 12 women per team:	$2 \times 12 = 24$
3 teams with 8 women per team:	$3 \times 8 = 24$
4 teams with 6 women per team:	$4 \times 6 = 24$
6 teams with 4 women per team:	$6 \times 4 = 24$
8 teams with 3 women per team:	$8 \times 3 = 24$
12 teams with 2 women per team:	$12 \times 2 = 24$
24 teams with 1 woman per team:	$24 \times 1 = 24.$

The natural numbers that are multiplied together resulting in a product of 24 are called *factors* of 24. Any natural number can be expressed as a product of two or more natural numbers. The natural numbers that are multiplied are called the **factors** of the product. Notice that a natural number may have many factors.

$$2 \times 12 = 24 \qquad 3 \times 8 = 24 \qquad 6 \times 4 = 24$$

Factors of 24   Factors of 24   Factors of 24

The numbers 1, 2, 3, 4, 6, 8, 12, and 24 are all factors of 24. Each of these numbers divides 24 without a remainder.

In general, let *a* and *b* represent natural numbers. We say that *a* is **divisible** by *b* if the operation of dividing *a* by *b* leaves a remainder of 0.

1 Determine divisibility.

A natural number is divisible by all of its factors. Thus, 24 is divisible by 1, 2, 3, 4, 6, 8, 12, and 24. Using the factor 8, we can express this divisibility in a number of ways:

24 is **divisible** by 8.

8 is a **divisor** of 24.

8 **divides** 24.

Mathematicians use a special notation to indicate divisibility.

DIVISIBILITY

If a and b are natural numbers, a is **divisible** by b if the operation of dividing a by b leaves a remainder of 0. This is the same as saying that b is a **divisor** of a, or b **divides** a. All three statements are symbolized by writing

$$b|a.$$

Using this new notation, we can write

$$12|24.$$

Twelve divides 24 because 24 divided by 12 leaves a remainder of 0. By contrast, 13 does not divide 24 because 24 divided by 13 does not leave a remainder of 0. The notation

$$13\!\!\not|\,24$$

means that 13 does not divide 24.

Table 5.1 shows some common rules for divisibility. Divisibility rules for 7 and 11 are difficult to remember and are not included in the table.

TABLE 5.1 RULES OF DIVISIBILITY

Divisible By	Test	Example
2	The last digit is 0, 2, 4, 6, or 8.	5,892,796 is divisible by 2 because the last digit is 6.
3	The sum of the digits is divisible by 3.	52,341 is divisible by 3 because the sum of the digits is $5 + 2 + 3 + 4 + 1 = 15$, and 15 is divisible by 3.
4	The last two digits form a number divisible by 4.	3,947,136 is divisible by 4 because 36 is divisible by 4.
5	The number ends in 0 or 5.	28,160 and 72,805 end in 0 and 5, respectively. Both are divisible by 5.
6	The number is divisible by both 2 and 3. (In other words, the number is even and the sum of its digits is divisible by 3.)	954 is divisible by 2 because it ends in 4. 954 is also divisible by 3 because the digit sum is 18, which is divisible by 3. Because 954 is divisible by both 2 and 3, it is divisible by 6.
8	The last three digits form a number that is divisible by 8.	593,777,832 is divisible by 8 because 832 is divisible by 8.
9	The sum of the digits is divisible by 9.	5346 is divisible by 9 because the sum of the digits, 18, is divisible by 9.
10	The last digit is 0.	998,746,250 is divisible by 10 because the number ends in 0.
12	The number is divisible by both 3 and 4. (In other words, the sum of the digits is divisible by 3 and the last two digits form a number divisible by 4.)	614,608,176 is divisible by 3 because the digit sum is 39, which is divisible by 3. It is also divisible by 4 because the last two digits form 76, which is divisible by 4. Because 614,608,176 is divisible by both 3 and 4, it is divisible by 12.

TECHNOLOGY

Calculators and Divisibility

You can use a calculator to verify each result in Example 1. Consider part (a):

$$4 \mid 3{,}754{,}086.$$

Divide 3,754,086 by 4 using the following keystrokes:

3754086 ÷ 4.

Press = or ENTER . The number displayed is 938521.5. This is not a natural number. The 0.5 shows that the division leaves a nonzero remainder. Thus, 4 does not divide 3,754,086. The given statement is false.

Now consider part (b):

$$9 \nmid 4{,}119{,}706{,}413.$$

Use your calculator to divide the number on the right by 9:

4119706413 ÷ 9.

Press = or ENTER . The display is 457745157. This is a natural number. The remainder of the division is 0, so 9 does divide 4,119,706,413. The given statement is false.

EXAMPLE 1   USING THE RULES OF DIVISIBILITY

Which one of the following statements is true?

a. $4 \mid 3{,}754{,}086$ **b.** $9 \nmid 4{,}119{,}706{,}413$ **c.** $8 \mid 677{,}840$

SOLUTION

a. $4 \mid 3{,}754{,}086$ states that 4 divides 3,754,086. Table 5.1 indicates that for 4 to divide a number, the last two digits must form a number that is divisible by 4. Because 86 is not divisible by 4, the given statement is false.

b. $9 \nmid 4{,}119{,}706{,}413$ states that 9 does *not* divide 4,119,706,413. Based on Table 5.1, if the sum of the digits is divisible by 9, then 9 does indeed divide this number. The sum of the digits is $4 + 1 + 1 + 9 + 7 + 0 + 6 + 4 + 1 + 3 = 36$, which is divisible by 9. Because 4,119,706,413 is divisible by 9, the given statement is false.

c. $8 \mid 667{,}840$ states that 8 divides 677,840. Table 5.1 indicates that for 8 to divide a number, the last three digits must form a number that is divisible by 8. Because 840 is divisible by 8, then 8 divides 677,840, and the given statement is true.

The statement given in part (c) is the only true statement.

 **Check 1 Point**   Which one of the following statements is true?

a. $8 \mid 48{,}324$ **b.** $6 \mid 48{,}324$ **c.** $4 \nmid 48{,}324$

Prime Factorization

By developing some other ideas of number theory, we will be able to solve the intramural league problem. We begin with the definition of a prime number.

> **PRIME NUMBERS**
>
> A **prime number** is a natural number greater than 1 that has only itself and 1 as factors.

Using this definition, we see that the number 7 is a prime number because it has only 1 and 7 as factors. Said in another way, 7 is prime because it is divisible by only 1 and 7. The first ten prime numbers are 2, 3, 5, 7, 11, 13, 17, 19, 23, and 29. Each number in this list has exactly two divisors, itself and 1. By contrast, 9 is not a prime number; in addition to being divisible by 1 and 9, it is also divisible by 3. The number 9 is an example of a *composite number*.

> **COMPOSITE NUMBERS**
>
> A **composite number** is a natural number greater than 1 that is divisible by a number other than itself and 1.

Using this definition, the first ten composite numbers are 4, 6, 8, 9, 10, 12, 14, 15, 16, and 18. Each number in this list has at least three divisors.

Every composite number can be expressed as the product of prime numbers. For example, the composite number 45 can be expressed as

$$45 = 3 \times 3 \times 5.$$

Note that 3 and 5 are prime numbers. Expressing a composite number as the product of prime numbers is called **prime factorization**. The prime factorization of 45 is

STUDY TIP

The number 2 is the only even prime number. Every other even number has at least three factors: 1, 2, and the number itself.

2   Write the prime factorization of a composite number.

$3 \times 3 \times 5$. The order in which we write these factors does not matter. This means that

$$45 = 3 \times 3 \times 5$$
$$\text{or } 45 = 5 \times 3 \times 3$$
$$\text{or } 45 = 3 \times 5 \times 3.$$

In Chapter 1, we defined a **theorem** as a statement that can be proved using deductive reasoning. The ancient Greeks proved that if the order of the factors is disregarded, there is only one prime factorization possible for any given composite number. This statement is called the **Fundamental Theorem of Arithmetic**.

THE FUNDAMENTAL THEOREM OF ARITHMETIC

Every composite number can be expressed as a product of prime numbers in one and only one way (if the order of the factors is disregarded).

One method used to find the prime factorization of a composite number is called a **factor tree**. To use this method, begin by selecting any two numbers whose product is the number to be factored. If one or both of the factors are not prime numbers, continue to factor each composite number. Stop when all numbers are prime.

EXAMPLE 2 | PRIME FACTORIZATION USING A FACTOR TREE

Find the prime factorization of 700.

SOLUTION   Start with any two numbers whose product is 700, such as 7 and 100. This forms the first branch of the tree. Continue factoring the composite number or numbers that result (in this case 100), branching until the end of each branch contains a prime number.

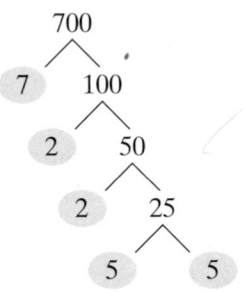

The prime factors are circled. Thus, the prime factorization of 700 is

$$700 = 7 \times 2 \times 2 \times 5 \times 5.$$

We can use exponents to show the repeated prime factors:

$$700 = 7 \times 2^2 \times 5^2.$$

Using a dot to indicate multiplication and arranging the factors from least to greatest, we can write

$$700 = 2^2 \cdot 5^2 \cdot 7.$$

Check
2
Point

Find the prime factorization of 120.

---

3 | Find the greatest common divisor of two numbers.

Greatest Common Divisor

The greatest common divisor of two or more natural numbers is the largest number that is a divisor (or factor) of all the numbers. For example, 8 is the greatest common divisor of 32 and 40 because it is the largest natural number that divides both 32 and 40. Some pairs of numbers have 1 as their greatest common divisor. Such number pairs are said to be **relatively prime**. For example, the greatest common divisor of 5 and 26 is 1. Thus, 5 and 26 are relatively prime.

The greatest common divisor can be found using prime factorizations.

FINDING THE GREATEST COMMON DIVISOR OF TWO OR MORE NUMBERS USING PRIME FACTORIZATIONS

To find the greatest common divisor of two or more numbers,

1. Write the prime factorization of each number.

2. Select each prime factor with the smallest exponent that is common to each of the prime factorizations.

3. Form the product of the numbers from step 2. The greatest common divisor is the product of these factors.

smaller ‡

EXAMPLE 3 | FINDING THE GREATEST COMMON DIVISOR

Find the greatest common divisor of 216 and 234.

SOLUTION

STEP 1. Write the prime factorization of each number. Begin by writing the prime factorizations of 216 and 234.

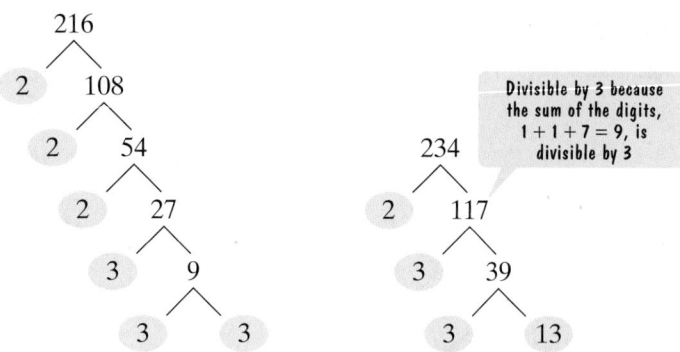

> Divisible by 3 because the sum of the digits, $1 + 1 + 7 = 9$, is divisible by 3

The factor tree at the left indicates that
$$216 = 2^3 \times 3^3.$$
The factor tree at the right indicates that
$$234 = 2 \times 3^2 \times 13.$$

STEP 2. Select each prime factor with the smallest exponent that is common to each of the prime factorizations. Look at the factorizations of 216 and 234 from step 1. Can you see that 2 is a prime number common to the factorizations of 216 and 234? Likewise, 3 is also a prime number common to the two factorizations. By contrast, 13 is a prime number that is not common to both factorizations.

$$216 = 2^3 \times 3^3$$
$$234 = 2 \times 3^2 \times 13$$

> **2 is a prime number common to both factorizations.**

> **3 is a prime number common to both factorizations.**

BLITZER BONUS

Simple Questions with No Answers

In number theory, a good problem is one that can be stated quite simply, but whose solution turns out to be particularly difficult, if not impossible.

In 1742, the mathematician Christian Goldbach (1690–1764) wrote a letter to Leonhard Euler (1707–1783) in which he proposed, without a proof, that every even number greater than 2 is the sum of two primes. For example,

Even number	Sum of two primes
$4 =$	$2 + 2$
$6 =$	$3 + 3$
$8 =$	$3 + 5$
$10 =$	$5 + 5$
$12 =$	$7 + 5$

and so on.

Two and a half centuries later, it is still not known if this conjecture is true or false. Inductively, it appears to be true; computer searches have written even numbers as large as one billion as the sum of two primes. Deductively, no mathematician has been able to prove that the conjecture is true.

$216 = 2^3 \times 3^3$

Smallest exponent on 2 is 1.   Smallest exponent on 3 is 2.

$234 = 2^1 \times 3^2 \times 13$

Now we need to use $216 = 2^3 \times 3^3$ and $234 = 2 \times 3^2 \times 13$ to determine which exponent is appropriate for 2 and which exponent is appropriate for 3. The appropriate exponent is the smallest exponent associated with the prime number in the factorizations. The exponents associated with 2 in the factorizations are 1 and 3, so we select 1. Therefore, one factor for the greatest common divisor is 2^1, or 2. The exponents associated with 3 in the factorizations are 2 and 3, so we select 2. Therefore, another factor for the greatest common divisor is 3^2.

STEP 3. Form the product of the numbers from step 2. The greatest common divisor is the product of these factors.

$$\text{Greatest common divisor} = 2 \times 3^2 = 2 \times 9 = 18$$

The greatest common divisor of 216 and 234 is 18.

Check Point 3   Find the greatest common divisor of 225 and 825.

4   Solve problems using the greatest common divisor.

EXAMPLE 4   SOLVING A PROBLEM USING THE GREATEST COMMON DIVISOR

For an intramural league, you need to divide 40 men and 24 women into all-male and all-female teams so that each team has the same number of people. What is the largest number of people that can be placed on a team?

SOLUTION   Because 40 men are to be divided into teams, the number of men on each team must be a divisor of 40. Because 24 women are to be divided into teams, the number of women placed on a team must be a divisor of 24. Although the teams are all-male and all-female, the same number of people must be placed on each team. The largest number of people that can be placed on a team is the largest number that will divide into 40 and 24 without a remainder. This is the greatest common divisor of 40 and 24.

To find the greatest common divisor of 40 and 24, begin with their prime factorizations.

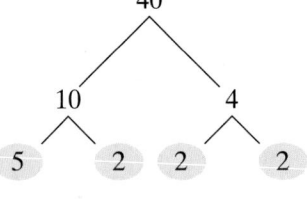

The factor trees indicate that

$$40 = 2^3 \times 5 \qquad \text{and} \qquad 24 = 2^3 \times 3.$$

We see that 2 is a prime number common to both factorizations. The exponents associated with 2 in the factorizations are 3 and 3, so we select 3.

$$\text{Greatest common divisor} = 2^3 = 2 \times 2 \times 2 = 8$$

The largest number of people that can be placed on a team is 8. Thus, the 40 men can form five teams with 8 men per team. The 24 women can form three teams with 8 women per team.

Check Point 4   A choral director needs to divide 192 men and 288 women into all-male and all-female singing groups so that each group has the same number of people. What is the largest number of people that can be placed in each singing group?

5 Find the least common multiple of two numbers.

BLITZER BONUS

Friendly Numbers

You probably do not describe numbers as sad, angry, happy, or friendly. But the ancient Greeks described two numbers as **friendly** if each was the sum of the other's factors, excluding the numbers themselves. The Greeks knew of only one such pair, 220 and 284. Factors of 220 have a sum of 284:

$1 + 2 + 4 + 5 + 10 + 11 + 20$
$+ 22 + 44 + 55 + 110 = 284.$

Factors of 284 have a sum of 220:

$1 + 2 + 4 + 71 + 142 = 220.$

In 1636, the French mathematician Pierre de Fermat discovered a second pair of friendly numbers, 17,296 and 18,416. By the middle of the nineteenth century, the number of known pairs of friendly numbers exceeded 60. Incredibly, the second-lowest pair of all had gone undiscovered. We still have unanswered questions about friendly numbers. All known friendly pairs consist of either two odd or two even numbers. Are pairs consisting of an odd and an even number possible? Why are all the known odd friendly numbers multiples of 3?

Least Common Multiple

The **least common multiple** of two or more natural numbers is the smallest natural number that is divisible by all of the numbers. One way to find the least common multiple is to make a list of the numbers that are divisible by each number. This list represents the **multiples** of each number. For example, if we wish to find the least common multiple of 15 and 20, we can list the sets of multiples of 15 and multiples of 20.

$\begin{cases} \text{Numbers Divisible by 15:} \\ \quad \text{Multiples of 15:} \qquad \{15, 30, 45, 60, 75, 90, 105, 120, \dots\} \end{cases}$

$\begin{cases} \text{Numbers Divisible by 20:} \\ \quad \text{Multiples of 20:} \qquad \{20, 40, 60, 80, 100, 120, 140, 160, \dots\} \end{cases}$

Some common multiples of 15 and 20 are 60 and 120. The least common multiple is 60. Equivalently, 60 is the smallest number that is divisible by both 15 and 20.

Sometimes a partial list of the multiples for each of two numbers does not reveal the smallest number that is divisible by both given numbers. A more efficient method for finding the least common multiple is to use prime factorizations.

FINDING THE LEAST COMMON MULTIPLE USING PRIME FACTORIZATIONS

To find the least common multiple of two or more numbers,

1. Write the prime factorization of each number.
2. Select every prime factor that occurs, raised to the greatest power to which it occurs, in these factorizations.
3. Form the product of the numbers from step 2. The least common multiple is the product of these factors.

big #

EXAMPLE 5 | FINDING THE LEAST COMMON MULTIPLE

Find the least common multiple of 144 and 300.

SOLUTION

STEP 1. Write the prime factorization of each number. Write the prime factorizations of 144 and 300.

$$144 = 2^4 \times 3^2$$
$$300 = 2^2 \times 3 \times 5^2$$

STEP 2. Select every prime factor that occurs, raised to the greatest power to which it occurs, in these factorizations. The prime factors that occur are 2, 3, and 5. The greatest exponent that appears on 2 is 4, so we select 2^4. The greatest exponent that appears on 3 is 2, so we select 3^2. The greatest exponent that occurs on 5 is 2, so we select 5^2. Thus, we have selected 2^4, 3^2, and 5^2.

STEP 3. Form the product of the numbers from step 2. The least common multiple is the product of these factors.

$$\text{Least common multiple} = 2^4 \times 3^2 \times 5^2 = 16 \times 9 \times 25 = 3600$$

The least common multiple of 144 and 300 is 3600. The smallest natural number divisible by 144 and 300 is 3600.

Check Point 5 Find the least common multiple of 18 and 30.

6   Solve problems using the least common multiple.

EXAMPLE 6   SOLVING A PROBLEM USING THE LEAST COMMON MULTIPLE

A movie theater runs its films continuously. One movie runs for 80 minutes and a second runs for 120 minutes. Both movies begin at 4:00 P.M. When will the movies begin again at the same time?

SOLUTION   The shorter movie lasts 80 minutes, or 1 hour, 20 minutes. It begins at 4:00, so it will be shown again at 5:20. The longer movie lasts 120 minutes, or 2 hours. It begins at 4:00, so it will be shown again at 6:00. We are asked to find when the movies will begin again at the same time. Therefore, we are looking for the least common multiple of 80 and 120. Find the least common multiple and then add this number of minutes to 4:00 P.M.

Begin with the prime factorizations of 80 and 120:

$$80 = 2^4 \times 5$$
$$120 = 2^3 \times 3 \times 5.$$

Now select each prime factor, with the greatest exponent from each factorization.

Least common multiple $= 2^4 \times 3 \times 5 = 16 \times 3 \times 5 = 240$

Therefore, it will take 240 minutes, or 4 hours, for the movies to begin again at the same time. By adding 4 hours to 4:00 P.M., they will start together again at 8:00 P.M.

STUDY TIP

Example 6 can also be solved by making a partial list of starting times for each movie.

Shorter Movie (Runs 1 hour, 20 minutes):

4:00,   5:20,   6:40,   8:00, …

Longer Movie (Runs 2 hours):

4:00,   6:00,   8:00, …

The list reveals that both movies start together again at 8:00 P.M.

Check 6 Point   A movie theater runs two documentary films continuously. One documentary runs for 40 minutes and a second documentary runs for 60 minutes. Both movies begin at 3:00 P.M. When will the movies begin again at the same time?

Exercise Set 5.1

Practice Exercises

Use rules of divisibility to determine whether each number given in Exercises 1–10 is divisible by

a. 2   **b.** 3   **c.** 4   **d.** 5   **e.** 6
f. 8   **g.** 9   **h.** 10   **i.** 12.

1. 6944   **2.** 7245   **3.** 21,408   **4.** 25,025
5. 26,428   **6.** 89,001   **7.** 374,832   **8.** 347,712
9. 6,126,120   **10.** 5,941,221

In Exercises 11–24, use a calculator to determine whether each statement is true or false. If the statement is true, explain why this is so using one of the rules of divisibility in Table 5.1.

11. 3|5958   **12.** 3|8142   **13.** 4|10,612

14. 4|15,984   **15.** 5|38,814   **16.** 5|48,659
17. 6|104,538   **18.** 6|163,944   **19.** 8|20,104
20. 8|28,096   **21.** 9|11,378   **22.** 9|23,772
23. 12|517,872   **24.** 12|785,172

In Exercises 25–44, find the prime factorization of each composite number.

25. 75   **26.** 45   **27.** 56   **28.** 48
29. 105   **30.** 180   **31.** 500   **32.** 360
33. 663   **34.** 510   **35.** 885   **36.** 999
37. 1440   **38.** 1280   **39.** 1996   **40.** 1575
41. 3675   **42.** 8316   **43.** 85,800   **44.** 30,600

In Exercises 45–56, find the greatest common divisor of the numbers.

45. 42 and 56 **46.** 25 and 70 **47.** 16 and 42

48. 66 and 90 **49.** 60 and 108 **50.** 96 and 212

51. 72 and 120 **52.** 220 and 400 **53.** 342 and 380

54. 224 and 430 **55.** 240 and 285 **56.** 150 and 480

In Exercises 57–68, find the least common multiple of the numbers.

57. 42 and 56 **58.** 25 and 70 **59.** 16 and 42

60. 66 and 90 **61.** 60 and 108 **62.** 96 and 212

63. 72 and 120 **64.** 220 and 400 **65.** 342 and 380

66. 224 and 430 **67.** 240 and 285 **68.** 150 and 480

Application Exercises

69. In Carl Sagan's novel *Contact*, Ellie Arroway, the book's heroine, has been working at SETI, the Search for Extraterrestrial Intelligence, listening to the crackle of the cosmos. One night, as the radio telescopes are turned toward Vega, they suddenly pick up strange pulses through the background noise. Two pulses are followed by a pause, then three pulses, five, seven,

$$11, \quad 13, \quad 17, \quad 19, \quad 23, \quad 29, \quad 31, \ldots$$

continuing through 97. Then it starts all over again. Ellie is convinced that only intelligent life could generate the structure in the sequence of pulses. "It's hard to imagine some radiating plasma sending out a regular set of mathematical signals like this." What is it about the structure of the pulses that the book's heroine recognizes as the sign of intelligent life? Asked in another way, what is significant about the number of pulses?

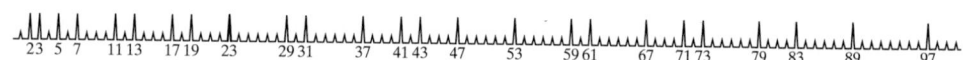

70. There are two species of insects, *Magicicada septendecim* and *Magicacada tredecim*, that live in the same environment. They have a life cycle of exactly 17 and 13 years, respectively. For all but their last year, they remain in the ground feeding on the sap of tree roots. Then, in their last year, they emerge en masse from the ground as fully formed cricketlike insects, taking over the forest in a single night. They chirp loudly, mate, eat, lay eggs, then die six weeks later.

(*Source*: Marcus du Sautoy, *The Music of the Primes*, Harper-Collins, 2003)

a. Suppose that the two species have life cycles that are not prime, say 18 and 12 years, respectively. List the set of multiples of 18 that are less than or equal to 216. List the set of multiples of 12 that are less than or equal to 216. Over a 216-year period, how many times will the two species emerge in the same year and compete to share the forest?

b. Recall that both species have evolved prime-number life cycles, 17 and 13 years, respectively. Find the least common multiple of 17 and 13. How often will the two species have to share the forest?

c. Compare your answers to parts (a) and (b). What explanation can you offer for each species having a prime number of years as the length of its life cycle?

71. A relief worker needs to divide 300 bottles of water and 144 cans of food into groups that each contain the same number of items. Also, each group must have the same type of item (bottled water or canned food). What is the largest number of relief supplies that can be put in each group?

72. A choral director needs to divide 180 men and 144 women into all-male and all-female singing groups so that each group has the same number of people. What is the largest number of people that can be placed in each singing group?

73. You have in front of you 310 five-dollar bills and 460 ten-dollar bills. Your problem: Place the five-dollar bills and the ten-dollar bills in stacks so that each stack has the same number of bills, and each stack contains only one kind of

bill (five-dollar or ten-dollar). What is the largest number of bills that you can place in each stack?

74. Harley collects sports cards. He has 360 football cards and 432 baseball cards. Harley plans to arrange his cards in stacks so that each stack has the same number of cards. Also, each stack must have the same type of card (football or baseball). Every card in Harley's collection is to be placed in one of the stacks. What is the largest number of cards that can be placed in each stack?

75. You and your brother both work the 4:00 P.M. to midnight shift. You have every sixth night off. Your brother has every tenth night off. Both of you were off on June 1. Your brother would like to see a movie with you. When will the two of you have the same night off again?

76. A movie theater runs its films continuously. One movie is a short documentary that runs for 40 minutes. The other movie is a full-length feature that runs for 100 minutes. Each film is shown in a separate theater. Both movies begin at noon. When will the movies begin again at the same time?

77. Two people are jogging around a circular track in the same direction. One person can run completely around the track in 15 minutes. The second person takes 18 minutes. If they both start running in the same place at the same time, how long will it take them to be together at this place if they continue to run?

78. Two people are in a bicycle race around a circular track. One rider can race completely around the track in 40 seconds. The other rider takes 45 seconds. If they both begin the race at a designated starting point, how long will it take them to be together at this starting point again if they continue to race around the track?

Writing in Mathematics

79. If a is a factor of c, what does this mean?

80. Why is 45 divisible by 5?

81. What does "a is divisible by b" mean?

82. Describe the difference between a prime number and a composite number.

83. What does the Fundamental Theorem of Arithmetic state?

84. What is the greatest common divisor of two or more natural numbers?

85. Describe how to find the greatest common divisor of two numbers.

86. What is the least common multiple of two or more natural numbers?

87. Describe how to find the least common multiple of two natural numbers.

88. The process of finding the greatest common divisor of two natural numbers is similar to finding the least common multiple of the numbers. Describe how the two processes differ.

89. What does the Blitzer Bonus on page 197 have to do with Gödel's discovery about mathematics and logic, described on page 108?

Critical Thinking Exercises

90. Write a four-digit natural number that is divisible by 4 and not by 8.

91. Find the greatest common divisor and the least common multiple of $2^{17} \cdot 3^{25} \cdot 5^{31}$ and $2^{14} \cdot 3^{37} \cdot 5^{30}$. Express answers in the same form as the numbers given.

92. A middle-age man observed that his present age was a prime number. He also noticed that the number of years in which his age would again be prime was equal to the number of years ago in which his age was prime. How old is the man?

93. A movie theater runs its films continuously. One movie runs for 85 minutes and a second runs for 100 minutes. The the-ater has a 15-minute intermission after each movie, at which point the movie is shown again. If both movies start at noon, when will the two movies start again at the same time?

Technology Exercises

Use the divisibility rules listed in Table 5.1 to answer the questions in Exercises 94–96. Then, using a calculator, perform the actual division to determine whether your answer is correct.

94. Is 67,234,096 divisible by 4?

95. Is 12,541,750 divisible by 3?

96. Is 48,201,651 divisible by 9?

Group Exercises

The following topics from number theory are appropriate for either individual or group research projects. A report should be given to the class on the researched topic. Useful references include liberal arts mathematics textbooks, books about numbers and number theory, books whose purpose is to excite the reader about mathematics, history of mathematics books, encyclopedias, and the World Wide Web.

97. Euclid and Number Theory

98. An Unsolved Problem from Number Theory

99. Perfect Numbers

100. Mersenne Primes

101. Deficient and Abundant Numbers

102. Formulas that Yield Primes

103. The Sieve of Eratosthenes

SECTION 5.2   THE INTEGERS; ORDER OF OPERATIONS

OBJECTIVES

1. Define the integers.

2. Graph integers on a number line.

3. Use the symbols $<$ and $>$.

4. Find the absolute value of an integer.

5. Perform operations with integers.

6. Use the order of operations agreement.

People are going to live longer in the twenty-first century. This will put added pressure on the Social Security and Medicare systems. How insecure is Social Security's future? In this section, we use subtraction and a set of numbers called the *integers* to describe numerically one aspect of the insecurity.

 Define the integers.

Defining the Integers

In Section 5.1, we applied some ideas of number theory to the set of natural, or counting, numbers:

$$\text{Natural numbers} = \{1, 2, 3, 4, 5, \dots\}.$$

When we combine the number 0 with the natural numbers, we obtain the set of **whole numbers**:

$$\text{Whole numbers} = \{0, 1, 2, 3, 4, 5, \dots\}.$$

The whole numbers do not allow us to describe certain everyday situations. For example, if the balance in your checking account is $30 and you write a check for $35, your checking account is overdrawn by $5. We can write this as −5, read *negative* 5. The set consisting of the natural numbers, 0, and the negatives of the natural numbers is called the set of **integers**.

$$\text{Integers} = \{\dots, \underbrace{-4, -3, -2, -1}, 0, \underbrace{1, 2, 3, 4, \dots}\}$$

$$\qquad\qquad\quad \text{Negative}\qquad\quad \text{Positive}$$
$$\qquad\qquad\quad \text{integers}\qquad\qquad \text{integers}$$

Notice that the term "positive integers" is another name for the natural numbers. The positive integers can be written in two ways:

1. Use a "+" sign. For example, +4 is "positive four."

2. Do not write any sign. For example, 4 is assumed to be "positive four."

The Number Line; The Symbols $<$ and $>$

The **number line** is a graph we use to visualize the set of integers, as well as sets of other numbers. The number line is shown in Figure 5.1.

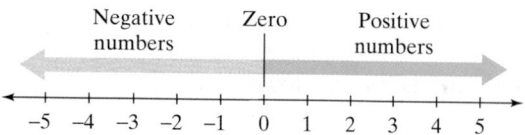

FIGURE 5.1 The number line

The number line extends indefinitely in both directions, shown by the arrows on the left and the right. Zero separates the positive numbers from the negative numbers on the number line. The positive integers are located to the right of 0, and the negative integers are located to the left of 0. **Zero is neither positive nor negative.** For every positive integer on a number line, there is a corresponding negative integer on the opposite side of 0.

Integers are graphed on a number line by placing a dot at the correct location for each number.

② Graph integers on a number line.

EXAMPLE 1 | GRAPHING INTEGERS ON A NUMBER LINE

Graph: **a.** −3   **b.** 4   **c.** 0.

SOLUTION   Place a dot at the correct location for each integer.

Check 1 Point   Graph:

       a. a. −4   **b.** 0   **c.** 3.

3 Use the symbols $<$ and $>$.

We will use the following symbols for comparing two integers:

$<$ means "is less than."

$>$ means "is greater than."

On the number line, the integers increase from left to right. The *lesser* of two integers is the one farther to the *left* on a number line. The *greater* of two integers is the one farther to the *right* on a number line.

Look at the number line in Figure 5.2. The integers 2 and 5 are graphed.

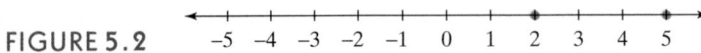

FIGURE 5.2

Observe that 2 is to the left of 5 on the number line. This means that 2 is less than 5:

$2 < 5$: 2 is less than 5 because 2 is to the *left* of 5 on the number line.

In Figure 5.2, we can also observe that 5 is to the right of 2 on the number line. This means that 5 is greater than 2:

$5 > 2$: 5 is greater than 2 because 5 is to the *right* of 2 on the number line.

STUDY TIP

The symbols $<$ and $>$ always point to the lesser of the two integers when the statement is true.

$2 < 5$ The symbol points to 2, the lesser number.

$5 > 2$ The symbol points to 2, the lesser number.

EXAMPLE 2   USING THE SYMBOLS $<$ AND $>$

Insert either $<$ or $>$ in the shaded area between the integers to make the statement true:

a. $-4 \blacksquare 3$ **b.** $-1 \blacksquare -4$ **c.** $-5 \blacksquare -2$ **d.** $0 \blacksquare -3$.

SOLUTION   The solution is illustrated by the number line in Figure 5.3.

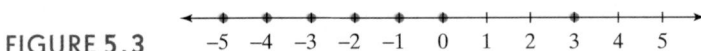

FIGURE 5.3

a. $-4 < 3$ (negative 4 is less than 3) because -4 is to the left of 3 on the number line.

b. $-1 > -4$ (negative 1 is greater than negative 4) because -1 is to the right of -4 on the number line.

c. $-5 < -2$ (negative 5 is less than negative 2) because -5 is to the left of -2 on the number line.

d. $0 > -3$ (zero is greater than negative 3) because 0 is to the right of -3 on the number line.

STUDY TIP

You can think of negative integers as amounts of money that you *owe*. It's better to owe less, so

$-1 > -4$.

Insert either $<$ or $>$ in the shaded area between the integers to make the statement true:

a. $6 \blacksquare -7$ **b.** $-8 \blacksquare -1$ **c.** $-25 \blacksquare -2$ **d.** $-14 \blacksquare 0$.

The symbols $<$ and $>$ are called **inequality symbols**. They may be combined with an equal sign, as shown in the following table.

Symbols	Meaning	Example	Explanation
$a \leq b$	a is less than or equal to b.	$3 \leq 7$	Because $3 < 7$
		$7 \leq 7$	Because $7 = 7$
$b \geq a$	b is greater than or equal to a.	$7 \geq 3$	Because $7 > 3$
		$-5 \geq -5$	Because $-5 = -5$

Find the absolute value of an integer.

Absolute Value

Absolute value describes distance from 0 on a number line. If *a* represents an integer, the symbol $|a|$, represents its absolute value, read "the absolute value of *a*." For example,

$$|-5| = 5.$$

The absolute value of -5 is 5 because -5 is 5 units from 0 on a number line.

ABSOLUTE VALUE

The **absolute value** of an integer *a*, denoted by $|a|$, is the distance from 0 to *a* on the number line. Because absolute value describes a distance, it is never negative.

EXAMPLE 3 FINDING ABSOLUTE VALUE

Find the absolute value:

   a. $|-3|$ **b.** $|5|$ **c.** $|0|$.

SOLUTION

   a. $|-3| = 3$ The absolute value of -3 is 3 because -3 is 3 units from 0.

   b. $|5| = 5$ 5 is 5 units from 0.

   c. $|0| = 0$ 0 is 0 units from itself.

Example 3 illustrates that the absolute value of a positive integer or 0 is the number itself. The absolute value of a negative integer, such as -3, is the number without the negative sign. Zero is the only real number whose absolute value is 0: $|0| = 0$. **The absolute value of any integer other than 0 is always positive.**

Find the absolute value:

   a. $|-8|$ **b.** $|6|$.

---

5 Perform operations with integers.

Addition of Integers

It has not been a good day! First, you lost your wallet with $30 in it. Then, you borrowed $10 to get through the day, which you somehow misplaced. Your loss of $30 followed by a loss of $10 is an overall loss of $40. This can be written

$$-30 + (-10) = -40.$$

The result of adding two or more numbers is called the **sum** of the numbers. The sum of -30 and -10 is -40.

   You can think of gains and losses of money to find sums. For example, to find $17 + (-13)$, think of a gain of $17 followed by a loss of $13. There is an overall gain of $4. Thus, $17 + (-13) = 4$. In the same way, to find $-17 + 13$, think of a loss of $17 followed by a gain of $13. There is an overall loss of $4, so $-17 + 13 = -4$.

Using gains and losses, we can develop the following rules for adding integers:

RULES FOR ADDITION OF INTEGERS

Rule	**Examples**

If the integers have the same sign,

1. Add their absolute values.
2. The sign of the sum is the same as the sign of the two numbers.

$-11 + (-15) = -26$

Use the common sign.

> Add absolute values:
> $11 + 15 = 26.$

If the integers have different signs,

1. Subtract the smaller absolute value from the larger absolute value.
2. The sign of the sum is the same as the sign of the number with the larger absolute value.

$-13 + 4 = -9$

Use the sign of the number with the greater absolute value.

> Subtract absolute values:
> $13 - 4 = 9.$

$13 + (-6) = 7$

Use the sign of the number with the greater absolute value.

> Subtract absolute values:
> $13 - 6 = 7.$

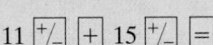
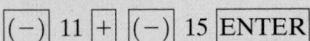
STUDY TIP

In addition to gains and losses of money, another good analogy for adding integers is temperatures above and below zero on a thermometer. Think of the thermometer as a number line standing straight up. For example,

$$-11 + (-15) = -26$$

> If it's 11 below zero and the temperature falls 15 degrees, it will then be 26 below zero.

$$-13 + 4 = -9$$

> If it's 13 below zero and the temperature rises 4 degrees, the new temperature will be 9 below zero.

$$13 + (-6) = 7.$$

> If it's 13 above zero and the temperature falls 6 degrees, it will then be 7 above zero.

Using the analogies of gains and losses of money or temperatures can make the formal rules for addition of integers easy to use.

Can you guess what number is displayed if you use a calculator to find a sum such as $18 + (-18)$? If you gain 18 and then lose 18, there is neither an overall gain or loss. Thus,

$$18 + (-18) = 0.$$

We call 18 and -18 **additive inverses**. Additive inverses have the same absolute value, but lie on opposite sides of zero on the number line. Thus, -7 is the additive inverse of 7, and 5 is the additive inverse of -5. In general, the sum of any integer and its additive inverse is 0:

$$a + (-a) = 0.$$

Subtraction of Integers

Time for a new computer! Your favorite model, which normally sells for $1500, has a price reduction of $600. The computer's reduced price, $900, can be expressed in two ways:

$$1500 - 600 = 900 \quad \text{or} \quad 1500 + (-600) = 900.$$

This means that

$$1500 - 600 = 1500 + (-600).$$

To subtract 600 from 1500, we add 1500 and the additive inverse of 600. Generalizing from this situation, we define subtraction as follows:

DEFINITION OF SUBTRACTION

For all integers a and b,

$$a - b = a + (-b).$$

In words, to subtract b from a, add the additive inverse of b to a. The result of subtraction is called the **difference**.

EXAMPLE 4   SUBTRACTING INTEGERS

Subtract:

a. $17 - (-11)$ **b.** $-18 - (-5)$ **c.** $-18 - 5$.

SOLUTION

a. $17 - (-11) = 17 + 11 = 28$

> Change the subtraction to addition.   Replace −11 with its additive inverse.

b. $-18 - (-5) = -18 + 5 = -13$

> Change the subtraction to addition.   Replace −5 with its additive inverse.

c. $-18 - 5 = -18 + (-5) = -23$

> Change the subtraction to addition.   Replace 5 with its additive inverse.

Check Point 4   Subtract:

a. $30 - (-7)$ **b.** $-14 - (-10)$ **c.** $-14 - 10.$

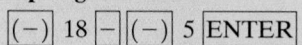
STUDY TIP

You can think of subtracting a negative integer as taking away a debt. Let's apply this analogy to $17 - (-11)$. Your checking account balance is $17 after an erroneous $11 charge was made against your account. When you bring this error to the bank's attention, they will take away the $11 debit and your balance will go up to $28:

$$17 - (-11) = 28.$$

Subtraction is used to solve problems in which the word *difference* appears. The difference between integers a and b is expressed as $a - b$.

EXAMPLE 5 | AN APPLICATION OF SUBTRACTION USING THE WORD *DIFFERENCE*

The bar graph in Figure 5.4 shows that in 1995, Social Security had an annual cash surplus of $233 billion. By 2020, this amount is expected to be a negative number—a deficit of $244 billion. What is the difference between the 1995 surplus and the projected 2020 deficit?

Social Security Annual Operating Surplus / Deficit

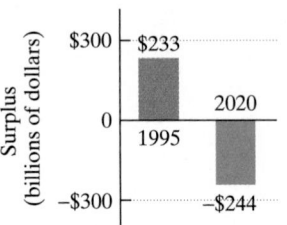

Source: U.S. Office of Management and Budget

FIGURE 5.4

SOLUTION

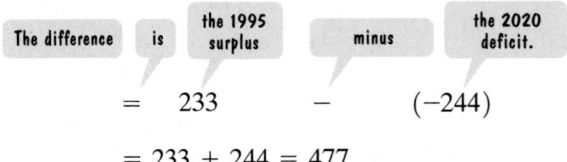

$$= \quad 233 \quad - \quad (-244)$$

$$= 233 + 244 = 477$$

The difference between the 1995 surplus and the projected 2020 deficit is $477 billion.

Check 5 Point The peak of Mount Everest is 8848 meters above sea level. The Marianas Trench, on the floor of the Pacific Ocean, is 10,915 meters below sea level. What is the difference in elevation between the peak of Mount Everest and the Marianas Trench?

Multiplication of Integers

Suppose that things go from bad to worse for Social Security, and the $244 deficit in 2020 doubles by the middle of the twenty-first century. The new deficit is $2(-\$244)$ or $-\$488$. Notice that multiplying a positive integer and a negative integer gives a negative integer. The result of the multiplication, -488, called the **product**, has the opposite sign of the first number in the multiplication. Using this observation, we would expect $-2(-244)$ to have the opposite sign as -2. Thus, the product is positive:

$$-2(-244) = 488.$$

These observations give us the following rules for multiplying integers:

RULES FOR MULTIPLYING INTEGERS

Rule

1. The product of two integers with the same sign is found by multiplying their absolute values. The product is positive.

2. The product of two integers with different signs is found by multiplying their absolute values. The product is negative.

3. The product of 0 and any integer is 0.

Examples

- $3(11) = 33$
- $-5(-9) = 45$

- $7(-10) = -70$
- $(-7)(11) = -77$

- $6(0) = 0$
- $0(-23) = 0$

Exponential Notation

We have seen that exponents are used to write repeated products. For example,

Exponent

$$3^2 = \underbrace{3 \cdot 3}_{\text{Two factors of 3}} = 9 \quad \text{and} \quad 5^3 = \underbrace{5 \cdot 5 \cdot 5}_{\text{Three factors of 5}} = 125.$$

Exponent

EXAMPLE 6 EVALUATING EXPONENTIAL EXPRESSIONS

Evaluate:

a. 4^2 **b.** $(-4)^2$ **c.** 6^3 **d.** $(-6)^3$ **e.** $(-2)^4$.

SOLUTION

a. $4^2 = 4 \cdot 4 = 16$

b. $(-4)^2 = (-4)(-4) = 16$

c. $6^3 = 6 \cdot 6 \cdot 6 = 36 \cdot 6 = 216$

d. $(-6)^3 = (-6)(-6)(-6) = 36(-6) = -216$

e. $(-2)^4 = (-2)(-2)(-2)(-2) = 4(-2)(-2) = (-8)(-2) = 16$

STUDY TIP

Do not confuse -2^4 and $(-2)^4$:

$-2^4 = -2 \cdot 2 \cdot 2 \cdot 2 = -16$

$(-2)^4 = (-2)(-2)(-2)(-2) = 16$.

In Example 6, notice the difference between the signs of the answers in parts (d) and (e). In general, if a negative number is raised to an odd power, the sign of the answer is negative. If a negative number is raised to an even power, the sign of the answer is positive.

Check Point 6 Evaluate:

a. 8^2 **b.** $(-8)^2$ **c.** 7^3 **d.** $(-7)^3$ **e.** $(-3)^4$.

Division of Integers

The result of dividing the integer a by the nonzero integer b is called the **quotient** of the numbers. We can write this quotient as $a \div b$ or $\frac{a}{b}$.

A relationship exists between multiplication and division. For example,

$$\frac{-12}{4} = -3 \text{ means that } 4(-3) = -12.$$

$$\frac{-12}{-4} = 3 \text{ means that } -4(3) = -12.$$

Because of the relationship between multiplication and division, the rules for obtaining the sign of a quotient are the same as those for obtaining the sign of a product.

TECHNOLOGY

Multiplying and Dividing on a Calculator

Example: $(-173)(-256)$

Scientific Calculator

173 $\boxed{+/\text{-}}$ $\boxed{\times}$ 256 $\boxed{+/\text{-}}$ $\boxed{=}$

Graphing Calculator

$\boxed{(-)}$ 173 $\boxed{\times}$ $\boxed{(-)}$ 256 $\boxed{\text{ENTER}}$

The number 44288 should be displayed.

Division is performed in the same manner, using $\boxed{\div}$ instead of $\boxed{\times}$. What happens when you divide by 0? Try entering

8 $\boxed{\div}$ 0

and pressing $\boxed{=}$ or $\boxed{\text{ENTER}}$.

 6 Use the order of operations agreement.

RULES FOR DIVIDING INTEGERS

Rule

1. The quotient of two integers with the same sign is found by dividing their absolute values. The quotient is positive.

2. The quotient of two integers with different signs is found by dividing their absolute values. The quotient is negative.

3. Zero divided by any nonzero integer is zero.

4. Division by 0 is undefined.

Examples

- $\dfrac{27}{9} = 3$

- $\dfrac{-45}{-3} = 15$

- $\dfrac{80}{-4} = -20$

- $\dfrac{-15}{5} = -3$

- $\dfrac{0}{-5} = 0$ (because $-5(0) = 0$)

- $\dfrac{-8}{0}$ is undefined (because 0 cannot be multiplied by an integer to obtain -8).

Order of Operations

Suppose that you want to find the value of $3 + 7 \cdot 5$. Which procedure shown below is correct?

$$3 + 7 \cdot 5 = 3 + 35 = 38 \quad \text{or} \quad 3 + 7 \cdot 5 = 10 \cdot 5 = 50$$

If you know the answer, you probably know certain rules, called the **order of operations**, to make sure that there is only one correct answer. One of these rules states that if a problem contains no parentheses, perform multiplication before addition. Thus, the procedure on the left is correct because the multiplication of 7 and 5 is done first. Then the addition is performed. The correct answer is 38.

Here are the rules for determining the order in which operations should be performed:

STUDY TIP

Here's a sentence to help remember the order of operations: "Please excuse my dear Aunt Sally."

Please	**P**arentheses
Excuse	**E**xponents
$\begin{cases}\textbf{M}\text{y} \\ \textbf{D}\text{ear}\end{cases}$	$\begin{cases}\textbf{M}\text{ultiplication} \\ \textbf{D}\text{ivision}\end{cases}$
$\begin{cases}\textbf{A}\text{unt} \\ \textbf{S}\text{ally}\end{cases}$	$\begin{cases}\textbf{A}\text{ddition} \\ \textbf{S}\text{ubtraction}\end{cases}$

ORDER OF OPERATIONS

1. Perform all operations within grouping symbols.

2. Evaluate all exponential expressions.

3. Do all multiplications and divisions in the order in which they occur, working from left to right.

4. Finally, do all additions and subtractions in the order in which they occur, working from left to right.

In the third step, be sure to do all multiplications and divisions *as they occur* from left to right. For example,

$$8 \div 4 \cdot 2 = 2 \cdot 2 = 4$$

Do the division first because it occurs first.

$$8 \cdot 4 \div 2 = 32 \div 2 = 16.$$

Do the multiplication first because it occurs first.

EXAMPLE 7   USING THE ORDER OF OPERATIONS

Simplify: $6^2 - 24 \div 2^2 \cdot 3 + 1$.

SOLUTION   There are no grouping symbols. Thus, we begin by evaluating exponential expressions. Then we multiply or divide. Finally, we add or subtract.

$6^2 - 24 \div 2^2 \cdot 3 + 1$

$= 36 - 24 \div 4 \cdot 3 + 1$ Evaluate exponential expressions: $6^2 = 6 \cdot 6 = 36$ and $2^2 = 2 \cdot 2 = 4$.

$= 36 - 6 \cdot 3 + 1$ Perform the multiplications and divisions from left to right. Start with $24 \div 4 = 6$.

$= 36 - 18 + 1$ Now do the multiplication: $6 \cdot 3 = 18$.

$= 18 + 1$ Finally, perform the additions and subtractions from left to right. Subtract: $36 - 18 = 18$.

$= 19$ Add: $18 + 1 = 19$.

 Check Point 7   Simplify: $7^2 - 48 \div 4^2 \cdot 5 + 2$.

EXAMPLE 8   USING THE ORDER OF OPERATIONS

Simplify: $(-6)^2 - (5 - 7)^2(-3)$.

SOLUTION   Because grouping symbols appear, we perform the operation within parentheses first.

$(-6)^2 - (5 - 7)^2(-3)$

$= (-6)^2 - (-2)^2(-3)$ Work inside parentheses first: $5 - 7 = 5 + (-7) = -2$.

$= 36 - 4(-3)$ Evaluate exponential expressions: $(-6)^2 = (-6)(-6) = 36$ and $(-2)^2 = (-2)(-2) = 4$.

$= 36 - (-12)$ Multiply: $4(-3) = -12$.

$= 48$ Subtract: $36 - (-12) = 36 + 12 = 48$.

 Check Point 8   Simplify: $(-8)^2 - (10 - 13)^2(-2)$.

Exercise Set 5.2

Practice Exercises

In Exercises 1–4, start by drawing a number line that shows integers from −5 to 5. Then graph each of the following integers on your number line.

1. 3   **2.** 5   **3.** −4   **4.** −2

In Exercises 5–12, insert either < or > in the shaded area between the integers to make the statement true.

5. −2 ▇ 7   **6.** −1 ▇ 13   **7.** −13 ▇ −2

8. −1 ▇ −13   **9.** 8 ▇ −50   **10.** 7 ▇ −9

11. −100 ▇ 0   **12.** 0 ▇ −300

In Exercises 13–18, find the absolute value.

13. $|-14|$ **14.** $|-16|$ **15.** $|14|$

16. $|16|$ **17.** $|-300,000|$ **18.** $|-1,000,000|$

In Exercises 19–30, find each sum.

19. $-7 + (-5)$ **20.** $-3 + (-4)$ **21.** $12 + (-8)$

22. $13 + (-5)$ **23.** $6 + (-9)$ **24.** $3 + (-11)$

25. $-9 + (+4)$ **26.** $-7 + (+3)$ **27.** $-9 + (-9)$

28. $-13 + (-13)$ **29.** $9 + (-9)$ **30.** $13 + (-13)$

In Exercises 31–42, perform the indicated subtraction.

31. $13 - 8$ **32.** $14 - 3$ **33.** $8 - 15$

34. $9 - 20$ **35.** $4 - (-10)$ **36.** $3 - (-17)$

37. $-6 - (-17)$ **38.** $-4 - (-19)$ **39.** $-12 - (-3)$

40. $-19 - (-2)$ **41.** $-11 - 17$ **42.** $-19 - 21$

In Exercises 43–52, find each product.

43. $6(-9)$ **44.** $5(-7)$ **45.** $(-7)(-3)$

46. $(-8)(-5)$ **47.** $(-2)(6)$ **48.** $(-3)(10)$

49. $(-13)(-1)$ **50.** $(-17)(-1)$ **51.** $0(-5)$

52. $0(-8)$

In Exercises 53–66, evaluate each exponential expression.

53. 5^2 **54.** 6^2 **55.** $(-5)^2$ **56.** $(-6)^2$

57. 4^3 **58.** 2^3 **59.** $(-5)^3$ **60.** $(-4)^3$

61. $(-5)^4$ **62.** $(-4)^4$ **63.** -3^4 **64.** -1^4

65. $(-3)^4$ **66.** $(-1)^4$

In Exercises 67–80, find each quotient, or, if applicable, state that the expression is undefined.

67. $\frac{-12}{4}$ **68.** $\frac{-40}{5}$ **69.** $\frac{21}{-3}$ **70.** $\frac{60}{-6}$

71. $\frac{-90}{-3}$ **72.** $\frac{-66}{-6}$ **73.** $\frac{0}{-7}$ **74.** $\frac{0}{-8}$

75. $\frac{-7}{0}$ **76.** $\frac{0}{0}$

77. $(-480) \div 24$ **78.** $(-300) \div 12$

79. $(465) \div (-15)$ **80.** $(-594) \div (-18)$

Use the order of operations to find the value of each expression in Exercises 81–98.

81. $7 + 6 \cdot 3$ **82.** $-5 + (-3) \cdot 8$

83. $(-5) - 6(-3)$ **84.** $-8(-3) - 5(-6)$

85. $6 - 4(-3) - 5$ **86.** $3 - 7(-1) - 6$

87. $3 - 5(-4 - 2)$ **88.** $3 - 9(-1 - 6)$

89. $(2 - 6)(-3 - 5)$ **90.** $9 - 5(6 - 4) - 10$

91. $3(-2)^2 - 4(-3)^2$ **92.** $5(-3)^2 - 2(-2)^3$

93. $(2 - 6)^2 - (3 - 7)^2$ **94.** $(4 - 6)^2 - (5 - 9)^3$

95. $6(3 - 5)^3 - 2(1 - 3)^3$

96. $-3(-6 + 8)^3 - 5(-3 + 5)^3$

97. $8^2 - 16 \div 2^2 \cdot 4 - 3$

98. $10^2 - 100 \div 5^2 \cdot 2 - (-3)$

Application Exercises

Temperatures sometimes fall below zero. A combination of low temperature and wind makes it feel colder than the actual temperature. The table shows how cold it feels when low temperatures are combined with different wind speeds.

WINDCHILL

Wind (mph)	Temperature (° F)											
	35	30	25	20	15	10	5	0	−5	−10	−15	−20
5	31	25	19	13	7	−1	−5	−11	−16	−22	−28	−34
10	27	21	15	9	3	−4	−10	−16	−22	−28	−35	−41
15	25	19	13	6	0	−7	−13	−19	−26	−32	−39	−45
20	24	17	11	4	−2	−9	−15	−22	−29	−35	−42	−48
25	23	16	9	3	−4	−11	−17	−24	−31	−37	−44	−51

Source: National Weather Service

Use the information from the table to solve Exercises 99–100.

99. Write a negative integer that indicates how cold the temperature feels when the temperature is 15° Fahrenheit and the wind is blowing at 20 miles per hour.

100. Write a negative integer that indicates how cold the temperature feels when the temperature is 10° Fahrenheit and the wind is blowing at 15 miles per hour.

101. The greatest temperature variation recorded in a day is 100 degrees in Browning, Montana, on January 23, 1916. The low temperature was −56°F. What was the high temperature?

102. In Spearfish, South Dakota, on January 22, 1943, the temperature rose 49 degrees in two minutes. If the initial temperature was −4°F, what was the high temperature?

103. The peak of Mount Kilimanjaro, the highest point in Africa, is 19,321 feet above sea level. Qattara Depression, Egypt, the lowest point in Africa, is 436 feet below sea level. What is the difference in elevation between the peak of Mount Kilimanjaro and the Qattara Depression?

104. The peak of Mount Whitney is 14,494 feet above sea level. Mount Whitney can be seen directly above Death Valley, which is 282 feet below sea level. What is the difference in elevation between these geographic locations?

The following table shows the amount of money, in millions of dollars, collected and spent by the U.S. government from 2000 through 2002.

Year	Money Collected	Money Spent
2000	2,025,200	1,788,800
2001	1,991,000	1,863,900
2002	1,946,100	2,052,300

Money is expressed in millions of dollars.
Source: Office of Management and Budget

Use the information from the table to solve Exercises 105–106.

105. In 2000, what was the difference between the amount of money collected and the amount spent? Was there a budget surplus or deficit in 2000?

106. Repeat Exercise 105 for 2002.

Do you enjoy cold weather? If so, try Fairbanks, Alaska. The average daily low temperature for each month in Fairbanks is shown in the bar graph. Use the graph to solve Exercises 107–110.

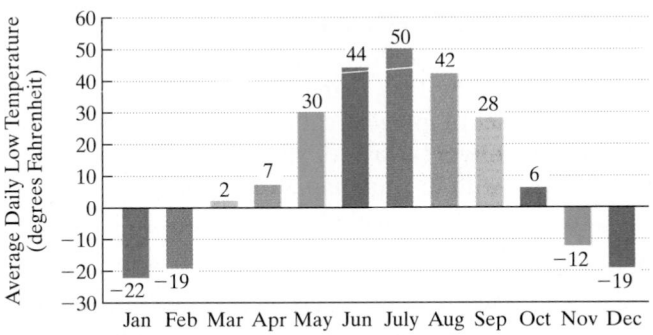

Each Month's Average Daily Low Temperature in Fairbanks, Alaska

Source: The Weather Channel Enterprises, Inc.

107. What is the difference between the average daily low temperatures for March and February?

108. What is the difference between the average daily low temperatures for October and November?

109. How many degrees warmer is February's average low temperature than January's average low temperature?

110. How many degrees warmer is November's average low temperature than December's average low temperature?

The graph shows the net profits and losses of the Ford Motor Company from 1998 through 2002. Use the graph to solve Exercises 111–112.

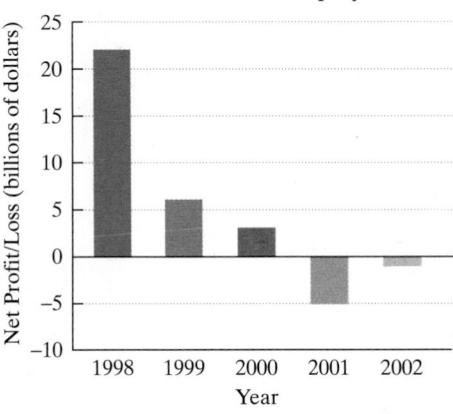

Net Profits and Losses of the Ford Motor Company

Source: Newsweek, June 16, 2003

111. According to *Newsweek*, "Ford swung from most profitable carmaker to biggest loser." Estimate the difference between Ford's profits in 1998 and its losses in 2002.

112. Estimate the difference between Ford's losses in 2002 and its losses in 2001.

Writing in Mathematics

113. How does the set of integers differ from the set of whole numbers?

114. Explain how to graph an integer on a number line.

115. If you are given two integers, explain how to determine which one is smaller.

116. Explain how to add integers.

117. Explain how to subtract integers.

118. Explain how to multiply integers.

119. Explain how to divide integers.

120. Describe what it means to raise a number to a power. In your description, include a discussion of the difference between -5^2 and $(-5)^2$.

121. Why is $\frac{0}{4}$ equal to 0, but $\frac{4}{0}$ is undefined?

Critical Thinking Exercises

In Exercises 122–123, insert one pair of parentheses to make each calculation correct.

122. $8 - 2 \cdot 3 - 4 = 10$ **123.** $8 - 2 \cdot 3 - 4 = 14$

Technology Exercises

Scientific calculators that have parentheses keys allow for the entry and computation of relatively complicated expressions in a single step. For example, the expression $15 + (10 - 7)^2$ can be evaluated by entering the following keystrokes:

$$15 \; \boxed{+} \; \boxed{(\!(} \; 10 \; \boxed{-} \; 7 \; \boxed{)\!)} \; \boxed{y^x} \; 2 \; \boxed{=} .$$

Find the value of each expression in Exercises 124–126 in a single step on your scientific calculator.

124. $8 - 2 \cdot 3 - 9$ **125.** $(8 - 2) \cdot (3 - 9)$

126. $5^3 + 4 \cdot 9 - (8 + 9 \div 3)$

Group Exercise

127. How do we first learn about numbers? At what age do we begin to understand what they mean? For this activity, group members should research the work of the innovative Swiss psychologist Jean Piaget (1896–1980) and present a group report on numbers and psychology.

OBJECTIVES

1. Define the rational numbers.

2. Reduce rational numbers.

3. Convert between mixed numbers and improper fractions.

4. Express rational numbers as decimals.

5. Express decimals in the form $\frac{a}{b}$.

6. Multiply and divide rational numbers.

7. Add and subtract rational numbers.

8. Apply the density property of rational numbers.

9. Solve problems involving rational numbers.

You are making eight dozen chocolate chip cookies for a large neighborhood block party. The recipe lists the ingredients needed to prepare five dozen cookies, such as $\frac{3}{4}$ cup sugar. How do you adjust the amount of sugar, as well as each of the other ingredients, given in the recipe?

Adapting a recipe to suit a different number of portions usually involves working with numbers that are not integers. For example, the number describing the amount of sugar, $\frac{3}{4}$ (cup), is not an integer, although it consists of the quotient of two integers, 3 and 4. Before returning to the problem of changing the size of a recipe, we study a new set of numbers consisting of the quotient of integers.

Defining the Rational Numbers

 Define the rational numbers.

If two integers are added, subtracted, or multiplied, the result is always another integer. This, however, is not always the case with division. For example, 10 divided by 5 is the integer 2. By contrast, 5 divided by 10 is $\frac{1}{2}$, and $\frac{1}{2}$ is not an integer. To permit divisions such as $\frac{5}{10}$, we enlarge the set of integers, calling the new collection the *rational numbers*. The set of **rational numbers** consists of all the numbers that can be expressed as a quotient of two integers, with the denominator not 0.

> **THE RATIONAL NUMBERS**
>
> The set of **rational numbers** is the set of all numbers which can be expressed in the form $\frac{a}{b}$, where a and b are integers and b is not equal to 0. The integer a is called the **numerator** and the integer b is called the **denominator**.

The following numbers are examples of rational numbers:

$$\frac{1}{2}, \frac{-3}{4}, 5, 0.$$

The integer 5 is a rational number because it can be expressed as the quotient of integers: $5 = \frac{5}{1}$. Similarly, 0 can be written as $\frac{0}{1}$.

In general, every integer a is a rational number because it can be expressed in the form $\frac{a}{1}$.

STUDY TIP

We know that the quotient of two numbers with different signs is a negative number. Thus,

$$\frac{-3}{4} = -\frac{3}{4} \quad \text{and} \quad \frac{3}{-4} = -\frac{3}{4}.$$

Reducing Rational Numbers

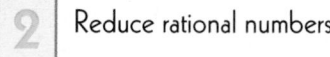

A rational number is **reduced to its lowest terms**, or **simplified**, when the numerator and denominator have no common divisors other than 1. Reducing rational numbers to lowest terms is done using the **Fundamental Principle of Rational Numbers**.

THE FUNDAMENTAL PRINCIPLE OF RATIONAL NUMBERS

If $\frac{a}{b}$ is a rational number and c is any number other than 0,

$$\frac{a \cdot c}{b \cdot c} = \frac{a}{b}.$$

The rational numbers $\frac{a}{b}$ and $\frac{a \cdot c}{b \cdot c}$ are called **equivalent fractions**.

When using the Fundamental Principle to reduce a rational number, the simplification can be done in one step by finding the greatest common divisor of the numerator and the denominator and using it for c. Thus, **to reduce a rational number to its lowest terms, divide both the numerator and the denominator by their greatest common divisor**.

For example, consider the rational number $\frac{12}{100}$. The greatest common divisor of 12 and 100 is 4. We reduce to lowest terms as follows:

$$\frac{12}{100} = \frac{3 \cdot 4}{25 \cdot 4} = \frac{3}{25} \quad \text{or} \quad \frac{12}{100} = \frac{12 \div 4}{100 \div 4} = \frac{3}{25}.$$

EXAMPLE 1 | REDUCING A RATIONAL NUMBER

Reduce $\frac{130}{455}$ to lowest terms.

SOLUTION Begin by finding the greatest common divisor of 130 and 455.

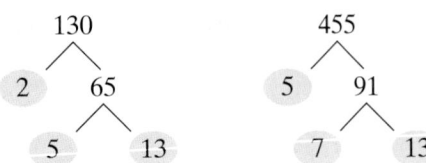

Thus, $130 = 2 \cdot 5 \cdot 13$, and $455 = 5 \cdot 7 \cdot 13$. The greatest common divisor is $5 \cdot 13$, or 65. Divide the numerator and the denominator of the given rational number by $5 \cdot 13$ or by 65.

$$\frac{130}{455} = \frac{2 \cdot 5 \cdot 13}{5 \cdot 7 \cdot 13} = \frac{2}{7} \quad \text{or} \quad \frac{130}{455} = \frac{130 \div 65}{455 \div 65} = \frac{2}{7}$$

There are no common divisors of 2 and 7 other than 1. Thus, the rational number $\frac{2}{7}$ is in its lowest terms.

Check Point 1   Reduce $\frac{72}{90}$ to lowest terms.

Mixed Numbers and Improper Fractions

A **mixed number** consists of the sum of an integer and a rational number, expressed without the use of an addition sign. Here is an example of a mixed number:

$$3\frac{4}{5}.$$ The integer is **3** and the rational number is $\frac{4}{5}$. $3\frac{4}{5}$ means $3 + \frac{4}{5}$.

An **improper fraction** is a rational number whose numerator is greater than its denominator. An example of an improper fraction is $\frac{19}{5}$.

③ Convert between mixed numbers and improper fractions.

The mixed number $3\frac{4}{5}$ can be converted to the improper fraction $\frac{19}{5}$ using the following procedure:

CONVERTING A POSITIVE MIXED NUMBER TO AN IMPROPER FRACTION

1. Multiply the denominator of the rational number by the integer and add the numerator to this product.
2. Place the sum in step 1 over the denominator in the mixed number.

EXAMPLE 2   CONVERTING FROM MIXED NUMBER TO IMPROPER FRACTION

Convert $3\frac{4}{5}$ to an improper fraction.

SOLUTION

$$3\frac{4}{5} = \frac{5 \cdot 3 + 4}{5}$$

Multiply the denominator by the integer and add the numerator.

Place the sum over the mixed number's denominator.

$$= \frac{15 + 4}{5} = \frac{19}{5}$$

Convert $2\frac{5}{8}$ to an improper fraction.

When converting a negative mixed number to an improper fraction, copy the negative sign and then follow the previous procedure. For example,

$$-2\frac{3}{4} = -\frac{4 \cdot 2 + 3}{4} = -\frac{8 + 3}{4} = -\frac{11}{4}.$$

Copy the negative sign from step to step and convert $2\frac{3}{4}$ to an improper fraction.

A positive improper fraction can be converted to a mixed number using the following procedure:

CONVERTING A POSITIVE IMPROPER FRACTION TO A MIXED NUMBER

1. Divide the denominator into the numerator. Record the quotient and the remainder.
2. Write the mixed number using the following form:

$$\text{quotient } \frac{\text{remainder}}{\text{original denominator}}.$$

integer part

rational number part

EXAMPLE 3 | CONVERTING FROM AN IMPROPER FRACTION TO A MIXED NUMBER

Convert $\frac{42}{5}$ to a mixed number.

SOLUTION

STEP 1. Divide the denominator into the numerator.

$$\begin{array}{r} 8 \\ 5\overline{)42} \\ 40 \\ \hline 2 \end{array}$$

quotient

remainder

STEP 2. Write the mixed number using quotient$\dfrac{\text{remainder}}{\text{original denominator}}$. Thus,

$$\frac{42}{5} = 8\frac{2}{5}.$$

remainder

original denominator

quotient

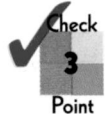 Check Point 3   Convert $\frac{5}{3}$ to a mixed number.

When converting a negative improper fraction to a mixed number, copy the negative sign and then follow the previous procedure. For example,

$$-\frac{29}{8} = -3\frac{5}{8}.$$

Copy the negative sign.

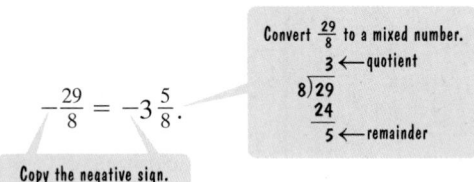

Convert $\frac{29}{8}$ to a mixed number.

$$\begin{array}{r} 3 \\ 8\overline{)29} \\ 24 \\ \hline 5 \end{array}$$

← quotient

← remainder

 4 Express rational numbers as decimals.

Tens	Ones	Tenths	Hundredths	Thousandths	Ten-Thousandths	Hundred-Thousandths
10	1	$\frac{1}{10}$	$\frac{1}{100}$	$\frac{1}{1000}$	$\frac{1}{10,000}$	$\frac{1}{100,000}$

decimal point

Rational Numbers and Decimals

We have seen that a rational number is the quotient of integers. Rational numbers can also be expressed as decimals. As shown in the place-value chart in the margin, it is convenient to represent rational numbers with denominators of 10, 100, 1000, and so on as decimals. For example,

$$\frac{7}{10} = 0.7, \quad \frac{3}{100} = 0.03, \quad \text{and} \quad \frac{8}{1000} = 0.008.$$

Any rational number $\frac{a}{b}$ can be expressed as a decimal by dividing the denominator, b, into the numerator, a.

EXAMPLE 4 EXPRESSING RATIONAL NUMBERS AS DECIMALS

Express each rational number as a decimal:

a. $\frac{5}{8}$ **b.** $\frac{7}{11}$.

SOLUTION In each case, divide the denominator into the numerator.

a.
$$
\begin{array}{r}
0.625 \\
8)\overline{5.000} \\
\underline{4\ 8} \\
20 \\
\underline{16} \\
40 \\
\underline{40} \\
0
\end{array}
$$

b.
$$
\begin{array}{r}
0.6363\ldots \\
11)\overline{7.0000\ldots} \\
\underline{6\ 6} \\
40 \\
\underline{33} \\
70 \\
\underline{66} \\
40 \\
\underline{33} \\
70 \\
\vdots
\end{array}
$$

In Example 4, the decimal for $\frac{5}{8}$, namely 0.625, stops and is called a **terminating decimal**. Other examples of terminating decimals are

$$\frac{1}{4} = 0.25, \qquad \frac{2}{5} = 0.4, \qquad \text{and} \qquad \frac{7}{8} = 0.875.$$

By contrast, the division process for $\frac{7}{11}$ results in $0.6363\ldots$, with the digits 63 repeating over and over indefinitely. To indicate this, write a bar over the digits that repeat. Thus,

$$\frac{7}{11} = 0.\overline{63}.$$

The decimal for $\frac{7}{11}$, $0.\overline{63}$, is called a **repeating decimal**. Other examples of repeating decimals are

$$\frac{1}{3} = 0.333\ldots = 0.\overline{3} \qquad \text{and} \qquad \frac{2}{3} = 0.666\ldots = 0.\overline{6}.$$

RATIONAL NUMBERS AND DECIMALS

Any rational number can be expressed as a decimal. The resulting decimal will either terminate (stop), or it will have a digit that repeats or a block of digits that repeat.

 Express decimals in the form $\frac{a}{b}$.

Tens	Ones	Tenths	Hundredths	Thousandths	Ten-Thousandths	Hundred-Thousandths
10	1	$\frac{1}{10}$	$\frac{1}{100}$	$\frac{1}{1000}$	$\frac{1}{10,000}$	$\frac{1}{100,000}$

decimal point

 **Check Point 4** Express each rational number as a decimal:

a. $\frac{3}{8}$ **b.** $\frac{5}{11}$.

Reversing Directions: Expressing Decimals as a Quotient of Two Integers

Terminating decimals can be expressed with denominators of 10, 100, 1000, 10,000, and so on. Use the place-value chart shown in the margin. The digits to the right of the decimal point are the numerator of the rational number. To find the denominator, observe the last digit to the right of the decimal point. The place-value of this digit will indicate the denominator.

EXAMPLE 5 EXPRESSING TERMINATING DECIMALS IN $\frac{a}{b}$ FORM

Express each terminating decimal as a quotient of integers:

a. 0.7 **b.** 0.49 **c.** 0.048.

SOLUTION

a. $0.7 = \frac{7}{10}$ because the 7 is in the tenths position.

b. $0.49 = \frac{49}{100}$ because the digit on the right, 9, is in the hundredths position.

c. $0.048 = \frac{48}{1000}$ because the digit on the right, 8, is in the thousandths position. Reducing to lowest terms, $\frac{48}{1000} = \frac{48 \div 8}{1000 \div 8} = \frac{6}{125}$.

 Check Point 5 Express each terminating decimal as a quotient of integers, reduced to lowest terms:

a. 0.9 **b.** 0.86 **c.** 0.053.

If you are given a rational number as a repeating decimal, there is a technique for expressing the number as a quotient of integers. Although we will not be discussing algebra until Chapter 6, the basic algebra used in this technique may be familiar to you. We begin by illustrating the technique with an example. Then we will summarize the steps in the procedure and apply them to another example.

EXAMPLE 6 EXPRESSING A REPEATING DECIMAL IN $\frac{a}{b}$ FORM

Express $0.\overline{6}$ as a quotient of integers.

SOLUTION
STEP 1. Let n equal the repeating decimal. Let $n = 0.\overline{6}$, so that $n = 0.66666\ldots$.

STEP 2. If there is one repeating digit, multiply both sides of the equation in step 1 by 10.

$$n = 0.66666\ldots \quad \text{This is the equation from step 1.}$$
$$10n = 10(0.66666\ldots) \quad \text{Multiply both sides by 10.}$$
$$10n = 6.66666\ldots \quad \text{Multiplying by 10 moves the decimal point one place to the right.}$$

STUDY TIP

In step 3, be sure to line up the decimal points before subtracting.

STEP 3. Subtract the equation in step 1 from the equation in step 2.

> Remember from algebra that n means $1n$. Thus, $10n - 1n = 9n$.

$$\begin{array}{r} 10n = 6.66666\ldots \quad \text{This is the equation from step 2.} \\ - \quad n = 0.66666\ldots \quad \text{This is the equation from step 1.} \\ \hline 9n = 6 \end{array}$$

STEP 4. Divide both sides of the equation in step 3 by the number in front of n and solve for n. We solve $9n = 6$ for n by dividing both sides by 9.

$$9n = 6 \quad \text{This is the equation from step 3.}$$
$$\frac{9n}{9} = \frac{6}{9} \quad \text{Divide both sides by 9.}$$
$$n = \frac{6}{9} = \frac{2}{3} \quad \begin{array}{l} \text{Reduce } \frac{6}{9} \text{ to lowest terms:} \\ \frac{6}{9} = \frac{2 \cdot 3}{3 \cdot 3} = \frac{2}{3}. \end{array}$$

We began the solution process with $n = 0.\overline{6}$, and now we have $n = \frac{2}{3}$. Therefore,

$$0.\overline{6} = \frac{2}{3}.$$

Here are the steps for expressing a repeating decimal as a quotient of integers. Assume that the repeating digit or digits begin directly to the right of the decimal point.

EXPRESSING A REPEATING DECIMAL AS A QUOTIENT OF INTEGERS

STEP 1. Let n equal the repeating decimal.

STEP 2. Multiply both sides of the equation in step 1 by 10 if one digit repeats, by 100 if two digits repeat, by 1000 if three digits repeat, and so on.

STEP 3. Subtract the equation in step 1 from the equation in step 2.

STEP 4. Divide both sides of the equation in step 3 by the number in front of n and solve for n.

 **Check Point 6**   Express $0.\overline{2}$ as a quotient of integers.

EXAMPLE 7 EXPRESSING A REPEATING DECIMAL IN $\frac{a}{b}$ FORM

Express $0.\overline{53}$ as a quotient of integers.

SOLUTION

STEP 1. Let n equal the repeating decimal. Let $n = 0.\overline{53}$, so that $n = 0.535353\ldots$.

STEP 2. If there are two repeating digits, multiply both sides of the equation in step 2 by 100.

$$n = 0.535353\ldots \qquad \text{This is the equation from step 1.}$$
$$100n = 100(0.535353\ldots) \qquad \text{Multiply both sides by 100.}$$
$$100n = 53.535353\ldots \qquad \text{Multiplying by 100 moves the decimal point two places to the right.}$$

STEP 3. Subtract the equation in step 1 from the equation in step 2.

$$
\begin{aligned}
100n &= 53.535353\ldots \qquad \text{This is the equation from step 2.}\\
-\ n &= \ \ 0.535353\ldots \qquad \text{This is the equation from step 1.}\\
\hline
99n &= 53
\end{aligned}
$$

STEP 4. Divide both sides of the equation in step 3 by the number in front of n and solve for n. We solve $99n = 53$ for n by dividing both sides by 99.

$$99n = 53 \qquad \text{This is the equation from step 3.}$$
$$\frac{99n}{99} = \frac{53}{99} \qquad \text{Divide both sides by 99.}$$
$$n = \frac{53}{99}$$

Because n equals $0.\overline{53}$ and n equals $\frac{53}{99}$,

$$0.\overline{53} = \frac{53}{99}.$$

 **Check Point 7**   Express $0.\overline{79}$ as a quotient of integers.

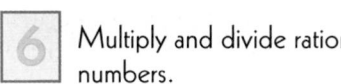

Multiply and divide rational numbers.

Multiplying and Dividing Rational Numbers

The product of two rational numbers is found as follows:

> ### MULTIPLYING RATIONAL NUMBERS
> The product of two rational numbers is the product of their numerators divided by the product of their denominators.
>
> If $\frac{a}{b}$ and $\frac{c}{d}$ are rational numbers, then $\frac{a}{b} \cdot \frac{c}{d} = \frac{a \cdot c}{b \cdot d}$.

EXAMPLE 8 | MULTIPLYING RATIONAL NUMBERS

Multiply.   If possible, reduce the product to its lowest terms:

a. $\frac{3}{8} \cdot \frac{5}{11}$ **b.** $\left(-\frac{2}{3}\right)\left(-\frac{9}{4}\right)$ **c.** $\left(3\frac{2}{3}\right)\left(1\frac{1}{4}\right)$.

SOLUTION

a. $\frac{3}{8} \cdot \frac{5}{11} = \frac{3 \cdot 5}{8 \cdot 11} = \frac{15}{88}$

b. $\left(-\frac{2}{3}\right)\left(-\frac{9}{4}\right) = \frac{(-2)(-9)}{3 \cdot 4} = \frac{18}{12} = \frac{3 \cdot 6}{2 \cdot 6} = \frac{3}{2}$ or $1\frac{1}{2}$

c. $\left(3\frac{2}{3}\right)\left(1\frac{1}{4}\right) = \frac{11}{3} \cdot \frac{5}{4} = \frac{11 \cdot 5}{3 \cdot 4} = \frac{55}{12}$ or $4\frac{7}{12}$

Check Point 8   Multiply.   If possible, reduce the product to its lowest terms:

a. $\frac{4}{11} \cdot \frac{2}{3}$ **b.** $\left(-\frac{3}{7}\right)\left(-\frac{14}{4}\right)$ **c.** $\left(3\frac{2}{5}\right)\left(1\frac{1}{2}\right)$.

STUDY TIP

> You can divide numerators and denominators by common factors *before* performing multiplication. Then multiply the remaining factors in the numerators and multiply the remaining factors in the denominators. For example,
>
> $$\frac{7}{15} \cdot \frac{20}{21} = \frac{\overset{1}{\cancel{7}}}{\underset{3}{\cancel{15}}} \cdot \frac{\overset{4}{\cancel{20}}}{\underset{3}{\cancel{21}}} = \frac{1 \cdot 4}{3 \cdot 3} = \frac{4}{9}.$$

Two numbers whose product is 1 are called **reciprocals**, or **multiplicative inverses**, of each other. Thus, the reciprocal of 2 is $\frac{1}{2}$ and the reciprocal of $\frac{1}{2}$ is 2 because $2 \cdot \frac{1}{2} = 1$. In general, if $\frac{c}{d}$ is a nonzero rational number, its reciprocal is $\frac{d}{c}$ because $\frac{c}{d} \cdot \frac{d}{c} = 1$.

Reciprocals are used to find the quotient of two rational numbers.

> ### DIVIDING RATIONAL NUMBERS
> The quotient of two rational numbers is the product of the first number and the reciprocal of the second number.
>
> If $\frac{a}{b}$ and $\frac{c}{d}$ are rational numbers and $\frac{c}{d}$ is not 0, then $\frac{a}{b} \div \frac{c}{d} = \frac{a}{b} \cdot \frac{d}{c} = \frac{a \cdot d}{b \cdot c}$.

EXAMPLE 9 DIVIDING RATIONAL NUMBERS

Divide.   If possible, reduce the quotient to its lowest terms:

a. $\frac{4}{5} \div \frac{1}{10}$ **b.** $-\frac{3}{5} \div \frac{7}{11}$ **c.** $4\frac{3}{4} \div 1\frac{1}{2}.$

SOLUTION

a. $\frac{4}{5} \div \frac{1}{10} = \frac{4}{5} \cdot \frac{10}{1} = \frac{4 \cdot 10}{5 \cdot 1} = \frac{40}{5} = 8$

b. $-\frac{3}{5} \div \frac{7}{11} = -\frac{3}{5} \cdot \frac{11}{7} = \frac{-3(11)}{5 \cdot 7} = -\frac{33}{35}$

c. $4\frac{3}{4} \div 1\frac{1}{2} = \frac{19}{4} \div \frac{3}{2} = \frac{19}{4} \cdot \frac{2}{3} = \frac{19 \cdot 2}{4 \cdot 3} = \frac{38}{12} = \frac{19 \cdot 2}{6 \cdot 2} = \frac{19}{6}$ or $3\frac{1}{6}$

 Divide.   If possible, reduce the quotient to its lowest terms:

Check Point 9

a. $\frac{9}{11} \div \frac{5}{4}$ **b.** $-\frac{8}{15} \div \frac{2}{5}$ **c.** $3\frac{3}{8} \div 2\frac{1}{4}.$

 Add and subtract rational numbers.

Adding and Subtracting Rational Numbers

Rational expressions with identical deonominators are added and subtracted using the following rules:

> **ADDING AND SUBTRACTING RATIONAL NUMBERS WITH IDENTICAL DENOMINATORS**
>
> The sum or difference of two rational numbers with identical denominators is the sum or difference of their numerators over the common denominator.
>   If $\frac{a}{b}$ and $\frac{c}{b}$ are rational numbers, then $\frac{a}{b} + \frac{c}{b} = \frac{a + c}{b}$ and $\frac{a}{b} - \frac{c}{b} = \frac{a - c}{b}.$

EXAMPLE 10 ADDING AND SUBTRACTING RATIONAL NUMBERS WITH IDENTICAL DENOMINATORS

Perform the indicated operations:

a. $\frac{3}{7} + \frac{2}{7}$ **b.** $\frac{11}{12} - \frac{5}{12}$ **c.** $-5\frac{1}{4} - \left(-2\frac{3}{4}\right).$

SOLUTION

a. $\frac{3}{7} + \frac{2}{7} = \frac{3 + 2}{7} = \frac{5}{7}$

b. $\frac{11}{12} - \frac{5}{12} = \frac{11 - 5}{12} = \frac{6}{12} = \frac{1 \cdot 6}{2 \cdot 6} = \frac{1}{2}$

c. $-5\frac{1}{4} - \left(-2\frac{3}{4}\right) = -\frac{21}{4} - \left(-\frac{11}{4}\right) = -\frac{21}{4} + \frac{11}{4} = \frac{-21 + 11}{4} = \frac{-10}{4} = -\frac{5}{2}$ or $-2\frac{1}{2}$

 Perform the indicated operations:

Check Point 10

a. $\frac{5}{12} + \frac{3}{12}$ **b.** $\frac{7}{4} - \frac{1}{4}$ **c.** $-3\frac{3}{8} - \left(-1\frac{1}{8}\right).$

   If the rational numbers to be added or subtracted have different denominators, we use the least common multiple of their denominators to rewrite the rational numbers. The least common multiple of the denominators is called the **least common denominator**.

   Rewriting rational numbers with a least common denominator is done using the Fundamental Principle of Rational Numbers, discussed at the beginning of this section. Recall that if $\frac{a}{b}$ is a rational number and c is a nonzero number, then

$$\frac{a}{b} = \frac{a}{b} \cdot \frac{c}{c} = \frac{a \cdot c}{b \cdot c}.$$

Multiplying the numerator and the denominator of a rational number by the same nonzero number is equivalent to multiplying by 1, resulting in an equivalent fraction.

EXAMPLE 11 ADDING RATIONAL NUMBERS WITH UNLIKE DENOMINATORS

Find the sum of $\frac{3}{4} + \frac{1}{6}$.

SOLUTION The smallest number divisible by both 4 and 6 is 12. Therefore, 12 is the least common multiple of 4 and 6, and will serve as the least common denominator. To obtain a denominator of 12, multiply the denominator and the numerator of the first rational number, $\frac{3}{4}$, by 3. To obtain a denominator of 12, multiply the denominator and the numerator of the second rational number, $\frac{1}{6}$, by 2.

$$\frac{3}{4} + \frac{1}{6} = \frac{3}{4} \cdot \frac{3}{3} + \frac{1}{6} \cdot \frac{2}{2}$$

Rewrite each rational number as an equivalent fraction with a denominator of 12. $\frac{3}{3} = 1$ and $\frac{2}{2} = 1$, and multiplying by 1 does not change a number's value.

$$= \frac{9}{12} + \frac{2}{12}$$

Multiply.

$$= \frac{11}{12}$$

Add numerators and put this sum over the least common denominator.

 Check Point 11   Find the sum of $\frac{1}{5} + \frac{3}{4}$.

If the least common denominator cannot be found by inspection, use prime factorizations of the denominators and the method for finding their least common multiple, discussed in Section 5.1.

EXAMPLE 12 SUBTRACTING RATIONAL NUMBERS WITH UNLIKE DENOMINATORS

Perform the indicated operation: $\frac{1}{15} - \frac{7}{24}$.

SOLUTION We need to first find the least common denominator, which is the least common multiple of 15 and 24. What is the smallest number divisible by both 15 and 24? The answer is not obvious, so we begin with the prime factorization of each number.

$$15 = 5 \cdot 3$$
$$24 = 8 \cdot 3 = 2^3 \cdot 3$$

The different factors are 5, 3, and 2. Using the greatest number of times each factor appears in any factorization, the least common multiple is $5 \cdot 3 \cdot 2^3 = 5 \cdot 3 \cdot 8 = 120$.

We will now express each rational number with a denominator of 120, which is the least common denominator. For the first rational number, $\frac{1}{15}$, 120 divided by 15 is 8. Thus, we will multiply the numerator and denominator by 8. For the second

TECHNOLOGY

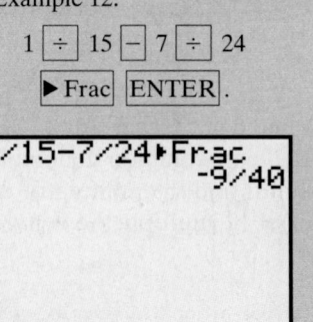

Here is a possible keystroke sequence on a graphing calculator for the subtraction problem in Example 12:

$$1 \boxed{\div} 15 \boxed{-} 7 \boxed{\div} 24$$

$$\boxed{\blacktriangleright \text{Frac}} \boxed{\text{ENTER}}.$$

```
1/15-7/24►Frac
            -9/40
```

The calculator display reads $-9/40$, serving as a check for our answer in Example 12.

8 Apply the density property of rational numbers.

rational number, $\frac{7}{24}$, 120 divided by 24 is 5. Thus, we will multiply the numerator and denominator by 5.

$$\frac{1}{15} - \frac{7}{24} = \frac{1}{15} \cdot \frac{8}{8} - \frac{7}{24} \cdot \frac{5}{5} \qquad \begin{array}{l}\text{Rewrite each rational number as an} \\ \text{equivalent fraction with a denomina-} \\ \text{tor of 120.}\end{array}$$

$$= \frac{8}{120} - \frac{35}{120} \qquad \text{Multiply.}$$

$$= \frac{8 - 35}{120} \qquad \begin{array}{l}\text{Subtract the numerators and put} \\ \text{this difference over the least com-} \\ \text{mon denominator.}\end{array}$$

$$= \frac{-27}{120} \qquad \text{Perform the subtraction.}$$

$$= \frac{-9 \cdot 3}{40 \cdot 3} \qquad \text{Reduce to lowest terms.}$$

$$= -\frac{9}{40}$$

 Perform the indicated operation: $\frac{3}{10} - \frac{7}{12}$.

Density of Rational Numbers

It is always possible to find a rational number between any two given rational numbers. Mathematicians express this idea by saying that the set of rational numbers is **dense**.

DENSITY OF THE RATIONAL NUMBERS

If r and t represent rational numbers, with $r < t$, then there is a rational number s such that s is between r and t:

$$r < s < t.$$

One way to find a rational number between two given rational numbers is to find the rational number halfway between them. Add the given rational numbers and divide their sum by 2, thereby finding the average of the numbers.

EXAMPLE 13 ILLUSTRATING THE DENSITY PROPERTY

Find a rational number halfway between $\frac{1}{2}$ and $\frac{3}{4}$.

SOLUTION First, add $\frac{1}{2}$ and $\frac{3}{4}$.

$$\frac{1}{2} + \frac{3}{4} = \frac{2}{4} + \frac{3}{4} = \frac{5}{4}$$

Next, divide this sum by 2.

$$\frac{5}{4} \div \frac{2}{1} = \frac{5}{4} \cdot \frac{1}{2} = \frac{5}{8}$$

The number $\frac{5}{8}$ is halfway between $\frac{1}{2}$ and $\frac{3}{4}$. Thus,

$$\frac{1}{2} < \frac{5}{8} < \frac{3}{4}.$$

STUDY TIP

The inequality $\frac{1}{2} < \frac{5}{8} < \frac{3}{4}$ is more obvious if all denominators are changed to 8:

$$\frac{4}{8} < \frac{5}{8} < \frac{6}{8}.$$

We can repeat the procedure of Example 13 and find a rational number halfway between $\frac{1}{2}$ and $\frac{5}{8}$. Repeated application of this procedure implies the following surprising result:

Between any two given rational numbers are *infinitely many* rational numbers.

 **Check 13 Point** Find a rational number halfway between $\frac{1}{3}$ and $\frac{1}{2}$.

 9 Solve problems involving rational numbers.

Problem Solving with Rational Numbers

A common application of rational numbers involves preparing food for a different number of servings than what the recipe gives. The amount of each ingredient can be found as follows:

Amount of ingredient needed

$$= \frac{\text{desired serving size}}{\text{recipe serving size}} \times \text{ingredient amount in the recipe.}$$

EXAMPLE 14 CHANGING THE SIZE OF A RECIPE

A chocolate-chip cookie recipe for five dozen cookies requires $\frac{3}{4}$ cup sugar. If you want to make eight dozen cookies, how much sugar is needed?

SOLUTION   Amount of sugar needed

$$= \frac{\text{desired serving size}}{\text{recipe serving size}} \times \text{sugar amount in recipe}$$

$$= \frac{8 \;\cancel{\text{dozen}}}{5 \;\cancel{\text{dozen}}} \times \frac{3}{4} \text{ cup}$$

The amount of sugar, in cups, needed is determined by multiplying the rational numbers:

$$\frac{8}{5} \times \frac{3}{4} = \frac{8 \cdot 3}{5 \cdot 4} = \frac{24}{20} = \frac{6 \cdot \cancel{4}}{5 \cdot \cancel{4}} = 1\frac{1}{5}.$$

Thus, $1\frac{1}{5}$ cups of sugar is needed. (Depending on the measuring cup you are using, you may need to round the sugar amount to $1\frac{1}{4}$ cups.)

 **Check 14 Point** A chocolate-chip cookie recipe for five dozen cookies requires two eggs. If you want to make seven dozen cookies, exactly how many eggs are needed? Now round your answer to a realistic number that does not involve fractional parts of an egg.

Exercise Set 5.3

Practice Exercises

In Exercises 1–12, reduce each rational number to its lowest terms.

1. $\frac{10}{15}$ **2.** $\frac{18}{45}$ **3.** $\frac{15}{18}$ **4.** $\frac{16}{64}$

5. $\frac{24}{42}$ **6.** $\frac{32}{80}$ **7.** $\frac{60}{108}$ **8.** $\frac{112}{128}$

9. $\frac{342}{380}$ **10.** $\frac{210}{252}$ **11.** $\frac{308}{418}$ **12.** $\frac{144}{300}$

In Exercises 13–18, convert each mixed number to an improper fraction.

13. $2\frac{3}{8}$ **14.** $2\frac{7}{9}$ **15.** $-7\frac{3}{5}$

16. $-6\frac{2}{5}$ **17.** $12\frac{7}{16}$ **18.** $11\frac{5}{16}$

In Exercises 19–24, convert each improper fraction to a mixed number.

19. $\frac{23}{5}$ **20.** $\frac{47}{8}$ **21.** $-\frac{76}{9}$

22. $-\frac{59}{9}$ **23.** $\frac{711}{20}$ **24.** $\frac{788}{25}$

In Exercises 25–36, express each rational number as a decimal.

25. $\frac{3}{4}$ **26.** $\frac{3}{5}$ **27.** $\frac{7}{20}$ **28.** $\frac{3}{20}$

29. $\frac{7}{8}$ **30.** $\frac{5}{16}$ **31.** $\frac{9}{11}$ **32.** $\frac{3}{11}$

33. $\frac{22}{7}$ **34.** $\frac{20}{3}$ **35.** $\frac{2}{7}$ **36.** $\frac{5}{7}$

In Exercises 37–48, express each terminating decimal as a quotient of integers. If possible, reduce to lowest terms.

37. 0.3   **38.** 0.9   **39.** 0.4   **40.** 0.6

41. 0.39   **42.** 0.59   **43.** 0.82   **44.** 0.64

45. 0.725   **46.** 0.625   **47.** 0.5399   **48.** 0.7006

In Exercises 49–56, express each repeating decimal as a quotient of integers. If possible, reduce to lowest terms.

49. $0.\overline{7}$ **50.** $0.\overline{1}$ **51.** $0.\overline{9}$ **52.** $0.\overline{3}$

53. $0.\overline{36}$ **54.** $0.\overline{81}$ **55.** $0.\overline{257}$ **56.** $0.\overline{529}$

In Exercises 57–92, perform the indicated operations. If possible, reduce the answer to its lowest terms.

57. $\frac{3}{8} \cdot \frac{7}{11}$ **58.** $\frac{5}{8} \cdot \frac{3}{11}$ **59.** $\left(-\frac{1}{10}\right)\left(\frac{7}{12}\right)$

60. $\left(-\frac{1}{8}\right)\left(\frac{5}{9}\right)$ **61.** $\left(-\frac{2}{3}\right)\left(-\frac{9}{4}\right)$ **62.** $\left(-\frac{5}{4}\right)\left(-\frac{6}{7}\right)$

63. $\left(3\frac{3}{4}\right)\left(1\frac{3}{5}\right)$ **64.** $\left(2\frac{4}{5}\right)\left(1\frac{1}{4}\right)$ **65.** $\frac{5}{4} \div \frac{3}{8}$

66. $\frac{5}{8} \div \frac{4}{3}$ **67.** $-\frac{7}{8} \div \frac{15}{16}$ **68.** $-\frac{13}{20} \div \frac{4}{5}$

69. $6\frac{3}{5} \div 1\frac{1}{10}$ **70.** $1\frac{3}{4} \div 2\frac{5}{8}$ **71.** $\frac{2}{11} + \frac{3}{11}$

72. $\frac{5}{13} + \frac{2}{13}$ **73.** $\frac{5}{6} - \frac{1}{6}$ **74.** $\frac{7}{12} - \frac{5}{12}$

75. $\frac{7}{12} - \left(-\frac{1}{12}\right)$ **76.** $\frac{5}{16} - \left(-\frac{5}{16}\right)$ **77.** $\frac{1}{2} + \frac{1}{5}$

78. $\frac{1}{3} + \frac{1}{5}$ **79.** $\frac{3}{4} + \frac{3}{20}$ **80.** $\frac{2}{5} + \frac{2}{15}$

81. $\frac{5}{24} + \frac{7}{30}$ **82.** $\frac{7}{108} + \frac{55}{144}$ **83.** $\frac{13}{18} - \frac{2}{9}$

84. $\frac{13}{15} - \frac{2}{45}$ **85.** $\frac{4}{3} - \frac{3}{4}$ **86.** $\frac{3}{2} - \frac{2}{3}$

87. $\frac{1}{15} - \frac{27}{50}$ **88.** $\frac{4}{15} - \frac{1}{6}$ **89.** $3\frac{3}{4} - 2\frac{1}{3}$

90. $3\frac{2}{3} - 2\frac{1}{2}$ **91.** $\left(\frac{1}{2} - \frac{1}{3}\right) \div \frac{5}{8}$ **92.** $\left(\frac{1}{2} + \frac{1}{4}\right) \div \left(\frac{1}{2} + \frac{1}{3}\right)$

In Exercises 93–98, find a rational number halfway between the two numbers in each pair.

93. $\frac{1}{4}$ and $\frac{1}{3}$ **94.** $\frac{2}{3}$ and $\frac{5}{6}$ **95.** $\frac{1}{2}$ and $\frac{2}{3}$

96. $\frac{3}{5}$ and $\frac{2}{3}$ **97.** $-\frac{2}{3}$ and $-\frac{5}{6}$ **98.** -4 and $-\frac{7}{2}$

Different operations with the same rational numbers usually result in different answers. Exercises 99–100 illustrate some curious exceptions.

99. Show that $\frac{13}{4} + \frac{13}{9}$ and $\frac{13}{4} \times \frac{13}{9}$ give the same answer.

100. Show that $\frac{169}{30} + \frac{13}{15}$ and $\frac{169}{30} \div \frac{13}{15}$ give the same answer.

Application Exercises

The circle graph shows the breakdown of the number of countries in the world that are free, partly free, or not free. Use the information in the graph to solve Exercises 101–102.

World's Countries by Status of Freedom

Source: Larry Berman and Bruce Murphy, *Approaching Democracy*, 4th Edition, Prentice Hall, 2003

101. What fractional part of the world's countries is free? Reduce the answer to its lowest terms.

102. What fractional part of the world's countries is not free? Reduce the answer to tis lowest terms.

In most societies, women say they prefer to marry men who are older than themselves, whereas men say they prefer women who are younger. Evolutionary psychologists attribute these preferences to female concern with a partner's material resources and male concern with a partner's fertility (Source: Davie M. Buss, Psychological Inquiry, 6, 1–30). The graph shows the preferred age in a mate in five selected countries. Use the information in the graph to solve Exercises 103–106. Express each answer as a mixed number.

Preferred Age in a Mate

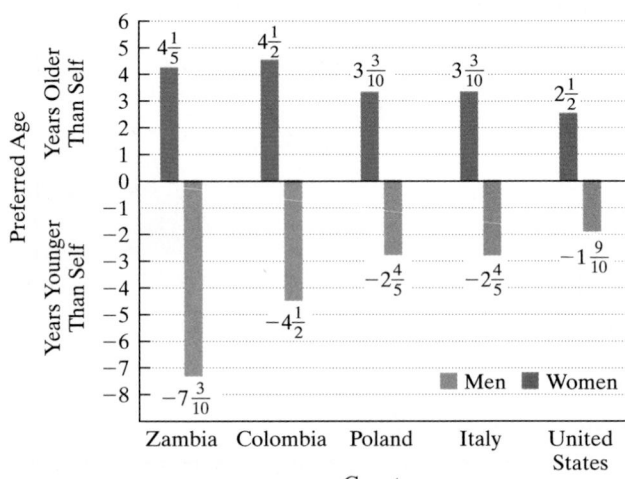

Source: Carole Wade and Carol Tavris, *Psychology*, 6th Edition, Prentice Hall, 2000

103. What is the difference between the preferred age in a mate for women in Italy and women in the United States?

104. What is the difference between the preferred age in a mate for women in Zambia and women in the United States?

105. What is the difference between the preferred age in a mate for men in Italy and men in the United States

106. What is the difference between the preferred age in a mate for men in Colombia and men in the United States?

107. A recipe calls for $\frac{3}{4}$ cup of sugar. How much is needed to make half of the recipe?

108. A recipe calls for $\frac{3}{4}$ cup of shortening. How much is needed to triple the recipe?

A mix for eight servings of instant potatoes requires $2\frac{2}{3}$ cups of water. Use this information to solve Exercises 109–110.

109. If you want to make 11 servings, how much water is needed?

110. If you want to make six servings, how much water is needed?

111. A franchise is owned by three people. The first owns $\frac{5}{12}$ of the business and the second owns $\frac{1}{4}$ of the business. What fractional part of the business is owned by the third person?

112. At a workshop on enhancing creativity, $\frac{1}{4}$ of the participants are musicians, $\frac{2}{5}$ are artists, $\frac{1}{10}$ are actors, and the remaining participants are writers. What fraction of the people attending the workshop are writers?

113. If you walk $\frac{3}{4}$ mile and then jog $\frac{2}{5}$ mile, what is the total distance covered? How much farther did you walk than jog?

114. Some companies pay people extra when they work more than a regular 40-hour work week. The overtime pay is often $1\frac{1}{2}$ times the regular hourly rate. This is called time and a half. A summer job for students pays $12 an hour and offers time and a half for the hours worked over 40. If a student works 46 hours during one week, what is the student's total pay before taxes?

115. A will states that $\frac{3}{5}$ of the estate is to be divided among relatives. Of the remaining estate, $\frac{1}{4}$ goes to the National Foundation for AIDS Research. What fraction of the estate goes to the National Foundation for AIDS Research?

116. The legend of a map indicates that 1 inch = 16 miles. If the distance on the map between two cities is $2\frac{3}{8}$ inches, how far apart are the cities?

Writing in Mathematics

117. What is a rational number?

118. Explain how to reduce a rational number to its lowest terms.

119. Explain how to convert from a mixed number to an improper fraction. Use $7\frac{2}{3}$ as an example.

120. Explain how to convert from an improper fraction to a mixed number. Use $\frac{47}{5}$ as an example.

121. Explain how to write a rational number as a decimal.

122. Explain how to write 0.083 as a quotient of integers.

123. Explain how to write $0.\overline{9}$ as a quotient of integers.

124. Explain how to multiply rational numbers. Use $\frac{5}{6} \cdot \frac{1}{2}$ as an example.

125. Explain how to divide rational numbers. Use $\frac{5}{6} \div \frac{1}{2}$ as an example.

126. Explain how to add rational numbers with different denominators. Use $\frac{5}{6} + \frac{1}{2}$ as an example.

127. What does it mean when we say that the set of rational numbers is dense?

128. Explain what is wrong with this statement. "If you'd like to save some money, I'll be happy to sell you my computer system for only $\frac{3}{2}$ of the price I originally paid for it."

Critical Thinking Exercises

129. Shown below is a short excerpt from "The Star-Spangled Banner." The time is $\frac{3}{4}$, which means that each measure must contain notes that add up to $\frac{3}{4}$. The values of the different notes tell musicians how long to hold each note.

Use vertical lines to divide this line of "The Star-Spangled Banner" into measures.

130. Use inductive reasoning to predict the addition problem and the sum that will appear in the fourth row. Then perform the arithmetic to verify your conjecture.

$$\frac{1}{1 \cdot 2} + \frac{1}{2 \cdot 3} = \frac{2}{3}$$

$$\frac{1}{1 \cdot 2} + \frac{1}{2 \cdot 3} + \frac{1}{3 \cdot 4} = \frac{3}{4}$$

$$\frac{1}{1 \cdot 2} + \frac{1}{2 \cdot 3} + \frac{1}{3 \cdot 4} + \frac{1}{4 \cdot 5} = \frac{4}{5}$$

Technology Exercises

131. Use a calculator to express the following rational numbers as decimals.

a. $\frac{197}{800}$ **b.** $\frac{4539}{3125}$ **c.** $\frac{7}{6250}$

132. Some calculators have a fraction feature. This feature allows you to perform operations with fractions and displays the answer as a fraction reduced to its lowest terms. If your calculator has this feature, use it to verify any five of the answers that you obtained in Exercises 57–92.

Group Exercise

133. Each member of the group should present an application of rational numbers. The application can be based on research or on how the group member uses rational numbers in his or her life. If you are not sure where to begin, ask yourself how your life would be different if fractions and decimals were concepts unknown to our civilization.

SECTION 5.4 THE IRRATIONAL NUMBERS

OBJECTIVES

1. Define the irrational numbers.
2. Simplify square roots.
3. Perform operations with square roots.
4. Rationalize the denominator.

Shown here is Renaissance artist Raphael Sanzio's (1483–1520) image of Pythagoras from *The School of Athens* mural. Detail of left side. Stanza della Segnatura, Vatican Palace, Vatican State.

Source: Scala/Art Resource, NY.

For the followers of the Greek mathematician Pythagoras in the sixth century B.C., numbers took on a life-and-death importance. The "Pythagorean Brotherhood" was a secret group whose members were convinced that properties of whole numbers were the key to understanding the universe. Members of the Brotherhood (which admitted women) thought that all numbers that were not whole numbers could be represented as the ratio of whole numbers. A crisis occurred for the Pythagoreans when they discovered the existence of a number that was not rational. Because the Pythagoreans viewed numbers with reverence and awe, the punishment for speaking about this number was death. However, a member of the Brotherhood revealed the secret of the number's existence. When he later died in a shipwreck, his death was viewed as punishment from the gods.

The triangle in Figure 5.5 led the Pythagoreans to the discovery of a number that could not be expressed as the quotient of integers. Based on their understanding of the relationship among the sides of this triangle, they knew that the length of the side shown in red had to be a number that, when squared, is equal to 2. The Pythagoreans discovered that this number seemed to be close to the rational numbers

$$\frac{14}{10}, \frac{141}{100}, \frac{1414}{1000}, \frac{14{,}141}{10{,}000}, \text{ and so on.}$$

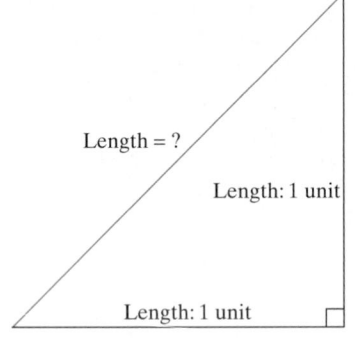

Length = ?
Length: 1 unit
Length: 1 unit

FIGURE 5.5

However, they were shocked to find that there is no quotient of integers whose square is equal to 2.

The positive number whose square is equal to 2 is written $\sqrt{2}$. We read this "the square root of 2," or "radical 2." The symbol $\sqrt{}$ is called the **radical sign**. The number under the radical sign, in this case 2, is called the **radicand**. The entire symbol $\sqrt{2}$ is called a **radical**.

Using deductive reasoning, mathematicians have proved that $\sqrt{2}$ cannot be represented as a quotient of integers. This means that there is no terminating or repeating decimal that can be multiplied by itself to give 2. We can, however, give a decimal approximation for $\sqrt{2}$. We use the symbol $\approx$, which means "is approximately equal to." Thus,

$$\sqrt{2} \approx 1.414214.$$

We can verify that this is only an approximation by multiplying 1.414214 by itself. The product is not exactly 2:

$$1.414214 \times 1.414214 = 2.00000123796.$$

1 Define the irrational numbers.

A number like $\sqrt{2}$, whose decimal representation does not come to an end and does not have a block of repeating digits, is an example of an **irrational number**.

THE IRRATIONAL NUMBERS

The set of **irrational numbers** is the set of numbers whose decimal representations are neither terminating nor repeating.

Perhaps the best known of all the irrational numbers is π (pi). This irrational number represents the distance around a circle (its circumference) divided by the diameter of the circle. In the *Star Trek* episode "Wolf in the Fold," Spock foils an evil computer by telling it to "compute the last digit in the value of π." Because π is an irrational number, there is no last digit in its decimal representation:

$$\pi = 3.14159265358979323846264433832795\ldots.$$

The nature of the irrational number π has fascinated mathematicians for centuries. Amateur and professional mathematicians have taken up the challenge of calculating π to more and more decimal places.

Computing π is the ultimate stress test for a computer. Brothers Gregory and David Chudnovsky put their homemade computer to the test, calculating π to over 8 billion digits. If printed in ordinary type, the number would stretch over 9600 miles!

TECHNOLOGY

You can obtain decimal approximations for irrational numbers using a calculator. For example, to approximate $\sqrt{2}$ use the following keystrokes:

Scientific Calculator	Graphing Calculator
2 $\boxed{\sqrt{}}$ or 2 $\boxed{\text{2ND}\atop\text{INV}}$ $\boxed{x^2}$	$\boxed{\sqrt{}}$ 2 $\boxed{\text{ENTER}}$

The display may read 1.41421356237, although your calculator may display more or fewer digits. Between which two integers would you graph $\sqrt{2}$ on a number line?

The U.N. building is designed with three golden rectangles.

Square Roots

The United Nations Building in New York was designed to represent its mission of promoting world harmony. Viewed from the front, the building looks like three rectangles stacked upon each other. In each rectangle, the width divided by the height is $\sqrt{5} + 1$ to 2, approximately 1.618 to 1. The ancient Greeks believed that such a rectangle, called a **golden rectangle**, was the most pleasing of all rectangles. The comparison 1.618 to 1 is approximate because $\sqrt{5}$ is an irrational number.

The **principal square root** of a nonnegative number n, written $\sqrt{n}$, is the positive number that when multiplied by itself gives n. Thus,

$$\sqrt{36} = 6 \text{ because } 6 \cdot 6 = 36$$

and

$$\sqrt{81} = 9 \text{ because } 9 \cdot 9 = 81.$$

Notice that both $\sqrt{36}$ and $\sqrt{81}$ are rational numbers because 6 and 9 are terminating decimals. Thus, **not all square roots are irrational**.

Numbers such as 36 and 81 are called *perfect squares*. A **perfect square** is a number that is the square of a whole number. The first few perfect squares are listed below.

0 $= 0^2$	**16** $= 4^2$	**64** $= 8^2$	**144** $= 12^2$
1 $= 1^2$	**25** $= 5^2$	**81** $= 9^2$	**169** $= 13^2$
4 $= 2^2$	**36** $= 6^2$	**100** $= 10^2$	**196** $= 14^2$
9 $= 3^2$	**49** $= 7^2$	**121** $= 11^2$	**225** $= 15^2$

The square root of a perfect square is a whole number. For example,

$$\sqrt{0} = 0, \sqrt{1} = 1, \sqrt{4} = 2, \sqrt{9} = 3, \sqrt{16} = 4, \sqrt{25} = 5, \sqrt{36} = 6,$$

and so on.

Simplifying Square Roots

A rule for simplifying square roots can be generalized by comparing $\sqrt{25 \cdot 4}$ and $\sqrt{25} \cdot \sqrt{4}$. Notice that

$$\sqrt{25 \cdot 4} = \sqrt{100} = 10 \quad \text{and} \quad \sqrt{25} \cdot \sqrt{4} = 5 \cdot 2 = 10.$$

Because we obtain 10 in both situations, the original radicals must be equal. That is,

$$\sqrt{25 \cdot 4} = \sqrt{25} \cdot \sqrt{4}.$$

This result is a particular case of the **product rule for square roots** that can be generalized as follows:

> **THE PRODUCT RULE FOR SQUARE ROOTS**
>
> If a and b represent nonnegative numbers, then
> $$\sqrt{ab} = \sqrt{a} \cdot \sqrt{b} \quad \text{and} \quad \sqrt{a} \cdot \sqrt{b} = \sqrt{ab}.$$
> The square root of a product is the product of the square roots.

Example 1 shows how the product rule is used to remove from the square root any perfect squares that occur as factors.

2 Simplify square roots.

STUDY TIP

There are no addition or subtraction rules for square roots:

$$\sqrt{a + b} \neq \sqrt{a} + \sqrt{b}$$
$$\sqrt{a - b} \neq \sqrt{a} - \sqrt{b}.$$

For example, if $a = 9$ and $b = 16$,

$$\sqrt{9 + 16} = \sqrt{25} = 5$$

and

$$\sqrt{9} + \sqrt{16} = 3 + 4 = 7.$$

Thus,

$$\sqrt{9 + 16} \neq \sqrt{9} + \sqrt{16}.$$

EXAMPLE 1   SIMPLIFYING SQUARE ROOTS

Simplify, if possible:

a. $\sqrt{75}$ **b.** $\sqrt{500}$ **c.** $\sqrt{17}$.

SOLUTION

a. $\sqrt{75} = \sqrt{25 \cdot 3}$ *25 is the largest perfect square that is a factor of 75.*

$\qquad = \sqrt{25} \cdot \sqrt{3}$ $\sqrt{ab} = \sqrt{a} \cdot \sqrt{b}$

$\qquad = 5\sqrt{3}$ *Write $\sqrt{25}$ as 5.*

b. $\sqrt{500} = \sqrt{100 \cdot 5}$ *100 is the largest perfect square factor of 500.*

$\qquad = \sqrt{100} \cdot \sqrt{5}$ $\sqrt{ab} = \sqrt{a} \cdot \sqrt{b}$

$\qquad = 10\sqrt{5}$ *Write $\sqrt{100}$ as 10.*

c. Because 17 has no perfect square factors (other than 1), $\sqrt{17}$ cannot be simplified.

Check Point 1

Simplify, if possible:

a. $\sqrt{12}$ **b.** $\sqrt{60}$ **c.** $\sqrt{55}$.

Perform operations with square roots.

Multiplying Square Roots

If a and b are nonnegative, then we can use the product rule

$$\sqrt{a} \cdot \sqrt{b} = \sqrt{a \cdot b}$$

to multiply square roots. The product of the square roots is the square root of the product. Once the square roots are multiplied, simplify the square root of the product when possible.

EXAMPLE 2   MULTIPLYING SQUARE ROOTS

Multiply:

a. $\sqrt{2} \cdot \sqrt{5}$ **b.** $\sqrt{7} \cdot \sqrt{7}$ **c.** $\sqrt{6} \cdot \sqrt{12}$.

SOLUTION

a. $\sqrt{2} \cdot \sqrt{5} = \sqrt{2 \cdot 5} = \sqrt{10}$
b. $\sqrt{7} \cdot \sqrt{7} = \sqrt{7 \cdot 7} = \sqrt{49} = 7$
c. $\sqrt{6} \cdot \sqrt{12} = \sqrt{6 \cdot 12} = \sqrt{72} = \sqrt{36 \cdot 2} = \sqrt{36} \cdot \sqrt{2} = 6\sqrt{2}$

 Multiply:

a. $\sqrt{3} \cdot \sqrt{10}$ **b.** $\sqrt{10} \cdot \sqrt{10}$ **c.** $\sqrt{6} \cdot \sqrt{2}$.

Dividing Square Roots

Another property for square roots involves division.

> ### THE QUOTIENT RULE FOR SQUARE ROOTS
> If a and b represent nonnegative real numbers and $b \neq 0$, then
> $$\frac{\sqrt{a}}{\sqrt{b}} = \sqrt{\frac{a}{b}} \quad \text{and} \quad \sqrt{\frac{a}{b}} = \frac{\sqrt{a}}{\sqrt{b}}.$$
> The quotient of two square roots is the square root of the quotient:

Once the square roots are divided, simplify the square root of the quotient when possible.

EXAMPLE 3   DIVIDING SQUARE ROOTS

Find the quotient:

a. $\dfrac{\sqrt{75}}{\sqrt{3}}$ **b.** $\dfrac{\sqrt{90}}{\sqrt{2}}$.

SOLUTION

a. $\dfrac{\sqrt{75}}{\sqrt{3}} = \sqrt{\dfrac{75}{3}} = \sqrt{25} = 5$

b. $\dfrac{\sqrt{90}}{\sqrt{2}} = \sqrt{\dfrac{90}{2}} = \sqrt{45} = \sqrt{9 \cdot 5} = \sqrt{9} \cdot \sqrt{5} = 3\sqrt{5}$

 Find the quotient:

a. $\dfrac{\sqrt{80}}{\sqrt{5}}$ **b.** $\dfrac{\sqrt{48}}{\sqrt{6}}$.

STUDY TIP

Because
$$\sqrt{7} \cdot \sqrt{7} = \sqrt{49} = 7,$$
it is possible to multiply irrational numbers and get a rational number for an answer.

BLITZER BONUS

Elephants and Irrational Numbers

The height of a fully grown elephant, measured from its foot to its shoulder, is approximately

$2\pi \cdot$ the diameter of its foot.

Source: David Blatner, *The Joy of π,* Walker and Company, 1997

Adding and Subtracting Square Roots

The number that multiplies a square root is called the square root's **coefficient**. For example, in $3\sqrt{5}$, 3 is the coefficient of the square root.

Square roots with the same radicand can be added or subtracted by adding or subtracting their coefficients:

$$a\sqrt{c} + b\sqrt{c} = (a + b)\sqrt{c} \qquad\qquad a\sqrt{c} - b\sqrt{c} = (a - b)\sqrt{c}$$

> Sum of coefficients times the common square root

> Difference of coefficients times the common square root

EXAMPLE 4 | ADDING AND SUBTRACTING SQUARE ROOTS

Add or subtract as indicated:

a. $7\sqrt{2} + 5\sqrt{2}$ **b.** $2\sqrt{5} - 6\sqrt{5}$ **c.** $3\sqrt{7} + 9\sqrt{7} - \sqrt{7}$.

SOLUTION

a.
$$\begin{aligned}
7\sqrt{2} + 5\sqrt{2} &= (7 + 5)\sqrt{2} \\
&= 12\sqrt{2}
\end{aligned}$$

b.
$$\begin{aligned}
2\sqrt{5} - 6\sqrt{5} &= (2 - 6)\sqrt{5} \\
&= -4\sqrt{5}
\end{aligned}$$

c.
$$\begin{aligned}
3\sqrt{7} + 9\sqrt{7} - \sqrt{7} &= 3\sqrt{7} + 9\sqrt{7} - 1\sqrt{7} \quad \text{Write } \sqrt{7} \text{ as } 1\sqrt{7}. \\
&= (3 + 9 - 1)\sqrt{7} \\
&= 11\sqrt{7}
\end{aligned}$$

 Check Point 4

Add or subtract as indicated:

a. $8\sqrt{3} + 10\sqrt{3}$ **b.** $4\sqrt{13} - 9\sqrt{13}$ **c.** $7\sqrt{10} + 2\sqrt{10} - \sqrt{10}$. ◼

In some situations, it is possible to add and subtract square roots that do not contain a common square root by first simplifying.

EXAMPLE 5 | ADDING AND SUBTRACTING SQUARE ROOTS BY FIRST SIMPLIFYING

Add or subtract as indicated:

a. $\sqrt{2} + \sqrt{8}$ **b.** $4\sqrt{50} - 6\sqrt{32}$.

SOLUTION

a.
$$\begin{aligned}
& \sqrt{2} + \sqrt{8} \\
&= \sqrt{2} + \sqrt{4 \cdot 2} && \text{Split 8 into two factors such that one is a perfect square.} \\
&= 1\sqrt{2} + 2\sqrt{2} && \sqrt{4 \cdot 2} = \sqrt{4} \cdot \sqrt{2} = 2\sqrt{2} \\
&= (1 + 2)\sqrt{2} && \text{Add coefficients and retain the common square root.} \\
&= 3\sqrt{2} && \text{Simplify.}
\end{aligned}$$

STUDY TIP

Sums or differences of square roots that cannot be simplified and that do not contain a common square root cannot be combined into one term by adding or subtracting coefficients. Some examples:

- $5\sqrt{3} + 3\sqrt{5}$ cannot be combined by adding coefficients. The square roots, $\sqrt{3}$ and $\sqrt{5}$, are different.

- $28 + 7\sqrt{3}$, or $28\sqrt{1} + 7\sqrt{3}$, cannot be combined by adding coefficients. The square roots, $\sqrt{1}$ and $\sqrt{3}$, are different.

b. $4\sqrt{50} - 6\sqrt{32}$

$= 4\sqrt{25\cdot 2} - 6\sqrt{16\cdot 2}$ 25 is the largest perfect square factor of 50 and 16 is the largest perfect square factor of 32.

$= 4\cdot 5\sqrt{2} - 6\cdot 4\sqrt{2}$ $\sqrt{25\cdot 2} = \sqrt{25}\sqrt{2} = 5\sqrt{2}$ and $\sqrt{16\cdot 2} = \sqrt{16}\sqrt{2} = 4\sqrt{2}$.

$= 20\sqrt{2} - 24\sqrt{2}$ Multiply.

$= (20 - 24)\sqrt{2}$ Subtract coefficients and retain the common square root.

$= -4\sqrt{2}$ Simplify.

Check Point 5   Add or subtract as indicated:

a. $\sqrt{3} + \sqrt{12}$ **b.** $4\sqrt{8} - 7\sqrt{18}$.

 Rationalize the denominator.

Rationalizing the Denominator

You can use a calculator to compare the approximate values for $\dfrac{1}{\sqrt{3}}$ and $\dfrac{\sqrt{3}}{3}$. The two approximations are the same. This is not a coincidence:

$$\frac{1}{\sqrt{3}} = \frac{1}{\sqrt{3}}\cdot \boxed{\frac{\sqrt{3}}{\sqrt{3}}} = \frac{\sqrt{3}}{\sqrt{9}} = \frac{\sqrt{3}}{3}.$$

> Any number divided by itself is 1. Multiplication by 1 does not change the value of $\dfrac{1}{\sqrt{3}}$.

This process involves rewriting a radical expression to remove the square root from the denominator without changing the value of the expression. The process is called **rationalizing the denominator**. If the denominator contains the square root of a natural number that is not a perfect square, **multiply the numerator and denominator by the smallest number that produces the square root of a perfect square in the denominator**.

EXAMPLE 6   RATIONALIZING DENOMINATORS

Rationalize the denominator:

a. $\dfrac{15}{\sqrt{6}}$ **b.** $\sqrt{\dfrac{3}{5}}$ **c.** $\dfrac{12}{\sqrt{8}}$.

SOLUTION

a. If we multiply numerator and denominator of $\dfrac{15}{\sqrt{6}}$ by $\sqrt{6}$, the denominator becomes $\sqrt{6}\cdot\sqrt{6} = \sqrt{36} = 6$. Therefore, we multiply by 1, choosing $\dfrac{\sqrt{6}}{\sqrt{6}}$ for 1.

$$\frac{15}{\sqrt{6}} = \frac{15}{\sqrt{6}}\cdot\frac{\sqrt{6}}{\sqrt{6}} = \frac{15\sqrt{6}}{\sqrt{36}} = \frac{15\sqrt{6}}{6} = \frac{5\sqrt{6}}{2}$$

> Multiply by 1.

> Simplify: $\dfrac{15}{6} = \dfrac{5\cdot 3}{2\cdot 3} = \dfrac{5}{2}$.

b. $\sqrt{\dfrac{3}{5}} = \dfrac{\sqrt{3}}{\sqrt{5}} = \dfrac{\sqrt{3}}{\sqrt{5}}\cdot\dfrac{\sqrt{5}}{\sqrt{5}} = \dfrac{\sqrt{15}}{\sqrt{25}} = \dfrac{\sqrt{15}}{5}$

> Multiply by 1.

STUDY TIP

You can rationalize the denominator of $\dfrac{12}{\sqrt{8}}$ by multiplying by $\dfrac{\sqrt{8}}{\sqrt{8}}$. However, it takes more work to simplify the result.

c. The *smallest* number that will produce a perfect square in the denominator of $\dfrac{12}{\sqrt{8}}$ is $\sqrt{2}$, because $\sqrt{8} \cdot \sqrt{2} = \sqrt{16} = 4$. We multiply by 1, choosing $\dfrac{\sqrt{2}}{\sqrt{2}}$ for 1.

$$\frac{12}{\sqrt{8}} = \frac{12}{\sqrt{8}} \cdot \frac{\sqrt{2}}{\sqrt{2}} = \frac{12\sqrt{2}}{\sqrt{16}} = \frac{12\sqrt{2}}{4} = 3\sqrt{2}$$

 **Check Point 6**   Rationalize the denominator:

a. $\dfrac{25}{\sqrt{10}}$ **b.** $\sqrt{\dfrac{2}{7}}$ **c.** $\dfrac{5}{\sqrt{18}}$.

Irrational Numbers and Other Kinds of Roots

Irrational numbers appear in roots other than square roots. The symbol $\sqrt[3]{}$ represents the **cube root** of a number. For example,

$$\sqrt[3]{8} = 2 \text{ because } 2 \cdot 2 \cdot 2 = 8 \quad \text{and} \quad \sqrt[3]{64} = 4 \text{ because } 4 \cdot 4 \cdot 4 = 64.$$

Although these cube roots are rational numbers, most cube roots are not. For example,

$$\sqrt[3]{217} \approx 6.0092 \text{ because } (6.0092)^3 \approx 216.995, \text{ not exactly } 217.$$

There is no end to the kinds of roots for numbers. For example, $\sqrt[4]{}$ represents the **fourth root** of a number. Thus, $\sqrt[4]{81} = 3$ because $3 \cdot 3 \cdot 3 \cdot 3 = 81$. Although the fourth root of 81 is rational, most fourth roots, fifth roots, and so on tend to be irrational.

Exercise Set 5.4

Practice Exercises

Evaluate each expression in Exercises 1–10.

1. $\sqrt{9}$ **2.** $\sqrt{16}$ **3.** $\sqrt{25}$ **4.** $\sqrt{49}$

5. $\sqrt{64}$ **6.** $\sqrt{100}$ **7.** $\sqrt{121}$ **8.** $\sqrt{144}$

9. $\sqrt{169}$ **10.** $\sqrt{225}$

In Exercises 11–16, use a calculator with a square root key to find a decimal approximation for each square root. Round the number displayed to the nearest **a.** *tenth,* **b.** *hundredth,* **c.** *thousandth.*

11. $\sqrt{173}$ **12.** $\sqrt{3176}$ **13.** $\sqrt{17{,}761}$

14. $\sqrt{779{,}264}$ **15.** $\sqrt{\pi}$ **16.** $\sqrt{2\pi}$

In Exercises 17–24, simplify the square root.

17. $\sqrt{20}$ **18.** $\sqrt{50}$ **19.** $\sqrt{80}$ **20.** $\sqrt{12}$

21. $\sqrt{250}$ **22.** $\sqrt{192}$ **23.** $7\sqrt{28}$ **24.** $3\sqrt{52}$

In Exercises 25–56, perform the indicated operation. Simplify the answer when possible.

25. $\sqrt{7} \cdot \sqrt{6}$ **26.** $\sqrt{19} \cdot \sqrt{3}$ **27.** $\sqrt{6} \cdot \sqrt{6}$

28. $\sqrt{5} \cdot \sqrt{5}$ **29.** $\sqrt{3} \cdot \sqrt{6}$ **30.** $\sqrt{12} \cdot \sqrt{2}$

31. $\sqrt{2} \cdot \sqrt{26}$ **32.** $\sqrt{5} \cdot \sqrt{50}$ **33.** $\dfrac{\sqrt{54}}{\sqrt{6}}$

34. $\dfrac{\sqrt{75}}{\sqrt{3}}$ **35.** $\dfrac{\sqrt{90}}{\sqrt{2}}$ **36.** $\dfrac{\sqrt{60}}{\sqrt{3}}$

37. $\dfrac{-\sqrt{96}}{\sqrt{2}}$ **38.** $\dfrac{-\sqrt{150}}{\sqrt{3}}$ **39.** $7\sqrt{3} + 6\sqrt{3}$

40. $8\sqrt{5} + 11\sqrt{5}$ **41.** $4\sqrt{13} - 6\sqrt{13}$

42. $6\sqrt{17} - 8\sqrt{17}$ **43.** $\sqrt{5} + \sqrt{5}$

44. $\sqrt{3} + \sqrt{3}$ **45.** $4\sqrt{2} - 5\sqrt{2} + 8\sqrt{2}$

46. $6\sqrt{3} + 8\sqrt{3} - 16\sqrt{3}$ **47.** $\sqrt{5} + \sqrt{20}$

48. $\sqrt{3} + \sqrt{27}$ **49.** $\sqrt{50} - \sqrt{18}$

50. $\sqrt{63} - \sqrt{28}$ **51.** $3\sqrt{18} + 5\sqrt{50}$

52. $4\sqrt{12} + 2\sqrt{75}$ **53.** $\dfrac{1}{4}\sqrt{12} - \dfrac{1}{2}\sqrt{48}$

54. $\dfrac{1}{5}\sqrt{300} - \dfrac{2}{3}\sqrt{27}$ **55.** $3\sqrt{75} + 2\sqrt{12} - 2\sqrt{48}$

56. $2\sqrt{72} + 3\sqrt{50} - \sqrt{128}$

In Exercises 57–66, rationalize the denominator.

57. $\dfrac{5}{\sqrt{3}}$ **58.** $\dfrac{12}{\sqrt{5}}$ **59.** $\dfrac{21}{\sqrt{7}}$

60. $\dfrac{30}{\sqrt{5}}$ **61.** $\dfrac{12}{\sqrt{30}}$ **62.** $\dfrac{15}{\sqrt{50}}$

63. $\dfrac{15}{\sqrt{12}}$ **64.** $\dfrac{13}{\sqrt{40}}$ **65.** $\sqrt{\dfrac{2}{5}}$ **66.** $\sqrt{\dfrac{5}{7}}$

67. In the Peanuts cartoon shown below, Woodstock appears to be working steps mentally. Fill in the missing steps that show how to go from $\dfrac{7\sqrt{2 \cdot 2 \cdot 3}}{6}$ to $\dfrac{7}{3}\sqrt{3}$.

PEANUTS reprinted by permission of United Feature Syndicate, Inc.

Application Exercises

Racing cyclists use the formula $s = 4\sqrt{r}$ to determine the maximum speed, in miles per hour, to turn a corner of radius r, in feet, without tipping over. Use this formula to solve Exercises 68–69.

68. What is the maximum speed that a cyclist could travel around a corner of radius 8 feet without tipping over? Write the answer in simplified radical form. Then use the simplified radical form and a calculator to express the answer to the nearest tenth.

69. What is the maximum speed that a cyclist could travel around a corner of radius 12 feet without tipping over? Write the answer in simplified radical form. Then use the simplified radical form and a calculator to express the answer to the nearest tenth.

The formula

$$t = \sqrt{\dfrac{x}{16}}$$

describes the time, t, in seconds, that it takes an object to fall x feet. Use this formula to solve Exercises 70–71.

70. How long will it take a ball dropped from the top of a 320-foot building to hit the ground? Write the answer in simplified radical form. Then use the simplified radical form and a calculator to express the answer to the nearest tenth of a second.

71. How long will it take a ball dropped from the top of a 640-foot building to hit the ground? Write the answer in simplified radical form. Then use the simplified radical form and a calculator to express the answer to the nearest tenth of a second.

The table shows the median, or average, heights for boys of various ages in the United States, from birth through 60 months, or five years. The data can be approximately described by the formula.

$$h = 2.9\sqrt{x} + 20.1$$

where h is the median height, in inches, of boys who are x months of age. Use the formula to solve Exercises 72–73.

BOYS' MEDIAN HEIGHTS

Age (months)	Height (inches)
0	20.5
6	27.0
12	30.8
18	32.9
24	35.0
36	37.5
48	40.8
60	43.4

Source: The Portable Pediatrician for Parents, by Laura Walther Nathanson, M.D., FAAP

72. According to the formula, what is the median height of boys who are 48 months, or four years, old? Use a calculator and round to the nearest tenth of an inch. How well does the formula describe the actual median height shown in the table?

73. According to the formula, what is the median height of boys who are 60 months, or five years, old? Use a calculator and round to the nearest tenth of an inch. How well does the formula describe the actual median height shown in the table?

The bar graph shows the percentage of U.S. households online from 1997 through 2001. The formula

$$P = 6.85\sqrt{t} + 19$$

approximately describes the percentage, P, of U.S. households online t years after 1997. Use the formula to solve Exercises 74–77.

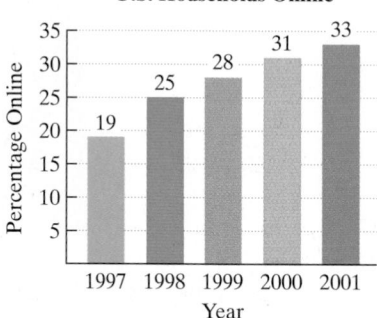

Percentage of U.S. Households Online

Source: Forrester Research

74. According to the formula, what was the percentage of households online in 1998—that is, 1 year after 1997? Answer the question by substituting 1 for *t*. How well does the formula describe the actual data for 1998 shown in the bar graph?

75. According to the formula, what was the percentage of households online in 2001—that is, 4 years after 1997? Answer the question by substituting 4 for *t*. How well does the formula describe the actual data for 2001 shown in the bar graph?

76. According to the formula, how many U.S. households will be online in 2005? Use a calculator and round to the nearest whole percent.

77. According to the formula, how many U.S. households will be online in 2010? Use a calculator and round to the nearest whole percent.

Writing in Mathematics

78. Describe the difference between a rational number and an irrational number.

79. Describe what is wrong with this statement: $\pi = \frac{22}{7}$.

80. Using $\sqrt{50}$, explain how to simplify a square root.

81. Describe how to multiply square roots.

82. Explain how to add square roots with the same radicand.

83. Explain how to add $\sqrt{3} + \sqrt{12}$.

84. Describe what it means to rationalize a denominator. Use $\dfrac{2}{\sqrt{5}}$ in your explanation.

Critical Thinking Exercises

85. Which one of the following is true?

    a. The product of any two irrational numbers is always an irrational number.

    b. $\sqrt{9} + \sqrt{16} = \sqrt{25}$

    c. $\sqrt{\sqrt{16}} = 2$

    d. $\dfrac{\sqrt{64}}{2} = \sqrt{32}$

In Exercises 86–88, insert either $<$ or $>$ in the shaded area between the numbers to make the statement true.

86. $\sqrt{2}$ ▪ 1.5 **87.** $-\pi$ ▪ -3.5 **88.** $-\dfrac{3.14}{2}$ ▪ $-\dfrac{\pi}{2}$

89. How does doubling a number affect its square root?

90. Between which two consecutive integers is $-\sqrt{47}$?

91. Simplify: $\sqrt{2} + \sqrt{\dfrac{1}{2}}$.

92. Create a counterexample to show the following statement is false: The difference between two irrational numbers is always an irrational number.

Group Exercises

The following topics related to irrational numbers are appropriate for either individual or group research projects. A report should be given to the class on the researched topic.

93. A History of How Irrational Numbers Developed

94. Pi: Its History, Applications, and Curiosities

95. Proving That $\sqrt{2}$ Is Irrational

96. Imaginary Numbers: Their History, Applications, and Curiosities

97. The Golden Rectangle in Art and Architecture

SECTION 5.5   REAL NUMBERS AND THEIR PROPERTIES

OBJECTIVES

1. Recognize subsets of the real numbers.

2. Recognize properties of real numbers.

Horror films offer the pleasure of vicarious terror, of being safely scared.

 Recognize subsets of the real numbers.

The Set of Real Numbers

The vampire legend is death as seducer; he/she sucks our blood to take us to a perverse immortality. The vampire resembles us, but appears only at night, hidden among mortals. In this section, you will find vampires in the world of numbers. Mathematicians even use the labels *vampire* and *weird* to describe sets of numbers. However, the label that appears most frequently is *real*. The union of the rational numbers and the irrational numbers is the set of **real numbers**.

The sets that make up the real numbers are summarized in Table 5.2. We refer to these sets as **subsets** of the real numbers, meaning that all elements in each subset are also elements in the set of real numbers.

Real numbers

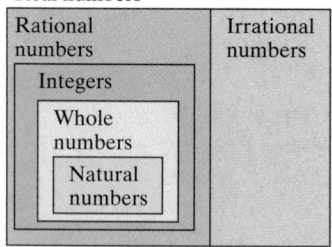

This diagram shows that every real number is rational or irrational.

TABLE 5.2 IMPORTANT SUBSETS OF THE REAL NUMBERS

Name	Description	Examples
Natural numbers	$\{1, 2, 3, 4, 5, \ldots\}$ These numbers are used for counting.	$2, 3, 5, 17$
Whole numbers	$\{0, 1, 2, 3, 4, 5, \ldots\}$ The set of whole numbers is formed by adding 0 to the set of natural numbers.	$0, 2, 3, 5, 17$
Integers	$\{\ldots, -5, -4, -3, -2, -1, 0, 1, 2, 3, 4, 5, \ldots\}$ The set of integers is formed by adding negatives of the natural numbers to the set of whole numbers.	$-17, -5, -3, -2,$ $0, 2, 3, 5, 17$
Rational numbers	The set of rational numbers is the set of all numbers which can be expressed in the form $\frac{a}{b}$, where a and b are integers, $b \neq 0$. The decimal representation of a rational number either terminates or repeats.	$-17 = \frac{-17}{1}, -5 = \frac{-5}{1}, -3, -2,$ $0, 2, 3, 5, 17,$ $\frac{2}{5} = 0.4,$ $\frac{-2}{3} = -0.6666\ldots = -0.\overline{6}$
Irrational numbers	This is the set of numbers whose decimal representations are neither terminating nor repeating. Irrational numbers cannot be expressed as a quotient of integers.	$\sqrt{2} \approx 1.414214$ $-\sqrt{3} \approx -1.73205$ $\pi \approx 3.142$ $-\frac{\pi}{2} \approx -1.571$

EXAMPLE 1 │ CLASSIFYING REAL NUMBERS

Consider the following set of numbers:

$$\left\{ -7, -\frac{3}{4}, 0, 0.\overline{6}, \sqrt{5}, \pi, 7.3, \sqrt{81} \right\}.$$

List the numbers in the set that are

a. natural numbers **b.** whole numbers
c. integers **d.** rational numbers
e. irrational numbers **f.** real numbers

SOLUTION

a. Natural numbers: The natural numbers are the numbers used for counting. The only natural number in the set is $\sqrt{81}$ because $\sqrt{81} = 9$. (9 multiplied by itself is 81.)

b. Whole numbers: The whole numbers consist of the natural numbers and 0. The elements of the set that are whole numbers are 0 and $\sqrt{81}$.

c. Integers: The integers consist of the natural numbers, 0, and the negatives of the natural numbers. The elements of the set that are integers are $\sqrt{81}$, 0, and -7.

$$\left\{-7, -\frac{3}{4}, 0, 0.\overline{6}, \sqrt{5}, \pi, 7.3, \sqrt{81}\right\}$$

The given set of numbers, repeated

d. Rational numbers: All numbers in the set that can be expressed as the quotient of integers are rational numbers. These include $-7\left(-7 = -\frac{7}{1}\right)$, $-\frac{3}{4}$, $0\left(0 = \frac{0}{1}\right)$, and $\sqrt{81}\left(\sqrt{81} = \frac{9}{1}\right)$. Furthermore, all numbers in the set that are terminating or repeating decimals are also rational numbers. These include $0.\overline{6}$ and 7.3.

e. Irrational numbers: The irrational numbers in the set are $\sqrt{5}$ ($\sqrt{5} \approx 2.236$) and π ($\pi \approx 3.14$). Both $\sqrt{5}$ and π are only approximately equal to 2.236 and 3.14, respectively. In decimal form, $\sqrt{5}$ and π neither terminate nor have blocks of repeating digits.

f. Real numbers: All the numbers in the given set are real numbers.

 **Check Point 1** Consider the following set of numbers:

$$\left\{-9, -1.3, 0, 0.\overline{3}, \frac{\pi}{2}, \sqrt{9}, \sqrt{10}\right\}.$$

List the numbers in the set that are

a. natural numbers   **b.** whole numbers

c. integers   **d.** rational numbers

e. irrational numbers   **f.** real numbers

BLITZER BONUS

Vampire Numbers

Like legendary vampires that lie concealed among humans, vampire numbers lie hidden within the set of real numbers, mostly undetected. By definition, vampire numbers have an even number of digits. Furthermore, they are the product of two numbers whose digits all survive, in scrambled form, in the vampire. For example, 1260, 1435, and 2187 are vampire numbers.

$$21 \times 60 = 1260 \qquad 35 \times 41 = 1435 \qquad 27 \times 81 = 2187$$

| The digits 2, 1, 6, and 0 lie scrambled in the vampire number. | The digits 3, 5, 4, and 1 lurk within the vampire number. | The digits 2, 7, 8, and 1 survive in the vampire number. |

As the real numbers grow increasingly larger, is it necessary to pull out a wooden stake with greater frequency? How often can you expect to find vampires hidden among the giants? And is it possible to find a weird vampire?

Here is a 40-digit vampire number that was discovered using a Pascal program on a personal computer:

$$98{,}765{,}432{,}198{,}765{,}432{,}198 \times 98{,}765{,}432{,}198{,}830{,}604{,}534 =$$

$$9{,}754{,}610{,}597{,}415{,}368{,}368{,}844{,}499{,}268{,}390{,}128{,}385{,}732.$$

Source: Clifford Pickover, *Wonders of Numbers*, Oxford University Press, 2001

---

2 Recognize properties of real numbers.

Properties of the Real Numbers

When you use your calculator to add two real numbers, you can enter them in any order. The fact that two real numbers can be added in any order is called the **commutative property of addition**. You probably use this property, as well as other properties of real numbers listed in Table 5.3, without giving it much thought. The properties of the real numbers are especially useful in algebra, as we shall see in Chapter 6.

TABLE 5.3 PROPERTIES OF THE REAL NUMBERS

Name	Meaning	Examples
Closure Property of Addition	The sum of any two real numbers is a real number.	$4\sqrt{2}$ is a real number and $5\sqrt{2}$ is a real number, so $4\sqrt{2} + 5\sqrt{2}$, or $9\sqrt{2}$, is a real number.
Closure Property of Multiplication	The product of any two real numbers is a real number.	10 is a real number and $\frac{1}{2}$ is a real number, so $10 \cdot \frac{1}{2}$, or 5, is a real number.
Commutative Property of Addition	Two real numbers can be added in any order. $a + b = b + a$	$13 + 7 = 7 + 13$
Commutative Property of Multiplication	Two real numbers can be multiplied in any order. $ab = ba$	$13 \cdot 7 = 7 \cdot 13$ $\sqrt{2} \cdot \sqrt{5} = \sqrt{5} \cdot \sqrt{2}$
Associative Property of Addition	If three real numbers are added, it makes no difference which two are added first. $(a + b) + c = a + (b + c)$	$(7 + 2) + 5 = 7 + (2 + 5)$ $9 + 5 = 7 + 7$ $14 = 14$
Associative Property of Multiplication	If three real numbers are multiplied, it makes no difference which two are multiplied first. $(a \cdot b) \cdot c = a \cdot (b \cdot c)$	$(7 \cdot 2) \cdot 5 = 7 \cdot (2 \cdot 5)$ $14 \cdot 5 = 7 \cdot 10$ $70 = 70$
Distributive Property of Multiplication over Addition	Multiplication distributes over addition. $a \cdot (b + c) = a \cdot b + a \cdot c$	$7(4 + \sqrt{3}) = 7 \cdot 4 + 7 \cdot \sqrt{3}$ $= 28 + 7\sqrt{3}$

STUDY TIP

Commutative: Changes *order*.
Associative: Changes *grouping*.

EXAMPLE 2   IDENTIFYING PROPERTIES OF THE REAL NUMBERS

Name the property illustrated:

a. $\sqrt{3} \cdot 7 = 7 \cdot \sqrt{3}$

b. $(4 + 7) + 6 = 4 + (7 + 6)$

c. $2(3 + \sqrt{5}) = 6 + 2\sqrt{5}$

d. $\sqrt{2} + (\sqrt{3} + \sqrt{7}) = \sqrt{2} + (\sqrt{7} + \sqrt{3})$.

SOLUTION

a. $\sqrt{3} \cdot 7 = 7 \cdot \sqrt{3}$ Commutative property of multiplication

b. $(4 + 7) + 6 = 4 + (7 + 6)$ Associative property of addition

c. $2(3 + \sqrt{5}) = 6 + 2\sqrt{5}$ Distributive property of multiplication over addition

d. $\sqrt{2} + (\sqrt{3} + \sqrt{7}) = \sqrt{2} + (\sqrt{7} + \sqrt{3})$ The only change between the left and the right sides is in the order that $\sqrt{3}$ and $\sqrt{7}$ are added. The order is changed from $\sqrt{3} + \sqrt{7}$ to $\sqrt{7} + \sqrt{3}$ using the commutative property of addition.

Check 2 Point

Name the property illustrated:

a. $(4 \cdot 7) \cdot 3 = 4 \cdot (7 \cdot 3)$

b. $3(\sqrt{5} + 4) = 3(4 + \sqrt{5})$

c. $3(\sqrt{5} + 4) = 3\sqrt{5} + 12$

d. $2(\sqrt{3} + \sqrt{7}) = (\sqrt{3} + \sqrt{7})2.$

BLITZER BONUS

The Associative Property and the English Language

In the English language, sentences can take on different meanings depending on the way the words are associated with commas. Here are two examples.

- Do not break your bread or roll in your soup.
  Do not break your bread, or roll in your soup.

- Woman, without her man, is nothing.
  Woman, without her, man is nothing.

Although the entire set of real numbers is closed with respect to addition and multiplication, some of the subsets of the real numbers do not satisfy the closure property for a given operation. If the operation results in just one answer that is not in the set, then the set is not closed for that operation.

EXAMPLE 3 | VERIFYING CLOSURE

a. Are the integers closed with respect to multiplication?

b. Are the irrational numbers closed with respect to multiplication?

c. Are the natural numbers closed with respect to division?

SOLUTION

a. Consider some examples of the multiplication of integers:

$$3 \cdot 2 = 6 \qquad 3(-2) = -6 \qquad -3(-2) = 6 \qquad -3 \cdot 0 = 0.$$

The product of any two integers is always a positive integer, a negative integer, or zero, which is an integer. Thus, the integers are closed under the operation of multiplication.

b. If we multiply two irrational numbers, must the product always be an irrational number? The answer is no. Here is an example:

$$\sqrt{7} \cdot \sqrt{7} = \sqrt{49} = 7.$$

Both irrational   Not an irrational number

This means that the irrational numbers are not closed under the operation of multiplication.

c. If we divide any two natural numbers, must the quotient always be a natural number? The answer is no. Here is an example:

$$4 \div 8 = \frac{1}{2}.$$

Both natural numbers   Not a natural number

Thus, the natural numbers are not closed under the operation of division.

a. Are the natural numbers closed with respect to multiplication?

b. Are the integers closed with respect to division?

The commutative property involves a change in order with no change in the final result. However, changing the order in which we subtract and divide real numbers can produce different answers. For example,

$$7 - 4 \neq 4 - 7 \quad \text{and} \quad 6 \div 2 \neq 2 \div 6.$$

Because the real numbers are not commutative with respect to subtraction and division, it is important that you enter numbers in the correct order when using a calculator to perform these operations.

The associative property does not hold for the operations of subtraction and division. The examples below show that if we change groupings when subtracting or dividing three numbers, the answer changes.

$$(6 - 1) - 3 \neq 6 - (1 - 3) \qquad (8 \div 4) \div 2 \neq 8 \div (4 \div 2)$$
$$5 - 3 \neq 6 - (-2) \qquad\qquad 2 \div 2 \neq 8 \div 2$$
$$2 \neq 8 \qquad\qquad\qquad 1 \neq 4$$

BLITZER BONUS

Beyond the Real Numbers

Only real numbers greater than or equal to zero have real number square roots. The square root of -1, $\sqrt{-1}$, is not a real number. This is because there is no real number that can be multiplied by itself that results in -1. Multiplying any real number by itself can never give a negative product. In the sixteenth century, mathematician Girolamo Cardano wrote that square roots of negative numbers would cause "mental tortures." In spite of these "tortures," mathematicians invented a new number, called i, to represent $\sqrt{-1}$. The number i is not a real number; it is called an **imaginary number**. Thus, $\sqrt{9} = 3$, $-\sqrt{9} = -3$, but $\sqrt{-9}$ is not a real number. However, $\sqrt{-9}$ is an imaginary number, represented by $3i$.

Exercise Set 5.5

Practice Exercises

In Exercises 1–4, list all numbers from the given set that are

a. *natural numbers* **b.** *whole numbers* **c.** *integers*

d. *rational numbers* **e.** *irrational numbers* **f.** *real numbers*

1. $\left\{-9, -\frac{4}{5}, 0, 0.25, \sqrt{3}, 9.2, \sqrt{100}\right\}$

2. $\left\{-7, -0.\overline{6}, 0, \sqrt{49}, \sqrt{50}\right\}$

3. $\left\{-11, -\frac{5}{6}, 0, 0.75, \sqrt{5}, \pi, \sqrt{64}\right\}$

4. $\left\{-5, -0.\overline{3}, 0, \sqrt{2}, \sqrt{4}\right\}$

5. Give an example of a whole number that is not a natural number.

6. Give an example of an integer that is not a whole number.

7. Give an example of a rational number that is not an integer.

8. Give an example of a rational number that is not a natural number.

9. Give an example of a number that is an integer, a whole number, and a natural number.

10. Give an example of a number that is a rational number, an integer, and a real number.

11. Give an example of a number that is an irrational number and a real number.

12. Give an example of a number that is a real number, but not an irrational number.

Complete each statement in Exercises 13–15 to illustrate the commutative property.

13. $3 + (4 + 5) = 3 + (5 + \underline{\quad})$

14. $\sqrt{5} \cdot 4 = 4 \cdot \underline{\quad}$

15. $9 \cdot (6 + 2) = 9 \cdot (2 + \underline{\quad})$

Complete each statement in Exercises 16–17 to illustrate the associative property.

16. $(3 + 7) + 9 = \underline{\quad} + (7 + \underline{\quad})$

17. $(4 \cdot 5) \cdot 3 = \underline{\quad} \cdot (5 \cdot \underline{\quad})$

Complete each statement in Exercises 18–20 to illustrate the distributive property.

18. $3 \cdot (6 + 4) = 3 \cdot 6 + 3 \cdot \underline{\quad}$

19. $\underline{\quad} \cdot (4 + 5) = 7 \cdot 4 + 7 \cdot 5$

20. $2 \cdot (\underline{\quad} + 3) = 2 \cdot 7 + 2 \cdot 3$

Use the distributive property to simplify the radical expressions in Exercises 21–28.

21. $5\left(6 + \sqrt{2}\right)$

22. $4\left(3 + \sqrt{5}\right)$

23. $\sqrt{7}\left(3 + \sqrt{2}\right)$

24. $\sqrt{6}\left(7 + \sqrt{5}\right)$

25. $\sqrt{3}\left(5 + \sqrt{3}\right)$

26. $\sqrt{7}\left(9 + \sqrt{7}\right)$

27. $\sqrt{6}\left(\sqrt{2} + \sqrt{6}\right)$

28. $\sqrt{10}\left(\sqrt{2} + \sqrt{10}\right)$

In Exercises 29–38, state the name of the property illustrated.

29. $6 + (-4) = (-4) + 6$

30. $11 \cdot (7 + 4) = 11 \cdot 7 + 11 \cdot 4$

31. $6 + (2 + 7) = (6 + 2) + 7$

32. $6 \cdot (2 \cdot 3) = 6 \cdot (3 \cdot 2)$

33. $(2 + 3) + (4 + 5) = (4 + 5) + (2 + 3)$

34. $7 \cdot (11 \cdot 8) = (11 \cdot 8) \cdot 7$

35. $2(-8 + 6) = -16 + 12$

36. $-8(3 + 11) = -24 + (-88)$

37. $\left(2\sqrt{3}\right) \cdot \sqrt{5} = 2\left(\sqrt{3} \cdot \sqrt{5}\right)$

38. $\sqrt{2}\pi = \pi\sqrt{2}$

In Exercises 39–43, use two numbers to show that

39. the natural numbers are not closed with respect to subtraction.

40. the natural numbers are not closed with respect to division.

41. the integers are not closed with respect to division.

42. the irrational numbers are not closed with respect to subtraction.

43. the irrational numbers are not closed with respect to multiplication.

Application Exercises

44. Are first putting on your left shoe and then putting on your right shoe commutative?

45. Are first getting undressed and then taking a shower commutative?

46. Give an example of two things that you do that are not commutative.

47. Give an example of two things that you do that are commutative.

48. Closure illustrates that a characteristic of a set is not necessarily a characteristic of all of its subsets. The real numbers are closed with respect to multiplication, but the irrational numbers, a subset of the real numbers, are not. Give an example of a set that is not mathematical that has a particular characteristic, but which has a subset without this characteristic.

Writing in Mathematics

49. What does it mean when we say that the rational numbers are a subset of the real numbers?

50. What does it mean if we say that a set is closed under a given operation?

51. State the commutative property of addition and give an example.

52. State the commutative property of multiplication and give an example.

53. State the associative property of addition and give an example.

54. State the associative property of multiplication and give an example.

55. State the distributive property of multiplication over addition and give an example.

56. Does $7 \cdot (4 \cdot 3) = 7 \cdot (3 \cdot 4)$ illustrate the commutative property or the associative property? Explain your answer.

Critical Thinking Exercises

57. Which one of the following statements is true?

   a. Every rational number is an integer.

   b. Some whole numbers are not integers.

   c. Some rational numbers are not positive.

   d. Irrational numbers cannot be negative.

58. Which one of the following statements is true?

   a. Subtraction is a commutative operation.

   b. $(24 \div 6) \div 2 = 24 \div (6 \div 2)$

   c. $7 \cdot a + 3 \cdot a = a \cdot (7 + 3)$

   d. $2 \cdot a + 5 = 5 \cdot a + 2$

SECTION 5.6 EXPONENTS AND SCIENTIFIC NOTATION

OBJECTIVES

1. Use properties of exponents.

2. Convert from scientific to decimal notation.

3. Convert from decimal to scientific notation.

4. Perform computations using scientific notation.

5. Solve applied problems using scientific notation.

We frequently encounter very large and very small numbers. Governments throughout the world are concerned about the billions of tons of carbon dioxide that the global population of 6 billion people release into the atmosphere each year. The national debt of the United States is about $6.8 trillion. A typical atom has a diameter of about one-ten-billionth of a meter. Exponents provide a way of putting these large and small numbers in perspective.

1   Use properties of exponents.

Properties of Exponents

We have seen that exponents are used to indicate repeated multiplication. For example,

$$2^4 \cdot 2^3$$

means that we are multiplying 4 factors of 2 and 3 factors of 2. We have a total of 7 factors of 2:

$$2^4 \cdot 2^3 = \underbrace{(2 \cdot 2 \cdot 2 \cdot 2)}_{\text{4 factors of 2}} \cdot \underbrace{(2 \cdot 2 \cdot 2)}_{\text{3 factors of 2}}.$$

Total: 7 factors of 2

Thus,

$$2^4 \cdot 2^3 = 2^7.$$

We can quickly find the exponent on the product, 7, by adding 4 and 3, the original exponents. Products of exponential expressions with the same base are found by adding exponents. For example,

$$3^2 \cdot 3^6 = 3^{2+6} = 3^8.$$

BLITZER BONUS

A Number with 369 Million Digits

The largest number that can be expressed with only three digits is

$$9^{(9^9)} \text{ or } 9^{387,420,489}.$$

The number begins with 428124773 ..., has 369 million digits, and would take around 70 years to read.

Properties of exponents allow us to perform operations with exponential expressions without having to write out long strings of factors. Three such properties are given in Table 5.4.

TABLE 5.4 PROPERTIES OF EXPONENTS

Property	Meaning	Examples
The Product Rule $b^m \cdot b^n = b^{m+n}$	Add exponents when multiplying with the same base. Use this sum as the exponent of the common base.	$9^6 \cdot 9^{12} = 9^{6+12} = 9^{18}$
The Power Rule $(b^m)^n = b^{m \cdot n}$	Multiply exponents when an exponential expression is raised to a power. Place the product of the exponents on the base and remove the parentheses.	$(3^4)^5 = 3^{4 \cdot 5} = 3^{20}$ $(5^3)^8 = 5^{3 \cdot 8} = 5^{24}$
The Quotient Rule $\dfrac{b^m}{b^n} = b^{m-n}$	Subtract exponents when dividing with the same base. Use this difference as the exponent of the common base.	$\dfrac{5^{12}}{5^4} = 5^{12-4} = 5^8$ $\dfrac{9^{40}}{9^5} = 9^{40-5} = 9^{35}$

The third property in Table 5.4, the quotient rule, can lead to a zero exponent when subtracting exponents. Here is an example:

$$\frac{4^3}{4^3} = 4^{3-3} = 4^0.$$

We can see what this zero exponent means by evaluating 4^3 in the numerator and the denominator:

$$\frac{4^3}{4^3} = \frac{4 \cdot 4 \cdot 4}{4 \cdot 4 \cdot 4} = \frac{64}{64} = 1.$$

This means that 4^0 must equal 1. This example illustrates the zero exponent rule.

THE ZERO EXPONENT RULE

If b is any real number other than 0,

$$b^0 = 1.$$

EXAMPLE 1 USING THE ZERO EXPONENT RULE

Use the zero exponent rule to simplify:

a. 7^0 **b.** π^0 **c.** $(-5)^0$ **d.** -5^0.

SOLUTION

a. $7^0 = 1$ **b.** $\pi^0 = 1$ **c.** $(-5)^0 = 1$ **d.** $-5^0 = -1$

Only 5 is raised to the 0 power.

Check **1** Point   Use the zero exponent rule to simplify:

a. 19^0 **b.** $(3\pi)^0$ **c.** $(-14)^0$ **d.** -14^0.

The quotient rule can result in a negative exponent. Consider, for example, $4^3 \div 4^5$:

$$\frac{4^3}{4^5} = 4^{3-5} = 4^{-2}.$$

We can see what this negative exponent means by evaluating the numerator and the denominator:

$$\frac{4^3}{4^5} = \frac{\cancel{4} \cdot \cancel{4} \cdot \cancel{4}}{\cancel{4} \cdot \cancel{4} \cdot \cancel{4} \cdot 4 \cdot 4} = \frac{1}{4^2}.$$

Notice that $\dfrac{4^3}{4^5}$ equals both 4^{-2} and $\dfrac{1}{4^2}$. This means that 4^{-2} must equal $\dfrac{1}{4^2}$. This example is a particular case of the negative exponent rule.

THE NEGATIVE EXPONENT RULE

If b is any real number other than 0 and m is a natural number,

$$b^{-m} = \frac{1}{b^m}.$$

NAMES OF LARGE NUMBERS

10^2	hundred
10^3	thousand
10^6	million
10^9	billion
10^{12}	trillion
10^{15}	quadrillion
10^{18}	quintillion
10^{21}	sextillion
10^{24}	septillion
10^{27}	octillion
10^{30}	nonillion
10^{100}	googol
10^{googol}	googoplex

John Scott "*Invoking the Googoplex*" 1981, water-colour on paper, 24×18 in. Photo courtesy/ The Gallery, Stratford, Ontario.

EXAMPLE 2 USING THE NEGATIVE EXPONENT RULE

Use the negative exponent rule to simplify:

a. 8^{-2} **b.** 5^{-3} **c.** 7^{-1}.

SOLUTION

a. $8^{-2} = \dfrac{1}{8^2} = \dfrac{1}{8 \cdot 8} = \dfrac{1}{64}$

b. $5^{-3} = \dfrac{1}{5^3} = \dfrac{1}{5 \cdot 5 \cdot 5} = \dfrac{1}{125}$

c. $7^{-1} = \dfrac{1}{7^1} = \dfrac{1}{7}$

Check **2** Point

Use the negative exponent rule to simplify:

a. 9^{-2} **b.** 6^{-3} **c.** 12^{-1}.

Powers of Ten

Exponents and their properties allow us to represent and compute with numbers that are large and small. For example, one billion, or 1,000,000,000 can be written as 10^9. In terms of exponents, 10^9 might not look very large, but consider this: If you can count to 200 in one minute and decide to count for 12 hours a day at this rate, it would take you in the region of 19 years, 9 days, 5 hours, and 20 minutes to count to 10^9!

Powers of 10 follow two basic rules:

1. **A positive exponent tells how many 0s follow the 1.** For example, 10^9 (one billion) is a 1 followed by nine zeros: 1,000,000,000. A googol, 10^{100}, is a 1 followed by one hundred zeros. (A googol far exceeds the number of protons, neutrons, and electrons in the universe.) A googol is a veritable pipsqueak compared to the googolplex, 10 raised to the googol power, or $10^{10^{100}}$; that's a 1 followed by a googol zeros. (If each zero in a googolplex were no larger than a grain of sand, there would not be enough room in the universe to represent the number.)

2. A negative exponent tells how many places there are to the right of the decimal point. For example, 10^{-9} (one billionth) has nine places to the right of the decimal point. The nine places contain eight 0s and the 1.

$$10^{-9} = 0.\underbrace{000000001}_{\text{nine places}}$$

BLITZER BONUS

Earthquakes and Powers of Ten

The earthquake that ripped through northern California on October 17, 1989, measured 7.1 on the Richter scale, killed more than 60 people, and injured more than 2400. Shown here is San Francisco's Marina district, where shock waves tossed houses off their foundations and into the street.

The Richter scale is misleading because it is not actually a 1 to 8, but rather a 1 to 10 million scale. Each level indicates a tenfold increase in magnitude from the previous level, making a 7.0 earthquake a million times greater than a 1.0 quake.

The following is a translation of the Richter scale:

Richter number (R)	Magnitude (10^{R-1})
1	$10^{1-1} = 10^0 = 1$
2	$10^{2-1} = 10^1 = 10$
3	$10^{3-1} = 10^2 = 100$
4	$10^{4-1} = 10^3 = 1000$
5	$10^{5-1} = 10^4 = 10,000$
6	$10^{6-1} = 10^5 = 100,000$
7	$10^{7-1} = 10^6 = 1,000,000$
8	$10^{8-1} = 10^7 = 10,000,000$

Scientific Notation

By the end of 2003, the national debt of the United States was about $6.8 trillion. This is the amount of money the government has had to borrow over the years, mostly by selling bonds, because it has spent more than it has collected in taxes. A stack of $1 bills equaling the national debt would rise to twice the distance from Earth to the moon. Because a trillion is 10^{12}, the national debt can be expressed as

$$6.8 \times 10^{12}.$$

The number 6.8×10^{12} is written in a form called *scientific notation*.

SCIENTIFIC NOTATION

A positive number is written in **scientific notation** when it is expressed in the form

$$a \times 10^n$$

where a is a number greater than or equal to 1 and less than 10 ($1 \leq a < 10$) and n is an integer.

It is customary to use the multiplication symbol, $\times$, rather than a dot, when writing a number in scientific notation.

Here are two examples of numbers in scientific notation:

- Each day, 2.6×10^7 pounds of dust from the atmosphere settle on Earth.
- The diameter of a hydrogen atom is 1.06×10^{-8} centimeter.

2 Convert from scientific to decimal notation.

We can use n, the exponent on the 10 in $a \times 10^n$, to change a number in scientific notation to decimal notation. If n is **positive**, move the decimal point in a to the **right** n places. If n is **negative**, move the decimal point in a to the **left** n places.

EXAMPLE 3   CONVERTING FROM SCIENTIFIC TO DECIMAL NOTATION

Write each number in decimal notation:

a. 2.6×10^7 **b.** 1.016×10^{-8}.

SOLUTION

a. We express 2.6×10^7 in decimal notation by moving the decimal point in 2.6 seven places to the right. We need to add six zeros.
$$2.6 \times 10^7 = 26{,}000{,}000$$

b. We express 1.016×10^{-8} in decimal notation by moving the decimal point in 1.016 eight places to the left. We need to add seven zeros to the right of the decimal point.
$$1.016 \times 10^{-8} = 0.00000001016$$

 Check
3
Point

Write each number in decimal notation:

a. 7.4×10^9 **b.** 3.017×10^{-6}.

3 Convert from decimal to scientific notation.

To convert a positive number from decimal notation to scientific notation, $a \times 10^n$, we reverse the procedure of Example 3.

- Move the decimal point in the given number to obtain a number greater than or equal to 1 and less than 10. This number is a in $a \times 10^n$.
- The number of places the decimal point moves gives the absolute value of n in $a \times 10^n$; n is positive if the given number is greater than or equal to 10 and n is negative if the given number is less than 1.

EXAMPLE 4   CONVERTING FROM DECIMAL NOTATION TO SCIENTIFIC NOTATION

Write each number in scientific notation:

a. 4,600,000   **b.** 0.00023.

SOLUTION

a. $4{,}600{,}000$ $=$ 4.6 $\times$ 10^6

| This number is greater than 10, so n is positive in $a \times 10^n$. | Move the decimal point in 4,600,000 to get $1 \leq a < 10$. | The decimal point moved 6 places from 4,600,000 to 4.6. |

b. 0.00023 $=$ 2.3 $\times$ 10^{-4}

| This number is less than 1, so n is negative in $a \times 10^n$. | Move the decimal point in 0.00023 to get $1 \leq a < 10$. | The decimal point moved 4 places from 0.00023 to 2.3. |

TECHNOLOGY

You can use your calculator's $\boxed{\text{EE}}$ (enter exponent) or $\boxed{\text{EXP}}$ key to convert from decimal to scientific notation. Here is how it's done for 0.00023:

Many Scientific Calculators

Keystrokes	Display
.00023 $\boxed{\text{EE}}$ $\boxed{=}$	$2.3 - 04$

Many Graphing Calculators

Use the mode setting for scientific notation.

Keystrokes	Display
.00023 $\boxed{\text{ENTER}}$	$2.3\text{E} - 4$

 Write each number in scientific notation:

a. 7,410,000,000 **b.** 0.000000092

Number of Unsolicited e-Mails Sent

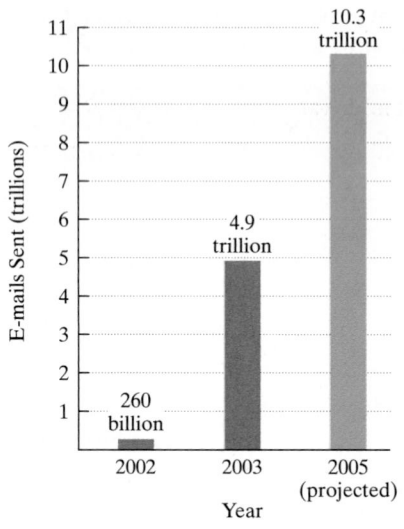

Source: Jupiter Research

FIGURE 5.6

EXAMPLE 5 EXPRESSING THE NUMBER OF UNSOLICITED E-MAILS IN SCIENTIFIC NOTATION

Driven by cheap technology and the promise of easy profit, spammers have evolved from pests to an invasive species that threatens to clog the inner workings of the Internet. In 2003, more than half the e-mails received by U.S. businesses were unsolicited, costing $8.9 billion in lost productivity. The bar graph in Figure 5.6 shows the number of unsolicited e-mails sent in 2002 and 2003, with projections for 2005. Express the number of unsolicited e-mails in 2002 in scientific notation.

SOLUTION Because a billion is 10^9, the number of unsolicited e-mails sent in 2002 can be expressed as

$$260 \times 10^9.$$

> This factor is not between 1 and 10, so the number is not in scientific notation.

The voice balloon indicates that we need to convert 260 to scientific notation.

$$260 \times 10^9 = (2.6 \times 10^2) \times 10^9 = 2.6 \times 10^{2+9} = 2.6 \times 10^{11}$$

There were 2.6×10^{11} unsolicited e-mails sent in 2002.

 In 2002, the U.S. budget deficit was $106.2 billion. Express this amount in scientific notation.

Computations with Scientific Notation

We use the product rule for exponents to multiply numbers in scientific notation:

$$(a \times 10^n) \times (b \times 10^m) = (a \times b) \times 10^{n+m}.$$

Add the exponents on 10 and multiply the other parts of the numbers separately.

EXAMPLE 6 MULTIPLYING NUMBERS IN SCIENTIFIC NOTATION

Multiply: $(3.4 \times 10^9)(2 \times 10^{-5})$. Write the product in decimal notation.

SOLUTION

$$
\begin{aligned}
(3.4 \times 10^9)(2 \times 10^{-5}) &= (3.4 \times 2) \times (10^9 \times 10^{-5}) && \text{Regroup factors.}\\
&= 6.8 \times 10^{9+(-5)} && \text{Add exponents on 10.}\\
&= 6.8 \times 10^4 && \text{Simplify.}\\
&= 68{,}000 && \text{Write the product in decimal notation.}
\end{aligned}
$$

 Multiply: $(1.3 \times 10^7)(4 \times 10^{-2})$. Write the product in decimal notation.

4 Perform computations using scientific notation.

TECHNOLOGY

$(3.4 \times 10^9)(2 \times 10^{-5})$ **on a Calculator:**

Many Scientific Calculators

3.4 $\boxed{\text{EE}}$ 9 $\boxed{\times}$ 2 $\boxed{\text{EE}}$ 5 $\boxed{+/-}$ $\boxed{=}$

Display: 6.8 04

Many Graphing Calculators

3.4 $\boxed{\text{EE}}$ 9 $\boxed{\times}$ 2 $\boxed{\text{EE}}$ $\boxed{(-)}$ 5 $\boxed{\text{ENTER}}$

Display: 6.8 E 4

We use the quotient rule for exponents to divide numbers in scientific notation:

$$\frac{a \times 10^n}{b \times 10^m} = \left(\frac{a}{b}\right) \times 10^{n-m}.$$

Subtract the exponents on 10 and divide the other parts of the numbers separately.

EXAMPLE 7   DIVIDING NUMBERS IN SCIENTIFIC NOTATION

Divide: $\dfrac{8.4 \times 10^{-7}}{4 \times 10^{-4}}$. Write the quotient in decimal notation.

SOLUTION

$$\frac{8.4 \times 10^{-7}}{4 \times 10^{-4}} = \left(\frac{8.4}{4}\right) \times \left(\frac{10^{-7}}{10^{-4}}\right) \qquad \text{Regroup factors.}$$

$$= 2.1 \times 10^{-7-(-4)} \qquad \text{Subtract exponents on 10.}$$

$$= 2.1 \times 10^{-7+4} \qquad \text{Simplify:} \quad -7 - (-4) = -7 + 4.$$

$$= 2.1 \times 10^{-3} \qquad \text{Simplify.}$$

$$= 0.0021 \qquad \text{Write the quotient in decimal notation.}$$

Divide: $\dfrac{6.9 \times 10^{-8}}{3 \times 10^{-2}}$. Write the quotient in decimal notation.

Multiplication and division involving very large or very small numbers can be performed by first converting each number to scientific notation.

EXAMPLE 8   USING SCIENTIFIC NOTATION TO MULTIPLY

Multiply: $0.00064 \times 9{,}400{,}000{,}000$. Express the product in **a.** scientific notation and **b.** decimal notation.

SOLUTION

a. $0.00064 \times 9{,}400{,}000{,}000$

$$= 6.4 \times 10^{-4} \times 9.4 \times 10^9 \qquad \text{Write each number in scientific notation.}$$

$$= (6.4 \times 9.4) \times (10^{-4} \times 10^9) \qquad \text{Regroup factors.}$$

$$= 60.16 \times 10^5 \qquad \text{Add exponents on 10: } -4 + 9 = 5$$

$$= (6.016 \times 10) \times 10^5 \qquad \text{Express 60.16 in scientific notation.}$$

$$= 6.016 \times 10^6 \qquad \text{Add exponents on 10:}$$
$$10^1 \times 10^5 = 10^{1+5} = 10^6.$$

b. The answer in decimal notation is obtained by moving the decimal point in 6.016 six places to the right. The product is 6,016,000.

Multiply: $0.0036 \times 5{,}200{,}000$. Express the product in **a.** scientific notation and **b.** decimal notation.

5 Solve applied problems using scientific notation.

Applications: Putting Numbers in Perspective

Due to tax cuts and spending increases, the United States began accumulating large deficits in the 1980s. To finance the deficit, the government had borrowed $6.8 trillion as of Novembre 2003. The graph in Figure 5.7 shows the national debt increasing over time.

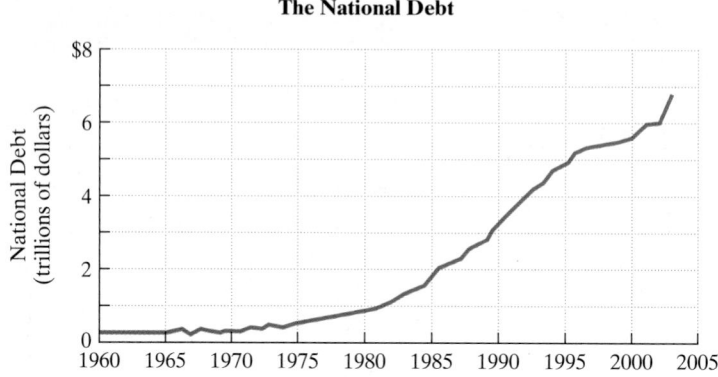

The National Debt

Source: Office of Management and Budget

FIGURE 5.7

Example 9 shows how we can use scientific notation to comprehend the meaning of a number such as 6.8 trillion.

EXAMPLE 9   THE NATIONAL DEBT

TECHNOLOGY

Here is the keystroke sequence for solving Example 9 using a calculator:

6.8 $\boxed{\text{EE}}$ 12 $\boxed{\div}$ 2.9 $\boxed{\text{EE}}$ 8.

The quotient is displayed by pressing $\boxed{=}$ on a scientific calculator and $\boxed{\text{ENTER}}$ on a graphing calculator. The answer can be displayed in scientific or decimal notation. Consult your manual.

As of November 2003, the national debt was $6.8 trillion, or 6.8×10^{12} dollars. At that time, the U.S. population was approximately 290,000,000 (290 million), or 2.9×10^8. If the national debt was evenly divided among every individual in the United States, how much would each citizen have to pay?

SOLUTION   The amount each citizen must pay is the total debt, 6.8×10^{12} dollars, divided by the number of citizens, 2.9×10^8.

$$\frac{6.8 \times 10^{12}}{2.9 \times 10^8} = \left(\frac{6.8}{2.9}\right) \times \left(\frac{10^{12}}{10^8}\right)$$

$$\approx 2.3 \times 10^{12-8}$$

$$= 2.3 \times 10^4$$

$$= 23,000$$

Every U.S. citizen would have to pay approximately $23,000 to the federal government to pay off the national debt. A family of three would owe $69,000.

Check 9 Point

In 2000, Americans spent 3.6×10^9 dollars on full-fat ice cream. At that time, the U.S. population was approximately 280 million, or 2.8×10^8. If ice cream spending is evenly divided, how much did each American spend?

BLITZER BONUS

Fermat's Last Theorem

Pierre de Fermat (1601–1665) was a lawyer who enjoyed studying mathematics. In a margin of one of his books, he claimed that no natural numbers satisfy

$$a^n + b^n = c^n$$

if n is greater than 2.

If $n = 2$, we can find numbers satisfying the equation. For example,

$$3^2 + 4^2 = 5^2.$$

However, Fermat claimed that no natural numbers satisfy

$$a^3 + b^3 = c^3, a^4 + b^4 = c^4,$$

and so on.

Fermat claimed to have a proof of his conjecture, but added, "The margin of my book is too narrow to write it down." Some believe that he never had a proof and his intent was to frustrate his colleagues.

In June 1993, 40-year-old Princeton math professor Andrew Wiles claimed that he discovered a proof of the theorem. Subsequent study revealed flaws, but Wiles corrected them. In 1995, his final proof served as a classic example of how great mathematicians accomplish great things: a combination of genius, hard work, frustration, and trial and error.

(Andrew Wiles 1953–)

Exercise Set 5.6

Practice Exercises

In Exercises 1–12, use properties of exponents to simplify each expression. First express the answer in exponential form. Then evaluate the expression.

1. $2^2 \cdot 2^3$ **2.** $3^3 \cdot 3^2$ **3.** $4 \cdot 4^2$ **4.** $5 \cdot 5^2$

5. $(2^2)^3$ **6.** $(3^3)^2$ **7.** $(1^4)^5$ **8.** $(1^3)^7$

9. $\dfrac{4^7}{4^5}$ **10.** $\dfrac{6^7}{6^5}$ **11.** $\dfrac{2^8}{2^4}$ **12.** $\dfrac{3^8}{3^4}$

In Exercises 13–24, use the zero and negative exponent rules to simplify each expression.

13. 3^0 **14.** 9^0 **15.** $(-3)^0$ **16.** $(-9)^0$

17. -3^0 **18.** -9^0 **19.** 2^{-2} **20.** 3^{-2}

21. 4^{-3} **22.** 2^{-3} **23.** 2^{-5} **24.** 2^{-6}

In Exercises 25–30, use properties of exponents to simplify each expression. First express the answer in exponential form. Then evaluate the expression.

25. $3^4 \cdot 3^{-2}$ **26.** $2^5 \cdot 2^{-2}$ **27.** $3^{-3} \cdot 3$

28. $2^{-3} \cdot 2$ **29.** $\dfrac{2^3}{2^7}$ **30.** $\dfrac{3^4}{3^7}$

In Exercises 31–46, express the number in decimal notation.

31. 2.7×10^2 **32.** 4.7×10^3 **33.** 9.12×10^5

34. 8.14×10^4 **35.** 8×10^7 **36.** 7×10^6

37. 1×10^5 **38.** 1×10^8 **39.** 7.9×10^{-1}

40. 8.6×10^{-1} **41.** 2.15×10^{-2} **42.** 3.14×10^{-2}

43. 7.86×10^{-4} **44.** 4.63×10^{-5} **45.** 3.18×10^{-6}

46. 5.84×10^{-7}

In Exercises 47–66, express the number in scientific notation.

47. 370   **48.** 530   **49.** 3600   **50.** 2700

51. 32,000   **52.** 64,000   **53.** 220,000,000

54. 370,000,000,000   **55.** 0.027   **56.** 0.014

57. 0.0037   **58.** 0.00083   **59.** 0.00000293

60. 0.000000647   **61.** 820×10^5 **62.** 630×10^8

63. 0.41×10^6 **64.** 0.57×10^9 **65.** 2100×10^{-9}

66. $97,000 \times 10^{-11}$

In Exercises 67–80, perform the indicated operation and express each answer in decimal notation.

67. $(2 \times 10^3)(3 \times 10^2)$ **68.** $(5 \times 10^2)(4 \times 10^4)$

69. $(2 \times 10^9)(3 \times 10^{-5})$ **70.** $(4 \times 10^8)(2 \times 10^{-4})$

71. $(4.1 \times 10^2)(3 \times 10^{-4})$ **72.** $(1.2 \times 10^3)(2 \times 10^{-5})$

73. $\dfrac{12 \times 10^6}{4 \times 10^2}$ **74.** $\dfrac{20 \times 10^{20}}{10 \times 10^{15}}$

75. $\dfrac{15 \times 10^4}{5 \times 10^{-2}}$ **76.** $\dfrac{18 \times 10^2}{9 \times 10^{-3}}$

77. $\dfrac{6 \times 10^3}{2 \times 10^5}$ **78.** $\dfrac{8 \times 10^4}{2 \times 10^7}$

79. $\dfrac{6.3 \times 10^{-6}}{3 \times 10^{-3}}$ **80.** $\dfrac{9.6 \times 10^{-7}}{3 \times 10^{-3}}$

In Exercises 81–90, perform the indicated operation by first expressing each number in scientific notation. Write the answer in scientific notation.

81. $(82,000,000)(3,000,000,000)$

82. $(94,000,000)(6,000,000,000)$

83. $(0.0005)(6,000,000)$ **84.** $(0.000015)(0.004)$

EXAMPLE 3 | USING THE FORMULA FOR THE GENERAL TERM OF AN ARITHMETIC SEQUENCE

Find the eighth term of the arithmetic sequence whose first term is 4 and whose common difference is −7.

SOLUTION   To find the eighth term, a_8, we replace n in the formula with 8, a_1 with 4, and d with −7.

$$a_n = a_1 + (n - 1)d$$
$$a_8 = 4 + (8 - 1)(-7) = 4 + 7(-7) = 4 + (-49) = -45$$

The eighth term is −45. We can check this result by writing the first eight terms of the sequence:

$$4, -3, -10, -17, -24, -31, -38, -45.$$

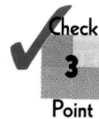 **Check Point 3**   Find the ninth term of the arithmetic sequence whose first term is 6 and whose common difference is −5.

EXAMPLE 4 | USING AN ARITHMETIC SEQUENCE TO DESCRIBE TEACHERS' EARNINGS

According to the National Education Association, teachers in the United States earned an average of $30,532 in 1990. This amount has increased by approximately $1472 per year.

a. Write a formula for the nth term of the arithmetic sequence that describes teachers' average earnings n years after 1989.

b. How much will U.S. teachers earn by the year 2010?

SOLUTION

a. We can express teachers' earnings by the following arithmetic sequence:

30,532,	32,004,	33,476,	34,948, ...
a_1: earnings in 1990, 1 year after 1989	a_2: earnings in 1991, 2 years after 1989	a_3: earnings in 1992, 3 years after 1989	a_4: earnings in 1993, 4 years after 1989

In this sequence, a_1, the first term, represents the amount teachers earned in 1990. Each subsequent year this amount increases by $1472, so $d = 1472$. We use the formula for the general term of an arithmetic sequence to write the nth term of the sequence that describes teachers' earnings n years after 1989.

$$a_n = a_1 + (n - 1)d$$ This is the formula for the general term of an arithmetic sequence.

$$a_n = 30,532 + (n - 1)1472$$ $a_1 = 30,532$ and $d = 1472$.

$$a_n = 30,532 + 1472n - 1472$$ Distribute 1472 to each term in parentheses.

$$a_n = 1472n + 29,060$$ Simplify.

Thus, teachers' earnings n years after 1989 can be described by $a_n = 1472n + 29,060$.

BLITZER BONUS

Fermat's Last Theorem

Pierre de Fermat (1601–1665) was a lawyer who enjoyed studying mathematics. In a margin of one of his books, he claimed that no natural numbers satisfy

$$a^n + b^n = c^n$$

if n is greater than 2.

If $n = 2$, we can find numbers satisfying the equation. For example,

$$3^2 + 4^2 = 5^2.$$

However, Fermat claimed that no natural numbers satisfy

$$a^3 + b^3 = c^3, a^4 + b^4 = c^4,$$

and so on.

Fermat claimed to have a proof of his conjecture, but added, "The margin of my book is too narrow to write it down." Some believe that he never had a proof and his intent was to frustrate his colleagues.

In June 1993, 40-year-old Princeton math professor Andrew Wiles claimed that he discovered a proof of the theorem. Subsequent study revealed flaws, but Wiles corrected them. In 1995, his final proof served as a classic example of how great mathematicians accomplish great things: a combination of genius, hard work, frustration, and trial and error.

(Andrew Wiles 1953–)

Exercise Set 5.6

Practice Exercises

In Exercises 1–12, use properties of exponents to simplify each expression. First express the answer in exponential form. Then evaluate the expression.

1. $2^2 \cdot 2^3$ **2.** $3^3 \cdot 3^2$ **3.** $4 \cdot 4^2$ **4.** $5 \cdot 5^2$

5. $(2^2)^3$ **6.** $(3^3)^2$ **7.** $(1^4)^5$ **8.** $(1^3)^7$

9. $\dfrac{4^7}{4^5}$ **10.** $\dfrac{6^7}{6^5}$ **11.** $\dfrac{2^8}{2^4}$ **12.** $\dfrac{3^8}{3^4}$

In Exercises 13–24, use the zero and negative exponent rules to simplify each expression.

13. 3^0 **14.** 9^0 **15.** $(-3)^0$ **16.** $(-9)^0$

17. -3^0 **18.** -9^0 **19.** 2^{-2} **20.** 3^{-2}

21. 4^{-3} **22.** 2^{-3} **23.** 2^{-5} **24.** 2^{-6}

In Exercises 25–30, use properties of exponents to simplify each expression. First express the answer in exponential form. Then evaluate the expression.

25. $3^4 \cdot 3^{-2}$ **26.** $2^5 \cdot 2^{-2}$ **27.** $3^{-3} \cdot 3$

28. $2^{-3} \cdot 2$ **29.** $\dfrac{2^3}{2^7}$ **30.** $\dfrac{3^4}{3^7}$

In Exercises 31–46, express the number in decimal notation.

31. 2.7×10^2 **32.** 4.7×10^3 **33.** 9.12×10^5

34. 8.14×10^4 **35.** 8×10^7 **36.** 7×10^6

37. 1×10^5 **38.** 1×10^8 **39.** 7.9×10^{-1}

40. 8.6×10^{-1} **41.** 2.15×10^{-2} **42.** 3.14×10^{-2}

43. 7.86×10^{-4} **44.** 4.63×10^{-5} **45.** 3.18×10^{-6}

46. 5.84×10^{-7}

In Exercises 47–66, express the number in scientific notation.

47. 370   **48.** 530   **49.** 3600   **50.** 2700

51. 32,000   **52.** 64,000   **53.** 220,000,000

54. 370,000,000,000   **55.** 0.027   **56.** 0.014

57. 0.0037   **58.** 0.00083   **59.** 0.00000293

60. 0.000000647   **61.** 820×10^5 **62.** 630×10^8

63. 0.41×10^6 **64.** 0.57×10^9 **65.** 2100×10^{-9}

66. $97,000 \times 10^{-11}$

In Exercises 67–80, perform the indicated operation and express each answer in decimal notation.

67. $(2 \times 10^3)(3 \times 10^2)$ **68.** $(5 \times 10^2)(4 \times 10^4)$

69. $(2 \times 10^9)(3 \times 10^{-5})$ **70.** $(4 \times 10^8)(2 \times 10^{-4})$

71. $(4.1 \times 10^2)(3 \times 10^{-4})$ **72.** $(1.2 \times 10^3)(2 \times 10^{-5})$

73. $\dfrac{12 \times 10^6}{4 \times 10^2}$ **74.** $\dfrac{20 \times 10^{20}}{10 \times 10^{15}}$

75. $\dfrac{15 \times 10^4}{5 \times 10^{-2}}$ **76.** $\dfrac{18 \times 10^2}{9 \times 10^{-3}}$

77. $\dfrac{6 \times 10^3}{2 \times 10^5}$ **78.** $\dfrac{8 \times 10^4}{2 \times 10^7}$

79. $\dfrac{6.3 \times 10^{-6}}{3 \times 10^{-3}}$ **80.** $\dfrac{9.6 \times 10^{-7}}{3 \times 10^{-3}}$

In Exercises 81–90, perform the indicated operation by first expressing each number in scientific notation. Write the answer in scientific notation.

81. $(82,000,000)(3,000,000,000)$

82. $(94,000,000)(6,000,000,000)$

83. $(0.0005)(6,000,000)$ **84.** $(0.000015)(0.004)$

85. $\dfrac{9,500,000}{500}$

86. $\dfrac{30,000}{0.0005}$

87. $\dfrac{0.00008}{200}$

88. $\dfrac{0.0018}{0.0000006}$

89. $\dfrac{480,000,000,000}{0.00012}$

90. $\dfrac{0.000000096}{16,000}$

Application Exercises

The graph shows the projected number of people in the United States ages 65 and over for the year 2000 and beyond. Use 10^6 for one million and the figures shown to solve Exercises 91–94. Express all answers in scientific notation.

U. S. Population Ages 65 and Over

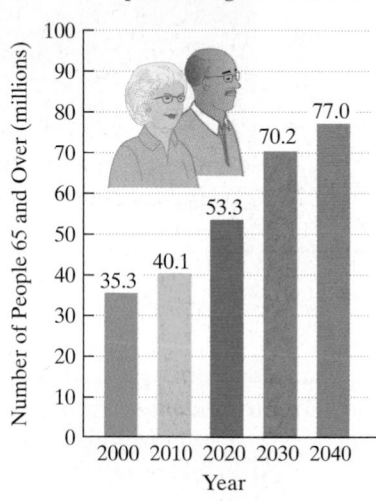

Source: U.S. Bureau of the Census

91. How many people 65 and over will there be in 2020?

92. How many people 65 and over will there be in 2030?

93. How many more people 65 and over will there be in 2040 than in 2000?

94. How many more people 65 and over will there be in 2030 than in 2000?

95. The United States government spends approximately $1.6 trillion per year. Use 10^{12} for one trillion and the graph to write the amount that it spends on Social Security in scientific notation.

Where the U.S. Government Spends Money

Interest on the national debt **22%**

Social Security **25%**

Health **21%**

Other (including education, agriculture, and transportation) **16%**

Defense (military) **16%**

Source: U.S. Office of Management and Budget

96. Americans say they lead active lives, but for the 205 million of us ages 18 and older, walking is often as strenuous as it gets. Use the graph to write the number of Americans whose lifestyle is not very active. Express the answer in scientific notation.

How Active Is Your Lifestyle?

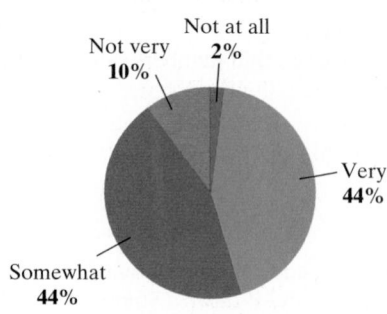

Not at all **2%**

Not very **10%**

Very **44%**

Somewhat **44%**

Source: Discovery Health Media

Use 10^{12} for one trillion and 2.8×10^8 for the U.S. population in 2000 to solve Exercises 97–99.

97. In 2000, the government collected approximately $1.9 trillion in taxes. What was the per capita tax burden, or the amount that each U.S. citizen paid in taxes? Round to the nearest hundred dollars.

98. In 2000, U.S. personal income was $8 trillion. What was the per capita income, or the income per U.S. citizen? Round to the nearest hundred dollars.

99. In the United States, we spend an average of $4000 per person each year on health care—the highest in the world. What do we spend each year on health care nationwide? Express the answer in scientific notation.

100. Approximately 2×10^4 people run in the New York City Marathon each year. Each runner runs a distance of 26 miles. Write the total distance covered by all the runners (assuming that each person completes the marathon) in scientific notation.

101. The mass of one oxygen molecule is 5.3×10^{-23} gram. Find the mass of 20,000 molecules of oxygen. Express the answer in scientific notation.

102. The mass of one hydrogen atom is 1.67×10^{-24} gram. Find the mass of 80,000 hydrogen atoms. Express the answer in scientific notation.

Writing in Mathematics

103. Explain the product rule for exponents. Use $2^3 \cdot 2^5$ in your explanation.

104. Explain the power rule for exponents. Use $(3^2)^4$ in your explanation.

105. Explain the quotient rule for exponents. Use $\dfrac{5^8}{5^2}$ in your explanation.

106. Explain the zero exponent rule and give an example.

107. Explain the negative exponent rule and give an example.

108. How do you know if a number is written in scientific notation?

109. Explain how to convert from scientific to decimal notation and give an example.

110. Explain how to convert from decimal to scientific notation and give an example.

111. Suppose you are looking at a number in scientific notation. Describe the size of the number you are looking at if the exponent on ten is **a.** positive **b.** negative **c.** zero.

112. Describe one advantage of expressing a number in scientific notation over decimal notation.

Critical Thinking Exercises

113. Which one of the following is true?
    a. $4^{-2} < 4^{-3}$
    b. $5^{-2} > 2^{-5}$
    c. $(-2)^4 = 2^{-4}$
    d. $5^2 \cdot 5^{-2} > 2^5 \cdot 2^{-5}$

114. Which one of the following is true?
    a. $534.7 = 5.347 \times 10^3$
    b. $\dfrac{8 \times 10^{30}}{4 \times 10^{-5}} = 2 \times 10^{25}$
    c. $(7 \times 10^5) + (2 \times 10^{-3}) = 9 \times 10^2$
    d. $(4 \times 10^3) + (3 \times 10^2) = 43 \times 10^2$

115. Give an example of a number for which there is no advantage to using scientific notation instead of decimal notation. Explain why this is the case.

116. The mad Dr. Frankenstein has gathered enough bits and pieces (so to speak) for $2^{-1} + 2^{-2}$ of his creature-to-be. Write a fraction that represents the amount of his creature that must still be obtained.

Technology Exercises

117. Use a calculator in a fraction mode to check your answers in Exercises 19–24.

118. Use a calculator to check any three of your answers in Exercises 31–46.

119. Use a calculator to check any three of your answers in Exercises 47–66.

120. Use a calculator with an $\boxed{\text{EE}}$ or $\boxed{\text{EXP}}$ key to check any four of your computations in Exercises 67–90. Display the result of the computation in scientific notation and decimal notation.

Group Exercises

121. *The Oddball's Oddball.*   In their March 29, 1999 issue on the century's greatest minds, *Time* magazine called Paul Erdös (1913–1996) "the most prolific and arguably the cleverest mathematician of the century." Mathematicians gathered to hear his fascinating ideas about integers, whole numbers, and primes. Erdös solved problems about the nature of numbers, although he could not figure out how to boil an egg or wash his underwear. Speaking of mathematicians in general and Erdös in particular, the writers of *Time* stated, "In a profession with no shortage of oddballs, he was the strangest." Group members should do research on Paul Erdös, presenting a report on his work with numbers and his eccentricities.

122. Fermat's most notorious theorem baffled the greatest minds for more than three centuries. In 1994, after ten years of work, Princeton University's Andrew Wiles proved Fermat's Last Theorem. *People* magazine put him on its list of "the 25 most intriguing people of the year," the Gap asked him to model jeans, and Barbara Walters chased him for an interview. "Who's Barbara Walters?" asked the bookish Wiles, who had somehow gone through life without a television.

    Using the 1993 PBS documentary "Solving Fermat: Andrew Wiles" or information about Andrew Wiles on the Internet, research and present a group seminar on what Wiles did to prove Fermat's Last Theorem, problems along the way, and how Wiles overcame them.

SECTION 5.7   ARITHMETIC AND GEOMETRIC SEQUENCES

OBJECTIVES

1. Write terms of an arithmetic sequence.
2. Use the formula for the general term of an arithmetic sequence.
3. Write terms of a geometric sequence.
4. Use the formula for the general term of a geometric sequence.

Sequences

Many creations in nature involve intricate mathematical designs, including a variety of spirals. For example, the arrangement of the individual florets in the head of a sunflower forms spirals. In some species, there are 21 spirals in the clockwise direction and 34 in the counterclockwise direction. The precise numbers depend on the species of sunflower: 21 and 34, or 34 and 55, or 55 and 89, or even 89 and 144.

This observation becomes even more interesting when we consider a sequence of numbers investigated by Leonardo of Pisa, also known as Fibonacci, an Italian mathematician of the thirteenth century. The **Fibonacci sequence** of numbers is an infinite sequence that begins as follows:

$$1, 1, 2, 3, 5, 8, 13, 21, 34, 55, 89, 144, 233 \ldots.$$

The first two terms are 1. Every term thereafter is the sum of the two preceding terms. For example, the third term, 2, is the sum of the first and second terms: $1 + 1 = 2$. The fourth term, 3, is the sum of the second and third terms: $1 + 2 = 3$, and so on. Did you know that the number of spirals in a daisy or a sunflower, 21 and 34, are two Fibonacci numbers? The number of spirals in a pinecone, 8 and 13, and a pineapple, 8 and 13, are also Fibonacci numbers.

We can think of a **sequence** as a list of numbers that are related to each other by a rule. The numbers in a sequence are called its **terms**. The letter a with a subscript is used to represent the terms of a sequence. Thus, a_1 represents the first term of the sequence, a_2 represents the second term, a_3 the third term, and so on. This notation is shown for the first six terms of the Fibonacci sequence:

$$1, \quad 1, \quad 2, \quad 3, \quad 5, \quad 8.$$

$$a_1 = 1 \quad a_2 = 1 \quad a_3 = 2 \quad a_4 = 3 \quad a_5 = 5 \quad a_6 = 8$$

Arithmetic Sequences

A formula that approximates the average annual salaries of major league baseball players generates the following data:

Year	1996	1997	1998	1999	2000	2001	2002
Salary	1,076,865	1,304,152	1,531,439	1,758,726	1,986,013	2,213,300	2,440,587

From 1996 to 1997, salaries increased by $\$1,304,152 - \$1,076,865 = \$227,287$. From 1997 to 1998, salaries increased by $\$1,531,439 - \$1,304,152 = \$227,287$. If we make these computations for each year, we find that the yearly salary increase is \$227,287. The sequence of annual salaries shows that each term after the first, 1,076,865, differs from the preceding term by a constant amount, namely 227,287. The sequence of annual salaries

$$1,076,865, \quad 1,304,152, \quad 1,531,439, \quad 1,758,726, \quad 1,986,013, \ldots$$

is an example of an *arithmetic sequence*.

DEFINITION OF AN ARITHMETIC SEQUENCE

An **arithmetic sequence** is a sequence in which each term after the first differs from the preceding term by a constant amount. The difference between consecutive terms is called the **common difference** of the sequence.

The common difference, d, is found by subtracting any term from the term that directly follows it. In the following examples, the common difference is found by subtracting the first term from the second term: $a_2 - a_1$.

Arithmetic Sequence	Common Difference
$1,076,865, 1,304,152, 1,531,439, 1,758,726, \ldots$	$d = 1,304,152 - 1,076,865 = 227,287$
$2, 6, 10, 14, 18, \ldots$	$d = 6 - 2 = 4$
$-2, -7, -12, -17, \ldots$	$d = -7 - (-2) = -5$

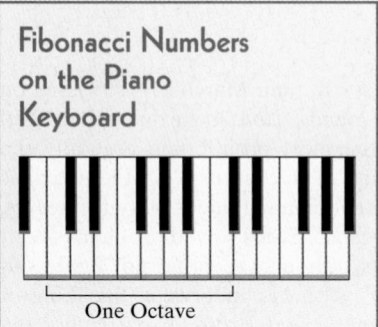

If the first term of an arithmetic sequence is a_1, each term after the first is obtained by adding d, the common difference, to the previous term.

 Write terms of an arithmetic sequence.

EXAMPLE 1 WRITING THE TERMS OF AN ARITHMETIC SEQUENCE

Write the first six terms of the arithmetic sequence with first term 6 and common difference 4.

SOLUTION   The first term is 6. The second term is $6 + 4$, or 10. The third term is $10 + 4$, or 14, and so on. The first six terms are

$$6, 10, 14, 18, 22, \text{ and } 26.$$

✓ Check 1 Point   Write the first six terms of the arithmetic sequence with first term 100 and common difference 20.

EXAMPLE 2 WRITING THE TERMS OF AN ARITHMETIC SEQUENCE

Write the first six terms of the arithmetic sequence with $a_1 = 5$ and $d = -2$.

SOLUTION   The first term, a_1, is 5. The common difference, d, is -2. To find the second term, we add -2 to 5, giving 3. For the next term, we add -2 to 3, and so on. The first six terms are

$$5, 3, 1, -1, -3, \text{ and } -5.$$

✓ Check 2 Point   Write the first six terms of the arithmetic sequence with first term 8 and common difference -3.

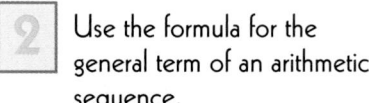 Use the formula for the general term of an arithmetic sequence.

The General Term of an Arithmetic Sequence

Consider an arithmetic sequence whose first term is a_1 and whose common difference is d. We are looking for a formula for the general term, a_n. Let's begin by writing the first six terms. The first term is a_1. The second term is $a_1 + d$. The third term is $a_1 + d + d$, or $a_1 + 2d$. Thus, we start with a_1 and add d to each successive term. The first six terms are

$$a_1, \quad a_1 + d, \quad a_1 + 2d, \quad a_1 + 3d, \quad a_1 + 4d, \quad a_1 + 5d.$$

a_1, first term a_2, second term a_3, third term a_4, fourth term a_5, fifth term a_6, sixth term

Using the pattern of the terms results in the following formula for the general or nth term of an arithmetic sequence:

GENERAL TERM OF AN ARITHMETIC SEQUENCE

The nth term (the general term) of an arithmetic sequence with first term a_1 and common difference d is

$$a_n = a_1 + (n - 1)d.$$

EXAMPLE 3 | USING THE FORMULA FOR THE GENERAL TERM OF AN ARITHMETIC SEQUENCE

Find the eighth term of the arithmetic sequence whose first term is 4 and whose common difference is −7.

SOLUTION   To find the eighth term, a_8, we replace n in the formula with 8, a_1 with 4, and d with −7.

$$a_n = a_1 + (n - 1)d$$
$$a_8 = 4 + (8 - 1)(-7) = 4 + 7(-7) = 4 + (-49) = -45$$

The eighth term is −45. We can check this result by writing the first eight terms of the sequence:

$$4, -3, -10, -17, -24, -31, -38, -45.$$

Check Point 3   Find the ninth term of the arithmetic sequence whose first term is 6 and whose common difference is −5.

EXAMPLE 4 | USING AN ARITHMETIC SEQUENCE TO DESCRIBE TEACHERS' EARNINGS

According to the National Education Association, teachers in the United States earned an average of $30,532 in 1990. This amount has increased by approximately $1472 per year.

a. Write a formula for the nth term of the arithmetic sequence that describes teachers' average earnings n years after 1989.

b. How much will U.S. teachers earn by the year 2010?

SOLUTION

a. We can express teachers' earnings by the following arithmetic sequence:

$$30{,}532, \qquad 32{,}004, \qquad 33{,}476, \qquad 34{,}948, \dots \; .$$

| a_1: earnings in 1990, 1 year after 1989 | a_2: earnings in 1991, 2 years after 1989 | a_3: earnings in 1992, 3 years after 1989 | a_4: earnings in 1993, 4 years after 1989 |

In this sequence, a_1, the first term, represents the amount teachers earned in 1990. Each subsequent year this amount increases by $1472, so $d = 1472$. We use the formula for the general term of an arithmetic sequence to write the nth term of the sequence that describes teachers' earnings n years after 1989.

$$a_n = a_1 + (n - 1)d \qquad \text{This is the formula for the general term of an arithmetic sequence.}$$
$$a_n = 30{,}532 + (n - 1)1472 \qquad a_1 = 30{,}532 \text{ and } d = 1472.$$
$$a_n = 30{,}532 + 1472n - 1472 \qquad \text{Distribute 1472 to each term in parentheses.}$$
$$a_n = 1472n + 29{,}060 \qquad \text{Simplify.}$$

Thus, teachers' earnings n years after 1989 can be described by $a_n = 1472n + 29{,}060$.

b. Now we need to find teachers' earnings in 2010. The year 2010 is 21 years after 1989: That is, $2010 - 1989 = 21$. Thus, $n = 21$. We substitute 21 for n in $a_n = 1472n + 29{,}060$.

$$a_{21} = 1472 \cdot 21 + 29{,}060 = 59{,}972$$

The 21st term of the sequence is 59,972. Therefore, U.S. teachers are predicted to earn an average of $59,972 by the year 2010.

 **Check 4 Point** According to the U.S. Census Bureau, new one-family houses sold for an average of $159,000 in 1995. This average sales price has increased by approximately $9700 per year.

a. Write a formula for the nth term of the arithmetic sequence that describes the average cost of new one-family houses n years after 1994.

b. How much will new one-family houses cost, on average, by the year 2010?

Geometric Sequences

Figure 5.8 shows a sequence in which the number of squares is increasing. From left to right, the number of squares is 1, 5, 25, 125, and 625. In this sequence, each term after the first, 1, is obtained by multiplying the preceding term by a constant amount, namely 5. This sequence of increasing number of squares is an example of a *geometric sequence*.

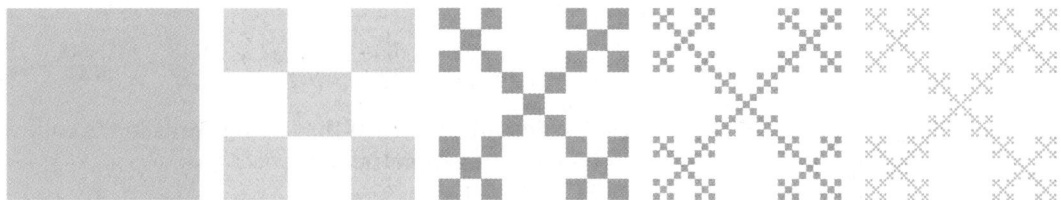

FIGURE 5.8 A geometric sequence of squares

DEFINITION OF A GEOMETRIC SEQUENCE

A **geometric sequence** is a sequence in which each term after the first is obtained by multiplying the preceding term by a fixed nonzero constant. The amount by which we multiply each time is called the **common ratio** of the sequence.

The common ratio, r, is found by dividing any term after the first term by the term that directly precedes it. In the examples below, the common ratio is found by dividing the second term by the first term: $\dfrac{a_2}{a_1}$.

Geometric sequence	Common ratio
$1, 5, 25, 125, 625, \ldots$	$r = \frac{5}{1} = 5$
$4, 8, 16, 32, 64, \ldots$	$r = \frac{8}{4} = 2$
$6, -12, 24, -48, 96, \ldots$	$r = \frac{-12}{6} = -2$
$9, -3, 1, -\frac{1}{3}, \frac{1}{9}, \ldots$	$r = \frac{-3}{9} = -\frac{1}{3}$

How do we write out the terms of a geometric sequence when the first term and the common ratio are known? We multiply the first term by the common ratio to get the second term, multiply the second term by the common ratio to get the third term, and so on.

 Write terms of a geometric sequence.

EXAMPLE 5 | WRITING THE TERMS OF A GEOMETRIC SEQUENCE

Write the first six terms of the geometric sequence with first term 6 and common ratio $\frac{1}{3}$.

SOLUTION The first term is 6. The second term is $6 \cdot \frac{1}{3}$, or 2. The third term is $2 \cdot \frac{1}{3}$, or $\frac{2}{3}$. The fourth term is $\frac{2}{3} \cdot \frac{1}{3}$, or $\frac{2}{9}$, and so on. The first six terms are

$$6, 2, \frac{2}{3}, \frac{2}{9}, \frac{2}{27}, \text{ and } \frac{2}{81}.$$

 **Check Point 5** Write the first six terms of the geometric sequence with first term 12 and common ratio $-\frac{1}{2}$.

The General Term of a Geometric Sequence

4 Use the formula for the general term of a geometric sequence.

Consider a geometric sequence whose first term is a_1 and whose common ratio is r. We are looking for a formula for the general term, a_n. Let's begin by writing the first six terms. The first term is a_1. The second term is $a_1 r$. The third term is $a_1 r \cdot r$, or $a_1 r^2$. The fourth term is $a_1 r^2 \cdot r$, or $a_1 r^3$, and so on. Starting with a_1 and multiplying each successive term by r, the first six terms are

$$a_1, \qquad a_1 r, \qquad a_1 r^2, \qquad a_1 r^3, \qquad a_1 r^4, \qquad a_1 r^5.$$

| a_1, first term | a_2, second term | a_3, third term | a_4, fourth term | a_5, fifth term | a_6, sixth term |

Using the pattern of the terms results in the following formula for the general or nth term of a geometric sequence:

GENERAL TERM OF A GEOMETRIC SEQUENCE

The nth term (the general term) of a geometric sequence with first term a_1 and common ratio r is

$$a_n = a_1 r^{n-1}.$$

EXAMPLE 6 | USING THE FORMULA FOR THE GENERAL TERM OF A GEOMETRIC SEQUENCE

Find the eighth term of the geometric sequence whose first term is -4 and whose common ratio is -2.

STUDY TIP

Be careful with the order of operations when evaluating

$$a_1 r^{n-1}.$$

First find r^{n-1}. Then multiply the result by a_1.

SOLUTION To find the eighth term, a_8, we replace n in the formula with 8, a_1 with -4, and r with -2.

$$a_n = a_1 r^{n-1}$$
$$a_8 = -4(-2)^{8-1} = -4(-2)^7 = -4(-128) = 512$$

The eighth term is 512. We can check this result by writing the first eight terms of the sequence:

$$-4, 8, -16, 32, -64, 128, -256, 512.$$

 **Check Point 6** Find the seventh term of the geometric sequence whose first term is 5 and whose common ratio is -3.

EXAMPLE 7 GEOMETRIC POPULATION GROWTH

The population of Florida from 1990 through 1997 is shown in the following table:

Year	1990	1991	1992	1993	1994	1995	1996	1997
Population in millions	12.94	13.20	13.46	13.73	14.00	14.28	14.57	14.86

a. Show that the population is increasing geometrically.

b. Write the general term for the geometric sequence describing population growth for Florida n years after 1989.

c. Estimate Florida's population, in millions, for the year 2000.

SOLUTION

a. First, we divide the population for each year by the population in the preceding year.

$$\frac{13.20}{12.94} \approx 1.02, \quad \frac{13.46}{13.20} \approx 1.02, \quad \frac{13.73}{13.46} \approx 1.02$$

Continuing in this manner, we will keep getting approximately 1.02. This means that the population is increasing geometrically with $r \approx 1.02$. In this situation, the common ratio is the growth rate, indicating that the population of Florida in any year shown in the table is approximately 1.02 times the population the year before.

b. The sequence of Florida's population growth is

$$12.94, 13.20, 13.46, 13.73, 14.00, 14.28, 14.57, 14.86, \ldots.$$

Because the population is increasing geometrically, we can find the general term of this sequence using

$$a_n = a_1 r^{n-1}.$$

In this sequence, $a_1 = 12.94$ and [from part (a)] $r \approx 1.02$. We substitute these values into the formula for the general term. This formula gives the general term for the geometric sequence describing Florida's population n years after 1989.

$$a_n = 12.94(1.02)^{n-1}$$

c. We can use the formula for the general term, a_n, in part (b) to estimate Florida's population for the year 2000. The year 2000 is 11 years after 1989—that is, $2000 - 1989 = 11$. Thus, $n = 11$. We substitute 11 for n in $a_n = 12.94(1.02)^{n-1}$:

$$a_{11} = 12.94(1.02)^{11-1} = 12.94(1.02)^{10} \approx 15.77.$$

The formula indicates that Florida had a population of approximately 15.77 million in the year 2000. According to the U.S. Census Bureau, Florida's population in 2000 was 15.98 million. Our geometric sequence describes the actual population fairly well.

Check 7 Point Write the general term for the geometric sequence

$$3, 6, 12, 24, 48, \ldots.$$

Then use the formula for the general term to find the eighth term.

Exercise Set 5.7

Practice Exercises

In Exercises 1–20, write the first six terms of the arithmetic sequence with the first term, a_1, and common difference, d.

1. $a_1 = 8, d = 2$

2. $a_1 = 5, d = 3$

3. $a_1 = 200, d = 20$

4. $a_1 = 300, d = 50$

5. $a_1 = -7, d = 4$

6. $a_1 = -8, d = 5$

7. $a_1 = -400, d = 300$

8. $a_1 = -500, d = 400$

9. $a_1 = 7, d = -3$

10. $a_1 = 9, d = -5$

11. $a_1 = 200, d = -60$

12. $a_1 = 300, d = -90$

13. $a_1 = \frac{5}{2}, d = \frac{1}{2}$

14. $a_1 = \frac{3}{4}, d = \frac{1}{4}$

15. $a_1 = \frac{3}{2}, d = \frac{1}{4}$

16. $a_1 = \frac{3}{2}, d = -\frac{1}{4}$

17. $a_1 = 4.25, d = 0.3$

18. $a_1 = 6.3, d = 0.25$

19. $a_1 = 4.5, d = -0.75$

20. $a_1 = 3.5, d = -1.75$

In Exercises 21–40, find the indicated term for the arithmetic sequence with first term, a_1, and common difference, d.

21. Find a_6, when $a_1 = 13, d = 4$.

22. Find a_{16}, when $a_1 = 9, d = 2$.

23. Find a_{50}, when $a_1 = 7, d = 5$.

24. Find a_{60}, when $a_1 = 8, d = 6$.

25. Find a_9, when $a_1 = -5, d = 9$.

26. Find a_{10}, when $a_1 = -8, d = 10$.

27. Find a_{200}, when $a_1 = -40, d = 5$.

28. Find a_{150}, when $a_1 = -60, d = 5$.

29. Find a_{10}, when $a_1 = -8, d = 10$.

30. Find a_{11}, when $a_1 = 10, d = -6$.

31. Find a_{60}, when $a_1 = 35, d = -3$.

32. Find a_{70}, when $a_1 = -32, d = 4$.

33. Find a_{12}, when $a_1 = 12, d = -5$.

34. Find a_{20}, when $a_1 = -20, d = -4$.

35. Find a_{90}, when $a_1 = -70, d = -2$.

36. Find a_{80}, when $a_1 = 106, d = -12$.

37. Find a_{12}, when $a_1 = 6, d = \frac{1}{2}$.

38. Find a_{14}, when $a_1 = 8, d = \frac{1}{4}$.

39. Find a_{50}, when $a_1 = 14, d = -0.25$.

40. Find a_{110}, when $a_1 = -12, d = -0.5$.

In Exercises 41–48, write a formula for the general term (the nth term) of each arithmetic sequence. Then use the formula for a_n to find a_{20}, the 20th term of the sequence.

41. $1, 5, 9, 13, \ldots$

42. $2, 7, 12, 17, \ldots$

43. $7, 3, -1, -5, \ldots$

44. $6, 1, -4, -9, \ldots$

45. $a_1 = 9, d = 2$

46. $a_1 = 6, d = 3$

47. $a_1 = -20, d = -4$

48. $a_1 = -70, d = -5$

In Exercises 49–70, write the first six terms of the geometric sequence with the first term, a_1, and common ratio, r.

49. $a_1 = 4, r = 2$

50. $a_1 = 2, r = 3$

51. $a_1 = 1000, r = 1$

52. $a_1 = 5000, r = 1$

53. $a_1 = 3, r = -2$

54. $a_1 = 2, r = -3$

55. $a_1 = 10, r = -4$

56. $a_1 = 20, r = -4$

57. $a_1 = 2000, r = -1$

58. $a_1 = 3000, r = -1$

59. $a_1 = -2, r = -3$

60. $a_1 = -4, r = -2$

61. $a_1 = -6, r = -5$

62. $a_1 = -8, r = -5$

63. $a_1 = \frac{1}{4}, r = 2$

64. $a_1 = \frac{1}{2}, r = 2$

65. $a_1 = \frac{1}{4}, r = \frac{1}{2}$

66. $a_1 = \frac{1}{5}, r = \frac{1}{2}$

67. $a_1 = -\frac{1}{16}, r = -4$

68. $a_1 = -\frac{1}{8}, r = -2$

69. $a_1 = 2, r = 0.1$

70. $a_1 = -1000, r = 0.1$

In Exercises 71–90, find the indicated term for the geometric sequence with first term, a_1, and common ratio, r.

71. Find a_7, when $a_1 = 4, r = 2$.

72. Find a_5, when $a_1 = 4, r = 3$.

73. Find a_{20}, when $a_1 = 2, r = 3$.

74. Find a_{20}, when $a_1 = 2, r = 2$.

75. Find a_{100}, when $a_1 = 50, r = 1$.

76. Find a_{200}, when $a_1 = 60, r = 1$.

77. Find a_7, when $a_1 = 5, r = -2$.

78. Find a_4, when $a_1 = 4, r = -3$.

79. Find a_{30}, when $a_1 = 2, r = -1$.

80. Find a_{40}, when $a_1 = 6, r = -1$.

81. Find a_6, when $a_1 = -2, r = -3$.

82. Find a_5, when $a_1 = -5, r = -2$.

83. Find a_8, when $a_1 = 6, r = \frac{1}{2}$.

84. Find a_8, when $a_1 = 12, r = \frac{1}{2}$.

85. Find a_6, when $a_1 = 18, r = -\frac{1}{3}$.

86. Find a_4, when $a_1 = 9, r = -\frac{1}{3}$.

87. Find a_{40}, when $a_1 = 1000, r = -\frac{1}{2}$.

88. Find a_{30}, when $a_1 = 8000, r = -\frac{1}{2}$.

89. Find a_8, when $a_1 = 1,000,000, r = 0.1$.

90. Find a_8, when $a_1 = 40,000, r = 0.1$.

In Exercises 91–98, write a formula for the general term (the nth term) of each geometric sequence. Then use the formula for a_n to find a_7, the seventh term of the sequence.

91. $3, 12, 48, 192, \ldots$

92. $3, 15, 75, 375, \ldots$

93. $18, 6, 2\frac{2}{3}, \ldots.$

94. $12, 6, 3, \frac{3}{2}, \ldots.$

95. $1.5, -3, 6, -12, \ldots$

96. $5, -1, \frac{1}{5}, -\frac{1}{25}, \ldots.$

97. $0.0004, -0.004, 0.04, -0.4, \ldots$

98. $0.0007, -0.007, 0.07, -0.7, \ldots$

Determine whether each sequence in Exercises 99–114 is arithmetic or geometric. In each sequence, find the next two terms.

99. $2, 6, 10, 14, \ldots$

100. $3, 8, 13, 18, \ldots$

101. $5, 15, 45, 135, \ldots$

102. $15, 30, 60, 120, \ldots$

103. $-7, -2, 3, 8, \ldots$

104. $-9, -5, -1, 3, \ldots$

105. $3, \frac{3}{2}, \frac{3}{4}, \frac{3}{8}, \ldots$

106. $6, 3, \frac{3}{2}, \frac{3}{4}, \ldots$

107. $\frac{1}{2}, 1, \frac{3}{2}, 2, \ldots$

108. $\frac{2}{3}, 1, \frac{4}{3}, \frac{5}{3}, \ldots$

109. $7, -7, 7, -7, \ldots$

110. $6, -6, 6, -6, \ldots$

111. $7, -7, -21, -35, \ldots$

112. $6, -6, -18, -30, \ldots$

113. $\sqrt{5}, 5, 5\sqrt{5}, 25, \ldots$

114. $\sqrt{3}, 3, 3\sqrt{3}, 9, \ldots$

Application Exercises

Use the formula for the nth term of an arithmetic sequence to solve Exercises 115–118.
The graph shows pounds of various food groups consumed per year by the average American. Exercises 115–116 involve developing arithmetic sequences that describe the data.

Per Capita Consumption of Various Food Groups

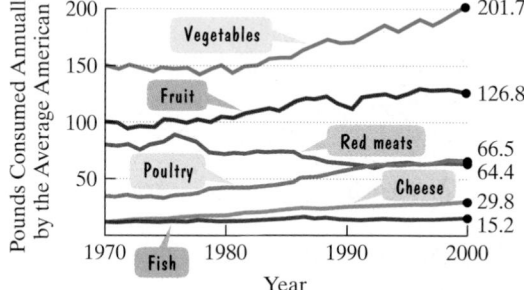

Source: U.S. Department of Agriculture

115. The graph shows that the average American consumed 150 pounds of vegetables in 1970. On average, this amount has increased by approximately 1.7 pounds per person per year.

   a. Write a formula for the nth term of the arithmetic sequence that describes pounds of vegetables consumed annually by the average American n years after 1969.

   b. How many pounds of vegetables will be consumed by the average American in 2006?

116. The graph shows that the average American consumed 100 pounds of fruit in 1970. On average, this amount has increased by approximately 0.9 pound per person per year.

   a. Write a formula for the nth term of the arithmetic sequence that describes pounds of fruit consumed annually by the average American n years after 1969.

   b. How many pounds of fruit will be consumed by the average American in 2006?

117. Company A pays $24,000 yearly with raises of $1600 per year. Company B pays $28,000 yearly with raises of $1000 per year. Which company will pay more in year 10? How much more?

118. Company A pays $23,000 yearly with raises of $1200 per year. Company B pays $26,000 yearly with raises of $800 per year. Which company will pay more in year 10? How much more?

Use the formula for the nth term of a geometric sequence to solve Exercises 119–124.
In Exercises 119–120, suppose you save $1 the first day of a month, $2 the second day, $4 the third day, and so on. That is, each day you save twice as much as you did the day before.

119. What will you put aside for savings on the fifteenth day of the month?

120. What will you put aside for savings on the thirtieth day of the month?

121. A professional baseball player signs a contract with a beginning salary of $3,000,000 for the first year with an annual increase of 4% per year beginning in the second year. That is, beginning in year 2, the athlete's salary will be 1.04 times what it was in the previous year. What is the athlete's salary for year 7 of the contract? Round to the nearest dollar.

122. You are offered a job that pays $30,000 for the first year with an annual increase of 5% per year beginning in the second year. That is, beginning in year 2, your salary will be 1.05 times what it was in the previous year. What can you expect to earn in your sixth year on the job? Round to the nearest dollar.

123. The population of California from 1990 through 1997 is shown in the following table.

Year	1990	1991	1992	1993
Population in millions	29.76	30.15	30.54	30.94

Year	1994	1995	1996	1997
Population in millions	31.34	31.75	32.16	32.58

   a. Divide the population for each year by the population in the preceding year. Round to three decimal places and show that the population of California is increasing geometrically.

   b. Write the general term of the geometric sequence describing population growth for California n years after 1989.

   c. Estimate California's population, in millions, for the year 2000. According to the U.S. Census Bureau, California's population in 2000 was 33.87 million. How well does your geometric sequence describe the actual population?

124. The population of Texas from 1990 through 1997 is shown in the following table.

Year	1990	1991	1992	1993
Population in millions	16.99	17.35	17.71	18.08

Year	1994	1995	1996	1997
Population in millions	18.46	18.85	19.25	19.65

    a. Divide the population for each year by the population in the preceding year. Round to three decimal places and show that the population of Texas is increasing geometrically.

    b. Write the general term of the geometric sequence describing population growth for Texas n years after 1989.

    c. Estimate Texas's population, in millions, for the year 2000. According to the U.S. Census Bureau, Texas's population in 2000 was 20.85 million. How well does your geometric sequence describe the actual population?

Writing in Mathematics

125. What is a sequence? Give an example with your description.

126. What is an arithmetic sequence? Give an example with your description.

127. What is the common difference in an arithmetic sequence?

128. What is a geometric sequence? Give an example with your description.

129. What is the common ratio in a geometric sequence?

130. If you are given a sequence that is arithmetic or geometric, how can you determine which type of sequence it is?

131. For the first 30 days of a flu outbreak, the number of students on your campus who become ill is increasing. Which is worse: The number of students with the flu is increasing arithmetically or is increasing geometrically? Explain your answer.

Critical Thinking Exercises

132. Which one of the following is true?

    a. The common difference for the arithmetic sequence given by $1, -1, -3, -5, \ldots$ is 2.

    b. The sequence $1, 4, 8, 13, 19, 26, \ldots$ is an arithmetic sequence.

    c. The nth term of an arithmetic sequence whose first term is a_1 and whose common difference is d is $a_n = a_1 + nd$.

    d. If the first term of an arithmetic sequence is 5 and the third term is -3, then the fourth term is -7.

133. Which one of the following is true?

    a. The sequence $2, 6, 24, 120, \ldots$ is an example of a geometric sequence.

    b. Adjacent terms in a geometric sequence have a common difference.

    c. A sequence that is not arithmetic must be geometric.

    d. If a sequence is geometric, we can write as many terms as we want by repeatedly multiplying by the common ratio.

134. A person is investigating two employment opportunities. They both have a beginning salary of $20,000 per year. Company A offers an increase of $1000 per year. Company B offers 5% more than during the preceding year. Which company will pay more in the sixth year?

Technology Exercises

In Exercises 135–136, use your calculator to determine the indicated term for the arithmetic sequence with first term, a_1, and common difference, d.

135. Find a_{327}, when $a_1 = 7721, d = -905$.

136. Find a_{3126}, when $a_1 = -3121, d = 698$.

In Exercises 137–138, use your calculator to determine the indicated term for the geometric sequence with first term, a_1, and common ratio, r.

137. Find a_{32}, when $a_1 = 7721, r = 5$.

138. Find a_{126}, when $a_1 = 196,200, r = 0.925$.

Group Exercise

139. Enough curiosities involving the Fibonacci sequence exist to warrant a flourishing Fibonacci Association. It publishes a quarterly journal. Do some research on the Fibonacci sequence by consulting the research department of your library or the Internet, and find one property that interests you. After doing this research, get together with your group to share these intriguing properties.

Chapter Summary, Review, and Test

SUMMARY

DEFINITIONS AND CONCEPTS   EXAMPLES

5.1 Number Theory; Prime and Composite Numbers

Divisibility

   a. Natural Numbers: $\mathbf{N} = \{1, 2, 3, 4, 5, \dots\}$

   b. $b|a$ (b divides a: a is divisible by b) for natural numbers a and b if the operation of dividing a by b leaves a remainder of 0.

   c. Rules of divisibility are given in Table 5.1 on page 194.   Ex. 1, p. 195

Prime and Composite Numbers

   a. A prime number is a natural number greater than 1 that has only itself and 1 as factors.

   b. A composite number is a natural number greater than 1 that is divisible by a number other than itself and 1.

   c. The Fundamental Theorem of Arithmetic: Every composite number can be expressed as a product of   Ex. 2, p. 196
prime numbers in one and only one way (if the order of the factors is disregarded).

Greatest Common Divisors and Least Common Multiples

   a. The greatest common divisor of two or more natural numbers is the largest number that is a divisor (or factor) of all the numbers.

   b. The procedure for finding the greatest common divisor is boxed on page 197.   Ex. 3, p. 197;
                                                                    Ex. 4, p. 198

   c. The least common multiple of two or more natural numbers is the smallest natural number that is divisible by all of the numbers.

   d. The procedure for finding the least common multiple is boxed on page 199.   Ex. 5, p. 199;
                                                                     Ex. 6, p. 200

5.2 The Integers; Order of Operations

Whole Numbers and Integers

   a. Whole Numbers $= \{0, 1, 2, 3, 4, 5, \dots\}$

   b. Integers $= \{\dots, -3, -2, -1, 0, 1, 2, 3, \dots\}$

The Symbols $<$ and $>$

   a. $a < b$ (a is less than b) means a is to the left of b on a number line.   Ex. 2, p. 204

   b. $a > b$ (a is greater than b) means a is to the right of b on a number line.

Absolute Value

   a. $|a|$, the absolute value of a, is the distance of a from 0 on a number line.   Ex. 3, p. 205

   b. The absolute value of a positive number is the number itself.

   c. The absolute value of 0 is 0: $|0| = 0$.

   d. The absolute value of a negative number is the number without the negative sign.
For example, $|-8| = 8$.

Order of Operations

   a. Perform all operations within grouping symbols.   Ex. 7, p. 211;

   b. Evaluate all exponential expressions.   Ex. 8, p. 211

   c. Do all multiplications and divisions from left to right.

   d. Do all additions and subtractions from left to right.

5.3 The Rational Numbers

Rational Numbers

   a. The set of rational numbers is the set of all numbers which can be expressed in the form $\frac{a}{b}$,
where a and b are integers and b is not equal to 0.

b. A rational number is reduced to its lowest terms, or simplified, by dividing both the numerator and the denominator by their greatest common divisor.

Ex. 1, p. 215

c. A mixed number consists of the sum of an integer and a rational number, expressed without the use of an addition sign. An improper fraction is a rational number whose numerator is greater than its denominator. Procedures for converting between these forms are given in the boxes on page 216.

Ex. 2, p. 216;;
Ex. 3, p. 217

d. Any rational number can be expressed as a decimal. The resulting decimal will either terminate (stop), or it will have a digit that repeats or a block of digits that repeat. The rational number $\frac{a}{b}$ is expressed as a decimal by dividing b into a.

Ex. 4, p. 218

e. To express a terminating decimal as a quotient of integers, the digits to the right of the decimal point are the numerator. The place-value of the last digit to the right of the decimal point is the denominator.

Ex. 5, p. 219

f. To express a repeating decimal as a quotient of integers, use the boxed procedure on page 220.

Ex. 6, p. 219;
Ex. 7, p. 220

Operations with Rational Numbers

a. The product of two rational numbers is the product of their numerators divided by the product of their denominators.

Ex. 8, p. 221

b. Two numbers whose product is 1 are called reciprocals, or multiplicative inverses, of each other. The quotient of two rational numbers is the product of the first number and the reciprocal of the second number.

Ex. 9, p. 222

c. The sum or difference of two rational numbers with identical denominators is the sum or difference of their numerators over the common denominator.

Ex. 10, p. 222

d. Add or subtract rational numbers with unlike denominators by first expressing each rational number with the least common denominator and then following part (c) above.

Ex. 11, p. 223;
Ex. 12, p. 223

Density of the Rational Numbers

a. Given any two rational numbers, there is always a rational number between them.

b. To find the rational number halfway between two rational numbers, add the rational numbers and divide their sum by 2.

Ex. 13, p. 224

5.4 The Irrational Numbers

Irrational Numbers

a. The set of irrational numbers is the set of numbers whose decimal representations are neither terminating nor repeating.

b. Examples of irrational numbers:

$\sqrt{2} \approx 1.414, \pi \approx 3.142$

Operations with Square Roots

a. Simplifying square roots: Use the product rule, $\sqrt{ab} = \sqrt{a} \cdot \sqrt{b}$, to remove from the square root any perfect squares that occur as factors.

Ex. 1, p. 230

b. Multiplying square roots: $\sqrt{a} \cdot \sqrt{b} = \sqrt{ab}$. The product of square roots is the square root of the product.

Ex. 2, p. 231

c. Dividing square roots: $\frac{\sqrt{a}}{\sqrt{b}} = \sqrt{\frac{a}{b}}$. The quotient of the square roots is the square root of the quotient.

Ex. 3, p. 232

d. Adding and subtracting square roots: If the radicals have the same radicand, add or subtract their coefficients. The answer is the sum or difference of the coefficients times the common square root. Addition or subtraction is sometimes possible by first simplifying the square roots.

Ex. 4, p. 232;
Ex. 5, p. 232

e. Rationalizing denominators: Multiply the numerator and denominator by the smallest number that produces a perfect square radicand in the denominator.

Ex. 6, p. 233

5.5 Real Numbers and Their Properties

The Real Numbers

a. The set of real numbers is obtained by combining the rational numbers with the irrational numbers.

Ex. 1, p. 237

b. The important subsets of the real numbers are summarized in Table 5.2 on page 237.

c. A diagram representing the relationships among the subsets of the real numbers is given to the left of Table 5.2 on page 237.

Properties of Real Numbers

a. Closure property of addition and multiplication

$$a + b \text{ and } a \cdot b \text{ are real numbers.}$$

Ex. 2, p. 239;
Ex. 3, p. 240

b. Commutative property of addition and multiplication

$$a + b = b + a \qquad a \cdot b = b \cdot a$$

c. Associative property of addition and multiplication

$$(a + b) + c = a + (b + c)$$
$$(a \cdot b) \cdot c = a \cdot (b \cdot c)$$

d. Distributive property of multiplication over addition

$$a \cdot (b + c) = a \cdot b + a \cdot c$$

5.6 Exponents and Scientific Notation

Properties of Exponents

a. Product rule: $b^m \cdot b^n = b^{m+n}$

b. Power rule: $(b^m)^n = b^{m \cdot n}$

Table 5.4, p. 244;
Ex. 1, p. 244;
Ex. 2, p. 245

c. Quotient rule: $\dfrac{b^m}{b^n} = b^{m-n}, b \neq 0$

d. Zero exponent rule: $b^0 = 1, b \neq 0$

e. Negative exponent rule: $b^{-m} = \dfrac{1}{b^m}$

Scientific Notation

a. A number in scientific notation is expressed as $a \times 10^n$, where $1 \leq a < 10$ and n is an integer.

b. Changing from Scientific to Decimal Notation: If n is positive, move the decimal point in a to the right n places. If n is negative, move the decimal point in a to the left n places.

Ex. 3, p. 247

c. Changing from Decimal to Scientific Notation: Move the decimal point in the given number to obtain a, where $1 \leq a < 10$. The number of places the decimal point moves gives the absolute value of n in $a \times 10^n$; n is positive if the number is greater than 10 and negative if the number is less than 1.

Ex. 4, p. 247;
Ex. 5, p. 248

d. The product and quotient rules for exponents are used to multiply and divide numbers in scientific notation.

Ex. 6, p. 248;
Ex. 7, p. 249;
Ex. 8, p. 249;
Ex. 9, p. 250

5.7 Arithmetic and Geometric Sequences

Arithmetic Sequence

a. Each term after the first differs from the preceding term by a constant, the common difference. Subtract any term from the term that directly follows to find the common difference.

Ex. 1, p. 255;
Ex. 2, p. 255

b. General term or nth term: $a_n = a_1 + (n - 1)d$. The first term is a_1 and the common difference is d.

Ex. 3, p. 256;
Ex. 4, p. 256

Geometric Sequence

a. Each term after the first is obtained by multiplying the preceding term by a constant, the common ratio. Divide any term after the first term by the term that directly precedes it to find the common ratio.

Ex. 5, p. 258

b. General term or nth term: $a_n = a_1 r^{n-1}$. The first term is a_1 and the common ratio is r.

Ex. 6, p. 258;
Ex. 7, p. 259

REVIEW EXERCISES

5.1

In Exercises 1 and 2, determine whether the number is divisible by each of the following numbers: 2, 3, 4, 5, 6, 8, 9, 10, and 12. If you are using a calculator, explain the divisibility shown by your calculator using one of the rules of divisibility.

1. 238,632

2. 421,153,470

In Exercises 3–5, find the prime factorization of each composite number.

3. 705

4. 960

5. 6825

In Exercises 6–8, find the greatest common divisor and the least common multiple of the numbers.

6. 30 and 48

7. 36 and 150

8. 216 and 254

9. For an intramural league, you need to divide 24 men and 60 women into all-male and all-female teams so that each team has the same number of people. What is the largest number of people that can be placed on a team?

10. The media center at a college runs videotapes of two lectures continuously. One videotape runs for 42 minutes and a second runs for 56 minutes. Both videotapes begin at 9:00 a.m. When will the videos of the two lectures begin again at the same time?

5.2

In Exercises 11–12, insert either $<$ or $>$ in the shaded area between the integers to make the statement true.

11. $-93 \blacksquare 17$

12. $-2 \blacksquare -200$

In Exercise 13–15, find the absolute value.

13. $|-860|$

14. $|53|$

15. $|0|$

Perform the indicated operations in Exercises 16–28.

16. $8 + (-11)$

17. $-6 + (-5)$

18. $-7 - 8$

19. $-7 - (-8)$

20. $(-9)(-11)$

21. $5(-3)$

22. $\dfrac{-36}{-4}$

23. $\dfrac{20}{-5}$

24. $-40 \div 5 \cdot 2$

25. $-6 + (-2) \cdot 5$

26. $6 - 4(-3 + 2)$

27. $28 \div (2 - 4^2)$

28. $36 - 24 \div 4 \cdot 3 - 1$

5.3

In Exercises 29–31, reduce each rational number to its lowest terms.

29. $\dfrac{40}{75}$

30. $\dfrac{36}{150}$

31. $\dfrac{165}{180}$

In Exercises 32–33, convert each mixed number to an improper fraction.

32. $5\frac{9}{11}$

33. $-3\frac{2}{7}$

In Exercises 34–35, convert each improper fraction to a mixed number.

34. $\dfrac{27}{5}$

35. $-\dfrac{17}{9}$

In Exercises 36–39, express each rational number as a decimal.

36. $\frac{4}{5}$

37. $\frac{3}{7}$

38. $\frac{5}{8}$

39. $\frac{9}{16}$

In Exercises 40–43, express each terminating decimal as a quotient of integers in lowest terms.

40. 0.6

41. 0.68

42. 0.588

43. 0.0084

In Exercises 44–46, express each repeating decimal as a quotient of integers in lowest terms.

44. $0.\overline{5}$

45. $0.\overline{34}$

46. $0.\overline{113}$

In Exercises 47–55 perform the indicated operations. Where possible, reduce the answer to lowest terms.

47. $\frac{3}{5} \cdot \frac{7}{10}$

48. $\left(3\frac{1}{3}\right)\left(1\frac{3}{4}\right)$

49. $\frac{4}{5} \div \frac{3}{10}$

50. $-1\frac{2}{3} \div 6\frac{2}{3}$

51. $\frac{2}{9} + \frac{4}{9}$

52. $\frac{7}{9} + \frac{5}{12}$

53. $\frac{3}{4} - \frac{2}{15}$

54. $\frac{1}{3} + \frac{1}{2} \cdot \frac{4}{5}$

55. $\frac{3}{8}\left(\frac{1}{2} + \frac{1}{3}\right)$

In Exercises 56–57, find a rational number halfway between the two numbers in each pair.

56. $\frac{1}{7}$ and $\frac{1}{8}$

57. $\frac{3}{4}$ and $\frac{3}{5}$

58. A recipe for coq au vin is meant for six people and requires $4\frac{1}{2}$ pounds of chicken. If you want to serve 15 people, how much chicken is needed?

59. The gas tank of a car is filled to its capacity. The first day, $\frac{1}{4}$ of the tank's gas is used for travel. The second day, $\frac{1}{3}$ of the tank's original amount of gas is used for travel. What fraction of the tank is filled with gas at the end of the second day?

5.4

In Exercises 60–63, simplify the square root.

60. $\sqrt{28}$

61. $\sqrt{72}$

62. $\sqrt{150}$

63. $\sqrt{300}$

In Exercises 64–72, perform the indicated operation. Simplify the answer when possible.

64. $\sqrt{6} \cdot \sqrt{8}$

65. $\sqrt{10} \cdot \sqrt{5}$

66. $\dfrac{\sqrt{24}}{\sqrt{2}}$

67. $\dfrac{\sqrt{27}}{\sqrt{3}}$

68. $\sqrt{5} + 4\sqrt{5}$

69. $7\sqrt{11} - 13\sqrt{11}$

70. $\sqrt{50} + \sqrt{8}$

71. $\sqrt{3} - 6\sqrt{27}$

72. $2\sqrt{18} + 3\sqrt{8}$

In Exercises 73–74, rationalize the denominator.

73. $\dfrac{30}{\sqrt{5}}$

74. $\sqrt{\dfrac{2}{3}}$

75. The expression $2\sqrt{5L}$ is used to estimate the speed of a car prior to an accident, in miles per hour, based on the length of its skid marks L, in feet. Find the speed of a car that left skid marks 40 feet long, and write the answer in simplified radical form. Then use your calculator to estimate the speed to the nearest mile per hour.

5.5

76. Consider the set

$$\left\{-17, -\tfrac{9}{13}, 0, 0.75, \sqrt{2}, \pi, \sqrt{81}\right\}.$$

List all numbers from the set that are **a.** natural numbers, **b.** whole numbers, **c.** integers, **d.** rational numbers, **e.** irrational numbers, **f.** real numbers.

77. Give an example of an integer that is not a natural number.

78. Give an example of a rational number that is not an integer.

79. Give an example of a real number that is not a rational number.

In Exercises 80–85, state the name of the property illustrated.

80. $3 + 17 = 17 + 3$ **81.** $(6 \cdot 3) \cdot 9 = 6 \cdot (3 \cdot 9)$

82. $\sqrt{3}\left(\sqrt{5} + \sqrt{3}\right) = \sqrt{15} + 3$

83. $(6 \cdot 9) \cdot 2 = 2 \cdot (6 \cdot 9)$

84. $\sqrt{3}\left(\sqrt{5} + \sqrt{3}\right) = \left(\sqrt{5} + \sqrt{3}\right)\sqrt{3}$

85. $(3 \cdot 7) + (4 \cdot 7) = (4 \cdot 7) + (3 \cdot 7)$

In Exercises 86–87, give on example to show that

86. The natural numbers are not closed with respect to division.

87. The whole numbers are not closed with respect to subtraction.

5.6

In Exercises 88–98, evaluate each expression.

88. $6 \cdot 6^2$ **89.** $2^3 \cdot 2^3$ **90.** $(2^2)^2$

91. $(3^3)^2$ **92.** $\dfrac{5^6}{5^4}$ **93.** 7^0

94. $(-7)^0$ **95.** 6^{-3} **96.** 2^{-4}

97. $\dfrac{7^4}{7^6}$ **98.** $3^5 \cdot 3^{-2}$

In Exercises 99–102, express the number in decimal notation.

99. 4.6×10^2 **100.** 3.74×10^4

101. 2.55×10^{-3} **102.** 7.45×10^{-5}

In Exercises 103–108, express the number in scientific notation.

103. 7520 **104.** $3{,}590{,}000$ **105.** 0.00725

106. 0.000000409 **107.** 420×10^{11} **108.** 0.97×10^{-4}

In Exercises 109–112, perform the indicated operation and express each answer in decimal notation.

109. $(3 \times 10^7)(1.3 \times 10^{-5})$ **110.** $(5 \times 10^3)(2.3 \times 10^2)$

111. $\dfrac{6.9 \times 10^3}{3 \times 10^5}$ **112.** $\dfrac{2.4 \times 10^{-4}}{6 \times 10^{-6}}$

In Exercises 113–116, perform the indicated operation by first expressing each number in scientific notation. Write the answer in scientific notation.

113. $(60{,}000)(540{,}000)$ **114.** $(91{,}000)(0.0004)$

115. $\dfrac{8{,}400{,}000}{4000}$ **116.** $\dfrac{0.000003}{0.00000006}$

117. If you earned $1 million per year ($10^6$), how long would it take to accumulate $1 billion dollars ($10^9$)?

118. If the population of the United States is 2.8×10^8 and each person spends about $150 per year going to the movies (or renting movies), express the total annual spending on movies in scientific notation.

119. The world's population is approximately 6.1×10^9 people. Current projections double this population in 40 years. Write the population 40 years from now in scientific notation.

5.7

In Exercises 120–122, write the first six terms of the arithmetic sequence with the first term, a_1, and common difference, d.

120. $a_1 = 7, d = 4$ **121.** $a_1 = -4, d = -5$

122. $a_1 = \frac{3}{2}, d = -\frac{1}{2}$

In Exercises 123–125, find the indicated term for the arithmetic sequence with first term, a_1, and common difference, d.

123. Find a_6, when $a_1 = 5, d = 3$.

124. Find a_{12}, when $a_1 = -8, d = -2$.

125. Find a_{14}, when $a_1 = 14, d = -4$.

In Exercises 126–127, write a formula for the general term (the nth term) of each arithmetic sequence. Then use the formula for a_n to find a_{20}, the 20th term of the sequence.

126. $-7, -3, 1, 5, \ldots$ **127.** $a_1 = 200, d = -20$

In Exercises 128–130, write the first six terms of the geometric sequence with the first term, a_1, and common ratio, r.

128. $a_1 = 3, r = 2$ **129.** $a_1 = \frac{1}{2}, r = \frac{1}{2}$

130. $a_1 = 16, r = -\frac{1}{2}$

In Exercises 131–133, find the indicated term for the geometric sequence with first term, a_1, and common ratio, r.

131. Find a_4, when $a_1 = 2, r = 3$.

132. Find a_6, when $a_1 = 16, r = \frac{1}{2}$.

133. Find a_5, when $a_1 = -3, r = 2$.

In Exercises 134–135, write a formula for the general term (the nth term) of each geometric sequence. Then use the formula for a_n to find a_8, the eighth term of the sequence.

134. $1, 2, 4, 8, \ldots$ **135.** $100, 10, 1, \frac{1}{10}, \ldots$

Determine whether each sequence in Exercises 136–139 is arithmetic or geometric. In each sequence, find the next two terms.

136. $4, 9, 14, 19, \ldots$ **137.** $2, 6, 18, 54, \ldots$

138. $1, \frac{1}{4}, \frac{1}{16}, \frac{1}{64}, \ldots$ **139.** $0, -7, -14, -21, \ldots$

140. The graph shows the changing pattern of work in the United States from 1900 through 2000. In 1900, 20% of the total labor force was comprised of white-collar workers. On average, this increased by approximately 0.52% per year since then.

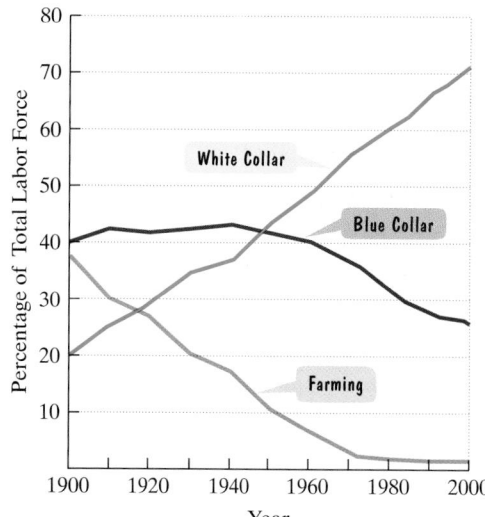

The Changing Pattern of Work in the United States

Source: U.S. Department of Labor

a. Write a formula for the nth term of the arithmetic sequence that describes the percentage of white-collar workers in the labor force n years after 1899.

b. Use the formula to predict the percentage of white-collar workers in the labor force by the year 2010.

141. The population of Iraq from 1998 through 2001 is shown in the following table.

Year	1998	1999	2000	2001
Population in Millions	19.96	20.72	21.51	22.33

Source: U.N. Population Division

a. Show that Iraq's population is increasing geometrically.

b. Write the general term of the geometric sequence describing population growth for Iraq n years after 1997.

c. Estimate Iraq's population, in millions, for the year 2008.

Group Exercise

142. *Putting Numbers into Perspective.* A large number can be put into perspective by comparing it with another number. For example, we put the $6.8 trillion national debt into perspective by comparing it to the number of U.S. citizens. The total distance covered by all the runners in the New York City Marathon (Exercise 100 in Exercise Set 5.6 on page 252) can be put into perspective by comparing this distance with, say, the distance from New York to San Francisco.

For this project, each group member should consult an almanac, a newspaper, or the World Wide Web to find a number greater than one million. Explain to other members of the group the context in which the large number is used. Express the number in scientific notation. Then put the number into perspective by comparing it with another number.

CHAPTER 5 TEST

1. Which of the numbers 2, 3, 4, 5, 6, 8, 9, 10, and 12 divide 391,248?

2. Find the prime factorization of 252.

3. Find the greatest common divisor and the least common multiple of 48 and 72.

Perform the indicated operations in Exercises 4–6.

4. $-6 - (5 - 12)$

5. $(-3)(-4) \div (7 - 10)$

6. $(6 - 8)^2(5 - 7)^3$

7. Express $\frac{7}{12}$ as a decimal.

8. Express $0.\overline{64}$ as a quotient of integers in lowest terms.

In Exercises 9–11, perform the indicated operations. Where possible, reduce the answer to its lowest terms.

9. $\left(-\frac{3}{7}\right) \div \left(-2\frac{1}{7}\right)$

10. $\frac{19}{24} - \frac{7}{40}$

11. $\frac{1}{2} - 8\left(\frac{1}{4} + 1\right)$

12. Find a rational number halfway between $\frac{1}{2}$ and $\frac{2}{3}$.

13. Multiply and simplify: $\sqrt{10} \cdot \sqrt{5}$.

14. Add: $\sqrt{50} + \sqrt{32}$.

15. Rationalize the denominator: $\dfrac{6}{\sqrt{2}}$.

16. List all the rational numbers in this set:
$$\left\{-7, -\frac{4}{5}, 0, 0.25, \sqrt{3}, \sqrt{4}, \frac{22}{7}, \pi\right\}.$$

In Exercises 17–18, state the name of the property illustrated.

17. $3(2 + 5) = 3(5 + 2)$

18. $6(7 + 4) = 6 \cdot 7 + 6 \cdot 4$

In Exercises 19–21, evaluate each expression.

19. $3^3 \cdot 3^2$

20. $\dfrac{4^6}{4^3}$

21. 8^{-2}

22. Multiply and express the answer in decimal notation.
$$(3 \times 10^8)(2.5 \times 10^{-5})$$

23. Divide by first expressing each number in scientific notation. Write the answer in scientific notation.
$$\frac{49,000}{0.007}$$

24. A human brain contains 3×10^{10} neurons and a gorilla brain contains 7.5×10^9 neurons. How many times as many neurons are in the brain of a human as in the brain of a gorilla?

25. Write the first six terms of the arithmetic sequence with first term, a_1, and common difference, d.
$$a_1 = 1, d = -5$$

26. Find a_9, the ninth term of the arithmetic sequence with the first term, a_1, and common difference, d.
$$a_1 = -2, d = 3$$

27. Write the first six terms of the geometric sequence with first term, a_1, and common ratio, r.
$$a_1 = 16, r = \frac{1}{2}$$

28. Find a_7, the seventh term of the geometric sequence with the first term, a_1, and common ratio, r.
$$a_1 = 5, r = 2$$

Algebra: Equations and Inequalities

Formulas like those that describe the height a child will attain as an adult are frequently obtained from actual data. Formulas can be used to describe what happened in the past and to make predictions about what might occur in the future. Knowing how to create and use formulas will help you recognize patterns, logic, and order in a world that can appear chaotic to the untrained eye. In many ways, algebra will provide you with a new way of looking at your world.

Sitting in the biology department office, you overhear two of the professors discussing the possible adult heights of their respective children. Looking at the blackboard that they've been writing on, you see that there are formulas that can estimate the height a child will attain as an adult. If the child is x years old and h inches tall, that child's adult height, H, in inches, is approximated by one of the following formulas:

$$\text{Girls:} \quad H = \frac{h}{0.00028x^3 - 0.0071x^2 + 0.0926x + 0.3524}$$

$$\text{Boys:} \quad H = \frac{h}{0.00011x^3 - 0.0032x^2 + 0.0604x + 0.3796}.$$

OBJECTIVES

1. Evaluate algebraic expressions.

2. Understand the vocabulary of algebraic expressions.

3. Simplify algebraic expressions.

4. Evaluate formulas.

Algebraic Expressions

Feeling attractive with a suntan that gives you a "healthy glow"? Think again. Direct sunlight is known to promote skin cancer. Although sunscreens protect you from burning, skin doctors are concerned with the long-term damage that results from the sun even without sunburn.

Algebra uses letters, such as x and y, to represent numbers. Such letters are called **variables**. For example, we can let x represent the number of minutes that a person can stay in the sun without burning with no sunscreen. With a number 6 sunscreen, exposure time without burning is six times as long, or 6 times x. This can be written $6 \cdot x$, but it is usually expressed as $6x$. Placing a number and a letter next to one another indicates multiplication.

Notice that $6x$ combines the number 6 and the variable x using the operation of multiplication. A combination of variables and numbers using the operations of addition, subtraction, multiplication, or division, as well as powers or roots, is called an **algebraic expression**. Here are some examples of algebraic expressions:

$$x + 6, \quad x^2 - 6, \quad 6x, \quad \frac{x}{6}, \quad 3x + 5, \quad \sqrt{x} + 7.$$

Evaluating Algebraic Expressions and Using the Order of Operations

1 Evaluate algebraic expressions.

Evaluating an algebraic expression means to find the value of the expression for a given value of the variable. For example, we can evaluate $6x$ (from the sunscreen example) when $x = 15$. We substitute 15 for x. We obtain $6 \cdot 15$, or 90. This means that if you can stay in the sun for 15 minutes without burning when you don't put on any lotion, then with a number 6 lotion, you can "cook" for 90 minutes with no visible damage.

Many algebraic expressions contain more than one operation. Evaluating an algebraic expression correctly involves carefully applying the order of operations agreement that we studied in Chapter 5. This agreement is summarized in the box at the top of the next page.

> **ORDER OF OPERATIONS**
>
> 1. First, perform all operations within grouping symbols.
> 2. Next, evaluate all exponential expressions.
> 3. Next, do all multiplications and divisions in the order in which they occur, working from left to right.
> 4. Finally, do all additions and subtractions in the order in which they occur, working from left to right.

EXAMPLE 1   EVALUATING AN ALGEBRAIC EXPRESSION

Evaluate $7x + 5$ for $x = 10$.

Solution   We substitute 10 for x and carry out the multiplication and addition.

$$7x + 5 \qquad \text{This is the given algebraic expression.}$$
$$= 7 \cdot 10 + 5 \qquad \text{Substitute 10 for x.}$$
$$= 70 + 5 \qquad \text{Multiply: } 7 \cdot 10 = 70.$$
$$= 75 \qquad \text{Add.}$$

 **Check Point 1**   Evaluate $6x + 9$ for $x = 12$.

EXAMPLE 2   EVALUATING AN ALGEBRAIC EXPRESSION

Evaluate $x^2 + 5x - 3$ for $x = -6$.

Solution   We substitute -6 for each of the two occurrences of x. Then we use the order of operations to evaluate the algebraic expression.

$$x^2 + 5x - 3 \qquad \text{This is the given algebraic expression.}$$
$$= (-6)^2 + 5(-6) - 3 \qquad \text{Substitute } -6 \text{ for each x.}$$
$$= 36 + 5(-6) - 3 \qquad \text{Evaluate the exponential expression: } (-6)^2 = (-6)(-6) = 36$$
$$= 36 + (-30) - 3 \qquad \text{Multiply: } 5(-6) = -30.$$
$$= 6 - 3 \qquad \text{Add and subtract from left to right. First add: } 36 + (-30) = 6.$$
$$= 3 \qquad \text{Subtract.}$$

 **Check Point 2**   Evaluate $x^2 + 4x - 7$ for $x = -5$.

EXAMPLE 3 EVALUATING AN ALGEBRAIC EXPRESSION

Evaluate $-2x^2 + 5xy - y^2$ for $x = 4$ and $y = 2$.

Solution   We substitute 4 for each x and 2 for each y. Then we use the order of operations to evaluate the algebraic expression.

$-2x^2 + 5xy - y^2$	This is the given algebraic expression.
$= -2 \cdot 4^2 + 5 \cdot 4 \cdot 2 - 2^2$	Substitute 4 for x and 2 for y.
$= -2 \cdot 16 + 5 \cdot 4 \cdot 2 - 4$	Evaluate the exponential expressions: $4^2 = 4 \cdot 4 = 16$ and $2^2 = 2 \cdot 2 = 4$.
$= -32 + 40 - 4$	Multiply: $-2 \cdot 16 = -32$ and $5 \cdot 4 \cdot 2 = 20 \cdot 2 = 40$.
$= 8 - 4$	Add and subtract from left to right. First add: $-32 + 40 = 8$.
$= 4$	Subtract.

 **Check Point 3**   Evaluate $-3x^2 + 4xy - y^2$ for $x = 5$ and $y = 6$.

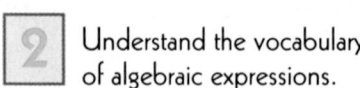 **2** Understand the vocabulary of algebraic expressions.

The Vocabulary of Algebraic Expressions

We have seen that an algebraic expression combines numbers and variables. Here is another example of an algebraic expression:

$$7x - 9y - 3.$$

The **terms** of an algebraic expression are those parts that are separated by addition. For example, we can rewrite $7x - 9y - 3$ as

$$7x + (-9y) + (-3).$$

This expression contains three terms, namely $7x$, $-9y$, and -3.

The numerical part of a term is called its **coefficient**. In the term $7x$, the 7 is the coefficient. In the term $-9y$, the -9 is the coefficient.

Coefficients of 1 and -1 are not written. Thus, the coefficient of x, meaning $1x$, is 1. Similarly, the coefficient of $-y$, meaning $-1y$, is -1.

A term that consists of just a number is called a **constant term**. The constant term of $7x - 9y - 3$ is -3.

Like terms are terms that have the same variables with the same exponents on the variables. Here are some examples of like terms:

$7x$ and $3x$	These terms have the same variable, x.
$4y$ and $9y$	These terms have the same variable, y.
$5x^2$ and $-3x^2$.	These terms have the same variable with the same exponent, x^2.

Constant terms are like terms. Thus, the constant terms 7 and -12 are like terms.

Simplifying Algebraic Expressions

3 Simplify algebraic expressions.

The properties of real numbers that we discussed in Chapter 5 can be applied to algebraic expressions.

PROPERTIES OF REAL NUMBERS

Property	Example
Commutative Property of Addition $a + b = b + a$	$4192 + 30.5x^2 = 30.5x^2 + 4192$
Commutative Property of Multiplication $ab = ba$	$x \cdot 6 = 6x$
Associative Property of Addition $(a + b) + c = a + (b + c)$	$3 + (8 + x) = (3 + 8) + x = 11 + x$
Associative Property of Multiplication $(ab)c = a(bc)$	$-2(3x) = (-2 \cdot 3)x = -6x$
Distributive Property $a(b + c) = ab + ac$	$5(3x + 7) = 5 \cdot 3x + 5 \cdot 7 = 15x + 35$
$a(b - c) = ab - ac$	$4(2x - 5) = 4 \cdot 2x - 4 \cdot 5 = 8x - 20$

The distributive property can be used to rewrite some algebraic expressions without parentheses.

EXAMPLE 4   APPLYING THE DISTRIBUTIVE PROPERTY

The optimum heart rate is the rate that a person should achieve during exercise for the exercise to be most beneficial. The algebraic expression

$$0.6(220 - a)$$

describes a person's optimum heart rate, in beats per minute, where a represents the age of the person, in years.

a. Use the distributive property to rewrite the algebraic expression without parentheses.

b. Use each form of the algebraic expression to determine the optimum heart rate for a 20-year-old runner.

Solution

a. $0.6(220 - a) = 0.6(220) - 0.6a$ Use the distributive property to remove parentheses.

$$= 132 - 0.6a$$ Multiply: $0.6(220) = 132$.

b. To determine the optimum heart rate for a 20-year-old runner, substitute 20 for a in each form of the algebraic expression.

Using $0.6(220 - a)$:

$0.6(220 - 20)$

$= 0.6(200)$

$= 120$

Using $132 - 0.6a$:

$132 - 0.6(20)$

$= 132 - 12$

$= 120$

Both forms of the algebraic expression indicate that the optimum heart rate for a 20-year-old runner is 120 beats per minute.

Check Point 4

Use each form of the algebraic expression in Example 5 to determine the optimum heart rate for a 40-year-old runner.

The distributive property (in reverse) can also be used to combine like terms such as $7x$ and $3x$:

$$7x + 3x = (7 + 3)x = 10x.$$

When combining like terms, you may find yourself leaving out the details of the distributive property. For example, you may simply write

$$7x + 3x = 10x.$$

It might be useful to think along these lines: Seven things plus three of the (same) things give ten of those things. To add like terms, add the coefficients and copy the common variable.

EXAMPLE 5   COMBINING LIKE TERMS

Combine like terms in each expression:

a. $12y + 9y$ **b.** $8x - 10x$ **c.** $x + 7x$ **d.** $3x^2 - 7x^2$.

Solution

a. $12y + 9y = (12 + 9)y = 21y$

b. $8x - 10x = (8 - 10)x = -2x$

c. $x + 7x = 1x + 7x = (1 + 7)x = 8x$

d. $3x^2 - 7x^2 = (3 - 7)x^2 = -4x^2$

STUDY TIP

You can work each part of Example 5 without writing the middle step. Add or subtract the coefficients of the terms. Use this result as the coefficient of the terms' common variable.

Check Point 5

Combine like terms in each expression:

a. $4y + 15y$ **b.** $7x - 20x$ **c.** $y + 99y$ **d.** $5x^2 - 15x^2$.

When an expression contains three or more terms, use the commutative and associative properties to group like terms. Then combine the like terms.

EXAMPLE 6   GROUPING AND COMBINING LIKE TERMS

Group and combine like terms in each expression:

a. $7x + 5 + 3x + 8$ **b.** $4x + 7y + 11 - 2x - 9y - 3$

Solution

a. The like terms are $7x$ and $3x$, as well as 5 and 8. Group the like terms and then combine them.

$$7x + 5 + 3x + 8$$

$$= \underbrace{7x + 3x}_{} + \underbrace{5 + 8}_{}$$

$$= \quad 10x \quad + \quad 13$$

b. The pairs of like terms are $4x$ and $-2x$, $7y$ and $-9y$, and 11 and -3. Group the like terms and then combine them.

$$4x + 7y + 11 - 2x - 9y - 3$$

$$= \underbrace{4x - 2x} + \underbrace{7y - 9y} + \underbrace{11 - 3}$$

$$= \quad 2x \quad + \quad -2y \quad + \quad 8$$

$$= \quad 2x \quad - \quad 2y \quad + \quad 8$$

 **Check 6 Point** Group and combine like terms in each expression:

a. $8x + 7 + 10x + 3$

b. $9x + 6y + 15 - 3x - 20y - 16$.

An algebraic expression is **simplified** when parentheses have been removed and like terms have been combined.

EXAMPLE 7 | SIMPLIFYING AN ALGEBRAIC EXPRESSION

Simplify: $5(3x - 7) - 6x$.

Solution

$$5(3x - 7) - 6x$$

$$= 5 \cdot 3x - 5 \cdot 7 - 6x \qquad \text{Use the distributive property to remove the parentheses.}$$

$$= 15x - 35 - 6x \qquad \text{Multiply.}$$

$$= (15x - 6x) - 35 \qquad \text{Group like terms.}$$

$$= 9x - 35 \qquad \text{Combine like terms.}$$

 **Check 7 Point** Simplify: $7(2x - 3) - 11x$.

It is not uncommon to see algebraic expressions with parentheses preceded by a negative sign or subtraction. An expression of the form $-(b + c)$ can be simplified as follows:

$$-(b + c) = -1(b + c) = (-1)b + (-1)c = -b + (-c) = -b - c.$$

Do you see a fast way to obtain the simplified expression on the right? **If a negative sign or a subtraction symbol appears outside parentheses, drop the parentheses and change the sign of every term within the parentheses.** For example,

$$-(3x^2 - 7x - 4) = -3x^2 + 7x + 4.$$

EXAMPLE 8 | SIMPLIFYING AN ALGEBRAIC EXPRESSION

Simplify: $2x - (3x - 4y + 6)$.

Solution

$$2x - (3x - 4y + 6)$$

$$= 2x - 3x + 4y - 6 \qquad \text{Drop parentheses and change the sign of each term in parentheses.}$$

$$= -x + 4y - 6 \qquad \text{Combine like terms: } 2x - 3x = -1x = -x.$$

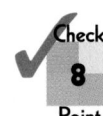

Simplify: $7x - (15x + 2y - 1)$.

Evaluate formulas.

Formulas and Mathematical Models

An **equation** is formed when an equal sign is placed between two algebraic expressions. One aim of algebra is to provide a compact, symbolic description of your world. These descriptions involve the use of **formulas**, statements of equality expressing a relationship among two or more variables. Many formulas give an approximate description of actual data and are called **mathematical models**. We say that such formulas **model**, or approximately describe, the data.

EXAMPLE 9   MODELING THE GRAY WOLF POPULATION

The formula

$$P = 0.72x^2 + 9.4x + 783$$

models the gray wolf population in the United States, P, x years after 1960. Use the formula to find the population in 1990. How well does the formula model the actual data shown in the bar graph in Figure 6.1?

**Gray Wolf Population
(to the Nearest Hundred)**

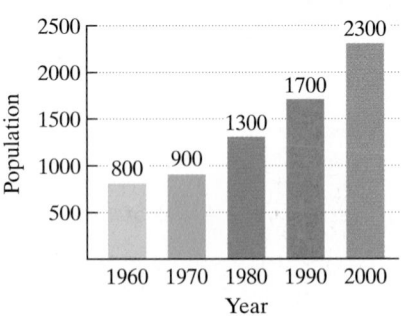

Source: U.S. Department of the Interior

FIGURE **6.1**

Solution   Because 1990 is 30 years after 1960, we substitute 30 for x in the given formula. Then we use the order of operations to find P, the gray wolf population in 1990.

$P = 0.72x^2 + 9.4x + 783$	This is the given mathematical model.
$P = 0.72(30)^2 + 9.4(30) + 783$	Replace each occurrence of x with 30.
$P = 0.72(900) + 9.4(30) + 783$	Evaluate the exponential expression: $30^2 = 30 \cdot 30 = 900$.
$P = 648 + 282 + 783$	Multiply from left to right: $0.72(900) = 648$ and $9.4(30) = 282$.
$P = 1713$	Add.

The formula indicates that in 1990, the gray wolf population in the United States was 1713. The number given in Figure 6.1 is 1700, so the formula models the data quite well.

Use the formula in Example 10 to find the gray wolf population in 2000. How well does the formula model the data in Figure 6.1?

Do mathematical models always provide an estimate that is a good approximation of actual data? No. Sometimes a mathematical model gives an estimate that is not a good approximation or is extended too far into the future, resulting in a prediction that does not make sense. In these cases, we say that **model break-down** has occurred.

BLITZER BONUS

Einstein's Famous Formula: $E = mc^2$

One of the most famous formulas in the world is $E = mc^2$, formulated by Albert Einstein. Einstein showed that any form of energy has mass and that mass itself is a form of energy. In this formula, E represents energy, in ergs, m represents mass, in grams, and c represents the speed of light. Because light travels at 30 billion centimeters per second, the formula indicates that 1 gram of mass will produce 900 billion ergs of energy.

Einstein's formula implies that the mass of a golf ball could provide the daily energy needs of the metropolitan Boston area. Mass and energy are equivalent, and the transformation of even a tiny amount of mass releases an enormous amount of energy. If this energy is released suddenly, a destructive force is unleashed, as in an atom bomb. When the release is gradual and controlled, the energy can be used to generate power.

The theoretical results implied by Einstein's formula $E = mc^2$ have not been realized because scientists have not yet developed a way of converting a mass completely to energy.

Exercise Set 6.1

Practice Exercises

In Exercises 1–30, evaluate the algebraic expression for the given value or values of the variables.

1. $5x + 7$; $x = 4$
2. $9x + 6$; $x = 5$
3. $-7x - 5$; $x = -4$
4. $-6x - 13$; $x = -3$
5. $x^2 + 4$; $x = 5$
6. $x^2 + 9$; $x = 3$
7. $x^2 - 6$; $x = -2$
8. $x^2 - 11$; $x = -3$
9. $x^2 + 4x$; $x = 10$
10. $x^2 + 6x$; $x = 9$
11. $8x^2 + 17$; $x = 5$
12. $7x^2 + 25$; $x = 3$
13. $x^2 - 5x$; $x = -11$
14. $x^2 - 8x$; $x = -5$
15. $x^2 + 5x - 6$; $x = 4$
16. $x^2 + 7x - 4$; $x = 6$
17. $2x^2 - 5x - 6$; $x = -3$
18. $3x^2 - 4x - 9$; $x = -5$
19. $-5x^2 - 4x - 11$; $x = -1$
20. $-6x^2 - 11x - 17$; $x = -2$
21. $4xy$; $x = -3, y = -1$
22. $6xy$; $x = -4, y = -1$
23. $x^2 - 4xy + y^2$; $x = -1, y = -3$
24. $x^2 - 6xy + y^2$; $x = -2, y = -1$
25. $3x^2 + 2xy + 5y^2$; $x = 2, y = 3$
26. $4x^2 + 3xy + 2y^2$; $x = 3, y = 2$
27. $-x^2 - 4xy + 3y^2$; $x = -1, y = -2$
28. $-x^2 - 3xy + 4y^2$; $x = -3, y = -1$
29. $x^3 - 5x^2 + 2$; $x = -4$
30. $x^3 - 4x^2 + 3$; $x = -5$

In Exercises 31–46, use the distributive property to rewrite each expression without parentheses.

31. $3(x + 5)$
32. $4(x + 6)$
33. $8(2x + 3)$
34. $9(2x + 5)$
35. $\frac{1}{3}(12 + 6r)$
36. $\frac{1}{4}(12 + 8r)$
37. $5(x + y)$
38. $7(x + y)$
39. $3(x - 2)$
40. $4(x - 5)$
41. $2(4x - 5)$
42. $6(3x - 2)$
43. $-4(2x - 3)$
44. $-3(4x - 5)$
45. $-3(-2x + 4)$
46. $-4(-3x + 2)$

In Exercises 47–76, simplify each algebraic expression.

47. $7x + 10x$
48. $5x + 13x$
49. $11x - 3x$
50. $14x - 5x$
51. $x + 9x$
52. $x + 12x$
53. $5y - 7y$
54. $8y - 11y$
55. $-5y + y$
56. $-6y + y$
57. $2x^2 + 5x^2$
58. $4x^2 + 6x^2$
59. $5x^2 - 8x^2$
60. $7x^2 - 10x^2$
61. $5x - 3 + 6x$
62. $8x - 7 + 10x$
63. $11y + 12 - 3y - 2$
64. $13y + 15 - 2y - 11$
65. $3x + 10y + 12 + 7x - 2y - 14$
66. $5x + 7y + 13 - 2x - 3y - 15$
67. $5(3x + 4) - 4$
68. $2(5x + 4) - 3$
69. $5(3x - 2) + 12x$
70. $2(5x - 1) + 14x$
71. $7(3y - 5) + 2(4y + 3)$
72. $4(2y - 6) + 3(5y + 10)$
73. $4x - (7x - 12y - 2)$
74. $5x - (9x - 17y - 3)$
75. $5(3y - 2) - (7y + 2)$
76. $4(5y - 3) - (6y + 3)$

Application Exercises

On the average, infant girls weigh 7 pounds at birth and gain 1.5 pounds for each month for the first six months. The formula

$$W = 1.5x + 7$$

models a baby girl's weight, W, in pounds, after x months, where x is less than or equal to 6. Use the formula to solve Exercises 77–78.

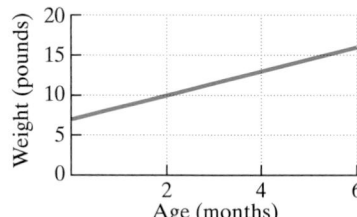

Average Weight for Infant Girls

77. What does an infant girl weigh after four months? Identify your computation as an appropriate point on the line graph.

78. What does an infant girl weigh after six months? Identify your computation as an appropriate point on the line graph.

The graph shows life expectancy in the United States by year of birth. In Exercises 79–80, you will need to refer to the data shown.

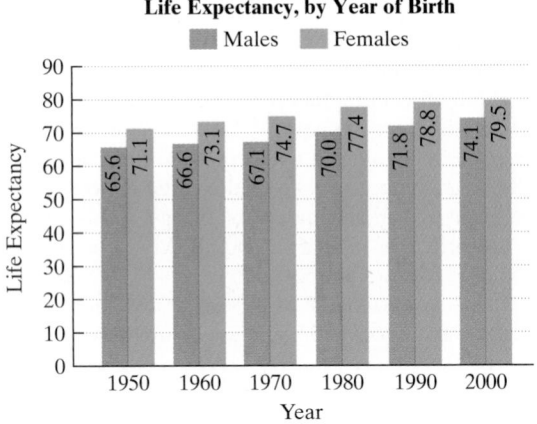

Life Expectancy, by Year of Birth

■ Males ■ Females

Source: U.S. Bureau of the Census

79. The data for U.S. women can be modeled by the formula

$$E = 0.18t + 71$$

where E represents life expectancy for women born t years after 1950. Use the formula to find the life expectancy for women born in 1990 and in 2000. How well does the formula model the data shown in the bar graph for each of these years?

80. The data for U.S. men can be modeled by the formula

$$E = 0.16t + 65$$

where E represents life expectancy for men born t years after 1950. Use the formula to find the life expectancy for men born in 1990 and in 2000. How well does the formula model the data shown in the bar graph for each of these years?

Bariatrics is the field of medicine that deals with the overweight. Bariatric surgery closes off a large part of the stomach. As a result, patients eat less and have a diminished appetite. Celebrities like pop singer Carnie Wilson and Blues Traveler harmonica player John Popper have become no-longer-larger-than-life walking billboards for the operation. The bar graph shows the number of bariatric surgeries from 1992 through 2002.

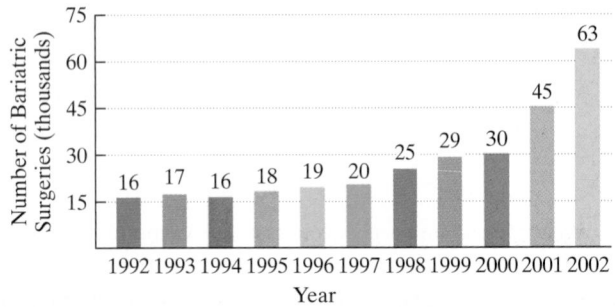

Bariatric Surgery in the U.S.

Source: American Society for Bariatric Surgery

The formula

$$N = 0.12x^3 - x^2 + 3x + 15$$

models the number of bariatric surgeries, N, in thousands, x years after 1992. Use this formula to solve Exercises 81–82.

81. According to the formula, how many bariatric surgeries were performed in 2002? How well does the formula model the data shown in the bar graph?

82. According to the formula, how many bariatric surgeries were performed in 1994? How well does the formula model the data shown in the bar graph?

Writing in Mathematics

83. What is an algebraic expression? Provide an example with your description.

84. What does it mean to evaluate an algebraic expression? Provide an example with your description.

85. What is a term? Provide an example with your description.

86. What are like terms? Provide an example with your description.

87. Explain how to add like terms. Give an example.

88. What does it mean to simplify an algebraic expression?

89. An algebra student incorrectly used the distributive property and wrote $3(5x + 7) = 15x + 7$. If you were that student's teacher, what would you say to help the student avoid this kind of error?

90. According to the mathematical models in Exercises 79–80, what will be the life expectancy of U.S. women and U.S. men born in 2050? What might occur by 2050 to cause these models to undergo model breakdown?

Critical Thinking Exercises

91. Which of the following is true?
  a. The term x has no coefficient.
  b. $5 + 3(x - 4) = 8(x - 4) = 8x - 32$
  c. $-x - x = -x + (-x) = 0$
  d. $x - 0.02(x + 200) = 0.98x - 4$

92. Which one of the following is true?
  a. $3 + 7x = 10x$
  b. $b \cdot b = 2b$
  c. $(3y - 4) - (8y - 1) = -5y - 3$
  d. $-4y + 4 = -4(y + 4)$

93. A business that manufactures small alarm clocks has weekly fixed costs of $5000. The average cost per clock for the business to manufacture x clocks is described by

$$\frac{0.5x + 5000}{x}.$$

  a. Find the average cost when $x = 100$, 1000, and 10,000.
  b. Like all other businesses, the alarm clock manufacturer must make a profit. To do this, each clock must be sold for at least 50¢ more than what it costs to manufacture. Due to competition from a larger company, the clocks can be sold for $1.50 each and no more. Our small manufacturer can only produce 2000 clocks weekly. Does this business have much of a future? Explain.

Technology Exercises

The United States has more people in prison, as well as more people in prison per capita, than any other western industrialized nation. The bar graph shows the number of inmates in U.S. state and federal prisons in eight selected years from 1985 through 2001.

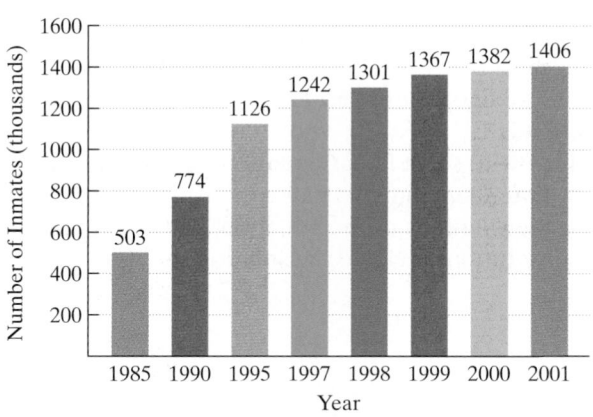

Number of Inmates in U.S. State and Federal Prisons

Source: U.S. Justice Department

The data in the graph can be modeled by each of the following formulas:

$$N = 59.5x + 505.12$$
$$N = -0.619x^2 + 69.52x + 485.28$$
$$N = -0.239x^3 + 5.288x^2 + 33.28x + 503.47.$$

In each formula, x represents the number of years after 1985 and N represents the number of inmates, in thousands. Use this information to solve Exercises 94–95.

94. Use a calculator to determine which of the three formulas serves as the best model for the inmate population in 2001.

95. Use a calculator to determine which, if any, of the three formulas serves as the best model for the inmate population for the period shown by the graph.

Group Exercise

96. By consulting textbooks from your major, algebra books, or liberal arts mathematics books, each group member should find an algebraic formula that he or she finds particularly intriguing. Group members should present these formulas to the entire class and illustrate their use.

SECTION 6.2   SOLVING LINEAR EQUATIONS

OBJECTIVES

1. Solve linear equations.

2. Solve linear equations containing fractions.

3. Solve for a variable in an equation or formula.

4. Identify equations with no solution or infinitely many solutions.

Average Hours Americans Work per Week

Source: U.S.A. Today

FIGURE 6.2

You would think that technological advances in the workplace have made it possible for us to work less and have more recreational time. Wrong. The bar graph in Figure 6.2 shows that the average number of hours that Americans work per week is gradually increasing. The data can be modeled by the formula

$$W = 0.3x + 46.6$$

where W represents the average number of hours that Americans work per week x years after 1980. So when will our workaholic nation average 55 hours of work per week? Substitute 55 for W in the formula:

$$55 = 0.3x + 46.6.$$

Now we need to find the value for x. This represents the number of years after 1980 when we will average 55 hours of work per week.

The use of algebra in a variety of everyday applications often involves *equations*. An **equation** consists of two algebraic expressions joined by an equal sign.

Thus $55 = 0.3x + 46.6$ is an example of an equation. The equal sign divides the equation into two parts, the left side and the right side:

$$\underbrace{55}_{\text{Left side}} = \underbrace{0.3x + 46.6}_{\text{Right side}}.$$

The two sides of an equation can be reversed, so we can express this equation as

$$0.3x + 46.6 = 55.$$

The form of this equation is $ax + b = c$, with $a = 0.3$, $b = 46.6$, and $c = 55$. Any equation in this form is called a **linear equation in one variable**. The exponent on the variable in such an equation is 1.

In this section, we will study how to solve linear equations. **Solving an equation** is the process of finding the set of numbers that will make the equation a true statement. These numbers are called the **solutions** of the equation, and we say that they **satisfy** the equation. The set of all solutions is called the **solution set** of the equation.

Solving Linear Equations

The idea in solving a linear equation is to get the variable (the letter) by itself on one side of the equal sign and a number by itself on the other side. For example, consider the equation $x - 3 = 8$. To get x by itself on the left side, add 3 to the left side, because $x - 3 + 3$ gives $x + 0$, or just x. You must then add 3 to the right side also. By doing this, we are using the **addition property of equality**.

THE ADDITION PROPERTY OF EQUALITY

The same real number (or algebraic expression) may be added to both sides of an equation without changing the solution set. This can be expressed symbolically as follows:

$$\text{If } a = b, \text{ then } a + c = b + c.$$

EXAMPLE 1	SOLVING AN EQUATION USING THE ADDITION PROPERTY

Solve the equation: $x - 3 = 8$.

Solution We can isolate the variable, x, by adding 3 to both sides of the equation.

$x - 3 = 8$	This is the given equation.
$x - 3 + 3 = 8 + 3$	Add 3 to both sides.
$x + 0 = 11$	This step is often done mentally and not listed.
$x = 11$	

Check

$x - 3 = 8$	This is the original equation.
$11 - 3 \stackrel{?}{=} 8$	Substitute 11 for x. The question mark over the equal sign indicates that we do not know yet if the statement is true.
$8 = 8$	This true statement indicates that 11 is the solution.

This verifies that the solution set is $\{11\}$. The braces, $\{\ \}$, show that we are representing a set.

Solve linear equations.

STUDY TIP

An equation is like a balanced scale—balanced because its two sides are equal. To maintain this balance, whatever is done to one side must also be done to the other side.

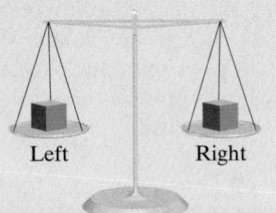

Left   Right

Consider $x - 3 = 8$

$x - 3$ 8

The scale is balanced if the left and right sides are equal.

Add 3 to the left side

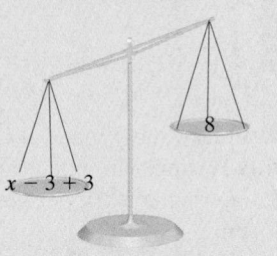

$x - 3 + 3$ 8

The scale is balanced by adding 3 to the right side.

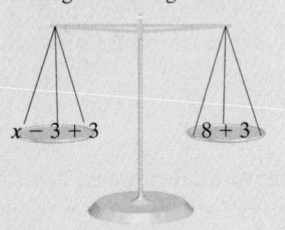

$x - 3 + 3$ $8 + 3$

Thus, $x = 11$.

 Solve the equation: $x - 5 = 12$.

A second property used to isolate the variable is the subtraction property:

THE SUBTRACTION PROPERTY OF EQUALITY

The same real number (or algebraic expression) may be subtracted from both sides of an equation without changing the solution set.

$$\text{If } a = b, \text{ then } a - c = b - c.$$

EXAMPLE 2 | SOLVING AN EQUATION USING THE SUBTRACTION PROPERTY

Solve the equation: $x + 7 = -15$.

Solution   We can isolate the variable, x, by subtracting 7 from both sides of the equation.

$$\begin{aligned}
x + 7 &= -15 &&\text{This is the given equation.}\\
x + 7 - 7 &= -15 - 7 &&\text{Subtract 7 from both sides.}\\
x &= -22
\end{aligned}$$

By substituting -22 for x in the given equation, we obtain the true statement $-15 = -15$. This verifies that the solution set is $\{-22\}$.

 Solve the equation: $x + 11 = -4$.

A third property used to isolate the variable is the multiplication property:

THE MULTIPLICATION PROPERTY OF EQUALITY

Both sides of an equation may be multiplied by the same nonzero real number (or algebraic expression) without changing the solution set.

$$\text{If } a = b \text{ and } c \neq 0, \text{ then } ac = bc.$$

EXAMPLE 3 | SOLVING AN EQUATION USING THE MULTIPLICATION PROPERTY

Solve the equation: $\dfrac{x}{5} = 9$.

Solution   We can isolate the variable, x, by multiplying both sides of the equation by 5.

$$\begin{aligned}
\frac{x}{5} &= 9 &&\text{This is the given equation.}\\
5 \cdot \frac{x}{5} &= 5 \cdot 9 &&\text{Multiply both sides by 5.}\\
1x &= 45 &&\text{Simplify: } 5 \cdot \frac{x}{5} = \left(5 \cdot \frac{1}{5}\right)x = 1x.\\
x &= 45 &&1x = x
\end{aligned}$$

By substituting 45 for x in the given equation, we obtain the true statement $9 = 9$. This verifies that the solution set is $\{45\}$.

 **Check Point 3**   Solve the equation: $\dfrac{x}{3} = 12$.

A fourth property used to isolate the variable is the division property:

THE DIVISION PROPERTY OF EQUALITY

Both sides of an equation may be divided by the same nonzero real number (or algebraic expression) without changing the solution set.

$$\text{If } a = b \text{ and } c \neq 0, \text{ then } \frac{a}{c} = \frac{b}{c}.$$

EXAMPLE 4   SOLVING AN EQUATION USING THE DIVISION PROPERTY

Solve the equation: $-6x = 30$.

Solution   We can isolate the variable, x, by dividing both sides of the equation by -6.

$$-6x = 30 \qquad \text{This is the given equation.}$$

$$\frac{-6x}{-6} = \frac{30}{-6} \qquad \text{Divide both sides by } -6.$$

$$1x = -5 \qquad \text{Simplify.}$$

$$x = -5 \qquad 1x = x$$

By substituting -5 for x in the given equation, we obtain the true statement $30 = 30$. This verifies that the solution set is $\{-5\}$.

 **Check Point 4**   Solve the equation: $-4x = 84$.

Example 5 illustrates that it is often necessary to use more than one property of equality to solve an equation.

EXAMPLE 5   SOLVING A LINEAR EQUATION

Solve and check: $2x + 3 = 17$.

Solution   Our goal is to get x by itself on the left side. First, get $2x$ by itself by subtracting 3 from both sides. Then we isolate x from $2x$ by dividing both sides of the equation by 2.

$$2x + 3 = 17 \qquad \text{This is the given equation.}$$

$$2x + 3 - 3 = 17 - 3 \qquad \text{Subtract 3 from both sides.}$$

$$2x = 14 \qquad \text{Simplify.}$$

$$\frac{2x}{2} = \frac{14}{2} \qquad \text{Divide both sides by 2.}$$

$$x = 7 \qquad \text{Simplify.}$$

STUDY TIP

If your proposed solution is incorrect, you will get a false statement when you check your answer. For example, 8 is not a solution of $2x + 3 = 17$. Look what happens when we substitute 8 for x:

$$2x + 3 = 17$$
$$2 \cdot 8 + 3 \overset{?}{=} 17$$
$$16 + 3 \overset{?}{=} 17$$
$$19 = 17, \text{ false}$$

Check   Now we check the proposed solution, 7, by replacing x with 7 in the original equation.

$$2x + 3 = 17 \qquad \text{This is the original equation.}$$
$$2 \cdot 7 + 3 \overset{?}{=} 17 \qquad \text{Substitute 7 for x.}$$
$$14 + 3 \overset{?}{=} 17 \qquad \text{Multiply: } 2 \cdot 7 = 14.$$
$$\boxed{\text{This statement is true.}} \quad 17 = 17 \qquad \text{Add: } 14 + 3 = 17.$$

Because the check results in a true statement, we conclude that the solution of the given equation is 7, or the solution set is $\{7\}$.

 **Check Point 5**   Solve and check: $4x + 5 = 29$.

EXAMPLE 6   SOLVING A LINEAR EQUATION BY FIRST SIMPLIFYING

Solve the equation: $2(x - 4) + 5x = -22$.

STUDY TIP

With increased practice in solving linear equations, try working some steps mentally. For example, consider

$$7x - 8 = -22,$$

the simplified equation on the right. Add 8 to both sides without listing all the steps:

$$7x = -14.$$

Now, divide both sides by 7:

$$x = -2.$$

Solution

$$2(x - 4) + 5x = -22 \qquad \text{This is the given equation.}$$
$$2x - 8 + 5x = -22 \qquad \text{Apply the distributive property on the left.}$$
$$7x - 8 = -22 \qquad \text{Combine like terms on the left: } 2x + 5x = 7x.$$
$$7x - 8 + 8 = -22 + 8 \qquad \text{Add 8 to both sides.}$$
$$7x = -14 \qquad \text{Simplify.}$$
$$\frac{7x}{7} = \frac{-14}{7} \qquad \text{Divide both sides by 7.}$$
$$x = -2 \qquad \text{Simplify.}$$

Check   Check by substituting the proposed solution, -2, into the original equation.

$$2(x - 4) + 5x = -22 \qquad \text{This is the original equation.}$$
$$2(-2 - 4) + 5(-2) \overset{?}{=} -22 \qquad \text{Substitute } -2 \text{ for x.}$$
$$2(-6) + 5(-2) \overset{?}{=} -22 \qquad \text{Simplify inside parentheses.}$$
$$-12 + (-10) \overset{?}{=} -22 \qquad \text{Multiply.}$$
$$-22 = -22 \qquad \text{This true statement indicates that } -2 \text{ is the solution.}$$

The solution set is $\{-2\}$.

 **Check Point 6**   Solve the equation: $6(x - 3) + 7x = -57$.

In Example 7, we solve a linear equation by subtracting the same algebraic expression from both sides.

STUDY TIP

You can solve

$$5x - 12 = 8x + 24$$

by collecting x-terms on the right and numbers on the left. To collect x-terms on the right, subtract $5x$ from both sides:

$$5x - 12 - 5x = 8x + 24 - 5x$$
$$-12 = 3x + 24.$$

To collect numbers on the left, subtract 24 from both sides:

$$-12 - 24 = 3x + 24 - 24$$
$$-36 = 3x.$$

Now isolate x by dividing both sides by 3:

$$\frac{-36}{3} = \frac{3x}{3}$$
$$-12 = x.$$

This is the same solution that we obtained in Example 7.

EXAMPLE 7 SOLVING A LINEAR EQUATION

Solve the equation: $5x - 12 = 8x + 24$.

Solution   As always, our goal is to isolate the variable, x. One way to do this is to collect all x-terms on the left and all numbers on the right. This is accomplished by subtracting $8x$ from both sides and adding 12 to both sides. Once this is done, we will have the term $-3x$ on the left. We isolate x by dividing both sides by -3. Here are the details.

$5x - 12 = 8x + 24$	This is the given equation.
$5x - 12 - 8x = 8x + 24 - 8x$	Subtract 8x from both sides.
$-3x - 12 = 24$	Simplify: 5x − 8x = −3x.
$-3x - 12 + 12 = 24 + 12$	Collect numbers on the right, adding 12 to both sides.
$-3x = 36$	Simplify.
$\dfrac{-3x}{-3} = \dfrac{36}{-3}$	Isolate x by dividing both sides by −3.
$x = -12$	Simplify.

Now do a check to verify that the solution set is $\{-12\}$.

Check Point 7   Solve the equation: $2x + 9 = 8x - 3$.

Let us summarize the major steps involved in solving a linear equation. Not all of these steps are necessary in every equation.

SOLVING A LINEAR EQUATION

1. Simplify the algebraic expression on each side.
2. Collect all the variable terms on one side and all the constant terms on the other side.
3. Isolate the variable and solve.
4. Check the proposed solution in the original equation.

EXAMPLE 8 SOLVING A LINEAR EQUATION

Solve the equation: $2(x - 3) - 17 = 13 - 3(x + 2)$.

Solution

STEP 1. Simplify the algebraic expression on each side.

$2(x - 3) - 17 = 13 - 3(x + 2)$	This is the given equation.
$2x - 6 - 17 = 13 - 3x - 6$	Use the distributive property.
$2x - 23 = -3x + 7$	Combine constant terms.

STEP 2. Collect variable terms on one side and constant terms on the other side. We will collect variable terms on the left by adding $3x$ to both sides. We will collect the numbers on the right by adding 23 to both sides.

$$2x - 23 + 3x = -3x + 7 + 3x \qquad \text{Add 3x to both sides.}$$
$$5x - 23 = 7 \qquad \text{Simplify.}$$
$$5x - 23 + 23 = 7 + 23 \qquad \text{Add 23 to both sides.}$$
$$5x = 30 \qquad \text{Simplify.}$$

STEP 3. Isolate the variable and solve. We isolate the variable, x, by dividing both sides by 5.

$$\frac{5x}{5} = \frac{30}{5} \qquad \text{Divide both sides by 5.}$$
$$x = 6 \qquad \text{Simplify.}$$

STEP 4. Check the proposed solution in the original equation. Substitute 6 for x in the original equation. You should obtain $-11 = -11$. This true statement verifies that the solution set is $\{6\}$.

 Check Point **8** Solve the equation: $4(2x + 1) - 29 = 3(2x - 5)$.

Linear Equations with Fractions

 2 Solve linear equations containing fractions.

Equations are easier to solve when they do not contain fractions. How do we remove fractions from an equation? We begin by multiplying each term of the equation by the least common denominator of the fractions in the equation. Example 9 shows how we "clear an equation of fractions."

EXAMPLE 9	SOLVING A LINEAR EQUATION INVOLVING FRACTIONS

Solve the equation: $\dfrac{3x}{2} = \dfrac{x}{5} - \dfrac{39}{5}$.

Solution The denominators are 2, 5, and 5. The smallest number that is divisible by 2, 5, and 5 is 10. We begin by multiplying both sides of the equation by 10, the least common denominator.

$$\frac{3x}{2} = \frac{x}{5} - \frac{39}{5} \qquad \text{This is the given equation.}$$

$$10 \cdot \frac{3x}{2} = 10\left(\frac{x}{5} - \frac{39}{5}\right) \qquad \text{Multiply both sides by 10.}$$

$$10 \cdot \frac{3x}{2} = 10 \cdot \frac{x}{5} - 10 \cdot \frac{39}{5} \qquad \text{Use the distributive property. Be sure to multiply all terms by 10.}$$

$$\overset{5}{\cancel{10}} \cdot \frac{3x}{\cancel{2}} = \overset{2}{\cancel{10}} \cdot \frac{x}{\cancel{5}} - \overset{2}{\cancel{10}} \cdot \frac{39}{\cancel{5}} \qquad \text{Divide out common factors in the multiplications.}$$

$$15x = 2x - 78 \qquad \text{Complete the multiplications. The fractions are now cleared.}$$

The compact, symbolic notation of algebra enables us to use a clear step-by-step method for solving equations, designed to avoid the confusion shown in Carnwath's painting.

Squeak Carnwath, *Equations*, 1981, oil on cotton canvas 96 in. h × 72 in. w.

At this point, we have an equation similar to those we have previously solved. Collect the variable terms on one side and the constant terms on the other side.

$$15x - 2x = 2x - 2x - 78 \qquad \text{Subtract 2x to get the variable terms on the left.}$$
$$13x = -78 \qquad \text{Simplify.}$$

Isolate x by dividing both sides by 13.

$$\frac{13x}{13} = \frac{-78}{13}$$ Divide both sides by 13.

$$x = -6$$ Simplify.

Check the proposed solution. Substitute -6 for x in the original equation. You should obtain $-9 = -9$. This true statement verifies that the solution set is $\{-6\}$.

Check Point 9 Solve the equation: $\dfrac{x}{4} = \dfrac{2x}{3} + \dfrac{5}{6}$.

EXAMPLE 10　AN APPLICATION: WORKING MORE AND MORE

The formula

$$W = 0.3x + 46.6$$

models the average number of hours that Americans work per week, W, x years after 1980. When will we average 55 hours of work per week?

Solution We are interested in when we will work 55 hours, so substitute 55 for W in the formula and solve for x, the number of years after 1980.

$W = 0.3x + 46.6$	This is the given formula.
$55 = 0.3x + 46.6$	Replace W with 55.
$55 - 46.6 = 0.3x + 46.6 - 46.6$	Subtract 46.6 from both sides.
$8.4 = 0.3x$	Simplify.
$\dfrac{8.4}{0.3} = \dfrac{0.3x}{0.3}$	Divide both sides by 0.3.
$28 = x$	Simplify.

The formula indicates that 28 years after 1980, or in 2008, we will average 55 hours of work per week.

Check Point 10 Use the formula in Example 10 to find when we will average 55.9 hours of work per week.

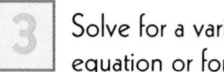 Solve for a variable in an equation or formula.

Solving for a Variable in an Equation or Formula

If an equation or formula contains two or more variables, it is often convenient to express one variable in terms of the others. For example, in the next chapter we will be studying graphing and we will need to solve an equation such as $3x + 7y = 14$ for the variable y. This is done by isolating y on one side of the equation. We solve for y using the same properties that we used to solve linear equations in one variable. This procedure is illustrated in Example 11.

EXAMPLE 11　SOLVING FOR A VARIABLE IN AN EQUATION

Solve the equation for y: $3x + 7y = 14$.

Solution First, isolate $7y$ on the left by subtracting $3x$ from both sides. Then, isolate y from $7y$ by dividing both sides of the equation by 7.

$$3x + 7y = 14 \qquad \text{This is the given equation.}$$

$$3x - 3x + 7y = -3x + 14 \qquad \text{Subtract 3x from both sides.}$$

$$7y = -3x + 14 \qquad \text{Simplify.}$$

$$\frac{7y}{7} = \frac{-3x + 14}{7} \qquad \text{Divide both sides by 7.}$$

$$y = \frac{-3x + 14}{7} \qquad \text{Simplify.}$$

We have now solved for y. We can express the answer by dividing each term in the numerator by 7.

$$y = \frac{-3x}{7} + \frac{14}{7}$$

$$y = -\frac{3}{7}x + 2$$

 **Check Point 11**   Solve the equation for y: $4x + 9y = 27$.

EXAMPLE 12 DEFERRED PAYMENT BUYING

The total price of an article purchased on a monthly deferred payment plan is described by the following formula:

$$T = D + pm.$$

In this formula, T is the total price, D is the down payment, p is the monthly payment, and m is the number of months one pays. Solve the formula for p.

Solution   First, isolate pm on the right by subtracting D from both sides. Then, isolate p from pm by dividing both sides of the formula by m.

> We need to isolate p.

$$T = D + pm \qquad \text{This is the given formula. We want p alone.}$$

$$T - D = D - D + pm \qquad \text{Isolate pm by subtracting D from both sides.}$$

$$T - D = pm \qquad \text{Simplify.}$$

$$\frac{T - D}{m} = \frac{pm}{m} \qquad \text{Now isolate p by dividing both sides by m.}$$

$$\frac{T - D}{m} = p \qquad \text{Simplify: } \frac{pm}{m} = \frac{p\,\cancel{m}}{\cancel{m}} = \frac{p}{1} = p.$$

The formula solved for p is $p = \dfrac{T - D}{m}$.

 **Check Point 12**   Solve the formula $T = D + pm$ for m.

4 Identify equations with no solution or infinitely many solutions.

Linear Equations with No Solution or Infinitely Many Solutions

Thus far, each equation that we have solved has had a single solution. However, some equations are not true for even one real number. By contrast, other equations are true for all real numbers.

If you attempt to solve an equation with no solution, you will eliminate the variable and obtain a false statement, such as $2 = 5$. If you attempt to solve an equation that is true for every real number, you will eliminate the variable and obtain a true statement, such as $4 = 4$.

EXAMPLE 13 | SOLVING A LINEAR EQUATION

Solve: $2x + 6 = 2(x + 4)$.

Solution

$2x + 6 = 2(x + 4)$	This is the given equation.
$2x + 6 = 2x + 8$	Use the distributive property.
$2x + 6 - 2x = 2x + 8 - 2x$	Subtract 2x from both sides.
$6 = 8$	Simplify. Keep reading; $6 = 8$ is not the solution.

The original equation results in the statement $6 = 8$, which is false for every value of x. Thus, the equation has no solution. The solution set for such an equation contains no elements and is called the empty set, written ∅.

 **Check Point 13** Solve: $3x + 7 = 3(x + 1)$.

EXAMPLE 14 | SOLVING A LINEAR EQUATION

Solve: $-3x + 5 + 5x = 4x - 2x + 5$.

Solution

$-3x + 5 + 5x = 4x - 2x + 5$	This is the given equation.
$2x + 5 = 2x + 5$	Combine like terms: $-3x + 5x = 2x$ and $4x - 2x = 2x$.
$2x + 5 - 2x = 2x + 5 - 2x$	Subtract 2x from both sides.
$5 = 5$	Simplify. Keep reading; $5 = 5$ is not the solution.

The original equation results in the statement $5 = 5$, which is true for every value of x. Thus, the solution set consists of the set of all real numbers. Try substituting any number of your choice for x in the original equation. You will obtain a true statement.

 **Check Point 14** Solve: $3(x - 1) + 9 = 8x + 6 - 5x$.

Exercise Set 6.2

Practice Exercises

In Exercises 1–42, solve each equation. Be sure to check your proposed solution by substituting it for the variable in the given equation.

1. $x - 7 = 3$

2. $x - 3 = -17$

3. $x + 5 = -12$

4. $x + 12 = -14$

5. $\dfrac{x}{3} = 4$

6. $\dfrac{x}{5} = 3$

7. $5x = 45$

8. $6x = 18$

9. $8x = -24$

10. $5x = -25$

11. $-8x = 2$

12. $-6x = 3$

13. $5x + 3 = 18$

14. $3x + 8 = 50$

15. $6x - 3 = 63$

16. $5x - 8 = 72$

17. $4x - 14 = -82$

18. $9x - 14 = -77$

19. $14 - 5x = -41$

20. $25 - 6x = -83$

21. $9(5x - 2) = 45$

22. $10(3x + 2) = 70$

23. $5x - (2x - 10) = 35$

24. $11x - (6x - 5) = 40$

25. $3x + 5 = 2x + 13$

26. $2x - 7 = 6 + x$

27. $8x - 2 = 7x - 5$

28. $13x + 14 = -5 + 12x$

29. $7x + 4 = x + 16$

30. $8x + 1 = x + 43$

31. $8y - 3 = 11y + 9$

32. $5y - 2 = 9y + 2$

33. $2(4 - 3x) = 2(2x + 5)$

34. $3(5 - x) = 4(2x + 1)$

35. $8(y + 2) = 2(3y + 4)$

36. $3(3y - 1) = 4(3 + 3y)$

37. $3(x + 1) = 7(x - 2) - 3$

38. $5x - 4(x + 9) = 2x - 3$

39. $5(2x - 8) - 2 = 5(x - 3) + 3$

40. $7(3x - 2) + 5 = 6(2x - 1) + 24$

41. $5(x - 2) - 2(2x + 1) = 2 + 5x$

42. $2(5x + 4) + 3(2x + 11) = 4x - 19$

Solve and check each equation in Exercises 43–50. Begin your work by rewriting each equation without fractions.

43. $\dfrac{x}{3} + \dfrac{x}{2} = \dfrac{5}{6}$

44. $\dfrac{x}{4} - 1 = \dfrac{x}{5}$

45. $\dfrac{x}{2} = 20 - \dfrac{x}{3}$

46. $\dfrac{x}{5} - \dfrac{1}{2} = \dfrac{x}{6}$

47. $\dfrac{3y}{4} - 3 = \dfrac{y}{2} + 2$

48. $y + \dfrac{1}{2} = 1 - \dfrac{y}{3}$

49. $\dfrac{3x}{5} - x = \dfrac{x}{10} - \dfrac{5}{2}$

50. $2x - \dfrac{2x}{7} = \dfrac{x}{2} + \dfrac{17}{2}$

In Exercises 51–60, solve each equation for y.

51. $3x + y = 6$

52. $5x + y = 7$

53. $x + 2y = 6$

54. $x + 4y = 8$

55. $2x + 6y = 12$

56. $3x + 9y = 18$

57. $-2x + 4y = 0$

58. $-3x + 9y = 0$

59. $2x - 3y = 5$

60. $9x - 4y = 7$

In Exercises 61–74, solve each formula for the specified variable. Do you recognize the formula? If so, what does it describe?

61. $A = LW$ for L

62. $D = RT$ for R

63. $A = \frac{1}{2}bh$ for b

64. $V = \frac{1}{3}Bh$ for B

65. $I = Prt$ for P

66. $C = 2\pi r$ for r

67. $E = mc^2$ for m

68. $V = \pi r^2 h$ for h

69. $y = mx + b$ for m

70. $P = C + MC$ for M

71. $A = \frac{1}{2}(a + b)$ for a

72. $A = \frac{1}{2}(a + b)$ for b

73. $S = P + Prt$ for r

74. $S = P + Prt$ for t

For Exercises 75–80, indicate whether each equation has no solution or is true for all real numbers. If neither is the case, solve for the variable.

75. $10x - 2(4 + 5x) = -8$

76. $9x - 3(-5 + 3x) = 15$

77. $10x - 2(4 + 5x) = 8$

78. $9x - 3(-5 + 3x) = -15$

79. $2(3x + 4) - 4 = 9x + 4 - 3x$

80. $3(x - 4) - 5 = -2x + 5x - 9$

Application Exercises

In Massachusetts, speeding fines are determined by the formula

$$F = 10(x - 65) + 50$$

where F is the cost, in dollars, of the fine if a person is caught driving x miles per hour for $x \geq 65$. Use this formula to solve Exercises 81–82.

81. If a fine comes to $250, how fast was that person driving?

82. If a fine comes to $400, how fast was that person driving?

The formula

$$\frac{c}{2} + 80 = 2F$$

models the relationship between temperature, F, in degrees Fahrenheit, and the number of cricket chirps per minute, c, for the snow tree cricket. Use this mathematical model to solve Exercises 83–84.

83. Find the number of chirps per minute at a temperature of 70°F.

84. Find the number of chirps per minute at a temperature of 80°F.

The formula

$$p = 15 + \frac{5d}{11}$$

describes the pressure of sea water, p, in pounds per square foot, at a depth of d feet below the surface. Use the formula to solve Exercises 85–86.

85. The record depth for breath-held diving, by Francisco Ferreras (Cuba) off Grand Bahama Island, on November 14, 1993, involved pressure of 201 pounds per square foot. To what depth did Ferreras descend on this ill-advised venture? (He was underwater for 2 minutes and 9 seconds!)

86. At what depth is the pressure 20 pounds per square foot?

Medical researchers have found that the desirable heart rate, R, in beats per minute, for beneficial exercise is approximated by the formulas

$$R = 143 - 0.65A \text{ for women}$$
$$R = 165 - 0.75A \text{ for men}$$

where A is the person's age. Use these formulas to solve Exercises 87–88.

87. If the desirable heart rate for a woman is 117 beats per minute, how old is she? How is the solution shown on the line graph?

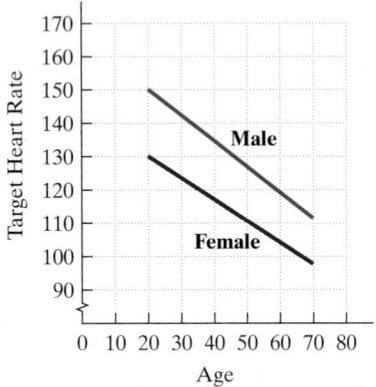

88. If the desirable heart rate for a man is 147 beats per minute, how old is he? How is the solution shown on the line graph?

The graph indicates a relationship between sense of humor and depression. Persons with a low sense of humor have higher levels of depression in response to negative life events than those with a high sense of humor.

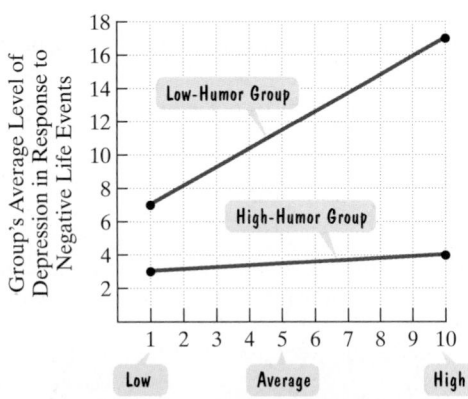

Sense of Humor and Depression

Source: Steven Davis and Joseph Palladino, *Psychology*, 3rd Edition, Prentice Hall, 2002

The data can be modeled by the formulas

$$D = \frac{10}{9}N + \frac{53}{9} \quad \text{Low humor}$$

$$D = \frac{1}{9}N + \frac{26}{9} \quad \text{High humor}$$

where N is the intensity of a negative life event $(1 \le N \le 10)$ *and D is the level of depression in response to that event. Use these formulas to solve Exercises 89–90.*

89. If the low-humor group averages a level of depression of 10 in response to a negative life event, what is the intensity of that event? How is the solution shown on the line graph?

90. If the high-humor group averages a level of depression of 3.5 in response to a negative life event, what is the intensity of that event? How is the solution shown on the line graph?

The graph shows the millions of welfare recipients in the United States who received cash assistance from 1994 through 2001. The data can be modeled by the formula

$$N = -1.4t + 14.7$$

in which N is the number of welfare recipients, in millions, t years after 1994. Use the formula to solve Exercises 91–92. Round to the nearest year.

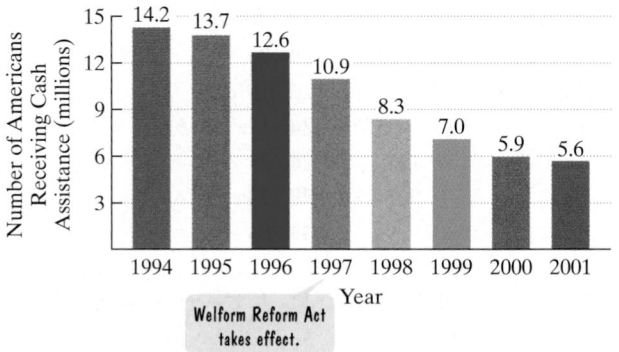

Welfare Recipients in the U.S.

Source: Department of Health and Human Services

91. When will the number of welfare recipients be reduced to 3 million?

92. According to the model, when will there be no welfare recipients? Does this prediction seem realistic? Explain your answer.

Writing in Mathematics

93. What is the solution set of an equation?

94. State the addition property of equality and give an example.

95. State the subtraction property of equality and give an example.

96. State the multiplication property of equality and give an example.

97. State the division property of equality and give an example.

98. How do you know if a linear equation has one solution, no solution, or infinitely many solutions?

99. What is the difference between solving an equation such as $2(y - 4) + 5y = 34$ and simplifying an algebraic expression such as $2(y - 4) + 5y$? If there is a difference, which topic should be taught first? Why?

100. Suppose that you solve $\frac{x}{5} - \frac{x}{2} = 1$ by multiplying both sides by 20, rather than the least common denominator of 5 and 2 (namely, 10). Describe what happens. If you get the correct solution, why do you think we clear the equation of fractions by multiplying by the *least* common denominator?

101. Suppose you are an algebra teacher grading the following solution on an examination:

Solve: $-3(x - 6) = 2 - x.$

Solution: $-3x - 18 = 2 - x$

$$-2x - 18 = 2$$

$$-2x = -16$$

$$x = 8.$$

You should note that 8 checks, and the solution set is $\{8\}$. The student who worked the problem therefore wants full credit. Can you find any errors in the solution? If full credit is 10 points, how many points should you give the student? Justify your position.

102. Although the formulas in Exercises 89–90 are correct, some people object to representing the variables with numbers, such as a 1-to-10 scale for the intensity of a negative life event. What might be their objection to quantifying the variables in this situation?

Critical Thinking Exercises

103. Which one of the following is true?

a. If $3x + 7 = 0$, then $x = \dfrac{7}{3}$.

b. Solving $A = LW$ for W gives $W = \dfrac{L}{A}$.

c. The equation $6x = 0$ has exactly one solution.

d. The final step in solving $x - b = 6x - c$ for x is $x = 6x - c + b$.

104. Write three equations whose solution set is $\{5\}$.

105. If x represents a number, write an English sentence about the number that results in an equation with no solution.

106. A woman's height, h, is related to the length of the femur, f (the bone from the knee to the hip socket), by the formula $f = 0.432h - 10.44$. Both h and f are measured in inches. A partial skeleton is found of a woman in which the femur is 16 inches long. Police find the skeleton in an area where a woman slightly over 5 feet tall has been missing for over a year. Can the partial skeleton be that of the missing woman? Explain.

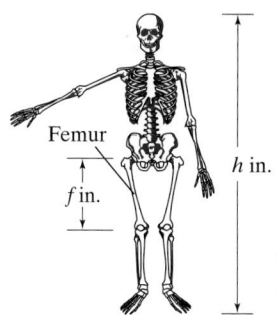

Group Exercise

107. In your group, describe the best procedure for solving the equation

$$0.47x + \frac{19}{4} = -0.2 + \frac{2}{5}x.$$

Use this procedure to actually solve the equation. Then compare procedures with other groups working on this problem. Which group devised the most streamlined method?

OBJECTIVE

1. Solve problems using linear equations.

The human race is undeniably becoming a faster race. Since the beginning of the past century, track-and-field records have fallen in everything from sprints to miles to marathons. The performance arc is clearly rising, but no one knows how much higher it can climb. At some point, even the best-trained body simply has to up and quit. The question is, just where is that point, and is it possible for athletes, trainers, and genetic engineers to push it higher? In this section, you will learn a problem-solving strategy that uses linear equations to determine if anyone will ever run a 3-minute mile.

Solve problems using linear equations.

A Strategy for Solving Word Problems Using Equations

Problem solving is an important part of algebra. The problems in this book are presented in English. We must *translate* from the ordinary language of English into the language of algebraic equations. To translate, however, we must understand the English prose and be familiar with the forms of algebraic language. Here are some general steps we will follow in solving word problems:

STRATEGY FOR SOLVING WORD PROBLEMS

STEP 1. Read the problem carefully at least twice. Attempt to state the problem in your own words and state what the problem is looking for. Let x (or any variable) represent one of the quantities in the problem.

STEP 2. If necessary, write expressions for any other unknown quantities in the problem in terms of x.

STEP 3. Write an equation in x that describes the verbal conditions of the problem.

STEP 4. Solve the equation and answer the problem's question.

STEP 5. Check the solution *in the original wording* of the problem, not in the equation obtained from the words.

Take great care with step 1. Reading a word problem is not the same as reading a newspaper. Reading the problem involves slowly working your way through its parts, making notes on what is given, and perhaps rereading the problem a few times. Only at this point should you let x represent one of the quantities.

The most difficult step in this process is step 3 because it involves translating verbal conditions into an algebraic equation. Translations of some commonly used English phrases are listed in Table 6.1. We choose to use x to represent the variable, but we can use any letter.

STUDY TIP

Cover the right column in Table 6.1 with a sheet of paper and attempt to formulate the algebraic expression in the column on your own. Then slide the paper down and check your answer. Work through the entire table in this manner.

TABLE 6.1 ALGEBRAIC TRANSLATIONS OF ENGLISH PHRASES

English Phrase	Algebraic Expression
Addition	
The sum of a number and 7	$x + 7$
Five more than a number; A number plus 5	$x + 5$
A number increased by 6; 6 added to a number	$x + 6$
Subtraction	
A number minus 4	$x - 4$
A number decreased by 5	$x - 5$
A number subtracted from 8	$8 - x$
The difference between a number and 6	$x - 6$
The difference between 6 and a number	$6 - x$
Seven less than a number	$x - 7$
Seven minus a number	$7 - x$
Nine fewer than a number	$x - 9$
Multiplication	
Five times a number	$5x$
The product of 3 and a number	$3x$
Two-thirds of a number (used with fractions)	$\frac{2}{3}x$
Seventy-five percent of a number (used with decimals)	$0.75x$
Thirteen multiplied by a number	$13x$

TABLE 6.1 *continued*

English Phrase	Algebraic Expression
A number multiplied by 13	$13x$
Twice a number	$2x$
Division	
A number divided by 3	$\frac{x}{3}$
The quotient of 7 and a number	$\frac{7}{x}$
The quotient of a number and 7	$\frac{x}{7}$
The reciprocal of a number	$\frac{1}{x}$
More than one operation	
The sum of twice a number and 7	$2x + 7$
Twice the sum of a number and 7	$2(x + 7)$
Three times the sum of 1 and twice a number	$3(1 + 2x)$
Nine subtracted from 8 times a number	$8x - 9$
Twenty-five percent of the sum of 3 times a number and 14	$0.25(3x + 14)$
Seven times a number, increased by 24	$7x + 24$
Seven times the sum of a number and 24	$7(x + 24)$

EXAMPLE 1 SOLVING A WORD PROBLEM

Nine subtracted from eight times a number is 39. Find the number.

Solution

STEP 1. Let x represent one of the quantities. Because we are asked to find a number, let

$$x = \text{the number.}$$

STEP 2. Represent other quantities in terms of x. There are no other unknown quantities to find, so we can skip this step.

STEP 3. Write an equation in x that describes the conditions.

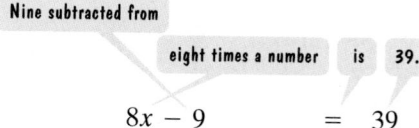

$$8x - 9 = 39$$

STEP 4. Solve the equation and answer the question.

$8x - 9 = 39$	This is the equation for the problem's conditions.
$8x - 9 + 9 = 39 + 9$	Add 9 to both sides.
$8x = 48$	Simplify.
$\dfrac{8x}{8} = \dfrac{48}{8}$	Divide both sides by 8.
$x = 6$	Simplify.

The number is 6.

STEP 5. Check the proposed solution in the original wording of the problem. "Nine subtracted from eight times a number is 39." The proposed number is 6. Eight times 6 is $8 \cdot 6$, or 48. Nine subtracted from 48 is $48 - 9$, or 39. The proposed solution checks in the problem's wording, verifying that the number is 6.

Four subtracted from six times a number is 68. Find the number.

Mile Records			
1886	4:12.3	1958	3:54.5
1923	4:10.4	1966	3:51.3
1933	4:07.6	1979	3:48.9
1945	4:01.3	1985	3:46.3
1954	3:59.4	1999	3:43.1

Source: U.S.A. Track and Field

BLITZER BONUS

A Poky Species

For a species that prides itself on its athletic prowess, human beings are a pretty poky group. Lions can sprint at up to 50 miles per hour; cheetahs move even faster, flooring it to a sizzling 70 miles per hour. But most humans—with our willowy spines and awkward, upright gait—would have trouble cracking 25 miles per hour with a tail wind, a flat track, and a good pair of running shoes.

EXAMPLE 2 WILL ANYONE EVER RUN A THREE-MINUTE MILE?

One yardstick for measuring how steadily—if slowly—athletic performance has improved is the mile run. In 1923, the record for the mile was a comparatively sleepy 4 minutes, 10.4 seconds. In 1954, Roger Bannister of Britain cracked the 4-minute mark, coming in at 3 minutes, 59.4 seconds. In the half-century since, about 0.3 second per year has been shaved off Bannister's record. If this trend continues, by which year will someone run a 3-minute mile?

Solution   In solving this problem, we will express time for the mile run in seconds. Our interest is in a time of 3 minutes, or 180 seconds.

STEP 1. Let x represent one of the quantities. The critical information in the problem is as follows:

• In 1954, the record was 3 minutes, 59.4 seconds, or 239.4 seconds.

• The record has decreased by 0.3 second per year since then.

We are interested in when the record will be 180 seconds. Let x = the number of years after 1954 when someone will run a 3-minute mile.

STEP 2. Represent other quantities in terms of x. There are no other unknown quantities to find, so we can skip this step.

STEP 3. Write an equation in x that describes the conditions.

The 1954 record time	decreased by	0.3 second per year for x years	equals	the 3-minute, or 180-second, mile.
239.4	−	0.3x	=	180

STEP 4. Solve the equation and answer the question.

$$239.4 - 0.3x = 180 \qquad \text{This is the equation for the problem's conditions.}$$

$$239.4 - 239.4 - 0.3x = 180 - 239.4 \qquad \text{Subtract 239.4 from both sides.}$$

$$-0.3x = -59.4 \qquad \text{Simplify.}$$

$$\frac{-0.3x}{-0.3} = \frac{-59.4}{-0.3} \qquad \text{Divide both sides by } -0.3.$$

$$x = 198 \qquad \text{Simplify.}$$

Using current trends, 198 years (gasp!) after 1954, or in 2152, someone will run a 3-minute mile.

STEP 5. Check the proposed solution in the original wording of the problem. The problem states that the record time should be 180 seconds. Do we obtain 180 seconds if we decrease the 1954 record time, 239.4 seconds, by 0.3 second per year for 198 years, our proposed solution?

$$239.4 - 0.3(198) = 239.4 - 59.4 = 180$$

This verifies that, using current trends, the 3-minute mile will be run 198 years after 1954.

Check 2 Point

Rent-a-Heap Agency charges $125 per week plus $0.20 per mile to rent a small car. How many miles can you travel for $335?

EXAMPLE 3 PET POPULATION

Americans love their pets. The number of cats in the United States exceeds the number of dogs by 7.5 million. The number of cats and dogs combined is 114.7 million. Determine the number of dogs and cats in the United States.

Solution

STEP 1. Let x represent one of the quantities. We know something about the number of cats: the cat population exceeds the dog population by 7.5 million. This means that there are 7.5 million more cats than dogs. We will let

x = the number (in millions) of dogs in the United States.

STEP 2. Represent other quantities in terms of x. The other unknown quantity is the number of cats. Because there are 7.5 million more cats than dogs, let

$x + 7.5$ = the number (in millions) of cats in the United States.

STEP 3. Write an equation in x that describes the conditions. The number of cats and dogs combined is 114.7 million.

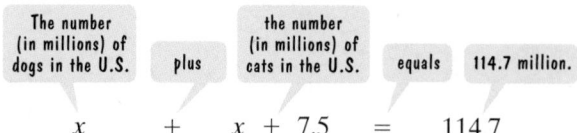

$$x + x + 7.5 = 114.7$$

STEP 4. Solve the equation and answer the question.

$x + x + 7.5 = 114.7$	This is the equation for the problem's conditions.
$2x + 7.5 = 114.7$	Combine like terms on the left side.
$2x + 7.5 - 7.5 = 114.7 - 7.5$	Subtract 7.5 from both sides.
$2x = 107.2$	Simplify.
$\dfrac{2x}{2} = \dfrac{107.2}{2}$	Divide both sides by 2.
$x = 53.6$	Simplify.

Because x represents the number (in millions) of dogs, there are 53.6 million dogs in the United States. Because $x + 7.5$ represents the number (in millions) of cats, there are $53.6 + 7.5$, or 61.1 million cats in the United States.

STEP 5. Check the proposed solution in the original wording of the problem. The problem states that the number of cats and dogs combined is 114.7 million. By adding 53.6 million, the dog population, and 61.1 million, the cat population, we do, indeed, obtain a sum of 114.7 million.

 Check 3 Point Two of the top-selling music albums of all time are *Jagged Little Pill* (Alanis Morissette) and *Saturday Night Fever* (Bee Gees). The Morissette album sold 5 million more copies than that of the Bee Gees. Combined, the two albums sold 27 million copies. Determine the number of sales for each of the albums.

U.S. Pet Population

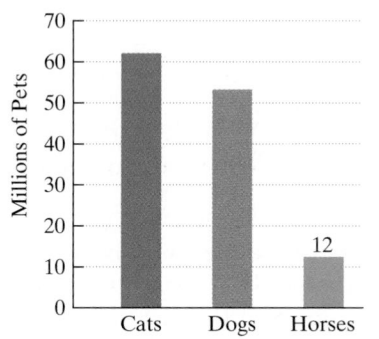

Source: American Veterinary Medical Association

EXAMPLE 4   SELECTING A LONG-DISTANCE CARRIER

You are choosing between two long-distance telephone plans. Plan A has a monthly fee of $20 with a charge of $0.05 per minute for all long-distance calls. Plan B has a monthly fee of $5 with a charge of $0.10 per minute for all long-distance calls. For how many minutes of long-distance calls will the costs for the two plans be the same?

Solution

STEP 1. Let x represent one of the quantities. Let

$$x = \text{the number of minutes of long-distance calls}$$

for which the two plans cost the same.

STEP 2. Represent other unknown quantities in terms of x. There are no other unknown quantities, so we can skip this step.

STEP 3. Write an equation in x that describes the conditions. The monthly cost for plan A is the monthly fee, $20, plus the per minute charge, $0.05, times the number of minutes of long-distance calls, x. The monthly cost for plan B is the monthly fee, $5, plus the per-minute charge, $0.10, times the number of minutes of long-distance calls, x.

The monthly cost for plan A	must equal	the monthly cost for plan B.
$20 + 0.05x$	$=$	$5 + 0.10x$

STEP 4. Solve the equation and answer the question.

$20 + 0.05x = 5 + 0.10x$	This is the equation for the problem's conditions.
$20 = 5 + 0.05x$	Subtract 0.05x from both sides.
$15 = 0.05x$	Subtract 5 from both sides.
$\dfrac{15}{0.05} = \dfrac{0.05x}{0.05}$	Divide both sides by 0.05.
$300 = x$	Simplify.

Because x represents the number of minutes of long-distance calls for the two plans to cost same, the costs will be the same with 300 minutes of long-distance calls.

STEP 5. Check the proposed solution in the original wording of the problem. The problem states that the costs for the two plans should be the same. Let's see if they are with 300 minutes of long distance calls.

$$\text{Cost for plan A} = \$20 + \$0.05(300) = \$20 + \$15 = \$35$$

Monthly fee   Per-minute charge

$$\text{Cost for plan B} = \$5 + \$0.10(300) = \$5 + \$30 = \$35$$

With 300 minutes, or 5 hours, of long-distance chatting, both plans cost $35 for the month. Thus, the proposed solution, 300 minutes, satisfies the problem's conditions.

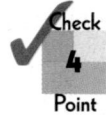 The costs for two different kinds of heating systems for a three-bedroom home are given below.

System	Cost to Install	Operating Cost/Year
Solar	$29,700	$150
Electric	$5000	$1100

After how many years will the total costs for solar heating and electric heating be the same? What will be the cost at that time?

Exercise Set 6.3

Practice Exercises

In Exercises 1–14, let x represent the number. Write each English phrase as an algebraic expression.

1. The sum of a number and 9
2. A number increased by 13
3. A number subtracted from 20
4. 13 less than a number
5. 8 decreased by 5 times a number
6. 14 less than the product of 6 and a number
7. The quotient of 15 and a number
8. The quotient of a number and 15
9. The sum of twice a number and 20
10. Twice the sum of a number and 20
11. 30 subtracted from 7 times a number
12. The quotient of 12 and a number, decreased by 3 times the number
13. Four times the sum of a number and 12
14. Five times the difference of a number and 6

In Exercises 15–34, let x represent the number. Use the given conditions to write an equation. Solve the equation and find the number.

15. A number increased by 40 is equal to 450. Find the number.
16. The sum of a number and 29 is 54. Find the number.
17. A number decreased by 13 is equal to 123. Find the number.
18. The difference between a number and 14 is 28. Find the number.
19. The product of 7 and a number is 91. Find the number.
20. The product of 8 and a number is 184. Find the number.
21. The quotient of a number and 18 is 6. Find the number.
22. The quotient of a number and 13 is 9. Find the number.
23. The sum of four and twice a number is 36. Find the number.
24. The sum of five and three times a number is 29. Find the number.
25. Seven subtracted from five times a number is 123. Find the number.
26. Eight subtracted from six times a number is 184. Find the number.
27. A number increased by 5 is two times the number. Find the number.
28. A number increased by 12 is four times the number. Find the number.
29. Twice the sum of four and a number is 36. Find the number.
30. Three times the sum of five and a number is 48. Find the number.

31. Nine times a number is 30 more than three times that number. Find the number.
32. Five more than four times a number is that number increased by 35. Find the number.
33. If the quotient of three times a number and five is increased by four, the result is 34. Find the number.
34. If the quotient of three times a number and 4 is decreased by three, the result is 9. Find the number.

Application Exercises

In Exercises 35–49, use the five-step strategy to solve each problem.

35. A car rental agency charges $200 per week plus $0.15 per mile to rent a car. How many miles can you travel in one week for $320?
36. A car rental agency charges $180 per week plus $0.25 per mile to rent a car. How many miles can you travel in one week for $395?

According to the National Center for Health Statistics, in 1990, 28% of babies in the United States were born to parents who were not married. Throughout the 1990s, this increased by approximately 0.6% per year. Use this information to solve Exercise 37–38.

37. If this trend continues, in which year will 37% of babies be born out of wedlock?
38. If this trend continues, in which year will 40% of babies be born out of wedlock?
39. Each day, the number of births in the world exceeds the number of deaths by 229 thousand. The combined number of births and deaths is 521 thousand. Determine the number of births and the number of deaths per day.

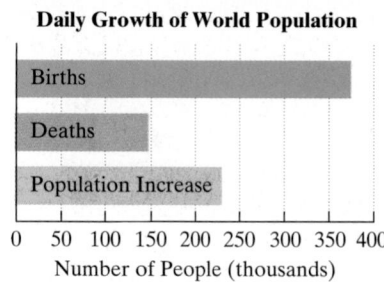

Daily Growth of World Population

Source: "Population Update" 2000

40. In 2001, the most populous countries in the world were China and India. In that year, China's population exceeded India's by 260 million. Combined, the two countries had a population of 2310 million. Determine the 2001 population for China and India. •

41. The graph shows the five costliest natural disasters in U.S. history. The cost of the Northridge, California, earthquake exceeded Hurricane Hugo by $5.5 billion, and the cost of Hurricane Andrew exceeded Hugo by $13 billion. The combined costs of the three natural disasters was $39.5 billion. Determine the cost of each.

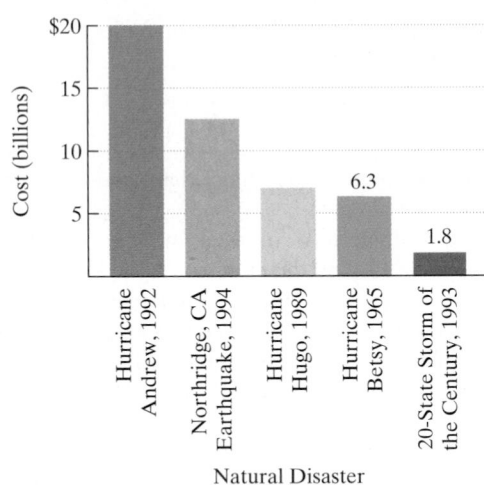

Costliest Natural Disasters in U.S. History

Source: Federal Emergency Management Agency

42. Commuters in one-third of the largest cities in the United States spend more than 40 hours per year, equivalent to at least one work week, sitting in traffic. The bar graph shows the number of hours in traffic per year for the average motorist in ten cities. The average motorist in Los Angeles spends 32 hours less than twice that of the average motorist in Miami stuck in traffic each year. Together, the average motorist in Miami and the average motorist in Los Angeles spend 139 hours per year in traffic. How many hours are wasted in traffic by the average motorist in Los Angeles and Miami?

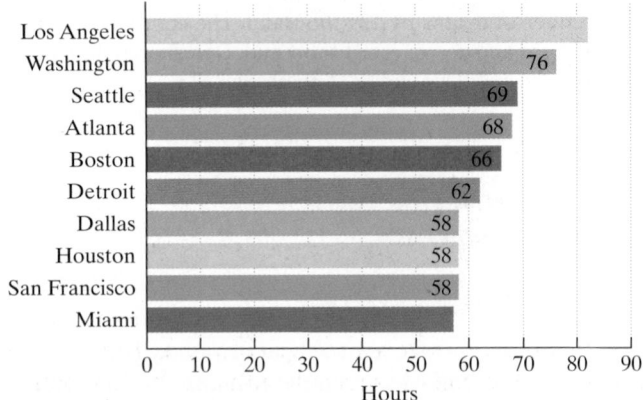

Hours in Traffic per Year for the Average Motorist

Source: Texas Transportation Institute

43. An automobile repair shop charged a customer $448, listing $63 for parts and the remainder for labor. If the cost of labor is $35 per hour, how many hours of labor did it take to repair the car?

44. A repair bill on a yacht came to $1603, including $532 for parts and the remainder for labor. If the cost of labor is $63 per hour, how many hours of labor did it take to repair the yacht?

45. You are choosing between two long-distance telephone plans. Plan A has a monthly fee of $15 with a charge of $0.08 per minute for all long-distance calls. Plan B has a monthly fee of $3 with a charge of $0.12 per minute for all long-distance calls. For how many minutes of long-distance calls will the costs for the two plans be the same?

46. You are choosing between two plans at a discount warehouse. Plan A offers an annual membership fee of $300 and you pay 70% of the manufacturer's recommended list price. Plan B offers an annual membership fee of $40 and you pay 90% of the manufacturer's recommended list price. How many dollars of merchandise would you have to purchase in a year to pay the same amount under both plans? What will be the cost for each plan?

47. The bus fare in a city is $1.25. People who use the bus have the option of purchasing a monthly coupon book for $21.00. With the coupon book, the fare is reduced to $0.50. Determine the number of times in a month the bus must be used so that the total monthly cost without the coupon book is the same as the total monthly cost with the coupon book.

48. A coupon book for a bridge costs $21 per month. The toll for the bridge is normally $2.50, but it is reduced to $1 for people who have purchased the coupon book. Determine the number of times in a month the bridge must be crossed so that the total monthly cost without the coupon book is the same as the total monthly cost with the coupon book.

49. A bookcase is to be constructed as shown in the figure. The length is to be 3 times the height. If 60 feet of lumber is available for the entire unit, find the length and height of the bookcase.

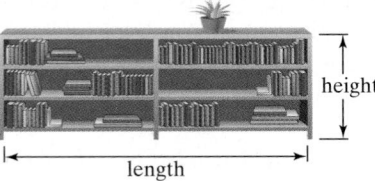

Writing in Mathematics

50. In your own words, describe a step-by-step approach for solving algebraic word problems.

51. Many students find solving linear equations much easier than solving algebraic word problems. Discuss some of the reasons why this is the case.

52. Did you have some difficulties solving some of the problems that were assigned in this exercise set? Discuss what you did if this happened to you. Did your course of action enhance your ability to solve algebraic word problems?

53. The mile records in Example 2 on page 294 are a yardstick for measuring how athletes are getting better and better. Do you think that there is a limit to human performance? Explain your answer. If so, when might we reach it?

Critical Thinking Exercises

54. An HMO pamphlet contains the following recommended weight for women: "Give yourself 100 pounds for the first 5 feet plus 5 pounds for every inch over 5 feet tall." Using this description, which height corresponds to an ideal weight of 135 pounds?

55. The rate for a particular international long distance telephone call is $0.55 for the first minute and $0.40 for each additional minute. Determine the length of a call that cost $6.95.

56. Every year, approximately 1760 Americans suffer spinal cord injuries due to falls. This represents 22% of the total number of Americans who suffer spinal cord injuries yearly. Determine the number of Americans who suffer spinal cord injuries each year. Then use the circle graph to find the number of Americans in each of the remaining four categories.

Causes of U.S. Spinal Cord Injuries

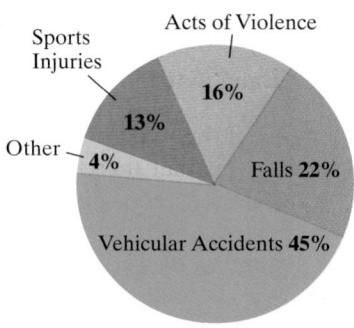

Source: U.S. News and World Report

57. In a film, the actor Charles Coburn plays an elderly "uncle" character criticized for marrying a woman when he is 3 times her age. He wittily replies, "Ah, but in 20 years time I shall only be twice her age." How old is the "uncle" and the woman?

58. Suppose that we agree to pay you 8¢ for every problem in this chapter that you solve correctly and fine you 5¢ for every problem done incorrectly. If at the end of 26 problems we do not owe each other any money, how many problems did you solve correctly?

Group Exercise

59. One of the best ways to learn how to solve a word problem in algebra is to design word problems of your own. Creating a word problem makes you very aware of precisely how much information is needed to solve the problem. You must also focus on the best way to present information to a reader and on how much information to give. As you write your problem, you gain skills that will help you solve problems created by others.

   The group should design five different word problems that can be solved using algebraic equations. All of the problems should be on different topics. For example, the group should not have more than one problem on finding a number. The group should turn in both the problems and their algebraic solutions.

SECTION 6.4   RATIO, PROPORTION, AND VARIATION

OBJECTIVES

1. Find ratios.

2. Solve proportions.

3. Solve problems using proportions.

4. Solve direct variation problems.

5. Solve inverse variation problems.

The possibility of seeing a blue whale, the largest mammal ever to grace the Earth, increases the excitement of gazing out over the ocean's swell of waves. Blue whales were hunted to near extinction in the last half of the nineteenth and the first half of the twentieth centuries. Using a method for estimating wildlife populations that we discuss in this section, by the mid-1960s it was determined that the world population of blue whales was less than 1000. This led the International Whaling Commission to ban the killing of blue whales to prevent their extinction. A dramatic increase in blue whale sightings indicates an ongoing increase in their population and the success of the killing ban.

Find ratios.

Ratios

A **ratio** compares quantities by division. For example, a group contains 60 women and 30 men. The ratio of women to men is $\frac{60}{30}$. We can express this ratio as a fraction reduced to lowest terms:

$$\frac{60}{30} = \frac{2 \cdot 30}{1 \cdot 30} = \frac{2}{1}.$$

This ratio can be expressed as $2:1$, or 2 to 1.

EXAMPLE 1 | FINDING RATIOS

The bar graph in Figure 6.3 shows the number of divorced people for every thousand married persons. Find the ratio of the number of divorced people for every thousand married persons in 1995 to the number in 1980.

Solution The ratio we must find shows the following information:

$$\frac{\text{The number of divorced persons per thousand married persons in 1995}}{\text{The number of divorced persons per thousand married persons in 1980}}.$$

First, we must use Figure 6.3 to find the number of divorced persons per thousand married persons in 1995. In 1995, the graph indicates that there were 160 divorced persons for every thousand married persons. Likewise, the graph shows that there were 100 divorced persons for every thousand married persons in 1980. We use 160 as the numerator of the ratio and 100 as the denominator. Thus, the ratio is

$$\frac{160}{100} = \frac{8 \cdot 20}{5 \cdot 20} = \frac{8}{5}, \text{ or } 8:5 \ (8 \text{ to } 5).$$

 **Check 1 Point** Use the bar graph in Figure 6.3 to find the ratio of the number of divorced people for every thousand married persons in 1995 to the number in 2000.

[bar graph in left margin]

Number of Divorced Persons per Thousand Married Persons

180
160
140
120
100
80
60
40
20
0

1970 1975 1980 1985 1990 1995 2000

Year

Source: U.S. Bureau of the Census

FIGURE 6.3

BLITZER BONUS

Ratios in Baseball

A baseball player's batting average is the ratio of the number of hits to the number of times at bat. For example, a player with 60 hits and 200 times at bat has a batting average of

$$\frac{60}{200} = \frac{3 \cdot 20}{10 \cdot 20} = \frac{3}{10}$$

or 0.300. Joe DiMaggio was the American League batting champion in 1939 with a 0.381 batting average. The leading batter of all time was left-handed hitter Ty Cobb. Cobb's batting average over 24 years (4191 hits and 11,429 times at bat) was 0.367.

Joe DiMaggio's (1914–1999) grace rivaled that of the great sailing ships, earning him the nickname the "Yankee Clipper."

Solve proportions.

Proportions

A **proportion** is a statement that says that two ratios are equal. If the ratios are $\dfrac{a}{b}$ and $\dfrac{c}{d}$, then the proportion is

$$\frac{a}{b} = \frac{c}{d}.$$

We can clear this equation of fractions by multiplying both sides by bd, the least common denominator:

$$\frac{a}{b} = \frac{c}{d} \qquad \text{This is the given proportion.}$$

$$bd \cdot \frac{a}{b} = bd \cdot \frac{c}{d} \qquad \text{Multiply both sides by } bd(b \neq 0 \text{ and } d \neq 0). \text{ Then simplify. On the}$$
$$\text{left, } \frac{\cancel{b}d}{1} \cdot \frac{a}{\cancel{b}} = da = ad. \text{ On the right, } \frac{b\cancel{d}}{1} \cdot \frac{c}{\cancel{d}} = bc.$$

$$ad = bc$$

We see that the following principle is true for any proportion:

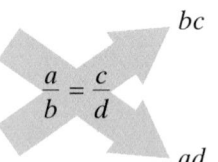

The cross-products principle: $ad = bc$

THE CROSS-PRODUCTS PRINCIPLE FOR PROPORTIONS

$$\text{If } \frac{a}{b} = \frac{c}{d}, \quad \text{then} \quad ad = bc. \quad (b \neq 0 \quad \text{and} \quad d \neq 0.)$$

The cross products ad and bc are equal.

For example, if $\frac{2}{3} = \frac{6}{9}$, we see that $2 \cdot 9 = 3 \cdot 6$, or $18 = 18$.

If three of the numbers in a proportion are known, the value of the missing quantity can be found by using the cross-products principle. This idea is illustrated in Example 2.

EXAMPLE 2 │ SOLVING PROPORTIONS

Solve each proportion for x:

a. $\dfrac{63}{x} = \dfrac{7}{5}$ **b.** $\dfrac{20}{x - 10} = \dfrac{30}{x}.$

Solution

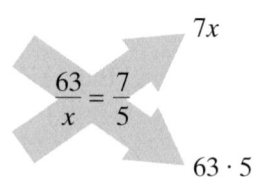

Cross products

a.
$$\frac{63}{x} = \frac{7}{5} \qquad \text{This is the given proportion.}$$
$$63 \cdot 5 = 7x \qquad \text{Apply the cross-products principle.}$$
$$315 = 7x \qquad \text{Simplify.}$$
$$\frac{315}{7} = \frac{7x}{7} \qquad \text{Divide both sides by 7.}$$
$$45 = x \qquad \text{Simplify.}$$

The solution set is $\{45\}$.

Check
$$\frac{63}{45} \overset{?}{=} \frac{7}{5} \qquad \text{Substitute 45 for } x \text{ in } \frac{63}{x} = \frac{7}{5}.$$
$$\frac{7 \cdot 9}{5 \cdot 9} \overset{?}{=} \frac{7}{5} \qquad \text{Reduce } \frac{63}{45} \text{ to lowest terms.}$$
$$\frac{7}{5} = \frac{7}{5} \qquad \text{This true statement verifies that the solution set is } \{45\}.$$

b. $\dfrac{20}{x-10} = \dfrac{30}{x}$ 　　　　This is the given proportion.

$20x = 30(x-10)$ 　　　Apply the cross-products principle.

$20x = 30x - 30 \cdot 10$ 　　Use the distributive property.

$20x = 30x - 300$ 　　　Simplify.

$20x - 30x = 30x - 300 - 30x$ 　　Subtract 30x from both sides.

$-10x = -300$ 　　　Simplify.

$\dfrac{-10x}{-10} = \dfrac{-300}{-10}$ 　　　Divide both sides by -10.

$x = 30$ 　　　Simplify.

Check

$\dfrac{20}{30-10} \stackrel{?}{=} \dfrac{30}{30}$ 　　Substitute 30 for x in $\dfrac{20}{x-10} = \dfrac{30}{x}$.

$\dfrac{20}{20} \stackrel{?}{=} \dfrac{30}{30}$ 　　Subtract: $30 - 10 = 20$.

$1 = 1$ 　　This true statement verifies that the solution is 30.

The solution set is $\{30\}$.

 Solve each proportion for x:

a. $\dfrac{10}{x} = \dfrac{2}{3}$ 　　　　**b.** $\dfrac{11}{910-x} = \dfrac{2}{x}$.

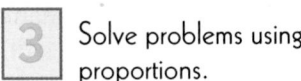 Solve problems using proportions.

Applications of Proportions

We now turn to practical application problems that can be solved using proportions. Here is a procedure for solving these problems:

SOLVING APPLIED PROBLEMS USING PROPORTIONS

1. Read the problem and represent the unknown quantity by x (or any letter).
2. Set up a proportion by listing the given ratio on one side and the ratio with the unknown quantity on the other side. Each respective quantity should occupy the same corresponding position on each side of the proportion.
3. Drop units and apply the cross-products principle.
4. Solve for x and answer the question.

EXAMPLE 3　APPLYING PROPORTIONS: CALCULATING TAXES

The property tax on a house whose assessed value is $65,000 is $825. Determine the property tax on a house with an assessed value of $180,000, assuming the same tax rate.

Solution

STEP 1. Represent the unknown by x. Let x = the tax on a $180,000 house.

STEP 2. Set up a proportion. We will set up a proportion comparing taxes to assessed value.

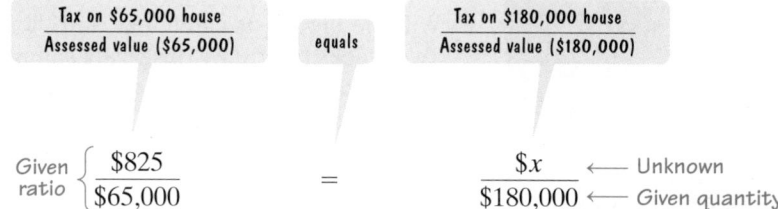

$$\underset{\text{ratio}}{\overset{\text{Given}}{}} \left\{ \frac{\$825}{\$65,000} \right. = \frac{\$x}{\$180,000} \quad \begin{matrix} \longleftarrow \text{Unknown} \\ \longleftarrow \text{Given quantity} \end{matrix}$$

STEP 3. Drop the units and apply the cross-products principle. We drop the dollar signs and begin to solve for x.

$$\frac{825}{65,000} = \frac{x}{180,000}$$

$$(825)(180,000) = 65,000x \qquad \text{Apply the cross-products principle.}$$

$$148,500,000 = 65,000x \qquad \text{Multiply.}$$

STEP 4. Solve for x and answer the question.

$$\frac{148,500,000}{65,000} = \frac{65,000x}{65,000} \qquad \text{Divide both sides by 65,000.}$$

$$2284.62 \approx x \qquad \text{Round the value of } x \text{ to the nearest cent.}$$

The property tax on the $180,000 house is approximately $2284.62.

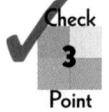

**Check
3
Point**

The tax on a house whose assessed value is $45,000 is $600. Determine the tax on a house with an assessed value of $112,500, assuming the same tax rate.

Sampling in Nature

The method that was used to estimate the blue whale population described in the section opener is called the **capture-recapture method**. Because it is impossible to count each individual animal within a population, wildlife biologists randomly catch and tag a given number of animals. Sometime later they recapture a second sample of animals and count the number of recaptured tagged animals. The total size of the wildlife population is then estimated using the following proportion:

$$\underset{\substack{\text{Initially} \\ \text{unknown} \\ (x) \longrightarrow}}{} \frac{\text{Original number of}}{\text{Total number}} = \frac{\text{Number of recaptured}}{\text{Number of animals}} \left. \cdot \right\} \begin{matrix} \text{Known} \\ \text{ratio} \end{matrix}$$

$$\frac{\text{tagged animals}}{\text{of animals in the}} = \frac{\text{tagged animals}}{\text{in second sample}}$$

$$\text{population}$$

Although this is called the capture-recapture method, it is not necessary to recapture animals in order to observe whether or not they are tagged. This could be done from a distance, with binoculars for instance.

EXAMPLE 4 | APPLYING PROPORTIONS: ESTIMATING WILDLIFE POPULATION

Wildlife biologists catch, tag, and then release 135 deer back into a wildlife refuge. Two weeks later they observe a sample of 140 deer, 30 of which are tagged. Assuming the ratio of tagged deer in the sample holds for all deer in the refuge, approximately how many deer are in the refuge?

Solution

STEP 1. Represent the unknown by x. Let x = the total number of deer in the refuge.

STEP 2. Set up a proportion.

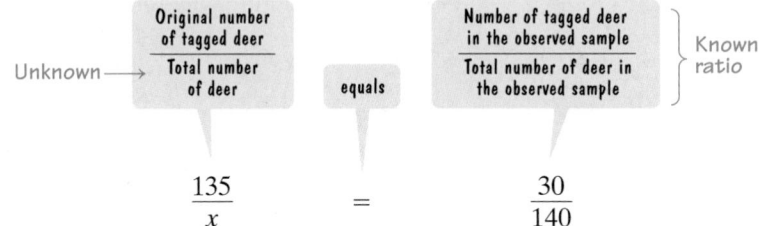

Unknown →

$$\frac{135}{x} = \frac{30}{140}$$

STEPS 3 and 4. Apply the cross-products principle, solve, and answer the question.

$$\frac{135}{x} = \frac{30}{140}$$ This is the proportion for the problem's conditions.

$$(135)(140) = 30x$$ Apply the cross-products principle.

$$18,900 = 30x$$ Multiply.

$$\frac{18,900}{30} = \frac{30x}{30}$$ Divide both sides by 30.

$$630 = x$$ Simplify.

There are approximately 630 deer in the refuge.

Check 4 Point   Wildlife biologists catch, tag, and then release 120 deer back into the wildlife refuge. Two weeks later they observe a sample of 150 deer, 25 of which are tagged. Assuming the ratio of tagged deer in the sample holds for all deer in the refuge, approximately how many deer are in the refuge?

---

④ Solve direct variation problems.

Direct Variation

In Example 3, we saw that as the assessed value on a house goes up, so do the taxes. When the assessed value is doubled, the taxes are doubled; when the assessed value is tripled, the taxes are tripled. Because of this, the tax on a house is said to **vary directly** as its assessed value. Direct variation situations can be solved using a proportion.

EXAMPLE 5 | SOLVING A DIRECT VARIATION PROBLEM

The amount of garbage created in a geographic area varies directly as the area's population. Dallas, Texas, has a population of 1.2 million and creates 38.4 million pounds of garbage each week. Find the amount of weekly garbage produced by New York City with a population of 8 million.

Solution

STEP 1. Represent the unknown by x. Let x = the weekly garbage produced by New York City (in millions of pounds).

STEP 2. Set up a proportion. We will set up a proportion comparing amount of garbage to population.

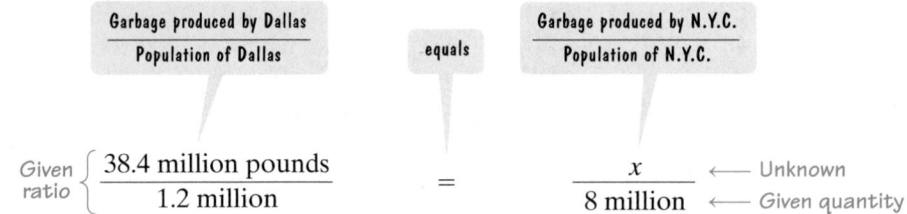

$$\text{Given ratio}\left\{\frac{38.4 \text{ million pounds}}{1.2 \text{ million}}\right. = \underset{\substack{\longleftarrow \text{ Unknown} \\ \longleftarrow \text{ Given quantity}}}{\frac{x}{8 \text{ million}}}$$

STEPS 3 and 4. Apply the cross-products principle, solve, and answer the question.

$$\frac{38.4}{1.2} = \frac{x}{8}$$ This is the proportion for the problem's conditions.

$$1.2x = (38.4)(8)$$ Apply the cross-products principle.

$$1.2x = 307.2$$ Multiply.

$$\frac{1.2x}{1.2} = \frac{307.2}{1.2}$$ Divide both sides by 1.2.

$$x = 256$$ Simplify.

The weekly garbage produced by New York City is 256 million pounds.

Check Point 5

The pressure of water on an object below the surface varies directly as its distance below the surface. If a submarine experiences a pressure of 25 pounds per square inch 60 feet below the surface, how much pressure will it experience 330 feet below the surface?

EXAMPLE 6 | SOLVING A DIRECT VARIATION PROBLEM

The distance that the sky divers shown in the photograph fall varies directly as the square of the time in which they fall. The sky divers fall 64 feet in 2 seconds. How far will they fall in 5 seconds?

Solution

STEP 1. Represent the unknown by *x*. Let x = the distance the sky divers fall in 5 seconds.

STEP 2. Set up a proportion. We will set up a proportion comparing the distance that the sky divers fall to the square of the time in which they fall.

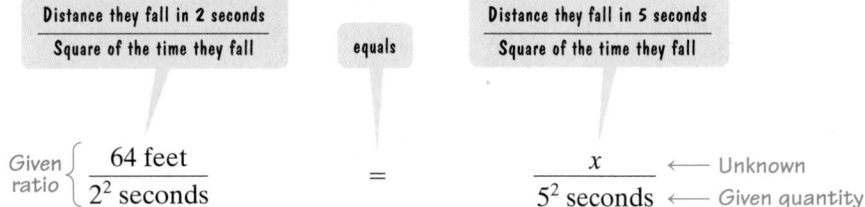

$$\text{Given ratio}\left\{\frac{64 \text{ feet}}{2^2 \text{ seconds}}\right. = \underset{\substack{\longleftarrow \text{ Unknown} \\ \longleftarrow \text{ Given quantity}}}{\frac{x}{5^2 \text{ seconds}}}$$

STEPS 3 and 4. Apply the cross-products principle, solve, and answer the question.

$$\frac{64}{2^2} = \frac{x}{5^2}$$ This is the proportion for the problem's conditions.

$$\frac{64}{4} = \frac{x}{25}$$ Square 2 and 5: $2^2 = 2 \cdot 2 = 4$ and $5^2 = 5 \cdot 5 = 25$.

$$4x = (64)(25)$$ Apply the cross-products principle.

$$4x = 1600$$ Multiply.

$$\frac{4x}{4} = \frac{1600}{4}$$ Divide both sides by 4.

$$x = 400$$ Simplify.

The sky divers will fall 400 feet in 5 seconds.

Check 6 Point The distance required to stop a car varies directly as the square of its speed. If 200 feet are required to stop a car traveling 60 miles per hour, how many feet are required to stop a car traveling 100 miles per hour? Round to the nearest foot.

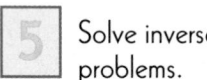 Solve inverse variation problems.

Inverse Variation

So far, we have looked at problems in which as one variable (or quantity) increases, so does the second one. However, in other problems one quantity increases as the other decreases, or vice versa. For example, if you drive to campus, consider the rate at which you drive and the time it takes you to get there. As your driving rate increases, the time it takes you to get to campus decreases. This is an example of *inverse variation*. When two quantities **vary inversely**, as one quantity increases the other decreases and vice versa.

The distance from Atlanta, Georgia, to Orlando, Florida, is 450 miles. If you average 45 miles per hour, the time for the drive is 10 hours. If you average 75 miles per hour, the time for the drive is 6 hours. The following proportion represents this situation in which time varies inversely as driving rate:

$$\frac{\text{time at the slower rate}}{\text{the faster rate}} = \frac{\text{time at the faster rate}}{\text{the slower rate}}$$

$$\frac{10 \text{ hours}}{75 \text{ miles per hour}} = \frac{6 \text{ hours}}{45 \text{ miles per hour}}.$$

The cross products are equal:

$$\frac{10 \cancel{\text{ hours}}}{1} \cdot \frac{45 \text{ miles}}{\cancel{\text{hour}}} = 10 \cdot 45 \text{ miles} = 450 \text{ miles}$$

$$\frac{75 \text{ miles}}{\cancel{\text{hour}}} \cdot \frac{6 \cancel{\text{ hours}}}{1} = 75 \cdot 6 \text{ miles} = 450 \text{ miles}.$$

Because the product of rate and time is distance, each cross product gives the distance from Atlanta to Orlando.

There is something unusual about the placement of quantities in a proportion in an inverse variation situation:

> These quantities correspond to each other.
> 6 hours is the travel time at 75 miles per hour.

$$\frac{10 \text{ hours}}{75 \text{ miles per hour}} = \frac{6 \text{ hours}}{45 \text{ miles per hour}}.$$

> These quantities correspond to each other.
> 10 hours is the travel time at 45 miles per hour.

In an inverse variation situation, corresponding values are not placed in the same ratio. *Placing corresponding values in opposite ratios* allows one variable to increase while the other decreases, or vice versa.

STUDY TIP

When you apply the cross-products principle in an inverse variation proportion, you multiply corresponding quantities.

SETTING UP A PROPORTION WHEN y VARIES INVERSELY AS x

$$\frac{\text{The first value for } y}{\text{The value for } x \text{ corresponding to the second value for } y} = \frac{\text{The second value for } y}{\text{The value for } x \text{ corresponding to the first value for } y}$$

Crude-Oil Price per Barrel

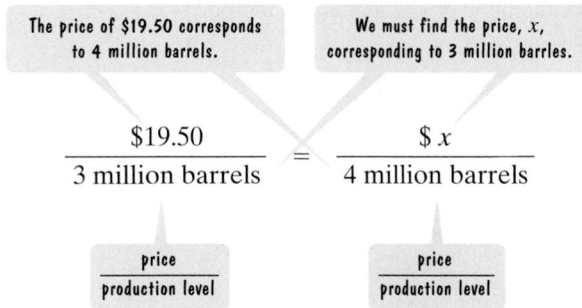

Source: U.S. Department of Commerce

FIGURE 6.4

EXAMPLE 7 SOLVING AN INVERSE VARIATION PROBLEM

Figure 6.4 shows the price per barrel for crude oil from 1990 through 2000. The price per barrel varies inversely as the supply. Suppose an OPEC nation sells oil for $19.50 per barrel when its daily production level is 4 million barrels. At what price will it sell oil if the daily production level is decreased to 3 million barrels?

Solution

STEP 1. Represent the unknown by x. Let x = the price of oil per barrel at a production level of 3 million barrels.

STEP 2. Set up a proportion. We will set up a proportion comparing price and production level. Because this is an inverse variation situation, we place corresponding values in opposite ratios.

> The price of $19.50 corresponds to 4 million barrels.

> We must find the price, x, corresponding to 3 million barrles.

$$\frac{\$19.50}{3 \text{ million barrels}} = \frac{\$x}{4 \text{ million barrels}}$$

> $\frac{\text{price}}{\text{production level}}$

> $\frac{\text{price}}{\text{production level}}$

STEPS 3 and 4. Apply the cross-products principle, solve, and answer the question.

$\dfrac{19.50}{3} = \dfrac{x}{4}$	This is the proportion for the problem's conditions.
$3x = (19.50)(4)$	Apply the cross-products principle.
$3x = 78$	Multiply.
$\dfrac{3x}{3} = \dfrac{78}{3}$	Divide both sides by 3.
$x = 26$	Divide.

The price will be $26 per barrel.

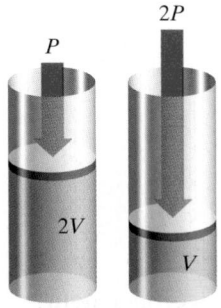

Doubling the pressure halves the volume.

✓Check
7
Point

When you use a spray can and press the value at the top, you decrease the pressure of the gas in the can. This decrease of pressure causes the volume of the gas in the can to increase. Because the gas needs more room than is provided in the can, it expands in spray form through the small hole near the valve. In general, if the temperature is constant, the pressure of a gas in a container varies inversely as the volume of the container. The pressure of a gas sample in a container whose volume is 8 cubic inches is 12 pounds per square inch. If the sample expands to a volume of 22 cubic inches, what is the new pressure of the gas?

Exercise Set 6.4

Practice Exercises

In Exercises 1–6, express each ratio as a fraction reduced to lowest terms.

1. 24 to 48   **2.** 14 to 49   **3.** 48 to 20

4. 24 to 15   **5.** 27 : 36   **6.** 25 : 40

In a class, there are 20 men and 10 women. Find each ratio in Exercises 7–10. First express the ratio as a fraction reduced to lowest terms. Then rewrite the ratio using the reduced fraction and a colon.

7. Find the ratio of the number of men to the number of women.

8. Find the ratio of the number of women to the number of men.

9. Find the ratio of the number of women to the number of students in the class.

10. Find the ratio of the number of men to the number of students in the class.

Solve each proportion in Exercises 11–22.

11. $\dfrac{24}{x} = \dfrac{12}{7}$ **12.** $\dfrac{56}{x} = \dfrac{8}{7}$

13. $\dfrac{x}{6} = \dfrac{18}{4}$ **14.** $\dfrac{x}{32} = \dfrac{3}{24}$

15. $\dfrac{x}{3} = -\dfrac{3}{4}$ **16.** $\dfrac{x}{2} = -\dfrac{1}{5}$

17. $\dfrac{-3}{8} = \dfrac{x}{40}$ **18.** $\dfrac{-3}{8} = \dfrac{6}{x}$

19. $\dfrac{x-2}{5} = \dfrac{3}{10}$ **20.** $\dfrac{x+4}{8} = \dfrac{3}{16}$

21. $\dfrac{y+10}{10} = \dfrac{y-2}{4}$ **22.** $\dfrac{2}{y-5} = \dfrac{3}{y+6}$

Application Exercises

The points in the graph show costs for private and public four-year colleges projected through the year 2017. According to these projections, your daughter's college education at a private four-year school could cost about $250,000. Use the graph to estimate each of the ratios in Exercises 23–24. First express the estimated ratio as a fraction reduced to lowest terms. Then rewrite the ratio using the reduced fraction and a colon.

Cost of a Four-Year College

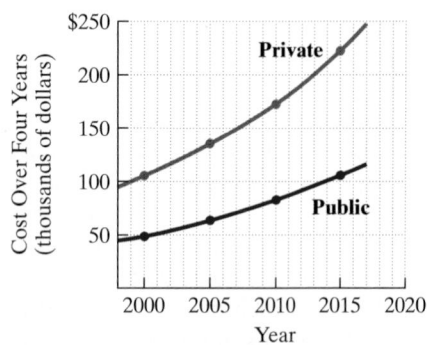

Source: U.S. Department of Education

23. The projected cost of a public four-year college to that of a private four-year college in 2010

24. The projected cost of a public four-year college to that of a private four-year college in 2015

Use a proportion to solve Exercises 25–30.

25. The tax on a property with an assessed value of $65,000 is $725. Find the tax on a property with an assessed value of $100,000.

26. The maintenance bill for a shopping center containing 180,000 square feet is $45,000. What is the bill for a store in the center that is 4800 square feet?

27. St. Paul Island in Alaska has 12 fur seal rookeries (breeding places). In 1961, to estimate the fur seal pup population in the Gorbath rookery, 4963 fur seal pups were tagged in early August. In late August, a sample of 900 pups was observed and 218 of these were found to have been previously tagged. Estimate the total number of fur seal pups in this rookery.

28. To estimate the number of bass in a lake, wildlife biologists tagged 50 bass and released them in the lake. Later they netted 108 bass and found that 27 of them were tagged. Approximately how many bass are in the lake?

29. The ratio of monthly child support to a father's yearly income is 1 : 40. How much should a father earning $38,000 annually pay in monthly child support?

30. A person who weighs 55 kilograms on earth weighs 8.8 kilograms on the moon. Find the moon weight of a person who weighs 90 kilograms on earth.

Exercises 31–40 involve direct and inverse variation. Use a proportion to solve each exercise.

31. Height varies directly as foot length. A person whose foot length is 10 inches is 67 inches tall. In 1951, photos of large footprints were published. Some believed that these footprints were made by the "Abominable Snowman." Each footprint was 23 inches long. If indeed they belonged to the Abominable Snowman, how tall is the critter?

32. A person's hair length varies directly as the number of years it has been growing. After 2 years, a person's hair length is 8 inches. The longest moustache on record was grown by Kalyan Sain of India. Sain grew his moustache for 17 years. How long was it?

33. The Mach number is a measurement of speed named after the man who suggested it, Ernst Mach (1838–1916). The speed of an aircraft varies directly as its Mach number. Shown here are two aircraft. Use the figures for the Concorde to determine the Blackbird's speed.

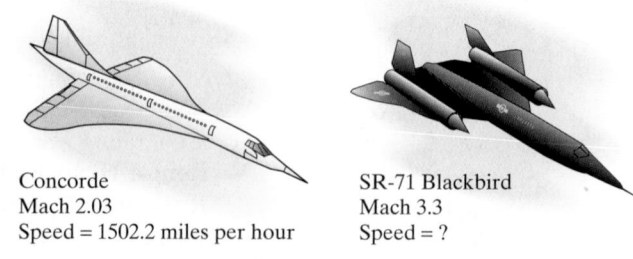

Concorde
Mach 2.03
Speed = 1502.2 miles per hour

SR-71 Blackbird
Mach 3.3
Speed = ?

34. Do you still own records, or are you strictly a CD person? Record (vinyl) owners claim that the quality of sound on good vinyl surpasses that of a CD, although this is up for debate. This, however, is not debatable: The number of revolutions a record makes as it is being played varies directly as the time that it is on the turntable. A record that lasted 3 minutes made 135 revolutions. If a record takes 2.4 minutes to play, how many revolutions does it make?

35. If all men had identical body types, their weight would vary directly as the cube of their height. Shown is Robert Wadlow, who reached a record height of 8 feet 11 inches (107 inches) before his death at age 22. If a man who is 5 feet 10 inches tall (70 inches) with the same body type as Mr. Wadlow weighs 170 pounds, what was Robert Wadlow's weight shortly before his death?

36. The distance that an object falls varies directly as the square of the time it has been falling. An object falls 144 feet in 3 seconds. Find how far it will fall in 7 seconds.

37. The time that it takes you to get to campus varies inversely as your driving rate. Averaging 20 miles per hour in terrible traffic, it takes you 1.5 hours to get to campus. How long would the trip take averaging 60 miles per hour?

38. The weight that can be supported by a 2-inch by 4-inch piece of pine (called a 2-by-4) varies inversely as its length. A 10-foot 2-by-4 can support 500 pounds. What weight can be supported by a 5-foot 2-by-4?

39. The volume of a gas in a container at a constant temperature varies inversely as the pressure. If the volume is 32 cubic centimeters at a pressure of 8 pounds, find the pressure when the volume is 40 cubic centimeters.

40. The current in a circuit varies inversely as the resistance. The current is 20 amperes when the resistance is 5 ohms. Find the current for a resistance of 16 ohms.

Writing in Mathematics

41. What is a ratio? Give an example with your description.

42. What is a proportion? Give an example with your description.

43. Explain how to solve a proportion. Illustrate your explanation with an example.

44. Explain the meaning of this statement: A company's monthly sales vary directly as its advertising budget.

45. Explain the meaning of this statement: A company's monthly sales vary inversely as the price of its product.

46. Explain the difference between setting up proportions for direct and inverse variation problems.

Critical Thinking Exercises

47. The front sprocket on a bicycle has 60 teeth and the rear sprocket has 20 teeth. For mountain biking, an owner needs a 5 : 1 front : rear ratio. If only one of the sprockets is to be replaced, describe the two ways in which this can be done.

48. My friend is 44 years old. My dog Phideaux is 7 years old. If Phideaux were human, he would be 56. Phideaux thinks my friend is another dog, which makes me wonder: If my friend were a dog, how old would my friend be?

49. In a hurricane, the wind pressure varies directly as the square of the wind velocity. If wind pressure is a measure of a hurricane's destructive capacity, what happens to this destructive power when the wind velocity doubles?

50. The illumination from a light source varies inversely as the square of the distance from the light source. If you raise a lamp from 15 inches to 30 inches over your desk, what happens to the illumination?

SECTION 6.5   SOLVING LINEAR INEQUALITIES

OBJECTIVES

1. Graph the solutions of an inequality on a number line.

2. Solve linear inequalities.

3. Solve applied problems using linear inequalities.

Rent-a-Heap, a car rental company, charges $125 per week plus $0.20 per mile to rent one of their cars. Suppose you are limited by how much money you can spend for the week: You can spend at most $335. If we let *x* represent the number of miles you drive the heap in a week, we can write an inequality that describes the given conditions:

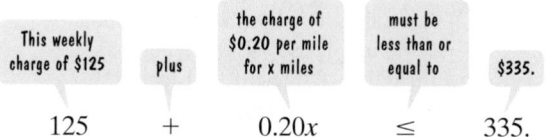

| This weekly charge of $125 | plus | the charge of $0.20 per mile for x miles | must be less than or equal to | $335. |

$$125 \quad + \quad 0.20x \quad \leq \quad 335.$$

Using the commutative property of addition, we can express this inequality as

$$0.20x + 125 \leq 335.$$

The form of this inequality is $ax + b \leq c$, with $a = 0.20$, $b = 125$, and $c = 335$. Any inequality in this form is called **a linear inequality in one variable**. The greatest exponent on the variable in such an inequality is 1. The symbol between $ax + b$ and c can be $\leq$ (is less than or equal to), $<$ (is less than), $\geq$ (is greater than or equal to), or $>$ (is greater than).

In this section, we will study how to solve linear inequalities such as $0.20x + 125 \leq 335$. **Solving an inequality** is the process of finding the set of numbers that make the inequality a true statement. These numbers are called the **solutions** of the inequality, and we say that they **satisfy** the inequality. The set of all solutions is called the **solution set** of the inequality. We begin by discussing how to graph and how to represent these solution sets.

---

1 Graph the solutions of an inequality on a number line.

Graphs of Inequalities

There are infinitely many solutions to the inequality $x < 3$, namely, all real numbers that are less than 3. Although we cannot list all the solutions, we can make a drawing on a number line that represents these solutions. Such a drawing is called the **graph of the inequality**.

Graphs of solutions to linear inequalities are shown on a number line by shading all points representing numbers that are solutions. *Open dots* (∘) indicate endpoints that are *not solutions* and *closed dots* (•) indicate endpoints that *are solutions*.

EXAMPLE 1 | GRAPHING INEQUALITIES

Graph the solutions of each inequality:

a. $x < 3$ **b.** $x \geq -1$ **c.** $-1 < x \leq 3$.

STUDY TIP

An inequality symbol points to the smaller number. Thus, another way to express $x < 3$ (x is less than 3) is $3 > x$ (3 is greater than x).

Solution

a. The solutions of $x < 3$ are all real numbers that are less than 3. They are graphed on a number line by shading all points to the left of 3. The open dot at 3 indicates that 3 is not a solution, but numbers such as 2.9999 and 2.6 are. The arrow shows that the graph extends indefinitely to the left.

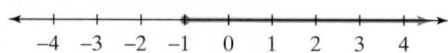

b. The solutions of $x \geq -1$ are all real numbers that are greater than or equal to -1. We shade all points to the right of -1 and the point for -1 itself. The closed dot at -1 shows that -1 is a solution of the given inequality. The arrow shows that the graph extends indefinitely to the right.

$$\begin{array}{cccccccccc} & -4 & -3 & -2 & -1 & 0 & 1 & 2 & 3 & 4 \end{array}$$

c. The inequality $-1 < x \leq 3$ is read "-1 is less than x *and* x is less than or equal to 3," or "x is greater than -1 *and* less than or equal to 3." The solutions of $-1 < x \leq 3$ are all real numbers between -1 and 3, not including -1 but including 3. The open dot at −1 indicates that −1 is not a solution. The closed dot at 3 shows that 3 is a solution. Shading indicates the other solutions.

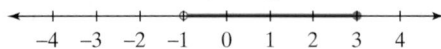

Check 1 Point Graph the solutions of each inequality:

     a. $x < 4$ **b.** $x \geq -2$ **c.** $-4 \leq x < 1$.

Solution Sets

The solutions of $x < 3$ are all real numbers that are less than 3. We can use the set concept introduced in Chapter 2 and state that the solution is the *set of all real numbers less than* 3. Using set-builder notation, the solution set of $x < 3$ is

$$\{x \mid x < 3\}.$$

We read this as "the set of all x such that x is less than 3." Solutions of inequalities should be expressed in set-builder notation.

Solving Linear Inequalities

 2 Solve linear inequalities.

Back to our question: How many miles can you drive your Rent-a-Heap car if you can spend at most \$335 per week? We answer the question by solving

$$0.20x + 125 \leq 335$$

for x. The solution procedure is nearly identical to that for solving

$$0.20x + 125 = 335.$$

Our goal is to get x by itself on the left side. We do this by subtracting 125 from both sides to isolate $0.20x$:

$$0.20x + 125 \leq 335$$
$$0.20x + 125 - 125 \leq 335 - 125$$
$$0.20x \leq 210.$$

Finally, we isolate x from $0.20x$ by dividing both sides of the inequality by 0.20:

$$\frac{0.20x}{0.20} \leq \frac{210}{0.20}$$
$$x \leq 1050.$$

With at most \$335 per week to spend, you can travel at most 1050 miles.

     We started with the inequality $0.20x + 125 \leq 335$ and obtained the inequality $x \leq 1050$ in the final step. Both of these inequalities have the same solution set, namely $\{x \mid x \leq 1050\}$. Inequalities such as these, with the same solution set, are said to be **equivalent**.

     We isolated x from $0.20x$ by dividing both sides of $0.20x \leq 210$ by 0.20, a positive number. Let's see what happens if we divide both sides of an inequality by a negative number. Consider the inequality $10 < 14$. Divide 10 and 14 by −2:

$$\frac{10}{-2} = -5 \quad \text{and} \quad \frac{14}{-2} = -7.$$

Because −5 lies to the right of −7 on the number line, −5 is greater than −7:

$$-5 > -7.$$

Notice that the direction of the inequality symbol is reversed:

$$10 < 14$$
$$-5 > -7.$$

STUDY TIP

English phrases such as "at least" and "at most" can be represented by inequalities.

English Sentence	Inequality
x is at least 5.	$x \geq 5$
x is at most 5.	$x \leq 5$
x is between 5 and 7.	$5 < x < 7$
x is no more than 5.	$x \leq 5$
x is no less than 5.	$x \geq 5$

In general, **when we multiply or divide both sides of an inequality by a negative number, the direction of the inequality symbol is reversed**. When we reverse the direction of the inequality symbol, we say that we change the *sense* of the inequality.

We can summarize this discussion with the following statement:

SOLVING LINEAR INEQUALITIES

The procedure for solving linear inequalities is the same as the procedure for solving linear equations, with one important exception: When multiplying or dividing both sides of the inequality by a negative number, reverse the direction of the inequality symbol, changing the sense of the inequality.

EXAMPLE 2   SOLVING A LINEAR INEQUALITY

Solve and graph the solution set: $4x - 7 \geq 5$.

Solution   Our goal is to get x by itself on the left side. We do this by getting $4x$ by itself, adding 7 to both sides.

$$4x - 7 \geq 5 \qquad \text{This is the given inequality.}$$
$$4x - 7 + 7 \geq 5 + 7 \qquad \text{Add 7 to both sides.}$$
$$4x \geq 12 \qquad \text{Simplify.}$$

Next, we isolate x from $4x$ by dividing both sides by 4. The inequality symbol stays the same because we are dividing by a positive number.

$$\frac{4x}{4} \geq \frac{12}{4} \qquad \text{Divide both sides by 4.}$$
$$x \geq 3 \qquad \text{Simplify.}$$

The solution set is $\{x \mid x \geq 3\}$. We read this as "the set of all x such that x is greater than or equal to 3." The graph of the solution set is shown as follows:

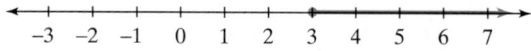

We cannot check all members of an inequality's solution set, but we can take a few values to get an indication of whether or not it is correct. In Example 2, we found that the solution set of $4x - 7 \geq 5$ is $\{x \mid x \geq 3\}$. Show that 3 and 4 satisfy the inequality, whereas 2 does not.

 **Check Point 2**   Solve and graph the solution set: $5x - 3 \leq 17$.

EXAMPLE 3   SOLVING LINEAR INEQUALITIES

Solve and graph the solution set:

a. $\frac{1}{3}x < 5$ **b.** $-3x < 21$.

Solution   In each case, our goal is to isolate x. In the first inequality, this is accomplished by multiplying both sides by 3. In the second inequality, we can do this by dividing both sides by -3.

a. $\dfrac{1}{3}x < 5$ This is the given inequality.

   $3 \cdot \dfrac{1}{3}x < 3 \cdot 5$ Isolate x by multiplying by 3 on both sides.
          The symbol $<$ stays the same because we are multiplying
          both sides by a positive number.

        $x < 15$ Simplify.

The solution set is $\{x \mid x < 15\}$. The graph of the solution set is shown as follows:

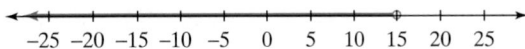

b. $-3x < 21$ This is the given inequality.

   $\dfrac{-3x}{-3} > \dfrac{21}{-3}$ Isolate x by dividing by -3 on both sides.
          The symbol $<$ must be reversed because we are dividing
          both sides by a negative number.

       $x > -7$ Simplify.

The solution set is $\{x \mid x > -7\}$. The graph of the solution set is shown as follows:

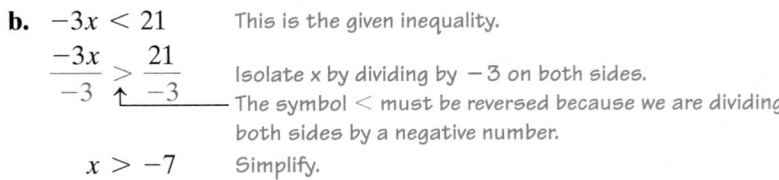

Check Point 3   Solve and graph the solution set:

a. $\dfrac{1}{4}x < 2$ **b.** $-6x < 18.$

EXAMPLE 4   SOLVING A LINEAR INEQUALITY

Solve and graph the solution set: $6x - 12 > 8x + 2.$

Solution   We will get x by itself on the left side. We begin by subtracting $8x$ from both sides so that the x-term appears on the left.

   $6x - 12 > 8x + 2$ This is the given inequality.
   $6x - 8x - 12 > 8x - 8x + 2$ Subtract 8x on both sides with the goal of
                   isolating x on the left.

      $-2x - 12 > 2$ Simplify.

Next, we get $-2x$ by itself, adding 12 to both sides.

   $-2x - 12 + 12 > 2 + 12$ Add 12 to both sides.

        $-2x > 14$ Simplify.

In the last step, we isolate x from $-2x$ by dividing both sides by -2. The direction of the inequality symbol must be reversed because we are dividing by a negative number.

      $\dfrac{-2x}{-2} < \dfrac{14}{-2}$ Divide both sides by -2 and
                   reverse the sense of the inequality.

        $x < -7$ Simplify.

The solution set is $\{x \mid x < -7\}$. The graph of the solution set is shown as follows:

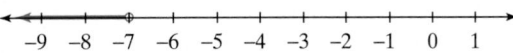

Check Point 4 Solve and graph the solution set: $7x - 3 > 13x + 33$.

EXAMPLE 5 | SOLVING A LINEAR INEQUALITY

Solve and graph the solution set:

$$2(x - 3) + 5x \le 8(x - 1).$$

Solution Begin by simplifying the algebraic expression on each side.

$$2(x - 3) + 5x \le 8(x - 1) \quad \text{This is the given inequality.}$$
$$2x - 6 + 5x \le 8x - 8 \quad \text{Use the distributive property.}$$
$$7x - 6 \le 8x - 8 \quad \text{Add like terms on the left.}$$

We will get x by itself on the left side. Subtract $8x$ from both sides.

$$7x - 8x - 6 \le 8x - 8x - 8$$
$$-x - 6 \le -8$$

Next, we get $-x$ by itself, adding 6 to both sides.

$$-x - 6 + 6 \le -8 + 6$$
$$-x \le -2$$

To isolate x, we must eliminate the negative sign in front of the x. Because $-x$ means $-1x$, we can do this by dividing both sides of the inequality by -1. This reverses the direction of the inequality symbol.

$$\frac{-x}{-1} \ge \frac{-2}{-1} \quad \text{Divide both sides by } -1 \text{ and reverse the sense of the inequality.}$$

$$x \ge 2 \quad \text{Simplify.}$$

The solution set is $\{x | x \ge 2\}$. The graph of the solution set is shown as follows:

$$\begin{array}{ccccccccccc} \leftarrow & | & | & | & | & | & | & \bullet & | & | & | & \rightarrow \\ & -5 & -4 & -3 & -2 & -1 & 0 & 1 & 2 & 3 & 4 & 5 \end{array}$$

Check Point 5 Solve and graph the solution set:

$$2(x - 3) - 1 \le 3(x + 2) - 14.$$

As you know, different professors may use different grading systems to determine your final course grade. Some professors require a final examination; others do not. In our next example, not only is a final exam required, but it also counts as two grades.

3 Solve applied problems using linear inequalities.

EXAMPLE 6 | AN APPLICATION: FINAL COURSE GRADE

To earn an A in a course, you must have a final average of at least 90%. On the first four examinations, you have grades of 86%, 88%, 92%, and 84%. If the final examination counts as two grades, what must you get on the final to earn an A in the course?

Solution We will use our five-step strategy for solving algebraic word problems.

STUDY TIP

You can solve

$$7x - 6 \le 8x - 8$$

by isolating x on the right side. Subtract $7x$ from both sides and add 8 to both sides:

$$7x - 6 - 7x \le 8x - 8 - 7x$$
$$-6 \le x - 8$$
$$-6 + 8 \le x - 8 + 8$$
$$2 \le x.$$

This last inequality means the same thing as

$$x \ge 2.$$

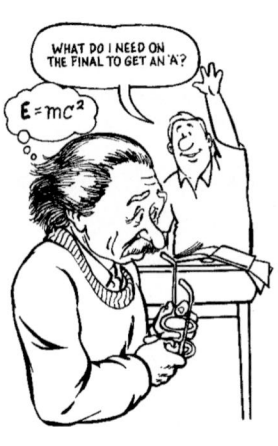

STEP 1 AND 2. **Represent unknown quantities in terms of *x*.** Let x = your grade on the final examination.

STEP 3. **Write an inequality in *x* that describes the conditions.** The average of the six grades is found by adding the grades and dividing the sum by 6.

$$\text{Average} = \frac{86 + 88 + 92 + 84 + x + x}{6}$$

Because the final counts as two grades, the x (your grade on the final examination) is added twice. This is also why the sum is divided by 6.

In order to get an A, your average must be at least 90. This means that your average must be greater than or equal to 90.

Your average	must be greater than or equal to	90
$\dfrac{86 + 88 + 92 + 84 + x + x}{6}$	$\geq$	90

STEP 4. **Solve the inequality and answer the problem's question.**

$$\frac{86 + 88 + 92 + 84 + x + x}{6} \geq 90 \qquad \text{This is the inequality for the given conditions.}$$

$$\frac{350 + 2x}{6} \geq 90 \qquad \text{Combine like terms in the numerator.}$$

$$6\left(\frac{350 + 2x}{6}\right) \geq 6(90) \qquad \text{Multiply both sides by 6, clearing fractions.}$$

$$350 + 2x \geq 540 \qquad \text{Multiply.}$$

$$350 + 2x - 350 \geq 540 - 350 \qquad \text{Subtract 350 from both sides.}$$

$$2x \geq 190 \qquad \text{Simplify.}$$

$$\frac{2x}{2} \geq \frac{190}{2} \qquad \text{Divide both sides by 2.}$$

$$x \geq 95 \qquad \text{Simplify.}$$

You must get at least 95% on the final examination to earn an A in the course.

STEP 5. **Check.** We can perform a partial check by computing the average with any grade that is at least 95. We will use 96. If you get 96% on the final examination, your average is

$$\frac{86 + 88 + 92 + 84 + 96 + 96}{6} = \frac{542}{6} = 90\frac{1}{3}.$$

Because $90\frac{1}{3} > 90$, you earn an A in the course.

Check Point 6

To earn a B in a course, you must have a final average of at least 80%. On the first three examinations, you have grades of 82%, 74%, and 78%. If the final examination counts as two grades, what must you get on the final to earn a B in the course?

Exercise Set 6.5

Practice Exercises

In Exercises 1–12, graph the solutions of each inequality on a number line.

1. $x > 6$

2. $x > -2$

3. $x < -4$

4. $x < 0$

5. $x \geq -3$

6. $x \geq -5$

7. $x \leq 4$

8. $x \leq 7$

9. $-2 < x \leq 5$

10. $-3 \leq x < 7$

11. $-1 < x < 4$

12. $-7 \leq x \leq 0$

Solve the inequalities in Exercises 13–54 and graph the solution sets.

13. $x - 3 > 2$

14. $x + 1 < 5$

15. $x + 4 \leq 9$

16. $x - 5 \geq 1$

17. $x - 3 < 0$

18. $x + 4 \geq 0$

19. $4x < 20$

20. $6x \geq 18$

21. $3x \geq -15$

22. $7x < -21$

23. $2x - 3 > 7$

24. $3x + 2 \leq 14$

25. $3x + 3 < 18$

26. $8x - 4 > 12$

27. $\frac{1}{2}x < 4$

28. $\frac{1}{2}x > 3$

29. $\frac{x}{3} > -2$

30. $\frac{x}{4} < -1$

31. $-3x < 15$

32. $-7x > 21$

33. $-3x \geq -15$

34. $-7x \leq -21$

35. $3x + 4 \leq 2x + 7$

36. $2x + 9 \leq x + 2$

37. $5x - 9 < 4x + 7$

38. $3x - 8 < 2x + 11$

39. $-2x - 3 < 3$

40. $14 - 3x > 5$

41. $3 - 7x \leq 17$

42. $5 - 3x \geq 20$

43. $-x < 4$

44. $-x > -3$

45. $5 - x \leq 1$

46. $3 - x \geq -3$

47. $2x - 5 > -x + 6$

48. $6x - 2 \geq 4x + 6$

49. $2x - 5 < 5x - 11$

50. $4x - 7 > 9x - 2$

51. $3(x + 1) - 5 < 2x + 1$

52. $4(x + 1) + 2 \geq 3x + 6$

53. $8x + 3 > 3(2x + 1) - x + 5$

54. $7 - 2(x - 4) < 5(1 - 2x)$

Application Exercises

The bar graph at the top of the next column shows the percentage of wages full-time workers pay in income tax in eight selected countries. (The percents shown are averages for single earners without children.) Let x represent the percentage of wages workers pay in income tax. In Exercises 55–60, write the name of the country or countries described by the given inequality.

55. $x \geq 30\%$

56. $x > 30\%$

57. $x < 20\%$

58. $x \leq 20\%$

59. $25\% \leq x < 40\%$

60. $5\% < x \leq 25\%$

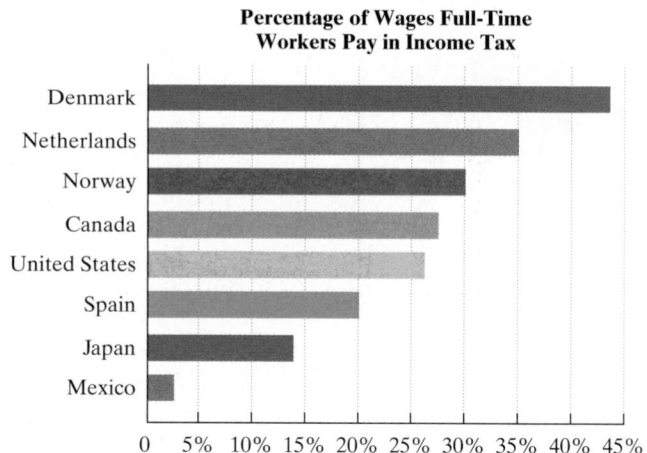

Percentage of Wages Full-Time Workers Pay in Income Tax

Source: The Washington Post

61. On three examinations, you have grades of 88, 78, and 86. There is still a final examination, which counts as one grade.

   a. In order to get an A, your average must be at least 90. If you get 100 on the final, compute your average and determine if an A in the course is possible.

   b. To earn a B in the course, you must have a final average of at least 80. What must you get on the final to earn a B in the course?

62. On two examinations, you have grades of 86 and 88. There is an optional final examination, which counts as one grade. You decide to take the final in order to get a course grade of A, meaning a final average of at least 90.

   a. What must you get on the final to earn an A in the course?

   b. By taking the final, if you do poorly, you might risk the B that you have in the course based on the first two exam grades. If your final average is less than 80, you will lose your B in the course. Describe the grades on the final that will cause this to happen.

The line graph on the next page shows the declining consumption of cigarettes in the United States. The data shown by the graph can be modeled by

$$N = 550 - 9x$$

where N is the number of cigarettes consumed, in billions, x years after 1988. Use this formula to solve Exercises 63–64.

63. Describe how many years after 1988 cigarette consumption will be less than 370 billion cigarettes each year. Which years are included in your description?

64. Describe how many years after 1988 cigarette consumption will be less than 325 billion cigarettes each year. Which years are included in your description?

65. A car can be rented from Continental Rental for $80 per week plus 25 cents for each mile driven. How many miles can you travel if you can spend at most $400 for the week?

66. A car can be rented from Basic Rental for $60 per week plus 50 cents for each mile driven. How many miles can you travel if you can spend at most $600 for the week?

**Consumption of
Cigarettes in the U.S.**

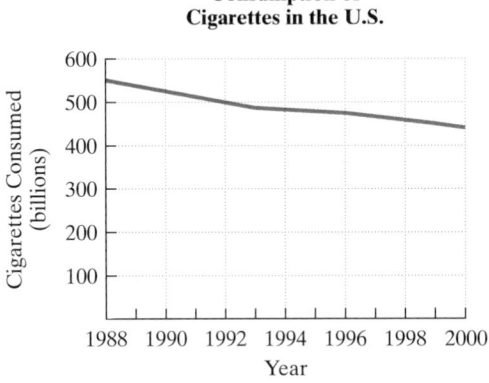

Source: Economic Research Service, USDA.

67. An elevator at a construction site has a maximum capacity of 3000 pounds. If the elevator operator weighs 245 pounds and each cement bag weighs 95 pounds, how many bags of cement can be safely lifted on the elevator in one trip?

68. An elevator at a construction site has a maximum capacity of 2800 pounds. If the elevator operator weighs 265 pounds and each cement bag weighs 65 pounds, how many bags of cement can be safely lifted on the elevator in one trip?

Writing in Mathematics

69. When graphing the solutions of an inequality, what is the difference between an open dot and a closed dot?

70. When solving an inequality, when is it necessary to change the direction of the inequality symbol? Give an example.

71. Describe ways in which solving a linear inequality is similar to solving a linear equation.

72. Describe ways in which solving a linear inequality is different than solving a linear equation.

73. Using current trends, future costs of Medicare can be modeled by $C = 18x + 250$, where x represents the number of years after 2000 and C represents the cost of Medicare, in billions of dollars. Use the formula to write a word problem that can be solved using a linear inequality. Then solve the problem.

Critical Thinking Exercises

74. A car can be rented from Basic Rental for $260 per week with no extra charge for mileage. Continental charges $80 per week plus 25 cents for each mile driven to rent the same car. How many miles should be driven in a week to make the rental cost for Basic Rental a better deal than Continental's?

75. A company manufactures and sells personalized stationery. The weekly fixed cost is $3000 and it cost $3.00 to produce each package of stationery. The selling price is $5.50 per package. How many packages of stationery must be produced and sold each week for the company to generate a profit?

Group Exercise

76. Each group member should research one situation that provides two different pricing options. These can involve areas such as public transportation options (with or without coupon books) or long-distance telephone plans or anything of interest. Be sure to bring in all the details for each option. At the group meeting, select the two pricing situations that are most interesting and relevant. Using each situation, write a word problem about selecting the better of the two options. The word problem should be one that can be solved using a linear inequality. The group should turn in the two problems and their solutions.

SECTION 6.6   SOLVING QUADRATIC EQUATIONS

OBJECTIVES

1. Multiply binomials using the FOIL method.

2. Factor trinomials.

3. Solve quadratic equations by factoring.

4. Solve quadratic equations using the quadratic formula.

© Bettmann Corbis

America is a nation of immigrants. Since 1820, over 40 million people have immigrated to the United States from all over the world. They chose to come for various reasons, such as to live in freedom, to practice religion freely, to escape poverty or oppression, and to make better lives for themselves and their children. As a result, a substantial percentage of the U.S. population is foreign-born. The graph in Figure 6.5 shows the percentage of foreign-born Americans for each decade of the twentieth century.

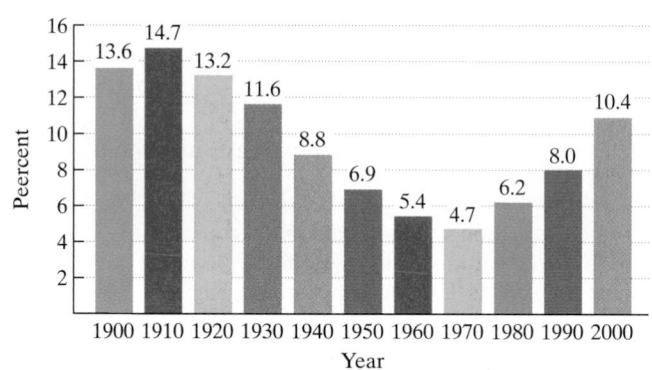

Percentage of U.S. Population That Was Foreign-Born, 1900-2000

Source: U.S. Census Bureau

FIGURE 6.5

The data shown in the graph can be modeled by the formula

$$P = 0.0021x^2 - 0.286x + 16.28$$

where P is the percentage of the U.S. population that was foreign-born x years after 1900. If the trend shown by the data continues, in which year will 20% of the U.S. population be foreign-born? To answer the question, it is necessary to substitute 20 for P in the formula and solve for x, the number of years after 1900:

$$20 = 0.0021x^2 - 0.286x + 16.28.$$

Do you see how this equation differs from a linear equation? This equation has a term in which the exponent on x is 2. Solving such an equation involves finding the set of numbers that will make the equation a true statement. In this section, we focus on two methods for solving equations in the form $ax^2 + bx + c = 0$: (1) factoring and (2) using a formula. Equations in this form are called **quadratic equations**. The greatest exponent on the variable in such an equation is 2.

Multiply binomials using the FOIL method.

Multiplying Two Binomials Using the FOIL Method

Before we learn about the first method, factoring, we need to consider the FOIL method for multiplying two binomials. A **binomial** is a simplified algebraic expression that contains two terms in which each exponent that appears on a variable is a whole number.

Examples of Binomials

$$x + 3, \quad x + 4, \quad 3x + 4, \quad 5x - 3$$

Two binomials can be quickly multiplied by using the FOIL method, in which F represents the product of the **first** terms in each binomial, O represents the product of the **outside** terms, I represents the product of the two **inside** terms, and L represents the product of the **last,** or second, terms in each binomial.

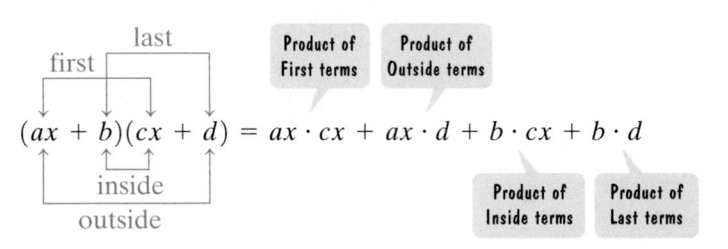

USING THE FOIL METHOD TO MULTIPLY BINOMIALS

$$(ax + b)(cx + d) = ax \cdot cx + ax \cdot d + b \cdot cx + b \cdot d$$

Once you have multiplied first, outside, inside, and last terms, combine all like terms.

EXAMPLE 1   USING THE FOIL METHOD

Multiply: $(x + 3)(x + 4)$.

Solution

$$\text{F : First terms} \quad = (x + 3)(x + 4) = x \cdot x = x^2$$
$$\text{O: Outside terms} = (x + 3)(x + 4) = x \cdot 4 = 4x$$
$$\text{I : Inside terms} \quad = (x + 3)(x + 4) = 3 \cdot x = 3x$$
$$\text{L : Last terms} \quad = (x + 3)(x + 4) = 3 \cdot 4 = 12$$

$$
\begin{aligned}
(x + 3)(x + 4) &= x \cdot x + x \cdot 4 + 3 \cdot x + 3 \cdot 4 \\
&= x^2 + 4x + 3x + 12 \\
&= x^2 + 7x + 12 \quad \textit{Combine like terms.}
\end{aligned}
$$

Check
1
Point

Multiply: $(x + 5)(x + 6)$.

EXAMPLE 2   USING THE FOIL METHOD

Multiply: $(3x + 4)(5x - 3)$.

Solution

$$
\begin{aligned}
(3x + 4)(5x - 3) &= 3x \cdot 5x + 3x(-3) + 4 \cdot 5x + 4(-3) \\
&= 15x^2 - 9x + 20x - 12 \\
&= 15x^2 + 11x - 12 \quad \textit{Combine like terms.}
\end{aligned}
$$

Check
2
Point   Multiply: $(7x + 5)(4x - 3)$.

2 Factor trinomials.

Factoring Trinomials Whose Leading Coefficient Is 1

The algebraic expression $x^2 + 7x + 12$ is called a trinomial. A **trinomial** is a simplified algebraic expression that contains three terms in which all variables have whole number exponents.

We can use the FOIL method to multiply two binomials to obtain the trinomial $x^2 + 7x + 12$:

Factored Form   **F   O   I   L   Trinomial Form**

$$(x + 3)(x + 4) = x^2 + 4x + 3x + 12 = x^2 + 7x + 12$$

Because the product of $x + 3$ and $x + 4$ is $x^2 + 7x + 12$, we call $x + 3$ and $x + 4$ the **factors** of $x^2 + 7x + 12$. **Factoring** an algebraic expression is the process of writing the expression as the product of two or more expressions. Thus, to factor $x^2 + 7x + 12$, we write

$$x^2 + 7x + 12 = (x + 3)(x + 4).$$

Now, we can make several important observations.

$x^2 + 7x + 12 = (x + 3)(x + 4)$ $x^2 + 7x + 12 = (x + 3)(x + 4)$ $x^2 + 7x + 12 = (x + 3)(x + 4)$

The first term of each
factor is x. The product
of the First terms
is $x \cdot x = x^2$.

3 and 4 are factors
of 12. The product of
the Last terms is
$3 \cdot 4 = 12$.

I: $3x$
O: $4x$

The sum of the Outside
and Inside products
is $4x + 3x = 7x$.

These observations provide us with a procedure for factoring $x^2 + bx + c$.

A STRATEGY FOR FACTORING $x^2 + bx + c$

1. Enter x as the first term of each factor.

$$(x \quad)(x \quad) = x^2 + bx + c$$

2. List pairs of factors of the constant c.

3. Try various combinations of these factors. Select the combination in which the sum of the *O*utside and *I*nside products is equal to bx.

$$(x + \square)(x + \square) = x^2 + bx + c$$

I

O

Sum of O + I

4. Check your work by multiplying the factors using the FOIL method. You should obtain the original trinomial.

If none of the possible combinations yield an *O*utside product and an *I*nside product whose sum is equal to bx, the trinomial cannot be factored using integers and is called **prime**.

EXAMPLE 3 FACTORING A TRINOMIAL IN $x^2 + bx + c$ FORM

Factor: $x^2 + 6x + 8$.

Solution

STEP 1. Enter x as the first term of each factor.

$$x^2 + 6x + 8 = (x \quad)(x \quad)$$

To find the second term of each factor, we must find two integers whose product is 8 and whose sum is 6.

STEP 2. List all pairs of factors of the constant, 8.

Factors of 8	8, 1	4, 2	−8, −1	−4, −2

STEP 3. Try various combinations of these factors. The correct factorization of $x^2 + 6x + 8$ is the one in which the sum of the *Outside* and *Inside* products is equal to $6x$. Here is a list of the possible factorizations:

Possible Factorizations of $x^2 + 6x + 8$	Sum of Outside and Inside Products (Should Equal $6x$)
$(x + 8)(x + 1)$	$x + 8x = 9x$
$(x + 4)(x + 2)$	$2x + 4x = 6x$
$(x − 8)(x − 1)$	$−x − 8x = −9x$
$(x − 4)(x − 2)$	$−2x − 4x = −6x$

This is the required middle term.

Thus, $x^2 + 6x + 8 = (x + 4)(x + 2)$.

Check this result by multiplying the right side using the FOIL method. You should obtain the original trinomial. Because of the commutative property, we can also say that

$$x^2 + 6x + 8 = (x + 2)(x + 4).$$

Check Point **3**   Factor: $x^2 + 5x + 6$.

EXAMPLE 4 FACTORING A TRINOMIAL IN $x^2 + bx + c$ FORM

Factor: $x^2 + 2x − 35$.

Solution

STEP 1. Enter x as the first term of each factor.

$$x^2 + 2x − 35 = (x \quad)(x \quad)$$

To find the second term of each factor, we must find two integers whose product is −35 and whose sum is 2.

STEP 2. List pairs of factors of the constant, − 35.

Factors of −35	35, −1	−35, 1	−7, 5	7, −5

STEP 3. Try various combinations of these factors. The correct factorization of $x^2 + 2x − 35$ is the one in which the sum of the *Outside* and *Inside* products is equal to $2x$. On the next page is a list of the possible factorizations.

STUDY TIP

To factor $x^2 + bx + c$ when c is positive, find two numbers with the same sign as the middle term.

$$x^2 + 6x + 8 = (x + 2)(x + 4)$$

Same signs

$$x^2 − 5x + 6 = (x − 3)(x − 2)$$

Same signs

Using this observation, it is not necessary to list the last two factorizations in step 3 on the right.

Possible Factorizations of $x^2 + 2x - 35$	Sum of Outside and Inside Products (Should Equal $2x$)
$(x - 1)(x + 35)$	$35x - x = 34x$
$(x + 1)(x - 35)$	$-35x + x = -34x$
$(x - 7)(x + 5)$	$5x - 7x = -2x$
$(x + 7)(x - 5)$	$-5x + 7x = 2x$

This is the required middle term.

Thus, $x^2 + 2x - 35 = (x + 7)(x - 5)$ or $(x - 5)(x + 7)$.

STEP 4. Verify the factorization using the FOIL method.

F   O   I   L

$$(x + 7)(x - 5) = x^2 - 5x + 7x - 35 = x^2 + 2x - 35$$

Because the product of the factors is the original polynomial, the factorization is correct.

Check Point 4   Factor: $x^2 + 3x - 10$.

STUDY TIP

To factor $x^2 + bx + c$ when c is negative, find two numbers with opposite signs whose sum is the coefficient of the middle term.

$x^2 + 2x - 35 = (x + 7)(x - 5)$

Negative   Opposite signs

Factoring Trinomials Whose Leading Coefficient Is Not 1

How do we factor a trinomial such as $3x^2 - 20x + 28$? Notice that the leading coefficient is 3. We must find two binomials whose product is $3x^2 - 20x + 28$. The product of the *First* terms must be $3x^2$:

$$(3x \quad)(x \quad).$$

From this point on, the factoring strategy is exactly the same as the one we use to factor trinomials whose leading coefficient is 1.

EXAMPLE 5 | FACTORING A TRINOMIAL

Factor: $3x^2 - 20x + 28$.

Solution

STEP 1. Find two *First* terms whose product is $3x^2$.

$$3x^2 - 20x + 28 = (3x \quad)(x \quad)$$

STEP 2. List all pairs of factors of the constant, 28. The number 28 has pairs of factors that are either both positive or both negative. Because the middle term, $-20x$, is negative, both factors must be negative. The negative factorizations of 28 are $-1(-28)$, $-2(-14)$, and $-4(-7)$.

STEP 3. Try various combinations of these factors. The correct factorization of $3x^2 - 20x + 28$ is the one in which the sum of the *Outside* and *Inside* products is equal to $-20x$. Here is a list of the possible factorizations:

Possible Factorizations of $3x^2 - 20x + 28$	Sum of Outside and Inside Products (Should Equal $-20x$)
$(3x - 1)(x - 28)$	$-84x - x = -85x$
$(3x - 28)(x - 1)$	$-3x - 28x = -31x$
$(3x - 2)(x - 14)$	$-42x - 2x = -44x$
$(3x - 14)(x - 2)$	$-6x - 14x = -20x$
$(3x - 4)(x - 7)$	$-21x - 4x = -25x$
$(3x - 7)(x - 4)$	$-12x - 7x = -19x$

This is the required middle term.

Thus,
$$3x^2 - 20x + 28 = (3x - 14)(x - 2) \quad \text{or} \quad (x - 2)(3x - 14).$$

STEP 4. Verify the factorization using the FOIL method.

$$\begin{aligned}
(3x - 14)(x - 2) &= \overset{F}{3x \cdot x} + \overset{O}{3x(-2)} + \overset{I}{(-14) \cdot x} + \overset{L}{(-14)(-2)} \\
&= 3x^2 - 6x - 14x + 28 \\
&= 3x^2 - 20x + 28
\end{aligned}$$

Because this is the trinomial we started with, the factorization is correct.

 **Check Point 5**   Factor: $5x^2 - 14x + 8$.

EXAMPLE 6   FACTORING A TRINOMIAL

Factor: $8y^2 - 10y - 3$.

Solution

STEP 1. Find two *First* terms whose product is $8y^2$.

$$8y^2 - 10y - 3 \overset{?}{=} (8y \qquad)(y \qquad)$$
$$8y^2 - 10y - 3 \overset{?}{=} (4y \qquad)(2y \qquad)$$

STEP 2. List all pairs of factors of the constant, -3. The possible factorizations are $1(-3)$ and $-1(3)$.

STEP 3. Try various combinations of these factors. The correct factorization of $8y^2 - 10y - 3$ is the one in which the sum of the *Outside* and *Inside* products is equal to $-10y$. Here is a list of the possible factorizations:

Possible Factorizations of $8y^2 - 10y - 3$	Sum of Outside and Inside Products (Should Equal $-10y$)
$(8y + 1)(y - 3)$	$-24y + y = -23y$
$(8y - 3)(y + 1)$	$8y - 3y = 5y$
$(8y - 1)(y + 3)$	$24y - y = 23y$
$(8y + 3)(y - 1)$	$-8y + 3y = -5y$
$(4y + 1)(2y - 3)$	$-12y + 2y = -10y$
$(4y - 3)(2y + 1)$	$4y - 6y = -2y$
$(4y - 1)(2y + 3)$	$12y - 2y = 10y$
$(4y + 3)(2y - 1)$	$-4y + 6y = 2y$

This is the required middle term.

Thus,

$$8y^2 - 10y - 3 = (4y + 1)(2y - 3) \quad \text{or} \quad (2y - 3)(4y + 1).$$

Show that this factorization is correct by multiplying the factors using the FOIL method. You should obtain the original trinomial.

 Factor: $6y^2 + 19y - 7$.

 3 Solve quadratic equations by factoring.

Solving Quadratic Equations by Factoring

We begin by restating the definition of a quadratic equation.

> ### DEFINITION OF A QUADRATIC EQUATION
> A **quadratic equation** in x is an equation that can be written in the **general form**
> $$ax^2 + bx + c = 0$$
> where a, b, and c are real numbers, with $a \neq 0$.

An example of a quadratic equation in general form is $x^2 - 7x + 10 = 0$. The coefficient of x^2 is 1 ($a = 1$), the coefficient of x is -7 ($b = -7$), and the constant term is 10 ($c = 10$). Notice that we can factor the left side of this equation.

$$x^2 - 7x + 10 = 0$$
$$(x - 5)(x - 2) = 0$$

If a quadratic equation has zero on one side and a factored trinomial on the other side, it can be solved using the **zero-product principle**:

> ### THE ZERO-PRODUCT PRINCIPLE
> If the product of two factors is zero, then one (or both) of the factors must have a value of zero.
> $$\text{If } A \cdot B = 0, \text{ then } A = 0 \text{ or } B = 0.$$

EXAMPLE 7 | SOLVING A QUADRATIC EQUATION USING THE ZERO-PRODUCT PRINCIPLE

Solve: $(x - 5)(x - 2) = 0$.

Solution   The product $(x - 5)(x - 2)$ is equal to zero. By the zero-product principle, the only way that this product can be zero is if at least one of the factors is zero. We set each individual factor equal to zero and solve each resulting equation for x.

$$(x - 5)(x - 2) = 0$$
$$x - 5 = 0 \quad \text{or} \quad x - 2 = 0$$
$$x = 5 \qquad\qquad x = 2$$

Check each proposed solution by substituting it for x in the original equation.

<table>
<tr><td align="center">Check 5:</td><td align="center">Check 2:</td></tr>
</table>

Check 5:	Check 2:
$(x - 5)(x - 2) = 0$	$(x - 5)(x - 2) = 0$
$(5 - 5)(5 - 2) \stackrel{?}{=} 0$	$(2 - 5)(2 - 2) \stackrel{?}{=} 0$
$0(3) \stackrel{?}{=} 0$	$-3(0) \stackrel{?}{=} 0$
$0 = 0, \quad$ true	$0 = 0, \quad$ true

The resulting true statements indicate that the solutions are 5 and 2. The solution set is $\{2, 5\}$.

 **Check Point 7**   Solve: $(x + 6)(x - 3) = 0$.

SOLVING A QUADRATIC EQUATION BY FACTORING

1. If necessary, rewrite the equation in the general form $ax^2 + bx + c = 0$, moving all terms to one side, thereby obtaining zero on the other side.
2. Factor.
3. Apply the zero-product principle, setting each factor equal to zero.
4. Solve the equations in step 3.
5. Check the solutions in the original equation.

EXAMPLE 8   USING FACTORING TO SOLVE A QUADRATIC EQUATION

Solve: $x^2 - 2x = 35$.

Solution

STEP 1. Move all terms to one side and obtain zero on the other side. Subtract 35 from both sides and write the equation in general form.

$$x^2 - 2x = 35$$
$$x^2 - 2x - 35 = 35 - 35$$
$$x^2 - 2x - 35 = 0$$

STEP 2. Factor. $(x - 7)(x + 5) = 0$

STEPS 3 and 4. Set each factor equal to zero and solve each resulting equation.

$$x - 7 = 0 \quad \text{or} \quad x + 5 = 0$$
$$x = 7 \qquad\qquad x = -5$$

STEP 5. Check the solutions in the original equation.

Check 7:	Check -5:
$x^2 - 2x = 35$	$x^2 - 2x = 35$
$7^2 - 2 \cdot 7 \stackrel{?}{=} 35$	$(-5)^2 - 2(-5) \stackrel{?}{=} 35$
$49 - 14 \stackrel{?}{=} 35$	$25 + 10 \stackrel{?}{=} 35$
$35 = 35, \quad$ true	$35 = 35, \quad$ true

The resulting true statements indicate that the solutions are 7 and -5. The solution set is $\{-5, 7\}$.

Check — Solve: $x^2 - 6x = 16.$

Point 8

EXAMPLE 9   SOLVING A QUADRATIC EQUATION BY FACTORING

Solve: $5x^2 - 33x + 40 = 0.$

Solution   All terms are already on the left and zero is on the other side. Thus, we can factor the trinomial on the left side. $5x^2 - 33x + 40$ factors as $(5x - 8)(x - 5)$.

$$5x^2 - 33x + 40 = 0 \qquad \text{This is the given quadratic equation.}$$

$$(5x - 8)(x - 5) = 0 \qquad \text{Factor.}$$

$$5x - 8 = 0 \quad \text{or} \quad x - 5 = 0 \qquad \text{Set each factor equal to zero.}$$

$$5x = 8 \qquad\qquad x = 5 \qquad \text{Solve the resulting equations.}$$

$$x = \frac{8}{5}$$

Check these values in the original equation to confirm that the solution set is $\left\{\frac{8}{5}, 5\right\}$.

 **Check** Solve: $2x^2 + 7x - 4 = 0.$

Point 9

---

4 Solve quadratic equations using the quadratic formula.

Solving Quadratic Equations Using the Quadratic Formula

The solutions of a quadratic equation cannot always be found by factoring. Some trinomials are difficult to factor, and others cannot be factored (that is, they are prime). However, there is a formula that can be used to solve all quadratic equations, whether or not they contain factorable trinomials. The formula is called the *quadratic formula*.

STUDY TIP

The entire numerator of the quadratic formula must be divided by $2a$. Always write the fraction bar all the way across the numerator.

$$x = \frac{-b \pm \sqrt{b^2 - 4ac}}{2a}$$

THE QUADRATIC FORMULA

The solutions of a quadratic equation in general form $ax^2 + bx + c = 0$, with $a \neq 0$, are given by the **quadratic formula**

$$x = \frac{-b \pm \sqrt{b^2 - 4ac}}{2a}.$$

x equals negative b, plus or minus the square root of $b^2 - 4ac$, all divided by 2a.

To use the quadratic formula, write the quadratic equation in general form if necessary. Then determine the numerical values for a (the coefficient of the squared term), b (the coefficient of the x-term), and c (the constant term). Substitute the values of a, b, and c in the quadratic formula and evaluate the expression. The $\pm$ sign indicates that there are two solutions of the equation.

To Die at Twenty

Can the equations

$$7x^5 + 12x^3 - 9x + 4 = 0$$

and

$$8x^6 - 7x^5 + 4x^3 - 19 = 0$$

be solved using a formula similar to the quadratic formula? The first equation has five solutions and the second has six solutions, but they cannot be found using a formula. How do we know? In 1832, a 20-year-old Frenchman, Evariste Galois, wrote a proof showing that there is no general formula to solve equations when the exponent on the variable is 5 or greater. Galois was jailed as a political activist several times while still a teenager. The day after his brilliant proof he fought a duel over a woman. The duel was a political setup. As he lay dying, Galois told his brother, Alfred, of the manuscript that contained his proof: "Mathematical manuscripts are in my room. On the table. Take care of my work. Make it known. Important. Don't cry, Alfred. I need all my courage—to die at twenty." (Our source is Leopold Infeld's biography of Galois, *Whom the Gods Love*. Some historians, however, dispute the story of Galois's ironic death the very day after his algebraic proof. Mathematical truths seem more reliable than historical ones!)

EXAMPLE 10 SOLVING A QUADRATIC EQUATION USING THE QUADRATIC FORMULA

Solve using the quadratic formula: $2x^2 + 9x - 5 = 0$.

Solution The given equation is in general form. Begin by identifying the values for a, b, and c.

$$2x^2 + 9x - 5 = 0$$

$a = 2$ $b = 9$ $c = -5$

Substituting these values into the quadratic formula and simplifying gives the equation's solutions.

$$x = \frac{-b \pm \sqrt{b^2 - 4ac}}{2a}$$ Use the quadratic formula.

$$x = \frac{-9 \pm \sqrt{9^2 - 4(2)(-5)}}{2(2)}$$ Substitute the values for a, b, and c: $a = 2$, $b = 9$, and $c = -5$.

$$= \frac{-9 \pm \sqrt{81 + 40}}{4}$$ $9^2 - 4(2)(-5) = 81 - (-40) = 81 + 40$

$$= \frac{-9 \pm \sqrt{121}}{4}$$ Add under the radical sign.

$$= \frac{-9 \pm 11}{4}$$ $\sqrt{121} = 11$

Now we will evaluate this expression to obtain the two solutions. On the left, we *add* 11 to -9. On the right, we *subtract* 11 from -9.

$$x = \frac{-9 + 11}{4} \quad \text{or} \quad x = \frac{-9 - 11}{4}$$

$$= \frac{2}{4} = \frac{1}{2} \qquad\qquad = \frac{-20}{4} = -5$$

The solution set is $\left\{-5, \frac{1}{2}\right\}$.

 Check Point 10   Solve using the quadratic formula: $8x^2 + 2x - 1 = 0$.

The quadratic equation in Example 10 has rational solutions, namely -5 and $\frac{1}{2}$. The equation can also be solved by factoring. Take a few minutes to do this now and convince yourself that you will arrive at the same two solutions.

Any quadratic equation that has rational solutions can be solved by factoring or using the quadratic formula. However, quadratic equations with irrational solutions cannot be solved by factoring. These equations can be readily solved using the quadratic formula.

EXAMPLE 11 | SOLVING A QUADRATIC EQUATION USING THE QUADRATIC FORMULA

Solve using the quadratic formula: $2x^2 = 4x + 1$.

Solution The quadratic equation must be in general form to identify the values for a, b, and c. To move all terms to one side and obtain zero on the right, we subtract $4x + 1$ from both sides. Then we can identify the values for a, b, and c.

$$2x^2 = 4x + 1 \qquad \text{This is the given equation.}$$
$$2x^2 - 4x - 1 = 0 \qquad \text{Subtract } 4x + 1 \text{ from both sides.}$$

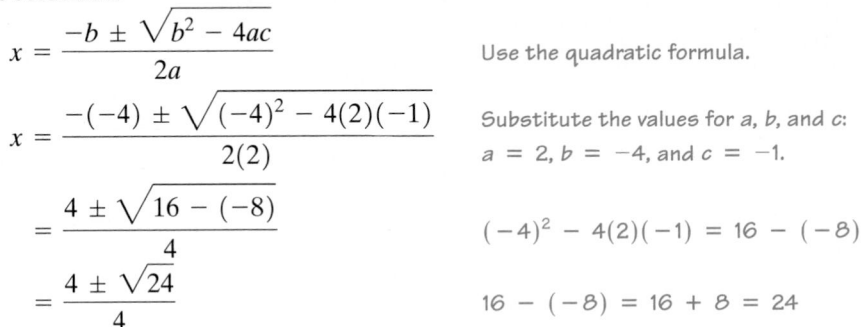

$$a = 2 \qquad b = -4 \qquad c = -1$$

Substituting these values into the quadratic formula and simplifying gives the equation's solutions.

$$x = \frac{-b \pm \sqrt{b^2 - 4ac}}{2a} \qquad \text{Use the quadratic formula.}$$

$$x = \frac{-(-4) \pm \sqrt{(-4)^2 - 4(2)(-1)}}{2(2)} \qquad \begin{aligned} &\text{Substitute the values for } a, b, \text{ and } c: \\ &a = 2, b = -4, \text{ and } c = -1. \end{aligned}$$

$$= \frac{4 \pm \sqrt{16 - (-8)}}{4} \qquad (-4)^2 - 4(2)(-1) = 16 - (-8)$$

$$= \frac{4 \pm \sqrt{24}}{4} \qquad 16 - (-8) = 16 + 8 = 24$$

The solutions are $\dfrac{4 + \sqrt{24}}{4}$ and $\dfrac{4 - \sqrt{24}}{4}$. These solutions are irrational numbers.

You can use a calculator to obtain a decimal approximation for each solution. However, in situations such as this that do not involve applications, it is best to leave the irrational solutions in radical form as exact answers. In some cases, we can simplify this radical form. Using methods for simplifying square roots discussed in Section 5.4, we can simplify $\sqrt{24}$:

$$\sqrt{24} = \sqrt{4 \cdot 6} = \sqrt{4}\,\sqrt{6} = 2\sqrt{6}.$$

Now we can use this result to simplify the two solutions. First, use the distributive property to factor out 2 from both terms in the numerator. Then, divide the numerator and the denominator by 2.

$$x = \frac{4 \pm \sqrt{24}}{4} = \frac{4 \pm 2\sqrt{6}}{4} = \frac{\overset{1}{2}(2 \pm \sqrt{6})}{\underset{2}{4}} = \frac{2 \pm \sqrt{6}}{2}$$

In simplified radical form, the equation's solution set is

$$\left\{ \frac{2 + \sqrt{6}}{2}, \frac{2 - \sqrt{6}}{2} \right\}.$$

Examples 10 and 11 illustrate that the solutions of quadratic equations can be rational or irrational numbers. In Example 10, the expression under the square root was 121, a perfect square ($\sqrt{121} = 11$), and we obtained rational solutions. In Example 11, this expression was 24, which is not a perfect square (although we simplified $\sqrt{24}$ to $2\sqrt{6}$), and we obtained irrational solutions. If the expression under the square root simplifies to a negative number, then the quadratic equation has **no real solution**.

 **Check Point 11** Solve using the quadratic formula: $2x^2 = 6x - 1$.

TECHNOLOGY

Using a Calculator to Approximate $\dfrac{4 + \sqrt{24}}{4}$:

Many Scientific Calculators

$(\!|\; 4 \;+\; 24 \;\boxed{\sqrt{}}\;)\; \div\; 4 \;=$

Many Graphing Calculators

$(\!|\; 4 \;+\; \boxed{\sqrt{}}\; 24 \;)\; \div\; 4 \;\boxed{\text{ENTER}}$

Correct to the nearest tenth,

$$\frac{4 + \sqrt{24}}{4} \approx 2.2.$$

STUDY TIP

Avoid these common errors by factoring the numerator *before* you divide.

Incorrect:

$$\frac{\overset{1}{4} \pm \sqrt{24}}{\underset{1}{4}} = 1 \pm \sqrt{24}$$

Incorrect:

$$\frac{4 \pm \overset{1}{2}\sqrt{6}}{\underset{2}{4}} = \frac{4 \pm \sqrt{6}}{2}$$

You cannot divide just one term in the numerator and the denominator by their greatest common divisor.

Exercise Set 6.6

Practice Exercises

Use FOIL to find the products in Exercises 1–8.

1. $(x + 3)(x + 5)$ **2.** $(x + 7)(x + 2)$

3. $(x - 5)(x + 3)$ **4.** $(x - 1)(x + 2)$

5. $(2x - 1)(x + 2)$ **6.** $(2x - 5)(x + 3)$

7. $(3x - 7)(4x - 5)$ **8.** $(2x - 9)(7x - 4)$

Factor the trinomials in Exercises 9–20, or state that the trinomial is prime. Check your factorization using FOIL multiplication.

9. $x^2 + 5x + 6$ **10.** $x^2 + 8x + 15$

11. $x^2 - 2x - 15$ **12.** $x^2 - 4x - 5$

13. $x^2 - 8x + 15$ **14.** $x^2 - 14x + 45$

15. $x^2 - 9x - 36$ **16.** $x^2 - x - 90$

17. $x^2 - 8x + 32$ **18.** $x^2 - 9x + 81$

19. $x^2 + 17x + 16$ **20.** $x^2 - 7x - 44$

Factor the trinomials in Exercises 21–32, or state that the trinomial is prime. Check your factorization using FOIL multiplication. Unlike the trinomials you factored in Exercises 9–20, the trinomials that follow have coefficients of squared terms that are not equal to 1.

21. $2x^2 + 7x + 3$ **22.** $3x^2 + 7x + 2$

23. $2x^2 - 17x + 30$ **24.** $5x^2 - 13x + 6$

25. $3x^2 - x - 2$ **26.** $2x^2 + 5x - 3$

27. $3x^2 - 25x - 28$ **28.** $3x^2 - 2x - 5$

29. $6x^2 - 11x + 4$ **30.** $6x^2 - 17x + 12$

31. $4x^2 + 16x + 15$ **32.** $8x^2 + 33x + 4$

In Exercises 33–36, solve each equation using the zero-product principle.

33. $(x - 8)(x + 3) = 0$ **34.** $(x + 11)(x - 5) = 0$

35. $(4x + 5)(x - 2) = 0$ **36.** $(x + 9)(3x - 1) = 0$

Solve the quadratic equations in Exercises 37–52 by factoring.

37. $x^2 + 8x + 15 = 0$ **38.** $x^2 + 5x + 6 = 0$

39. $x^2 - 2x - 15 = 0$ **40.** $x^2 + x - 42 = 0$

41. $x^2 - 4x = 21$ **42.** $x^2 + 7x = 18$

43. $x^2 + 9x = -8$ **44.** $x^2 - 11x = -10$

45. $x^2 - 12x = -36$ **46.** $x^2 - 14x = -49$

47. $2x^2 = 7x + 4$ **48.** $3x^2 = x + 4$

49. $5x^2 + x = 18$ **50.** $3x^2 - 4x = 15$

51. $x(6x + 23) + 7 = 0$ **52.** $x(6x + 13) + 6 = 0$

Solve the equations in Exercises 53–72 using the quadratic formula.

53. $x^2 + 8x + 15 = 0$ **54.** $x^2 + 8x + 12 = 0$

55. $x^2 + 5x + 3 = 0$ **56.** $x^2 + 5x + 2 = 0$

57. $x^2 + 4x = 6$ **58.** $x^2 + 2x = 4$

59. $x^2 + 4x - 7 = 0$ **60.** $x^2 + 4x + 1 = 0$

61. $x^2 - 3x = 18$ **62.** $x^2 - 3x = 10$

63. $6x^2 - 5x - 6 = 0$ **64.** $9x^2 - 12x - 5 = 0$

65. $x^2 - 2x - 10 = 0$ **66.** $x^2 + 6x - 10 = 0$

67. $x^2 - x = 14$ **68.** $x^2 - 5x = 10$

69. $6x^2 + 6x + 1 = 0$ **70.** $3x^2 = 5x - 1$

71. $4x^2 = 12x - 9$ **72.** $9x^2 + 6x + 1 = 0$

Application Exercises

The formula

$$N = \frac{t^2 - t}{2}$$

describes the number of football games, N, that must be played in a league with t teams if each team is to play every other team once. Use this information to solve Exercises 73–74.

73. If a league has 36 games scheduled, how many teams belong to the league, assuming that each team plays every other team once?

74. If a league has 45 games scheduled, how many teams belong to the league, assuming that each team plays every other team once?

The bar graph in Figure 6.5 on page 318 shows the percentage of the U.S. population that was foreign-born for each decade of the twentieth century. The percentage, P, of the U.S. population that was foreign-born x years after 1900 can be modeled by the formula

$$P = 0.0021x^2 - 0.286x + 16.28.$$

Use the formula to solve Exercises 75–76.

75. In which year will 20% of the U.S. population be foreign-born? Use a calculator and round to the nearest year.

76. In which year will 25% of the U.S. population be foreign-born? Use a calculator and round to the nearest year.

A driver's age has something to do with his or her chance of getting into a fatal car crash. The bar graph shows the number of fatal vehicle crashes per 100 million miles driven for drivers of various age groups. For example, 25-year-old drivers are involved in 4.1 fatal crashes per 100 million miles driven. Thus, when a group of 25-year-old Americans have driven a total of 100 million miles, approximately 4 have been in accidents in which someone died.

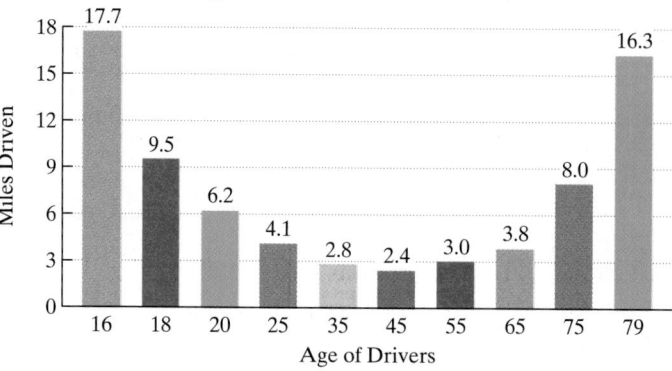

Age of U.S. Drivers and Fatal Crashes

Source: Insurance Institute for Highway Safety

The number of fatal vehicle crashes per 100 million miles, N, for drivers of age x can be modeled by the formula

$$N = 0.013x^2 - 1.19x + 28.24.$$

Use the formula to solve Exercises 77–78. Use a calculator and round answers to the nearest year.

77. What age groups are expected to be involved in 3 fatal crashes per 100 million miles driven? How well does the formula model the trend in the actual data shown in the bar graph on the previous page?

78. What age groups are expected to be involved in 10 fatal crashes per 100 million miles driven? How well does the formula model the trend in the actual data shown in the bar graph on the previous page?

Writing in Mathematics

79. Explain how to multiply two binomials using the FOIL method. Give an example with your explanation.

80. Explain how to factor $x^2 - 5x + 6$.

81. Explain how to solve a quadratic equation by factoring. Use the equation $x^2 + 6x + 8 = 0$ in your explanation.

82. Explain how to solve a quadratic equation using the quadratic formula. Use the equation $x^2 + 6x + 8 = 0$ in your explanation.

83. Describe the trend shown by the data for the percentage of foreign-born Americans in the graph in Figure 6.5 on page 318. Do you believe that this trend is likely to continue or might something occur that would make it impossible to extend the model into the future? Explain your answer.

Critical Thinking Exercises

84. The radicand of the quadratic formula, $b^2 - 4ac$, can be used to determine whether $ax^2 + bx + c = 0$ has solutions that are rational, irrational, or not real numbers. Explain how this works. Is it possible to determine the kinds of answers that one will obtain to a quadratic equation without actually solving the equation? Explain.

In Exercises 85–86, find all positive integers b so that the trinomial can be factored.

85. $x^2 + bx + 15$

86. $x^2 + 4x + b$

87. Factor: $x^{2n} + 20x^n + 99$.

88. A person throws a rock upward from the edge of an 80-foot cliff. The height, h, in feet, of the rock above the water at the bottom of the cliff after t seconds is described by the formula

$$h = -16t^2 + 64t + 80.$$

How long will it take for the rock to reach the water?

Technology Exercise

89. If you have access to a calculator that solves quadratic equations, consult the owner's manual to determine how to use this feature. Then use your calculator to solve any five of the equations in Exercises 37–72.

Chapter Summary, Review, and Test

SUMMARY

DEFINITIONS AND CONCEPTS

EXAMPLES

6.1 Algebraic Expressions and Formulas

a. An algebraic expression combines variables and numbers using addition, subtraction, multiplication, division, powers, or roots.

b. Evaluating an algebraic expression means finding its value for a given value of the variable.

Ex. 1, p. 271;
Ex. 2, p. 271;
Ex. 3, p. 272

c. Terms of an algebraic expression are separated by addition. Like terms have the same variables with the same exponents on the variables. To add or subtract like terms, add or subtract the coefficients and copy the common variable.

Ex. 5, p. 274;
Ex. 6, p. 274

d. An algebraic expression is simplified when parentheses have been removed (using the distributive property) and like terms have been combined.

Ex. 7, p. 275;
Ex. 8, p. 275

e. Formulas are statements of equality expressing a relationship among two or more variables. Formulas that approximately describe data are called mathematical models.

Ex. 9, p. 276

6.2 Solving Linear Equations

a. A linear equation in one variable can be written in the form $ax + b = c$. The greatest exponent on the variable is 1.

b. Solving a linear equation is the process of finding the set of numbers that will make the equation a true statement. These numbers are the solutions.

c. A step-by-step procedure for solving a linear equation is given in the box on page 284.

Ex. 6, p. 283;
Ex. 7, p. 284;
Ex. 8, p. 284

d. If an equation contains fractions, multiply each term of the equation by the least common denominator of the fractions in the equation, thereby clearing the fractions.

Ex. 9, p. 285

e. If a false statement (such as $-6 = 7$) is obtained in solving an equation, the equation has no solution. The solution set is $\varnothing$, the empty set.

Ex. 13, p. 288

f. If a true statement (such as $-6 = -6$) is obtained, the equation has infinitely many solutions. The solution set is the set of all real numbers.

Ex. 14, p. 288

6.3 Applications of Linear Equations

a. A step-by-step strategy for solving word problems using linear equations is given in the box on page 292.

Ex. 1, p. 293;
Ex. 2, p. 294;
Ex. 3, p. 295;
Ex. 4, p. 296

b. Algebraic translations of English phrases are given in Table 6.1 on page 292.

6.4 Ratio, Proportion, and Variation

a. The ratio of a to b is written $\dfrac{a}{b}$, or $a:b$.

Ex. 1, p. 300

b. A proportion is a statement in the form $\dfrac{a}{b} = \dfrac{c}{d}$.

c. The cross-products principle states that if $\dfrac{a}{b} = \dfrac{c}{d}$, then $ad = bc$.

Ex. 2, p. 301

d. A step-by-step procedure for solving applied problems using proportions is given in the box on page 302.

Ex. 3, p. 302;
Ex. 4, p. 304

e. If two quantities vary directly, as one increases, so does the other. As one decreases, so does the other. Proportions can be used to solve direct variation problems.

Ex. 5, p. 304;
Ex. 6, p. 305

f. If two quantities vary inversely, as one increases, the other decreases. As one decreases, the other increases. Proportions, as explained in the box on page 306, can be used to solve inverse variation problems.

Ex. 7, p. 307

6.5 Solving Linear Inequalities

A procedure for solving a linear inequality is given in the box on page 312. Remember to reverse the direction of the inequality symbol when multiplying or dividing both sides of an inequality by a negative number.

Ex. 2, p. 312;
Ex. 3, p. 312;
Ex. 4, p. 313;
Ex. 5, p. 314

6.6 Solving Quadratic Equations

a. The general form of a quadratic equation is $ax^2 + bx + c = 0, a \neq 0$.

Ex. 8, p. 325;
Ex. 9, p. 326
Ex. 10, p. 327;
Ex. 11, p. 328

b. Some quadratic equations can be solved using factoring and the zero-product principle. A step-by-step procedure is given in the box on page 325.

c. All quadratic equations in general form can be solved using the quadratic formula:

$$x = \frac{-b \pm \sqrt{b^2 - 4ac}}{2a}.$$

REVIEW EXERCISES

6.1

In Exercises 1–3, evaluate the algebraic expression for the given value of the variable.

1. $6x + 9$; $x = 4$

2. $4x^2 - 3x + 2$; $x = 5$

3. $7x^2 + 4x - 5$; $x = -2$

In Exercises 4–6, simplify each algebraic expression.

4. $7x + 9 - 12 - x$ **5.** $6(5x + 3) - 20$

6. $4(7x - 1) + 11x$

7. Suppose you can stay in the sun for x minutes without burning when you don't put on any lotion. The algebraic expression $9x$ describes how long you can tan without burning with a number 9 lotion. Evaluate the expression when $x = 15$. Describe what the answer means in practical terms.

8. Suppose that a store is selling all computers at 25% off the regular price. If x is the regular price, the algebraic expression $x - 0.25x$ describes the sale price. Evaluate the expression when $x = 2400$. Describe what the answer means in practical terms.

9. The bar graph shows the cost of Medicare, in billions of dollars, projected through 2005. The data can be modeled by the formula

$$N = 1.2x^2 + 15.2x + 181.4$$

where N represents Medicare spending, in billions of dollars, x years after 1995. According to the formula, what will be the cost of Medicare, in billions of dollars, in 2005? How well does the formula describe the cost for that year shown by the bar graph?

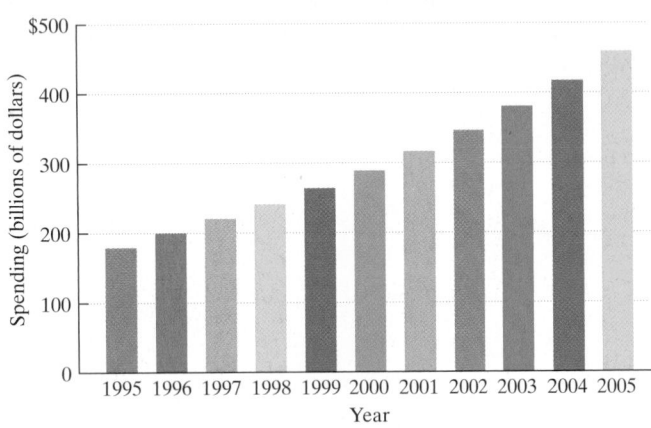

Medicare Spending

Source: Congressional Budget Office

6.2

Solve each equation in Exercises 10–14.

10. $4x + 9 = 33$ **11.** $5x - 3 = x + 5$

12. $3(x + 4) = 5x - 12$

13. $2(x - 2) + 3(x + 5) = 2x - 2$

14. $\dfrac{2x}{3} = \dfrac{x}{6} + 1$

In Exercises 15–16, solve each equation for y.

15. $3x + y = 9$ **16.** $4x + 2y = 16$

In Exercises 17–20, solve each formula for the specified variable.

17. $D = RT$, for T **18.** $P = 2l + 2w$, for w

19. $A = \dfrac{1}{2}bh$, for h **20.** $A = \dfrac{B + C}{2}$, for B

21. The formula $y = 420x + 720$ describes the amount of money, y, in millions of dollars, lost to credit card fraud worldwide x years after 1989. In how many years after 1989 did losses amount to 4080 million dollars? In which year was that?

6.3

In Exercises 22–25, let x represent the number. Write each English phrase as an algebraic expression.

22. 17 decreased by 9 times a number

23. 3 times a number, decreased by 7

24. 8 more than 5 times a number

25. The quotient of 6 and a number, increased by triple the number

In Exercises 26–29, let x represent the number. Use the given conditions to write an equation. Solve the equation and find the number.

26. Seventeen decreased by 4 times a number is 5. Find the number.

27. The product of 5 and a number, increased by 2, is 22 more than the number. Find the number.

28. When 7 times a number is decreased by 1, the result is 9 more than 5 times the number. Find the number.

29. Eight times the difference between a number and 5 is 56. Find the number.

In Exercises 30–33, use an equation to solve each problem.

30. In 2000, the average weekly salary for workers in the United States was $567. If this amount is increasing by $15 yearly, in how many years after 2000 will the average salary reach $702. In which year will that be?

31. In New York City, a fitness trainer earns $22,870 more per year than a preschool teacher. The yearly average salaries for fitness trainers and preschool teachers combined is $79,030. Determine the average yearly salary of a fitness trainer and a preschool teacher in New York City. (*Source: Time*, April 14, 2003)

32. On average, the number of unhealthy air days per year in Los Angeles exceeds three times that of New York City by 48 days. If Los Angeles and New York combined have 268 unhealthy air days per year, determine the number of unhealthy days for the two cities. (*Source:* Environmental Protection Agency)

33. You are choosing between two long-distance telephone plans. One plan has a monthly fee of $15 with a charge of $0.05 per minute. The other plan has a monthly fee of $5 with a charge of $0.07 per minute. For how many minutes of long-distance calls will the costs for the two plans be the same?

6.4

34. The bar graph indicates countries where ten or more languages have become extinct. Use the graph to find each of the following ratios. First express the ratio as a fraction reduced to lowest terms. Then rewrite the ratio using the reduced fraction and a colon.

a. The number of extinct languages in Brazil to that of the United States

b. The number of extinct languages in Australia and India combined to that of Colombia

Countries Where Ten or More Languages Have Become Extinct (Number of Languages)

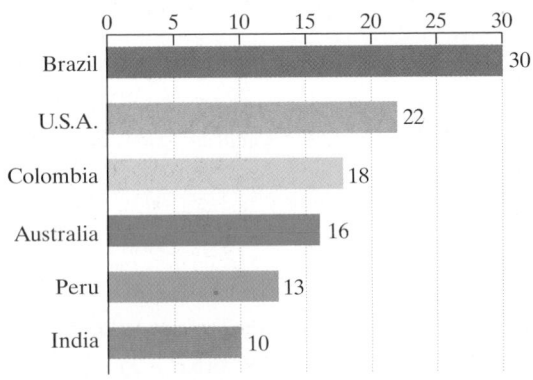

Source: Grimes

Solve each proportion in Exercises 35–38.

35. $\dfrac{3}{x} = \dfrac{15}{25}$

36. $\dfrac{-7}{5} = \dfrac{91}{x}$

37. $\dfrac{x+2}{3} = \dfrac{4}{5}$

38. $\dfrac{5}{x+7} = \dfrac{3}{x+3}$

Use a proportion to solve Exercises 39–40.

39. If a school board determines that there should be 3 teachers for every 50 students, how many teachers are needed for an enrollment of 5400 students?

40. To determine the number of trout in a lake, a conservationist catches 112 trout, tags them, and returns them to the lake. Later, 82 trout are caught, and 32 of them are found to be tagged. How many trout are in the lake?

Exercises 41–43 involve direct and inverse variation. Use a proportion to solve each exercise.

41. An electric bill varies directly as the amount of electricity used. The bill for 1400 kilowatts of electricity is $98. What is the bill for 2200 kilowatts of electricity?

42. The distance that a body falls from rest varies directly as the square of the time of the fall. If skydivers fall 144 feet in 3 seconds, how far will they fall in 10 seconds?

43. The time it takes to drive a certain distance varies inversely as the rate of travel. If it takes 4 hours at 50 miles per hour to drive the distance, how long will it take at 40 miles per hour?

6.5

Solve the inequalities in Exercises 44–49 and graph the solution sets.

44. $2x - 5 < 3$

45. $\dfrac{x}{2} > -4$

46. $3 - 5x \le 18$

47. $4x + 6 < 5x$

48. $6x - 10 \ge 2(x+3)$

49. $4x + 3(2x - 7) \le x - 3$

50. To pass a course, a student must have an average on three examinations of at least 60. If a student scores 42 and 74 on the first two tests, what must be earned on the third test to pass the course?

6.6

Use FOIL to find the products in Exercises 51–52.

51. $(x+9)(x-5)$

52. $(4x-7)(3x+2)$

Factor the trinomials in Exercises 53–58, or state that the trinomial is prime.

53. $x^2 - x - 12$

54. $x^2 - 8x + 15$

55. $x^2 + 2x + 3$

56. $3x^2 - 17x + 10$

57. $6x^2 - 11x - 10$

58. $3x^2 - 6x - 5$

Solve the quadratic equations in Exercises 59–62 by factoring.

59. $x^2 + 5x - 14 = 0$

60. $x^2 - 4x = 32$

61. $2x^2 + 15x - 8 = 0$

62. $3x^2 = -21x - 30$

Solve the quadratic equations in Exercises 63–66 using the quadratic formula.

63. $x^2 - 4x + 3 = 0$

64. $x^2 - 5x = 4$

65. $2x^2 + 5x - 3 = 0$

66. $3x^2 - 6x = 5$

In a national park, alligators are the subjects of a species protection program. The line graph shows the alligator population in the park during the first 12 years that the program is in effect. The formula

$$P = -10x^2 + 475x + 3500$$

models the park's alligator population, P, after x years of the protection program, where $0 \le x \le 12$. Use the formula to solve Exercises 67–68.

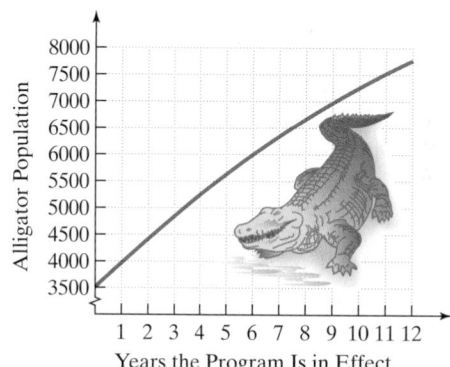

67. What was the alligator population three years after the program was in effect? Identify this information as a point on the line graph.

68. According to the formula, after how many years of the program was the park's alligator population up to 7250? Solve the resulting quadratic equation by the method of your choice. Then identify your solution as a point on the line graph.

Group Exercise

69. Throughout this chapter, we have seen examples of formulas and mathematical models that predict future values for things that are changing. These changing quantities included the gray wolf population (Example 9, page 276), life expectancy for men and women (Exercises 79–80, page 278), the number of bariatric surgeries (Exercises 81–82, page 278), the average number of hours Americans work per week (Example 10, page 286), the number of welfare recipients (Exercises 91–92, page 290), cigarette consumption (Exercises 63–64, page 316), the percentage of foreign-born Americans (Exercises 75–76, page 329), and Medicare spending (Exercise 9, page 332). Group members should select one of these formulas and

a. Use the formula to predict a future value for the changing quantity.

b. Use the formula to write and solve a word problem. The solution should require a linear equation, a linear inequality, or a quadratic equation.

c. Describe changes that might take place in the future that will affect the accuracy of the prediction made by the formula.

CHAPTER 6 TEST

1. Evaluate $5x^2 - 7x - 2$ when $x = -3$.

2. Simplify: $5(3x - 2) + 7x$.

3. The formula $S = 91t + 164$ models the average annual salary, S, of major league baseball players, in thousands of dollars, t years after 1984. What was the average annual salary in 1994?

In Exercises 4 and 5, solve each equation.

4. $8x - 5(x - 2) = x + 26$

5. $3(2x - 4) = 9 - 3(x + 1)$

6. The formula $N = 3.5x + 58$ models the average mortgage loan, N, in thousands of dollars, x years after 1980. How many years after 1980 will the average mortgage loan be $142 thousand? In which year will that be?

7. Solve for y: $2x + 4y = 8$.

8. Solve for P: $L = \dfrac{P - 2W}{2}$.

In Exercises 9–11, use an equation to solve each problem.

9. The product of 5 and a number, decreased by 9, is 310. What is the number?

10. At the time they took office, Ronald Reagan and James Buchanan were among the oldest U.S. presidents. Reagan was 4 years older than Buchanan. The sum of their ages was 134. Determine Reagan's age and Buchanan's age at the time each man took office.

11. A long distance telephone plan has a monthly fee of $15.00 and a rate of $0.05 per minute. How many minutes can you chat long distance in a month for a total cost, including the $15.00, of $45.00?

12. In a class, there are 15 men and 10 women. Find the ratio of the number of women to the number of students in the class. First express the ratio as a fraction reduced to lowest terms. Then rewrite the ratio using a second method.

In Exercises 13 and 14, solve each proportion.

13. $\dfrac{5}{8} = \dfrac{x}{12}$

14. $\dfrac{x + 5}{8} = \dfrac{x + 2}{5}$

15. Park rangers catch, tag, and release 200 tule elk back into a wildlife refuge. Two weeks later they observe a sample of 150 elk, of which 5 are tagged. Assuming that the ratio of tagged elk in the sample holds for all elk in the refuge, how many elk are there in the park?

16. The pressure of water on an object below the surface varies directly as its distance below the surface. If a submarine experiences a pressure of 25 pounds per square inch 60 feet below the surface, how much pressure will it experience 330 feet below the surface?

17. The amount of current flowing in an electrical circuit varies inversely as the resistance in the circuit. When the resistance in a particular circuit is 5 ohms, the current is 42 amperes. What is the current when the resistance is 4 ohms?

Solve the inequalities in Exercises 18–19 and graph the solution sets.

18. $6 - 9x \geq 33$

19. $4x - 2 > 2(x + 6)$

20. A student has grades on three examinations of 76, 80, and 72. What must the student earn on a fourth examination in order to have an average of at least 80?

21. Use FOIL to find this product: $(2x - 5)(3x + 4)$.

22. Factor: $2x^2 - 9x + 10$.

23. Solve by factoring: $x^2 + 5x = 36$.

24. Solve using the quadratic formula: $2x^2 + 4x = -1$.

Algebra: Graphs, Functions, and Linear Systems

We have seen that the language of algebra provides a compact symbolic description of your world. There is even a formula that models the number of car accidents as a function of a driver's age. (We'll have lots to say about functions in this chapter.) Do you have any idea about the age at which drivers get into the least number of car accidents? The answer to this question will unfold as we study a technique for making pictures of formulas.

In this chapter, we view formulas as functions and see how to represent them as pictures we call graphs. Functions and their graphs provide powerful tools for describing your world and solving some of its problems.

It's been one of those days! Traffic is really backed up on the highway. Finally, you see the source of the traffic jam—a minor fender-bender. Still stuck in traffic, you notice that the driver appears to be quite young. This might seem like a strange observation. After all, what does a driver's age have to do with his or her chances of getting into an accident?

This problem appears as Example 8 in Section 7.3.

| SECTION 7.1 | GRAPHING AND FUNCTIONS |

OBJECTIVES

1. Plot ordered pairs in the rectangular coordinate system.
2. Graph equations.
3. Use $f(x)$ notation.
4. Graph functions.
5. Use the vertical line test.
6. Obtain information about a function from its graph.

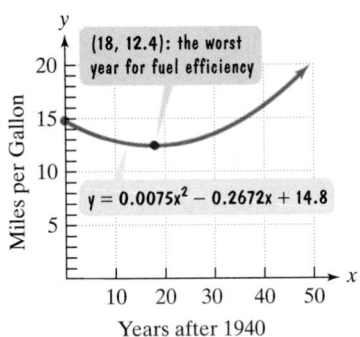

FIGURE 7.1 A picture of a formula

A picture, as they say, is worth a thousand words. Have you seen pictures of gas-guzzling cars from the 1950s, with their huge fins and overstated designs? The worst year for automobile fuel efficiency was 1958, when U.S. cars averaged a dismal 12.4 miles per gallon.

There is a formula that models fuel efficiency of U.S. cars over time. The formula is

$$y = 0.0075x^2 - 0.2672x + 14.8.$$

The variable x represents the number of years after 1940. The variable y represents the average number of miles per gallon for U.S. automobiles. Looking at the formula does not make it obvious that 1958, 18 years after 1940, was the worst year for fuel efficiency. However, if we could somehow make a picture of the formula, such as the one shown in Figure 7.1, the lowest point on the picture would reveal approximately 1958 as the year in which gas-guzzling cars averaged less than 13 miles per gallon. The shape of the graph also shows decreasing fuel efficiency from 1940 through 1958 and increasing fuel efficiency after 1958. In this section, we will be making pictures of formulas. We can use these pictures to visualize the behavior of the variables in the formulas.

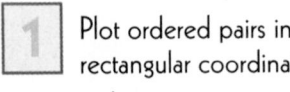

Plot ordered pairs in the rectangular coordinate system.

Points and Ordered Pairs

Making a picture of an equation with two variables, such as the formula for automobile fuel efficiency over time, is accomplished using the **rectangular coordinate system**. The rectangular coordinate system is based on the work of the French mathematician René Descartes (1596–1650). It is also called the **Cartesian coordinate system**.

Descartes used two number lines that intersect at right angles at their zero points, as shown in Figure 7.2. The horizontal number line is the **x-axis**. The vertical number line is the **y-axis**. The point of intersection of these axes is their zero points, called the **origin**. Positive numbers are shown to the right and above the origin. Negative numbers are shown to the left and below the origin. The axes divide the plane into four quarters, called **quadrants**. The points located on the axes are not in any quadrant.

Each point in the rectangular coordinate system corresponds to an **ordered pair** of real numbers, (x, y). Examples of such pairs are $(4, 2)$ and $(-5, -3)$. The first number in each pair, called the **x-coordinate**, denotes the distance and direction from the origin along the x-axis. The second number, called the **y-coordinate**, denotes vertical distance and direction along a line parallel to the y-axis or along the y-axis itself.

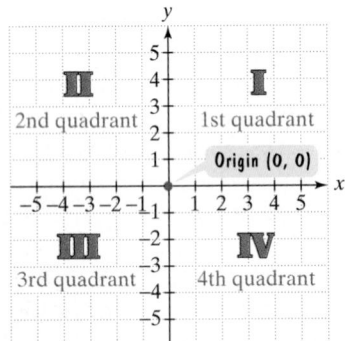

FIGURE 7.2 The rectangular coordinate system

Figure 7.3 shows how we **plot**, or locate, the points corresponding to the ordered pairs $(4, 2)$ and $(-5, -3)$. We plot $(4, 2)$ by going 4 units from 0 to the right along the x-axis. Then we go 2 units up parallel to the y-axis. We plot $(-5, -3)$ by going 5 units from 0 to the left along the x-axis and 3 units down parallel to the y-axis. The phrase "the point corresponding to the ordered pair $(-5, -3)$" is often abbreviated as "the point $(-5, -3)$."

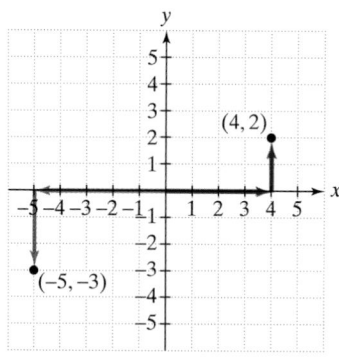

FIGURE 7.3 Plotting $(4, 2)$ and $(-5, -3)$

EXAMPLE 1 | PLOTTING POINTS IN THE RECTANGULAR COORDINATE SYSTEM

Plot the points: $A(-3, 5)$, $B(2, -4)$, $C(5, 0)$, $D(-5, -3)$, $E(0, 4)$, and $F(0, 0)$.

Solution   See Figure 7.4. We move from the origin and plot the points in the following way:

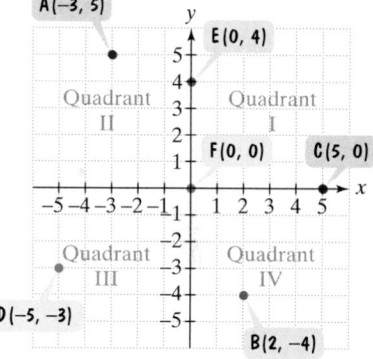

$A(-3, 5)$: 3 units left, 5 units up

$B(2, -4)$: 2 units right, 4 units down

$C(5, 0)$: 5 units right, 0 units up or down

$D(-5, -3)$: 5 units left, 3 units down

$E(0, 4)$: 0 units right or left, 4 units up

$F(0, 0)$: 0 units right or left, 0 units up or down.

FIGURE 7.4 Plotting points

 Check 1 Point

Plot the points:

$$A(-2, 4), B(4, -2), C(-3, 0), \text{ and } D(0, -3).$$

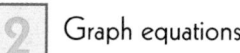 Graph equations.

Graphing Equations

The rectangular coordinate system allows us to visualize relationships between two variables by connecting any equation in two variables with a geometric figure. Consider, for example, the following equation in two variables: $y = x^2 - 4$. A **solution** to this equation is an ordered pair of real numbers with the following property: When the x-coordinate is substituted for x and the y-coordinate is substituted for y in the equation, we obtain a true statement.

We can generate as many ordered-pair solutions as desired to $y = x^2 - 4$ by substituting numbers for x and then finding the values for y. For example, if we let $x = 3$, then $y = 3^2 - 4 = 9 - 4 = 5$. The ordered pair $(3, 5)$ is a solution to the equation $y = x^2 - 4$. We also say that $(3, 5)$ **satisfies** the equation. The **graph of the equation** is the set of all points whose coordinates satisfy the equation.

One method for graphing an equation such as $y = x^2 - 4$ is the **point-plotting method**. First, we find several ordered pairs that are solutions to the equation.

Next, we plot these ordered pairs as points in the rectangular coordinate system. Finally, we connect the points with a smooth curve or line. This often gives us a picture of all ordered pairs that satisfy the equation.

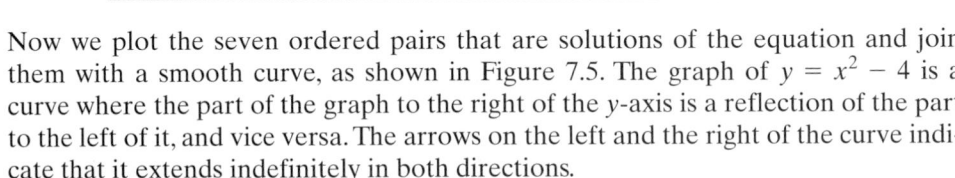

Graph $y = x^2 - 4$. Select integers for x, starting with -3 and ending with $3 (-3 \leq x \leq 3)$.

Solution For each value of x, we find the corresponding value for y.

x	$y = x^2 - 4$	Ordered Pair (x, y)
-3	$y = (-3)^2 - 4 = 9 - 4 = 5$	$(-3, 5)$
-2	$y = (-2)^2 - 4 = 4 - 4 = 0$	$(-2, 0)$
-1	$y = (-1)^2 - 4 = 1 - 4 = -3$	$(-1, -3)$
0	$y = 0^2 - 4 = 0 - 4 = -4$	$(0, -4)$
1	$y = 1^2 - 4 = 1 - 4 = -3$	$(1, -3)$
2	$y = 2^2 - 4 = 4 - 4 = 0$	$(2, 0)$
3	$y = 3^2 - 4 = 9 - 4 = 5$	$(3, 5)$

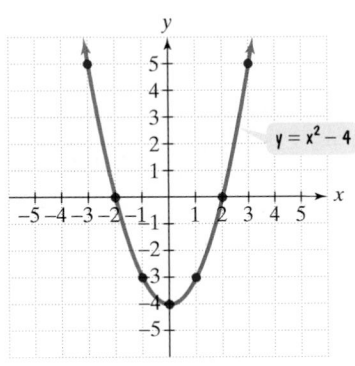

FIGURE 7.5 The graph of $y = x^2 - 4$

Now we plot the seven ordered pairs that are solutions of the equation and join them with a smooth curve, as shown in Figure 7.5. The graph of $y = x^2 - 4$ is a curve where the part of the graph to the right of the y-axis is a reflection of the part to the left of it, and vice versa. The arrows on the left and the right of the curve indicate that it extends indefinitely in both directions.

Check Point 2 Graph $y = x^2 - 1$. Select integers for x, $-3 \leq x \leq 3$.

Part of the beauty of the rectangular coordinate system is that it allows us to "see" formulas and visualize the solution to a problem. This idea is demonstrated in Example 3.

EXAMPLE 3 AN APPLICATION USING THE RECTANGULAR COORDINATE SYSTEM

The toll to a bridge costs $2.50. Commuters have the option of purchasing a monthly coupon book for $21.00. With the coupon book, the toll is reduced to $1.00. The monthly cost, y, of using the bridge x times can be described by the following formulas:

Without the coupon book:

$$y = 2.5x$$ The monthly cost, y, is $2.50 times the number of times, x, that the bridge is used.

With the coupon book:

$$y = 21 + 1 \cdot x$$ The monthly cost, y, is $21 for the book plus $1 times
$$y = 21 + x.$$ the number of times, x, that the bridge is used.

a. Let $x = 0, 2, 4, 10, 12, 14,$ and 16. Make a table of values showing seven solutions for each of the equations.

b. Graph the equations in the same rectangular coordinate system.

c. What are the coordinates of the intersection point for the two graphs? Interpret the coordinates in practical terms.

Solution

a. A table of values showing seven solutions for each equation follows. For each value of x, we find the corresponding value for y.

Without the Coupon Book

x	$y = 2.5x$	(x, y)
0	$y = 2.5(0) = 0$	$(0, 0)$
2	$y = 2.5(2) = 5$	$(2, 5)$
4	$y = 2.5(4) = 10$	$(4, 10)$
10	$y = 2.5(10) = 25$	$(10, 25)$
12	$y = 2.5(12) = 30$	$(12, 30)$
14	$y = 2.5(14) = 35$	$(14, 35)$
16	$y = 2.5(16) = 40$	$(16, 40)$

With the Coupon Book

x	$y = 21 + x$	(x, y)
0	$y = 21 + 0 = 21$	$(0, 21)$
2	$y = 21 + 2 = 23$	$(2, 23)$
4	$y = 21 + 4 = 25$	$(4, 25)$
10	$y = 21 + 10 = 31$	$(10, 31)$
12	$y = 21 + 12 = 33$	$(12, 33)$
14	$y = 21 + 14 = 35$	$(14, 35)$
16	$y = 21 + 16 = 37$	$(16, 37)$

b. Now we are ready to graph the two equations. Because the x- and y-coordinates are nonnegative, it is only necessary to use the origin, the positive portions of the x- and y-axes, and the first quadrant of the rectangular coordinate system. The x-coordinates begin at 0 and end at 16. We will let each tick-mark on the x-axis represent two units. However, the y-coordinates begin at 0 and get as large as 40 in the equation that gives monthly cost without the coupon book. So that our y-axis does not get too long, we will let each tick mark on the y-axis represent five units. Using this setup and the two tables of values, we construct the graphs of $y = 2.5x$ and $y = 21 + x$, shown in Figure 7.6.

c. The graphs intersect at $(14, 35)$. This means that if the bridge is used 14 times in a month, the total monthly cost without the coupon book is the same as the total monthly cost with the coupon book, namely $35.

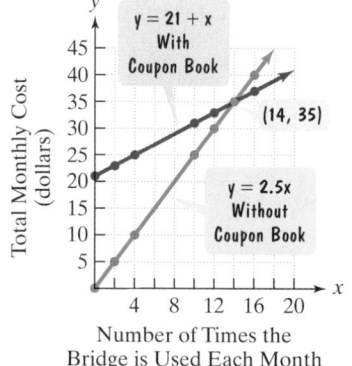

FIGURE 7.6 Options for a toll

In Figure 7.6, look at the two graphs to the right of the intersection point $(14, 35)$. The red graph of $y = 21 + x$ lies below the blue graph of $y = 2.5x$. This means that if the bridge is used more than 14 times in a month ($x > 14$), the (red) monthly cost, y, with the coupon book is cheaper than the (blue) monthly cost, y, without the coupon book.

Check Point 3   The toll to a bridge costs $2.00. If you use the bridge x times in a month, the monthly cost, y, is $y = 2x$. With a $10 coupon book, the toll is reduced to $1.00. The monthly cost, y, of using the bridge x times in a month with the coupon book is $y = 10 + x$.

a. Let $x = 0, 2, 4, 6, 8, 10$, and 12. Make tables of values showing seven solutions of $y = 2x$ and seven solutions of $y = 10 + x$.

b. Graph the equations in the same rectangular coordinate system.

c. What are the coordinates of the intersection point for the two graphs? Interpret the coordinates in practical terms.

3 Use $f(x)$ notation.

Functions

Reconsider one of the equations from Example 3, $y = 2.5x$. Recall that this equation describes the monthly cost, y, of using the bridge x times, with a toll cost of $2.50 each time the bridge is used. The monthly cost, y, depends on the number of times the bridge is used, x. For each value of x, there is one and only one value of y. If an equation in two variables (x and y) yields precisely one value of y for each value of x, we say that y is a **function** of x.

The notation $y = f(x)$ indicates that the variable y is a function of x. The notation $f(x)$ is read "f of x."

For example, the formula for the cost of the bridge

$$y = 2.5x$$

can be expressed in function notation as

$$f(x) = 2.5x.$$

We read this as "f of x is equal to $2.5x$." If, say, x equals 10 (meaning that the bridge is used 10 times), we can find the corresponding value of y (monthly cost) using the equation $f(x) = 2.5x$.

$$f(x) = 2.5x$$
$$f(10) = 2.5(10) \quad \text{To find f(10), read "f of 10," replace x with 10.}$$
$$= 25$$

Because $f(10) = 25$ (f of 10 equals 25), this means that if the bridge is used 10 times in a month, the total monthly cost is $25.

Table 7.1 compares our previous notation with the new notation of functions.

TABLE 7.1 FUNCTION NOTATION

"y Equals" Notation	"$f(x)$ Equals" Notation
$y = 2.5x$	$f(x) = 2.5x$
If $x = 10$, $y = 2.5(10) = 25$.	$f(10) = 2.5(10) = 25$ f of 10 equals 25.

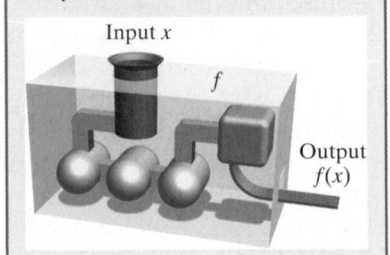
We now return to our formula for fuel efficiency of U.S. cars over time.

EXAMPLE 4   USING FUNCTION NOTATION

The formula $y = 0.0075x^2 - 0.2672x + 14.8$ models the average number of miles per gallon of U.S. automobiles, y, as a function of time, x, in which x is the number of years after 1940.

a. Express the equation in function notation.

b. Find and interpret $f(18)$.

Solution

a. We can express $y = 0.0075x^2 - 0.2672x + 14.8$ in function notation by replacing y with $f(x)$. The formula in function notation is

$$f(x) = 0.0075x^2 - 0.2672x + 14.8.$$

b. We can find $f(18)$, f of 18, by replacing each occurrence of x with 18. The arithmetic gets somewhat "messy," so it is probably a good idea to use a calculator.

$f(x) = 0.0075x^2 - 0.2672x + 14.8$ This is the given formula in function notation.

$f(18) = 0.0075(18)^2 - 0.2672(18) + 14.8$ Replace each x with 18.

$= 0.0075(324) - 0.2672(18) + 14.8$ Use the order of operations, evaluating the exponential expression.

$= 2.43 - 4.8096 + 14.8$ Perform the multiplications.

$= 12.4204$ Subtract and add as indicated.

We see that $f(18) = 12.4204$—that is, f of 18 is 12.4204. Because 18 represents the number of years after 1940, this means that in 1958, U.S. automobiles averaged about 12.4 miles per gallon of gasoline. Figure 7.1 on page 336 shows that $(18, 12.4)$ is the lowest point on the graph of the function. Thus, automobile fuel efficiency was at its very worst in 1958, a dismal 12.4 miles per gallon.

Check 4 Point   The formula $y = -0.05x^2 + 4.2x - 26$ models the percentage of coffee drinkers, y, who are x years old who become irritable if they do not have coffee at their regular time.

    a. Express the equation in function notation.

    b. Find and interpret $f(20)$.

  Graph functions.

Graphing Functions

The **graph of a function** is the graph of its ordered pairs. In our next example, we will graph two functions. It would be awkward to call both of them f. We will call one function f and the other g. These are the letters most frequently used to name functions.

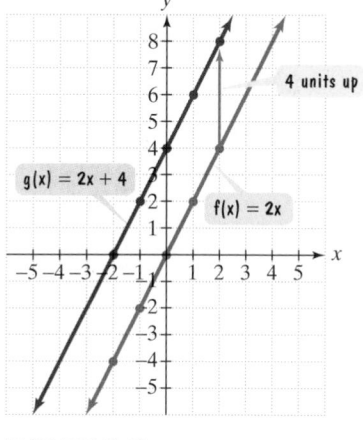

FIGURE 7.7

EXAMPLE 5   GRAPHING FUNCTIONS

Graph the functions $f(x) = 2x$ and $g(x) = 2x + 4$ in the same rectangular coordinate system. Select integers for x, $-2 \le x \le 2$.

Solution   For each function, we use the suggested values for x to create a table of some of the coordinates. These tables are shown below. Now, we plot the five points in each table and connect them, as shown in Figure 7.7. The graph of each function is a straight line. Do you see a relationship between the two graphs? The graph of g is the graph of f shifted vertically up 4 units.

x	$f(x) = 2x$	(x, y) or $(x, f(x))$
-2	$f(-2) = 2(-2) = -4$	$(-2, -4)$
-1	$f(-1) = 2(-1) = -2$	$(-1, -2)$
0	$f(0) = 2 \cdot 0 = 0$	$(0, 0)$
1	$f(1) = 2 \cdot 1 = 2$	$(1, 2)$
2	$f(2) = 2 \cdot 2 = 4$	$(2, 4)$

x	$g(x) = 2x + 4$	(x, y) or $(x, g(x))$
-2	$g(-2) = 2(-2) + 4 = 0$	$(-2, 0)$
-1	$g(-1) = 2(-1) + 4 = 2$	$(-1, 2)$
0	$g(0) = 2 \cdot 0 + 4 = 4$	$(0, 4)$
1	$g(1) = 2 \cdot 1 + 4 = 6$	$(1, 6)$
2	$g(2) = 2 \cdot 2 + 4 = 8$	$(2, 8)$

Choose x.   Compute f(x) by evaluating f at x.   Form the pair.   Choose x.   Compute g(x) by evaluating g at x.   Form the pair.

Check 5 Point   Graph the functions $f(x) = 2x$ and $g(x) = 2x - 3$ in the same rectangular coordinate system. Select integers for x, $-2 \le x \le 2$. How is the graph of g related to the graph of f?

5 Use the vertical line test.

TECHNOLOGY

A graphing calculator is a powerful tool that quickly generates the graph of an equation in two variables. Here is the graph of $y = x^2 - 4$ that we drew by hand in Figure 7.5 on page 338.

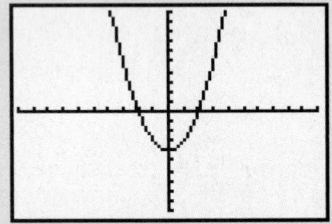

What differences do you notice between this graph and the graph we drew by hand? This graph seems a bit "jittery." Arrows do not appear on the left and right ends of the graph. Furthermore, numbers are not given along the axes. For the graph shown above, the *x*-axis extends from −10 to 10 and the *y*-axis also extends from −10 to 10. The distance represented by each consecutive tick mark is one unit. We say that the **viewing rectangle**, is [−10, 10, 1] by [−10, 10, 1].

To graph an equation in *x* and *y* using a graphing calculator, enter the equation and specify the size of the viewing rectangle. The size of the viewing rectangle sets minimum and maximum values for both the *x*- and *y*-axes. Enter these values, as well as the values between consecutive tick marks, on the respective axes. The [−10, 10, 1] by [−10, 10, 1] viewing rectangle used above is called the **standard viewing rectangle**.

The Vertical Line Test

Not every graph in the rectangular coordinate system is the graph of a function. The definition of a function specifies that no value of *x* can be paired with two or more different values of *y*. Consequently, if a graph contains two or more different points with the same first coordinate, the graph cannot represent a function. This is illustrated in Figure 7.8. Observe that points sharing a common first coordinate are vertically above or below each other.

This observation is the basis of a useful test for determining whether a graph defines *y* as a function of *x*. The test is called the **vertical line test**.

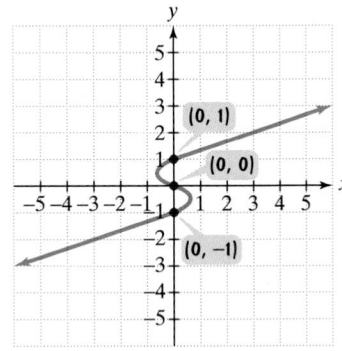

FIGURE 7.8 *y* is not a function of *x* because 0 is paired with three values of *y*, namely, 1, 0, and −1.

THE VERTICAL LINE TEST FOR FUNCTIONS

If any vertical line intersects a graph in more than one point, the graph does not define *y* as a function of *x*.

EXAMPLE 6 | USING THE VERTICAL LINE TEST

Use the vertical line test to identify graphs in which *y* is a function of *x*.

a. 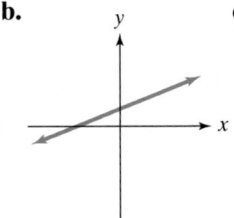   **b.**   **c.** 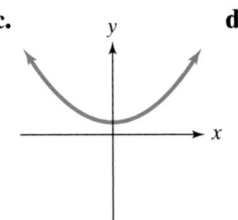   **d.**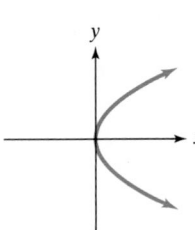

Solution *y* is a function of *x* for the graphs in (b) and (c).

a. 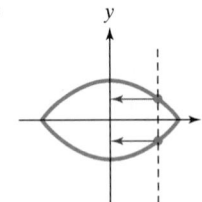   **b.** 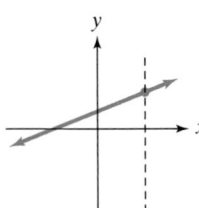   **c.** 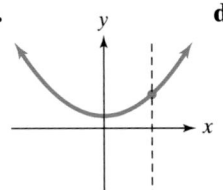   **d.**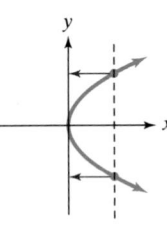

y **is not a function** of *x*. Two values of *y* correspond to an *x*-value.   *y* **is a function** of *x*.   *y* **is a function** of *x*.   *y* **is not a function** of *x*. Two values of *y* correspond to an *x*-value.

Check Point 6   Use the vertical line test to identify graphs in which *y* is a function of *x*.

a. 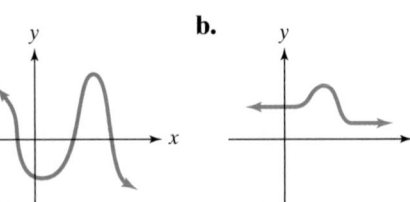   **b.**   **c.**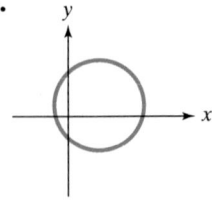

6	Obtain information about a function from its graph.

Obtaining Information from Graphs

Example 7 illustrates how to obtain information about a function from its graph.

EXAMPLE 7   ANALYZING THE GRAPH OF A FUNCTION

The function

$$f(x) = -0.016x^2 + 0.93x + 8.5$$

models the average number of paid vacation days each year, $f(x)$, for full-time workers at medium to large U.S. companies after x years. The graph of f is shown for $x \geq 1$ in Figure 7.9.

a. Explain why f represents the graph of a function.

b. Use the graph to find a reasonable estimate of $f(5)$.

c. Describe the general trend shown by the graph.

Solution

a. No vertical line intersects the graph of f more than once. By the vertical line test, f represents the graph of a function.

b. To find $f(5)$, or f of 5, we locate 5 on the x-axis. The figure shows the point on the graph of f for which 5 is the first coordinate. From this point, we look to the y-axis to find the corresponding y-coordinate. A reasonable estimate of the y-coordinate is 13. Thus, $f(5) \approx 13$. After 5 years, a worker can expect approximately 13 paid vacation days.

c. The graph of f is rising from left to right. This shows that paid vacation days increase as time with the company increases. However, the rate of increase is slowing down as the graph moves to the right. This means that the increase in paid vacation days takes place more slowly the longer an employee is with the company.

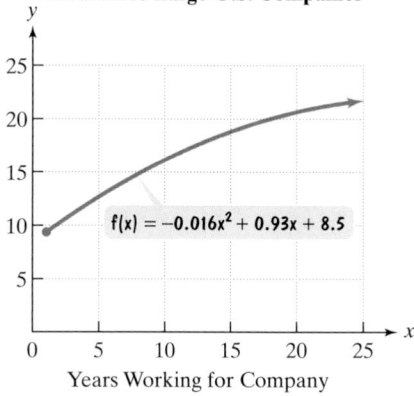

Average Number of Paid Vacation Days for Full-Time Workers at Medium to Large U.S. Companies

$f(x) = -0.016x^2 + 0.93x + 8.5$

Years Working for Company

Source: Bureau of Labor Statistics

FIGURE 7.9

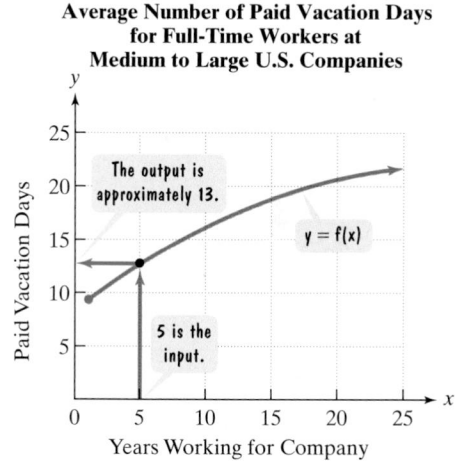

Average Number of Paid Vacation Days for Full-Time Workers at Medium to Large U.S. Companies

The output is approximately 13.

$y = f(x)$

5 is the input.

Years Working for Company

Check 7 Point Use the graph of f in Figure 7.9 to find a reasonable estimate of $f(10)$.

Exercise Set 7.1

Practice Exercises

In Exercises 1–20, plot the given point in a rectangular coordinate system.

1. $(1, 4)$ **2.** $(2, 5)$ **3.** $(-2, 3)$ **4.** $(-1, 4)$

5. $(-3, -5)$ **6.** $(-4, -2)$ **7.** $(4, -1)$ **8.** $(3, -2)$

9. $(-4, 0)$ **10.** $(-5, 0)$ **11.** $(0, -3)$ **12.** $(0, -4)$

13. $(0, 0)$ **14.** $\left(-3, -1\frac{1}{2}\right)$ **15.** $\left(-2, -3\frac{1}{2}\right)$ **16.** $(-5, -2.5)$

17. $(3.5, 4.5)$ **18.** $(2.5, 3.5)$

19. $(1.25, -3.25)$ **20.** $(2.25, -4.25)$

Graph each equation in Exercises 21–30. Select integers for x, $-3 \leq x \leq 3$.

21. $y = x^2 - 2$ **22.** $y = x^2 + 2$

23. $y = x - 2$ **24.** $y = x + 2$

25. $y = 2x + 1$ **26.** $y = 2x - 4$

27. $y = -\frac{1}{2}x$ **28.** $y = -\frac{1}{2}x + 2$

29. $y = x^3$ **30.** $y = x^3 - 1$

In Exercises 31–40, find f of each given value of x.

31. $f(x) = x - 4$ **a.** $f(8)$ **b.** $f(1)$

32. $f(x) = x - 6$ **a.** $f(9)$ **b.** $f(2)$

33. $f(x) = 3x - 2$ **a.** $f(7)$ **b.** $f(0)$

34. $f(x) = 4x - 3$ **a.** $f(7)$ **b.** $f(0)$

35. $f(x) = x^2 + 1$ **a.** $f(2)$ **b.** $f(-2)$

36. $f(x) = x^2 + 4$ **a.** $f(3)$ **b.** $f(-3)$

37. $f(x) = 3x^2 + 5$ **a.** $f(4)$ **b.** $f(-1)$

38. $f(x) = 2x^2 - 4$ **a.** $f(5)$ **b.** $f(-1)$

39. $f(x) = 2x^2 + 3x - 1$ **a.** $f(3)$ **b.** $f(-4)$

40. $f(x) = 3x^2 + 4x - 2$ **a.** $f(2)$ **b.** $f(-1)$.

In Exercises 41–48, evaluate $f(x)$ for the given values for x. Then use the ordered pairs $(x, f(x))$ from your table to graph the function.

41. $f(x) = x^2 - 1$

x	$f(x) = x^2 - 1$
-2	
-1	
0	
1	
2	

42. $f(x) = x^2 + 1$

x	$f(x) = x^2 + 1$
-2	
-1	
0	
1	
2	

43. $f(x) = x - 1$

x	$f(x) = x - 1$
-2	
-1	
0	
1	
2	

44. $f(x) = x + 1$

x	$f(x) = x + 1$
-2	
-1	
0	
1	
2	

45. $f(x) = (x - 2)^2$

x	$f(x) = (x - 2)^2$
0	
1	
2	
3	
4	

46. $f(x) = (x + 1)^2$

x	$f(x) = (x + 1)^2$
-3	
-2	
-1	
0	
1	

47. $f(x) = x^3 + 1$

x	$f(x) = x^3 + 1$
-3	
-2	
-1	
0	
1	

48. $f(x) = (x + 1)^3$

x	$f(x) = (x + 1)^3$
-3	
-2	
-1	
0	
1	

For Exercises 49–56, use the vertical line test to identify graphs in which y is a function of x.

49.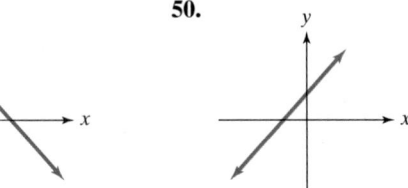

50.

51.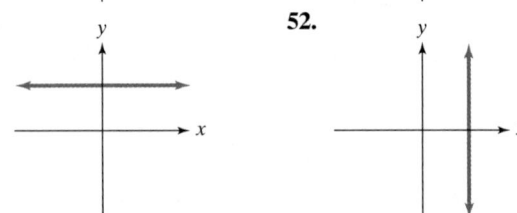

52.

53.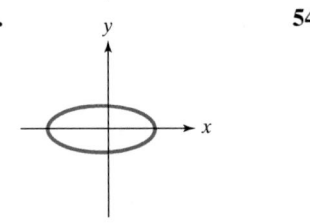

54.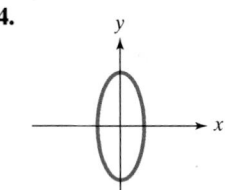

55.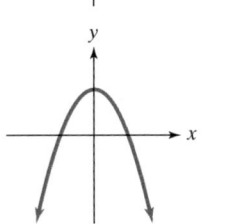

56.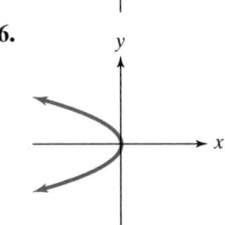

Application Exercises

A football is thrown by a quarterback to a receiver. The points in the figure show the height of the football, in feet, above the ground in terms of its distance, in yards, from the quarterback. Use this information to solve Exercises 57–62.

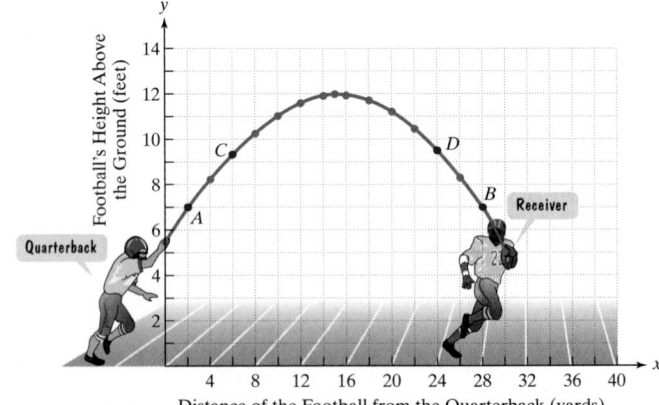

57. Find the coordinates of point A. Then interpret the coordinates in terms of the information given.

58. Find the coordinates of point B. Then interpret the coordinates in terms of the information given.

59. Estimate the coordinates of point C.

60. Estimate the coordinates of point D.

61. What is the football's maximum height? What is its distance from the quarterback when it reaches its maximum height?

62. What is the football's height when it is caught by the receiver? What is the receiver's distance from the quarterback when he catches the football?

The function $f(x) = 0.76x + 171.4$ models the cholesterol level of an American man as a function of his age, x, in years. Use the function to solve Exercises 63–64.

63. Find and interpret $f(20)$.   **64.** Find and interpret $f(50)$.

The Food Stamp Program is America's first line of defense against hunger for millions of families. Over half of all participants are children; one out of six is a low-income older adult. Exercises 65–68 involve the number of participants in the program from 1990 through 2000.
    The function

$$f(x) = -\frac{1}{2}x^2 + 4x + 19$$

models the number of people, $f(x)$, in millions, receiving food stamps x years after 1990. Use the function to solve Exercises 65–66.

65. Find and interpret $f(4)$.

66. Find and interpret $f(0)$ and $f(8)$.

The graph of the function in Exercises 65–66 is shown. Use the graph to solve Exercises 67–68.

Number of People Receiving Food Stamps

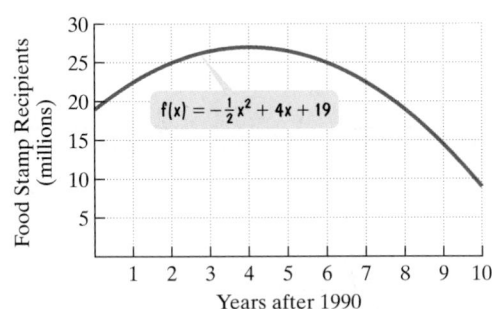

Years after 1990

Source: New York Times

67. Identify your solution in Exercise 65 as a point on the graph. Describe what is significant about this point.

68. Identify your solutions in Exercise 66 as two points on the graph. Then describe the trend shown by the graph.

The figure shows the graph of the divorce rate for the United States, f, measured in divorces per thousand persons ages 15 and older, as a function of time x, where x is a year from 1890 through 2001. Use the graph to solve Exercises 69–74.

The Divorce Rate for the U.S.

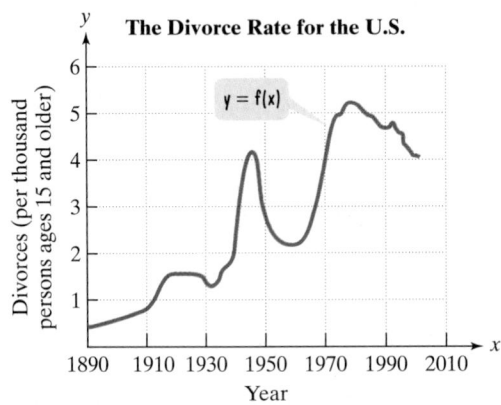

Year

Source: U.S. Department of Health and Human Services

69. Use the graph to find a reasonable estimate of $f(1970)$. What does this mean in terms of the variables in this situation?

70. Use the graph to find a reasonable estimate of $f(2001)$. What does this mean in terms of the variables in this situation?

71. In which year for $1940 \le x \le 1970$ did the divorce rate in the U.S. reach a maximum? What is a reasonable estimate of the divorce rate for that year?

72. In which year for $1970 \le x \le 2001$ did the divorce rate in the U.S. reach a maximum? What is a reasonable estimate of the divorce rate for that year?

73. Explain why f represents the graph of a function.

74. Describe the general trend shown by the graph.

75. Although the level of air pollution varies from day to day and from hour to hour, during the summer the level of air pollution is a function of the time of day. The function

$$f(x) = 0.1x^2 - 0.4x + 0.6$$

models the level of air pollution, $f(x)$, in parts per million (ppm), x hours after 9 A.M.

a. Construct a table of values using integers for x, $0 \le x \le 5$, and graph the function for $0 \le x \le 5$.

b. Researchers have determined that a level of 0.3 ppm of pollutants in the air can be hazardous to your health. Based on the graph, at which time of day should runners exercise to avoid unsafe air?

Writing in Mathematics

76. What is the rectangular coordinate system?

77. Explain how to plot a point in the rectangular coordinate system. Give an example with your explanation.

78. Explain why $(5, -2)$ and $(-2, 5)$ do not represent the same ordered pair.

79. Explain how to graph an equation in the rectangular coordinate system.

80. What is a function?

81. Explain how the vertical line test is used to determine whether a graph represents a function.

Critical Thinking Exercises

82. Researchers at Yale University have suggested that levels of passion and commitment are functions of time. Based on the shapes of the graphs shown, which do you think depicts passion and which represents commitment? Explain how you arrived at your answer.

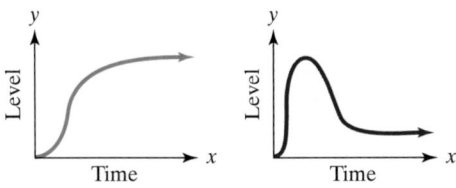

83. Consider a function defined by

$$f(1) = 1$$
$$f(2) = 1 - 2 = -1$$
$$f(3) = 1 - 2 + 3 = 2$$
$$f(4) = 1 - 2 + 3 - 4 = -2$$
$$f(5) = 1 - 2 + 3 - 4 + 5 = 3$$
$$f(6) = 1 - 2 + 3 - 4 + 5 - 6 = -3.$$

a. What is the formula for $f(x)$ if x is even?

b. What is the formula for $f(x)$ if x is odd?

c. Find $f(20) + f(40) + f(65)$.

84. Reexamine Figure 7.6 on page 339. Explain why it is more realistic to graph each equation using only the red and blue points shown rather than connecting them, respectively, with lines.

Technology Exercises

85. Use a graphing calculator to verify the graphs that you drew by hand in Exercises 41–48.

86. The function

$$f(x) = -0.00002x^3 + 0.008x^2 - 0.3x + 6.95$$

models the number of annual physician visits, $f(x)$, by a person of age x. Graph the function in a $[0, 100, 5]$ by $[0, 40, 2]$ viewing rectangle. What does the shape of the graph indicate about the relationship between one's age and the number of annual physician visits? Use the ⬛TRACE or minimum function capability to find the coordinates of the minimum point on the graph of the function. What does this mean?

Group Exercise

87. In Exercise 82, passion and commitment are graphed over time. For this activity, you will be creating a graph of a particular experience that involved your feelings of love, anger, sadness, or any other emotion you choose. The horizontal axis should be labeled time and the vertical axis the emotion you are graphing. You will not be using your algebra skills to create your graph: however, you should try to make the graph as precise as possible. You may use negative numbers on the vertical axis, if appropriate. After each group member has created a graph, pool together all of the graphs and study them to see if there are any similarities in the graphs for a particular emotion or for all emotions.

SECTION 7.2   LINEAR FUNCTIONS AND THEIR GRAPHS

OBJECTIVES

1. Use intercepts to graph a linear equation.

2. Calculate slope.

3. Use the slope and y-intercept to graph a line.

4. Graph horizontal or vertical lines.

5. Interpret slope and y-intercept in applied situations.

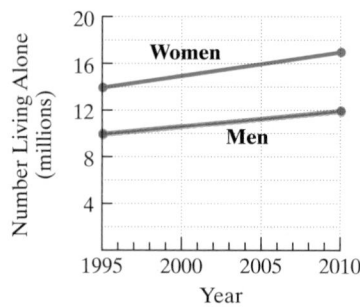

Number of People in the U.S. Living Alone

Source: Forrester Research

FIGURE 7.10

A best guess at the look of our nation in the next decades indicates that the number of men and women living alone will increase each year. Figure 7.10 shows that by 2010, approximately 12 million men and 17 million women will be living alone.

The data points in Figure 7.10 fall along straight lines. The data for the women can be modeled by

$$y = 0.2x + 14$$

where y represents the number of women living alone, in millions, x years after 1995. Using function notation, we can rewrite the equation as

$$f(x) = 0.2x + 14.$$

A function such as this, whose graph is a straight line, is called a **linear function**. In this section, we will study linear functions and their graphs.

① Use intercepts to graph a linear equation.

Graphing Using Intercepts

There is another way that we can write the equation

$$y = 0.2x + 14.$$

We will collect the x- and y-terms on the left side. This is done by subtracting $0.2x$ from both sides:

$$-0.2x + y = 14.$$

All equations of the form $Ax + By = C$ are straight lines when graphed, as long as A and B are not both zero. Such equations are called **linear equations in two variables**. We can quickly obtain the graph for equations in this form by finding the points where the graph crosses the x-axis and the y-axis. The x-coordinate of the point where the graph crosses the x-axis is called the **x-intercept**. The y-coordinate of the point where the graph crosses the y-axis is called the **y-intercept**.

LOCATING INTERCEPTS

To locate the x-intercept, set $y = 0$ and solve the equation for x.

To locate the y-intercept, set $x = 0$ and solve the equation for y.

An equation of the form $Ax + By = C$ can be graphed by finding the x- and y-intercepts, plotting the intercepts, and drawing a straight line through these points. When graphing using intercepts, it is a good idea to use a third point, a checkpoint, before drawing the line. A checkpoint can be obtained by selecting a value for x, other than 0, and finding the corresponding value for y. The checkpoint should lie on the same line as the x- and y-intercepts. If it does not, recheck your work and find the error.

EXAMPLE 1   USING INTERCEPTS TO GRAPH A LINEAR EQUATION

Graph: $3x + 2y = 6$.

Solution

Find the x-intercept by letting $y = 0$ and solving for x.	**Find the y-intercept by letting $x = 0$ and solving for y.**
$3x + 2y = 6$	$3x + 2y = 6$
$3x + 2 \cdot 0 = 6$	$3 \cdot 0 + 2y = 6$
$3x = 6$	$2y = 6$
$x = 2$	$y = 3$

The x-intercept is 2, so the line passes through the point $(2, 0)$. The y-intercept is 3, so the line passes through the point $(0, 3)$.

For our checkpoint, we will let $x = 1$ and find the corresponding value for y.

$3x + 2y = 6$	This is the given equation.
$3 \cdot 1 + 2y = 6$	Substitute 1 for x.
$3 + 2y = 6$	Simplify.
$2y = 3$	Subtract 3 from both sides.
$y = \frac{3}{2}$	Divide both sides by 2.

The checkpoint is the ordered pair $\left(1, \frac{3}{2}\right)$, or $(1, 1.5)$.

The three points in Figure 7.11 lie along the same line. Drawing a line through the three points results in the graph of $3x + 2y = 6$.

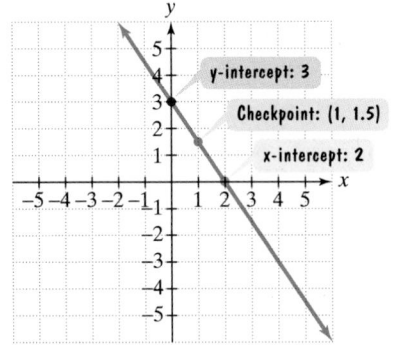

FIGURE 7.11 The graph of $3x + 2y = 6$

Check **1** Point Graph: $2x + 3y = 6$.

 2 Calculate slope.

Slope

Mathematicians have developed a useful measure of the steepness of a line, called the **slope** of the line. Slope compares the vertical change (the **rise**) to the horizontal change (the **run**) when moving from one fixed point to another along the line. To calculate the slope of a line, we use a ratio that compares the change in y (the rise) to the change in x (the run).

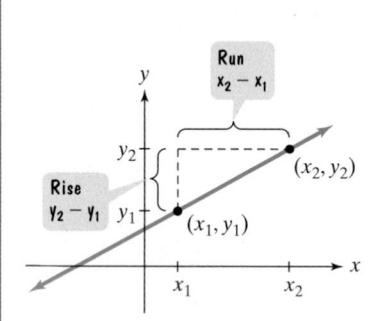

DEFINITION OF SLOPE

The **slope** of the line through the distinct points (x_1, y_1) and (x_2, y_2) is

$$\frac{\text{Change in } y}{\text{Change in } x} = \frac{\text{Rise}}{\text{Run}}$$

$$= \frac{y_2 - y_1}{x_2 - x_1}.$$

where $x_2 - x_1 \neq 0$.

It is common notation to let the letter m represent the slope of a line. The letter m is used because it is the first letter of the French verb *monter*, meaning to rise, or to ascend.

EXAMPLE 2 USING THE DEFINITION OF SLOPE

Find the slope of the line passing through each pair of points:

a. $(-3, -1)$ and $(-2, 4)$ **b.** $(-3, 4)$ and $(2, -2)$.

Solution

a. Let $(x_1, y_1) = (-3, -1)$ and $(x_2, y_2) = (-2, 4)$. We obtain a slope of

$$m = \frac{\text{Change in } y}{\text{Change in } x} = \frac{y_2 - y_1}{x_2 - x_1} = \frac{4 - (-1)}{-2 - (-3)} = \frac{5}{1} = 5.$$

The situation is illustrated in Figure 7.12. The slope of the line is 5, indicating that there is a vertical change, a rise, of 5 units for each horizontal change, a run, of 1 unit. The slope is positive, and the line rises from left to right.

b. Let $(x_1, y_1) = (-3, 4)$ and $(x_2, y_2) = (2, -2)$. The slope of the line shown in Figure 7.13 is computed as follows:

$$m = \frac{-2 - 4}{2 - (-3)} = \frac{-6}{5} = -\frac{6}{5}.$$

The slope of the line is $-\frac{6}{5}$. For every vertical change of -6 units (6 units down), there is a corresponding horizontal change of 5 units. The slope is negative, and the line falls from left to right.

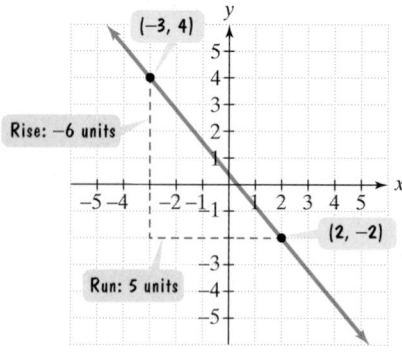

FIGURE 7.12 Visualizing a slope of 5

FIGURE 7.13 Visualizing a slope of $-\frac{6}{5}$

Check **2** Point Find the slope of the line passing through each pair of points:

a. $(-3, 4)$ and $(-4, -2)$

b. $(4, -2)$ and $(-1, 5)$.

STUDY TIP

When computing slope, it makes no difference which point you call (x_1, y_1) and which point you call (x_2, y_2). In Example 2(a), we obtained a slope of 5 by letting $(x_1, y_1) = (-3, -1)$ and $(x_2, y_2) = (-2, 4)$. If we let $(x_1, y_1) = (-2, 4)$ and $(x_2, y_2) = (-3, -1)$, the slope is still 5:

$$m = \frac{y_2 - y_1}{x_2 - x_1} = \frac{-1 - 4}{-3 - (-2)} = \frac{-5}{-1} = 5.$$

However, you should not subtract in one order in the numerator $(y_2 - y_1)$ and then in a different order in the denominator $(x_1 - x_2)$. The slope is *not*

$$\frac{-1 - 4}{-2 - (-3)} = \frac{-5}{1} = -5. \quad \text{Incorrect!}$$

Example 2 illustrates that a line with a positive slope is rising from left to right and a line with a negative slope is falling from left to right. By contrast, a horizontal line neither rises nor falls and has a slope of zero. A vertical line has no horizontal change, so $x_2 - x_1 = 0$ in the formula for slope. Because we cannot divide by zero, the slope of a vertical line is undefined. This discussion is summarized in Table 7.2.

TABLE 7.2 POSSIBILITIES FOR A LINE'S SLOPE

Positive Slope	Negative Slope	Zero Slope	Undefined Slope
Line rises from left to right.	Line falls from left to right.	Line is horizontal.	Line is vertical.

3 Use the slope and y-intercept to graph a line.

The Slope-Intercept Form of a Line

We opened this section with the linear function

$$y = 0.2x + 14.$$

Recall that y represents the number of U.S. women living alone, in millions, x years after 1995. A linear equation such as this, in the form $y = mx + b$, is said to be in **slope-intercept form**.

SLOPE-INTERCEPT FORM OF THE EQUATION OF A LINE

The equation $y = mx + b$ or, equivalently, $f(x) = mx + b$ is called the **slope-intercept form of a line**. The coefficient of x is the slope of the line, and the constant term is its y-intercept.

If a line's equation is written with y isolated on one side, we can use the y-intercept and the slope to obtain its graph.

GRAPHING $y = mx + b$ **USING THE SLOPE AND y-INTERCEPT**

1. Plot the point containing the y-intercept on the y-axis. This is the point $(0, b)$.

2. Obtain a second point using the slope, m. Write m as a fraction, and use rise over run, starting at the point containing the y-intercept, to plot this point.

3. Use a straightedge to draw a line through the two points. Draw arrowheads at the ends of the line to show that the line continues indefinitely in both directions.

EXAMPLE 3 ░ GRAPHING BY USING THE SLOPE AND y-INTERCEPT

Graph the linear function $y = \dfrac{2}{3}x + 2$ by using the slope and y-intercept.

Solution The equation of the linear function is in the form $y = mx + b$. We can find the slope, m, by identifying the coefficient of x. We can find the y-intercept, b, by identifying the constant term.

$$y = \frac{2}{3}x + 2$$

The slope is $\frac{2}{3}$. The y-intercept is 2.

Now that we have identified the slope and the y-intercept, we use the three-step procedure to graph the equation.

STEP 1. Plot the point containing the y-intercept on the y-axis. The y-intercept is 2. We plot $(0, 2)$, shown in Figure 7.14.

STEP 2. Obtain a second point using the slope, m. Write m as a fraction, and use rise over run, starting at the point containing the y-intercept, to plot this point. The slope, $\frac{2}{3}$, is already written as a fraction:

$$m = \frac{2}{3} = \frac{\text{Rise}}{\text{Run}}.$$

We plot the second point on the line by starting at $(0, 2)$, the first point. Based on the slope, we move 2 units *up* (the rise) and 3 units to the *right* (the run). This puts us at a second point on the line, $(3, 4)$, shown in Figure 7.14.

STEP 3. Use a straightedge to draw a line through the two points. The graph of $y = \frac{2}{3}x + 2$ is shown in Figure 7.14.

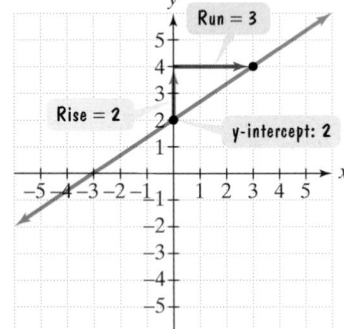

FIGURE 7.14 The graph of $y = \frac{2}{3}x + 2$

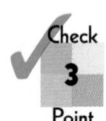

Check Point 3 Graph the linear function $y = \dfrac{3}{5}x + 1$ by using the slope and y-intercept.

Earlier in this section, we considered linear functions whose equations are of the form $Ax + By = C$. We used x- and y-intercepts, as well as checkpoints, to graph these functions. It is also possible to obtain the graphs by using the slope and y-intercept. To do this, begin by solving $Ax + By = C$ for y. This will put the equation in slope-intercept form. Then use the three-step procedure shown in the box above to graph the equation. This is illustrated in Example 4.

EXAMPLE 4 GRAPHING BY USING THE SLOPE AND y-INTERCEPT

Graph the linear function $2x + 5y = 0$ by using the slope and y-intercept.

Solution We put the equation in slope-intercept form by solving for y.

$$2x + 5y = 0 \qquad \text{This is the given equation.}$$

$$2x - 2x + 5y = -2x + 0 \qquad \text{Subtract 2x from both sides.}$$

$$5y = -2x + 0 \qquad \text{Simplify.}$$

$$\frac{5y}{5} = \frac{-2x + 0}{5} \qquad \text{Divide both sides by 5.}$$

$$y = \frac{-2x}{5} + \frac{0}{5} \qquad \text{Divide each term in the numerator by 5.}$$

$$y = -\frac{2}{5}x + 0 \qquad \text{Simplify. Equivalently, } f(x) = -\tfrac{2}{5}x + 0.$$

Now that the equation is in slope-intercept form, we can use the slope and y-intercept to obtain its graph. Examine the slope-intercept form:

$$y = -\frac{2}{5}x + 0.$$

slope: $-\frac{2}{5}$ y-intercept: 0

Note that the slope is $-\frac{2}{5}$ and the y-intercept is 0. Use the y-intercept to plot $(0,0)$ on the y-axis. Then locate a second point by using the slope.

$$m = -\frac{2}{5} = \frac{-2}{5} = \frac{\text{Rise}}{\text{Run}}$$

Because the rise is -2 and the run is 5, move *down* 2 units and to the *right* 5 units, starting at the point $(0, 0)$. This puts us at a second point on the line, $(5, -2)$. The graph of $2x + 5y = 0$ is the line drawn through these points, shown in Figure 7.15.

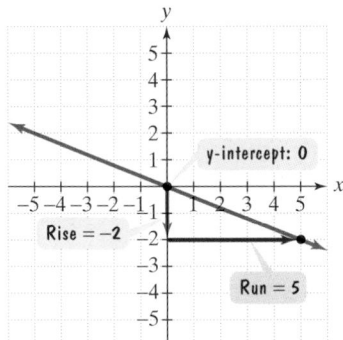

FIGURE 7.15 The graph of $2x + 5y = 0$, or $y = -\frac{2}{5}x$

If you try graphing $2x + 5y = 0$ by using intercepts, you will find that the x-intercept is 0 and the y-intercept is 0. This means that the graph passes through the origin. A second point must be found to graph the line. In Example 4, the line's slope gave us the second point.

Check Point 4 Graph the linear function $3x + 4y = 0$ by using the slope and y-intercept.

4 Graph horizontal or vertical lines.

Horizontal and Vertical Lines

If a line is horizontal, its slope is zero: $m = 0$. Thus, the equation $y = mx + b$ becomes $y = b$, where b is the y-intercept. All horizontal lines have equations of the form $y = b$.

EXAMPLE 5 GRAPHING A HORIZONTAL LINE

Graph $y = -4$ in the rectangular coordinate system.

Solution All points on the graph of $y = -4$ have a value of y that is always -4. No matter what the x-coordinate is, the y-coordinate for every point on the line is -4. Let us select three of the possible values for x: -2, 0, and 3. Using these values of x, three of the points on the graph $y = -4$ are $(-2, -4)$, $(0, -4)$, and $(3, -4)$. Plot each of these points. Drawing a line that passes through the three points gives the horizontal line shown in Figure 7.16.

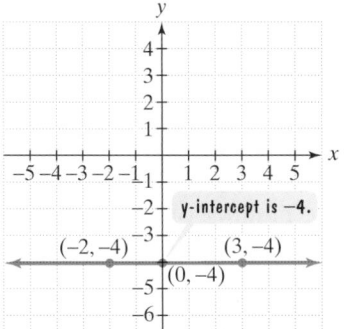

FIGURE 7.16 The graph of $y = -4$

Check
5
Point

Graph $y = 3$ in the rectangular coordinate system.

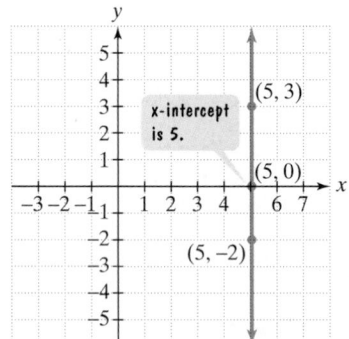

FIGURE 7.17 The graph of $x = 5$

Next, let's see what we can discover about the graph of an equation of the form $x = a$ by looking at an example.

EXAMPLE 6 | GRAPHING A VERTICAL LINE

Graph $x = 5$ in the rectangular coordinate system.

Solution All points on the graph of $x = 5$ have a value of x that is always 5. No matter what the y-coordinate is, the corresponding x-coordinate for every point on the line is 5. Let us select three of the possible values of y: $-2, 0$, and 3. Using these values of y, three of the points on the graph of $x = 5$ are $(5, -2)$, $(5, 0)$, and $(5, 3)$. Plot each of these points. Drawing a line that passes through the three points gives the vertical line shown in Figure 7.17.

HORIZONTAL AND VERTICAL LINES

The graph of $y = b$ or $f(x) = b$ is a horizontal line. The y-intercept is b.

The graph of $x = a$ is a vertical line. The x-intercept is a.

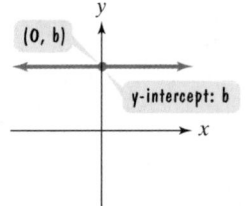

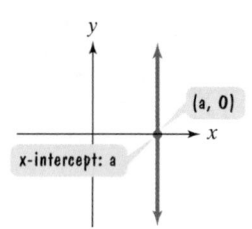

Check
6
Point

Graph $x = -2$ in the rectangular coordinate system.

The graph of a vertical line does not pass the vertical line test. One of the vertical lines that can be drawn will coincide with the given vertical line. This line will certainly intersect the graph more than once; it will intersect it in infinitely many points! Thus, a vertical line does not define y as a function of x. All other lines are graphs of functions.

Applications of Linear Functions

5 Interpret slope and y-intercept in applied situations.

Slope is defined as the ratio of a change in y to a corresponding change in x. Our next example shows how slope can be interpreted as a **rate of change** in an applied situation.

EXAMPLE 7 | SLOPE AS A RATE OF CHANGE

Shown, again, in Figure 7.18, are the line graphs for the number of U.S. men and women living alone, projected through 2010. Find the slope of the line segment for the women. Describe what the slope represents.

Solution We let x represent a year and y the number of women living alone in that year. The two points shown on the line segment for women have the following coordinates:

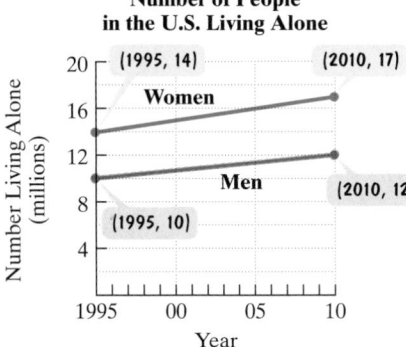

Number of People in the U.S. Living Alone

Source: Forrester Research

FIGURE 7.18

(1995, 14)   and   (2010, 17).

In 1995, 14 million U.S. women lived alone.

In 2010, 17 million U.S. women are projected to live alone.

Now we compute the slope:

$$m = \frac{\text{Change in } y}{\text{Change in } x} = \frac{17 - 14}{2010 - 1995}$$

The unit in the numerator is million people.

The unit in the denominator is year.

$$= \frac{3}{15} = \frac{1}{5} = \frac{0.2 \text{ million people}}{\text{year}}.$$

The slope indicates that the number of U.S. women living alone is projected to increase by 0.2 million each year. The rate of change is 0.2 million women per year.

Check Point 7 Use the graph in Figure 7.18 to find the slope of the line segment for the men. Express the slope correct to two decimal places and describe what it represents.

If an equation in slope-intercept form models relationships between variables, then the slope and *y*-intercept have physical interpretations. For the equation $y = mx + b$, the *y*-intercept, *b*, tells us what is happening to *y* when *x* is 0. If *x* represents time, the *y*-intercept describes the value of *y* at the beginning, or when time equals 0. The slope represents the rate of change in *y* per unit change in *x*.

Using these ideas, we can develop a model for the data for women living alone, shown in Figure 7.18. We let *x* = the number of years after 1995. At the beginning of our data, or 0 years after 1995, 14 million women lived alone. Thus, *b* = 14. In Example 7, we found that *m* = 0.2 (rate of change is 0.2 million women per year). An equation of the form $y = mx + b$ that models the data is

$$y = 0.2x + 14$$

where *y* is the number, in millions, of U.S. women living alone *x* years after 1995.

BLITZER BONUS

Railroads and Highways

The steepest part of Mt. Washington Cog Railway in New Hampshire has a 37% grade. This is equivalent to a slope of $\frac{37}{100}$. For every horizontal change of 100 feet, the railroad ascends 37 feet vertically. Engineers denote slope by grade, expressing slope as a percentage.

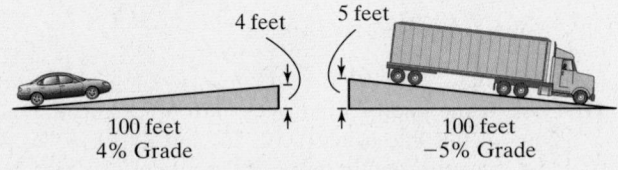

4 feet

5 feet

100 feet
4% Grade

100 feet
−5% Grade

Railroad grades are usually less than 2%, although in the mountains they may go as high as 4%. The grade of the Mt. Washington Cog Railway is phenomenal, making it necessary for locomotives to *push* single cars up its steepest part.

A Mount Washington Cog Railway locomotive pushing a single car up the steepest part of the railroad. The locomotive is about 120 years old.

Exercise Set 7.2

Practice Exercises

In Exercises 1–8, use the x- and y-intercepts to graph each linear equation.

1. $x - y = 3$

2. $x + y = 4$

3. $3x - 4y = 12$

4. $2x - 5y = 10$

5. $2x + y = 6$

6. $x + 3y = 6$

7. $5x = 3y - 15$

8. $3x = 2y + 6$

In Exercises 9–20, calculate the slope of the line passing through the given points. If the slope is undefined, so state. Then indicate whether the line rises, falls, is horizontal, or is vertical.

9. $(2, 6)$ and $(3, 5)$

10. $(4, 2)$ and $(3, 4)$

11. $(-2, 1)$ and $(2, 2)$

12. $(-1, 3)$ and $(2, 4)$

13. $(-2, 4)$ and $(-1, -1)$

14. $(6, -4)$ and $(4, -2)$

15. $(5, 3)$ and $(5, -2)$

16. $(3, -4)$ and $(3, 5)$

17. $(2, 0)$ and $(0, 8)$

18. $(3, 0)$ and $(0, -9)$

19. $(5, 1)$ and $(-2, 1)$

20. $(-2, 3)$ and $(1, 3)$

In Exercises 21–32, graph each linear function using the slope and y-intercept.

21. $y = 2x + 3$

22. $y = 2x + 1$

23. $y = -2x + 4$

24. $y = -2x + 3$

25. $y = \frac{1}{2}x + 3$

26. $y = \frac{1}{2}x + 2$

27. $f(x) = \frac{2}{3}x - 4$

28. $f(x) = \frac{3}{4}x - 5$

29. $y = -\frac{3}{4}x + 4$

30. $y = -\frac{2}{3}x + 5$

31. $f(x) = -\frac{5}{3}x$

32. $f(x) = -\frac{4}{3}x$

In Exercises 33–40,

    a. *Put the equation in slope-intercept form by solving for y.*
    b. *Identify the slope and the y-intercept.*
    c. *Use the slope and y-intercept to graph the line.*

33. $3x + y = 0$

34. $2x + y = 0$

35. $3y = 4x$

36. $4y = 5x$

37. $2x + y = 3$

38. $3x + y = 4$

39. $7x + 2y = 14$

40. $5x + 3y = 15$

In Exercises 41–48, graph each horizontal or vertical line.

41. $y = 4$

42. $y = 2$

43. $y = -2$

44. $y = -3$

45. $x = 2$

46. $x = 4$

47. $x + 1 = 0$

48. $x + 5 = 0$

Application Exercises

The graph shows the projected online shopping per U.S. online household through 2004. Use the information provided by the graph to solve Exercises 49–50.

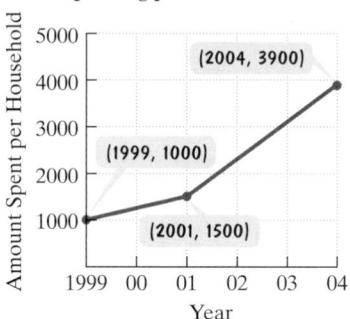

Online Spending: Yearly Spending per Online Household

Source: Forrester Research

49. Find the slope of the line passing through (1999, 1000) and (2001, 1500). What does this represent in terms of the increase in online shopping per year?

50. Find the slope of the line passing through (2001, 1500) and (2004, 3900). What does this represent in terms of the increase in online shopping per year?

If talk about a federal budget surplus sounded too good to be true, that's because it probably was. The Congressional Budget Office's estimates for 2010 range from a $1.2 trillion budget surplus to a $286 billion deficit. Use the information provided by the Congressional Budget Office graphs to solve Exercises 51–52.

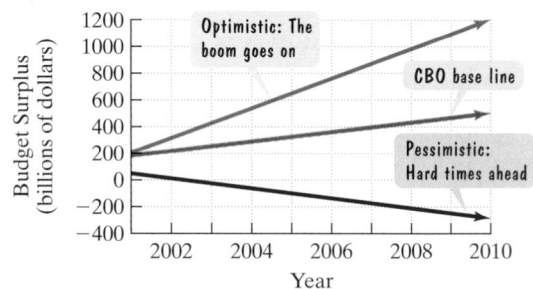

Federal Budget Projections

Source: Congressional Budget Office

51. Look at the line that indicates hard times ahead. Find the slope of this line using (2001, 50) and (2010, −286). What does this represent in terms of the rate of decrease in budget surplus per year?

52. Look at the line that indicates the boom goes on. Find the slope of this line using (2001, 200) and (2010, 1200). What does this represent in terms of the rate of increase in the budget surplus per year?

53. Horrified at the cost the last time you needed a prescription drug? The graph shows that the cost of the average retail prescription has been rising steadily since 1991.

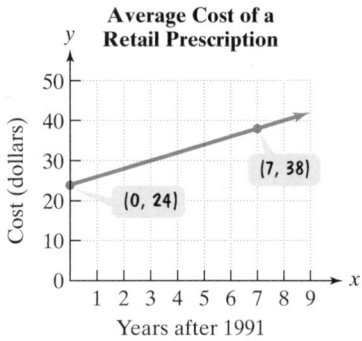

Average Cost of a Retail Prescription

Source: Newsweek

a. According to the graph, what is the *y*-intercept? Describe what this represents in this situation.

b. Use the coordinates of the two points shown to compute the slope. What does this mean about the cost of the average retail prescription?

c. Use the *y*-intercept from part (a) and the slope from part (b) to write an equation that models the cost of the average retail prescription, *y*, *x* years after 1991. Write your model in *y* = *mx* + *b* form.

d. Use your model from part (c) to predict the cost of the average retail prescription in 2005.

54. Films may not be getting any better, but in this era of moviegoing, the number of screens available for new films and the classics has exploded. The points in the graph show the number of screens in the United States from 1995 through 2000. Also shown is a line that passes through or near the points.

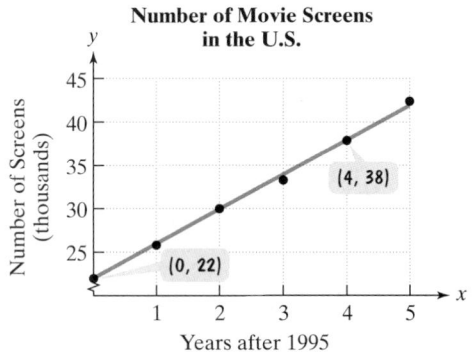

Number of Movie Screens in the U.S.

Source: Motion Picture Association of America

a. According to the graph, what is the *y*-intercept? Describe what this represents in this situation.

b. Use the coordinates of the two points shown to compute the slope. What does this mean about the number of movie screens?

c. Use the *y*-intercept from part (a) and the slope from part (b) to write an equation that models the number of screens, *y*, in thousands, *x* years after 1995.

d. Use your model from part (c) to predict the number of screens, in thousands, in 2005.

Writing in Mathematics

55. Describe how to find the *x*-intercept of a linear equation.

56. Describe how to find the *y*-intercept of a linear equation.

57. What is the slope of a line?

58. Describe how to calculate the slope of a line passing through two points.

59. Describe how to graph a line using the slope and *y*-intercept. Provide an original example with your description.

60. What does it mean if the slope of a line is 0?

61. What does it mean if the slope of a line is undefined?

62. What is the least number of points needed to graph a line? How many should actually be used? Explain.

63. Explain why the *y*-values can be any number for the equation *x* = 5. How is this shown in the graph of the equation?

Critical Thinking Exercises

In Exercises 64–65, find the coefficients that must be placed in each shaded area so that the equation's graph will be a line with the specified intercepts.

64. ▓ *x* + ▓ *y* = 10   *x*-intercept = 5;   *y*-intercept = 2

65. ▓ *x* + ▓ *y* = 12   *x*-intercept = −2;   *y*-intercept = 4

Technology Exercises

Use a graphing calculator to graph each equation in Exercises 66–69. Then use the $\boxed{\text{TRACE}}$ *feature to trace along the line and find the coordinates of two points. Use these points to verify the line's slope.*

66. $y = 2x + 4$

67. $y = -3x + 6$

68. $y = -\frac{1}{2}x - 5$

69. $y = \frac{3}{4}x - 2$

70. Use a graphing utility to verify any three of your hand-drawn graphs in Exercises 21–32.

71. Use a graphing utility to verify any three of your hand-drawn graphs in Exercises 33–40. Solve the equation for *y* before entering it.

| SECTION 7.3 | QUADRATIC FUNCTIONS AND THEIR GRAPHS |

OBJECTIVES

1. Graph parabolas.
2. Solve applied problems based on knowing a parabola's vertex.

In the opening to this chapter, we mentioned that a driver's age has something to do with his or her chances of getting into a car accident. The equation

$$y = 0.4x^2 - 36x + 1000$$

models the number of accidents, y, per 50 million miles driven, as a function of a driver's age, x, in years, where x is an age between 16 and 74, inclusive ($16 \leq x \leq 74$). Can you see how this function differs from the linear functions studied in the previous section? The greatest exponent on x is 2. In a linear function, the greatest exponent on the variable is 1. The equation

$$y = 0.4x^2 - 36x + 1000$$

is of the form $y = ax^2 + bx + c$. An equation of this form, in which $a \neq 0$ and the greatest exponent on the variable is 2, is called a **quadratic function**. Using function notation, we can also express this equation as

$$f(x) = 0.4x^2 - 36x + 1000.$$

| **EXAMPLE 1** | USING A QUADRATIC FUNCTION |

Use the quadratic function described above to find and interpret $f(20)$.

Solution   We can find $f(20)$, f of 20, by replacing each x with 20.

$f(x) = 0.4x^2 - 36x + 1000$	This is the given quadratic function.
$f(20) = 0.4(20)^2 - 36(20) + 1000$	Replace each x with 20.
$\quad\quad = 0.4(400) - 36(20) + 1000$	Use the order of operations, evaluating the exponential expression first.
$\quad\quad = 160 - 720 + 1000$	Perform the multiplications.
$\quad\quad = 440$	Subtract and add as indicated.

We see that $f(20) = 440$—that is, f of 20 is 440. In practical terms, this indicates that 20-year-olds have 440 accidents per 50 million miles driven.

Check
1
Point

Use the quadratic function described above to find and interpret $f(30)$.

Graph parabolas.

Graphs of Quadratic Functions

The graph of the quadratic function $y = ax^2 + bx + c$, or $f(x) = ax^2 + bx + c$, is called a **parabola**. Parabolas are shaped like cups, as shown in Figure 7.19. If the coefficient of x^2 (the value of a in $ax^2 + bx + c$) is positive, the parabola opens upward. If the coefficient of x^2 is negative, the parabola opens downward. The **vertex** (or turning point) of the parabola is the lowest point on the graph when it opens upward and the highest point on the graph when it opens downward.

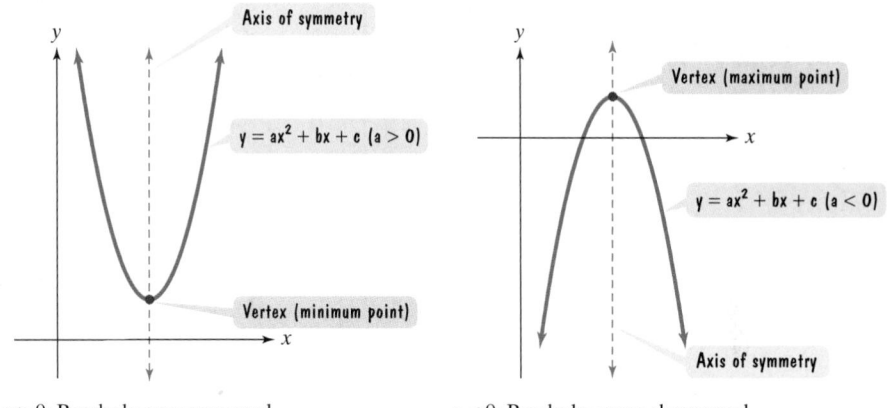

$a > 0$: Parabola opens upward. $a < 0$: Parabola opens downward.

FIGURE 7.19 Characteristics of graphs of quadratic functions

Look at the unusual image of the word "mirror" shown below. The artist, Scott Kim, has created the image so that the two halves of the whole are mirror images of each other. A parabola shares this kind of symmetry, in which a line through the vertex divides the figure in half. Parabolas are symmetric with respect to this line, called the **axis of symmetry**. If a parabola is folded along its axis of symmetry, the two halves match exactly. The movements of gymnasts, divers, and swimmers can approximate this kind of symmetry.

EXAMPLE 2   USING POINT PLOTTING TO GRAPH A PARABOLA

Consider the quadratic function $y = x^2 + 4x + 3$.

a. Is the graph a parabola that opens upward or downward?

b. Use point plotting to graph the parabola.

Solution

a. To determine whether a parabola opens upward or downward, we begin by identifying a, the coefficient of x^2. The following equation shows the values for $a, b,$ and c in $y = x^2 + 4x + 3$. Notice that we wrote x^2 as $1x^2$.

$$y = 1x^2 + 4x + 3$$

a, the coefficient of x^2, is 1.   b, the coefficient of x, is 4.   c, the constant term, is 3.

When a is greater than 0, the parabola opens upward. When a is less than 0, the parabola opens downward. Because $a = 1$, which is greater than 0, the parabola opens upward.

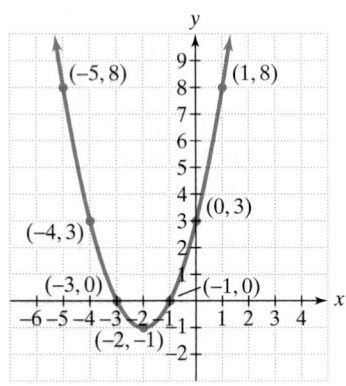

FIGURE 7.20 The graph of
$y = x^2 + 4x + 3$

b. To use point plotting to graph the parabola, we first make a table of x- and y-coordinates.

x	$y = x^2 + 4x + 3$	(x, y)
-5	$y = (-5)^2 + 4(-5) + 3 = 8$	$(-5, 8)$
-4	$y = (-4)^2 + 4(-4) + 3 = 3$	$(-4, 3)$
-3	$y = (-3)^2 + 4(-3) + 3 = 0$	$(-3, 0)$
-2	$y = (-2)^2 + 4(-2) + 3 = -1$	$(-2, -1)$
-1	$y = (-1)^2 + 4(-1) + 3 = 0$	$(-1, 0)$
0	$y = 0^2 + 4(0) + 3 = 3$	$(0, 3)$
1	$y = 1^2 + 4(1) + 3 = 8$	$(1, 8)$

Then we plot the points and connect them with a smooth curve. The graph of $y = x^2 + 4x + 3$ is shown in Figure 7.20.

 **Check Point 2**   Consider the quadratic function $y = x^2 - 6x + 8$.

a. Is the graph a parabola that opens upward or downward?

b. Use point plotting to graph the parabola. Select integers for x, $0 \leq x \leq 6$.

A number of points are important when graphing a quadratic function. These points, labeled in Figure 7.21, are the x-intercepts (although not every parabola has two x-intercepts), the y-intercept, and the vertex. Let's see how we can locate these points.

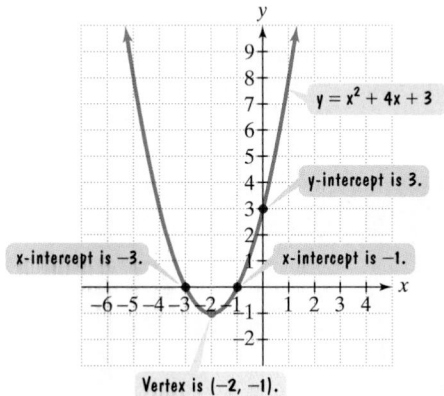

FIGURE 7.21 Useful points in graphing a parabola

Finding a Parabola's x-Intercepts

At each point where a parabola crosses the x-axis, the value of y equals 0. Thus, the x-intercepts can be found by replacing y with 0 in $y = ax^2 + bx + c$. Use factoring or the quadratic formula to solve the resulting quadratic equation for x.

EXAMPLE 3   FINDING A PARABOLA'S x-INTERCEPTS

Find the x-intercepts for the parabola whose equation is $y = x^2 + 4x + 3$.

Solution   Replace y with 0 in $y = x^2 + 4x + 3$. We obtain $0 = x^2 + 4x + 3$, or $x^2 + 4x + 3 = 0$. We can solve this equation by factoring.

$$x^2 + 4x + 3 = 0$$
$$(x + 3)(x + 1) = 0$$
$$x + 3 = 0 \quad \text{or} \quad x + 1 = 0$$
$$x = -3 \qquad\qquad x = -1$$

Thus, the x-intercepts are -3 and -1. The parabola passes through $(-3, 0)$ and $(-1, 0)$, shown in Figure 7.21.

 **Check Point 3**   Find the x-intercepts for the parabola whose equation is $y = x^2 - 6x + 8$.

Finding a Parabola's *y*-Intercept

At the point where a parabola crosses the *y*-axis, the value of *x* equals 0. Thus, the *y*-intercept can be found by replacing *x* with 0 in $y = ax^2 + bx + c$. Simple arithmetic will produce a value for *y*, which is the *y*-intercept.

EXAMPLE 4   FINDING A PARABOLA'S *y*-INTERCEPT

Find the *y*-intercept for the parabola whose equation is $y = x^2 + 4x + 3$.

Solution   Replace *x* with 0 in $y = x^2 + 4x + 3$.

$$y = 0^2 + 4 \cdot 0 + 3 = 0 + 0 + 3 = 3$$

The *y*-intercept is 3. The parabola passes through $(0, 3)$, shown in Figure 7.21 on the previous page.

 **Check Point 4**   Find the *y*-intercept for the parabola whose equation is $y = x^2 - 6x + 8$.

Finding a Parabola's Vertex

Keep in mind that a parabola's vertex is its turning point. The *x*-coordinate of the vertex for the parabola in Figure 7.21, −2, is midway between the *x*-intercepts, −3 and −1. However, not every parabola has two *x*-intercepts. There is a formula that can be used to determine the vertex of a parabola regardless of how many *x*-intercepts it has.

STUDY TIP

Using function notation

$$f(x) = ax^2 + bx + c,$$

a parabola's vertex can be expressed as

$$\left(\frac{-b}{2a}, f\left(\frac{-b}{2a}\right)\right).$$

THE VERTEX OF A PARABOLA

For a parabola whose equation is $y = ax^2 + bx + c$

1. The *x*-coordinate of the vertex is $\dfrac{-b}{2a}$.

2. The *y*-coordinate of the vertex is found by substituting the *x*-coordinate into the parabola's equation and evaluating *y*.

EXAMPLE 5   FINDING A PARABOLA'S VERTEX

Find the vertex for the parabola whose equation is $y = x^2 + 4x + 3$.

Solution   In the equation $y = x^2 + 4x + 3$, $a = 1$ and $b = 4$.

$$x\text{-coordinate of vertex} = \frac{-b}{2a} = \frac{-4}{2 \cdot 1} = \frac{-4}{2} = -2$$

To find the *y*-coordinate of the vertex, we substitute −2 for *x* in $y = x^2 + 4x + 3$ and then evaluate.

$$y\text{-coordinate of vertex} = (-2)^2 + 4(-2) + 3 = 4 + (-8) + 3 = -1$$

The vertex is $(-2, -1)$, shown in Figure 7.21 on the previous page.

 **Check Point 5**   Find the vertex for the parabola whose equation is $y = x^2 - 6x + 8$.

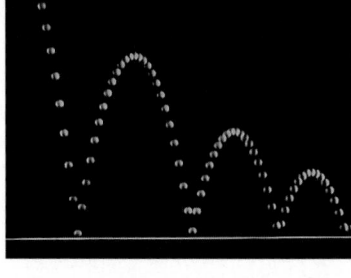

© Bernice Abbot/Commerce Graphics Ltd., Inc.

A Strategy for Graphing Quadratic Functions

Here is a general procedure to sketch the graph of the quadratic function $y = ax^2 + bx + c$ or $f(x) = ax^2 + bx + c$:

GRAPHING QUADRATIC FUNCTIONS

The graph of $y = ax^2 + bx + c$ or $f(x) = ax^2 + bx + c$, called a parabola, can be graphed using the following steps:

1. Determine whether the parabola opens upward or downward. If $a > 0$, it opens upward. If $a < 0$, it opens downward.

2. Determine the vertex of the parabola. The x-coordinate is $\dfrac{-b}{2a}$.

   The y-coordinate is found by substituting the x-coordinate in the parabola's equation.

3. Find any x-intercepts by replacing y or $f(x)$ with 0. Solve the resulting quadratic equation for x.

4. Find the y-intercept by replacing x with 0.

5. Plot the intercepts and the vertex.

6. Connect these points with a smooth U-shaped curve.

EXAMPLE 6 | GRAPHING A PARABOLA

Graph the quadratic function: $y = x^2 - 2x - 3$.

Solution   We can graph this function by following the steps in the box.

STEP 1. Determine how the parabola opens. Note that a, the coefficient of x^2, is 1. Thus, $a > 0$; this positive value tells us that the parabola opens upward.

STEP 2. Find the vertex. We know that the x-coordinate of the vertex is $\dfrac{-b}{2a}$. Let's identify the numbers a, b, and c in the given equation, which is in the form $y = ax^2 + bx + c$.

$$y = x^2 - 2x - 3$$

$$a = 1 \qquad b = -2 \qquad c = -3$$

Now we substitute the values of a and b into the equation for the x-coordinate:

$$x\text{-coordinate of vertex} = \frac{-b}{2a} = \frac{-(-2)}{2(1)} = \frac{2}{2} = 1.$$

The x-coordinate of the vertex is 1. We substitute 1 for x in the equation $y = x^2 - 2x - 3$ to find the y-coordinate:

$$y\text{-coordinate of vertex} = 1^2 - 2 \cdot 1 - 3 = 1 - 2 - 3 = -4.$$

The vertex is $(1, -4)$.

STEP 3. Find the x-intercepts. Replace y with 0 in $y = x^2 - 2x - 3$. We obtain $0 = x^2 - 2x - 3$ or $x^2 - 2x - 3 = 0$. We can solve this equation by factoring.

$$x^2 - 2x - 3 = 0$$
$$(x - 3)(x + 1) = 0$$
$$x - 3 = 0 \quad \text{or} \quad x + 1 = 0$$
$$x = 3 \qquad\qquad x = -1$$

Parabola

Suspension bridge

Parabola

Arch bridge

Cables hung between structures to form suspension bridges often form parabolas. Arches constructed of steel and concrete, whose main purpose is strength, are usually parabolic in shape.

The x-intercepts are 3 and -1. The parabola passes through $(3, 0)$ and $(-1, 0)$.

STEP 4. Find the y-intercept. Replace x with 0 in $y = x^2 - 2x - 3$:

$$y = 0^2 - 2 \cdot 0 - 3 = 0 - 0 - 3 = -3.$$

The y-intercept is -3. The parabola passes through $(0, -3)$.

STEPS 5 AND 6. Plot the intercepts and the vertex. Connect these points with a smooth curve. The intercepts and the vertex are shown as the four labeled points in Figure 7.22. Also shown is the graph of the quadratic function, obtained by connecting the points with a smooth curve. Can you see that the axis of symmetry is the vertical line whose equation is $x = 1$?

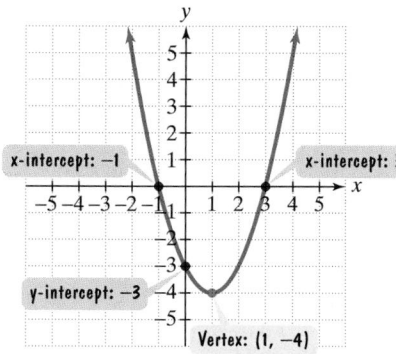

x-intercept: −1   x-intercept: 3
y-intercept: −3
Vertex: (1, −4)

FIGURE 7.22 The graph of $y = x^2 - 2x - 3$

Check Point 6

Graph the quadratic function: $y = x^2 + 6x + 5$.

EXAMPLE 7   GRAPHING A PARABOLA

Graph the quadratic function: $f(x) = -x^2 + 4x - 1$.

Solution

STEP 1. Determine how the parabola opens. Note that a, the coefficient of x^2, is -1: $a = -1$. Thus, $a < 0$; this negative value tells us that the parabola opens downward.

STEP 2. Find the vertex. The x-coordinate of the vertex is $\dfrac{-b}{2a}$.

$$f(x) = -x^2 + 4x - 1$$

$a = -1$ $b = 4$ $c = -1$

$$x\text{-coordinate of vertex} = \frac{-b}{2a} = \frac{-4}{2(-1)} = \frac{-4}{-2} = 2$$

Substitute 2 for x in $f(x) = -x^2 + 4x - 1$ to find the y-coordinate:

$$y\text{-coordinate of vertex} = f(2) = -2^2 + 4 \cdot 2 - 1 = -4 + 8 - 1 = 3.$$

The vertex is $(2, 3)$.

STEP 3. Find the x-intercepts. Replace $f(x)$ with 0 in $f(x) = -x^2 + 4x - 1$. We obtain $0 = -x^2 + 4x - 1$ or $-x^2 + 4x - 1 = 0$. This equation cannot be solved by factoring. We will use the quadratic formula to solve it.

$$a = -1, \quad b = 4, \quad c = -1$$

$$x = \frac{-b \pm \sqrt{b^2 - 4ac}}{2a} = \frac{-4 \pm \sqrt{4^2 - 4(-1)(-1)}}{2(-1)} = \frac{-4 \pm \sqrt{16 - 4}}{-2}$$

$$x = \frac{-4 + \sqrt{12}}{-2} \approx 0.3 \quad \text{or} \quad x = \frac{-4 - \sqrt{12}}{-2} \approx 3.7$$

The x-intercepts are approximately 0.3 and 3.7. The parabola passes through $(0.3, 0)$ and $(3.7, 0)$.

STEP 4. Find the y-intercept. Replace x with 0 in $f(x) = -x^2 + 4x - 1$.

$$f(0) = -0^2 + 4 \cdot 0 - 1 = -1$$

The y-intercept is -1. The parabola passes through $(0, -1)$.

STEPS 5 AND 6. Plot the intercepts and the vertex. Connect these points with a smooth curve. The intercepts and the vertex are shown as the four labeled points in Figure 7.23. Also shown is the graph of the quadratic function, obtained by connecting the points with a smooth curve. The parabola's axis of symmetry is the vertical line whose equation is $x = 2$.

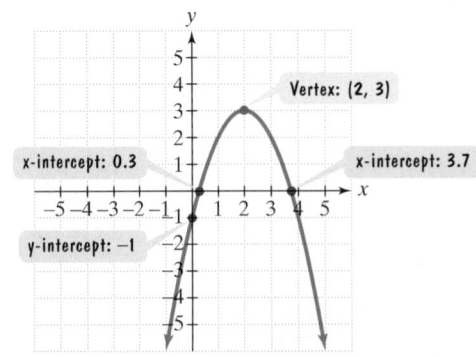

FIGURE 7.23 The graph of $f(x) = -x^2 + 4x - 1$

 Check Point **7**

Graph the quadratic function: $f(x) = -x^2 - 2x + 5$.

Applications of Quadratic Functions

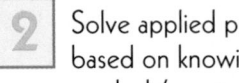 **2** Solve applied problems based on knowing a parabola's vertex.

We opened this section with a formula that described accidents as a function of age. We saw that 20-year-olds have 440 car accidents per 50 million miles driven. Do you have a guess about the age of a driver who has the least number of accidents per 50 million miles driven? We answer this question in Example 8.

EXAMPLE 8 CAR ACCIDENTS AS A FUNCTION OF AGE

We have seen that the equation

$$y = 0.4x^2 - 36x + 1000$$

models the number of accidents, y, per 50 million miles driven, as a function of a driver's age, x, in years, where $16 \leq x \leq 74$. The graph of the function is shown in Figure 7.24 on the next page. Find the coordinates of the vertex and describe what this represents in practical terms.

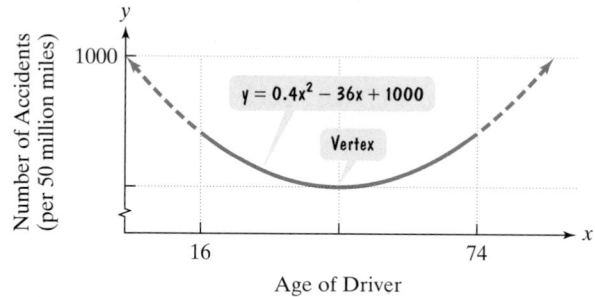

FIGURE 7.24

Solution   We begin by identifying the numbers a, b, and c in the given model.

$$y = 0.4x^2 - 36x + 1000$$

$$a = 0.4 \quad b = -36 \qquad c = 1000$$

$$x\text{-coordinate of vertex} = \frac{-b}{2a} = \frac{-(-36)}{2(0.4)} = \frac{36}{0.8} = 45$$

Substitute 45 for x in $y = 0.4x^2 - 36x + 1000$ to find the y-coordinate.

$$y\text{-coordinate of vertex} = 0.4(45)^2 - 36(45) + 1000 = 810 - 1620 + 1000 = 190$$

The vertex is $(45, 190)$. Because the coefficient of x^2, 0.4, is positive, the parabola opens upward and the vertex, $(45, 190)$, is the lowest point on the graph. In practical terms, this indicates that 45-year-olds have the least number of car accidents, 190 per 50 million miles driven. Because the vertex is a minimum point, drivers both younger and older than 45 have more than 190 accidents per 50 million miles driven.

Check 8 Point   The function

$$y = -0.5x^2 + 4x + 19$$

models the number of people, y, in millions, receiving food stamps x years after 1990. The graph of the resulting parabola is shown in Figure 7.25. Find the coordinates of the vertex and describe what this represents in practical terms.

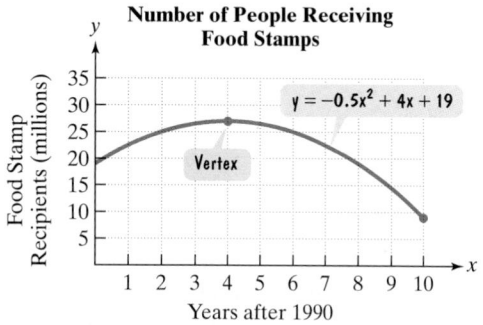

Source: New York Times

FIGURE 7.25

Exercise Set 7.3

Practice Exercises

In Exercises 1–4, determine whether the parabola whose equation is given opens upward or downward.

1. $y = x^2 - 4x + 3$ **2.** $y = x^2 - 6x + 5$

3. $f(x) = -2x^2 + x + 6$ **4.** $f(x) = -2x^2 - 4x + 6$

In Exercises 5–10, find the x-intercepts for the parabola whose equation is given. If the x-intercepts are irrational numbers, round your answers to the nearest tenth.

5. $y = x^2 - 4x + 3$ **6.** $y = x^2 - 6x + 5$

7. $y = -x^2 + 8x - 12$ **8.** $y = -x^2 - 2x + 3$

9. $f(x) = x^2 + 2x - 4$ **10.** $f(x) = x^2 + 8x + 14$

In Exercises 11–18, find the y-intercept for the parabola whose equation is given.

11. $y = x^2 - 4x + 3$ **12.** $y = x^2 - 6x + 5$

13. $y = -x^2 + 8x - 12$ **14.** $y = -x^2 - 2x + 3$

15. $y = x^2 + 2x - 4$ **16.** $y = x^2 + 8x + 14$

17. $f(x) = x^2 + 6x$ **18.** $f(x) = x^2 + 8x$

In Exercises 19–24, find the vertex for the parabola whose equation is given. Is the vertex a maximum point or a minimum point?

19. $y = x^2 - 4x + 3$ **20.** $y = x^2 - 6x + 5$

21. $y = 2x^2 + 4x - 6$ **22.** $y = -2x^2 - 4x - 2$

23. $f(x) = x^2 + 6x$ **24.** $f(x) = x^2 + 8x$

In Exercises 25–36, graph the quadratic function whose equation is given.

25. $y = x^2 + 8x + 7$ **26.** $y = x^2 + 10x + 9$

27. $y = x^2 - 2x - 8$ **28.** $y = x^2 + 4x - 5$

29. $y = -x^2 + 4x - 3$ **30.** $y = -x^2 + 2x + 3$

31. $y = x^2 - 1$ **32.** $y = x^2 - 4$

33. $f(x) = x^2 + 2x + 1$ **34.** $f(x) = x^2 - 2x + 1$

35. $f(x) = -2x^2 + 4x + 5$ **36.** $f(x) = -3x^2 + 6x - 2$

Application Exercises

The quadratic function

$$f(x) = 0.036x^2 - 2.8x + 58.14$$

models the number of deaths per year per thousand people, $f(x)$, for people who are x years old, $40 \le x \le 60$. Use this function to solve Exercises 37–38.

37. Find and interpret $f(50)$. **38.** Find and interpret $f(40)$.

The graph of the quadratic function in Exercises 37–38 is shown in the figure. Use the graph to solve Exercises 39–42.

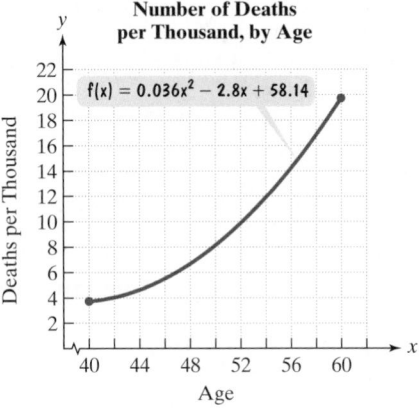

Number of Deaths per Thousand, by Age

$f(x) = 0.036x^2 - 2.8x + 58.14$

Source: U.S. Department of Health and Human Services

39. Identify your answer from Exercise 37 as a point on the graph.

40. Identify your answer from Exercise 38 as a point on the graph.

41. At which age, to the nearest year, do approximately 16 people per thousand die each year?

42. At which age, to the nearest year, do approximately 6 people per thousand die each year?

Use the formula for the vertex of a parabola to solve Exercises 43–46.

43. Fireworks are launched into the air. The quadratic function

$$y = -16x^2 + 200x + 4$$

models the fireworks' height, y, in feet, x seconds after they are launched. When should the fireworks explode so that they go off at the greatest height? What is that height?

44. A football is thrown by a quarterback to a receiver 40 yards away. The quadratic function

$$y = -0.025x^2 + x + 5$$

models the football's height above the ground, y, in feet, when it is x yards from the quarterback. How many yards from the quarterback does the football reach its greatest height? What is that height?

45. One would think that the more avocado trees planted per acre, the higher the yield of avocados. However, this is not the case. When more than a certain number of trees are present per acre, they tend to crowd one another and the yield per tree drops. The function $y = -0.01x^2 + 0.8x$ models the yield, y, in bushels of avocados per tree, when x trees are planted per acre. The graph of the resulting parabola was obtained with a graphing utility and is shown here. Find the number of trees that should be planted per acre to produce the maximum yield. How many bushels of avocados per tree is the maximum yield?

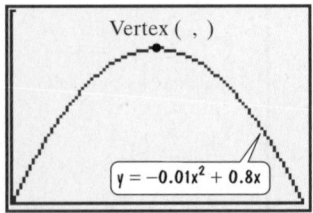

Vertex (,)

$y = -0.01x^2 + 0.8x$

46. You have 120 feet of fencing to enclose a rectangular plot that borders on a river. The figure shows that you do not fence the side along the river. The function

$$A(x) = x(120 - 2x)$$

gives the area of the plot, $A(x)$, in square feet, in terms of its width, x, in feet. Find the width and the length of the plot that you should enclose to obtain the maximum area. What is the largest area that can be enclosed?

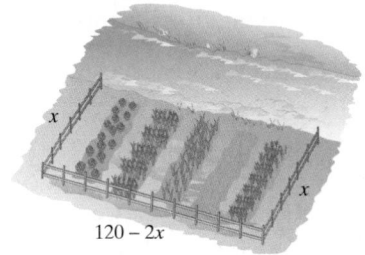

120 − 2x

Writing in Mathematics

47. What is a quadratic function?

48. What is a parabola? Describe its shape.

49. Explain how to decide whether a parabola opens upward or downward.

50. Describe how to find a parabola's x-intercepts. Use $y = x^2 + x - 6$ as an example.

51. Describe how to find a parabola's y-intercept. Use $y = x^2 + x - 6$ as an example.

52. Describe how to find a parabola's vertex. Use $y = x^2 - 6x + 8$ as an example.

53. A parabola that opens downward has its vertex at $(1, 5)$. Describe as much as you can about this parabola based on this information. Include in your discussion the number of x-intercepts (if any) for the parabola.

54. The quadratic function

$$y = -0.018x^2 + 1.93x - 25.34$$

models the miles per gallon, y, of a Ford Taurus driven at x miles per hour. Suppose that you own a Ford Taurus. Describe how you can use this formula to save money.

Critical Thinking Exercises

In Exercises 55–57, the value of a in $y = ax^2 + bx + c$ and the vertex of the parabola are given. How many x-intercepts does the parabola have? Explain how you arrived at this number.

55. $a = -2$; vertex at $(4, 8)$ **56.** $a = 1$; vertex at $(2, 0)$

57. $a = 3$; vertex at $(3, 1)$

58. A parabola has x-intercepts at 3 and 7, a y-intercept at -21, and $(5, 4)$ for its vertex. Write the parabola's equation.

Technology Exercise

59. Use a graphing calculator to graph the quadratic functions that you graphed by hand in Exercises 25–36. Describe similarities and differences between the graphs obtained by hand and those that appear in the calculator's viewing window.

Group Exercise

60. This activity is for a group of students who are familiar with graphing calculators. The group should prepare a presentation to the entire class that will amaze people with what graphing calculators can do. Show how functions are graphed. Take some of the topics studied in our two algebra chapters and show how the calculator does the things that we have been doing by hand. If possible, show how models are obtained from actual data. Make the presentation as entertaining as possible. *Note*: Your professor will need to have the graphing calculator equipped with an attachment so that the calculator display can be shown on a large screen using an overhead projector.

SECTION 7.4   EXPONENTIAL FUNCTIONS

OBJECTIVES

1. Graph exponential functions.

2. Solve applied problems using exponential functions.

The space shuttle *Challenger* exploded approximately 73 seconds into flight on January 28, 1986. The tragedy involved damage to O-rings, which were used to seal the connections between different sections of the shuttle engines. The number of O-rings damaged increases dramatically as temperature falls.

The function

$$f(x) = 13.49(0.967)^x - 1$$

models the number of ○-rings expected to fail when the temperature is $x°$F. Can you see how this function is different from linear and quadratic functions? The variable x is in the exponent. Functions whose equations contain a variable in the exponent are called **exponential functions**. Many real-life situations, including population growth, growth of epidemics, radioactive decay, and other changes that involve rapid increase or decrease, can be described using exponential functions.

DEFINITION OF THE EXPONENTIAL FUNCTION

The **exponential function** f **with base** b is defined by

$$y = b^x \quad \text{or} \quad f(x) = b^x$$

where b is a positive constant other than $1\,(b > 0$ and $b \neq 1)$ and x is any real number.

---

1 Graph exponential functions.

Graphing Exponential Functions

We begin by looking at the graph of an exponential function with a base, b, greater than 1.

EXAMPLE 1   GRAPHING AN EXPONENTIAL FUNCTION

Graph: $f(x) = 2^x$.

Solution   We start by selecting numbers for x and finding the corresponding values for $f(x)$.

x	$f(x) = 2^x$
-3	$f(-3) = 2^{-3} = \frac{1}{8}$
-2	$f(-2) = 2^{-2} = \frac{1}{4}$
-1	$f(-1) = 2^{-1} = \frac{1}{2}$
0	$f(0) = 2^0 = 1$
1	$f(1) = 2^1 = 2$
2	$f(2) = 2^2 = 4$
3	$f(3) = 2^3 = 8$

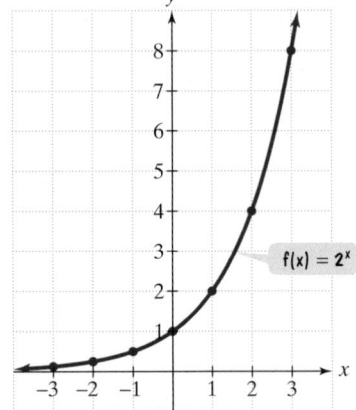

FIGURE 7.26 The graph of $f(x) = 2^x$

We plot these points, connecting them with a smooth curve. Figure 7.26 shows the graph of $f(x) = 2^x$.

All exponential functions of the form $y = b^x$, or $f(x) = b^x$, where b is a number greater than 1, have the shape of the graph shown in Figure 7.26. The graph approaches, but never touches, the negative portion of the x-axis.

Check Point 1   Graph: $f(x) = 3^x$.

In our next example, we graph an exponential function of the form $f(x) = b^x$, where $0 < b < 1$.

EXAMPLE 2   GRAPHING AN EXPONENTIAL FUNCTION

Graph: $f(x) = \left(\frac{1}{2}\right)^x$.

Solution   We begin by selecting numbers for x and finding the corresponding values for $f(x)$. We compute the values for $f(x)$ by noting that

$$f(x) = \left(\frac{1}{2}\right)^x = (2^{-1})^x = 2^{-x}.$$

x	$f(x) = \left(\frac{1}{2}\right)^x$ or 2^{-x}
-3	$f(-3) = 2^{-(-3)} = 2^3 = 8$
-2	$f(-2) = 2^{-(-2)} = 2^2 = 4$
-1	$f(-1) = 2^{-(-1)} = 2^1 = 2$
0	$f(0) = 2^{-0} = 2^0 = 1$
1	$f(1) = 2^{-1} = \dfrac{1}{2^1} = \dfrac{1}{2}$
2	$f(2) = 2^{-2} = \dfrac{1}{2^2} = \dfrac{1}{4}$
3	$f(3) = 2^{-3} = \dfrac{1}{2^3} = \dfrac{1}{8}$

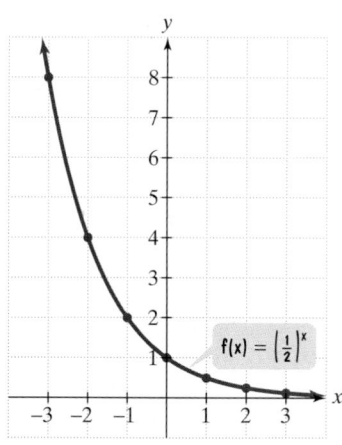

FIGURE 7.27 The graph of $f(x) = \left(\dfrac{1}{2}\right)^x$

We plot these points, connecting them with a smooth curve. Figure 7.27 shows the graph of $f(x) = \left(\frac{1}{2}\right)^x$.

All exponential functions of the form $y = b^x$, or $f(x) = b, 0 < b < 1$, have the shape of the graph shown in Figure 7.27. The graph approaches, but never touches, the positive portion of the x-axis.

 **Check Point 2**   Graph: $f(x) = \left(\frac{1}{3}\right)^x$.

Applications of Exponential Functions

2 Solve applied problems using exponential functions.

EXAMPLE 3   EVALUATING AN EXPONENTIAL FUNCTION

The exponential function $f(x) = 13.49(0.967)^x - 1$ describes the number of O-rings expected to fail, $f(x)$, when the temperature is $x°$F. On the morning the *Challenger* was launched, the temperature was 31°F, colder than any previous experience. Find the number of O-rings expected to fail at this temperature.

Solution   Because the temperature was 31°F, substitute 31 for x and evaluate the function at 31.

$$f(x) = 13.49(0.967)^x - 1 \quad \text{This is the given function.}$$
$$f(31) = 13.49(0.967)^{31} - 1 \quad \text{Substitute 31 for } x$$

Use a scientific or graphing calculator to evaluate $f(31) = 13.49(0.967)^{31} - 1$. Press the following keys on your calculator to do this:

Scientific calculator: 13.49 $\times$ .967 y^x 31 $-$ 1 $=$

Graphing calculator: 13.49 $\times$ .967 $\wedge$ 31 $-$ 1 ENTER .

The display should be approximately 3.7668627.

$$f(31) = 13.49(0.967)^{31} - 1 \approx 3.8 \approx 4$$

Thus, four ○-rings are expected to fail at a temperature of 31°F.

 **Check 3 Point** Use the function in Example 3 to find the number of ○-rings expected to fail at a temperature of 60°F. Round to the nearest whole number.

EXAMPLE 4 | THE CHERNOBYL DISASTER

Horses were born with eight deformed legs, pigs with no eyes, and eggs contained several yolks. This was part of the grotesque aftermath of the 1986 explosion at the Chernobyl nuclear power plant in the former Soviet Union. Nearby cities were abandoned and 335,000 people were permanently displaced from their homes. The explosion sent about 1000 kilograms of radioactive cesium-137 into the atmosphere. The function

$$f(x) = 1000(0.5)^{x/30}$$

describes the amount of cesium-137, $f(x)$, in kilograms, remaining in Chernobyl x years after 1986. If even 100 kilograms of cesium-137 remain in Chernobyl's atmosphere, the area is considered unsafe for human habitation. Will people be able to live in the area 80 years after the accident?

The spread of radioactive fallout from the Chernobyl accident is expected to result in at least 100,000 deaths from cancer in the Northern Hemisphere.

Solution   We substitute 80 for x in the given function. If the resulting value of $f(x)$ is 100 or greater, the area will still be unsafe for human habitation.

$f(x) = 1000(0.5)^{x/30}$ *This is the given function.*

$f(80) = 1000(0.5)^{80/30}$ *Substitute 80 for x.*

≈ 157 *Use a calculator.*

After 80 years, 157 kilograms of cesium-137 will remain in the atmosphere. Because this exceeds 100, even by the year 2066 the Chernobyl area will be unsafe for human habitation.

 **Check 4 Point** Use the function in Example 4 to determine whether people will be able to live in the cities surrounding Chernobyl 90 years after the explosion at the power plant.

The Role of e in Applied Exponential Functions

An irrational number, symbolized by the letter e, appears as the base in many applied exponential functions. This irrational number is approximately equal to 2.72. More accurately,

$$e \approx 2.71828\ldots.$$

The number e is called the **natural base**. The function $f(x) = e^x$ is called the **natural exponential function**.

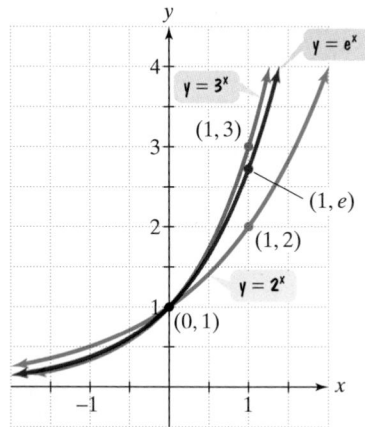

FIGURE 7.28 Graphs of three exponential functions

Use a scientific or graphing calculator with an $\boxed{e^x}$ key to evaluate e to various powers. For example, to find e^2, press the following keys on most calculators:

Scientific calculator: $2\ \boxed{e^x}$

Graphing calculator: $\boxed{e^x}\ 2\ \boxed{\text{ENTER}}$.

The display is approximately 7.389.

$$e^2 \approx 7.389$$

The number e lies between 2 and 3. Because $2^2 = 4$ and $3^2 = 9$, it makes sense that e^2, approximately 7.389, lies between 4 and 9.

Because $2 < e < 3$, the graph of $y = e^x$ is between the graphs of $y = 2^x$ and $y = 3^x$, shown in Figure 7.28.

EXAMPLE 5   WORLD POPULATION

In a report entitled *Resources and Man*, the U.S. National Academy of Sciences concluded that a world population of 10 billion "is close to (if not above) the maximum that an intensely managed world might hope to support with some degree of comfort and individual choice." At the time the report was issued in 1969, the world population was approximately 3.6 billion, with a growth rate of 2% per year. The function

$$f(x) = 3.6e^{0.02x}$$

models world population, $f(x)$, in billions, x years after 1969. Use the function to find world population in the year 2020. Is there cause for alarm?

Solution   Because 2020 is 51 years after 1969, we substitute 51 for x in $f(x) = 3.6e^{0.02x}$:

$$f(51) = 3.6e^{0.02(51)}.$$

Perform this computation on your calculator.

Scientific Calculator: $3.6\ \boxed{\times}\ \boxed{(}\ .02\ \boxed{\times}\ 51\ \boxed{)}\ \boxed{e^x}\ \boxed{=}$

Graphing Calculator: $3.6\ \boxed{\times}\ \boxed{e^x}\ \boxed{(}\ .02\ \boxed{\times}\ 51\ \boxed{)}\ \boxed{\text{ENTER}}$

The display should be approximately 9.9835012. Thus,

$$f(51) = 3.6e^{0.02(51)} \approx 9.98.$$

This indicates that world population in the year 2020 will be approximately 9.98 billion. Because this number is quite close to 10 billion, the given function suggests that there may be cause for alarm.

World population in 2000 was approximately 6 billion, but the growth rate was no longer 2%. It has slowed down to 1.3%. Using this current growth rate, exponential functions now predict a world population of 7.8 billion in the year 2020. Experts think the population may stabilize at 10 billion after 2200 if the deceleration in growth rate continues.

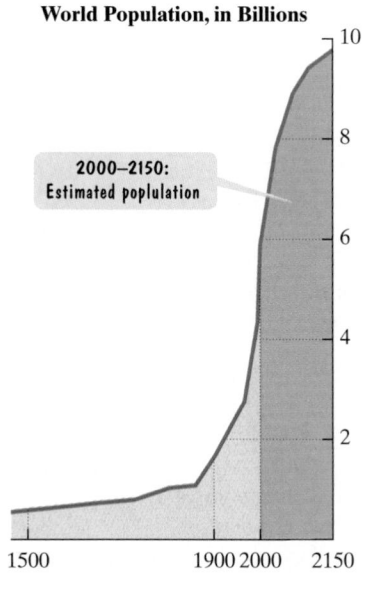

World Population, in Billions

2000–2150: Estimated population

Source: U.N. Population Division

Check 5 Point   The function $f(x) = 6e^{0.013x}$ models world population, $f(x)$, in billions, x years after 2000 subject to a growth rate of 1.3% annually. Use the function to find world population in 2050.

Exercise Set 7.4

Practice Exercises

In Exercises 1–10, graph the exponential function whose equation is given. Start by using −2, −1, 0, 1, and 2 for x and find the corresponding values for f(x).

1. $f(x) = 3^x$

2. $f(x) = 4^x$

3. $f(x) = 2^{x+1}$

4. $f(x) = 2^{x-1}$

5. $f(x) = 3^{x-1}$

6. $f(x) = 3^{x+1}$

7. $f(x) = \left(\frac{1}{3}\right)^x$

8. $f(x) = \left(\frac{1}{4}\right)^x$

9. $f(x) = \left(\frac{1}{2}\right)^{x-1}$

10. $f(x) = \left(\frac{1}{2}\right)^{x+1}$

Application Exercises

Use a calculator with a $\boxed{y^x}$ key or a $\boxed{\wedge}$ key to solve Exercises 11–14.

11. India is currently one of the world's fastest-growing countries. By 2040, the population of India will be larger than the population of China; by 2050, nearly one-third of the world's population will live in these two countries alone. The exponential function $f(x) = 574(1.026)^x$ models the population of India, $f(x)$, in millions, x years after 1974.

    a. Substitute 0 for x and, without using a calculator, find India's population in 1974.

    b. Substitute 27 for x and use your calculator to find India's population in the year 2001 as predicted by this function.

    c. Find India's population in the year 2028 as predicted by this function.

    d. Find India's population in the year 2055 as predicted by this function.

    e. What appears to be happening to India's population every 27 years?

12. In California, where 38% of fatal traffic crashes involve drunk drivers, it is illegal to drive with a blood alcohol concentration of 0.08 or higher. Medical research indicates that the risk of having a car accident increases exponentially as the concentration of alcohol in the blood increases. The risk is modeled by the function

$$y = 6(351,512)^x$$

where x is the blood alcohol concentration and y, given as a percent, is the risk of having a car accident. Find the risk of a car accident for a blood alcohol concentration of 0.17. Round to the nearest percent.

The formula $S = C(1 + r)^t$ models inflation, where C = the value today, r = the annual inflation rate (in decimal form), and S = the inflated value t years from now. Use this formula to solve Exercises 13–14.

13. If the inflation rate is 6%, how much will a house now worth $65,000 be worth in 10 years?

14. If the inflation rate is 3%, how much will a house now worth $110,000 be worth in 5 years?

Use a calculator with an $\boxed{e^x}$ key to solve Exercises 15–18.

15. The graph at the top of the next column shows the average annual salary, in millions of dollars, for each of ten groups of major-league baseball players, arranged from the lowest-paid 10% (group 1) to the highest-paid 10% (group 10).

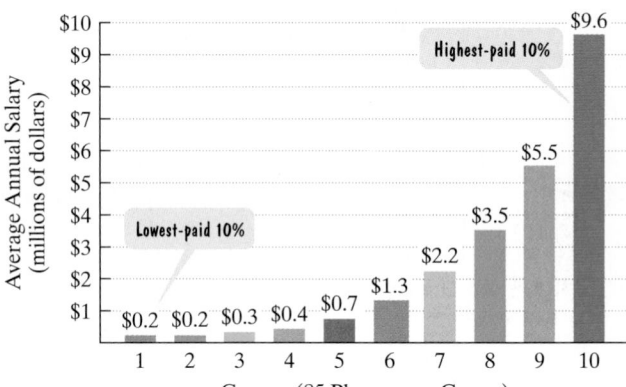

Salaries of Major-League Baseball Players

Source: Newsweek

    a. The data can be modeled by the exponential function

$$f(x) = 0.08e^{0.47x}$$

which describes the average annual salary, $f(x)$, in millions of dollars, for group x. According to the model, what is the average annual salary for the players in group 9, the next-to-highest paid group of players? Round to the nearest tenth of a million. How well does the function model the average salary shown for this group?

    b. According to the data in the bar graph, what percentage of all pay is earned by the top 20% of players? Round to the nearest percent.

16. In college, we study large volumes of information—information that, unfortunately, we do not often retain for very long. The function

$$f(x) = 80e^{-0.5x} + 20$$

describes the percentage of information, $f(x)$, that a particular person remembers x weeks after learning the information.

    a. Substitute 0 for x and, without using a calculator, find the percentage of information remembered at the moment it is first learned.

    b. Substitute 1 for x and find the percentage of information that is remembered after 1 week.

    c. Find the percentage of information that is remembered after 4 weeks.

    d. Find the percentage of information that is remembered after one year (52 weeks).

The function

$$f(x) = \frac{90}{1 + 270e^{-0.122x}}$$

models the percentage, $f(x)$, of people x years old with some coronary heart disease. Use this function to solve Exercises 17–18. Round answers to the nearest tenth of a percent.

17. Evaluate $f(30)$ and describe what this means in practical terms.

18. Evaluate $f(70)$ and describe what this means in practical terms.

Writing in Mathematics

19. What is an exponential function?

20. Look at the graph of $y = 2^x$ (Figure 7.26 on page 366) and explain why the graph defines y as a function of x.

21. What is the natural exponential function?

22. In 2000, world population was 6 billion with an annual growth rate of 1.3%. Discuss two factors that would cause this growth rate to slow down over the next ten years.

Critical Thinking Exercises

23. Explain why the graph of $y = 2^x$ does not cross the x-axis.

24. The graphs labeled (a)–(d) in the accompanying figure represent $y = 3^x$, $y = 5^x$, $y = \left(\frac{1}{3}\right)^x$, and $y = \left(\frac{1}{5}\right)^x$, but not necessarily in that order. Which is which? Describe the process that enables you to make your decision.

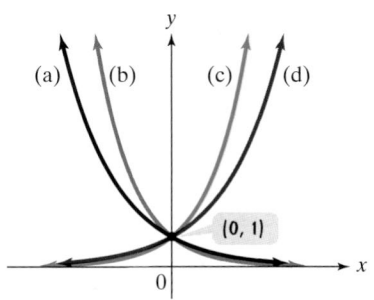

Technology Exercises

25. Use a graphing calculator to graph the exponential functions that you graphed by hand in Exercises 1–10. Describe similarities and differences between the graphs obtained by hand and those that appear in the calculator's viewing window.

26. Graph $y = 13.49(0.967)^x - 1$, the function for the number of O-rings expected to fail at $x°$F, in a $[0, 90, 10]$ by $[0, 20, 5]$ viewing rectangle. If NASA engineers had used this function and its graph, is it likely they would have allowed the *Challenger* to be launched when the temperature was $31°$F? Explain.

Group Exercise

27. Group members should consult the current edition of *The World Almanac* or an equivalent reference, including the Internet. Select data of interest to the group that can be modeled with an exponential function. Then consult a person who is familiar with graphing calculators who can use the exponential regression feature of the calculator to show you how to obtain a function that models your data. Once you have this function, each group member should make one prediction based on the exponential function, and then discuss a consequence of this prediction. What factors might change the accuracy of each prediction?

SECTION 7.5   SYSTEMS OF LINEAR EQUATIONS

OBJECTIVES

1. Decide whether an ordered pair is a solution of a linear system.

2. Solve linear systems by graphing.

3. Solve linear systems by substitution.

4. Solve linear systems by addition.

5. Identify systems that do not have exactly one ordered-pair solution.

6. Solve problems using systems of linear equations.

Key West residents Brian Goss (left), George Wallace, and Michael Mooney (right) hold on to each other as they battle 90-mph winds along Houseboat Row in Key West, Fla., on Friday, Sept. 25, 1998. The three had sought shelter behind a Key West hotel as Hurricane Georges descended on the Florida Keys but were forced to seek other shelter when the storm conditions became too rough. Hundreds of people were killed by the storm when it swept through the Caribbean.

Real-world problems often involves thousands of equations, sometimes containing a million variables. Problems ranging from scheduling airline flights to controlling traffic flow to routing phone calls over the nation's communication network often require solutions in a matter of moments. AT&T's domestic long-distance network involves 800,000 variables! Meteorologists describing atmospheric conditions surrounding a hurricane must solve problems involving thousands of equations rapidly and efficiently. The difference between a two-hour warning and a two-day warning is a life-and-death issue for thousands of people in the path of one of nature's most destructive forces.

Although we will not be solving 800,000 equations with 800,000 variables, we will turn our attention to two equations with two variables, such as

$$2x - 3y = -4$$
$$2x + y = 4.$$

The methods that we consider for solving such problems provide the foundation for solving far more complex systems with many variables.

Systems of Linear Equations and Their Solutions

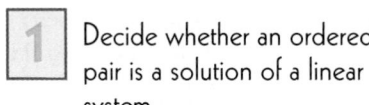 Decide whether an ordered pair is a solution of a linear system.

We have seen that all equations in the form $Ax + By = C$ are straight lines when graphed. Two such equations, such as those listed above, are called a **system of linear equations** or a **linear system**. A **solution to a system of linear equations** is an ordered pair that satisfies all equations in the system. For example, $(3, 4)$ satisfies the system

$$x + y = 7 \quad \text{(3 + 4 is, indeed, 7.)}$$
$$x - y = -1. \quad \text{(3 − 4 is, indeed, −1.)}$$

Thus, $(3, 4)$ satisfies both equations and is a solution of the system. The solution can be described by saying that $x = 3$ and $y = 4$. The solution can also be described using set notation. The solution set to the system is $\{(3, 4)\}$—that is, the set consisting of the ordered pair $(3, 4)$.

A system of linear equations can have exactly one solution, no solution, or infinitely many solutions. We begin with systems having exactly one solution.

EXAMPLE 1	DETERMINING WHETHER AN ORDERED PAIR IS A SOLUTION OF A SYSTEM

Determine whether $(1, 2)$ is a solution of the system:

$$2x - 3y = -4$$
$$2x + y = 4.$$

Solution Because 1 is the x-coordinate and 2 is the y-coordinate of $(1, 2)$, we replace x with 1 and y with 2.

$$
\begin{array}{ll}
2x - 3y = -4 & 2x + y = 4 \\
2(1) - 3(2) \stackrel{?}{=} -4 & 2(1) + 2 \stackrel{?}{=} 4 \\
2 - 6 \stackrel{?}{=} -4 & 2 + 2 \stackrel{?}{=} 4 \\
-4 = -4, \quad \text{true} & 4 = 4, \quad \text{true}
\end{array}
$$

The pair $(1, 2)$ satisfies both equations: It makes each equation true. Thus, the pair is a solution of the system.

 Check Point 1

Determine whether $(-4, 3)$ is a solution of the system:

$$x + 2y = 2$$
$$x - 2y = 6.$$

② Solve linear systems by graphing.

Solving Linear Systems by Graphing

The solution to a system of linear equations can be found by graphing both of the equations in the same rectangular coordinate system. For a system with one solution, **the coordinates of the point of intersection of the lines is the system's solution.**

EXAMPLE 2   SOLVING A LINEAR SYSTEM BY GRAPHING

Solve by graphing:

$$x + 2y = 2$$
$$x - 2y = 6.$$

Solution   We find the solution by graphing both $x + 2y = 2$ and $x - 2y = 6$ in the same rectangular coordinate system. We will use intercepts to graph each equation.

$x + 2y = 2$	$x - 2y = 6$
x-intercept: Set $y = 0$.	**x-intercept: Set $y = 0$.**
$x + 2 \cdot 0 = 2$	$x - 2 \cdot 0 = 6$
$x = 2$	$x = 6$
Graph $(2, 0)$.	Graph $(6, 0)$.
y-intercept: Set $x = 0$.	**y-intercept: Set $x = 0$.**
$0 + 2y = 2$	$0 - 2y = 6$
$2y = 2$	$-2y = 6$
$y = 1$	$y = -3$
Graph $(0, 1)$.	Graph $(0, -3)$.

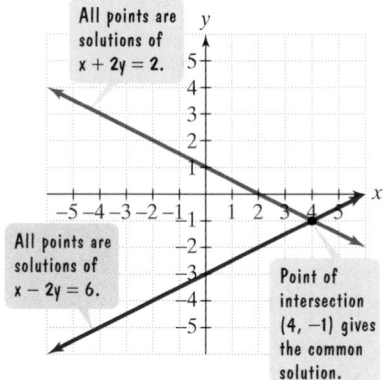

FIGURE 7.29 Visualizing a system's solution

The system is graphed in Figure 7.29. To ensure that the graph is accurate, check the coordinates of the intersection point, $(4, -1)$, in both equations. We replace x with 4 and y with -1.

$x + 2y = 2$	$x - 2y = 6$
$4 + 2(-1) \overset{?}{=} 2$	$4 - 2(-1) \overset{?}{=} 6$
$4 + (-2) \overset{?}{=} 2$	$4 - (-2) \overset{?}{=} 6$
$2 = 2,$ true	$4 + 2 \overset{?}{=} 6$
	$6 = 6,$ true

The pair $(4, -1)$ satisfies both equations—that is, it makes each equation true. This verifies that the system's solution set is $\{(4, -1)\}$.

 Solve by graphing:

$$2x + 3y = 6$$
$$2x + y = -2.$$

 ③ Solve linear systems by substitution.

Solving Linear Systems by the Substitution Method

Finding the solution to a linear system by graphing equations may not be easy to do. For example, a solution of $\left(-\frac{2}{3}, \frac{157}{29}\right)$ would be difficult to "see" as an intersection point of two lines.

Let's consider a method that does not depend on finding a system's solution visually: the substitution method. This method involves converting the system to one equation in one variable by an appropriate substitution.

EXAMPLE 3 SOLVING A SYSTEM BY SUBSTITUTION

Solve by the substitution method:

$$y = -x - 1$$
$$4x - 3y = 24.$$

Solution

STEP 1. Solve either of the equations for one variable in terms of the other. This step has already been done for us. The first equation, $y = -x - 1$, has y solved in terms of x.

STEP 2. Substitute the expression from step 1 into the other equation. We substitute the expression $-x - 1$ for y in the other equation:

$$y = \boxed{-x - 1} \qquad 4x - 3\,\boxed{y} = 24. \quad \text{Substitute } -x - 1 \text{ for } y.$$

This gives us an equation in one variable, namely

$$4x - 3(-x - 1) = 24.$$

The variable y has been eliminated.

STEP 3. Solve the resulting equation containing one variable.

$$4x - 3(-x - 1) = 24 \qquad \text{This is the equation containing one variable.}$$
$$4x + 3x + 3 = 24 \qquad \text{Apply the distributive property.}$$
$$7x + 3 = 24 \qquad \text{Combine like terms.}$$
$$7x = 21 \qquad \text{Subtract 3 from both sides.}$$
$$x = 3 \qquad \text{Divide both sides by 7.}$$

TECHNOLOGY

A graphing calculator can be used to solve the system in Example 3. Graph each equation and use the intersection feature. The calculator displays the solution $(3, -4)$ as $x = 3, y = -4$.

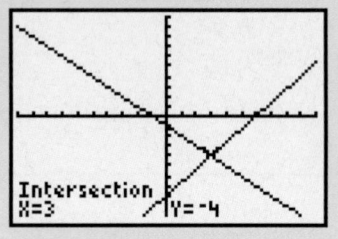

Intersection
X=3   Y=-4

STEP 4. Back-substitute the obtained value into the equation from step 1. We now know that the x-coordinate of the solution is 3. To find the y-coordinate, we back-substitute the x-value into the equation from step 1,

$$y = -x - 1.$$

Substitute 3 for x.

$$y = -3 - 1 = -4$$

With $x = 3$ and $y = -4$, the proposed solution is $(3, -4)$.

STEP 5. Check the proposed solution in both of the system's given equations. Replace x with 3 and y with -4.

$$y = -x - 1 \qquad\qquad 4x - 3y = 24$$
$$-4 \overset{?}{=} -3 - 1 \qquad\qquad 4(3) - 3(-4) \overset{?}{=} 24$$
$$-4 = -4, \quad \text{true} \qquad\qquad 12 + 12 \overset{?}{=} 24$$
$$24 = 24, \quad \text{true}$$

The pair $(3, -4)$ satisfies both equations. The system's solution set is $\{(3, -4)\}$.

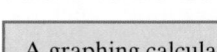 **Check Point 3** Solve by the substitution method:

$$y = 3x - 7$$
$$5x - 2y = 8.$$

Before considering additional examples, let's summarize the steps used in the substitution method.

STUDY TIP

In step 1, if possible, solve for a variable whose coefficient is 1 or −1 to avoid working with fractions.

SOLVING LINEAR SYSTEMS BY SUBSTITUTION

1. Solve either of the equations for one variable in terms of the other. (If one of the equations is already in this form, you can skip this step.)
2. Substitute the expression found in step 1 into the other equation. This will result in an equation in one variable.
3. Solve the equation obtained in step 2.
4. Back-substitute the value found from step 3 into the equation from step 1. Simplify and find the value of the remaining variable.
5. Check the proposed solution in both of the system's given equations.

EXAMPLE 4 | SOLVING A SYSTEM BY SUBSTITUTION

Solve by the substitution method:

$$5x - 4y = 9$$
$$x - 2y = -3.$$

Solution

STEP 1. Solve either of the equations for one variable in terms of the other. We begin by isolating one of the variables in either of the equations. By solving for x in the second equation, which has a coefficient of 1, we can avoid fractions.

$$x - 2y = -3 \qquad \text{This is the second equation in the given system.}$$
$$x = 2y - 3 \qquad \text{Solve for x by adding 2y to both sides.}$$

STEP 2. Substitute the expression from step 1 into the other equation. We substitute $2y - 3$ for x in the first equation.

$$x = \boxed{2y - 3} \qquad 5\,\boxed{x} - 4y = 9$$

This gives us an equation in one variable, namely

$$5(2y - 3) - 4y = 9.$$

The variable x has been eliminated.

STEP 3. Solve the resulting equation containing one variable.

$$5(2y - 3) - 4y = 9 \qquad \text{This is the equation containing one variable.}$$
$$10y - 15 - 4y = 9 \qquad \text{Apply the distributive property.}$$
$$6y - 15 = 9 \qquad \text{Combine like terms.}$$
$$6y = 24 \qquad \text{Add 15 to both sides.}$$
$$y = 4 \qquad \text{Divide both sides by 6.}$$

STEP 4. Back-substitute the obtained value into the equation from step 1. Now that we have the y-coordinate of the solution, we back-substitute 4 for y in the equation $x = 2y - 3$.

$$x = 2y - 3 \qquad \text{Use the equation obtained in step 1.}$$
$$x = 2(4) - 3 \qquad \text{Substitute 4 for y.}$$
$$x = 8 - 3 \qquad \text{Multiply.}$$
$$x = 5 \qquad \text{Subtract.}$$

With $x = 5$ and $y = 4$, the proposed solution is $(5, 4)$.

4 Solve linear systems by addition.

STEP 5. Check. Take a moment to show that $(5, 4)$ satisfies both given equations. The solution set is $\{(5, 4)\}$.

 **Check Point 4** Solve by the substitution method:

$$3x + 2y = -1$$
$$x - y = 3.$$

Solving Linear Systems by the Addition Method

The substitution method is most useful if one of the given equations has an isolated variable. A third, and frequently the easiest, method for solving a linear system is the addition method. Like the substitution method, the addition method involves eliminating a variable and ultimately solving an equation containing only one variable. However, this time we eliminate a variable by adding the equations. In Example 5, the x-terms are eliminated when the two equations are added.

EXAMPLE 5   SOLVING A SYSTEM BY THE ADDITION METHOD

Solve by the addition method:

$$3x - 4y = 11$$
$$-3x + 2y = -7.$$

Solution   Because the coefficients of x in the two equations differ only in sign, we can add the two left sides and add the two right sides, thereby eliminating the x-terms.

$$
\begin{array}{rcr}
3x - 4y & = & 11 \\
-3x + 2y & = & -7 \\
\hline
-2y & = & 4 \\
y & = & -2
\end{array}
$$

Add: Solve for y, dividing both sides by -2.

Now we can back-substitute -2 for y into one of the original equations to find x. We will use both equations to show that we obtain the same value for x in either case.

Using the First Equation:	**Using the Second Equation:**
$3x - 4y = 11$	$-3x + 2y = -7$
$3x - 4(-2) = 11$	$-3x + 2(-2) = -7$ Replace y with -2.
$3x + 8 = 11$	$-3x - 4 = -7$ Solve for x.
$3x = 3$	$-3x = -3$
$x = 1$	$x = 1$

Thus, $x = 1$ and $y = -2$. The proposed solution, $(1, -2)$, can be shown to satisfy both equations in the system. Consequently, the solution set is $\{(1, -2)\}$.

 **Check Point 5** Solve by the addition method:

$$2x - 5y = 26$$
$$-2x + 9y = -42.$$

When we use the addition method, we want to obtain two equations whose sum is an equation containing only one variable. The key step is to obtain, for one of the variables, coefficients that differ only in sign. To do this, we may need to multiply one or both equations by some nonzero number so that the coefficients of one of the variables, x or y, become opposites. Then when the two equations are added, this variable is eliminated.

EXAMPLE 6   SOLVING A SYSTEM BY THE ADDITION METHOD

Solve by the addition method:

$$3x + 2y = 48$$
$$9x - 8y = -24.$$

Solution   We must rewrite one or both equations in equivalent forms so that the coefficients of the same variable (either x or y) are opposites of each other. Consider the x-terms in each equation, that is, $3x$ and $9x$. To eliminate x, we can multiply each term of the first equation by -3 and then add the equations.

$$
\begin{array}{ll}
3x + 2y = 48 & \xrightarrow{\text{Multiply by } -3.} \quad -9x - 6y = -144 \\
9x - 8y = -24 & \xrightarrow{\text{No change}} \quad 9x - 8y = -24 \\
& \text{Add:} \quad\quad\quad\quad -14y = -168 \\
& y = 12 \quad \text{Solve for } y, \text{dividing} \\
& \text{both sides by } -14.
\end{array}
$$

Thus, $y = 12$. We back-substitute this value into either one of the given equations. We'll use the first one.

$$
\begin{array}{ll}
3x + 2y = 48 & \text{This is the first equation in the given system.} \\
3x + 2(12) = 48 & \text{Substitute 12 for } y. \\
3x + 24 = 48 & \text{Multiply.} \\
3x = 24 & \text{Subtract 24 from both sides.} \\
x = 8 & \text{Divide both sides by 3.}
\end{array}
$$

Finally, we verify that the ordered pair $(8, 12)$ satisfies both equations in the system.

$$
\begin{array}{lll}
3x + 2y = 48 & 9x - 8y = -24 & \text{These are the given equations.} \\
3(8) + 2(12) \overset{?}{=} 48 & 9(8) - 8(12) \overset{?}{=} -24 & \text{Replace } x \text{ with 8 and } y \text{ with 12.} \\
24 + 24 \overset{?}{=} 48 & 72 - 96 \overset{?}{=} -24 & \\
48 = 48, \quad \text{true} & -24 = -24, \quad \text{true} &
\end{array}
$$

Because $(8, 12)$ satisfies both equations, the system's solution set is $\{(8, 12)\}$.

   **Check Point 6**   Solve by the addition method:

$$4x + 5y = 3$$
$$2x - 3y = 7.$$

SOLVING LINEAR SYSTEMS BY ADDITION

1. If necessary, rewrite both equations in the form $Ax + By = C$.
2. If necessary, multiply either equation or both equations by appropriate nonzero numbers so that the sum of the x-coefficients or the sum of the y-coefficients is 0.
3. Add the equations in step 2. The sum is an equation in one variable.
4. Solve the equation in one variable.
5. Back-substitute the value obtained in step 4 into either of the given equations and solve for the other variable.
6. Check the solution in both of the original equations.

EXAMPLE 7   SOLVING A SYSTEM BY THE ADDITION METHOD

Solve by the addition method:

$$7x = 5 - 2y$$
$$3y = 16 - 2x.$$

STEP 1: Rewrite both equations in the form $Ax + By = C$. We first arrange the system so that variable terms appear on the left and constants appear on the right. We obtain

$$7x + 2y = 5 \qquad \text{Add 2y to both sides in the first equation.}$$
$$2x + 3y = 16. \qquad \text{Add 2x to both sides in the second equation.}$$

STEP 2: If necessary, multiply either equation or both equations by appropriate numbers so that the sum of the x-coefficients or the sum of the y-coefficients is 0. We can eliminate x or y. Let's eliminate y by multiplying the first equation by 3 and the second equation by -2.

$$
\begin{array}{ll}
7x + 2y = 5 & \xrightarrow{\text{Multiply by 3.}} \quad 21x + 6y = 15 \\
2x + 3y = 16 & \xrightarrow{\text{Multiply by } -2.} \quad -4x - 6y = -32
\end{array}
$$

STEP 3: Add the equations. $\qquad\qquad \text{Add: } \overline{17x + 0y = -17}$
$$17x = -17$$

STEP 4: Solve the equation in one variable. We solve $17x = -17$ by dividing both sides by 17.

$$\frac{17x}{17} = \frac{-17}{17} \qquad \text{Divide both sides by 17.}$$

$$x = -1 \qquad \text{Simplify.}$$

STEP 5: Back-substitute and find the value for the other variable. We can back-substitute -1 for x into either one of the given equations. We'll use the second one.

$$3y = 16 - 2x \qquad \text{This is the second equation in the given system.}$$
$$3y = 16 - 2(-1) \qquad \text{Substitute } -1 \text{ for x.}$$
$$3y = 16 + 2 \qquad \text{Multiply.}$$
$$3y = 18 \qquad \text{Add.}$$
$$y = 6 \qquad \text{Divide both sides by 3.}$$

With $x = -1$ and $y = 6$, the proposed solution is $(-1, 6)$.

STEP 6: Check. Take a moment to show that $(-1, 6)$ satisfies both given equations. The solution is $(-1, 6)$ and the solution set is $\{(-1, 6)\}$.

Check Point 7   Solve by the addition method:

$$3x = 2 - 4y$$
$$5y = -1 - 2x.$$

5 Identify systems that do not have exactly one ordered-pair solution.

Linear Systems Having No Solution or Infinitely Many Solutions

Because the graph of a linear equation is a straight line, there are three possibilities for the number of solutions to a system of two linear equations.

THE NUMBER OF SOLUTIONS TO A SYSTEM OF TWO LINEAR EQUATIONS

The number of solutions to a system of two linear equations in two variables is given by one of the following: (See Figure 7.30.)

Number of Solutions	What This Means Graphically
Exactly one ordered-pair solution	The two lines intersect at one point.
No solution	The two lines are parallel.
Infinitely many solutions	The two lines are identical.

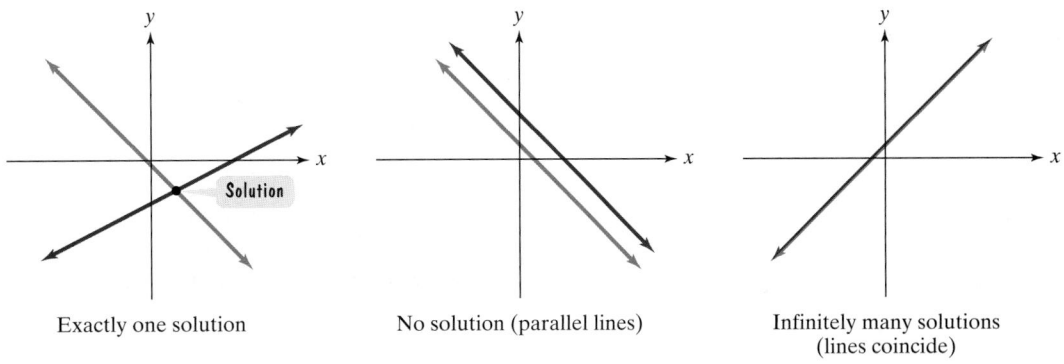

Exactly one solution   No solution (parallel lines)   Infinitely many solutions (lines coincide)

FIGURE 7.30 Possible graphs for a system of two linear equations in two variables

---

EXAMPLE 8 | A SYSTEM WITH NO SOLUTION

Solve the system:

$$4x + 6y = 12$$
$$6x + 9y = 12.$$

Solution   Because no variable is isolated, we will use the addition method. To obtain coefficients of x that differ only in sign, we multiply the first equation by 3 and the second equation by -2.

$$
\begin{array}{ll}
4x + 6y = 12 & \xrightarrow{\text{Multiply by 3.}} \quad 12x + 18y = 36 \\
6x + 9y = 12 & \xrightarrow{\text{Multiply by } -2.} \quad -12x - 18y = -24 \\
& \qquad\qquad\text{Add:} \qquad\qquad 0 = 12
\end{array}
$$

There are no values of x and y for which 0 = 12.

The false statement $0 = 12$ indicates that the system has no solution. The solution set is the empty set, $\varnothing$.

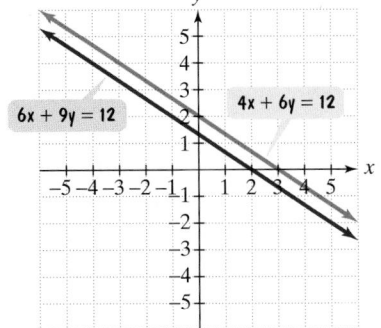

FIGURE 7.31 The graph of a system with no solution

The lines corresponding to the two equations in Example 8 are shown in Figure 7.31. The lines are parallel and have no point of intersection.

Check Point 8   Solve the system:

$$x + 2y = 4$$
$$3x + 6y = 13.$$

EXAMPLE 9   A SYSTEM WITH INFINITELY MANY SOLUTIONS

Solve the system:

$$y = 3 - 2x$$
$$4x + 2y = 6.$$

Solution   Because the variable y is isolated in the first equation, we can use the substitution method. We substitute the expression for y, in the second equation.

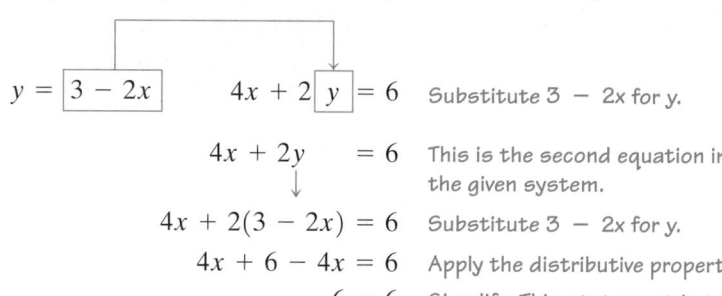

$y = \boxed{3 - 2x}$ $4x + 2\boxed{y} = 6$	Substitute $3 - 2x$ for y.
$4x + 2y \quad = 6$	This is the second equation in the given system.
$4x + 2(3 - 2x) = 6$	Substitute $3 - 2x$ for y.
$4x + 6 - 4x = 6$	Apply the distributive property.
$6 = 6$	Simplify. This statement is true for all values of x and y.

In our final step, both variables have been eliminated, and the resulting statement $6 = 6$ is true. This true statement indicates that the system has infinitely many solutions. The solution set consists of all points (x, y) lying on either of the coinciding lines, $y = 3 - 2x$ or $4x + 2y = 6$, as shown in Figure 7.32.

We express the solution set for the system in one of two equivalent ways:

$$\{(x, y) \mid y = 3 - 2x\}$$ The set of all ordered pairs (x, y) such that $y = 3 - 2x$
or $\{(x, y) \mid 4x + 2y = 6\}$.   The set of all ordered pairs (x, y) such that $4x + 2y = 6$

FIGURE 7.32 The graph of a system with infinitely many solutions

---

Check Point 9   Solve the system:

$$y = 4x - 4$$
$$8x - 2y = 8.$$

LINEAR SYSTEMS HAVING NO SOLUTION OR INFINITELY MANY SOLUTIONS

If both variables are eliminated when a system of linear equations is solved by substitution or addition, one of the following is true:

1. There is no solution if the resulting statement is false.
2. There are infinitely many solutions if the resulting statement is true.

---

6   Solve problems using systems of linear equations.

Problem Solving

When we solved word problems in Section 6.3, we let x represent a quantity that was unknown. Problems in this section involve *two* unknown quantities. We will let x and y represent these quantities. We then translate from the verbal conditions of the problem to a *system* of linear equations.

EXAMPLE 10 | CHOLESTEROL AND HEART DISEASE

About 15 million hamburgers are eaten every day in the United States.

The verdict is in. After years of research, the nation's health experts agree that high cholesterol in the blood is a major contributor to heart disease. Thus, cholesterol intake should be limited to 300 milligrams or less each day. Fast foods provide a cholesterol carnival. All together, two McDonald's Quarter Pounders and three Burger King Whoppers with cheese contain 520 milligrams of cholesterol. Three Quarter Pounders and one Whopper with cheese exceed the suggested daily cholesterol intake by 53 milligrams. Determine the cholesterol content in each item.

Solution

STEP 1. Use variables to represent the unknown quantities. Let x represent the cholesterol content, in milligrams, of a Quarter Pounder and y the cholesterol content, in milligrams, of a Whopper with cheese.

STEP 2. Write a system of equations describing the problem's conditions.

The amount of cholesterol in 2 Quarter Pounders	plus	the amount of cholesterol in 3 Whoppers with cheese	is	520 mg.
$2x$	$+$	$3y$	$=$	520

The amount of cholesterol in 3 Quarter Pounders	plus	the amount of cholesterol in 1 Whopper with cheese	is	the suggested daily limit plus 53 mg.
$3x$	$+$	y	$=$	$300 + 53$

STEP 3. Solve the system and answer the problem's question. The system

$$2x + 3y = 520$$
$$3x + y = 353$$

can be solved by substitution or addition. We will use addition; if we multiply the second equation by -3, adding equations will eliminate y.

$$
\begin{array}{lll}
2x + 3y = 520 & \xrightarrow{\text{No change}} & 2x + 3y = 520 \\
3x + y = 353 & \xrightarrow{\text{Multiply by } -3} & -9x - 3y = -1059 \\
& \text{Add: } & -7x = -539 \\
& & x = \dfrac{-539}{-7} = 77
\end{array}
$$

Because x represents the cholesterol content of a Quarter Pounder, we see that a Quarter Pounder contains 77 milligrams of cholesterol. Now we can find y, the cholesterol content of a Whopper with cheese. We do so by back-substituting 77 for x in either of the system's equations.

$$3x + y = 353 \qquad \text{We'll use the second equation.}$$
$$3(77) + y = 353 \qquad \text{Back-substitute 77 for x.}$$
$$231 + y = 353 \qquad \text{Multiply.}$$
$$y = 122 \qquad \text{Subtract 231 from both sides.}$$

Because $x = 77$ and $y = 122$, a Quarter Pounder contains 77 milligrams of cholesterol and a Whopper with cheese contains 122 milligrams of cholesterol.

STEP 4. Check the proposed answers in the original wording of the problem. Two Quarter Pounders and three Whoppers with cheese contain

$$2(77 \text{ mg}) + 3(122 \text{ mg}) = 520 \text{ mg}$$

which checks with the given conditions. Furthermore, three Quarter Pounders and one Whopper with cheese contain

$$3(77 \text{ mg}) + 1(122 \text{ mg}) = 353 \text{ mg}$$

which does exceed the daily limit of 300 milligrams by 53 milligrams.

 Check 10 Point

How do the Quarter Pounder and Whopper with cheese measure up in the calorie department? Actually, not too well. Two Quarter Pounders and three Whoppers with cheese provide 2607 calories. Even one of each provides enough calories to bring tears to Jenny Craig's eyes—9 calories in excess of what is allowed on a 1000-calorie-a-day diet. Find the caloric content of each item.

Exercise Set 7.5

Practice Exercises

In Exercises 1–4, determine whether the given ordered pair is a solution of the system.

1. $(2, 3)$
$x + 3y = 11$
$x - 5y = -13$

2. $(-3, 5)$
$9x + 7y = 8$
$8x - 9y = -69$

3. $(2, 5)$
$2x + 3y = 17$
$x + 4y = 16$

4. $(8, 5)$
$5x - 4y = 20$
$3y = 2x + 1$

In Exercises 5–12, solve each system by graphing. Check the coordinates of the intersection point in both equations.

5. $x + y = 6$
$x - y = 2$

6. $x + y = 2$
$x - y = 4$

7. $2x - 3y = 6$
$4x + 3y = 12$

8. $4x + y = 4$
$3x - y = 3$

9. $y = x + 5$
$y = -x + 3$

10. $y = x + 1$
$y = 3x - 1$

11. $y = -x - 1$
$4x - 3y = 24$

12. $y = 3x - 4$
$2x + y = 1$

In Exercises 13–24, solve each system by the substitution method. Be sure to check all proposed solutions.

13. $x + y = 4$
$y = 3x$

14. $x + y = 6$
$y = 2x$

15. $x + 3y = 8$
$y = 2x - 9$

16. $2x - 3y = -13$
$y = 2x + 7$

17. $x + 3y = 5$
$4x + 5y = 13$

18. $y = 2x + 7$
$2x - y = -15$

19. $2x - y = -5$
$x + 5y = 14$

20. $2x + 3y = 11$
$x - 4y = 0$

21. $2x - y = 3$
$5x - 2y = 10$

22. $-x + 3y = 10$
$2x + 8y = -6$

23. $x + 8y = 6$
$2x + 4y = -3$

24. $-4x + y = -11$
$2x - 3y = 5$

In Exercises 25–36, solve each system by the addition method. Be sure to check all proposed solutions.

25. $x + y = 1$
$x - y = 3$

26. $x + y = 6$
$x - y = -2$

27. $2x + 3y = 6$
$2x - 3y = 6$

28. $3x + 2y = 14$
$3x - 2y = 10$

29. $x + 2y = 2$
$-4x + 3y = 25$

30. $2x - 7y = 2$
$3x + y = -20$

31. $4x + 3y = 15$
$2x - 5y = 1$

32. $3x - 7y = 13$
$6x + 5y = 7$

33. $3x - 4y = 11$
$2x + 3y = -4$

34. $2x + 3y = -16$
$5x - 10y = 30$

35. $2x = 3y - 4$
$-6x + 12y = 6$

36. $5x = 4y - 8$
$3x + 7y = 14$

In Exercises 37–44, solve by the method of your choice. Identify systems with no solution and systems with infinitely many solutions, using set notation to express their solution sets.

37. $x = 9 - 2y$
$x + 2y = 13$

38. $6x + 2y = 7$
$y = 2 - 3x$

39. $y = 3x - 5$
$21x - 35 = 7y$

40. $9x - 3y = 12$
$y = 3x - 4$

41. $3x - 2y = -5$
$4x + y = 8$

42. $2x + 5y = -4$
$3x - y = 11$

43. $x + 3y = 2$
$3x + 9y = 6$

44. $4x - 2y = 2$
$2x - y = 1$

In Exercises 45–48, let x represent one number and let y represent the other number. Use the given conditions to write a system of equations. Solve the system and find the numbers.

45. The sum of two numbers is 7. If one number is subtracted from the other, their difference is −1. Find the numbers.

46. The sum of two numbers is 2. If one number is subtracted from the other, their difference is 8. Find the numbers.

47. Three times a first number decreased by a second number is 1. The first number increased by twice the second number is 12. Find the numbers.

48. The sum of three times a first number and twice a second number is 8. If the second number is subtracted from twice the first number, the result is 3. Find the numbers.

Application Exercises

The graph below shows the calories in some favorite fast foods. Use the information in Exercises 49–50 to find the exact caloric content of the specified foods.

Calories in Some Favorite Fast Foods

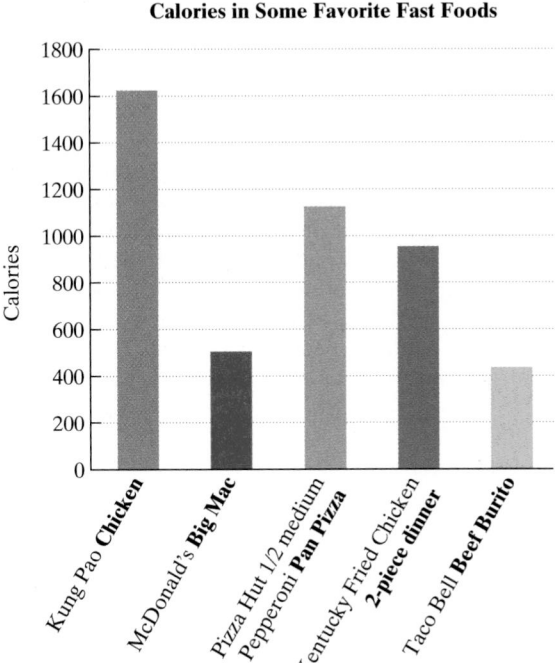

Source: Center for Science in the Public Interest

49. One pan pizza and two beef burritos provide 1980 calories. Two pan pizzas and one beef burrito provide 2670 calories. Find the caloric content of each item.

50. One Kung Pao chicken and two Big Macs provide 2620 calories. Two Kung Pao chickens and one Big Mac provide 3740 calories. Find the caloric content of each item.

51. In the United States, deaths from car accidents, per 100,000 persons, are decreasing at a faster rate than deaths from gunfire, shown by the blue and red lines that model the data points in the figure.

Annual Deaths in the U.S. from Car Accidents and Gunfire

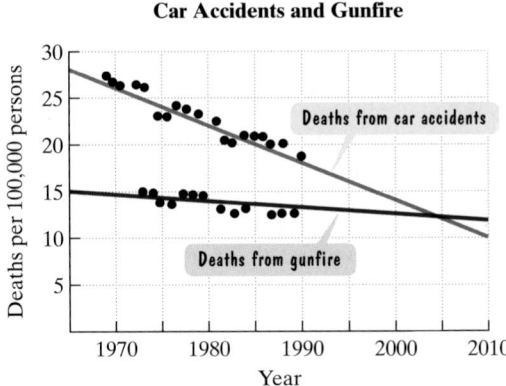

Source: Journal of the American Medical Association

The function $y = -0.4x + 28$ models deaths from car accidents, y, per 100,000 persons, x years after 1965. The function $0.07x + y = 15$ models deaths from gunfire, y, per 100,000 persons, x years after 1965. Use these models to project when the number of deaths from gunfire will equal the number of deaths from car accidents. Round to the nearest year. How many annual deaths, per 100,000 persons, will there be from each at that time? Describe how this is illustrated by the lines in the figure shown.

52. As consumption of carbonated soft drinks in the United States declines, there is an emerging market for alternative beverages, including juices, teas, bottled waters, and sports drinks. The line graphs show the projected U.S. beverage market share for carbonated and noncarbonated alternative beverages through 2010.

U.S. Beverage Market Share

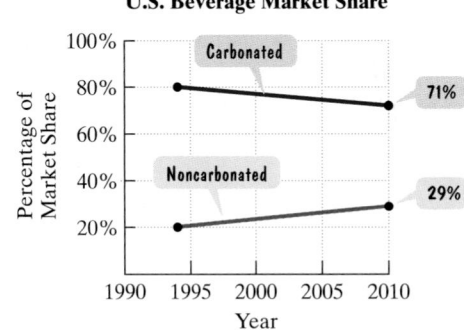

Source: Beverage Digest

The market share for carbonated beverages can be modeled by $M = -0.57x + 80$ and for noncarbonated beverages by $M = 0.57x + 20$. In both models, x is the number of years after 1994 and M is the market share, measured as a percentage of all sales. Use these models to determine when carbonated and noncarbonated beverages will have equal shares of the market. Round to the nearest year.

The graphs show average weekly earnings of full-time wage and salary workers 25 and older, by educational attainment. Exercises 53–54 involve the information in these graphs.

Average Weekly Earnings by Educational Attainment

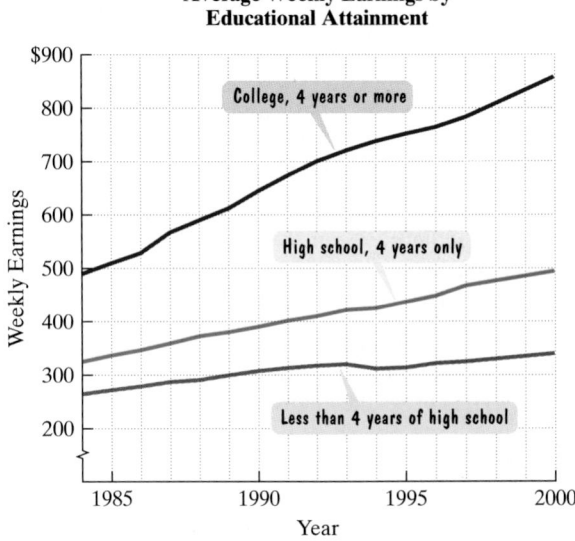

Source: U.S. Bureau of Labor Statistics

(The graph for Exercises 53–54 are at the bottom of page 383.)

53. In 1985, college graduates averaged $508 in weekly earnings. This amount has increased by approximately $25 in weekly earnings per year. By contrast, in 1985, high school graduates averaged $345 in weekly earnings. This amount has only increased by approximately $9 in weekly earnings per year. How many years after 1985 will college graduates be earning twice as much per week as high school graduates? In which year will this occur? What will be the weekly earnings for each group at that time?

54. In 1985, college graduates averaged $508 in weekly earnings. This amount has increased by approximately $25 in weekly earnings per year. By contrast, in 1985, people with less than four years of high school averaged $270 in weekly earnings. This amount has only increased by approximately $4 in weekly earnings per year. How many years after 1985 will college graduates be earning three times as much per week as people with less than four years of high school? (Round to the nearest whole number.) In which year will this occur? What will be the weekly earnings for each group at that time?

Writing in Mathematics

55. What is a system of linear equations? Provide an example with your description.

56. What is the solution to a system of linear equations?

57. Explain how to solve a system of equations using graphing.

58. Explain how to solve a system of equations using the substitution method. Use $y = 3 - 3x$ and $3x + 4y = 6$ to illustrate your explanation.

59. Explain how to solve a system of equations using the addition method. Use $3x + 5y = -2$ and $2x + 3y = 0$ to illustrate your explanation.

60. What is the disadvantage to solving a system of equations using the graphing method?

61. When is it easier to use the addition method rather than the substitution method to solve a system of equations?

62. When using the addition or substitution method, how can you tell whether a system of linear equations has infinitely many solutions? What is the relationship between the graphs of the two equations?

63. When using the addition or substitution method, how can you tell whether a system of linear equations has no solution? What is the relationship between the graphs of the two equations?

64. The law of supply and demand states that, in a free market economy, a commodity tends to be sold at its equilibrium price. At this price, the amount that the seller will supply is the same amount that the consumer will buy. Explain how systems of equations can be used to determine the equilibrium price.

Critical Thinking Exercises

65. Write a word problem that can be solved by translating to a system of linear equations. Then solve the problem.

66. Write a system of equations having $\{(-2, 7)\}$ as a solution set. (More than one system is possible.)

67. One apartment is directly above a second apartment. The resident living downstairs calls his neighbor living above him and states, "If one of you is willing to come downstairs, we'll have the same number of people in both apartments." The upstairs resident responds, "We're all too tired to move. Why don't one of you come up here? Then we will have twice as many people up here as you've got down there." How many people are in each apartment?

Technology Exercises

68. Verify your solutions to Exercises 5–12 using a graphing calculator to graph the two equations in the system in the same viewing rectangle. After entering the two equations, one as y_1 and the other as y_2, and graphing them, use the $\boxed{\text{TRACE}}$ and $\boxed{\text{ZOOM}}$ features to find the coordinates of the intersection point. (It may first be necessary to solve an equation for y before entering it.) Many graphing utilities have a special $\boxed{\text{INTERSECTION}}$ feature that displays the coordinates of the intersection point once the equations are graphed. Consult your manual.

Group Exercise

69. a. In solving

$$3x + 5y = 26$$
$$y = 2x$$

by substitution, a student found that $x = 2$. At that point the student asserted that the system's solution is $x = 2$. What is the error?

b. In solving

$$y = 4 - x$$
$$2x + 2y = 8$$

by substitution, a student obtained $0 = 0$, giving the solution as $(0, 0)$. What is the error?

c. In your group, discuss other common errors that can occur in solving linear systems by any of the three methods presented in this section. Give specific examples of these kinds of errors. What suggestions can group members offer for avoiding these errors?

SECTION 7.6   LINEAR INEQUALITIES IN TWO VARIABLES

OBJECTIVES
1. Graph a linear inequality.
2. Graph a system of linear inequalities.

Had a good workout lately? If so, could you tell if you were overdoing it or not pushing yourself hard enough? In this section, we will use systems of inequalities in two variables to help you establish a target zone for your workouts.

Linear Inequalities in Two Variables and Their Solutions

We have seen that equations in the form $Ax + By = C$ are straight lines when graphed. If we change the $=$ sign to $>$, $<$, $\geq$, or $\leq$, we obtain a **linear inequality in two variables**. Some examples of linear inequalities in two variables are

$$x + y < 2 \quad \text{and} \quad 3x - 5y \geq 15.$$

A **solution of an inequality in two variables**, x and y, is an ordered pair of real numbers with the following property: When the x-coordinate is substituted for x and the y-coordinate is substituted for y in the inequality, we obtain a true statement. For example, $(3, 1)$ is a solution of the inequality $x + y > 2$. When 3 is substituted for x and 1 is substituted for y, we obtain the true statement $3 + 1 > 2$, or $4 > 2$. Because there are infinitely many pairs of numbers that have a sum greater than 2, the inequality $x + y > 2$ has infinitely many solutions. Each ordered pair solution is said to **satisfy** the inequality. Thus, $(3, 1)$ satisfies the inequality $x + y > 2$.

1 Graph a linear inequality.

The Graph of a Linear Inequality in Two Variables

We know that the graph of an equation in two variables is the set of all points whose coordinates satisfy the equation. Similarly, the **graph of an inequality in two variables** is the set of all points whose coordinates satisfy the inequality.

Let's use Figure 7.33 to get an idea of what the graph of a linear inequality in two variables looks like. Part of the figure shows the graph of the linear equation $x + y = 2$. The line divides the points in the rectangular coordinate system into three sets. First, there is the set of points along the line, satisfying $x + y = 2$. Next, there is the set of points in the green region above the line. Points in the green region satisfy the linear inequality $x + y > 2$. Finally, there is the set of points in the pink region below the line. Points in the pink region satisfy the linear inequality $x + y < 2$.

A **half-plane** is the set of all the points on one side of a line. In Figure 7.33, the green region is a half-plane. The pink region is also a half-plane. A half-plane is the graph of a linear inequality that involves $>$ or $<$. The graph of an inequality that involves $\geq$ or $\leq$ is a half-plane and a line. A solid line is used to show that the line is part of the graph. A dashed line is used to show that a line is not part of a graph.

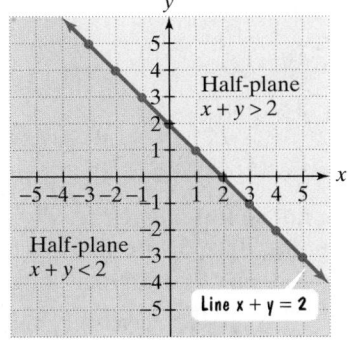

FIGURE 7.33

GRAPHING A LINEAR INEQUALITY IN TWO VARIABLES

1. Replace the inequality symbol with an equal sign and graph the corresponding linear equation. Draw a solid line if the original inequality contains a ≤ or ≥ symbol. Draw a dashed line if the original inequality contains a < or > symbol.
2. Choose a test point in one of the half-planes that is not on the line. Substitute the coordinates of the test point into the inequality.
3. If a true statement results, shade the half-plane containing this test point. If a false statement results, shade the half-plane not containing this test point.

EXAMPLE 1 GRAPHING A LINEAR INEQUALITY IN TWO VARIABLES

Graph: $3x - 5y \geq 15$.

Solution

STEP 1. Replace the inequality symbol by = and graph the linear equation. We need to graph $3x - 5y = 15$. We can use intercepts to graph this line.

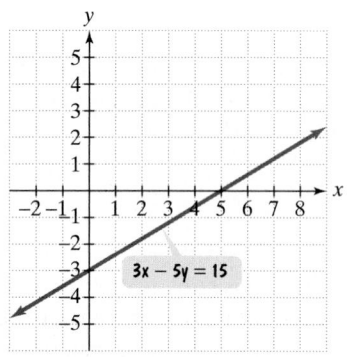

FIGURE 7.34 Preparing to graph $3x - 5y \geq 15$

We set $y = 0$ to find the x-intercept.	We set $x = 0$ to find the y-intercept.
$3x - 5y = 15$	$3x - 5y = 15$
$3x - 5 \cdot 0 = 15$	$3 \cdot 0 - 5y = 15$
$3x = 15$	$-5y = 15$
$x = 5$	$y = -3$

The x-intercept is 5, so the line passes through $(5, 0)$. The y-intercept is -3, so the line passes through $(0, -3)$. Using the intercepts, the line is shown in Figure 7.34 as a solid line. This is because the inequality $3x - 5y \geq 15$ contains a ≥ symbol, in which equality is included.

STEP 2. Choose a test point in one of the half-planes that is not on the line. Substitute its coordinates into the inequality. The line $3x - 5y = 15$ divides the plane into three parts—the line itself and two half-planes. The points in one half-plane satisfy $3x - 5y > 15$. The points in the other half-plane satisfy $3x - 5y < 15$. We need to find which half-plane belongs to the solution. To do so, we test a point from either half-plane. The origin, $(0, 0)$, is the easiest point to test.

$$3x - 5y \geq 15 \quad \text{This is the given inequality.}$$

$$3 \cdot 0 - 5 \cdot 0 \overset{?}{\geq} 15 \quad \text{Test } (0, 0) \text{ by substituting 0 for } x \text{ and 0 for } y.$$

$$0 - 0 \overset{?}{\geq} 15 \quad \text{Multiply.}$$

$$0 \geq 15 \quad \text{This statement is false.}$$

STEP 3. If a false statement results, shade the half-plane not containing the test point. Because 0 is not greater than or equal to 15, the test point, $(0, 0)$, is not part of the solution set. Thus, the half-plane below the solid line $3x - 5y = 15$ is part of the solution set. The solution set is the line and the half-plane that does not contain the point $(0, 0)$, indicated by shading this half-plane. The graph is shown using green shading and a blue line in Figure 7.35.

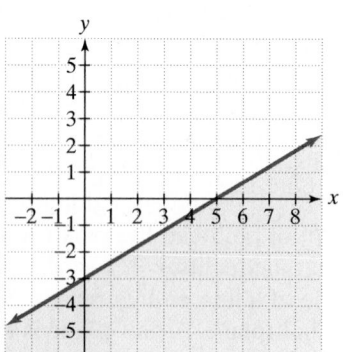

FIGURE 7.35 The graph of $3x - 5y \geq 15$

Check 1 Point   Graph: $2x - 4y \geq 8$.

When graphing a linear inequality, test a point that lies in one of the half-planes and *not on the line dividing the half-planes*. The test point $(0, 0)$ is convenient because it is easy to calculate when 0 is substituted for each variable. However, if $(0, 0)$ lies on the dividing line and not in a half-plane, a different test point must be selected.

EXAMPLE 2	GRAPHING A LINEAR INEQUALITY IN TWO VARIABLES

Graph: $y > -\dfrac{2}{3}x$.

Solution

STEP 1. Replace the inequality symbol by = and graph the linear equation. We need to graph $y = -\frac{2}{3}x$. We can use the slope and the y-intercept to graph this linear function.

$$y = -\frac{2}{3}x + 0$$

Slope $= \dfrac{-2}{3} = \dfrac{\text{rise}}{\text{run}}$ y-intercept $= 0$

The y-intercept is 0, so the line passes through $(0, 0)$. Using the y-intercept and the slope, the line is shown in Figure 7.36 as a dashed line. This is because the inequality $y > -\frac{2}{3}x$ contains a $>$ symbol, in which equality is not included.

STEP 2. Choose a test point in one of the half-planes that is not on the line. Substitute its coordinates into the inequality. We cannot use $(0, 0)$ as a test point because it lies on the line and not in a half-plane. Let's use $(1, 1)$, which lies in the half-plane above the line.

$$y > -\frac{2}{3}x \qquad \text{This is the given inequality.}$$

$$1 \overset{?}{>} -\frac{2}{3} \cdot 1 \qquad \text{Test } (1, 1) \text{ by substituting 1 for } x \text{ and 1 for } y.$$

$$1 > -\frac{2}{3} \qquad \text{This statement is true.}$$

STEP 3. If a true statement results, shade the half-plane containing the test point. Because 1 is greater than $-\frac{2}{3}$, the test point, $(1, 1)$, is part of the solution set. All the points on the same side of the line $y = -\frac{2}{3}x$ as the point $(1, 1)$ are members of the solution set. The solution set is the half-plane that contains the point $(1, 1)$, indicated by shading this half-plane. The graph is shown using green shading and a dashed blue line in Figure 7.36.

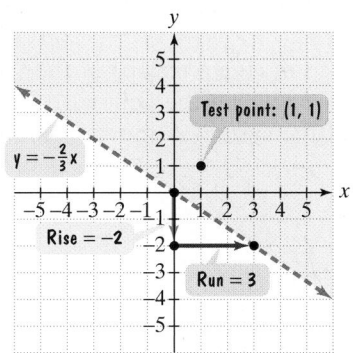
FIGURE 7.36 The graph of $y > -\frac{2}{3}x$

Test point: (1, 1)

$y = -\frac{2}{3}x$

Rise = -2

Run = 3

Check 2 Point   Graph: $y > -\dfrac{3}{4}x$.

You can graph inequalities in the form $y > mx + b$ or $y < mx + b$ without using test points. The inequality symbol indicates which half-plane to shade.

• If $y > mx + b$, shade above the line $y = mx + b$.
• If $y < mx + b$, shade below the line $y = mx + b$.

 2 Graph a system of linear inequalities.

Systems of Linear Inequalities

We have seen that a system of linear equations can be solved by graphing the individual equations in the system and then finding the intersection of the lines. To graph a system of inequalities, we graph each inequality and then find the intersection of the graphs.

EXAMPLE 3   GRAPHING A SYSTEM OF LINEAR INEQUALITIES

Graph the system:

$$2x - y < 4$$
$$x + y \geq -1.$$

Solution   We begin by graphing $2x - y < 4$. Because the inequality contains a $<$ symbol, we graph $2x - y = 4$ as a dashed line. (If $x = 0$, then $y = -4$, and if $y = 0$, then $x = 2$. The x-intercept is 2, and the y-intercept is -4.) Because $(0, 0)$ makes the inequality $2x - y < 4$ true, we shade the half-plane containing $(0, 0)$, shown in yellow in Figure 7.37.

Now we graph $x + y \geq -1$ in the same rectangular coordinate system. Because the inequality contains a $\geq$ symbol, in which equality is included, we graph $x + y = -1$ as a solid line. (If $x = 0$, then $y = -1$, and if $y = 0$, then $x = -1$. The x-intercept and y-intercept are both -1.) Because $(0, 0)$ makes the inequality true, we shade the half-plane containing $(0, 0)$. This is shown in Figure 7.38 using green vertical shading above the solid red line.

The graph of the system is shown by the intersection (the overlap) of the two half-planes. This is shown in Figure 7.38 as the region in which the yellow shading and the green vertical shading overlap. The graph of the system is shown again in Figure 7.39.

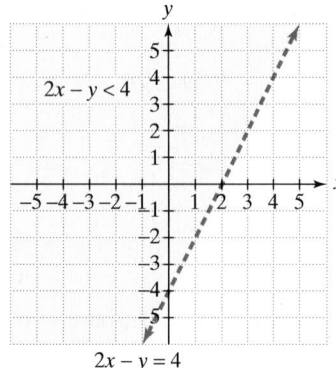

FIGURE 7.37 The graph of $2x - y < 4$

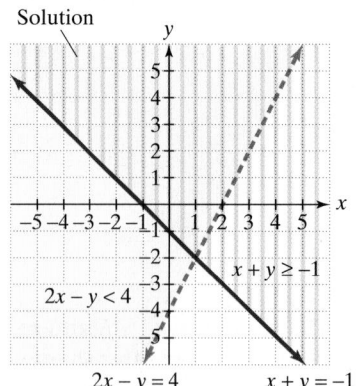

FIGURE 7.38 Adding the graph of $x + y \geq -1$

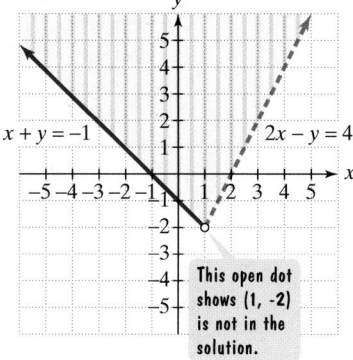

FIGURE 7.39 The graph of $2x - y < 4$ and $x + y \geq -1$

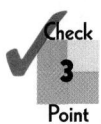 **Check Point 3**   Graph the system:

$$x + 2y > 4$$
$$2x - 3y \leq -6.$$

It is not necessary to use test points when graphing inequalities involving vertical or horizontal lines.

For the Vertical Line $x = a$:
- If $x > a$, shade to the right of $x = a$.
- If $x < a$, shade to the left of $x = a$.

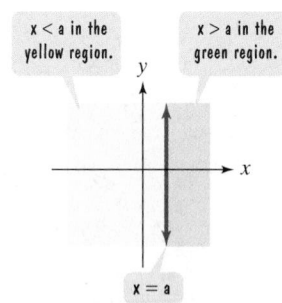

For the Horizontal Line $y = b$:
- If $y > b$, shade above $y = b$.
- If $y < b$, shade below $y = b$.

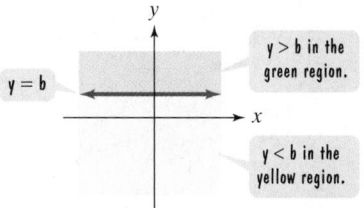

EXAMPLE 4 | GRAPHING A SYSTEM OF LINEAR INEQUALITIES

Graph the system:

$$x \leq 4$$
$$y > -2.$$

Solution We begin by graphing $x \leq 4$. Because equality is included in $\leq$, we graph $x = 4$ as a solid blue vertical line. The graph of $x < 4$ is the half-plane to the left of the blue line. See Figure 7.40 in which the half-plane is shown using yellow shading.

Now we add the graph of $y > -2$. Because equality is not included in $>$, we graph $y = -2$ as a dashed red horizontal line. The graph of $y > -2$ is the half-plane above the red line. See Figure 7.41 in which the half-plane is shown using green vertical shading.

The graph of the system is shown by the intersection of the two half-planes. This is shown in Figure 7.41 as the region in which the yellow shading and the green vertical shading overlap. The graph of the system is shown again in Figure 7.42. The point of intersection of the two lines, $(4, -2)$, is not part of the system's graph because it does not satisfy the inequality $y > -2$.

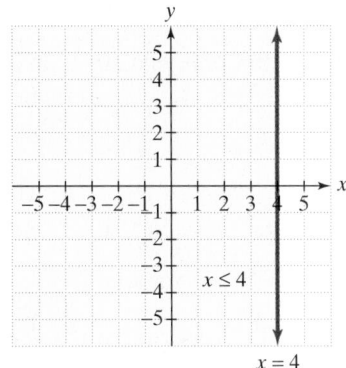

FIGURE 7.40 The graph of $x \leq 4$

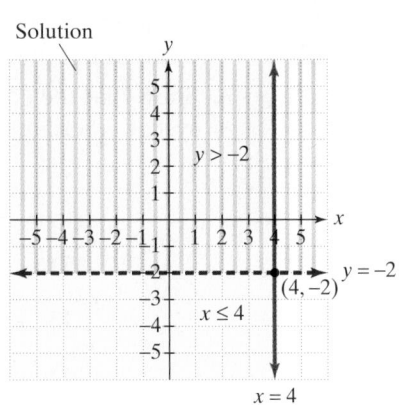

FIGURE 7.41 Adding the graph of $y > -2$

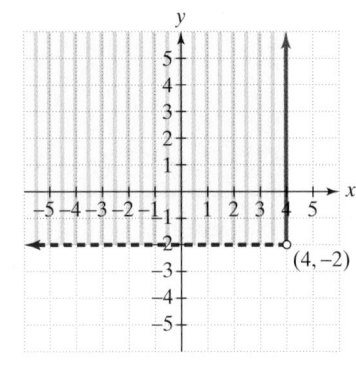

FIGURE 7.42 The graph of $x \leq 4$ and $y > -2$

✓ **Check Point 4** Graph the system:

$$x < 3$$
$$y \geq -1.$$

EXAMPLE 5 INEQUALITIES AND AEROBIC EXERCISE

For people between ages 10 and 70, inclusive, the target zone for aerobic exercise is given by the following system of inequalities in which a represents one's age and p is one's pulse rate:

$$2a + 3p \geq 450$$
$$a + p \leq 190.$$

The graph of this target zone is shown in Figure 7.43. Find your age. The line segments on the top and bottom of the shaded region indicate upper and lower limits for your pulse rate, in beats per minute, when engaging in aerobic exercise.

a. What are the coordinates of point A and what does this mean in terms of age and pulse rate?

b. Show that the coordinates of point A satisfy each inequality in the system.

FIGURE 7.43

Solution

a. Point A has coordinates $(20, 160)$. This means that a pulse rate of 160 beats per minute is within the target zone for a 20-year-old person engaged in aerobic exercise.

b. We can show that $(20, 160)$ satisfies each inequality by substituting 20 for a and 160 for p.

$2a + 3p \geq 450$	$a + p \leq 190$
Is $2(20) + 3(160) \geq 450$?	Is $20 + 160 \leq 190$?
$40 + 480 \geq 450$	$180 \leq 190$, true
$520 \geq 450$, true	

The pair $(20, 160)$ makes each inequality true, so it satisfies each inequality in the system.

✓ **Check Point 5** Identify a point other than A in the target zone in Figure 7.43.

a. What are the coordinates of this point and what does this mean in terms of age and pulse rate?

b. Show that the coordinates of the point satisfy each inequality in the system in Example 5.

Exercise Set 7.6

Practice Exercises

Graph each linear inequality in Exercises 1–22.

1. $x + y \geq 2$

2. $x - y \leq 1$

3. $3x - y \geq 6$

4. $3x + y \leq 3$

5. $2x + 3y > 12$

6. $2x - 5y < 10$

7. $5x + 3y \leq -15$

8. $3x + 4y \leq -12$

9. $2y - 3x > 6$

10. $2y - x > 4$

11. $y > \dfrac{1}{3}x$

12. $y > \dfrac{1}{4}x$

13. $y \leq 3x + 2$

14. $y \leq 2x - 1$

15. $y < -\dfrac{1}{4}x$

16. $y < -\dfrac{1}{3}x$

17. $x \leq 2$

18. $x \leq -4$

19. $y > -4$

20. $y > -2$

21. $y \geq 0$

22. $x \geq 0$

Graph each system of linear inequalities in Exercises 23–38.

23. $3x + 6y \leq 6$
$2x + y \leq 8$

24. $x - y \geq 4$
$x + y \leq 6$

25. $2x + y < 3$
$x - y > 2$

26. $x + y < 4$
$4x - 2y < 6$

27. $2x + y < 4$
$x - y > 4$

28. $2x - y < 3$
$x + y < 6$

29. $x \geq 2$
$y \leq 3$

30. $x \geq 4$
$y \leq 2$

31. $x \leq 5$
$y > -3$

32. $x \leq 3$
$y > -1$

33. $x - y \leq 1$
$x \geq 2$

34. $4x - 5y \geq -20$
$x \geq -3$

35. $y > 2x - 3$
$y < -x + 6$

36. $y < -2x + 4$
$y < x - 4$

37. $x + 2y \leq 4$
$y \geq x - 3$

38. $x + y \leq 4$
$y \geq 2x - 4$

Application Exercises

The figure shows three kinds of regions—deserts, grasslands, and forests—that result from various ranges of temperature, T, and precipitation, P.

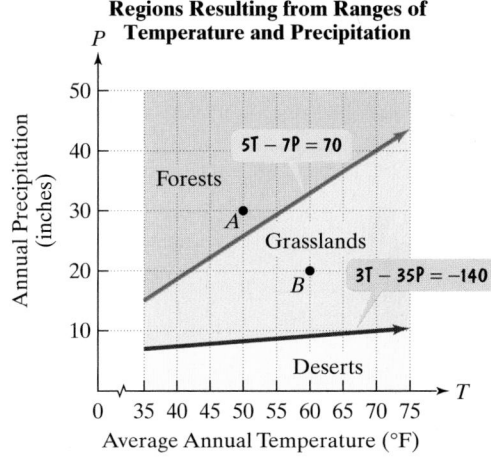

Regions Resulting from Ranges of Temperature and Precipitation

Source: A. Miller and J. Thompson, *Elements of Meteorology*

Systems of inequalities can be used to describe where forests, grasslands, and deserts occur. Because these regions occur when the average annual temperature, T, is 35°F or greater, each system contains the inequality $T \geq 35$.

Forests occur if	**Grasslands occur if**	**Deserts occur if**
$T \geq 35$	$T \geq 35$	$T \geq 35$
$5T - 7P < 70.$	$5T - 7P \geq 70$	$3T - 35P > -140.$
	$3T - 35P \leq -140.$	

Use this information to solve Exercises 39–40.

39. Find the coordinates of point *A*. Then show that these coordinates satisfy each inequality in the system that describes where forests occur.

40. Find the coordinates of point *B*. Then show that these coordinates satisfy each inequality in the system that describes where grasslands occur.

41. Many elevators have a capacity of 2000 pounds.

a. If a child averages 50 pounds and an adult 150 pounds, write an inequality that describes when *x* children and *y* adults will cause the elevator to be overloaded.

b. Graph the inequality. Because *x* and *y* must be positive, limit the graph to quadrant I and its boundary only.

c. Select an ordered pair satisfying the inequality. What are its coordinates and what do they represent in this situation?

42. A patient is not allowed to have more than 330 milligrams of cholesterol per day from a diet of eggs and meat. Each egg provides 165 milligrams of cholesterol. Each ounce of meat provides 110 milligrams of cholesterol.

a. Write an inequality that describes the patient's dietary restrictions for *x* eggs and *y* ounces of meat.

b. Graph the inequality. Because *x* and *y* must be positive, limit the graph to quadrant I and its boundary only.

c. Select an ordered pair satisfying the inequality. What are its coordinates and what do they represent in this situation?

The graph of a linear inequality in two variables is a region in the rectangular coordinate system. Regions in coordinate systems have numerous applications. For example, the regions in the following two graphs indicate whether a person is overweight, borderline overweight, or normal weight.

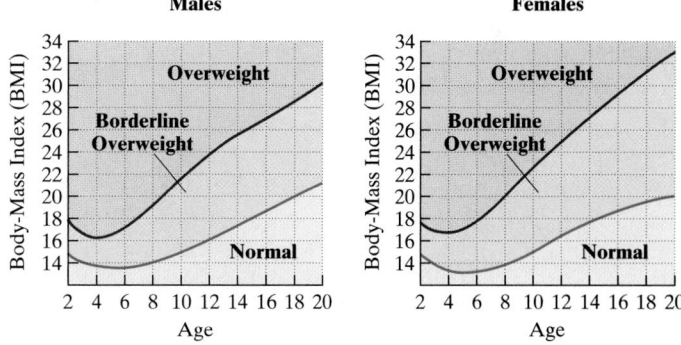

Source: Centers for Disease Control and Prevention

The horizontal axis shows a person's age. The vertical axis shows that person's body-mass index (BMI), computed using the following formula:

$$\text{BMI} = \frac{703W}{H^2}.$$

The variable W represents weight, in pounds. The variable H represents height, in inches. Use this information to solve Exercises 43–44.

43. A man is 20 years old, 72 inches (6 feet) tall, and weighs 200 pounds.

a. Compute the man's BMI. Round to the nearest tenth.

b. Use the man's age and his BMI to locate this information as a point in the coordinate system for males. Is this person overweight, borderline overweight, or normal weight?

44. (*Refer back to the graphs and formula on page 391.*) A girl is 10 years old, 50 inches (4 feet, 2 inches) tall, and weighs 100 pounds.

   a. Compute the girl's BMI. Round to the nearest tenth.

   b. Use the girl's age and her BMI to locate this information as a point in the coordinate system for females. Is this person overweight, borderline overweight, or normal weight?

Writing in Mathematics

45. What is a half-plane?

46. What does a dashed line mean in the graph of an inequality?

47. Write a paragraph explaining how to graph $2x - 3y < 6$.

48. Compare the graphs of $3x - 2y > 6$ and $3x - 2y \leq 6$. Discuss similarities and differences between the graphs.

49. Describe how to solve a system of linear inequalities.

Critical Thinking Exercises

50. Which one of the following is true?

   a. The graph of $4x - y > 2$ is the half-plane above the line whose equation is $4x - y = 2$.

   b. The graph of $x < -2$ is the region to the left of the vertical line whose equation is $x = 2$.

   c. The graph of $-\frac{5}{12}x + \frac{1}{4}y \geq \frac{3}{4}$ includes a boundary line whose y-intercept is 3.

   d. The graph of a half-plane in rectangular coordinates is the graph of a function.

51. Graph the system of inequalities:

$$x + 2y \geq 8$$
$$x - y \leq 2$$
$$y \leq 4.$$

52. Promoters of a rock concert must sell at least 25,000 tickets priced at $35 and $50 per ticket. Furthermore, the promoters must take in at least $1,025,000 in ticket sales. Find and graph a system of inequalities that describes all possibilities for selling the $35 tickets and the $50 tickets.

Technology Exercises

Graphing calculators can be used to shade regions in the rectangular coordinate system, thereby graphing an inequality in two variables. Read the section of the user's manual for your graphing utility that describes how to shade a region. Then use your graphing calculator to graph the inequalities in Exercises 53–54.

53. $y \leq 4x + 4$ **54.** $3x - 2y \geq 6$

55. Does your graphing calculator have any limitations in terms of graphing inequalities? If so, what are they?

56. Use a graphing calculator with a $\boxed{\text{SHADE}}$ feature to verify any five of the graphs that you drew by hand in Exercises 1–22.

57. Use a graphing calculator with a $\boxed{\text{SHADE}}$ feature to verify any five of the graphs that you drew by hand for the systems in Exercises 23–38.

SECTION 7.7   LINEAR PROGRAMMING

OBJECTIVES

1. Write an objective function describing a quantity that must be maximized or minimized.

2. Use inequalities to describe limitations in a situation.

3. Use linear programming to solve problems.

West Berlin children at Tempelhof airport watch fleets of U.S. airplanes bringing in supplies to circumvent the Russian blockade. The airlift began June 28, 1948 and continued for 15 months.

The Berlin Airlift (1948–1949) was an operation by the United States and Great Britain in response to military action by the former Soviet Union: Soviet troops closed all roads and rail lines between West Germany and Berlin, cutting off supply routes to the city. The Allies used a mathematical technique developed during World War II to maximize the amount of supplies transported. During the 15-month airlift, 278,228 flights provided basic necessities to blockaded Berlin, saving one of the world's great cities.

    In this section, we will look at an important application of systems of linear inequalities. Such systems arise in **linear programming**, a method for solving problems in which a particular quantity that must be maximized or minimized is limited by other factors. Linear programming is one of the most widely used tools in management science. It helps businesses allocate resources to manufacture products in a way that will maximize profit. Linear programming accounts for more than 50% and perhaps as much as 90% of all computing time used for management decisions in business. The Allies used linear programming to save Berlin.

Objective Functions in Linear Programming

① Write an objective function describing a quantity that must be maximized or minimized.

Many problems involve quantities that must be maximized or minimized. Businesses are interested in maximizing profit. An operation in which bottled water and medical kits are shipped to earthquake victims needs to maximize the number of victims helped by this shipment. An **objective function** is an algebraic expression in two or more variables describing a quantity that must be maximized or minimized.

EXAMPLE 1 | WRITING AN OBJECTIVE FUNCTION

Bottled water and medical supplies are to be shipped to victims of an earthquake by plane. Each container of bottled water will serve 10 people and each medical kit will aid 6 people. Let x represent the number of bottles of water to be shipped and y the number of medical kits. Write the objective function that describes the number of people that can be helped.

Solution   Because each bottle of water serves 10 people and each medical kit aids 6 people, we have

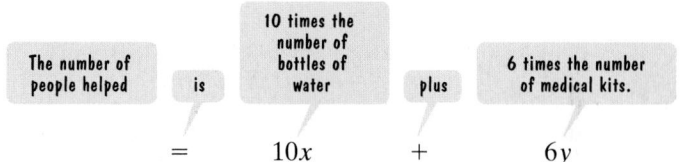

$$= \qquad 10x \qquad + \qquad 6y$$

Using z to represent the objective function, we have

$$z = 10x + 6y.$$

Unlike the functions that we have seen so far, the objective function is an equation in three variables. For a value of x and a value of y, there is one and only one value of z. Thus, z is a function of x and y.

A company manufactures bookshelves and desks for computers. Let x represent the number of bookshelves manufactured daily and y the number of desks manufactured daily. The company's profits are $25 per bookshelf and $55 per desk. Write the objective function that describes the company's total daily profit, z, from x bookshelves and y desks. (Check Points 2 through 4 are also related to this situation, so keep track of your answers.)

Constraints in Linear Programming

② Use inequalities to describe limitations in a situation.

Ideally, the number of earthquake victims helped in Example 1 should increase without restriction so that every victim receives water and medical kits. However, the planes that ship these supplies are subject to weight and volume restrictions. In linear programming problems, such restrictions are called **constraints**. Each constraint is expressed as a linear inequality. The list of constraints forms a system of linear inequalities.

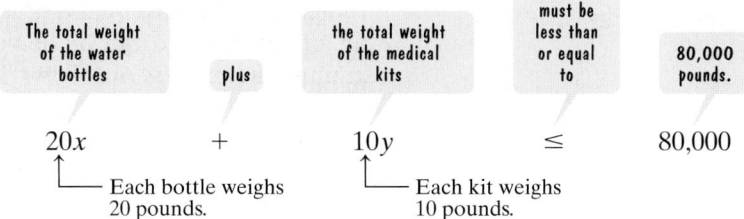

Each plane can carry no more than 80,000 pounds. The bottled water weighs 20 pounds per container and each medical kit weighs 10 pounds. Let x represent the number of bottles of water to be shipped and y the number of medical kits. Write an inequality that describes this constraint.

Solution   Because each plane can carry no more than 80,000 pounds, we have

The total weight of the water bottles	plus	the total weight of the medical kits		must be less than or equal to	80,000 pounds.
$20x$	$+$	$10y$		$\leq$	$80{,}000$

Each bottle weighs 20 pounds.   Each kit weighs 10 pounds.

The plane's weight constraint is described by the inequality
$$20x + 10y \leq 80{,}000.$$

Check 2 Point   To maintain high quality, the company in Check Point 1 should not manufacture more than 80 bookshelves and desks per day. Write an inequality that describes this constraint.

In addition to a weight constraint on its cargo, each plane has a limited amount of space in which to carry supplies. Example 3 demonstrates how to express this constraint.

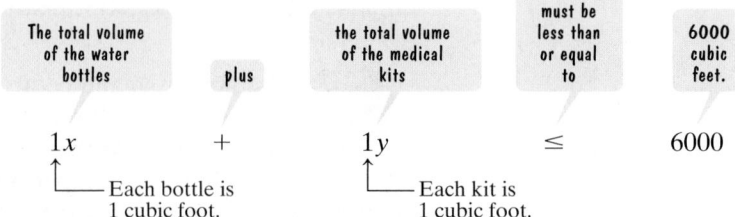

The total volume of supplies that a plane carries cannot exceed 6000 cubic feet. Each water bottle is 1 cubic foot and each medical kit also has a volume of 1 cubic foot. With x still representing the number of water bottles and y the number of medical kits, write an inequality that describes this second constraint.

Solution   Because each plane can carry a volume of supplies that does not exceed 6000 cubic feet, we have

The total volume of the water bottles	plus	the total volume of the medical kits		must be less than or equal to	6000 cubic feet.
$1x$	$+$	$1y$		$\leq$	6000

Each bottle is 1 cubic foot.   Each kit is 1 cubic foot.

The plane's volume constraint is described by the inequality $x + y \leq 6000$.

In summary, here's what we have described so far in this aid-to-earthquake-victims situation:

$$z = 10x + 6y$$ This is the objective function describing the number of people helped with x bottles of water and y medical kits.

$$20x + 10y \leq 80{,}000$$
$$x + y \leq 6000.$$ These are the constraints based on each plane's weight and volume limitations.

 **Check 3 Point** To meet customer demand, the company in Check Point 1 must manufacture between 30 and 80 bookshelves per day, inclusive. Furthermore, the company must manufacture at least 10 and no more than 30 desks per day. Write an inequality that describes each of these sentences. Then summarize what you have described about this company by writing the objective function for its profits, and the three constraints.

Solving Problems with Linear Programming

The goal in the earthquake situation described previously is to maximize the number of victims who can be helped, subject to the planes' weight and volume constraints. The process of solving this problem is called *linear programming*, based on a theorem that was proven during World War II.

---

3 Use linear programming to solve problems.

> **SOLVING A LINEAR PROGRAMMING PROBLEM**
>
> Let $z = ax + by$ be an objective function that depends on x and y. Furthermore, z is subject to a number of constraints on x and y. If a maximum or minimum value of z exists, it can be determined as follows:
>
> 1. Graph the system of inequalities representing the constraints.
> 2. Find the value of the objective function at each corner, or **vertex**, of the graphed region. The maximum and minimum of the objective function occur at one or more of the corner points.

EXAMPLE 4 SOLVING A LINEAR PROGRAMMING PROBLEM

Determine how many bottles of water and how many medical kits should be sent on each plane to maximize the number of earthquake victims who can be helped.

Solution   We must maximize $z = 10x + 6y$ subject to the constraints:

$$20x + 10y \leq 80{,}000$$
$$x + y \leq 6000.$$

STEP 1. Graph the system of inequalities representing the constraints. Because x (the number of bottles of water per plane) and y (the number of medical kits per plane) must be nonnegative, we need to graph the system of inequalities in quadrant I and its boundary only. To graph the inequality $20x + 10y \leq 80{,}000$, we graph the equation $20x + 10y = 80{,}000$ as a solid blue line (Figure 7.44). Setting $y = 0$, the x-intercept is 4000 and setting $x = 0$, the y-intercept is 8000. Using $(0, 0)$ as a test point, the inequality is satisfied, so we shade below the blue line, as shown in yellow in Figure 7.44. Now we graph $x + y \leq 6000$ by first graphing $x + y = 6000$ as a solid red line. Setting $y = 0$, the x-intercept is 6000. Setting $x = 0$, the y-intercept is 6000. Using $(0, 0)$ as a test point, the inequality is satisfied, so we shade below the red line, as shown using green vertical shading in Figure 7.44.

We use the addition method to find where the lines $20x + 10y = 80{,}000$ and $x + y = 6000$ intersect.

$$
\begin{array}{ll}
20x + 10y = 80{,}000 & \xrightarrow{\text{No change}} \quad 20x + 10y = 80{,}000 \\
x + y = 600 & \xrightarrow{\text{Multiply by } -10.} \quad -10x - 10y = -60{,}000 \\
& \qquad\qquad\text{Add:} \quad \underline{10x = 20{,}000} \\
& \qquad\qquad\qquad\qquad x = 2000
\end{array}
$$

Back-substituting 2000 for x in $x + y = 6000$, we find $y = 4000$, so the intersection point is $(2000, 4000)$.

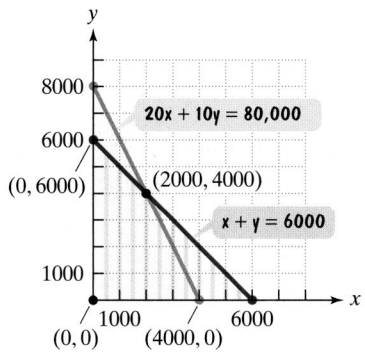

FIGURE 7.44 The region in quadrant I representing the constraints
$20x + 10y \leq 80{,}000$
$x + y \leq 6000$

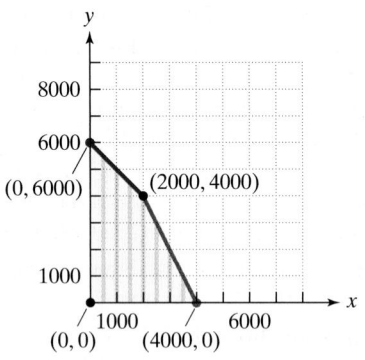

FIGURE 7.45

The system of inequalities representing the constraints is shown by the region in which the yellow shading and the green vertical shading overlap in Figure 7.44. The graph of the system of inequalities is shown again in Figure 7.45. The red and blue line segments are included in the graph.

STEP 2. Find the value of the objective function at each corner of the graphed region. The maximum and minimum of the objective function occur at one or more of the corner points. We must evaluate the objective function, $z = 10x + 6y$, at the four corners, or vertices, of the region in Figure 7.45.

Corner (x, y)	Objective Function $z = 10x + 6y$
$(0, 0)$	$z = 10(0) + 6(0) = 0$
$(4000, 0)$	$z = 10(4000) + 6(0) = 40{,}000$
$(2000, 4000)$	$z = 10(2000) + 6(4000) = 44{,}000 \leftarrow$ maximum
$(0, 6000)$	$z = 10(0) + 6(6000) = 36{,}000$

Thus, the maximum value of z is 44,000 and this occurs when $x = 2000$ and $y = 4000$. In practical terms, this means that the maximum number of earthquake victims who can be helped with each plane shipment is 44,000. This can be accomplished by sending 2000 water bottles and 4000 medical kits per plane.

Check 4 Point

For the company in Check Points 1–3, how many bookshelves and how many desks should be manufactured per day to obtain a maximum daily profit? What is the maximum daily profit?

Exercise Set 7.7

Practice Exercises

In Exercises 1–4, find the value of the objective function at each corner of the graphed region. What is the maximum value of the objective function? What is the minimum value of the objective function?

1. Objective Function
$z = 5x + 6y$

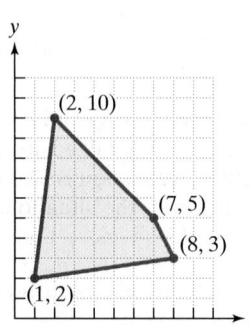

2. Objective Function
$z = 3x + 2y$

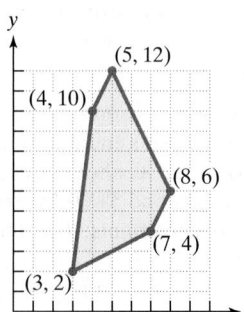

3. Objective Function
$z = 40x + 50y$

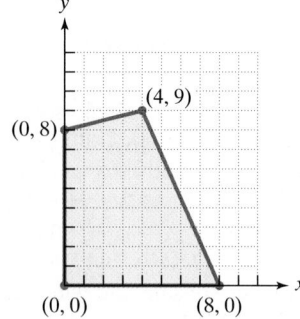

4. Objective Function
$z = 30x + 45y$

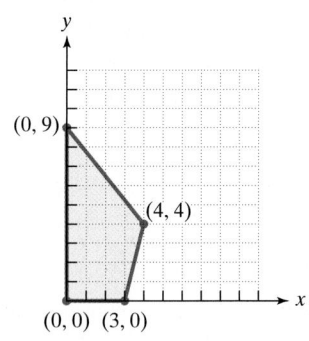

In Exercises 5–8, an objective function and a system of inequalities representing constraints are given.

    a. *Graph the system of inequalities representing the constraints.*

b. *Find the value of the objective function at each corner of the graphed region.*

c. *Use the values in part (b) to determine the maximum and minimum values of the objective function and the values of x and y for which they occur.*

5. Objective Function

$z = x + y$

Constraints

$x \le 6$

$y \ge 1$

$2x - y \ge -1$

6. Objective Function

$z = 3x - 2y$

Constraints

$x \ge 1$

$x \le 5$

$y \ge 2$

$x - y \ge -3$

7. Objective Function

$z = 6x + 10y$

Constraints

$x + y \le 12$

$x + 2y \le 20$

$\left. \begin{array}{l} x \ge 0 \\ y \ge 0 \end{array} \right\}$ Quadrant 1 and its boundary

8. Objective Function

$z = x + 3y$

Constraints

$x + y \ge 2$

$x \le 6$

$y \le 5$

$\left. \begin{array}{l} x \ge 0 \\ y \ge 0 \end{array} \right\}$ Quadrant 1 and its boundary

Application Exercises

9. a. A student earns $10 per hour for tutoring and $7 per hour as a teacher's aid. Let $x =$ the number of hours each week spent tutoring and $y =$ the number of hours each week spent as a teacher's aid. Write the objective function that describes total weekly earnings.

b. The student is bound by the following constraints:

- To have enough time for studies, the student can work no more 20 hours per week.

- The tutoring center requires that each tutor spend at least three hours per week tutoring.

- The tutoring center requires that each tutor spend no more than eight hours per week tutoring.

Write a system of three inequalities that describes these constraints.

c. Graph the system of inequalities in part (b). Use only the first quadrant and its boundary, because x and y are nonnegative.

d. Evaluate the objective function for total weekly earnings at each of the four vertices of the graphed region. [The vertices should occur at $(3, 0)$, $(8, 0)$, $(3, 17)$, and $(8, 12)$.]

e. Complete the missing portions of this statement: The student can earn the maximum amount per week by tutoring for _____ hours per week and working as a teacher's aid for _____ hours per week. The maximum amount that the student can earn each week is $_____.

10. A television manufacturer makes console and wide-screen televisions. The profit per unit is $125 for the console televisions and $200 for the wide-screen televisions.

a. Let $x =$ the number of consoles manufactured in a month and $y =$ the number of wide-screens manufactured in a month. Write the objective function that describes the total monthly profit.

b. The manufacturer is bound by the following constraints:

- Equipment in the factory allows for making at most 450 console televisions in one month.

- Equipment in the factory allows for making at most 200 wide-screen televisions in one month.

- The cost to the manufacturer per unit is $600 for the console televisions and $900 for the wide-screen televisions. Total monthly costs cannot exceed $360,000.

Write a system of three inequalities that describes these constraints.

c. Graph the system of inequalities in part (b). Use only the first quadrant and its boundary, because x and y must both be nonnegative.

d. Evaluate the objective function for total monthly profit at each of the five vertices of the graphed region. [The vertices should occur at $(0, 0)$, $(0, 200)$, $(300, 200)$, $(450, 100)$, and $(450, 0)$.]

e. Complete the missing portions of this statement: The television manufacturer will make the greatest profit by manufacturing _____ console televisions each month and _____ wide-screen televisions each month. The maximum monthly profit is $_____.

Use the two steps for solving a linear programming problem, given in the box on page 395, to solve the problems in Exercises 11–14.

11. Food and clothing are shipped to victims of a natural disaster. Each carton of food will feed 12 people, while each carton of clothing will help 5 people. Each 20-cubic-foot box of food weighs 50 pounds and each 10-cubic-foot box of clothing weighs 20 pounds. The commercial carriers transporting food and clothing are bound by the following constraints:

- The total weight per carrier cannot exceed 19,000 pounds.

- The total volume must be no more than 8000 cubic feet.

How many cartons of food and clothing should be sent with each plane shipment to maximize the number of people who can be helped?

12. A manufacturer produces two models of mountain bicycles. The times (in hours) required for assembling and painting each model are given in the following table:

	Model A	Model B
Assembling	5	4
Painting	2	3

The maximum total weekly hours available in the assembly department and the paint department are 200 hours and 108 hours, respectively. The profits per unit are $25 for model A and $15 for model B. How many of each type should be produced to maximize profit?

13. A theater is presenting a program on drinking and driving for students and their parents. The proceeds will be donated to a local alcohol information center. Admission is $2.00 for parents and $1.00 for students. However, the situation has two constraints: The theater can hold no more than 150 people and every two parents must bring at least one student. How many parents and students should attend to raise the maximum amount of money?

14. On June 24, 1948, the former Soviet Union blocked all land and water routes through East Germany to Berlin. A gigantic airlift was organized using American and British planes to bring food, clothing, and other supplies to the more than 2 million people in West Berlin. The cargo capacity was 30,000 cubic feet for an American plane and 20,000 cubic feet for a British plane. To break the Soviet blockade, the Western Allies had to maximize cargo capacity, but were subject to the following restrictions:

- No more than 44 planes could be used.

- The larger American planes required 16 personnel per flight, double that of the requirement for the British planes. The total number of personnel available could not exceed 512.

- The cost of an American flight was $9000 and the cost of a British flight was $5000. Total weekly costs could not exceed $300,000.

Find the number of American and British planes that were used to maximize cargo capacity.

Writing in Mathematics

15. What kinds of problems are solved using the linear programming method?

16. What is an objective function in a linear programming problem?

17. What is a constraint in a linear programming problem? How is a constraint represented?

18. In your own words, describe how to solve a linear programming problem.

19. Describe a situation in your life in which you would like to maximize something, but are limited by at least two constraints. Can linear programming be used in this situation? Explain your answer.

Critical Thinking Exercise

20. Suppose that you inherit $10,000. The will states how you must invest the money. Some (or all) of the money must be invested in stocks and bonds. The requirements are that at least $3000 be invested in bonds, with expected returns of $0.08 per dollar, and at least $2000 be invested in stocks, with expected returns of $0.12 per dollar. Because the stocks are medium risk, the final stipulation requires that the investment in bonds should never be less than the investment in stocks. How should the money be invested so as to maximize your expected returns?

Group Exercises

21. Group members should choose a particular field of interest. Research how linear programming is used to solve problems in that field. If possible, investigate the solution of a specific practical problem. Present a report on your findings, including the contributions of George Dantzig, Narendra Karmarkar, and L. G. Khachion to linear programming.

22. Members of the group should interview a business executive who is in charge of deciding the product mix for a business. How are production policy decisions made? Are other methods used in conjunction with linear programming? What are these methods? What sort of academic background, particularly in mathematics, does this executive have? Write a group report addressing these questions, emphasizing the role of linear programming for the business.

Chapter Summary, Review, and Test

SUMMARY

DEFINITIONS AND CONCEPTS	EXAMPLES

7.1 Graphing and Functions

a. The rectangular coordinate system is formed using two number lines that intersect at right angles at their zero points. See Figure 7.2 on page 336. The horizontal line is the x-axis and the vertical line is the y-axis. Their point of intersection, $(0, 0)$, is the origin. — Ex. 1, p. 337; Ex. 2, p. 338

b. If an equation in x and y yields one value of y for each value of x, then y is a function of x, indicated by writing $f(x)$ for y. — Ex. 4, p. 340; Ex. 5, p. 341

c. The Vertical Line Test: If any vertical line intersects a graph in more than one point, the graph does not define y as a function of x. — Ex. 6, p. 342; Ex. 7, p. 343

7.2 Linear Functions and Their Graphs

a. A function whose graph is a straight line is a linear function.

b. The graph of $Ax + By = C$, a linear equation in two variables, is a straight line. The line can be graphed using intercepts and a checkpoint. To locate the x-intercept, set $y = 0$ and solve for x. To locate the y-intercept, set $x = 0$ and solve for y. — Ex. 1, p. 347

c. The slope of the line through (x_1, y_1) and (x_2, y_2) is — Ex. 2, p. 348; Ex. 7, p. 352

$$m = \frac{\text{Rise}}{\text{Run}} = \frac{y_2 - y_1}{x_2 - x_1} = \frac{y_1 - y_2}{x_1 - x_2}.$$

d. The equation $y = mx + b$ is the slope-intercept form of a line, in which m is the slope and b is the y-intercept.

Ex. 3, p. 350;
Ex. 4, p. 351

e. Horizontal and Vertical Lines

   1. The graph of $y = b$ is a horizontal line that intersects the y-axis at $(0, b)$.

Ex. 5, p. 351

   2. The graph of $x = a$ is a vertical line that intersects the x-axis at $(a, 0)$.

Ex. 6, p. 352

7.3 Quadratic Functions and Their Graphs

a. An equation of the form $y = ax^2 + bx + c$ or $f(x) = ax^2 + bx + c$, $a \neq 0$, is called a quadratic function.

b. The graph of a quadratic function is called a parabola.

Ex. 2, p. 357

c. The x-coordinate of a parabola's vertex is $x = \frac{-b}{2a}$. The y-coordinate is found by substituting the x-coordinate in the parabola's equation. The vertex, or turning point, is the low point when the graph opens upward and the high point when the graph opens downward.

Ex. 5, p. 359;
Ex. 8, p. 362

d. The six steps for graphing a parabola are given in the box on page 360.

Ex. 6, p. 360;
Ex. 7, p. 361

7.4 Exponential Functions

a. The exponential function f with base b is defined by $y = b^x$ or $f(x) = b^x$, $b > 0$ and $b \neq 1$.

b. All exponential functions of the form $y = b^x$, where $b > 1$, have the shape of the graph in Figure 7.26 on page 366.

Ex. 1, p. 366

c. All exponential functions of the form $y = b^x$, where $0 < b < 1$, have the shape of the graph in Figure 7.27 on page 367.

Ex. 2, p. 367

d. The irrational number e, $e \approx 2.72$, appears in many applied exponential functions.

Ex. 5, p. 369

7.5 Systems of Linear Equations

a. Two equations in the form $Ax + By = C$ are called a system of linear equations. A solution to the system is an ordered pair that satisfies both equations in the system.

Ex. 1, p. 372

b. Linear systems can be solved by graphing, substitution (see the box on page 375), and addition (see the box on page 377).

Exs. 2–7,
pp. 373–378

c. Some linear systems have no solution: others have infinitely many solutions. For details, see the box on page 380.

Ex. 8, p. 379;
Ex. 9, p. 380

7.6 Linear Inequalities in Two Variables

a. A linear inequality in two variables can be written in the form $Ax + By > C$, $Ax + By \geq C$, $Ax + By < C$, or $Ax + By \leq C$.

b. The procedure for graphing a linear inequality in two variables is given in the box on page 386.

Ex. 1, p. 386;
Ex. 2, p. 387

c. Graphing Systems of Linear Inequalities

   1. Graph each inequality in the system in the same rectangular coordinate system.

Ex. 3, p. 388;
Ex. 4, p. 389

   2. Find the intersection of the individual graphs.

7.7 Linear Programming

a. An objective function is an algebraic expression in three variables describing a quantity to be maximized or minimized.

Ex. 1, p. 393

b. Constraints are restrictions, expressed as linear inequalities.

Ex. 2, p. 394;
Ex. 3, p. 394

c. Steps for solving a linear programming problem are given in the box on page 395.

Ex. 4, p. 395

REVIEW EXERCISES

7.1

In Exercises 1–4, plot the given point in a rectangular coordinate system.

1. $(2, 5)$

2. $(-4, 3)$

3. $(-5, -3)$

4. $(2, -5)$

Graph each equation in Exercises 5–7. Let $x = -3, -2, -1, 0, 1, 2$, and 3.

5. $y = 2x - 2$ **6.** $y = x^2 - 3$ **7.** $y = x$

8. If $f(x) = 4x + 11$, find $f(-2)$.

9. If $f(x) = -7x + 5$, find $f(-3)$.

10. If $f(x) = 3x^2 - 5x + 2$, find $f(4)$.

11. If $f(x) = -3x^2 + 6x + 8$, find $f(-4)$.

In Exercises 12–13, evaluate $f(x)$ for the given values of x. Then use the ordered pairs $(x, f(x))$ from your table to graph the function.

12. $f(x) = \frac{1}{2}x$

x	$f(x) = \frac{1}{2}x$
-6	
-4	
-2	
0	
2	
4	
6	

13. $f(x) = x^2 - 2$

x	$f(x) = x^2 - 2$
-2	
-1	
0	
1	
2	

In Exercises 14–16, use the vertical line test to identify graphs, in which y is a function of x.

14.

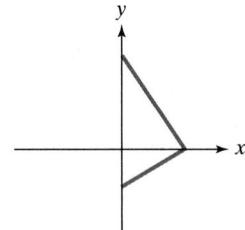

15.

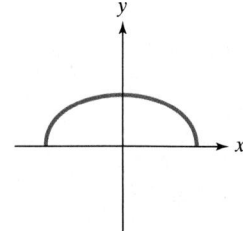

16.

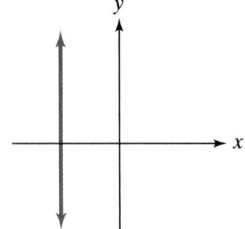

17. The figure shows the percentage of the U.S. population, $f(x)$, made up of Jewish Americans as a function of time, x, where x is the number of years after 1900.

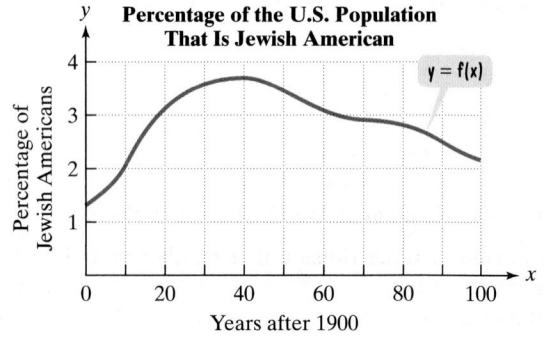

Percentage of the U.S. Population That Is Jewish American

$y = f(x)$

Years after 1900

Source: American Jewish Yearbook

a. Use the graph to find a reasonable estimate of $f(60)$. What does this mean in terms of the variables in this situation?

b. In which year did the percentage of Jewish Americans in the U.S. population reach a maximum? What is a reasonable estimate of the percentage for that year?

c. In which year was the percentage of Jewish Americans in the U.S. population at a minimum? What is a reasonable estimate of the percentage for that year?

d. Explain why f represents the graph of a function.

e. Describe the general trend shown by the graph.

7.2

In Exercises 18–20, use the x- and y-intercepts to graph each linear equation.

18. $2x + y = 4$ **19.** $2x - 3y = 6$

20. $5x - 3y = 15$

In Exercises 21–24, calculate the slope of the line passing through the given points. If the slope is undefined, so state. Then indicate whether the line rises, falls, is horizontal, or is vertical.

21. $(3, 2)$ and $(5, 1)$ **22.** $(-1, 2)$ and $(-3, -4)$

23. $(-3, 4)$ and $(6, 4)$ **24.** $(5, 3)$ and $(5, -3)$

In Exercises 25–28, graph each linear function using the slope and y-intercept.

25. $y = 2x - 4$ **26.** $y = -\frac{2}{3}x + 5$

27. $f(x) = \frac{3}{4}x - 2$ **28.** $y = \frac{1}{2}x + 0$

In Exercises 29–31, **a.** *Put the equation in slope-intercept form by solving for y;* **b.** *Identify the slope and the y-intercept; and* **c.** *Use the slope and y-intercept to graph the line.*

29. $2x + y = 0$ **30.** $3y = 5x$

31. $3x + 2y = 4$

In Exercises 32–34, graph each horizontal or vertical line.

32. $x = 3$ **33.** $y = -4$

34. $x + 2 = 0$

35. The graph shows that sales of light trucks, including SUVs, outpaced sales of cars in recent years.

Millions of Cars and SUVs Sold in the U.S.

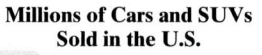

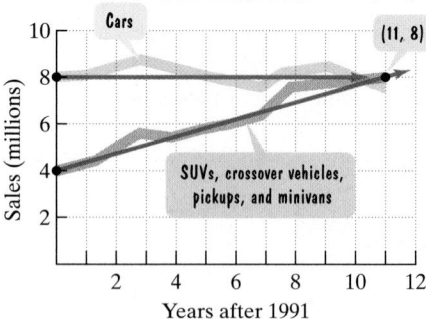

Cars

(11, 8)

SUVs, crossover vehicles, pickups, and minivans

Sales (millions)

Years after 1991

Source: National Automobile Dealers Association

a. Shown in the graph is a line that passes through or near the data for the SUVs. What is the line's y-intercept? Describe what this represents in this situation.

b. Use the coordinates of the two points shown to compute the slope. What does this mean about the millions of SUVs sold?

c. Use the y-intercept from part (a) and the slope from part (b) to write an equation that models the number of SUVs (and other light trucks) sold, y, in millions, x years after 1991.

d. Use your model from part (c) to predict SUV sales, to the nearest tenth of a million, in 2013.

e. Shown in the graph is a horizontal line that passes through or near the data for the cars. Write the equation of this horizontal line that approximates the millions of cars sold for the period shown.

7.3

In Exercises 36–38,

a. *Determine whether the parabola whose equation is given opens upward or downward.*

b. *Find the parabola's x-intercepts. If they are irrational, round to the nearest tenth.*

c. *Find the parabola's y-intercept.*

d. *Find the parabola's vertex.*

e. *Graph the parabola.*

36. $y = x^2 - 6x - 7$ **37.** $y = -x^2 - 2x + 3$

38. $f(x) = -3x^2 + 6x + 1$

39. The quadratic function

$$y = -0.05x^2 + 4.2x - 26$$

models the percentage of coffee drinkers, y, who are x years old who become irritable if they do not have coffee at their regular time. The graph of the resulting parabola is shown. Find the coordinates of the vertex. Then describe what the vertex represents in practical terms.

**What? No Coffee! Percentage of
Coffee Drinkers Who Become Irritable**

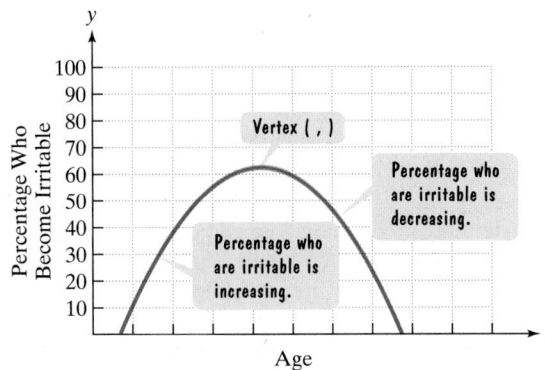

Source: LMK Associates

7.4

In Exercises 40–42, graph the exponential function whose equation is given. Start by using −2, −1, 0, 1, and 2 for x and find the corresponding values for y.

40. $f(x) = 2^x$ **41.** $f(x) = 4^{x-1}$ **42.** $f(x) = \left(\frac{1}{2}\right)^{2x}$

43. Use a calculator with an $\boxed{y^x}$ key or a $\boxed{\wedge}$ key to solve this exercise.

The amount of carbon dioxide in the atmosphere, measured in parts per million, has been increasing as a result of the burning of oil and coal. The buildup of gases and particles traps heat and raises the planet's temperature, a phenomenon called the greenhouse effect. Carbon dioxide accounts for about half of the warming. The function $f(x) = 364(1.005)^x$ projects carbon dioxide concentration, $f(x)$, in parts per million, x years after 2000. According to the model, what will be the concentration in the year 2086? Round to the nearest part per million. How does this compare with the preindustrial level of 280 parts per million?

44. Use a calculator with an $\boxed{e^x}$ key to solve this exercise. According to the U.S. Bureau of the Census, in 1990 there were 22.4 million residents of Hispanic origin living in the United States. By 2000, the number had increased to 35.3 million. The exponential growth function $f(x) = 22.4e^{0.045x}$ describes the U.S. Hispanic population, $f(x)$, in millions, x years after 1990. Find the Hispanic resident population of the United States in the year 2010 as predicted by this function. Round to the nearest tenth of a million.

7.5

In Exercises 45–47, solve each system by graphing. Check the coordinates of the intersection point in both equations.

45. $x + y = 5$
    $3x - y = 3$

46. $2x - y = -1$
    $x + y = -5$

47. $y = -x + 5$
    $2x - y = 4$

In Exercises 48–50, solve each system by the substitution method.

48. $2x + 3y = 2$
    $x = 3y + 10$

49. $y = 4x + 1$
    $3x + 2y = 13$

50. $x + 4y = 14$
    $2x - y = 1$

In Exercises 51–53, solve each system by the addition method.

51. $x + 2y = -3$
    $x - y = -12$

52. $2x - y = 2$
    $x + 2y = 11$

53. $5x + 3y = 1$
    $3x + 4y = -6$

In Exercises 54–56, solve by the method of your choice. Identify systems with no solution and systems with infinitely many solutions, using set notation to express their solution sets.

54. $y = -x + 4$
    $3x + 3y = -6$

55. $3x + y = 8$
    $2x - 5y = 11$

56. $3x - 2y = 6$
    $6x - 4y = 12$

Use a system of equations to solve Exercises 57–58.

57. The sum of three times a first number and eight times a second number is −1. When twice the second number is subtracted from the first number, the result is −5. What are the numbers?

58. Health experts agree that cholesterol intake should be limited to 300 mg or less each day. Three ounces of shrimp and 2 ounces of scallops contain 156 mg of cholesterol. Five ounces of shrimp and 3 ounces of scallops contain 45 mg of cholesterol less than the suggested maximum daily intake. Determine the cholesterol content in an ounce of each item.

7.6

Graph each inequality in Exercises 59–65.

59. $x - 3y \leq 6$

60. $2x + 3y \geq 12$

61. $2x - 7y > 14$

62. $y > \dfrac{3}{5}x$

63. $y \leq -\dfrac{1}{2}x + 2$

64. $x \leq 2$

65. $y > -3$

Graph each system of linear inequalities in Exercises 66–71.

66. $3x - y \leq 6$
    $x + y \geq 2$

67. $x + y < 4$
    $x - y < 4$

68. $x \leq 3$
    $y > -2$

69. $4x + 6y \leq 24$
    $y > 2$

70. $x + y \leq 6$
    $y \geq 2x - 3$

71. $y < -x + 4$
    $y > x - 4$

7.7

72. Find the value of the objective function $z = 2x + 3y$ at each corner of the graphed region shown. What is the maximum value of the objective function? What is the minimum value of the objective function?

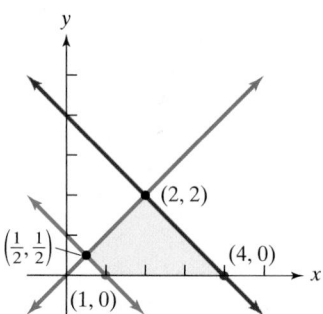

73. Consider the objective function $z = 2x + 3y$ and the following constraints:

$$x \leq 6,\ y \leq 5,\ x + y \geq 2,\ \underbrace{x \geq 0,\ y \geq 0}_{\substack{\text{Quadrant 1 and} \\ \text{its boundary}}}$$

a. Graph the system of inequalities representing the constraints.

b. Find the value of the objective function at each corner of the graphed region.

c. Use the values in part (b) to determine the maximum and minimum values of the objective function and the values of x and y for which they occur.

74. A paper manufacturing company converts wood pulp to writing paper and newsprint. The profit on a unit of writing paper is $500 and the profit on a unit of newsprint is $350.

a. Let x represent the number of units of writing paper produced daily. Let y represent the number of units of newsprint produced daily. Write the objective function that models total daily profit.

b. The manufacturer is bound by the following constraints:

- Equipment in the factory allows for making at most 200 units of paper (writing paper and newsprint) in a day.

- Regular customers require at least 10 units of writing paper and at least 80 units of newsprint daily.

Write a system of inequalities that models these constraints.

c. Graph the inequalities in part (b). Use only the first quadrant and its boundary, because x and y must both be nonnegative. (*Suggestion:* Let each unit along the x- and y-axes represent 20.)

d. Evaluate the objective profit function at each of the three vertices of the graphed region.

e. Complete the missing portions of this statement: The company will make the greatest profit by producing _____ units of writing paper and _____ units of newsprint each day. The maximum daily profit is $_____.

Group Exercise

75. Figure 7.24 on page 363 and the accompanying quadratic function show that the number of car accidents is a function of a driver's age. In Exercises 37–38 on page 364, we saw that deaths per thousand are also a function of age. For this activity, group members should list as many characteristics as possible that are functions of a person's age. For each characteristic, group members should create a graph. The horizontal axis should be labeled *age* and the vertical axis the characteristic you are graphing. Is the shape of the graph linear, quadratic, exponential, or something quite different? Group members should consider what happens to the characteristic from childhood to old age as they create each graph. Are there similarities in some of the graphs even though they describe different characteristics?

CHAPTER 7 TEST

1. Graph $y = (x - 1)^2$. Let $x = -1, 0, 1, 2,$ and 3.

2. If $f(x) = 3x^2 - 7x - 5$, find $f(-2)$.

3. Use the x- and y-intercepts to graph $4x - 2y = -8$.

4. Find the slope of the line passing through $(-3, 4)$ and $(-5, -2)$.

5. Strong demand plus higher fuel and labor costs are driving up the price of flying. The graph shows the national averages for one-way fares. Also shown is a line that models the data.

National Averages for One-Way Airline Fares: Business Travel

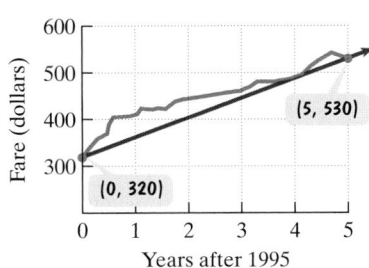

Source: American Express

Use the two points whose coordinates are shown in the voice balloons to find the slope of the line that models the average one-way fare, y, in dollars, x years after 1995. Describe what this means in terms of airline fares.

In Exercises 6–7, graph each linear function using the slope and y-intercept.

6. $y = \frac{2}{3}x - 1$

7. $f(x) = -2x + 3$

8. For the function $f(x) = x^2 - 2x - 8$,
   a. Determine whether the parabola opens upward or downward.
   b. Find the vertex.
   c. Find the x-intercepts.
   d. Find the y-intercept.
   e. Graph the parabola.

9. Use the vertex of a quadratic function to solve this exercise.

   A batter hits a baseball into the air. The function
   $$y = -16x^2 + 96x + 3$$
   models the baseball's height above the ground, y, in feet, x seconds after it was hit. When does the baseball reach its maximum height? What is that height?

10. Graph $f(x) = 3^x$. Use $-2, -1, 0, 1,$ and 2 for x and find the corresponding values for y.

11. Use a calculator with an $\boxed{e^x}$ key to solve this exercise. The function
   $$f(x) = 6e^{12.77x}$$
   models the risk of having a car accident, $f(x)$, as a percent, when driving with a blood alcohol concentration of x. In most states it is illegal to drive with a blood alcohol concentration of 0.08 or higher. Find the risk of a car accident for a blood alcohol concentration of 0.08. Round to the nearest percent.

12. Solve by graphing:
   $$x + y = 6$$
   $$4x - y = 4.$$

13. Solve by substitution:
   $$x = y + 4$$
   $$3x + 7y = -18.$$

14. Solve by addition:
   $$5x + 4y = 10$$
   $$3x + 5y = -7.$$

15. A travel agent offers two package vacation plans. The first plan costs $360 and includes 3 days at a hotel and a rental car for 2 days. The second plan costs $500 and includes 4 days at a hotel and a rental car for 3 days. The daily charge for the hotel is the same under each plan, as is the daily charge for the car. Find the cost per day for the hotel and for the car.

Graph each linear inequality in Exercises 16–17.

16. $3x - 2y < 6$

17. $y \le \frac{1}{2}x - 1$

18. Graph the system of linear inequalities:
   $$2x - y \le 4$$
   $$2x - y > -1.$$

19. Find the value of the objective function $z = 3x + 2y$ at each corner of the graphed region shown. What is the maximum value of the objective function? What is the minimum value of the objective function?

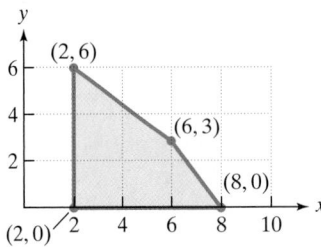

20. Find the maximum value of the objective function $z = 3x + 5y$ subject to the following constraints: $x + y \le 6, x \ge 2, y \ge 0$.

21. A manufacturer makes two types of jet skis, regular and deluxe. The profit on a regular jet ski is $200 and the profit on the deluxe model is $250. To meet customer demand, the company must manufacture at least 50 regular jet skis per week and at least 75 deluxe models. To maintain high quality, the total number of both models of jet skis manufactured by the company should not exceed 150 per week. How many jet skis of each type should be manufactured per week to obtain maximum profit? What is the maximum weekly profit?

Consumer Mathematics and Financial Management

Managing your personal finances is not an easy task. At one time, it did not involve much thought and planning: Everything was purchased using cash, and banks were the only reliable place for savings. Now we face numerous choices. Money invested may increase in value over time. If $1000 is not invested, the amount of goods it can buy diminishes over time. If it's spent, it's gone, although what it's spent on may be life-affirming in ways that cannot be measured in dollars and cents. However, we also know that if the $1000 is invested wisely, its value may increase considerably in just a few years.

We all want a wonderful life with fulfilling work, good health, and loving relationships. And let's be honest: Financial security wouldn't hurt! Achieving this goal depends on understanding basic ideas about savings, loans, and investments. A solid understanding of the topics in this chapter can pay, literally, by making your financial goals a reality. And how can that small investment now be worth a million dollars in the future? The chapter concludes with an essay that answers this question.

Writing the checks to cover your monthly bills gets you thinking about the future. Last night you heard an infomercial selling the idea that if you invest a small amount of money while you are young, you will be a millionaire by the time you are 50 and can retire early. How can this be true?

This problem appears in the Blitzer Bonus on How to Make Your Child a Millionaire in Section 8.6.

SECTION 8.1   PERCENT

OBJECTIVES

1. Express a fraction as a percent.
2. Express a decimal as a percent.
3. Express a percent as a decimal.
4. Use the percent formula.
5. Solve applied problems involving percent.
6. Investigate some of the ways percent can be abused.

"And if elected, it is my solemn pledge to cut your taxes by 10% for each of my first three years in office, for a total cut of 30%."

Did you know that one of the most common ways that you are given numerical information is with percents? This section will provide you with the tools to make sense of the politician's promise, as we present the uses, and abuses, of percent.

Basics of Percent

Percents are the result of expressing numbers as a part of 100. The word *percent* means *per hundred*. For example, the circle graph in Figure 8.1 shows that 57 out of every 100 single-family houses have three bedrooms. Thus, $\frac{57}{100} = 57\%$, indicating that 57% of the houses have three bedrooms. The percent sign, %, is used to indicate the number of parts out of one hundred parts.

A fraction can be expressed as a percent using the following procedure:

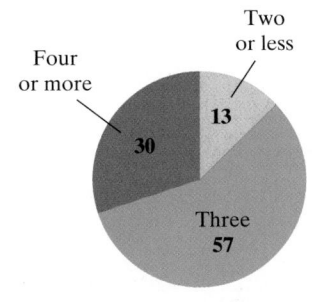

Source: U.S. Census Bureau and HUD

FIGURE 8.1 Number of bedrooms in privately owned single-family U.S. houses per 100 houses

---

EXPRESSING A FRACTION AS A PERCENT

1. Divide the numerator by the denominator.
2. Multiply the quotient by 100. This is done by moving the decimal point in the quotient two places to the right.
3. Add a percent sign.

---

1 Express a fraction as a percent.

EXAMPLE 1   EXPRESSING A FRACTION AS A PERCENT

Express $\frac{5}{8}$ as a percent.

Solution

STEP 1. **Divide the numerator by the denominator.**
$$5 \div 8 = 0.625$$

STEP 2. **Multiply the quotient by 100.**
$$0.625 \times 100 = 62.5$$

STEP 3. **Add a percent sign.**
$$62.5\%$$

Thus, $\frac{5}{8} = 62.5\%$.

✓ **Check 1 Point**   Express $\frac{1}{8}$ as a percent.

2 Express a decimal as a percent.

Our work in Example 1 shows that $0.625 = 62.5\%$. This illustrates the procedure for expressing a decimal number as a percent.

STUDY TIP

Dictionaries indicate that the word *percentage* has the same meaning as the word *percent*. Use the word that sounds best in the circumstance.

> **EXPRESSING A DECIMAL NUMBER AS A PERCENT**
> 1. Move the decimal point two places to the right.
> 2. Add a percent sign.

EXAMPLE 2   EXPRESSING A DECIMAL AS A PERCENT

Express 0.47 as a percent.

Solution

Move decimal point two places right.

$$0.47 \,\% \quad \text{Add a percent sign.}$$

Thus, $0.47 = 47\%$.

✓ **Check 2 Point**   Express 0.023 as a percent.

3 Express a percent as a decimal.

We reverse the procedure of Example 2 to express a percent as a decimal number.

> **EXPRESSING A PERCENT AS A DECIMAL NUMBER**
> 1. Move the decimal point two places to the left.
> 2. Remove the percent sign.

EXAMPLE 3   EXPRESSING PERCENTS AS DECIMALS

Express each percent as a decimal:

   a. 19%   **b.** 180%.

Solution   Use the two steps in the box.

   a. $19\% = 19.\% = 0.19\,\%$ The percent sign is removed.

   The decimal point starts at the far right.

   The decimal point is moved two places to the left.

   Thus, $19\% = 0.19$.

   b. $180\% = 1.80\,\% = 1.80$ or 1.8.

Check Point 3 Express each percent as a decimal:

a. 67% **b.** 250%.

If a fraction is part of a percent, as in $\frac{1}{4}$%, begin by expressing the fraction as a decimal, retaining the percent sign. Then, use the two steps in the box to express the percent as a decimal number. For example,

$$\frac{1}{4}\% = 0.25\% = 00.25\,\% = 0.0025.$$

 Use the percent formula.

A Formula Involving Percent

Percents are useful in comparing two numbers. To compare the number A to the number B using a percent P, the following formula is used:

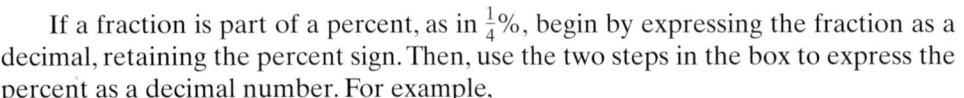

| A | is | P percent | of | B. |

$$A = P \cdot B$$

In the formula

$$A = PB$$

B = the base number, P = the percent (in decimal form), and A = the number compared to B.

Causes of U.S. Spinal Cord Injuries

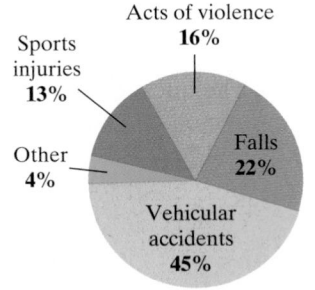

Source: U.S. News and World Report

FIGURE 8.2

EXAMPLE 4 USING THE PERCENT FORMULA: WHAT IS P PERCENT OF B?

Each year, approximately 8000 Americans suffer spinal cord injuries. The circle graph in Figure 8.2 shows that 22% of these injuries are due to falls. Approximately how many spinal cord injuries are due to falls each year?

Solution We use the formula $A = PB$: A is P percent of B. The number of injuries due to falls, A in the formula, is 22% of 8000.

| What | is | 22% | of | 8000? |

$$A = 0.22 \cdot 8000 \qquad \text{Express 22\% as 0.22.}$$
$$A = 1760 \qquad \text{Multiply.}$$

Thus, 1760 is 22% of 8000. This means that 1760 Americans suffer spinal cord injuries due to falls each year.

Check Point 4 Each year, approximately 8000 Americans suffer spinal cord injuries. Use the information in Figure 8.2 to determine approximately how many spinal cord injuries are due to sports injuries each year.

TABLE 8.1 LEADING CAUSES OF DEATH IN THE U.S. IN 2002

	No. of Deaths
1 Heart disease	710,760
2 Cancer	553,091
3 Stroke	167,661
4 Respiratory diseases	122,009
5 Accidents	97,900

Source: Department of Health and Human Services

EXAMPLE 5 | USING THE PERCENT FORMULA: A IS P PERCENT OF WHAT?

The five leading causes of death in the United States for 2002 are given in Table 8.1. If approximately 30% of all deaths were caused by heart disease, find the total number of deaths in the United States in 2002.

Solution We use the formula $A = PB$: A is P percent of B. The number of deaths caused by heart disease, 710,760, is 30% of the total number of deaths, represented by B in the formula.

> 710,760   is   30%   of   what?

$$710,760 = 0.3 \cdot B \qquad \text{Express 30\% as 0.3.}$$

$$\frac{710,760}{0.3} = \frac{0.3B}{0.3} \qquad \text{Divide both sides by 0.3.}$$

$$2,369,200 = B \qquad \text{Simplify.}$$

Thus, 710,760 is 30% of 2,369,200. The total number of deaths in the United States in 2002 was approximately 2,369,200.

 **Check 5 Point**   9 is 60% of what?

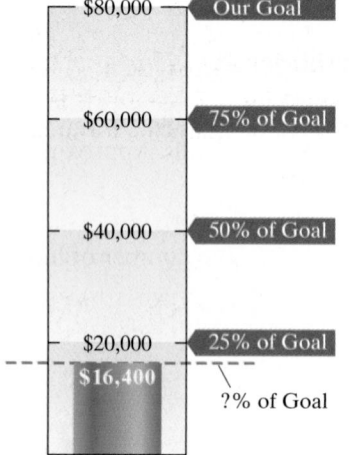

$80,000 — Our Goal
$60,000 — 75% of Goal
$40,000 — 50% of Goal
$20,000 — 25% of Goal
$16,400 — ?% of Goal

Raising money for a charity

EXAMPLE 6 | USING THE PERCENT FORMULA: A IS WHAT PERCENT OF B?

A charity has raised $16,400, with a goal of raising $80,000. What percent of the goal has been raised?

Solution We use the formula $A = PB$: A is P percent of B. The question we must answer is, "$16,400 is what percent of $80,000?" We are interested in finding P in the formula.

> $16,400   is   what percent   of   80,000?

$$16,400 = P \cdot 80,000$$

$$\frac{16,400}{80,000} = \frac{P \cdot 80,000}{80,000} \qquad \text{Divide both sides by 80,000.}$$

$$0.205 = P \qquad \text{Simplify.}$$

We express 0.205 as a percent by moving the decimal point two places to the right and adding a percent sign: $0.205 = 20.5\%$. Thus, $16,400 is 20.5% of $80,000. We see that 20.5% of the charity's goal has been raised.

 **Check 6 Point**   1.3 is what percent of 26?

5 Solve applied problems involving percent.

Applications of Percent

A common application of percent involves **sales tax** collected by states, counties, and cities on sales of items to customers. The sales tax is a percent of the cost of an item, implying the following formula:

> Sales tax amount = tax rate × item's cost.

EXAMPLE 7 | PERCENTS AND SALES TAX

Suppose that the local sales tax rate is 7.5% and you purchase a bicycle for $394.

a. How much tax is paid?

b. What is the bicycle's total cost?

Solution

a. Sales tax amount = tax rate × item's cost

$$= 7.5\% \times \$394 = 0.075 \times \$394 = \$29.55$$

The tax paid is $29.55.

b. The bicycle's total cost is the purchase price, $394, plus the sales tax, $29.55.

$$\text{Total cost} = \$394.00 + \$29.55 = \$423.55$$

The bicycle's total cost is $423.55.

 Check 7 Point Suppose that the local sales tax, rate is 6% and you purchase a computer for $1260.

a. How much tax is paid?

b. What is the computer's total cost?

None of us is thrilled about sales tax, but we do like buying things that are *on sale*. Businesses reduce prices, or **discount**, to attract customers and to reduce inventory. The discount rate is a percent of the original price, implying the following formula:

> Discount amount = discount rate × original price.

EXAMPLE 8 | PERCENTS AND SALE PRICE

A computer with an original price of $1460 is on sale at 15% off.

a. What is the discount amount?

b. What is the computer's sale price?

Solution

a. Discount amount = discount rate × original price

$$= 15\% \times \$1460 = 0.15 \times \$1460 = \$219$$

The discount amount is $219.

b. The computer's sale price is the original price, $1460, minus the discount amount, $219.

$$\text{Sale price} = \$1460 - \$219 = \$1241$$

The computer's sale price is $1241.

TECHNOLOGY

A calculator is useful in this chapter. The keystroke sequence that gives the sale price in Example 8 is

1460 − .15 × 1460.

Press = on a scientific calculator or ENTER on a graphing calculator to display the answer, 1241.

Check 8 Point   A CD player with an original price of $380 is on sale at 35% off.

a. What is the discount amount?

b. What is the CD player's sale price?

Percents are used for comparing changes, such as increases or decreases in sales, population, prices, and production. If a quantity changes, its **percent increase** or its **percent decrease** can be found as follows:

FINDING PERCENT INCREASE OR PERCENT DECREASE

1. Find the fraction for the percent increase or the percent decrease:

$$\frac{\text{amount of increase}}{\text{original amount}} \quad \text{or} \quad \frac{\text{amount of decrease}}{\text{original amount}}.$$

2. Find the percent increase or the percent decrease by expressing the fraction in step 1 as a percent.

EXAMPLE 9   FINDING PERCENT INCREASE AND DECREASE

In 2000, world population was approximately 6 billion. Figure 8.3 shows world population projections through the year 2150. The data are from the United Nations Family Planning Program and are based on optimistic or pessimistic expectations for successful control of human population growth.

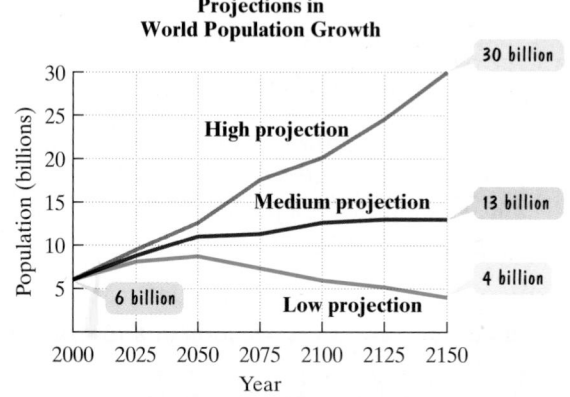

Source: United Nations

FIGURE 8.3

a. Find the percent increase in world population from 2000 to 2150 using the high projection data.

b. Find the percent decrease in world population from 2000 to 2150 using the low projection data.

Solution

a. Use the data shown on the blue, high-projection, graph.

$$\text{Percent increase} = \frac{\text{amount of increase}}{\text{original amount}}$$

$$= \frac{30 - 6}{6} = \frac{24}{6} = 4 = 400\%$$

The projected percent increase in world population is 400%.

b. Use the data shown on the green, low-projection, graph.

$$\text{Percent decrease} = \frac{\text{amount of decrease}}{\text{original amount}}$$

$$= \frac{6 - 4}{6} = \frac{2}{6} = \frac{1}{3} = 0.33\frac{1}{3} = 33\frac{1}{3}\%$$

The projected percent decrease in world population is $33\frac{1}{3}\%$.

In Example 9, we expressed the percent decrease as $33\frac{1}{3}\%$ because of the familiar conversion $\frac{1}{3} = 0.33\frac{1}{3}$. However, in many situations, rounding is needed. We suggest that you round to the nearest tenth of a percent. Carry the division in the fraction for percent increase or decrease to four places after the decimal point. Then round the decimal to three places, or to the nearest thousandth. Expressing this rounded decimal as a percent gives percent increase or decrease to the nearest tenth of a percent.

a. If 6 is increased to 10, find the percent increase.

b. If 10 is decreased to 6, find the percent decrease.

EXAMPLE 10 FINDING PERCENT DECREASE

A jacket regularly sells for $135.00. The sale price is $60.75. Find the percent decrease of the sale price from the regular price.

Solution $\text{Percent decrease} = \dfrac{\text{amount of decrease}}{\text{original amount}}$

$$= \frac{135.00 - 60.75}{135} = \frac{74.25}{135} = 0.55 = 55\%$$

The percent decrease of the sale price from the regular price is 55%. This means that the sale price of the jacket is 55% lower than the regular price.

A television regularly sells for $940. The sale price is $611. Find the percent decrease of the sale price from the regular price.

Abuses of Percent

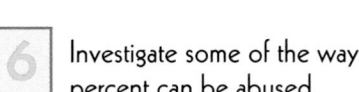

In our next examples, we look at a few of the many ways that percent can be used incorrectly. Confusion often arises when percent increase (or decrease) refers to a changing quantity that is itself a percent.

EXAMPLE 11 PERCENTS OF PERCENTS

John Tesh, while he was still coanchoring "Entertainment Tonight," reported that the PBS series "The Civil War" had an audience of 13% versus the usual 4% PBS audience, "an increase of more than 300%." Did Tesh report the percent increase correctly?

Solution   We begin by finding the percent increase.

$$\text{Percent increase} = \frac{\text{amount of increase}}{\text{original amount}}$$

$$= \frac{13\% - 4\%}{4\%} = \frac{9\%}{4\%} = \frac{9}{4} = 2.25 = 225\%$$

The percent increase for PBS was 225%. This is not more than 300%, so Tesh did not report the percent increase correctly.

Check
11
Point

An episode of a television series had an audience of 12% versus its usual 10%. What was the percent increase for this episode?

EXAMPLE 12 PROMISES OF A POLITICIAN

A politician states, "If you elect me to office, I promise to cut your taxes for each of my first three years in office by 10% each year, for a total reduction of 30%." Evaluate the accuracy of the politician's statement.

Solution   To make things simple, let's assume that a taxpayer paid $100 in taxes in the year previous to the politician's election. A 10% reduction during year 1 is 10% of $100.

$$10\% \text{ of previous year tax} = 10\% \text{ of } \$100 = 0.10 \times \$100 = \$10$$

With a 10% reduction the first year, the taxpayer will pay only $100 − $10, or $90, in taxes during the politician's first year in office.

The following table shows how we calculate the new, reduced tax for each of the first three years in office:

Year	Tax paid the year before	10% reduction	Taxes paid this year
1	$100	$0.10 \times \$100 = \$10$	$\$100 - \$10 = \$90$
2	$ 90	$0.10 \times \ \ \$90 = \$9$	$\$90 - \$9 = \$81$
3	$ 81	$0.10 \times \ \ \$81 = \$8.10$	$\$81 - \$8.10 = \$72.90$

Now, we determine the percent decrease in taxes over the three years.

$$\text{Percent decrease} = \frac{\text{amount of decrease}}{\text{original amount}}$$

$$= \frac{\$100 - \$72.90}{\$100} = \frac{\$27.10}{\$100} = \frac{27.1}{100} = 0.271 = 27.1\%$$

The taxes decline by 27.1%, not by 30%. The politician is ill-informed in saying that three consecutive 10% cuts add up to a total tax cut of 30%. In our calculation, which serves as a counterexample to the promise, the total tax cut is only 27.1%.

Check
12
Point

Suppose you paid $1200 in taxes. During year 1, taxes decrease by 20%. During year 2, taxes increase by 20%.

a. What do you pay in taxes for year 2?

b. How do your taxes for year 2 compare with what you originally paid, namely $1200? If the taxes are not the same, find the percent increase or decrease.

Exercise Set 8.1

Practice Exercises

In Exercises 1–10, express each fraction as a percent.

1. $\frac{2}{5}$ **2.** $\frac{3}{5}$ **3.** $\frac{1}{4}$ **4.** $\frac{3}{4}$

5. $\frac{3}{8}$ **6.** $\frac{7}{8}$ **7.** $\frac{1}{40}$ **8.** $\frac{3}{40}$

9. $\frac{9}{80}$ **10.** $\frac{13}{80}$

In Exercises 11–20, express each decimal as a percent.

11. 0.59   **12.** 0.96   **13.** 0.3844   **14.** 0.003

15. 2.87   **16.** 9.83   **17.** 14.87   **18.** 19.63

19. 100   **20.** 95

In Exercises 21–34, express each percent as a decimal.

21. 72%   **22.** 38%   **23.** 43.6%   **24.** 6.25%

25. 130%   **26.** 260%   **27.** 2%   **28.** 6%

29. $\frac{1}{2}$%   **30.** $\frac{3}{4}$%   **31.** $\frac{5}{8}$%   **32.** $\frac{1}{8}$%

33. $62\frac{1}{2}$%   **34.** $87\frac{1}{2}$%

Use the percent formula, $A = PB$: A is P percent of B, to solve Exercises 35–46.

35. What is 3% of 200?   **36.** What is 8% of 300?

37. What is 18% of 40?   **38.** What is 16% of 90?

39. 3 is 60% of what?   **40.** 8 is 40% of what?

41. 24% of what number is 40.8?

42. 32% of what number is 51.2?

43. 3 is what percent of 15?

44. 18 is what percent of 90?

45. What percent of 2.5 is 0.3?

46. What percent of 7.5 is 0.6?

Application Exercises

A recent Time/CNN telephone poll included never-married single women between the ages of 18 and 49 and never-married single men between the ages of 18 and 49. The circle graphs show the results for one of the questions in the poll. Use this information to solve Exercises 47–48.

**If You Couldn't Find the Perfect Mate,
Would You Marry Someone Else?**

Single women   Single men

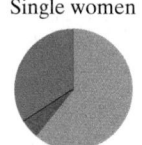   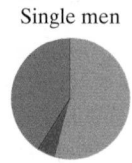

	Yes	No	Not sure
Single women	34%	61%	5%
Single men	41%	54%	5%

Source: Time, August 28, 2000

47. There were 1200 single women who participated in the poll. How many stated they would marry someone other than the perfect mate?

48. There were 1200 single men who participated in the poll. How many stated they would marry someone other than the perfect mate?

49. In 2000, 4% of the Hispanic population in the United States was Cuban. If the Cuban-American population at that time was 1,242,000, what was the U.S. Hispanic population in 2000?

50. In 2001, there were approximately 612,000 inmates in U.S. jails. At that time, jails were operating at 90% of their rated capacity. Find the capacity, in number of inmates, of U.S. jails in 2001.
(*Source:* U.S. Justice Department)

51. A charity has raised $7500, with a goal of raising $60,000. What percent of the goal has been raised?

52. A charity has raised $225,000, with a goal of raising $500,000. What percent of the goal has been raised?

The circle graph shows the average number of hours a typical group of 400 Americans sleep each night. Use this information to solve Exercises 53–54.

**Average Number of Hours 400
Americans Sleep Each Night**

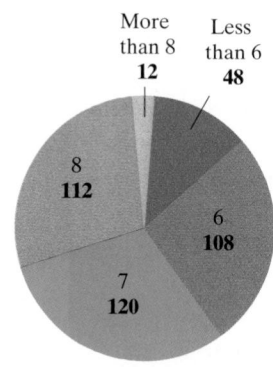

Source: American Medical Association

53. What percent of Americans sleep 8 hours or more per night?

54. What percent of Americans sleep 6 hours or less per night?

55. A restaurant bill came to $60. If 15% of this bill was left as a tip, how much was the tip?

56. If income tax is $3502 plus 28% of taxable income over $23,000, how much is the income tax on a taxable income of $35,000?

57. Suppose that the local sales tax rate is 6% and you purchase a car for $16,800.
   a. How much tax is paid?
   b. What is the car's total cost?

58. Suppose that the local sales tax rate is 7% and you purchase a graphing calculator for $96.
   a. How much tax is paid?
   b. What is the calculator's total cost?

59. An exercise machine with an original price of $860 is on sale at 12% off.
   a. What is the discount amount?
   b. What is the exercise machine's sale price?

60. A dictionary that normally sells for $16.50 is on sale at 40% off.
   a. What is the discount amount?
   b. What is the dictionary's sale price?

Although average annual earnings in the United States declined from 2001 to 2002, some workers gained while others lost.

Gaining

	2001 Salary	2002 Salary
Dieticians	$32,600	$89,200
Pharmacists	$67,100	$87,400

Losing

	2001 Salary	2002 Salary
Dentists	$86,400	$72,100
Musicians, Composers	$92,300	$39,600

Source: *Time*

Use this information to solve Exercises 61–64. Round answers to the nearest tenth of a percent.

61. Find the percent increase in annual earnings for dieticians from 2001 to 2002.

62. Find the percent increase in annual earnings for pharmacists from 2001 to 2002.

63. Find the percent decrease in annual earnings for dentists from 2001 to 2002.

64. Find the percent decrease in annual earnings for musicians and composers from 2001 to 2002.

65. A sofa regularly sells for $840. The sale price is $714. Find the percent decrease of the sale price from the regular price.

66. A FAX machine regularly sells for $380. The sale price is $266. Find the percent decrease of the sale price from the regular price.

67. Suppose that you have $10,000 in a rather risky investment recommended by your financial advisor. During the first year, your investment decreases by 30% of its original value. During the second year, your investment increases by 40% of its first-year value. Your advisor tells you that there must have been a 10% overall increase of your original $10,000 investment. Is your financial advisor using percentages properly? If not, what is your actual percent gain or loss of your original $10,000 investment?

68. The price of a color printer is reduced by 30% of its original price. When it still does not sell, its price is reduced by 20% of the reduced price. The salesperson informs you that there has been a total reduction of 50%. Is the salesperson using percentages properly? If not, what is the actual percent reduction from the original price?

69. The graph at the top of the next column shows the extent of socialized medicine in selected countries. Based on the data shown, can you conclude that the government of Belgium spends more money on health expenditures than the government of the United States? Explain your answer.

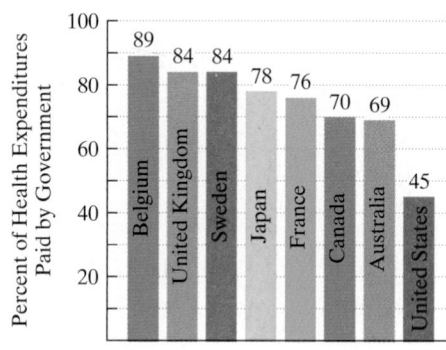

Percent of Health Expenditures Paid by Governments of Selected Countries

Source: The World Bank

Writing in Mathematics

70. What is a percent?

71. Describe how to express a decimal number as a percent and give an example.

72. Describe how to express a percent as a decimal number and give an example.

73. What does the percent formula, $A = PB$, describe? Give an example of how the formula is used.

74. Describe how to find percent increase and give an example.

75. Describe two ways in which you use percents in your daily life.

Critical Thinking Exercises

76. Which one of the following is true?
   a. If your weight increases by 5% this year and 5% next year, the increase in weight over two years is 10%.
   b. A grade point average that increases by 10% this semester and decreases by 10% next semester will be unchanged after two semesters.
   c. If $\frac{1}{10}$ of a person's salary is spent for clothing, $\frac{1}{3}$ for food, and $\frac{1}{5}$ for rent, the percent of the salary that is left is $36\frac{2}{3}\%$.
   d. If the amount that a person spends on rent increases from 20% to 30% of that person's income, the percent increase is 10%.

77. A condominium is taxed based on its $78,500 value. The tax rate is $3.40 for every $100 of value. If the tax is paid before March 1, 3% of the normal tax is given as a discount. How much tax is paid if the condominium owner takes advantage of the discount?

78. In January, each of 60 people purchased a $200 washing machine. In February, 10% fewer customers purchased the same washing machine that had increased in price by 20%. What was the change in sales from January to February?

Group Exercise

79. Each person should find two examples in which percents are used in news reports. Share these examples with other members of the group. If possible, try to find an example in which percent is used incorrectly.

SECTION 8.2   SIMPLE INTEREST

OBJECTIVES

1. Calculate simple interest.

2. Use the future value formula.

3. Use the simple interest formula on discounted loans.

In 1626, Peter Minuit convinced the Wappinger Indians to sell him Manhattan Island for $24. If the native Americans had put the $24 into a bank account at a 5% interest rate compounded monthly, by the year 2000 there would be well over $3 billion in the account!

Although you may not yet understand terms such as *interest rate* and *compounded monthly*, one thing seems clear: Money in certain savings accounts grows in remarkable ways. You, too, can take advantage of such accounts with astonishing results. In the next two sections, we will show you how.

[1] Calculate simple interest.

Simple Interest

Interest is the dollar amount that we get paid for lending money or pay for borrowing money. When we deposit money in a savings institution, the institution pays us interest for its use. When we borrow money, interest is the price we pay for the privilege of using the money until we repay it.

The amount of money that we deposit or borrow is called the **principal**. For example, if you deposit $2000 in a savings account, then $2000 is the principal. The amount of interest depends on the principal, the interest **rate**, which is given as a percent and varies from bank to bank, and the length of time for which the money is deposited. In this section, the rate is assumed to be per year.

Simple interest involves interest calculated only on the principal. The following formula is used to find simple interest:

CALCULATING SIMPLE INTEREST

$$\text{Interest} = \text{principal} \times \text{rate} \times \text{time}$$
$$I = Prt$$

The rate, r, is expressed as a decimal when calculating simple interest.

EXAMPLE 1   CALCULATING SIMPLE INTEREST FOR A YEAR

You deposit $2000 in a savings account at Hometown Bank, which has a rate of 6%. Find the interest at the end of the first year.

Solution   To find the interest at the end of the first year, we use the simple interest formula.

$$I = Prt = (2000)(0.06)(1) = 120$$

| Principal, or amount deposited, is $2000. | Rate is 6% = 0.06. | Time is 1 year. |

At the end of the first year, the interest is $120. You can withdraw the $120 interest, and you still have $2000 in the savings account.

 **Check 1 Point**   You deposit $3000 in a savings account at Yourtown Bank, which has a rate of 5%. Find the interest at the end of the first year.

EXAMPLE 2 CALCULATING SIMPLE INTEREST FOR MORE THAN A YEAR

A student took out a simple interest loan for $1800 for two years at a rate of 8% to purchase a used car. What is the interest on the loan?

Solution   To find the interest on the loan, we use the simple interest formula.

$$I = Prt = (1800)(0.08)(2) = 288$$

| Principal, or amount borrowed, is $1800. | Rate is 8% = 0.08. | Time is 2 years. |

The interest on the loan is $288.

 **Check 2 Point**   A student took out a simple interest loan for $2400 for two years at a rate of 7%. What is the interest on the loan?

2 Use the future value formula.

Future Value: Principal Plus Interest

When a loan is repaid, the interest is added to the original principal to find the total amount due. In Example 2, at the end of two years the student will have to repay

$$\text{principal} + \text{interest} = \$1800 + \$288 = \$2088.$$

In general, if a principal P is borrowed at a simple interest rate r, then after t years the amount due, A, can be determined as follows:

$$A = P + I = P + Prt = P(1 + rt).$$

The amount due, A, is called the **future value** of the loan. The principal borrowed now, P, is also known as the loan's **present value**.

CALCULATING FUTURE VALUE FOR SIMPLE INTEREST

The future value, A, of P dollars at simple interest rate r (as a decimal) for t years is given by

$$A = P(1 + rt).$$

STUDY TIP

Throughout this chapter, keep in mind that all given rates are assumed to be *per year*, unless otherwise stated.

EXAMPLE 3 CALCULATING FUTURE VALUE

A loan of $1060 has been made at 6.5% for three months. Find the loan's future value.

Solution   The amount borrowed, or principal, P, is $1060. The rate, r, is 6.5%, or 0.065. The time, t, is given as 3 months. We need to express the time in years because the rate is understood to be 6.5% per year. Because 3 months is $\frac{3}{12}$ of a year, $t = \frac{3}{12} = \frac{1}{4} = 0.25$.
   The loan's future value, or the total amount due after three months, is

$$A = P(1 + rt) = 1060[1 + (0.065)(0.25)] \approx \$1077.23.$$

Rounded to the nearest cent, the loan's future value is $1077.23.

 **Check 3 Point**   A loan of $2040 has been made at 7.5% for four months. Find the loan's future value.

   Simple interest is used for many short-term loans, including automobile and consumer loans. Imagine that a short-term loan is taken for 125 days. The time of the loan is $\frac{125}{365}$ because there are 365 days in a year. However, the **Banker's rule** allows financial institutions to use 360 in the denominator of such a fraction, claiming that this simplifies the interest calculation. The time, t, for a 125-day short-term loan is

$$\frac{125 \text{ days}}{360 \text{ days}} = \frac{125}{360}.$$

   Consumer groups question whether a year should be simplified to 360 days. Why? Compare the values for time, t, for a 125-day short-term loan using denominators of 360 and 365.

$$\frac{125}{360} \approx 0.347 \qquad \frac{125}{365} \approx 0.342$$

The denominator of 360 benefits the bank by resulting in a greater period of time for the loan, and consequently more interest.
   The formula for future value, $A = P(1 + rt)$, has four variables. If we are given values for any three of these variables, we can solve for the fourth.

EXAMPLE 4 DETERMINING A SIMPLE INTEREST RATE

You borrow $2500 from a friend and promise to pay back $2655 in six months. What simple interest rate will you pay?

Solution   We use the formula for future value, $A = P(1 + rt)$. You borrow $2500: $P = 2500$. You will pay back $2655, so this is the future value: $A = 2655$. You will do this in six months, which must be expressed in years: $t = \frac{6}{12} = \frac{1}{2} = 0.5$. To determine the simple interest rate you will pay, we solve the future value formula for r.

$$A = P(1 + rt) \qquad \text{This is the formula for future value.}$$
$$2655 = 2500[1 + r(0.5)] \qquad \text{Substitute the given values.}$$
$$2655 = 2500 + 1250r \qquad \text{Use the distributive property.}$$
$$155 = 1250r \qquad \text{Subtract 2500 from both sides.}$$
$$\frac{155}{1250} = \frac{1250r}{1250} \qquad \text{Divide both sides by 1250.}$$
$$r = 0.124 = 12.4\% \qquad \text{Express } \frac{155}{1250} \text{ as a percent.}$$

You will pay a simple interest rate of 12.4%.

 **Check Point 4** You borrow $5000 from a friend and promise to pay back $6800 in two years. What simple interest rate will you pay?

EXAMPLE 5   DETERMINING A PRESENT VALUE

You plan to save $2000 for a trip to Europe in two years. You decide to purchase a certificate of deposit (CD) from your bank that pays a simple interest rate of 4%. How much must you put in this CD now in order to have the $2000 in two years?

Solution   We use the formula for future value, $A = P(1 + rt)$. We are interested in finding the principal, P, or the present value.

$$A = P(1 + rt)$$ This is the formula for future value.

$$2000 = P[1 + (0.04)(2)]$$ A(future value) = $2000, r(interest rate) = 0.04, and t = 2 (you want $2000 in two years).

$$2000 = 1.08P$$ Simplify: 1 + (0.04)(2) = 1.08.

$$\frac{2000}{1.08} = \frac{1.08P}{1.08}$$ Divide both sides by 1.08.

$$P \approx 1851.852$$ Simplify.

To make sure you will have enough money for the vacation, let's round this principal *up* to $1851.86. Thus, you should put $1851.86 in the CD now to have $2000 in two years.

 **Check Point 5** How much should you put an an investment paying a simple interest rate of 8% if you need $4000 in six months?

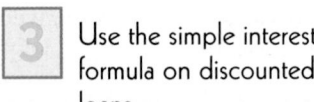 Use the simple interest formula on discounted loans.

Discounted Loans

Some lenders collect the interest from the amount of the loan at the time that the loan is made. This type of loan is called a **discounted loan**. The interest that is deducted from the loan is the **discount**.

EXAMPLE 6   A DISCOUNTED LOAN

You borrow $10,000 on a 10% discounted loan for a period of 8 months.

a. What is the loan's discount?

b. Determine the net amount of money you receive.

c. What is the loan's actual interest rate?

Solution

a. Because the loan's discount is the deducted interest, we use the simple interest formula.

$$I = Prt = (10,000)(0.10)\left(\tfrac{2}{3}\right) \approx 666.67$$

Principal is $10,000.   Rate is 10% = 0.10.   Time is 8 months = $\frac{8}{12}$ = $\frac{2}{3}$ year.

The loan's discount is $666.67.

b. The net amount that you receive is the amount of the loan, $10,000, minus the discount, $666.67:

$$10,000 - 666.67 = 9333.33.$$

Thus, you receive $9333.33.

c. We can calculate the loan's actual interest rate, rather than the stated 10%, by using the simple interest formula.

$$I = Prt \qquad \text{This is the simple interest formula.}$$

$$666.67 = (9333.33)r\left(\frac{2}{3}\right) \qquad \begin{array}{l} I \text{ (interest)} = \$666.67, P \text{ (principal,} \\ \text{or the net amount received)} \\ = \$9333.33, \text{ and } t = \frac{2}{3} \text{ year.} \end{array}$$

$$666.67 = 6222.22r \qquad \text{Simplify: } 9333.33\left(\frac{2}{3}\right) = 6222.22.$$

$$r = \frac{666.67}{6222.22} \approx 0.107 = 10.7\% \qquad \begin{array}{l} \text{Solve for } r \text{ and express } r \text{ to} \\ \text{the nearest tenth of a percent.} \end{array}$$

The actual rate of interest on the 10% discounted loan is approximately 10.7%.

Check Point 6 You borrow $5000 on a 12% discounted loan for a period of two years. Determine: **a.** the loan's discount; **b.** the net amount you receive; **c.** the actual interest rate.

Exercise Set 8.2

Practice Exercises

In Exercises 1–8, the principal P is borrowed at simple interest rate r for a period of time t. Find the simple interest owed for the use of the money. Assume 360 days in a year and round answers to the nearest cent.

1. $P = \$4000, r = 6\%, t = 1$ year

2. $P = \$7000, r = 5\%, t = 1$ year

3. $P = \$180, r = 3\%, t = 2$ years

4. $P = \$260, r = 4\%, t = 3$ years

5. $P = \$5000, r = 8.5\%, t = 9$ months

6. $P = \$18,000, r = 7.5\%, t = 18$ months

7. $P = \$15,500, r = 11\%, t = 90$ days

8. $P = \$12,600, r = 9\%, t = 60$ days

In Exercises 9–14, the principal P is borrowed at simple interest rate r for a period of time t. Find the loan's future value, A, or the total amount due at time t. Round answers to the nearest cent.

9. $P = \$3000, r = 7\%, t = 2$ years

10. $P = \$2000, r = 6\%, t = 3$ years

11. $P = \$26,000, r = 9.5\%, t = 5$ years

12. $P = \$24,000, r = 8.5\%, t = 6$ years

13. $P = \$9000, r = 6.5\%, t = 8$ months

14. $P = \$6000, r = 4.5\%, t = 9$ months

In Exercises 15–20, the principal P is borrowed and the loan's future value, A, at time t is given. Determine the loan's simple interest rate, r, to the nearest tenth of a percent.

15. $P = \$2000, A = \$2150, t = 1$ year

16. $P = \$3000, A = \$3180, t = 1$ year

17. $P = \$5000, A = \$5900, t = 2$ years

18. $P = \$10,000, A = \$14,060, t = 2$ years

19. $P = \$2300, A = \$2840, t = 9$ months

20. $P = \$1700, A = \$1820, t = 6$ months

In Exercises 21–26, determine the present value, P, you must invest to have the future value, A, at simple interest rate r after time t. Round answers up to the nearest cent.

21. $A = \$6000, r = 8\%, t = 2$ years

22. $A = \$8500, r = 7\%, t = 3$ years

23. $A = \$14,000, r = 9.5\%, t = 6$ years

24. $A = \$16,000, r = 11.5\%, t = 5$ years

25. $A = \$5000, r = 14.5\%, t = 9$ months

26. $A = \$2000, r = 12.6\%, t = 8$ months

Exercises 27–30 involve discounted loans. In each exercise, determine

   a. the loan's discount.

   b. the net amount of money you receive.

   c. the loan's actual interest rate, to the nearest tenth of a percent.

27. You borrow $2000 on a 7% discounted loan for a period of 8 months.

28. You borrow $3000 on an 8% discounted loan for a period of 9 months.

29. You borrow $12,000 on a 6.5% discounted loan for a period of two years.

30. You borrow $20,000 on an 8.5% discounted loan for a period of three years.

Application Exercises

31. In order to start a small business, a student takes out a simple interest loan for $4000 for 9 months at a rate of 8.25%.

   a. How much interest must the student pay?

   b. Find the future value of the loan.

32. In order to pay for baseball uniforms, a school takes out a simple interest loan for $20,000 for 7 months at a rate of 12%.

   a. How much interest must the school pay?

   b. Find the future value of the loan.

33. You borrow $1400 from a friend and promise to pay back $2000 in two years. What simple interest rate, to the nearest tenth of a percent, will you pay?

34. Treasury bills (T-bills) can be purchased from the U.S. Treasury Department. You buy a T-bill for $981.60 that pays $1000 in 13 weeks. What simple interest rate, to the nearest tenth of a percent, does this T-bill earn?

35. A bank offers a CD that pays a simple interest rate of 6.5%. How much must you put in this CD now in order to have $3000 for a home-entertainment center in two years? Round *up* to the nearest cent.

36. A bank offers a CD that pays a simple interest rate of 5.5%. How much must you put in this CD now in order to have $8000 for a kitchen remodeling project in two years? Round *up* to the nearest cent.

Writing in Mathematics

37. Explain how to calculate simple interest.

38. What is the future value of a loan and how is it determined?

39. What is a discounted loan? How is the net amount of money received from such a loan determined?

Critical Thinking Exercises

40. Use the future value formula to show that the time required for an amount of money P to double in value to $2P$ is given by

$$t = \frac{1}{r}.$$

41. You deposit $5000 in an account that earns 5.5% simple interest.

   a. Express the future value in the account as a linear function of time, t.

   b. Determine the slope of the function in part (a) and describe what this means. Use the phrase "rate of change" in your description.

OBJECTIVES

1. Use compound interest formulas.

2. Calculate present value.

3. Understand and compute effective annual yield.

So, how did the present value of Manhattan in 1626—that is, the $24 paid to the native Americans—attain a future value of over $3 billion in 2000, 374 years later, at a mere 5% interest rate? After all, the future value on $24 for 374 years at 5% simple interest is

$$A = P(1 + rt) = 24[1 + (0.05)(374)] = 472.80$$

or a paltry $472.80 compared to over $3 billion. To understand this dramatic difference in future value, we turn to the concept of *compound interest*.

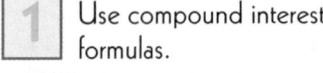
Use compound interest formulas.

Compound Interest

Compound interest is interest computed on the original principal as well as on any accumulated interest. Many savings accounts pay compound interest. For example, suppose you deposit $1000 in a savings account at a rate of 5%. Table 8.2 shows how the investment grows if the interest earned is automatically added on to the principal.

TABLE 8.2 COMPOUND INTEREST ON $1000

Year	Starting Balance	Interest Earned: $I = Prt$	New Balance
1	$1000	$1000 \times 0.05 \times 1 = \$50$	$1050
2	$1050	$1050 \times 0.05 \times 1 = \$52.50$	$1102.50
3	$1102.50	$1102.50 \times 0.05 \times 1 \approx \$55.13$	$1157.63

A faster way to determine the amount, A, in an account subject to compound interest is to use the following formula and a calculator:

> ### CALCULATING THE AMOUNT IN AN ACCOUNT FOR COMPOUND INTEREST PAID ONCE A YEAR
>
> If P dollars are deposited at rate r, in decimal form, subject to compound interest, then the amount, A, of money in the account after t years is given by
>
> $$A = P(1 + r)^t.$$
>
> The amount A is called the account's **future value** and the principal P is called its **present value**.

EXAMPLE 1   USING THE COMPOUND INTEREST FORMULA

You deposit $2000 in a savings account at Hometown Bank, which has a rate of 6%.

a. Find the amount, A, of money in the account after 3 years subject to compound interest.

b. Find the interest.

Solution

a. The amount deposited, or principal, P, is $2000. The rate, r, is 6%, or 0.06. The time of the deposit, t, is three years. The amount in the account after three years is

$$A = P(1 + r)^t = 2000(1 + 0.06)^3 = 2000(1.06)^3 \approx 2382.03.$$

Rounded to the nearest cent, the amount in the savings account after three years in $2382.03.

b. Because the amount in the account is $2382.03 and the original principal is $2000, the interest is $2382.03 − $2000, or $383.03.

TECHNOLOGY

Here are the calculator keystrokes to compute $2000(1.06)^3$:

Many Scientific Calculators

2000 $\boxed{\times}$ 1.06 $\boxed{y^x}$ 3 $\boxed{=}$

Many Graphing Calculators

2000 $\boxed{\times}$ 1.06 $\boxed{\wedge}$ 3 $\boxed{\text{ENTER}}$.

Check Point 1 You deposit $1000 in a savings account at a bank that has a rate of 4%.

a. Find the amount, A, of money in the account after 5 years subject to compound interest. Round to the nearest cent.

b. Find the interest.

Compound Interest Paid More Than Once a Year

The period of time between two interest payments is called the **compounding period**. When compound interest is paid once per year, the compounding period is one year. We say that the interest is **compounded annually**.

Most savings institutions have plans in which interest is paid more than once per year. If compound interest is paid twice per year, the compounding period is six months. We say that the interest is **compounded semiannually**. When compound interest is paid four times per year, the compounding period is three months and the interest is said to be **compounded quarterly**. Some plans allow for monthly compounding or daily compounding.

In general, when compound interest is paid n times per year, we say that there are **n compounding periods per year**. The following formula is used to calculate the amount in an account subject to compound interest with n compounding periods per year:

CALCULATING THE AMOUNT IN AN ACCOUNT FOR COMPOUND INTEREST PAID n TIMES A YEAR

If P dollars are deposited at rate r, in decimal form, subject to compound interest paid n times per year, then the amount, A, of money in the account after t years is given by

$$A = P\left(1 + \frac{r}{n}\right)^{nt}.$$

A is the account's **future value** and the principal P is its **present value**.

EXAMPLE 2 | USING THE COMPOUND INTEREST FORMULA

TECHNOLOGY

Here are the calculator keystrokes to compute

$$7500\left(1 + \frac{0.06}{12}\right)^{12 \cdot 5}:$$

Many Scientific Calculators

7500 $\times$ ((1 $+$ .06 $\div$ 12))
y^x ((12 $\times$ 5)) $=$

Many Graphing Calculators

7500 ((1 $+$ .06 $\div$ 12))
$\wedge$ ((12 $\times$ 5)) ENTER .

You deposit $7500 in a savings account that has a rate of 6%. The interest is compounded monthly.

a. How much money will you have after five years?

b. Find the interest after five years.

Solution

a. The amount deposited, or principal, P, is $7500. The rate, r, is 6%, or 0.06. Because interest is compounded monthly, there are 12 compounding periods per year, so $n = 12$. The time of the deposit, t, is five years. The amount in the account after five years is

$$A = P\left(1 + \frac{r}{n}\right)^{nt} = 7500\left(1 + \frac{0.06}{12}\right)^{12 \cdot 5} = 7500(1.005)^{60} \approx 10{,}116.38.$$

Rounded to the nearest cent, you will have $10,116.38 after five years.

b. Because the amount in the account is $10,116.38 and the original principal is $7500, the interest after five years is $10,116.38 − $7500, or $2616.38.

Check 2 Point

You deposit $4200 in a savings account that has a rate of 4%. The interest is compounded quarterly.

a. How much money will you have after ten years? Round to the nearest cent.

b. Find the interest after ten years.

Some banks use **continuous compounding**, where the compounding periods increase infinitely (compounding interest every trillionth of a second, every quadrillionth of a second, etc.). Although continuous compounding sounds terrific, it yields only a fraction of a percent more interest over a year than daily compounding.

2   Calculate present value.

Planning for the Future with Compound Interest

Just as we did in Section 8.2, we can determine P, the principal or present value, that should be deposited now in order to have a certain amount, A, in the future. If an account earns compound interest, the amount of money that should be invested today to obtain a future value of A dollars can be determined by solving the compound interest formula for P:

CALCULATING PRESENT VALUE

If A dollars are to be accumulated in t years in an account that pays rate r compounded n times per year, then the present value, P, that needs to be invested now is given by

$$P = \frac{A}{\left(1 + \dfrac{r}{n}\right)^{nt}}.$$

EXAMPLE 3   CALCULATING PRESENT VALUE

How much money should be deposited in an account today that earns 8% compounded monthly so that it will accumulate to $20,000 in five years?

Solution   The amount we need today, or the present value, is determined by the present value formula. Because the interest is compounded monthly, $n = 12$. Furthermore, A (the future value) = $20,000, r (the rate) = 8% = 0.08, and t (time in years) = 5.

$$P = \frac{A}{\left(1 + \dfrac{r}{n}\right)^{nt}} = \frac{20,000}{\left(1 + \dfrac{0.08}{12}\right)^{12 \cdot 5}} \approx 13,424.21$$

Approximately $13,424.21 should be invested today in order to accumulate to $20,000 in five years.

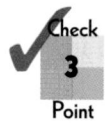 Check 3 Point

How much money should be deposited in an account today that earns 6% compounded weekly so that it will accumulate to $10,000 in eight years?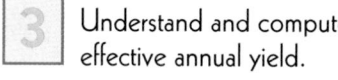

3   Understand and compute effective annual yield.

Effective Annual Yield

A common problem in financial planning is selecting the best investment from two or more investments. For example, is an investment that pays 8.25% interest compounded quarterly better than one that pays 8.3% interest compounded semiannually? One way to answer the question is to compare their *effective rates*, also called their *effective annual yields*.

EFFECTIVE ANNUAL YIELD

The **effective annual yield**, or the **effective rate**, is the simple interest rate that produces the same amount of money in account at the end of one year as when the account is subjected to compound interest at a stated rate.

EXAMPLE 4 UNDERSTANDING EFFECTIVE ANNUAL YIELD

You deposit $4000 in an account that pays 8% interest compounded monthly.

a. Find the future value after one year.

b. Use the future value formula for simple interest to determine the effective annual yield.

Solution

a. We use the compound interest formula to find the account's future value after one year.

$$A = P\left(1 + \frac{r}{n}\right)^{nt} = 4000\left(1 + \frac{0.08}{12}\right)^{12 \cdot 1} \approx 4332.00$$

Principal is $4000.   Stated rate is 8% = 0.08.   Monthly compounding: n = 12   Time is one year: t = 1

Rounded to the nearest cent, the future value after one year is $4332.00.

b. **The effective annual yield**, or effective rate, **is a simple interest rate.** We use the future value formula for simple interest to determine the simple interest rate that produces a future value of $4332 for a $4000 deposit after one year.

$$A = P(1 + rt)$$ This is the future value formula for simple interest.

$$4332 = 4000(1 + r \cdot 1)$$ Substitute the given values.

$$4332 = 4000 + 4000r$$ Use the distributive property.

$$332 = 4000r$$ Subtract 4000 from both sides.

$$\frac{332}{4000} = \frac{4000r}{4000}$$ Divide both sides by 4000.

$$r = \frac{332}{4000} = 0.083 = 8.3\%$$ Express r as a percent.

The effective annual yield, or effective rate, is 8.3%. This means that money invested at 8.3% simple interest earns the same amount in one year as money invested at 8% interest compounded monthly.

In Example 4, the stated 8% rate is called the **nominal rate**. The 8.3% rate is the effective rate and is a simple interest rate.

 **Check 4 Point** You deposit $6000 in an account that pays 10% interest compounded monthly.

a. Find the future value after one year.

b. Determine the effective annual yield.

Generalizing the procedure of Example 4 and Check Point 4 gives a formula for effective annual yield:

CALCULATING EFFECTIVE ANNUAL YIELD

Suppose that an investment has a nominal interest rate, r, in decimal form, and pays compound interest n times per year. The investment's effective annual yield, Y, is given by

$$Y = \left(1 + \frac{r}{n}\right)^n - 1.$$

TECHNOLOGY

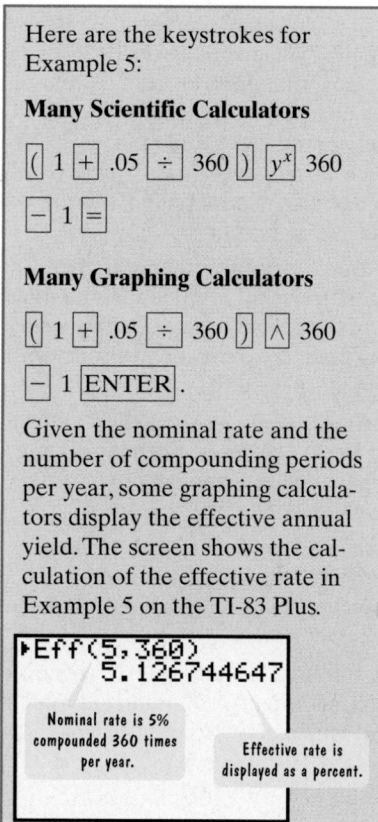

Here are the keystrokes for Example 5:

Many Scientific Calculators

$(1 + .05 ÷ 360) \; y^x \; 360$

$- 1 =$

Many Graphing Calculators

$(1 + .05 ÷ 360) \; \wedge \; 360$

$- 1$ ENTER.

Given the nominal rate and the number of compounding periods per year, some graphing calculators display the effective annual yield. The screen shows the calculation of the effective rate in Example 5 on the TI-83 Plus.

▸Eff(5,360)
        5.126744647

Nominal rate is 5% compounded 360 times per year.

Effective rate is displayed as a percent.

EXAMPLE 5 CALCULATING EFFECTIVE ANNUAL YIELD

A passbook savings account has a nominal rate of 5%. The interest is compounded daily. Find the account's effective annual yield. (Assume 360 days in a year.)

Solution   The rate, r, is 5%, or 0.05. Because interest is compounded daily and we assume 360 days in a year, $n = 360$. The account's effective annual yield is

$$Y = \left(1 + \frac{r}{n}\right)^n - 1 = \left(1 + \frac{0.05}{360}\right)^{360} - 1 \approx 0.0513 = 5.13\%.$$

The effective annual yield is 5.13%. Thus, money invested at 5.13% simple interest earns the same amount of interest in one year as money invested at 5% interest, the nominal rate, compounded daily.

Check 5 Point   What is the effective annual yield of an account paying 8% compounded quarterly?

$$\left(1 + \frac{.08}{4}\right)^4 - 1 =$$

The effective annual yield is often included in the information about investments or loans. Because it's the true interest rate you're earning or paying, it's the number you should pay attention to. **If you are selecting the best investment from two or more investments, the best choice is the account with the greatest effective annual yield.** However, there are differences in the types of accounts that you need to take into consideration. Some pay interest from the day of deposit to the day of withdrawal. Other accounts start paying interest the first day of the month that follows the day of deposit. Some savings institutions stop paying interest if the balance in the account falls below a certain amount.

When borrowing money, the effective rate or effective annual yield is usually called the **annual percentage rate**. If all other factors are equal and you are borrowing money, select the option with the least annual percentage rate. In the next section on installment buying, we will discuss annual percentage rates in more detail.

Exercise Set 8.3

Practice Exercises

In Exercises 1–12, the principal represents an amount of money deposited in a savings account subject to compound interest at the given rate.

  a. *Use the formula $A = P(1 + r)^t$ for one compounding period per year or $A = P(1 + \frac{r}{n})^{nt}$ for n compounding periods per year to find how much money there will be in the account after the given number of years. (Assume 360 days in a year.)*

  b. *Find the interest earned.*

Round answers to the nearest cent.

	Principal	Rate	Compounded	Time
1.	$10,000	4%	annually	2 years
2.	$8000	6%	annually	3 years
3.	$3000	5%	semiannually	4 years
4.	$4000	4%	semiannually	5 years
5.	$9500	6%	quarterly	5 years
6.	$2500	8%	quarterly	6 years
7.	$4500	4.5%	monthly	3 years
8.	$2500	6.5%	monthly	4 years
9.	$1500	8.5%	daily	2.5 years
10.	$1200	8.5%	daily	3.5 years
11.	$20,000	4.5%	daily	20 years
12.	$25,000	5.5%	daily	20 years

Solve Exercises 13–16 using the present value formula:

$$P = \frac{A}{\left(1 + \dfrac{r}{n}\right)^{nt}}.$$

Round answers up to the nearest cent.

13. How much money should be deposited today in an account that earns 6% compounded semiannually so that it will accumulate to $10,000 in three years?

14. How much money should be deposited today in an account that earns 7% compounded semiannually so that it will accumulate to $12,000 in four years?

15. How much money should be deposited today in an account that earns 9.5% compounded monthly so that it will accumulate to $10,000 in three years?

16. How much money should be deposited today in an account that earns 10.5% compounded monthly so that it will accumulate to $22,000 in four years?

17. You deposit $10,000 in an account that pays 4.5% interest compounded quarterly.
    a. Find the future value after one year.
    b. Use the future value formula for simple interest to determine the effective annual yield.

18. You deposit $12,000 in an account that pays 6.5% interest compounded quarterly.
    a. Find the future value after one year.
    b. Use the future value formula for simple interest to determine the effective annual yield.

Solve Exercises 19–24 using the effective annual yield formula:

$$Y = \left(1 + \frac{r}{n}\right)^{n} - 1.$$

A passbook savings account has a rate of 6%. Find the effective annual yield, rounded to the nearest tenth of a percent, if the interest is compounded

19. semiannually.   20. quarterly.
21. monthly.
22. daily. (Assume 360 days in a year.)
23. 1000 times per year.
24. 100,000 times per year.

In Exercises 25–28, determine the effective annual yield, to the nearest tenth of a percent, for each investment. Then select the better investment. Assume 360 days in a year.

25. 8% compounded monthly; 8.25% compounded annually
26. 5% compounded monthly; 5.25% compounded quarterly
27. 5.5% compounded semiannually; 5.4% compounded daily
28. 7% compounded annually; 6.85% compounded daily

Application Exercises

29. In 1626, Peter Minuit convinced the Wappinger Indians to sell him Manhattan Island for $24. If the Native Americans had put the $24 into a bank account paying compound interest at a 5% rate, how much would the investment be worth in the year 2000 ($t = 374$ years) if interest were compounded **a.** monthly? **b.** 360 times per year?

30. In 1777, Jacob DeHaven loaned George Washington's army $450,000 in gold and supplies. Due to a disagreement over the method of repayment (gold versus Continental money), DeHaven was never repaid, dying penniless. In 1989, his descendants sued the U.S. government over the 212-year-old debt. If the DeHavens used an interest rate of 6% and daily compounding (the rate offered by the Continental Congress in 1777), how much money did the DeHaven family demand in their suit? (*Hint:* Use the compound interest formula with $n = 360$ and $t = 212$ years.)

31. At the time of her grandson's birth, a grandmother deposits $10,000 in an account that pays 9% compounded monthly. What will be the value of the account at the child's twenty-first birthday, assuming that no other deposits or withdrawals are made during this period?

32. Parents wish to have $80,000 available for a child's education. If the child is now 5 years old, how much money must be set aside at 6% compounded semiannually to meet their financial goal when the child is 18?

33. A 30-year-old worker plans to retire at age 65. He believes that $500,000 is needed to retire comfortably. How much should be deposited now at 9% compounded monthly to meet the $500,000 retirement goal?

34. In 2001, First Internet Bank of Indiana offered a money market account paying 4.5% interest compounded monthly. NetBank offered a money market account paying 4.4% interest compounded daily. (*Source:* www.bankrate.com) Which account is the better investment?

*In this section, we saw how to find the future value of money invested at a fixed interest rate that is compounded. Instead of investing money and then watching it grow, many people invest smaller amounts at regular intervals. An **annuity** is a sequence of equal deposits made at equal time periods.*

If P is the amount of each deposit, n is the number of deposits per year, and r is the annual interest rate, in decimal form, compounded n times per year, the future value, A, of the annuity after t years is

$$A = P\frac{\left(1 + \dfrac{r}{n}\right)^{nt} - 1}{\dfrac{r}{n}}.$$

Use this formula to solve Exercises 35–36.

35. To save for retirement, you decide to deposit $1000 into an IRA at the end of each year. If the interest rate is rate is 7% per year compounded annually, find the value of the IRA after 30 years.

36. You decide to deposit $100 at the end of each month into an account paying 8% interest compounded monthly to save for your child's education. How much will you save over 16 years?

Writing in Mathematics

37. Describe the difference between simple and compound interest.

38. Give two examples that illustrate the difference between a compound interest problem involving future value and a compound interest problem involving present value.

39. What is effective annual yield?

40. Explain how to select the best investment from two or more investments.

41. Shown are figures in an actual advertisement for Great Western Bank. Explain the meaning of the given percents, as well as the statement below these percents.

9.21%	**8.85**%
Yield	*Rate*

Interest compounded monthly.

Critical Thinking Exercises

42. A depositor opens a new savings account with $6000 at 5% compounded semiannually. At the beginning of year 3, an additional $4000 is deposited. At the end of six years, what is the balance in the account?

43. A depositor opens a money market account with $5000 at 8% compounded monthly. After two years, $1500 is withdrawn from the account to buy a new computer. A year later, $2000 is put in the account. What will be the ending balance if the money is kept in the account for another three years?

44. Use the future value formulas for simple and compound interest in one year to derive the formula for effective annual yield.

Group Exercises

45. This activity is a group research project intended for four or five people. Present your research in a seminar on the history of interest and banking. The seminar should last about 30 minutes. Address the following questions:

When was interest first charged on loans? How was lending money for a fee opposed historically? What is usury? What connection did banking and interest rates play in the historic European rivalries between Christians and Jews? When and where were some of the highest interest rates charged? What were the rates? Where does the word "interest" come from? What is the origin of the word "shylock"? What is the difference between usury and interest in modern times? What is the history of a national bank in the United States?

46. Group members should conduct a survey of banks, savings and loan companies, and credit unions in the area. Report to the class on the different kinds of savings accounts available and the interest rates they pay. What are the methods of payment?

SECTION 8.4   INSTALLMENT BUYING

OBJECTIVES

1. Determine the amount financed, the installment price, and the finance charge for a fixed loan.

2. Determine the APR.

3. Compute unearned interest and the payoff amount for a loan paid off early.

4. Find the interest, the balance due, and the minimum monthly payment for credit card loans.

5. Calculate interest on credit cards using three methods.

Do you buy products with a credit card? Your card lets you use a product while paying for it. However, the costs associated with such cards contributed to nearly 1.5 million U.S. consumers filing for bankruptcy in 2001. If you use a credit card, you are engaging in **installment buying**, in which you repay a loan for the cost of a product on a monthly basis. A loan that you pay off with weekly or monthly payments, or payments in some other time period, is called an **installment loan**. The advantage of an installment loan is that the consumer gets to use a product immediately. In this section, we will see that the disadvantage is that it can add a substantial amount to the cost of a purchase. When it comes to installment buying, consumer beware!

BLITZER BONUS

The Most Expensive Credit Card

The American Express Centurion card is offered by invitation only to selected individuals with a minimum yearly salary of $215,000. The card has an annual fee of $1000 and no spending limits.

1 Determine the amount financed, the installment price, and the finance charge for a fixed loan.

Fixed Installment Loans

You decide to purchase a pick-up truck that costs $9345. You can finance the truck by paying $300 at the time of purchase, called the **down payment**, and $194.38 per month for 60 months. A loan like this that has a schedule for paying a fixed amount each period is called a **fixed installment loan**. We begin with three terms associated with such loans.

THE VOCABULARY OF FIXED INSTALLMENT LOANS

The **amount financed** is what the consumer borrows:

$$\text{Amount financed} = \text{Cash price} - \text{Down payment}.$$

The **total installment price** is the sum of all monthly payments plus the down payment:

$$\text{Total installment price} = \text{Total of all monthly payments} + \text{Down payment}.$$

The **finance charge** is the interest on the installment loan:

$$\text{Finance charge} = \text{Total installment price} - \text{Cash price}.$$

EXAMPLE 1 | BUYING A PICK-UP TRUCK USING A FIXED INSTALLMENT LOAN

The cost of a pick-up truck is $9345. We can finance the truck by paying $300 down and $194.38 per month for 60 months.

a. Determine the amount financed.

b. Determine the total installment price.

c. Determine the finance charge.

Solution

a. Amount financed = Cash price − Down payment

$$= \$9345 - \$300$$
$$= \$9045$$

The amount financed is $9045.

b. The total installment price is obtained by adding the total of all monthly payments and the down payment. Because we are paying $194.38 per month for 60 months, the total of all monthly payments is

$$\frac{\$194.38}{\text{month}} \times 60 \text{ months} = \$194.38 \times 60.$$

Total installment price = Total of all monthly payments + Down payment

$$= \$194.38 \times 60 + \$300$$
$$= \$11{,}662.80 + \$300$$
$$= \$11{,}962.80$$

The total installment price is $11,962.80.

c. Finance charge = Total installment price − Cash price

= $11,962.80 − $9345

= $2617.80

The finance charge for the loan is $2617.80.

 **Check 1 Point** The cost of a new car is $14,000. You can finance the car by paying $280 down and $315 per month for 60 months.

a. Determine the amount financed.

b. Determine the total installment price.

c. Determine the finance charge.

You will be using these figures in Check Points 2, 3, and 4, so keep track of your answers.

 Determine the APR.

STUDY TIP

Do not confuse effective annual yield with annual percentage rate. Effective annual yield applies to *interest earned* on invested money. By contrast, annual percentage rate applies to *interest owed* on loans.

In Example 1, the finance charge for the loan, $2617.80, is the **interest** paid to finance the truck. What interest rate are we paying when we pay $2617.80 interest over 60 months? The interest rate per year is called the **annual percentage rate**, abbreviated **APR**. In 1969, the Federal Reserve Board established the **Truth-in-Lending Act**. It requires lending institutions to inform borrowers in writing of a loan's APR. **When comparing two or more loans with different terms, the loan with the lowest APR is the one that charges the least interest.** For many people, an important factor in deciding on a fixed installment loan is the cash down payment that is required. If you do not have a lot of cash at the time of purchase, you might select a loan with a smaller down payment even though it has a not-so-desirable APR.

The APR for a fixed installment loan can be determined using a table similar to the abbreviated version shown in Table 8.3. Here are the steps involved in using an APR table. These steps are illustrated in Example 2.

STEPS IN USING AN APR TABLE

1. Compute the finance charge per $100 financed:

$$\frac{\text{Finance charge}}{\text{Amount financed}} \times \$100.$$

2. Look in the row corresponding to the number of payments to be made and find the entry closest to the value in step 1.

3. Find the APR at the top of the column in which the entry from step 2 is found. (This is the APR rounded to the nearest $\frac{1}{2}$%.)

TABLE 8.3 ANNUAL PERCENTAGE RATE (APR) FOR MONTHLY PAYMENT LOANS

Number of Monthly Payments	Annual Percentage Rate (APR)												
	10.0%	10.5%	11.0%	11.5%	12.0%	12.5%	13.0%	13.5%	14.0%	14.5%	15.0%	15.5%	16.0%
	(Finance charge per $100 of amount financed)												
6	$2.94	$3.08	$3.23	$3.38	$3.53	$3.68	$3.83	$3.97	$4.12	$4.27	$4.42	$4.57	$4.72
12	5.50	5.78	6.06	6.34	6.62	6.90	7.18	7.46	7.74	8.03	8.31	8.59	8.88
18	8.10	8.52	8.93	9.35	9.77	10.19	10.61	11.03	11.45	11.87	12.29	12.72	13.14
24	10.75	11.30	11.86	12.42	12.98	13.54	14.10	14.66	15.23	15.80	16.37	16.94	17.51
30	13.43	14.13	14.83	15.54	16.24	16.95	17.66	18.38	19.10	19.81	20.54	21.26	21.99
36	16.16	17.01	17.86	18.71	19.57	20.43	21.30	22.17	23.04	23.92	24.80	25.68	26.57
48	21.74	22.90	24.06	25.23	26.40	27.58	28.77	29.97	31.17	32.37	33.59	34.81	36.03
60	27.48	28.96	30.45	31.96	33.47	34.99	36.52	38.06	39.61	41.17	42.74	44.32	45.91

EXAMPLE 2 DETERMINING THE APR

In Example 1, we found that the amount financed for the truck was $9045 and the finance charge was $2617.80. The borrower financed the truck with 60 monthly payments. Use Table 8.3 to determine the APR.

Solution

STEP 1. Find the finance charge per $100 of the amount financed.

$$\text{Finance charge per } \$100 \text{ financed} = \frac{\text{Finance charge}}{\text{Amount financed}} \times \$100$$

$$= \frac{\$2617.80}{\$9045.00} \times \$100$$

$$\approx \$28.94$$

This means that the borrower pays $28.94 interest for each $100 being financed.

STEP 2. Look in the row corresponding to the number of payments to be made (60) and find the entry closest to the value in step 1 ($28.94). Table 8.3 on page 429 presents three types of information: the number of monthly payments, the APR, and the finance charge per $100 financed. There are 60 monthly payments, so look for 60 in the left column. Then move across to the right until you find the amount closest to the finance charge per $100 financed, namely $28.94. This amount, $28.96, is circled in the table on page 429.

STEP 3. Find the APR at the top of the column in which the entry from step 2 is found. Take a second look at the circled entry, $28.96. At the top of this column is 10.5%. Thus, the APR for the fixed installment truck loan is approximately 10.5%.

Use Table 8.3 to determine the APR for the car loan described in Check Point 1.

STUDY TIP

Use Table 8.3 for this example as follows.

10.5%
↓   ↑
60 → (28.96)

---

3 | Compute unearned interest and the payoff amount for a loan paid off early.

A fixed installment loan can be paid off early so that a borrower need not pay the entire finance charge. The amount by which the finance charge is reduced is called the **unearned interest**. The interest saved by paying off the loan early can be calculated by one of two methods, the **actuarial method** or the **rule of 78**. The actuarial method uses the APR table, and the rule of 78 does not. The Truth-in-Lending Act requires that the method for calculating the reduction of the finance charge be disclosed at the time the loan is signed.

Here are two formulas for computing unearned interest:

STUDY TIP

Keep in mind that a loan's finance charge is the same thing as the loan's interest.

METHODS FOR COMPUTING UNEARNED INTEREST

Unearned interest is the amount by which a loan's finance charge is reduced when the loan is paid off early.

Actuarial Method	**Rule of 78**
$$u = \frac{kRV}{100 + V}$$	$$u = \frac{k(k+1)}{n(n+1)} \times F$$
u = unearned interest	u = unearned interest
k = remaining number of scheduled payments (excluding current payment)	k = remaining number of scheduled payments (excluding current payment)
R = regular monthly payment	n = original number of payments
V = finance charge per $100 (from the APR table) for a loan with the same APR and k monthly payments	F = original finance charge

Pick-up truck
cost: $9345
Financing: $300 + $194.38 per month for 60 months
Amount financed: $9045
Total installment price: $11,962.80
Finance charge (interest): $2617.80
Finance charge per $100 financed: $28.94
APR: 10.5%.

The total amount due on the day that a loan is paid off early is called the **payoff amount**. Unearned interest is subtracted in computing the payoff amount. Examples 3 and 4 illustrate how the payoff amount is computed using the actuarial method in Example 3 and the rule of 78 in Example 4. Both examples refer back to the pick-up truck in Example 1. Shown in the margin is a summary of given and computed amounts for the truck.

EXAMPLE 3   EARLY PAYOFF: THE ACTUARIAL METHOD

You got a big bonus at work—and a raise. Instead of making the thirty-sixth payment on your truck, you decide to pay the remaining balance and terminate the loan.

a. Use the actuarial method to determine how much interest will be saved (the unearned interest, u) by repaying the loan early.

b. Find the payoff amount.

Solution

a. We will use the formula $u = \dfrac{kRV}{100 + V}$ to find u, the unearned interest. We need values for k, R, and V. The current payment is payment number 36. There are 60 payments total. We subtract the number of the current payment from the total number of payments to find the number of remaining payments, k: $k = 60 - 36 = 24$. The value of R, the regular monthly payment, is given: $R = \$194.38$. To find the value for the finance charge per $100, V, use the APR table (Table 8.3 on page 429). In the Number of Monthly Payments Column, find the number of remaining payments, 24, and then look to the right until you reach the column headed by 10.5%, the APR. This row and column intersect at 11.30. Thus, $V = 11.30$. This is the finance charge per $100 for a loan with the same APR and k ($k = 24$) monthly payments. We substitute these values into the actuarial method formula.

$$u = \frac{kRV}{100 + V} = \frac{24 \times 194.38 \times 11.30}{100 + 11.30} \approx 473.64$$

Approximately $473.64 will be saved in interest by terminating the loan early using the actuarial method.

b. The payoff amount, the amount due on the day of the loan's termination, is determined as follows:

Payoff amount	=	payment number 36	plus	total of remaining payments after payment 36	minus	interest saved (unearned interest).

$$= \$194.38 \quad + \quad 24 \times \$194.38 \quad - \quad \$473.64$$

$$= \$4385.86.$$

The payment needed to terminate the loan at the end of 36 months is $4385.86.

Check 3 Point

Instead of making the twenty-fourth payment on the car loan described in Check Points 1 and 2, you decide to pay the remaining balance and terminate the loan.

a. Use the actuarial method to determine how much interest will be saved (the unearned interest, u) by repaying the loan early.

b. Find the payoff amount.

Pick-up truck
cost: $9345
Financing: $300 + $194.38 per month for 60 months
Amount financed: $9045
Total installment price: $11,962.80
Finance charge (interest): $2617.80
Finance charge per $100 financed: $28.94
APR: 10.5%.

EXAMPLE 4   EARLY PAYOFF: THE RULE OF 78

The loan in Example 3 is paid off at the time of the thirty-sixth monthly payment. Use the rule of 78 to find

a. the unearned interest.

b. the payoff amount.

Solution

a. We will use the formula

$$u = \frac{k(k + 1)}{n(n + 1)} \times F$$

to find u, the unearned interest. We need values for k, n, and F. From Example 3, k, the number of remaining payments, is 24: $k = 24$. The original number of payments, n, is 60: $n = 60$. The summary of figures for the pick-up truck shown in the margin lists the finance charge: $F = \$2617.80$. We substitute these values into the rule of 78 formula.

$$u = \frac{k(k + 1)}{n(n + 1)} \times F = \frac{24(24 + 1)}{60(60 + 1)} \times \$2617.80 \approx \$429.15$$

By the rule of 78, the borrower will save $429.15 in interest.

b. Next, we determine the payoff amount.

Payoff amount	=	payment number 36	plus	total of remaining payments after payment 36	minus	interest saved (unearned interest).

$$= \quad \$194.38 \quad + \quad 24 \times \$194.38 \quad - \quad \$429.15$$

$$= \$4430.35$$

The payoff amount is $4430.35.

In Example 3, we found that $473.64 is saved in interest using the actuarial method, while in Example 4, we found that $429.15 is saved in interest using the rule of 78. These examples illustrate that the actuarial method is more beneficial to the borrower than the rule of 78 because it results in greater savings in interest at the loan's termination. For loans of 61 months or longer, only the actuarial method is allowed. Also, some states have shorter loan periods that do not allow the rule of 78.

 **Check 4 Point** The loan in Check Point 3 is paid off at the time of the twenty-fourth monthly payment. Use the rule of 78 to find

a. the unearned interest.

b. the payoff amount.

 Find the interest, the balance due, and the minimum monthly payment for credit card loans.

Open-End Installment Loans

Using a credit card is an example of an open-end installment loan, commonly called **revolving credit**. Open-end loans differ from fixed installment loans in that there is no schedule for paying a fixed amount each period. Credit card loans require users to make only a minimum monthly payment that depends on the unpaid balance and the interest rate. Credit cards have high interest rates compared to other kinds of loans. The interest on credit cards is computed using the simple interest formula $I = Prt$. However, r represents the *monthly* interest rate and t is time in months rather than in years. A typical interest rate is 1.57% monthly. This is equivalent to a yearly rate of $12 \times 1.57\%$, or 18.84%. With such a high APR, credit card balances should be paid off as quickly as possible.

Most credit card customers are billed every month. A typical billing period is May 1 through May 31, but it can also run from, say, May 5 through June 4. Customers receive a statement, called an **itemized billing**, that contains the unpaid balance on the first day of the billing period, the total balance owed on the last day of the billing period, a list of purchases and cash advances made during the billing period, the date of the last day of the billing period, the payment due date, and the minimum payment required.

Customers who make a purchase during the billing period and pay the entire amount of the purchase by the payment due date are not charged interest. By contrast, customers who make cash advances using their credit cards must pay interest from the day the money is advanced until the day it is repaid.

One method for calculating interest on credit cards is the **unpaid balance method**. Interest is calculated on the unpaid balance on the first day of the billing period less payments and credits.

EXAMPLE 5   BALANCE DUE ON A CREDIT CARD

A particular VISA card calculates interest using the unpaid balance method. The monthly interest rate is 1.3% on the unpaid balance on the first day of the billing period less payments and credits. Here are some of the details in the May 1–May 31 itemized billing:

May 1 Unpaid Balance: $1350
Payment Received May 8: $250
Purchases Charged to the VISA Account: Airline tickets: $375, Books: $57, Meals: $65
Last Day of the Billing Period: May 31
Payment Due Date: June 9

a. Find the interest due on the payment due date.

b. Find the total balance owed on the last day of the billing period.

c. This credit card requires a $10 minimum monthly payment if the total balance owed on the last day of the billing period is less than $360. Otherwise, the minimum monthly payment is $\frac{1}{36}$ of the balance owed on the last day of the billing period, rounded to the nearest whole dollar. What is the minimum monthly payment due by June 9?

Solution

a. The monthly interest rate is 1.3% on the unpaid balance on the first day of the billing period, $1350, less payments and credits, $250. The interest due is computed using $I = Prt$.

$$I = Prt = (\$1350 - \$250) \times 0.013 \times 1 = \$1100 \times 0.013 \times 1 = \$14.30$$

Time, t, is measured in months, and $t = 1$ month.

The interest due on the payment due date is $14.30.

b. The total balance owed on the last day of the billing period is determined as follows:

Balance owed = unpaid balance on the first day of the billing period minus payments plus interest plus charges for three items during billing period.

= $1100 + $14.30 + $375 + $57 + $65

= $1611.30

The total balance owed on the last day of the billing period is $1611.30.

Credit-Card Debt per U.S. Household

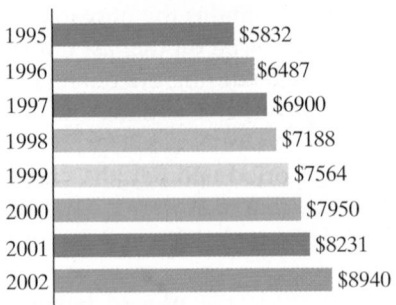

Year	Debt
1995	$5832
1996	$6487
1997	$6900
1998	$7188
1999	$7564
2000	$7950
2001	$8231
2002	$8940

Source: Cardweb.Com, Inc.

---

5 Calculate interest on credit cards using three methods.

BLITZER BONUS

The Dark Side of Credit Cards

- The average APR is a steep 18.84% and increasing each year.
- Half the credit card companies charge a late fee of more than $18.
- Some credit card companies charge their cardholders a fee for not charging during a six-month period.
- Some credit card companies charge a fee if you cancel your account.
- In addition to high interest rates, many credit card companies charge annual fees.

---

c. Because the balance owed on the last day of the billing period, $1611.30, exceeds $360, the customer must pay a minimum of $\frac{1}{36}$ of the balance owed.

$$\text{Minimum monthly payment} = \frac{\text{balance owed}}{36} = \frac{\$1611.30}{36} \approx \$45$$

Rounded to the nearest whole dollar, the minimum monthly payment due by June 9 is $45.

Check Point 5 A credit card calculates interest using the unpaid balance method. The monthly interest rate is 1.6% on the unpaid balance on the first day of the billing period less payments and credits. Here are some of the details in the May 1–May 31 itemized billing:

May 1 Unpaid Balance: $4720
Payment Received May 8: $1000
Purchases Charged to the Account: Computer: $1025, Meals: $45
Last Day of the Billing Period: May 31
Payment Due Date: June 9
Answer parts (a) through (c) in Example 5 using this information.

Methods for calculating interest on credit cards vary, and the interest can differ on credit cards that show the same APR. The three methods for calculating interest are summarized in the following box:

METHODS FOR CALCULATING INTEREST ON CREDIT CARDS

For all three methods, $I = Prt$, where r is the monthly rate and t is one month.

Unpaid balance method:

The principal, P, is the balance on the first day of the billing period less payments and credits.

Previous balance method:

The principal, P, is the unpaid balance on the first day of the billing period.

Average daily balance method:

The principal, P, is the **average daily balance**. This is determined by adding the unpaid balances for each day in the billing period and dividing by the number of days in the billing period.

EXAMPLE 6 | COMPARING METHODS FOR CALCULATING INTEREST

A credit card has a monthly rate of 1.75%. (The APR is 21%.) In the January 1–January 31 itemized billing, the January 1 unpaid balance is $2500. A payment of $1000 was received on January 8. There are no purchases or cash advances in this billing period. The payment due date is February 9. Find the interest due on this date using each of the three methods for calculating credit card interest.

Solution Because the monthly rate is 1.75%, for all three methods $I = Prt = P \times 0.0175 \times 1$ month. The principal, P, is different for each method.

a. The Unpaid Balance Method

The principal, P, is the unpaid balance on the first day of the billing period, $2500, less payments and credits, $1000. The interest is

$$I = Prt = (\$2500 - \$1000) \times 0.0175 \times 1 = \$1500 \times 0.0175 \times 1 = \$26.25.$$

The interest due on the payment due date is $26.25.

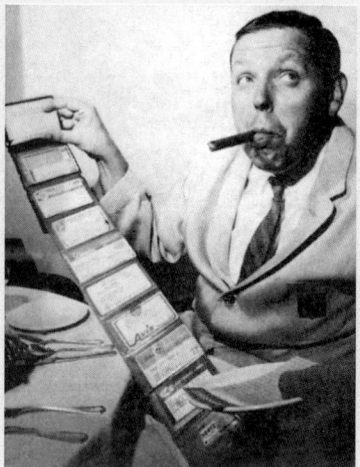
b. The Previous Balance Method

The principal, P, is the unpaid balance on the first day of the billing period, $2500. The interest is

$$I = Prt = \$2500 \times 0.0175 \times 1 = \$43.75.$$

The interest due on the payment due date is $43.75.

c. The Average Daily Balance Method

The principal, P, is the average daily balance. Add the unpaid balances for each day in the billing period and divide by the number of days in the billing period, 31. The unpaid balance on the first day of the billing period is $2500, and a $1000 payment is recorded on January 8. The sum of the balances owed for each day of the billing period is $2500 added for each of the first 7 days, expressed as $2500 × 7, plus $1500 added for each of the remaining 24 days, expressed as $1500 × 24.

Average daily balance

$$= \frac{\text{Sum of the unpaid balances for each day in the billing period}}{\text{Number of days in the billing period}}$$

$$= \frac{\$2500(7) + \$1500(24)}{31} \approx \$1725.81$$

The average daily balance serves as the principal. The interest is

$$I = Prt = \$1725.81 \times 0.0175 \times 1 \approx \$30.20.$$

The interest due on the payment due date is $30.20.

Most credit cards use the average daily balance method to determine interest due. Calculating the average daily balance by hand can be quite tedious when there are numerous transactions during a billing period. Credit card customers who are charged interest by the average daily balance method will find the average daily balance provided on monthly statements.

Check Point 6 A credit card has a monthly rate of 1.8%. In the January 1–January 31 itemized billing, the January 1 unpaid balance is $6800. A payment of $500 was received on January 8. There are no purchases or cash advances in this billing period. The payment due date is February 9. Find the interest due on this date using each of the three methods for calculating credit card interest.

Exercise Set 8.4

Practice and Application Exercises

1. The cost of a sports utility vehicle is $27,000. We can finance this by paying $5000 down and $410 per month for 60 months. Determine **a.** the amount financed; **b.** the total installment price; **c.** the finance charge.

2. The cost of a computer is $2450. We can finance this by paying $550 down and $94.50 per month for 24 months. Determine **a.** the amount financed; **b.** the total installment price; **c.** the finance charge.

3. The cost of a washer-dryer is $1100. We can finance this by paying $100 down and $110 per month for 12 months. Determine **a.** the amount financed; **b.** the total installment price; **c.** the finance charge.

4. The cost of a used car is $5675. We can finance this by paying $1223 down and $125 per month for 48 months. Determine **a.** the amount financed; **b.** the total installment price; **c.** the finance charge.

5. You plan to pay for a computer in 12 equal monthly payments. The finance charge per $100 financed is $6.90. Use Table 8.3 on page 429 to find the APR for this loan.

6. You plan to pay for a refrigerator in 18 equal monthly payments. The finance charge per $100 financed is $12.72. Use Table 8.3 on page 429 to find the APR for this loan.

7. The finance charge per $100 financed for a computer that is paid off in 24 equal monthly payments is $15.80. Use Table 8.3 to find the APR for this loan.

8. The finance charge per $100 financed for a refrigerator that is paid off in 12 monthly payments is $8.59. Use Table 8.3 to find the APR for this loan.

9. A used car is financed for $4450 over 48 months. If the total finance charge is $1279, find the APR for this loan.

10. A desk is financed for $1200 over 30 months. If the total finance charge is $264, find the APR for this loan.

11. The cash price for furniture for all rooms of a three-bedroom house is $17,500. The furniture can be financed by paying $500 down and $360.55 per month for 60 months.
    a. Determine the amount financed.
    b. Determine the total installment price.
    c. Determine the finance charge.
    d. What is the APR for this loan?

12. The cost of a Blazer is $18,000, which can be financed by paying $600 down and $385 per month for 60 months.
    a. Determine the amount financed.
    b. Determine the total installment price.
    c. Determine the finance charge.
    d. What is the APR for this loan?

13. In Exercise 11, instead of making the twenty-fourth payment, the borrower decides to pay the remaining balance and terminate the loan for the furniture.
    a. Use the actuarial method to determine how much interest will be saved by repaying the loan early.
    b. By the actuarial method, what is the total amount due on the day of the loan's termination?
    c. Use the rule of 78 to determine how much interest will be saved by repaying the loan early.
    d. By the rule of 78, what is the total amount due on the day of the loan's termination?

14. In Exercise 12, instead of making the twenty-fourth payment, the borrower decides to pay the remaining balance and terminate the loan for the Blazer.
    a. Use the actuarial method to determine how much interest will be saved by repaying the loan early.
    b. By the actuarial method, what is the total amount due on the day of the loan's termination?
    c. Use the rule of 78 to determine how much interest will be saved by repaying the loan early.
    d. By the rule of 78, what is the total amount due on the day of the loan's termination?

15. A particular VISA card calculates interest using the unpaid balance method. The monthly interest rate is 1.3% on the unpaid balance on the first day of the billing period less payments and credits. Here are some of the details in the May 1–May 31 itemized billing:

    May 1 Unpaid Balance: $950
    Payment Received May 8: $100
    Purchases Charged to the VISA Account: clothing, $85 and car repair, $67
    Last Day of the Billing Period: May 31
    Payment Due Date: June 9

a. Find the interest due on the payment due date.
b. Find the total balance owed on the last day of the billing period.
c. This credit card requires a $10 minimum monthly payment if the total balance owed on the last day of the billing period is less than $360. Otherwise, the minimum monthly payment is $\frac{1}{36}$ of the balance owed on the last day of the billing period, rounded to the nearest whole dollar. What is the minimum monthly payment due by June 9?

16. A particular credit card calculates interest using the unpaid balance method. The monthly interest rate is 1.75% on the unpaid balance on the first day of the billing period less payments and credits. Here are some of the details in the September 1–September 30 itemized billing:

    September 1 Unpaid Balance: $425
    Payment Received September 6: $75
    Purchases Charged to the Account: groceries, $45 and clothing, $77
    Last Day of the Billing Period: September 30
    Payment Due Date: October 9

a. Find the interest due on the payment due date.
b. Find the total balance owed on the last day of the billing period.
c. Terms for this credit card are shown in the following table. What is the minimum monthly payment due by October 9?

New Balance	Minimum Payment
$0.01 to $10.00	No payment due
10.01 to 200.00	$10.00
200.01 to 250.00	15.00
250.01 to 300.00	20.00
300.01 to 350.00	25.00
350.01 to 400.00	30.00
400.01 to 450.00	35.00
450.01 to 500.00	40.00
Over $500.00	$\frac{1}{10}$ of new balance

17. A credit card has a monthly rate of 1.5% and uses the average daily balance method for calculating interest. Here are some of the details in the April 1–April 30 itemized billing:

    April 1 Unpaid Balance: $445.59
    Payment Received April 5: $110
    Purchases Charged to the Account: $278.06
    Average Daily Balance: $330.90
    Last Day of the Billing Period: April 30
    Payment Due Date: May 9

a. Find the interest due on the payment due date.
b. Find the total balance owed on the last day of the billing period.
c. Terms for this credit card are given in Exercise 16. What is the minimum monthly payment due by May 9?

18. A credit card has a monthly rate of 1.8% and uses the average daily balance method for calculating interest. Here are some of the details in the December 1–December 31 itemized billing:

> December 1 Unpaid Balance: $220
> Payment Received December 7: $60
> Purchases Charged to the Account: $90
> Average Daily Balance: $205.60
> Last Day of the Billing Period: December 31
> Payment Due Date: January 9

  a. Find the interest due on the payment due date.

  b. Find the total balance owed on the last day of the billing period.

  c. Terms for this credit card are given in Exercise 16. What is the minimum monthly payment due on January 9?

19. A credit card has a monthly rate of 1.5%. In the September 1–September 30 itemized billing, the September 1 unpaid balance is $3000. A payment of $2500 was received on September 6. There are no purchases or cash advances in this billing period. The payment due date is October 9. Find the interest due on this date using

  a. the unpaid balance method.

  b. the previous balance method.

  c. the average daily balance method.

20. A credit card has a monthly rate of 2.2%. In the October 1–October 31 itemized billing, the October 1 unpaid balance is $2000. A payment of $400 was received on October 6. There are no purchases or cash advances in this billing period. The payment due date is November 9. Find the interest due on this date using

  a. the unpaid balance method.

  b. the previous balance method.

  c. the average daily balance method.

Writing in Mathematics

21. Describe the difference between a fixed installment loan and an open-end installment loan.

22. For a fixed installment loan, what is the total installment price?

23. For a fixed installment loan, how is the total finance charge determined?

24. What is the APR?

25. What are the two methods for computing unearned interest when a loan is paid off early? Describe how the payoff amount is determined regardless of which method is used in the computation.

26. Name and describe each of the three methods for calculating interest on credit cards.

27. For a credit card billing period, describe how the average daily balance is determined. Why is this computation somewhat tedious when done by hand?

28. Which method for calculating interest on credit cards is most beneficial to the borrower and which is least beneficial? Explain why this is so.

29. A Sears Revolving Charge Card has a monthly rate of 1.75%. The interest is a minimum of 50¢ if the average daily balance is $28.50 or less. Explain how this policy is beneficial to Sears.

Critical Thinking Exercises

30. Which one of the following is true?

  a. The finance charge on a fixed installment loan is the cash price minus the total installment price.

  b. It is to a borrower's advantage to have unearned interest computed by the rule of 78 rather than by the actuarial method.

  c. It is not necessary to know the number of days in a credit card billing period to determine the average daily balance.

  d. If a credit card has a 2.2% monthly interest rate, the annual rate exceeds 25%.

Use estimation and not calculation to select the most reasonable value in Exercises 31–32.

31. If you purchase a $1400 item, put $200 down, and pay the balance in 30 monthly installments, what is a reasonable estimate of the monthly payment?

  a. $35   **b.** $47   **c.** $70   **d.** $100

32. A reasonable estimate of the monthly interest on an average daily balance of $359.58 at a 1.3% monthly rate is

  a. $50   **b.** $10   **c.** $5   **d.** $2

33. A bank bills its credit card holders on the first of each month for each itemized billing. The card provides a 20-day period in which to pay the bill before charging interest. If the card holder wants to buy an expensive gift for a September 30 wedding but can't pay for it until November 5, explain how this can be done without adding an interest charge.

34. A $1500 computer can be purchased using a credit card that charges a monthly rate of 1.5% using the unpaid balance method. The borrower is considering one of the following options:

> Option A: Make a credit card payment of $300 at the end of each month for five months and pay off the balance at the end of the sixth month.
> Option B: Make a credit card payment of $300 plus the month's interest at the end of each month for five months.

How much is saved in interest using option B?

Technology Exercises

35. Set up a spreadsheet with one column each for the balance, number of days at that balance, payments, charges, and the product of each balance and the number of days at that balance. The balance on John and Jane Doe's credit card on July 5, their billing date, was $375.80. For the period ending August 4, they had the following transactions:

> July 13, Payment: $150.00
> July 15, Charge: Computer Store, $74.35
> July 23, Charge: Clothing, $123.50
> July 29, Charge: Restaurant, $42.50

  a. Use the summation capabilities of the spreadsheet to determine the number of days in the billing period and the average daily balance.

  b. Find the finance charge (the interest) that is owed on August 4. Assume the monthly rate is 1.3%.

  c. Find the balance that is owed on August 4.

36. Set up a spreadsheet with one column each for the balance, number of days at that balance, payments, charges, and the product of each balance and the number of days at that balance. The balance on the Does' credit card on May 12, their billing date, was $378.50. For the period ending June 11, they had the following transactions:

May 13, Charge: Toys, $129.79
May 15, Payment: $50.00
May 18, Charge: Clothing, $135.85
May 29, Charge: Housewares, $37.63

a. Use the summation capabilities of the spreadsheet to determine the number of days in the billing period and the average daily balance.

b. Find the finance charge (the interest) that is owed on June 11. Assume the monthly rate is 0.75%.

c. Find the balance that is owed on June 11.

Group Exercise

37. Credit card trouble can compound quickly, and many people get into financial trouble as a result. As a group, present ten guidelines for avoiding credit card trouble. Assume that these guidelines are to be distributed to all students at your college by the Student Government Association.

SECTION 8.5   THE COST OF HOME OWNERSHIP

OBJECTIVES

1. Understand mortgage options.

2. Compute interest costs for a mortgage.

3. Determine monthly mortgage payments using a formula.

4. Prepare a partial loan amortization schedule.

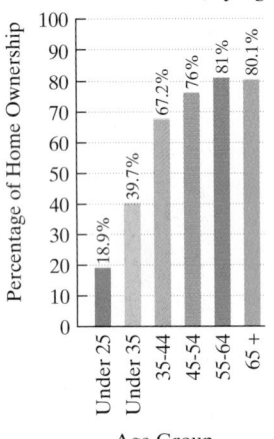

Percentage of People in the United States Who Are Homeowners, by Age

Source: U.S. Census Bureau

FIGURE 8.4

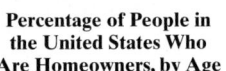

1   Understand mortgage options.

Figure 8.4 shows the percentage of Americans who are homeowners, by age. For most people, buying a home is a major lifetime decision and their largest purchase ever.

Many factors must be considered before deciding to purchase a home. Two major questions are "Does this house have what I require?" and, if it does, "Can I afford it?" In this section, we focus on the cost of home ownership, a centerpiece of the American dream.

Mortgages

A **mortgage** is a long-term loan (perhaps up to 30 or even 40 years) for the purpose of buying a home, and for which the property is pledged as security for payment. If payments are not made on the loan, the lender gets to take possession of the property. The **down payment** is the portion of the sale price of the home that the buyer initially pays to the seller. The minimum required down payment is computed as a percentage of the sale price. For example, suppose you decide to buy a $120,000 home that requires you to pay the seller 10% of the sale price. You must pay 10% of $120,000, which is $0.10 \times $120,000$, or $12,000, to the seller. Thus, $12,000 is the down payment. The **amount of the mortgage** is the difference between the sale price and the down payment. For your $120,000 home, the amount of the mortgage is $120,000 − $12,000, or $108,000.

There are many types of lenders and types of mortgages to choose from. Many banks, savings associations, mortgage companies, and credit unions provide mortgage loans. Some companies, called **mortgage brokers**, offer to find you a mortgage lender willing to make you a loan. Computer loan origination systems, or CLOs, are computer terminals sometimes available in real estate offices to help buyers sort through the various types of loans offered by different lenders.

Monthly payments for a mortgage depend on the amount of the mortgage, or the principal, the interest rate, and the time of the mortgage. Mortgages can have a fixed interest rate or a variable interest rate. **Fixed-rate mortgages** have the same monthly payment during the entire time of the loan. **Variable-rate mortgages**, also known as **adjustable-rate mortgages** (ARMs), have payment amounts that change from time to time depending on changes in the interest rate. ARMS are less predictable than fixed-rate mortgages. They start out at lower rates than fixed-rate mortgages. Caps limit how high rates can go over the term of the loan.

There are many other mortgage options. One option involves a short-term loan in which only the interest is paid, followed by a final payment (called a "balloon payment") for the entire loan amount. This kind of mortgage might appeal to a buyer who plans to sell the house in a short period of time. Some buyers are eligible for a mortgage insured through the Federal Housing Administration (FHA) or guaranteed by the Department of Veteran Affairs or similar programs operated by cities or states. These programs require a smaller down payment than standard loans.

Computation Involved with Buying a Home

Although monthly payments for a mortgage depend on the amount of the mortgage, the time of the loan, and the interest rate, the rate is not the only cost of a mortgage. Most lending institutions require the buyer to pay one or more **points** at the time of closing—that is, the time at which the mortgage begins. A point is a one-time charge that equals 1% of the loan amount. For example, two points means that the buyer must pay 2% of the loan amount at closing. Often, a buyer can pay fewer points in exchange for a higher interest rate or more points for a lower rate. A document, called the **Truth-in-Lending Disclosure Statement**, shows the buyer the APR for the mortgage. The APR takes into account the interest rate and points.

A monthly mortgage payment is used to repay the principal plus interest. In addition, lending institutions can require that part of the monthly payment be deposited into an **escrow account**, an account used by the lender to pay real estate taxes and insurance. Interest is paid to the buyer on the amount in an escrow account.

Table 8.4 gives the monthly mortgage payment for each $1000 of mortgage at various interest rates for mortgage loans of 15, 20, 25, 30, 35, and 40 years. The entries in the table do not include amounts deposited monthly into an escrow account.

TABLE 8.4 MONTHLY PAYMENT PER $1000 OF MORTGAGE, INCLUDING PRINCIPAL AND INTEREST

Rate %	Number of Years					
	15	20	25	30	35	40
6.5	$8.71	$7.46	$6.75	$6.32	$6.04	$5.85
7	8.99	7.75	7.07	6.65	6.39	6.21
7.5	9.28	8.06	7.39	6.99	6.74	6.58
8	9.56	8.36	7.72	7.34	7.10	6.95
8.5	9.85	8.68	8.05	7.69	7.47	7.33
9	10.15	9.00	8.40	8.05	7.84	7.72
9.5	10.45	9.33	8.74	8.41	8.22	8.11
10	10.75	9.66	9.09	8.70	8.60	8.50
10.5	11.06	9.98	9.44	9.15	8.98	8.89
11	11.37	10.32	9.80	9.52	9.37	9.28
11.5	11.69	10.66	10.16	9.90	9.76	9.68
12	12.01	11.01	10.53	10.29	10.16	10.08
12.5	12.33	11.36	10.90	10.67	10.55	10.49

Compute interest costs for a mortgage.

BLITZER BONUS

An Easy Way to Find the Cheapest Mortgage

When people learn all the major costs of getting a mortgage, they often feel like they're drowning in an ocean of rates, points, and formulas. Assume you'll be in your new home for five years, the average length of time most people stay. An easy way to shop for the best mortgage is to ask six different lenders for the total *dollar costs* of

1. up-front fees (such as appraisal and credit reports)
2. closing costs, including points
3. interest for the first five years.

Then add up the total cost of these three items at each lender and compare the various deals. Don't let lenders get away with saying, "Then, of course, there are certain other fees and charges." You want them all now—period.

EXAMPLE 1 | COMPUTING THE TOTAL COST OF INTEREST OVER THE LIFE OF A FIXED-RATE MORTGAGE

The price of a home is $195,000. The bank requires a 10% down payment and two points at the time of closing. The cost of the home is financed with a 30-year fixed-rate mortgage at 7.5%.

a. Find the required down payment.
b. Find the amount of the mortgage.
c. How much must be paid for the two points at closing?
d. Find the monthly payment (excluding escrowed taxes and insurance).
e. Find the total interest paid over 30 years.

Solution

a. The required down payment is 10% of $195,000 or

$$0.10 \times \$195,000 = \$19,500.$$

b. The amount of the mortgage is the difference between the price of the home and the down payment.

$$\begin{array}{ccc} \text{Amount of the mortgage} & = & \text{sale price} & - & \text{down payment} \end{array}$$

$$= \$195,000 - \$19,500$$

$$= \$175,500$$

c. To find the cost of two points on a mortgage of $175,500, find 2% of $175,500.

$$0.02 \times \$175,500 = \$3510$$

The down payment ($19,500) is paid to the seller and the cost of two points ($3510) is paid to the lending institution.

d. We need to find the monthly mortgage payment. To do so, we will use Table 8.4 on page 439, which shows the monthly payment *per $1000 of the mortgage amount*. For a mortgage of 30 years at 7.5%, the monthly payment *per $1000* (including the interest payment and the principal payment) is $6.99. This amount is circled in Table 8.4. Take a moment to turn back a page and verify this figure.

We can multiply this amount, $6.99, by the number of thousands of dollars in the mortgage amount to find the monthly payment for the entire mortgage amount. We divide the mortgage amount, $175,500, by $1000 to find the number of thousands of dollars in the mortgage amount.

$$\frac{\$175,500}{\$1000} = 175.5$$

The monthly mortgage payment is found by multiplying the number of thousands of dollars in the mortgage, 175.5, by the amount shown in Table 8.4.

$$\$6.99 \times 175.5 \approx \$1227.00$$

The monthly mortgage payment for principal and interest is approximately $1227.00. (Keep in mind that this payment does not include escrowed taxes and insurance.)

e. The total cost of interest over 30 years is equal to the difference between the total of all monthly payments and the amount of the mortgage. The total of all monthly payments is equal to the amount of the monthly payment multiplied by the number of payments. We found the amount of each monthly payment in (d): $1227. The number of payments is equal to the number of months in a year, 12, multiplied by the number of years in the mortgage, 30:

12 × 30 = 360. Thus, the total of all monthly payments = $1227 × 360. Now we can calculate the interest over 30 years.

$$\underbrace{\text{Total interest paid}} = \underbrace{\text{total of all monthly payments}} \quad \underbrace{\text{minus}} \quad \underbrace{\text{amount of the mortgage.}}$$

$$= \$1227 \quad \times \quad 360 \quad - \quad \$175,500$$

$$= \$441,720 \quad - \quad \$175,500 \quad = \quad \$266,220$$

The total interest paid over 30 years is approximately $266,220.

 **Check 1 Point**   The price of a home is $240,000. The bank requires a 10% down payment and three points at the time of closing. The cost of the home is financed with a 30-year fixed-rate mortgage at 6.5%. Use this information to solve parts (a) through (e) in Example 1.

 3 Determine monthly mortgage payments using a formula.

Computing Monthly Mortgage Payments Using a Formula

A loan is said to be **amortized** if both principal and interest are paid by a sequence of equal payments over equal time periods. Although the monthly mortgage payment needed to amortize a loan can be found using Table 8.4, it can also be determined using the following formula:

STUDY TIP

An **annuity** is a sequence of equal payments made at equal time periods. (See Exercise Set 8.3, Exercises 35–36.) Thus, a mortgage is an example of an annuity.

MONTHLY MORTGAGE PAYMENTS

The monthly mortgage payment, *PMT*, required to repay a loan of *PV* dollars paid *n* times per year over *t* years at an annual rate *r* is given by

$$PMT = PV \dfrac{\dfrac{r}{n}}{1 - \left(1 + \dfrac{r}{n}\right)^{-nt}}.$$

EXAMPLE 2 | USING THE MONTHLY MORTGAGE FORMULA

In Example 1, the $175,500 mortgage was financed with a 30-year fixed rate at 7.5%. The total interest paid over 30 years was approximately $266,220.

a. Use the monthly mortgage formula to find the monthly payment if the time of the mortgage is reduced to 15 years.

b. Find the total interest paid over 15 years.

c. How much interest is saved by reducing the mortgage from 30 to 15 years?

Solution

a. We are interested in finding the monthly payment for a $175,500 mortgage at 7.5% amortized over 15 years.

PV, the mortgage amount, is $175,500.   Fixed rate, r, is 7.5%.   12 payments per year   The mortgage time, t, is 15 years.

$$PMT = PV \dfrac{\dfrac{r}{n}}{1 - \left(1 + \dfrac{r}{n}\right)^{-nt}} = 175,500 \dfrac{\dfrac{0.075}{12}}{1 - \left(1 + \dfrac{0.075}{12}\right)^{-12(15)}} \approx 1627$$

The monthly mortgage payment is approximately $1627.

b. Now we calculate the interest over 15 years. There are 12×15, or 180, monthly payments.

$$\text{Total interest paid} = \underbrace{\text{total of all monthly payments}} \quad \text{minus} \quad \underbrace{\text{amount of the mortgage}}$$

$$= \quad \$1627 \times 180 - \$175{,}500$$

$$= \quad \$117{,}360$$

The total interest paid over 15 years is approximately $117,360.

c. Now we calculate the interest saved by reducing the mortgage from 30 to 15 years.

$$\text{Interest saved} = \underbrace{\text{total interest over 30 years}} \quad \text{minus} \quad \underbrace{\text{total interest over 15 years}}$$

$$= \quad \$266{,}220 - \$117{,}360$$

$$= \quad \$148{,}860$$

Thus, $148,860 is saved when the mortgage is amortized over 15 years.

 **Check Point 2** Use the monthly mortgage formula to find the monthly payment for the mortgage in Check Point 1 if the time of the mortgage is reduced to 15 years. How much interest is saved when the mortgage is amortized over 15 years?

TECHNOLOGY

Some graphing calculators have a special FINANCE menu. The screen shows the result of Example 2(a) on the TI-83 Plus.

Total number of payments ⟶	N=180
Interest rate ⟶	I%=7.5
Mortgage amount (present value) ⟶	PV=175500
Calculated mortgage payment ⟶	PMT=-1626.9066
Future value, 0, when loan is amortized ⟶	FV=0
Payments per year ⟶	P/Y=12
Compoundings per year ⟶	C/Y=12
	PMT:**END** BEGIN

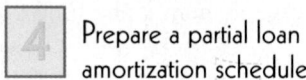 Prepare a partial loan amortization schedule.

Loan Amortization Schedules

In Example 2, were you surprised at the interest saved when the mortgage was amortized over 15 years rather than over 30 years? What adds to the interest cost is the long period over which the loan is financed. Although each payment is the same, with each successive payment the interest portion decreases and the principal portion increases. The interest is computed using the simple interest formula $I = Prt$. The principal, P, is equal to the balance of the loan, which changes each month. The rate, r, is the annual interest rate of the mortgage loan. Because a payment is made each month, the time, t, is

$$\frac{1 \text{ month}}{12 \text{ months}} = \frac{1 \;\text{month}}{12 \;\text{months}}$$

or $\frac{1}{12}$ of a year.

A document showing important information about the status of the mortgage is called a **loan amortization schedule**. Typically, the document includes the number of the most recent payment and those of any previous monthly payments, the interest for each monthly payment, the principal payment for each monthly payment, and the balance of the loan.

EXAMPLE 3 | PREPARING A LOAN AMORTIZATION SCHEDULE

Prepare a loan amortization schedule for the first two months of the mortgage loan shown in the following table:

LOAN AMORTIZATION SCHEDULE

Annual % rate: 9.5%			
Amount of Mortgage: $130,000		Monthly payment: $1357.50	
Number of Monthly Payments: 180		Term: Years 15, Months 0	
Payment Number	**Interest Payment**	**Principal Payment**	**Balance of Loan**
1			
2			

Solution   We begin with payment number 1.

$$\text{Interest for the month} = Prt = \$130,000 \times 0.095 \times \tfrac{1}{12} \approx \$1029.17$$

$$\text{Principal payment} = \text{Monthly payment} - \text{Interest payment}$$

$$= \$1357.50 - \$1029.17 = \$328.33$$

$$\text{Balance of loan} = \text{Principal balance} - \text{Principal payment}$$

$$= \$130,000 - \$328.33 = \$129,671.67$$

Now, starting with a loan balance of $129,671.67, we repeat these computations for the second month.

$$\text{Interest for the month} = Prt = \$129,671.67 \times 0.095 \times \tfrac{1}{12} = \$1026.57$$

$$\text{Principal payment} = \text{Monthly payment} - \text{Interest payment}$$

$$= \$1357.50 - \$1026.57 = \$330.93$$

$$\text{Balance of loan} = \text{Principal balance} - \text{Principal payment}$$

$$= \$129,671.67 - \$330.93 = \$129,340.74$$

The results of these computations are included in Table 8.5, a partial loan amortization schedule. By using the simple interest formula month-to-month on the loan's balance, a complete loan amortization schedule for all 180 payments can be calculated.

TABLE 8.5 LOAN AMORTIZATION SCHEDULE

Annual % rate: 9.5%			
Amount of Mortgage: $130,000		Monthly payment: $1357.50	
Number of Monthly Payments: 180		Term: Years 15, Months 0	
Payment Number	**Interest Payment**	**Principal Payment**	**Balance of Loan**
1	$1029.17	$328.33	$129,671.67
2	1026.57	330.93	129,340.74
3	1023.96	333.54	129,007.22
4	1021.32	336.18	128,671.04
30	944.82	412.68	118,931.35
31	941.55	415.95	118,515.52
125	484.62	872.88	60,340.84
126	477.71	879.79	59,461.05
179	21.26	1336.24	1347.74
180	9.76	1347.74	

Many lenders supply a loan amortization schedule like the one in Example 3. Such a schedule shows how the buyer pays slightly less in interest and more in principal for each payment over the entire life of the loan.

Check 3 Point Prepare a loan amortization schedule for the first two months of the mortgage loan shown in the following table:

Annual % rate: 7.0%			
Amount of Mortgage: $200,000		Monthly payment: $1550.00	
Number of Monthly Payments: 240		Term: Years 20, Months 0	
Payment Number	**Interest Payment**	**Principal Payment**	**Balance of Loan**
1			
2			

Exercise Set 8.5

Practice and Application Exercises

In Exercises 1–7, use Table 8.4 on page 439 or the formula in the box on page 441 to determine the monthly mortgage payment, rounded to the nearest dollar. (It will be necessary to use the formula for exercises in which the interest rates are not given in the table.)

1. The price of a home is $125,000. The bank requires a 20% down payment and three points at the time of closing. The cost of the home is financed with a 30-year fixed-rate mortgage at 7%.
   a. Find the required down payment.
   b. Find the amount of the mortgage.
   c. How much must be paid for the three points at closing?
   d. Find the monthly payment (excluding escrowed taxes and insurance).
   e. Find the total cost of interest over 30 years.

2. The price of a condominium is $85,000. The bank requires a 5% down payment and one point at the time of closing. The cost of the condominium is financed with a 30-year fixed-rate mortgage at 8%.
   a. Find the required down payment.
   b. Find the amount of the mortgage.
   c. How much must be paid for the one point at closing?
   d. Find the monthly payment (excluding escrowed taxes and insurance).
   e. Find the total cost of interest over 30 years.

3. The price of a small cabin is $40,000. The bank requires a 5% down payment. The buyer is offered two mortgage options: 20-year fixed at 8% or 30-year fixed at 8%. Calculate the amount of interest paid for each option. How much does the buyer save in interest with the 20-year option?

4. The price of a home is $160,000. The bank requires a 15% down payment. The buyer is offered two mortgage options: 15-year fixed at 8% or 30-year fixed at 8%. Calculate the amount of interest paid for each option. How much does the buyer save in interest with the 15-year option?

5. In terms of paying less in interest, which is more economical for a $100,000 mortgage: a 30-year fixed-rate at 8% or a 20-year fixed-rate at 7.5%?

6. In terms of paying less in interest, which is more economical for a $90,000 mortgage: a 30-year fixed-rate at 8% or a 15-year fixed-rate at 7.5%?

7. According to the *Guinness Book of World Records*, the most expensive private house ever built is the Hearst Ranch in San Simeon, California. It was built 1922–1939 for William Randolph Hearst (1863–1951), at a cost of over $30 million. Suppose that you do very well in your career and decide to treat yourself to this 100-room home, complete with a 104-foot-long heated pool, an 83-foot-long assembly hall, and a garage for 25 limousines. The house is yours for a "steal"—namely $80 million, with 30% down and a 30-year fixed-rate mortgage at 6.3%. Find the total cost of interest over 30 years. (By the way, if you do get the roof of the Hearst Ranch over your head, you may need some help to keep things in tip-top shape. Hearst originally maintained the house with 60 servants.)

The unpaid balance of a mortgage loan is equal to the present value of the remaining payments. The unpaid balance, PV, is given by

$$PV = PMT \dfrac{1 - \left(1 + \dfrac{r}{n}\right)^{-nt}}{\dfrac{r}{n}}$$

where PMT is the monthly mortgage payment, r is the annual interest rate, n is the number of payments per year, and t is the number of years remaining in the mortgage. Use this formula to solve Exercises 8–11.

8. The price of a home is $180,000. The bank requires a 10% down payment. After the down payment, the balance is financed with a 30-year fixed-rate mortgage at 6.3%.
   a. Determine the monthly mortgage payment.
   b. Determine the unpaid balance after 10 years.
   c. Determine the unpaid balance after 20 years.

9. The price of a home is $220,000. The bank requires a 10% down payment. After the down payment, the balance is financed with a 30-year fixed-rate mortgage at 6.4%.

   a. Determine the monthly mortgage payment.

   b. Determine the unpaid balance after 10 years.

   c. Determine the unpaid balance after 20 years.

*The difference between the appraised value of your home and your unpaid mortgage balance is called **equity**. With a home equity loan, you borrow against the amount of its equity. In Exercises 10–11 a bank offers a home equity loan called an 80% LTV (loan-to-value). The maximum amount available on this loan is*

   (80% *of home's value*) − (*unpaid balance on mortgage*).

10. You bought a home five years ago for $100,000 and put 5% down on a 30-year fixed mortgage at 6.5%. No points were charged at the time of closing.

   a. Determine the monthly payment needed to amortize the loan.

   b. Determine the unpaid balance on the mortgage after five years.

   c. After five years, the value of your home appreciates to $125,000. What is the maximum amount you can receive on a home equity loan?

11. You bought a home ten years ago for $120,000 and put 5% down on a 30-year fixed mortgage at 7%. No points were charged at the time of closing.

   a. Determine the monthly payment needed to amortize the loan.

   b. Determine the unpaid balance on the mortgage after ten years.

   c. After ten years, the value of your home appreciates to $250,000. What is the maximum amount you can receive on a home equity loan?

Writing in Mathematics

12. What is a mortgage?

13. What is a down payment?

14. How is the amount of a mortgage determined?

15. Describe why a buyer would select a 30-year fixed-rate mortgage instead of a 15-year fixed-rate mortgage if interest rates are $\frac{1}{4}$% to $\frac{1}{2}$% lower on a 15-year mortgage.

16. Describe one advantage and one disadvantage of an adjustable-rate mortgage over a fixed-rate mortgage.

17. What is the worst thing that can happen with a balloon-payment mortgage?

18. In terms of current interest rates for mortgage loans, describe the type of mortgage that best suits your needs at this time. What features of the mortgage are compatible with your vocational goals and your projected future earnings?

19. What is a loan amortization schedule?

20. Describe one advantage and one disadvantage of home ownership over renting.

Critical Thinking Exercises

21. Refer to Exercise 9 and write a function that approximates the unpaid mortgage balance after x payments.

22. Use the formula for monthly mortgage payments, given in the box on page 441, to derive the formula for the unpaid balance of a mortgage, given in the directions for Exercises 8–11.

Group Exercise

23. Group members are employed by a company that works with local banks. A client with a monthly income of $4000 needs a loan to purchase a $125,000 home. This client has $16,000 in savings that can be used as a down payment. Find the current rates available from two banks for both fixed- and adjustable-rate mortgages. Analyze the offerings and then write a report for this client summarizing the best mortgage options, along with the advantages and disadvantages of each option.

SECTION 8.6   INVESTING IN STOCKS, BONDS, AND MUTUAL FUNDS

OBJECTIVES

1. Understand stocks, bonds, and mutual funds as investments.

2. Read stock tables.

3. Calculate costs involved in trading stock.

Within a week after the horrific terrorist attacks launched September 11, 2001, the stock market reopened, swooned, and then revived. Even as fires burned in downtown Manhattan and soldiers headed off to war, investors believed that the economy was on the right track.

Then came 2002's summer of corporate scandals. Story after story of crooked CEOs gave new meaning to the traditional business attire of pinstripes. The scandals started with Enron, continued with WorldCom, and extended to the diva of domesticity herself, Martha Stewart. Scamming corporate criminals had accomplished what Osama bin Laden could not, causing the financial markets to decline significantly. In this section, we provide you with information about various kinds of investments so that you can make informed choices in managing your finances and achieving your financial goals at a time when the perception of deception is so widespread.

1 Understand stocks, bonds, and mutual funds as investments.

Investments

When you deposit money into a bank account, you are making a **cash investment**. Bank accounts up to $100,000 are insured by the federal government, so there is no risk of losing the principal you've invested. The account's interest rate guarantees a certain percent increase in your investment, called its **return**. For example, if you deposit $7500 in a savings account that has a rate of 4% compounded annually, the annual return is 4%. There are other kinds of investments that are riskier, meaning that it is possible to lose all or part of your principal. These investments include **stocks** and **bonds**. We begin by discussing these investment categories.

Stocks

Investors purchase **stock**, shares of ownership in a company. The shares indicate the percent of ownership. For example, if a company has issued a total of one million shares and an investor owns 20,000 of these shares, that investor owns

$$\frac{20,000 \text{ shares}}{1,000,000 \text{ shares}} = 0.02$$

or 2% of the company. Any investor who owns some percentage of the company is called a **shareholder**.

Buying or selling stock is referred to as **trading**. Shares of stock need both a seller and a buyer to be traded. Stocks are traded on a **stock exchange**. The price of a share of stock is determined by the law of supply and demand. If a company is prospering, investors will be willing to pay a good price for its stock, and so the stock price goes up. If the company does not do well, investors may decide to sell, and the stock price goes down. Stock prices indicate the performance of the companies they represent, as well as the state of the national and global economies.

Stock exchanges provide an orderly place for determining the price of a share of stock. The New York Stock Exchange on Wall Street in Manhattan is the largest stock exchange in the world. Only **stockbrokers** are allowed on the floor of the exchange. Investors buy and sell stock through stockbrokers, who charge a commission for their services.

Americans now have more money invested in stocks than in their homes. Stock ownership grew dramatically in the last decade of the twentieth century. Today, over 43% of adult Americans are stock owners. Many people with computers and access to the Internet use the world of online investing to buy and sell stocks. With the click of a mouse, industry research suggesting companies to invest in is now available to online investors. A note of caution: Some of the information on the Web may be inaccurate or posted for fraudulent reasons.

Investors make money with stock by selling a stock for more money than they paid for it. Over the past 75 years, the return on stock has been 11% per year on average. However, the ability of anyone to call which way the market is going over the short term—short being defined as any period less than five years—is pretty bad. At an 11% rate of return, compounded annually, the NASDAQ would not return to its March 2000 high of 5048 until 2010. But who says we can expect 11% any time soon?

If you make money by selling stock for more than you paid for it, you have a **capital gain** on the sale of stock. When more and more average Americans were investing and making money in the 1990s, the federal government cut the capital-gains tax rate in 1997.

BLITZER BONUS

Online Trading

Internet, or online brokerages, allow people to trade stocks directly without a broker for commissions as low as $5 per trade. It's wise to be cautious about making investments using the Internet. The rapidly changing prices of stocks can cause an online trader to see a huge profit and an even bigger loss in a very short amount of time. Just like any investment, online trading should be thoroughly investigated before you commit your hard-earned money. Many of the largest online brokers offer extensive Web sites. You may be able to access many features of these sites without even investing.

BLITZER BONUS

The Dow of History

The Dow Jones Industrial Average (DJIA) is an index of 30 U.S. stocks. It is computed by adding up stock prices and dividing by the number of stocks. The DJIA is the best known market indicator in the world. A **bull market** requires a 30% rise in the DJIA over 50 calendar days or a 13% rise over 155 calendar days. A **bear market** requires a 30% drop in the DJIA over 50 calendar days or a 13% decline over 145 calendar days.

The DJIA does not like surprises, dropping sharply to past crises. What happens afterward depends on the situation. When the Japanese attacked Pearl Harbor, the market sold off for several months. In other crises—Eisenhower's heart attack, the Cuban missile crisis, and the assassination of President Kennedy—the market recovered quickly.

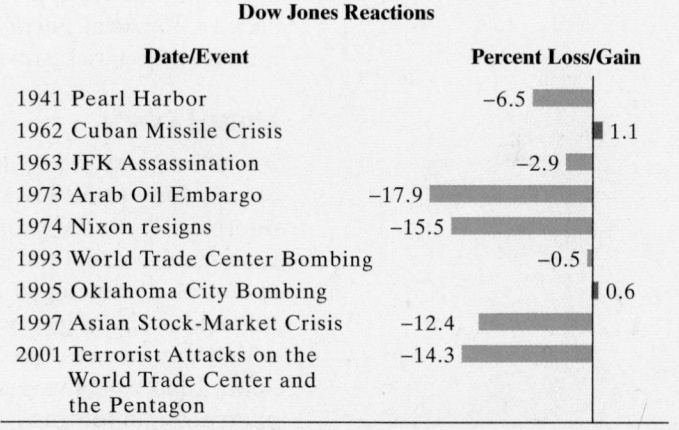

Dow Jones Reactions

Date/Event	Percent Loss/Gain
1941 Pearl Harbor	−6.5
1962 Cuban Missile Crisis	1.1
1963 JFK Assassination	−2.9
1973 Arab Oil Embargo	−17.9
1974 Nixon resigns	−15.5
1993 World Trade Center Bombing	−0.5
1995 Oklahoma City Bombing	0.6
1997 Asian Stock-Market Crisis	−12.4
2001 Terrorist Attacks on the World Trade Center and the Pentagon	−14.3

Source: Ned Davis Research

Ned Davis Research looked at 28 crises dating back to the fall of France to the Nazis and found that in 25 of those cases, an initial market decline turned to solid gains within six months. The average initial decline in the DJIA was 4.6%, followed by a rally of 12.1%. On September 11, 2001, a new date of infamy, icons of American vigor and affluence were reduced to rubble. Our national psyche was devastated, but based on the Dow of history, for the long-range investor, this may not be a catastrophic event.

HOW MARKETS BOUNCE BACK

Major News Event	Dow's Change During Crisis	Dow's Change Six Months Later
U.S. Bombing of Cambodia (April 29–May 26, 1970)	▼ −14.4%	▲ 20.7%
Arab Oil Embargo (Oct. 18–Dec. 5, 1973)	▼ −17.9%	▲ 7.2%
Gulf War Ultimatum (Dec. 24, 1990–Jan. 16, 1991)	▼ −4.3%	▲ 18.7%
World Trade Center Bombing (Feb. 26–27, 1993)	▼ −0.5%	▲ 8.5%
Asian Stock-Market Crisis (Oct. 7–27, 1997)	▼ −12.4%	▲ 25.0%
Average (of these events and 23 others)	▼ −4.6%	▲ **12.1%**

Investors can make money while they own stock if a company distributes all or part of its profits to shareholders as **dividends**. Each share of stock is paid the same dividend, so the amount of money received by an investor depends on the number of shares owned. Some companies do not pay dividends, reinvesting all profits within the company.

Bonds

People who buy stock become part owners in a company. In order to raise money and not dilute the ownership of current stockholders, companies sell **bonds**. People who buy a bond are **lending money** to the company from which they buy the bond. Bonds are a commitment from a company to pay the price an investor pays for the bond at the time it was purchased, called the **face value**, along with interest payments at a given rate.

There are many reasons for issuing bonds. A company might need to raise money for research on a drug that has the potential for curing AIDS, so it issues bonds. The U.S. Treasury Department issues 30-year bonds at a fixed 7% annual rate to borrow money to cover possible federal deficits. Local governments often issue bonds to borrow money to build schools, parks, and libraries.

Bonds are traded like stock, and their price is a function of supply and demand. If a company goes bankrupt, bondholders are the first to claim the company's assets. They make their claims before the stockholders, even though (unlike stockholders) they do not own a share of the company. Buying and selling bonds is frequently done through online investing.

Generally speaking, investing in bonds is less risky than investing in stocks, although the return is lower. A listing of all the investments that a person holds is called a **financial portfolio**. Most financial advisors recommend a portfolio with a mixture of low-risk and high-risk investments, called a **diversified portfolio**.

Mutual Funds

It is not an easy job to determine which stocks and bonds to buy or sell, or when to do so. Many small investors have decided that they do not have the time to stay informed about the progress of corporations, even with the help of online industry research. Instead, they invest in a **mutual fund**. A mutual fund is a group of stocks and/or bonds managed by a professional investor. When you purchase shares in a mutual fund, you give your money to the **fund manager**. Your money is combined with the money of other investors in the mutual fund. The fund manager invests this pool of money, buying and selling shares of stocks and bonds to obtain the maximum possible returns.

Investors in mutual funds own a small portion of many different companies, which may protect them against the poor performance of a single company. In the United States, there are more mutual funds that invest in stocks than there are stocks listed on the New York Stock Exchange. A total of $2.2 trillion reside in stock mutual funds, and this is increasing by $20 billion a month. New funds, some specializing in high-risk investments, others in low-risk investments, some investing in U.S. markets only, others investing in foreign markets, are forming at the rate of three per day.

Newspapers publish ratings from 1 (worst) to 5 (best) of mutual fund performance based on whether the fund manager is doing a good job with its investors' money. Two numbers are given. The first number compares the performance of the mutual fund to a large group of similar funds. The second number compares the performance to funds that are nearly identical. The best rating a fund manager can receive is 5/5; the worst is 1/1.

2 Read stock tables.

Reading Stock Tables

Daily newspapers and online services give current stock prices and other information about stocks. We will use Citigroup stock to learn how to read these daily stock tables. Look at the following newspaper listing of Citigroup stock.

52-Week High	Low	Stock	SYM	Div	Yld %	PE	Vol 100s	Hi	Lo	Close	Net Chg
77	28.50	Citigroup	C	.84	1.1	29	64981	76	74.44	74.56	−0.94

The headings indicate the meanings of the numbers across the row.

The heading **52-Week High** refers to the *highest price* at which Citigroup stock traded during the past 52 weeks. The highest price was $77.00 per share. This means that during the past 52 weeks an investor was willing to pay $77.00 for a share of stock. Notice that 77 represents a quantity in dollars, although the stock table does not show the dollar sign.

52-Week High
77

The heading **52-Week Low** refers to the *lowest price* that Citigroup stock reached during the past 52 weeks. This price is $28.50.

52-Week Low
28.50

The heading **Stock** is the *company name*, Citigroup. The heading **SYM** is the *symbol* the company uses for trading. Citigroup uses the symbol C.

Stock	SYM
Citigroup	C

The heading **Div** refers to *dividends* paid per share to stockholders last year. Citigroup paid a dividend of $0.84 per share. Once again, the dollar symbol does not appear in the table. Thus, if you owned 100 shares, you received a dividend of $0.84 × 100, or $84.00.

Div
.84

The heading **Yld %** stands for *percent yield*. In this case, the percent yield is 1.1%. (The stock table does not show the percent sign.) This means that the dividends alone give investors an annual return of 1.1%. This is much lower than interest rates offered by most banks. However, this percent does not take into account the fact that Citigroup stock prices might rise. If an investor sells shares for more than what they were paid for, the gain will probably make Citigroup stock a much better investment than a bank account.

Yld %
1.1

In order to understand the meaning of the heading PE, we need to understand some of the other numbers in the table. We will return to this column.

The heading **Vol 100s** stands for *sales volume in hundreds*. This is the number of shares traded yesterday, in hundreds. The number in the table is 64,981. This means that yesterday, a total of 64,981 × 100, or 6,498,100 shares of Citigroup were traded.

Vol 100s
64981

The heading **Hi** stands for the *highest price* at which Citigroup stock traded *yesterday*. This number is 76. Yesterday, Citigroup's highest trading price was $76 a share.

Hi
76

The heading **Lo** stands for the *lowest price* at which Citigroup stock traded *yesterday*. This number is 74.44. Yesterday, Citigroup's lowest trading price was $74.44 a share.

Lo
74.44

The heading **Close** stands for the *price* at which shares traded *when the stock exchange closed yesterday*. This number is 74.56. Thus, the the price at which shares of Citigroup traded when the stock exchange closed yesterday was $74.56 per share. This is called yesterday's **closing price**.

Close
74.56

The heading **Net Chg** stands for *net change*. This is the change in price from the market close two days ago to yesterday's market close. This number is −0.94. Thus, the price of a share of Citigroup stock went down by $0.94. For some stock listings, the notation . . . appears under Net Chg. This means that there was *no change in price* for a share of stock from the market close two days ago to yesterday's market close.

Net Chg
−0.94

Now, we are ready to return to the heading **PE**, standing for the *price-to-earnings ratio*.

PE
29

$$\text{PE ratio} = \frac{\text{Yesterday's closing price per share}}{\text{Annual earnings per share}}$$

This can be expressed as

$$\text{Annual earnings per share} = \frac{\text{Yesterday's closing price per share}}{\text{PE ratio}}.$$

Close
74.56

PE
29

The PE ratio for Citigroup is given to be 29. Yesterday's closing price per share was 74.56. We can substitute these numbers into the formula to find annual earnings per share:

$$\text{Annual earnings per share} = \frac{74.56}{29} \approx 2.57.$$

The annual earnings per share for Citigroup are $2.57. The PE ratio, 29, tells us that yesterday's closing price per share, $74.56, is 29 times greater than the earnings per share, $2.57.

EXAMPLE 1   READING STOCK TABLES

52-Week High	Low	Stock	SYM	Div	Yld %	PE	Vol 100s	Hi	Lo	Close	Net Chg
42.38	22.50	Disney	DIS	.21	.6	43	115900	32.50	31.25	32.50	...

Use the stock table for Disney to answer the following questions.

a. What were the high and low prices for the past 52 weeks?

b. If you owned 3000 shares of Disney stock last year, what dividend did you receive?

c. What is the annual return for dividends alone? How does this compare to a bank account offering a 3.5% interest rate?

d. How many shares of Disney were traded yesterday?

e. What were the high and low prices for Disney shares yesterday?

f. What was the price at which Disney shares traded when the stock exchange closed yesterday?

g. What does . . . in the net change column mean?

h. Compute Disney's annual earnings per share using

$$\text{Annual earnings per share} = \frac{\text{Yesterday's closing price per share}}{\text{PE ratio}}.$$

Solution

a. We find the high price for the past 52 weeks by looking under the heading **High**. The price is listed in dollars, given as 42.38. Thus, the high price for a share of stock for the past 52 weeks was $42.38. We find the low price for the past 52 weeks by looking under the heading **Low**. This price is also listed in dollars, given as 22.50. Thus, the low price for a share of Disney stock for the past 52 weeks was $22.50.

b. We find the dividend paid for a share of Disney stock last year by looking under the heading **Div**. The price is listed in dollars, given as .21. Thus, Disney paid a dividend of $0.21 per share to stockholders last year. If you owned 3000 shares, you received a dividend of $0.21 × 3000, or $630.

c. We find the annual return for dividends alone by looking under the heading **Yld** %, standing for percent yield. The number in the table, .6, is a percent. This means that the dividends alone give Disney investors an annual return of 0.6%. This is much lower than a bank account paying a 3.5% interest rate. However, if Disney shares increase in value, the gain might make Disney stock a better investment than the bank account.

d. We find the number of shares of Disney traded yesterday by looking under the heading **Vol 100s**, standing for sales volume in hundreds. The number in the table is 115900. This means that yesterday, a total of 115,900 × 100, or 11,590,000 shares, were traded.

BLITZER BONUS

Black Monday

On Monday, October 19, 1987, the Dow Jones Industrial Average plunged 508 points, losing 22.6% of its value—almost double that of the 1929 crash that set off the Great Depression. Analysts pointed to one factor above all: the then-deplorable state of the U.S. economy, with its massive budget and trade deficits. On the Thursday following "Black Monday" (as it was instantly dubbed), President Reagan announced plans for a deficit cutting conference with Congress. America's spending spree, related to its military buildup, came to an end.

After the fall, an anxious crowd gathered on Wall Street.

e. We find the high and low prices for Disney shares yesterday by looking under the headings **Hi** and **Lo**. Both prices are listed in dollars, given as 32.50 and 31.25. Thus, the high and low prices for Disney shares yesterday were $32.50 and $31.25, respectively.

f. We find the price at which Disney shares traded when the stock exchange closed yesterday by looking under the heading **Close**. The price is listed in dollars, given as 32.50. Thus, when the stock exchange closed yesterday, the price of a share of Disney stock was $32.50.

g. The . . . under **Net Chg** means that there was no change in price in Disney stock from the market close two days ago to yesterday's market close. In part (f), we found that the price of a share of Disney stock at yesterday's close was $32.50, so the price at the market close two days ago was also $32.50.

h. We are now ready to use

$$\text{Annual earnings per share} = \frac{\text{Yesterday's closing price per share}}{\text{PE ratio}}$$

to compute Disney's annual earnings per share. We found that yesterday's closing price per share was $32.50. We find the PE ratio under the heading **PE**. The given number is 43. Thus,

$$\text{Annual earnings per share} = \frac{\$32.50}{43} \approx \$0.76.$$

The annual earnings per share for Disney are $0.76. The PE ratio, 43, tells us that yesterday's closing price per share, $32.50, is 43 times greater than the earnings per share, approximately $0.76.

 **Check 1 Point** Use the stock table for Coca Cola to solve parts (a) through (h) in Example 1.

52-Week High	Low	Stock	Div	Yld %	PE	Vol 100s	Hi	Lo	Close	Net Chg
63.38	42.37	CocaCl	.72	1.5	37	72032	49.94	48.33	49.50	+0.03

 Calculate costs involved in trading stock.

Costs Involved in Trading Stock

Frank and Ernest reprinted by permission of Newspaper Enterprise Association, Inc.

A broker must be used to buy or sell stock. The broker has representatives at the stock exchange who carry out an investor's order. Brokers charge commissions for their service. They make most of their money by persuading customers to trade into and out of individual stocks. Commissions are competitive and vary among brokers.

Brokers also charge commissions for buying and selling bonds. Online investing also involves paying commissions, although these are usually at a discount. Investors must also pay fees for subscribing to online research offering advice about which stocks and bonds to buy and sell and when to do so. There are management fees for mutual funds, although many funds trade shares with no commission charge.

EXAMPLE 2 SELLING STOCK

Suppose that you bought 400 shares of Citigroup stock at the 52-week low, $28.50 per share, and sold at the 52-week high, $77.00 per share.

a. Ignoring dividends, how much money did you make on this transaction?

b. If the broker charges 2.5% of the total sale price, find the broker's commission.

Solution

a. The difference between the high price per share, $77.00, and the low price per share, $28.50, is $77.00 − $28.50, or $48.50. Because you bought at the low price and sold at the high price, you made $48.50 per share. For 400 shares, you made $48.50 × 400, or $19,400.

b. You are selling the stock at $77.00 per share. The total sale price for 400 shares is $77.00 × 400, or $30,800. The broker charges 2.5% of this price:

$$0.025 \times \$30,800 = \$770.00.$$

The broker's commission is $770.00.

Check 2 Point
Use this information to answer the two questions in Example 2. You buy 500 shares of Blockbuster stock at $6.88 per share and sell them at $19.16 per share.

STUDY TIP

Many brokerage firms on the Internet charge fixed fees per trade, rather than a percentage of the total sale.

BLITZER BONUS

How to Make Your Child a Millionaire

People in charge of pension funds and retirement accounts invest the money collected by members in stocks. Based on an average annual return of 10%, *The Wall Street Journal* article on the right explains how to make your child a millionaire at age 65.

Make a Child a Millionaire
BY JONATHAN CLEMENTS
Staff Reporter of THE WALL STREET JOURNAL

Thanks a million.

Even if you haven't got a lot of money, you can easily give $1 million or more to your children, grandchildren or favorite charity. All it takes is a small initial investment and a lot of time.

Suppose your 16-year-old daughter plans to take a summer job that will pay her at least $2000. Because she has earned income, she can open an individual retirement account. If you would like to help fund her retirement, Kenneth Klegon, a financial planner in Lansing, Mich., suggests giving her $2000 to set up the IRA. He then advises doing the same in each of the next five years, so that your daughter stashes away a total of $12,000.

Result? If the money is invested in stocks, and stocks deliver their historical average annual return of 10%, your daughter will have more than $1 million by the time she turns 65.

Exercise Set 8.6

Practice and Application Exercises

Exercises 1 and 2 refer to the stock tables for Goodyear (the tire company) and JC Penney (the department store) given below. In each exercise, use the stock table to answer the following questions. Where necessary, round dollar amounts to the nearest cent.

a. What were the high and low prices for a share for the past 52 weeks?

b. If you owned 700 shares of this stock last year, what dividend did you receive?

c. What is the annual return for the dividends alone? How does this compare to a bank offering a 3% interest rate?

d. How many shares of this company's stock were traded yesterday?

e. What were the high and low prices for a share yesterday?

f. What was the price at which a share traded when the stock exchange closed yesterday?

g. What was the change in price for a share of stock from the market close two days ago to yesterday's market close?

h. Compute the company's annual earnings per share using

$$\text{Annual earnings per share} = \frac{\text{Yesterday's closing price per share}}{\text{PE ratio}}.$$

1.

52-Week High	Low	Stock	SYM	Div	Yld %	PE	Vol 100s	Hi	Lo	Close	Net Chg
73.25	45.44	Goodyear	GT	1.20	2.2	17	5915	56.38	54.38	55.50	+1.25

2.

52-Week High	Low	Stock	SYM	Div	Yld %	PE	Vol 100s	Hi	Lo	Close	Net Chg
78.34	35.38	Penney JC	JCP	2.18	4.7	22	7473	48.19	46.63	46.88	−1.31

3. Suppose that you bought 250 shares of Sears stock at its 52-week low, $39.06 per share, and sold at the 52-week high, $65.00 per share.

a. Ignoring dividends, how much did you make on this transaction?

b. If the broker charges 2.5% of the total sale price, find the broker's commission.

4. Suppose that you bought 350 shares of Walmart stock at its 52-week low, $36.75 per share, and sold at the 52-week high, $72.44 per share.

a. Ignoring dividends, how much did you make on this transaction?

b. If the broker charges 2.5% of the total sale price, find the broker's commission.

5. An investor purchased 400 shares of stock for $37.50 per share.

a. Find the cost of the stock.

b. If the broker charges 2% of the cost of the stock, find the broker's commission.

6. An investor purchased 600 shares of stock for $56.25 per share.

a. Find the cost of the stock.

b. If the broker charges 2% of the cost of the stock, find the broker's commission.

7. An investor purchased 240 shares of stock for $17.75 per share.

a. Find the cost of the stock.

b. The broker charges 2.5% of the cost of the stock plus 12.5¢ per share for 40 of the shares purchased. Find the broker's commission.

8. An investor purchased 370 shares of stock for $22.75 per share.

a. Find the cost of the stock.

b. The broker charges 2.5% of the cost of the stock plus 12.5¢ per share for 70 of the shares purchased. Find the broker's commission.

Writing in Mathematics

9. What is stock?

10. Describe how to find the percent ownership that a shareholder has in a company.

11. Describe the two ways that investors make money with stock.

12. What is a bond? Describe the difference between a stock and a bond.

13. What is a mutual fund? Describe two advantages to investing in mutual funds.

14. Using a recent newspaper, copy the stock table for a company of your choice. Then explain the meaning of the numbers in the columns.

15. If an investor sees that the dividends for a stock have a lower annual rate than those for a no-risk bank account, should the stock be sold and the money placed in the bank account? Explain your answer.

Use the following investments to answer Exercises 16–19.

> *Investment 1: 1000 shares of IBM stock*
> *Investment 2: A 5-year bond with a 22% interest rate issued by a small company that is testing and planning to sell delicious, nearly zero-calorie desserts*
> *Investment 3: A 30-year U.S. treasury bond at a fixed 7% annual rate*

16. Which of these investments has the greatest risk? Explain why.

17. Which of these investments has the least risk? Explain why.

18. Which of these investments is likely to have the greatest return? Explain why.

19. If you could be given one of these investments as a gift, which one would you choose? Explain why.

Critical Thinking Exercise

20. Use mathematics to show that the *Wall Street Journal* article on page 452 is correct. In particular, use one or more formulas to show that the $12,000 will be worth more than $1 million by the time the teenager turns 65.

Group Exercises

21. Each group should have a newspaper with current stock quotations. Choose nine stocks that group members think would make good investments. Imagine that you invest $1000 in each of these nine investments. Check the value of your stock each day over the next five weeks and then sell

the nine stocks after five weeks. What is the group's profit or loss over the five-week period? Compare this figure with the profit or loss of other groups in your class for this activity.

22. This activity is a group research project intended for four or five people. Use the research to present a seminar on investments. The seminar is intended to last about 30 minutes and should result in an interesting and informative presentation to the entire class. The seminar should include investment considerations, how to read the bond section of the newspaper, how to read the mutual fund section, and higher risk investments.

23. Group members have inherited $1 million. However, the group cannot spend any of the money for ten years. As a group, determine how to invest this money in order to maximize the money you will make over ten years. The money can be invested in as many ways as the group decides. Explain each investment decision. What are the risks involved in this investment plan?

Chapter Summary, Review, and Test

SUMMARY

DEFINITIONS AND CONCEPTS

EXAMPLES

8.1 Percent

a. Percent means per hundred. Thus, $97\% = \frac{97}{100}$.

b. To express a fraction as a percent, divide the numerator by the denominator, move the decimal point in the quotient two places to the right, and add a percent sign.

Ex. 1, p. 405

c. To express a decimal number as a percent, move the decimal point two places to the right and add a percent sign.

Ex. 2, p. 406

d. To express a percent as a decimal number, move the decimal point two places to the left and remove the percent sign.

Ex. 3, p. 406

e. The percent formula, $A = PB$, means A is P percent of B.

Ex. 4, p. 407;
Ex. 5, p. 408;
Ex. 6, p. 408

f. Sales tax amount = tax rate × item's cost

Ex. 7, p. 409

g. Discount amount = discount rate × original price

Ex. 8, p. 409

h. The fraction for percent increase (or decrease) is

$$\frac{\text{amount of increase (or decrease)}}{\text{original amount}}.$$

Ex. 9, p. 410;
Ex. 10, p. 411

Find the percent increase (or decrease) by expressing this fraction as a percent.

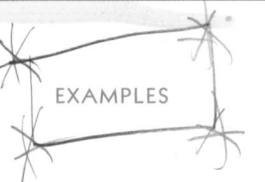

8.2 Simple Interest

a. Interest is the price we are paid for lending money or pay for borrowing money. The amount deposited or borrowed is the principal. The charge for interest, given as a percent, is the rate, assumed to be per year.

b. Simple interest involves interest calculated only on the principal and is computed using $I = Prt$.

Ex. 1, p. 415;
Ex. 2, p. 416

c. The future value, A, of P dollars at simple interest rate r for t years is $A = P(1 + rt)$.

Ex. 3, p. 417;
Ex. 4, p. 417;
Ex. 5, p. 418

d. Discounted loans deduct the interest, called the discount, from the loan amount at the time the loan is made.

Ex. 6, p. 418

8.3 Compound Interest

a. Compound interest involves interest computed on the original principal as well as on any accumulated interest. The amount in an account for one compounding period per year is $A = P(1 + r)^t$. For n compoundings per year, the amount is $A = P\left(1 + \frac{r}{n}\right)^{nt}$.

b. Calculating Present Value

If A dollars are to be accumulated in t years in an account that pays rate r compounded n times per year, then the present value, P, that needs to be invested now is given by

$$P = \frac{A}{\left(1 + \dfrac{r}{n}\right)^{nt}}.$$

c. Effective Annual Yield

Effective annual yield is defined in the box on page 423. The effective annual yield, Y, for an account that pays rate r compounded n times per year is given by

$$Y = \left(1 + \frac{r}{n}\right)^n - 1.$$

8.4 Installment Buying

a. A fixed installment loan is paid off with a series of equal periodic payments. An open-end installment loan is paid off with variable monthly payments. Credit card loans are open-end installment loans.

b. The terms of fixed installment loans—the amount financed, the total installment price, and the finance charge—are explained in the box on page 428.

c. The interest rate per year on a loan is called the annual percentage rate (APR). The box on page 429 shows how to find the APR for a fixed installment loan.

d. Unearned interest is the amount by which the original finance charge is reduced when a fixed installment loan is paid off early. The two methods for computing unearned interest—the actuarial method and the rule of 78—are explained in the box on page 430.

e. Open-end installment loans such as credit cards calculate interest using the simple interest formula $I = Prt$. The methods for determining the interest, the unpaid balance method, the previous balance method, and the average daily balance method, are described in the box on page 434.

8.5 The Cost of Home Ownership

a. A mortgage is a long-term loan for the purpose of buying a home, and for which the property is pledged as security for payment. The term of the mortgage is the number of years until final payoff. The down payment is the portion of the sale price of the home that the buyer initially pays. The amount of the mortgage is the difference between the sale price and the down payment.

b. Fixed-rate mortgages have the same monthly payment during the entire time of the loan. Variable-rate mortgages, or adjustable-rate mortgages, have payment amounts that change from time to time depending on changes in the interest rate.

c. A point is a one-time charge that equals 1 percent of the amount of a mortgage loan.

d. The monthly mortgage payment, PMT, to repay a loan of PV dollars paid n times per year over t years at annual rate r is

$$PMT = PV \frac{\dfrac{r}{n}}{1 - \left(1 + \dfrac{r}{n}\right)^{-nt}}.$$

Table 8.4 on page 439 can also be used to determine the PMT for certain rates.

e. Amortizing the loan is the process of making regular payments on the principal and interest until the loan is paid off. A document containing the payment number, payment toward the interest, payment toward the principal, and balance of the loan is called a loan amortization schedule. Such a schedule shows how the buyer pays slightly less in interest and more in principal for each payment over the entire life of the loan.

8.6 Investing in Stocks, Bonds, and Mutual Funds

a. The return of an investment is the percent increase in the investment.

b. Investors purchase stock, shares of ownership in a company. The shares indicate the percent of ownership. Trading refers to buying and selling stock. Investors make money by selling a stock for more money than they paid for it. They can also make money while they own stock if a company distributes all or part of its profits as dividends. Each share of stock is paid the same dividend.

c. Investors purchase a bond, lending money to the company from which they purchase the bond. The company commits itself to pay the price an investor pays for the bond at the time it was purchased, called its face value, along with interest payments at a given rate.

d. A portfolio is a listing of all the investments that a person holds.

e. A mutual fund is a group of stocks and/or bonds managed by a professional investor, called the fund manager. Investors in mutual funds own a small portion of many different companies, protecting them against the poor performance of a single company.

f. Reading stock tables is explained on pages 448–450.　　Ex. 1, p. 450

g. Costs involved in trading stock include commissions to brokers and management fees to fund managers of mutual funds.　　Ex. 2, p. 452

REVIEW EXERCISES

8.1

In Exercises 1–3, express each fraction as a percent.

1. $\frac{4}{5}$　　　　**2.** $\frac{1}{8}$　　　　**3.** $\frac{3}{4}$

In Exercises 4–6, express each decimal as a percent.

4. 0.72　　　　**5.** 0.0035　　　　**6.** 4.756

In Exercises 7–12, express each percent as a decimal.

7. 65%　　　　**8.** 99.7%　　　　**9.** 150%

10. 3%　　　　**11.** 0.65%　　　　**12.** $\frac{1}{4}$%

13. What is 8% of 120?　　　**14.** 90 is 45% of what?

15. 36 is what percent of 75?

The circle graph shows the results of one of the questions in a telephone poll taken for Time/CNN during 2002's summer of corporate scandals.

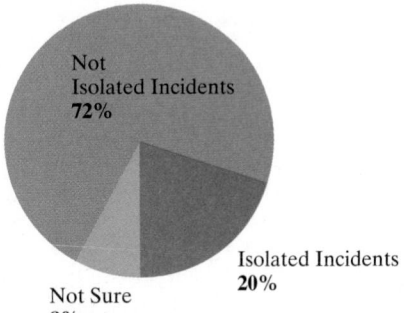

Which of the following best describes your view of the recent accounting scandals involving major American corporations?

Not Isolated Incidents 72%

Isolated Incidents 20%

Not Sure 8%

Source: Time/CNN

Use the graph to solve Exercises 16–17.

16. If 1440 people replied that the accounting scandals were not isolated incidents, find the total number of people polled.

17. Use your result from Exercise 16 to answer the following question: How many people felt that the accounting scandals were isolated incidents?

18. If $35,000 is invested in stock and there is a total of $50,000 invested, what percent of the invested money is in stock?

19. Suppose that the local sales tax rate is 6% and you purchase a backpack for $24.
　a. How much tax is paid?
　b. What is the backpack's total cost?

20. A television with an original price of $850 is on sale at 35% off.
　a. What is the discount amount?
　b. What is the television's sale price?

21. A college that had 40 students for each lecture course increased the number to 45 students. What is the percent increase in the number of students in a lecture course?

22. A dictionary regularly sells for $56.00. The sale price is $36.40. Find the percent decrease of the sale price from the regular price.

23. Consider the following statement:

My portfolio fell 10% last year, but then it rose 10% this year, so at least I recouped my losses.

Is this statement true? In particular, suppose you invested $10,000 in the stock market last year. How much money would be left in your portfolio with a 10% fall and then a 10% rise? If there is a loss, what is the percent decrease, to the nearest tenth of a percent, in your portfolio?

8.2

In Exercises 24–28, find the simple interest. (Assume 360 days in a year.)

	Principal	Rate	Time
24.	$6000	3%	1 year
25.	$8400	5%	6 years
26.	$20,000	8%	9 months
27.	$36,000	15%	60 days

28. In order to pay for tuition and books, a student borrows $3500 for four months at 10.5% interest.
   a. How much interest must the student pay?
   b. Find the future value of the loan.

In Exercises 29–33, use the formula for future value at simple interest, $A = P(1 + rt)$, to find the missing quantity. Round dollar amounts to the nearest cent and rates to the nearest tenth of a percent.

29. $A = ?, P = \$12,000, r = 8.2\%, t = 9$ months

30. $A = \$5750, P = \$5000, r = ?, t = 2$ years

31. $A = \$16,000, P = ?, r = 6.5\%, t = 3$ years

32. You plan to buy a $12,000 sailboat in four years. How much should you invest now, at 7.3% simple interest, to have enough for the boat in four years? (Round up to the nearest cent.)

33. You borrow $1500 from a friend and promise to pay back $1800 in six months. What simple interest rate will you pay?

34. You borrow $1800 on a 7% discounted loan for a period of 9 months.
   a. What is the loan's discount?
   b. Determine the net amount of money you will receive.
   c. What is the loan's actual interest rate, to the nearest tenth of a percent?

8.3

In Exercises 35–37, the principal represents an amount of money deposited in a savings account that provides the lender compound interest at the given rate.

   a. *Use the formula $A = P(1 + r)^t$ for one compounding period per year or $A = P\left(1 + \frac{r}{n}\right)^{nt}$ for n compounding periods per year to find how much money, to the nearest cent, there will be in the account after the given number of years.*
   b. *Find the interest earned.*

Principal	Rate	Compoundings per Year	Time
35. $7000	3%	1	5 years
36. $30,000	2.5%	4	10 years
37. $2500	4%	12	20 years

Solve Exercises 38–39 using the present value formula

$$P = \frac{A}{\left(1 + \frac{r}{n}\right)^{nt}}.$$

Round answers to the nearest cent.

38. How much money should parents deposit today in an account that earns 10% compounded monthly so that it will accumulate to $100,000 in 18 years for their child's college education?

39. How much money should be deposited today in an account that earns 5% compounded quarterly so that it will accumulate to $75,000 in 35 years for retirement?

40. You deposit $2000 in an account that pays 6% interest compounded quarterly.
   a. Find the future value, to the nearest cent, after one year.
   b. Use the future value formula for simple interest to determine the effective annual yield. Round to the nearest tenth of a percent.

Use the effective annual yield formula

$$Y = \left(1 + \frac{r}{n}\right)^n - 1$$

to solve Exercises 41–42.

41. What is the effective annual yield, to the nearest hundredth of a percent, of an account paying 5.5% compounded quarterly? What does your answer mean?

42. Which investment is the better choice: 6.25% compounded monthly or 6.3% compounded annually?

8.4

43. The cost of a new car is $16,500. We can finance the purchase by paying $500 down and $350 per month for 60 months.
   a. Determine the amount financed.
   b. Determine the total installment price.
   c. Determine the finance charge.
   d. Use Table 8.3 on page 429 to find the APR for this loan.

44. Use the actuarial method formula

$$u = \frac{kRV}{100 + V}$$

described in the box on page 430 to solve this problem. In Exercise 43, instead of making the forty-eighth payment, the borrower decides to pay the remaining balance and terminate the loan for the car.
   a. How much interest is saved by repaying the loan early?
   b. What is the total amount due on the day of the loan's termination?

45. Use the rule of 78 formula

$$u = \frac{k(k + 1)}{n(n + 1)} \times F$$

described in the box on page 430 to solve this problem. In Exercise 43, instead of making the forty-eighth payment, the borrower decides to pay the remaining balance and terminate the loan for the car.
   a. How much interest is saved by repaying the loan early?
   b. What is the total amount due on the day of the loan's termination?

46. Describe the difference between the answers in Exercises 44 and 45. Which method saves the borrower more money?

47. The terms of a particular credit card are based on the unpaid balance method. The monthly interest rate is 1.5% on the unpaid balance on the first day of the billing period less payments and credits. Here are some of the details in the June 1–June 30 itemized billing:

   June 1 Unpaid Balance: $1300

   Payment Received June 4: $200

   Purchases Charged to the Account: airline ticket, $380; car repair, $120; groceries, $140

   Last Day of the Billing Period: June 30

   Due Date: July 9

a. Find the interest due on the payment due date.

b. Find the total balance owed on the last day of the billing period.

c. Terms for the credit card require a $10 minimum monthly payment if the balance due is less than $360. Otherwise, the minimum monthly payment is $\frac{1}{36}$ of the balance due, rounded to the nearest whole dollar. What is the minimum monthly payment due by July 9?

48. A credit card has a monthly rate of 1.8%. In the March 1– March 31 itemized billing, the March 1 unpaid balance is $3600. A payment of $2000 was received on March 6. There are no purchases or cash advances in this billing period. The payment due date is April 9. Find the interest due on this date using

a. the unpaid balance method.

b. the previous balance method.

c. the average daily balance method.

8.5

49. The price of a home is $145,000. The bank requires a 20% down payment and two points at the time of closing. The cost of the home is financed with a 30-year fixed-rate mortgage at 7%.

a. Find the required down payment.

b. Find the amount of the mortgage.

c. How much must be paid for the two points at closing?

d. Use Table 8.4 on page 439 or the formula for monthly mortgage payments on page 441 to find the monthly payment (excluding escrowed taxes and insurance).

e. Find the total cost of interest over 30 years.

50. Use Table 8.4 on page 439 or the formula on page 441 to solve this problem. In terms of paying less in interest, which is more economical for a $70,000 mortgage: a 30-year fixed-rate at 8.5% or a 20-year fixed-rate at 8%? Discuss one advantage and one disadvantage for each mortgage option.

51. Prepare an amortization schedule for the first two months of the following mortgage loan.

AMORTIZATION SCHEDULE

Annual % rate: 7.5%			
Amount of Mortgage: $80,000		Monthly payment: $559.20	
Number of Monthly Payments: 360		Term: Years 30, Months 0	

Payment Number	Interest Payment	Principal Payment	Balance of Loan
1			
2			

52. You need a loan of $100,000 to buy a home. Here are your options:

Option A: 30-year fixed-rate at 8.5% with no closing costs and no points

Option B: 30-year fixed-rate at 7.5% with closing costs of $1300 and three points

Determine your monthly payments for each option and discuss how you would decide between the two options.

8.6

Exercises 53–60, refer to the stock table for Harley Davidson (the motorcycle company). Where necessary, round dollar amounts to the nearest cent.

52-Week					Yld	
High	Low	Stock	SYM	Div	%	PE
64.06	26.13	Harley Dav	HDI	.16	.3	41

Vol 100s	Hi	Lo	Close	Net Chg
5458	61.25	59.25	61	+1.75

53. What were the high and low prices for a share for the past 52 weeks?

54. If you owned 900 shares of this stock last year, what dividend did you receive?

55. What is the annual return for the dividends alone?

56. How many shares of this company's stock were traded yesterday?

57. What were the high and low prices for a share yesterday?

58. What was the price at which a share traded when the stock exchange closed yesterday?

59. What was the change in price for a share of stock from the market close two days ago to yesterday's market close?

60. Compute the company's annual earnings per share using

$$\text{Annual earnings per share} = \frac{\text{Yesterday's closing price per share}}{\text{PE ratio}}.$$

61. Suppose that you bought 600 shares of Pepsi stock at $27.50 per share and sold them for $43.75 per share.

a. Ignoring dividends, how much did you make on this transaction?

b. If the broker charges 2.5% of the total sale price, find the broker's commission.

62. Explain the difference between investing in a stock and investing in a bond.

63. What is a mutual fund? Why might a person want to invest in one?

Group Exercise

64. Group members have $12,000 to invest. The Blitzer Bonus on page 452 suggests one way that the investment can be worth over $1 million, but this will take about 45 years. Your group task is to determine strategies for investing this money in order to make it worth over $1 million in a shorter period of time. The money can be invested in as many ways as the group decides. Explain each investment decision, using formulas to show that the $12,000 will be worth more than $1 million in a shorter period than 45 years. What are the risks involved in each investment plan?

CHAPTER 8 TEST

1. What is 7.4% of 260?
2. 48 is 30% of what?
3. 18 is what percent of 80?
4. A CD player with an original price of $120 is on sale at 15% off.
   a. What is the amount of the discount?
   b. What is the sale price of the CD player?
5. A club that had 10 members increased its membership to 12. Find the percent increase in the club's membership.

Use the appropriate formulas to solve Exercises 6–11. Unless otherwise stated, round dollar amounts to the nearest cent and rates to the nearest tenth of a percent.

6. Find the future value and the amount of interest earned if $2000 is deposited for five years at 6% compounded monthly.
7. You borrow $2400 for three months at 12% simple interest. Find the amount of interest paid and the future value of the loan.
8. How much money should be deposited today in an account that earns 10% compounded semiannually so that it will accumulate to $100,000 in 20 years? (In this exercise, round up to the nearest cent.)
9. You borrow $2000 from a friend and promise to pay back $3000 in two years. What simple interest rate will you pay?
10. In six months, you want to have $7000 worth of remodeling done to your home. How much should you invest now, at 9% simple interest, to have enough money for the project? (Round up to the nearest cent.)
11. Find the effective annual yield, to the nearest hundredth of a percent, of an account paying 4.5% compounded quarterly. What does your answer mean?

In Exercises 12–17, the cost of a new car is $16,000, which can be financed by paying $3000 down and $300 per month for 60 months.

12. Determine the amount financed.
13. Determine the total installment price.
14. Determine the finance charge.
15. Use Table 8.3 on page 429 to find the APR for this loan.
16. Use the rule of 78 formula

$$u = \frac{k(k+1)}{n(n+1)} \times F$$

to solve this problem. Instead of making the thirty-sixth payment, the borrower decides to pay the remaining balance and terminate the loan for the car. How much interest is saved by repaying the loan early?

17. Use the answer from Exercise 16 to find the total amount due on the day of the loan's termination.

Use this information to solve Exercises 18–20. The terms of a particular credit card are based on the unpaid balance method. The monthly interest rate is 2% on the unpaid balance on the first day of the billing period less payments and credits. Here are some of the details in the November 1–November 30 itemized billing:

November 1 Unpaid Balance: $880
Payment Received November 5: $100
Purchases Charged to the Account: clothing, $350; gasoline, $70; groceries, $120
Last Day of the Billing Period: November 30
Due Date: December 8

18. Find the interest due on the payment due date.
19. Find the total balance owed on the last day of the billing period.
20. Terms for the credit card require a $10 minimum monthly payment if the balance due is less than $360. Otherwise, the minimum monthly payment is $\frac{1}{36}$ of the balance due, rounded to the nearest whole dollar. What is the minimum monthly payment due by December 8?
21. A credit card has a monthly rate of 1.6%. In the September 1–September 30 itemized billing, the September 1 unpaid balance is $2400. A payment of $1500 was received on September 4. There are no purchases or cash advances in this billing period. The payment due date is October 10. Find the interest due on this date using the average daily balance method.

Use this information to solve Exercises 22–26. The price of a home is $120,000. The bank requires a 10% down payment and two points at the time of closing. The cost of the home is financed with a 30-year fixed-rate mortgage at 8.5%.

22. Find the required down payment.
23. Find the amount of the mortgage.
24. How much must be paid for the two points at closing?
25. Use Table 8.4 on page 439 or the formula on page 441 to find the monthly payment (excluding escrowed taxes and insurance).
26. Find the total cost of interest over 30 years.
27. You need a loan of $100,000 to buy a home. Here are your options:

   Option A: 30-year fixed-rate at $7\frac{1}{2}\%$ with closing costs of $1200 and one point

   Option B: 15-year fixed-rate at 7% with closing costs of $1500 and three points

   Use Table 8.4 on page 439 or the formula on page 441 to determine your monthly payment for each option. Then discuss how you would decide between the two options.

Use the stock table for AT&T to solve Exercises 28–30.

| 52-Week | | | | | Yld | |
High	Low	Stock	SYM	Div	%	PE
26.50	24.25	AT & T	PNS	2.03	7.9	18

| Vol | | | | Net |
100s	Hi	Lo	Close	Chg
961	25.75	25.50	25.75	+0.13

28. What were the high and low prices for a share yesterday?
29. If you owned 1000 shares of this stock last year, what dividend did you receive?
30. Suppose that you bought 600 shares of AT&T, paying the price per share at which a share traded when the stock exchange closed yesterday. If the broker charges 2.5% of the price paid for all 600 shares, find the broker's commission.

Measurement

You use measurements all the time to describe things: the distance you travel to school, the temperature of your room, the size of your house or apartment, and your weight gain after a vacation with lots of great food. We live in a country that still uses the units of the English system, whereas most of the world now uses metric measurements. Having to use unfamiliar units of measure can be extremely disorienting when you travel outside the country.

In this chapter, we explore ways of measuring things in our English system, as well as in the metric system. We will also return to find a state with lots of room to spread out. Knowing how units of measure are used to describe your world will help you make decisions based on understanding the numbers and units you encounter every day.

You are feeling crowded in. Perhaps it would be a good time to move to a state with more elbow room. But which state? You can look up the population of each state, but that does not take into account the amount of land the population occupies. How is this land measured and how can you use this measure to select a place where wildlife outnumber humans?

This problem appears as Example 2 in Section 9.2.

SECTION 9.1   MEASURING LENGTH; THE METRIC SYSTEM

OBJECTIVES

1. Use dimensional analysis to change units of measurement.
2. Understand and use metric prefixes.
3. Convert units within the metric system.
4. Use dimensional analysis to change to and from the metric system.

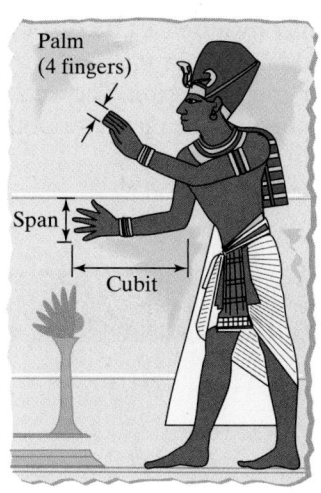

Linear units of measure were originally based on parts of the body. The Egyptians used the palm (equal to four fingers), the span (a handspan), and the cubit (length of forearm).

Have you seen either of the *Jurassic Park* films? The popularity of these movies reflects our fascination with dinosaurs and their incredible size. From end to end, the largest dinosaur from the Jurassic period, which lasted from 208 to 146 million years ago, was about 88 feet. To **measure** an object such as a dinosaur is to assign a number to its size. The number representing its measure from end to end is called its **length**. Measurements are used to describe properties of length, area, volume, weight, and temperature. Over the centuries, people have developed systems of measurement that are now accepted in most of the world.

Length

Every measurement consists of two parts: a number and a unit of measure. For example, if the length of a dinosaur is 88 feet, the number is 88 and the unit of measure is the foot. Many different units are commonly used in measuring length. The foot is from a system of measurement called the **English system**, which is generally used in the United States. In this system of measurement, length is expressed in such units as inches, feet, yards, and miles.

The result obtained from measuring length is called a **linear measurement** and is stated in **linear units**.

LINEAR UNITS OF MEASURE: THE ENGLISH SYSTEM

12 inches (in.) = 1 foot (ft)
3 feet = 1 yard (yd)
36 inches = 1 yard
5280 feet = 1 mile (mi)

Because many of us are familiar with the measures in the box, we find it simple to change from one measure to another, say from feet to inches. We know that there are 12 inches in a foot. To convert from 5 feet to a measure in inches, we multiply by 12. Thus, 5 feet = 5 × 12 inches = 60 inches.

Another procedure to convert from one unit of measurement to another is called **dimensional analysis**. Dimensional analysis uses **unit fractions**. A unit fraction has two properties: The numerator and denominator contain different units, and the value of the unit fraction is 1. Here are some examples of unit fractions:

$$\frac{12 \text{ in.}}{1 \text{ ft}}; \quad \frac{1 \text{ ft}}{12 \text{ in.}}; \quad \frac{3 \text{ ft}}{1 \text{ yd}}; \quad \frac{1 \text{ yd}}{3 \text{ ft}}; \quad \frac{5280 \text{ ft}}{1 \text{ mi}}; \quad \frac{1 \text{ mi}}{5280 \text{ ft}}.$$

In each unit fraction, the numerator and denominator are equal, making the value of the fraction 1.

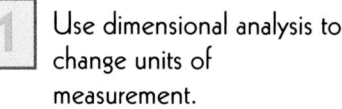

Use dimensional analysis to change units of measurement.

Let's see how to convert 5 feet to inches using a unit fraction.

$$5 \text{ ft} = ? \text{ in.}$$

We need to eliminate feet and introduce inches. The unit we need to introduce, inches, must appear in the numerator of the fraction. The unit we need to eliminate, feet, must appear in the denominator. Therefore, we choose the unit fraction with inches in the numerator and feet in the denominator. The units divide out as follows:

 $$5 \text{ ft} = \frac{5 \text{ ft}}{1} \cdot \frac{12 \text{ in.}}{1 \text{ ft}} = 5 \cdot 12 \text{ in.} = 60 \text{ in.}$$

unit fraction

DIMENSIONAL ANALYSIS

To convert a measurement to a different unit, multiply by a unit fraction (or by unit fractions). The given unit of measurement should appear in the denominator of the unit fraction so that this unit cancels upon multiplication. The unit of measurement that needs to be introduced should appear in the numerator of the fraction so that this unit will be retained upon multiplication.

EXAMPLE 1   USING DIMENSIONAL ANALYSIS TO CHANGE UNITS OF MEASUREMENT

Convert:

a. 40 inches to feet

b. 13,200 feet to miles

c. 9 inches to yards.

Solution

a. Because we want to convert 40 inches to feet, feet should appear in the numerator and inches in the denominator. We use the unit fraction

$$\frac{1 \text{ ft}}{12 \text{ in.}}$$

and proceed as follows:

$$40 \text{ in.} = \frac{40 \text{ in.}}{1} \cdot \frac{1 \text{ ft}}{12 \text{ in.}} = \frac{40}{12} \text{ ft} = 3\frac{1}{3} \text{ ft or } 3.\overline{3} \text{ ft.}$$

This period ends the sentence and is not part of the abbreviated unit.

b. To convert 13,200 feet to miles, miles should appear in the numerator and feet in the denominator. We use the unit fraction

$$\frac{1 \text{ mi}}{5280 \text{ ft}}$$

and proceed as follows:

$$13,200 \text{ ft} = \frac{13,200 \text{ ft}}{1} \cdot \frac{1 \text{ mi}}{5280 \text{ ft}} = \frac{13,200}{5280} \text{ mi} = 2.5 \text{ mi.}$$

STUDY TIP

The abbreviations for units of measurement are written without the period, such as ft for feet, except for the abbreviation for inch (in.).

c. To convert 9 inches to yards, yards should appear in the numerator and inches in the denominator. We use the unit fraction

$$\frac{1 \text{ yd}}{36 \text{ in.}}$$

and proceed as follows:

$$9 \text{ in.} = \frac{9 \ \cancel{\text{in.}}}{1} \cdot \frac{1 \text{ yd}}{36 \ \cancel{\text{in.}}} = \frac{9}{36} \text{ yd} = \frac{1}{4} \text{ yd or } 0.25 \text{ yd.}$$

Check 1 Point Convert:

  a. 78 inches to feet

  b. 17,160 feet to miles

  c. 3 inches to yards.

 Understand and use metric prefixes.

Length and the Metric System

Although the English system of measurement is most commonly used in the United States, most industrialized countries use the metric system of measurement. One of the advantages of the metric system is that units are based on powers of ten, making it much easier than the English system to change from one unit of measure to another.

The basic unit for linear measure in the metric system is the meter (m). A meter is slightly longer than a yard, approximately 39.37 inches. Prefixes are used to denote a multiple or part of a meter. Table 9.1 summarizes the more commonly used metric prefixes and their meanings.

TABLE 9.1 COMMONLY USED METRIC PREFIXES

Prefix	Symbol	Meaning
kilo	k	$1000 \times$ base unit
hecto	h	$100 \times$ base unit
deka	da	$10 \times$ base unit
deci	d	$\frac{1}{10}$ of base unit
centi	c	$\frac{1}{100}$ of base unit
milli	m	$\frac{1}{1000}$ of base unit

STUDY TIP

Prefixes containing an "i" (deci, centi, milli) are all fractional parts of one unit.

The prefixes kilo, centi, and milli are used more frequently than hecto, deka, and deci. Table 9.2 applies all six prefixes to the meter. The first part of the symbol indicates the prefix and the second part (m) indicates meter.

TABLE 9.2 COMMONLY USED UNITS OF LINEAR MEASURE IN THE METRIC SYSTEM

Symbol	Unit	Meaning
km	kilometer	1000 meters
hm	hectometer	100 meters
dam	dekameter	10 meters
m	meter	1 meter
dm	decimeter	0.1 meter
cm	centimeter	0.01 meter
mm	millimeter	0.001 meter

Kilometer is pronounced kil'-oh-met-er with the accent on the **FIRST** syllable. If pronounced correctly, kilometers should sound something like "kill all meters."

kilodollar

hectodollar

dekadollar

dollar

decidollar   centidollar

In the metric system, the kilometer is used to measure distances comparable to those measured in miles in the English system. One kilometer is approximately 0.6 mile, and one mile is approximately 1.6 kilometers.

Metric units of centimeters and millimeters are used to measure what the English system measures in inches. Figure 9.1 shows that a centimeter is less than half an inch; there are 2.54 centimeters in an inch. The smaller markings on the bottom scale are millimeters. A millimeter is approximately the thickness of a dime. The length of a bee or a fly may be measured in millimeters.

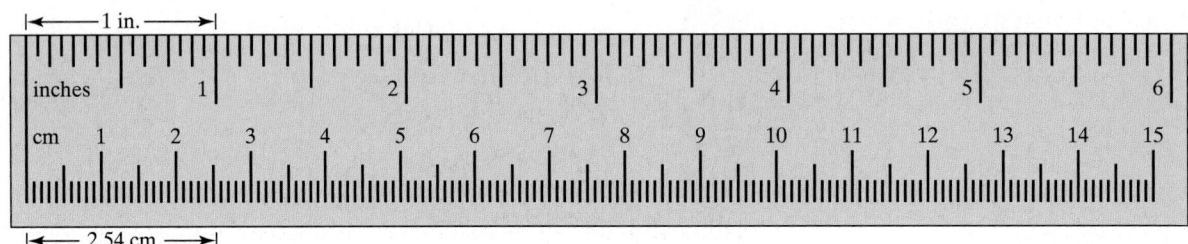

FIGURE 9.1

Those of us born in the United States have a good sense of what a length in the English system tells us about an object. An 88-foot dinosaur is huge, about 15 times the length of a 6-foot man. But what sense can we make of knowing that a whale is 25 meters long? The following lengths and the given approximations can help give you a feel for metric units of linear measure.

(1 meter ≈ 39 inches   1 kilometer ≈ 0.62 mile)

Item	Approximate Length
Width of lead in pencil	2 mm or 0.08 in.
Width of an adult's thumb	2 cm or 0.8 in.
Height of adult male	1.8 m or 6 ft
Typical room height	2.5 m or 8.3 ft
Length of medium-size car	5 m or 16.7 ft
Height of Empire State Building	381 m or 1270 ft
Average depth of ocean	4 km or 2.5 mi
Length of Manhattan Island	18 km or 11.25 mi
Distance from New York City to San Francisco	4800 km or 3000 mi
Radius of Earth	6378 km or 3986 mi
Distance from Earth to the moon	384,401 km or 240,251 mi

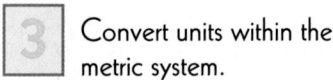

Convert units within the metric system.

Although dimensional analysis can be used to convert from one unit to another within the metric system, there is an easier, faster way to accomplish this conversion. The procedure is based on the observation that successively smaller units involve division by 10 and successively larger units involve multiplication by 10.

BLITZER BONUS

The First Meter

The French first defined the meter in 1791, calculating its length as a romantic one ten-millionth of a line running from the Equator through Paris to the North Pole. Today's meter, officially accepted in 1983, is equal to the length of the path traveled by light in a vacuum during the time interval of 1/299,794,458 of a second.

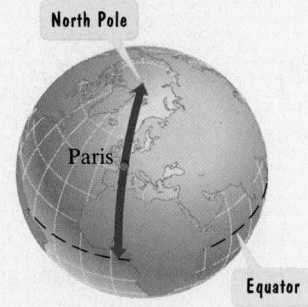

1 meter = 1/10,000,000 of the distance from the North Pole to the Equator on the meridian through Paris.

CHANGING UNITS WITHIN THE METRIC SYSTEM

Use the following chart to find equivalent measures of length:

$\times$ 10

Multiply by 10 for each step to the right.

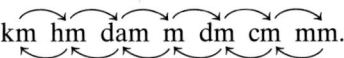

km hm dam m dm cm mm.

Divide by 10 for each step to the left.

$\div$ 10

1. To change from a larger unit to a smaller unit (moving to the right in the diagram), multiply by 10 for each step to the right. Thus, move the decimal point in the given quantity one place to the right for each smaller unit until the desired unit is reached.

2. To change from a smaller unit to a larger unit (moving to the left in the diagram), divide by 10 for each step to the left. Thus, move the decimal point in the given quantity one place to the left for each larger unit until the desired unit is reached.

STUDY TIP

The following sentence provides a way to remember the metric units for length from largest to smallest.

Kitty , how does my dog catch mice ?

km   hm dam m   dm cm mm

EXAMPLE 2 CHANGING UNITS WITHIN THE METRIC SYSTEM

a. Convert 504.7 meters to kilometers.
b. Convert 27 meters to centimeters.
c. Convert 704 mm to hm.
d. Convert 9.71 dam to dm.

Solution

a. To convert from meters to kilometers, we start at meters and move three steps to the left to obtain kilometers:

km hm dam m dm cm mm.

Hence, we move the decimal point three places to the left:

$$504.7 \text{ m} = 0.5047 \text{ km}$$

Thus, 504.7 meters converts to 0.5047 kilometers. Changing from a smaller unit of measurement (meter) to a larger unit of measurement (kilometer) results in an answer with a smaller number of units.

b. To convert from meters to centimeters, we start at meters and move two steps to the right to obtain centimeters:

$$\text{km} \quad \text{hm} \quad \text{dam} \quad \text{m} \quad \overset{\frown}{\text{dm}} \quad \overset{\frown}{\text{cm}} \quad \text{mm}.$$

Hence, we move the decimal point two places to the right:

$$27 \text{ m} = 2700 \text{ cm}.$$

Thus, 27 meters converts to 2700 centimeters. Changing from a larger unit of measurement (meter) to a smaller unit of measurement (centimeter) results in an answer with a larger number of units.

c. To convert from mm (millimeters) to hm (hectometers), we start at mm and move five steps to the left to obtain hm:

$$\text{km} \quad \text{hm} \quad \text{dam} \quad \text{m} \quad \text{dm} \quad \text{cm} \quad \text{mm}.$$

Hence, we move the decimal point five places to the left:

$$704 \text{ mm} = 0.00704 \text{ hm}.$$

d. To convert from dam (dekameters) to dm (decimeters), we start at dam and move two places to the right to obtain dm:

$$\text{km} \quad \text{hm} \quad \text{dam} \quad \overset{\frown}{\text{m}} \quad \overset{\frown}{\text{dm}} \quad \text{cm} \quad \text{mm}.$$

Hence, we move the decimal point two places to the right:

$$9.71 \text{ dam} = 971 \text{ dm}.$$

In Example 2(b), we showed that 27 meters converts to 2700 centimeters. This is the average length of the California blue whale, the longest of the great whales. Blue whales can have lengths that exceed 30 meters, making them over 100 feet long.

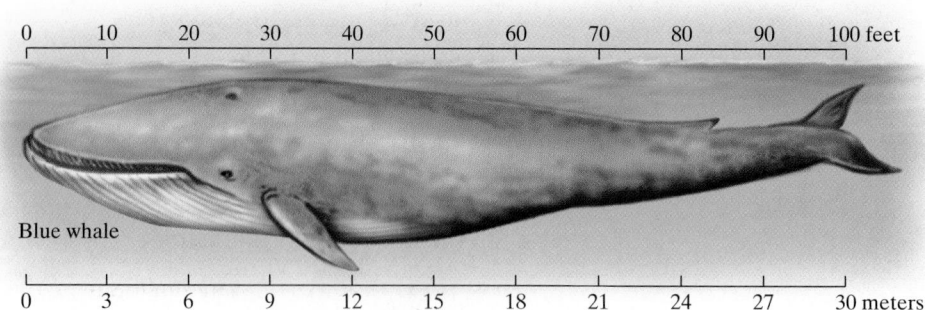

Check 2 Point

a. Convert 8000 meters to kilometers.

b. Convert 53 meters to millimeters.

c. Convert 604 cm to hm.

d. Convert 6.72 dam to cm.

BLITZER BONUS

Viruses and Metric Prefixes

Viruses are measured in attometers. An attometer is one quintillionth of a meter, or 10^{-18} meter, symbolized am. If a virus measures 1 am, you can place 10^{15} of them across a penciled 1 millimeter dot. If you were to enlarge each of these viruses to the size of the dot, they would stretch far into space, almost reaching Saturn.

Here is a list of all twenty metric prefixes. When applied to the meter, they range from the yottameter (10^{24} meters) to the yoctometer (10^{-24} meters).

LARGER THAN THE BASIC UNIT

Prefix	Symbol	Power of Ten	English Name
yotta-	Y	+24	septillion
zetta-	Z	+21	sextillion
exa-	E	+18	quintillion
peta-	P	+15	quadrillion
tera-	T	+12	trillion
giga-	G	+9	billion
mega-	M	+6	million
kilo-	k	+3	thousand
hecto-	h	+2	hundred
deca-	da	+1	ten

SMALLER THAN THE BASIC UNIT

Prefix	Symbol	Power of Ten	English Name
deci-	d	−1	tenth
centi-	c	−2	hundredth
milli-	m	−3	thousandth
micro-	μ	−6	millionth
nano-	n	−9	billionth
pico-	p	−12	trillionth
femto-	f	−15	quadrillionth
atto-	a	−18	quintillionth
zepto-	z	−21	sextillionth
yocto-	y	−24	septillionth

 4 Use dimensional analysis to change to and from the metric system.

Although dimensional analysis is not necessary when changing units within the metric system, it is a useful tool when converting to and from the metric system. Some approximate conversions are given in Table 9.3.

TABLE 9.3 APPROXIMATE ENGLISH AND METRIC EQUIVALENTS

1 inch (in.)	= 2.54 centimeters (cm)
1 foot (ft)	= 30.48 centimeters (cm)
1 yard (yd)	= 0.9 meter (m)
1 mile (mi)	= 1.6 kilometers (km)

These conversions are exact.

These conversions are approximate.

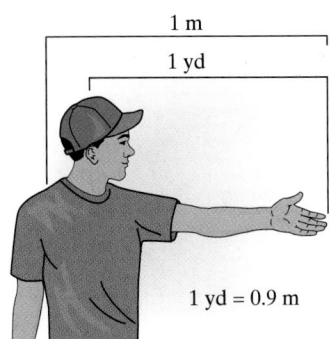

1 m

1 yd

1 yd = 0.9 m

EXAMPLE 3 | USING DIMENSIONAL ANALYSIS TO CHANGE TO AND FROM THE METRIC SYSTEM

a. Convert 8 inches to centimeters.

b. Convert 125 miles to kilometers.

c. Convert 26,800 millimeters to inches.

Solution

a. To convert 8 inches to centimeters, we use a unit fraction with centimeters in the numerator and inches in the denominator:

$$\frac{2.54 \text{ cm}}{1 \text{ in.}}.$$
    Table 9.3 shows that 1 in. = 2.54 cm.

We proceed as follows.

$$8 \text{ in.} = \frac{8 \text{ in.}}{1} \cdot \frac{2.54 \text{ cm}}{1 \text{ in.}} = 8(2.54) \text{ cm} = 20.32 \text{ cm}$$

b. To convert 125 miles to kilometers, we use a unit fraction with kilometers in the numerator and miles in the denominator:

$$\frac{1.6 \text{ km}}{1 \text{ mi}}. \qquad \textit{Table 9.3 shows that 1 mi} = \textit{1.6 km.}$$

Thus,

$$125 \text{ mi} = \frac{125 \text{ mi}}{1} \cdot \frac{1.6 \text{ km}}{1 \text{ mi}} = 125(1.6) \text{ km} = 200 \text{ km}.$$

c. To convert 26,800 millimeters to inches, we observe that Table 9.3 has only a conversion factor between inches and centimeters. We begin by changing millimeters to centimeters:

$$26,800 \text{ mm} = 2680.0 \text{ cm}.$$

Now we need to convert 2680 centimeters to inches. We use a unit fraction with inches in the numerator and centimeters in the denominator:

$$\frac{1 \text{ in.}}{2.54 \text{ cm}}.$$

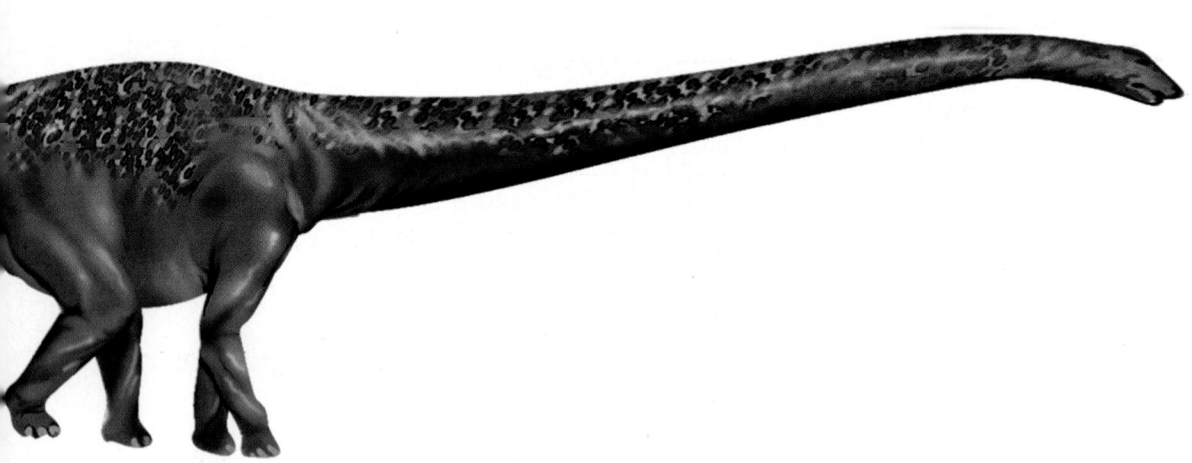

Thus,

$$26,800 \text{ mm} = 2680 \text{ cm} = \frac{2680 \text{ cm}}{1} \cdot \frac{1 \text{ in.}}{2.54 \text{ cm}} = \frac{2680}{2.54} \text{ in.} \approx 1055 \text{ in.}$$

This measure is equivalent to about 88 feet, the length of the largest dinosaur from the Jurassic period. The diplodocus, a plant eater, was 26.8 meters, approximately 88 feet, long.

Check
3
Point

a. Convert 8 feet to centimeters.

b. Convert 20 meters to yards.

c. Convert 30 meters to inches.

So far, we have used dimensional analysis to change units of length. Dimensional analysis may also be used to convert other kinds of measures, such as speed.

EXAMPLE 4   USING DIMENSIONAL ANALYSIS

a. The speed limit on many highways in the United States is 55 miles per hour (mi/hr). How many kilometers per hour (km/hr) is this?

b. If a high-speed train in Japan is capable of traveling at 200 kilometers per hour, how many miles per hour is this?

Solution

a. To change miles per hour to kilometers per hour, we need to concentrate on changing miles to kilometers, so we need a unit fraction with kilometers in the numerator and miles in the denominator:

$$\frac{1.6 \text{ km}}{1 \text{ mi}}.$$

Thus,

$$\frac{55 \text{ mi}}{\text{hr}} = \frac{55 \text{ mi}}{\text{hr}} \cdot \frac{1.6 \text{ km}}{1 \text{ mi}} = 55(1.6)\frac{\text{km}}{\text{hr}} = 88 \text{ km/hr}.$$

This shows that 55 miles per hour is equivalent to 88 kilometers per hour.

b. To change 200 kilometers per hour to miles per hour, we must convert kilometers to miles. We need a unit fraction with miles in the numerator and kilometers in the denominator:

$$\frac{1 \text{ mi}}{1.6 \text{ km}}.$$

Thus,

$$\frac{200 \text{ km}}{\text{hr}} = \frac{200 \text{ km}}{\text{hr}} \cdot \frac{1 \text{ mi}}{1.6 \text{ km}} = \frac{200 \text{ mi}}{1.6 \text{ hr}} = 125 \text{ mi/hr}.$$

A train capable of traveling at 200 kilometers per hour can therefore travel at 125 miles per hour.

 **Check Point 4**   A road in Europe has a speed limit of 60 kilometers per hour. How many miles per hour is this?

Exercise Set 9.1

Practice Exercises

In Exercises 1–16, use dimensional analysis to convert the quantity to the indicated units. If necessary, round the answer to two decimal places.

1. 30 in. to ft

2. 100 in. to ft

3. 30 ft to in.

4. 100 ft to in.

5. 6 in. to yd

6. 21 in. to yd

7. 6 yd to in.

8. 21 yd to in.

9. 6 yd to ft

10. 12 yd to ft

11. 6 ft to yd

12. 12 ft to yd

13. 23,760 ft to mi

14. 19,800 ft to mi

15. 0.75 mi to ft

16. 0.25 mi to ft

In Exercises 17–26, use the diagram in the box on page 465 to convert the given measurement to the unit indicated.

17. 5 m to cm

18. 8 dam to m

19. 16.3 hm to m

20. 0.37 hm to m

21. 317.8 cm to hm

22. 8.64 hm to cm

23. 0.023 mm to m

24. 0.00037 km to cm

25. 2196 mm to dm

26. 71 dm to km

In Exercises 27–44, use the following English and metric equivalents, along with dimensional analysis, to convert the given measurement to the unit indicated.

Approximate English and Metric Equivalents

$$1 \text{ in.} = 2.54 \text{ cm}$$
$$1 \text{ ft} = 30.48 \text{ cm}$$
$$1 \text{ yd} = 0.9 \text{ m}$$
$$1 \text{ mi} = 1.6 \text{ km}$$

27. 14 in. to cm   **28.** 26 in. to cm

29. 14 cm to in.   **30.** 26 cm to in.

31. 265 mi to km   **32.** 776 mi to km

33. 265 km to mi   **34.** 776 km to mi

35. 12 m to yd   **36.** 20 m to yd

37. 14 dm to in.   **38.** 1.2 dam to in.

39. 160 in. to dam   **40.** 180 in. to hm

41. 5 ft to m   **42.** 8 ft to m

43. 5 m to ft   **44.** 8 m to ft

Use 1 mi = 1.6 km to solve Exercises 45–48.

45. Express 96 kilometers per hour in miles per hour.

46. Express 104 kilometers per hour in miles per hour.

47. Express 45 miles per hour in kilometers per hour.

48. Express 50 miles per hour in kilometers per hour.

Application Exercises

In Exercises 49–58, selecting from millimeter, meter, and kilometer, determine the best unit of measure to express the given length.

49. A person's height

50. The length of a football field

51. The length of a bee

52. The distance from New York City to Washington, D.C.

53. The distance around a one-acre lot

54. The length of a car

55. The width of a book

56. The altitude of an airplane

57. The diameter of a screw

58. The width of a human foot

In Exercises 59–66, select the best estimate for the measure of the given item.

59. The length of a pen
    a. 30 cm   **b.** 19 cm   **c.** 19 mm

60. The length of this page
    a. 2.5 mm   **b.** 25 mm   **c.** 250 mm

61. The height of a skyscraper
    a. 325 m   **b.** 32.5 km   **c.** 325 km   **d.** 3250 km

62. The length of a pair of pants
    a. 700 cm   **b.** 70 cm   **c.** 7 cm

63. The height of a room
    a. 4 mm   **b.** 4 cm   **c.** 4 m   **d.** 4 dm

64. The length of a rowboat
    a. 4 cm   **b.** 4 dm   **c.** 4 m   **d.** 4 dam

65. The width of an electric cord
    a. 4 mm   **b.** 4 cm   **c.** 4 dm   **d.** 4 m

66. The dimensions of a piece of typing paper
    a. 22 mm by 28 mm   **b.** 22 cm by 28 cm
    c. 22 dm by 28 dm   **d.** 22 m by 28 m

67. A baseball diamond measures 27 meters along each side. If a batter scored two home runs in a game, how many kilometers did the batter run?

68. If you jog six times around a track that is 700 meters long, how many kilometers have you covered?

69. The distance from the Earth to the sun is about 93 million miles. What is this distance in kilometers?

70. The distance from New York City to Los Angeles is 4690 kilometers. What is the distance in miles?

Writing in Mathematics

71. Describe the two parts of a measurement.

72. Describe how to use dimensional analysis to convert 20 inches to feet.

73. Describe advantages of the metric system over the English system.

74. Explain how to change units within the metric system.

75. You jog 500 meters in a given period of time. The next day, you jog 500 yards over the same time period. On which day was your speed faster? Explain your answer.

76. What kind of difficulties might arise if the United States immediately eliminated all units of measure in the English system and replaced the system by the metric system?

77. The United States is the only Westernized country that does not use the metric system as its primary system of measurement. What reasons might be given for continuing to use the English system?

Critical Thinking Exercises

In Exercises 78–82, convert to an appropriate metric unit so that the numerical expression in the given measure does not contain any zeros.

78. 6000 cm   **79.** 900 m

80. 7000 dm   **81.** 11,000 mm

82. 0.0002 km

83. Use the unit fractions

$$\frac{36 \text{ in.}}{1 \text{ yd}} \quad \text{and} \quad \frac{2.54 \text{ cm}}{1 \text{ in.}}$$

to convert 5 yards to centimeters.

84. Use the unit fractions

$$\frac{5280 \text{ ft}}{1 \text{ mi}}, \quad \frac{12 \text{ in.}}{1 \text{ ft}}, \quad \text{and} \quad \frac{2.54 \text{ cm}}{1 \text{ in.}}$$

to convert 30 miles to kilometers.

Group Exercise

85. Jurassic Math: A Matter of Scale

Given only a list of numbers describing the height and length of a dinosaur, would you be able to make realistic judgments about how this historic creature would fit into your world? Could a Tyrannosaurus Rex peer in the window of a three-story building? Could a Stegosaurus fit in your classroom? How many parking spaces would a Brontosaurus fill?

a. In a small group, select a dinosaur to study. Write the approximate dimensions of the dinosaur in terms of height and length in an uncommon unit of measurement, such as inches, centimeters, or a fraction of a mile. Ask several people to make a comparative estimate of the size of the dinosaur from these measurements. For example, you might ask, "Is it bigger than a car?" or "Is it taller than this house?" You may tell the person if the dinosaur walked on two or four legs, but do not reveal the name of the dinosaur. Record the answers from several sources and compare the results with the rest of your group. How accurate were the answers?

b. Conduct another set of interviews, this time giving the information in a more typical form, such as feet or yards. How does the accuracy of the new estimates compare to the ones in part (a)? Share your results with other groups in the class. Does the actual size of the dinosaur seem to affect the estimates? In other words, does the size of larger dinosaurs seem harder to estimate accurately than smaller ones or vice versa?

SECTION 9.2   MEASURING AREA AND VOLUME

OBJECTIVES

1. Use square units to measure area.

2. Use dimensional analysis to change units for area.

3. Use cubic units to measure volume.

4. Use English and metric units to measure capacity.

Are you feeling a bit crowded in? Although there are more people on the East coast of the United States than there are bears, there are places in the Northwest where bears outnumber humans. The most densely populated state is New Jersey, with 1042 people per square mile. The least densely populated state is Alaska, with one person per square mile.

A square mile is one way of measuring the **area** of a state. A state's area is the region within its boundaries. Its **population density** is its population divided by its area. In this section, we discuss methods for measuring both area and volume.

1 Use square units to measure area.

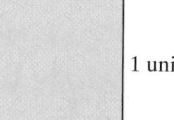

Measuring Area

In order to measure a region that is enclosed by boundaries, we begin by selecting a **square unit**. A square unit is a square, each of whose sides is one unit in length, illustrated in Figure 9.2. The region in Figure 9.2 is said to have an area of **one square unit**. The side of the square can be 1 inch, 1 centimeter, 1 meter, 1 foot, or 1 of any linear unit of measure. The corresponding units of area are the square inch (in.^{2}), the square centimeter (cm^2), the square meter (m^2), the square foot (ft^2),

FIGURE 9.2 One square unit

and so on. Figure 9.3 illustrates 1 square inch and 1 square centimeter, drawn to actual size.

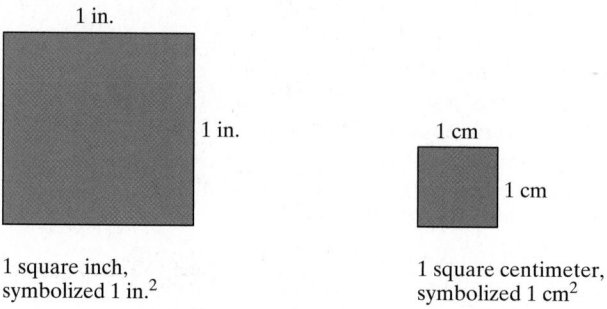

FIGURE 9.3 Common units of measurement for area, drawn to actual size

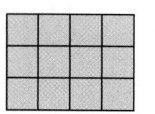

Square Unit

FIGURE 9.4

EXAMPLE 1 | MEASURING AREA

What is the area of the region shown in Figure 9.4?

Solution We can determine the area of the region by counting the number of square units contained within the region. There are 12 such units. Therefore, the area of the region is 12 square units.

Check 1 Point What is the area of the region represented by the first two rows in Figure 9.4?

Although there are 12 inches in one foot and 3 feet in one yard, these numerical relationships are not the same for square units:

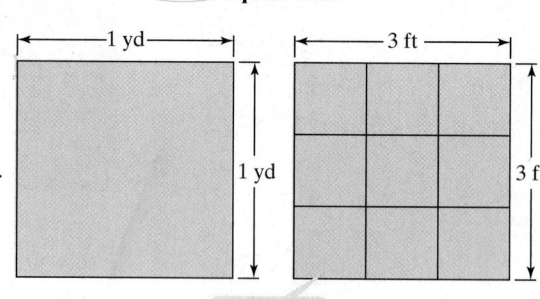

One Square Foot

Area is:
1 ft · 1 ft = 1 ft^2

1 square inch

Area is:
12 in. · 12 in. = 144 in.^{2}

Because the two figures have the same area,
1 ft^2 = 144 in.^{2}

One Square Yard

Area is:
1 yd · 1 yd = 1 yd^2

1 square foot

Area is:
3 ft · 3 ft = 9 ft^2

Because the two figures have the same area,
1 yd^2 = 9 ft^2

STUDY TIP

A small plot of land is usually measured in square feet, rather than a small fraction of an acre. Curiously, square yards are rarely used. However, square yards are commonly used for carpet and flooring measures.

SQUARE UNITS OF MEASURE: THE ENGLISH SYSTEM

1 square foot (ft^2)   = 144 square inches (in.^{2})

1 square yard (yd^2) = 9 square feet (ft^2)

1 acre (a)   = 43,560 ft^2 or 4840 yd^2

1 square mile (mi^2) = 640 acres

EXAMPLE 2 USING SQUARE UNITS TO COMPUTE POPULATION DENSITY

After Alaska, Wyoming is the least densely populated state. The population of Wyoming is 494,423 and its area is 97,818 square miles. What is Wyoming's population density?

Solution We compute the population density by dividing Wyoming's population by its area.

$$\text{population density} = \frac{\text{population}}{\text{area}} = \frac{494{,}423 \text{ people}}{97{,}818 \text{ square miles}}$$

Using a calculator and rounding to the nearest tenth, we obtain a population density of 5.1 people per square mile. This means that there is an average of only 5.1 people for each square mile of area.

Check 2 Point The population of Berkeley, California, is 103,000 and its area is 10.5 square miles. What is Berkeley's population density?

2 Use dimensional analysis to change units for area.

One theory is that using square units to measure area has its origin in the weaving of fabric.

In Section 9.1, we saw that in most other countries, the system of measurement that is used is the metric system. In the metric system, the square centimeter is used instead of the square inch. The square meter replaces the square foot and the square yard.

The English system uses the acre and the square mile to measure large land areas. The metric system uses the hectare (symbolized ha and pronounced "hectair"). A hectare is about the area of two football fields placed side by side, approximately 2.5 acres. One square mile of land consists of approximately 260 hectares. Just as the hectare replaces the acre, the square kilometer is used instead of the square mile. One square kilometer is approximately 0.38 square mile.

Some basic approximate conversions for square units of area are given in Table 9.4.

TABLE 9.4 ENGLISH AND METRIC EQUIVALENTS FOR AREA

1 square inch (in.^{2})	= 6.5 square centimeters (cm^2)
1 square foot (ft^2)	= 0.09 square meter (m^2)
1 square yard (yd^2)	= 0.8 square meter (m^2)
1 square mile (mi^2)	= 2.6 square kilometers (km^2)
1 acre	= 0.4 hectare (ha)

EXAMPLE 3 USING DIMENSIONAL ANALYSIS ON UNITS OF AREA

A property in Italy is advertised at $545,000 for 6.8 hectares.

a. Find the area of the property in acres.

b. What is the price per acre?

Solution

a. Using Table 9.4, we see that 1 acre = 0.4 hectare. To convert 6.8 hectares to acres, we use a unit fraction with acres in the numerator and hectares in the denominator.

$$6.8 \text{ ha} = \frac{6.8 \text{ ha}}{1} \cdot \frac{1 \text{ acre}}{0.4 \text{ ha}} = \frac{6.8}{0.4} \text{ acres} = 17 \text{ acres}$$

The area of the property is 17 acres.

b. The price per acre is the total price, $545,000, divided by the number of acres, 17.

$$\text{price per acre} = \frac{\$545,000}{17 \text{ acres}} \approx \$32,059/\text{acre}$$

The price is approximately $32,059 per acre.

 **Check 3 Point** A property in Northern California is on the market at $415,000 for 1.8 acres.

    a. Find the area of the property in hectares.

    b. What is the price per hectare?

 Use cubic units to measure volume.

Measuring Volume

A shoe box and a basketball are examples of three-dimensional figures. **Volume** refers to the amount of space occupied by such figures. In order to measure this space, we begin by selecting a *cubic unit*. Two such cubic units are shown in Figure 9.5.

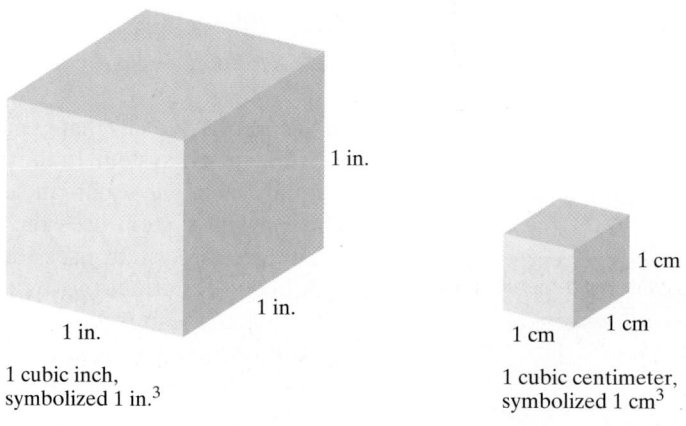

1 in.

1 in.

1 in.

1 cubic inch, symbolized 1 in.³

1 cm

1 cm

1 cm

1 cubic centimeter, symbolized 1 cm³

FIGURE 9.5 Common units of measurement for volume

The edges of a cube all have the same length. Other cubic units used to measure volume include 1 cubic foot (1 ft^3) and 1 cubic meter (1 m^3). One way to measure the volume of a solid is to calculate the number of cubic units contained in its interior.

EXAMPLE 4   MEASURING VOLUME

What is the volume of the solid shown in Figure 9.6?

Solution We can determine the volume of the solid by counting the number of cubic units contained within the region. Because we have drawn a solid three-dimensional figure on a flat two-dimensional page, some of the small cubic units in the back, left are hidden. The figures below show how the cubic units are used to fill the inside of the solid.

Cubic unit   Volume = ?

FIGURE 9.6

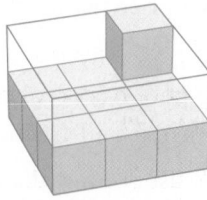 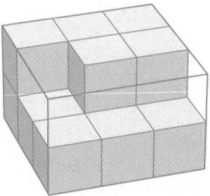

Do these figures help you to see that there are 18 cubic units inside the solid? The volume of the solid is 18 cubic units.

Check 4 Point What is the volume of the region represented by the bottom row of blocks in Figure 9.6?

We have seen that there are 3 feet in a yard, but 9 square feet in a square yard. Neither of these relationships holds for cubic units. Figure 9.7 illustrates that there are 27 cubic feet in a cubic yard. Furthermore, there are 1728 cubic inches in a cubic foot.

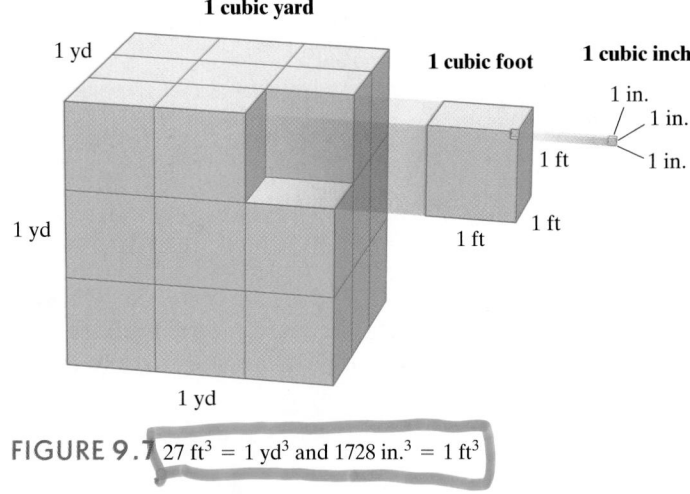

FIGURE 9.7 $27 \text{ ft}^3 = 1 \text{ yd}^3$ and $1728 \text{ in.}^3 = 1 \text{ ft}^3$

4 Use English and metric units to measure capacity.

The measure of volume also includes the amount of fluid that a three-dimensional object can hold. This is often called the object's **capacity**. For example, we often refer to the capacity, in gallons, of a gas tank. A cubic yard has a capacity of about 200 gallons and a cubic foot has a capacity of about 7.48 gallons.

Table 9.5 contains information about standard units of capacity in the English system.

TABLE 9.5 ENGLISH UNITS FOR CAPACITY

2 pints (pt) = 1 quart (qt)	
4 quarts	= 1 gallon (gal)
1 gallon	= 128 ounces (oz)
1 cup (c)	= 8 ounces

Volume in Cubic Units	Capacity
1 cubic yard	about 200 gallons
1 cubic foot	about 7.48 gallons
231 cubic inches	about 1 gallon

EXAMPLE 5 VOLUME AND CAPACITY IN THE ENGLISH SYSTEM

A swimming pool has a volume of 22,500 cubic feet. How many gallons of water does the pool hold?

Solution We use the fact that 1 cubic foot has a capacity of about 7.48 gallons to set up our unit fraction:

$$\frac{7.48 \text{ gal}}{1 \text{ ft}^3}.$$

We use this unit fraction to find the capacity of the 22,500 cubic feet.

$$22{,}500 \text{ ft}^3 = \frac{22{,}500 \text{ ft}^3}{1} \cdot \frac{7.48 \text{ gal}}{1 \text{ ft}^3} = 22{,}500(7.48) \text{ gal} = 168{,}300 \text{ gal}$$

The pool holds approximately 168,300 gallons of water.

 **Check Point 5** A pool has a volume of 10,000 cubic feet. How many gallons of water does the pool hold?

As we have come to expect, things are simpler when the metric system is used to measure capacity. The basic unit is the **liter**, symbolized by L. A liter is slightly larger than a quart.

$$1 \text{ liter} \approx 1.0567 \text{ quarts}$$

The standard metric prefixes are used to denote a multiple or part of a liter. Table 9.6 applies these prefixes to the liter.

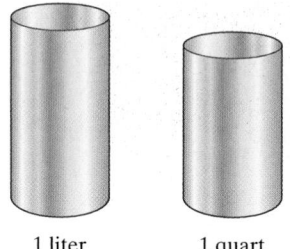

1 liter   1 quart

TABLE 9.6 UNITS OF CAPACITY IN THE METRIC SYSTEM

Symbol	Unit	Meaning
kL	kiloliter	1000 liters
hL	hectoliter	100 liters
daL	dekaliter	10 liters
L	liter	1 liter ≈ 1.06 quarts
dL	deciliter	0.1 liter
cL	centiliter	0.01 liter
mL	milliliter	0.001 liter

The following list should help give you a feel for capacity in the metric system.

Item	Capacity
Average cup of coffee	250 mL
12-ounce can of soda	355 mL
One quart of fruit juice	0.95 L
One gallon of milk	3.78 L
Average gas capacity of a car (about 18.5 gallons)	70 L

Figure 9.8 shows a 1 liter container filled with water. The water in the liter container will fit exactly into the box shown to its right. The volume of this box is 1000 cubic centimeters, or equivalently, 1 cubic decimeter. Thus,

$$1000 \text{ cm}^3 = 1 \text{ dm}^3 = 1 \text{ L}.$$

Table 9.7 expands on this relationship between volume and capacity in the metric system.

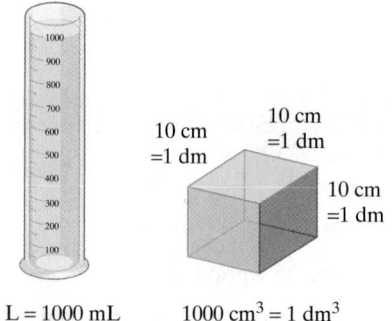

1 L = 1000 mL   1000 cm³ = 1 dm³

FIGURE 9.8

Measuring Dosages of Medicine

Table 9.7 indicates that

$$1 \text{ cm}^3 = 1 \text{ mL}.$$

Dosages of medicine are measured using cubic centimeters, or milliliters as they are also called. (In the United States, cc, rather than cm³, denotes cubic centimeters.) When taking medication by mouth, such as cough syrup, there is an easy conversion between the English and metric systems:

$$1 \text{ teaspoon} = 5 \text{ cc}.$$

TABLE 9.7 VOLUME AND CAPACITY IN THE METRIC SYSTEM

Volume in Cubic Units		Capacity
1 cm³	=	1 mL
1 dm³ = 1000 cm³	=	1 L
1 m³	=	1 kL

EXAMPLE 6 | VOLUME AND CAPACITY IN THE METRIC SYSTEM

An aquarium has a volume of 36,000 cubic centimeters. How many liters of water does the aquarium hold?

Solution   We use the fact that 1000 cubic centimeters corresponds to a capacity of 1 liter to set up our unit fraction:

$$\frac{1 \text{ L}}{1000 \text{ cm}^3}.$$

We use this unit fraction to find the capacity of the 36,000 cubic centimeters.

$$36,000 \text{ cm}^3 = \frac{36,000 \text{ cm}^3}{1} \cdot \frac{1 \text{ L}}{1000 \text{ cm}^3} = \frac{36,000}{1000} \text{ L} = 36 \text{ L}$$

The aquarium holds 36 liters of water.

 A fish pond has a volume of 220,000 cubic centimeters. How many liters of water does the pond hold?

Exercise Set 9.2

Practice Exercises

In Exercises 1–4, use the given figure to find its area in square units.

1.    **2.**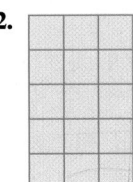

3. 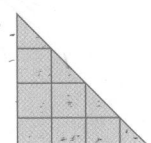   **4.**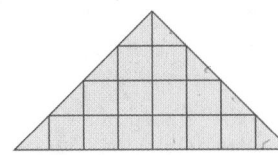

In Exercises 5–12, use Table 9.4 on page 473, along with dimensional analysis, to convert the given square unit to the square unit indicated. Where necessary, round answers to two decimal places.

5. 14 cm² to in.²   **6.** 20 m² to ft²

7. 30 m² to yd²   **8.** 14 mi² to km²

9. 10.2 ha to acres   **10.** 20.6 ha to acres

11. 14 in.² to cm²   **12.** 20 in.² to cm²

In Exercises 13–14, use the given figure to find its volume in cubic units.

13. 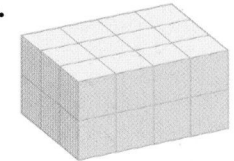   **14.**

In Exercises 15–22, use Table 9.5 on page 475, along with dimensional analysis, to convert the given unit to the unit indicated. Where necessary, round answers to two decimal places.

15. 10,000 ft³ to gal   **16.** 25,000 ft³ to gal

17. 8 yd³ to gal   **18.** 35 yd³ to gal

19. 2079 in.³ to gal   **20.** 6237 in.³ to gal

21. 2700 gal to yd³   **22.** 1496 gal to ft³

In Exercises 23–32, use Table 9.7 on page 477, along with dimensional analysis, to convert the given unit to the unit indicated.

23. $45,000 \text{ cm}^3$ to L

24. $75,000 \text{ cm}^3$ to L

25. 17 cm^3 to mL

26. 19 cm^3 to mL

27. 1.5 L to cm^3

28. 4.5 L to cm^3

29. 150 mL to cm^3

30. 250 mL to cm^3

31. 12 kL to dm^3

32. 16 kL to dm^3

Application Exercises

33. The population of Montana is 904,433 and its area is 147,046 square miles. Find Montana's population density to the nearest tenth.

34. The population of Utah is 2,269,789 and its area is 84,904 square miles. Find Utah's population density to the nearest tenth.

35. A property that measures 8 hectares is for sale.

 a. How large is the property in acres?

 b. If the property is selling for $250,000, what is the price per acre?

36. A property that measures 100 hectares is for sale.

 a. How large is the property in acres?

 b. If the property is selling for $350,000, what is the price per acre?

In Exercises 37–40, selecting from square centimeters, square meters, or square kilometers, determine the best unit of measure to express the area of the object described.

37. The top of a desk

38. A dollar bill

39. A national park

40. The wall of a room

In Exercises 41–44, select the best estimate for the measure of the area of the object described.

41. The area of the floor of a room

 a. 25 cm^2 **b.** 25 m^2 **c.** 25 km^2

42. The area of a television screen

 a. 2050 mm^2 **b.** 2050 cm^2 **c.** 2050 dm^2

43. The area of the face of a small coin

 a. 6 mm^2 **b.** 6 cm^2 **c.** 6 dm^2

44. The area of a parcel of land in a large metropolitan area on which a house can be built

 a. 900 cm^2 **b.** 900 m^2 **c.** 900 ha

45. A swimming pool has a volume of 45,000 cubic feet. How many gallons of water does the pool hold?

46. A swimming pool has a volume of 66,000 cubic feet. How many gallons of water does the pool hold?

47. A container of grapefruit juice has a volume of 4000 cubic centimeters. How many liters of juice does the container hold?

48. An aquarium has a volume of 17,500 cubic centimeters. How many liters of water does the aquarium hold?

Writing in Mathematics

49. Describe how area is measured. Explain why linear units cannot be used.

50. New Mexico has a population density of 12.5 people per square mile. Describe what this means. Use an almanac or the Internet to compare this population density to that of the state in which you are now living.

51. Describe the difference between the following problems: How much fencing is needed to enclose a garden? How much fertilizer is needed for the garden?

52. Describe how volume is measured. Explain why linear or square units cannot be used.

53. For a swimming pool, what is the difference between the following units of measure: cubic feet and gallons? For each unit, write a sentence about the pool that makes the use of the unit appropriate.

54. If there are 10 decimeters in a meter, explain why there are not 10 cubic decimeters in a cubic meter.

Critical Thinking Exercises

55. Singapore has the highest population density of any country: 46,690 people per 1000 hectares. How many people are there per square mile?

56. Nebraska has a population density of 20.5 people per square mile and a population of 1,578,417. What is the area of Nebraska?

57. A high population density is a condition common to extremely poor and extremely rich locales. Explain why this is so.

58. Although Alaska is the least densely populated state, over 90% of its land is protected federal property that is off-limits to settlement. A resident of Anchorage, Alaska, might feel hemmed in. In terms of "elbow room," what other important factor must be considered when calculating a state's population density?

59. Does an adult's body contain approximately 6.5 liters, kiloliters, or milliliters of blood? Explain your answer.

60. Is the volume of a coin approximately 1 cubic centimeter, 1 cubic millimeter, or 1 cubic decimeter? Explain your answer.

Group Exercise

61. Discovering how a particular measure came into common use requires a look at the history of mathematics and sometimes the history of everyday life. Most of the shorter measures we use for length came from the measure of different parts of the human body. Ancient Egyptians, Mesopotamians,

Greeks, and Romans all had measures based on the width of a finger or the span of a palm. For example, the smallest Roman measure, the width of a finger, was named *uncia*, the Latin word for one-twelfth, which has survived to this day in the form of "inch."

Listed in the chart are various units for measuring length and area. For this project, you will select a measure and determine how it came into its current usage. Bring this information to class and share it with other students in the group. Pool the research and prepare a presentation on your findings.

UNITS FOR MEASURING LENGTH AND AREA

statute mile	nautical mile
furlong	arc
foot	rod
inch	cubit
yard	span
meter	fathom
acre	hand

SECTION 9.3   MEASURING WEIGHT AND TEMPERATURE

OBJECTIVES

1. Apply metric prefixes to units of weight.
2. Convert units of weight within the metric system.
3. Use relationships between volume and weight within the metric system.
4. Use dimensional analysis to change units of weight to and from the metric system.
5. Understand temperature scales.

You are watching CNN International on cable television. The temperature in Honolulu, Hawaii, is reported as 30°C. Are Honolulu's tourists running around in winter jackets? In this section, we will make sense of Celsius temperature readings, as we discuss methods for measuring temperature and weight.

Measuring Weight

You step on the scale at the doctor's office to check your weight, discovering that you are 150 pounds. Compare this to your weight on the moon: 25 pounds. Why the difference? **Weight** is the measure of the gravitational pull on an object. The gravitational pull on the moon is only about one-sixth the gravitational pull on Earth. Although your weight varies depending on the force of gravity, your mass is exactly the same in all locations. **Mass** is a measure of the quantity of matter in an object, determined by its molecular structure. On Earth, as your weight increases, so does your mass. In this section, we will treat weight and mass as equivalent and refer strictly to weight.

UNITS OF WEIGHT: THE ENGLISH SYSTEM

16 ounces (oz) = 1 pound (lb)

2000 pounds (lb) = 1 ton (T)

The basic metric unit of weight is the **gram** (g), used for very small objects such as a coin, a candy bar, or a teaspoon of salt. A nickel has a weight of about 5 grams.

1 Apply metric prefixes to units of weight.

As with meters, prefixes are used to denote a multiple or part of a gram. Table 9.8 applies the common metric prefixes to the gram. The first part of the symbol indicates the prefix and the second part (g) indicates gram.

TABLE 9.8 COMMONLY USED UNITS OF WEIGHT IN THE METRIC SYSTEM

Symbol	Unit	Meaning
kg	kilogram	1000 grams
hg	hectogram	100 grams
dag	dekagram	10 grams
g	gram	1 gram
dg	decigram	0.1 gram
cg	centigram	0.01 gram
mg	milligram	0.001 gram

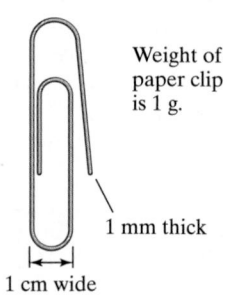

Weight of paper clip is 1 g.

1 mm thick

1 cm wide

In the metric system, the kilogram is the comparable unit to the pound in the English system. **A weight of 1 kilogram is approximately 2.2 pounds**. Thus, an average man has a weight of about 75 kilograms. Objects that we measure in pounds are measured in kilograms in most countries.

A milligram, equivalent to 0.001 gram, is an extremely small unit of weight and is used extensively in the pharmaceutical industry. If you look at the label on a bottle of tablets, you will see that the amounts of different substances in each tablet are expressed in milligrams.

The weight of a very heavy object is expressed in terms of the metric tonne (t), which is equivalent to 1000 kilograms, or about 2200 pounds. This is 10 percent more than the English ton (T) of 2000 pounds.

We change units of weight within the metric system exactly the same way that we changed units of length.

2 Convert units of weight within the metric system.

CHANGING UNITS OF WEIGHT WITHIN THE METRIC SYSTEM

Use the following diagram to find equivalent measures of weight:

Multiply by 10 for each step to the right.

kg hg dag g dg cg mg.

Divide by 10 for each step to the left.

EXAMPLE 1   CHANGING UNITS WITHIN THE METRIC SYSTEM

a. Convert 8.7 dg to mg.

b. Convert 950 mg to g.

Solution

a. To convert from dg (decigrams) to mg (milligrams), we start at dg and move two steps to the right:

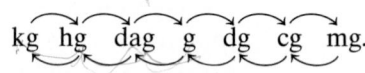

kg hg dag g dg cg mg.

Hence, we move the decimal point two places to the right:

$$8.7 \text{ dg} = 870 \text{ mg}.$$

b. To convert from mg (milligrams) to g (grams), we start at mg and move three steps to the left:

kg hg dag g dg cg mg.

Hence, we move the decimal point three places to the left:

$$950 \text{ mg} = 0.950 \text{ g}.$$

Check 1 Point

a. Convert 4.2 dg to mg.

b. Convert 620 cg to g.

 3 Use relationships between volume and weight within the metric system.

We have seen a convenient relationship in the metric system between volume and capacity:

$$1000 \text{ cm}^3 = 1 \text{ dm}^3 = 1 \text{ L}.$$

This relationship can be extended to include weight based on the following:

One kilogram of water has a volume of 1 liter.

Thus,

$$1000 \text{ cm}^3 = 1 \text{ dm}^3 = 1 \text{ L} = 1 \text{ kg}.$$

Table 9.9 shows the relationships between volume and weight of water in the metric system.

TABLE 9.9 VOLUME AND WEIGHT OF WATER IN THE METRIC SYSTEM

Volume		Capacity		Weight
1 cm^3	$=$	1 mL	$=$	1 g
$1 \text{ dm}^3 = 1000 \text{ cm}^3$	$=$	1 L	$=$	1 kg
1 m^3	$=$	1 kL	$=$	$1000 \text{ kg} = 1 \text{ t}$

EXAMPLE 2   VOLUME AND WEIGHT IN THE METRIC SYSTEM

An aquarium holds 0.25 m^3 of water. How much does the water weigh?

Solution   We use the fact that 1 m^3 of water $= 1000 \text{ kg}$ of water to set up our unit fraction:

$$\frac{1000 \text{ kg}}{1 \text{ m}^3}.$$

Thus,

$$0.25 \text{ m}^3 = \frac{0.25 \text{ m}^3}{1} \cdot \frac{1000 \text{ kg}}{1 \text{ m}^3} = 250 \text{ kg}.$$

The water weighs 250 kilograms.

 An aquarium holds 0.145 m³ of water. How much does the water weigh?

A problem like Example 2 involves more awkward computation in the English system. For example, if you know the aquarium's volume in cubic feet, you must also know that 1 cubic foot of water weighs about 62.5 pounds to determine the water's weight.

Dimensional analysis is a useful tool when converting units of weight between the English and metric systems. Some basic conversions are given in Table 9.10.

TABLE 9.10 WEIGHT: ENGLISH AND METRIC EQUIVALENTS

1 ounce (oz) = 28 grams (g)
1 pound (lb) = 0.45 kilogram (kg)
1 ton (T) = 0.9 tonne (t)

4 Use dimensional analysis to change units of weight to and from the metric system.

EXAMPLE 3   USING DIMENSIONAL ANALYSIS

a. Convert 160 pounds to kilograms.

b. Convert 300 grams to ounces.

Solution

a. To convert 160 pounds to kilograms, we use a unit fraction with kilograms in the numerator and pounds in the denominator:

$$\frac{0.45 \text{ kg}}{1 \text{ lb}}.$$ Table 9.10 shows that 1 lb = 0.45 g.

Thus,

$$160 \text{ lb} = \frac{160 \text{ lb}}{1} \cdot \frac{0.45 \text{ kg}}{1 \text{ lb}} = 160(0.45) \text{ kg} = 72 \text{ kg}.$$

b. To convert 300 grams to ounces, we use a unit fraction with ounces in the numerator and grams in the denominator:

$$\frac{1 \text{ oz}}{28 \text{ g}}.$$ Table 9.10 shows that 1 oz = 28 g.

Thus,

$$300 \text{ g} = \frac{300 \text{ g}}{1} \cdot \frac{1 \text{ oz}}{28 \text{ g}} = \frac{300}{28} \text{ oz} \approx 10.7 \text{ oz}.$$

 **a.** A man weighs 186 pounds. Convert his weight to kilograms.

b. For each kilogram of weight, 1.2 milligrams of a drug is to be given. What dosage should a 186-pound man be given?

5 Understand temperature scales.

Measuring Temperature

You'll be leaving the cold of winter for a vacation to Hawaii. CNN International reports a temperature in Hawaii of 30°C. Should you pack a winter coat?

The idea of changing from Celsius readings—or is it Centigrade?—to familiar Fahrenheit readings can be disorienting. Reporting a temperature of 30°C doesn't

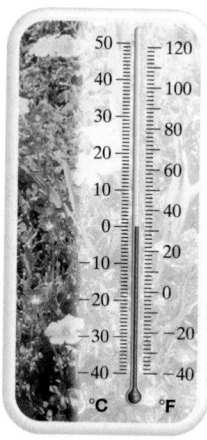

FIGURE 9.9 The Celsius scale is on the left and the Fahrenheit scale is on the right.

have the same impact as the Fahrenheit equivalent of 86 degrees (don't pack the coat). Why these annoying temperature scales?

The **Fahrenheit temperature scale**, the one we are accustomed to, was established in 1714 by the German physicist Gabriel Daniel Fahrenheit. He took a mixture of salt and ice, then thought to be the coldest possible temperature, and called it 0 degrees. He called the temperature of the human body 96 degrees, dividing the space between 0 and 96 into 96 parts. Fahrenheit was wrong about body temperature. It was later found to be 98.6 degrees. On his scale, water froze (without salt) at 32 degrees and boiled at 212 degrees. The symbol ° was used to replace the word "degree".

Twenty years later, the Swedish scientist Anders Celsius introduced another temperature scale. He set the freezing point of water at 0° and its boiling point at 100°, dividing the space into 100 parts. Degrees were called centigrade until 1948, when the name was officially changed to honor its inventor. However, "centigrade" is still commonly used in the United States.

Figure 9.9 shows a thermometer that measures temperatures in both degrees Celsius (°C is the scale on the left) and degrees Fahrenheit (°F is the scale on the right). The thermometer should help orient you if you need to know what a temperature in °C means. For example, if it is 40°C, find the horizontal line representing this temperature on the left. Now read across to the °F scale on the right. The reading is above 100°, indicating heat wave conditions.

The following formulas can be used to convert from one temperature scale to the other:

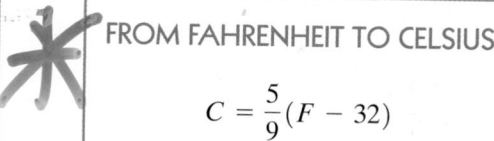

FROM CELSIUS TO FAHRENHEIT	FROM FAHRENHEIT TO CELSIUS
$F = \dfrac{9}{5}C + 32$	$C = \dfrac{5}{9}(F - 32)$

EXAMPLE 4   CONVERTING FROM CELSIUS TO FAHRENHEIT

The bills from your European vacation have you feeling a bit feverish, so you decide to take your temperature. The thermometer reads 37°. Should you panic?

Solution   Use the formula

$$F = \frac{9}{5}C + 32$$

to convert 37°C from °C to °F. Substitute 37 for C in the formula and find the value of F.

$$F = \frac{9}{5}(37) + 32 = 66.6 + 32 = 98.6$$

No need to panic! Your temperature is 98.6°F, which is perfectly normal.

Check 4 Point   Convert 50°C from °C to °F.

EXAMPLE 5 | CONVERTING FROM FAHRENHEIT TO CELSIUS

The temperature on a warm spring day is 77°F. Find the equivalent temperature on the Celsius scale.

Solution   Use the formula

$$C = \frac{5}{9}(F - 32)$$

to convert 77°F from °F to °C. Substitute 77 for F in the formula and find the value of C.

$$C = \frac{5}{9}(77 - 32) = \frac{5}{9}(45) = 25$$

Thus, 77°F is equivalent to 25°C.

 **Check 5 Point**   Convert 59°F from °F to °C.

Because temperature is a measure of heat, scientists do not find negative temperatures meaningful in their work. In 1948, the British physicist Lord Kelvin introduced a third temperature scale. He put 0 degrees at absolute zero, the coldest possible temperature, at which there is no heat and molecules stop moving. Figure 9.10 illustrates the three temperature scales.

Lake Baikal, Siberia, is one of the coldest places on Earth, reaching −76°F (−60°C) in winter. The lowest temperature possible is absolute zero. Scientists have cooled atoms to a few millionths of a degree above absolute zero.

BLITZER BONUS

Running a 5 K Race?

A 5 K race means a race at 5 Kelvins. This is a race so cold that no one would be able to move because all the participants would be frozen solid! The proper symbol for a race five kilometers long is a 5 km race.

Kelvin	Celsius	Fahrenheit	
373.15 K	100° C	212° F	Water boils
273.15 K	0° C	32° F	Water freezes
0 K	−273.15° C	−459.67° F	Absolute zero

FIGURE 9.10 The three temperature scales

Figure 9.10 shows that water freezes at 273.15 K (read "K" or "Kelvins," not "degrees Kelvin") and boils at 373.15 K. The Kelvin scale is the same as the Celsius scale, except in its starting (zero) point. This makes it easy to go back and forth from Celsius to Kelvin.

FROM CELSIUS TO KELVIN	FROM KELVIN TO CELSIUS
$K = C + 273.15$	$C = K - 273.15$

Kelvin's scale was embraced by the scientific community. Today, it is the final authority, as scientists define Celsius and Fahrenheit in terms of Kelvins.

Exercise Set 9.3

Practice Exercises

In Exercises 1–10, convert the given unit of weight to the unit indicated.

1. 7.4 dg to mg
2. 6.9 dg to mg
3. 870 mg to g
4. 640 mg to g
5. 8 g to cg
6. 7 g to cg
7. 18.6 kg to g
8. 0.37 kg to g
9. 0.018 mg to g
10. 0.029 mg to g

In Exercises 11–18, use Table 9.9 on page 481 to convert the given measurement of water to the unit indicated.

11. 0.05 m^3 to kg
12. 0.02 m^3 to kg
13. 4.2 kg to cm^3
14. 5.8 kg to cm^3
15. 1100 m^3 to t
16. 1500 t to m^3
17. 0.04 kL to g
18. 0.03 kL to g

In Exercises 19–30, use the following equivalents, along with dimensional analysis, to convert the given unit to the unit indicated. When necessary, round answers to two decimal places.

$$16 \text{ oz} = 1 \text{ lb}$$
$$2000 \text{ lb} = 1 \text{ T}$$
$$1 \text{ oz} = 28 \text{ g}$$
$$1 \text{ lb} = 0.45 \text{ kg}$$
$$1 \text{ T} = 0.9 \text{ t}$$

19. 36 oz to lb
20. 26 oz to lb
21. 36 oz to g
22. 26 oz to g
23. 540 lb to kg
24. 220 lb to kg
25. 80 lb to g
26. 150 lb to g
27. 540 kg to lb
28. 220 kg to lb
29. 200 t to T
30. 100 t to T

In Exercises 31–38, convert the given Celsius temperature to its equivalent temperature on the Fahrenheit scale. Where appropriate, round to the nearest tenth of a degree.

31. 10°C
32. 20°C
33. 35°C
34. 45°C
35. 57°C
36. 98°C
37. −5°C
38. −10°C

In Exercises 39–50, convert the given Fahrenheit temperature to its equivalent temperature on the Celsius scale. Where appropriate, round to the nearest tenth of a degree.

39. 68°F
40. 86°F
41. 41°F
42. 50°F
43. 72°F
44. 90°F
45. 23°F
46. 14°F
47. 350°F
48. 475°F
49. −22°F
50. −31°F

Application Exercises

In Exercises 51–57, selecting from milligram, gram, kilogram, and tonne, determine the best unit of measure to express the given item's weight.

51. A bee
52. This book
53. A tablespoon of salt
54. A Boeing 747
55. A stacked washer-dryer
56. A pen
57. An adult male

In Exercises 58–64, select the best estimate for the weight of the given item.

58. A newborn infant's weight
  a. 3000 kg   **b.** 300 kg   **c.** 30 kg   **d.** 3 kg
59. The weight of a nickel
  a. 5 kg   **b.** 5 g   **c.** 5 mg
60. A person's weight
  a. 60 kg   **b.** 60 g   **c.** 60 dag
61. The weight of a box of cereal
  a. 0.5 kg   **b.** 0.5 g   **c.** 0.5 t
62. The weight of a glass of water
  a. 400 dg   **b.** 400 g   **c.** 400 dag   **d.** 400 hg
63. The weight of a regular-size car
  a. 1500 dag   **b.** 1500 hg   **c.** 1500 kg   **d.** 15,000 kg
64. The weight of a bicycle
  a. 140 kg   **b.** 140 hg   **c.** 140 dag   **d.** 140 g

65. Six items purchased at a grocery store weigh 14 kilograms. One of the items is detergent weighing 720 grams. What is the total weight, in kilograms, of the other five items?

66. If a nickel weighs 5 grams, how many nickels are there in 4 kilograms of nickels?

67. If the cost to mail a letter is 37 cents for mail weighing up to one ounce and 23 cents for each additional ounce or fraction of an ounce, find the cost of mailing a letter that weighs 85 grams.

68. Using the information given below the pictured finback whale, estimate the weight, in tons and kilograms, of the killer whale.

Killer whale

Weight: 50 T or 45,360 kg

Finback whale

69. Which is more economical: purchasing the economy size of a detergent at 3 kilograms for $3.15 or purchasing the regular size at 720 grams for 60¢?

Exercises 70–71 ask you to determine drug dosage by a patient's weight. Use the fact that 1 lb = 0.45 kg.

70. For each kilogram of a person's weight, 20 milligrams of a drug is to be given. What dosage should be given to a person who weighs 200 pounds?

71. For each kilogram of a person's weight, 2.5 milligrams of a drug is to be given. What dosage should be given to a child who weighs 80 pounds.

The label on a bottle of Emetrol ("for food or drink indiscretions") reads

> *Each 5 mL teaspoonful contains glucose, 1.87 g; levulose, 1.87 g; and phosphoric acid, 21.5 mg.*

Use this information to solve Exercises 72–73.

72. a. Find the amount of glucose in the recommended dosage of two teaspoons.

  b. If the bottle contains 4 ounces, find the quantity of glucose in the bottle. (1 oz = 30 mL)

73. a. Find the amount of phosophroic acid in the recommended dosage of two teaspoons.

  b. If the bottle contains 4 ounces, find the quantity of phosphoric acid in the bottle. (1 oz = 30 mL)

In Exercises 74–77, select the best estimate of the Celsius temperature of

74. A very hot day.
  a. 85°C   **b.** 65°C   **c.** 35°C   **d.** 20°C

75. A warm winter day in Washington, D.C.
  a. 10°C   **b.** 30°C   **c.** 50°C   **d.** 70°C

76. A setting for a home thermostat.
  a. 20°C   **b.** 40°C   **c.** 60°C   **d.** 80°C

77. The oven temperature for cooking a roast.
  a. 80°C   **b.** 100°C   **c.** 175°C   **d.** 350°C

78. Key West, Florida, has the highest average yearly temperature of any city in the United States: 25.7°C. What is this temperature, in °F, to the nearest tenth of a degree?

79. Plateau Station, Antarctica, has the lowest average yearly temperature of any place on Earth: −7°C. What is this temperature, in °F, to the nearest tenth of a degree?

Writing in Mathematics

80. Describe the difference between weight and mass.

81. Explain how to use dimensional analysis to convert 200 pounds to kilograms.

82. Why do you think that countries using the metric system prefer the Celsius scale over the Fahrenheit scale?

83. Describe in words how to convert from Celsius to Fahrenheit.

84. Describe in words how to convert from Fahrenheit to Celsius.

85. If you decide to travel outside the United States, which one of the two temperature conversion formulas should you take? Explain your answer.

Critical Thinking Exercises

86. Which one of the following is true?
  a. A 4-pound object weighs more than a 2000-gram object.
  b. A 100-milligram object weighs more than a 2-ounce object.
  c. A 50-gram object weighs more than a 2-ounce object.
  d. A 10-pound object weighs more than a 4-kilogram object.

87. Which one of the following is true?
  a. Flour selling at 3¢ per gram is a better buy than flour selling at 55¢ per pound.
  b. The measures

$$32{,}600 \text{ g,} \quad 32.1 \text{ kg,} \quad 4 \text{ lb,} \quad 36 \text{ oz}$$

  are arranged in order, from greatest to least weight.
  c. If you are taking aspirin to relieve cold symptoms, a reasonable dose is 2 kilograms four times a day.
  d. A large dog weighs about 350 kilograms.

88. Start with the conversion formula from Fahrenheit to Celsius. Use algebra to obtain the conversion formula from Celsius to Fahrenheit.

Group Exercise

89. Present a group report on the current status of the metric system in the United States. At present, does it appear that the United States will convert to the metric system? Who supports the conversion and who opposes it? Summarize each side's position. Give examples of how our current system of weights and measures is an economic liability. What are the current obstacles to metric conversion?

Chapter Summary, Review, and Test

SUMMARY

DEFINITIONS AND CONCEPTS	EXAMPLES

9.1 Measuring Length; The Metric System

a. The result obtained from measuring length is called a linear measurement, stated in linear units.

b. Linear Units: The English System

$$12 \text{ in.} = 1 \text{ ft}, 3 \text{ ft} = 1 \text{ yd}, 36 \text{ in.} = 1 \text{ yd}, 5280 \text{ ft} = 1 \text{ mi}$$

c. Dimensional Analysis

Ex. 1, p. 462

Multiply the given measurement by a unit fraction with the unit of measurement that needs to be introduced in the numerator and the unit of measurement that needs to be eliminated in the denominator.

d. Linear Units: The Metric System

The basic unit is the meter (m), approximately 39 inches.

$$1 \text{ km} = 1000 \text{ m}, 1 \text{ hm} = 100 \text{ m}, 1 \text{ dam} = 10 \text{ m},$$
$$1 \text{ dm} = 0.1 \text{ m}, 1 \text{ cm} = 0.01 \text{ m}, 1 \text{ mm} = 0.001 \text{ m}$$

e. Changing Linear Units within the Metric System

Ex. 2, p. 465

$$\times 10$$

km hm dam m dm cm mm

$$\div 10$$

f. English and metric equivalents for length are given in Table 9.3 on page 467.

Ex. 3, p. 467;
Ex. 4, p. 469

9.2 Measuring Area and Volume

a. The area measure of a plane region is the number of square units contained in the given region.

Ex. 1, p. 472

b. English and metric equivalents for area are given in Table 9.4 on page 473.

Ex. 3, p. 473

c. The volume measure of a three-dimensional figure is the number of cubic units contained in its interior.

Ex. 4, p. 474

d. Capacity refers to the amount of fluid that a three-dimensional object can hold. English units for capacity include pints, quarts, and gallons: 2 pt = 1 qt; 4 qt = 1 gal; one cubic yard has a capacity of about 200 gallons. See Table 9.5 on page 475.

Ex. 5, p. 475

e. The basic unit for capacity in the metric system is the liter (L). One liter is about 1.06 quarts. Prefixes for the liter are the same as throughout the metric system.

f. Table 9.7 on page 477 shows relationships between volume and capacity in the metric system.

Ex. 6, p. 477

$$1 \text{ dm}^3 = 1000 \text{ cm}^3 = 1 \text{ L}$$

9.3 Measuring Weight and Temperature

a. Weight: The English System

$$16 \text{ oz} = 1 \text{ lb}, 2000 \text{ lb} = 1 \text{ T}$$

b. Units of Weight: The Metric System

The basic unit is the gram (g).

$$1000 \text{ grams } (1 \text{ kg}) \approx 2.2 \text{ lb},$$
$$1 \text{ kg} = 1000 \text{ g}, 1 \text{ hg} = 100 \text{ g}, 1 \text{ dag} = 10 \text{ g},$$
$$1 \text{ dg} = 0.1 \text{ g}, 1 \text{ cg} = 0.01 \text{ g}, 1 \text{ mg} = 0.001 \text{ g}$$

c. Changing Units of Weight within the Metric System   Ex. 1, p. 480

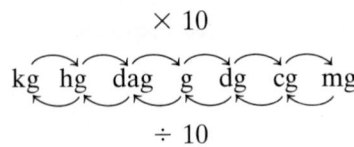

$\times$ 10

kg hg dag g dg cg mg

$\div$ 10

d. One kilogram of water has a volume of 1 liter. Table 9.9 on page 481 shows relationships between volume and weight in the metric system.   Ex. 2, p. 481

e. English and metric equivalents for weight are given in Table 9.10 on page 482.   Ex. 3, p. 482

f. Temperature Scales: Celsius to Fahrenheit: $F = \frac{9}{5}C + 32$ Ex. 4, p. 483

g. Temperature Scales: Fahrenheit to Celsius: $C = \frac{5}{9}(F - 32)$ Ex. 5, p. 484

REVIEW EXERCISES

9.1

In Exercises 1–4, use dimensional analysis to convert the quantity to the indicated unit.

1. 69 in. to ft

2. 9 in. to yd

3. 21 ft to yd

4. 13,200 ft to mi

In Exercises 5–10, convert the given linear measurement to the metric unit indicated.

5. 22.8 m to cm

6. 7 dam to m

7. 19.2 hm to m

8. 144 cm to hm

9. 0.5 mm to m

10. 18 cm to mm

In Exercises 11–16, use the given English and metric equivalents, along with dimensional analysis, to convert the given measurement to the unit indicated. Where necessary, round answers to two decimal places.

11. 23 in. to cm

12. 19 cm to in.

13. 330 mi to km

14. 600 km to mi

| 1 in. = 2.54 cm |
| 1 ft = 30.48 cm |
| 1 yd = 0.9 m |
| 1 mi = 1.6 km |

15. 14 m to yd

16. 12 m to ft

17. Express 45 kilometers per hour in miles per hour.

18. Express 60 miles per hour in kilometers per hour.

19. Arrange from smallest to largest: 0.024 km, 2400 m, 24,000 cm.

20. If you jog six times around a track that is 800 meters long, how many kilometers have you covered?

9.2

21. Use the given figure to find its area in square units.

22. The population of California is 33,870,000, making it the U.S. state with the largest population. If the area of California is 163,707 square miles, find California's population density to the nearest tenth. Describe what this means.

23. Given 1 acre = 0.4 hectare, use dimensional analysis to find the size of a property in acres measured at 7.2 hectares.

24. Using 1 ft^2 = 0.09 m^2, convert 30 m^2 to ft^2.

25. Using 1 mi^2 = 2.6 km^2, convert 12 mi^2 to km^2.

26. Which one of the following is a reasonable measure for the area of a flower garden in a person's yard?

a. 100 m^2 **b.** 0.4 ha   **c.** 0.01 km^2

27. Use the given figure to find its volume in cubic units.

28. A swimming pool has a volume of 33,600 cubic feet. Given that 1 cubic foot has a capacity of about 7.48 gallons, how many gallons of water does the pool hold?

29. An aquarium has a volume of 76,000 cubic centimeters. How many liters of water does the aquarium hold?

30. The capacity of a one-quart container of juice is approximately

a. 0.1 kL   **b.** 0.5 L   **c.** 1 L   **d.** 1 mL

31. There are 3 feet in a yard. Explain why there are not 3 square feet in a square yard. If helpful, illustrate your explanation with a diagram.

32. Explain why the area of California could not be measured in cubic miles.

9.3

In Exercises 33–36, convert the given unit of weight to the unit indicated.

33. 12.4 dg to mg

34. 12 g to cg

35. 0.012 mg to g

36. 450 mg to kg

In Exercises 37–38, use Table 9.9 on page 481 to convert the given measurement of water to the unit or units indicated.

37. 50 kg to cm³

38. 4 kL to dm³ to g

39. Using 1 lb = 0.45 kg, convert 210 pounds to kilograms.

40. Using 1 oz = 28 g, convert 392 grams to ounces.

41. If you are interested in your weight in the metric system, would it be best to report it in milligrams, grams, or kilograms? Explain why the unit you selected would be most appropriate. Explain why the other two units are not the best choice for reporting your weight.

42. Given 16 oz = 1 lb, use dimensional analysis to convert 36 ounces to pounds.

In Exercises 43–44, select the best estimate for the weight of the given item.

43. A dollar bill:
  a. 1 g   **b.** 10 g   **c.** 1 kg   **d.** 4 kg

44. A hamburger:
  a. 3 kg   **b.** 1 kg   **c.** 200 g   **d.** 5 g

In Exercises 45–49, convert the given Celsius temperature to its equivalent temperature on the Fahrenheit scale.

45. 15°C   **46.** 100°C   **47.** 5°C

48. 0°C   **49.** −25°C

In Exercises 50–55, convert the given Fahrenheit temperature to its equivalent temperature on the Celsius scale.

50. 59°F   **51.** 41°F   **52.** 212°F

53. 98.6°F   **54.** 0°F   **55.** 14°F

56. Is a decrease of 15° Celsius more or less than a decrease of 15° Fahrenheit? Explain your answer.

Group Exercise

57. If you could select any place in the world, where would you like to live? Look up the population and area of your ideal place, and compute its population density. Group members should share places to live and population densities. What trend, if any, does the group observe?

CHAPTER 9 TEST

1. Change 807 mm to hm.

2. Given 1 inch = 2.54 centimeters, use dimensional analysis to change 635 centimeters to inches.

3. If you jog eight times around a track that is 600 meters long, how many kilometers have you covered?

In Exercises 4–6, write the most reasonable metric unit for length in each blank. Select from mm, cm, m, and km.

4. A human thumb is 20____ wide.

5. The height of the table is 45____.

6. The towns are 60____ apart.

7. If 1 mile = 1.6 kilometers, express 80 miles per hour in kilometers per hour.

8. How many times greater is a square yard than a square foot?

9. Spain has a population of 39,133,966 and an area of 194,896 square miles. Find Spain's population density to the nearest tenth. Describe what this means.

10. Given 1 acre = 0.4 hectare, use dimensional analysis to find the area of a property measured at 18 hectares.

11. The area of a dollar bill is approximately
  a. 10 cm²   **b.** 100 cm²   **c.** 1000 cm²   **d.** 1 m²

12. There are 10 decimeters in a meter. Explain why there are not 10 cubic decimeters in a cubic meter. How many times greater is a cubic meter than a cubic decimeter?

13. A swimming pool has a volume of 10,000 cubic feet. Given that 1 cubic foot has a capacity of about 7.48 gallons, how many gallons of water does the pool hold?

14. The capacity of a pail used to wash floors is approximately
  a. 3 L   **b.** 12 L   **c.** 80 L   **d.** 2 kL

15. Change 137 g to kg.

16. Using 1 lb = 0.45 kg, convert 90 pounds to kilograms.

In Exercises 17–18, write the most reasonable metric unit for weight in each blank. Select from mg, g, and kg.

17. My suitcase weighs 20____.

18. I took a 350____ aspirin.

19. Convert 30°C to Fahrenheit.

20. Convert 176°F to Celsius.

21. Comfortable room temperature is approximately
  a. 70°C   **b.** 50°C   **c.** 30°C   **d.** 20°C

Geometry

Geometry is the study of the space you live in and the shapes that surround you. You're even made of it! The human lung consists of nearly 300 spherical air sacs, geometrically designed to provide the greatest surface area within the limited volume of our bodies. Viewed in this way, geometry becomes an intimate experience.

For thousands of years, people have studied geometry in some form to obtain a better understanding of the world in which they live. A study of the shape of your world will provide you with many practical applications and perhaps help to increase your appreciation of its beauty.

On a summer job, you are delivering three large cylindrical tanks filled with water to a remote country house. Your truck weighs 12 tons and you are worried about the extra weight of the tanks on these poorly paved, winding roads. Up ahead you see a warning sign at the entrance to a wobbly looking small bridge: Weight Not to Exceed 20 Tons. Knowing that making an error would not be a good idea, you pull over to compute the weight of your truck. Your boss told you every cubic inch of water that you are carrying weighs about 0.036 pound. You need to compute the volume of the tanks and then determine the combined weight of the tanks and your truck. What else do you need to know to decide whether to cross the bridge or go home without pay?

This problem appears as Exercise 71 on page 556.

SECTION 10.1   POINTS, LINES, PLANES, AND ANGLES

OBJECTIVES

1. Understand points, lines, and planes as the basis of geometry.

2. Solve problems involving angle measures.

3. Solve problems involving angles formed by parallel lines and transversals.

1 Understand points, lines, and planes as the basis of geometry.

The San Francisco Museum of Modern Art was constructed in 1996 to illustrate how art and architecture can enrich one another. The exterior involves geometric shapes, symmetry, and unusual facades. Although there are no windows, natural light streams in through a truncated cylindrical skylight that crowns the building. The architect worked with a scale model of the museum at the site and observed how light hit it during different times of the day. These observations were used to cut the cylindrical skylight at an angle that maximizes sunlight entering the interior.

Angles play a critical role in creating modern architecture. They are also fundamental in the study of geometry. The word "geometry" means "earth measure." Because it involves the mathematics of shapes, geometry connects mathematics to art and architecture. It also has many practical applications. You can use geometry at home when you buy a carpet, build a fence, tile a floor, or determine whether a piece of furniture will fit through your doorway. In this chapter, we look at the shapes that surround us and their applications.

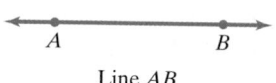

Point *A*

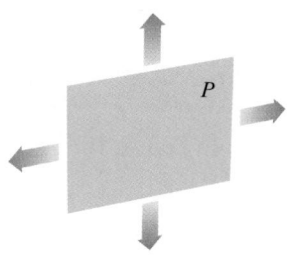

Line *AB*

Plane *P*

FIGURE 10.1 Representing a point, a line, and a plane

Points, Lines, and Planes

Points, lines, and planes make up the basis of all geometry. Stars in the night sky look like points of light. Long stretches of overhead power lines that appear to extend endlessly look like lines. The top of a flat table resembles part of a plane. However, stars, power lines, and table tops only approximate points, lines, and planes. Points, lines, and planes do not exist in the physical world. Representations of these forms are shown in Figure 10.1. A **point**, represented as a small dot, has no length, width, or thickness. No object in the real world has zero size. A **line**, connecting two points along the shortest possible path, has no thickness and extends infinitely in both directions. However, no familiar everyday object is infinite in length. A **plane** is a flat surface with no thickness and no boundaries. This page resembles a plane, although it does not extend indefinitely and it does have thickness.

A line may be named using any two of its points. In Figure 10.2(a), line AB can be symbolized $\overrightarrow{AB}$ or $\overleftrightarrow{BA}$. Any point on the line divides the line into three parts—the point and two **half-lines**. Figure 10.2(b) illustrates half-line AB, symbolized $\overset{\circ}{AB}$. The open circle above the A in the symbol and below the A the diagram indicates that point A is not included in the half-line. A **ray** is a half-line with its endpoint included. Figure 10.2(c) illustrates ray AB, symbolized $\overrightarrow{AB}$. The closed dot below the A in the diagram shows that point A is included in the ray. A portion of a line joining two points and including the endpoints is called a **line segment**. Figure 10.2(d) illustrates line segment AB, symbolized $\overline{AB}$ or $\overline{BA}$.

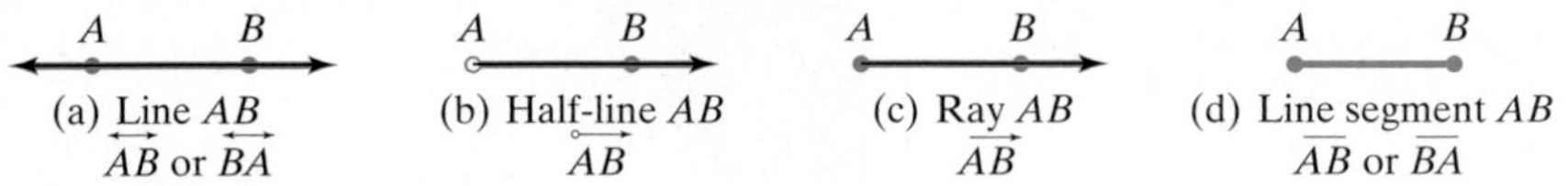

A B	A B	A B	A B
(a) Line AB	(b) Half-line AB	(c) Ray AB	(d) Line segment AB
$\overleftrightarrow{AB}$ or $\overleftrightarrow{BA}$	$\overset{\circ}{AB}$	$\overrightarrow{AB}$	$\overline{AB}$ or $\overline{BA}$

FIGURE **10.2** Lines, half-lines, rays, and line segments

Ray

Ray

FIGURE **10.3** Clock with hands forming an angle

Angles

An **angle**, symbolized $\measuredangle$, is formed by the union of two rays that have a common endpoint. One ray is called the **initial side** and the other the **terminal side**.

A rotating ray is often a useful way to think about angles. The ray in Figure 10.3 rotates from 12 to 2. The ray pointing to 12 is the initial side and the ray pointing to 2 is the terminal side. The common endpoint of an angle's initial side and terminal side is the **vertex** of the angle.

Figure 10.4 shows an angle. The common endpoint of the two rays, B, is the vertex. The two rays that form the angle, $\overrightarrow{BA}$ and $\overrightarrow{BC}$, are the sides. The four ways of naming the angle are shown to the right of Figure 10.4.

FIGURE **10.4** An angle: two rays with a common endpoint

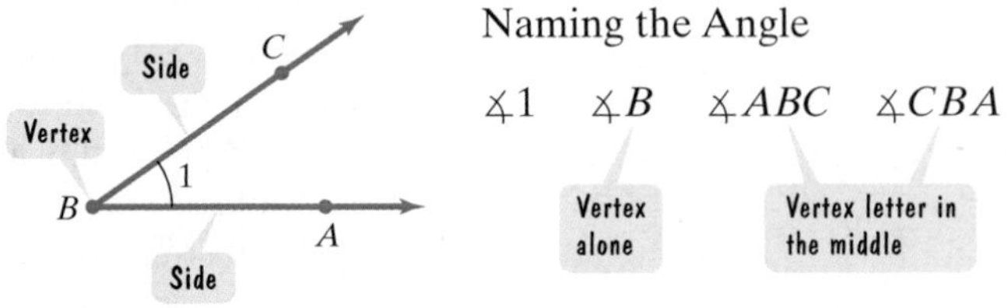

Side   C

Vertex

1

B

Side

A

Naming the Angle

$\measuredangle 1$ $\measuredangle B$ $\measuredangle ABC$ $\measuredangle CBA$

Vertex alone

Vertex letter in the middle

 Solve problems involving angle measures.

Measuring Angles Using Degrees

Angles are measured by determining the amount of rotation from the initial side to the terminal side. One way to measure angles is in **degrees**, symbolized by a small, raised circle °. Think of the hour hand of a clock. From 12 noon to 12 midnight, the hour hand moves around in a complete circle. By definition, the ray has rotated through 360 degrees, or 360°. Using 360° as the amount of rotation of a ray back onto itself, **a degree, 1°, is $\frac{1}{360}$ of a complete rotation**.

> ### EXAMPLE 1   USING DEGREE MEASURE

The hour hand of a clock moves from 12 to 2 o'clock, shown in Figure 10.3. Through how many degrees does it move?

Solution   We know that one complete rotation is 360°. Moving from 12 to 2 o'clock is $\frac{2}{12}$, or $\frac{1}{6}$, of a complete revolution. Thus, the hour hand moves

$$\frac{1}{6} \times 360° = \frac{360°}{6} = 60°$$

in going from 12 to 2 o'clock.

A complete 360° rotation

Check 1 Point The hour hand of a clock moves from 12 to 1 o'clock. Through how many degrees does it move?

Figure 10.5 shows angles classified by their degree measurement. An **acute angle** measures less than 90° [see Figure 10.5(a)]. A **right angle**, one-quarter of a complete rotation, measures 90° [Figure 10.5(b)]. Examine the right angle—do you see a small square at the vertex? This symbol is used to indicate a right angle. An **obtuse angle** measures more than 90°, but less than 180° [Figure 10.5(c)]. Finally, a **straight angle**, one-half a complete rotation, measures 180° [Figure 10.5(d)]. The two rays in a straight angle form a straight line.

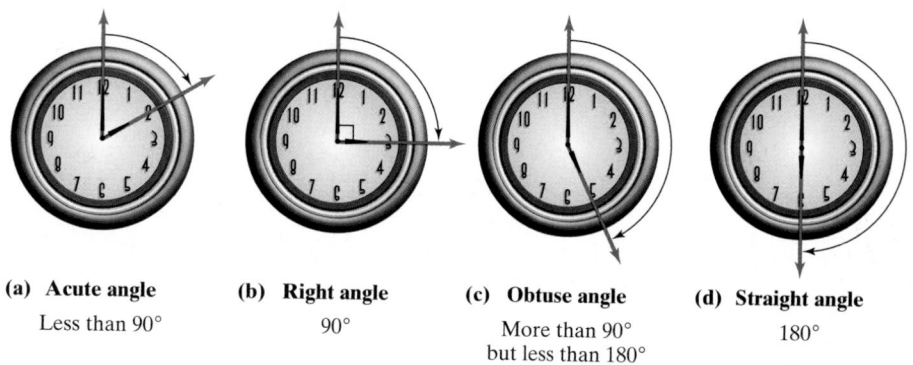

(a) Acute angle   **(b) Right angle**   **(c) Obtuse angle**   **(d) Straight angle**
Less than 90°   90°   More than 90°   180°
                                              but less than 180°

FIGURE 10.5 Classifying angles by their degree measurement

Figure 10.6 illustrates a **protractor**, used for finding the degree measure of an angle. As shown in the figure, we measure an angle by placing the center point of the protractor on the vertex of the angle and the straight side of the protractor along one side of the angle. The measure of ∡ABC is then read as 50°. Observe that the measure is not 130° because the angle is obviously less than 90°. We indicate the angle's measure by writing m∡$ABC = 50°$, read "the measure of angle ABC is 50°."

FIGURE 10.6 Using a protractor to measure an angle:
m∡$ABC = 50°$

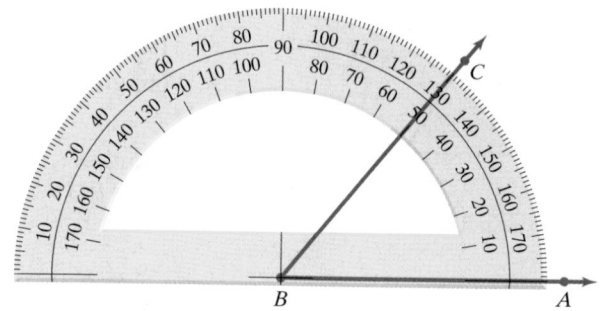

Two angles whose measures have a sum of 90° are called **complementary angles**. For example, angles measuring 70° and 20° are complementary angles because 70° + 20° = 90°. For angles such as those measuring 70° and 20°, each angle is the **complement** of the other: The 70° angle is the complement of the 20° angle, and the 20° angle is the complement of the 70° angle. **The measure of the complement can be found by subtracting the angle's measure from 90°**. For example, we can find the complement of a 25° angle by subtracting 25° from 90°: 90° − 25° = 65°. Thus, an angle measuring 65° is the complement of one measuring 25°.

FIGURE 10.7

EXAMPLE 2 | ANGLE MEASURES AND COMPLEMENTS

Use Figure 10.7 to find $m\angle DBC$.

Solution The measure of $\angle DBC$ is not yet known. It is shown as ?° in Figure 10.7. The acute angles $\angle ABD$, which measures 62°, and $\angle DBC$ form a right angle, indicated by the square at the vertex. This means that the measures of the acute angles add up to 90°. Thus, $\angle DBC$ is the complement of the angle measuring 62°. The measure of $\angle DBC$ is found by subtracting 62° from 90°:

$$m\angle DBC = 90° - 62° = 28°.$$

The measure of $\angle DBC$ can also be found using an algebraic approach.

$m\angle ABD + m\angle DBC = 90°$	The sum of the measures of complementary angles is 90°.
$62° + m\angle DBC = 90°$	We are given $m\angle ABD = 62°$.
$m\angle DBC = 90° - 62° = 28°$	Subtract 62° from both sides of the equation.

✓ **Check Point 2** In Figure 10.7, let $m\angle DBC = 19°$. Find $m\angle DBA$.

Two angles whose measures have a sum of 180° are called **supplementary angles**. For example, angles measuring 110° and 70° are supplementary angles because 110° + 70° = 180°. For angles such as those measuring 110° and 70°, each angle is the **supplement** of the other: the 110° angle is the supplement of the 70° angle, and the 70° angle is the supplement of the 110° angle. **The measure of the supplement can be found by subtracting the angle's measure from 180°.** For example, we can find the supplement of a 25° angle by subtracting 25° from 180°: 180° − 25° = 155°. Thus, an angle measuring 155° is the supplement of one measuring 25°.

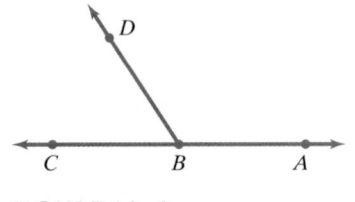

FIGURE 10.8

EXAMPLE 3 | ANGLE MEASURES AND SUPPLEMENTS

Figure 10.8 shows that $\angle ABD$ and $\angle DBC$ are supplementary angles. If $m\angle ABD$ is 66° greater than $m\angle DBC$, find the measure of each angle.

Solution Let $m\angle DBC = x$. Because $m\angle ABD$ is 66° greater than $m\angle DBC$, then $m\angle ABD = x + 66°$. We are given that these angles are supplementary.

$m\angle DBC + m\angle ABD = 180°$	The sum of the measures of supplementary angles is 180°.
$x + (x + 66°) = 180°$	Substitute the variable expressions for the measures.
$2x + 66° = 180°$	Combine like terms: $x + x = 2x$.
$2x = 114°$	Subtract 66° from both sides.
$x = 57°$	Divide both sides by 2.

Thus, $m\angle DBC = 57°$ and $m\angle ABD = 57° + 66° = 123°$.

Check 3 Point   In Figure 10.8, if $m\angle ABD$ is 88° greater than $m\angle DBC$, find the measure of each angle.

Figure 10.9 illustrates a highway sign that warns of a railroad crossing. When two lines intersect, the opposite angles formed are called **vertical angles**.

In Figure 10.10, there are two pairs of vertical angles. Angles 1 and 3 are vertical angles. Angles 2 and 4 are also vertical angles.

We can use Figure 10.10 to show that vertical angles have the same measure. Let's concentrate on angles 1 and 3, each denoted by one tick mark. Can you see that each of these angles is supplementary to angle 2?

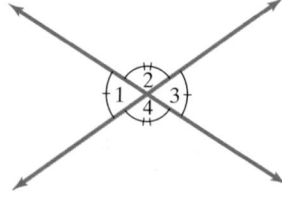

FIGURE 10.10

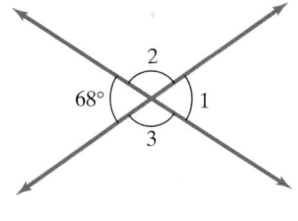

FIGURE 10.9

$$m\angle 1 + m\angle 2 = 180°$$ The sum of the measures of supplementary angles is 180°.

$$m\angle 2 + m\angle 3 = 180°$$

$$m\angle 1 + m\angle 2 = m\angle 2 + m\angle 3$$ Substitute $m\angle 2 + m\angle 3$ for 180° in the first equation.

$$m\angle 1 = m\angle 3$$ Subtract $m\angle 2$ from both sides.

Using a similar approach, we can show that $m\angle 2 = m\angle 4$, each denoted by two tick marks in Figure 10.10.

Vertical angles have the same measure.

EXAMPLE 4   USING VERTICAL ANGLES

Figure 10.11 shows that the angle on the left measures 68°. Find the measures of the other three angles.

Solution   Angle 1 and the angle measuring 68° are vertical angles. Because vertical angles have the same measure,

$$m\angle 1 = 68°.$$

Angle 2 and the angle measuring 68° form a straight angle and are supplementary. Because their measures add up to 180°,

$$m\angle 2 = 180° - 68° = 112°.$$

Angle 2 and angle 3 are also vertical angles, so they have the same measure. Because the measure of angle 2 is 112°,

$$m\angle 3 = 112°.$$

FIGURE 10.11

Check 4 Point   In Figure 10.11, assume that the angle on the left measures 57°. Find the measures of the other three angles.

3   Solve problems involving angles formed by parallel lines and transversals.

Parallel Lines

Parallel lines are lines that lie in the same plane and have no points in common. If two different lines in the same plane are not parallel, they have a single point in common and are called **intersecting lines**. If the lines intersect at an angle of 90°, they are called **perpendicular lines**.

If we intersect a pair of parallel lines with a third line, called a **transversal**, eight angles are formed, as shown in Figure 10.12 on the bottom of the page. Certain pairs of these angles have special names, as well as special properties. These names and properties are summarized in Table 10.1.

TABLE 10.1 NAMES OF ANGLE PAIRS FORMED BY A TRANSVERSAL INTERSECTING PARALLEL LINES

Name	Description	Sketch	Angle Pairs Described	Property
Alternate interior angles	Interior angles that do not have a common vertex on alternate sides of the transversal		∡3 and ∡6 ∡4 and ∡5	Alternate interior angles have the same measure. $m\angle 3 = m\angle 6$ $m\angle 4 = m\angle 5$
Alternate exterior angles	Exterior angles that do not have a common vertex on alternate sides of the transversal		∡1 and ∡8 ∡2 and ∡7	Alternate exterior angles have the same measure. $m\angle 1 = m\angle 8$ $m\angle 2 = m\angle 7$
Corresponding angles	One interior and one exterior angle on the same side of the transversal		∡1 and ∡5 ∡2 and ∡6 ∡3 and ∡7 ∡4 and ∡8	Corresponding angles have the same measure. $m\angle 1 = m\angle 5$ $m\angle 2 = m\angle 6$ $m\angle 3 = m\angle 7$ $m\angle 4 = m\angle 8$

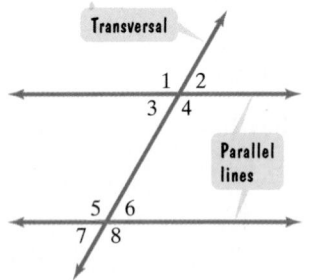

FIGURE 10.12

When two parallel lines are cut by a transversal, the following relationships are true:

PARALLEL LINES AND ANGLE PAIRS

If parallel lines are cut by a transversal,

- alternate interior angles have the same measure,
- alternate exterior angles have the same measure, and
- corresponding angles have the same measure.

Conversely, if an alternate interior angle pair or an alternate exterior angle pair or a pair of corresponding angles have the same measure, then the lines are parallel.

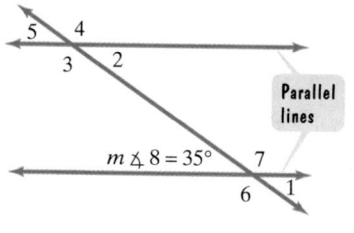

FIGURE 10.13

EXAMPLE 5 | FINDING ANGLE MEASURES WHEN PARALLEL LINES ARE INTERSECTED BY A TRANSVERSAL

In Figure 10.13, two parallel lines are intersected by a transversal. One of the angles (∡8) has a measure of 35°. Find the measure of each of the other seven angles.

Solution Look carefully at Figure 10.13 and fill in the angle measures as you read each line in this solution.

$m∡1 = 35°$ ∡8 and ∡1 are vertical angles and vertical angles have the same measure.

$m∡6 = 180° − 35° = 145°$ ∡8 and ∡6 are supplementary.

$m∡7 = 145°$ ∡6 and ∡7 are vertical angles, so they have the same measure.

$m∡2 = 35°$ ∡8 and ∡2 are alternate interior angles, so they have the same measure.

$m∡3 = 145°$ ∡7 and ∡3 are alternate interior angles. Thus, they have the same measure.

$m∡5 = 35°$ ∡8 and ∡5 are corresponding angles. Thus, they have the same measure.

$m∡4 = 180° − 35° = 145°$ ∡4 and ∡5 are supplementary.

STUDY TIP

There is more than one way to solve Example 6. For example, once you know that $m∡2 = 35°$, $m∡5$ is also 35° because ∡2 and ∡5 are vertical angles.

✓ Check 5 Point

In Figure 10.13, assume that $m∡8 = 29°$. Find the measure of each of the other seven angles.

Exercise Set 10.1

Practice Exercises

1. The hour hand of a clock moves from 12 to 5 o'clock. Through how many degrees does it move?
2. The hour hand of a clock moves from 12 to 4 o'clock. Through how many degrees does it move?
3. The hour hand of a clock moves from 1 to 4 o'clock. Through how many degrees does it move?
4. The hour hand of a clock moves from 1 to 7 o'clock. Through how many degrees does it move?

In Exercises 5–10, use the protractor to find the measure of each angle. Then classify the angle as acute, right, straight, or obtuse.

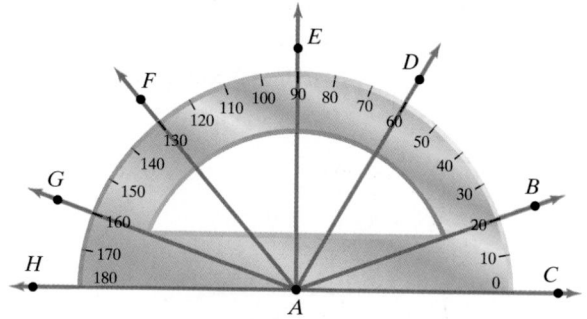

5. ∡CAB   6. ∡CAF   7. ∡HAB

8. ∡HAF   9. ∡CAH   10. ∡HAE

In Exercises 11–14, find the measure of the angle in which ?° appears.

11.
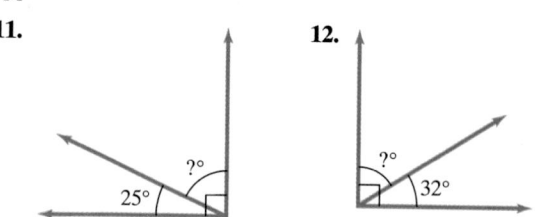

12.

13.
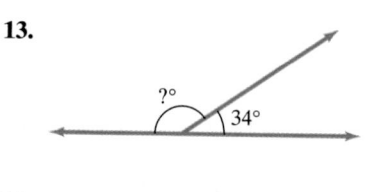

14.

In Exercises 15–20, find the measure of the complement and the supplement of each angle.

15. 48°   16. 52°   17. 89°

18. 1°   19. 37.4°   20. $15\frac{1}{3}°$

In Exercises 21–24, use an algebraic equation to find the measures of the two angles described. Begin by letting x represent the degree measure of the angle's complement or its supplement.

21. The measure of the angle is 12° greater than its complement.

22. The measure of the angle is 56° greater than its complement.

23. The measure of the angle is three times greater than its supplement.

24. The measure of the angle is 81° more than twice that of its supplement.

In Exercises 25–28, find the measures of angles 1, 2, and 3.

25.
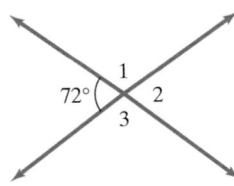

26.
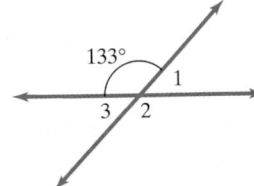

27.
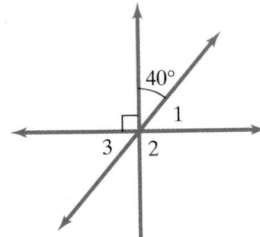

28.
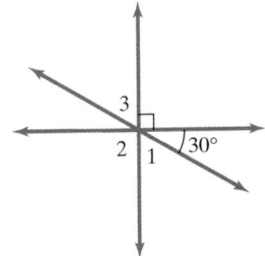

The figures for Exercises 29–30 show two parallel lines intersected by a transversal. One of the angle measures is given. Find the measure of each of the other seven angles.

29.
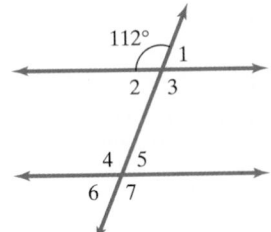

30.
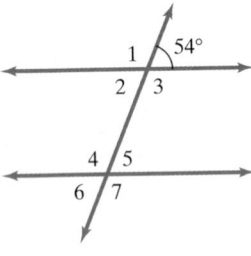

The figures for Exercises 31–32 show two parallel lines intersected by more than one transversal. Two of the angle measures are given. Find the measures of angles 1, 2, and 3.

31.
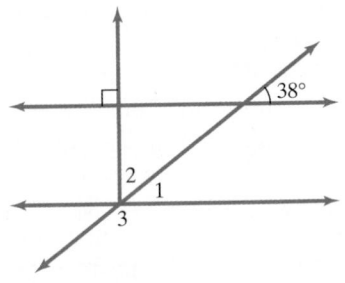

32.
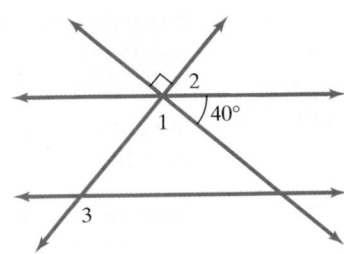

Application Exercises

33. Look around the room you are in. Describe a feature that approximates
  a. a point.   **b.** a line segment.   **c.** a plane.
  d. perpendicular lines.   **e.** parallel lines.

34. Describe a feature of the room you are in where an angle can be found. Classify the angle by its size.

Writing in Mathematics

35. Describe the differences among lines, half lines, rays, and line segments.

36. What is an angle and what determines its size?

37. Describe each type of angle: acute, right, obtuse, and straight.

38. What are complementary angles? Describe how to find the measure of an angle's complement.

39. What are supplementary angles? Describe how to find the measure of an angle's supplement.

40. Describe the difference between perpendicular and parallel lines.

41. If two parallel lines are intersected by a transversal, describe the location of the alternate interior angles, the alternate exterior angles, and the corresponding angles.

42. Describe everyday objects that approximate points, lines, and planes.

43. If a transversal is perpendicular to one of two parallel lines, must it be perpendicular to the other parallel line as well? Explain your answer.

Critical Thinking Exercises

44. Use the figure shown to select a pair of complementary angles from the following choices:

  a. ∡1 and ∡4   **b.** ∡3 and ∡6
  c. ∡2 and ∡5   **d.** ∡1 and ∡5.

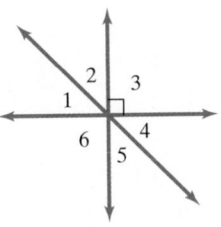

45. If $m\angle AGB = m\angle BGC$, and $m\angle CGD = m\angle DGE$, find $m\angle BGD$.

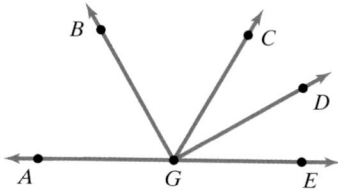

Group Exercises

46. Group members should find pictures of interesting shapes in architecture. Bring in a picture of the building that you find interesting so that you can share it with other group members. In each case, the group should discuss what they observe about the angles that are formed in the geometric shape of the building.

47. Have you ever noticed that we use the vocabulary of angles in everyday speech? Here is an example:

My opinion about the metric system took a 180° turn.

Group members should explain what this means. Then members should give examples of the vocabulary of angles in everyday use. Describe what the everyday meaning has to do with angles and their measures. Can the group give examples in which the vocabulary of angles takes on a different meaning in everyday speech?

SECTION 10.2   TRIANGLES

OBJECTIVES

1. Solve problems involving angle relationships in triangles.

2. Solve problems involving similar triangles.

3. Solve problems using the Pythagorean Theorem.

Jasper Johns "*0 Through 9*," 1961. Oil on canvas, 137 × 105 cm. The Saatchi Collection, courtesy of the Leo Castelli Gallery. (c) Jasper Johns/ Licensed by VAGA, New York, NY

In Chapter 1, we defined deductive reasoning as the process of proving a specific conclusion from one or more general statements. A conclusion that is proved to be true through deductive reasoning is called a **theorem**. The Greek mathematician Euclid, who lived more than 2000 years ago, used deductive reasoning. In his 13-volume book, the *Elements*, Euclid proved over 465 theorems about geometric figures. Euclid's work established deductive reasoning as a fundamental tool of mathematics.

A **triangle** is a geometric figure that has three sides, all of which lie on a flat surface or plane. If you start at any point along the triangle and trace along the entire figure exactly once, you will end at the same point at which you started. Because the beginning point and ending point are the same, the triangle is called a **closed** geometric figure. Euclid used parallel lines to prove one of the most important properties of triangles: The sum of the measures of the three angles of any triangle is 180°. Here is how he did it. He began with the following general statement:

EUCLID'S ASSUMPTION ABOUT PARALLEL LINES

Given a line and a point not on the line, one and only one line can be drawn through the given point parallel to the given line.

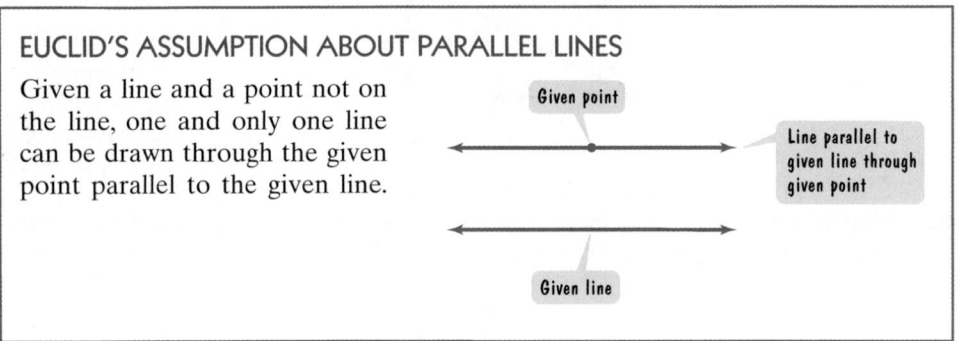

In Figure 10.14, triangle *ABC* represents any triangle. Using the general assumption given above, we draw a line through point *B* parallel to line *AC*.

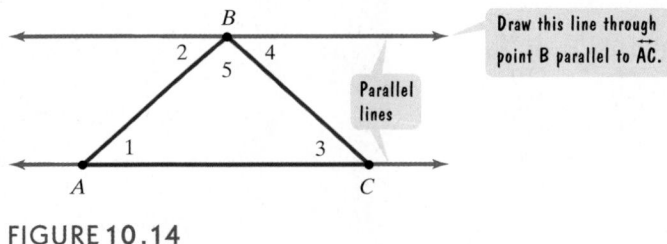

FIGURE 10.14

Because the lines are parallel, alternate interior angles have the same measure.

$$m\angle 1 = m\angle 2 \quad \text{and} \quad m\angle 3 = m\angle 4$$

Also observe that angles 2, 5, and 4 form a straight angle.

$$m\angle 2 + m\angle 5 + m\angle 4 = 180°$$

Because $m\angle 1 = m\angle 2$, replace $m\angle 2$ with $m\angle 1$. Because $m\angle 3 = m\angle 4$, replace $m\angle 4$ with $m\angle 3$.

$$m\angle 1 + m\angle 5 + m\angle 3 = 180°$$

Because $\angle 1$, $\angle 5$, and $\angle 3$ are the three angles of the triangle, this last equation shows that the measures of the triangle's three angles have a 180° sum.

THE ANGLES OF A TRIANGLE

The sum of the measures of the three angles of any triangle is 180°.

Solve problems involving angle relationships in triangles.

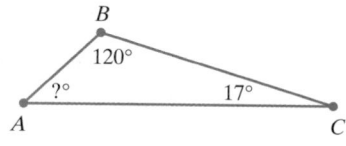

FIGURE 10.15

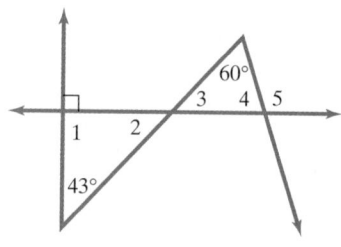

FIGURE 10.16

EXAMPLE 1   USING ANGLE RELATIONSHIPS IN TRIANGLES

Find the measure of angle A for triangle ABC in Figure 10.15.

Solution   Because $m\angle A + m\angle B + m\angle C = 180°$, we obtain

$$m\angle A + 120° + 17° = 180°.$$ The sum of the measures of a triangle's three angles is 180°.

$$m\angle A + 137° = 180°$$ Simplify: $120° + 17° = 137°$.

$$m\angle A = 180° - 137°$$ Find the measure of A by subtracting 137° from both sides of the equation.

$$m\angle A = 43°$$ Simplify.

Check 1 Point   In Figure 10.15, suppose that $m\angle B = 116°$ and $m\angle C = 15°$. Find $m\angle A$.

EXAMPLE 2   USING ANGLE RELATIONSHIPS IN TRIANGLES

Find the measures of angles 1 through 5 in Figure 10.16.

Solution   Because $\angle 1$ is supplementary to the right angle, $m\angle 1 = 90°$.
$m\angle 2$ can be found using the fact that the sum of the measures of the angles of a triangle is 180°.

$$m\angle 1 + m\angle 2 + 43° = 180°$$ The sum of the measures of a triangle's three angles is 180°.

$$90° + m\angle 2 + 43° = 180°$$ We previously found that $m\angle 1 = 90°$.

$$m\angle 2 + 133° = 180°$$ Simplify: $90° + 43° = 133°$.

$$m\angle 2 = 180° - 133°$$ Subtract 133° from both sides.

$$m\angle 2 = 47°$$ Simplify.

$m\angle 3$ can be found using the fact that vertical angles have equal measures. $m\angle 3 = m\angle 2$. Thus, $m\angle 3 = 47°$.
$m\angle 4$ can be found using the fact that the sum of the measures of the angles of a triangle is 180°. Refer to the triangle at the top of Figure 10.16.

$$m\angle 3 + m\angle 4 + 60° = 180°$$ The sum of the measures of a triangle's three angles is 180°.

$$47° + m\angle 4 + 60° = 180°$$ We previously found that $m\angle 3 = 47°$.

$$m\angle 4 + 107° = 180°$$ Simplify: $47° + 60° = 107°$.

$$m\angle 4 = 180° - 107°$$ Subtract 107° from both sides.

$$m\angle 4 = 73°$$ Simplify.

Finally, we can find $m\angle 5$ by observing that angles 4 and 5 form a straight angle.

$$m\angle 4 + m\angle 5 = 180°$$ A straight angle measures 180°.

$$73° + m\angle 5 = 180°$$ We previously found that $m\angle 4 = 73°$.

$$m\angle 5 = 180° - 73°$$ Subtract 73° from both sides.

$$m\angle 5 = 107°$$ Simplify.

Check 2 Point   In Figure 10.16, suppose that the angle shown to measure 43° measures, instead, 36°. Further suppose that the angle shown to measure 60° measures, instead, 58°. Under these new conditions, find the measures of angles 1 through 5 in the figure.

Triangles can be described using characteristics of their angles or their sides.

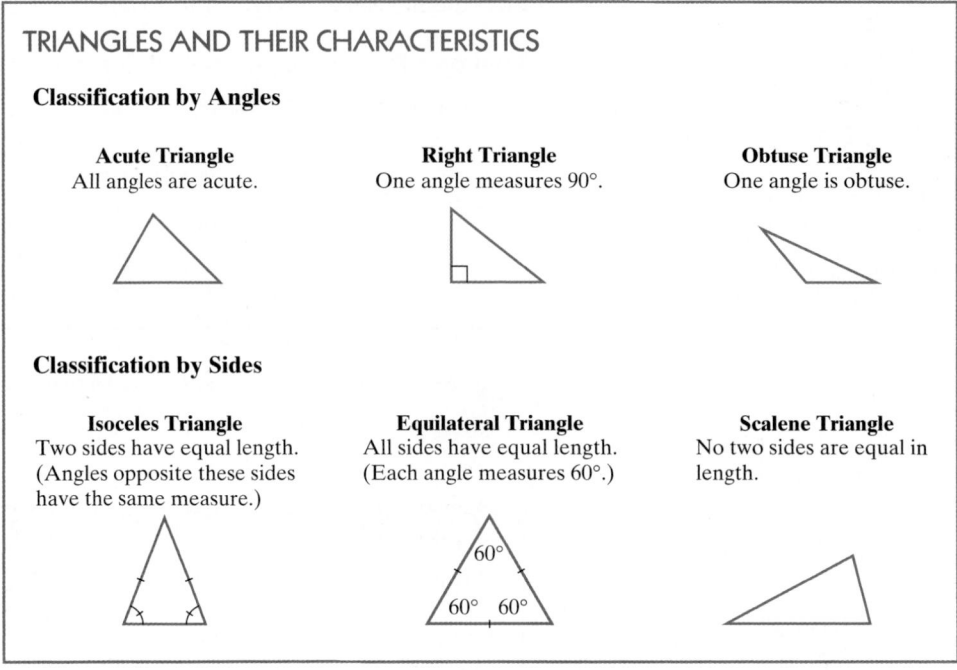

TRIANGLES AND THEIR CHARACTERISTICS

Classification by Angles

Acute Triangle
All angles are acute.

Right Triangle
One angle measures 90°.

Obtuse Triangle
One angle is obtuse.

Classification by Sides

Isoceles Triangle
Two sides have equal length.
(Angles opposite these sides
have the same measure.)

Equilateral Triangle
All sides have equal length.
(Each angle measures 60°.)

Scalene Triangle
No two sides are equal in
length.

 Solve problems involving similar triangles.

Pedestrian crossing

Similar Triangles

Shown in the margin is an international road sign. This sign is shaped just like the actual sign, although its size is smaller. Figures that have the same shape, but not the same size, are used in **scale drawings**. A scale drawing always pictures the exact shape of the object that the drawing represents. Architects, engineers, landscape gardeners, and interior decorators use scale drawings in planning their work.

Figures that have the same shape, but not necessarily the same size, are called **similar figures**. In Figure 10.17, triangles ABC and DEF are similar. Angles A and D measure the same number of degrees and are called **corresponding angles**. Angles C and F are corresponding angles, as are angles B and E. Angles with the same number of tick marks in Figure 10.17 are the corresponding angles.

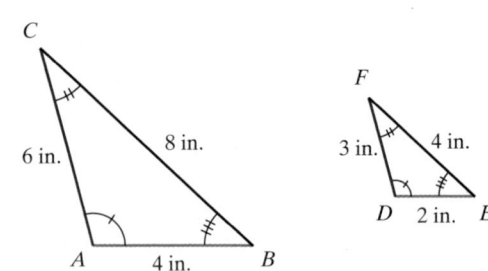

FIGURE **10.17**

The sides opposite the corresponding angles are called **corresponding sides**. Thus, $\overline{CB}$ and $\overline{FE}$ are corresponding sides. $\overline{AB}$ and $\overline{DE}$ are also corresponding sides, as are $\overline{AC}$ and $\overline{DF}$. Corresponding angles measure the same number of degrees, but corresponding sides may or may not be the same length. For the triangles in Figure 10.17, each side in the smaller triangle is half the length of the corresponding side in the larger triangle.

Triangles in Construction Work

The triangle is used in most types of construction work, including bridges, radio towers, airplane wings, and pitched roofs. Its use in construction gives an object the quality of firmness, or stiffness, resulting in rigid, strong structures.

The triangles in Figure 10.17 illustrate what it means to be **similar triangles**. **Corresponding angles have the same measure and the ratios of the lengths of the corresponding sides are equal**.

$$\frac{\text{length of } \overline{AC}}{\text{length of } \overline{DF}} = \frac{6 \text{ in.}}{3 \text{ in.}} = \frac{2}{1}; \quad \frac{\text{length of } \overline{CB}}{\text{length of } \overline{FE}} = \frac{8 \text{ in.}}{4 \text{ in.}} = \frac{2}{1}; \quad \frac{\text{length of } \overline{AB}}{\text{length of } \overline{DE}} = \frac{4 \text{ in.}}{2 \text{ in.}} = \frac{2}{1}$$

In similar triangles, the lengths of the corresponding sides are proportional. Thus,

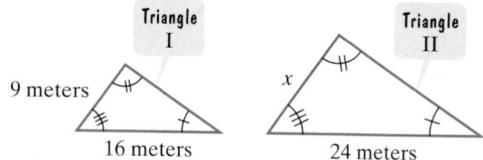

AC represents the length of $\overline{AC}$, DF the length of $\overline{DF}$, and so on.

$$\frac{AC}{DF} = \frac{CB}{FE} = \frac{AB}{DE}.$$

If we know that two triangles are similar, we can set up a proportion to solve for the length of an unknown side.

EXAMPLE 3 │ USING SIMILAR TRIANGLES

Triangles I and II in Figure 10.18 are similar. Find the missing length, x.

FIGURE 10.18

Solution Because the triangles are similar, their corresponding sides are proportional.

Left side of △ I

Bottom side of △ I

$$\frac{9}{x} = \frac{16}{24}$$

Corresponding side on left of △ II

Corresponding side on bottom of △ II

We solve this equation for x by applying the cross-products principle for proportions that we discussed in Section 6.4.

$$16x = 9 \cdot 24 \qquad \text{Apply the cross-products principle.}$$
$$16x = 216 \qquad \text{Multiply: } 9 \cdot 24 = 216.$$
$$\frac{16x}{16} = \frac{216}{16} \qquad \text{Divide both sides by 16.}$$
$$x = 13.5 \qquad \text{Simplify.}$$

The missing length, x, is 13.5 meters.

✓ **Check Point 3** The similar triangles in the figure are positioned so that corresponding parts are in the same position. Find the missing length, x.

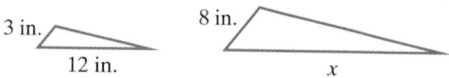

How can we quickly determine whether two triangles are similar? **If the measures of two angles of one triangle are equal to those of two angles of a second triangle, then the two triangles are similar.** If the triangles are similar, then their corresponding sides are proportional.

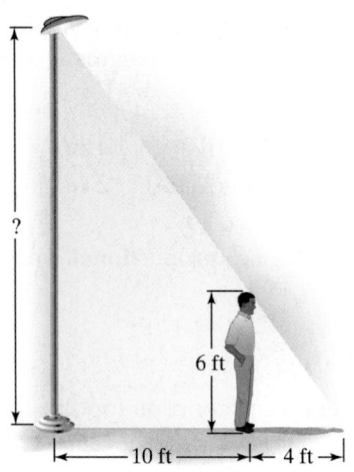

FIGURE 10.19

EXAMPLE 4 | PROBLEM SOLVING USING SIMILAR TRIANGLES

A man who is 6 feet tall is standing 10 feet from the base of a lamppost (see Figure 10.19). The man's shadow has a length of 4 feet. How tall is the lamppost?

Solution The drawing in Figure 10.20 makes the similarity of the triangles easier to see. The large triangle with the lamppost on the left and the small triangle with the man on the left both contain 90° angles. They also share an angle. Thus, two angles of the large triangle are equal in measure to two angles of the small triangle. This means that the triangles are similar and their corresponding sides are proportional. We begin by letting x represent the height of the lamppost, in feet. Because corresponding sides of the two similar triangles are proportional,

Left side of big △ → $\frac{x}{6}$ ← Corresponding side on left of small △

Bottom side of big △ → $= \frac{14}{4}$. ← Corresponding side on bottom of small △

We solve for x by applying the cross-products principle.

$$4x = 6 \cdot 14 \quad \text{Apply the cross-products principle.}$$

$$4x = 84 \quad \text{Multiply: } 6 \cdot 14 = 84.$$

$$\frac{4x}{4} = \frac{84}{4} \quad \text{Divide both sides by 4.}$$

$$x = 21 \quad \text{Simplify.}$$

The lamppost is 21 feet tall.

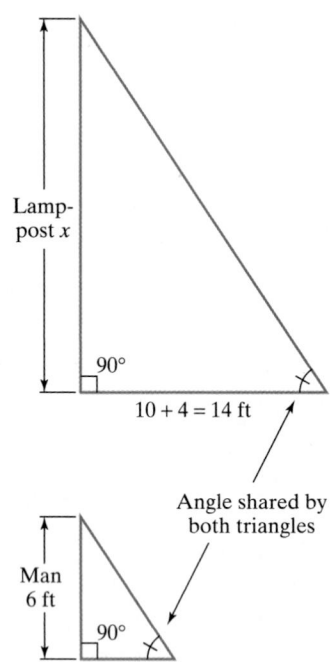

FIGURE 10.20

Check 4 Point Find the height of the lookout tower using the figure that lines up the top of the tower with the top of a stick that is 2 yards long.

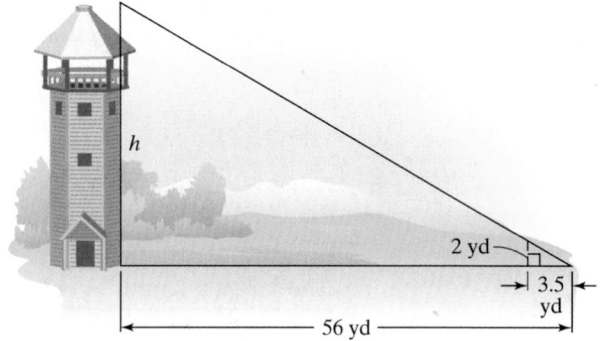

---

3 Solve problems using the Pythagorean Theorem.

The Pythagorean Theorem

The ancient Greek philosopher and mathematician Pythagoras (approximately 582–500 B.C.) founded a school whose motto was "All is number." Pythagoras is best remembered for his work with the **right triangle**, a triangle with one angle measuring 90°. The side opposite the 90° angle is called the **hypotenuse**. The other sides are called **legs**. Pythagoras found that if he constructed squares on each of the legs, as well as a larger square on the hypotenuse, the sum of the areas of the smaller squares is equal to the area of the larger square. This is illustrated in Figure 10.21.

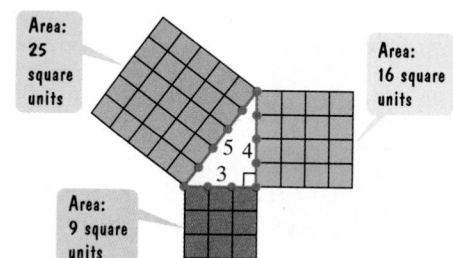

FIGURE 10.21 The area of the large square equals the sum of the areas of the smaller squares.

This relationship is usually stated in terms of the lengths of the three sides of a right triangle and is called the **Pythagorean Theorem**.

THE PYTHAGOREAN THEOREM

The sum of the squares of the lengths of the legs of a right triangle equals the square of the length of the hypotenuse.

If the legs have lengths a and b and the hypotenuse has length c, then

$$a^2 + b^2 = c^2.$$

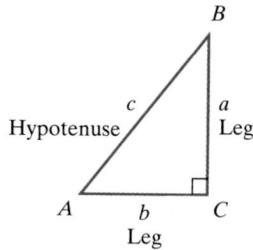

EXAMPLE 5   USING THE PYTHAGOREAN THEOREM

Find the length of the hypotenuse c in the right triangle shown in Figure 10.22.

Solution   Let $a = 9$ and $b = 12$. Substituting these values into $c^2 = a^2 + b^2$ enables us to solve for c.

$c^2 = a^2 + b^2$ Use the symbolic statement of the Pythagorean Theorem.

$c^2 = 9^2 + 12^2$ Let $a = 9$ and $b = 12$.

$c^2 = 81 + 144$ $9^2 = 9 \cdot 9 = 81$ and $12^2 = 12 \cdot 12 = 144$.

$c^2 = 225$ Add.

$c = \sqrt{225} = 15$ Solve for c by taking the positive square root of 225.

The length of the hypotenuse is 15 feet.

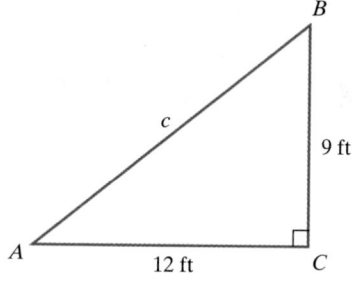

FIGURE 10.22

Check **5** Point   Find the length of the hypotenuse in a right triangle whose legs have lengths 7 feet and 24 feet.

FIGURE 10.23 A right triangle is formed by the television's height, width, and diagonal.

EXAMPLE 6 | USING THE PYTHAGOREAN THEOREM

In a 25-inch television set, the length of the screen's diagonal is 25 inches. If the screen's height is 15 inches, what is its width?

Solution Figure 10.23 shows a right triangle that is formed by the height, width, and diagonal. We can find w, the screen's width, using the Pythagorean Theorem.

(Leg)²	plus	(Leg)²	equals	(Hypotenuse)²

$$w^2 \quad + \quad 15^2 \quad = \quad 25^2$$

This is the equation resulting from the Pythagorean Theorem.

$$w^2 + 225 = 625$$
Square 15 and 25.

$$w^2 = 400$$
Subtract 225 from both sides.

$$w = \sqrt{400}$$
Take the square root of 400.

$$w = 20$$
Simplify.

The screen's width is 20 inches.

Check Point 6 What is the width of a 15-inch television set whose height is 9 inches?

BLITZER BONUS

The Universality of Mathematics

Is mathematics discovered or invented? The "Pythagorean" Theorem, credited to Pythagoras, was known in China, from land surveying, and in Egypt, from pyramid building, centuries before Pythagoras was born. The same mathematics is often discovered/invented by independent researchers separated by time, place, and culture.

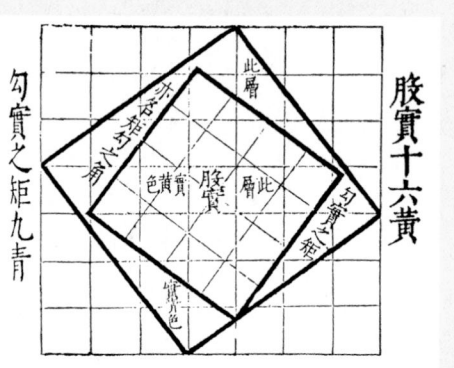

This diagram of the Pythagorean Theorem is from a Chinese manuscript dated as early as 1200 B.C.

Exercise Set 10.2

Practice Exercises

In Exercises 1–4, find the measure of angle A for the triangle shown.

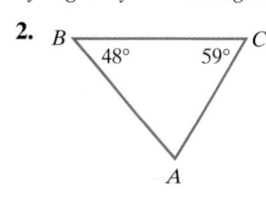

1.

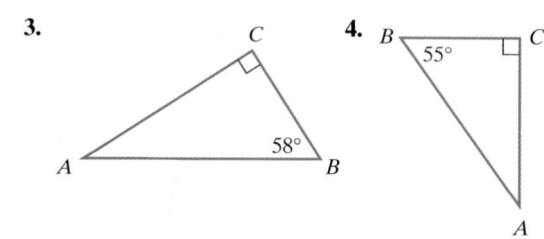

2. **3.** **4.**

In Exercises 5–6, find the measures of angles 1 through 5 in the figure shown.

5.

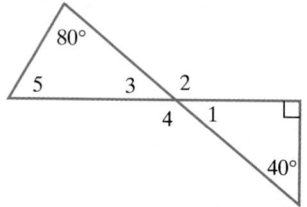

6.

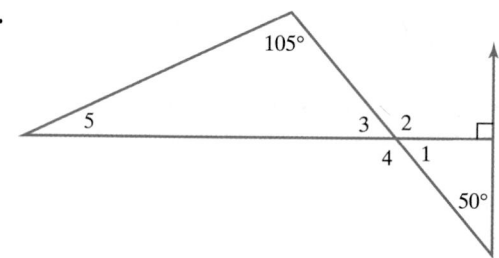

We have seen that isosceles triangles have two sides of equal length. The angles opposite these sides have the same measure. In Exercises 7–8, use this information to help find the measure of each numbered angle.

7.

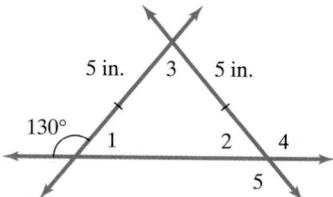

8.

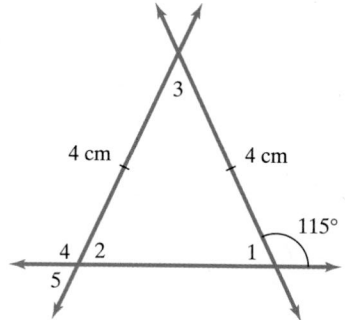

In Exercises 9–10, lines l and m are parallel. Find the measure of each numbered angle.

9.

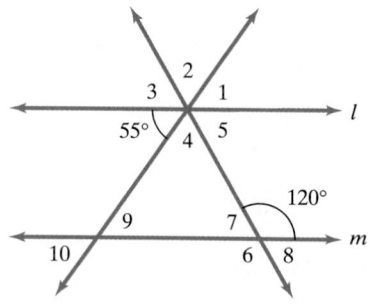

10.

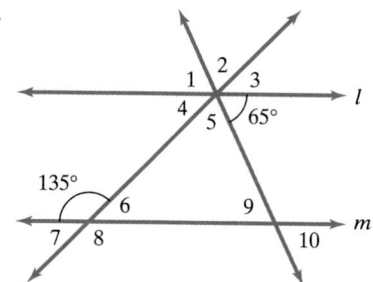

In Exercises 11–16, use similar triangles and the fact that corresponding sides are proportional to find the missing length, x.

11.

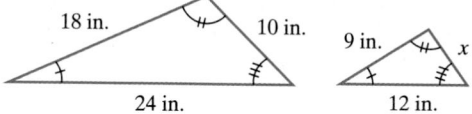

12.

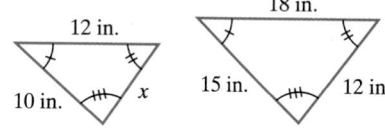

13.

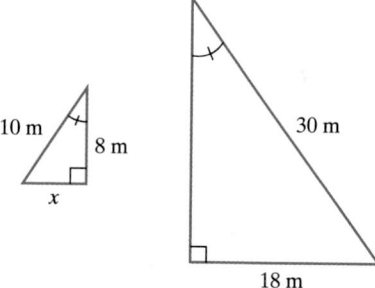

14.

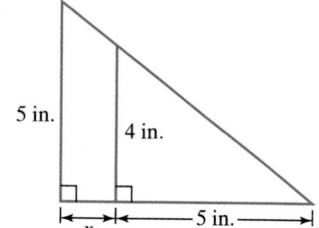

15.

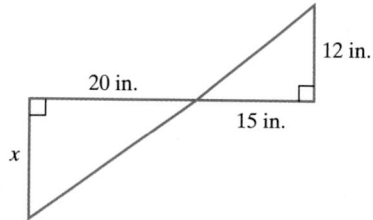

16.

In Exercises 17–19, △ABC and △ADE are similar. Find the length of the indicated side.

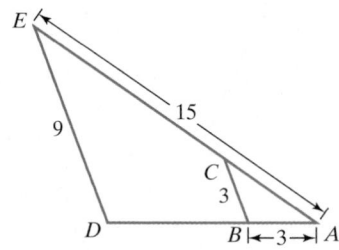

17. $\overline{CA}$ **18.** $\overline{DB}$ **19.** $\overline{DA}$

20. In the diagram for Exercises 17–19, suppose that you are not told that △ABC and △ADE are similar. Instead, you are given that $\overleftrightarrow{ED}$ and $\overleftrightarrow{CB}$ are parallel. Under these conditions, explain why the triangles must be similar.

In Exercises 21–26, use the Pythagorean Theorem to find the missing length in each right triangle. Use your calculator to find square roots, rounding, if necessary, to the nearest tenth.

21.
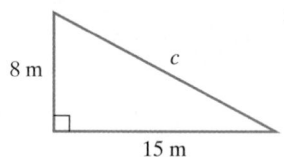

22.
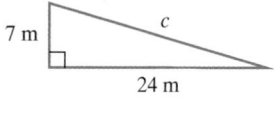

23.
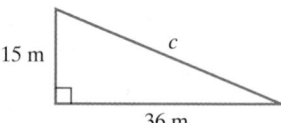

24.
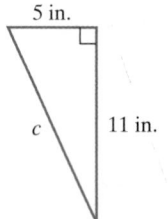

25.
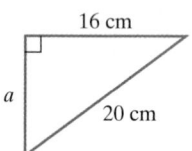

26.
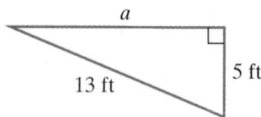

Application Exercises

27. A surveyor measured a triangular plot of ground, measuring the three angles as 46.1°, 58.3°, and 75.9°. Are these measurements correct? If not, how much of an error is there?

Use similar triangles to solve Exercises 28–29.

28. A tree casts a shadow 12 feet long. At the same time, a vertical rod 8 feet high casts a shadow that is 6 feet long. How tall is the tree?

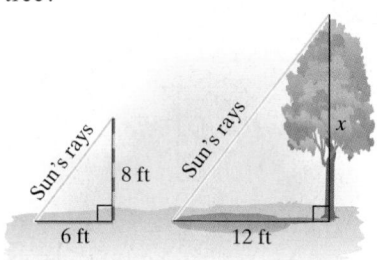

29. A person who is 5 feet tall is standing 80 feet from the base of a tree, and the tree casts an 86-foot shadow. The person's shadow is 6 feet in length. What is the tree's height?

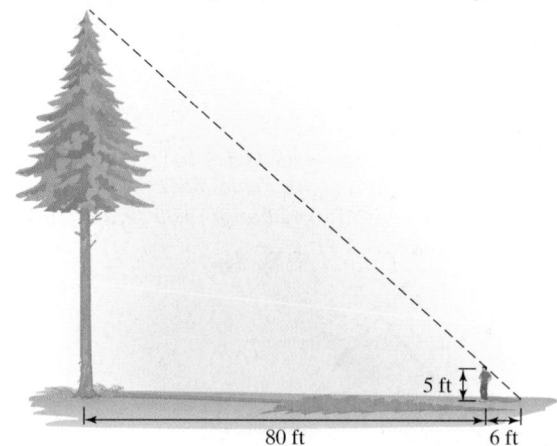

Use the Pythagorean Theorem to solve Exercises 30–35. Use your calculator to find square roots, rounding, if necessary, to the nearest tenth.

30. Find the length of the ladder.

31. How high is the airplane above the ground?

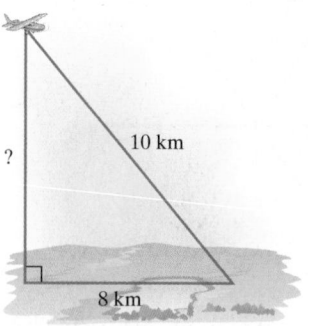

32. A baseball diamond is actually a square with 90-foot sides. What is the distance from home plate to second base?

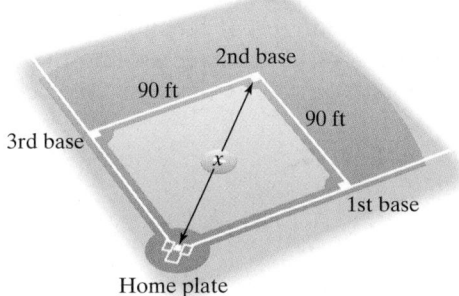

33. A 20-foot ladder is 15 feet from the house. How far up the house does the ladder reach?

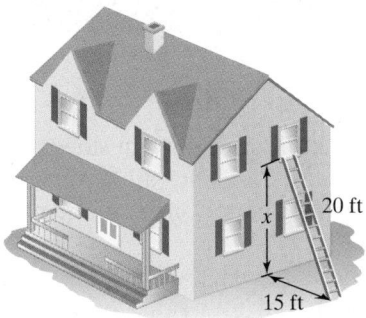

34. An 8-foot tree is to be supported by two wires that extend from the top of the tree to a point on the ground located 15 feet from the base of the tree. Find the total length of the two support wires.

35. If construction costs are $150,000 per kilometer, find the cost of building the new road in the figure shown.

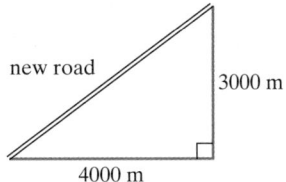

Writing in Mathematics

36. If the measures of two angles of a triangle are known, explain how to find the measure of the third angle.

37. Can a triangle contain two right angles? Explain your answer.

38. What general assumption did Euclid make about a point and a line in order to prove that the sum of the measures of the angles of a triangle is 180°?

39. What are similar triangles?

40. If the ratio of the corresponding sides of two similar triangles is 1 to 1 $\left(\frac{1}{1}\right)$, what must be true about the triangles?

41. What are corresponding angles in similar triangles?

42. Describe how to identify the corresponding sides in similar triangles.

43. In your own words, state the Pythagorean Theorem.

44. In the 1939 movie *The Wizard of Oz*, upon being presented with a Th.D. (Doctor of Thinkology), the Scarecrow proudly exclaims, "The sum of the square roots of any two sides of an isosceles triangle is equal to the square root of the remaining side." Did the Scarecrow get the Pythagorean Theorem right? In particular, describe four errors in the Scarecrow's statement.

Critical Thinking Exercises

45. Find the measure of angle R.

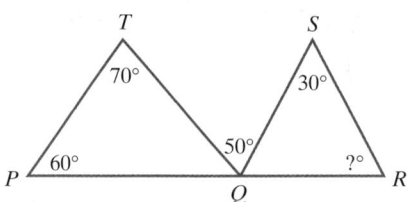

46. What is the length of $\overline{AB}$ in the accompanying figure?

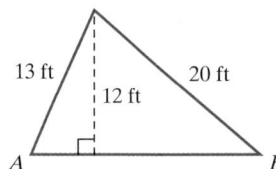

47. A builder is adding a 16-foot by 20-foot rectangular room to a house. Once the floor is constructed, it is found that the length of its diagonal is $24\frac{1}{4}$ feet. Is the floor squared off properly so that once the walls are constructed they will form a right angle?

Group Exercises

48. Group members should write and solve two practical problems that can be solved using similar triangles. The problems should be different from those in this section.

49. Group members should write and solve two practical problems that can be solved using the Pythagorean Theorem. As in Exercise 48, be sure that the problems are different from those in this section.

SECTION 10.3	POLYGONS, QUADRILATERALS, AND PERIMETER

OBJECTIVES

1. Name certain polygons according to the number of sides.

2. Recognize the characteristics of certain quadrilaterals.

3. Solve problems involving a polygon's perimeter.

4. Find the sum of the measures of a polygon's angles.

You have just purchased a beautiful plot of land in the country, shown in Figure 10.24. In order to have more privacy, you decide to put fencing along each of its four sides. The cost of this project depends on the distance around the four outside edges of the plot, called its **perimeter**, as well as the cost for each foot of fencing.

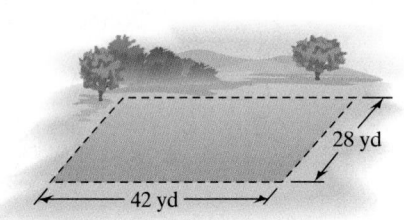

FIGURE 10.24

 Name certain polygons according to the number of sides.

Your plot of land is a geometric figure: It has four straight sides that are line segments. The plot is on level ground, so that the four sides all lie on a flat surface, or plane. The plot is an example of a *polygon*. Any closed shape in the plane formed by three or more line segments is a **polygon**.

A polygon is named according to the number of sides it has. We know that a three-sided polygon is called a **triangle**. A four-sided polygon is called a **quadrilateral**.

A polygon whose sides are all the same length and whose angles all have the same measure is called a **regular polygon**. Table 10.2 provides the names of six polygons. Also shown are illustrations of regular polygons.

TABLE 10.2 ILLUSTRATIONS OF REGULAR POLYGONS

Name	Picture
Triangle 3 sides	
Quadrilateral 4 sides	
Pentagon 5 sides	

Name	Picture
Hexagon 6 sides	
Heptagon 7 sides	
Octagon 8 sides	

 Recognize the characteristics of certain quadrilaterals.

Quadrilaterals

The plot of land in Figure 10.24 is a four-sided polygon, or a quadrilateral. However, when you first looked at the figure, perhaps you thought of the plot as a rectangular field. A **rectangle** is a special kind of quadrilateral in which both pairs of opposite sides are parallel, have the same measure, and whose angles are right angles. Table 10.3 presents some special quadrilaterals and their characteristics.

BLITZER BONUS

Bees and Regular Hexagons

Bees use honeycombs to store honey and house larvae. They construct honey storage cells from wax. Each cell has the shape of a regular hexagon. The cells fit together perfectly, preventing dirt or predators from entering. Squares or equilateral triangles would fit equally well, but regular hexagons provide the largest storage space for the amount of wax used.

TABLE 10.3 TYPES OF QUADRILATERALS

Name	Characteristics	Representation
Parallelogram	Quadrilateral in which both pairs of opposite sides are parallel and have the same measure. Opposite angles have the same measure.	
Rhombus	Parallelogram with all sides having equal length.	
Rectangle	Parallelogram with four right angles. Because a rectangle is a parallelogram, opposite sides are parallel and have the same measure.	
Square	A rectangle with all sides having equal length. Each angle measures 90°, and the square is a regular quadrilateral.	
Trapezoid	A quadrilateral with exactly one pair of parallel sides.	

Perimeter

The **perimeter**, P, of a polygon is the sum of the lengths of its sides. Perimeter is measured in linear units, such as inches, feet, yards, meters, or kilometers.

Example 1 involves the perimeter of a rectangle. Because perimeter is the sum of the lengths of the sides, the perimeter of the rectangle shown in Figure 10.25 is $l + w + l + w$. This can be expressed as

$$P = 2l + 2w.$$

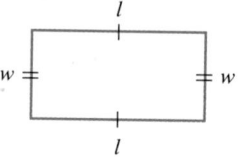

FIGURE 10.25 A rectangle with length l and width w

3 Solve problems involving a polygon's perimeter.

EXAMPLE 1 · AN APPLICATION OF PERIMETER

The rectangular field we discussed at the beginning of this section (see Figure 10.24 on page 510) has a length of 42 yards and a width of 28 yards. If fencing costs $5.25 per foot, find the cost to enclose the field with fencing.

Solution We begin by finding the perimeter of the rectangle in yards. Using 3 ft = 1 yd and dimensional analysis, we express the perimeter in feet. Finally, we multiply the perimeter, in feet, by $5.25 because the fencing costs $5.25 per foot.

The length, l, is 42 yards and the width, w, is 28 yards. The perimeter of the rectangle is determined using the formula $P = 2l + 2w$.

$$P = 2l + 2w = 2 \cdot 42 \text{ yd} + 2 \cdot 28 \text{ yd} = 84 \text{ yd} + 56 \text{ yd} = 140 \text{ yd}$$

Because 3 ft = 1 yd, we use the unit fraction $\frac{3 \text{ ft}}{1 \text{ yd}}$ to convert from yards to feet.

$$140 \text{ yd} = \frac{140 \text{ yd}}{1} \cdot \frac{3 \text{ ft}}{1 \text{ yd}} = 140 \cdot 3 \text{ ft} = 420 \text{ ft}$$

BLITZER BONUS

Decorating with Geometry

When your bathroom wall is tiled, tiles are fitted together to cover the wall without leaving any gaps, so that water cannot get through. Tiling the plane means covering the entire plane without leaving any space empty or overlapping any tiles. Shown here is a tiling using regular polygons. If you look at a soccer ball, you will see that the entire surface is filled with regular pentagons and regular hexagons. Designs formed by tiling the plane have been featured in art and architecture for centuries.

 Find the sum of the measures of a polygon's angles.

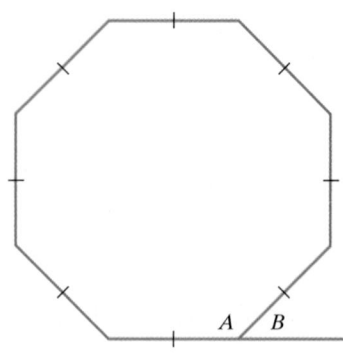

FIGURE 10.27 A regular octagon

The perimeter of the rectangle is 420 feet. Now we are ready to find the cost of the fencing. We multiply 424 feet by $5.25, the cost per foot.

$$\text{Cost} = \frac{420 \text{ feet}}{1} \cdot \frac{\$5.25}{\text{foot}} = 420(\$5.25) = \$2205$$

The cost to enclose the rectangle with fencing is $2205.

 **Check Point 1**   A rectangular field has a length of 50 yards and a width of 30 yards. If fencing costs $6.50 per foot, find the cost to enclose the rectangle with fencing.

The Sum of the Measures of a Polygon's Angles

We know that the sum of the measures of the three angles of any triangle is 180°. We can use inductive reasoning to find the sum of the measures of the angles of any polygon. Start by drawing line segments from a single point where two sides meet so that nonoverlapping triangles are formed. This is done in Figure 10.26.

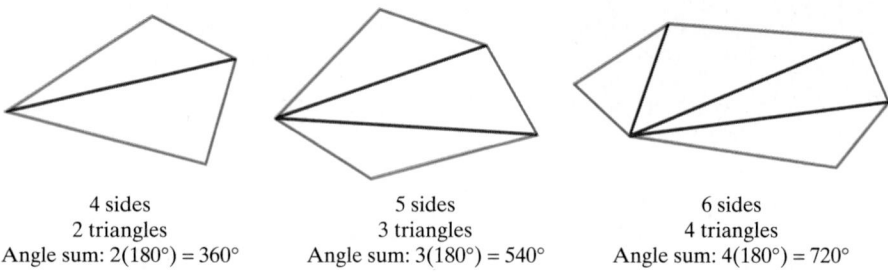

4 sides
2 triangles
Angle sum: 2(180°) = 360°

5 sides
3 triangles
Angle sum: 3(180°) = 540°

6 sides
4 triangles
Angle sum: 4(180°) = 720°

FIGURE 10.26

In each case, the number of triangles is two less than the number of sides of the polygon. Thus, for an n-sided polygon, there are $n - 2$ triangles. Because each triangle has an angle-measure sum of 180°, the sum of the measures for the angles in the $n - 2$ triangles is $(n - 2)180°$. Thus, the sum of the measures of the angles of an n-sided polygon is $(n - 2)180°$.

THE ANGLES OF A POLYGON

The sum of the measures of the angles of a polygon with n sides is

$$(n - 2)180°.$$

EXAMPLE 2 | USING THE FORMULA FOR THE ANGLES OF A POLYGON

a. Find the sum of the measures of the angles of an octagon.

b. Figure 10.27 shows a regular octagon. Find the measure of angle A.

c. Find the measure of exterior angle B.

Solution

a. An octagon has eight sides. Using the formula $(n - 2)180°$ with $n = 8$, we can find the sum of the measures of its eight angles.
The sum of the measures of an octagon's angles

$$= (n - 2)180°$$
$$= (8 - 2)180°$$
$$= 6 \cdot 180°$$
$$= 1080°.$$

b. Examine the regular octagon in Figure 10.27. Note that all eight sides have the same length. Likewise, all eight angles have the same degree measure. Angle A is one of these eight angles. We find its measure by taking the sum of the measures of all eight angles, $1080°$, and dividing by 8.

$$m\angle A = \frac{1080°}{8} = 135°$$

c. Because $\angle B$ is the supplement of $\angle A$,

$$m\angle B = 180° - 135° = 45°.$$

Check Point 2

a. Find the sum of the measures of the angles of a 12-sided polygon.

b. Find the measure of an angle of a regular 12-sided polygon.

Exercise Set 10.3

Practice Exercises

In Exercises 1–4, use the number of sides to name the polygon.

1.

2.

3.
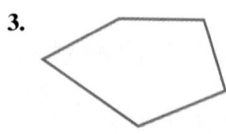

4.
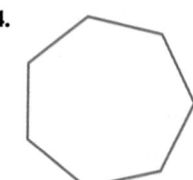

Use these quadrilaterals to solve Exercises 5–10.

a.
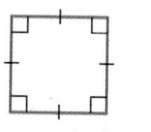

b.
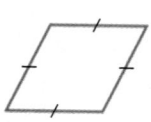

c.
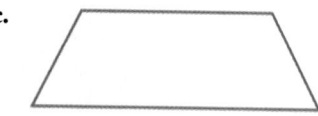

d.
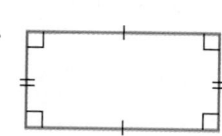

e.

5. Which of these quadrilaterals have opposite sides that are parallel? Name these quadrilaterals.

6. Which of these quadrilaterals have sides of equal length that meet at a vertex? Name these quadrilaterals.

7. Which of these quadrilaterals have right angles? Name these quadrilaterals.

8. Which of these quadrilaterals do not have four sides of equal length? Name these quadrilaterals.

9. Which of these quadrilaterals is not a parallelogram? Name this quadrilateral.

10. Which of these quadrilaterals is/are a regular polygon? Name this/these quadrilateral(s).

In Exercises 11–20, find the perimeter of the figure named and shown. Express the perimeter using the same unit of measure that appears on the given side or sides.

11. Rectangle

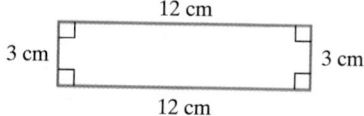

12 cm

3 cm   3 cm

12 cm

12. Parallelogram

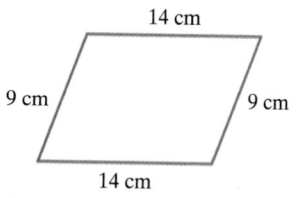

14 cm

9 cm   9 cm

14 cm

13. Rectangle

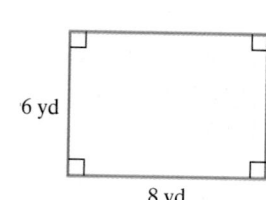

6 yd

8 yd

14. Rectangle

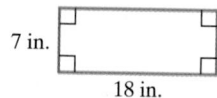

7 in.

18 in.

15. Square

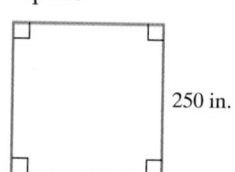

250 in.

16. Square

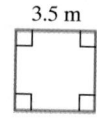

3.5 m

17. Triangle

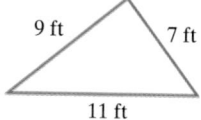

9 ft   7 ft

11 ft

18. Triangle

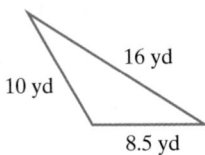

16 yd
10 yd
8.5 yd

19. Equilateral triangle

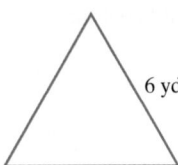

6 yd

20. Regular hexagon

4 mm

In Exercises 21–24, find the perimeter of the figure shown. Express the perimeter using the same unit of measure that appears on the given side or sides.

21.

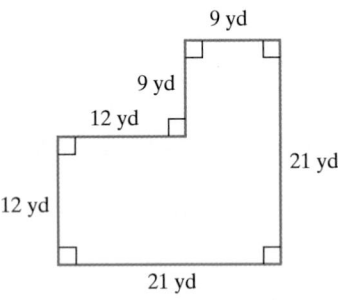

9 yd
9 yd
12 yd
21 yd
12 yd
21 yd

22.

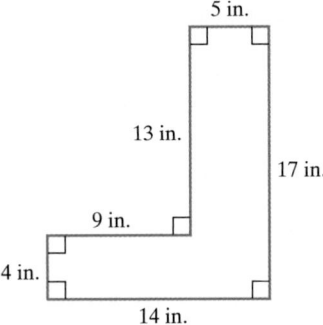

5 in.
13 in.
17 in.
9 in.
4 in.
14 in.

23.

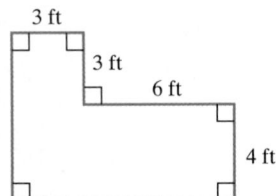

3 ft
3 ft
6 ft
4 ft

24.

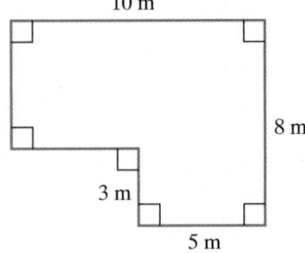

10 m
8 m
3 m
5 m

25. Find the sum of the measures of the angles of a five-sided polygon.

26. Find the sum of the measures of the angles of a six-sided polygon.

27. Find the sum of the measures of the angles of a quadrilateral.

28. Find the sum of the measures of the angles of a heptagon.

In Exercises 29–30, each figure shows a regular polygon. Find the measure of angle A and angle B.

29.

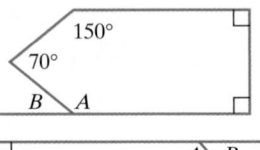

B A

30.

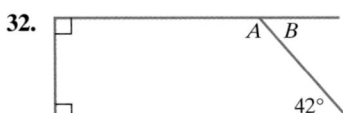

B A

In Exercises 31–32, **a.** *Find the sum of the measures of the angles for the figure given;* **b.** *Find the measure of angle A and angle B.*

31.

150°
70°
B A

32.

A B
42°

Application Exercises

For Exercises 33–40, use the road signs shown.

Stop at intersection

One-way traffic

No entry

Yield

Deer crossing

33. Which sign has a shape that is not a polygon?

34. Describe the shape of the STOP sign as specifically as possible.

35. Which signs have the shape of a regular polygon?

36. Which sign(s) has/have the shape of a polygon that is not regular?

37. Describe the shape of the YIELD sign as specifically as possible.

38. Describe the shape of the DEER CROSSING sign as specifically as possible.

39. Which sign is shaped like a polygon with a 60° angle?

40. Which sign has obtuse angles?

41. A school playground is in the shape of a rectangle 400 feet long and 200 feet wide. If fencing costs $14 per yard, what will it cost to place fencing around the playground?

42. A rectangular field is 70 feet long and 30 feet wide. If fencing costs $8 per yard, how much will it cost to enclose the field?

43. One side of a square flower bed is 8 feet long. How many plants are needed if they are to be spaced 8 inches apart around the outside of the bed?

44. What will it cost to place baseboard around the region shown if the baseboard costs $0.25 per foot? No baseboard is needed for the 2-foot doorway.

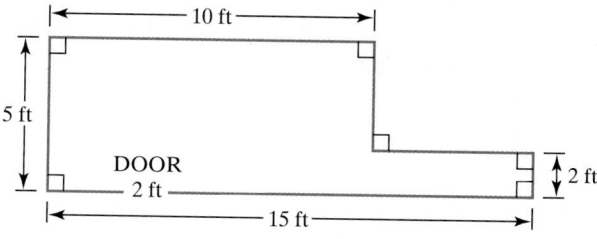

Writing in Mathematics

45. What is a polygon?

46. Explain why rectangles and rhombuses are also parallelograms.

47. Explain why every square is a rectangle, a rhombus, a parallelogram, a quadrilateral, and a polygon.

48. Explain why a square is a regular polygon, but a rhombus is not.

49. Using words only, describe how to find the perimeter of a rectangle.

50. Describe a practical situation in which you needed to apply the concept of a geometric figure's perimeter.

51. Describe how to find the measure of an angle of a regular pentagon.

Critical Thinking Exercises

In Exercises 52–53, write an algebraic expression that represents the perimeter of the figure shown.

52.

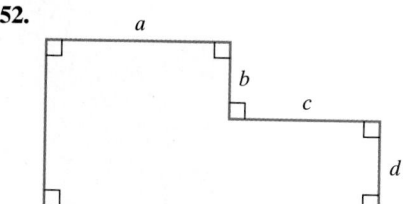

53.

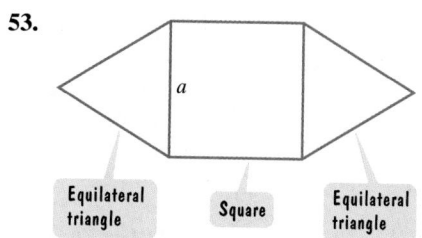

54. Find $m\angle 1$ in the figure shown.

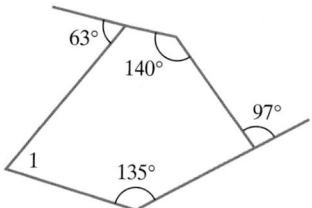

Group Exercise

55. Tiling the plane, in a mathematical sense, is very similar to tiling a floor. Tiling, or tessellating, the plane means covering the entire plane with endless repetitions of a basic design, without leaving any space empty. Present a group report on tiling the plane. Include a discussion of cultures that have used tessellations on fabrics, wall coverings, baskets, rugs, and pottery, with examples. Include the Alhambra, a fourteenth-century palace in Granada, Spain, in the presentation, as well as works by the artist M. C. Escher. Finally, be sure to discuss practical applications of tiling the plane.

SECTION 10.4   AREA AND CIRCUMFERENCE

OBJECTIVES

1. Use area formulas to compute the areas of plane regions and solve applied problems.
2. Use formulas for a circle's circumference and area.

The size of a house is described in square feet. But how do you know from the real estate ad whether the 1200-square-foot home with the backyard pool is large enough to warrant a visit? Faced with hundreds of ads, you need some way to sort out the best bets. What does 1200 square feet mean and how is this area determined? In this section, we discuss how to compute the areas of plane regions.

1 Use area formulas to compute the areas of plane regions and solve applied problems.

Square unit of measure

FIGURE 10.28 The area of the region on the left is 12 square units.

Formulas for Area

In Section 9.2, we saw that the area of a two-dimensional figure is the number of square units, such as square inches or square miles, it takes to fill the interior of the figure. For example, Figure 10.28 shows that there are 12 square units contained within the rectangular region. The area of the region is 12 square units. Notice that the area can be determined in the following manner:

Distance across   Distance down

$$4 \text{ units} \times 3 \text{ units} = 4 \times 3 \times \text{units} \times \text{units}$$
$$= 12 \text{ square units.}$$

The area of a rectangular region, usually referred to as the area of a rectangle, is the product of the distance across (length) and the distance down (width).

AREA OF A RECTANGLE AND A SQUARE

The area, A, of a rectangle with length l and width w is given by the formula

$$A = lw.$$

The area, A, of a square with one side measuring s linear units is given by the formula

$$A = s^2.$$

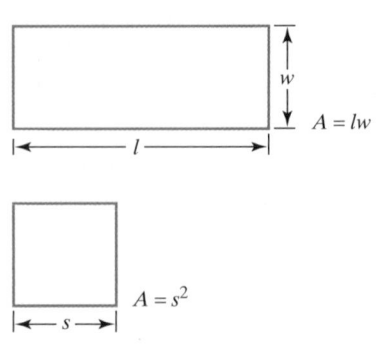

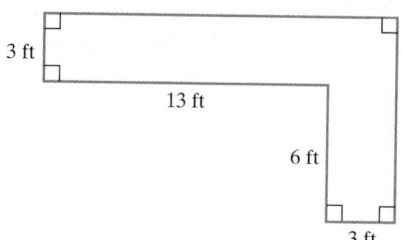

FIGURE 10.29

EXAMPLE 1 SOLVING AN AREA PROBLEM

You decide to cover the path shown in Figure 10.29 with bricks.

a. Find the area of the path.

b. If the path requires four bricks for every square foot, how many bricks are needed for the project?

Solution

a. Because we have a formula for the area of a rectangle, we begin by drawing a dashed line that divides the path into two rectangles. One way of doing this is shown on the right. We then use the length and width of

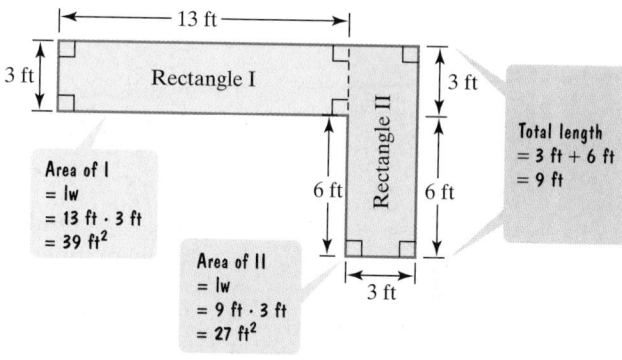

each rectangle to find its area. The computations for area are shown in the green and blue voice balloons.

The area of the path is found by adding the areas of the two rectangles.

$$\text{Area of path} = 39 \text{ ft}^2 + 27 \text{ ft}^2 = 66 \text{ ft}^2$$

b. The path requires 4 bricks per square foot. The number of bricks needed for the project is the number of square feet in the path, its area, times 4.

$$\text{Number of bricks needed} = 66 \text{ ft}^2 \cdot \frac{4 \text{ bricks}}{\text{ft}^2} = 66 \cdot 4 \text{ bricks} = 264 \text{ bricks}$$

Thus, 264 bricks are needed for the project.

 **Check 1 Point** Find the area of the path described in Example 1, rendered below as a green region, by first measuring off a large rectangle as shown. The area of the path is the area of the large rectangle (the blue and green regions combined) minus the area of the blue rectangle. Do you get the same answer as we did in Example 1(a)?

In Section 9.2, we saw that although there are 3 linear feet in a linear yard, there are 9 square feet in a square yard. If a problem requires measurement of area in square yards and the linear measures are given in feet, to avoid errors first convert feet to yards. Then apply the area formula. This idea is illustrated in Example 2.

EXAMPLE 2 SOLVING AN AREA PROBLEM

What will it cost to carpet a rectangular floor measuring 12 feet by 15 feet if the carpet costs $18.50 per square yard?

Solution   We begin by converting the linear measures from feet to yards.

$$12 \text{ ft} = \frac{12 \text{ ft}}{1} \cdot \frac{1 \text{ yd}}{3 \text{ ft}} = \frac{12}{3} \text{ yd} = 4 \text{ yd}$$

$$15 \text{ ft} = \frac{15 \text{ ft}}{1} \cdot \frac{1 \text{ yd}}{3 \text{ ft}} = \frac{15}{3} \text{ yd} = 5 \text{ yd}$$

Next, we find the area of the rectangular floor in square yards.

$$A = lw = 5 \text{ yd} \cdot 4 \text{ yd} = 20 \text{ yd}^2$$

Finally, we find the cost of the carpet by multiplying the cost per square yard, $18.50, by the number of square yards in the floor, 20.

$$\text{Cost of carpet} = \frac{\$18.50}{\text{yd}^2} \cdot \frac{20 \text{ yd}^2}{1} = \$18.50(20) = \$370$$

It will cost $370 to carpet the floor.

 **Check Point 2**   What will it cost to carpet a rectangular floor measuring 18 feet by 21 feet if the carpet costs $16 per square yard?

We can use the formula for the area of a rectangle to develop formulas for areas of other polygons. We begin with a parallelogram, a quadrilateral with opposite sides equal and parallel. The **height** of a parallelogram is the perpendicular distance between two of the parallel sides. Height is denoted by h in Figure 10.30. The **base**, denoted by b, is the length of either of these parallel sides.

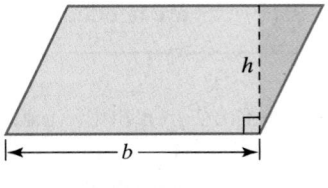

FIGURE **10.30**

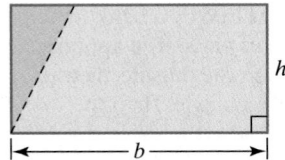

FIGURE **10.31**

In Figure 10.31, the red triangular region has been cut off from the right of the parallelogram and attached to the left. The resulting figure is a rectangle with length b and width h. Because bh is the area of the rectangle, it also represents the area of the parallelogram.

STUDY TIP

The height of a parallelogram is the perpendicular distance between two of the parallel sides. It is *not* the length of a side.

AREA OF A PARALLELOGRAM

The area, A, of a parallelogram with height h and base b is given by the formula

$$A = bh.$$

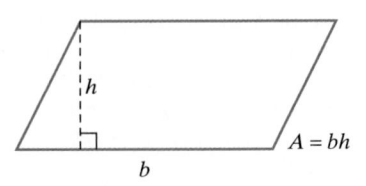

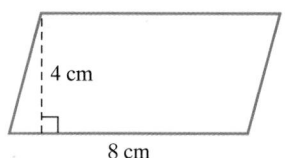

FIGURE 10.32

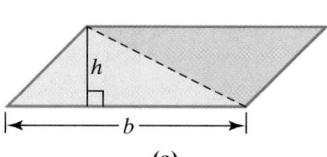

(a)

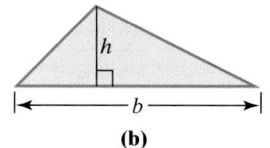

(b)

FIGURE 10.33

EXAMPLE 3 USING THE FORMULA FOR A PARALLELOGRAM'S AREA

Find the area of the parallelogram in Figure 10.32.

Solution   As shown in the figure, the base is 8 centimeters and the height is 4 centimeters. Thus, $b = 8$ and $h = 4$.

$$A = bh$$
$$A = 8 \text{ cm} \cdot 4 \text{ cm} = 32 \text{ cm}^2$$

The area is 32 cm².

Check 3 Point   Find the area of a parallelogram with a base of 10 inches and a height of 6 inches.

Figure 10.33 demonstrates how we can use the formula for the area of a parallelogram to obtain a formula for the area of a triangle. The area of the parallelogram in Figure 10.33(a) is given by $A = bh$. The diagonal shown in the parallelogram divides it into two triangles with the same size and shape. This means that the area of each triangle is one-half that of the parallelogram. Thus, the area of the triangle in Figure 10.33(b) is given by $A = \frac{1}{2}bh$.

AREA OF A TRIANGLE

The area, A, of a triangle with height h and base b is given by the formula

$$A = \frac{1}{2}bh.$$

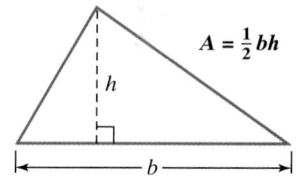

EXAMPLE 4 USING THE FORMULA FOR A TRIANGLE'S AREA

Find the area of each triangle in Figure 10.34.

Solution

a. In Figure 10.34(a), the base is 16 meters and the height is 10 meters, so $b = 16$ and $h = 10$. We do not need the 10.5 meters or the 14 meters to find the area. The area of the triangle is

$$A = \frac{1}{2}bh = \frac{1}{2} \cdot 16 \text{ m} \cdot 10 \text{ m} = 80 \text{ m}^2.$$

The area is 80 square meters.

b. In Figure 10.34(b), the base is 12 inches. The base line needs to be extended to draw the height. However, we still use 12 inches for b in the area formula. The height, h, is given to be 9 inches. The area of the triangle is

$$A = \frac{1}{2}bh = \frac{1}{2} \cdot 12 \text{ in.} \cdot 9 \text{ in.} = 54 \text{ in.}^2.$$

The area of the triangle is 54 square inches.

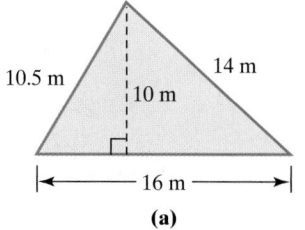

(a)

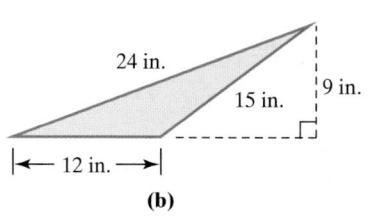

(b)

FIGURE 10.34

Check Point 4 A sailboat has a triangular sail with a base of 12 feet and a height of 5 feet. Find the area of the sail.

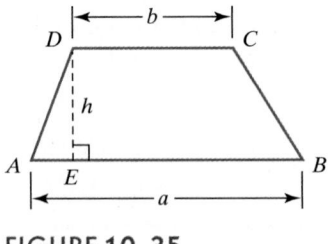

FIGURE 10.35

The formula for the area of a triangle can be used to obtain a formula for the area of a trapezoid. Consider the trapezoid shown in Figure 10.35. The lengths of the two parallel sides, called the **bases**, are represented by a (the lower base) and b (the upper base). The trapezoid's height, denoted by h, is the perpendicular distance between the two parallel sides.

In Figure 10.36, we have drawn line segment BD, dividing the trapezoid into two triangles, shown in yellow and red. The area of the trapezoid is the sum of the areas of these triangles.

Area of trapezoid	=	Area of yellow △	Plus	Area of red △

$$A = \frac{1}{2}ah + \frac{1}{2}bh$$

$$= \frac{1}{2}h(a + b) \quad \text{Factor out } \frac{1}{2}h.$$

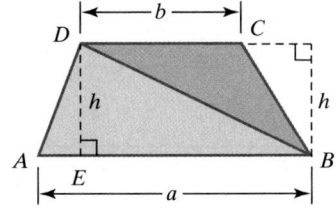

FIGURE 10.36

AREA OF A TRAPEZOID

The area, A, of a trapezoid with parallel bases a and b and height h is given by the formula

$$A = \frac{1}{2}h(a + b).$$

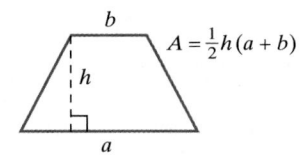

$A = \frac{1}{2}h(a + b)$

EXAMPLE 5   FINDING THE AREA OF A TRAPEZOID

Find the area of the trapezoid in Figure 10.37.

FIGURE 10.37

Solution   The height, h, is 13 feet. The lower base, a, is 46 feet, and the upper base, b, is 32 feet. We do not use the 17-foot and 19-foot sides in finding the trapezoid's area.

$$A = \frac{1}{2}h(a + b) = \frac{1}{2} \cdot 13 \text{ ft} \cdot (46 \text{ ft} + 32 \text{ ft})$$

$$= \frac{1}{2} \cdot 13 \text{ ft} \cdot 78 \text{ ft} = 507 \text{ ft}^2$$

The area of the trapezoid is 507 square feet.

Check Point 5 Find the area of a trapezoid with bases of length 20 feet and 10 feet and height 7 feet.

2   Use formulas for a circle's circumference and area.

Circles

It's a good idea to know your way around a circle. Clocks, angles, maps, and compasses are based on circles. Circles occur everywhere in nature: in ripples on water, patterns on a butterfly's wings, and cross sections of trees. Some consider the circle to be the most pleasing of all shapes.

The point at which a pebble hits a flat surface of water becomes the center of a number of circular ripples.

A **circle** is a set of points in the plane equally distant from a given point, its **center**. Figure 10.38 shows two circles. The **radius** (plural: radii), r, is a line segment from the center to any point on the circle. For a given circle, all radii have the same length. The **diameter**, d, is a line segment through the center whose endpoints both lie on the circle. For a given circle, all diameters have the same length. In any circle, the **length of the diameter is twice the length of the radius**.

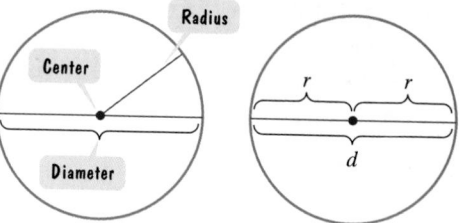

FIGURE 10.38

The words "radius" and "diameter" refer to both the line segments in Figure 10.38 as well as to their linear measures. The distance around a circle (its perimeter) is called its **circumference**, C. For all circles, if you divide the circumference by the diameter, or by twice the radius, you will get the same number. This ratio is the irrational number π and is approximately equal to 3.14:

$$\frac{C}{d} = \pi \quad \text{or} \quad \frac{C}{2r} = \pi.$$

Thus,

$$C = \pi d \quad \text{or} \quad C = 2\pi r.$$

FINDING THE DISTANCE AROUND A CIRCLE

The circumference, C, of a circle with diameter d and radius r is

$$C = \pi d \quad \text{or} \quad C = 2\pi r.$$

When computing a circle's circumference by hand, round π to 3.14. When using a calculator, use the $\boxed{\pi}$ key, which gives the value of π rounded to approximately 11 decimal places. In either case, calculations involving π give approximate answers. These answers can vary slightly depending on how π is rounded. The symbol $\approx$ (is approximately equal to) will be written in these calculations.

EXAMPLE 6 FINDING A CIRCLE'S CIRCUMFERENCE

Find the circumference of the circle in Figure 10.39.

Solution   The diameter is 40 yards, so we use the formula for circumference with d in it.

$$C = \pi d = \pi(40 \text{ yd}) = 40\pi \text{ yd} \approx 125.7 \text{ yd}$$

The distance around the circle is approximately 125.7 yards.

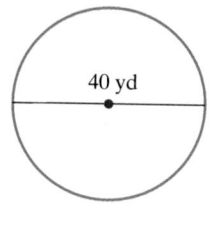

40 yd

FIGURE 10.39

Check 6 Point

Find the circumference of a circle whose diameter measures 10 inches. Express the answer in terms of π and then round to the nearest tenth of an inch.

FIGURE 10.40

EXAMPLE 7 USING THE CIRCUMFERENCE FORMULA

How much trim, to the nearest tenth of a foot, is needed to go around the window shown in Figure 10.40?

Solution The trim covers the 6-foot bottom of the window, the two 8-foot sides, and the half-circle (called a semicircle) on top. The length needed is

$$6 \text{ ft} + 8 \text{ ft} + 8 \text{ ft} + \text{circumference of the semicircle.}$$

The circumference of the semicircle is half the circumference of a circle whose diameter is 6 feet.

Circumference of semicircle

$$\text{Circumference of semicircle} = \tfrac{1}{2}\pi d$$
$$= \tfrac{1}{2}\pi(6 \text{ ft}) = 3\pi \text{ ft} \approx 9.4 \text{ ft}$$

Rounding the circumference to the nearest tenth (9.4 feet), the length of trim that is needed is approximately

$$6 \text{ ft} + 8 \text{ ft} + 8 \text{ ft} + 9.4 \text{ ft}$$

or 31.4 feet.

Check Point 7 In Figure 10.40, suppose that the dimensions are 10 feet and 12 feet for the window's bottom and side, respectively. How much trim, to the nearest tenth of a foot, is needed to go around the window?

We also use π to find the area of a circle in square units.

> **FINDING THE AREA OF A CIRCLE**
> The area, A, of a circle with radius r is
> $$A = \pi r^2.$$

EXAMPLE 8 PROBLEM SOLVING USING THE FORMULA FOR A CIRCLE'S AREA

Which one of the following is the better buy: a large pizza with a 16-inch diameter for $15.00 or a medium pizza with an 8-inch diameter for $7.50?

Solution The better buy is the pizza with the lower price per square inch. The radius of the large pizza is $\tfrac{1}{2} \cdot 16$ inches, or 8 inches, and the radius of the medium pizza is $\tfrac{1}{2} \cdot 8$ inches, or 4 inches. The area of the surface of each circular pizza is determined using the formula for the area of a circle.

$$\text{Large pizza: } A = \pi r^2 = \pi(8 \text{ in.})^2 = 64\pi \text{ in.}^2 \approx 201 \text{ in.}^2$$
$$\text{Medium pizza: } A = \pi r^2 = \pi(4 \text{ in.})^2 = 16\pi \text{ in.}^2 \approx 50 \text{ in.}^2$$

TECHNOLOGY

You can use your calculator to obtain the price per square inch for each pizza in Example 8. The price per square inch for the large pizza, $\dfrac{15}{64\pi}$, is approximated by one of the following sequences of keystrokes:

Many Scientific Calculators

15 ÷ (64 × π) =

Many Graphing Calculators

15 ÷ (64 π) ENTER .

For each pizza, the price per square inch is found by dividing the price by the area:

$$\text{Price per square inch for large pizza} = \frac{\$15.00}{64\pi \text{ in.}^2} \approx \frac{\$15.00}{201 \text{ in.}^2} \approx \frac{\$0.07}{\text{in.}^2}$$

$$\text{Price per square inch for medium pizza} = \frac{\$7.50}{16\pi \text{ in.}^2} \approx \frac{\$7.50}{50 \text{ in.}^2} = \frac{\$0.15}{\text{in.}^2}.$$

The large pizza is the better buy.

In Example 8, did you at first think that the price per square inch would be the same for the large and the medium pizzas? After all, the radius of the large pizza is twice that of the medium pizza, and the cost of the large is twice that of the medium. However, the large pizza's area, 64π square inches, is *four times the area* of the medium pizza's, 16π square inches. Doubling the radius of a circle increases its area by four times the original amount. In general, if the radius of a circle is increased by k times its original linear measure, the area is multiplied by k^2. The same principle is true for any two-dimensional figure: If the shape of the figure is kept the same while linear dimensions are increased k times, the area of the larger, similar, figure is k^2 times greater than the area of the original figure.

Check Point 8 Which one of the following is the better buy: a large pizza with an 18-inch diameter for $20.00 or a medium pizza with a 14-inch diameter for $14.00?

Exercise Set 10.4

Practice Exercises

In Exercises 1–14, use the formulas developed in this section to find the area of each figure.

1.
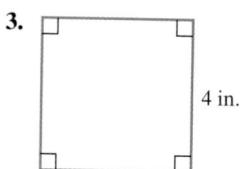
3 m
6 m

(Note: figure 1 shows 3 m and 6 m)

1.
6 m, 3 m

2.
4 ft, 3 ft

3.
4 in., 4 in.

4.
3 cm, 3 cm

5.
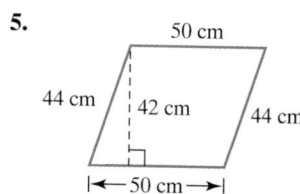
50 cm
44 cm   42 cm   44 cm
50 cm

6.
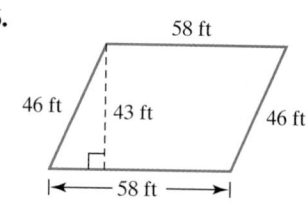
58 ft
46 ft   43 ft   46 ft
58 ft

7.
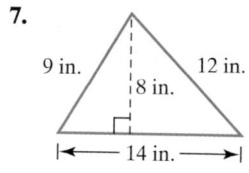
9 in.   12 in.
8 in.
14 in.

8.
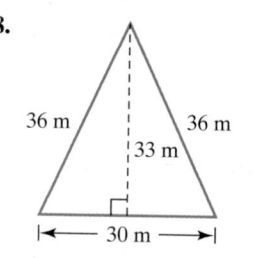
36 m   36 m
33 m
30 m

9.
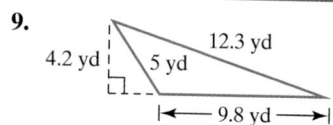
4.2 yd   5 yd   12.3 yd
9.8 yd

10.
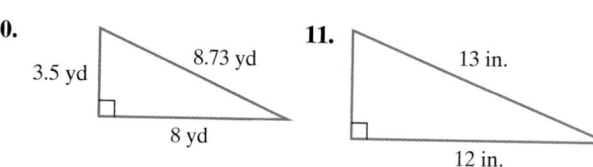
3.5 yd   8.73 yd
8 yd

11.
13 in.
12 in.

12.
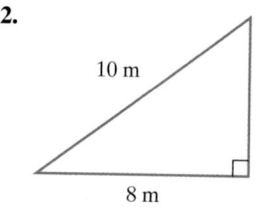
10 m
8 m

13.
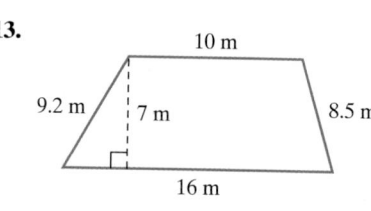
10 m
9.2 m   7 m   8.5 m
16 m

14.
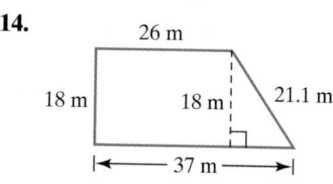
26 m
18 m   18 m   21.1 m
37 m

In Exercises 15–18, find the circumference and area of each circle.
Express answers in terms of π and then round to the nearest tenth.

15.

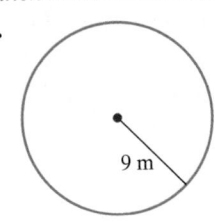

4 cm

16.

9 m

17.

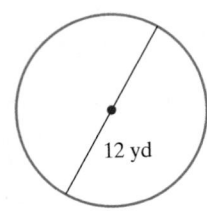

12 yd

18.

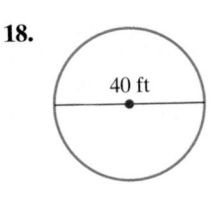

40 ft

Find the area of each figure in Exercises 19–24. Where necessary,
express answers in terms of π and then round to the nearest tenth.

19.

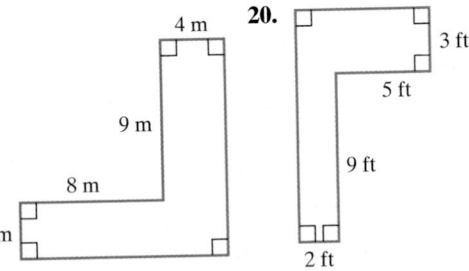

4 m
9 m
8 m
3 m

20.

3 ft
5 ft
9 ft
2 ft

21.

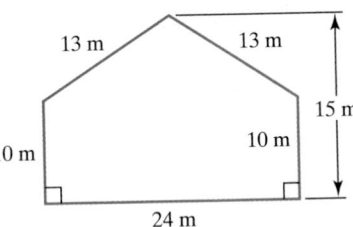

13 m   13 m
15 m
10 m   10 m
24 m

22.

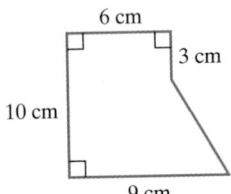

6 cm
3 cm
10 cm
9 cm

23.

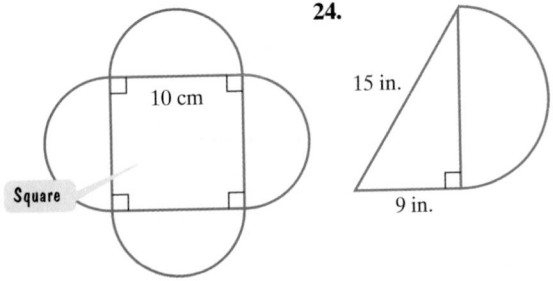

10 cm
Square

24.

15 in.
9 in.

Application Exercises

25. What will it cost to carpet a rectangular floor measuring 9 feet by 21 feet if the carpet costs $26.50 per square yard?

26. A plastering contractor charges $18 per square yard. What is the cost of plastering 60 feet of wall in a house with a 9-foot ceiling?

27. A rectangular kitchen floor measures 12 feet by 15 feet. A stove on the floor has a rectangular base measuring 3 feet by 4 feet, and a refrigerator covers a rectangular area of the floor measuring 4 feet by 5 feet. How many square feet of tile will be needed to cover the kitchen floor not counting the area used by the stove and the refrigerator?

28. A rectangular room measures 12 feet by 15 feet. The entire room is to be covered with rectangular tiles that measure 3 inches by 2 inches. If the tiles are sold at ten for 30¢, what will it cost to tile the room?

29. The lot in the figure shown, except for the house, shed, and driveway, is lawn. One bag of lawn fertilizer costs $25.00 and covers 4000 square feet.

   a. Determine the minimum number of bags of fertilizer needed for the lawn.

   b. Find the total cost of the fertilizer.

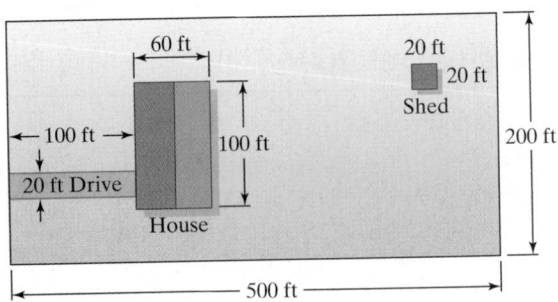

60 ft
20 ft
20 ft
Shed
100 ft
100 ft
200 ft
20 ft Drive
House
500 ft

30. Taxpayers with an office in their home may deduct a percentage of their home-related expenses. This percentage is based on the ratio of the office's area to the area of the home. A taxpayer with an office in a 2200 square-foot home maintains a 20 foot by 16 foot office. If the yearly electric bills for the home come to $4800, how much of this is deductible?

31. You are planning to paint the house whose dimensions are shown in the figure.

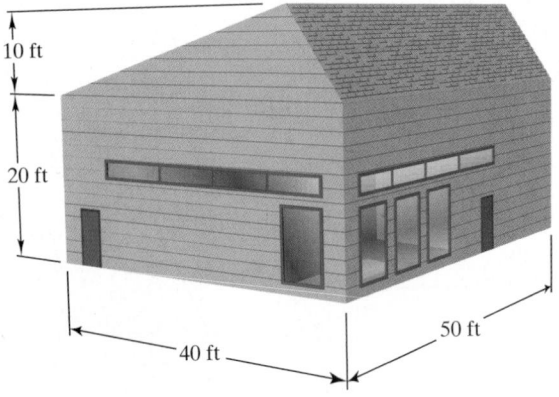

10 ft
20 ft
40 ft
50 ft

a. How many square feet will you need to paint? (There are four windows, each 8 ft by 5 ft; two windows, each 30 ft by 2 ft; and two doors, each 80 in. by 36 in., that do not require paint.)

b. The paint that you have chosen is available in gallon cans only. Each can covers 500 square feet. If you want to use two coats of paint, how many cans will you need for the project?

c. If the paint you have chosen sells for $26.95 per gallon, what will it cost to paint the house?

The diagram shows the floor plan for a one-story home. Use the given measurements to solve Exercises 32–34. (A calculator will be helpful in performing the necessary computations.)

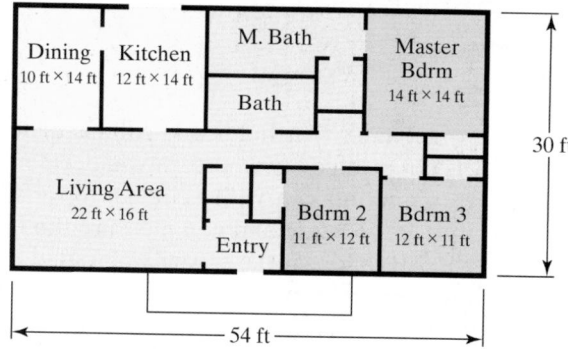

32. If construction costs $95 per square foot, find the cost of building the home.

33. If carpet costs $17.95 per square yard and is available in whole square yards only, find the cost of carpeting the three bedroom floors.

34. If ceramic tile costs $26.95 per square yard and is available in whole square yards only, find the cost of installing ceramic tile on the kitchen and dining room floors.

In Exercises 35–36, express the required calculation in terms of π and then round to the nearest tenth.

35. How much fencing is required to enclose a circular garden whose radius is 20 meters?

36. A circular rug is 6 feet in diameter. How many feet of fringe is required to edge this rug?

37. How many plants spaced every 6 inches are needed to surround a circular garden with a 30-foot radius?

38. A stained glass window is to be placed in a house. The window consists of a rectangle, 6 feet high by 3 feet wide, with a semicircle at the top. Approximately how many feet of stripping, to the nearest tenth of a foot, will be needed to frame the window?

39. Which one of the following is a better buy: a large pizza with a 14-inch diameter for $12.00 or a medium pizza with a 7-inch diameter for $5.00?

40. Which one of the following is a better buy: a large pizza with a 16-inch diameter for $12.00 or two small pizzas, each with a 10-inch diameter, for $12.00?

Writing in Mathematics

41. Using the formula for the area of a rectangle, explain how the formula for the area of a parallelogram ($A = bh$) is obtained.

42. Using the formula for the area of a parallelogram ($A = bh$), explain how the formula for the area of a triangle ($A = \frac{1}{2}bh$) is obtained.

43. Using the formula for the area of a triangle, explain how the formula for the area of a trapezoid is obtained.

44. Explain why a circle is not a polygon.

45. Describe the difference between the following problems: How much fencing is needed to enclose a circular garden? How much fertilizer is needed for a circular garden?

Critical Thinking Exercises

46. You need to enclose a rectangular region with 200 feet of fencing. Experiment with different lengths and widths to determine the maximum area you can enclose. Which quadrilateral encloses the most area?

47. Suppose you know the cost for building a rectangular deck measuring 8 feet by 10 feet. If you decide to increase the dimensions to 12 feet by 15 feet, by how much will the cost increase?

48. A rectangular swimming pool measures 14 feet by 30 feet. The pool is surrounded on all four sides by a path that is 3 feet wide. If the cost to resurface the path is $2 per square foot, what is the total cost of resurfacing the path?

49. A proposed oil pipeline will cross 16.8 miles of national forest. The width of the land needed for the pipeline is 200 feet. If the U.S. Forest Service charges the oil company $32 per acre, calculate the total cost. (1 mile = 5280 feet and 1 acre = 43,560 square feet.)

50. Find the area of the shaded region in terms of π.

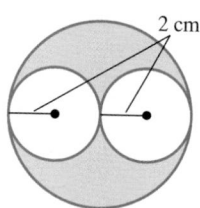

SECTION 10.5   VOLUME

OBJECTIVES

1. Use volume formulas to compute the volumes of three-dimensional figures and solve applied problems.
2. Compute the surface area of a three-dimensional figure.

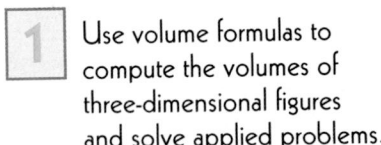

1 Use volume formulas to compute the volumes of three-dimensional figures and solve applied problems.

Your contractor promised to install a water tank that holds 500 gallons of water. Upon delivery, you notice that capacity is not printed anywhere, so you decide to do some measuring. The tank is shaped like a giant tuna can, with a circular top and bottom. You measure the radius of each circle to be 3 feet and you measure the tank's height to be 2 feet 4 inches. You know that 500 gallons is the capacity of a solid figure with a volume of about 67 cubic feet. Now you need some sort of method to compute the volume of the water tank. In this section, we discuss how to compute the volumes of various solid, three-dimensional figures. Using a formula you will learn in the section, you can determine whether the evidence indicates you can win a case against the contractor if it goes to court. (See Exercise 33 in Exercise Set 10.5.)

Formulas for Volume

In Section 9.2, we saw that **volume** refers to the amount of space occupied by a solid object, determined by the number of cubic units it takes to fill the interior of that object. For example, Figure 10.41 shows that there are 18 cubic units contained within the box. The volume of the box, called a **rectangular solid**, is 18 cubic units. The box has a length of 3 units, a width of 3 units, and a height of 2 units. The volume, 18 cubic units, may be determined by finding the product of the length, the width, and the height:

$$\text{Volume} = 3 \text{ units} \cdot 3 \text{ units} \cdot 2 \text{ units} = 18 \text{ units}^3.$$

In general, the volume, V, of a rectangular solid is the product of its length, l, its width, w, and its height, h:

$$V = lwh.$$

If the length, width, and height are the same, the rectangular solid is called a **cube**. Formulas for these boxlike shapes are given below.

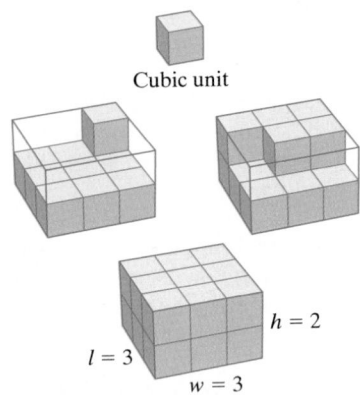

Cubic unit

FIGURE 10.41   Volume = 18 cubic units

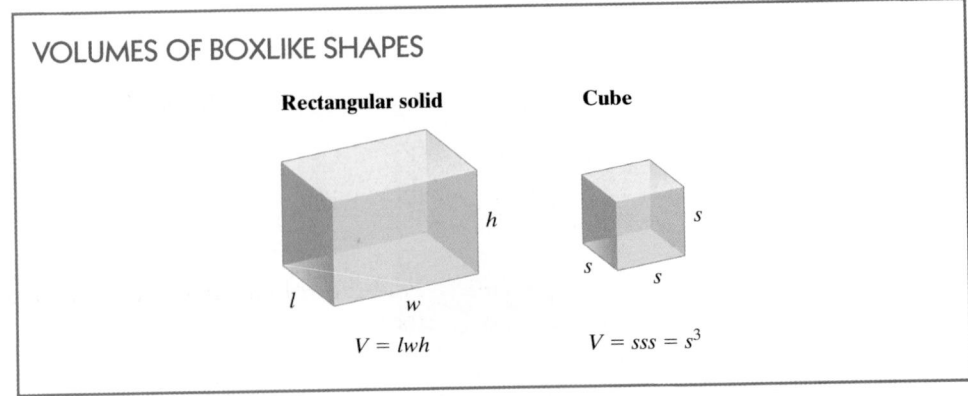

VOLUMES OF BOXLIKE SHAPES

Rectangular solid

h

l w

$V = lwh$

Cube

s

s s

$V = sss = s^3$

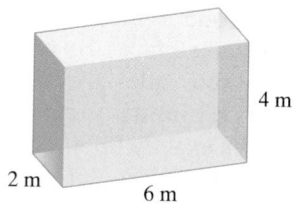

4 m

2 m   6 m

FIGURE 10.42

EXAMPLE 1   FINDING THE VOLUME OF A RECTANGULAR SOLID

Find the volume of the rectangular solid in Figure 10.42.

Solution   As shown in the figure, the length is 6 meters, the width is 2 meters, and the height is 4 meters. Thus, $l = 6$, $w = 2$, and $h = 4$.

$$V = lwh = 6 \text{ m} \cdot 2 \text{ m} \cdot 4 \text{ m} = 48 \text{ m}^3$$

The volume of the rectangular solid is 48 cubic meters.

 Check 1 Point   Find the volume of a rectangular solid with length 5 feet, width 3 feet, and height 7 feet.

In Section 9.2, we saw that although there are 3 feet in a yard, there are 27 cubic feet in a cubic yard. If a problem requires measurement of volume in cubic yards and the linear measures are given in feet, to avoid errors first convert feet to yards. Then apply the volume formula. This idea is illustrated in Example 2.

EXAMPLE 2   SOLVING A VOLUME PROBLEM

You are about to begin work on a swimming pool in your yard. The first step is to have a hole dug that is 90 feet long, 60 feet wide, and 6 feet deep. You will use a truck that can carry 10 cubic yards of dirt and charges $12 per load. How much will it cost you to have all the dirt hauled away?

Solution   We begin by converting feet to yards:

$$90 \text{ ft} = \frac{90 \text{ ft}}{1} \cdot \frac{1 \text{ yd}}{3 \text{ ft}} = \frac{90}{3} \text{ yd} = 30 \text{ yd}.$$

Similarly, 60 ft = 20 yd and 6 ft = 2 yd. Next, we find the volume of dirt that needs to be dug out and hauled off.

$$V = lwh = 30 \text{ yd} \cdot 20 \text{ yd} \cdot 2 \text{ yd} = 1200 \text{ yd}^3$$

Now, we find the number of loads that the truck needs to haul off all the dirt. Because the truck carries 10 cubic yards, divide the number of cubic yards of dirt by 10.

$$\text{Number of truckloads} = \frac{1200 \text{ yd}^3}{\dfrac{10 \text{ yd}^3}{\text{trip}}} = \frac{1200 \text{ yd}^3}{1} \cdot \frac{\text{trip}}{10 \text{ yd}^3} = \frac{1200}{10} \text{ trips} = 120 \text{ trips}$$

Because the truck charges $12 per trip, the cost to have all the dirt hauled away is the number of trips, 120, times the cost per trip, $12.

$$\text{Cost to haul all dirt away} = \frac{120 \text{ trips}}{1} \cdot \frac{\$12}{\text{trip}} = 120(\$12) = \$1440$$

The dirt-hauling phase of the pool project will cost you $1440.

 Check 2 Point   Find the volume, in cubic yards, of a cube whose edges each measure 6 feet.

FIGURE 10.43 The volume of a pyramid is $\frac{1}{3}$ the volume of a rectangular solid having the same base and the same height.

The Transamerica Tower's 3678 windows take cleaners one month to wash. Its foundation is sunk 15.5 m (52 ft) into the ground and is designed to move with earth tremors.

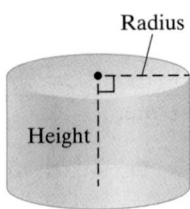

FIGURE 10.44

A rectangular solid is an example of a **polyhedron**, a solid figure bounded by polygons. A rectangular solid is bounded by six rectangles, called faces. By contrast, a **pyramid** is a polyhedron whose base is a polygon and whose sides are triangles. Figure 10.43 shows a pyramid with a rectangular base drawn inside a rectangular solid. The contents of three pyramids with rectangular bases exactly fill a rectangular solid of the same base and height. Thus, the formula for the volume of the pyramid is $\frac{1}{3}$ that of the rectangular solid.

VOLUME OF A PYRAMID

The volume, V, of a pyramid is given by the formula

$$V = \frac{1}{3}Bh$$

where B is the area of the base and h is the height (the perpendicular distance from the top to the base).

Pyramid

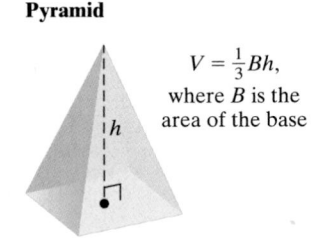

$V = \frac{1}{3}Bh$, where B is the area of the base

EXAMPLE 3 USING THE FORMULA FOR A PYRAMID'S VOLUME

Capped with a pointed spire on top of its 48 stories, the Transamerica Tower in San Francisco is a pyramid with a square base. The pyramid is 256 meters (853 feet) tall. Each side of the square base has a length of 52 meters. Although San Franciscans disliked it when it opened in 1972, they have since accepted it as part of the skyline. Find the volume of the building.

Solution First find the area of the square base, represented as B in the volume formula. Because each side of the square base is 52 meters, the area of the square base is

$$B = 52 \text{ m} \cdot 52 \text{ m} = 2704 \text{ m}^2.$$

The area of the square base is 2704 square meters. Because the pyramid is 256 meters tall, its height, h, is 256 meters. Now we apply the formula for the volume of a pyramid:

$$V = \frac{1}{3}Bh = \frac{1}{3} \cdot \frac{2704 \text{ m}^2}{1} \cdot \frac{256 \text{ m}}{1} = \frac{2704 \cdot 256}{3} \text{ m}^3 \approx 230{,}741 \text{ m}^3.$$

The volume of the building is approximately 230,741 cubic meters.

The San Francisco pyramid is relatively small compared to the Great Pyramid outside Cairo, Egypt. Built about 2550 B.C. by a labor force of 100,000, the Great Pyramid is approximately 11 times the volume of San Francisco's pyramid.

Check Point 3 A pyramid is 4 feet tall. Each side of the square base has a length of 6 feet. Find the pyramid's volume.

Not every three-dimensional figure is a polyhedron. Take, for example, the right circular cylinder shown in Figure 10.44. Its shape should remind you of a soup can or a stack of coins. The right circular cylinder is so named because the top and bottom are circles, and the side forms a right angle with the top and bottom. The formula for the volume of a right circular cylinder is given as follows:

VOLUME OF A RIGHT CIRCULAR CYLINDER

The volume, V, of a right circular cylinder is given by the formula

$$V = \pi r^2 h$$

where r is the radius of the circle at either end and h is the height.

Right circular cylinder

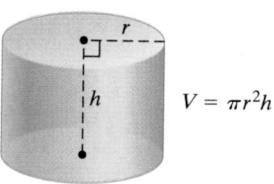

$V = \pi r^2 h$

EXAMPLE 4 FINDING THE VOLUME OF A CYLINDER

Find the volume of the cylinder in Figure 10.45.

Solution In order to find the cylinder's volume, we need both its radius and its height. Because the diameter is 20 yards, the radius is half this length, or 10 yards. The height of the cylinder is given to be 9 yards. Thus, $r = 10$ and $h = 9$. Now we apply the formula for the volume of a cylinder.

$$V = \pi r^2 h = \pi (10 \text{ yd})^2 \cdot 9 \text{ yd} = 900\pi \text{ yd}^3 \approx 2827 \text{ yd}^3$$

The volume of the cylinder is approximately 2827 cubic yards.

 **Check 4 Point** Find the volume, to the nearest cubic inch, of a cylinder with a diameter of 8 inches and a height of 6 inches.

20 yd

9 yd

FIGURE 10.45

Figure 10.46 shows a **right circular cone** inside a cylinder, sharing the same circular base as the cylinder. The height of the cone, the perpendicular distance from the top to the circular base, is the same as that of the cylinder. Three such cones can occupy the same amount of space as the cylinder. Therefore, the formula for the volume of the cone is $\frac{1}{3}$ the volume of the cylinder.

FIGURE 10.46

VOLUME OF A CONE

The volume, V, of a right circular cone that has height h and radius r is given by the formula

$$V = \frac{1}{3}\pi r^2 h.$$

Cone

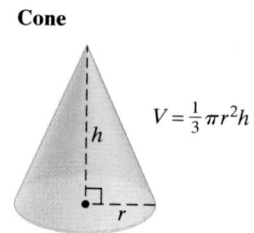

$V = \frac{1}{3}\pi r^2 h$

EXAMPLE 5 FINDING THE VOLUME OF A CONE

Find the volume of the cone in Figure 10.47.

Solution The radius of the cone is 7 meters and the height is 10 meters. Thus, $r = 7$ and $h = 10$. Now we apply the formula for the volume of a cone.

$$V = \frac{1}{3}\pi r^2 h = \frac{1}{3}\pi (7 \text{ m})^2 \cdot 10 \text{ m} = \frac{490\pi}{3} \text{ m}^3 \approx 513 \text{ m}^3$$

The volume of the cone is approximately 513 cubic meters.

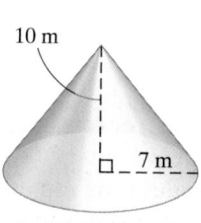

10 m

7 m

FIGURE 10.47

Find the volume, to the nearest cubic inch, of a cone with a radius of 4 inches and a height of 6 inches.

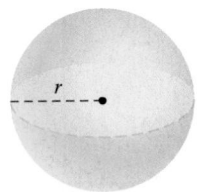

FIGURE 10.48

Figure 10.48 shows a *sphere*. Its shape may remind you of a basketball. The Earth is not a perfect sphere, but it's close. A **sphere** is the set of points in space equally distant from a given point, its **center**. The **radius** is the line segment from the center to any point on the sphere. The word *radius* is also used to refer to the length of this line segment. A sphere's volume can be found by using π and its radius.

VOLUME OF A SPHERE

The volume, V, of a sphere of radius r is given by the formula

$$V = \frac{4}{3}\pi r^3.$$

Sphere

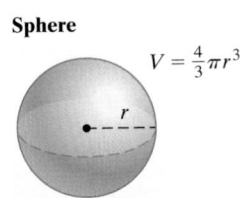

$V = \frac{4}{3}\pi r^3$

EXAMPLE 6 | APPLYING VOLUME FORMULAS

An ice cream cone is 5 inches deep and has a radius of 1 inch. A spherical scoop of ice cream also has a radius of 1 inch. (See Figure 10.49.) If the ice cream melts into the cone, will it overflow?

Solution The ice cream will overflow if the volume of the ice cream, a sphere, is greater than the volume of the cone. Find the volume of each.

$$V_{\text{cone}} = \frac{1}{3}\pi r^2 h = \frac{1}{3}\pi(1 \text{ in.})^2 \cdot 5 \text{ in.} = \frac{5\pi}{3} \text{ in.}^3 \approx 5 \text{ in.}^3$$

$$V_{\text{sphere}} = \frac{4}{3}\pi r^3 \quad = \frac{4}{3}\pi(1 \text{ in.})^3 = \frac{4\pi}{3} \text{ in.}^3 \approx 4 \text{ in.}^3$$

The volume of the spherical scoop of ice cream is less than the volume of the cone, so there will be no overflow.

A basketball has a radius of 4.5 inches. If the ball is filled with 350 cubic inches of air, is this enough air to fill it completely?

FIGURE 10.49

 2 Compute the surface area of a three-dimensional figure.

Surface Area

In addition to volume, we can also measure the area of the outer surface of a three-dimensional object, called its **surface area**. Like area, surface area is measured in square units. For example, the surface area of the rectangular solid in Figure 10.50 is the sum of the areas of the six outside rectangles of the solid.

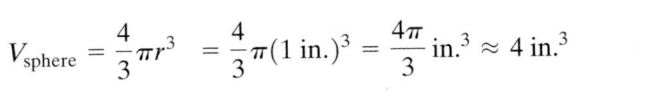

$$\text{Surface Area} = lw + lw + lh + lh + wh + wh$$

| Areas of top and bottom rectangles | Areas of front and back rectangles | Areas of rectangles on left and right sides |

$$= \quad 2lw \quad + \quad 2lh \quad + \quad 2wh$$

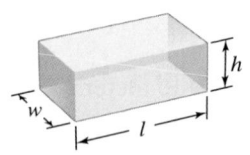

FIGURE 10.50

Formulas for the surface area, abbreviated SA, of three-dimensional figures are given in Table 10.4.

TABLE 10.4 COMMON FORMULAS FOR SURFACE AREA

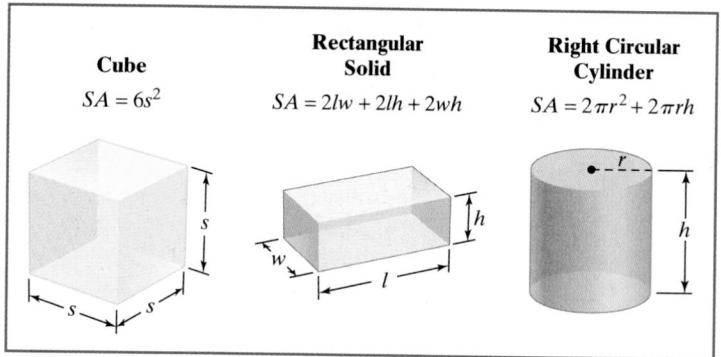

Cube	Rectangular Solid	Right Circular Cylinder
$SA = 6s^2$	$SA = 2lw + 2lh + 2wh$	$SA = 2\pi r^2 + 2\pi rh$

EXAMPLE 7 | FINDING THE SURFACE AREA OF A SOLID

Find the surface area of the rectangular solid in Figure 10.51.

Solution   As shown in the figure, the length is 8 yards, the width is 5 yards, and the height is 3 yards. Thus, $l = 8$, $w = 5$, and $h = 3$.

$$SA = 2lw + 2lh + 2wh$$
$$= 2 \cdot 8 \text{ yd} \cdot 5 \text{ yd} + 2 \cdot 8 \text{ yd} \cdot 3 \text{ yd} + 2 \cdot 5 \text{ yd} \cdot 3 \text{ yd}$$
$$= 80 \text{ yd}^2 + 48 \text{ yd}^2 + 30 \text{ yd}^2 = 158 \text{ yd}^2$$

The surface area is 158 square yards.

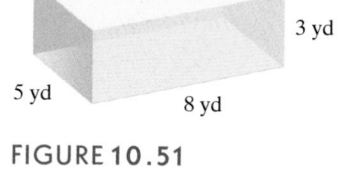

3 yd
5 yd
8 yd

FIGURE 10.51

Check 7 Point

If the length, width, and height shown in Figure 10.51 are each doubled, find the surface area of the resulting rectangular solid.

Exercise Set 10.5

Practice Exercises

In Exercises 1–20, find the volume of each figure. If necessary, express answers in terms of π and then round to the nearest whole number.

1.

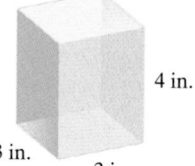

4 in.
3 in.
3 in.

2.

3 cm
3 cm
5 cm

3.

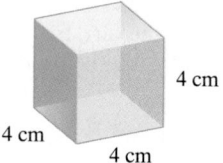

4 cm
4 cm
4 cm

4.

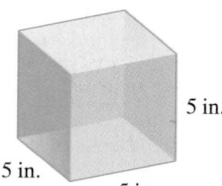

5 in.
5 in.
5 in.

5.

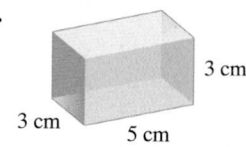

$h = 15$ yd
5 yd   7 yd

6.

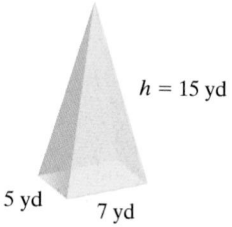

$h = 20$ yd
8 yd   15 yd

7.

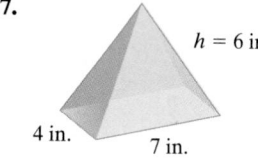

$h = 6$ in.
4 in.   7 in.

8.

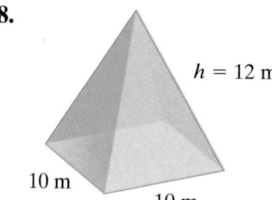

$h = 12$ m
10 m   10 m

9.

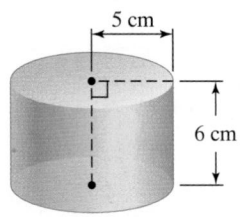

5 cm

6 cm

10.

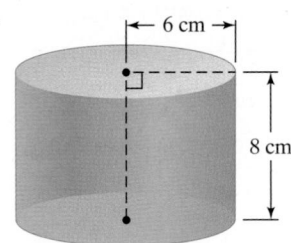

6 cm

8 cm

11.

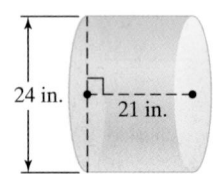

24 in.

21 in.

12.

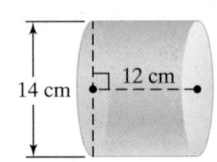

14 cm

12 cm

13.

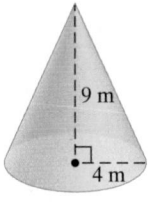

9 m

4 m

14.

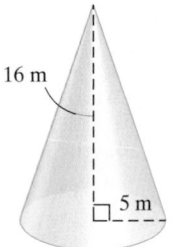

16 m

5 m

15.

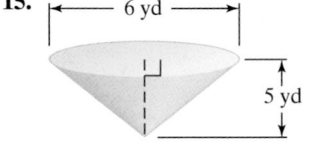

6 yd

5 yd

16.

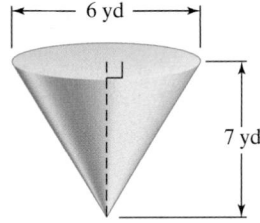

6 yd

7 yd

17.

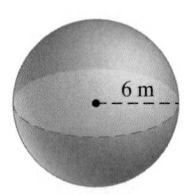

6 m

18.

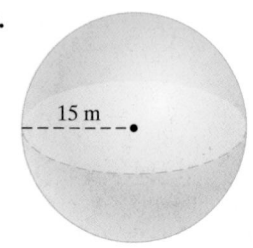

15 m

19.

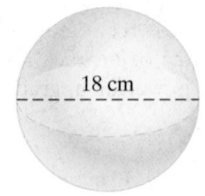

18 cm

20.

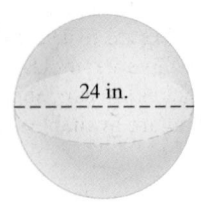

24 in.

In Exercises 21–24, find the surface area of each figure.

21.

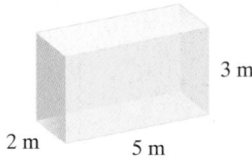

3 m

2 m

5 m

22.

3 m

4 m

6 m

23.

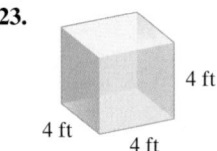

4 ft

4 ft

4 ft

24.

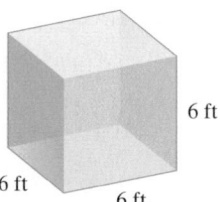

6 ft

6 ft

6 ft

In Exercises 25–26, use two formulas for volume to find the volume of each figure. Express answers in terms of π and then round to the nearest whole number.

25.

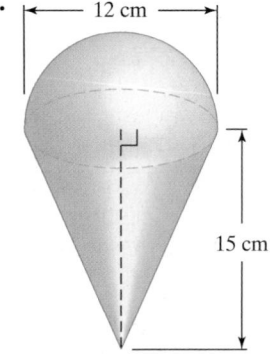

12 cm

15 cm

26.

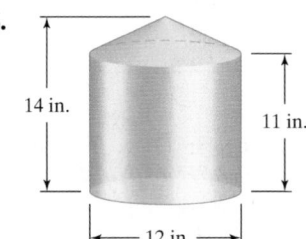

14 in.

11 in.

12 in.

Application Exercises

27. A building contractor is to dig a foundation 12 feet long, 9 feet wide, and 6 feet deep for a toll booth. The contractor pays $10 per load for trucks to remove the dirt. Each truck holds 6 cubic yards. What is the cost to the contractor to have all the dirt hauled away?

28. What is the cost of concrete for a walkway that is 15 feet long, 8 feet wide, and 9 inches deep if the concrete costs $30 per cubic yard?

29. A furnace is designed to heat 10,000 cubic feet. Will this furnace be adequate for a 1400-square-foot house with a 9-foot ceiling?

30. A water reservoir is shaped like a rectangular solid with a base that is 50 yards by 30 yards, and a vertical height of 20 yards. At the start of a three-month period of no rain, the reservoir was completely full. At the end of this period, the height of the water was down to 6 yards. How much water was used in the three-month period?

31. The Great Pyramid outside Cairo, Egypt, has a square base measuring 756 feet on a side and a height of 480 feet.

   a. What is the volume of the Great Pyramid, in cubic yards?

   b. The stones used to build the Great Pyramid were limestone blocks with an average volume of 1.5 cubic yards. How many of these blocks were needed to construct the Great Pyramid?

32. Although the Eiffel Tower in Paris is not a solid pyramid, its shape approximates that of a pyramid with a square base measuring 120 feet on a side and a height of 980 feet. If it were a solid pyramid, what would be the Eiffel Tower's volume, in cubic yards?

33. You are about to sue your contractor who promised to install a water tank that holds 500 gallons of water. You know that 500 gallons is the capacity of a tank that holds 67 cubic feet. The cylindrical tank has a radius of 3 feet and a height of 2 feet 4 inches. Does the evidence indicate you can win the case against the contractor if it goes to court?

34. Two cylindrical cans of soup sell for the same price. One can has a diameter of 6 inches and a height of 5 inches. The other has a diameter of 5 inches and a height of 6 inches. Which can contains more soup and, therefore, is the better buy?

35. A circular backyard pool has a diameter of 24 feet and is 4 feet deep. One cubic foot of water has a capacity of approximately 7.48 gallons. If water costs $2 per thousand gallons, how much, to the nearest dollar, will it cost to fill the pool?

36. The tunnel under the English Channel that connects England and France is the world's longest tunnel. There are actually three separate tunnels built side by side. Each is a half-cylinder that is 50,000 meters long and 4 meters high. How many cubic meters of dirt had to be removed to build the tunnel?

Writing in Mathematics

37. Explain the following analogy:

In terms of formulas used to compute volume, a pyramid is to a rectangular solid just as a cone is to a cylinder.

38. Explain why a cylinder is not a polyhedron.

Critical Thinking Exercises

39. What happens to the volume of a sphere if its radius is doubled?

40. A scale model of a car is constructed so that its length, width, and height are each $\frac{1}{10}$ the length, width, and height of the actual car. By how many times does the volume of the car exceed its scale model?

In Exercises 41–42, find the volume of the darkly shaded region. If necessary, round to the nearest whole number.

41.

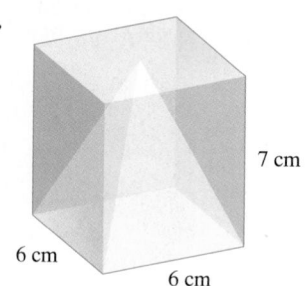

7 cm

6 cm

6 cm

42.

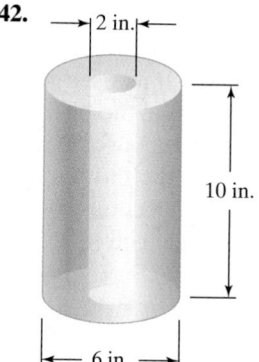

2 in.

10 in.

6 in.

43. Find the surface area of the figure shown.

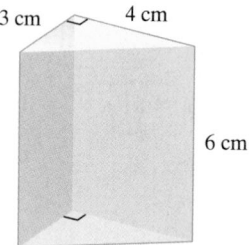

3 cm   4 cm

6 cm

Group Exercise

44. People who love cars should be members of this group. The group will prepare a presentation to the class on how the power of a car engine is measured. One person, who has worked on car engines, should explain the details of how they work, including cylinders, pistons, fuel, and so on, in the discussion. In explaining how the power of the engine is measured, be sure to discuss horsepower, cubic centimeters, and liters. Give examples of the measures of engines for various makes of cars, including more powerful upgrade options that are available.

OBJECTIVES

1. Use the lengths of the sides of a right triangle to find trigonometric ratios.

2. Use trigonometric ratios to find missing parts of right triangles.

3. Use trigonometric ratios to solve applied problems.

Ang Rita Sherpa climbed Mount Everest eight times, all without the use of bottled oxygen.

Mountain climbers have forever been fascinated by reaching the top of Mount Everest, sometimes with tragic results. The mountain, on Asia's Tibet-Nepal border, is Earth's highest, peaking at an incredible 29,035 feet. The heights of mountains can be found using **trigonometry**. The word "trigonometry" means "measurement of triangles." Trigonometry is used in navigation, building, and engineering. For centuries, Muslims used trigonometry and the stars to navigate across the Arabian desert to Mecca, the birthplace of the prophet Muhammad, the founder of Islam. The ancient Greeks used trigonometry to record the locations of thousands of stars and worked out the motion of the moon relative to Earth. Today, trigonometry is used to study the structure of DNA, the master molecule that determines how we grow from a single cell to a complex, fully developed adult.

Ratios in Right Triangles

The right triangle forms the basis of trigonometry. If either acute angle of a right triangle stays the same size, the shape of the triangle does not change even if it is made larger or smaller. Because of properties of similar triangles, this means that the ratios of certain lengths stay the same regardless of the right triangle's size. These ratios have special names and are defined in terms of the **side opposite** an acute angle, the **side adjacent** to the acute angle, and the **hypotenuse**. In Figure 10.52, the length of the hypotenuse, the side opposite the 90° angle, is represented by c. The length of the side opposite angle A is represented by a. The length of the side adjacent to angle A is represented by b.

The three fundamental trigonometric ratios, **sine** (abbreviated sin), **cosine** (abbreviated cos), and **tangent** (abbreviated tan) are defined as ratios of the lengths of the sides of a right triangle. In the box that follows, when a side of a triangle is mentioned, we are referring to the *length* of that side.

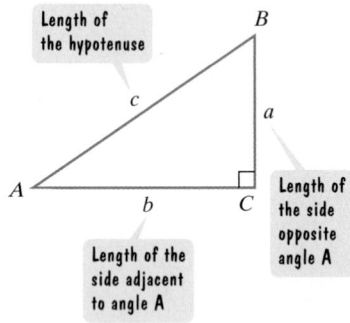

Length of the hypotenuse

Length of the side adjacent to angle A

Length of the side opposite angle A

FIGURE 10.52 Naming a right triangle's sides from the point of view of an acute angle

1 Use the lengths of the sides of a right triangle to find trigonometric ratios.

TRIGONOMETRIC RATIOS

Let A represent an acute angle of a right triangle, with right angle C, shown in Figure 10.52. For angle A, the trigonometric ratios are defined as follows:

sine of A
$$\sin A = \frac{\text{side opposite angle } A}{\text{hypotenuse}} = \frac{a}{c}$$

cosine of A
$$\cos A = \frac{\text{side adjacent to angle } A}{\text{hypotenuse}} = \frac{b}{c}$$

tangent of A
$$\tan A = \frac{\text{side opposite angle } A}{\text{side adjacent to angle } A} = \frac{a}{b}.$$

STUDY TIP

The word

SOHCAHTOA (pronounced: so-cah-tow-ah)

is a way to remember the definitions of the three trigonometric ratios.

S   O H   C   A H   T   O A
↑   opp   ↑   adj   ↑   opp
     hyp   hyp   adj
Sine   Cosine   Tangent

"<u>S</u>ome <u>O</u>ld <u>H</u>og <u>C</u>ame <u>A</u>round <u>H</u>ere and <u>T</u>ook <u>O</u>ur <u>A</u>pples."

EXAMPLE 1 | BECOMING FAMILIAR WITH THE TRIGONOMETRIC RATIOS

Find the sine, cosine, and tangent of A in Figure 10.53.

Solution   We begin by finding the measure of hypotenuse c using the Pythagorean Theorem.

$$c^2 = a^2 + b^2 = 5^2 + 12^2 = 25 + 144 = 169$$
$$c = \sqrt{169} = 13$$

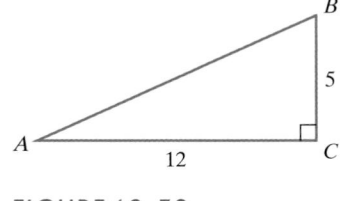

FIGURE 10.53

Now, we apply the definitions of the trigonometric ratios.

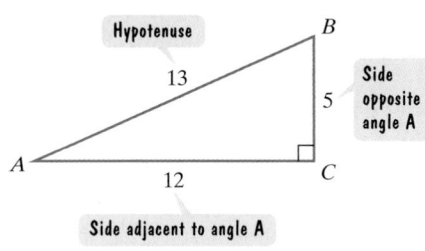

$$\sin A = \frac{\text{side opposite angle } A}{\text{hypotenuse}} = \frac{5}{13}$$

$$\cos A = \frac{\text{side adjacent to angle } A}{\text{hypotenuse}} = \frac{12}{13}$$

$$\tan A = \frac{\text{side opposite angle } A}{\text{side adjacent to angle } A} = \frac{5}{12}$$

 Check 1 Point   Find the sine, cosine, and tangent of A in the figure shown.

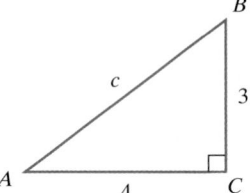

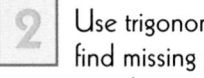

2 Use trigonometric ratios to find missing parts of right triangles.

A scientific or graphing calculator in the degree mode will give you decimal approximations for the trigonometric ratios of any angle. For example, to find an approximation for tan 37°, the tangent of 37°, a keystroke sequence similar to one of the following can be used:

Many Scientific Calculators: 37 | TAN |

Many Graphing Calculators: | TAN | 37 | ENTER | .

The tangent of 37°, rounded to four decimal places, is 0.7536.

If we are given the length of one side and the measure of an acute angle of a right triangle, we can use trigonometry to solve for the length of either of the other two sides. Example 2 illustrates how this is done.

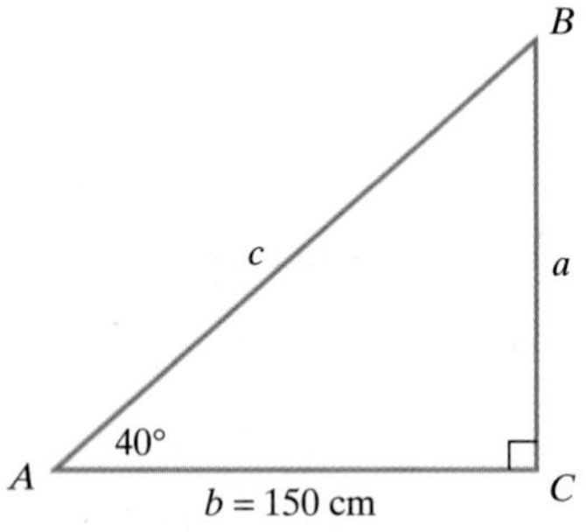

FIGURE 10.54

TECHNOLOGY

Here is the keystroke sequence for 150 tan 40°:

Many Scientific Calculators

150 ⊠ 40 TAN ⊟

Many Graphing Calculators

150 TAN 40 ENTER .

TECHNOLOGY

Here is the keystroke sequence for $\dfrac{150}{\cos 40°}$:

Many Scientific Calculators

150 ⊟ 40 COS ⊟

Many Graphing Calculators

150 ⊟ COS 40 ENTER .

3 Use trigonometric ratios to solve applied problems.

EXAMPLE 2   FINDING A MISSING LEG OF A RIGHT TRIANGLE

Find a in the right triangle in Figure 10.54.

Solution   We need to identify a trigonometric ratio that will make it possible to find a. Because we have a known angle, 40°, an unknown opposite side, a, and a known adjacent side, 150 cm, we use the tangent ratio.

$$\tan 40° = \frac{a}{150}$$

Side opposite the 40° angle

Side adjacent to the 40° angle

Now we solve for a by multiplying both sides by 150.

$$a = 150 \tan 40° \approx 126$$

The tangent ratio reveals that a is approximately 126 centimeters.

✓ Check 2 Point   In Figure 10.54, let $m \angle A = 62°$ and $b = 140$ cm. Find a to the nearest centimeter.

EXAMPLE 3   FINDING A MISSING HYPOTENUSE OF A RIGHT TRIANGLE

Find c in the right triangle in Figure 10.54.

Solution   In Example 2, we found a: $a \approx 126$. Because we are given that $b = 150$, it is possible to find c using the Pythagorean Theorem: $c^2 = a^2 + b^2$. However, if we made an error in computing a, we will perpetuate our mistake using this approach.

Instead, we will use the quantities given and identify a trigonometric ratio that will make it possible to find c. Refer to Figure 10.54. Because we have a known angle, 40°, a known adjacent side, 150 cm, and an unknown hypotenuse, c, we use the cosine ratio.

$$\cos 40° = \frac{150}{c}$$

Side adjacent to the 40° angle

Hypotenuse

$c \cos 40° = 150$ Multiply both sides by c.

$c = \dfrac{150}{\cos 40°}$ Divide both sides by cos 40°.

$c \approx 196$ Use a calculator.

The cosine ratio reveals that the hypotenuse is approximately 196 centimeters.

✓ Check 3 Point   In Figure 10.54, let $m \angle A = 62°$ and $b = 140$ cm. Find c to the nearest centimeter.

Applications of the Trigonometric Ratios

Trigonometry was first developed to measure heights and distances that are inconvenient or impossible to measure. These applications often involve the angle made with an imaginary horizontal line. As shown in Figure 10.55, an angle formed by a horizontal line and the line of sight to an object that is above the horizontal line is called the **angle of elevation**. The angle formed by a horizontal line and the line of sight to an object that is below the horizontal line is called the **angle of depression**. Transits and sextants are instruments used to measure such angles.

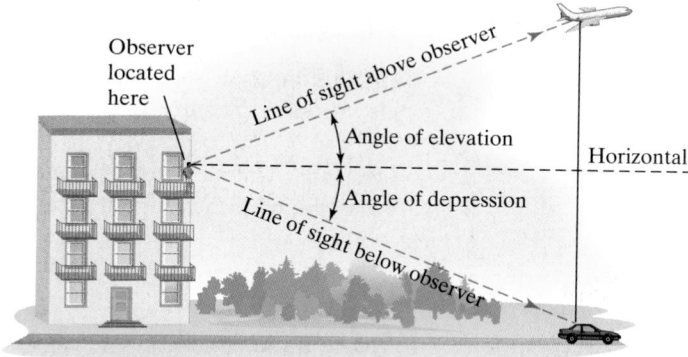

FIGURE 10.55

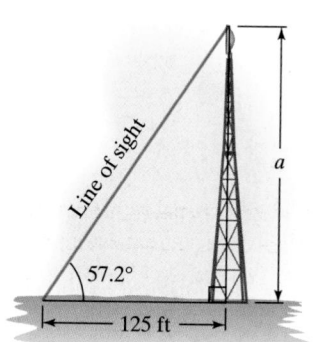

FIGURE 10.56 Determining height without using direct measurement

EXAMPLE 4	PROBLEM SOLVING USING AN ANGLE OF ELEVATION

From a point on level ground 125 feet from the base of a tower, the angle of elevation is 57.2°. Approximate the height of the tower to the nearest foot.

Solution   A sketch is shown in Figure 10.56, where a represents the height of the tower. In the right triangle, we have a known angle, an unknown opposite side, and a known adjacent side. Therefore, we use the tangent ratio.

$$\tan 57.2° = \frac{a}{125}$$

> Side opposite the 57.2° angle
>
> Side adjacent to the 57.2° angle

We solve for a by multiplying both sides of this equation by 125:

$$a = 125 \tan 57.2° \approx 194.$$

The tower is approximately 194 feet high.

Check 4 Point   From a point on level ground 80 feet from the base of the Eiffel Tower, the angle of elevation is 85.4°. Approximate the height of the Eiffel Tower to the nearest foot.

If the measures of two sides of a right triangle are known, the measures of the two acute angles can be found using the **inverse trigonometric keys** on a calculator. For example, suppose that $\sin A = 0.866$. We can find the measure of angle A by using the **inverse sine** key, usually labeled $\boxed{\text{SIN}^{-1}}$.

Many Scientific Calculators: **Many Graphing Calculators:**

.866 $\boxed{\text{2nd}}$ $\boxed{\text{SIN}^{-1}}$ $\boxed{\text{2nd}}$ $\boxed{\text{SIN}^{-1}}$ .866 $\boxed{\text{ENTER}}$

> Finding A in sin A = 0.866

The display should show approximately 59.99°, which can be rounded to 60°. Thus, if $\sin A = 0.866$, then $m\angle A \approx 60°$.

EXAMPLE 5 DETERMINING THE ANGLE OF ELEVATION

A building that is 21 meters tall casts a shadow 25 meters long. Find the angle of elevation of the sun.

Solution   The situation is illustrated in Figure 10.57. We are asked to find $m \angle A$. We begin with the tangent ratio.

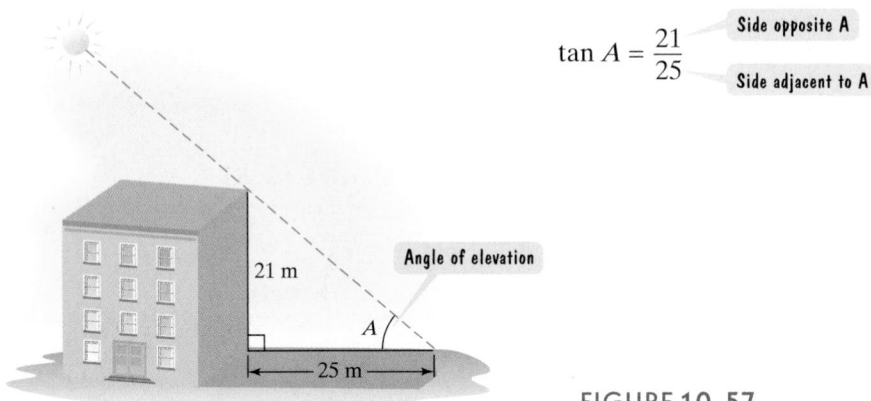

$$\tan A = \frac{21}{25}$$

Side opposite A

Side adjacent to A

21 m

Angle of elevation

A

25 m

FIGURE 10.57

We use the **inverse tangent** key, $\boxed{\text{TAN}^{-1}}$, to find A.

Finding A in
$\tan A = \frac{21}{25}$

Many Scientific Calculators:

$\boxed{(}$ $\boxed{21}$ $\boxed{\div}$ $\boxed{25}$ $\boxed{)}$ $\boxed{2^{\text{nd}}}$ $\boxed{\text{TAN}^{-1}}$

Many Graphing Calculators:

$\boxed{2^{\text{nd}}}$ $\boxed{\text{TAN}^{-1}}$ $\boxed{(}$ $\boxed{21}$ $\boxed{\div}$ $\boxed{25}$ $\boxed{)}$ $\boxed{\text{ENTER}}$

The display should show approximately 40. Thus, the angle of elevation of the sun is approximately 40°.

Check 5 Point   A flagpole that is 14 meters tall casts a shadow 10 meters long. Find the angle of elevation of the sun to the nearest degree.

BLITZER BONUS

The Mountain Man

In the 1930s, a *National Geographic* team headed by Brad Washburn used trigonometry to create a map of the 5000-square-mile region of the Yukon, near the Canadian border. The team started with aerial photography. By drawing a network of angles on the photographs, the approximate locations of the major mountains and their rough heights were determined. The expedition then spent three months on foot to find the exact heights. Team members established two base points a known distance apart, one directly under the mountain's peak. By measuring the angle of elevation from one of the base points to the peak, the tangent ratio was used to determine the peak's height. The Yukon expedition was a major advance in the way maps are made.

Exercise Set 10.6

Practice Exercises

In Exercises 1–8, use the given right triangles to find ratios, in reduced form, for sin A, cos A, and tan A.

1.
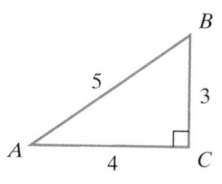

2.
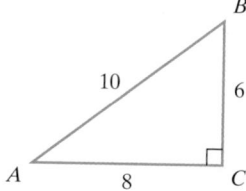

3.
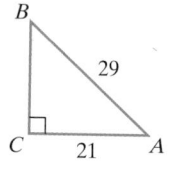

4.
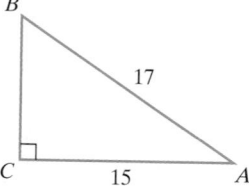

5.
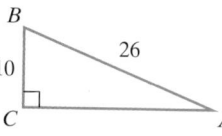

6.
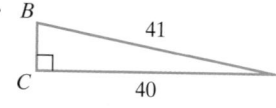

7.
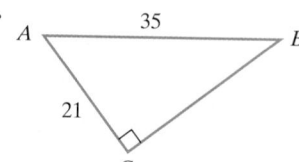

8.
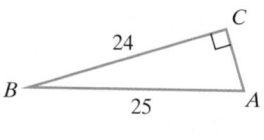

In Exercises 9–18, find the measure of the side of the right triangle whose length is designated by a lowercase letter. Round answers to the nearest whole number.

9.
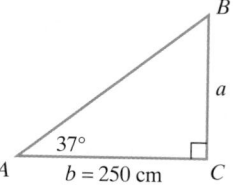

10.
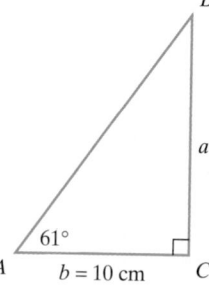

11.
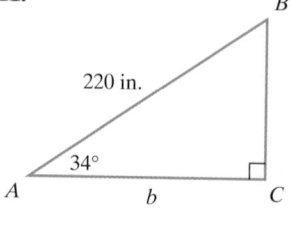

12.
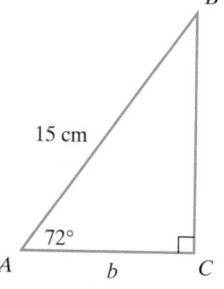

13.
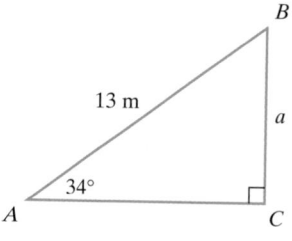

14.
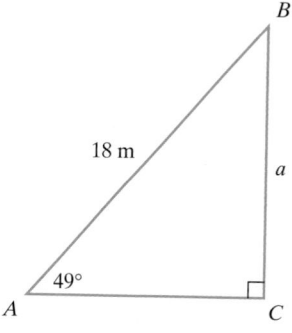

15.
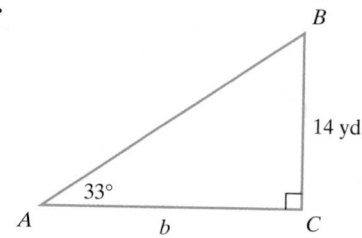

16.
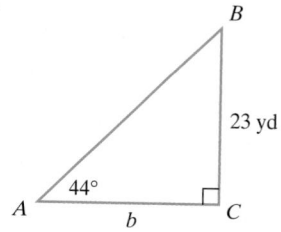

17.
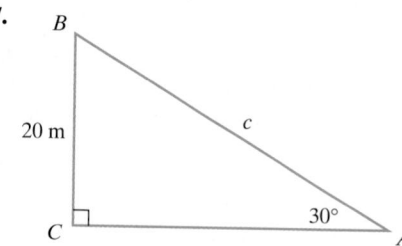

18.
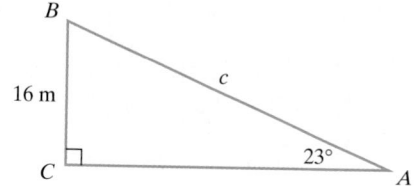

In Exercises 19–22, find the measures of the parts of the right triangle that are not given. Round all answers to the nearest whole number.

19.

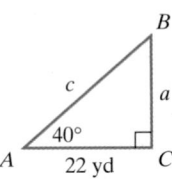

20.

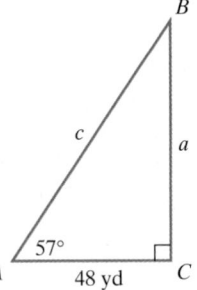

21.

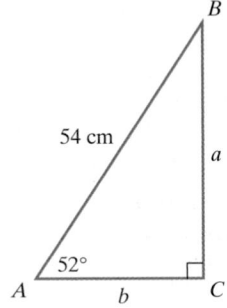

22.

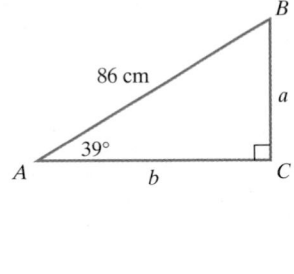

In Exercises 23–26, use the inverse trigonometric keys on a calculator to find the measure of angle A, rounded to the nearest whole degree.

23.

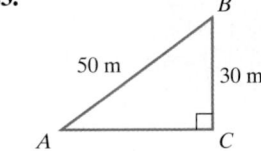

24.

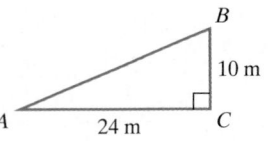

25.

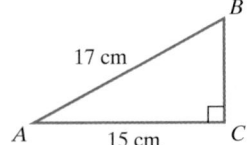

26.

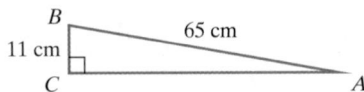

Application Exercises

27. To find the distance across a lake, a surveyor took the measurements in the figure shown. Use these measurements to determine how far it is across the lake. Round to the nearest yard.

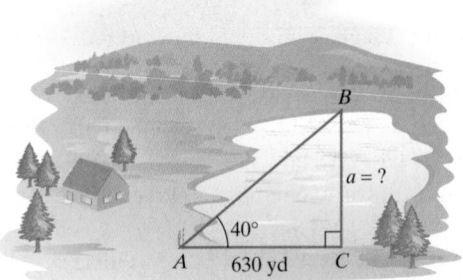

28. At a certain time of day, the angle of elevation of the sun is 40°. To the nearest foot, find the height of a tree whose shadow is 35 feet long.

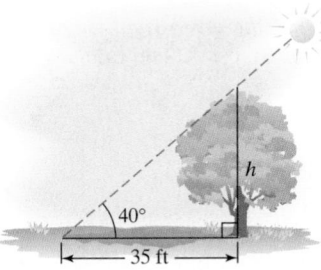

29. A plane rises from take-off and flies at an angle of 10° with the horizontal runway. When it has gained 500 feet, find the distance, to the nearest foot, the plane has flown.

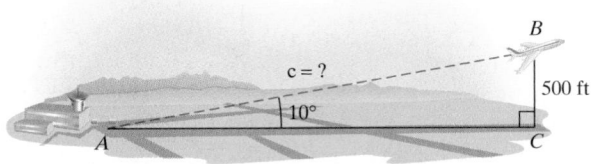

30. A road is inclined at an angle of 5°. After driving 5000 feet along this road, find the driver's increase in altitude. Round to the nearest foot.

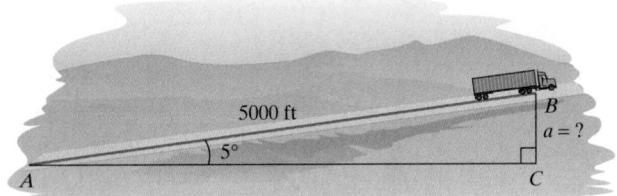

31. The tallest television transmitting tower in the world is in North Dakota. From a point on level ground 5280 feet (one mile) from the base of the tower, the angle of elevation is 21.3°. Approximate the height of the tower to the nearest foot.

32. From a point on level ground 30 yards from the base of a building, the angle of elevation is 38.7°. Approximate the height of the building to the nearest foot.

33. The Statue of Liberty is approximately 305 feet tall. If the angle of elevation of a ship to the top of the statue is 23.7°, how far, to the nearest foot, is the ship from the statue's base?

34. A 200-foot cliff drops vertically into the ocean. If the angle of elevation of a ship to the top of the cliff is 22.3°, how far off shore, to the nearest foot, is the ship?

35. A tower that is 125 feet tall casts a shadow 172 feet long. Find the angle of elevation of the sun to the nearest degree.

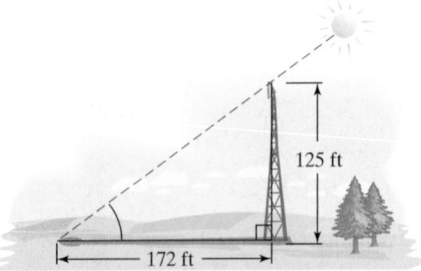

36. The Washington Monument is 555 feet high. If you stand one quarter of a mile, or 1320 feet, from the base of the monument and look to the top, find the angle of elevation to the nearest degree.

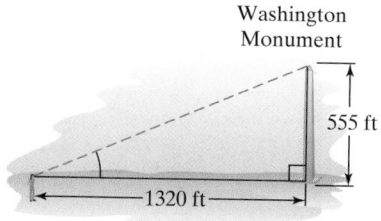

Washington
Monument

555 ft

1320 ft

37. A helicopter hovers 1000 feet above a small island. The figure shows that the angle of depression from the helicopter to point P is 36°. How far off the coast, to the nearest foot, is the island?

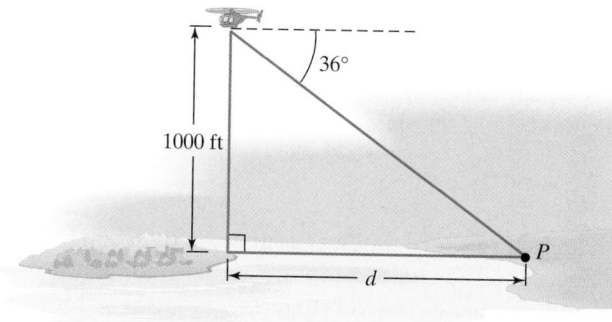

36°

1000 ft

d

P

38. A police helicopter is flying at 800 feet. A stolen car is sighted at an angle of depression of 72°. Find the distance of the stolen car, to the nearest foot, from a point directly below the helicopter.

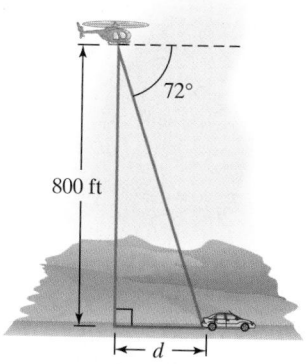

72°

800 ft

d

39. A wheelchair ramp is to be built beside the steps to the campus library. Find the angle of elevation of the 23-foot ramp, to the nearest tenth of a degree, if its final height is 6 feet.

40. A kite flies at a height of 30 feet when 65 feet of string is out. If the string is in a straight line, find the angle that it makes with the ground. Round to the nearest tenth of a degree.

Writing in Mathematics

41. If you are given the lengths of the sides of a right triangle, describe how to find the sine of either acute angle.

42. Describe one similarity and one difference between the sine ratio and the cosine ratio in terms of the sides of a right triangle.

43. If one of the acute angles of a right triangle is 37°, explain why the sine ratio does not increase as the size of the triangle increases.

44. If the measure of one of the acute angles and the hypotenuse of a right triangle are known, describe how to find the measure of the remaining parts of the triangle.

45. Describe what is meant by an angle of elevation and an angle of depression.

46. Give an example of an applied problem that can be solved using one or more trigonometric ratios. Be as specific as possible.

47. Use a calculator to find each of the following: sin 32° and cos 58°; sin 17° and cos 73°; sin 50° and cos 40°; sin 88° and cos 2°. Describe what you observe. Based on your observations, what do you think the *co* in *cosine* stands for?

48. Stonehenge, the famous "stone circle" in England, was built between 2750 B.C. and 1300 B.C. using solid stone blocks weighing over 99,000 pounds each. It required 550 people to pull a single stone up a ramp inclined at a 9° angle. Describe how right triangle trigonometry can be used to determine the distance the 550 workers had to drag a stone in order to raise it to a height of 30 feet.

Critical Thinking Exercises

49. Explain why the sine or cosine of an acute angle cannot be greater than or equal to 1.

50. Describe what happens to the tangent of an angle as the measure of the angle gets close to 90°. What happens at 90°?

51. From the top of a 250-foot lighthouse, a plane is sighted overhead and a ship is observed directly below the plane. The angle of elevation of the plane is 22° and the angle of depression of the ship is 35°. Find **a.** the distance of the ship from the lighthouse; **b.** the plane's height above the water. Round to the nearest foot.

52. Sighting the top of a building, a surveyor measured the angle of elevation to be 22°. The transit is 5 feet above the ground and 300 feet from the building. Find the building's height. Round to the nearest foot.

SECTION 10.7   BEYOND EUCLIDEAN GEOMETRY

OBJECTIVE

1. Gain an understanding of some of the general ideas of other kinds of geometries.

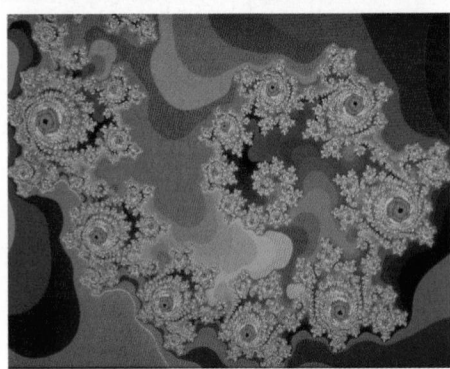

Think of your favorite natural setting. During the last quarter of the twentieth century, mathematicians developed a new kind of geometry that uses the computer to produce the diverse forms of nature that we see around us. These forms are not polygons, polyhedrons, circles, or cylinders. In this section, we move beyond Euclidean geometry to explore ideas which have extended geometry beyond the boundaries first laid down by the ancient Greek scholars.

---

1 Gain an understanding of some of the general ideas of other kinds of geometries.

The Geometry of Graphs

In the early 1700s, the city of Königsberg, Germany, was connected by seven bridges, shown in Figure 10.58. Many people in the city were interested in finding if it were possible to walk through the city so as to cross each bridge exactly once. After a few trials, you may be convinced that the answer is no. However, it is not easy to prove your answer by trial and error because there are a large number of ways of taking such a walk.

STUDY TIP

A more detailed presentation on the geometry of graphs, called *graph theory*, can be found in Chapter 15.

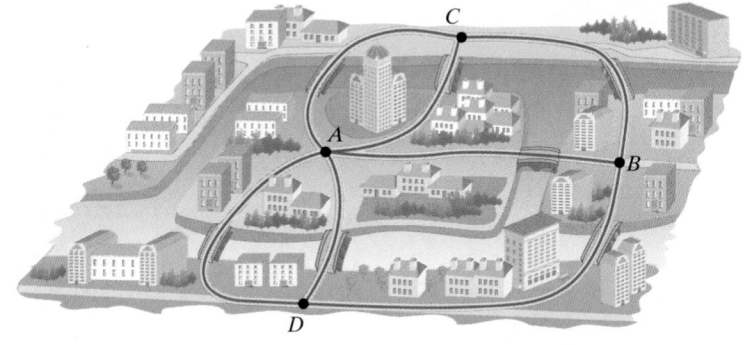

FIGURE 10.58

Ethane

Graph theory is used to show how atoms are linked to form molecules.

The problem was taken to the Swiss mathematician Leonhard Euler (1707–1783). In the year 1736, Euler proved that it is not possible to stroll through the city and cross each bridge exactly once. His solution of the problem opened up a new kind of geometry called **graph theory**. Graph theory is now used to design city streets, analyze traffic patterns, and find the most efficient routes for public transportation.

Euler solved the problem by introducing the following definitions:

A **vertex** is a point. An **edge** is a line segment or curve that starts and ends at a vertex.

Vertices and edges form a **graph**. A vertex with an odd number of attached edges is an **odd vertex**. A vertex with an even number of attached edges is an **even vertex**.

Using these definitions, Figure 10.59 shows that the graph in the Königsberg bridge problem has four odd vertices.

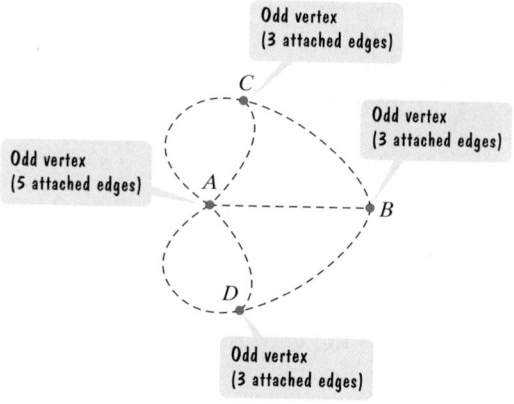

FIGURE 10.59

A graph is **traversable** if it can be traced without lifting the pencil from the paper and without tracing an edge more than once. Euler proved the following rules of traversability in solving the problem:

RULES OF TRAVERSABILITY

1. A graph with all even vertices is traversable. One can start at any vertex and end where one began.
2. A graph with two odd vertices is traversable. One must start at either of the odd vertices and finish at the other.
3. A graph with more than two odd vertices is not traversable.

Because the graph in the Königsberg bridge problem has four odd vertices, it cannot be traversed.

EXAMPLE 1 | TO TRAVERSE OR NOT TO TRAVERSE?

Consider the graph in Figure 10.60.
 a. Is this graph traversable?
 b. If it is, describe a path that will traverse it.

Solution
 a. Begin by determining if each vertex is even or odd.

FIGURE 10.60

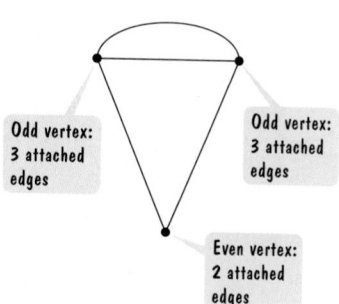

Because this graph has two odd vertices, by Euler's second rule, it is traversable.

End •   • Start

FIGURE 10.61

b. In order to find a path that traverses this graph, we again use Euler's second rule. It states that we must start at either of the odd vertices and finish at the other. One of two possible paths is shown in Figure 10.61.

Check
1
Point

Create a graph with two even and two odd vertices. Then describe a path that will traverse it.

Graph theory is used to solve many kinds of practical problems. The first step is to create a graph that represents, or models, the problem. For example, consider a UPS driver trying to find the best way to deliver packages around town. By modeling the delivery locations as vertices and roads between locations as edges, graph theory reveals the driver's most efficient path. Graphs are powerful tools because they illustrate the important aspects of a problem and leave out unnecessary detail.

Topology

A branch of modern geometry called **topology** looks at shapes in a completely new way. In Euclidean geometry, shapes are rigid and unchanging. In topology, shapes can be twisted, stretched, bent, and shrunk. Total flexibility is the rule.

A topologist does not know the difference between a doughnut and a coffee cup! This is because, topologically speaking, there is no difference. They both have one hole, and in this strange geometry of transformations, a doughnut can be flattened, pulled, and pushed to form a coffee cup.

In topology, objects are classified according to the number of holes in them, called their **genus**. The genus gives the largest number of complete cuts that can be made in the object without cutting the object into two pieces. Objects with the same genus are **topologically equivalent**.

The three shapes shown below (sphere, cube, irregular blob) have the same genus: 0. No complete cuts can be made without cutting these objects into two pieces.

STUDY TIP

The genus of an object is the same as the number of holes in the object.

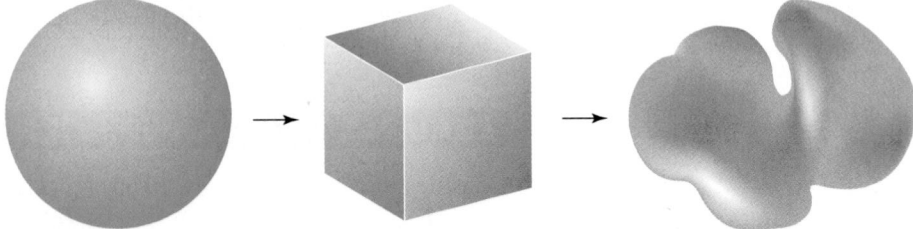

How can a doughnut be transformed into a coffee cup? It can be stretched and deformed to make a bowl part of the surface. Both the doughnut and the coffee cup have genus 1. One complete cut can be made without cutting these objects into two pieces.

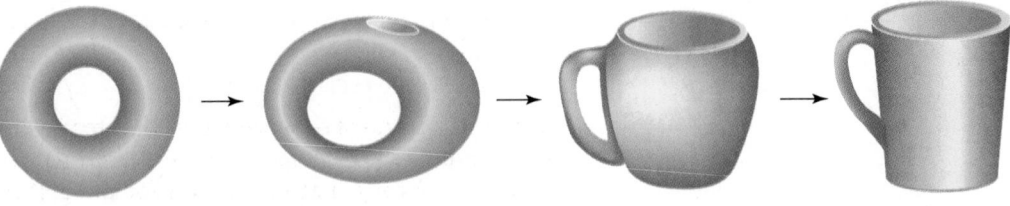

To a topologist, a lump with two holes in it is no different than a sugar bowl with two handles. Both have genus 2. Two complete cuts can be made without cutting these objects into two pieces.

Shown below is a transformation that results in a most unusual topological figure.

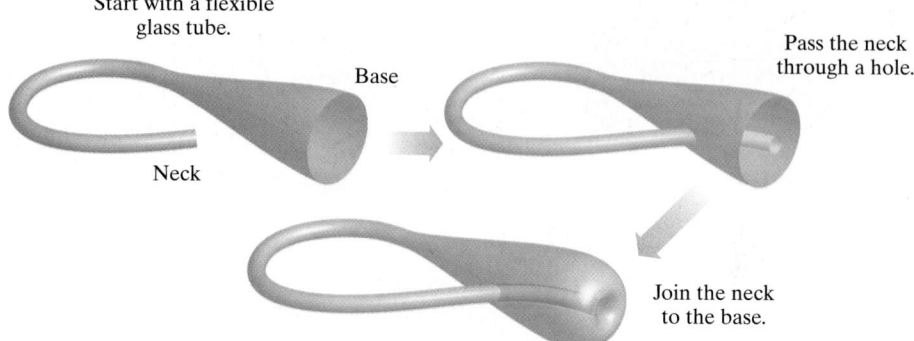

The figure that results is called a **Klein bottle**. You couldn't use one to carry water on a long hike. Because the inside surface loops back on itself to merge with the outside, it has neither an outside nor an inside and cannot hold water. A Klein bottle passes through itself without the existence of a hole, which is impossible in three-dimensional space. A true Klein bottle is visible only when generated on a computer.

A Klein bottle generated on a computer

Topology is more than an excursion in geometric fantasy. Topologists study knots and ways in which they are topologically equivalent. Knot theory is used to identify viruses and understand the ways in which they invade our cells. Such an understanding is a first step in finding vaccines for viruses ranging from the common cold to HIV.

Non-Euclidean Geometries

Imagine this: Earth is squeezed and compressed to the size of a golf ball. It becomes a **black hole**, a region in space where everything seems to disappear. Its pull of gravity is now so strong that the space near the golf ball can be thought of as a funnel with the black hole sitting at the bottom. Any object that comes near the funnel will be pulled into it by the immense gravity of the golf-ball-sized Earth. Then the object disappears! Although this sounds like science fiction, some physicists believe that there are billions of black holes in space.

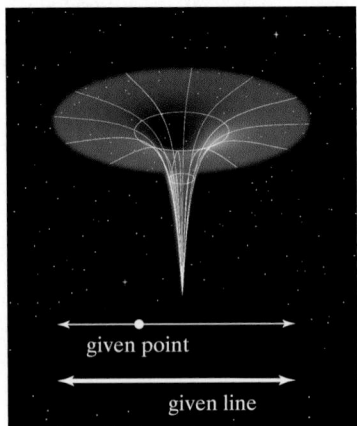

FIGURE 10.62

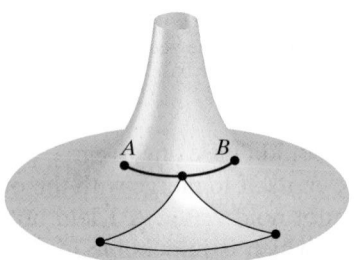

FIGURE 10.63

Is there a geometry that describes this unusual phenomenon? Yes, although it's not the geometry of Euclid. Euclidean geometry is based on the assumption that given a point not on a line, there is one line that can be drawn through the point parallel to the given line, shown as the thinner white line in Figure 10.62. We used this assumption to prove that the sum of the measures of the three angles of any triangle is 180°.

Hyperbolic geometry was developed independently by the Russian mathematician Nikolay Lobachevsky (1792–1856) and the Hungarian mathematician Janos Bolyai (1802–1860). It is based on the assumption that given a point not on a line, there are an infinite number of lines that can be drawn through the point parallel to the given line. The shapes in this non-Euclidean geometry are not represented on a plane. Instead, they are drawn on a funnel-like surface, called a *pseudosphere*, shown in Figure 10.63. On such a surface, the shortest distance between two points is a curved line. A triangle's sides are composed of arcs, and the sum of the measures of its angles is less than 180°. The shape of the pseudosphere looks like the distorted space near a black hole.

Once the ice had been broken by Lobachevsky and Bolyai, mathematicians were stimulated to set up other non-Euclidean geometries. The best known of these was **elliptic geometry**, proposed by the German mathematician Bernhard Riemann (1826–1866). Riemann began his geometry with the assumption that there are no parallel lines. As shown in Figure 10.64, elliptic geometry is on a sphere, and the sum of the measures of the angles of a triangle is greater than 180°.

Riemann's elliptic geometry was used by Albert Einstein in his theory of the universe. One aspect of the theory states that if you set off on a journey through space and keep going in the same direction, you will eventually come back to your

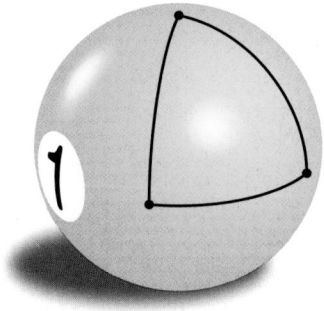

FIGURE 10.64

BLITZER BONUS

Colorful Puzzles

How many different colors are needed to color a plane map so that any two regions that share a common border are colored differently? Even though no map could be found that required more than four colors, a proof that four colors would work for every map remained elusive to mathematicians for over 100 years. The conjecture that at most four colors are needed, stated in 1852, was proved by American mathematicians Kenneth Appel and Wolfgang Haken of the University of Illinois in 1976. Their proof translated maps into graphs that represented each region as a vertex and each shared common border as an edge between corresponding vertices. The proof relied heavily on computer computations that checked approximately 1500 special cases, requiring 1200 hours of computer time. Although there are recent rumors about a flaw in the proof, no one has found a proof that does not enlist the use of computer calculations.

If you take a strip of paper, give it a single half-twist, and paste the ends together, the result is a one-sided surface called a Möbius strip. A map on a Möbius strip requires six colors to make sure that no adjacent areas are the same color.

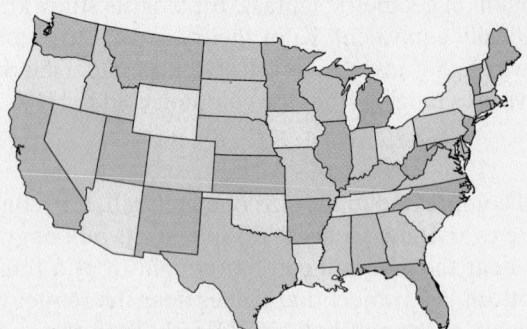

At most four colors are needed on a plane map.

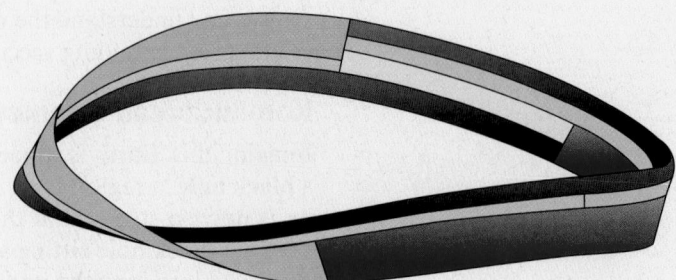

At most six colors are needed on a map drawn on a Möbius strip.

starting point. The same thing happens on the surface of a sphere. This means that space itself is "curved," although the idea is difficult to visualize. It is interesting that Einstein used ideas of non-Euclidean geometry more than 50 years after the system was logically developed. Mathematics often moves ahead of our understanding of the physical world.

The Dutch artist M. C. Escher (1898–1972) is known for combining art and geometry. Many of his images are based on non-Euclidean geometry. In the woodcut shown on the right, things enlarge as they approach the center and shrink proportionally as they approach

M. C. Escher, *Circle Limit III* © 2002 Cordon Art B.V. – Baarn, Holland. Maurits Cornelis Escher (1898–1972).

the boundary. The shortest distance between two points is a curved line, and the triangles that are formed look like those from hyperbolic geometry.

Table 10.5 Compares Euclidean geometry with hyperbolic and elliptic geometry.

TABLE 10.5 COMPARING THREE SYSTEMS OF GEOMETRY

Euclidean Geometry Euclid (300 B.C.)	Hyperbolic Geometry Lobachevsky, Bolyai (1830)	Elliptic Geometry Riemann (1850)
Given a point not on a line, there is one and only one line through the point parallel to the given line.	Given a point not on a line, there are an infinite number of lines through the point that do not intersect the given line.	There are no parallel lines.
Geometry is on a plane:	Geometry is on a pseudosphere:	Geometry is on a sphere:
The sum of the measures of the angles of a triangle is 180°.	The sum of the measures of the angles of a triangle is less than 180°.	The sum of the measures of the angles of a triangle is greater than 180°.

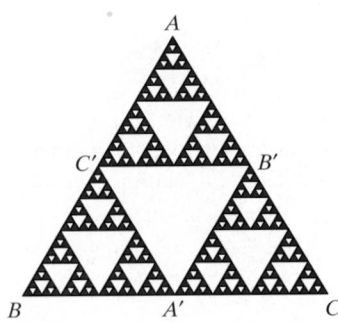

FIGURE 10.65

"I love Euclidean geometry, but it is quite clear that it does not give a reasonable presentation of the world. Mountains are not cones, clouds are not spheres, trees are not cylinders, neither does lightning travel in a straight line. Almost everything around us is non-Euclidean."

BENOIT MANDELBROT

Fractal Geometry

How can geometry describe nature's complexity in objects such as ferns, coastlines, or the human circulatory system? Any magnified portion of a fern repeats much of the pattern of the whole fern, as well as new and unexpected patterns. This property is called **self-similarity**.

Magnified portion

Through the use of computers, a new geometry of natural shapes, called **fractal geometry**, has arrived. The word "fractal" is from the Latin word *fractus*, meaning "broken up," or "fragmented." The fractal shape shown in Figure 10.65 is, indeed, broken up. This shape was obtained by repeatedly subtracting triangles from within triangles. From the midpoints of the sides of triangle ABC, remove triangle $A'B'C'$. Repeat the process for the three remaining triangles, then with each of the triangles left after that, and so on ad infinitum.

The process of repeating a rule again and again to create a self-similar fractal like the broken-up triangle is called **iteration**. The self-similar fractal shape in Figure 10.66 was generated on a computer by iteration that involved subtracting pyramids within pyramids. Computers can generate fractal shapes by carrying out thousands or millions of iterations.

In 1975, Benoit Mandelbrot (1924–), the mathematician who developed fractal geometry, published *The Fractal Geometry of Nature*. This book of beautiful computer graphics shows images that look like actual mountains, coastlines, underwater coral gardens, and flowers, all imitating nature's forms. Many of the realistic-looking landscapes that you see in movies are actually computer-generated fractal images.

FIGURE 10.66 Daryl H. Hepting and Allan N. Snider, "Desktop Tetrahedron" 1990. Computer generated at the University of Regina in Canada

These images are computer-generated fractals.

Exercise Set 10.7

Practice Exercises

For each graph in Exercises 1–6, **a.** *Is the graph traversable?* **b.** *If it is, describe a path that will traverse it.*

1.

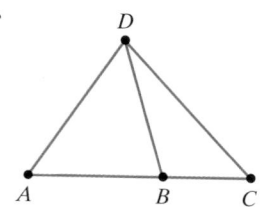

2.

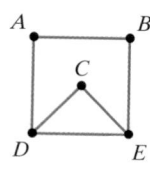

3.

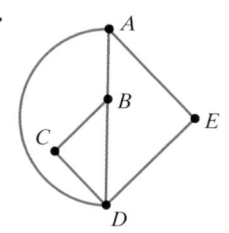

4.

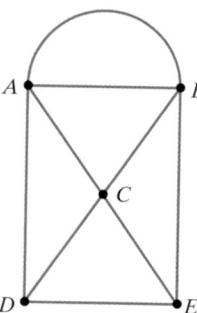

5.

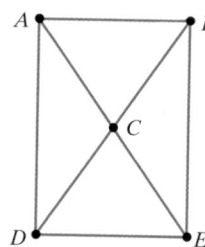

6.

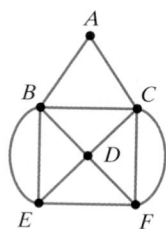

In Exercises 7–10, give the genus of each object.

7.

Pretzel

8.

Pitcher

9.

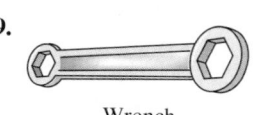

Wrench

10.

Button

11. In Exercises 7–10, which objects are topologically equivalent?

12. Draw (or find and describe) an object of genus 4 or more.

13. What does the figure shown illustrate about angles in a triangle when the triangle is drawn on a sphere?

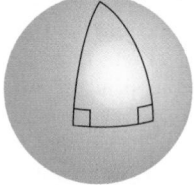

14. The figure shows a quadrilateral that was drawn on the surface of a sphere. Angles C and D are obtuse angles. What does the figure illustrate about the sum of the measures of the four angles in a quadrilateral in elliptic geometry?

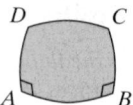

In Exercises 15–17, describe whether each figure shown exhibits self-similarity.

15.

16.

17.

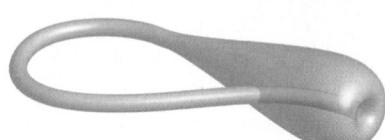

Application Exercises

The model shows the way that carbon atoms (red) and hydrogen atoms (blue) link together to form a propane molecule. Use the model to answer Exercises 18–20.

Propane

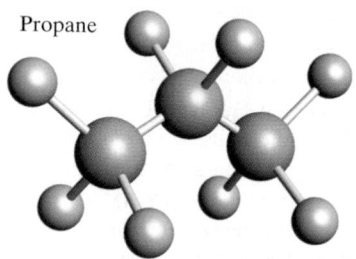

(In Exercises 18–20, be sure to refer to the figure on the bottom of page 549.)

18. Using each of the three red carbon atoms as a vertex and each of the eight black hydrogen atoms as a vertex, draw a graph for the propane molecule.

19. Determine if each vertex of the network is even or odd.

20. Is the network traversable?

The figure on the left shows the floor plan of a four-room house. By representing rooms as vertices, the outside, E, as a vertex, and doors as edges, the figure on the right is a graph that models the floor plan. Use these figures to solve Exercises 21–24.

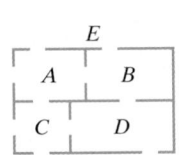 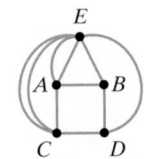

21. Use the floor plan to determine how many doors connect the outside, *E*, to room *A*. How is this shown in the graph?

22. Use the floor plan to determine how many doors connect the outside, *E*, to room *C*. How is this shown in the graph?

23. Is the graph traversable? Explain your answer.

24. If the graph is traversable, determine a path to walk through every room, using each door only once.

25. Consult an Internet site devoted to fractals and download two fractal images that imitate nature's forms. How is self-similarity shown in each image?

Writing in Mathematics

26. What is a graph?

27. What does it mean if a graph is traversable?

28. How do you determine whether or not a graph is traversable?

29. Describe one way in which topology is different than Euclidean geometry.

30. What is the genus of an object?

31. Describe how a rectangular solid and a sphere are topologically equivalent.

32. State the assumption that Euclid made about parallel lines that was altered in both hyperbolic and elliptic geometry.

33. How does hyperbolic geometry differ from Euclidean geometry?

34. How does elliptic geometry differ from Euclidean geometry?

35. What is self-similarity? Describe an object in nature that has this characteristic.

36. Some suggest that nature produces its many forms through a combination of both iteration and a touch of randomness. Describe what this means.

37. Find an Internet site devoted to fractals. Use the site to write a paper on a specific use of fractals.

38. Did you know that laid end to end, the veins, arteries, and capillaries of your body would reach over 40,000 miles? However, your vascular system occupies a very small fraction of your body's volume. Describe a self-similar object in nature, such as your vascular system, with enormous length but relatively small volume.

39. Describe a difference between the shapes of man-made objects and objects that occur in nature.

40. Explain what this short poem by Jonathan Swift has to do with fractal images:

> So Nat'ralists observe, A Flea
> Hath Smaller Fleas that on him prey
> and these have smaller Fleas to bite'em
> And so proceed, ad infinitum

41. In Tom Stoppard's play *Arcadia*, the characters talk about mathematics, including ideas from fractal geometry. Read the play and write a paper on one of the ideas that the characters discuss that is related to something presented in this section.

Group Exercises

42. This activity is suggested for two or three people. Research some of the practical applications of topology. Present the results of your research in a seminar to the entire class. You may also want to include a discussion of some of topology's more unusal figures, such as the Klein bottle and the Möbius strip.

43. Research non-Euclidean geometry and plan a seminar based on your group's research. Each group member should research one of the following five areas:

   a. Present an overview of the history of the people who developed non-Euclidean geometry. Who first used the term and why did he never publish his work?

   b. Present an overview of the connection between Saccheri quadrilaterals and non-Euclidean geometry. Describe the work of Girolamo Saccheri.

   c. Describe how Albert Einstein applied the ideas of Gauss and Riemann. Discuss the notion of curved space and a fourth dimension.

   d. Present examples of the work of M. C. Escher that provide ways of visualizing hyperbolic and elliptic geometry.

   e. Describe how non-Euclidean geometry changed the direction of subsequent research in mathematics.

   After all research has been completed, the group should plan the order in which each group member will speak. Each person should plan on taking about five minutes for his or her portion of the presentation.

44. Albert Einstein's theory of general relativity is concerned with the structure, or the geometry, of the universe. In order to describe the universe, Einstein discovered that he needed four variables: three variables to locate an object in space and a fourth variable describing time. This system is known as space-time.

   Because we are three-dimensional beings, how can we imagine four dimensions? One interesting approach to visualizing four dimensions is to consider an analogy of a two-dimensional being struggling to understand three dimensions. This approach first appeared in a book called *Flatland* by Edwin Abbott, written around 1884.

   Flatland describes an entire race of beings who are two dimensional, living on a flat plane, unaware of the existence of anything outside their universe. A house in Flatland would look like a blueprint or a line drawing to us. If we were to draw a closed circle around Flatlanders, they would be imprisoned in a cell with no way to see out or escape because there is no way to move up and over the circle.

For a two-dimensional being moving only on a plane, the idea of up would be incomprehensible. We could explain that up means moving in a new direction, perpendicular to the two dimensions they know, but it would be similar to telling us we can move in the fourth dimension by traveling perpendicular to our three dimensions.

Group members should obtain copies of or excerpts from Edwin Abbott's *Flatland*. Once all group members have read the story, the following questions are offered for group discussion.

a. How does the sphere, the visitor from the third dimension, reflect the same narrow perspective as the Flatlanders?

b. What are some of the sociological problems raised in the story?

c. What happens when we have a certain way of seeing the world that is challenged by coming into contact with something quite different? Be as specific as possible, citing either personal examples or historical examples.

d. How are A. Square's difficulties in visualizing three dimensions similar to those of a three-dimensional dweller trying to visualize four dimensions?

e. How does the author reflect the overt sexism of his time?

f. What "upward not northward" ideas do you hold that, if shared, would result in criticism, rejection, or a fate similar to that of the narrator of *Flatland*?

Chapter Summary, Review, and Test

SUMMARY

DEFINITIONS AND CONCEPTS	EXAMPLES

10.1 Points, Lines, Planes, and Angles

a. Line AB ($\overleftrightarrow{AB}$ or $\overleftrightarrow{BA}$), half-line AB ($\overrightarrow{AB}$), ray AB ($\overrightarrow{AB}$), and line segment AB ($\overline{AB}$ or $\overline{BA}$) are represented in Figure 10.2 on page 492.

b. Angles are measured in degrees. A degree, $1°$, is $\frac{1}{360}$ of a complete rotation. Acute angles measure less than $90°$, right angles $90°$, obtuse angles more than $90°$ but less than $180°$, and straight angles $180°$.   Ex. 1, p. 492

c. Complementary angles are two angles whose measures have a sum of $90°$ and supplementary angles a sum of $180°$.   Ex. 2, p. 494; Ex. 3, p. 494

d. Vertical angles have the same measure.   Ex. 4, p. 495

e. If parallel lines are cut by a transversal, alternate interior angles, alternate exterior angles, and corresponding angles have the same measure.   Ex. 5, p. 497

10.2 Triangles

a. The sum of the measures of the three angles of any triangle is $180°$.   Ex. 1, p. 501;

b. Triangles can be classified by angles (acute right, obtuse) or by sides (isosceles, equilateral, scalene). See the box on page 502.   Ex. 2, p. 501

c. Similar triangles have the same shape, but not necessarily the same size. Corresponding angles have the same measure and corresponding sides are proportional. If the measures of two angles of one triangle are equal to those of two angles of a second triangle, then the two triangles are similar.   Ex. 3, p. 503; Ex. 4, p. 504

d. The Pythagorean Theorem: The sum of the squares of the lengths of the legs of a right triangle equals the square of the length of the hypotenuse.   Ex. 5, p. 505; Ex. 6, p. 506

10.3 Polygons, Quadrilaterals, and Perimeter

a. A polygon is a closed geometric figure in a plane formed by three or more line segments. Names of some polygons are given in Table 10.2 on page 510. A regular polygon is one whose sides are all the same length and whose angles all have the same measure. The sum of the measures of the angles of an n-sided polygon is $(n - 2)180°$.   Ex. 2, p. 512

b. Quadrilaterals are four-sided polygons. The sum of the measures of the angles is $360°$. Types of quadrilaterals and their characteristics are given in Table 10.3 on page 511.

c. The perimeter of a polygon is the sum of the lengths of its sides.   Ex. 1, p. 511

10.4 Area and Circumference

a. Formulas for Area
Rectangle: $A = lw$; Square: $A = s^2$; Parallelogram: $A = bh$;
Triangle: $A = \frac{1}{2}bh$; Trapezoid: $A = \frac{1}{2}h(a + b)$

Ex. 1–Ex. 5,
pp. 517–520

b. Circles
Circumference: $C = 2\pi r$ or $C = \pi d$;
Area: $A = \pi r^2$

Ex. 6, p. 521;
Ex. 7, p. 522;
Ex. 8, p. 522

10.5 Volume

a. Formulas for Volume
Rectangular Solid: $V = lwh$; Cube: $V = s^3$; Pyramid: $V = \frac{1}{3}Bh$;
Cylinder: $V = \pi r^2 h$; Cone: $V = \frac{1}{3}\pi r^2 h$;
Sphere: $V = \frac{4}{3}\pi r^3$

Ex. 1–Ex. 6,
pp. 527–530

b. Formulas for surface area are given in Table 10.4 on page 531.

Ex. 7, p. 531

10.6 Right Triangle Trigonometry

a. Trigonometric ratios, $\sin A$, $\cos A$, and $\tan A$, are defined in the box on page 534.

Ex. 1, p. 535

b. Given one side and the measure of an acute angle of a right triangle, the trigonometric ratios can be used to find the measures of the other parts of the right triangle.

Ex. 2, p. 536;
Ex. 3, p. 536;
Ex. 4, p. 537

c. Given the measures of two sides of a right triangle, the measures of the acute angles can be found using the inverse trigonometric ratio keys $\boxed{\text{SIN}^{-1}}$, $\boxed{\text{COS}^{-1}}$, and $\boxed{\text{TAN}^{-1}}$, on a calculator.

Ex. 5, p. 538

10.7 Beyond Euclidean Geometry

a. A vertex is a point. An edge is a line segment or curve that starts and ends at a vertex. Vertices and edges form a graph. A vertex with an odd number of edges is odd; one with an even number of edges is even. A traversable graph is one that can be traced without removing the pencil from the paper and without tracing an edge more than once. A graph with all even vertices or one with two odd vertices is traversable. With more than two odd vertices, the graph cannot be traversed.

Ex. 1, p. 543

b. In topology, shapes are twisted, stretched, bent, and shrunk. The genus of an object refers to the number of holes in the object. Objects with the same genus are topologically equivalent.

c. Non-Euclidean geometries, including hyperbolic geometry and elliptic geometry, are outlined in Table 10.5 on page 547.

d. Fractal geometry includes self-similar forms; magnified portions of such forms repeat much of the pattern of the whole form.

REVIEW EXERCISES

10.1

In the figure shown, lines l and m are parallel. In Exercises 1–7, match each term with the numbered angle or angles in the figure.

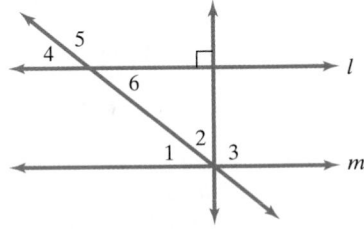

1. right angle

2. obtuse angle

3. vertical angles

4. alternate interior angles

5. corresponding angles

6. the complement of ∡1

7. the supplement of ∡6

In Exercises 8–9, find the measure of the angle in which ?° appears.

8.

115°
?°

9.

?°
41°

10. If an angle measures 73°, find the measure of its complement.

11. If an angle measures 46°, find the measure of its supplement.

12. In the figure shown, find the measures of angles 1, 2, and 3.

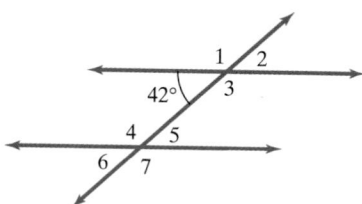

13. In the figure shown, two parallel lines are intersected by a transversal. One of the angle measures is given. Find the measure of each of the other seven angles.

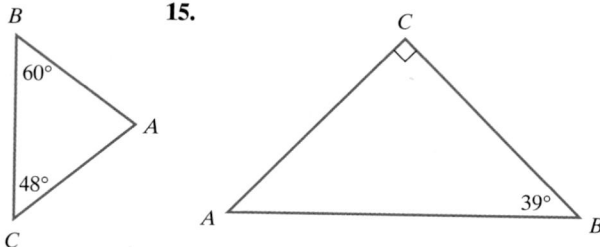

10.2

In Exercises 14–15, find the measure of angle A for the triangle shown.

14.

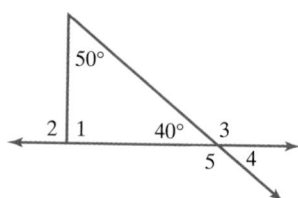

15.

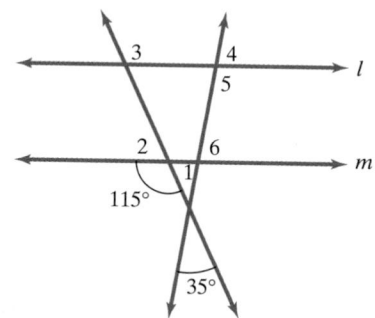

16. Find the measures of angles 1 through 5 in the figure shown.

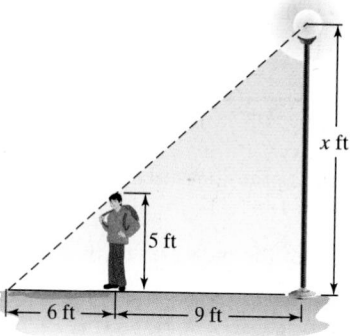

17. In the figure shown, lines *l* and *m* are parallel. Find the measure of each numbered angle.

In Exercises 18–19, use similar triangles and the fact that corresponding sides are proportional to find the length of each side marked with an x.

18.

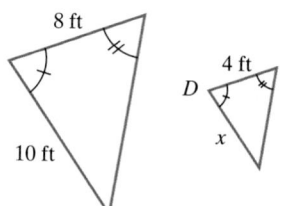

19.

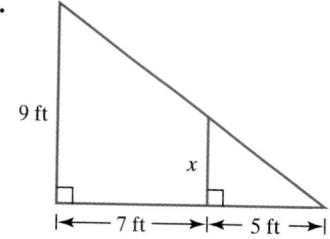

In Exercises 20–22, use the Pythagorean Theorem to find the missing length in each right triangle. Round, if necessary, to the nearest tenth.

20.   **21.**

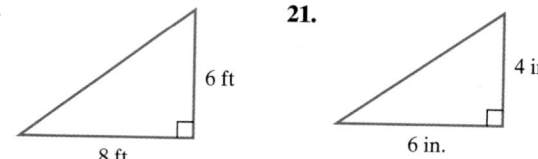

22.

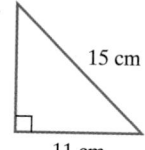

23. Find the height of the lamppost in the figure.

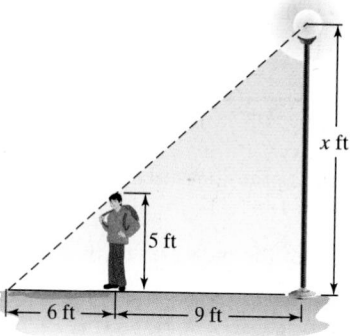

24. How far away from the building in the figure shown is the bottom of the ladder?

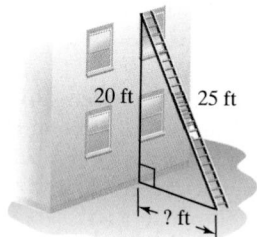

25. A vertical pole is to be supported by three wires. Each wire is 13 yards long and is anchored 5 yards from the base of the pole. How far up the pole will the wires be attached?

10.3

26. Write the names of all quadrilaterals that always have four right angles.

27. Write the names of all quadrilaterals with four sides always having the same measure.

28. Write the names of all quadrilaterals that do not always have four angles with the same measure.

In Exercises 29–31, find the perimeter of the figure shown. Express the perimeter using the same unit of measure that appears in the figure.

29.

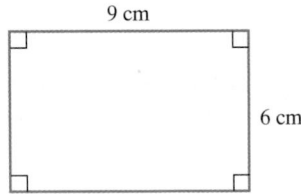

30.

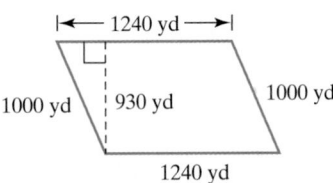

31.

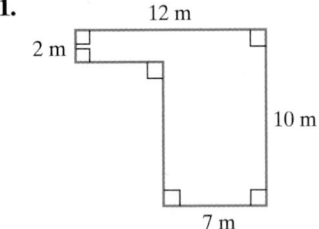

32. Find the sum of the measures of the angles of a 12-sided polygon.

33. Find the sum of the measures of the angles of an octagon.

34. The figure shown is a regular polygon. Find the measure of angle 1 and angle 2.

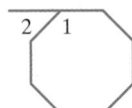

35. A carpenter is installing a baseboard around a room that has a length of 35 feet and a width of 15 feet. The room has four doorways and each doorway is 3 feet wide. If no baseboard is to be put across the doorways and the cost of the baseboard is $1.50 per foot, what is the cost of installing the baseboard around the room?

10.4

In Exercises 36–39, find the area of each figure.

36.

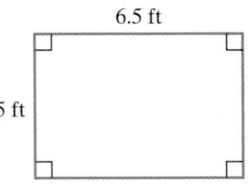

37.

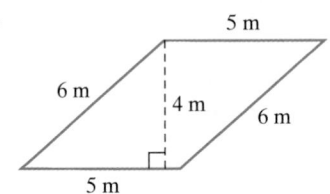

38.

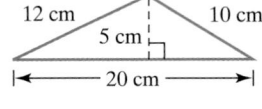

39.

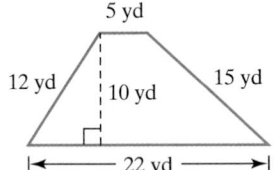

40. Find the circumference and the area of a circle with a diameter of 20 meters. Express answers in terms of π and then round to the nearest tenth.

In Exercises 41–42, find the area of each figure.

41.

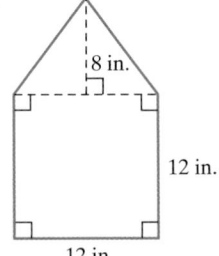

42.

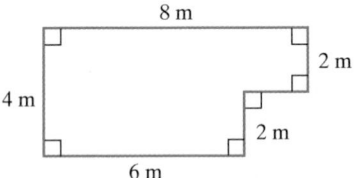

43. What will it cost to carpet a rectangular floor measuring 15 feet by 21 feet if the carpet costs $22.50 per square yard?

44. What will it cost to cover a rectangular floor measuring 40 feet by 50 feet with square tiles that measure 2 feet on each side if a package of 10 tiles costs $13 per package?

45. How much fencing, to the nearest whole yard, is needed to enclose a circular garden that measures 10 yards across?

10.5

In Exercises 46–50, find the volume of each figure. Where necessary, express answers in terms of π and then round to the nearest whole number.

46.

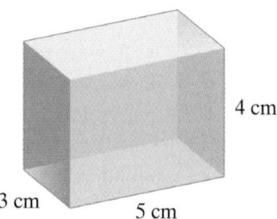

4 cm

3 cm   5 cm

47.

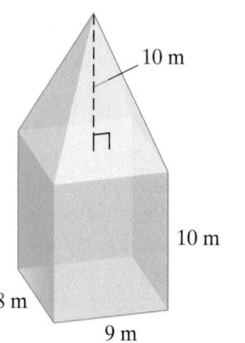

10 m

10 m

8 m

9 m

48.

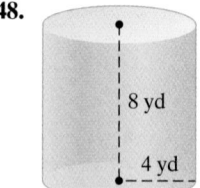

8 yd

4 yd

49.

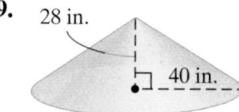

28 in.

40 in.

50.

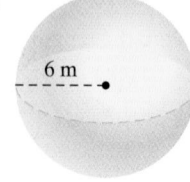

6 m

51. Find the surface area of the figure shown.

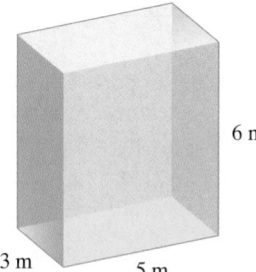

6 m

3 m   5 m

52. A train is being loaded with shipping boxes. Each box is 8 meters long, 4 meters wide, and 3 meters high. If there are 50 shipping boxes, how much space is needed?

53. An Egyptian pyramid has a square base measuring 145 meters on each side. If the height of the pyramid is 93 meters, find its volume.

54. What is the cost of concrete for a walkway that is 27 feet long, 4 feet wide, and 6 inches deep if the concrete is $40 per cubic yard?

10.6

55. Use the right triangle shown to find ratios, in reduced form, for sin A, cos A, and tan A.

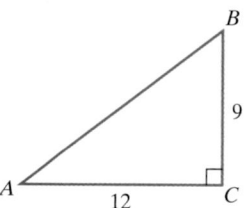

B

9

A   12   *C*

In Exercises 56–58, find the measure of the side of the right triangle whose length is designated by a lowercase letter. Round answers to the nearest whole number.

56.

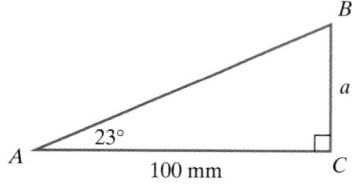

B

a

23°

A   100 mm   *C*

57.

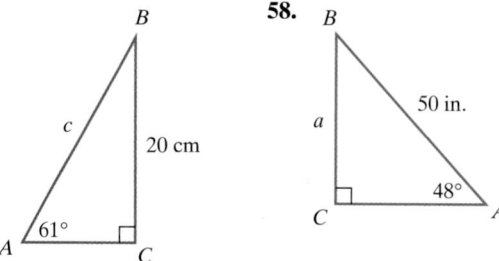

B

c   20 cm

61°

A   *C*

58.

B

50 in.

a

48°

C   *A*

59. Find the measure of angle A in the right triangle shown. Round to the nearest whole degree.

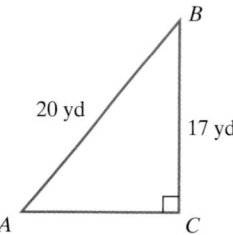

B

20 yd   17 yd

A   *C*

60. A hiker climbs for a half mile (2640 feet) up a slope whose inclination is 17°. How many feet of altitude, to the nearest foot, does the hiker gain?

61. To find the distance across a lake, a surveyor took the measurements in the figure shown. What is the distance across the lake? Round to the nearest meter.

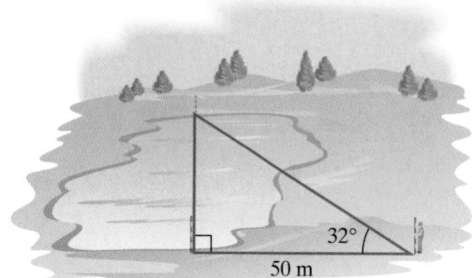

32°

50 m

62. When a six-foot pole casts a four-foot shadow, what is the angle of elevation of the sun? Round to the nearest whole degree.

10.7

For each graph in Exercises 63–64, determine whether it is traversable. If it is, describe a path that will traverse it.

63.
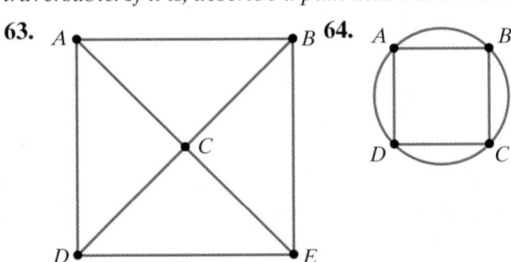

64.
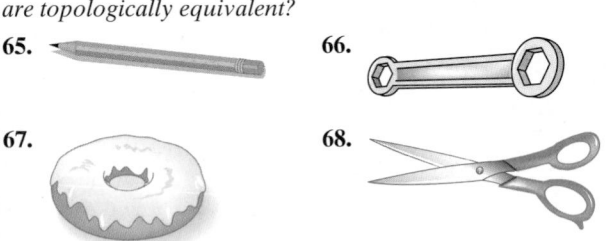

In Exercises 65–68, give the genus of each object. Which objects are topologically equivalent?

65.
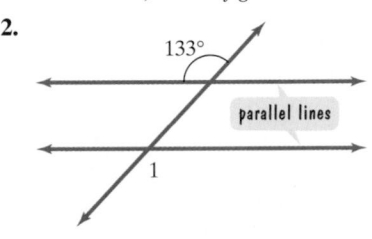

66.
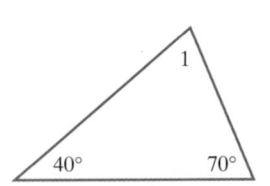

67.
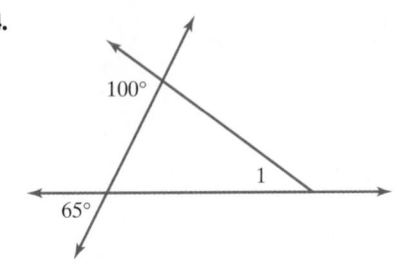

68.

69. State Euclid's assumption about parallel lines that no longer applies in hyperbolic and elliptic geometry.

70. What is self-similarity? Describe an object in nature that has this characteristic.

Group Exercise

71. a. As a group, write a problem that is based on the situation that opened this chapter. Group members will need to assign realistic values, in inches, to the radius and height of each cylindrical tank. You can decide if each tank is the same size, or if they differ in size. Use these assigned numbers, the formula for the volume of a cylinder, and the information given to the left of the photograph on page 490 to determine if it is safe to cross the bridge.

b. We have seen that geometry has many practical applications. Now, group members need to take this one step further. Describe as many situations as possible in which a knowledge of geometry can prevent a disaster (or at least a serious mishap) from happening. Be as creative as possible. It is not necessary to assign specific numbers in each situation or to solve the problem, as you were asked to do in part (a).

CHAPTER 10 TEST

1. If an angle measures 54°, find the measure of its complement and supplement.

In Exercises 2–4, use the figure shown to find the measure of angle 1.

2.
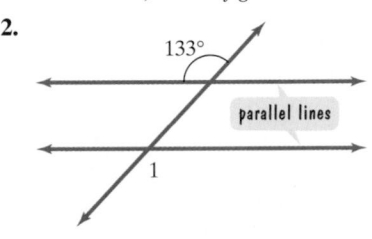

133°

parallel lines

1

3.

1

40°   70°

4.

100°

1

65°

5. The triangles in the figure are similar. Find the length of the side marked with an *x*.

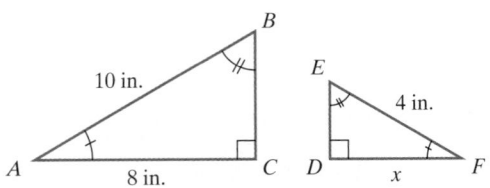

6. A vertical pole is to be supported by three wires. Each wire is 26 feet long and is anchored 24 feet from the base of the pole. How far up the pole should the wires be attached?

7. Find the sum of the measures of the angles of a ten-sided polygon.

8. Find the perimeter of the figure shown.

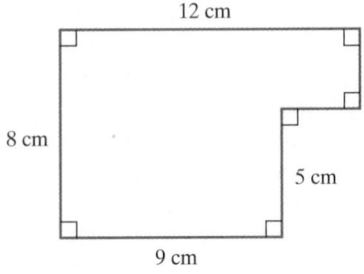

12 cm

8 cm

5 cm

9 cm

9. Which one of the following names a quadrilateral in which the sides that meet at each vertex have the same measure?

a. rectangle   **b.** parallelogram

c. trapezoid   **d.** rhombus

In Exercises 10–11, find the area of each figure.

10.

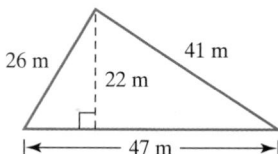

11.

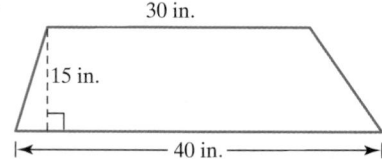

12. The right triangle shown has one leg of length 5 centimeters and a hypotenuse of length 13 centimeters.

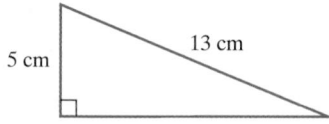

a. Find the length of the other leg.

b. What is the perimeter of the triangle?

c. What is the area of the triangle?

13. Find the circumference and area of a circle with a diameter of 40 meters. Express answers in terms of π and then round to the nearest tenth.

14. A rectangular floor measuring 8 feet by 6 feet is to be completely covered with square tiles measuring 8 inches on each side. How many tiles are needed to completely cover the floor?

In Exercises 15–17, find the volume of each figure. If necessary, express the answer in terms of π and then round to the nearest whole number.

15.   **16.**

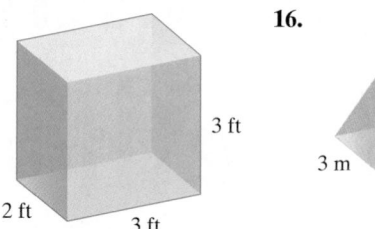

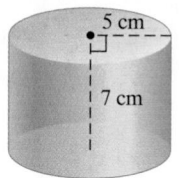

17.

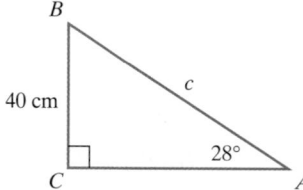

18. Find the measure, to the nearest whole number, of the side of the right triangle whose length is designated by c.

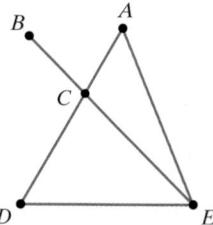

19. At a certain time of day, the angle of elevation of the sun is $34°$. If a building casts a shadow measuring 104 feet, find the height of the building to the nearest foot.

20. Determine if the graph shown is traversable. If it is, describe a path that will traverse it.

21. Describe a difference between the shapes of fractal geometry and those of Euclidean geometry.

Counting Methods and Probability Theory

ife is filled with risks and uncertainty. We purchase insurance to compensate ourselves for the possibility of an unpleasant event. Weather forecasters tell us the chance of rain tomorrow, and on the basis of the information they provide we decide what to wear when we go out. Stock market investments are placed according to the probabilities of future performances of different companies.

The mathematics of uncertainty and risk is called probability theory. In this chapter, you will see how probability theory gives you the tools for making decisions, such as whether or not to list an expensive house. In this sense, the mathematics of uncertainty will give you a way to peer into the future, enabling you to see things that would otherwise be invisible.

You have a plan to help pay for college. It took a lot of hard work, but you got your real estate license. After a few months of making no money, a great opportunity has come your way. You have the chance to list a relatively expensive house. If it sells during your listing period, you can make enough money to pay for at least a year of college. However, the listing will cost you quite a bit of money for advertising and providing food for other realtors during open showings. If the house does not sell after four months, you lose the listing, along with your out-of-pocket money to cover these expenses. Such a loss will leave you short of money for the fall semester. What should you do?

This problem appears as Example 4 in Section 11.8.

SECTION 11.1   THE FUNDAMENTAL COUNTING PRINCIPLE

OBJECTIVE

1. Use the Fundamental Counting Principle to determine the number of possible outcomes in a given situation.

Have you ever imagined what your life would be like if you won the lottery? What changes would you make? Before you fantasize about becoming a person of leisure with a staff of obedient elves, think about this: The probability of winning top prize in the lottery is about the same as the probability of being struck by lightning. There are millions of possible number combinations in lottery games, and only one way of winning the grand prize. Determining the probability of winning involves calculating the chance of getting the winning combination from all possible outcomes. In this section, we begin preparing for the surprising world of probability by looking at methods for counting possible outcomes.

| 1 | Use the Fundamental Counting Principle to determine the number of possible outcomes in a given situation. |

The Fundamental Counting Principle with Two Groups of Items

It's early morning, you're groggy, and you have to select something to wear for your 8 A.M. class. (What *were* you thinking when you signed up for a class at that hour?!) Fortunately, your "lecture wardrobe" is rather limited—just two pairs of jeans to choose from (one blue, one black) and three T-shirts to choose from (one beige, one yellow, and one blue). Your early-morning dilemma is illustrated in Figure 11.1.

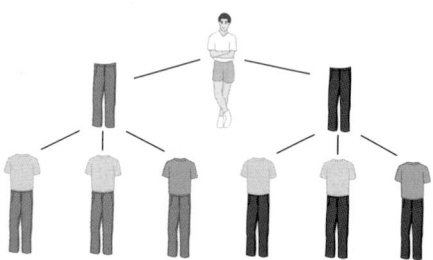

FIGURE 11.1 Selecting a wardrobe

The **tree diagram**, so named because of its branches, shows that you can form six different outfits from your two pairs of jeans and three T-shirts. Each pair of jeans can be combined with one of three T-shirts. Notice that the total number of outfits can be obtained by multiplying the number of choices for the jeans, 2, by the number of choices for the T-shirts, 3:

$$2 \cdot 3 = 6.$$

We can generalize this idea to any two groups of items—not just jeans and T-shirts—with the **Fundamental Counting Principle**.

THE FUNDAMENTAL COUNTING PRINCIPLE

If you can choose one item from a group of M items and a second item from a group of N items, then the total number of two-item choices is $M \cdot N$.

EXAMPLE 1   APPLYING THE FUNDAMENTAL COUNTING PRINCIPLE

The Greasy Spoon Restaurant offers 6 appetizers and 14 main courses. In how many ways can a person order a two-course meal?

Solution   Choosing from one of 6 appetizers and one of 14 main courses, the total number of two-course meals is

$$6 \cdot 14 = 84.$$

A person can order a two-course meal in 84 different ways.

 A restaurant offers 10 appetizers and 15 main courses. In how many ways can you order a two-course meal?

EXAMPLE 2   APPLYING THE FUNDAMENTAL COUNTING PRINCIPLE

This is the semester that you will take your required psychology and social science courses. Because you decide to register early, there are 15 sections of psychology from which you can choose. Furthermore, there are 9 sections of social science that are available at times that do not conflict with those for psychology. In how many ways can you create two-course schedules that satisfy the psychology-social science requirement?

Solution   The number of ways that you can satisfy the requirement is found by multiplying the number of choices for each course. You can choose your psychology course from 15 sections and your social science course from 9 sections. For both courses you have

$$15 \cdot 9, \text{ or } 135$$

choices. Thus, you can satisfy the psychology-social science requirement in 135 ways.

 Rework Example 2 given that the number of sections of psychology and nonconflicting sections of social science each decrease by 5.

The Fundamental Counting Principle with More Than Two Groups of Items

Whoops! You forgot something in choosing your lecture wardrobe—shoes! You have two pairs of sneakers to choose from—one black and one red, for that extra fashion flair! Your possible outfits, including sneakers, are shown in Figure 11.2.

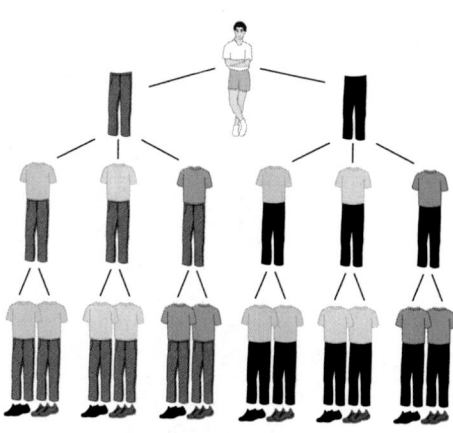

FIGURE 11.2 Increasing wardrobe selections

The number of possible ways of playing the first four moves on each side in a game of chess is 318,979,564,000.

The tree diagram shows that you can form 12 outfits from your two pairs of jeans, three T-shirts, and two pairs of sneakers. Notice that the number of outfits can be obtained by multiplying the number of choices for jeans, 2, the number of choices for T-shirts, 3, and the number of choices for sneakers, 2:

$$2 \cdot 3 \cdot 2 = 12.$$

Unlike your earlier dilemma, you are now dealing with *three* groups of items. The Fundamental Counting Principle can be extended to determine the number of possible outcomes in situations in which there are three or more groups of items.

THE FUNDAMENTAL COUNTING PRINCIPLE

The number of ways in which a series of successive things can occur is found by multiplying the number of ways in which each thing can occur.

For example, if you own 30 pairs of jeans, 20 T-shirts, and 12 pairs of sneakers, you have

$$30 \cdot 20 \cdot 12 = 7200$$

choices for your wardrobe!

EXAMPLE 3   OPTIONS IN PLANNING A COURSE SCHEDULE

Next semester you are planning to take three courses—math, English, and humanities. Based on time blocks and highly recommended professors, there are 8 sections of math, 5 of English, and 4 of humanities that you find suitable. Assuming no scheduling conflicts, how many different three-course schedules are possible?

Solution This situation involves making choices with three groups of items.

Math	English	Humanities
8 choices	5 choices	4 choices

We use the Fundamental Counting Principle to find the number of three-course schedules. Multiply the number of choices for each of the three groups.

$$8 \cdot 5 \cdot 4 = 160$$

Thus, there are 160 different three-course schedules.

 **Check 3 Point** A pizza can be ordered with two choices of size (medium or large), three choices of crust (thin, thick, or regular), and five choices of toppings (ground beef, sausage, pepperoni, bacon, or mushrooms). How many different one-topping pizzas can be ordered?

EXAMPLE 4 CAR OF THE FUTURE

Car manufacturers are now experimenting with lightweight three-wheel cars, designed for one person, and considered ideal for city driving. Intrigued? Suppose you could order such a car with a choice of 9 possible colors, with or without air conditioning, electric or gas powered, and with or without an onboard computer. In how many ways can this car be ordered with regard to these options?

Solution This situation involves making choices with four groups of items.

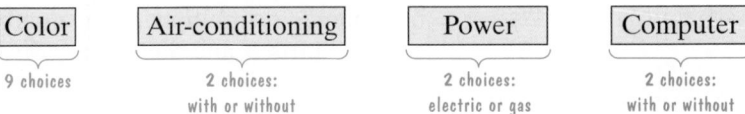

Color	Air-conditioning	Power	Computer
9 choices	2 choices: with or without	2 choices: electric or gas	2 choices: with or without

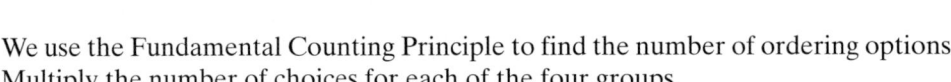

We use the Fundamental Counting Principle to find the number of ordering options. Multiply the number of choices for each of the four groups.

$$9 \cdot 2 \cdot 2 \cdot 2 = 72$$

Thus, the car can be ordered in 72 different ways.

Source: Corbin Motors (www.Corbinmotors.com)

Check 4 Point The car in Example 4 is now available in 10 possible colors. The options involving air conditioning, power, and an onboard computer still apply. Furthermore, the car is available with or without a global positioning system (for pinpointing your location at every moment). In how many ways can this car be ordered in terms of these options?

EXAMPLE 5 A MULTIPLE-CHOICE TEST

You are taking a multiple-choice test that has ten questions. Each of the questions has four answer choices, with one correct answer per question. If you select one of these four choices for each question and leave nothing blank, in how many ways can you answer the questions?

Solution This situation involves making choices with ten questions.

Question 1	Question 2	Question 3	· · ·	Question 9	Question 10
4 choices	4 choices	4 choices		4 choices	4 choices

We use the Fundamental Counting Principle to determine the number of ways that you can answer the questions on the test. Multiply the number of choices, 4, for each of the ten questions.

$$4 \cdot 4 \cdot 4 \cdot 4 \cdot 4 \cdot 4 \cdot 4 \cdot 4 \cdot 4 \cdot 4 = 4^{10} = 1{,}048{,}576$$

Thus, you can answer the questions in 1,048,576 different ways.

Are you surprised that there are over one million ways of answering a ten-question multiple-choice test? Of course, there is only one way to answer the test and receive a perfect score. The probability of guessing your way into a perfect score involves calculating the chance of getting a perfect score, just one way, from all 1,048,576 possible outcomes. In short, prepare for the test and do not rely on guessing!

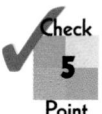 **Check 5 Point** You are taking a multiple-choice test that has six questions. Each of the questions has three answer choices, with one correct answer per question. If you select one of these three choices for each question and leave nothing blank, in how many ways can you answer the questions?

BLITZER BONUS

Running Out of Telephone Numbers

By the year 2020, portable telephones used for business and pleasure will all be videophones. At that time, U.S. population is expected to be 323 million. Faxes, beepers, cell phones, computer phone lines, and business lines may result in certain areas running out of phone numbers. Solution: Add more digits!

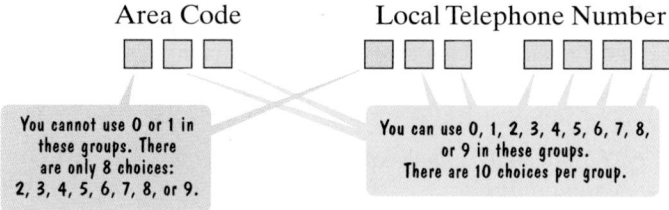
EXAMPLE 6 TELEPHONE NUMBERS IN THE UNITED STATES

Telephone numbers in the United States begin with three-digit area codes followed by seven-digit local telephone numbers. Area codes and local telephone numbers cannot begin with 0 or 1. How many different telephone numbers are possible?

Solution This situation involves making choices with ten groups of items.

Area Code   Local Telephone Number

□ □ □   □ □ □   □ □ □ □

You cannot use 0 or 1 in these groups. There are only 8 choices: 2, 3, 4, 5, 6, 7, 8, or 9.

You can use 0, 1, 2, 3, 4, 5, 6, 7, 8, or 9 in these groups. There are 10 choices per group.

Here are the choices for each of the ten groups of items:

Area Code   **Local Telephone Number**

$\boxed{8}\;\boxed{10}\;\boxed{10}$ $\boxed{8}\;\boxed{10}\;\boxed{10}\quad\boxed{10}\;\boxed{10}\;\boxed{10}\;\boxed{10}$.

We use the Fundamental Counting Principle to determine the number of different telephone numbers that are possible. The total number of telephone numbers possible is

$$8 \cdot 10 \cdot 10 \cdot 8 \cdot 10 \cdot 10 \cdot 10 \cdot 10 \cdot 10 \cdot 10 = 6{,}400{,}000{,}000.$$

There are six billion four hundred million different telephone numbers that are possible.

 **Check 6 Point** An electronic gate can be opened by entering five digits on a keypad containing the digits $0, 1, 2, 3, \ldots, 8, 9$. How many different keypad sequences are possible if the digit 0 cannot be used as the first digit?

Exercise Set 11.1

Practice and Application Exercises

Solve Exercises 1–6 using the Fundamental Counting Principle with two groups of items.

1. A restaurant offers 8 appetizers and 10 main courses. In how many ways can a person order a two-course meal?

2. The model of the car you are thinking of buying is available in nine different colors and three different styles (hatchback, sedan, or station wagon). In how many ways can you order the car?

3. A popular brand of pen is available in three colors (red, green, or blue) and four writing tips (bold, medium, fine, or micro). How many different choices of pens do you have with this brand?

4. In how many ways can a casting director choose a female lead and a male lead from five female actors and six male actors?

5. A student is planning a two-part trip. The first leg of the trip is from San Francisco to New York, and the second leg is from New York to Paris. From San Francisco to New York, travel options include airplane, train, or bus. From New York to Paris, the options are limited to airplane or ship. In how many ways can the two-part trip be made?

6. For a temporary job between semesters, you are painting the parking spaces for a new shopping mall with a letter of the alphabet and a single digit from 1 to 9. The first parking space is A1 and the last parking space is Z9. How many parking spaces can you paint with distinct labels?

Solve Exercises 7–22 using the Fundamental Counting Principle with three or more groups of items.

7. An ice cream store sells two drinks (sodas or milk shakes), in four sizes (small, medium, large, or jumbo), and five flavors (vanilla, strawberry, chocolate, coffee, or pistachio). In how many ways can a customer order a drink?

8. A pizza can be ordered with three choices of size (small, medium, or large), four choices of crust (thin, thick, crispy, or regular), and six choices of toppings (ground beef, sausage, pepperoni, bacon, mushrooms, or onions). How many one-topping pizzas can be ordered?

9. A restaurant offers the following limited lunch menu.

Main Course	Vegetables	Beverages	Desserts
Ham	Potatoes	Coffee	Cake
Chicken	Peas	Tea	Pie
Fish	Green beans	Milk	Ice cream
Beef		Soda	

If one item is selected from each of the four groups, in how many ways can a meal be ordered? Describe two such orders.

10. An apartment complex offers apartments with four different options, designated by A through D.

A	B	C	D
one bedroom	one bathroom	first floor	lake view
two bedrooms	two bathrooms	second floor	golf course view
three bedrooms			no special view

How many apartment options are available? Describe two such options.

11. Shoppers in a large shopping mall are categorized as male or female, over 30 or 30 and under, and cash or credit card shoppers. In how many ways can the shoppers be categorized?

12. There are three highways from city A to city B, two highways from city B to city C, and four highways from city C to city D. How many different highway routes are there from city A to city D?

13. A person can order a new car with a choice of six possible colors, with or without air conditioning, with or without automatic transmission, with or without power windows, and with or without a CD player. In how many different ways can a new car be ordered with regard to these options?

14. A car model comes in nine colors, with or without air conditioning, with or without a sun roof, with or without automatic transmission, and with or without antilock brakes. In how many ways can the car be ordered with regard to these options?

15. You are taking a multiple-choice test that has five questions. Each of the questions has three answer choices, with one correct answer per question. If you select one of these three choices for each question and leave nothing blank, in how many ways can you answer the questions?

16. You are taking a multiple-choice test that has eight questions. Each of the questions has three answer choices, with one correct answer per question. If you select one of these three choices for each question and leave nothing blank, in how many ways can you answer the questions?

17. In the original plan for area codes in 1945, the first digit could be any number from 2 through 9, the second digit was either 0 or 1, and the third digit could be any number except 0. With this plan, how many different area codes are possible?

18. The local seven-digit telephone numbers in Inverness, California, have 669 as the first three digits. How many different telephone numbers are possible in Inverness?

19. License plates in a particular state display two letters followed by three numbers, such as AT-887 or BB-013. How many different license plates can be manufactured for this state?

20. How many different four-letter radio station call letters can be formed if the first letter must be W or K?

21. A stock can go up, go down, or stay unchanged. How many possibilities are there if you own seven stocks?

22. A social security number contains nine digits, such as 074-66-7795. How many different social security numbers can be formed?

Writing in Mathematics

23. Explain the Fundamental Counting Principle.

24. Figure 11.2 on page 561 shows that a tree diagram can be used to find the total number of outfits. Describe one advantage of using the Fundamental Counting Principle rather than a tree diagram.

25. Write an original problem that can be solved using the Fundamental Counting Principle. Then solve the problem.

Critical Thinking Exercises

26. How many four-digit odd numbers are there? Assume that the digit on the left cannot be 0.

27. In order to develop a more appealing hamburger, a franchise used taste tests with 12 different buns, 30 sauces, 4 types of lettuce, and 3 types of tomatoes. If the taste test was done at one restaurant by one tester who took 10 minutes to eat each hamburger, approximately how long would it take the tester to eat all possible hamburgers?

Group Exercise

28. The group should select real-world situations where the Fundamental Counting Principle can be applied. These can involve the number of possible student ID numbers on your campus, the number of possible phone numbers in your community, the number of meal options at a local restaurant, the number of ways a person in the group can select outfits for class, the number of ways a condominium can be purchased in a nearby community, and so on. Once situations have been selected, group members should determine in how many ways each part of the task can be done. Group members will need to obtain menus, find out about telephone-digit requirements in the community, count shirts, pants, shoes in closets, visit condominium sales offices, and so on. Once the group reassembles, apply the Fundamental Counting Principle to determine the number of available options in each situation. Because these numbers may be quite large, use a calculator

---

SECTION 11.2   PERMUTATIONS

OBJECTIVES

1. Use the Fundamental Counting Principle to count permutations.

2. Evaluate factorial expressions.

3. Use the permutations formula.

4. Find the number of permutations of duplicate items.

Ladies and gentlemen: Please give a huge round of applause for

U2!   'N Sync!   Aerosmith!   The Rolling Stones!

You are in charge of planning one of the most anticipated concert tours of the decade. All four of these groups will appear in concert, which will be seen in a pay-per-view cable special by millions of people throughout the world. One of your jobs is to determine the order in which the groups will perform. Each group will perform once. How many different ways can you put together this four-group concert?

You are familiar with the work of all these musicians *and* you know the Fundamental Counting Principle! (Who could be better qualified for this job?) You can choose any of the four groups as the first performer. Once you've chosen the first group, you'll have three groups left to choose from for the second performer. You'll then have two groups left to choose from for the third performance. After the first three performers are determined, you'll have only one group left for the final appearance in the concert. This situation can be shown as follows:

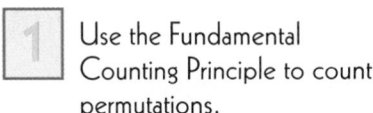

Use the Fundamental Counting Principle to count permutations.

First Group to Perform	Second Group to Perform	Third Group to Perform	Last Group to Perform
4 choices	3 choices	2 choices	1 choice

We use the Fundamental Counting Principle to find the number of ways you can put together the concert. Multiply the choices:

$$4 \cdot 3 \cdot 2 \cdot 1 = 24.$$

Thus, there are 24 different ways to arrange the concert. One of the 24 possible arrangements is

Aerosmith–'N Sync–U2–The Rolling Stones.

Such an ordered arrangement is called a *permutation* of the four rock groups.

A **permutation** is an ordered arrangement of items that occurs when

- No item is used more than once. (Each rock group performs exactly once.)
- The order of arrangement makes a difference. (It will make a difference in terms of how the concert is received if the Rolling Stones are the first group or the last group to perform.)

EXAMPLE 1 | COUNTING PERMUTATIONS

Based on their long-standing contribution to rock music, you decide that the Rolling Stones should be the last group to perform at the four-group U2, 'N Sync, Aerosmith, Rolling Stones concert. Given this decision, in how many ways can you put together the concert?

Solution You can now choose any one of the three groups, U2, 'N Sync, or Aerosmith, as the opening act. Once you've chosen the first group, you'll have two groups left to choose from for the second performance. You'll then have just one group left to choose for the third performance. There is also just one choice for the closing act—the Rolling Stones. This situation can be shown as follows:

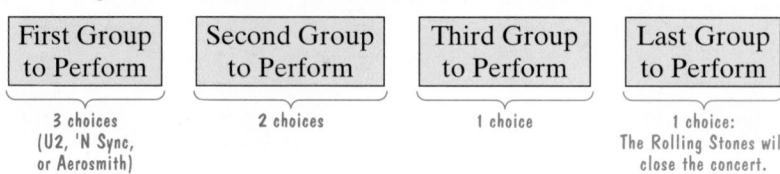

| First Group to Perform | Second Group to Perform | Third Group to Perform | Last Group to Perform |
| 3 choices (U2, 'N Sync, or Aerosmith) | 2 choices | 1 choice | 1 choice: The Rolling Stones will close the concert. |

We use the Fundamental Counting Principle to find the number of ways you can put together the concert. Multiply the choices:

$$3 \cdot 2 \cdot 1 \cdot 1 = 6.$$

Thus, there are six different ways to arrange the concert if the Rolling Stones are the final group to perform.

Check 1 Point For the concert in Example 1, suppose that U2 is to be the opening act and the Rolling Stones are to be the last group to perform. In how many ways can you put together the concert?

EXAMPLE 2 | COUNTING PERMUTATIONS

You need to arrange seven of your favorite books along a small shelf. How many different ways can you arrange the books, assuming that the order of the books makes a difference to you?

Solution You may choose any of the seven books for the first position on the shelf. This leaves six choices for second position. After the first two positions are filled, there are five books to choose from for third position, four choices left for the fourth position, three choices left for the fifth position, then two choices for the sixth position, and only one choice for the last position. This situation can be shown as follows:

The Rolling Stones

BLITZER BONUS

How to Pass the Time for $2\frac{1}{2}$ Million Years

If you were to arrange 15 different books on a shelf and it took you one minute for each permutation, the entire task would take 2,487,965 years.

Source: Isaac Asimov's Book of Facts

First Shelf Position	Second Shelf Position	Third Shelf Position	Fourth Shelf Position	Fifth Shelf Position	Sixth Shelf Position	Seventh Shelf Position
7 choices	6 choices	5 choices	4 choices	3 choices	2 choices	1 choice

We use the Fundamental Counting Principle to find the number of ways you can arrange the seven books along the shelf. Multiply the choices:

$$7 \cdot 6 \cdot 5 \cdot 4 \cdot 3 \cdot 2 \cdot 1 = 5040.$$

Thus, you can arrange the books in 5040 ways. There are 5040 different possible permutations.

 **Check Point 2** In how many ways can you arrange five books along a shelf, assuming that the order of the books makes a difference?

 2 Evaluate factorial expressions.

FACTORIALS FROM 0 THROUGH 20

0!	1
1!	1
2!	2
3!	6
4!	24
5!	120
6!	720
7!	5040
8!	40,320
9!	362,880
10!	3,628,800
11!	39,916,800
12!	479,001,600
13!	6,227,020,800
14!	87,178,291,200
15!	1,307,674,368,000
16!	20,922,789,888,000
17!	355,687,428,096,000
18!	6,402,373,705,728,000
19!	121,645,100,408,832,000
20!	2,432,902,008,176,640,000

As n increases, $n!$ grows very rapidly. Factorial growth is more explosive than exponential growth discussed in Chapter 7.

Factorial Notation

The product in Example 2,

$$7 \cdot 6 \cdot 5 \cdot 4 \cdot 3 \cdot 2 \cdot 1$$

is given a special name and symbol. It is called 7 **factorial**, and written 7!. Thus,

$$7! = 7 \cdot 6 \cdot 5 \cdot 4 \cdot 3 \cdot 2 \cdot 1.$$

In general, if n is a positive integer, then $n!$ (*n factorial*) is the product of all positive integers from n down through 1. For example,

$$1! = 1$$
$$2! = 2 \cdot 1 = 2$$
$$3! = 3 \cdot 2 \cdot 1 = 6$$
$$4! = 4 \cdot 3 \cdot 2 \cdot 1 = 24$$
$$5! = 5 \cdot 4 \cdot 3 \cdot 2 \cdot 1 = 120$$
$$6! = 6 \cdot 5 \cdot 4 \cdot 3 \cdot 2 \cdot 1 = 720.$$

FACTORIAL NOTATION

If n is a positive integer, the notation $n!$ (read "n factorial") is the product of all positive integers from n down through 1.

$$n! = n(n - 1)(n - 2) \cdots (3)(2)(1)$$

0! (zero factorial), by definition, is 1.

$$0! = 1$$

EXAMPLE 3 USING FACTORIAL NOTATION

Evaluate the following factorial expressions without using the factorial key on your calculator:

a. $\dfrac{8!}{5!}$ **b.** $\dfrac{26!}{21!}$ **c.** $\dfrac{500!}{499!}$.

TECHNOLOGY

Most calculators have a key or menu item for calculating factorials. Here are the keystrokes for finding 9!:

Many Scientific Calculators:

9 =

Many Graphing Calculators:

9 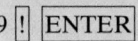 ENTER.

Solution

a. We can evaluate the numerator and the denominator of $\frac{8!}{5!}$. However, it is easier to use the following simplification:

$$\frac{8!}{5!} = \frac{8 \cdot 7 \cdot 6 \cdot \boxed{5 \cdot 4 \cdot 3 \cdot 2 \cdot 1}}{\boxed{5 \cdot 4 \cdot 3 \cdot 2 \cdot 1}} = \frac{8 \cdot 7 \cdot 6 \cdot \boxed{5!}}{\boxed{5!}} = \frac{8 \cdot 7 \cdot 6 \cdot \cancel{5!}}{\cancel{5!}} = 8 \cdot 7 \cdot 6 = 336.$$

b. Rather than write out 26!, the numerator of $\frac{26!}{21!}$, as the product of all integers from 26 down to 1, we can express 26! as

$$26! = 26 \cdot 25 \cdot 24 \cdot 23 \cdot 22 \cdot 21!.$$

In this way, we can cancel 21! in the numerator and the denominator of the given expression.

$$\frac{26!}{21!} = \frac{26 \cdot 25 \cdot 24 \cdot 23 \cdot 22 \cdot 21!}{21!} = \frac{26 \cdot 25 \cdot 24 \cdot 23 \cdot 22 \cdot \cancel{21!}}{\cancel{21!}}$$
$$= 26 \cdot 25 \cdot 24 \cdot 23 \cdot 22 = 7{,}893{,}600$$

c. In order to cancel identical factorials in the numerator and the denominator of $\frac{500!}{499!}$, we can express 500! as $500 \cdot 499!$.

$$\frac{500!}{499!} = \frac{500 \cdot 499!}{499!} = \frac{500 \cdot \cancel{499!}}{\cancel{499!}} = 500$$

 Evaluate without using a calculator's factorial key:

a. $\frac{9!}{6!}$ **b.** $\frac{16!}{11!}$ **c.** $\frac{100!}{99!}$.

 Use the permutations formula.

A Formula for Permutations

You are the coach of a little league baseball team. There are 13 players on the team (and lots of parents hovering in the background, dreaming of stardom for their little "Barry Bonds"). You need to choose a batting order having 9 players. The order makes a difference, because, for instance, if bases are loaded and "Little Barry" is fourth or fifth at bat, his possible home run will drive in three additional runs. How many batting orders can you form?

You can choose any of 13 players for the first person at bat. Then you will have 12 players from which to choose the second batter, then 11 from which to choose the third batter, and so on. The situation can be shown as follows:

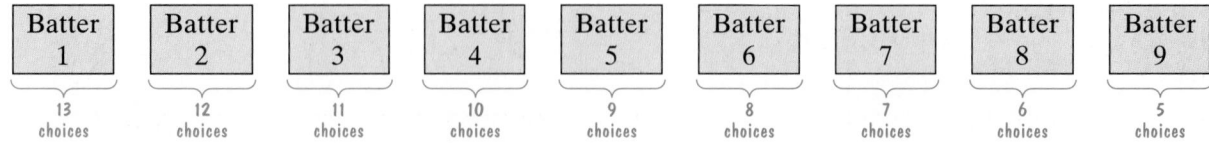

The total number of batting orders is

$$13 \cdot 12 \cdot 11 \cdot 10 \cdot 9 \cdot 8 \cdot 7 \cdot 6 \cdot 5 = 259{,}459{,}200.$$

Nearly 260 million batting orders are possible for your 13-player little league team. Each batting order is a permutation because the order of the batters makes a difference. The number of permutations of 13 players taken 9 at a time is 259,459,200.

We can obtain a formula for finding the number of permutations by rewriting our computation:

$13 \cdot 12 \cdot 11 \cdot 10 \cdot 9 \cdot 8 \cdot 7 \cdot 6 \cdot 5$

$$= \frac{13 \cdot 12 \cdot 11 \cdot 10 \cdot 9 \cdot 8 \cdot 7 \cdot 6 \cdot 5 \cdot \boxed{4 \cdot 3 \cdot 2 \cdot 1}}{\boxed{4 \cdot 3 \cdot 2 \cdot 1}} = \frac{13!}{4!} = \frac{13!}{(13 - 9)!}.$$

Thus, the number of permutations of 13 things taken 9 at a time is $\frac{13!}{(13 - 9)!}$. The special notation $_{13}P_9$ is used to replace the phrase "the number of permutations of 13 things taken 9 at a time." Using this new notation, we can write

$$_{13}P_9 = \frac{13!}{(13 - 9)!}.$$

The numerator of this expression is the factorial of the number of items, 13 team members: 13!. The denominator is also a factorial. It is the factorial of the difference between the number of items, 13, and the number of items in each permutation, 9 batters: $(13 - 9)!$.

The notation $_nP_r$ means the **number of permutations of n things taken r at a time**. We can generalize from the situation in which 9 batters were taken from 13 players. By generalizing, we obtain the following formula for the number of permutations if r items are taken from n items:

PERMUTATIONS OF n THINGS TAKEN r AT A TIME

The number of possible permutations if r items are taken from n items is

$$_nP_r = \frac{n!}{(n - r)!}.$$

EXAMPLE 4   USING THE FORMULA FOR PERMUTATIONS

You and 19 of your friends have decided to form an Internet marketing consulting firm. The group needs to choose three officers—a CEO, an operating manager, and a treasurer. In how many ways can those offices be filled?

Solution   Your group is choosing $r = 3$ officers from a group of $n = 20$ people (you and 19 friends). The order in which the officers are chosen matters because the CEO, the operating manager, and the treasurer each have different responsibilities. Thus, we are looking for the number of permutations of 20 things taken 3 at a time. We use the formula

$$_nP_r = \frac{n!}{(n - r)!}$$

with $n = 20$ and $r = 3$.

$$_{20}P_3 = \frac{20!}{(20 - 3)!} = \frac{20!}{17!} = \frac{20 \cdot 19 \cdot 18 \cdot 17!}{17!} = \frac{20 \cdot 19 \cdot 18 \cdot \cancel{17!}}{\cancel{17!}} = 20 \cdot 19 \cdot 18 = 6840$$

Thus, there are 6840 different ways of filling the three offices.

A corporation has seven members on its board of directors. In how many different ways can it elect a president, vice-president, secretary, and treasurer?

EXAMPLE 5 USING THE FORMULA FOR PERMUTATIONS

You are working for The Sitcom Television Network. Your assignment is to help set up the television schedule for Monday evenings between 7 and 10 P.M. You need to schedule a show in each of six 30-minute time blocks, beginning with 7 to 7:30 and ending with 9:30 to 10:00. You can select from among the following situation comedies: "Home Improvement," "Seinfeld," "Mad About You," "Cheers," "Friends," "Frasier," "All in the Family," "I Love Lucy," "M*A*S*H," "The Larry Sanders Show," "The Jeffersons," "Married with Children," and "Happy Days." How many different programming schedules can be arranged?

Solution   You are choosing $r = 6$ situation comedies from a collection of $n = 13$ classic sitcoms. The order in which the programs are aired matters. Family-oriented comedies have higher ratings when aired in earlier time blocks, such as 7 to 7:30. By contrast, comedies with adult themes do better in later time blocks. In short, we are looking for the number of permutations of 13 things taken 6 at a time. We use the formula

$$_nP_r = \frac{n!}{(n-r)!}$$

with $n = 13$ and $r = 6$.

$$_{13}P_6 = \frac{13!}{(13-6)!} = \frac{13!}{7!} = \frac{13 \cdot 12 \cdot 11 \cdot 10 \cdot 9 \cdot 8 \cdot 7!}{7!} = 13 \cdot 12 \cdot 11 \cdot 10 \cdot 9 \cdot 8 = 1,235,52\blacksquare$$

There are 1,235,520 different programming schedules that can be arranged.

 **Check 5 Point**   How many different programming schedules can be arranged by choosing 5 situation comedies from a collection of 9 classic sitcoms?

 Find the number of permutations of duplicate items.

Permutations of Duplicate Items

The number of permutations of the letters in the word SET is 3!, or 6. The six permutations are

SET,   STE,   EST,   ETS,   TES,   TSE.

Are there also six permutations of the letters in the name ANA? The answer is no. Unlike SET, with three distinct letters, ANA contains three letters, of which the two As are duplicates. If we rearrange the letters just as we did with SET, we obtain

ANA,   AAN,   NAA,   NAA,   ANA,   AAN.

Without the use of color to distinguish between the two As, there are only three distinct permutations: ANA, AAN, NAA.

There is a formula for finding the number of distinct permutations when duplicate items exist:

PERMUTATIONS OF DUPLICATE ITEMS

The number of permutations of n items, where p items are identical, q items are identical, r items are identical, and so on, is given by

$$\frac{n!}{p!q!r!\cdots}.$$

For example, ANA contains three letters ($n = 3$), where two of the letters are identical ($p = 2$). The number of distinct permutations is

$$\frac{n!}{p!} = \frac{3!}{2!} = \frac{3 \cdot 2!}{2!} = 3.$$

We saw that the three distinct permutations are ANA, AAN, and NAA.

TECHNOLOGY

Parentheses are necessary to enclose the factorials in the denominator when using a calculator to find

$$\frac{11!}{4!4!2!}.$$

EXAMPLE 6 USING THE FORMULA FOR PERMUTATIONS OF DUPLICATE ITEMS

In how many distinct ways can the letters of the word MISSISSIPPI be arranged?

Solution The word contains 11 letters ($n = 11$), where four Is are identical ($p = 4$), four Ss are identical ($q = 4$), and 2 Ps are identical ($r = 2$). The number of distinct permutations is

$$\frac{n!}{p!q!r!} = \frac{11!}{4!4!2!} = \frac{11 \cdot 10 \cdot 9 \cdot 8 \cdot 7 \cdot 6 \cdot 5 \cdot 4!}{4!4 \cdot 3 \cdot 2 \cdot 1 \cdot 2 \cdot 1} = 34,650$$

There are 34,650 distinct ways the letters in the word MISSISSIPPI can be arranged.

Check **6** Point   In how many ways can the letters of the word OSMOSIS be arranged?

Exercise Set 11.2

Practice and Application Exercises

Use the Fundamental Counting Principle to solve Exercises 1–12.

1. Six performers are to present their comedy acts on a weekend evening at a comedy club. How many different ways are there to schedule their appearances?

2. Five singers are to perform on a weekend evening at a night club. How many different ways are there to schedule their appearances?

3. In the *Cambridge Encyclopedia of Language* (Cambridge University Press, 1987), author David Crystal presents five sentences that make a reasonable paragraph regardless of their order. The sentences are as follows:

   Mark had told him about the foxes.
   John looked out of the window.
   Could it be a fox?
   However, nobody had seen one for months.
   He thought he saw a shape in the bushes.

   In how many different orders can the five sentences be arranged?

4. In how many different ways can a police department arrange eight suspects in a police lineup if each lineup contains all eight people?

5. As in Exercise 1, six performers are to present their comedy acts on a weekend evening at a comedy club. One of the performers insists on being the last stand-up comic of the evening. If this performer's request is granted, how many different ways are there to schedule the appearances?

6. As in Exercise 2, five singers are to perform at a night club. One of the singers insists on being the last performer of the evening. If this singer's request is granted, how many different ways are there to schedule the appearances?

7. You need to arrange nine of your favorite books along a small shelf. How many different ways can you arrange the books, assuming that the order of the books makes a difference to you?

8. You need to arrange ten of your favorite photographs on the mantel above a fireplace. How many ways can you arrange the photographs, assuming that the order of the pictures makes a difference to you?

In Exercises 9–10, use the five sentences that are given in Exercise 3.

9. How many different five-sentence paragraphs can be formed if the paragraph begins with "He thought he saw a shape in the bushes" and ends with "John looked out of the window"?

10. How many different five-sentence paragraphs can be formed if the paragraph begins with "He thought he saw a shape in the bushes" followed by "Mark had told him about the foxes"?

11. A television programmer is arranging the order that five movies will be seen between the hours of 6 P.M. and 4 A.M. Two of the movies have a G rating, and they are to be shown in the first two time blocks. One of the movies is rated NC-17, and it is to be shown in the last of the time blocks, from 2 A.M. until 4 A.M. Given these restrictions, in how many ways can the five movies be arranged during the indicated time blocks?

12. A camp counselor and six campers are to be seated along a picnic bench. In how many ways can this be done if the counselor must be seated in the middle and a camper who has a tendency to engage in food fights must sit to the counselor's immediate left?

In Exercises 13–32, evaluate each factorial expression.

13. $\dfrac{9!}{6!}$ **14.** $\dfrac{12!}{10!}$ **15.** $\dfrac{29!}{25!}$

16. $\dfrac{31!}{28!}$ **17.** $\dfrac{19!}{11!}$ **18.** $\dfrac{17!}{9!}$

19. $\dfrac{600!}{599!}$ **20.** $\dfrac{700!}{699!}$ **21.** $\dfrac{104!}{102!}$

22. $\dfrac{106!}{104!}$ **23.** $7! - 3!$ **24.** $6! - 3!$

25. $(7 - 3)!$ **26.** $(6 - 3)!$ **27.** $\left(\dfrac{12}{4}\right)!$

28. $\left(\dfrac{45}{9}\right)!$ **29.** $\dfrac{7!}{(7 - 2)!}$ **30.** $\dfrac{8!}{(8 - 5)!}$

31. $\dfrac{13!}{(13 - 3)!}$ **32.** $\dfrac{17!}{(17 - 3)!}$

In Exercises 33–40, use the formula for $_nP_r$ to evaluate each expression.

33. $_9P_4$ **34.** $_7P_3$

35. $_8P_5$ **36.** $_{10}P_4$

37. $_6P_6$ **38.** $_9P_9$

39. $_8P_0$ **40.** $_6P_0$

Use the formula for $_nP_r$ to solve Exercises 41–48.

41. A club with ten members is to choose three officers—president, vice-president, and secretary-treasurer. If each office is to be held by one person and no person can hold more than one office, in how many ways can those offices be filled?

42. A corporation has seven members on its board of directors. In how many different ways can it elect a president, vice-president, secretary, and treasurer?

43. For a segment of a radio show, a disc jockey can play 7 records. If there are 13 records to select from, in how many ways can the program for this segment be arranged?

44. Suppose you are asked to list, in order of preference, the three best movies you have seen this year. If you saw 20 movies during the year, in how many ways can the three best be chosen and ranked?

45. In a race in which six automobiles are entered and there are no ties, in how many ways can the first three finishers come in?

46. In a production of *West Side Story*, eight actors are considered for the male roles of Tony, Riff, and Bernardo. In how many ways can the director cast the male roles?

47. Nine bands have volunteered to perform at a benefit concert, but there is only enough time for five of the bands to play. How many lineups are possible?

48. How many arrangements can be made using four of the letters of the word COMBINE if no letter is to be used more than once?

Use the formula for the number of permutations of duplicate items to solve Exercises 49–56.

49. In how many distinct ways can the letters of the word DALLAS be arranged?

50. In how many distinct ways can the letters of the word SCIENCE be arranged?

51. How many distinct permutations can be formed using the letters of the word TALLAHASSEE?

52. How many distinct permutations can be formed using the letters of the word TENNESSEE?

53. In how many ways can the digits in the number 5,446,666 be arranged?

54. In how many ways can the digits in the number 5,432,435 be arranged?

In Exercises 55–56, a signal can be formed by running different colored flags up a pole, one above the other.

55. Find the number of different signals consisting of eight flags that can be made using three white flags, four red flags, and one blue flag.

56. Find the number of different signals consisting of nine flags that can be made using three white flags, five red flags, and one blue flag.

Writing in Mathematics

57. What is a permutation?

58. Explain how to find $n!$, where n is a positive integer.

59. Explain the best way to evaluate $\dfrac{900!}{899!}$ without a calculator.

60. Describe what $_nP_r$ represents.

61. Write a word problem that can be solved by evaluating $5!$.

62. Write a word problem that can be solved by evaluating $_7P_3$.

63. If 24 permutations can be formed using the letters in the word BAKE, why can't 24 permutations also be formed using the letters in the word BABE? How is the number of permutations in BABE determined?

Critical Thinking Exercises

64. Ten people board an airplane that has 12 aisle seats. In how many ways can they be seated if they all select aisle seats?

65. Six horses are entered in a race. If two horses are tied for first place, and there are no ties among the other four horses, in how many ways can the six horses cross the finish line?

66. Performing at a concert are eight rock bands and eight jazz groups. How many ways can the program be arranged if the first, third, and eighth performers are jazz groups?

67. Five men and five women line up at a checkout counter in a store. In how many ways can they line up if the first person in line is a woman, and the people in line alternate woman, man, woman, man, and so on?

68. How many four-digit odd numbers less than 6000 can be formed using the digits 2, 4, 6, 7, 8, and 9?

69. Express $_nP_{n-2}$ without using factorials.

SECTION 11.3   COMBINATIONS

OBJECTIVES

1. Distinguish between permutation and combination problems.

2. Solve problems involving combinations using the combinations formula.

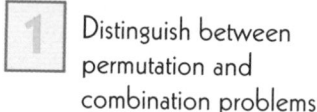

1 Distinguish between permutation and combination problems.

As the twentieth century drew to a close, *Time* magazine presented a series of special issues on the most influential people of the century. In their issue on heroes and icons (June 14, 1999), they discussed a number of people whose careers became more profitable after their tragic deaths, including Marilyn Monroe, James Dean, Jim Morrison, Kurt Cobain, and Selena.

Imagine that you ask your friends the following question: "Of these five people, which three would you select to be included in a documentary featuring the best of their work?" You are not asking your friends to rank their three favorite artists in any kind of order—they should merely select the three to be included in the documentary.

One friend answers, "Jim Morrison, Kurt Cobain, and Selena." Another responds, "Selena, Kurt Cobain, and Jim Morrison." These two people have the same artists in their group of selections, even if they are named in a different order. We are interested *in which artists are named, not the order in which they are named* for the documentary. Because the items are taken without regard to order, this is not a permutation problem. No ranking of any sort is involved.

Later on, you ask your roommate which three artists she would select for the documentary. She names Marilyn Monroe, James Dean, and Selena. Her selection is different from those of your two other friends because different entertainers are cited.

Mathematicians describe the group of artists given by your roommate as a *combination*. A **combination** of items occurs when

- The items are selected from the same group (the five stars who died young and tragically).

- No item is used more than once. (You may adore Selena, but your three selections cannot be Selena, Selena, and Selena).

- The order of items makes no difference. (Morrison, Cobain, Selena is the same group in the documentary as Selena, Cobain, Morrison.)

Do you see the difference between a permutation and a combination? A permutation is an ordered arrangement of a given group of items. A combination is a group of items taken without regard to their order. **Permutation** problems involve situations in which **order matters**. **Combination** problems involve situations in which the **order** of items **makes no difference**.

Marilyn Monroe, Actress (1927–1962)

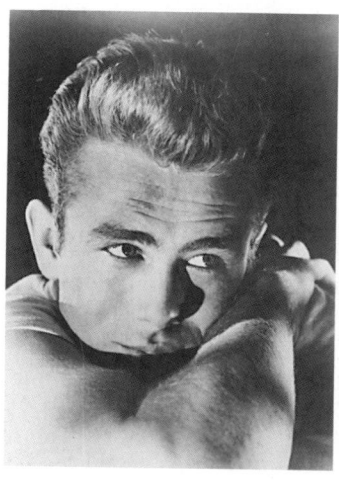

James Dean, Actor (1931–1955)

Jim Morrison, Musician and Lead Singer of The Doors (1943–1971)

Kurt Cobain, Musician and Front Man for Nirvana (1967–1994)

Selena, Musician of Tejano Music (1971–1995)

EXAMPLE 1 · DISTINGUISHING BETWEEN PERMUTATIONS AND COMBINATIONS

For each of the following problems, determine whether the problem is one involving permutations or combinations. (It is not necessary to solve the problem.)

a. Six students are running for student government president, vice-president, and treasurer. The student with the greatest number of votes becomes the president, the second highest vote-getter becomes vice-president, and the student who gets the third largest number of votes will be treasurer. How many different outcomes are possible for these three positions?

b. Six people are on the board of supervisors for your neighborhood park. A three-person committee is needed to study the possibility of expanding the park. How many different committees could be formed from the six people?

c. Baskin-Robbins offers 31 different flavors of ice cream. One of their items is a bowl consisting of three scoops of ice cream, each a different flavor. How many such bowls are possible?

Solution

a. Students are choosing three student government officers from six candidates. The order in which the officers are chosen makes a difference because each of the offices (president, vice-president, treasurer) is different. Order matters. This is a problem involving permutations.

b. A three-person committee is to be formed from the six-person board of supervisors. The order in which the three people are selected does not matter because they are not filling different roles on the committee. Because order makes no difference, this is a problem involving combinations.

c. A three-scoop bowl of three different flavors is to be formed from Baskin-Robbin's 31 flavors. The order in which the three scoops of ice cream are put into the bowl is irrelevant. A bowl with chocolate, vanilla, and strawberry is exactly the same as a bowl with vanilla, strawberry, and chocolate. Different orderings do not change things, and so this is a problem involving combinations.

 Check 1 Point

For each of the following problems, determine whether the problem is one involving permutations or combinations. (It is not necessary to solve the problem.)

a. How many ways can you select 6 free videos from a list of 200 videos?

b. In a race in which there are 50 runners and no ties, in how many ways can the first three finishers come in?

 Solve problems involving combinations using the combinations formula.

A Formula for Combinations

We have seen that the notation $_nP_r$ means the number of permutations of n things taken r at a time. Similarly, the notation $_nC_r$ **means the number of combinations of n things taken r at a time**.

We can develop a formula for $_nC_r$ by comparing permutations and combinations. Consider the letters A, B, C, and D. The number of permutations of these four letters taken three at a time is

$$_4P_3 = \frac{4!}{(4-3)!} = \frac{4!}{1!} = \frac{4 \cdot 3 \cdot 2 \cdot 1}{1} = 24.$$

Here are the 24 permutations:

ABC,	ABD,	ACD,	BCD,
ACB,	ADB,	ADC,	BDC,
BAC,	BAD,	CAD,	CBD,
BCA,	BDA,	CDA,	CDB,
CAB,	DAB,	DAC,	DBC,
CBA,	DBA,	DCA,	DCB.

> This column contains only one combination, **ABC.**

> This column contains only one combination, **ABD.**

> This column contains only one combination, **ACD.**

> This column contains only one combination, **BCD.**

Because the order of items makes no difference in determining combinations, each column of six permutations represents one combination. There are a total of four combinations:

ABC,   ABD,   ACD,   BCD.

Thus, $_4C_3 = 4$: The number of combinations of 4 things taken 3 at a time is 4. With 24 permutations and only four combinations, there are 6, or 3!, times as many permutations as there are combinations.

In general, there are $r!$ times as many permutations of n things taken r at a time as there are combinations of n things taken r at a time. Thus, we find the number of combinations of n things taken r at a time by dividing the number of permutations of n things taken r at a time by $r!$.

$$_nC_r = \frac{_nP_r}{r!} = \frac{\dfrac{n!}{(n-r)!}}{r!} = \frac{n!}{(n-r)!r!}$$

STUDY TIP

The number of combinations if r items are taken from n items cannot be found using the Fundamental Counting Principle and requires the use of the formula shown on the right.

COMBINATIONS OF n THINGS TAKEN r AT A TIME

The number of possible combinations if r items are taken from n items is

$$_nC_r = \frac{n!}{(n-r)!r!}.$$

EXAMPLE 2 USING THE FORMULA FOR COMBINATIONS

A three-person committee is needed to study ways of improving public transportation. How many committees could be formed from the eight people on the board of supervisors?

Solution The order in which the three people are selected does not matter. This is a problem of selecting $r = 3$ people from a group of $n = 8$ people. We are looking for the number of combinations of eight things taken three at a time. We use the formula

$$_nC_r = \frac{n!}{(n-r)!r!}$$

TECHNOLOGY

Graphing calculators have a menu item for calculating combinations, usually labeled $_nC_r$. For example, to find $_8C_3$, the keystrokes on most graphing calculators are

$$8 \boxed{_nC_r} \; 3 \; \boxed{\text{ENTER}}.$$

If you are using a scientific calculator, check your manual to see whether there is a menu item for calculating combinations.

If you use your calculator's factorial key to find $\frac{8!}{5!3!}$, be sure to enclose the factorials in the denominator with parentheses

$$8 \boxed{!} \; \boxed{\div} \; \boxed{(} \; 5 \boxed{!} \; \boxed{\times} \; 3 \boxed{!} \; \boxed{)}$$

pressing $\boxed{=}$ or $\boxed{\text{ENTER}}$ to obtain the answer.

with $n = 8$ and $r = 3$.

$$_8C_3 = \frac{8!}{(8-3)!3!} = \frac{8!}{5!3!} = \frac{8 \cdot 7 \cdot 6 \cdot 5!}{5! \cdot 3 \cdot 2 \cdot 1} = \frac{8 \cdot 7 \cdot 6 \cdot 5!}{5! \cdot 3 \cdot 2 \cdot 1} = 56$$

Thus, 56 committees of three people each can be formed from the eight people on the board of supervisors.

✔ **Check Point 2** You volunteer to pet-sit for your friend who has seven different animals. How many different pet combinations are possible if you take three of the seven pets?

EXAMPLE 3 | USING THE FORMULA FOR COMBINATIONS

In poker, a person is dealt 5 cards from a standard 52-card deck. The order in which you are dealt the 5 cards does not matter. How many different 5-card poker hands are possible?

Solution Because the order in which the 5 cards are dealt does not matter, this is a problem involving combinations. We are looking for the number of combinations of $n = 52$ cards drawn $r = 5$ cards at a time. We use the formula

$$_nC_r = \frac{n!}{(n-r)!r!}$$

with $n = 52$ and $r = 5$.

$$_{52}C_5 = \frac{52!}{(52-5)!5!} = \frac{52!}{47!5!} = \frac{52 \cdot 51 \cdot 50 \cdot 49 \cdot 48 \cdot 47!}{47! \cdot 5 \cdot 4 \cdot 3 \cdot 2 \cdot 1} = 2,598,960$$

Thus, there are 2,598,960 different 5-card poker hands possible. It surprises many people that more than 2.5 million 5-card hands can be dealt from a mere 52 cards.

FIGURE 11.3 A royal flush

If you are a card player, it does not get any better than to be dealt the 5-card poker hand shown in Figure 11.3. This hand is called a *royal flush*. It consists of an ace, king, queen, jack, and 10, all of the same suit: all hearts, all diamonds, all clubs, or all spades. The probability of being dealt a royal flush involves calculating the number of ways of being dealt such a hand: just 4 of all 2,598,960 possible hands. In the next section, we move from counting possibilities to computing probabilities.

✔ **Check Point 3** How many different 4-card hands can be dealt from a deck that has 16 different cards?

There are situations in which both the formula for combinations and the Fundamental Counting Principle are used together. Let's say that the U.S. Senate, with 100 members, consists of 54 Republicans and 46 Democrats. We want to form a committee of 3 Republicans and 2 Democrats.

100 Senate Members

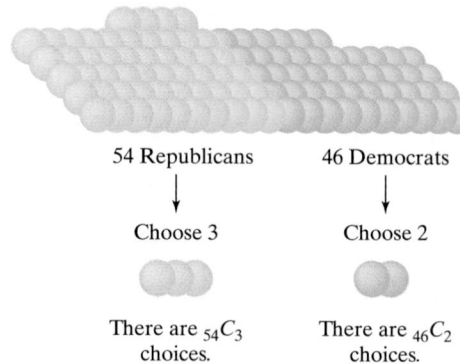

54 Republicans   46 Democrats

Choose 3   Choose 2

There are $_{54}C_3$ There are $_{46}C_2$
choices.   choices.

By the Fundamental Counting Principle, the number of ways of choosing 3 Republicans and 2 Democrats is given by

$$_{54}C_3 \cdot _{46}C_2.$$

In Example 4, we develop these ideas in more detail.

EXAMPLE 4 | USING THE FORMULA FOR COMBINATIONS AND THE FUNDAMENTAL COUNTING PRINCIPLE

The U.S. Senate of the 104th Congress consisted of 54 Republicans and 46 Democrats. How many committees can be formed if each committee must have 3 Republicans and 2 Democrats?

Solution   The order in which the members are selected does not matter. Thus, this is a problem involving combinations.

We begin with the number of ways of selecting 3 Republicans out of 54 Republicans without regard to order. We are looking for the number of combinations of $n = 54$ people taken $r = 3$ people at a time. We use the formula

$$_nC_r = \frac{n!}{(n-r)!r!}$$

with $n = 54$ and $r = 3$.

$$_{54}C_3 = \frac{54!}{(54-3)!3!} = \frac{54!}{51!3!} = \frac{54 \cdot 53 \cdot 52 \cdot \cancel{51!}}{\cancel{51!} \cdot 3 \cdot 2 \cdot 1} = \frac{54 \cdot 53 \cdot 52}{3 \cdot 2 \cdot 1} = 24{,}804$$

There are 24,804 choices for forming 3-member Republican committees.

Next, we find the number of ways of selecting 2 Democrats out of 46 Democrats without regard to order. We are looking for the number of combinations of $n = 46$ people taken $r = 2$ people at a time. Once again, we use the formula

$$_nC_r = \frac{n!}{(n-r)!r!}.$$

This time, $n = 46$ and $r = 2$.

$$_{46}C_2 = \frac{46!}{(46-2)!2!} = \frac{46!}{44!2!} = \frac{46 \cdot 45 \cdot \cancel{44!}}{\cancel{44!} \cdot 2 \cdot 1} = 1035$$

There are 1035 choices for forming 2-member Democratic committees.

We use the Fundamental Counting Principle to find the number of committees that can be formed:

$$_{54}C_3 \cdot _{46}C_2 = 24{,}804 \cdot 1035 = 25{,}672{,}140.$$

Thus, 25,672,140 committees can be formed.

Check 4 Point The U.S. Senate of the 107th Congress consisted of 50 Democrats, 49 Republicans, and one Independent. How many committees can be formed if each committee must have 3 Democrats and 2 Republicans?

Exercise Set 11.3

Practice Exercises

In Exercises 1–10, does the problem involve permutations or combinations? Explain your answer. (It is not necessary to solve the problem.)

1. In a race in which six automobiles are entered and there are no ties, in how many ways can the first three finishers come in?

2. A book club offers a choice of 8 books from a list of 40. In how many ways can a member make a selection?

3. A medical researcher needs 6 people to test the effectiveness of an experimental drug. If 13 people have volunteered for the test, in how many ways can 6 people be selected?

4. Fifty people purchase raffle tickets. Three winning tickets are selected at random. If first prize is $1000, second prize is $500, and third prize is $100, in how many different ways can the prizes be awarded?

5. From a club of 20 people, in how many ways can a group of three members be selected to attend a conference?

6. Fifty people purchase raffle tickets. Three winning tickets are selected at random. If each prize is $500, in how many different ways can the prizes be awarded?

7. How many different four-letter passwords can be formed from the letters A, B, C, D, E, F, and G if no repetition of letters is allowed?

8. Nine comedy acts will perform over two evenings. Five of the acts will perform on the first evening. How many ways can the schedule for the first evening be made?

9. Using 15 flavors of ice cream, how many cones with three different flavors can you create if it is important to you which flavor goes on the top, middle, and bottom?

10. A sample of four laser printers from a shipment of 50 will be selected and tested. How many ways are there to make this selection?

In Exercises 11–26, use the formula for $_nC_r$ to evaluate each expression.

11. $_6C_5$ 12. $_8C_7$ 13. $_9C_5$ 14. $_{10}C_6$

15. $_{11}C_4$ 16. $_{12}C_5$ 17. $_8C_1$ 18. $_7C_1$

19. $_7C_7$ 20. $_4C_4$ 21. $_{30}C_3$ 22. $_{25}C_4$

23. $_5C_0$ 24. $_6C_0$ 25. $\dfrac{_7C_3}{_5C_4}$ 26. $\dfrac{_{10}C_3}{_6C_4}$

Application Exercises

Use the formula for $_nC_r$ to solve Exercises 27–36.

27. An election ballot asks voters to select three city commissioners from a group of six candidates. In how many ways can this be done?

28. A four-person committee is to be elected from an organization's membership of 11 people. How many different committees are possible?

29. Of 12 possible books, you plan to take 4 with you on vacation. How many different collections of 4 books can you take?

30. There are 14 standbys who hope to get seats on a flight, but only 6 seats are available on the plane. How many different ways can the 6 people be selected?

31. You volunteer to help drive children at a charity event to the zoo, but you can fit only 8 of the 17 children present in your van. How many different groups of 8 children can you drive?

32. How many different 4-card hands can be dealt from a deck that has 16 different cards?

33. Baskin-Robbins offers 31 different flavors of ice cream. One of their items is a bowl consisting of three scoops of ice cream, each a different flavor. How many such bowls are possible?

34. Of the 100 people in the U.S. Senate, 18 serve on the Foreign Relations Committee. How many ways are there to select Senate members for this committee (assuming party affiliation is not a factor in selection)?

35. To win at LOTTO in the state of Florida, one must correctly select 6 numbers from a collection of 53 numbers (1 through 53). The order in which the selection is made does not matter. How many different selections are possible?

36. To win in the New York State lottery, one must correctly select 6 numbers from 59 numbers. The order in which the selection is made does not matter. How many different selections are possible?

Use the formula for $_nC_r$ and the Fundamental Counting Principle to solve Exercises 37–40.

37. In how many ways can a committee of four men and five women be formed from a group of seven men and seven women?

38. How many different committees can be formed from 5 professors and 15 students if each committee is made up of 2 professors and 10 students?

39. The U.S. Senate of the 106th Congress consisted of 55 Republicans and 45 Democrats. How many committees can be formed if each committee must have 4 Republicans and 3 Democrats?

40. A mathematics exam consists of 10 multiple-choice questions and 5 open-ended problems in which all work must be shown. If an examinee must answer 8 of the multiple-choice questions and 3 of the open-ended problems, in how many ways can the questions and problems be chosen?

Writing in Mathematics

41. What is a combination?

42. Explain how to distinguish between permutation and combination problems.

43. Write a word problem that can be solved by evaluating $_7C_3$.

Critical Thinking Exercises

44. Write a word problem that can be solved by evaluating $_{10}C_3 \cdot _7C_2$.

45. A 6/53 lottery involves choosing 6 of the numbers from 1 through 53 and a 5/36 lottery involves choosing 5 of the numbers from 1 through 36. The order in which the numbers are chosen does not matter. Which lottery is easier to win? Explain your answer.

46. If the number of permutations of n objects taken r at a time is six times the number of combinations of n objects taken r at a time, determine the value of r. Is there enough information to determine the value of n? Why or why not?

47. In a group of 20 people, how long will it take each person to shake hands with each of the other persons in the group, assuming that it takes three seconds for each shake and only 2 people can shake hands at a time? What if the group is increased to 40 people?

48. A sample of 4 telephones is selected from a shipment of 20 phones. There are 5 defective telephones in the shipment. How many of the samples of 4 phones do not include any of the defective ones?

SECTION 11.4   FUNDAMENTALS OF PROBABILITY

OBJECTIVES

1. Compute theoretical probability.

2. Compute empirical probability.

TABLE 11.1 THE HOURS OF SLEEP AMERICANS GET ON A TYPICAL NIGHT

Hours of Sleep	Number of Americans, in millions
4 or less	11.36
5	25.56
6	71
7	85.2
8	76.68
9	8.52
10 or more	5.68

Total: 284

Source: Discovery Health Media

How many hours of sleep do you typically get each night? Table 11.1 indicates that 71 million out of 284 million Americans are getting six hours of sleep on a typical night. The *probability* of an American getting six hours of sleep on a typical night is $\frac{71}{284}$. This fraction can be reduced to $\frac{1}{4}$, or expressed as 0.25, or 25%. Thus, 25% of Americans get six hours of sleep each night.

We find a probability by dividing one number by another. Probabilities are assigned to an *event*, such as getting six hours of sleep on a typical night. Events that are certain to occur are assigned probabilities of 1, or 100%. For example, the probability that a given individual will eventually die is 1. Regrettably, taxes and death are always certain! By contrast, if an event cannot occur, its probability is 0. For example, the probability that Elvis will return from the dead and serenade us with one final reprise of "Heartbreak Hotel" is 0.

Probabilities of events are expressed as numbers ranging from 0 to 1, or 0% to 100%. The closer the probability of a given event is to 1, the more likely it is that the event will occur. The closer the probability of a given event is to 0, the less likely it is that the event will occur.

100% or 1 — Certain

Likely

50% or $\frac{1}{2}$ — 50-50 Chance

Unlikely

0% or 0 — Impossible

Possible Values for Probabilities

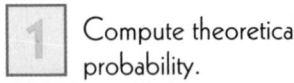

1 Compute theoretical probability.

Theoretical Probability

You toss a coin. Although it is equally likely to land either heads up, denoted by H, or tails up, denoted by T, the actual outcome is uncertain. Any occurrence for which the outcome is uncertain is called an **experiment**. Thus, tossing a coin is an example of an experiment. The set of all possible outcomes of an experiment is the **sample space** of the experiment, denoted by S. The sample space for the coin-tossing experiment is

$$S = \{H, T\}.$$

Lands heads up   Lands tails up

An **event**, denoted by E, is any subset of a sample space. For example, the subset $E = \{T\}$ is the event of landing tails up when a coin is tossed.

Theoretical probability applies to situations like this, in which the sample space only contains equally-likely outcomes, all of which are known. To calculate the theoretical probability of an event, we divide the number of outcomes resulting in the event by the total number of outcomes in the sample space.

COMPUTING THEORETICAL PROBABILITY

If an event E has $n(E)$ equally-likely outcomes and its sample space S has $n(S)$ equally-likely outcomes, the **theoretical probability** of event E, denoted by $P(E)$, is

$$P(E) = \frac{\text{number of outcomes in event } E}{\text{total number of possible outcomes}} = \frac{n(E)}{n(S)}.$$

How can we use this formula to compute the probability of a coin landing tails up? We use the following sets:

$$E = \{T\} \qquad S = \{H, T\}.$$

This is the event of landing tails up.   This is the sample space with all equally-likely outcomes.

The probability of a coin landing tails up is

$$P(E) = \frac{\text{number of outcomes that result in tails up}}{\text{total number of possible outcomes}} = \frac{n(E)}{n(S)} = \frac{1}{2}.$$

Theoretical probability applies to many games of chance, including dice rolling, lotteries, card games, and roulette. We begin with rolling a die. Figure 11.4 illustrates that when a die is rolled, there are six equally-likely possible outcomes. The sample space can be shown as

$$S = \{1, 2, 3, 4, 5, 6\}.$$

FIGURE 11.4 Outcomes when a die is rolled

EXAMPLE 1   COMPUTING THEORETICAL PROBABILITY

A die is rolled once. Find the probability of rolling

a. a 3.   **b.** an even number.   **c.** a number less than 5.

d. a number less than 10.   **e.** a number greater than 6.

BLITZER BONUS

Alcohol and the Probability of a Car Accident

In California, where 38% of fatal traffic crashes involve drinking drivers, it is illegal to drive with a blood alcohol concentration of 0.08 or higher. At these levels, drivers may be arrested and charged with driving under the influence.

Medical research indicates that the probability of having a car accident increases exponentially as the concentration of alcohol in the blood increases. The probability is modeled by

$$P = 6e^{12.77x}$$

where x is the blood alcohol concentration and P, given as a percent, is the probability of having a car accident.

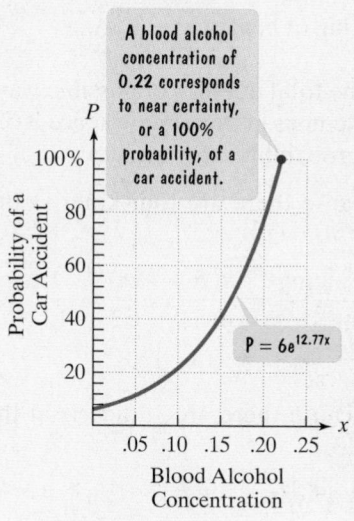

A blood alcohol concentration of 0.22 corresponds to near certainty, or a 100% probability, of a car accident.

Solution   The sample space is $S = \{1, 2, 3, 4, 5, 6\}$ with $n(S) = 6$. We will use 6, the total number of possible outcomes, in the denominator of each probability fraction.

a. The phrase "rolling a 3" describes the event $E = \{3\}$. This event can occur in one way: $n(E) = 1$.

$$P(3) = \frac{\text{number of outcomes that result in 3}}{\text{total number of possible outcomes}} = \frac{n(E)}{n(S)} = \frac{1}{6}$$

The probability of rolling a 3 is $\frac{1}{6}$.

b. The phrase "rolling an even number" describes the event $E = \{2, 4, 6\}$. This event can occur in three ways: $n(E) = 3$.

$$P(\text{even number}) = \frac{\text{number of outcomes that result in an even number}}{\text{total number of possible outcomes}} = \frac{n(E)}{n(S)} = \frac{3}{6} = \frac{1}{2}$$

The probability of rolling an even number is $\frac{1}{2}$.

c. The phrase "rolling a number less than 5" describes the event $E = \{1, 2, 3, 4\}$. This event can occur in four ways: $n(E) = 4$.

$$P(\text{less than 5}) = \frac{\text{number of outcomes that are less than 5}}{\text{total number of possible outcomes}} = \frac{n(E)}{n(S)} = \frac{4}{6} = \frac{2}{3}$$

The probability of rolling a number less than 5 is $\frac{2}{3}$.

d. The phrase "rolling a number less than 10" describes the event $E = \{1, 2, 3, 4, 5, 6\}$. This event can occur in six ways: $n(E) = 6$. Can you see that all of the possible outcomes are less than 10? This event is certain to occur.

$$P(\text{less than 10}) = \frac{\text{number of outcomes that are less than 10}}{\text{total number of possible outcomes}} = \frac{n(E)}{n(S)} = \frac{6}{6} = 1$$

The probability of any certain event is 1.

e. The phrase "rolling a number greater than 6" describes an event that cannot occur, or the empty set. Thus, $E = \varnothing$ and $n(E) = 0$.

$$P(\text{greater than 6}) = \frac{\text{number of outcomes that are greater than 6}}{\text{total number of possible outomes}} = \frac{n(E)}{n(S)} = \frac{0}{6} = 0$$

The probability of an event that cannot occur is 0.

In Example 1, there are six possible outcomes, each with a probability of $\frac{1}{6}$:

$$P(1) = \frac{1}{6} \quad P(2) = \frac{1}{6} \quad P(3) = \frac{1}{6} \quad P(4) = \frac{1}{6} \quad P(5) = \frac{1}{6} \quad P(6) = \frac{1}{6}.$$

The sum of these probabilities is 1: $\frac{1}{6} + \frac{1}{6} + \frac{1}{6} + \frac{1}{6} + \frac{1}{6} + \frac{1}{6} = 1$. In general, **the sum of the theoretical probabilities of all possible outcomes in the sample space is 1**.

Check Point 1   A die is rolled once. Find the probability of rolling

a. a 2.

b. a number less than 4.

c. a number greater than 7.

d. a number less than 7.

Our next example involves a standard 52-card bridge deck, illustrated in Figure 11.5. The deck has four suits: Hearts and diamonds are red, and clubs and spades are black. Each suit has 13 different face values—A(ace), 2, 3, 4, 5, 6, 7, 8, 9, 10. J(jack), Q(queen), and K(king). Jacks, queens, and kings are called **picture cards** or **face cards**.

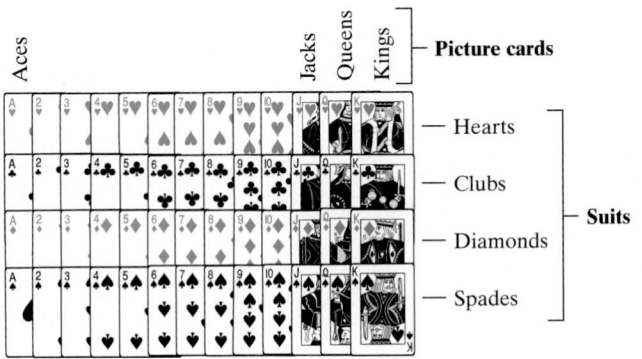

FIGURE 11.5 A standard 52-card bridge deck

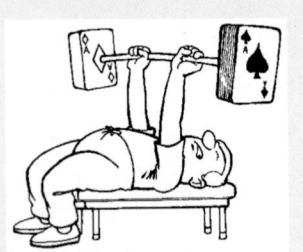
EXAMPLE 2   PROBABILITY AND A DECK OF 52 CARDS

You are dealt one card from a standard 52-card deck. Find the probability of being dealt

a. a king.   **b.** a heart.   **c.** the king of hearts.

Solution   Because there are 52 cards in the deck, the total number of possible ways of being dealt a single card is 52. The number of outcomes in the sample space is 52: $n(S) = 52$. We use 52 as the denominator of each probability fraction.

a. Let E be the event of being dealt a king. Because there are four kings in the deck, this event can occur in four ways: $n(E) = 4$.

$$P(\text{king}) = \frac{\text{number of outcomes that result in a king}}{\text{total number of possible outcomes}} = \frac{n(E)}{n(S)} = \frac{4}{52} = \frac{1}{13}$$

The probability of being dealt a king is $\frac{1}{13}$.

b. Let E be the event of being dealt a heart. Because there are 13 hearts in the deck, this event can occur in 13 ways: $n(E) = 13$.

$$P(\text{heart}) = \frac{\text{number of outcomes that result in a heart}}{\text{total number of possible outcomes}} = \frac{n(E)}{n(S)} = \frac{13}{52} = \frac{1}{4}$$

The probability of being dealt a heart is $\frac{1}{4}$.

c. Let E be the event of being dealt the king of hearts. Because there is only one card in the deck that is the king of hearts, this event can occur in just one way: $n(E) = 1$.

$$P(\text{king of hearts}) = \frac{\text{number of outcomes that result in the king of hearts}}{\text{total number of possible outcomes}} = \frac{n(E)}{n(S)} =$$

The probability of being dealt the king of hearts is $\frac{1}{52}$.

Check Point 2   You are dealt one card from a standard 52-card deck. Find the probability of being dealt

a. an ace.   **b.** a red card.   **c.** a red king.

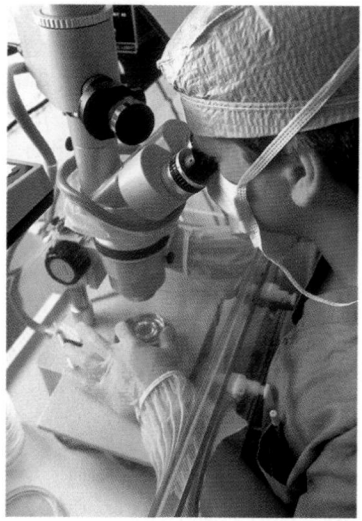

Genetic engineering offers some hope for cystic fibrosis patients. Geneticists can now isolate the *c* gene, the gene responsible for the disease. It is hoped that it will soon be possible to replace the defective *c* gene with a normal *C* gene.

Probabilities play a valuable role in the science of genetics. Example 3 deals with cystic fibrosis, an inherited lung disease occurring in about 1 out of every 2000 births among Caucasians and in about 1 out of every 250,000 births among non-Caucasians.

EXAMPLE 3   PROBABILITIES IN GENETICS

Each person carries two genes that are related to the absence or presence of the disease cystic fibrosis. Most Americans have two normal genes for this trait and are unaffected by cystic fibrosis. However, 1 in 25 Americans carries one normal gene and one defective gene. If we use *c* to represent a defective gene and *C* a normal gene, such a carrier can be designated as *Cc*. Thus, *CC* is a person who neither carries nor has cystic fibrosis, *Cc* is a carrier who is not actually sick, and *cc* is a person sick with the disease. Table 11.2 shows the four equally-likely outcomes for a child's genetic inheritance from two parents who are both carrying one cystic fibrosis gene. One copy of each gene is passed on to the child from the parents.

TABLE 11.2 CYSTIC FIBROSIS AND GENETIC INHERITANCE

		Second Parent	
		C	*c*
First	*C*	*CC*	*Cc*
Parent	*c*	*cC*	*cc*

> Shown in the table are the four possibilities for a child whose parents each carry one cystic fibrosis gene.

If each parent carries one cystic fibrosis gene, what is the probability that their child will have cystic fibrosis?

Solution   Table 11.2 shows that there are four equally likely outcomes. The sample space is $S = \{CC, Cc, cC, cc\}$ and $n(S) = 4$. The phrase "will have cystic fibrosis" describes only the *cc* child. Thus, $E = \{cc\}$ and $n(E) = 1$.

$$P(\text{cystic fibrosis}) = \frac{\text{number of outcomes that result in cystic fibrosis}}{\text{total number of possible outcomes}} = \frac{n(E)}{n(S)} = \frac{1}{4}$$

If each parent carries one cystic fibrosis gene, the probability that their child will have cystic fibrosis is $\frac{1}{4}$.

Check Point 3   Use the table in Example 3 to solve this exercise. If each parent carries one cystic fibrosis gene, find the probability that their child will be a carrier of the disease who is not actually sick.

2 Compute empirical probability.

Empirical Probability

Theoretical probability is based on a set of equally-likely outcomes and the number of elements in the set. By contrast, *empirical probability* applies to situations in which we observe the frequency of occurrence of an event. We use the following formula to compute the empirical probability of an event:

COMPUTING EMPIRICAL PROBABILITY

The empirical probability of event *E* is

$$P(E) = \frac{\text{observed number of times } E \text{ occurs}}{\text{total number of observed occurrences}}.$$

Arab-American Faiths

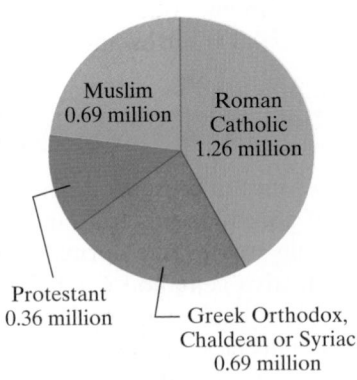

Muslim 0.69 million

Roman Catholic 1.26 million

Protestant 0.36 million

Greek Orthodox, Chaldean or Syriac 0.69 million

Source: Arab American Institute

FIGURE 11.6

EXAMPLE 4 COMPUTING EMPIRICAL PROBABILITY

There are approximately 3 million Arab Americans in the United States. The circle graph in Figure 11.6 shows religious affiliations of Arab Americans. If an Arab American is selected at random, find the empirical probability of selecting a Catholic.

Solution The probability of selecting a Catholic is the observed number of Arab Americans who are Catholic, 1.26 (million), divided by the total number of Arab Americans, 3 (million).

P (selecting a Catholic from the Arab-American population)

$$= \frac{\text{number of Arab Americans who are Catholic}}{\text{total number of Arab Americans}} = \frac{1.26}{3.00} = \frac{126}{300} = 0.42$$

The empirical probability of selecting a Catholic from the Arab-American population is $\frac{126}{300}$, or 0.42. Equivalently, 42% of Arab Americans are Catholic.

Check Point 4 If an Arab American is selected at random, find the empirical probability of selecting a Muslim.

In certain situations, we can establish a relationship between the two kinds of probability. Consider, for example, a coin that is equally likely to land heads or tails. Such a coin is called a **fair coin**. Empirical probability can be used to determine whether a coin is fair. Suppose we toss a coin 10, 50, 100, 1000, 10,000, and 100,000 times. We record the number of heads observed, shown in Table 11.3. For each of the six cases in the table, the empirical probability of heads is determined by dividing the number of heads observed by the number of tosses.

TABLE 11.3 EMPIRICAL PROBABILITIES OF HEADS AS THE NUMBER OF TOSSES INCREASES

Number of Tosses	Number of Heads Observed	Empirical Probability of Heads, or $P(H)$
10	4	$P(H) = \frac{4}{10} = 0.4$
50	27	$P(H) = \frac{27}{50} = 0.54$
100	44	$P(H) = \frac{44}{100} = 0.44$
1000	530	$P(H) = \frac{530}{1000} = 0.53$
10,000	4851	$P(H) = \frac{4851}{10,000} = 0.4851$
100,000	49,880	$P(H) = \frac{49,880}{100,000} = 0.4988$

A pattern is exhibited by the empirical probabilities in the right-hand column of Table 11.3. As the number of tosses increases, the empirical probabilities tend to get closer to 0.5, the theoretical probability. These results give us no reason to suspect that the coin is not fair.

Table 11.3 illustrates an important principle when observing uncertain outcomes such as the event of a coin landing on heads. As an experiment is repeated more and more times, the empirical probability of an event tends to get closer to the theoretical probability of that event. This principle is known as the **law of large numbers**.

Exercise Set 11.4

Practice and Application Exercises

Exercises 1–54 involve theoretical probability. Use the theoretical probability formula to solve each exercise. Express each probability as a fraction reduced to lowest terms.

In Exercises 1–10, a die is rolled. The set of equally-likely outcomes is $\{1, 2, 3, 4, 5, 6\}$. Find the probability of rolling

1. a 4.
2. a 5.
3. an odd number.
4. a number greater than 3.
5. a number less than 3.
6. a number greater than 4.
7. a number less than 7.
8. a number less than 8.
9. a number greater than 7.
10. a number greater than 8.

In Exercises 11–20, you are dealt one card from a standard 52-card deck. Find the probability of being dealt

11. a queen.
12. a jack.
13. a club.
14. a diamond.
15. a picture card.
16. a card greater than 3 and less than 7.
17. the queen of spades.
18. the ace of clubs.
19. a diamond and a spade.
20. a card with a green heart.

In Exercises 21–26, a fair coin is tossed two times in succession. The set of equally-likely outcomes is $\{HH, HT, TH, TT\}$. Find the probability of getting

21. two heads.
22. two tails.
23. the same outcome on each toss.
24. different outcomes on each toss.
25. a head on the second toss.
26. at least one head.

In Exercises 27–34, you select a family with three children. If M represents a male child and F a female child, the set of equally-likely outcomes for the children's genders is $\{MMM, MMF, MFM, MFF, FMM, FMF, FFM, FFF\}$. Find the probability of selecting a family with

27. exactly one female child.
28. exactly one male child.
29. exactly two male children.
30. exactly two female children.

31. at least one male child.
32. at least two female children.
33. four male children.
34. fewer than four female children.

In Exercises 35–40, a single die is rolled twice. The 36 equally-likely outcomes are shown as follows:

		Second Roll					
		⚀	⚁	⚂	⚃	⚄	⚅
First Roll	⚀	(1, 1)	(1, 2)	(1, 3)	(1, 4)	(1, 5)	(1, 6)
	⚁	(2, 1)	(2, 2)	(2, 3)	(2, 4)	(2, 5)	(2, 6)
	⚂	(3, 1)	(3, 2)	(3, 3)	(3, 4)	(3, 5)	(3, 6)
	⚃	(4, 1)	(4, 2)	(4, 3)	(4, 4)	(4, 5)	(4, 6)
	⚄	(5, 1)	(5, 2)	(5, 3)	(5, 4)	(5, 5)	(5, 6)
	⚅	(6, 1)	(6, 2)	(6, 3)	(6, 4)	(6, 5)	(6, 6)

Find the probability of getting

35. two even numbers.
36. two odd numbers.
37. two numbers whose sum is 5.
38. two numbers whose sum is 6.
39. two numbers whose sum exceeds 12.
40. two numbers whose sum is less than 13.

Use the spinner shown to answer Exercises 41–48. Assume that it is equally probable that the pointer will land on any one of the ten colored regions. If the pointer lands on a borderline, spin again.

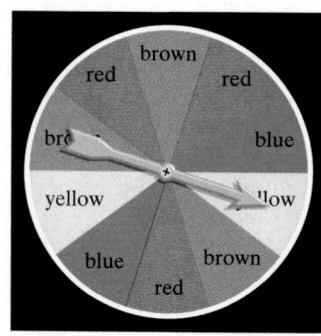

Find the probability that the spinner lands in

41. a red region.
42. a yellow region.
43. a blue region.
44. a brown region.
45. a region that is red or blue.
46. a region that is yellow or brown.
47. a region that is red and blue.
48. a region that is yellow and brown.

Exercises 49–54 deal with sickle cell anemia, an inherited disease in which red blood cells become distorted and deprived of oxygen. Approximately 1 in every 500 African-American infants is born with the disease; only 1 in 160,000 white infants has the disease. A person with two sickle cell genes will have the disease, but a person with only one sickle cell gene will have a mild, nonfatal anemia called sickle cell trait. (Approximately 8%–10% of the African-American population has this trait.)

		Second Parent	
		S	**s**
First	**S**	SS	Ss
Parent	**s**	sS	ss

If we use s to represent a sickle cell gene and S a healthy gene, the table shows the four possibilities for the children of two Ss parents. Each parent has only one sickle cell gene, so each has the relatively mild sickle cell trait. Find the probability that these parents give birth to a child who

49. has sickle cell anemia.　　**50.** has sickle cell trait.

51. is healthy.

In Exercises 52–54, use the following table that shows the four possibilities for the children of one healthy, SS, parent, and one parent with sickle cell trait, Ss.

		Second Parent (with Sickle Cell Trait)	
		S	**s**
Healthy	**S**	SS	Ss
First Parent	**S**	SS	Ss

Find the probability that these parents give birth to a child who

52. has sickle cell anemia.　　**53.** has sickle cell trait.

54. is healthy.

Exercises 55–64 involve empirical probability. First express answers as fractions. Then express probabilities as decimals, rounded to the nearest thousandth, if necessary.

Use the table showing the number of people who regularly participate in various forms of exercise, based on a survey of 2000 Americans, to solve Exercises 55–58.

NUMBER OF PEOPLE WHO REGULARLY PARTICIPATE IN VARIOUS FORMS OF EXERCISE IN A SURVEY OF 2000 PEOPLE

Forms of Exercise	Number of People
Walking/hiking	1140
Weight training	320
Running/jogging	280
Biking	240
Aerobics	240
Exercise machines	220

Source: Discovery Health Media

Find the probability that a randomly selected American participates in

55. weight training.　　**56.** running/jogging.

57. biking.　　**58.** walking/hiking.

The table shows the breakdown of the 89 thousand single parents on active duty in the U.S. military in 2002. All numbers are in thousands and rounded to the nearest thousand. Use the data in the table to solve Exercises 59–64.

SINGLE PARENTS ON ACTIVE DUTY IN THE U.S. MILITARY, IN THOUSANDS

	Army	**Navy**	**Marine Corps**	**Air Force**	**Total**	
Male	26	23	5	12	66	Total male: 26 + 23 + 5 + 12 = 66
Female	10	6	1	6	23	Total female: 10 + 6 + 1 + 6 = 23
Total	36	29	6	18	89	

Total Army　Total Navy　Total Marines　Total Air Force　Total on active duty

Source: U.S. Defense Department

Find the probability that a randomly selected single parent in the U.S. military is

59. female.　　**60.** male.

61. in the Army.　　**62.** in the Navy.

63. a woman in the Air Force.

64. a man in the Marine Corps.

Writing in Mathematics

65. What is the sample space of an experiment? What is an event?

66. How is the theoretical probability of an event computed?

67. Describe the difference between theoretical probability and empirical probability.

68. Give an example of an event whose probability must be determined empirically rather than theoretically.

69. Use the definition of theoretical probability to explain why the probability of an event that cannot occur is 0.

70. Use the definition of theoretical probability to explain why the probability of an event that is certain to occur is 1.

71. Write a probability word problem whose answer is one of the following fractions: $\frac{1}{6}$ or $\frac{1}{4}$ or $\frac{1}{3}$.

72. The president of a large company with 10,000 employees is considering mandatory cocaine testing for every employee. The test that would be used is 90% accurate, meaning that it will detect 90% of the cocaine users who are tested, and that 90% of the nonusers will test negative. This also means that the test gives 10% false positive. Suppose that 1% of the employees actually use cocaine. Find the probability that someone who tests positive for cocaine use is, indeed, a user.

Hint: Find the following probability fraction:

$$\frac{\text{the number of employees who test positive and are cocaine users}}{\text{the number of employes who test positive}}.$$

This fraction is given by

$$\frac{90\% \text{ of } 1\% \text{ of } 10,000}{\text{the number who test positive who actually use cocaine plus the number who test positive who do not use cocaine}}.$$

What does this probability indicate in terms of the percentage of employees who test positive who are not actually users? Discuss these numbers in terms of the issue of mandatory drug testing. Write a paper either in favor of or against mandatory drug testing, incorporating the actual percentage accuracy for such tests.

Critical Thinking Exercises

73. The target in the figure shown contains four squares. If a dart thrown at random hits the target, find the probability that it will land in an orange region.

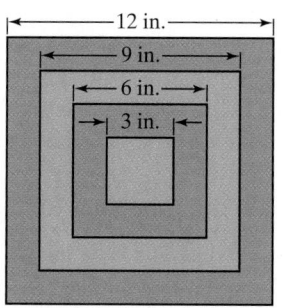

74. Some three-digit numbers, such as 101 and 313, read the same forward and backward. If you select a number from all three-digit numbers, find the probability that it will read the same forward and backward.

Group Exercise

75. There is a third type of probability, called **subjective probability**. Subjective probability states a probability value using an estimate or an educated guess, utilizing opinions and inexact information. For example, a sports journalist might say that there is a 60% probability that the Yankees will win the pennant next year. A seismologist might state that there is a 70% probability that a major earthquake will occur along the San Andreas fault within the next ten years.

Each group member should give an example of how subjective probability is used in everyday life, or of how he or she uses this probability.

SECTION 11.5	PROBABILITY WITH THE FUNDAMENTAL COUNTING PRINCIPLE, PERMUTATIONS, AND COMBINATIONS

OBJECTIVES

1. Compute probabilities with permutations.

2. Compute probabilities with combinations.

PROBABILITY OF DYING AT ANY GIVEN AGE

Age	Probability of Male Death	Probability of Female Death
10	0.00013	0.00010
20	0.00140	0.00050
30	0.00153	0.00050
40	0.00193	0.00095
50	0.00567	0.00305
60	0.01299	0.00792
70	0.03473	0.01764
80	0.07644	0.03966
90	0.15787	0.11250
100	0.26876	0.23969
110	0.39770	0.39043

Source: George Shaffner, *The Arithmetic of Life and Death*

George Tooker (b. 1920) "*Mirror II*" 1963, egg tempera on gesso panel, 20 × 20 in., 1968.4. Gift of R. H. Donnelley Erdman (PA 1956). Addison Gallery of American Art, Phillips Academy, Andover, Massachusetts. All Rights Reserved.

According to actuarial tables, there is no year in which death is as likely as continued life, at least until the age of 115. Until that age, the probability of dying in any one year ranges from a low of 0.00009 for a girl at age 11 to a high of 0.465 for either gender at age 114. For a healthy 30-year-old, how does the probability of dying this year compare to the probability of winning the top prize in a state lottery game? In this section, we provide the surprising answer to this question, as we study probability with the Fundamental Counting Principle, permutations, and combinations.

1 Compute probabilities with permutations.

Probability with Permutations

We return to our concert tour with U2, 'N Sync, Aerosmith, and the Rolling Stones. Now, the two surviving members of the Beatles have agreed to join the tour! You really have your work cut out for you. In which order should the five groups in the concert perform? Because order makes a difference, this is a permutation situation. Example 1 is based on this scenario, and uses permutations to solve a probability problem.

EXAMPLE 1 | PROBABILITY AND PERMUTATIONS

The five groups in the tour agree to determine the order of performance based on a random selection. Each band's name is written on one of five cards. The cards are placed in a hat and then five cards are drawn, one at a time. The order in which the cards are drawn determines the order in which the bands perform. What is the probability of the Rolling Stones performing fourth and the Beatles last?

Solution We begin by applying the definition of probability to this situation.

P(Rolling Stones fourth, Beatles last)

$$= \frac{\text{number of permutations with Rolling Stones fourth, Beatles last}}{\text{total number of possible permutations}}$$

We can use the Fundamental Counting Principle to find the total number of possible permutations. This represents the number of ways you can put together the concert.

First Group to Perform	Second Group to Perform	Third Group to Perform	Fourth Group to Perform	Last Group to Perform
5 choices	4 choices	3 choices	2 choices	1 choice

There are $5 \cdot 4 \cdot 3 \cdot 2 \cdot 1$, or 120, possible permutations. Equivalently, the five groups can perform in 120 different orders.

We can also use the Fundamental Counting Principle to find the number of permutations with the Rolling Stones performing fourth and the Beatles performing last. You can choose any one of the three groups, U2, 'N Sync, or Aerosmith, as the opening act. This leaves two choices for the second group to perform, and only one choice for the third group to perform. There is only one choice for the fourth group—the Rolling Stones, and one choice for the closing act—the Beatles:

First Group to Perform	Second Group to Perform	Third Group to Perform	Fourth Group to Perform	Last Group to Perform
3 choices (U2, 'NSync, or Aerosmith)	2 choices	1 choice	1 choice: The Rolling Stones	1 choice: The Beatles

Thus, there are $3 \cdot 2 \cdot 1 \cdot 1 \cdot 1$, or 6 possible permutations. Equivalently, there are 6 lineups with the Rolling Stones performing fourth and the Beatles last.

Now we can return to our probability fraction.

P(Rolling Stones fourth, Beatles last)

$$= \frac{\text{number of permutations with Rolling Stones fourth, Beatles last}}{\text{total number of possible permutations}}$$

$$= \frac{6}{120} = \frac{1}{20}$$

The probability of the Rolling Stones performing fourth and the Beatles last is $\frac{1}{20}$.

Check 1 Point Use the information given in Example 1 to find the probability of U2 performing first, the Rolling Stones performing fourth, and the Beatles last.

Probability with Combinations

2 Compute probabilities with combinations.

If your state has a lottery drawing each week, the probability that someone will win the top prize is relatively high. If there is no winner this week, it is virtually certain that eventually someone will be graced with millions of dollars. So, why are you so unlucky compared to this undisclosed someone? In Example 2, we provide an answer to this question.

EXAMPLE 2 | PROBABILITY AND COMBINATIONS: WINNING THE LOTTERY

Florida's lottery game, LOTTO, is set up so that each player chooses six different numbers from 1 to 53. If the six numbers chosen match the six numbers drawn randomly, the player wins (or shares) the top cash prize. (As of this writing, the top cash prize has ranged from $7 million to $106.5 million.) With one LOTTO ticket, what is the probability of winning this prize?

Solution Because the order of the six numbers does not matter, this is a situation involving combinations. We begin with the formula for probability.

$$P(\text{winning}) = \frac{\text{number of ways of winning}}{\text{total number of possible combinations}}$$

We can use the combinations formula

$$_nC_r = \frac{n!}{(n-r)!\,r!}$$

to find the total number of possible combinations. We are selecting $r = 6$ numbers from a collection of $n = 53$ numbers.

$$_{53}C_6 = \frac{53!}{(53-6)!\,6!} = \frac{53!}{47!\,6!} = \frac{53\cdot52\cdot51\cdot50\cdot49\cdot48\cdot47!}{47!\cdot6\cdot5\cdot4\cdot3\cdot2\cdot1} = 22{,}957{,}480$$

There are nearly 23 million number combinations possible in LOTTO! If a person buys one LOTTO ticket, that person has selected only one combination of the six numbers. With one LOTTO ticket, there is only one way of winning.

Now we can return to our probability fraction.

$$P(\text{winning}) = \frac{\text{number of ways of winning}}{\text{total number of possible combinations}} = \frac{1}{22{,}957{,}480} \approx 0.0000000436$$

The probability of winning the top prize with one LOTTO ticket is $\frac{1}{22{,}957{,}480}$, or about 1 in 23 million.

In 2001, Americans spent nearly 18 billion dollars on lotteries set up by revenue-hungry states. If a person buys, say 5000 different tickets in Florida's LOTTO, that person has selected 5000 different combinations of the six numbers. The probability of winning is

$$\frac{5000}{22{,}957{,}480} \approx 0.000218.$$

The chances of winning top prize are about 218 in a million. At $1 per LOTTO ticket, it is highly probable that our LOTTO player will be $5000 poorer.

State lotteries keep 50 cents on the dollar, resulting in $10 billion a year for public funding.
© Damon Higgins/The Palm Beach Post

BLITZER BONUS

Comparing the Probability of Dying to the Probability of Winning Florida's LOTTO

As a healthy nonsmoking 30-year-old, your probability of dying this year is approximately 0.001. Divide this probability by the probability of winning LOTTO with one ticket:

$$\frac{0.001}{0.0000000436} \approx 22{,}936.$$

A healthy 30-year-old is nearly 23,000 times more likely to die this year than to win Florida's lottery.

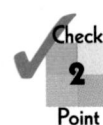

Check 2 Point People lose interest when they do not win at games of chance, including Florida's LOTTO. With drawings twice weekly instead of once, the game described in Example 2 was added to bring back lost players and increase ticket sales. The original LOTTO was set up so that each player chose six different numbers from 1 to 49, rather than from 1 to 53, with a lottery drawing only once a week. With one LOTTO ticket, what was the probability of winning the top cash prize in Florida's original LOTTO? Express the answer as a fraction and as a decimal correct to ten places.

EXAMPLE 3   PROBABILITY AND COMBINATIONS

A club consists of five men and seven women. Three members are selected at random to attend a conference. Find the probability that the selected group consists of

a. three men.

b. one man and two women.

Solution   The order in which the three people are selected does not matter, so this is a problem involving combinations.

12 Club Members

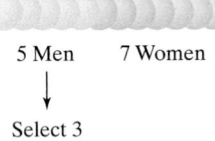

5 Men   7 Women

↓

Select 3

a. We begin with the probability of selecting three men.

$$P(3 \text{ men}) = \frac{\text{number of ways of selecting 3 men}}{\text{total number of possible combinations}}$$

First, we consider the denominator of the probability fraction. We are selecting $r = 3$ people from a total group of $n = 12$ people (five men and seven women). The total number of possible combinations is

$$_{12}C_3 = \frac{12!}{(12-3)!3!} = \frac{12!}{9!3!} = \frac{12 \cdot 11 \cdot 10 \cdot 9!}{9! \cdot 3 \cdot 2 \cdot 1} = 220.$$

Thus, there are 220 possible three-person selections.

Next, we consider the numerator of the probability fraction. We are interested in the number of ways of selecting three men from five men. We are selecting $r = 3$ men from a total group of $n = 5$ men. The number of possible combinations of three men is

$$_{5}C_3 = \frac{5!}{(5-3)!3!} = \frac{5!}{2!3!} = \frac{5 \cdot 4 \cdot 3!}{2 \cdot 1 \cdot 3!} = 10.$$

Thus, there are 10 ways of selecting three men from five men. Now we can fill in the numbers in the numerator and the denominator of our probability fraction.

$$P(3 \text{ men}) = \frac{\text{number of ways of selecting 3 men}}{\text{total number of possible combinations}} = \frac{10}{220} = \frac{1}{22}$$

The probability that the group selected to attend the conference consists of three men is $\frac{1}{22}$.

12 Club Members

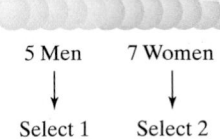

5 Men   7 Women

↓   ↓

Select 1   Select 2

b. We set up the fraction for the probability that the selected group consists of one man and two women.

$$P(1 \text{ man, } 2 \text{ women}) = \frac{\text{number of ways of selecting 1 man and 2 women}}{\text{total number of possible combinations}}$$

The denominator of this fraction is the same as the denominator in part (a). The total number of possible combinations is found by selecting $r = 3$ people from $n = 12$ people: $_{12}C_3 = 220$.

Next, we move to the numerator of the probability fraction. The number of ways of selecting $r = 1$ man from $n = 5$ men is

$$_5C_1 = \frac{5!}{(5-1)!1!} = \frac{5!}{4!1!} = \frac{5 \cdot 4!}{4! \cdot 1} = 5.$$

The number of ways of selecting $r = 2$ women from $n = 7$ women is

$$_7C_2 = \frac{7!}{(7-2)!2!} = \frac{7!}{5!2!} = \frac{7 \cdot 6 \cdot 5!}{5! \cdot 2 \cdot 1} = 21.$$

By the Fundamental Counting Principle, the number of ways of selecting 1 man and 2 women is

$$_5C_1 \cdot {}_7C_2 = 5 \cdot 21 = 105.$$

Now we can fill in the numbers in the numerator and the denominator of our probability fraction.

$$P(1 \text{ man, 2 women}) = \frac{\text{number of ways of selecting 1 man and 2 women}}{\text{total number of possible combinations}} = \frac{_5C_1 \cdot {}_7C_2}{_{12}C_3} = \frac{105}{220} = \frac{21}{44}$$

The probability that the group selected to attend the conference consists of one man and two women is $\frac{21}{44}$.

Check Point 3 A club consists of six men and four women. Three members are selected at random to attend a conference. Find the probability that the selected group consists of

a. three men.

b. two men and one woman.

Exercise Set 11.5

Practice and Application Exercises

1. Martha, Lee, Nancy, Paul, and Armando have all been invited to a dinner party. They arrive randomly and each person arrives at a different time.
   a. In how many ways can they arrive?
   b. In how many ways can Martha arrive first and Armando last?
   c. Find the probability that Martha will arrive first and Armando last.

2. Three men and three women line up at a checkout counter in a store.
   a. In how many ways can they line up?
   b. In how many ways can they line up if the first person in line is a woman, and then the line alternates by gender—that is a woman, a man, a woman, a man, and so on?
   c. Find the probability that the first person in line is a woman and the line alternates by gender.

3. Six stand-up comics, A, B, C, D, E, and F, are to perform on a single evening at a comedy club. The order of performance is determined by random selection. Find the probability that

   a. Comic E will perform first.
   b. Comic C will perform fifth and comic B will perform last.
   c. The comedians will perform in the following order: D, E, C, A, B, F.
   d. Comic A or comic B will perform first.

4. Seven performers, A, B, C, D, E, F, and G, are to appear in a fund raiser. The order of performance is determined by random selection. Find the probability that
   a. D will perform first.
   b. E will perform sixth and B will perform last.
   c. They will perform in the following order: C, D, B, A, G, F, E.
   d. F or G will perform first.

5. A group consists of four men and five women. Three people are selected to attend a conference.
   a. In how many ways can three people be selected from this group of nine?
   b. In how many ways can three women be selected from the five women?
   c. Find the probability that the selected group will consist of all women.

6. A political discussion group consists of five Democrats and six Republicans. Four people are selected to attend a conference.

　a. In how many ways can four people be selected from this group of eleven?

　b. In how many ways can four Republicans be selected from the six Republicans?

　c. Find the probability that the selected group will consist of all Republicans.

7. To play the California lottery, a person has to correctly select 6 out of 51 numbers, paying $1 for each six-number selection. If the six numbers picked are the same as the ones drawn by the lottery, mountains of money are bestowed. What is the probability that a person with one combination of six numbers will win? What is the probability of winning if 100 different lottery tickets are purchased?

8. A state lottery is designed so that a player chooses five numbers from 1 to 30 on one lottery ticket. What is the probability that a player with one lottery ticket will win? What is the probability of winning if 100 different lottery tickets are purchased?

9. A box contains 25 transistors, 6 of which are defective. If 6 are selected at random, find the probability that

　a. all are defective.

　b. none are defective.

10. A committee of five people is to be formed from six lawyers and seven teachers. Find the probability that

　a. all are lawyers.

　b. none are lawyers.

11. A city council consists of six Democrats and four Republicans. If a committee of three people is selected, find the probability of selecting one Democrat and two Republicans.

12. A parent-teacher committee consisting of four people is to be selected from fifteen parents and five teachers. Find the probability of selecting two parents and two teachers.

Exercises 13–18 involve a deck of 52 cards. If necessary, refer to the picture of a deck of cards, Figure 11.5 on page 582.

13. A poker hand consists of five cards.

　a. Find the total number of possible five-card poker hands.

　b. A diamond flush is a five-card hand consisting of all diamonds. Find the number of possible diamond flushes.

　c. Find the probability of being dealt a diamond flush.

14. A poker hand consists of five cards.

　a. Find the total number of possible five-card poker hands.

　b. Find the number of ways in which four aces can be selected.

　c. Find the number of ways in which one king can be selected.

　d. Use the Fundamental Counting Principle and your answers from parts (b) and (c) to find the number of ways of getting four aces and one king.

　e. Find the probability of getting a poker hand of four aces and one king.

15. If you are dealt 3 cards from a shuffled deck of 52 cards, find the probability that all 3 cards are picture cards.

16. If you are dealt 4 cards from a shuffled deck of 52 cards, find the probability that all 4 are hearts.

17. If you are dealt 4 cards from a shuffled deck of 52 cards, find the probability of getting two queens and two kings.

18. If you are dealt 4 cards from a shuffled deck of 52 cards, find the probability of getting three jacks and one queen.

Writing in Mathematics

19. If people understood the mathematics involving probabilities and lotteries, as you now do, do you think they would continue to spend hundreds of dollars per year on lottery tickets? Explain your answer.

20. Write and solve an original problem involving probability and permutations.

21. Write and solve an original problem involving probability and combinations whose solution requires $\dfrac{_{14}C_{10}}{_{20}C_{10}}$.

Critical Thinking Exercises

22. An apartment complex offers apartments with four different options, designated by A through D. There are an equal number of apartments with each combination of options.

A	B	C	D
one bedroom two bedrooms three bedrooms	one bathroom two bathrooms	first floor second floor	lake view golf course view no special view

If there is only one apartment left, what is the probability that it is precisely what a person is looking for, namely two bedrooms, two bathrooms, first floor, and a lake or golf course view?

23. Reread Exercise 7. How much must a person spend so that the probability of winning the California lottery is $\frac{1}{2}$?

24. Suppose that it is a week in which the cash prize in Florida's LOTTO is promised to exceed $50 million. If a person purchases 22,957,480 tickets in LOTTO at $1 per ticket (all possible combinations), isn't this a guarantee of winning the lottery? Because the probability in this situation is 1, what's wrong with doing this?

25. The digits 1, 2, 3, 4, and 5 are randomly arranged to form a three-digit number. (Digits are not repeated.) Find the probability that the number is even and greater than 500.

26. In a five-card poker hand, what is the probability of being dealt exactly one ace and no picture cards?

Group Exercise

27. Research and present a group report on state lotteries. Include answers to some or all of the following questions. As always, make the report interesting and informative. Which states do not have lotteries? Why not? How much is spent per capita on lotteries? What are some of the lottery games? What is the probability of winning top prize in these games? What income groups spend the greatest amount of money on lotteries? If your state has a lottery, what does it do with the money it makes? Is the way the money is spent what was promised when the lottery first began

SECTION 11.6   EVENTS INVOLVING *NOT* AND *OR*; ODDS

OBJECTIVES

1. Find the probability that an event will not occur.

2. Find the probability of one event or a second event occurring.

3. Understand and use odds.

You take your first trip to London. You are surprised to learn that the British gamble on everything. Shops with bookmakers are available everywhere for placing bets. In such a shop, you overhear a conversation about turning up the king of hearts in a deck of cards. You are expecting to hear something about a probability of $\frac{1}{52}$. Instead, you hear the phrase "51 to 1 against." Are you having difficulty understanding the British accents or did you miss something in your liberal arts math course? Whatever happened to probability?

No, it's not the accent. There are several ways to express the likelihood of an event. For example, we can discuss the probability of an event. We can also discuss *odds against* an event and *odds in favor* of an event. In this section, we expand our knowledge of probability and explain the meaning of odds.

1 Find the probability that an event will not occur.

Probability of an Event Not Occurring

A survey (source: Penn, Schoen, and Berland) asked 500 Americans to rate their health. Of those surveyed, 270 rated their health as good/excellent. This means that $500 - 270$, or 230, people surveyed did not rate their health as good/excellent. Notice that

$$P(\text{good/excellent}) + P(\text{not good/excellent}) = \frac{270}{500} + \frac{230}{500} = \frac{500}{500} = 1.$$

In general, because the sum of the probabilities of all possible outcomes in any situation is 1,

$$P(E) + P(\text{not } E) = 1.$$

We now solve this equation for $P(\text{not } E)$, the probability that event E will not occur, by subtracting $P(E)$ from both sides. The resulting formula is given in the following box:

THE PROBABILITY OF AN EVENT NOT OCCURRING

The probability that an event E will not occur is equal to 1 minus the probability that it will occur.

$$P(\text{not } E) = 1 - P(E)$$

STUDY TIP

You can work Example 1 without using the formula for $P(\text{not } E)$. Here is how it's done:

$$P(\text{not a queen})$$

$$= \frac{\begin{array}{c}\text{number of ways a}\\ \text{non-queen can occur}\end{array}}{\text{total number of outcomes}}$$

$$= \frac{48}{52} \quad \boxed{\begin{array}{c}\textbf{With 4 queens,}\\ \textbf{52 − 4 = 48 cards}\\ \textbf{are not queens.}\end{array}}$$

$$= \frac{4 \cdot 12}{4 \cdot 13} = \frac{12}{13}.$$

EXAMPLE 1 THE PROBABILITY OF AN EVENT NOT OCCURRING

If you are dealt one card from a standard 52-card deck, find the probability that you are not dealt a queen.

Solution Because

$$P(\text{not } E) = 1 - P(E)$$

then

$$P(\text{not a queen}) = 1 - P(\text{queen}).$$

There are four queens in a deck of 52 cards. The probability of being dealt a queen is $\frac{4}{52} = \frac{1}{13}$.

Thus,

$$P(\text{not a queen}) = 1 - P(\text{queen}) = 1 - \frac{1}{13} = \frac{13}{13} - \frac{1}{13} = \frac{12}{13}.$$

The probability that you are not dealt a queen is $\frac{12}{13}$.

 **Check Point 1** If you are dealt one card from a standard 52-card deck, find the probability that you are not dealt a diamond.

EXAMPLE 2 THE PROBABILITY OF AN EVENT NOT OCCURRING

The graph in Figure 11.7 shows the distribution, by branch and gender, of the 1.37 million, or 1370 thousand, active-duty personnel in the U.S. military in 2001. Numbers are given in thousands and rounded to the nearest ten thousand. If one person is randomly selected from the U.S. military and the distribution is the same as it was in 2001, find the probability that this person is not in the Army.

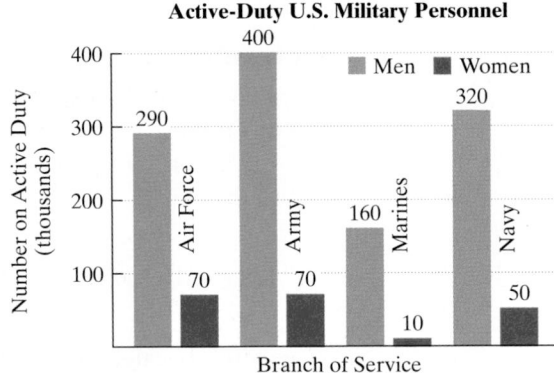

Source: U.S. Defense Department

FIGURE 11.7

Solution We begin by finding the probability that the selected person *is* in the Army.

$$P(\text{Army}) = \frac{\text{number of people in the Army}}{\text{total number of people in the U.S. military}}$$

$$= \frac{400 + 70}{1370} \quad \boxed{\begin{array}{l}\text{The graph shows 400 thousand men}\\ \text{and 70 thousand women in the Army.}\end{array}}$$

$$= \frac{470}{1370} = \frac{47}{137} \quad \boxed{\begin{array}{l}\text{This number was given, but can be}\\ \text{obtained by adding the eight numbers}\\ \text{above the bars.}\end{array}}$$

Thus,

$$P(\text{not in Army}) = 1 - P(\text{Army}) = 1 - \frac{47}{137} = \frac{137}{137} - \frac{47}{137} = \frac{90}{137}.$$

The probability that a person selected from the U.S. military is not in the Army is $\frac{90}{137}$.

 **Check Point 2**   Use the graph in Figure 11.7. If one person is randomly selected from the U.S. military, find the probability that this person is not in the Marines.

Or **Probabilities with Mutually Exclusive Events**

 Find the probability of one event or a second event occurring.

Suppose that you randomly select one card from a deck of 52 cards. Let A be the event of selecting a king and B be the event of selecting a queen. Only one card is selected, so it is impossible to get both a king and a queen. The outcomes of selecting a king and a queen cannot occur simultaneously. They are called *mutually exclusive events*.

MUTUALLY EXCLUSIVE EVENTS

If it is impossible for events A and B to occur simultaneously, the events are said to be **mutually exclusive**.

In general, if A and B are mutually exclusive events, the probability that either A or B will occur is determined by adding their individual probabilities.

OR PROBABILITIES WITH MUTUALLY EXCLUSIVE EVENTS

If A and B are mutually exclusive events, then

$$P(A \text{ or } B) = P(A) + P(B).$$

EXAMPLE 3 | THE PROBABILITY OF EITHER OF TWO MUTUALLY EXCLUSIVE EVENTS OCCURRING

If one card is randomly selected from a deck of cards, what is the probability of selecting a king or a queen?

Solution   We find the probability that either of these mutually exclusive events will occur by adding their individual probabilities.

$$P(\text{king or queen}) = P(\text{king}) + P(\text{queen}) = \frac{4}{52} + \frac{4}{52} = \frac{8}{52} = \frac{2}{13}$$

The probability of selecting a king or a queen is $\frac{2}{13}$.

 **Check Point 3**   If you roll a single, six-sided die, what is the probability of getting either a 4 or a 5?

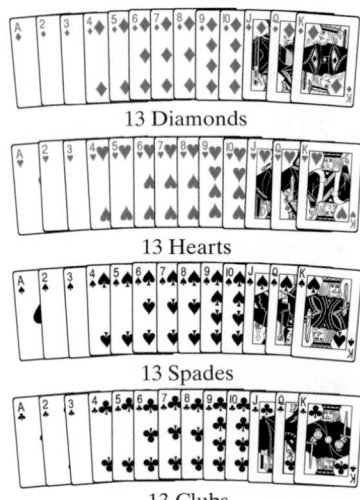

13 Diamonds

13 Hearts

13 Spades

13 Clubs

FIGURE 11.8 A deck of 52 cards

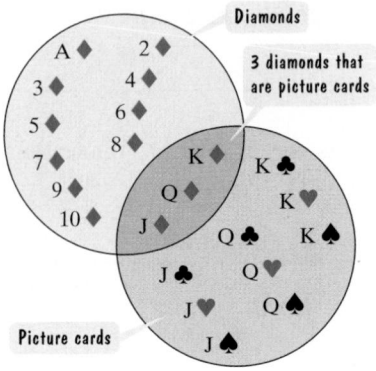

FIGURE 11.10

Or **Probabilities with Events That Are Not Mutually Exclusive**

Consider the deck of 52 cards shown in Figure 11.8. Suppose that these cards are shuffled and you randomly select one card from the deck. What is the probability of selecting a diamond or a picture card (jack, queen, king)? Begin by adding their individual probabilities.

$$P(\text{diamond}) + P(\text{picture card}) = \frac{13}{52} + \frac{12}{52}$$

There are 13 diamonds in the deck of 52 cards.

There are 12 picture cards in the deck of 52 cards.

However, this sum is not the probability of selecting a diamond or a picture card. The problem is that there are three cards that are *simultaneously* diamonds and picture cards, shown in Figure 11.9. The events of selecting a diamond and selecting a picture card are not mutually exclusive. It is possible to select a card that is both a diamond and a picture card.

The situation is illustrated in the Venn diagram in Figure 11.10. Why can't we find the probability of selecting a diamond or a picture card by adding their individual probabilities? The Venn diagram shows that three of the cards, the three diamonds that are picture cards, get counted twice when we add the individual probabilities. First the three cards get counted as diamonds, and then they get counted as picture cards. In order to avoid the error of counting the three cards twice, we need to subtract the probability of getting a diamond and a picture card, $\frac{3}{52}$, as follows:

FIGURE 11.9 Three diamonds are picture cards.

$P(\text{diamond or picture card})$

$$= P(\text{diamond}) + P(\text{picture card}) - P(\text{diamond and picture card})$$

$$= \frac{13}{52} + \frac{12}{52} - \frac{3}{52} = \frac{13 + 12 - 3}{52} = \frac{22}{52} = \frac{11}{26}.$$

Thus, the probability of selecting a diamond or a picture card is $\frac{11}{26}$.

In general, if A and B are events that are not mutually exclusive, the probability that A or B will occur is determined by adding their individual probabilities and then subtracting the probability that A and B occur simultaneously.

OR PROBABILITIES WITH EVENTS THAT ARE NOT MUTUALLY EXCLUSIVE

If A and B are not mutually exclusive events, then

$$P(A \text{ or } B) = P(A) + P(B) - P(A \text{ and } B).$$

| EXAMPLE 4 | AN *OR* PROBABILITY WITH EVENTS THAT ARE NOT MUTUALLY EXCLUSIVE |

In a group of 25 baboons, 18 enjoy grooming their neighbors, 16 enjoy screeching wildly, while 10 enjoy grooming their neighbors and screeching wildly. If one baboon is selected at random from the group, find the probability that it enjoys grooming its neighbors or screeching wildly.

Solution   It is possible for a baboon in the group to enjoy both grooming its neighbors and screeching wildly. Ten of the brutes are given to engage in both activities. These events are not mutually exclusive.

$$P\left(\begin{array}{c}\text{grooming}\\\text{or screeching}\end{array}\right) = P(\text{grooming}) + P(\text{screeching}) - P\left(\begin{array}{c}\text{grooming}\\\text{and screeching}\end{array}\right)$$

$$= \frac{18}{25} + \frac{16}{25} - \frac{10}{25}$$

| 18 of the 25 baboons enjoy grooming. | 16 of the 25 baboons enjoy screeching. | 10 of the 25 baboons enjoy both. |

$$= \frac{18 + 16 - 10}{25} = \frac{24}{25}$$

The probability that a baboon in the group enjoys grooming its neighbors or screeching wildly is $\frac{24}{25}$.

 **Check Point 4**   In a group of 50 students, 23 take math, 11 take psychology, and 7 take both math and psychology. If one student is selected at random, find the probability that the student takes math or psychology.

| EXAMPLE 5 | AN *OR* PROBABILITY WITH EVENTS THAT ARE NOT MUTUALLY EXCLUSIVE |

Figure 11.11 illustrates a spinner. It is equally probable that the pointer will land on any one of the eight regions, numbered 1 through 8. If the pointer lands on a borderline, spin again. Find the probability that the pointer will stop on an even number or on a number greater than 5.

Solution   It is possible for the pointer to land on a number that is both even and greater than 5. Two of the numbers, 6 and 8, are even and greater than 5. These events are not mutually exclusive. The probability of landing on a number that is even or greater than 5 is calculated as follows:

$$P\left(\begin{array}{c}\text{even or}\\\text{greater than 5}\end{array}\right) = P(\text{even}) + P(\text{greater than 5}) - P\left(\begin{array}{c}\text{even and}\\\text{greater than 5}\end{array}\right)$$

$$= \frac{4}{8} + \frac{3}{8} - \frac{2}{8}$$

| Four of the eight numbers, 2, 4, 6, and 8, are even. | Three of the eight numbers, 6, 7, and 8, are greater than 5. | Two of the eight numbers, 6 and 8, are even and greater than 5. |

$$= \frac{4 + 3 - 2}{8} = \frac{5}{8}$$

The probability that the pointer will stop on an even number or a number greater than 5 is $\frac{5}{8}$.

FIGURE 11.11 It is equally probable that the pointer will land on any one of the eight regions.

 Use Figure 11.11 on the previous page to find the probability that the pointer will stop on an odd number or a number less than 5.

EXAMPLE 6 | AN *OR* PROBABILITY WITH REAL-WORLD DATA

Earlier in this section, we saw a graph showing the distribution, by branch and gender, of active-duty personnel in the U.S. military. The data are shown again in Table 11.4. If one person is randomly selected from the U.S. military, find the probability that this person is in the Army or is a woman.

TABLE 11.4 ACTIVE-DUTY U.S. MILITARY PERSONNEL, IN THOUSANDS

	Air Force	Army	Marine Corps	Navy	Total
Male	290	400	160	320	1170
Female	70	70	10	50	200
Total	360	470	170	370	1370

Total Air Force — Total Army — Total Marines — Total Navy — Total on active duty

Total male: 290 + 400 + 160 + 320 = 1170

Total female: 70 + 70 + 10 + 50 = 200

Source: U.S. Defense Department

Solution It is possible to select a person who is both in the Army and is a woman. Thus, these events are not mutually exclusive.

$$P(\text{Army or woman}) = P(\text{Army}) + P(\text{woman}) - P(\text{Army and woman})$$

$$= \frac{470}{1370} + \frac{200}{1370} - \frac{70}{1370}$$

Of the 1370 (thousand) personnel, 470 are in the Army — 400 men and 70 women.

Of the 1370 personnel, 200 are women — 70 Air Force + 70 Army + 10 Marines + 50 Navy.

Of the 1370 personnel, 70 are Army women.

$$= \frac{470 + 200 - 70}{1370} = \frac{600}{1370} = \frac{60}{137}$$

The probability that a person selected from the U.S. military is in the Army or is a woman is $\frac{60}{137}$.

 Use Table 11.4. If one person is randomly selected from the U.S. military, find the probability that this person is in the Navy or is a man.

Understand and use odds.

Odds

If we know the probability of an event E, we can also speak of the *odds in favor*, or the *odds against*, the event. The following definitions link together the concepts of odds and probabilities:

PROBABILITY TO ODDS

If $P(E)$ is the probability of an event E occurring, then

1. The **odds in favor of E** are found by taking the probability that E will occur and dividing by the probability that E will not occur.

$$\text{Odds in favor of } E = \frac{P(E)}{P(\text{not } E)}$$

2. The **odds against E** are found by taking the probability that E will not occur and dividing by the probability that E will occur.

$$\text{Odds against } E = \frac{P(\text{not } E)}{P(E)}$$

The odds against E can also be found by reversing the ratio representing the odds in favor of E.

EXAMPLE 7   FROM PROBABILITY TO ODDS

You roll a single, six-sided die.

a. Find the odds in favor of rolling a 2.

b. Find the odds against rolling a 2.

Solution   Let E represent the event of rolling a 2. In order to determine odds, we must first find the probability of E occurring and the probability of E not occurring. With $S = \{1, 2, 3, 4, 5, 6\}$ and $E = \{2\}$, we see that

$$P(E) = \frac{1}{6}$$

$$\text{and}\quad P(\text{not } E) = 1 - \frac{1}{6} = \frac{6}{6} - \frac{1}{6} = \frac{5}{6}.$$

Now we are ready to construct the ratios for the odds in favor of E and the odds against E.

STUDY TIP

When computing odds, the denominators of the two probabilities will always divide out.

a.
$$\text{Odds in favor of } E(\text{rolling a 2}) = \frac{P(E)}{P(\text{not } E)} = \frac{\dfrac{1}{6}}{\dfrac{5}{6}} = \frac{1}{6} \cdot \frac{6}{5} = \frac{1}{5}$$

The odds in favor of rolling a 2 are $\frac{1}{5}$. The ratio $\frac{1}{5}$ is usually written $1:5$ and is read "1 to 5." Thus, the odds in favor of rolling a 2 are 1 to 5.

b. Now that we have the odds in favor of rolling a 2, namely $\frac{1}{5}$ or $1:5$, we can find the odds against rolling a 2 by reversing this ratio. Thus,

$$\text{Odds against } E(\text{rolling a 2}) = \frac{5}{1} \quad \text{or} \quad 5:1.$$

The odds against rolling a 2 are 5 to 1.

Check 7 Point

You are dealt one card from a 52-card deck.

a. Find the odds in favor of getting a red queen.

b. Find the odds against getting a red queen.

EXAMPLE 8   FROM PROBABILITY TO ODDS

The winner of a raffle will receive a new sports utility vehicle. If 500 raffle tickets were sold and you purchased ten tickets, what are the odds against your winning the car?

Solution   Let E represent the event of winning the SUV. Because you purchased ten tickets and 500 tickets were sold,

$$P(E) = \frac{10}{500} = \frac{1}{50} \quad \text{and} \quad P(\text{not } E) = 1 - \frac{1}{50} = \frac{50}{50} - \frac{1}{50} = \frac{49}{50}.$$

Now we are ready to construct the ratio for the odds against E (winning the SUV).

$$\text{Odds against } E = \frac{P(\text{not } E)}{P(E)} = \frac{\dfrac{49}{50}}{\dfrac{1}{50}} = \frac{49}{50} \cdot \frac{50}{1} = \frac{49}{1}$$

The odds against winning the SUV are 49 to 1, written 49 : 1.

Check 8 Point

The winner of a raffle will receive a two-year scholarship to the college of his or her choice. If 1000 raffle tickets were sold and you purchased five tickets, what are the odds against your winning the scholarship?

Odds enable us to play and bet fairly on games. For example, we have seen that the odds in favor of getting 2 when you roll a die are 1 to 5. Suppose this is a gaming situation and you bet $1 on a 2 turning up. In terms of your bet, there is one favorable outcome, rolling 2, and five unfavorable outcomes, rolling 1, 3, 4, 5, or 6. The odds in favor of getting 2, 1 to 5, compares the number of favorable outcomes, one, to the number of unfavorable outcomes, five.

Using odds in a gaming situation where money is waged, we can determine if the game is *fair*. If the odds in favor of an event E are a to b, the **game is fair** if a bet of $\$a$ is lost if event E does not occur, but a win of $\$b$ (as well as returning the bet of $\$a$) is realized if event E does occur. For example, the odds in favor of getting 2 on a die roll are 1 to 5. If you bet $1 on a 2 turning up and the game is fair, you you should win $5 (and have your bet of $1 returned) if a 2 turns up.

Now that we know how to convert from probability to odds, let's see how to convert from odds to probability. Suppose that the odds in favor of event E occurring are a to b. This means that

$$\frac{P(E)}{P(\text{not } E)} = \frac{a}{b} \quad \text{or} \quad \frac{P(E)}{1 - P(E)} = \frac{a}{b}.$$

By solving the equation on the right for $P(E)$, we obtain the following formula for converting from odds to probability:

BLITZER BONUS

House Odds

The odds given at horse races and at all games of chance are usually *odds against*. The horse in Example 9, with odds of winning at 2 to 5, has odds against winning at 5 to 2. At a horse race, the odds on this particular horse are given simply as 5 to 2. These **house odds** tell a gambler what the payoff is on a bet. For every $2 bet on the horse, the gambler would win $5 if the horse won, in addition to having the $2 bet returned.

ODDS TO PROBABILITY

If the odds in favor of event E are a to b, then the probability of the event is given by

$$P(E) = \frac{a}{a+b}.$$

EXAMPLE 9 | FROM ODDS TO PROBABILITY

The odds in favor of a particular horse winning a race are 2 to 5. What is the probability that this horse will win the race?

Solution   Because odds in favor, a to b, means a probability of $\frac{a}{a+b}$, then odds in favor, 2 to 5, means a probability of

$$\frac{2}{2+5} = \frac{2}{7}.$$

The probability that this horse will win the race is $\frac{2}{7}$.

Check
9
Point

The odds against a particular horse winning a race are 15 to 1. Find the odds in favor of the horse winning the race and the probability of the horse winning the race.

Exercise Set 11.6

Practice and Application Exercises

In Exercises 1–6, you are dealt one card from a 52-card deck. Find the probability that you are not dealt

1. an ace. **2.** a 3. **3.** a heart. **4.** a club. **5.** a picture card. **6.** a red picture card.

In 5-card poker, played with a standard 52-card deck, $_{52}C_5$, or 2,598,960, different hands are possible. The probability of being dealt various hands is the number of different ways they can occur divided by 2,598,960. Shown in Exercises 7–10 are various types of poker hands and their probabilities. In each exercise, find the probability of not being dealt this type of hand.

Type of Hand	Illustration	Number of Ways the Hand Can Occur	Probability
7. Straight flush: 5 cards with consecutive numbers, all in the same suit (excluding royal flush)		36	$\dfrac{36}{2{,}598{,}960}$
8. Four of a kind: 4 cards with the same number, plus 1 additional card		624	$\dfrac{624}{2{,}598{,}960}$
9. Full house: 3 cards of one number and 2 cards of a second number		3744	$\dfrac{3744}{2{,}598{,}960}$
10. Flush: 5 cards of the same suit (excluding royal flush and straight flush)		5108	$\dfrac{5108}{2{,}598{,}960}$

The graph shows the probability of cardiovascular disease, by age and gender. Use the information in the graph to solve Exercises 11–12. Express all probabilities as decimals, estimated to two decimal places.

Probability of Cardiovascular Disease, by Age and Gender

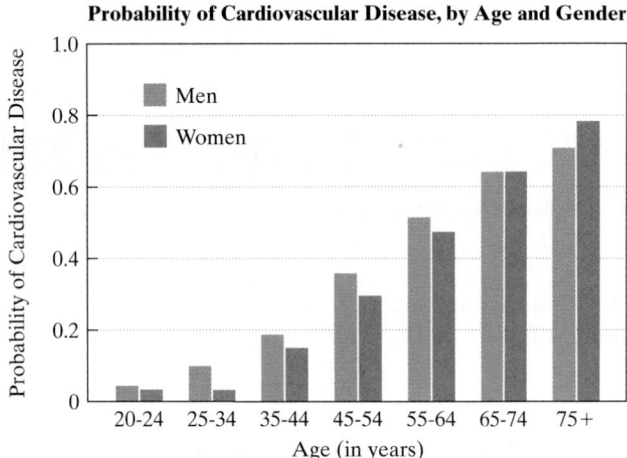

Source: American Heart Association

11. a. What is the probability that a randomly selected man between the ages of 25 and 34 has cardiovascular disease?

  b. What is the probability that a randomly selected man between the ages of 25 and 34 does not have cardiovascular disease?

12. a. What is the probability that a randomly selected woman, 75 or older, has cardiovascular disease?

  b. What is the probability that a randomly selected woman, 75 or older, does not have cardiovascular disease?

Exercises 13–22 involve probabilities with mutually exclusive events.

In 1900, acute illnesses like pneumonia and tuberculosis caused most deaths in the United States. Now our longer life spans increase the probability of developing, and eventually dying from, chronic conditions. The bar graph shows five causes of death and the percentage of all deaths in the United States attributed to each cause. Use the information shown to solve Exercises 13–14. Express all probabilities as decimals to three decimal places.

Causes of Death in the United States

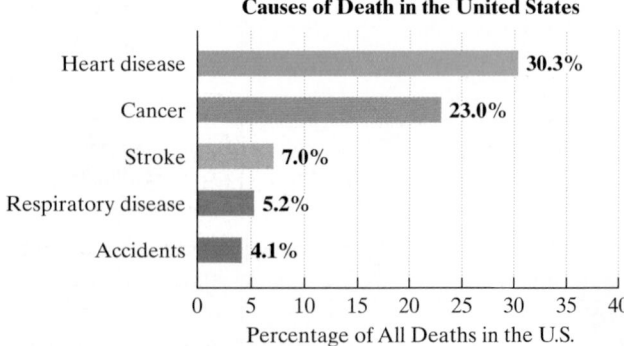

Source: U.S. Department of Health and Human Services

13. What is the probability of dying from heart disease or cancer?

14. What is the probability of dying from heart disease or a stroke?

In Exercises 15–20, you randomly select one card from a 52-card deck. Find the probability of selecting

15. a 2 or a 3.

16. a 7 or an 8.

17. a red 2 or a black 3.

18. a red 7 or a black 8.

19. the 2 of hearts or the 3 of spades.

20. the 7 of hearts or the 8 of spades.

21. The mathematics faculty at a college consists of 8 professors, 12 associate professors, 14 assistant professors, and 10 instructors. If one faculty member is randomly selected, find the probability of choosing a professor or an instructor.

22. A political discussion group consists of 30 Republicans, 25 Democrats, 8 Independents, and 4 members of the Green party. If one person is randomly selected from the group, find the probability of choosing an Independent or a Green.

Exercises 23–38 involve probabilities with events that are not mutually exclusive.

In Exercises 23–24, a single die is rolled. Find the probability of rolling

23. an even number or a number less than 5.

24. an odd number or a number less than 4.

In Exercises 25–28, you are dealt one card from a 52-card deck. Find the probability that you are dealt

25. a 7 or a red card.

26. a 5 or a black card.

27. a heart or a picture card.

28. a card greater than 2 and less than 7, or a diamond.

In Exercises 29–32, it is equally probable that the pointer on the spinner shown will land on any one of the eight regions, numbered 1 through 8. If the pointer lands on a borderline, spin again.

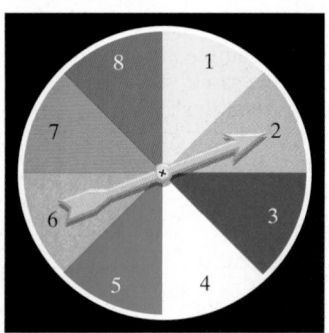

Find the probability that the pointer will stop on

29. an odd number or a number less than 6.

30. an odd number or a number greater than 3.

31. an even number or a number greater than 5.

32. an even number or a number less than 4.

Use this information to solve Exercises 33–36. The mathematics department of a college has 8 male professors, 11 female professors, 14 male teaching assistants, and 7 female teaching assistants. If a person is selected at random from the group, find the probability that the selected person is

33. a professor or a male.

34. a professor or a female.

35. a teaching assistant or a female.

36. a teaching assistant or a male.

37. In a class of 50 students, 29 are Democrats, 11 are business majors, and 5 of the business majors are Democrats. If one student is randomly selected from the class, find the probability of choosing a Democrat or a business major.

38. A student is selected at random from a group of 200 students in which 135 take math, 85 take English, and 65 take both math and English. Find the probability that the selected student takes math or English.

The table shows the educational attainment of the U.S. population, ages 25 and over. Use the data in the table, expressed in millions, to solve Exercises 39–46.

EDUCATIONAL ATTAINMENT, IN MILLIONS, OF THE U.S. POPULATION, AGES 25 AND OVER

	Less Than 4 Years High School	4 Years High School Only	Some College (Less than 4 years)	4 Years College (or More)	Total
Male	14	25	20	23	82
Female	15	31	24	22	92
Total	29	56	44	45	174

Source: U.S. Census Bureau

Find the probability that a randomly selected American, aged 25 and over

39. has not completed four years (or more) of college.

40. has not completed four years of high school.

41. has completed four years of high school only or less than four years of college.

42. has completed less than four years of high school or four years of high school only.

43. has completed four years of high school only or is a man.

44. has completed four years of high school only or is a woman.

Find the odds in favor and the odds against a randomly selected American, aged 25 and over, with

45. four years (or more) of college.

46. less than four years of high school.

In Exercises 47–50, a single die is rolled. Find the odds

47. in favor of rolling a number greater than 2.

48. in favor of rolling a number less than 5.

49. against rolling a number greater than 2.

50. against rolling a number less than 5.

The circle graphs show the percentage of children in the United States whose parents are college graduates in one-parent households and two-parent households. Use the information shown to solve Exercises 51–52.

Percentage of U.S. Children Whose Parents Are College Graduates

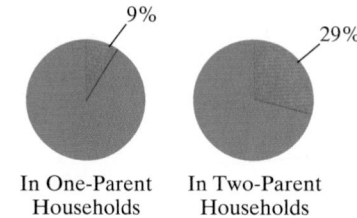

In One-Parent Households   In Two-Parent Households

Source: U.S. Census Bureau

51. a. What are the odds in favor of a child in a one-parent household having a parent who is a college graduate?

   b. What are the odds against a child in a one-parent household having a parent who is a college graduate?

52. a. What are the odds in favor of a child in a two-parent household having parents who are college graduates?

   b. What are the odds against a child in a two-parent household having parents who are college graduates?

In Exercises 53–62, one card is randomly selected from a deck of cards. Find the odds

53. in favor of drawing a heart.

54. in favor of drawing a picture card.

55. in favor of drawing a red card.

56. in favor of drawing a black card.

57. against drawing a 9.

58. against drawing a 5.

59. against drawing a black king.

60. against drawing a red jack.

61. against drawing a spade greater than 3 and less than 9.

62. against drawing a club greater than 4 and less than 10.

63. The winner of a raffle will receive a 21-foot outboard boat. If 1000 raffle tickets were sold and you purchased 20 tickets, what are the odds against your winning the boat?

64. The winner of a raffle will receive a 30-day all-expense-paid trip throughout Europe. If 5000 raffle tickets were sold and you purchased 30 tickets, what are the odds against your winning the trip?

65. According to the book *What Are the Chances* by Siskin, Staller, and Rorvik, the probability of an American being the victim of a serious crime at some point in his or her lifetime is $\frac{1}{20}$. What are the odds in favor of being the victim of a serious crime?

66. One in four Americans has a high cholesterol level. What are the odds in favor of randomly selecting an American with a high cholesterol level?

67. If you are given odds of 3 to 4 in favor of winning a bet, what is the probability of winning the bet?

68. If you are given odds of 3 to 7 in favor of winning a bet, what is the probability of winning the bet?

69. Based on his skills in basketball, it was computed that when Michael Jordan shot a free throw, the odds in favor of his making it were 21 to 4. Find the probability that when Michael Jordan shot a free throw, he missed it. Out of every 100 free throws he attempted, on the average how many did he make?

70. The odds in favor of a person who is alive at age 20 still being alive at age 70 are 193 to 270. Find the probability that a person who is alive at age 20 will still be alive at age 70.

Writing in Mathematics

71. Explain how to find the probability of an event not occurring. Give an example.

72. What are mutually exclusive events? Give an example of two events that are mutually exclusive.

73. Explain how to find *or* probabilities with mutually exclusive events. Give an example.

74. Give an example of two events that are not mutually exclusive.

75. Explain how to find *or* probabilities with events that are not mutually exclusive. Give an example.

76. Explain how to find the odds in favor of an event if you know the probability that the event will occur.

77. Explain how to find the probability of an event if you know the odds in favor of that event.

Critical Thinking Exercises

78. In Exercise 37, find the probability of choosing **a.** a Democrat who is not a business major; **b.** a student who is neither a Democrat nor a business major.

79. On New Year's Eve, the probability of a person driving while intoxicated or having a driving accident is 0.35. If the probability of driving while intoxicated is 0.32 and the probability of having a driving accident is 0.09, find the probability of a person having a driving accident while intoxicated.

80. The formula for converting from odds to probability is given in the box on page 601. Read the paragraph on the bottom of page 600 that precedes this box and derive the formula.

Group Exercise

81. Group members should check the Web for government statistics about transportation safety. Group members should use these numbers to write and solve problems involving *not* and *or*. For example, what is the probability of *not* dying in an airline accident? What is the probability of dying in a car accident *or* an airline accident? Be sure to specify time periods in your problems. If the group writes a probability problem on airline safety, does it seem like a good idea to spend $10 to insure your life on an airplane flight

---

SECTION 11.7　　EVENTS INVOLVING *AND*; CONDITIONAL PROBABILITY

OBJECTIVES

1. Find the probability of one event and a second event occurring.

2. Compute conditional probabilities.

You are considering a job offer in South Florida. You were thrilled by images of Miami on MTVs "The Real World." The job offer is just what you wanted, and you are excited about living in the midst of Miami's tropical diversity. However, there is just one thing: the risk of hurricanes. If you expect to stay in Miami ten years and buy a home, what is the probability that South Florida will be hit by a hurricane at least once in the next ten years?

In this section, we look at the probability that an event occurs at least once by expanding our discussion of probability to events involving *and*.

1   Find the probability of one event and a second event occurring.

And Probabilities with Independent Events

Considers tossing a fair coin two times in succession. The outcome of the first toss, heads or tails, does not affect what happens when you toss the coin a second time. For example, the occurrence of tails on the first toss does not make tails more likely or less likely to occur on the second toss. The repeated toss of a coin produces *independent events* because the outcome of one toss does not affect the outcome of others.

> INDEPENDENT EVENTS
>
> Two events are **independent events** if the occurrence of either of them has no effect on the probability of the other.

When a fair coin is tossed two times in succession, the set of equally-likely outcomes is

{heads heads, heads tails, tails heads, tails tails}.

We can use this set to find the probability of getting heads on the first toss and heads on the second toss:

$$P(\text{heads and heads}) = \frac{\text{number of ways two heads can occur}}{\text{total number of possible outcomes}} = \frac{1}{4}.$$

We can also determine the probability of two heads, $\frac{1}{4}$, without having to list all the equally-likely outcomes. The probability of heads on the first toss is $\frac{1}{2}$. The probability of heads on the second toss is also $\frac{1}{2}$. The product of these probabilities, $\frac{1}{2} \cdot \frac{1}{2}$, results in the probability of two heads, namely $\frac{1}{4}$. Thus,

$$P(\text{heads and heads}) = P(\text{heads}) \cdot P(\text{heads}).$$

In general, if two events are independent, we can calculate the probability of the first occurring and the second occurring by multiplying their probabilities.

> *AND* PROBABILITIES WITH INDEPENDENT EVENTS
>
> If A and B are independent events, then
> $$P(A \text{ and } B) = P(A) \cdot P(B).$$

EXAMPLE 1   INDEPENDENT EVENTS ON A ROULETTE WHEEL

Figure 11.12 shows a U.S. roulette wheel that has 38 numbered slots (1 through 36, 0, and 00). Of the 38 compartments, 18 are black, 18 are red, and two are green. A play has the dealer spin the wheel and a small ball in opposite directions. As the ball slows to a stop, it can land with equal probability on any one of the 38 numbered slots. Find the probability of red occurring on two consecutive plays.

Solution   The wheel has 38 equally-likely outcomes and 18 are red. Thus, the probability of red occurring on a play is $\frac{18}{38}$, or $\frac{9}{19}$. The result that occurs on each play is independent of all previous results. Thus,

$$P(\text{red and red}) = P(\text{red}) \cdot P(\text{red}) = \frac{9}{19} \cdot \frac{9}{19} = \frac{81}{361} \approx 0.224.$$

The probability of red occurring on two consecutive plays is $\frac{81}{361}$.

FIGURE 11.12 A U.S. roulette wheel

Some roulette players incorrectly believe that if red occurs on two consecutive plays, then another color is "due." Because the events are independent, the outcomes of previous spins have no effect on any other spins.

Check 1 Point Find the probability of green occurring on two consecutive plays on a roulette wheel.

The *and* rule for independent events can be extended to cover three or more events. Thus, if *A*, *B*, and *C* are independent events, then

$$P(A \text{ and } B \text{ and } C) = P(A) \cdot P(B) \cdot P(C).$$

EXAMPLE 2   INDEPENDENT EVENTS IN A FAMILY

The picture in the margin shows a family that has had nine girls in a row. Find the probability of this occurrence.

Solution   If two or more events are independent, we can find the probability of them all occurring by multiplying their probabilities. The probability of a baby girl is $\frac{1}{2}$, so the probability of nine girls in a row is $\frac{1}{2}$ used as a factor nine times.

$$P(\text{nine girls in a row}) = \frac{1}{2} \cdot \frac{1}{2} \cdot \frac{1}{2} \cdot \frac{1}{2} \cdot \frac{1}{2} \cdot \frac{1}{2} \cdot \frac{1}{2} \cdot \frac{1}{2} \cdot \frac{1}{2}$$
$$= \left(\frac{1}{2}\right)^9 = \frac{1}{512}$$

The probability of a run of nine girls in a row is $\frac{1}{512}$. (If another child is born into the family, this event is independent of the other nine, and the probability of a girl is still $\frac{1}{2}$.)

Check 2 Point Find the probability of a family having four boys in a row.

TABLE 11.5 THE SAFFIR/SIMPSON HURRICANE SCALE

Category	Winds (Miles per Hour)
1	74–95
2	96–110
3	111–130
4	131–155
5	>155

Now let us return to the hurricane problem that opened this section. The Saffir/Simpson scale assigns numbers 1 through 5 to measure the disaster potential of a hurricane's winds. Table 11.5 describes the scale. According to the National Hurricane Center, the probability that South Florida will be hit by a category 1 hurricane or higher in any single year is $\frac{5}{19}$, or approximately 0.26. In Example 3, we explore the risks of living in "Hurricane Alley."

EXAMPLE 3   HURRICANES AND PROBABILITIES

If the probability that South Florida will be hit by a hurricane in any single year is $\frac{5}{19}$,

a. What is the probability that South Florida will be hit by a hurricane in three consecutive years?

b. What is the probability that South Florida will not be hit by a hurricane in the next ten years?

Solution

a. The probability that South Florida will be hit by a hurricane in three consecutive years is

P(hurricane and hurricane and hurricane)

$$= P(\text{hurricane}) \cdot P(\text{hurricane}) \cdot P(\text{hurricane}) = \frac{5}{19} \cdot \frac{5}{19} \cdot \frac{5}{19} = \frac{125}{6859} \approx 0.018.$$

b. We will first find the probability that South Florida will not be hit by a hurricane in any single year.

$$P(\text{no hurricane}) = 1 - P(\text{hurricane}) = 1 - \frac{5}{19} = \frac{14}{19} \approx 0.737$$

The probability of not being hit by a hurricane in a single year is $\frac{14}{19}$. Therefore, the probability of not being hit by a hurricane ten years in a row is $\frac{14}{19}$ used as a factor ten times.

$$P(\text{no hurricanes for ten years})$$

$$= P\left(\begin{array}{c}\text{no hurricane}\\ \text{for year 1}\end{array}\right) \cdot P\left(\begin{array}{c}\text{no hurricane}\\ \text{for year 2}\end{array}\right) \cdot P\left(\begin{array}{c}\text{no hurricane}\\ \text{for year 3}\end{array}\right) \cdot \ldots \cdot P\left(\begin{array}{c}\text{no hurricane}\\ \text{for year 10}\end{array}\right)$$

$$= \frac{14}{19} \cdot \frac{14}{19} \cdot \frac{14}{19} \cdot \ldots \cdot \frac{14}{19}$$

$$= \left(\frac{14}{19}\right)^{10} \approx (0.737)^{10} \approx 0.047$$

The probability that South Florida will not be hit by a hurricane in the next ten years is approximately 0.047.

STUDY TIP

When solving probability problems, begin by deciding whether to use the *or* formulas or the *and* formulas.

- *Or* problems usually have the word *or* in the statement of the problem. These problems involve only one selection.

Example:

If one person is selected, find the probability of selecting a man or a Canadian.

- *And* problems often do not have the word *and* in the statement of the problem. These problems involve more than one selection.

Example:

If two people are selected, find the probability that both are men.

Now we are ready to answer your question:

What is the probability that South Florida will be hit by a hurricane at least once in the next ten years?

Because $P(\text{not } E) = 1 - P(E)$,

$$P(\text{no hurricane for ten years}) = 1 - P(\text{at least one hurricane in ten years}).$$

In our logic chapter, we saw that the negation of "at least one" is "no."

Equivalently,

$$P(\text{at least one hurricane in ten years}) = 1 - P(\text{no hurricane for ten years})$$
$$= 1 - 0.047 = 0.953.$$

With a probability of 0.953, it is nearly certain that South Florida will be hit by a hurricane at least once in the next ten years.

> ### THE PROBABILITY OF AN EVENT HAPPENING AT LEAST ONCE
>
> $$P(\text{event happening at least once}) = 1 - P(\text{event does not happen})$$

 **Check Point 3** If the probability that South Florida will be hit by a hurricane in any single year is $\frac{5}{19}$,

a. What is the probability that South Florida will be hit by a hurricane in four consecutive years?

b. What is the probability that South Florida will not be hit by a hurricane in the next four years?

c. What is the probability that South Florida will be hit by a hurricane at least once in the next four years?

Express all probabilities as fractions and as decimals rounded to three places. ▪

5 chocolate-covered cherries lie within the 20 pieces.

Once a chocolate-covered cherry is selected, only 4 chocolate-covered cherries lie within the remaining 19 pieces.

And Probabilities with Dependent Events

Chocolate lovers, please help yourself! There are 20 mouth-watering tidbits to select from. What's that? You want 2? And you prefer chocolate-covered cherries? The problem is that there are only 5 chocolate-covered cherries, and it's impossible to tell what is inside each piece. They're all shaped exactly alike. At any rate, reach in, select a piece, enjoy, choose another piece, eat, and be well. There is nothing like savoring a good piece of chocolate in the midst of all this chit-chat about probability and hurricanes.

Another question? You want to know what your chances are of selecting 2 chocolate-covered cherries? Well, let's see. Five of the 20 pieces are chocolate-covered cherries, so the probability of getting one of them on your first selection is $\frac{5}{20}$, or $\frac{1}{4}$. Now, suppose that you did choose a chocolate-covered cherry on your first pick. Eat it slowly; there's no guarantee that you'll select your favorite on the second selection. There are now only 19 pieces of chocolate left. Only 4 are chocolate-covered cherries. The probability of getting a chocolate-covered cherry on your second try is 4 out of 19, or $\frac{4}{19}$. This is a different probability than the $\frac{1}{4}$ probability on your first selection. Selecting a chocolate-covered cherry the first time changes what is in the candy box. The probability of what you select the second time *is* affected by the outcome of the first event. For this reason, we say that these are *dependent events*.

DEPENDENT EVENTS

Two events are **dependent events** if the occurrence of one of them has an effect on the probability of the other.

The probability of selecting two chocolate-covered cherries in a row can be found by multiplying the $\frac{1}{4}$ probability on the first selection by the $\frac{4}{19}$ probability on the second selection:

P(chocolate-covered cherry and chocolate-covered cherry)

$$= P(\text{chocolate-covered cherry}) \cdot P\left(\begin{array}{c} \text{chocolate-covered cherry} \\ \text{given that one was selected} \end{array}\right)$$

$$= \frac{1}{4} \cdot \frac{4}{19} = \frac{1}{19}.$$

The probability of selecting two chocolate-covered cherries in a row is $\frac{1}{19}$. This is a special case of finding the probability that each of two dependent events occurs.

AND PROBABILITIES WITH DEPENDENT EVENTS

If A and B are dependent events, then

$$P(A \text{ and } B) = P(A) \cdot P(B \text{ given that } A \text{ has occurred}).$$

EXAMPLE 4   DEPENDENT EVENTS WITH YOUR COUSINS

Good news: You won a free trip to Madrid and can take two people with you, all expenses paid. Bad news: Ten of your cousins have appeared out of nowhere and are begging you to take them. You write each cousin's name on a card, place the cards in a hat, and select one name. Then you select a second name without replacing the first card. If three of your ten cousins speak Spanish, find the probability of selecting two Spanish-speaking cousins.

STUDY TIP

Example 4 can also be solved using the combinations formula.

P(two Spanish speakers)

$$= \frac{\text{number of ways of selecting 2 Spanish-speaking cousins}}{\text{number of ways of selecting 2 cousins}}$$

$$= \frac{{}_3C_2}{{}_{10}C_2}$$

> 2 Spanish speakers selected from 3 Spanish-speaking cousins

> 2 cousins selected from 10 cousins

$$= \frac{3}{45} = \frac{1}{15}$$

Solution Because $P(A \text{ and } B) = P(A) \cdot P(B \text{ given that } A \text{ has occurred})$, then

P(two Spanish-speaking cousins)

$= P$(speaks Spanish and speaks Spanish)

$= P$(speaks Spanish) $\cdot P\left(\begin{array}{c}\text{speaks Spanish given that a Spanish-speaking} \\ \text{cousin was selected first}\end{array}\right)$

$$= \frac{3}{10} \cdot \frac{2}{9}$$

> There are ten cousins, three of whom speak Spanish.

> After picking a Spanish-speaking cousin, there are nine cousins left, two of whom speak Spanish.

$$= \frac{6}{90} = \frac{1}{15} \approx 0.067.$$

The probability of selecting two Spanish-speaking cousins is $\frac{1}{15}$.

 **Check Point 4** You are dealt two cards from a 52-card deck. Find the probability of getting two kings.

The multiplication rule for dependent events can be extended to cover three or more dependent events. For example, in the case of three such events,

$P(A \text{ and } B \text{ and } C)$

$= P(A) \cdot P(B \text{ given that } A \text{ occurred}) \cdot P(C \text{ given that } A \text{ and } B \text{ occurred}).$

> **EXAMPLE 5** AN *AND* PROBABILITY WITH THREE DEPENDENT EVENTS

Three people are randomly selected, one person at a time, from five freshmen, two sophomores, and four juniors. Find the probability that the first two people selected are freshmen and the third is a junior.

Solution

P(first two are freshmen and the third is a junior)

$= P$(freshman) $\cdot P\left(\begin{array}{c}\text{freshman given that a} \\ \text{freshman was selected first}\end{array}\right) \cdot P\left(\begin{array}{c}\text{junior given that a freshman was} \\ \text{selected first and a freshman was} \\ \text{selected second}\end{array}\right)$

$$= \frac{5}{11} \cdot \frac{4}{10} \cdot \frac{4}{9}$$

> There are 11 people, five of whom are freshmen.

> After picking a freshman, there are 10 people left, four of whom are freshmen.

> After the first two selections, 9 people are left, four of whom are juniors.

$$= \frac{8}{99}$$

The probability that the first two people selected are freshmen and the third is a junior is $\frac{8}{99}$.

 **Check Point 5** You are dealt three cards from a 52-card deck. Find the probability of getting three hearts.

2 Compute conditional probabilities.

Conditional Probability

We have seen that for any two dependent events A and B,

$$P(A \text{ and } B) = P(A) \cdot P(B \text{ given that } A \text{ occurs}).$$

The probability of B given that A occurs is called *conditional probability*, denoted by $P(B|A)$.

CONDITIONAL PROBABILITY

The probability of event B, assuming that the event A has already occurred, is called the **conditional probability** of B, given A. This probability is denoted by $P(B|A)$.

It is helpful to think of the conditional probability $P(B|A)$ as the **probability that event B occurs if the sample space is restricted to the outcomes associated with event A.**

EXAMPLE 6 FINDING CONDITIONAL PROBABILITY

A letter is randomly selected from the letters of the English alphabet. Find the probability of selecting a vowel, given that the outcome is a letter that precedes h.

Solution We are looking for

$$P(\text{vowel}|\text{letter precedes h}).$$

This is the probability of a vowel if the sample space is restricted to the set of letters that precede h. Thus, the sample space is given by

$$S = \{a, b, c, d, e, f, g\}.$$

There are seven possible outcomes in the sample space. We can select a vowel from this set in one of two ways: a or e. Therefore, the probability of selecting a vowel, given that the outcome is a letter that precedes h, is $\frac{2}{7}$.

$$P(\text{vowel}|\text{letter precedes h}) = \frac{2}{7}$$

Check Point 6

You are dealt one card from a 52-card deck. Find the probability of getting a heart, given that the card you were dealt is a red card.

EXAMPLE 7 CONDITIONAL PROBABILITIES WITH REAL-WORLD DATA

Table 11.6 shows the differences in political ideology per 100 males and per 100 females in the 2000 U.S. presidential election. Assuming that these numbers are representative of all Americans and one American is randomly selected, find the probability that the person

a. is liberal, given that the person is female.

b. is male, given that the person is conservative.

TABLE 11.6 DIFFERENCES IN POLITICAL IDEOLOGY PER 100 MALES AND 100 FEMALES IN THE 2000 U.S. ELECTION

	Liberal	**Moderate**	**Conservative**
Male	16	45	39
Female	20	55	25

Source: Center for Political Studies, University of Michigan

Solution

a. We begin with the probability that the selected person is liberal, given female:

$$P(\text{liberal}|\text{female}).$$

This is the probability of selecting a liberal if the data are restricted to the females:

	Liberal	Moderate	Conservative
Female	20	55	25

Within the restricted data, there are 20 liberals and 20 + 55 + 25, or 100 women. Thus,

$$P(\text{liberal}|\text{female}) = \frac{20}{100} = \frac{1}{5}.$$

The probability that the selected person is liberal, given female, is $\frac{1}{5}$.

b. Now, we find the probability that the selected person is male, given conservative:

$$P(\text{male}|\text{conservative}).$$

This is the probability of selecting a male if the data are restricted to the conservatives:

	Conservative
Male	39
Female	25

Within the restricted data, there are 39 males and 39 + 25, or 64 conservatives. Thus,

$$P(\text{male}|\text{conservative}) = \frac{39}{64}.$$

The probability that the selected person is male, given conservative, is $\frac{39}{64}$.

Check Point 7   Use Table 11.6 to solve this exercise. If one American is randomly selected, find the probability that the person

a. is conservative, given that the person is male.

b. is female, given that the person is liberal.

We have seen that $P(B|A)$ is the probability that event B occurs if the sample space is restricted to event A. Thus,

$$P(B|A) = \frac{\text{number of outcomes of } B \text{ that are in the restricted sample space } A}{\text{number of outcomes in the restricted sample space } A}.$$

This can be stated in terms of the following formula:

A FORMULA FOR CONDITIONAL PROBABILITY

$$P(B|A) = \frac{n(B \cap A)}{n(A)} = \frac{\text{number of outcomes common to } B \text{ and } A}{\text{number of outcomes in } A}$$

Exercise Set 11.7

Practice and Application Exercises

Exercises 1–26 involve probabilities with independent events.

Use the spinner shown to solve Exercises 1–10. It is equally probable that the pointer will land on any one of the six regions. If the pointer lands on a borderline, spin again. If the pointer is spun twice, find the probability it will land on

1. green and then red.
2. yellow and then green.
3. yellow and then yellow.
4. red and then red.
5. a color other than red each time.
6. a color other than green each time.

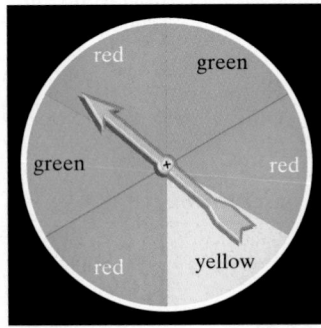

If the pointer is spun three times, find the probability it will land on

7. green and then red and then yellow.
8. red and then red and then green.
9. red every time.
10. green every time.

In Exercises 11–14, a single die is rolled twice. Find the probability of rolling

11. a 2 the first time and a 3 the second time.
12. a 5 the first time and a 1 the second time.
13. an even number the first time and a number greater than 2 the second time.
14. an odd number the first time and a number less than 3 the second time.

In Exercises 15–20, you draw one card from a 52-card deck. Then the card is replaced in the deck, the deck is shuffled, and you draw again. Find the probability of drawing

15. a picture card the first time and a heart the second time.
16. a jack the first time and a club the second time.
17. a king each time.
18. a 3 each time.
19. a red card each time.
20. a black card each time.

21. If you toss a fair coin six times, what is the probability of getting all heads?
22. If you toss a fair coin seven times, what is the probability of getting all tails?

In Exercises 23–24, a coin is tossed and a die is rolled. Find the probability of getting

23. a head and a number greater than 4.
24. a tail and a number less than 5.
25. The probability that South Florida will be hit by a major hurricane (category 4 or 5) in any single year is $\frac{1}{16}$. (*Source:* National Hurricane Center)
    a. What is the probability that South Florida will be hit by a major hurricane two years in a row?
    b. What is the probability that South Florida will be hit by a major hurricane in three consecutive years?
    c. What is the probability that South Florida will not be hit by a major hurricane in the next ten years?
    d. What is the probability that South Florida will be hit by a major hurricane at least once in the next ten years?
26. The probability that a region prone to flooding will flood in any single year is $\frac{1}{10}$.
    a. What is the probability of a flood two years in a row?
    b. What is the probability of flooding in three consecutive years?
    c. What is the probability of no flooding for ten consecutive years?
    d. What is the probability of flooding at least once in the next ten years?

When making two or more selections from populations with large numbers, such as the population of Americans ages 45 to 65, we assume that each selection is independent of every other selection. In Exercises 27–28, assume that the selections are independent events.

The graph shows how Americans 45 to 65 rate their health. Use the information shown to solve Exercises 27–28.

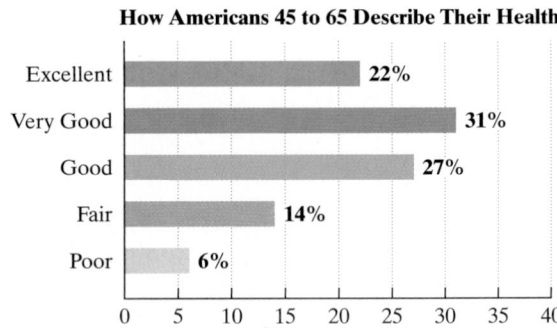

How Americans 45 to 65 Describe Their Health

Source: Newsweek

27. If four Americans ages 45 to 65 are selected at random, find the probability that all four rate their health as excellent.
28. If four Americans ages 45 to 65 are selected at random, find the probability that all four rate their health as poor.

Exercises 29–44 involve probabilities with dependent events.

In Exercises 29–32, we return to our box of chocolates. There are 30 chocolates in the box, all identically shaped. Five are filled with coconut, 10 with caramel, and 15 are solid chocolate. You randomly select one piece, eat it, and then select a second piece. Find the probability of selecting

29. two solid chocolates in a row.

30. two caramel-filled chocolates in a row.

31. a coconut-filled chocolate followed by a caramel-filled chocolate.

32. a coconut-filled chocolate followed by a solid chocolate.

In Exercises 33–38, consider a political discussion group consisting of 5 Democrats, 6 Republicans, and 4 Independents. Suppose that two group members are randomly selected, in succession, to attend a political convention. Find the probability of selecting

33. two Democrats.

34. two Republicans.

35. an Independent and then a Republican.

36. an Independent and then a Democrat.

37. no Independents.

38. no Democrats.

In Exercises 39–44, an ice chest contains six cans of apple juice, eight cans of grape juice, four cans of orange juice, and two cans of mango juice. Suppose that you reach into the container and randomly select three cans in succession. Find the probability of selecting

39. three cans of apple juice.

40. three cans of grape juice.

41. a can of grape juice, then a can of orange juice, then a can of mango juice.

42. a can of apple juice, then a can of grape juice, then a can of orange juice.

43. no grape juice.

44. no apple juice.

In Exercises 45–52, the numbered disks shown are placed in a box and one disk is selected at random.

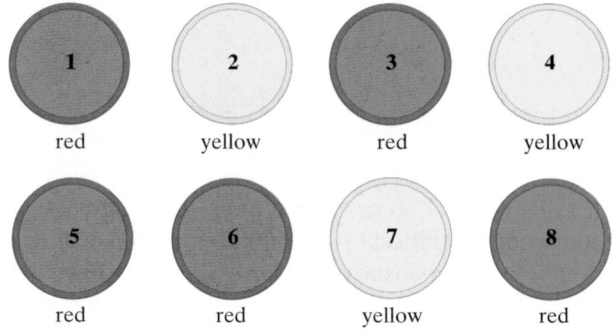

Find the probability of selecting

45. a 3, given that a red disk is selected.

46. a 7, given that a yellow disk is selected.

47. an even number, given that a yellow disk is selected.

48. an odd number, given that a red disk is selected.

49. a red disk, given that an odd number is selected.

50. a yellow disk, given that an odd number is selected.

51. a red disk, given that the number selected is at least 5.

52. a yellow disk, given that the number selected is at most 3

When women turn 40, their gynecologists typically remind them that it is time to undergo mammography screening for breast cancer. The data in the table are based on 100,000 U.S. women, ages 40 to 50, who participated in mammography screening. Use the data to solve Exercises 53–58. Express probabilities as fractions and as decimals to three decimal places.

	Breast Cancer	**No Breast Cancer**
Positive Mammogram	720	6944
Negative Mammogram	80	92,256

Source: Gerd Gigerenzer, *Calculated Risks*, Simon and Schuster, 2002

Find the probability that a woman ages 40 to 50

53. has breast cancer.

54. does not have breast cancer.

55. has a positive mammogram, given that she has breast cancer.

56. has a positive mammogram, given that she does not have breast cancer.

57. has breast cancer, given that she has a positive mammogram.

58. does not have breast cancer, given that she has a positive mammogram.

The table shows differences in political ideology, by education, for a random sample of U.S. voters. (The ratios for each group's ideologies are sourced from voting patterns in the 2000 U.S. election. The frequencies shown are hypothetical.) Use the data to solve Exercises 59–72. Express probabilites as simplified fractions.

	Liberal	**Moderate**	**Conservative**
High School only	7	35	13
College	10	15	20

Find the probability that a randomly selected person from this group

59. is not liberal.

60. is not conservative.

61. is liberal or moderate.

62. is moderate or conservative.

63. is conservative or attended college.

64. is liberal or has only a high school education.

65. is conservative, given only a high school education.

66. is moderate, given college attendance.

67. attended high school only, given a conservative ideology.

68. attended college, given a moderate ideology.

Suppose that two people are randomly selected, in succession, from this group. Find the probability of selecting

69. two liberals.

70. two conservatives.

71. a moderate and then a conservative.

72. a moderate and then a liberal.

Writing in Mathematics

73. Explain how to find *and* probabilities with independent events. Give an example.

74. Explain how to find *and* probabilities with dependent events. Give an example.

75. What does $P(B|A)$ mean? Give an example.

In Exercises 76–80, write a probability problem involving the word "and" whose solution results in the probability fractions shown.

76. $\frac{1}{2} \cdot \frac{1}{2}$ **77.** $\frac{1}{6} \cdot \frac{1}{6} \cdot \frac{1}{6}$ **78.** $\frac{1}{2} \cdot \frac{1}{6}$

79. $\frac{13}{52} \cdot \frac{12}{51}$ **80.** $\frac{1}{4} \cdot \frac{3}{5}$

Critical Thinking Exercises

81. If the probability of being hospitalized during a year is 0.1, find the probability that no one in a family of five will be hospitalized in a year.

82. If a single die is rolled five times, what is the probability it lands on 2 on the first, third, and fourth rolls, but not on any of the other rolls?

83. a. If two people are selected at random, the probability that they do not have the same birthday (day and month) is $\frac{365}{365} \cdot \frac{364}{365}$. Explain why this is so. (Ignore leap years and assume 365 days in a year.)

b. If three people are selected at random, find the probability that they all have different birthdays.

c. If three people are selected at random, find the probability that at least two of them have the same birthday.

d. If 20 people are selected at random, find the probability that at least 2 of them have the same birthday.

e. How large a group is needed to give a 0.5 chance of at least two people having the same birthday?

84. Nine cards numbered from 1 through 9 are placed into a box and two cards are selected without replacement. Find the probability that both numbers selected are odd, given that their sum is even.

Group Exercises

85. Do you live in an area prone to catastrophes, such as earthquakes, fires, tornadoes, hurricanes, or floods? If so, research the probability of this catastrophe occurring in a single year. Group members should then use this probability to write and solve a problem similar to Exercise 25 in this exercise set.

86. Group members should use the table for Exercises 59–72 to write and solve four probability problems different than those in the exercises. Two should involve *or* (one with events that are mutually exclusive and one with events that are not), one should involve *and*—that is, events in succession—and one should involve conditional probability.

SECTION 11.8 EXPECTED VALUE

OBJECTIVES

1. Compute expected value.

2. Use expected value to solve applied problems.

3. Use expected value to determine the average payoff or loss in a game of chance.

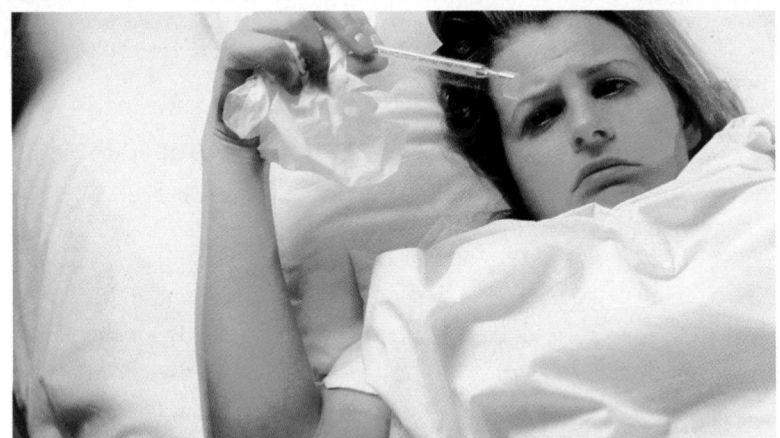

| 1 | Compute expected value.

Would you be willing to spend $50 a year for an insurance policy that pays $200,000 if you become too ill to continue your education? It is unlikely that this will occur. Insurance companies make money by compensating us for events that have a low probability. If one in every 5000 students needs to quit college due to serious illness, the probability of this event is $\frac{1}{5000}$. Multiplying the amount of the claim, $200,000, by its probability, $\frac{1}{5000}$, tells the insurance company what to expect on average for each policy:

$$\$200,000 \times \frac{1}{5000} = \$40.$$

Amount of the claim   Probability of paying the claim

TABLE 11.7 OUTCOMES AND PROBABILITIES FOR THE NUMBER OF GIRLS IN A THREE-CHILD FAMILY

Outcome: Number of Girls	Probability
0	$\frac{1}{8}$
1	$\frac{3}{8}$
2	$\frac{3}{8}$
3	$\frac{1}{8}$

Over the long run, the insurance company can expect to pay $40 for each policy it sells. By selling the policy for $50, the expected profit is $10 per policy. If 400,000 students choose to take out this insurance, the company can expect to make 400,000 × $10, or $4,000,000.

Expected value is a mathematical way to use probabilities to determine what to expect in various situations over the long run. Expected value is used to determine premiums on insurance policies, weigh the risks versus the benefits of alternatives in business ventures, and indicate to a player of any game of chance what will happen if the game is played repeatedly.

The standard way to find expected value is to multiply each possible outcome by its probability, and then add these products. We use the letter E to represent expected value.

EXAMPLE 1   COMPUTING EXPECTED VALUE

Find the expected value for the outcome of the roll of a fair die.

Solution   The outcomes are 1, 2, 3, 4, 5, and 6, each with a probability of $\frac{1}{6}$. The expected value, E, is computed by multiplying each outcome by its probability and then adding these products.

$$E = 1 \cdot \frac{1}{6} + 2 \cdot \frac{1}{6} + 3 \cdot \frac{1}{6} + 4 \cdot \frac{1}{6} + 5 \cdot \frac{1}{6} + 6 \cdot \frac{1}{6}$$

$$= \frac{1 + 2 + 3 + 4 + 5 + 6}{6} = \frac{21}{6} = 3.5$$

The expected value of the roll of a fair die is 3.5. This means that if the die is rolled repeatedly, there are an average of 3.5 dots per roll over the long run. This expected value cannot occur on a single roll of the die. However, it is a long-run average of the various outcomes that can occur when a fair die is rolled.

 **Check 1 Point**   It is equally probable that a pointer will land on any one of four regions, numbered 1 through 4. Find the expected value for where the pointer will stop.

EXAMPLE 2   COMPUTING EXPECTED VALUE

Find the expected value for the number of girls for a family with three children.

Solution   A family with three children can have 0, 1, 2, or 3 girls. There are eight ways these outcomes can occur.

No girls   : Boy Boy Boy   **One way**
One girl   : Girl Boy Boy, Boy Girl Boy, Boy Boy Girl   **Three ways**
Two girls   : Girl Girl Boy, Girl Boy Girl, Boy Girl Girl   **Three ways**
Three girls : Girl Girl Girl   **One way**

Table 11.7 shows the probabilities for 0, 1, 2, and 3 girls.

The expected value, E, is computed by multiplying each outcome by its probability and then adding these products.

$$E = 0 \cdot \frac{1}{8} + 1 \cdot \frac{3}{8} + 2 \cdot \frac{3}{8} + 3 \cdot \frac{1}{8} = \frac{0 + 3 + 6 + 3}{8} = \frac{12}{8} = \frac{3}{2} = 1.5$$

TABLE 11.8 OUTCOMES AND PROBABILITIES FOR THE NUMBER OF HEADS

Number of Heads	Probability
0	$\frac{1}{16}$
1	$\frac{4}{16}$
2	$\frac{6}{16}$
3	$\frac{4}{16}$
4	$\frac{1}{16}$

 Use expected value to solve applied problems.

The expected value is 1.5. This means that if we record the number of girls in many different three-child families, the average number of girls for all these families will be 1.5. In a three-child family, half the children are expected to be girls, so the expected value of 1.5 is consistent with this observation.

 **Check Point 2** A fair coin is tossed four times in succession. Table 11.8 show the probabilities for the different number of heads that can arise. Find the expected value for the number of heads.

Applications of Expected Value

Empirical probabilities can be determined in many situations by examining what has occurred in the past. For example, an insurance company can tally various claim amounts over many years. If 15% of these amounts are for a $2000 claim, then the probability of this claim amount is 0.15. By studying sales of similar houses in a particular area, a realtor can determine the probability that he or she will sell a listed house, another agent will sell the house, or the listed house will remain unsold. Once probabilities have been assigned to all possible outcomes, expected value can indicate what is expected to happen in the long run. These ideas are illustrated in Examples 3 and 4.

EXAMPLE 3   DETERMINING AN INSURANCE PREMIUM

TABLE 11.9 PROBABILITIES FOR AUTO CLAIMS

Amount of Claim (to the nearest $2000)	Probability
$0	0.70
$2000	0.15
$4000	0.08
$6000	0.05
$8000	0.01
$10,000	0.01

An automobile insurance company has determined the probabilities for various claim amounts for drivers ages 16 through 21, shown in Table 11.9.

a. Calculate the expected value and describe what this means in practical terms.

b. How much should the company charge as an average premium so that it does not lose or gain money on its claim costs?

Solution

a. The expected value, E, is computed by multiplying each outcome by its probability, and then adding these products.

$$E = \$0(0.70) + \$2000(0.15) + \$4000(0.08) + \$6000(0.05)$$
$$+ \$8000(0.01) + \$10,000(0.01)$$
$$= \$0 + \$300 + \$320 + \$300 + \$80 + \$100$$
$$= \$1100$$

The expected value is $1100. This means that in the long run the average cost of a claim is $1100. The insurance company should expect to pay $1100 per car insured to people in the 16–21 age group.

b. At the very least, the amount that the company should charge as an average premium for each person in the 16–21 group is $1100. In this way, it will not lose or gain money on its claims costs. It's quite probable that the company will charge more, moving from break-even to profit.

 **Check Point 3** Work Example 3 again if the probabilities for claims of $0 and $10,000 are reversed. Thus, the probability of a $0 claim is 0.01 and the probability of a $10,000 claim is 0.70.

Business decisions are interpreted in terms of dollars and cents. In these situations, **expected value is calculated by multiplying the gain or loss for each possible outcome by its probability. The sum of these products is the expected value.**

EXAMPLE 4 EXPECTATION IN A BUSINESS DECISION

You are a realtor considering listing a $500,000 house. The cost of advertising and providing food for other realtors during open showings is anticipated to cost you $5000. The house is quite unusual, and you are given a four-month listing. If the house is unsold after four months, you lose the listing and receive nothing. You anticipate that the probability you sell your own listed house is 0.3, the probability that another agent sells your listing is 0.2, and the probability that the house is unsold after 4 months is 0.5. If you sell your own listed house, the commission is a hefty $30,000. If another realtor sells your listing, the commission is $15,000. The bottom line: You will not take the listing unless you anticipate earning at least $6000. Should you list the house?

Solution   Shown in the margin is a summary of the amounts of money and probabilities that will determine your decision. The expected value in this situation is the sum of each income possibility times its probability. The expected value represents the amount you can anticipate earning if you take the listing. If the expected value is not at least $6000, you should not list the house.

The possible incomes listed in the margin, $30,000, $15,000, and $0, do not take into account your $5000 costs. Because of these costs, each amount needs to be reduced by $5000. Thus, you can gain $30,000 − $5000, or $25,000, or you can gain $15,000 − $5000, or $10,000. Because $0 − $5000 = −$5000, you can also lose $5000. Table 11.10 summarizes possible outcomes if you take the listing, and their respective probabilities.

TABLE 11.10 GAINS, LOSSES, AND PROBABILITIES FOR LISTING A $500,000 HOUSE

Outcome	Gain or Loss	Probability
Sells house	$25,000	0.3
Another agent sells house	$10,000	0.2
House doesn't sell	−$5000	0.5

The expected value, E, is computed by multiplying each gain or loss in Table 11.10 by its probability, and then adding these results.

$$E = \$25{,}000(0.3) + \$10{,}000(0.2) + (-\$5000)(0.5)$$
$$= \$7500 + \$2000 + (-\$2500) = \$7000$$

You can expect to earn $7000 by listing the house. Because the expected value exceeds $6000, you should list the house.

Check
4
Point

The SAT is a multiple-choice test. Each question has five possible answers. The test taker must select one answer for each question or not answer the question. One point is awarded for each correct response and $\frac{1}{4}$ point is subtracted for each wrong answer. No points are added or subtracted for answers left blank. Table 11.11 summarizes the information for the outcomes of a random guess on an SAT question. Find the expected point value of a random guess. Is there anything to gain or lose on average by guessing? Explain your answer.

TABLE 11.11 GAINS AND LOSSES FOR GUESSING ON THE SAT

Outcome	Gain or Loss	Probability
Guess correctly	1	$\frac{1}{5}$
Guess incorrectly	$-\frac{1}{4}$	$\frac{4}{5}$

3 Use expected value to determine the average payoff or loss in a game of chance.

Expected Value and Games of Chance

Expected value can be interpreted as the average payoff in a contest or game when either is played a large number of times. **To find the expected value of a game, multiply the gain or loss for each possible outcome by its probability. Then add the products.**

EXAMPLE 5 | EXPECTED VALUE AS AVERAGE PAYOFF

A game is played using one die. If the die is rolled and shows 1, 2, or 3, the player wins nothing. If the die shows 4 or 5, the player wins $3. If the die shows 6, the player wins $9. If there is a charge of $1 to play the game, what is the game's expected value? Describe what this means in practical terms.

Solution Because there is a charge of $1 to play the game, a player who wins $9 gains $9 − $1, or $8. A player who wins $3 gains $3 − $1, or $2. If the player gets $0, there is a loss of $1 because $0 − $1 = −$1. The outcomes for the die, with their respective gains, losses, and probabilities, are summarized in Table 11.12.

TABLE 11.12 GAINS, LOSSES, AND PROBABILITIES IN A GAME OF CHANCE

Outcome	Gain or Loss	Probability
1, 2, or 3	−$1	$\frac{3}{6}$
4 or 5	$2	$\frac{2}{6}$
6	$8	$\frac{1}{6}$

Expected value, E, is computed by multiplying each gain or loss in Table 11.10 by its probability and then adding these results.

$$E = (-\$1)\left(\frac{3}{6}\right) + \$2\left(\frac{2}{6}\right) + \$8\left(\frac{1}{6}\right)$$

$$= \frac{-\$3 + \$4 + \$8}{6} = \frac{\$9}{6} = \$1.50$$

The expected value is $1.50. This means that in the long run, a player can expect to win an average of $1.50 for each game played. However, this does not mean that the player will win $1.50 on any single game. It does mean that if the game is played repeatedly, then, in the long run, the player should expect to win about $1.50 per play on the average. If 1000 games are played, one could expect to win $1500. However, if only three games are played, one's net winnings can range between −$3 and $24, even though the expected winnings are $1.50(3), or $4.50.

 **Check Point 5** A charity is holding a raffle and sells 1000 raffle tickets for $2 each. One of the tickets will be selected to win a grand prize of $1000. Two other tickets will be selected to win consolation prizes of $50 each. Fill in the missing column in Table 11.13. Then find the expected value if you buy one raffle ticket. Describe what this means in practical terms. What can you expect to happen if you purchase five tickets?

TABLE 11.13 GAINS, LOSSES, AND PROBABILITIES IN A RAFFLE

Outcome	Gain or Loss	Probability
Win Grand Prize		$\frac{1}{1000}$
Win Consolation Prize		$\frac{2}{1000}$
Win Nothing		$\frac{997}{1000}$

FIGURE 11.13 A U.S. roulette wheel

Unlike the game in Example 5, games in gambling casinos are set up so that players will lose in the long run. These games have negative expected values. Such a game is roulette, French for "little wheel." We first saw the roulette wheel in Section 11.7. It is shown again in Figure 11.13. Recall that the wheel has 38 numbered slots (1 through 36, 0, and 00). In each play of the game, the dealer spins the wheel and a small ball in opposite directions. The ball is equally likely to come to rest in any one of the slots, which are colored black, red, or green. Gamblers can place a number of different bets in roulette. Example 6 illustrates one gambling option.

EXAMPLE 6 **EXPECTED VALUE AND ROULETTE**

One way to bet in roulette is to place $1 on a single number. If the ball lands on that number, you are awarded $35 and get to keep the $1 that you paid to play the game. If the ball lands on any one of the other 37 slots, you are awarded nothing and the $1 that you bet is collected. Find the expected value for playing roulette if you bet $1 on number 20. Describe what this means.

Solution Table 11.14 contains the two outcomes of interest: the ball landing on the winning number 20 and the ball landing elsewhere (in any one of the other 37 slots). The outcomes, their respective gains, losses, and probabilities, are summarized in the table.

TABLE 11.14 PLAYING ONE NUMBER WITH A 35 TO 1 PAYOFF IN ROULETTE

Outcome	Gain or Loss	Probability
Ball lands on 20	$35	$\frac{1}{38}$
Ball does not land on 20	−$1	$\frac{37}{38}$

Expected value, E, is computed by multiplying each gain or loss in Table 11.14 by its probability and then adding these results.

$$E = \$35\left(\frac{1}{38}\right) + (-\$1)\left(\frac{37}{38}\right) = \frac{\$35 - \$37}{38} = \frac{-\$2}{38} \approx -\$0.05$$

The expected value is approximately −$0.05. This means that in the long run, a player can expect to lose about 5¢ for each game played. If 2000 games are played, one could expect to lose $100.

In the game of one-spot keno, a card is purchased for $1. It allows a player to choose one number from 1 to 80. A dealer then chooses twenty numbers at random. If the player's number is among those chosen, the player is paid $3.20, but does not get to keep the $1 paid to play the game. Find the expected value of a $1 bet. Describe what this means.

Exercise Set 11.8

Practice and Application Exercises

In Exercises 1–2, the numbers that each pointer can land on and their respective probabilities are shown. Compute the expected value for the number on which each pointer lands.

1.

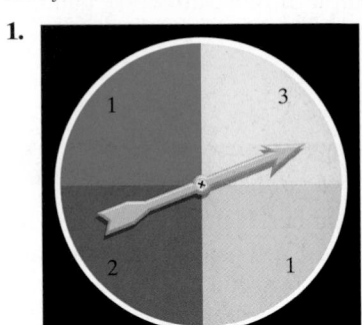

Outcome	Probability
1	$\frac{1}{2}$
2	$\frac{1}{4}$
3	$\frac{1}{4}$

2.

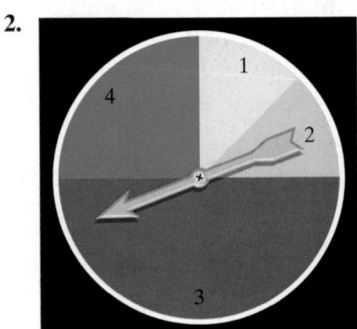

Outcome	Probability
1	$\frac{1}{8}$
2	$\frac{1}{8}$
3	$\frac{1}{2}$
4	$\frac{1}{4}$

The tables in Exercises 3–4 show claims and their probabilities for an insurance company.

   a. *Calculate the expected value and describe what this means in practical terms.*

   b. *How much should the company charge as an average premium so that it breaks even on its claim costs?*

   c. *How much should the company charge to make a profit of $50 per policy?*

3. PROBABILITIES FOR HOMEOWNERS' INSURANCE CLAIMS

Amount of claim (to the nearest $50,000)	Probability
$0	0.65
$50,000	0.20
$100,000	0.10
$150,000	0.03
$200,000	0.01
$250,000	0.01

4. PROBABILITIES FOR MEDICAL INSURANCE CLAIMS

Amount of claim (to the nearest $20,000)	Probability
$0	0.70
$20,000	0.20
$40,000	0.06
$60,000	0.02
$80,000	0.01
$100,000	0.01

5. An architect is considering bidding for the design of a new museum. The cost of drawing plans and submitting a model is $10,000. The probability of being awarded the bid is 0.1, and anticipated profits are $100,000, resulting in a possible gain of this amount minus the $10,000 cost for plans and a model. What is the expected value in this situation? Describe what this value means.

6. A construction company is planning to bid on a building contract. The bid costs the company $1500. The probability that the bid is accepted is $\frac{1}{5}$. If the bid is accepted, the company will make $40,000 minus the cost of the bid. Find the expected value in this situation. Describe what this value means.

7. It is estimated that there are 27 deaths for every 10 million people who use airplanes. A company that sells flight insurance provides $100,000 in case of death in a plane crash. A policy can be purchased for $1. Calculate the expected value and thereby determine how much the insurance company can make over the long run for each policy that it sells.

8. A 25-year old can purchase a one-year life insurance policy for $10,000 at a cost of $100. Past history indicates that the probability of a person dying at age 25 is 0.002. Determine the company's expected gain per policy.

Exercises 9–10 are related to the SAT, described in Check Point 4 on page 617.

9. Suppose that you can eliminate one of the possible five answers. Modify the two probabilities shown in the final column in Table 11.11 on page 617 by finding the probabilities of guessing correctly and guessing incorrectly under these circumstances. What is the expected point value of a random guess? Is it advantageous to guess under these circumstances?

10. Suppose that you can eliminate two of the possible five answers. Modify the two probabilities shown in the final column in Table 11.11 on page 617 by finding the probabilities of guessing correctly and guessing incorrectly under these circumstances. What is the expected point value of a random guess? Is it advantageous to guess under these circumstances?

11. A store specializing in mountain bikes is to open in one of two malls. If the first mall is selected, the store anticipates a yearly profit of $300,000 if successful and a yearly loss of $100,000 otherwise. The probability of success is $\frac{1}{2}$. If the second mall is selected, it is estimated that the yearly profit will be $200,000 if successful; otherwise, the annual loss will be $60,000. The probability of success at the second mall is $\frac{3}{4}$. Which mall should be chosen in order to maximize the expected profit?

12. An oil company is considering two sites on which to drill, described as follows:

Site A: Profit if oil is found: $80 million
           Loss if no oil is found: $10 million
           Probability of finding oil: 0.2
Site B: Profit if oil is found: $120 million
           Loss if no oil is found: $18 million
           Probability of finding oil: 0.1

Which site has the larger expected profit? By how much?

13. In a product liability case, a company can settle out of court for a loss of $350,000, or go to trial, losing $700,000 if found guilty and nothing if found not guilty. Lawyers for the company estimate the probability of a not-guilty verdict to be 0.8.
   a. Find the expected value of the amount the company can lose by taking the case to court.
   b. Should the company settle out of court?

14. A service that repairs air conditioners sells maintenance agreements for $80 a year. The average cost for repairing an air conditioner is $350 and 1 in every 100 people who purchase maintenance agreements have air conditioners that require repair. Find the service's expected profit per maintenance agreement.

Exercises 15–19 involve computing expected values in games of chance.

15. A game is played using one die. If the die is rolled and shows 1, the player wins $5. If the die shows any number other than 1, the player wins nothing. If there is a charge of $1 to play the game, what is the game's expected value? What does this value mean?

16. A game is played using one die. If the die is rolled and shows 1, the player wins $1; if 2, the player wins $2; if 3, the player wins $3. If the die shows 4, 5, or 6, the player wins nothing. If there is a charge of $1.25 to play the game, what is the game's expected value? What does this value mean?

17. Another option in a roulette game (see Example 6 on page 619) is to bet $1 on red. (There are 18 red compartments, 18 black compartments, and two compartments that are neither red nor black.) If the ball lands on red, you get to keep the $1 that you paid to play the game and you are awarded $1. If the ball lands elsewhere, you are awarded nothing and the $1 that you bet is collected. Find the expected value for playing roulette if you bet $1 on red. Describe what this number means.

18. The spinner on a wheel of fortune can land with an equal chance on any one of ten regions. Three regions are red, four are blue, two are yellow, and one is green. A player wins $4 if the spinner stops on red and $2 if it stops on green. The player loses $2 if it stops on blue and $3 if it stops on yellow. What is the expected value? What does this mean if the game is played ten times?

19. For many years, organized crime ran a numbers game that is now run legally by many state governments. The player selects a three-digit number from 000 to 999. There are 1000 such numbers. A bet of $1 is placed on a number, say number 115. If the number is selected, the player wins $500. If any other number is selected, the player wins nothing. Find the expected value for this game and describe what this means.

Writing in Mathematics

20. What does the expected value for the outcome of the roll of a fair die represent?

21. Explain how to find the expected value for the number of girls for a family with two children. What is the expected value?

22. How do insurance companies use expected value to determine what to charge for a policy?

23. Describe a situation in which a business can use expected value.

24. If the expected value of a game is negative, what does this mean? Also describe what it means for a positive and a zero expected value.

25. The expected value for purchasing a ticket in a raffle is −$0.75. Describe what this means. Will a person who purchases a ticket lose $0.75?

Critical Thinking Exercises

26. A popular state lottery is the 5/35 lottery, played in Arizona, Connecticut, Illinois, Iowa, Kentucky, Maine, Massachusetts, New Hampshire, South Dakota, and Vermont. In Arizona's version of the game, prizes are set: First prize is $50,000, second prize is $500, and third prize is $5. To win first prize, you must select all five of the winning numbers, numbered from 1 to 35. Second prize is awarded to players who select any four of the five winning numbers, and third prize is awarded to players who select any three of the winning numbers. The cost to purchase a lottery ticket is $1. Find the expected value of Arizona's "Fantasy Five" game, and describe what this means in terms of buying a lottery ticket over the long run.

27. Refer to the probabilities of dying at any given age on page 587 to solve this exercise. A 20-year-old woman wants to purchase a $200,000 one-year life insurance policy. What should the insurance company charge the woman for the policy if it wants an expected profit of $60?

Group Exercise

28. This activity is a group research project intended for people interested in games of chance at casinos. The research should culminate in a seminar on games of chance and their expected values. The seminar is intended to last about 30 minutes and should result in an interesting and informative presentation made to the entire class.

Each member of the group should research a game available at a typical casino. Describe the game to the class and compute its expected value. After each member has done this, so that class members now have an idea of those games with the greatest and smallest house advantages, a final group member might want to research and present ways for currently treating people whose addiction to these games has caused their lives to swirl out of control.

Chapter Summary, Review, and Test

SUMMARY

DEFINITIONS AND CONCEPTS

EXAMPLES

11.1 The Fundamental Counting Principle

The number of ways in which a series of successive things can occur is found by multiplying the number of ways in which each thing can occur.

Exs. 1–6,
pp. 560–563

11.2 Permutations

a. A permutation from a group of items occurs when no item is used more than once and the order of arrangement makes a difference.

Ex. 1, p. 566;
Ex. 2, p. 566

b. Factorial Notation

Ex. 3, p. 567

$$n! = n(n-1)(n-2)\cdots(3)(2)(1) \text{ and } 0! = 1$$

c. Permutations Formula

The number of permutations possible if r items are taken from n items is $_nP_r = \dfrac{n!}{(n-r)!}$.

Ex. 4, p. 569;
Ex. 5, p. 570

d. Permutations of Duplicate Items

Ex. 6, p. 571

The number of permutations of n items, where p items are identical, q items are identical, r items are identical, and so on, is

$$\frac{n!}{p!q!r!\cdots}.$$

11.3 Combinations

a. A combination from a group of items occurs when no item is used more than once and the order of items makes no difference.

Ex. 1, p. 574

b. Combinations Formula

The number of combinations possible if r items are taken from n items is $_nC_r = \dfrac{n!}{(n-r)!r!}$.

Ex. 2, p. 575;
Ex. 3, p. 576;
Ex. 4, p. 577

11.4 Fundamentals of Probability

a. Theoretical probability applies to experiments in which the set of all equally-likely outcomes, called the sample space, is known. An event is any subset of the sample space.

b. The theoretical probability of event E with sample space S is

Ex. 1, p. 580;
Ex. 2, p. 582;
Ex. 3, p. 583

$$P(E) = \frac{\text{number of outcomes in } E}{\text{total number of possible outcomes}} = \frac{n(E)}{n(S)}.$$

c. Empirical probability applies to situations in which we observe the frequency of the occurrence of an event.

d. The empirical probability of event E is

Ex. 4, p. 584;

$$P(E) = \frac{\text{observed number of times } E \text{ occurs}}{\text{total number of observed occurrences}}.$$

11.5 Probability with the Fundamental Counting Principle, Permutations, and Combinations

a. Probability of a permutation

Ex. 1, p. 588

$$= \frac{\text{the number of ways the permutation can occur}}{\text{total number of possible permutations}}.$$

b. Probability of a combination

Ex. 2, p. 589;
Ex. 3, p. 590

$$= \frac{\text{the number of ways the combination can occur}}{\text{total number of possible combinations}}.$$

11.6 Events Involving *Not* and *Or*

a. $P(\text{not } E) = 1 - P(E)$

b. If it is impossible for events A and B to occur simultaneously, the events are mutually exclusive.

c. If A and B are mutually exclusive events, then
$$P(A \text{ or } B) = P(A) + P(B).$$

d. If A and B are not mutually exclusive events, then

$$P(A \text{ or } B) = P(A) + P(B) - P(A \text{ and } B).$$

Odds

a. Probability to Odds

1. Odds in favor of $E = \dfrac{P(E)}{P(\text{not } E)}$

2. Odds against $E = \dfrac{P(\text{not } E)}{P(E)}$

b. Odds to Probability

If odds in favor of E are a to b then

$$P(E) = \frac{a}{a + b}.$$

11.7 Events Involving *And*

a. Two events are independent if the occurrence of either of them has no effect on the probability of the other.

b. If A and B are independent events,

$$P(A \text{ and } B) = P(A) \cdot P(B).$$

c. The probability of a succession of independent events is the product of each of their probabilities.

d. Two events are dependent if the occurrence of one of them has an effect on the probability of the other.

e. If A and B are dependent events,

$$P(A \text{ and } B) = P(A) \cdot P(B \text{ given that } A \text{ has occurred}).$$

f. The multiplication rule for dependent events can be extended to cover three or more dependent events. In the case of three such events,

$$P(A \text{ and } B \text{ and } C)$$
$$= P(A) \cdot P(B \text{ given } A \text{ occurred}) \cdot P(C \text{ given } A \text{ and } B \text{ occurred}).$$

Conditional Probability

The conditional probability of B, given A, written $P(B|A)$, is the probability of B if the sample space is restricted to A.

$$P(B|A) = \frac{n(B \cap A)}{n(A)} = \frac{\text{number of outcomes common to } B \text{ and } A}{\text{number of outcomes in } A}$$

11.8 Expected Value

a. Expected value, E, is found by multiplying every possible outcome by its probability and then adding these products.

b. In situations involving business decisions, expected value is calculated by multiplying the gain or loss for each possible outcome by its probability. The sum of these products is the expected value.

c. In a game of chance, expected value is the average payoff when the game is played a large number of times. To find the expected value of a game, multiply the gain or loss for each possible outcome by its probability. Then add the products.

REVIEW EXERCISES

11.1

1. A restaurant offers 20 appetizers and 40 main courses. In how many ways can a person order a two-course meal?

2. A popular brand of pen comes in red, green, blue, or black ink. The writing tip can be chosen from extra bold, bold, regular, fine, or micro. How many different choices of pens do you have with this brand?

3. In how many ways can first and second prize be awarded in a contest with 100 people, assuming that each prize is awarded to a different person?

4. You are answering three multiple-choice questions. Each question has five answer choices, with one correct answer per question. If you select one of these five choices for each question and leave nothing blank, in how many ways can you answer the questions?

5. A stock can go up, go down, or stay unchanged. How many possibilities are there if you own five stocks?

6. A person can purchase a condominium with a choice of five kinds of carpeting, with or without a pool, with or without a porch, and with one, two, or three bedrooms. How many different options are there for the condominium?

11.2

7. Six acts are scheduled to perform in a variety show. How many different ways are there to schedule their appearances?

8. In how many ways can five airplanes line up for departure on a runway?

9. You need to arrange seven of your favorite books along a small shelf. Although you are not arranging the books by height, the tallest of the books is to be placed at the left end and the shortest of the books at the right end. How many different ways can you arrange the books?

In Exercises 10–13, evaluate each factorial expression.

10. $\dfrac{16!}{14!}$ 11. $\dfrac{800!}{799!}$

12. $5! - 3!$ 13. $\dfrac{11!}{(11 - 3)!}$

In Exercises 14–15, use the formula for $_nP_r$ to evaluate each expression.

14. $_{10}P_6$ 15. $_{100}P_2$

Use the formula for $_nP_r$ to solve Exercises 16–17.

16. A club with 15 members is to choose four officers—president, vice-president, secretary, and treasurer. In how many ways can these offices be filled?

17. Suppose you are asked to list, in order of preference, the five favorite CDs you purchased in the past 12 months. If you bought 20 CDs over this time period, in how many ways can the five favorite be ranked?

Use the formula for the number of permutations of duplicate items to solve Exercises 18–19.

18. In how many distinct ways can the letters of the word TORONTO be arranged?

19. In how many ways can the digits in the number 335,557 be arranged?

11.3

In Exercises 20–22, does the problem involve permutations or combinations? Explain your answer. (It is not necessary to solve the problem.)

20. How many different 4-card hands can be dealt from a 52-card deck?

21. How many different ways can a director select from 20 male actors to cast the roles of Mark, Roger, Angel, and Collins in the musical *Rent*?

22. How many different ways can a director select 4 actors from a group of 20 actors to attend a workshop on performing in rock musicals?

In Exercises 23–24, use the formula for $_nC_r$ to evaluate each expression.

23. $_{11}C_7$ 24. $_{14}C_5$

Use the formula for $_nC_r$ to solve Exercises 25–28.

25. An election ballot asks voters to select four city commissioners from a group of ten candidates. In how many ways can this be done?

26. How many different 5-card hands can be dealt from a deck that has only hearts (13 different cards)?

27. From the 20 CDs that you've bought during the past year, you plan to take 3 with you on vacation. How many different sets of three CDs can you take?

28. A political discussion group consists of 12 Republicans and 8 Democrats. In how many ways can 5 Republicans and 4 Democrats be selected to attend a conference on politics and social issues?

11.4

In Exercises 29–32, a die is rolled. Find the probability of rolling

29. a 6.

30. a number less than 5.

31. a number less than 7.

32. a number greater than 6.

In Exercises 33–37, you are dealt one card from a 52-card deck. Find the probability of being dealt

33. a 5.

34. a picture card.

35. a card greater than 4 and less than 8.

36. a 4 of diamonds.

37. a red ace.

In Exercises 38–40, suppose that you reach into a bag and randomly select one piece of candy from 15 chocolates, 10 caramels, and 5 peppermints. Find the probability of selecting

38. a chocolate.

39. a caramel.

40. a peppermint.

41. Tay-Sachs disease occurs in 1 of every 3600 births among Jews from central and eastern Europe, and in 1 in 600,000 births in other populations. The disease causes abnormal accumulation of certain fat compounds in the spinal cord and brain, resulting in paralysis, blindness, and mental impairment. Death generally occurs before the age of five. If we use t to represent a Tay-Sachs gene and T a healthy gene, the table below shows the four possibilities for the children of one healthy, TT, parent, and one parent who carries the disease, Tt, but is not sick.

a. Find the probability that a child of these parents will be a carrier without the disease.

b. Find the probability that a child of these parents will have the disease.

		Second Parent	
		T	t
First	T	TT	Tt
Parent	T	TT	Tt

Exercises 42–43 involve empirical probabilities. Express each probability as a fraction. Then use a calculator to express the probability in decimal form, rounded to two decimal places.
The circle graph shows the Hispanic population in the United States, by origin. Find the probability that

42. a person randomly selected from the U.S. Hispanic population is Cuban American.

43. a person randomly selected from the U.S. Hispanic population is Mexican American.

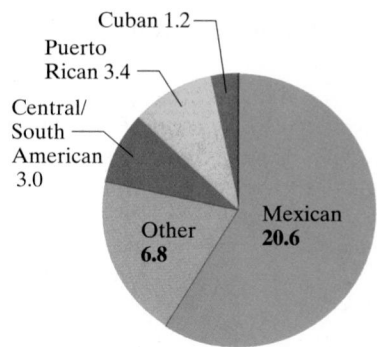

U.S Hispanic Population, in Millions

Cuban 1.2
Puerto Rican 3.4
Central/South American 3.0
Other 6.8
Mexican 20.6

Source: U.S. Census Bureau

11.5

44. If cities A, B, C, and D are visited in random order, each city visited once, find the probability that city D will be visited first, city B second, city A third, and city C last.

In Exercises 45–48, suppose that six singers are being lined up to perform at a charity. Call the singers A, B, C, D, E, and F. The order of performance is determined by writing each singer's name on one of six cards, placing the cards in a hat, and then drawing one card at a time. The order in which the cards are drawn determines the order in which the singers perform. Find the probability that

45. singer C will perform last.

46. singer B will perform first and singer A will perform last.

47. the singers will perform in the following order: F, E, A, D, C, B.

48. the performance will begin with singer A or C.

49. A lottery game is set up so that each player chooses five different numbers from 1 to 20. If the five numbers match the five numbers drawn in the lottery, the player wins (or shares) the top cash prize. What is the probability of winning the prize

a. with one lottery ticket?

b. with 100 different lottery tickets?

50. A committee of four people is to be selected from six Democrats and four Republicans. Find the probability that

a. all are Democrats.

b. two are Democrats and two are Republicans.

51. If you are dealt 3 cards from a shuffled deck of red cards (26 different cards), find the probability of getting exactly 2 picture cards.

11.6

In Exercises 52–56, a die is rolled. Find the probability of

52. not rolling a 5.

53. not rolling a number less than 4.

54. rolling a 3 or a 5.

55. rolling a number less than 3 or greater than 4.

56. rolling a number less than 5 or greater than 2.

In Exercises 57–62, you draw one card from a 52-card deck. Find the probability of

57. not drawing a picture card.

58. not drawing a diamond.

59. drawing an ace or a king.

60. drawing a black 6 or a red 7.

61. drawing a queen or a red card.

62. drawing a club or a picture card.

In Exercises 63–68, it is equally probable that the pointer on the spinner shown will land on any one of the six regions, numbered 1 through 6, and colored as shown. If the pointer lands on a borderline, spin again. Find the probability of

63. not stopping on 4.

64. not stopping on yellow.

65. not stopping on red.

66. stopping on red or yellow.

67. stopping on red or an even number.

68. stopping on red or a number greater than 3.

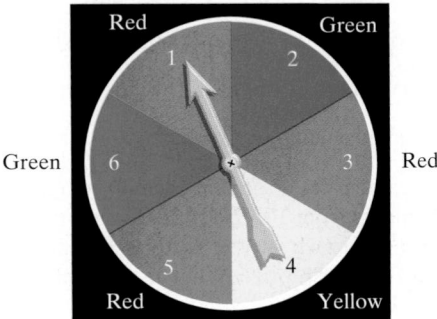

Use this information to solve Exercises 69–70. At a workshop on police work and the African American community, there are 50 African American male police officers, 20 African American female police officers, 90 white male police officers, and 40 white female police officers. If one police officer is selected at random from the people at the workshop, find the probability that the selected person is

69. African American or male.

70. female or white.

Suppose that a survey of 350 college students is taken. Each student is asked the type of college attended (public or private) and the family's income level (low, middle, high). Use the data in the table to solve Exercises 71–74. Express probabilities as simplified fractions.

	Public	**Private**	**Total**
Low	120	20	140
Middle	110	50	160
High	22	28	50
Total	252	98	350

Find the probability that a randomly selected student in the survey

71. attends a public college.

72. is not from a high-income family.

73. is from a middle-income or a high-income family.

74. attends a private college or is from a high-income family.

75. One card is randomly selected from a deck of 52 cards. Find the odds in favor and the odds against getting a queen.

76. The winner of a raffle will receive a two-year scholarship to any college of the winner's choice. If 2000 raffle tickets were sold and you purchased 20 tickets, what are the odds against your winning the scholarship?

77. The odds in favor of a candidate winning an election are given at 3 to 1. What is the probability that this candidate will win the election?

11.7

Use the spinner shown to solve Exercises 78–82. It is equally likely that the pointer will land on any one of the six regions, numbered 1 through 6, and colored as shown. If the pointer lands on a borderline, spin again. If the pointer is spun twice, find the probability it will land on

78. yellow and then red.　　**79.** 1 and then 3.

80. yellow both times.

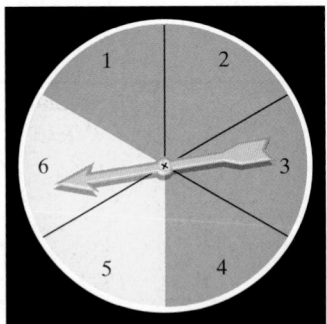

If the pointer is spun three times, find the probability it will land on

81. yellow and then 4 and then an odd number.

82. red every time.

83. What is the probability of a family having five boys born in a row?

84. The probability of a flood in any given year in a region prone to flooding is 0.2.

　a. What is the probability of a flood two years in a row?

　b. What is the probability of a flood for three consecutive years?

　c. What is the probability of no flooding for four consecutive years?

　d. What is the probability of a flood at least once in the next four years?

In Exercises 85–86, two students are selected from a group of four psychology majors, three business majors, and two music majors. The two students are to meet with the campus cafeteria manager to voice the group's concerns about food prices and quality. One student is randomly selected and leaves for the cafeteria manager's office. Then, a second student is selected. Find the probability of selecting

85. a music major and then a psychology major.

86. two business majors.

87. A final visit to the box of chocolates: It's now grown to a box of 50, of which 30 are solid chocolate, 15 are filled with jelly, and 5 are filled with cherries. The story is still the same: They all look alike. You select a piece, eat it, select a second piece, eat it, and help yourself to a final sugar rush. Find the probability of selecting a solid chocolate followed by two cherry-filled chocolates.

88. A single die is tossed. Find the probability that the tossed die shows 5, given that the outcome is an odd number.

89. A letter is randomly selected from the letters of the English alphabet. Find the probability of selecting a vowel, given that the outcome is a letter that precedes k.

90. The numbers shown below are each written on a colored chip. The chips are placed into a bag and one chip is selected at random. Find the probability of selecting

   a. an odd number, given that a red chip is selected

   b. a yellow chip, given that the number selected is at least 3

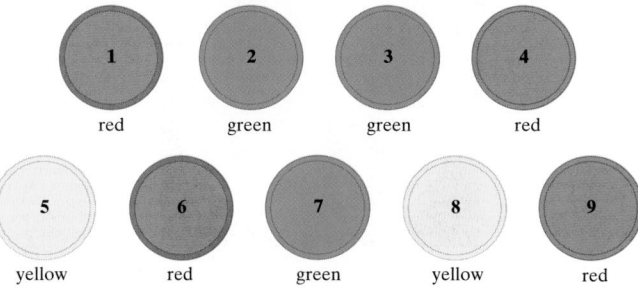

The data in the table are based on 145 Americans tested for tuberculosis. Use the data to solve Exercises 91–98. Express probabilities as simplified fractions.

	TB	No TB
Positive Screening Test	9	11
Negative Screening Test	1	124

Source: Deborah J. Bennett, *Randomness*, Harvard University Press, 1998

Find the probability that a randomly selected person from this group

91. does not have TB.

92. tests positive.

93. does not have TB or tests positive.

94. does not have TB, given a positive test.

95. has a positive test, given no TB.

96. has TB, given a negative test.

Suppose that two people are randomly selected, in succession, from this group. Find the probability of selecting

97. two people with TB.

98. two people with positive screening tests.

11.8

99. The numbers that the pointer can land on and their respective probabilities are shown. Compute the expected value for the number on which the pointer lands.

Outcome	Probability
1	$\frac{1}{4}$
2	$\frac{1}{8}$
3	$\frac{1}{8}$
4	$\frac{1}{4}$
5	$\frac{1}{4}$

100. The table shows claims and their probabilities for an insurance company.

LIFE INSURANCE FOR AN AIRLINE FLIGHT

Amount of Claim	Probability
$0	0.9999995
$1,000,000	0.0000005

   a. Calculate the expected value and describe what this value means.

   b. How much should the company charge to make a profit of $9.50 per policy?

101. A construction company is planning to bid on a building contract. The bid costs the company $3000. The probability that the bid is accepted is $\frac{1}{4}$. If the bid is accepted, the company will make $30,000 minus the cost of the bid. Find the expected value in this situation. Describe what this value means.

102. A game is played using a fair coin that is tossed twice. The sample space is $\{HH, HT, TH, TT\}$. If exactly one head occurs, the player wins $5, and if exactly two tails occur, the player also wins $5. For any other outcome, the player receives nothing. There is a $4 charge to play the game. What is the expected value? What does this value mean?

Group Exercise

103. We opened the chapter with a decision involving whether or not to accept a real estate listing. In Example 4 on page 617 we saw that it would be worth the risk to accept the listing. In this exercise, each group member should present a situation related to what he or she is doing now or plans to do in the future in which risk and decision making come into play. List all the possible outcomes and assign to each outcome the most realistic probability that you can. The group should select three of these situations. For each, the group should write and solve a problem involving expected value, similar to Example 4 on page 617.

CHAPTER 11 TEST

1. A person can purchase a particular model of a new car with a choice of ten colors, with or without automatic transmission, with or without four-wheel drive, with or without air conditioning, and with two, three, or four radio-CD speakers. How many different options are there for this model of the car?

2. Four acts are scheduled to perform in a variety show. How many different ways are there to schedule their appearances?

3. In how many ways can seven airplanes line up for a departure on a runway if the plane with the greatest number of passengers must depart first?

4. A human resource manager has 11 applicants to fill three different positions. Assuming that all applicants are equally qualified for any of the three positions, in how many ways can this be done?

5. From the ten books that you've recently bought but not read, you plan to take four with you on vacation. How many different sets of four books can you take?

6. In how many distinct ways can the letters of the word ATLANTA be arranged?

In Exercises 7–9, one student is selected at random from a group of 12 freshmen, 16 sophomores, 20 juniors, and 2 seniors. Find the probability that the person selected is

7. a freshman.

8. not a sophomore.

9. a junior or a senior.

10. If you are dealt one card from a 52-card deck, find the probability of being dealt a card greater than 4 and less than 10.

11. Seven movies (A, B, C, D, E, F, and G) are being scheduled for showing. The order of showing is determined by random selection. Find the probability that film C will be shown first, film A next-to-last, and film E last.

12. A lottery game is set up so that each player chooses six different numbers from 1 to 15. If the six numbers match the six numbers drawn in the lottery, the player wins (or shares) the top cash prize. What is the probability of winning the prize with 50 different lottery tickets?

In Exercises 13–14, it is equally probable that the pointer on the spinner shown will land on any one of the eight colored regions. If the pointer lands on a borderline, spin again.

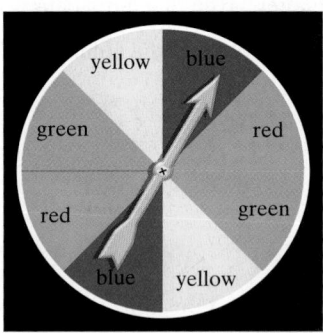

13. If the spinner is spun once, find the probability that the pointer will land on red or blue.

14. If the spinner is spun twice, find the probability that the pointer lands on red on the first spin and blue on the second spin.

15. A region is prone to flooding once every 20 years. The probability of flooding in any one year is $\frac{1}{20}$. What is the probability of flooding for three consecutive years?

16. One card is randomly selected from a deck of 52 cards. Find the probability of selecting a black card or a picture card.

17. A group of students consists of 10 male freshmen, 15 female freshmen, 20 male sophomores, and 5 female sophomores. If one person is randomly selected from the group, find the probability of selecting a freshman or a female.

18. A box contains five red balls, six green balls, and nine yellow balls. Suppose you select one ball at random from the box and do not replace it. Then you randomly select a second ball. Find the probability that both balls selected are red.

19. A quiz consisting of four multiple-choice questions has four available options (a, b, c, or d) for each question. If a person guesses at every question, what is the probability of answering *all* questions correctly?

20. A group is comprised of 20 men and 15 women. If one person is randomly selected from the group, find the odds against the person being a man.

21. The odds against a candidate winning an election are given at 1 to 4.

    a. What are the odds in favor of the candidate winning?

    b. What is the probability that the candidate will win the election?

A class is collecting data on eye color and gender. They organize the data they collected into the table shown. Numbers in the table represent the number of students from the class that belong to each of the categories. Use the data to solve Exercises 22–26. Express probabilities as simplified fractions.

	Brown	**Blue**	**Green**
Male	22	18	10
Female	18	20	12

Find the probability that a randomly selected student from this class

22. does not have brown eyes.

23. has brown eyes or blue eyes.

24. is female or has green eyes.

25. is male, given the student has blue eyes.

26. If two people are randomly selected, in succession, from the students in this class, find the probability that they both have green eyes.

27. An architect is considering bidding for the design of a new theater. The cost of drawing plans and submitting a model is $15,000. The probability of being awarded the bid is 0.2, and anticipated profits are $80,000, resulting in a possible gain of this amount minus the $15,000 cost for plans and models. What is the expected value if the architect decides to bid for the design? Describe what this value means.

28. A game is played by selecting one bill at random from a bag that contains ten $1 bills, five $2 bills, three $5 bills, one $10 bill, and one $100 bill. The player gets to keep the selected bill. There is a $20 charge to play the game. What is the expected value? What does this value mean?

Statistics

Statisticians collect numerical data from subgroups of populations to find out everything imaginable about the population as a whole, including whom they favor in an election, what they watch on TV, how much money they make, what worries them, and how they feel about the quality of their education. In the twenty-first century, statistics is big business. Because statisticians both record and influence our behavior, it is important to distinguish between good and bad methods for collecting, presenting, and interpreting data. In this chapter, you will gain an understanding of where data come from and how these numbers are used to make decisions.

This course is wonderful! You are sitting in a large auditorium classroom, enjoying every second of the lecture. You are energized by your professor's knowledge, delivery, and sense of humor. This is truly what college is all about. Are you the only one so enthused about this course? Well over 100 students are in the room. Is there any way to use the opinions of a small subgroup to find out if a substantial number of all the students share your opinion?

This problem is discussed following the boxed definition of random samples in Section 12.1.

SECTION 12.1   SAMPLING, FREQUENCY DISTRIBUTIONS, AND GRAPHS

OBJECTIVES

1. Describe the population whose properties are to be analyzed.

2. Select an appropriate sampling technique.

3. Organize and present data.

4. Identify deceptions in visual displays of data.

"M∗A∗S∗H" took place in the early 1950s, during the Korean War. By the final episode, the show had lasted four times as long as the Korean War.

At the end of the twentieth century, there were 94 million households in the United States with television sets. The television program viewed by the greatest percentage of such households in that century was the final episode of "M∗A∗S∗H." Over 50 million American households watched this program.

Numerical information, such as the information about the top three TV shows of the twentieth century, shown in Table 12.1, is called **data**. The word **statistics** is often used when referring to data. However, statistics has a second meaning: Statistics is also a method for collecting, organizing, analyzing, and interpreting data, as well as drawing conclusions based on the data. This methodology divides statistics into two main areas. **Descriptive statistics** is concerned with collecting, organizing, summarizing, and presenting data. **Inferential statistics** has to do with making generalizations about and drawing conclusions from the data collected.

TABLE 12.1 TV PROGRAMS WITH THE GREATEST U.S. AUDIENCE VIEWING PERCENTAGE OF THE TWENTIETH CENTURY

Program	Total Households	Viewing Percentage
1. "M∗A∗S∗H" Feb. 28, 1983	50,150,000	60.2%
2. "Dallas" Nov. 21, 1980	41,470,000	53.3%
3. "Roots" Part 8 Jan. 30, 1977	36,380,000	51.1%

Source: Nielsen Media Research

Populations and Samples

1 Describe the population whose properties are to be analyzed.

Consider the set of all American TV households. Such a set is called the *population*. In general, a **population** is the set containing all the people or objects whose properties are to be described and analyzed by the data collector.

BLITZER BONUS

A Sampling Fiasco

In 1936, the *Literary Digest* mailed out over ten million ballots to voters throughout the country. The results poured in, and the magazine predicted a landslide victory for Republican Alf Landon over Democrat Franklin Roosevelt. However, the prediction of the *Literary Digest* was wrong. Why? The mailing lists the editors used included people from their own subscriber list, directories of automobile owners, and telephone books. As a result, its sample was anything but random. It excluded most of the poor, who were unlikely to subscribe to the *Literary Digest*, or to own a car or telephone in the heart of the Depression. Prosperous people in 1936 were more likely to be Republican than the poor. Thus, although the sample was massive, it included a higher percentage of affluent individuals than the population as a whole did. A victim of both the Depression and the 1936 sampling fiasco, the *Literary Digest* folded in 1937.

The population of American TV households is huge. At the time of the "M∗A∗S∗H" conclusion, there were nearly 84 million such households. Did over 50 million American TV households really watch the final episode of "M∗A∗S∗H"? A friendly phone call to each household ("So, how are you? What's new? Watch any good television last night? If so, what?") is, of course, absurd. A **sample**, which is a subset or subgroup of the population, is needed. In this case, it would be appropriate to have a sample of a few thousand TV households to draw conclusions about the population of all TV households.

EXAMPLE 1   POPULATIONS AND SAMPLES

A group of hotel owners in a large city decide to conduct a survey among citizens of the city to discover their opinions about casino gambling.

a. Describe the population.

b. One of the hotel owners suggests obtaining a sample by surveying all the people at six of the largest nightclubs in the city on a Saturday night. Each person will be asked to express his or her opinion on casino gambling. Does this seem like a good idea?

Solution

a. The population is the set containing all the citizens of the city.

b. Questioning people at six of the city's largest nightclubs is a terrible idea. The nightclub subset is probably more likely to have a positive attitude toward casino gambling than the population of all the city's citizens.

A city government wants to conduct a survey among the city's homeless to discover their opinions about required residence in city shelters from midnight until 6 A.M.

a. Describe the population.

b. A city commissioner suggests obtaining a sample by surveying all the homeless people at the city's largest shelter on a Sunday night. Does this seem like a good idea? Explain your answer.

Random Sampling

There is a way to use a small sample to make generalizations about a large population: Guarantee that every member of the population has an equal chance to be selected for the sample. Surveying people at six of the city's largest nightclubs does not provide this guarantee. Unless it can be established that all citizens of the city frequent these clubs, which seems unlikely, this sampling scheme does not permit each citizen an equal chance of selection.

> RANDOM SAMPLES
>
> A **random sample** is a sample obtained in such a way that every element in the population has an equal chance of being selected for the sample.

Suppose that you are elated with the quality of one of your courses. Although it's an auditorium section with 120 students, you feel that the professor is lecturing right to you. During a wonderful lecture, you look around the auditorium to see if any of the other students are sharing your enthusiasm. Based on body language, it's hard to tell. You really want to know the opinion of the population of 120 students taking this course. You think about asking students to grade the course on an A to F scale, anticipating a unanimous A. You cannot survey everyone. Eureka! Suddenly you have an idea on how to take a sample. Place cards numbered from 1 through 120, one number per card, in a box. Because the course has assigned seating by number, each numbered card corresponds to a student in the class. Reach in and randomly select six cards. Each card, and therefore each student, has an equal chance of being selected. Then use the opinions about the course from the six randomly selected students to generalize about the course opinion for the entire 120-student population.

Your idea is precisely how random samples are obtained. In random sampling, each element in the population must be identified and assigned a number. The numbers are generally assigned in order. The way to sample from the larger numbered population is to generate random numbers using a computer or calculator. Each numbered element from the population that corresponds to one of the generated random numbers is selected for the sample.

Call-in polls on radio and television are not reliable because those polled do not represent the larger population. A person who calls in is likely to have feelings about an issue that are consistent with the politics of the show's host. For a poll to be accurate, the sample must be chosen randomly from the larger population. The A.C. Nielsen Company uses a random sample of approximately 5000 TV households to measure the percentage of households tuned in to a television program.

2 Select an appropriate sampling technique.

EXAMPLE 2   SELECTING AN APPROPRIATE SAMPLING TECHNIQUE

We return to the hotel owners in the large city who are interested in how the city's citizens feel about casino gambling. Which of the following would be the most appropriate way to select a random sample?

a. Randomly survey people who live in the oceanfront condominiums in the city.

b. Survey the first 200 people whose names appear in the city's telephone directory.

c. Randomly select neighborhoods of the city and then randomly survey people within the selected neighborhoods.

Solution   Keep in mind that the population is the set containing all the city's citizens. A random sample must give each citizen an equal chance of being selected.

a. Randomly selecting people who live in the city's oceanfront condominiums is not a good idea. Many hotels lie along the oceanfront, and the oceanfront property owners might object to the traffic and noise as a result of casino gambling. Furthermore, this sample does not give each citizen of the city an equal chance of being selected.

b. If the hotel owners survey the first 200 names in the city's telephone directory, all citizens do not have an equal chance of selection. For example, individuals whose last name begins with a letter toward the end of the alphabet have no chance of being selected.

c. Randomly selecting neighborhoods of the city and then randomly surveying people within the selected neighborhood is an appropriate technique. Using this method, each citizen has an equal chance of being selected.

In summary, given the three options, the sampling technique in part (c) is the most appropriate.

Surveys and polls involve data from a sample of some population. Regardless of the sampling technique used, the sample should exhibit characteristics typical of those possessed by the target population. This type of sample is called a **representative sample**.

Explain why the sampling technique described in Check Point 1(b) on page 632 is not a random sample. Then describe an appropriate way to select a random sample of the city's homeless.

Organize and present data.

Frequency Distributions

After data have been collected from a sample of the population, the next task facing the statistician is to present the data in a condensed and manageable form. In this way, the data can be more easily interpreted.

Suppose, for example, that researchers are interested in determining the age at which adolescent males show the greatest rate of physical growth. A random sample of 35 ten-year-old boys is measured for height and then remeasured each year until they reach 18. The age of maximum yearly growth for each subject is as follows:

12, 14, 13, 14, 16, 14, 14, 17, 13, 10, 13, 18, 12, 15, 14, 15, 15, 14, 14, 13, 15, 16, 15, 12, 13, 16, 11, 15, 12, 13, 12, 11, 13, 14, 14.

A piece of data is called a **data item**. This list of data has 35 data items. Some of the data items are identical. Two of the data items are 11 and 11. Thus, we can say that the **data value** 11 occurs twice. Similarly, because five of the data items are 12, 12, 12, 12, and 12, the data value 12 occurs five times.

Collected data can be presented using a **frequency distribution**. Such a distribution consists of two columns. The data values are listed in one column. Numerical data are generally listed from smallest to largest. The adjacent column is labeled **frequency** and indicates the number of times each value occurs.

TABLE 12.2 A FREQUENCY DISTRIBUTION FOR A BOY'S AGE OF MAXIMUM YEARLY GROWTH

Age of Maximum Growth	Number of Boys (Frequency)
10	1
11	2
12	5
13	7
14	9
15	6
16	3
17	1
18	1
Total:	$n = 35$

> 35 is the sum of the frequencies.

EXAMPLE 3 CONSTRUCTING A FREQUENCY DISTRIBUTION

Construct a frequency distribution for the data of the age of maximum yearly growth for 35 boys:

12, 14, 13, 14, 16, 14, 14, 17, 13, 10, 13, 18, 12, 15, 14, 15, 15, 14, 14, 13, 15, 16, 15, 12, 13, 16, 11, 15, 12, 13, 12, 11, 13, 14, 14.

Solution It is difficult to determine trends in the data above in its current format. Perhaps we can make sense of the data by organizing it into a frequency distribution. Let us create two columns. One lists all possible data values, from smallest (10) to largest (18). The other column indicates the number of times the value occurs in the sample. The frequency distribution is shown in Table 12.2.

The frequency distribution indicates that one subject had maximum growth at age 10, two at age 11, five at age 12, seven at age 13, and so on. The maximum growth for most of the subjects occurred between the ages of 12 and 15. Nine boys experienced maximum growth at age 14, more than at any other age within the sample. The sum of the frequencies, 35, is equal to the original number of data items.

The trend shown by the frequency distribution in Table 12.2 indicates that the number of boys who attain their maximum yearly growth at a given age increases until age 14 and decreases after that. This trend is not evident in the data in its original format.

Construct a frequency distribution for the data showing final course grades for students in a precalculus course, listed alphabetically by student name in a grade book:

F, A, B, B, C, C, B, C, A, A, C, C, D, C, B, D, C, C, B, C.

A frequency distribution that lists all possible data items can be quite cumbersome when there are many such items. For example, consider the following data items. These are statistics test scores for a class of 40 students.

82	47	75	64	57	82	63	93
76	68	84	54	88	77	79	80
94	92	94	80	94	66	81	67
75	73	66	87	76	45	43	56
57	74	50	78	71	84	59	76

It's difficult to determine how well the group did when the grades are displayed like this. Because there are so many data items, one way to organize these data so that the results are more meaningful is to arrange the grades into groups, or **classes**, based on something that interests us. Many grading systems assign an A to grades in the 90–100 class, B to grades in the 80–89 class, C to grades in the 70–79 class, and so on. These classes provide one way to organize the data.

Looking at the 40 statistics test scores, we see that they range from a low of 43 to a high of 94. We can use classes that run from 40 through 49, 50 through 59, 60 through 69, and so on up to 90 through 99, to organize the scores. In Example 4, we go through the data and tally each item into the appropriate class. This method for organizing data is called a **grouped frequency distribution**.

EXAMPLE 4 | CONSTRUCTING A GROUPED FREQUENCY DISTRIBUTION

Use the classes 40–49, 50–59, 60–69, 70–79, 80–89, and 90–99 to construct a grouped frequency distribution for the 40 test scores above.

Solution   We use the 40 given scores and tally the number of scores in each class.

Test Scores (Class)	Tally	Number of Students (Frequency)
40–49	|||	3
50–59	||||		6
60–69	||||		6
70–79	|||| ||||		11
80–89	|||| ||||	9
90–99	||||	5

> The second score in the list, 47, is shown as the first tally in this row.

> The first score in the list, 82, is shown as the first tally in this row.

TABLE 12.3

Class	Frequency
40–49	3
50–59	6
60–69	6
70–79	11
80–89	9
90–99	5
Total: $n = 40$	

> 40, the sum of the frequencies, is the number of data items.

Omitting the tally column results in the grouped frequency distribution in Table 12.3. The distribution shows that the greatest frequency of students scored in the 70–79 class. The number of students decreases in classes that contain successively lower and higher scores. The sum of the frequencies, 40, is equal to the original number of data items.

TABLE 12.3, repeated

Class	Frequency
40–49	3
50–59	6
60–69	6
70–79	11
80–89	9
90–99	5
Total:	$n = 40$

The leftmost number in each class of a grouped frequency distribution is called the **lower class limit**. For example, in Table 12.3, the lower limit of the first class is 40 and the lower limit of the third class is 60. The rightmost number in each class is called the **upper class limit**. In Table 12.3, 49 and 69 are the upper limits for the first and third classes, respectively. Notice that if we take the difference between any two lower class limits, we get the same number:

$$50 - 40 = 10, \ 60 - 50 = 10, \ 70 - 60 = 10, \ 80 - 70 = 10, \ 90 - 80 = 10.$$

The number 10 is called the **class width**.

When setting up class limits, each class, with the possible exception of the first or last, should have the same width. Because each data item must fall into exactly one class, it is sometimes helpful to vary the width of the first or last class to allow for items that fall far above or below most of the data.

 **Check 4 Point** Use the classes in Table 12.3 to construct a grouped frequency distribution for the following 37 exam scores:

73	58	68	75	94	79	96	79
87	83	89	52	99	97	89	58
95	77	75	81	75	73	73	62
69	76	77	71	50	57	41	98
77	71	69	90	75.			

TABLE 12.2, repeated
A FREQUENCY DISTRIBUTION FOR A BOY'S AGE OF MAXIMUM YEARLY GROWTH

Age of Maximum Growth	Number of Boys (Frequency)
10	1
11	2
12	5
13	7
14	9
15	6
16	3
17	1
18	1
Total:	$n = 35$

Histograms and Frequency Polygons

Take a second look at the frequency distribution for the age of a boy's maximum yearly growth, repeated in Table 12.2. A bar graph with bars that touch can be used to visually display the data. Such a graph is called a **histogram**. Figure 12.1 illustrates a histogram that was constructed using the frequency distribution in Table 12.2. A series of rectangles whose heights represent the frequencies are placed next to each other. For example, the height of the bar for the data value 10, shown in Figure 12.1, is 1. This corresponds to the frequency for 10 given in Table 12.2. The higher the bar, the more frequent the score. The break along the horizontal axis, symbolized by ⌇, eliminates listing the ages 1 through 9.

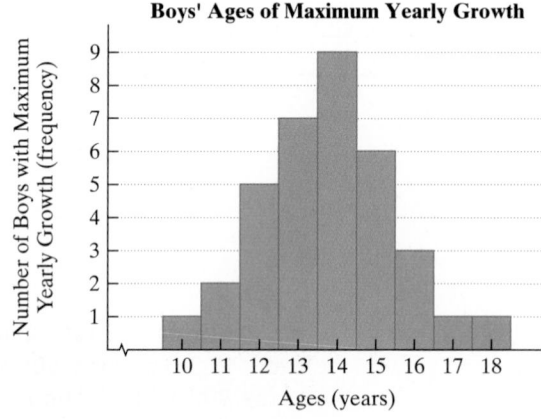

FIGURE 12.1 A histogram for a boy's age of maximum yearly growth

A line graph called a **frequency polygon** can also be used to visually convey the information shown in Figure 12.1. The axes are labeled just like those in a histogram. Thus, the horizontal axis shows data values and the vertical axis shows frequencies. Once a histogram has been constructed, it's fairly easy to draw a frequency polygon. Figure 12.2 shows a histogram with a dot at the top of each rectangle at its midpoint. Connect each of these midpoints with a straight line. To complete the frequency polygon at both ends, the lines should be drawn down to touch the horizontal axis. The completed frequency polygon is shown in Figure 12.3.

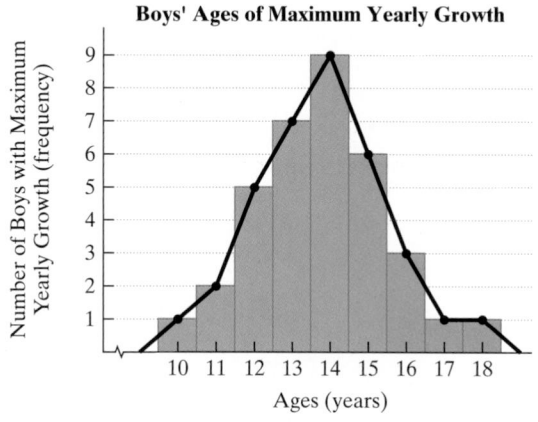

FIGURE 12.2 A histogram with a superimposed frequency polygon

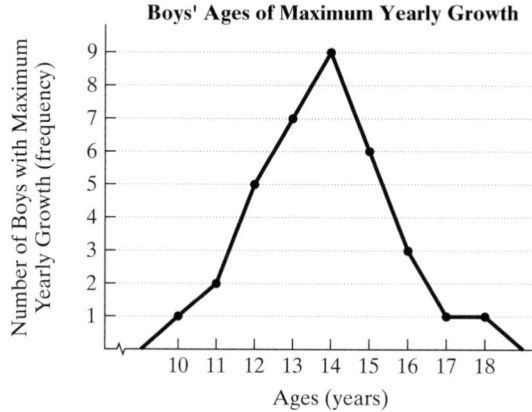

FIGURE 12.3 A frequency polygon

Stem-and-Leaf Plots

A unique way of displaying data uses a tool called a **stem-and-leaf plot**. Example 5 illustrates how we sort the data, revealing the same visual impression created by a histogram.

EXAMPLE 5 CONSTRUCTING A STEM-AND-LEAF PLOT

Use the data showing statistics test scores for 40 students to construct a stem-and-leaf plot:

82	47	75	64	57	82	63	93
76	68	84	54	88	77	79	80
94	92	94	80	94	66	81	67
75	73	66	87	76	45	43	56
57	74	50	78	71	84	59	76.

Solution The plot is constructed by separating each data item into two parts. The first part is the *stem*. The **stem** consists of the tens' digit. For example, the stem for the score of 82 is 8. The second part is the *leaf*. The **leaf** consists of the units' digit for a given value. For the score of 82, the leaf is 2. The possible stems for the 40 scores are 4, 5, 6, 7, 8, and 9, entered in the left column of the plot.

Begin by entering each data item in the first row:

82   47   75   64   57   82   63   93.

Entering **8 2:**		**Adding** **4 7:**		**Adding** **7 5:**		**Adding** **6 4:**	
Stems	Leaves	Stems	Leaves	Stems	Leaves	Stems	Leaves
4		4	7	4	7	4	7
5		5		5		5	
6		6		6		6	4
7		7		7	5	7	5
8	2	8	2	8	2	8	2
9		9		9		9	

Adding **5 7:**		**Adding** **8 2:**		**Adding** **6 3:**		**Adding** **9 3:**	
Stems	Leaves	Stems	Leaves	Stems	Leaves	Stems	Leaves
4	7	4	7	4	7	4	7
5	7	5	7	5	7	5	7
6	4	6	4	6	4 3	6	4 3
7	5	7	5	7	5	7	5
8	2	8	2 2	8	2 2	8	2 2
9		9		9		9	3

We continue in this manner and enter all the data items. Figure 12.4 shows the completed stem-and-leaf plot. If you turn the page so that the left margin is on the bottom and facing you, the visual impression created by the enclosed leaves is the same as that created by a histogram. An advantage over the histogram is that the stem-and-leaf plot preserves exact data items. The enclosed leaves extend farthest to the right when the stem is 7. This shows that the greatest frequency of students scored in the 70s.

Tens' digit   Units' digit

Stems	Leaves
4	7 5 3
5	7 4 6 7 0 9
6	4 3 8 6 7 6
7	5 6 7 9 5 3 6 4 8 1 6
8	2 2 4 8 0 0 1 7 4
9	3 4 2 4 4

FIGURE 12.4 A stem-and-leaf plot displaying 40 test scores

Construct a stem-and-leaf plot for the data in Check Point 4.

Deceptions in Visual Displays of Data

4   Identify deceptions in visual displays of data.

Benjamin Disraeli, Queen Victoria's prime minister, stated that there are "lies, damned lies, and statistics." The problem is not that statistics lie, but rather that liars use statistics. Graphs can be used to distort the underlying data, making it difficult for the viewer to learn the truth. One potential source of misunderstanding is the scale on the vertical axis used to draw the graph. This scale is important because it lets a researcher "inflate" or "deflate" a trend. For example, both graphs in Figure 12.5 present identical data for the average math SAT scores from 1996 through 2001. The graph in on the left stretches the scale on the vertical axis to create an overall impression of SAT scores that are increasing dramatically over time. The graph in on the right compresses the scale on the vertical axis to create an impression of SAT scores that are holding steady over time.

Average Math SAT Scores

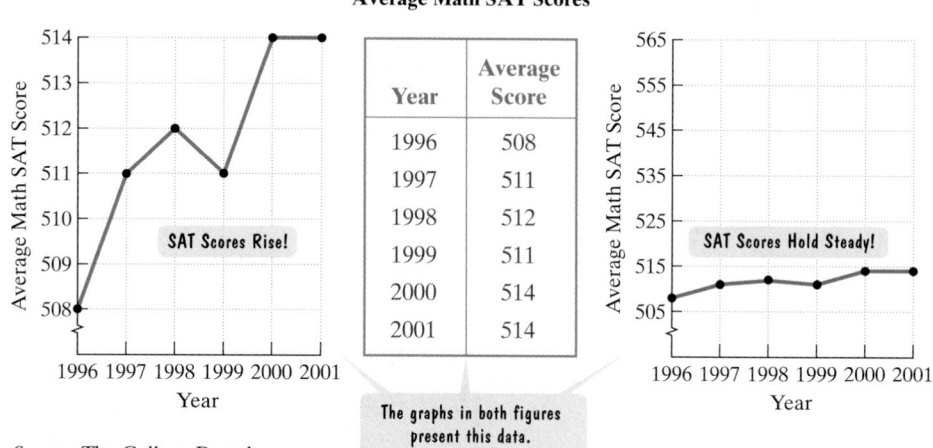

Year	Average Score
1996	508
1997	511
1998	512
1999	511
2000	514
2001	514

The graphs in both figures present this data.

FIGURE 12.5   *Source:* The College Board

We now look at two graphs that, although visually appealing, are misleading. Both graphs are discussed in *The Visual Display of Quantitative Information* by Edward Tufte. We begin with a graph that Tufte cites from the *New York Times*, August 9, 1978. Shown in Figure 12.6, the graph creates the impression that Congress has set forth unreasonable fuel economy standards for automobiles. The line representing 18 miles per gallon for 1978 is 0.6 inch long, so the line representing 27.5 miles per gallon for 1985 should be about 0.92 inch long.

$$\frac{\text{inches}}{\text{miles per gallon}} : \frac{0.6}{18} = \frac{x}{27.5}$$

$$18x = 0.6(27.5)$$

$$x = \frac{0.6(27.5)}{18} \approx 0.92$$

Rather than 0.92 inch, the line is more than 5 inches long. Because this line represents $27\frac{1}{2}$ miles per gallon, its length compared to that of the short line for 1978 creates the impression that fuel standards for 1985 are unreasonable.

This line, representing 18 miles per gallon in 1978, is 0.6 inches long.

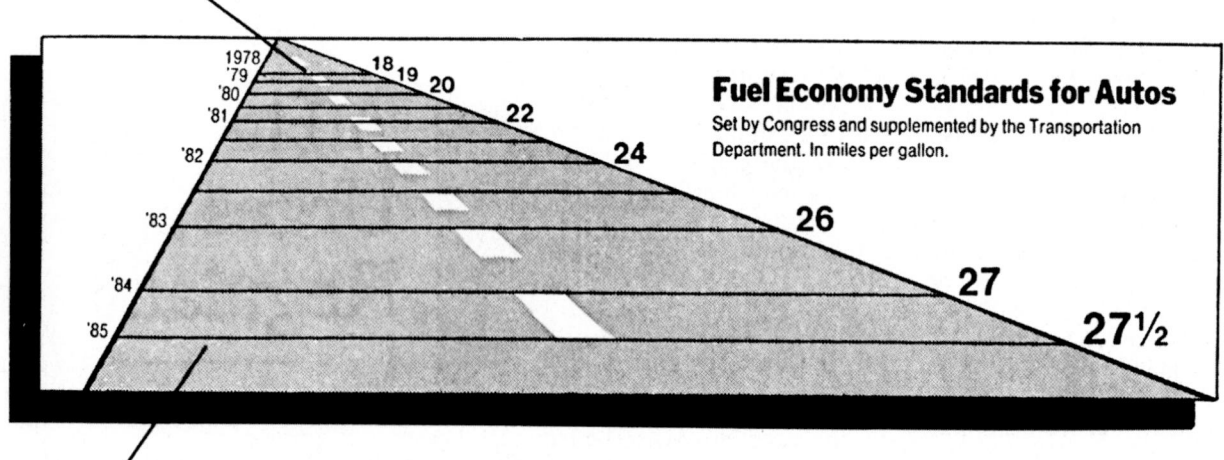

This line, representing 27.5 miles per gallon in 1985, is 5.3 inches long.

Source: New York Times, August 9, 1978, p. D-2. Reprinted by permission.

FIGURE 12.6

Required Fuel Economy Standards: New Cars Built from 1978 through 1985

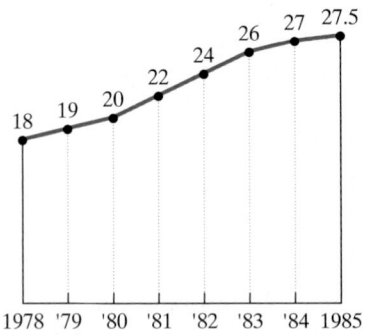

FIGURE 12.7

Do you notice anything else that is unusual about the graph in Figure 12.6? The dates on the left are all the same size. The numbers on the right are different sizes. The largest number in the figure is $27\frac{1}{2}$, which also exaggerates the severity of the 1985 mileage standard. Take a moment to turn back the page and verify these observations. Figure 12.7 shows how the data can be presented in a nondeceptive manner.

The graph in Figure 12.8 varies both the length and width of each dollar bill as the purchasing power of the bill diminishes. The 1958 dollar was worth $1.00. By contrast, the 1978 dollar is worth 44¢. Thus, the 1978 dollar should be drawn to be slightly less than half the size of the 1958 dollar. If the area of the dollar were drawn to reflect its purchasing power, the 1978 dollar would be approximately twice as large as the one shown in the visual display.

FIGURE 12.8

Examples of graphs that contain visual frauds are endless. We need to become aware of ways that visual displays can be deceptive. With this knowledge, we can avoid interpreting potentially deceptive graphs incorrectly.

BLITZER BONUS

Creating an Inaccurate Picture by Leaving Something Out

On Monday, October 19, 1987, the Dow Jones Industrial Average plunged 508 points, losing 22.6% of its value. The graph shown on the left, which appeared in a major newspaper following "Black Monday" (as it was instantly dubbed), creates the impression that the Dow average had been "bullish" from 1972 through 1987, increasing throughout this period. The graph creates this inaccurate picture by leaving something out. The graph on the right illustrates that the stock market rose and fell sharply over these years. The impressively smooth curve on the left was obtained by plotting only three of the data points. By ignoring most of the data, increases and decreases are not accounted for and the actual behavior of the market over the 15 years leading to "Black Monday" is inaccurately conveyed.

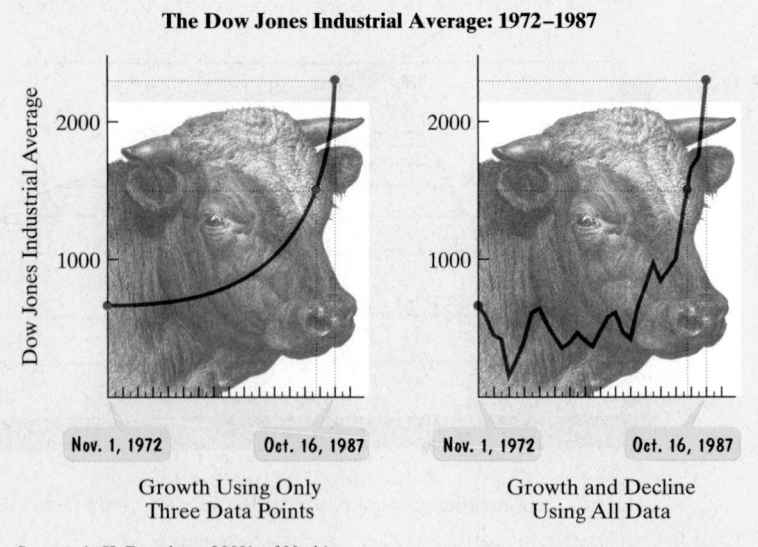

Source: A. K. Dewdney, *200% of Nothing*

Exercise Set 12.1

Practice and Application Exercises

1. "The Man Poll" of 1014 randomly selected American men ages 18 and older was taken by *Newsweek* June 2–4, 2003. The data shown below are from the poll.

IN GENERAL, HOW WOULD YOU DESCRIBE YOUR HEALTH?

	Age		
	18–34	35–55	>56
Excellent	23%	19%	15%
Very good	37	39	32
Good	28	28	28
Fair	10	10	16
Poor	2	4	8
Don't know	0	0	1

a. Describe the population and the sample of this poll.

b. For each man, what variable is measured? Are the data quantitative (numerical) or qualitative (nonnumerical categories)?

2. The *American Association of Nurse Anesthetists Journal* (Feb. 2000) published the results of a study on the use of herbal medicines before surgery. Of 500 surgical patients who were randomly selected for the study, 51% used herbal or alternative medicines prior to surgery.

a. Describe the population and the sample of this study.

b. Is the sample representative of the population? Explain your answer.

c. For each patient, what variable is measured? Are the data quantitative (numerical) or qualitative (nonnumerical categories)?

3. The government of a large city needs to determine whether the city's residents will support the construction of a new jail. The government decides to conduct a survey of a sample of the city's residents. Which one of the following procedures would be most appropriate for obtaining a sample of the city's residents?

a. Survey a random sample of the employees and inmates at the old jail.

b. Survey every fifth person who walks into City Hall on a given day.

c. Survey a random sample of persons within each geographic region of the city.

d. Survey the first 200 people listed in the city's telephone directory.

4. The city council of a large city needs to know whether its residents will support the building of three new schools. The council decides to conduct a survey of a sample of the city's residents. Which procedure would be most appropriate for obtaining a sample of the city's residents?

a. Survey a random sample of teachers who live in the city.

b. Survey 100 individuals who are randomly selected from a list of all people living in the state in which the city in question is located.

c. Survey a random sample of persons within each neighborhood of the city.

d. Survey every tenth person who enters City Hall on a randomly selected day.

A questionnaire was given to students in an introductory statistics class during the first week of the course. One question asked, "How stressed have you been in the last $2\frac{1}{2}$ weeks, on a scale of 0 to 10, with 0 being not at all stressed and 10 being as stressed as possible?" The students' responses are shown in the frequency distribution. Use this frequency distribution to solve Exercises 5–8.

Stress Rating	Frequency
0	2
1	1
2	3
3	12
4	16
5	18
6	13
7	31
8	26
9	15
10	14

Source: Journal of Personality and Social Psychology, 69, 1102–1112

5. Which stress rating describes the greatest number of students? How many students responded with this rating?

6. Which stress rating describes the least number of students? How many responded with this rating?

7. How many students were involved in this study?

8. How many students had a stress rating of 8 or more?

9. A random sample of 30 college students is selected. Each student is asked how much time he or she spent on homework during the previous week. The following times (in hours) are obtained:

16, 24, 18, 21, 18, 16, 18, 17, 15, 21, 19, 17, 17, 16, 19, 18,

15, 15, 20, 17, 15, 17, 24, 19, 16, 20, 16, 19, 18, 17.

Construct a frequency distribution for the data.

10. A random sample of 30 male college students is selected. Each student is asked his height (to the nearest inch). The heights are as follows:

72, 70, 68, 72, 71, 71, 71, 69, 73, 71, 73, 75, 66, 67, 75

74, 73, 71, 72, 67, 72, 68, 67, 71, 73, 71, 72, 70, 73, 70.

Construct a frequency distribution for the data.

A college professor had students keep a diary of their social interactions for a week. Excluding family and work situations, the number of social interactions of ten minutes or longer over the week is shown in the following grouped frequency distribution. Use this information to solve Exercises 11–18.

Number of Social Interactions	Frequency
0–4	12
5–9	16
10–14	16
15–19	16
20–24	10
25–29	11
30–34	4
35–39	3
40–44	3
45–49	3

Source: Society for Personality and Social Psychology

11. Identify the lower class limit for each class.

12. Identify the upper class limit for each class.

13. What is the class width?

14. How many students were involved in this study?

15. How many students had at least 30 social interactions for the week?

16. How many students had at most 14 social interactions for the week?

17. Among the classes with the greatest frequency, which class has the least number of social interactions?

18. Among the classes with the smallest frequency, which class has the least number of social interactions?

19. As of 2003, the following are the ages, in chronological order, at which U.S. presidents were inaugurated:

57, 61, 57, 57, 58, 57, 61, 54, 68, 51, 49, 64, 50, 48, 65, 52, 56, 46, 54, 49, 50, 47, 55, 55, 54, 42, 51, 56, 55, 51, 54, 51, 60, 62, 43, 55, 56, 61, 52, 69, 64, 46, 54

Source: Time Almanac, 2003

Construct a grouped frequency distribution for the data. Use 41–45 for the first class and use the same width for each subsequent class.

20. The IQ scores of 70 students enrolled in a liberal arts course at a college are as follows:

102, 100, 103, 86, 120, 117, 111, 101, 93, 97, 99, 95, 95, 104, 104, 105, 106, 109, 109, 89, 94, 95, 99, 99, 103, 104, 105, 109, 110, 114, 124, 123, 118, 117, 116, 110, 114, 114, 96, 99, 103, 103, 104, 107, 107, 110, 111, 112, 113, 117, 115, 116, 100, 104, 102, 94, 93, 93, 96, 96, 111, 116, 107, 109, 105, 106, 97, 106, 107, 108.

Construct a grouped frequency distribution for the data. Use 85–89 for the first class and use the same width for each subsequent class.

21. Construct a histogram and a frequency polygon for the data involving stress ratings in Exercises 5–8.

22. Construct a histogram and a frequency polygon for the data in Exercise 9.

23. Construct a histogram and a frequency polygon for the data in Exercise 10.

24. The histogram shows the distribution of starting salaries (rounded to the nearest thousand dollars) for college graduates based on a random sample of recent graduates.

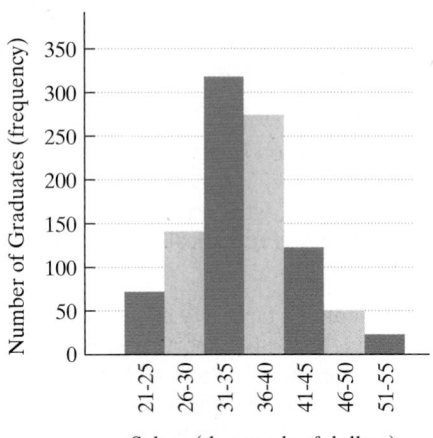

Starting Salaries of Recent College Graduates

Salary (thousands of dollars)

Which one of the following is true according to the graph?

a. The graph is based on a sample of approximately 500 recent college graduates.

b. More college graduates had starting salaries in the $41,000–$45,000 range than in the $26,000–$30,000 range.

c. If the sample is truly representative, then for a group of 400 college graduates, we can expect about 28 of them to have starting salaries in the $21,000–$25,000 range.

d. The percentage of starting salaries falling above those shown by any rectangular bar is equal to the percentage of starting salaries falling below that bar.

25. The frequency polygon shows a distribution of IQ scores. Which one of the following is true based upon the graph?

a. The graph is based on a sample of approximately 50 people.

b. More people had an IQ score of 100 than any other IQ score, and as the deviation from 100 increases or decreases, the scores fall off in a symmetrical manner.

c. More people had an IQ score of 110 than a score of 90.

d. The percentage of scores above any IQ score is equal to the percentage of scores below that score.

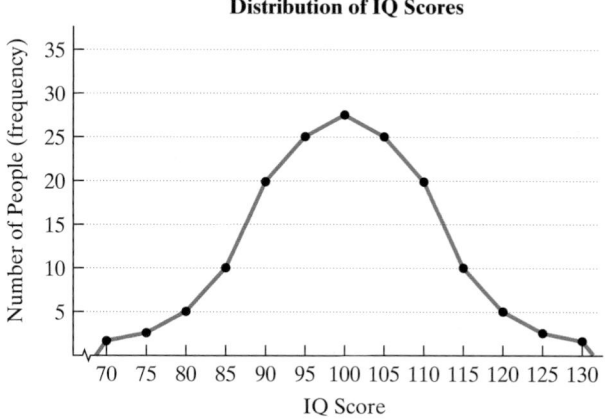

Distribution of IQ Scores

26. Construct a stem-and-leaf plot for the data in Exercise 19 showing the ages at which U.S. presidents were inaugurated.

27. A random sample of 40 college professors is selected from all professors at a university. The following list gives their ages:

63, 48, 42, 42, 38, 59, 41, 44, 45, 28, 54, 62, 51, 44, 63,

66, 59, 46, 51, 28, 37, 66, 42, 40, 30, 31, 48, 32, 29, 42,

63, 37, 36, 47, 25, 34, 49, 30, 35, 50.

Construct a stem-and-leaf plot for the data. What does the shape of the display reveal about the ages of the professors?

28. In "Ages of Oscar-Winning Best Actors and Actresses" (*Mathematics Teacher* magazine) by Richard Brown and Gretchen Davis, the stem-and-leaf plots shown compare the ages of actors and actresses for 30 winners of the Oscar at the time they won the award.

Actors	Stems	Actresses
	2	146667
98753221	3	00113344455778
88767543322100	4	11129
6651	5	
210	6	011
6	7	4
	8	0

a. What is the age of the youngest actor to win an Oscar?

b. What is the age difference between the oldest and the youngest actress to win an Oscar?

c. What is the oldest age shared by two actors to win an Oscar?

d. What differences do you observe between the two stem-and-leaf plots? What explanations can you offer for these differences?

29. A study by the American Association of University Women concluded that "Only 29% of high school girls are happy with themselves, compared to 66% of elementary school girls." The actual data involved 3000 high school girls, 29% of whom responded "always true" to the statement "I am happy the way I am." Most answered "sort of true" or "sometimes true." Do you think the study's conclusion is accurate or misleading? Explain your answer.

30. The circle graph shows the results of a study of U.S. children living in poverty (*Source:* National Center for Children in Poverty, cited in *Population Today*, 1995). The researchers titled this graph "Majority of Children in Poverty Live with Parents Who Work." Do you think this title is accurate or misleading? Explain your answer.

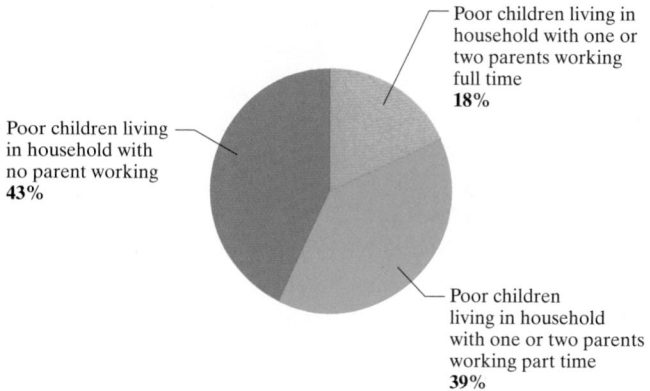

31. The graph shows world population, in billions, from 2000 B.C. through 2000. Describe what is misleading about the scale used along the horizontal axis. Does this cause the rapid increase in world population in the twentieth century to appear more or less dramatic?

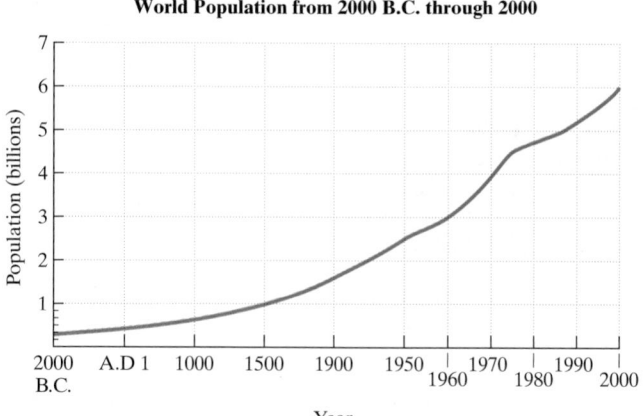

World Population from 2000 B.C. through 2000

32. The graph shown was cited in *The Visual Display of Quantitative Information* by Edward Tufte. Describe what is misleading about the size of the barrels. (*Note:* The abbreviation bbl. stands for barrel.)

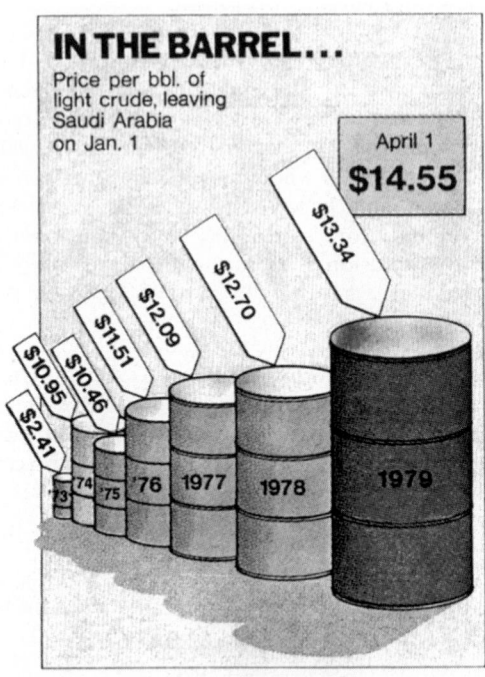

Source: Time, April 9, 1979, p. 57; © 1979 Time Inc. Reprinted by permission.

Writing in Mathematics

33. What is a population? What is a sample?

34. Describe what is meant by a random sample.

35. Suppose you are interested in whether or not the students at your college would favor a grading system in which students may receive final grades of A+, A, A−, B+, B, B−, C+, C, C−, and so on. Describe how you might obtain a random sample of 100 students from the entire student population.

36. For Exercise 35, would questioning every fifth student as he or she is leaving the campus library until 100 students are interviewed be a good way to obtain a random sample? Explain your answer.

37. What is a frequency distribution?

38. What is a histogram?

39. What is a frequency polygon?

40. Describe how to construct a frequency polygon from a histogram.

41. Describe how to construct a stem-and-leaf plot from a set of data.

42. Describe two ways that graphs can be misleading.

Critical Thinking Exercise

43. Construct a grouped frequency distribution for the following data, showing the length, in miles, of the 25 longest rivers in the United States. Use five classes that have the same width.

2540	2340	1980	1900	1900
1460	1450	1420	1310	1290
1280	1240	1040	990	926
906	886	862	800	774
743	724	692	659	649

Source: U.S. Department of the Interior

Group Exercises

44. The classic book on distortion using statistics is *How to Lie with Statistics* by Darrell Huff. This activity is designed for five people. Each person should select two chapters from Huff's book and then present to the class the common methods of statistical manipulation and distortion that Huff discusses.

45. Each group member should find one example of a graph that presents data with integrity and one example of a graph that is misleading. Use newspapers, magazines, the Internet, books, and so forth. Once graphs have been collected, each member should share his or her graphs with the entire group. Be sure to explain why one graph depicts quantitative data in a forthright manner and how the other graph deliberately misleads the viewer.

SECTION 12.2   MEASURES OF CENTRAL TENDENCY

OBJECTIVES

1. Determine the mean for a data set.

2. Determine the median for a data set.

3. Determine the mode for a data set.

4. Determine the midrange for a data set.

© NAS. Reprinted with permission of North America Syndicate.

Singers Ricky Martin and Jennifer Lopez, television journalist Soledad O'Brien, boxer Oscar De La Hoya, Internet entrepreneur Carlos Cardona, actor John Leguizamo, and political organizer Luigi Crespo are members of a generation of young Hispanics who are changing the way America looks, feels, thinks, eats, dances,

Median Age of U.S. Whites, African Americans, and Hispanics

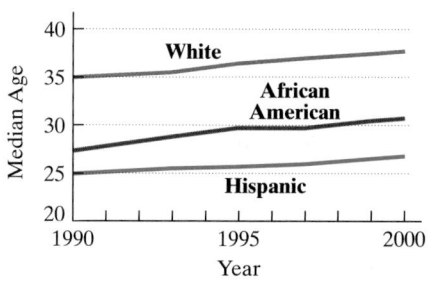

Source: U.S. Census Bureau

FIGURE 12.9

and votes. There are 31 million Hispanics in the United States, pumping $300 billion a year into the economy. By 2050, the Hispanic population is expected to reach 96 million—an increase of more than 200%. And Hispanics are younger than the rest of the nation: a third are under 18. The graph in Figure 12.9 shows the *median* age of whites, African Americans, and Americans of Hispanic origin.

The *what* age? Perhaps you have seen the word "median" in the presentation of numerical information. The median age for U.S. Hispanics in 2000 was 27: The number of Hispanics who were older than 27 is the same as the number who were younger than 27. The single number 27 represents what is "average" or "typical" of the age of Hispanic Americans. In statistics, such a value is known as a **measure of central tendency** because it is generally located toward the center of a distribution. Four such measures are discussed in this section: the mean, the median, the mode, and the midrange. Each measure of central tendency is calculated in a different way. Thus, it is better to use a specific term (mean, median, mode, or midrange) than to use the generic descriptive term "average."

1 Determine the mean for a data set.

The Mean

By far the most commonly used measure of central tendency is the *mean*. The **mean** is obtained by adding all the data items and then dividing the sum by the number of items. The Greek letter sigma, Σ, called a **symbol of summation**, is used to indicate the sum of data items. The notation Σx, read "the sum of *x*," means to add all the data items in a given data set. We can use this symbol to give a formula for calculating the mean.

> **THE MEAN**
> The **mean** is the sum of the data items divided by the number of items.
>
> $$\text{Mean} = \frac{\Sigma x}{n}$$
>
> where Σx represents the sum of all the data items and *n* represents the number of items.

EXAMPLE 1   CALCULATING THE MEAN

Table 12.4 shows the ten youngest male singers in the United States to have a number 1 single. Find the mean age of these male singers at the time of their number 1 single.

TABLE 12.4 YOUNGEST U.S. MALE SINGERS TO HAVE A NUMBER 1 SINGLE

Artist/Year	Title	Age
Stevie Wonder, 1963	*Fingertips*	13
Donny Osmond, 1971	*Go Away Little Girl*	13
Michael Jackson, 1972	*Ben*	14
Laurie London, 1958	*He's Got the Whole World in His Hands*	14
Paul Anka, 1957	*Diana*	16
Brian Hyland, 1960	*Itsy Bitsy Teenie Weenie Yellow Polkadot Bikini*	16
Shaun Cassidy, 1977	*Da Doo Ron Ron*	17
Bobby Vee, 1961	*Take Good Care of My Baby*	18
Usher, 1998	*Nice & Slow*	19
Andy Gibb, 1977	*I Just Want to Be Your Everything*	19

Source: Russell Ash, *The Top 10 of Everything*

BLITZER BONUS

The Mean American Guy

According to researchers, "Robert" is 31 years old, 5 feet 10 inches, 172 pounds, commutes to work, works 6.1 hours daily, and sleeps 7.7 hours.

Solution   We find the mean by adding the ages and dividing this sum by 10, the number of data items.

$$\text{Mean} = \frac{\Sigma x}{n} = \frac{13 + 13 + 14 + 14 + 16 + 16 + 17 + 18 + 19 + 19}{10} = \frac{159}{10} = 15.9$$

The mean age of the ten youngest singers to have a number 1 single is 15.9.

One and only one mean can be calculated for any group of numerical data. The mean may or many not be one of the actual data items. In Example 1, the mean was 15.9, although no data item is 15.9.

Check 1 Point   Find the mean for each group of data items:

    a. 10, 20, 30, 40, 50

    b. 3, 10, 10, 10, 117.

In Example 1, some of the data items were identical. We can use multiplication when computing the mean for these identical items:

$$\text{Mean} = \frac{13 + 13 + 14 + 14 + 16 + 16 + 17 + 18 + 19 + 19}{10}$$

$$= \frac{13 \cdot 2 + 14 \cdot 2 + 16 \cdot 2 + 17 \cdot 1 + 18 \cdot 1 + 19 \cdot 2}{10}$$

The data values 13, 14, 16, and 19 each have a frequency of 2.

When many data values occur more than once and a frequency distribution is used to organize the data, we can use the following formula to calculate the mean:

CALCULATING THE MEAN FOR A FREQUENCY DISTRIBUTION

$$\text{Mean} = \frac{\Sigma x f}{n}$$

where

    x represents each data value.

    f represents the frequency of that data value.

    $\Sigma x f$ represents the sum of all the products obtained by multiplying each data value by its frequency.

    n represents the *total frequency* of the distribution.

TABLE 12.5 STUDENTS' STRESS-LEVEL RATINGS

Stress Rating x	Frequency f
0	2
1	1
2	3
3	12
4	16
5	18
6	13
7	31
8	26
9	15
10	14

Source: Journal of Personality and Social Psychology, 69, 1102–1112.

EXAMPLE 2 CALCULATING THE MEAN FOR A FREQUENCY DISTRIBUTION

In the previous exercise set, we mentioned a questionnaire given to students in an introductory statistics class during the first week of the course. One question asked, "How stressed have you been in the last $2\frac{1}{2}$ weeks, on a scale of 0 to 10, with 0 being not at all stressed and 10 being as stressed as possible?" Table 12.5 shows the students' responses. Use this frequency distribution to find the mean of the stress-level ratings.

Solution   We use the formula

$$\text{Mean} = \frac{\Sigma x f}{n}.$$

First, we must find xf, obtained by multiplying each data value, x, by its frequency, f. Then, we need to find the sum of these products, Σxf. We can use the frequency distribution to organize these computations. Add a third column in which each data value is multiplied by its frequency. This column, shown below, is headed xf. Then, find the sum of the values, Σxf, in this column.

x	f	xf
0	2	$0 \cdot 2 = 0$
1	1	$1 \cdot 1 = 1$
2	3	$2 \cdot 3 = 6$
3	12	$3 \cdot 12 = 36$
4	16	$4 \cdot 16 = 64$
5	18	$5 \cdot 18 = 90$
6	13	$6 \cdot 13 = 78$
7	31	$7 \cdot 31 = 217$
8	26	$8 \cdot 26 = 208$
9	15	$9 \cdot 15 = 135$
10	14	$10 \cdot 14 = 140$

Totals: $n = 151$ $\Sigma xf = 975$

> This value, the sum of the numbers in the second column, is the total frequency of the distribution.

> Σxf is the sum of the numbers in the third column.

Now, substitute these values into the formula for the mean. Remember that n is the *total frequency* of the distribution, or 151.

$$\text{Mean} = \frac{\Sigma xf}{n} = \frac{975}{151} \approx 6.46$$

The mean of the 0 to 10 stress-level ratings is approximately 6.46. Notice that the mean is greater than 5, the middle of the 0 to 10 scale.

Check Point 2   Find the mean for the data items in the frequency distribution. (In order to save space, we've written the frequency distribution horizontally.)

Score, x	30	33	40	50
Frequency, f	3	4	4	1

The Median

② Determine the median for a data set.

We have seen that the median age for U.S. Hispanics in 2000 was 27. To find this value, researchers begin with a random sample of all U.S. Hispanics. The data items—that is, the ages for sampled Hispanics—are arranged in order, from youngest to oldest. The **median** age is the data item in the middle of this set of ranked, or ordered, data.

> **THE MEDIAN**
>
> To find the **median** of a group of data items,
>
> 1. Arrange the data items in order, from smallest to largest.
> 2. If the number of data items is odd, the median is the data item in the middle of the list.
> 3. If the number of data items is even, the median is the mean of the two middle data items.

Colombian-born actor John Leguizamo is a member of a generation of young Hispanics shaping American pop culture.

EXAMPLE 3   FINDING THE MEDIAN

Find the median for each of the following groups of data:

a. 84, 90, 98, 95, 88

b. 68, 74, 7, 13, 15, 25, 28, 59, 34, 47.

Solution

a. Arrange the data items in order, from smallest to largest. The number of data items in the list, five, is odd. Thus, the median is the middle number.

84, 88, 90, 95, 98

Middle data
item

The median is 90. Notice that two data items lie above 90 and two data items lie below 90.

b. Arrange the data items in order, from smallest to largest. The number of data items in the list, ten, is even. Thus, the median is the mean of the two middle data items.

7, 13, 15, 25, 28, 34, 47, 59, 68, 74

Middle data items
are 28 and 34.

$$\text{Median} = \frac{28 + 34}{2} = \frac{62}{2} = 31$$

The median is 31. Five data items lie above 31 and five data items lie below 31.

7   13   15   25   28 | 34   47   59   68   74

Five data items lie below 31.   Five data items lie above 31.

Median is 31.

 **Check Point 3**   Find the median for each of the following groups of data:

a. 28, 42, 40, 25, 35

b. 72, 61, 85, 93, 79, 87.

If a relatively long list of data items is arranged in order, it may be difficult to identify the item or items in the middle. In cases like this, the median can be found by determining its position in the list of items.

POSITION OF THE MEDIAN

If n data items are arranged in order, from smallest to largest, the median is the value in the

$$\frac{n + 1}{2}$$

position.

EXAMPLE 4 FINDING THE MEDIAN USING THE POSITION FORMULA

Listed below are the points scored per season by the 13 top point scorers in the National Football League. Find the median points scored per season for the top 13 scorers.

$$144, 144, 145, 145, 145, 146, 147, 149, 150, 155, 161, 164, 176$$

Solution The data items are arranged from smallest to largest. There are 13 data items, so $n = 13$. The median is the value in the

$$\frac{n + 1}{2} \text{ position} = \frac{13 + 1}{2} \text{ position} = \frac{14}{2} \text{ position} = \text{seventh position}.$$

We find the median by selecting the data item in the seventh position.

Position 3	Position 4		Position 7

$$144, \quad 144, \quad 145, \quad 145, \quad 145, \quad 146, \quad 147, \quad 149, \quad 150, \quad 155, \quad 161, \quad 164, \quad 176$$

Position 1	Position 2		Position 5	Position 6

The median is 147. Notice that six data items lie above 147 and six data items lie below it. The median points scored per season for the top 13 scorers in the National Football League is 147.

 Check Point 4   Find the median for the following group of data items:

$$1, 2, 2, 2, 3, 3, 3, 3, 3, 5, 6, 7, 7, 10, 11, 13, 19, 24, 26.$$

EXAMPLE 5 FINDING THE MEDIAN USING THE POSITION FORMULA

Listed below are the salaries, to the nearest tenth of a million dollars, for the 14 members of the Los Angeles Lakers basketball team in the year 2000. Find the team's median salary.

SALARIES FOR THE LOS ANGELES LAKERS

John Celestand	$0.3 million	Travis Knight	$3.1 million
Tyronne Lue	$0.7 million	Rick Fox	$4.2 million
John Salley	$0.8 million	Derek Fisher	$4.3 million
Brian Shaw	$1.0 million	Glen Rice	$4.5 million
Devean George	$1.0 million	Robert Horry	$5.0 million
A. C. Green	$2.0 million	Kobe Bryant	$11.8 million
Ron Harper	$2.1 million	Shaquille O'Neal	$17.1 million

Solution The data items are arranged from smallest to largest. There are 14 data items, so $n = 14$. The median is the value in the

$$\frac{n + 1}{2} \text{ position} = \frac{14 + 1}{2} \text{ position} = \frac{15}{2} \text{ position} = 7.5 \text{ position}.$$

This means that the median is the mean of the data items in positions 7 and 8.

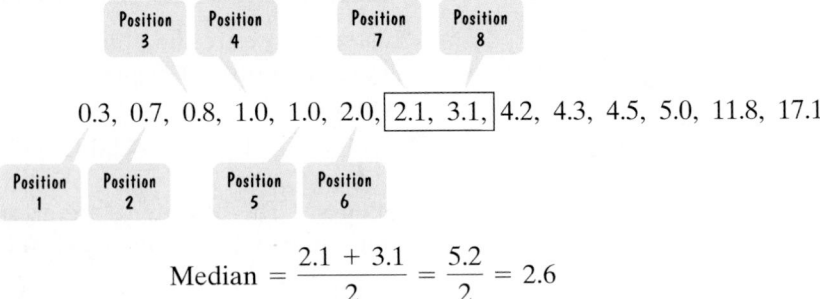

$$\text{Median} = \frac{2.1 + 3.1}{2} = \frac{5.2}{2} = 2.6$$

The team's median salary is $2.6 million.

Check 5 Point Before Barry Bonds, Mark McGwire, and Sammy Sosa, New York Yankee outfielder Roger Maris held the record for the most home runs in one season. Listed below are Maris's home run totals for each of his ten years in the American League. Find Maris's median number of home runs.

$$8, 13, 14, 16, 23, 26, 28, 33, 39, 61$$

When individual data items are listed from smallest to largest, you can find the median by identifying the item or items in the middle or by using the $\frac{n+1}{2}$ formula for its position. However, the formula for the position of the median is useful when data items are organized in a frequency distribution.

EXAMPLE 6 | FINDING THE MEDIAN FOR A FREQUENCY DISTRIBUTION

The frequency distribution for the stress-level ratings of 151 students is repeated below using a horizontal format. Find the median stress-level rating.

Stress rating

Number of college students

x	0	1	2	3	4	5	6	7	8	9	10	
f	2	1	3	12	16	18	13	31	26	15	14	Total: $n = 151$

Solution There are 151 data items, so $n = 151$. The median is the value in the

$$\frac{n+1}{2} \text{ position} = \frac{151+1}{2} \text{ position} = \frac{152}{2} \text{ position} = 76\text{th position}.$$

We find the median by selecting the data item in the 76th position. The frequency distribution indicates that the data items begin with

$$0, 0, 1, 2, 2, 2, \ldots.$$

We can write them all out and then select the median, the 76th data item. A more efficient way to proceed is to count down the frequency column in the distribution until we identify the 76th data item:

x	f
0	2
1	1
2	3
3	12
4	16
5	18
6	13
7	31
8	26
9	15
10	14

We count down the frequency column.

1, 2

3

4, 5, 6

7, 8, 9, 10, 11, 12, 13, 14, 15, 16, 17, 18

19, 20, 21, 22, 23, 24, 25, 26, 27, 28, 29, 30, 31, 32, 33, 34

35, 36, 37, 38, 39, 40, 41, 42, 43, 44, 45, 46, 47, 48, 49, 50, 51, 52,

53, 54, 55, 56, 57, 58, 59, 60, 61, 62, 63, 64, 65

66, 67, 68, 69, 70, 71, 72, 73, 74, 75, 76

Stop counting. We've reached the 76th data item.

The 76th data item is 7. The median stress-level rating is 7.

Check Point 6   Find the median for the following frequency distribution.

Age at presidential inauguration

Number of U.S. presidents assuming office in the 20th century with the given age

x	42	43	46	51	52	54	55	56	60	61	64	69
f	1	1	1	3	1	2	2	2	1	2	1	1

Statisticians generally use the median, rather than the mean, when reporting income. Why? Our next example will help to answer this question.

EXAMPLE 7 COMPARING THE MEDIAN AND THE MEAN

Five employees in the assembly section of a television manufacturing company earn salaries of $19,700, $20,400, $21,500, $22,600, and $23,000 annually. The section manager has an annual salary of $95,000.

a. Find the median annual salary for the six people.

b. Find the mean annual salary for the six people.

Solution

a. To compute the median, first arrange the salaries in order:

$19,700, $20,400, $21,500, $22,600, $23,000, $95,000.

Because the list contains an even number of data items, six, the median is the mean of the two middle items.

$$\text{Median} = \frac{\$21,500 + \$22,600}{2} = \frac{\$44,100}{2} = \$22,050$$

The median annual salary is $22,050.

b. We find the mean annual salary by adding the six annual salaries and dividing by 6.

$$\text{Mean} = \frac{\$19,700 + \$20,400 + \$21,500 + \$22,600 + \$23,000 + \$95,000}{6}$$

$$= \frac{\$202,200}{6} = \$33,700$$

The mean annual salary is $33,700.

In Example 7, the median annual salary is $22,050 and the mean annual salary is $33,700. Why such a big difference between these two measures of central tendency? The relatively high annual salary of the section manager, $95,000, pulls the mean salary to a value considerably higher than the median salary. When one or more data items are much greater than the other items, these extreme values can greatly influence the mean. In cases like this, the median is often more representative of the data.

This is why the median, rather than the mean, is used to summarize the incomes, by gender and race, shown in Figure 12.10. Because no one can earn less than $0, the distribution of income must come to an end at $0 for each of these six groups. By contrast, there is no upper limit on income on the high side. In the United States, the wealthiest 5% of the population earn about 21% of the total income. The relatively few people with very high annual incomes tend to pull the mean income to a value considerably greater than the median income. Reporting mean incomes in Figure 12.10 would inflate the numbers shown, making them nonrepresentative of the millions of workers in each of the six groups.

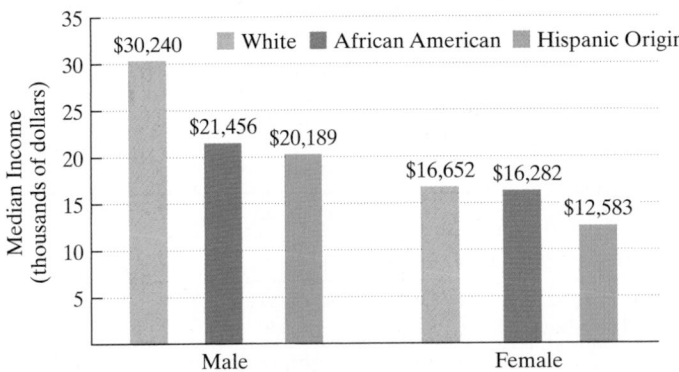

U.S. Median Income in 2001, by Gender and Race

Source: U.S. Census Bureau

FIGURE 12.10

 Check 7 Point In Example 5 on page 649, we found the median salary in the year 2000 for the Los Angeles Lakers. Use the 14 data items shown to find the team's mean salary. How does the mean compare to the $2.6 million median? Describe why one of the measures of central tendency is so much greater than the other.

The Mode

3 | Determine the mode for a data set.

Let's take one final look at the frequency distribution for the stress-level ratings of 151 college students.

Stress rating x	0	1	2	3	4	5	6	7	8	9	10
Number of college students f	2	1	3	12	16	18	13	31	26	15	14

7 is the stress rating with the greatest frequency.

The data value that occurs most often in this distribution is 7, the stress rating for 31 of the 151 students. We call 7 the *mode* of this distribution.

> ### THE MODE
>
> The **mode** is the data value that occurs most often in a data set. If no data items are repeated, then the data set has no mode. If more than one data value has the highest frequency, then each of these data values is a mode.

EXAMPLE 8 FINDING THE MODE

Find the mode for the following group of data:

$$7, 2, 4, 7, 8, 10.$$

Solution   The number 7 occurs more often than any other. Therefore, 7 is the mode.

 Check 8 Point   Find the mode for the following group of data:
$$8, 6, 2, 4, 6, 8, 10, 8.$$

Be aware that a data set might not have a mode. For example, no data item in 2, 1, 4, 5, 3 is repeated, so this data group has no mode. By contrast, 3, 3, 4, 5, 6, 6 has two data values with the highest frequency, namely 3 and 6. Each of these data values is a mode and the data set is said to be **bimodal**.

The Midrange

 4 Determine the midrange for a data set.

Table 12.6 shows the ten hottest cities in the United States. Because temperature is constantly changing, you might wonder how the mean temperatures shown in the table are obtained.

First, we need to find a representative daily temperature. This is obtained by adding the lowest and highest temperatures for the day and then dividing this sum by 2. Next, we take the representative daily temperatures for all 365 days, add them, and divide the sum by 365. These are the mean temperatures that appear in Table 12.6.

Representative daily temperature,

$$\frac{\text{lowest daily temperature} + \text{highest daily temperature}}{2},$$

is an example of a measure of central tendency called the *midrange*.

TABLE 12.6 TEN HOTTEST U.S. CITIES

City	Mean Temperature
Key West, FL	77.8°
Miami, FL	75.9°
West Palm Beach, FL	74.7°
Fort Myers, FL	74.4°
Yuma, AZ	74.2°
Brownsville, TX	73.8°
Phoenix, AZ	72.6°
Vero Beach, FL	72.4°
Orlando, FL	72.3°
Tampa, FL	72.3°

Source: National Oceanic and Atmospheric Administration

THE MIDRANGE

The **midrange** is found by adding the lowest and highest data values and dividing the sum by 2.

$$\text{Midrange} = \frac{\text{lowest data value} + \text{highest data value}}{2}$$

EXAMPLE 9 FINDING THE MIDRANGE

The best paid state governor is in New York, earning $179,000 annually. The worst paid is the governor of Nebraska, earning $65,000 annually. Find the midrange for annual salaries of U.S. governors.

Solution

$$\text{Midrange} = \frac{\text{lowest annual salary} + \text{highest annual salary}}{2}$$

$$= \frac{\$65,000 + \$179,000}{2} = \frac{\$244,000}{2} = \$122,000$$

The midrange for annual salaries of U.S. governors is $122,000.

We can find the mean annual salary of U.S. governors by adding up the salaries of all 50 governors and then dividing the sum by 50. It is much faster to calculate the midrange, which is often used as an estimate for the mean.

 **Check 9 Point** The best paid state attorney general is in New York, earning $151,500 annually. The worst paid is the state attorney general of Connecticut, earning $60,000 annually. Find the midrange for annual salaries of state attorneys general.

EXAMPLE 10 FINDING THE FOUR MEASURES OF CENTRAL TENDENCY

Suppose your six exam grades in a course are

$$52, 69, 75, 86, 86, \text{ and } 92.$$

Compute your final course grade (90–100 = A, 80–89 = B, 70–79 = C, 60–69 = D, below 60 = F) using the

a. mean. **b.** median. **c.** mode. **d.** midrange.

Solution

a. The mean is the sum of the data items divided by the number of items, 6.

$$\text{Mean} = \frac{52 + 69 + 75 + 86 + 86 + 92}{6} = \frac{460}{6} \approx 76.67$$

Using the mean, your final course grade is C.

b. The six data items, 52, 69, 75, 86, 86 and 92, are arranged in order. Because the number of data items is even, the median is the mean of the two middle items.

$$\text{Median} = \frac{75 + 86}{2} = \frac{161}{2} = 80.5$$

Using the median, your final course grade is B.

c. The mode is the data value that occurs most frequently. Because 86 occurs most often, the mode is 86. Using the mode, your final course grade is B.

d. The midrange is the mean of the lowest and highest data values.

$$\text{Midrange} = \frac{52 + 92}{2} = \frac{144}{2} = 72$$

Using the midrange, your final course grade is C.

 **Check 10 Point** *Consumer Reports* magazine gave the following data for the number of calories in a meat hot dog for each of 17 brands:

$$173, 191, 182, 190, 172, 147, 146, 138, 175, 136, 179,$$
$$153, 107, 195, 135, 140, 138.$$

Find the mean, median, mode, and midrange for the number of calories in a meat hot dog for the 17 brands. If necessary, round answers to the nearest tenth of a calorie.

Exercise Set 12.2

Practice Exercises

In Exercises 1–8, find the mean for each group of data items.

1. 7, 4, 3, 2, 8, 5, 1, 3

2. 11, 6, 4, 0, 2, 1, 12, 0, 0

3. 91, 95, 99, 97, 93, 95

4. 100, 100, 90, 30, 70, 100

5. 100, 40, 70, 40, 60

6. 1, 3, 5, 10, 8, 5, 6, 8

7. 1.6, 3.8, 5.0, 2.7, 4.2, 4.2, 3.2, 4.7, 3.6, 2.5, 2.5

8. 1.4, 2.1, 1.6, 3.0, 1.4, 2.2, 1.4, 9.0, 9.0, 1.8

In Exercises 9–12, find the mean for the data items in the given frequency distribution.

9.

Score x	Frequency f
1	1
2	3
3	4
4	4
5	6
6	5
7	3
8	2

10.

Score x	Frequency f
1	2
2	4
3	5
4	7
5	6
6	4
7	3

11.

Score x	Frequency f
1	1
2	1
3	2
4	5
5	7
6	9
7	8
8	6
9	4
10	3

12.

Score x	Frequency f
1	3
2	4
3	6
4	8
5	9
6	7
7	5
8	2
9	1
10	1

In Exercises 13–20, find the median for each group of data items.

13. 7, 4, 3, 2, 8, 5, 1, 3

14. 11, 6, 4, 0, 2, 1, 12, 0, 0

15. 91, 95, 99, 97, 93, 95

16. 100, 100, 90, 30, 70, 100

17. 100, 40, 70, 40, 60

18. 1, 3, 5, 10, 8, 5, 6, 8

19. 1.6, 3.8, 5.0, 2.7, 4.2, 4.2, 3.2, 4.7, 3.6, 2.5, 2.5

20. 1.4, 2.1, 1.6, 3.0, 1.4, 2.2, 1.4, 9.0, 9.0, 1.8

Find the median for the data items in the frequency distribution in

21. Exercise 9

22. Exercise 10

23. Exercise 11

24. Exercise 12

In Exercises 25–32, find the mode for each group of data items. If there is no mode, so state.

25. 7, 4, 3, 2, 8, 5, 1, 3

26. 11, 6, 4, 0, 2, 1, 12, 0, 0

27. 91, 95, 99, 97, 93, 95

28. 100, 100, 90, 30, 70, 100

29. 100, 40, 70, 40, 60

30. 1, 3, 5, 10, 8, 5, 6, 8

31. 1.6, 3.8, 5.0, 2.7, 4.2, 4.2, 3.2, 4.7, 3.6, 2.5, 2.5

32. 1.4, 2.1, 1.6, 3.0, 1.4, 2.2, 1.4, 9.0, 9.0, 1.8

Find the mode for the data items in the frequency distribution in

33. Exercise 9

34. Exercise 10

35. Exercise 11

36. Exercise 12

In Exercises 37–44, find the midrange for each group of data items.

37. 7, 4, 3, 2, 8, 5, 1, 3

38. 11, 6, 4, 0, 2, 1, 12, 0, 0

39. 91, 95, 99, 97, 93, 95

40. 100, 100, 90, 30, 70, 100

41. 100, 40, 70, 40, 60

42. 1, 3, 5, 10, 8, 5, 6, 8

43. 1.6, 3.8, 5.0, 2.7, 4.2, 4.2, 3.2, 4.7, 3.6, 2.5, 2.5

44. 1.4, 2.1, 1.6, 3.0, 1.4, 2.2, 1.4, 9.0, 9.0, 1.8

Find the midrange for the data items in the frequency distribution in

45. Exercise 9

46. Exercise 10

47. Exercise 11

48. Exercise 12

Application Exercises

Exercises 49–52 present data on a variety of topics. For each data set described in boldface, find the

    a. *mean.*

    b. *median.*

    c. *mode (or state that there is no mode).*

    d. *midrange.*

49. Ages of the Justices of the United States Supreme Court in 2003

Rehnquist (79), Stevens (83), O'Connor (73), Scalia (67), Kennedy (67), Souter (64), Thomas (55), Ginsburg (70), Breyer (65)

50. The Number of Home Runs Hit by Babe Ruth in His 15 Years with the New York Yankees (1920–1934)

54, 59, 35, 41, 46, 25, 47, 60, 54, 46, 49, 46, 41, 34, 22

51. Ten Highest-Paid Baseball Players

Player	Team	Length of Contract	Average Annual Salary in Millions of Dollars
Alex Rodriguez	Texas Rangers	2001–10	$25.2
Manny Ramirez	Boston Red Sox	2001–08	$20.0
Derek Jeter	New York Yankees	2001–10	$18.9
Barry Bonds	San Francisco Giants	2002–06	$18.0
Sammy Sosa	Chicago Cubs	2002–05	$18.0
Jason Giambi	New York Yankees	2002–08	$17.1
Jeff Bagwell	Houston Astros	2002–06	$17.0
Carlos Delgado	Toronto Blue Jays	2001–04	$17.0
Todd Helton	Colorado Rockies	2003–11	$15.7
Roger Clemens	New York Yankees	2001–02	$15.5

Source: Sports Illustrated

52. Number of Social Interactions of College Students

In Exercise Set 12.1, we presented a grouped frequency distribution showing the number of social interactions of ten minutes or longer over a one-week period for a group of college students. (These interactions excluded family and work situations.) Use the frequency distribution shown to solve this exercise. (This distribution was obtained by replacing the classes in the grouped frequency distribution previously shown with the midpoints of the classes.)

Social interactions in a week

x	2	7	12	17	22	27	32	37	42	47
f	12	16	16	16	10	11	4	3	3	3

Number of college students

The weights (to the nearest five pounds) of 40 randomly selected male college students are organized in a histogram with a superimposed frequency polygon. Use the graph to answer Exercises 53–56.

53. Find the mean weight.   **54.** Find the median weight.

55. Find the modal weight.   **56.** Find the midrange weight.

Weights of 40 Male College Students

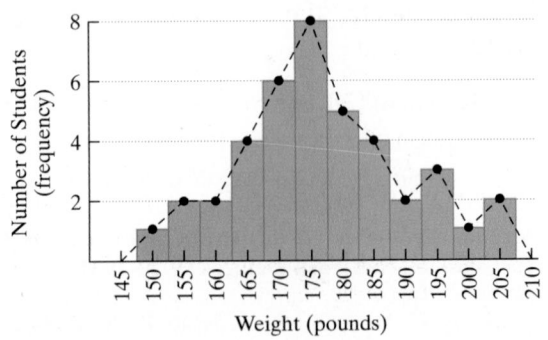

57. In Example 5 on page 649, we found the median salary in the year 2000 for the Los Angeles Lakers to be $2.6 million. Delete the salaries for Shaquille O'Neal and Kobe Bryant and find the mean for the remaining 12 data items. How does this mean compare to the team's median salary?

58. An advertisement for a speed-reading course claimed that the "average" reading speed for people completing the course was 1000 words per minute. Shown below are the actual data for the reading speeds per minute for a sample of 24 people who completed the course.

1000	900	800	1000	900	850
650	1000	1050	800	1000	850
700	750	800	850	900	950
600	1100	950	700	750	650

a. Find the mean, median, mode, and midrange. (If you prefer, first organize the data in a frequency distribution.)

b. Which measure of central tendency was given in the advertisement?

c. Which measure of central tendency is the best indicator of the "average" reading speed in this situation? Explain your answer.

59. In one common system for finding a grade-point average, or GPA,

$$A = 4, B = 3, C = 2, D = 1, F = 0.$$

The GPA is calculated by multiplying the number of credit hours for a course and the number assigned to each grade, and then adding these products. Then divide this sum by the total number of credit hours. Because each course grade is weighted according to the number of credits of the course, GPA is called a *weighted mean*. Calculate the GPA for this transcript:

Sociology: 3 cr. A; Biology: 3.5 cr. C; Music: 1 cr. B; Math: 4 cr. B; English: 3 cr. C.

Writing in Mathematics

60. What is the mean and how is it obtained?

61. What is the median and how is it obtained?

62. What is the mode and how is it obtained?

63. What is the midrange and how is it obtained?

Use the median sales prices of existing single-family homes to answer Exercises 64–66.

MEDIAN SALES PRICE OF EXISTING SINGLE-FAMILY HOMES

Year	United States	Northeast	Midwest	South	West
1999	$195,800	$249,200	$187,000	$173,500	$222,000
2000	$207,200	$277,400	$199,100	$179,300	$239,500

Source: U.S. Department of Housing and Urban Development

64. The median sales price in 2000 for the entire country was $207,200. Explain what this means.

65. Explain why median sales prices, rather than mean sales prices, are reported.

66. Without using specific numbers, write two statements that summarize the trends shown by the median sales prices.

67. The "average" income in the United States can be given by the mean or the median.

  a. Which measure would be used in anti-U.S. propaganda? Explain your answer.

  b. Which measure would be used in pro-U.S. propaganda? Explain your answer.

68. In a class of 40 students, 21 have examination scores of 77%. Which measure or measures of central tendency can you immediately determine? Explain your answer.

69. You read an article that states "of the 411 players in the National Basketball Association, only 138 make more than the average salary of $3.12 million." Is $3.12 million the mean or the median salary? Explain, your answer.

70. A student's parents promise to pay for next semester's tuition if an A average is earned in chemistry. With examination grades of 97%, 97%, 75%, 70%, and 55%, the student reports that an A average has been earned. Which measure of central tendency is the student reporting as the average? How is this student misrepresenting the course performance with statistics?

71. According to the National Oceanic and Atmospheric Administration, the coldest city in the United States is International Falls, Minnesota, with a mean temperature of 36.8°. Explain how this mean is obtained.

Critical Thinking Exercises

72. Give an example of a set of six examination grades (from 0 to 100) with each of the following characteristics:

  a. The mean and the median have the same value, but the mode has a different value.

  b. The mean and the mode have the same value, but the median has a different value.

  c. The mean is greater than the median.

  d. The mode is greater than the mean.

  e. The mean, median, and mode have the same value.

  f. The mean and mode have values of 72.

73. On an examination given to 30 students, no student scored below the mean. Describe how this occurred.

Group Exercises

74. Select a characteristic, such as shoe size or height, for which each member of the group can provide a number. Choose a characteristic of genuine interest to the group. For this characteristic, organize the data collected into a frequency distribution and a graph. Compute the mean, median, mode and midrange. Discuss any differences among these values. What happens if the group is divided (men and women, or people under a certain age and people over a certain age) and these measures of central tendency are computed for each of the subgroups? Attempt to use measures of central tendency to discover something interesting about the entire group or the subgroups.

75. A recent book on spotting bad statistics and learning to think critically about these influential numbers is *Damn Lies and Statistics* by Joel Best (University of California Press, 2001). This activity is designed for six people. Each person should select one chapter from Best's book. The group report should include examples of the use, misuse, and abuse of statistical information. Explain exactly how and why bad statistics emerge, spread, and come to shape policy debates. What specific ways does Best recommend to detect bad statistics?

SECTION 12.3   MEASURES OF DISPERSION

OBJECTIVES

1. Determine the range for a data set.

2. Determine the standard deviation for a data set.

When you think of Houston, Texas and Honolulu, Hawaii, do balmy temperatures come to mind? Both cities have a mean temperature of 75°. However, the mean temperature does not tell the whole story. The temperature in Houston differs seasonally from a low of about 40° in January to a high of close to 100° in July and August. By contrast, Honolulu's temperature varies less throughout the year, usually ranging between 60° and 90°.

Measures of dispersion are used to describe the spread of data items in a data set. Two of the most common measures of dispersion, the *range* and the *standard deviation*, are discussed in this section.

1 Determine the range for a data set.

The Range

A quick but rough measure of dispersion is the **range**, the difference between the highest and lowest data values in a data set. For example, if Houston's hottest annual temperature is 103° and its coldest annual temperature is 33°, the range in temperature is

$$103° - 33°, \quad \text{or} \quad 70°.$$

If Honolulu's hottest day is 89° and its coldest day 61°, the range in temperature is

$$89° - 61°, \quad \text{or} \quad 28°.$$

THE RANGE

The **range**, the difference between the highest and lowest data values in a data set, indicates the total spread of the data.

Range = highest data value − lowest data value

EXAMPLE 1 COMPUTING THE RANGE

The average person living in the United States watches about 1551 hours of television per year. That's equal to nearly 65 straight days, or 18% of one year. Figure 12.11 shows the five countries with the most televisions per 1000 people. Find the range of televisions per 1000 people for the five countries with the most televisions per capita.

Solution

$$\text{Range} = \text{highest data value} - \text{lowest data value}$$
$$= 806 - 597 = 209$$

The range is 209 televisions per 1000 people.

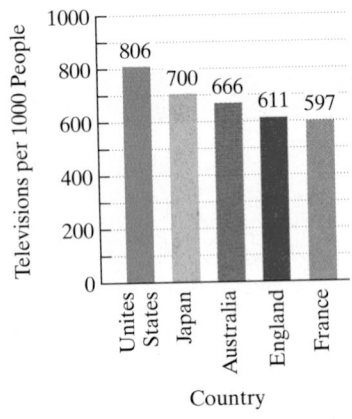

Countries With the Most Televisions per Capita

Source: International Telecommunications Union

FIGURE 12.11

Find the range for the following group of data items:

$$4, 2, 11, 7.$$

The Standard Deviation

A second measure of dispersion, and one that is dependent on *all* of the data items, is called the **standard deviation**. The standard deviation is found by determining how much each data item differs from the mean.

In order to compute the standard deviation, it is necessary to find by how much each data item deviates from the mean. First compute the mean. Then subtract the mean from each data item. Example 2 shows how this is done. In Example 3, we will use this skill to actually find the standard deviation.

EXAMPLE 2 PREPARING TO FIND THE STANDARD DEVIATION; FINDING DEVIATIONS FROM THE MEAN

Find the deviations from the mean for the five data items 806, 700, 666, 611, and 597, shown in Figure 12.11.

Solution First, calculate the mean.

$$\text{Mean} = \frac{\Sigma x}{n} = \frac{806 + 700 + 666 + 611 + 597}{5} = \frac{3380}{5} = 676$$

The mean for the five countries with the most televisions per capita is 676 televisions per 1000 people. Now, let's find out by how much each of the five data items in Figure 12.11 differs from 676, the mean. For the United States, with 806 televisions per 1000 people, the computation is shown as follows:

$$\text{Deviation from mean} = \text{data item} - \text{mean}$$

$$= 806 - 676 = 130.$$

This indicates that the number of televisions per 1000 people in the United States exceeds the mean by 130.

The computation for France, with 597 televisions per 1000 people, is given by

$$\text{Deviation from mean} = \text{data item} - \text{mean}$$

$$= 597 - 676 = -79.$$

This indicates that the number of televisions per 1000 people in France is 79 below the mean.

The deviations from the mean for each of the five given data items are shown in Table 12.7.

TABLE 12.7 DEVIATIONS FROM THE MEAN

Data item	Deviation: Data item − mean
806	806 − 676 = 130
700	700 − 676 = 24
666	666 − 676 = −10
611	611 − 676 = −65
597	597 − 676 = −79

 Check Point 2

Compute the mean for the following group of data items:

$$2, 4, 7, 11.$$

Then find the deviations from the mean for the four data items. Organize your work in table form just like Table 12.7. Keep track of these computations. You will be using them in Check Point 3.

The sum of the deviations for a set of data is always zero. For the deviations shown in Table 12.7,

$$130 + 24 + (-10) + (-65) + (-79) = 154 + (-154) = 0.$$

This shows that we cannot find a measure of dispersion by finding the mean of the deviations, because this value is always zero. However, a kind of average of the deviations from the mean, called the **standard deviation**, can be computed. We do so by squaring each deviation and later introducing a square root in the computation. On the next page are the details on how to find the standard deviation for a set of data.

Determine the standard deviation for a data set.

COMPUTING THE STANDARD DEVIATION FOR A DATA SET

1. Find the mean of the data items.
2. Find the deviation of each data item from the mean:

$$\text{data item} - \text{mean}.$$

3. Square each deviation:

$$(\text{data item} - \text{mean})^2.$$

4. Sum the squared deviations:

$$\Sigma(\text{data item} - \text{mean})^2.$$

5. Divide the sum in step 4 by $n - 1$, where n represents the number of data items:

$$\frac{\Sigma(\text{data item} - \text{mean})^2}{n - 1}.$$

6. Take the square root of the quotient in step 5. This value is the standard deviation for the data set.

$$\text{Standard deviation} = \sqrt{\frac{\Sigma(\text{data item} - \text{mean})^2}{n - 1}}$$

The computation of the standard deviation can be organized using a table with three columns:

Data item	Deviation: Data item − mean	(Deviation)^2: (Data item − mean)^2

In Example 2, we worked out the first two columns of such a table. Let's continue working with the data for the countries with the most televisions per capita and compute the standard deviation.

EXAMPLE 3 COMPUTING THE STANDARD DEVIATION

Figure 12.11, showing the five countries with the most televisions per capita, is repeated in the margin. Find the standard deviation for the number of televisions per 1000 people.

Solution

STEP 1. Find the mean. From our work in Example 2, the mean is 676.

STEP 2. Find the deviation of each data item from the mean: data item − mean. This, too, was done in Example 2 for each of the five data items.

STEP 3. Square each deviation: (data item − mean)^2. We square each of the numbers in the (data item − mean) column, shown in Table 12.8. Notice that squaring the difference always results in a positive number.

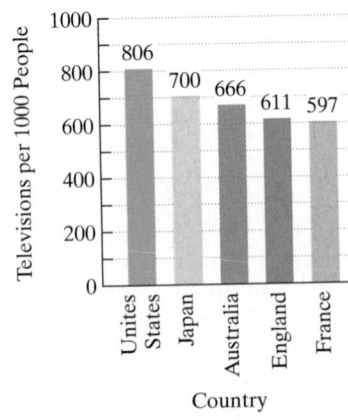

Countries With the Most Televisions per Capita

FIGURE 12.11, repeated

TECHNOLOGY

TABLE 12.8 COMPUTING THE STANDARD DEVIATION

Data item	Deviation: Data item − mean	(Deviation)²: (Data item − mean)²
806	806 − 676 = 130	$130^2 = 130 \cdot 130 = 16{,}900$
700	700 − 676 = 24	$24^2 = 24 \cdot 24 = 576$
666	666 − 676 = −10	$(-10)^2 = (-10)(-10) = 100$
611	611 − 676 = −65	$(-65)^2 = (-65)(-65) = 4225$
597	597 − 676 = −79	$(-79)^2 = (-79)(-79) = 6241$
Totals:	0	28,042

The sum of the deviations for a set of data is always zero.

Adding the five numbers in the third column gives the sum of the squared deviations:
$\Sigma(\text{data item} - \text{mean})^2$.

STEP 4. Sum the squared deviations: $\Sigma(\text{data item} - \text{mean})^2$. This step is shown in Table 12.8. The squares in the third column were added.

STEP 5. Divide the sum in step 4 by $n - 1$, where n represents the number of data items. The number of data items is 5 so we divide by 4.

$$\frac{\Sigma(\text{data item} - \text{mean})^2}{n - 1} = \frac{28{,}042}{5 - 1} = \frac{28{,}042}{4} = 7010.5$$

STEP 6. The standard deviation is the square root of the quotient in step 5.

$$\text{Standard deviation} = \sqrt{\frac{\Sigma(\text{data item} - \text{mean})^2}{n - 1}} = \sqrt{7010.5} \approx 83.73$$

The standard deviation for the number of televisions per 1000 people for the five countries with the most televisions per capita is approximately 83.73.

Check Point 3 Find the standard deviation for the group of data items in Check Point 2 on page 659. Round to two decimal places.

Example 4 illustrates that as the spread of data items increases, the standard deviation gets larger.

EXAMPLE 4 COMPUTING THE STANDARD DEVIATION

Find the standard deviation of the data items in each of the samples shown below.

Sample A	Sample B
17, 18, 19, 20, 21, 22, 23	5, 10, 15, 20, 25, 30, 35

Solution Begin by finding the mean for each sample.
Sample A:

$$\text{Mean} = \frac{17 + 18 + 19 + 20 + 21 + 22 + 23}{7} = \frac{140}{7} = 20$$

Data items, repeated

Sample A

17, 18, 19, 20, 21, 22, 23

Sample B

5, 10, 15, 20, 25, 30, 35

Sample B:

$$\text{Mean} = \frac{5 + 10 + 15 + 20 + 25 + 30 + 35}{7} = \frac{140}{7} = 20$$

Although both samples have the same mean, the data items in sample B are more spread out. Thus, we would expect sample B to have the greater standard deviation. The computation of the standard deviation requires that we find $\Sigma(\text{data item} - \text{mean})^2$, shown in Table 12.9.

TABLE 12.9 COMPUTING STANDARD DEVIATIONS FOR TWO SAMPLES

Sample A			Sample B		
Data item	**Deviation:** Data item − mean	**(Deviation)²:** (Data item − mean)²	Data item	**Deviation:** Data item − mean	**(Deviation)²:** (Data item − mean)²
17	$17 - 20 = -3$	$(-3)^2 = 9$	5	$5 - 20 = -15$	$(-15)^2 = 225$
18	$18 - 20 = -2$	$(-2)^2 = 4$	10	$10 - 20 = -10$	$(-10)^2 = 100$
19	$19 - 20 = -1$	$(-1)^2 = 1$	15	$15 - 20 = -5$	$(-5)^2 = 25$
20	$20 - 20 = 0$	$0^2 = 0$	20	$20 - 20 = 0$	$0^2 = 0$
21	$21 - 20 = 1$	$1^2 = 1$	25	$25 - 20 = 5$	$5^2 = 25$
22	$22 - 20 = 2$	$2^2 = 4$	30	$30 - 20 = 10$	$10^2 = 100$
23	$23 - 20 = 3$	$3^2 = 9$	35	$35 - 20 = 15$	$15^2 = 225$

Totals: $\quad\quad\quad\quad\quad \Sigma(\text{data item} - \text{mean})^2 = 28 \quad\quad\quad\quad\quad \Sigma(\text{data item} - \text{mean})^2 = 700$

Each sample contains seven data items, so we compute the standard deviation by dividing the sums in Table 12.9, 28 and 700, by $7 - 1$, or 6. Then we take the square root of each quotient.

$$\text{Standard deviation} = \sqrt{\frac{\Sigma(\text{data item} - \text{mean})^2}{n - 1}}$$

Sample A:

$$\text{Standard deviation} = \sqrt{\frac{28}{6}} \approx 2.16$$

Sample B:

$$\text{Standard deviation} = \sqrt{\frac{700}{6}} \approx 10.80$$

Sample A has a standard deviation of approximately 2.16, and sample B has a standard deviation of approximately 10.80. The scores in sample B are more spread out than those in sample A.

 Check
4
Point

Find the standard deviation of the data items in each of the samples shown below. Round to two decimal places.

Sample A: 73, 75, 77, 79, 81, 83

Sample B: 40, 44, 92, 94, 98, 100

Figure 12.12 on the next page illustrates four sets of data items organized in histograms. From left to right, the data items are

Figure 12.12(a): 4, 4, 4, 4, 4, 4, 4

Figure 12.12(b): 3, 3, 4, 4, 4, 5, 5

Figure 12.12(c): 3, 3, 3, 4, 5, 5, 5

Figure 12.12(d): 1, 1, 1, 4, 7, 7, 7.

Each data set has a mean of 4. However, as the spread of data items increases, the standard deviation gets larger. Observe that when all the data items are the same, the standard deviation is 0.

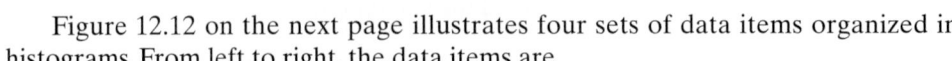

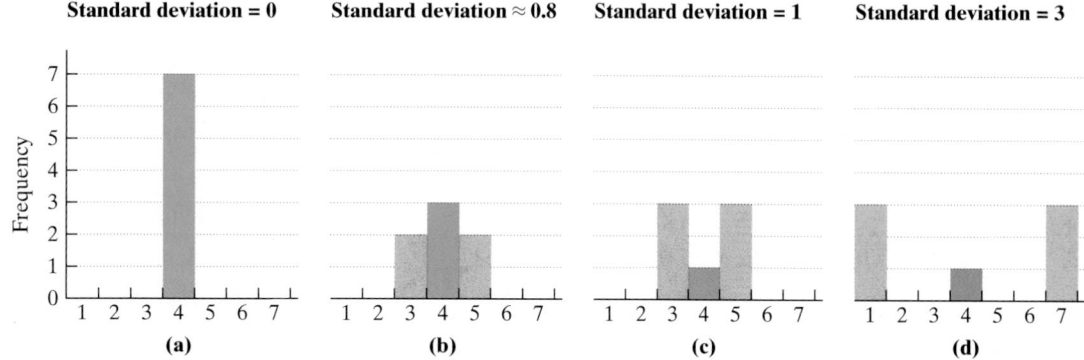

FIGURE 12.12 The standard deviation gets larger with increased dispersion among data items. In each case, the mean is 4.

EXAMPLE 5 | INTERPRETING STANDARD DEVIATION

Two fifth grade classes have nearly identical mean scores on an aptitude test, but one class has a standard deviation three times that of the other. All other factors being equal, which class is easier to teach, and why?

Solution   The class with the smaller standard deviation is easier to teach because there is less variation among student aptitudes. Course work can be aimed at the average student without too much concern that the work will be too easy for some or too difficult for others. By contrast, the class with greater dispersion poses a greater challenge. By teaching to the average student, the students whose scores are significantly above the mean will be bored; students whose scores are significantly below the mean will be confused.

Check 5 Point

Shown below are the means and standard deviations of the yearly returns on three investments over the 50 years from 1951 through 2000.

Investment	Mean Yearly Interest	Standard Deviation
Treasury Bills	5.3%	2.9%
Stocks	14.6%	16.3%

a. Use the means to determine which investment provides the greater yearly return.

b. Use the standard deviations to determine which investment has the greater risk. Explain your answer.

Exercise Set 12.3

Practice Exercises

In Exercises 1–6, find the range for each group of data items.

1. 1, 2, 3, 4, 5

2. 16, 17, 18, 19, 20

3. 7, 9, 9, 15

4. 11, 13, 14, 15, 17

5. 3, 3, 4, 4, 5, 5

6. 3, 3, 3, 4, 5, 5, 5

In Exercises 7–10, a group of data items and their mean are given.

  a. *Find the deviation from the mean for each of the data items.*

  b. *Find the sum of the deviations in part (a).*

7. 3, 5, 7, 12, 18, 27; Mean = 12

8. 84, 88, 90, 95, 98; Mean = 91

9. 29, 38, 48, 49, 53, 77; Mean = 49

10. 60, 60, 62, 65, 65, 65, 66, 67, 70, 70; Mean = 65

*In Exercises 11–16, find **a.** the mean; **b.** the deviation from the mean for each data item; and **c.** the sum of the deviations in part (b).*

11. 85, 95, 90, 85, 100

12. 94, 62, 88, 85, 91

13. 146, 153, 155, 160, 161

14. 150, 132, 144, 122

15. 2.25, 3.50, 2.75, 3.10, 1.90

16. 0.35, 0.37, 0.41, 0.39, 0.43

In Exercises 17–26, find the standard deviation for each group of data items. Round answers to two decimal places.

17. 1, 2, 3, 4, 5

18. 16, 17, 18, 19, 20

19. 7, 9, 9, 15

20. 11, 13, 14, 15, 17

21. 3, 3, 4, 4, 5, 5

22. 3, 3, 3, 4, 5, 5, 5

23. 1, 1, 1, 4, 7, 7, 7

24. 6, 6, 6, 6, 7, 7, 7, 4, 8, 3

25. 9, 5, 9, 5, 9, 5, 9, 5

26. 6, 10, 6, 10, 6, 10, 6, 10

In Exercises 27–28, compute the mean, range, and standard deviation for the data items in each of the three samples. Then describe one way in which the samples are alike and one way in which they are different.

27. Sample A: 6, 8, 10, 12, 14, 16, 18

Sample B: 6, 7, 8, 12, 16, 17, 18

Sample C: 6, 6, 6, 12, 18, 18, 18

28. Sample A: 8, 10, 12, 14, 16, 18, 20

Sample B: 8, 9, 10, 14, 18, 19, 20

Sample C: 8, 8, 8, 14, 20, 20, 20

Application Exercises

The graph shows U.S. voting age population and voter turnout from 1990 through 2000. Use the data shown to solve Exercises 29–30.

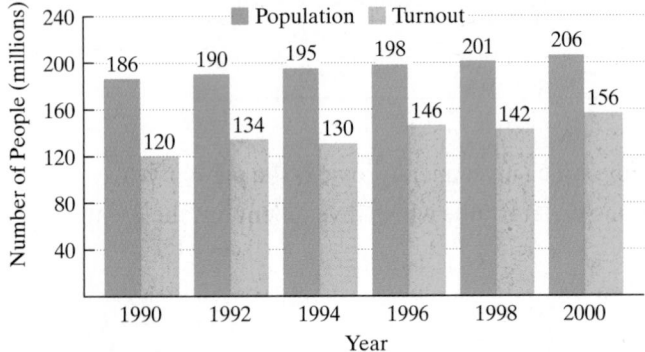

U.S. Voting Age Population and Voter Turnout: 1990–2000

Source: Federal Election Commission

29. Find the mean and the standard deviation for U.S. voting age population from 1990 through 2000.

30. Find the mean and the standard deviation for U.S. voter turnout from 1990 through 2000.

31. Use the data shown in the next column to find the mean and the standard deviation for the number of vetoes overridden for U.S. presidents from Roosevelt through Clinton.

President	Number of Vetoes Overridden
F. D. Roosevelt	9
Truman	12
Eisenhower	2
Kennedy	0
L. Johnson	0
Nixon	7
Ford	12
Carter	2
Reagan	9
Bush	1
Clinton	2

Source: Senate Library

Writing in Mathematics

32. Describe how to find the range of a data set.

33. Describe why the range might not be the best measure of dispersion.

34. Describe how the standard deviation is computed.

35. Describe what the standard deviation reveals about a data set.

36. If a set of test scores has a standard deviation of zero, what does this mean about the scores?

37. Two classes took a statistics test. Both classes had a mean score of 73. The scores of class A had a standard deviation of 5 and those of class B had a standard deviation of 10. Discuss the difference between the two classes' performance on the test.

38. A sample of cereals indicates a mean potassium content per serving of 93 milligrams and a standard deviation of 2 milligrams. Write a description of what this means for a person who knows nothing about statistics.

39. Over a one-month period, stock A had a mean daily closing price of 124.7 and a standard deviation of 12.5. By contrast, stock B had a mean daily closing price of 78.2 and a standard deviation of 6.1. Which stock was more volatile? Explain your answer.

Critical Thinking Exercises

40. Which one of the following is true?

a. If the same number is added to each data item in a set of data, the standard deviation does not change.

b. If each number in a data set is multiplied by 4, the standard deviation is doubled.

c. It is possible for a set of scores to have a negative standard deviation.

d. Data sets with different means cannot have the same standard deviation.

41. Describe a situation in which a relatively large standard deviation is desirable.

42. If a set of test scores has a large range but a small standard deviation, describe what this means about students' performance on the test.

43. Use the data 1, 2, 3, 5, 6, 7. Without actually computing the standard deviation, which of the following best approximates the standard deviation?

a. 2   **b.** 6   **c.** 10   **d.** 20

44. Use the data 0, 1, 3, 4, 4, 6. Add 2 to each of the numbers. How does this affect the mean? How does this affect the standard deviation?

Group Exercises

45. As a follow-up to Group Exercise 74 on page 657, the group should reassemble and compute the standard deviation for each data set whose mean you previously determined. Does the standard deviation tell you anything new or interesting about the entire group or subgroups that you did not discover during the previous group activity?

46. Group members should consult a current almanac or the Internet and select intriguing data. The group's function is to use statistics to tell a story. Once "intriguing" data are identified, as a group

a. Summarize the data. Use words, frequency distributions, and graphic displays.

b. Compute measures of central tendency and dispersion, using these statistics to discuss the data.

SECTION 12.4   THE NORMAL DISTRIBUTION

OBJECTIVES

1. Recognize characteristics of normal distributions.

2. Understand the 68–95–99.7 Rule.

3. Find scores at a specified standard deviation from the mean.

4. Use the 68–95–99.7 Rule.

5. Convert a data item to a z-score.

6. Understand and use percentiles.

7. Use and interpret margins of error.

8. Recognize distributions that are not normal.

Mean Adult Heights

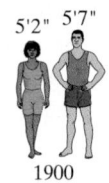   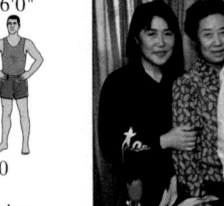 

5'2" 5'7"   5'5" 5'10"   5'7" 6'0"

1900   2000   2050

Source: National Center for Health Statistics

Our heights are on the rise! In one million B.C., the mean height for men was 4 feet 6 inches. The mean height for women was 4 feet 2 inches. Because of improved diets and medical care, the mean height for men is now 5 feet 10 inches and for women it is 5 feet 5 inches. Mean adult heights are expected to plateau by 2050.

Suppose that a researcher selects a random sample of 100 adult men, measures their heights, and constructs a histogram. The graph is shown in Figure 12.13. Figure 12.13 illustrates what happens as the sample size increases. In Figure 12.13(c), if you were to fold the graph down the middle, the left side would fit the right side. As we move out from the middle, the heights of the bars are the same to the left and right. Such a histogram is called **symmetric**. As the sample size increases, so does the graph's symmetry. If it were possible to measure the heights of all adult males, the entire population, the histogram would approach what is called the **normal distribution**, shown in Figure 12.13(d). This distribution is also called the **bell curve** or the **Gaussian distribution**, named for the German mathematician Carl Friedrich Gauss (1777–1855).

> **1** Recognize characteristics of normal distributions.

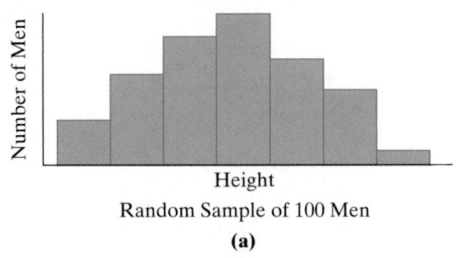

Height
Random Sample of 100 Men
(a)

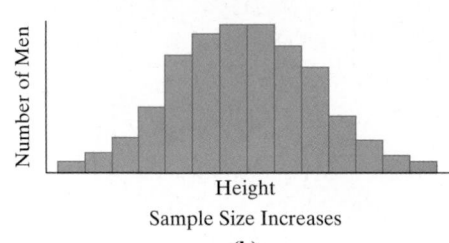

Height
Sample Size Increases
(b)

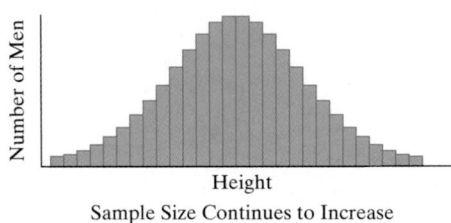

Height
Sample Size Continues to Increase
(c)

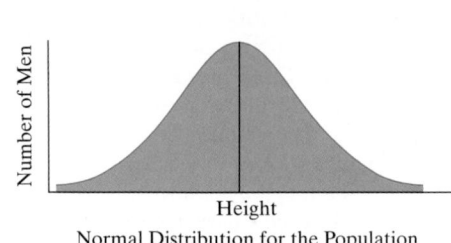

Height
Normal Distribution for the Population
(d)

FIGURE **12.13**   Heights of adult males

Figure 12.13(d) illustrates that the normal distribution is bell shaped and symmetric about a vertical line through its center. Furthermore, **the mean, median, and mode** of a normal distribution **are all equal** and located at the center of the distribution.

The shape of the normal distribution depends on the mean and the standard deviation. Figure 12.14 illustrates three normal distributions with the same mean, but different standard deviations. As the standard deviation increases, the distribution becomes more dispersed, or spread out, but retains its symmetric bell shape.

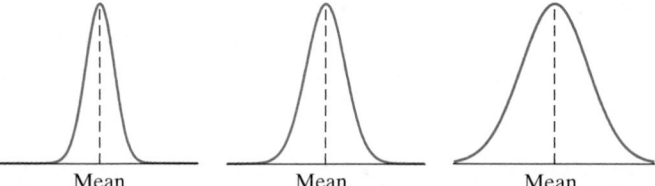

FIGURE 12.14

The normal distribution provides a wonderful model for all kinds of phenomena because many sets of data items closely resemble this population distribution. Examples include heights and weights of adult males, intelligence quotients, SAT scores, prices paid for a new car model, and life spans of light bulbs. In these distributions, the data items tend to cluster around the mean. The more an item differs from the mean, the less likely it is to occur.

The normal distribution is used to make predictions about an entire population using data from a sample. In this section, we focus on the characteristics and applications of the normal distribution.

2 | Understand the 68–95–99.7 Rule.

The Standard Deviation and z-Scores in Normal Distributions

The standard deviation plays a crucial role in the normal distribution, summarized by the **68–95–99.7 Rule**. This rule is illustrated in Figure 12.15.

THE 68–95–99.7 RULE FOR THE NORMAL DISTRIBUTION

1. Approximately 68% of the data items fall within 1 standard deviation of the mean (in both directions).
2. Approximately 95% of the data items fall within 2 standard deviations of the mean.
3. Approximately 99.7% of the data items fall within 3 standard deviations of the mean.

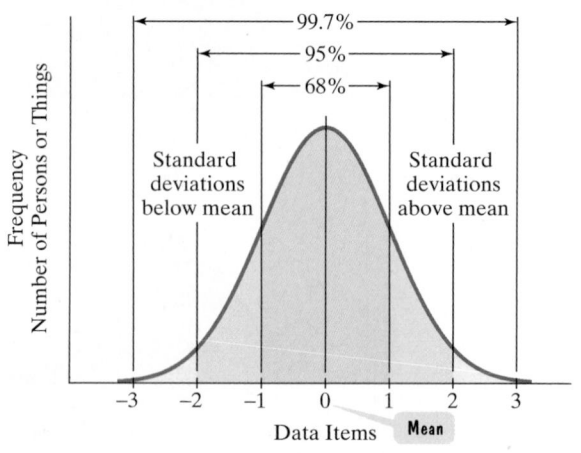

FIGURE 12.15

Figure 12.15 illustrates that a very small percentage of the data in a normal distribution lies more than 3 standard deviations above or below the mean. As we move from the mean, the curve falls rapidly, and then more and more gradually, toward the horizontal axis. The tails of the curve approach, but never touch, the horizontal axis, although they are quite close to the axis at 3 standard deviations from the mean. The range of the population normal distribution is infinite. No matter how far out from the mean we move, there is always the probability (although very small) of a data item occurring even farther out.

The SAT is a test that has been taken by many college-bound seniors since 1941. Scores on each part have a mean of 500 and a standard deviation of 100. Although there can be no scores less than 200 or more than 800, the scores can be modeled by the normal distribution.

3 Find scores at a specified standard deviation from the mean.

| **EXAMPLE 1** | FINDING SCORES AT A SPECIFIED STANDARD DEVIATION FROM THE MEAN |

With a mean of 500 and a standard deviation of 100, find the SAT score that is

a. 2 standard deviations above the mean.

b. 3 standard deviations below the mean.

Solution

a. First, let us find the SAT score that is 2 standard deviations above the mean.

$$\text{Score} = \text{mean} + 2 \cdot \text{standard deviation}$$
$$= 500 + 2 \cdot 100 = 500 + 200 = 700$$

An SAT score of 700 is 2 standard deviations above the mean.

b. Next, let us find the SAT score that is 3 standard deviations below the mean.

$$\text{Score} = \text{mean} - 3 \cdot \text{standard deviation}$$
$$= 500 - 3 \cdot 100 = 500 - 300 = 200$$

An SAT score of 200 is 3 standard deviations below the mean.

 **Check Point 1** The distribution of heights of young men is approximately normal with a mean of 70 inches and a standard deviation of 2.5 inches. Find the height that is

a. 3 standard deviations above the mean.

b. 2 standard deviations below the mean.

The distribution of SAT scores is illustrated as a population normal distribution in Figure 12.16.

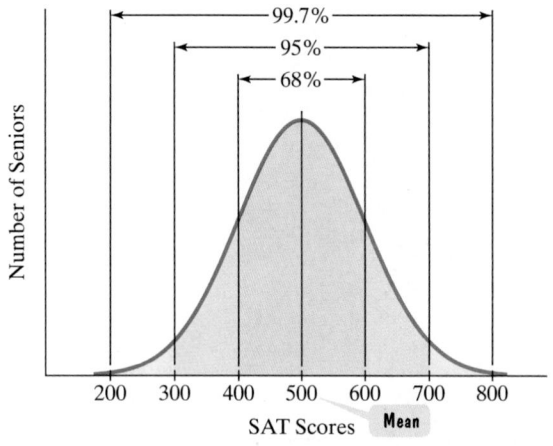

FIGURE 12.16

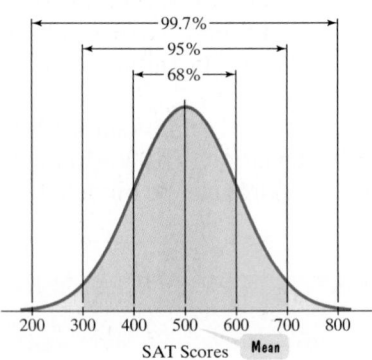

4 Use the 68–95–99.7 Rule.

Normal Distribution of SAT Scores

FIGURE **12.16,** repeated

EXAMPLE 2 | USING THE 68–95–99.7 RULE

Use the distribution of SAT scores in Figure 12.16 to find the percentage of seniors who score

a. between 400 and 600.   **b.** between 500 and 600.   **c.** above 700.

Solution

a. The 68–95–99.7 Rule states that approximately 68% of the data items fall within 1 standard deviation, 100, of the mean, 500.

$$\text{mean} - 1 \cdot \text{standard deviation} = 500 - 1 \cdot 100 = 400$$

$$\text{mean} + 1 \cdot \text{standard deviation} = 500 + 1 \cdot 100 = 600$$

Figure 12.16 shows that 68% of the seniors score between 400 and 600.

b. The percentage of seniors who score between 500 and 600 is not given directly in Figure 12.16 or 12.17. Because of the distribution's symmetry, the percentage who score between 400 and 500 is the same as the percentage who score between 500 and 600. Figure 12.17 indicates that 68% score between 400 and 600. Thus, half of 68%, or 34%, of seniors score between 500 and 600.

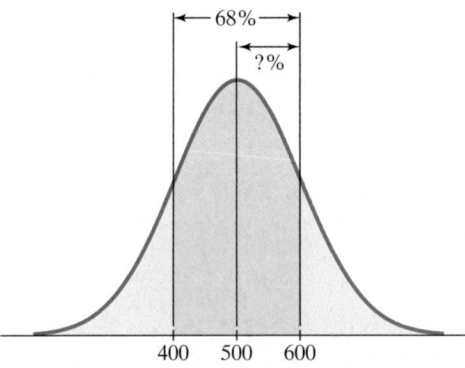

FIGURE **12.17** What percentage score between 500 and 600?

c. The percentage of seniors who score above 700 is not given directly in Figure 12.16 or 12.18. A score of 700 is 2 standard deviations, $2 \cdot 100$, above the mean, 500. The 68–95–99.7 Rule states that approximately 95% of the data items fall within 2 standard deviations of the mean. Thus, approximately $100\% - 95\%$, or 5%, of the data items are farther than 2 standard deviations from the mean. The 5% of the data items are represented by the two shaded green regions in Figure 12.18. Because of the distribution's symmetry, half of 5%, or 2.5%, of the data items are more than 2 standard deviations above the mean. This means that 2.5% of seniors score above 700.

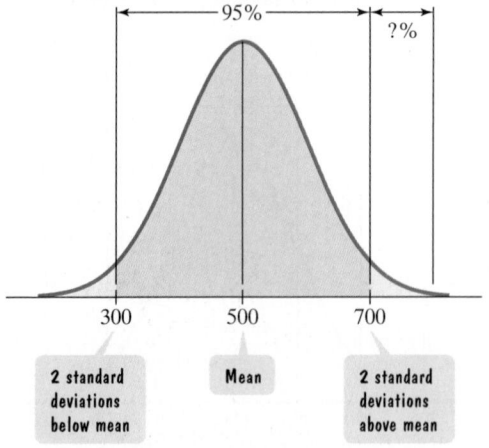

FIGURE **12.18** What percentage score above 700?

 **Check Point 2** Use the distribution of SAT scores in Figure 12.16 to find the percentage of seniors who score

    a. between 300 and 700.

    b. between 500 and 700.

    c. above 600.

Because the SAT has a mean of 500 and a standard deviation of 100, a score of 700 lies 2 standard deviations above the mean. In a normal distribution, a *z*-**score** describes how many standard deviations a particular data item lies above or below the mean. Thus, the *z*-score for the data item 700 is 2.

The following formula can be used to express a data item in a normal distribution as a *z*-score:

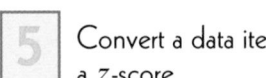 **5** Convert a data item to a *z*-score.

COMPUTING *z*-SCORES

A *z*-score describes how many standard deviations a data item in a normal distribution lies above or below the mean. The *z*-score can be obtained using

$$z\text{-score} = \frac{\text{data item} - \text{mean}}{\text{standard deviation}}.$$

Data items above the mean have positive *z*-scores. Data items below the mean have negative *z*-scores. The *z*-score for the mean is 0.

EXAMPLE 3 **COMPUTING *z*-SCORES**

The mean weight of newborn infants is 7 pounds and the standard deviation is 0.8 pound. The weights of newborn infants are normally distributed. Find the *z*-score for a weight of

  a. 9 pounds.   **b.** 7 pounds.   **c.** 6 pounds.

Solution We compute the *z*-score for each weight by using the *z*-score formula. The mean is 7 and the standard deviation is 0.8.

  a. The *z*-score for a weight of 9 pounds, written z_9, is

$$z_9 = \frac{\text{data item} - \text{mean}}{\text{standard deviation}} = \frac{9 - 7}{0.8} = \frac{2}{0.8} = 2.5.$$

The *z*-score of a data item greater than the mean is always positive. A 9-pound infant is a chubby little tyke, with a weight that is 2.5 standard deviations above the mean.

  b. The *z*-score for a weight of 7 pounds is

$$z_7 = \frac{\text{data item} - \text{mean}}{\text{standard deviation}} = \frac{7 - 7}{0.8} = \frac{0}{0.8} = 0.$$

The *z*-score for the mean is always 0. A 7-pound infant is right at the mean, deviating 0 pounds above or below it.

  c. The *z*-score for a weight of 6 pounds is

$$z_6 = \frac{\text{data item} - \text{mean}}{\text{standard deviation}} = \frac{6 - 7}{0.8} = \frac{-1}{0.8} = -1.25.$$

The *z*-score of a data item less than the mean is always negative. A 6-pound infant's weight is 1.25 standard deviations below the mean.

Figure 12.19 shows the normal distribution of weights of newborn infants. The horizontal axis is labeled in terms of weights and z-scores.

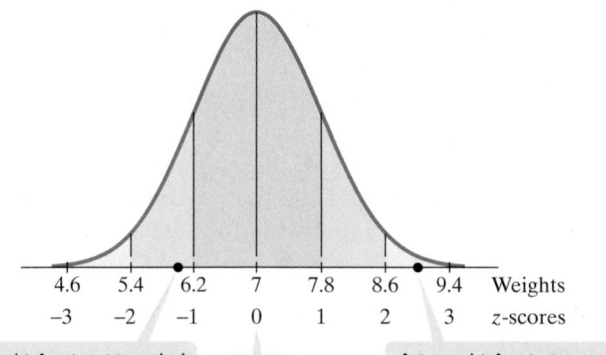

Normal Distribution of Weights of Newborn Infants

Weights	4.6	5.4	6.2	7	7.8	8.6	9.4
z-scores	–3	–2	–1	0	1	2	3

A 6-pound infant is 1.25 standard deviations below the mean.

Mean

A 9-pound infant is 2.5 standard deviations above the mean.

FIGURE 12.19 Infants' weights are normally distributed.

Check 3 Point The length of horse pregnancies from conception to birth is normally distributed with a mean of 336 days and a standard deviation of 3 days. Find the z-score for a horse pregnancy of
a. 342 days.   **b.** 336 days.   **c.** 333 days.

In Example 4, we consider two normally distributed sets of test scores, in which a higher score generally indicates a better result. To compare scores on two different tests in relation to the mean on each test, we can use z-scores. The better score is the item with the greater z-score.

EXAMPLE 4 | USING AND INTERPRETING z-SCORES

A student scores 70 on an arithmetic test and 66 on a vocabulary test. The scores for both tests are normally distributed. The arithmetic test has a mean of 60 and a standard deviation of 20. The vocabulary test has a mean of 60 and a standard deviation of 2. On which test did the student have the better score?

Solution To answer the question, we need to find the student's z-score on each test, using

$$z = \frac{\text{data item} - \text{mean}}{\text{standard deviation}}.$$

The arithmetic test has a mean of 60 and a standard deviation of 20.

$$z\text{-score for } 70 = z_{70} = \frac{70 - 60}{20} = \frac{10}{20} = 0.5$$

The vocabulary test has a mean of 60 and a standard deviation of 2.

$$z\text{-score for } 66 = z_{66} = \frac{66 - 60}{2} = \frac{6}{2} = 3$$

The arithmetic score, 70, is half a standard deviation above the mean, whereas the vocabulary score, 66, is 3 standard deviations above the mean. The student did much better than the mean on the vocabulary test.

Understand and use percentiles.

Check 4 Point We have seen that the SAT has a mean of 500 and a standard deviation of 100. The ACT (American College Test) has a mean of 18 and a standard deviation of 6. Both tests measure the same kind of ability, with scores that are normally distributed. Suppose that you score 550 on the SAT and 24 on the ACT. On which test did you have the better score?

EXAMPLE 5   UNDERSTANDING z-SCORES

Intelligence quotients (IQs) are normally distributed with a mean of 100 and a standard deviation of 15. What is the IQ corresponding to a z-score of -1.5?

Solution   The negative sign in -1.5 tells us that the score in question is $1\frac{1}{2}$ standard deviations below the mean.

$$\text{Score} = \text{mean} - 1.5 \cdot \text{standard deviation}$$
$$= 100 - 1.5 \cdot 15 = 100 - 22.5 = 77.5$$

The IQ corresponding to a z-score of -1.5 is 77.5.

Check 5 Point Use the information in Example 5 to find the IQ corresponding to a z-score of 2.5.

z-scores and Percentiles

A z-score measures a data item's position in a normal distribution. Another measure of a data item's position is its **percentile**. Percentiles are often associated with scores on standardized tests. If a score is in the 45th percentile, this means that 45% of the scores are less than this score. If a score is in the 95th percentile, this indicates that 95% of the scores are less that this score.

> **PERCENTILES**
>
> If n% of the items in a distribution are less than a particular data item, we say that the data item is in the **nth percentile** of the distribution.

EXAMPLE 6   INTERPRETING PERCENTILE

A student scored in the 93rd percentile on the SAT. What does this mean?

Solution   By scoring in the 93rd percentile, the student did better than about 93% of all those who took the exam.

STUDY TIP

A score in the 93rd percentile does *not* mean that 93% of the answers are correct. Nor does it mean that the score was 93%.

Check 6 Point A student scored in the 62nd percentile on the SAT. What does this mean?

Table 12.10 gives a percentile interpretation for z-scores. For example, the corresponding percentile for a data item with a z-score of 2 is 97.72. A student with this score on a test whose results are normally distributed outperformed 97.72% of those who took the test.

In a normal distribution, the mean, median, and mode all have a corresponding z-score of 0. Table 12.10 shows that the percentile for a z-score of 0 is 50.00. Thus, 50% of the data items in a normal distribution are less than the mean, median, and mode. Fifty percent of the data items are greater than or equal to the mean, median, and mode.

Table 12.10 can be used to find the percentage of data items that are less than any data item in a normal distribution. Begin by converting the data item to a z-score. Then, use the table to find the percentile for this z-score. This percentile is the percentage of data items that are less than the data item in question.

TABLE 12.10 z-SCORES AND PERCENTILES

z-score	Percentile	z-score	Percentile	z-score	Percentile	z-score	Percentile
−4.0	0.003	−1.0	15.87	0.0	50.00	1.1	86.43
−3.5	0.02	−0.95	17.11	0.05	51.99	1.2	88.49
−3.0	0.13	−0.90	18.41	0.10	53.98	1.3	90.32
−2.9	0.19	−0.85	19.77	0.15	55.96	1.4	91.92
−2.8	0.26	−0.80	21.19	0.20	57.93	1.5	93.32
−2.7	0.35	−0.75	22.66	0.25	59.87	1.6	94.52
−2.6	0.47	−0.70	24.20	0.30	61.79	1.7	95.54
−2.5	0.62	−0.65	25.78	0.35	63.68	1.8	96.41
−2.4	0.82	−0.60	27.43	0.40	65.54	1.9	97.13
−2.3	1.07	−0.55	29.12	0.45	67.36	2.0	97.72
−2.2	1.39	−0.50	30.85	0.50	69.15	2.1	98.21
−2.1	1.79	−0.45	32.64	0.55	70.88	2.2	98.61
−2.0	2.28	−0.40	34.46	0.60	72.57	2.3	98.93
−1.9	2.87	−0.35	36.32	0.65	74.22	2.4	99.18
−1.8	3.59	−0.30	38.21	0.70	75.80	2.5	99.38
−1.7	4.46	−0.25	40.13	0.75	77.34	2.6	99.53
−1.6	5.48	−0.20	42.07	0.80	78.81	2.7	99.65
−1.5	6.68	−0.15	44.04	0.85	80.23	2.8	99.74
−1.4	8.08	−0.10	46.02	0.90	81.59	2.9	99.81
−1.3	9.68	−0.05	48.01	0.95	82.89	3.0	99.87
−1.2	11.51	0.0	50.00	1.0	84.13	3.5	99.98
−1.1	13.57					4.0	99.997

EXAMPLE 7 FINDING THE PERCENTAGE OF DATA ITEMS LESS THAN A GIVEN DATA ITEM

According to the Department of Health and Education, cholesterol levels are normally distributed. For men between 18 and 24 years, the mean is 178.1 (measured in milligrams per 100 milliliters) and the standard deviation is 40.7. What percentage of men in this age range have a cholesterol level less than 239.15?

Solution If you are familiar with your own cholesterol level, you probably recognize that a level of 239.15 is fairly high for a young man. Because of this, we would expect most young men to have a level less than 239.15. Let's see if this is so. Table 12.10 requires that we use z-scores. We compute the z-score for a 239.15 cholesterol level by using the z-score formula.

$$z_{239.15} = \frac{\text{data item} - \text{mean}}{\text{standard deviation}} = \frac{239.15 - 178.1}{40.7} = \frac{61.05}{40.7} = 1.5$$

A man between 18 and 24 with a 239.15 cholesterol level is 1.5 standard deviations above the mean, illustrated in Figure 12.20(a). The question mark indicates that we must find the percentage of men with a cholesterol level less than $z = 1.5$, the z-score for a 239.15 cholesterol level. Table 12.10 gives this percentage as a percentile. Find 1.5 in the z-score column in the right portion of the table. The percentile given to the right of 1.5 is 93.32. Thus, 93.32% of men between 18 and 24 have a cholesterol level less than 239.15, shown in Figure 12.20(b).

A Portion of Table 12.10

z-score	Percentile
1.4	91.92
1.5	93.32
1.6	94.52

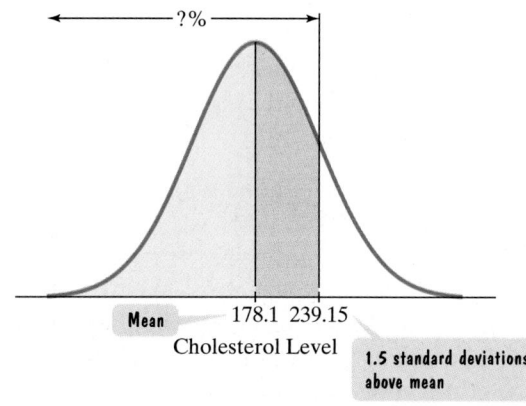

FIGURE 12.20 (a)

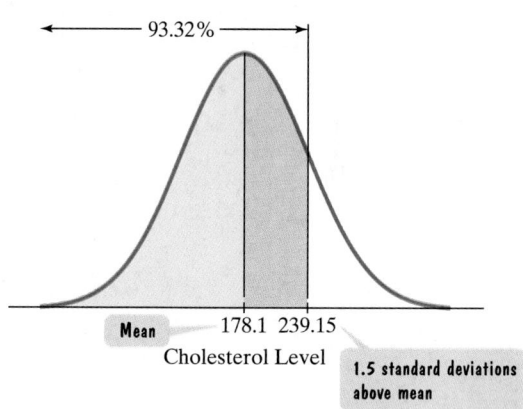

FIGURE 12.20 (b)

 **Check Point 7** The distribution of heights of young women is approximately normal with a mean of 65 inches and a standard deviation of 2.5 inches. What percentage of young women are shorter than 68 inches?

The normal distribution accounts for all data items, meaning 100% of the scores. This means that Table 12.10 can also be used to find the percentage of data items that are greater than any data item in a normal distribution. Use the percentile in the table to determine the percentage of data items less than the data item in question. Then subtract this percentage from 100% to find the percentage of data items greater than the item in question. In using this technique, we will treat the phrases "greater than" and "greater than or equal to" as equivalent.

A Portion of Table 12.10

z-score	Percentile
0.45	67.36
0.50	69.15
0.55	70.88

EXAMPLE 8 | FINDING THE PERCENTAGE OF DATA ITEMS GREATER THAN A GIVEN DATA ITEM

Lengths of pregnancy of women are normally distributed with a mean of 266 days and a standard deviation of 16 days. What percentage of children are born from pregnancies lasting more than 274 days?

Solution Table 12.10 requires that we use z-scores. We compute the z-score for a 274-day pregnancy by using the z-score formula.

$$z_{274} = \frac{\text{data item} - \text{mean}}{\text{standard deviation}} = \frac{274 - 266}{16} = \frac{8}{16} = 0.5$$

A 274-day pregnancy is 0.5 standard deviation above the mean. Table 12.10 gives the percentile corresponding to 0.50 as 69.15. This means that 69.15% of pregnancies last less than 274 days, illustrated in Figure 12.21. We must find the number of pregnancies lasting more than 274 days by subtracting 69.15% from 100%.

$$100\% - 69.15\% = 30.85\%$$

Thus, 30.85% of children are born from pregnancies lasting more than 274 days.

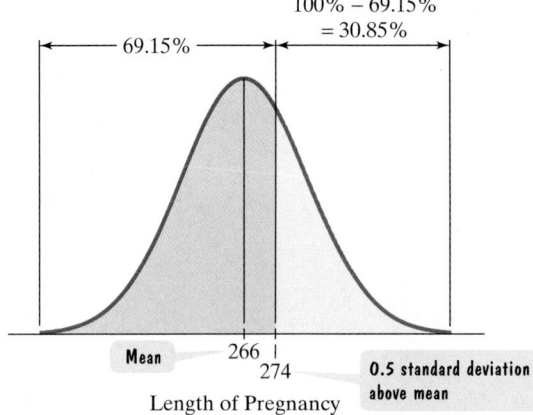

FIGURE 12.21

 **Check 8 Point** The Wechsler Adult Intelligence Scale (WAIS) is an IQ test. Scores on the WAIS for the 20–34 age group are normally distributed with a mean of 110 and a standard deviation of 25. What percentage of IQ scores for this age group are greater than 145?

We have seen how Table 12.10 is used to find the percentage of data items that are less than or greater than any given item. The table can also be used to find the percentage of data items *between* two given items. Because the percentile for each item is the percentage of data items less than the given item, the percentage of data between the two given items is found by subtracting the lesser percent from the greater percent. This is illustrated in Figure 12.22.

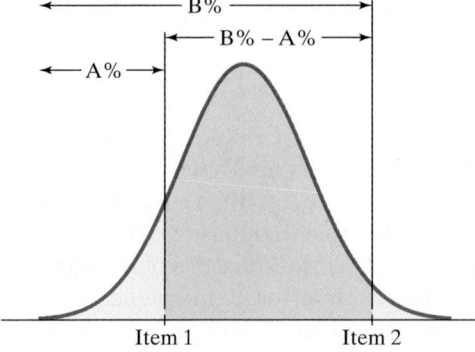

FIGURE 12.22 The percentile for data item 1 is A. The percentile for data item 2 is B. The percentage of data items between item 1 and item 2 is B% − A%.

FINDING THE PERCENTAGE OF DATA ITEMS BETWEEN TWO GIVEN ITEMS IN A NORMAL DISTRIBUTION

1. Convert each given data item to a z-score:

$$z = \frac{\text{data item} - \text{mean}}{\text{standard deviation}}.$$

2. Use Table 12.10 to find the percentile corresponding to each z-score in step 1.

3. Subtract the lesser percentile from the greater percentile and attach a % sign.

EXAMPLE 9 FINDING THE PERCENTAGE OF DATA ITEMS BETWEEN TWO GIVEN DATA ITEMS

According to the National Federation of State High School Associations, the amount of time that high school students work at jobs each week is normally distributed with a mean of 10.7 hours and a standard deviation of 11.2 hours. What percentage of high school students work between 5.1 and 38.7 hours each week?

Solution

STEP 1. Convert each given data item to a z-score.

$$z_{5.1} = \frac{\text{data item} - \text{mean}}{\text{standard deviation}} = \frac{5.1 - 10.7}{11.2} = \frac{-5.6}{11.2} = -0.5$$

$$z_{38.7} = \frac{\text{data item} - \text{mean}}{\text{standard deviation}} = \frac{38.7 - 10.7}{11.2} = \frac{28}{11.2} = 2.5$$

A Portion of Table 12.10

z-score	Percentile
−0.55	29.12
−0.50	30.85
−0.45	32.64

A Portion of Table 12.10

z-score	Percentile
2.4	99.18
2.5	99.38
2.6	99.53

STEP 2. Use Table 12.10 to find the percentile corresponding to these z-scores. The percentile given to the right of −0.50 is 30.85. This means that 30.85% of high school students work less than 5.1 hours each week.

Table 12.10 also gives the percentile corresponding to $z = 2.5$. Find 2.5 in the z-score column in the far-right portion of the table. The percentile given to the right of 2.5 is 99.38. This means that 99.38% of high school students work less than 38.7 hours per week.

STEP 3. Subtract the lesser percentile from the greater percentile and attach a % sign. Subtracting percentiles, we obtain

$$99.38 - 30.85 = 68.53.$$

Thus, 68.53% of high school students work between 5.1 and 38.7 hours each week. The solution is illustrated in Figure 12.23.

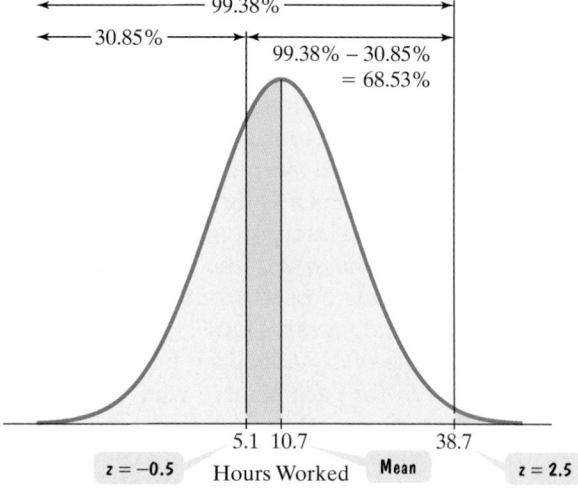

FIGURE 12.23

Check 9 Point The distribution of heights of young men is approximately normal with a mean of 70 inches and a standard deviation of 2.5 inches. What percentage of young men have heights between 67 inches and 74 inches?

Our work in Examples 7 through 9 is summarized as follows:

COMPUTING PERCENTAGE OF DATA ITEMS FOR NORMAL DISTRIBUTIONS

Description of Percentage	Graph	Computation of Percentage
Percentage of data items less than a given data item with $z = b$		Use the table percentile for $z = b$ and add a % sign.
Percentage of data items greater than a given data item with $z = a$		Subtract the table percentile for $z = a$ from 100 and add a % sign.
Percentage of data items between two given data items with $z = a$ and $z = b$		Subtract the table percentile for $z = a$ from the table percentile for $z = b$ and add a % sign.

---

7 Use and interpret margins of error.

Polls and Margins of Error

When you were between the ages of 6 and 14, how would you have responded to this question:

What is bad about being a kid?

In a random sample of 1172 children ages 6 through 14, 17% of the children responded, "Getting bossed around." The problem is that this is a single random sample. Do 17% of kids in the entire population of children ages 6 through 14 think that getting bossed around is a bad thing?

Statisticians use properties of the normal distribution to estimate the probability that a result obtained from a single sample reflects what is truly happening in the population. If you look at the results of a poll like the one shown in the margin, you will observe that a *margin of error* is reported. Surveys and opinion polls often give a margin of error. Let's use our understanding of the normal distribution to see how to calculate and interpret margins of error.

Suppose that p% of the population of children ages 6 through 14 hold the opinion that getting bossed around is a bad thing about being a kid. Instead of taking only one random sample of 1172 children, we repeat the process of selecting a random sample of 1172 children hundreds of times. Then, we calculate the percentage of children for each sample who think being bossed around is bad. With random sampling, we expect to find the percentage in many of the samples close to p%, with relatively few samples having percentages far from p%. Figure 12.24 shows that the percentages of children from the hundreds of samples form a normal distribution.

What Is Bad About Being a Kid?

Kids Say	
Getting bossed around	17%
School, homework	15%
Can't do everything I want	11%
Chores	9%
Being grounded	9%

Source: Penn, Schoen, and Berland using 1172 interviews with children ages 6 to 14 from May 14 to June 1, 1999, Margin of error: ±2.9%

Note the margin of error.

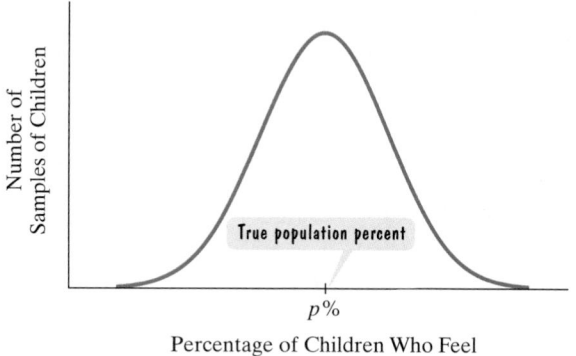

Number of
Samples of Children

True population percent

$p\%$

Percentage of Children Who Feel
Being Bossed Around is Bad

FIGURE 12.24 Percentage of children who feel being bossed around is bad

The mean of this distribution is the actual population percent, $p\%$, and is the most frequent result from the samples.

Mathematicians have shown that the standard deviation of a normal distribution of samples like the one in Figure 12.24 is approximately $\dfrac{1}{2\sqrt{n}}$, where n is the sample size. Using the 68–95–99.7 Rule, approximately 95% of the samples have a percentage within 2 standard deviations of the true population percentage, $p\%$:

$$2 \text{ standard deviations} = 2 \cdot \frac{1}{2\sqrt{n}} = \frac{1}{\sqrt{n}}.$$

If we use a single random sample of size n, there is a 95% probability that the percent obtained will lie within two standard deviations, or $\dfrac{1}{\sqrt{n}}$, of the true population percent. We can be 95% confident that the true population percent lies between

$$\text{the sample percent} - \frac{1}{\sqrt{n}}$$

and

$$\text{the sample percent} + \frac{1}{\sqrt{n}}.$$

We call $\pm \dfrac{1}{\sqrt{n}}$ the **margin of error**.

MARGIN OF ERROR IN A SURVEY

If a statistic is obtained from a random sample of size n, there is a 95% probability that it lies within $\dfrac{1}{\sqrt{n}}$ of the true population statistic, where $\pm \dfrac{1}{\sqrt{n}}$ is called the **margin of error**.

EXAMPLE 10 USING AND INTERPRETING MARGIN OF ERROR

In a random sample of 1172 children ages 6 through 14, 17% of the children said getting bossed around is a bad thing about being a kid.

a. Verify the margin of error that was given for this survey.

b. Write a statement about the percentage of children in the population who feel that getting bossed around is a bad thing about being a kid.

What Is Bad About Being a Kid?

Kids Say	
Getting bossed around	17%
School, homework	15%
Can't do everything I want	11%
Chores	9%
Being grounded	9%

Source: Penn, Schoen, and Berland using 1172 interviews with children ages 6 to 14 from May 14 to June 1, 1999, Margin of error: ±2.9%

Solution

a. The sample size is $n = 1172$. The margin of error is

$$\pm \frac{1}{\sqrt{n}} = \pm \frac{1}{\sqrt{1172}} \approx \pm 0.029 = \pm 2.9\%.$$

b. There is a 95% probability that the true population percentage lies between

$$\text{the sample percent} - \frac{1}{\sqrt{n}} = 17\% - 2.9\% = 14.1\%$$

and

$$\text{the sample percent} + \frac{1}{\sqrt{n}} = 17\% + 2.9\% = 19.9\%.$$

We can be 95% confident that between 14.1% and 19.9% of all children feel that getting bossed around is a bad thing about being a kid.

 Check Point 10

Following the terrorist attacks on September 11, 2001, a telephone poll of 1082 randomly selected adult Americans taken for Time/CNN asked the following question:

If it becomes necessary to tighten airport security in order to reduce chances of other hijackings, do you think that raising airplane ticket prices $50 to pay for increased security is acceptable?

64% of those polled stated that this was acceptable.

a. Find the margin of error for this survey. Round to the nearest percent.

b. Write a statement about the percentage of American adults in the population who found this an acceptable way to increase security.

BLITZER BONUS

A CAVEAT GIVING A TRUE PICTURE OF A POLL'S ACCURACY

Unlike the precise calculation of a poll's margin of error, certain polling imperfections cannot be determined exactly. One problem is that people do not always respond to polls honestly and accurately. Some people are embarrassed to say "undecided," so they make up an answer. Other people may try to respond to questions in the way they think will make the pollster happy, just to be "nice." Perhaps the following caveat, applied to the poll in Example 10, would give the public a truer picture of its accuracy:

The poll results are 14.1% to 19.9% at the 95% confidence level, but it's only under ideal conditions that we can be 95% confident that the true numbers are within 2.9% of the poll's results. The true error span is probably greater than 2.9% due to limitations that are inherent in this and every poll, but, unfortunately, this additional error amount cannot be calculated precisely. Warning: Five percent of the time—that's one time out of 20—the error will be greater than 2.9%. We remind readers of the poll that things occurring "only" 5% of the time do, indeed, happen.

We suspect that the public would tire of hearing this.

Other Kinds of Distributions

8 Recognize distributions that are not normal.

The histogram in Figure 12.25 represents the frequencies of the ages of women interviewed by Kinsey and his colleagues in their study of female sexual behavior. This distribution is not symmetric. The greatest frequency of women interviewed was in the 16–20 age range. The bars get shorter and shorter after this. The shorter bars fall on the right, indicating that relatively few older women were included in Kinsey's interviews.

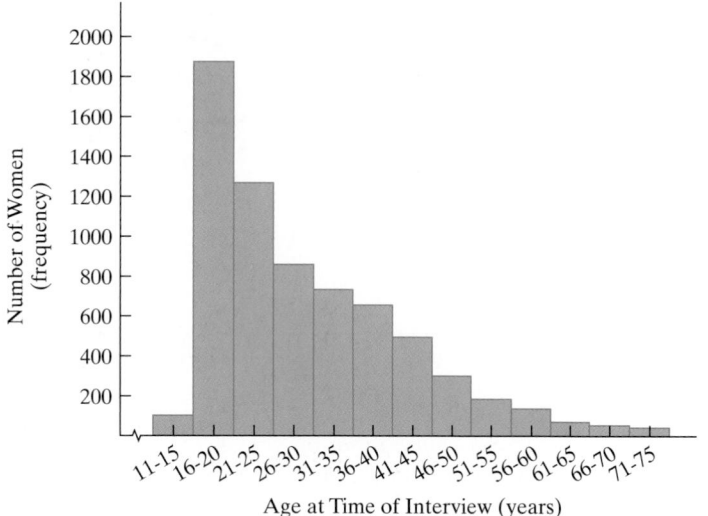

FIGURE 12.25 Histogram of the ages of females interviewed by Kinsey and his associates

Although the normal distribution is the most important of all distributions in terms of analyzing data, not all data can be approximated by this symmetric distribution with its mean, median, and mode all having the same value.

In our discussion of measures of central tendency, we mentioned that the median, rather than the mean, is used to summarize income. Figure 12.26 illustrates the population distribution of weekly earnings in the United States. There is no upper limit on weekly earnings. The relatively few people with very high weekly incomes tend to pull the mean income to a value greater than the median. The most frequent income, the mode, occurs toward the low end of the data items. The mean, median, and mode do not have the same value, and a normal distribution is not an appropriate model for describing weekly earnings in the United States.

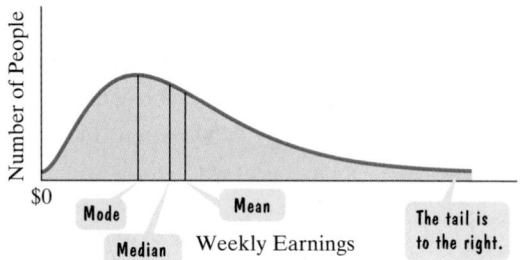

FIGURE 12.26 Skewed to the right

The distribution in Figure 12.26 is called a *skewed distribution*. A distribution of data is **skewed** if a large number of data items are piled up at one end or the other, with a "tail" at the opposite end. In the distribution of weekly earnings in Figure 12.26, the tail is to the right. Such a distribution is said to be **skewed to the right**.

By contrast to the distribution of weekly earnings, the distribution in Figure 12.27 has more data items at the high end of the scale than at the low end. The tail of this distribution is to the left. The distribution is said to be **skewed to the left**. In many colleges, an example of a distribution skewed to the left is based on the student ratings of faculty teaching performance. Most professors are given rather high ratings, while only a few are rated terrible. These low ratings pull the value of the mean lower than the median.

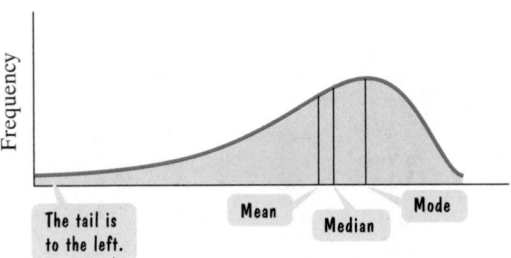

FIGURE 12.27 Skewed to the left

Exercise Set 12.4

Practice and Application Exercises

The scores on a test are normally distributed with a mean of 100 and a standard deviation of 20. In Exercises 1–10, find the score that is

1. 1 standard deviation above the mean.
2. 2 standard deviations above the mean.
3. 3 standard deviations above the mean.
4. $1\frac{1}{2}$ standard deviations above the mean.
5. $2\frac{1}{2}$ standard deviations above the mean.
6. 1 standard deviation below the mean.
7. 2 standard deviations below the mean.
8. 3 standard deviations below the mean.
9. one-half a standard deviation below the mean.
10. $2\frac{1}{2}$ standard deviations below the mean.

Not everyone pays the same price for the same model of a car. The figure illustrates a normal distribution for the prices paid for a particular model of a new car. The mean is $17,000 and the standard deviation is $500.

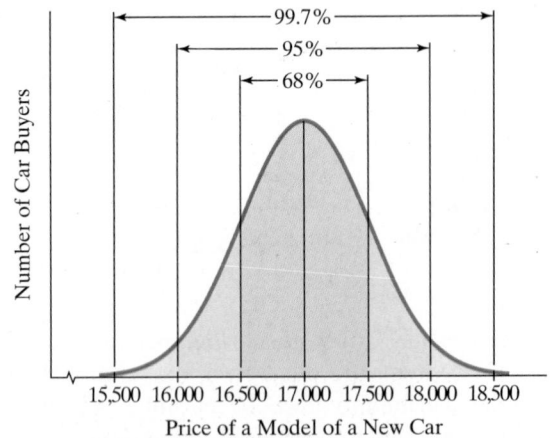

Price of a Model of a New Car

In Exercises 11–22, use the 68–95–99.7 Rule, illustrated in the figure, to find the percentage of buyers who paid

11. between $16,500 and $17,500.
12. between $16,000 and $18,000.
13. between $17,000 and $17,500.
14. between $17,000 and $18,000.
15. between $16,000 and $17,000.
16. between $16,500 and $17,000.
17. between $15,500 and $17,000.
18. between $17,000 and $18,500.
19. more than $17,500.
20. more than $18,000.
21. less than $16,000.
22. less than $16,500.

Scores on the GRE (Graduate Record Examination) are normally distributed with a mean of 530 and a standard deviation of 128. In Exercises 23–32, use the 68–95–99.7 Rule to find the percentage of people taking the test who score

23. between 274 and 786.
24. between 402 and 658.
25. between 274 and 530.
26. between 402 and 530.
27. above 658.
28. above 786.
29. below 274.
30. below 402.
31. above 914.
32. below 146.

A set of data items is normally distributed with a mean of 60 and a standard deviation of 8. In Exercises 33–48, convert each data item to a z-score.

33. 68	34. 76	35. 84	36. 92
37. 64	38. 72	39. 74	40. 78
41. 60	42. 100	43. 52	44. 44
45. 48	46. 40	47. 34	48. 30

Lengths of pregnancy of women are normally distributed with a mean of 266 days and a standard deviation of 16 days. In Exercises 49–56, find the z-score for a pregnancy lasting

49. 290 days. **50.** 294 days. **51.** 302 days. **52.** 318 days.

53. 258 days. **54.** 254 days. **55.** 242 days. **56.** 226 days.

57. A student scores 230 on a math test and 540 on a reading comprehension test. The math test has a mean of 200 and a standard deviation of 10. The reading comprehension test has a mean of 500 and a standard deviation of 15. If the data for both tests are normally distributed, on which test did the student have the better score?

58. A student scores 72 on a grammar test and 255 on a vocabulary test. The grammar test has a mean of 84 and a standard deviation of 10. The vocabulary test has a mean of 300 and a standard deviation of 30. If the data for both tests are normally distributed, on which test did the student have the better score?

A set of data items is normally distributed with a mean of 400 and a standard deviation of 50. In Exercises 59–66, find the data item in this distribution that corresponds to the given z-score.

59. $z = 2$ **60.** $z = 3$

61. $z = 1.5$ **62.** $z = 2.5$

63. $z = -3$ **64.** $z = -2$

65. $z = -2.5$ **66.** $z = -1.5$

Use Table 12.10 on page 672 to solve Exercises 67–82. In Exercises 67–74, find the percentage of data items in a normal distribution that lie a. below and b. above the given z-score.

67. $z = 0.6$ **68.** $z = 0.8$

69. $z = 1.2$ **70.** $z = 1.4$

71. $z = -0.7$ **72.** $z = -0.4$

73. $z = -1.2$ **74.** $z = -1.8$

In Exercises 75–82, find the percentage of data items in a normal distribution that lie between

75. $z = 0.2$ and $z = 1.4$ **76.** $z = 0.3$ and $z = 2.1$

77. $z = 1$ and $z = 3$ **78.** $z = 2$ and $z = 3$

79. $z = -1.5$ and $z = 1.5$ **80.** $z = -1.2$ and $z = 1.2$

81. $z = -2$ and $z = -0.5$ **82.** $z = -2.2$ and $z = -0.3$

We have seen that SAT scores are normally distributed with a mean of 500 and a standard deviation of 100. In Exercises 83–92, begin by converting any given score or scores into z-scores. Then use Table 12.10 on page 672 to find the percentage of seniors who score

83. below 650. **84.** below 680.

85. above 560. **86.** above 590.

87. above 380. **88.** above 360.

89. between 640 and 710. **90.** between 660 and 740.

91. between 440 and 560. **92.** between 420 and 580.

The weights for 12-month-old baby boys are normally distributed with a mean of 22.5 pounds and a standard deviation of 2.2 pounds. In Exercises 93–96, use Table 12.10 on page 672 to find the percentage of 12-month-old baby boys who weigh

93. more than 25.8 pounds.

94. more than 23.6 pounds.

95. between 19.2 and 21.4 pounds.

96. between 18.1 and 19.2 pounds.

97. Penn, Schoen, and Berland conducted interviews with 397 randomly selected parents. Twenty-six percent of the parents said that crime is a bad thing about being a kid.

   a. Find the margin of error for this survey.

   b. Write a statement about the percentage of parents in the population who feel that crime is a bad thing about being a kid.

98. Penn, Schoen, and Berland conducted interviews with 397 randomly selected parents. Five percent of the parents said that drugs are a bad thing about being a kid.

   a. Find the margin of error for this survey.

   b. Write a statement about the percentage of parents in the population who feel that drugs are a bad thing about being a kid.

99. Using a random sample of 4000 TV households, Nielsen Media Research found that 60.2% watched the final episode of "M*A*S*H."

   a. Find the margin of error in this percent.

   b. Write a statement about the percentage of TV households in the population who tuned into the final episode of "M*A*S*H."

100. Using a random sample of 4000 TV households, Nielsen Media Research found that 51.1% watched "Roots," Part 8.

   a. Find the margin of error in this percent.

   b. Write a statement about the percentage of TV households in the population who tuned into "Roots," Part 8.

101. In 1997, Nielsen Media Research increased their random sample to 5000 TV households. By how much, to the nearest tenth of a percent, did this improve their margin of error over that in Exercises 99 and 100?

102. If Nielsen Media Research were to increase their random sample from 5000 to 10,000 TV households, by how much, to the nearest tenth of a percent, would this improve their margin of error?

The histograms shown below indicate the frequencies of the number of syllables per word for randomly selected words in English and Japanese. Use the histograms to solve Exercises 103–106.

103. Is either histogram symmetric? Explain your answer.

104. Can the number of syllables in English or Japanese be described using the normal distribution? Explain your answer.

105. Describe one way in which English and Japanese are similar.

106. Describe one way in which English and Japanese are different.

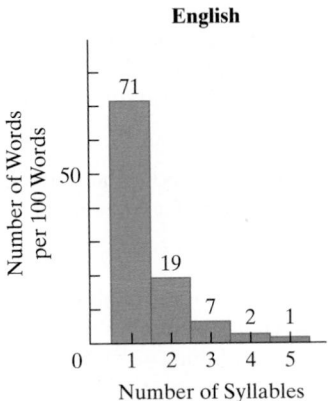

English

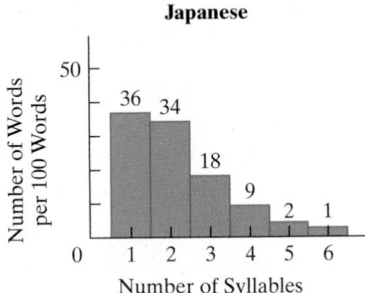

Japanese

Writing in Mathematics

107. What is a symmetric histogram?

108. Describe the normal distribution and discuss some of its properties.

109. Describe the 68–95–99.7 Rule.

110. Describe how to determine the z-score for a data item in a normal distribution.

111. What does a z-score measure?

112. Give an example of both a commonly occurring and an infrequently occurring z-score. Explain how you arrived at these examples.

113. Describe when a z-score is negative.

114. If you score in the 73rd percentile, what does this mean?

115. Two students have scores with the same percentile, but for different administrations of the SAT. Does this mean that the students have the same score on the SAT? Explain your answer.

116. Give an example of a phenomenon that is normally distributed. Explain why. (Try to be creative and not use one of the distributions discussed in this section.) Estimate what the mean and the standard deviation might be and describe how you determined these estimates.

117. Give an example of a phenomenon that is not normally distributed and explain why.

Critical Thinking Exercises

118. Find two z-scores so that 40% of the data in the distribution lies between them. (More than one answer is possible.)

119. A woman insists that she will never marry a man as short or shorter than she, knowing that only one man in 400 falls into this category. Assuming a mean height of 69 inches for men with a standard deviation of 2.5 inches (and a normal distribution), approximately how tall is the woman?

120. The placement test for a college has scores that are normally distributed with a mean of 500 and a standard deviation of 100. If the college accepts only the top 10% of examinees, what is the cutoff score on the test for admission?

Group Exercise

121. For this activity, group members will conduct interviews with a random sample of students on campus. Each student is to be asked. "What is the worst thing about being a student?" One response should be recorded for each student.

    a. Each member should interview enough students so that there are at least 50 randomly selected students in the sample.

    b. After all responses have been recorded, the group should organize the four most common answers. For each answer, compute the percentage of students in the sample who felt that this is the worst thing about being a student.

    c. Find the margin of error for your survey.

    d. For each of the four most common answers, write a statement about the percentage of all students on your campus who feel that this is the worst thing about being a student.

SECTION 12.5   SCATTER PLOTS, CORRELATION, AND REGRESSION LINES

OBJECTIVES

1. Make a scatter plot for a table of data items.

2. Interpret information given in a scatter plot.

3. Compute the correlation coefficient.

4. Write the equation of the regression line.

5. Use a sample's correlation coefficient to determine whether there is a correlation in the population.

Surprised by the number of people smoking cigarettes in movies and television shows made in the 1940s and 1950s? At that time, there was little awareness of the relationship between tobacco use and numerous diseases. Cigarette smoking was seen as a healthy way to relax and help digest a hearty meal. Then, in 1964, an equation changed everything. To understand the mathematics behind this turning point in public health, we need to explore situations involving data collected on two variables.

Up to this point in the chapter, we have studied situations in which data sets involve a single variable, such as SAT scores, weights, cholesterol levels, and lengths of pregnancies. By contrast, the 1964 study involved data collected on two variables from 11 countries—annual cigarette consumption for each adult male and deaths per million males from lung cancer. In this section, we consider situations in which there are two data items for each randomly selected person or thing. Our interest is in determining whether or not there is a relationship between the two variables and, if so, the strength of that relationship.

1 | Make a scatter plot for a table of data items.

Scatter Plots and Correlation

Is there a relationship between education and prejudice? With increased education, does a person's level of prejudice tend to decrease? Notice that we are interested in two quantities—years of education and level of prejudice. For each person in our sample, we will record the number of years of school completed and the score on a test measuring prejudice. Higher scores on this 1 to 10 test indicate greater prejudice. Using x to represent years of education and y to represent scores on a test measuring prejudice, Table 12.11 shows these two quantities for a random sample of ten people.

TABLE 12.11 RECORDING TWO QUANTITIES IN A SAMPLE OF TEN PEOPLE

Respondent	A	B	C	D	E	F	G	H	I	J
Years of education (x)	12	5	14	13	8	10	16	11	12	4
Score on prejudice test (y)	1	7	2	3	5	4	1	2	3	10

When two data items are collected for every person or object in a sample, the data items can be visually displayed using a *scatter plot*. A **scatter plot** is a collection of data points, one data point per person or object. We can make a scatter plot of the data in Table 12.11 by drawing a horizontal axis to represent years of education and a vertical axis to represent scores on a test measuring prejudice. We then represent each of the ten respondents with a single point on the graph. For example, the dot for respondent A is located to represent 12 years of education on the horizontal axis and 1 on the prejudice test on the vertical axis. Plotting each of the ten pieces of data in a rectangular coordinate system results in the scatter plot shown in Figure 12.28.

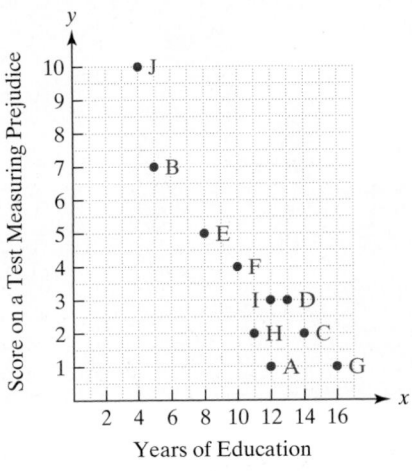

FIGURE 12.28 A scatter plot for education-prejudice data

A scatter plot like the one in Figure 12.28 can be used to determine whether two quantities are related. If there is a clear relationship, the quantities are said to be **correlated**. The scatter plot shows a downward trend among the data points, although there are a few exceptions. People with increased education tend to have a lower score on the test measuring prejudice. **Correlation** is used to determine if there is a relationship between two variables and, if so, the strength and direction of that relationship.

Correlation and Causal Connections

Correlations can often be seen when data items are displayed on a scatter plot. Although the scatter plot in Figure 12.28 indicates a correlation between education and prejudice, we cannot conclude that increased education causes a person's level of prejudice to decrease. There are at least three possible explanations:

1. The correlation between increased education and decreased prejudice is simply a coincidence.
2. Education usually involves classrooms with a variety of different kinds of people. Increased exposure to diversity in the classroom setting, which accompanies increased levels of education, might be an underlying cause for decreased prejudice.
3. Education, the process of acquiring knowledge, requires people to look at new ideas and see things in different ways. Thus, education causes one to be more tolerant and less prejudiced.

Establishing that one thing causes another is extremely difficult, even if there is a strong correlation between these things. For example, as the air temperature increases, there is an increase in the number of people stung by jellyfish at the beach. This does not mean that an increase in air temperature causes more people to be stung. It might mean that because it is hotter, more people go into the water. With an increased number of swimmers, more people are likely to be stung. In short, correlation is not necessarily causation.

Regression Lines and Correlation Coefficients

Figure 12.29 on the next page shows the scatter plot for the education-prejudice data. Also shown is a straight line that seems to approximately "fit" the data points fairly well. Most of the data points lie either near or on this line. A line that best fits the data points in a scatter plot is called a **regression line**. The regression line is the particular line in which the spread of the data points around it is as small as possible.

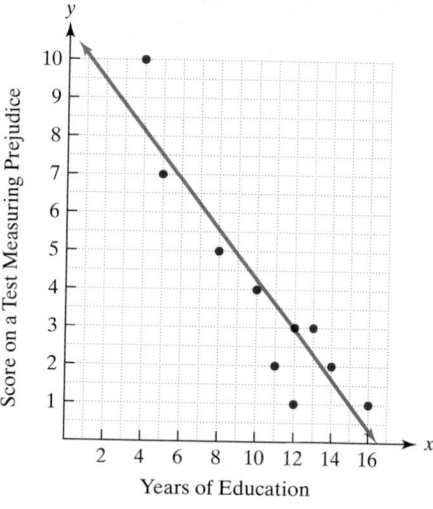

FIGURE 12.29 A scatter plot with a regression line

<table>
<tr><td>2</td><td>Interpret information given in a scatter plot.</td></tr>
</table>

A measure that is used to describe the strength and direction of a relationship between variables whose data points lie on or near a line is called the **correlation coefficient**, designated by r. Figure 12.30 shows scatter plots and correlation coefficients. Variables are **positively correlated** if they tend to increase or decrease together, as in Figure 12.30 (a), (b), and (c). By contrast, variables are **negatively correlated** if one variable tends to decrease while the other increases, as in Figure 12.30 (e), (f), and (g). Figure 12.30 illustrates that a correlation coefficient, r, is a number between -1 and 1, inclusive. Figure 12.30(a) shows a value of 1. This indicates a **perfect positive correlation** in which all points in the scatter plot lie precisely on the regression line that rises from left to right. Figure 12.30(g) shows a value of -1. This indicates a **perfect negative correlation** in which all points in the scatter plot lie precisely on the regression line that falls from left to right.

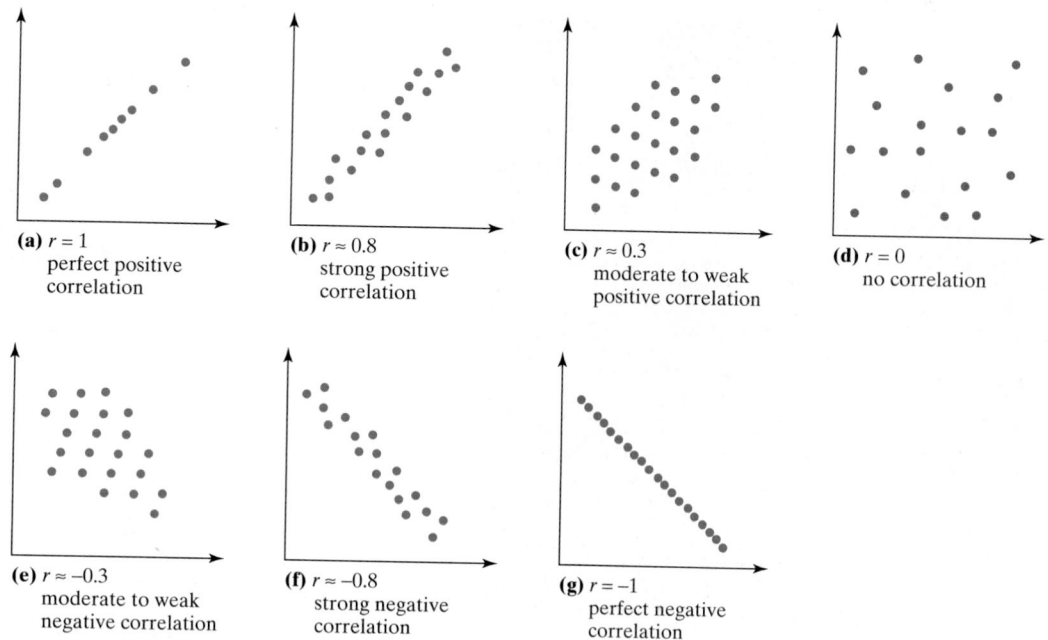

FIGURE 12.30 Scatter plots and correlation coefficients

(a) $r = 1$
perfect positive correlation

(b) $r \approx 0.8$
strong positive correlation

(c) $r \approx 0.3$
moderate to weak positive correlation

(d) $r = 0$
no correlation

(e) $r \approx -0.3$
moderate to weak negative correlation

(f) $r \approx -0.8$
strong negative correlation

(g) $r = -1$
perfect negative correlation

Take another look at Figure 12.30. If r is between 0 and 1, as in (b) and (c), the two variables are positively correlated, but not perfectly. Although all the data points will not lie on the regression line, as in (a), an increase in one variable tends to be accompanied by an increase in the other.

Negative correlations are also illustrated in Figure 12.30. If r is between 0 and -1, as in (e) and (f), the two variables are negatively correlated, but not perfectly.

Although all the data points will not lie on the regression line, as in (g), an increase in one variable tends to be accompanied by a decrease in the other.

EXAMPLE 1 | INTERPRETING A CORRELATION COEFFICIENT

In a 1971 study involving 232 subjects, researchers found a relationship between the subjects' level of stress and how often they became ill. The correlation coefficient in this study was 0.32. Does this indicate a strong relationship between stress and illness?

Solution The correlation coefficient $r = 0.32$ means that as stress increases, frequency of illness also tends to increase. However, 0.32 is only a moderate correlation, illustrated in Figure 12.30(c) on page 685. There is not, based on this study, a strong relationship between stress and illness. In this study, the relationship is somewhat weak.

In a 1996 study involving obesity in mothers and daughters, researchers found a relationship between a high body-mass index for the girls and their mothers. (Body-mass index is a measure of weight relative to height. People with a high body-mass index are overweight or obese.) The correlation in this study was 0.51. Does this indicate a weak relationship between the body-mass index of daughters and the body-mass index of their mothers?

BLITZER BONUS

Missing America

Mary Katherine Campbell, Miss America 1922

Angela Perez Baraquio, Miss America 2000

Here she is, Miss America, the icon of American beauty. Always thin, she is becoming more so. The scatter plot in the figure shows Miss America's body-mass index, a ratio comparing weight divided by the square of height. Two lines are also shown: a line that passes near the data points and a horizontal line representing the World Health Organization's cutoff point for undernutrition. The intersection point indicates that in approximately 1978, Miss America reached this cutoff. There she goes: If the trend continues, Miss America's body-mass index could reach zero in about 320 years.

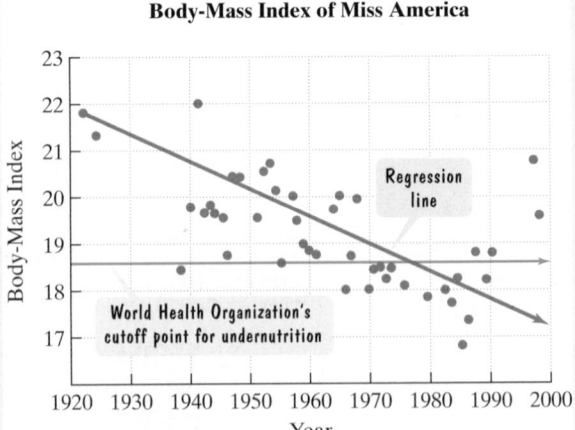

Source: John Hopkins School of Public Health

BLITZER BONUS

Beneficial Uses of Correlation Coefficients

- A Florida study showed a high positive correlation between the number of powerboats and the number of manatee deaths. Many of these deaths were seen to be caused by boats' propellers gashing into the manatees' bodies. Based on this study, Florida set up coastal sanctuaries where power boats are prohibited so that these large gentle mammals that float just below the water's surface could thrive.

- In 1986, researchers studied how psychiatric patients readjusted to their community after their release from a mental hospital. A moderate positive correlation ($r = 0.38$) was found between patients' attractiveness and their postdischarge social adjustment. The better-looking patients were better off. The researchers suggested that physical attractiveness plays a role in patients' readjustment to community living because good-looking people tend to be treated better by others than homely people are.

How to Obtain the Correlation Coefficient and the Equation of the Regression Line

The easiest way to find the correlation coefficient and the equation of the regression line is to use a graphing or statistical calculator. Graphing calculators have statistical menus that enable you to enter the x and y data items for the variables. Based on this information, you can instruct the calculator to display a scatter plot, the equation of the regression line, and the correlation coefficient.

We can also compute the correlation coefficient and the equation of the regression line by hand using formulas. First, we compute the correlation coefficient.

COMPUTING THE CORRELATION COEFFICIENT BY HAND

The following formula is used to calculate the correlation coefficient, r:

$$r = \frac{n(\Sigma xy) - (\Sigma x)(\Sigma y)}{\sqrt{n(\Sigma x^2) - (\Sigma x)^2}\sqrt{n(\Sigma y^2) - (\Sigma y)^2}}.$$

In the formula,

$$n = \text{the number of data points, } (x, y)$$
$$\Sigma x = \text{the sum of the } x\text{-values}$$
$$\Sigma y = \text{the sum of the } y\text{-values}$$
$$\Sigma xy = \text{the sum of the product of } x \text{ and } y \text{ in each pair}$$
$$\Sigma x^2 = \text{the sum of the squares of the } x\text{-values}$$
$$\Sigma y^2 = \text{the sum of the squares of the } y\text{-values}$$
$$(\Sigma x)^2 = \text{the square of the sum of the } x\text{-values}$$
$$(\Sigma y)^2 = \text{the square of the sum of the } y\text{-values}$$

When computing the correlation coefficient by hand, organize your work in five columns:

$$\underline{x \quad y \quad xy \quad x^2 \quad y^2}$$

Find the sum of the numbers in each column. Then, substitute these values into the formula for r. Example 2 illustrates computing the correlation coefficient for the education-prejudice test data.

EXAMPLE 2   COMPUTING THE CORRELATION COEFFICIENT

Shown below are the data involving the number of years of school, x, completed by ten randomly selected people and their scores on a test measuring prejudice, y. Recall that higher scores on the measure of prejudice (1 to 10) indicate greater levels of prejudice. Determine the correlation coefficient between years of education and scores on a prejudice test.

Respondent	A	B	C	D	E	F	G	H	I	J
Years of education (x)	12	5	14	13	8	10	16	11	12	4
Score on prejudice test (y)	1	7	2	3	5	4	1	2	3	10

TECHNOLOGY

Graphing Calculators, Scatter Plots, and Regression Lines

You can use a graphing calculator to display a scatter plot and the regression line. After entering the x and y data items for years of education and scores on a prejudice test, the calculator shows the scatter plot of the data and the regression line.

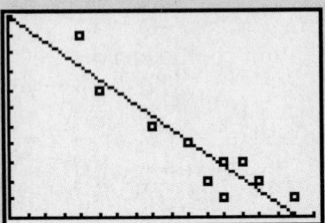

Also displayed below is the regression line's equation and the correlation coefficient, r. The slope shown below is approximately -0.69. The negative slope reinforces the fact that there is a negative correlation between the variables in Example 2.

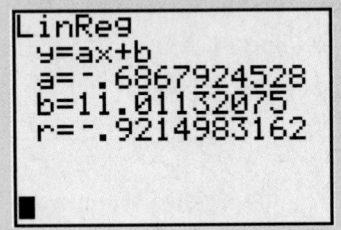

```
LinReg
 y=ax+b
 a=-.6867924528
 b=11.01132075
 r=-.9214983162
 ■
```

"Common wisdom among statisticians is that at least 5% of all data points are corrupted, either when they are initially recorded or when they are entered into the computer."

JESSICA UTTS,
Statistician

Solution   As suggested, organize the work in five columns.

x	y	xy	x^2	y^2
12	1	12	144	1
5	7	35	25	49
14	2	28	196	4
13	3	39	169	9
8	5	40	64	25
10	4	40	100	16
16	1	16	256	1
11	2	22	121	4
12	3	36	144	9
4	10	40	16	100
$\Sigma x = 105$	$\Sigma y = 38$	$\Sigma xy = 308$	$\Sigma x^2 = 1235$	$\Sigma y^2 = 218$

Add all values in the x-column.   Add all values in the y-column.   Add all values in the xy-column.   Add all values in the x²-column.   Add all values in the y²-column.

We use these five sums to calculate the correlation coefficient.

Another value in the formula for r that we have not yet determined is n, the number of data points (x, y). Because there are ten items in the x-column and ten items in the y-column, the number of data points (x, y) is ten. Thus, $n = 10$.

In order to calculate r, we also need to find the square of the sum of the x-values and the y-values:

$$(\Sigma x)^2 = (105)^2 = 11{,}025 \quad \text{and} \quad (\Sigma y)^2 = (38)^2 = 1444.$$

We are ready to determine the value for r.

$$
\begin{aligned}
r &= \frac{n(\Sigma xy) - (\Sigma x)(\Sigma y)}{\sqrt{n(\Sigma x^2) - (\Sigma x)^2}\sqrt{n(\Sigma y^2) - (\Sigma y)^2}} \\[2mm]
&= \frac{10(308) - 105(38)}{\sqrt{10(1235) - 11{,}025}\sqrt{10(218) - 1444}} \\[2mm]
&= \frac{-910}{\sqrt{1325}\sqrt{736}} \\[2mm]
&\approx -0.92
\end{aligned}
$$

This value for r is fairly close to -1 and indicates a strong negative correlation. This means that the more education a person has, the less prejudiced that person is (based on scores on the test measuring levels of prejudice).

 Check 2 Point

Is there a relationship between alcohol from moderate wine consumption and heart disease death rate? The table on the next page gives data from 19 developed countries. Using a calculator, determine the correlation coefficient between these variables. Round to two decimal places.

France

Country	A	B	C	D	E	F	G
Liters of alcohol from drinking wine, per person, per year (x)	2.5	3.9	2.9	2.4	2.9	0.8	9.1
Deaths from heart disease, per 100,000 people per year (y)	211	167	131	191	220	297	71

U.S.

Country	H	I	J	K	L	M	N	O	P	Q	R	S
(x)	0.8	0.7	7.9	1.8	1.9	0.8	6.5	1.6	5.8	1.3	1.2	2.7
(y)	211	300	107	167	266	227	86	207	115	285	199	172

Source: New York Times, December 28, 1994

Once we have determined that two variables are related, we can use the equation of the regression line to determine the exact relationship. Here is the formula for writing the equation of the line that best fits the data:

Write the equation of the regression line.

WRITING THE EQUATION OF THE REGRESSION LINE BY HAND

The equation of the regression line is

$$y = mx + b$$

where

$$m = \frac{n(\Sigma xy) - (\Sigma x)(\Sigma y)}{n(\Sigma x^2) - (\Sigma x)^2} \quad \text{and} \quad b = \frac{\Sigma y - m(\Sigma x)}{n}.$$

EXAMPLE 3 WRITING THE EQUATION OF THE REGRESSION LINE

a. Shown, again, in Figure 12.29 is the scatter plot and the regression line for the data in Example 2. Use the data to find the equation of the regression line that relates years of education and scores on a prejudice test.

b. Approximately what score on the test can be anticipated by a person with nine years of education?

Solution

a. We use the sums obtained in Example 3. We begin by computing m.

$$m = \frac{n(\Sigma xy) - (\Sigma x)(\Sigma y)}{n(\Sigma x^2) - (\Sigma x)^2} = \frac{10(308) - 105(38)}{10(1235) - (105)^2} = \frac{-910}{1325} \approx -0.69$$

With a negative correlation coefficient, it makes sense that the slope of the regression line is negative. This line falls from left to right, indicating a negative correlation.

Now, we find the y-intercept, b.

$$b = \frac{\Sigma y - m(\Sigma x)}{n} = \frac{38 - (-0.69)(105)}{10} = \frac{110.45}{10} \approx 11.05$$

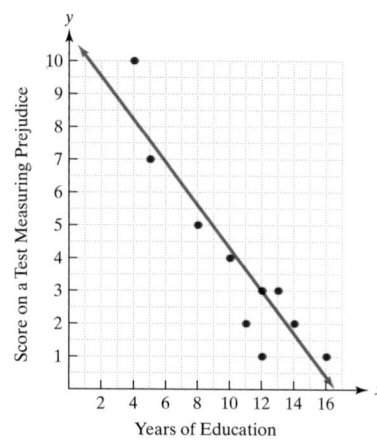

FIGURE **12.29**, repeated

TABLE 12.12 VALUES FOR DETERMINING CORRELATIONS IN A POPULATION

n	$\alpha = 0.05$	$\alpha = 0.01$
4	0.950	0.990
5	0.878	0.959
6	0.811	0.917
7	0.754	0.875
8	0.707	0.834
9	0.666	0.798
10	0.632	0.765
11	0.602	0.735
12	0.576	0.708
13	0.553	0.684
14	0.532	0.661
15	0.514	0.641
16	0.497	0.623
17	0.482	0.606
18	0.468	0.590
19	0.456	0.575
20	0.444	0.561
22	0.423	0.537
27	0.381	0.487
32	0.349	0.449
37	0.325	0.418
42	0.304	0.393
47	0.288	0.372
52	0.273	0.354
62	0.250	0.325
72	0.232	0.302
82	0.217	0.283
92	0.205	0.267
102	0.195	0.254

The larger the sample size, n, the smaller is the value of r needed for a correlation in the population.

5 Use a sample's correlation coefficient to determine whether there is a correlation in the population.

Therefore, the equation of the regression line is

$$y = -0.69x + 11.05$$

This is $y = mx + b$ with $m = -0.69$ and $b = 11.05$.

where x represents the number of years of education and y represents the score on the prejudice test.

b. To anticipate the score on the prejudice test for a person with nine years of education, substitute 9 for x in the regression line's equation.

$$y = -0.69x + 11.05$$
$$y = -0.69(9) + 11.05 = 4.84$$

A person with nine years of education is anticipated to have a score close to 5 on the prejudice test.

> **Check Point 3** Use the data in Check Point 2 on page 689 to find the equation of the regression line. Use the equation to predict the heart disease death rate in a country where adults average 10 liters of alcohol per person per year.

The Level of Significance of r

In Example 2, we found a strong negative correlation between education and prejudice, computing the correlation coefficient, r, to be -0.92. However, the sample size ($n = 10$) was relatively small. With such a small sample, can we truly conclude that a correlation exists in the population? Or could it be that education and prejudice are not related? Perhaps the results we obtained were simply due to sampling error and chance.

Mathematicians have identified values to determine whether r, the correlation coefficient for a sample, can be attributed to a relationship between variables in the population. These values are shown in the second and third columns of Table 12.12. They depend on the sample size, n, listed in the left column. If $|r|$, the absolute value of the correlation coefficient computed for the sample, is greater than the value given in the table, a correlation exists between the variables in the population. The column headed $\alpha = 0.05$ denotes a **significance level of 5%**, meaning that there is a 0.05 probability that, when the statistician says the variables are correlated, they are actually not related in the population. The column on the right, headed $\alpha = 0.01$, denotes a **significance level of 1%**, meaning that there is a 0.01 probability that, when the statistician says the variables are correlated, they are actually not related in the population. Values in the $\alpha = 0.01$ column are greater than those in the $\alpha = 0.05$ column. Because of the possibility of sampling error, there is always a probability that when we say the variables are related, there is actually not a correlation in the population from which the sample was randomly selected.

EXAMPLE 4 | DETERMINING A CORRELATION IN THE POPULATION

In Example 2, we computed $r = -0.92$ for $n = 10$. Can we conclude that there is a negative correlation between education and prejudice in the population?

Solution Begin by taking the absolute value of the calculated correlation coefficient.

$$|r| = |-0.92| = 0.92$$

Now, look to the right of $n = 10$ in Table 12.12. Because 0.92 is greater than both of these values (0.632 and 0.765), we may conclude that a correlation does exist between education and prejudice in the population. (There is a probability of at most 0.01 that the variables are not really correlated in the population and our results could be attributed to chance.)

 **Check 4 Point** If you worked Check Point 2 correctly, you should have found that $r \approx -0.84$ for $n = 19$. Can you conclude that there is a negative correlation between moderate wine consumption and heart disease death rate?

BLITZER BONUS

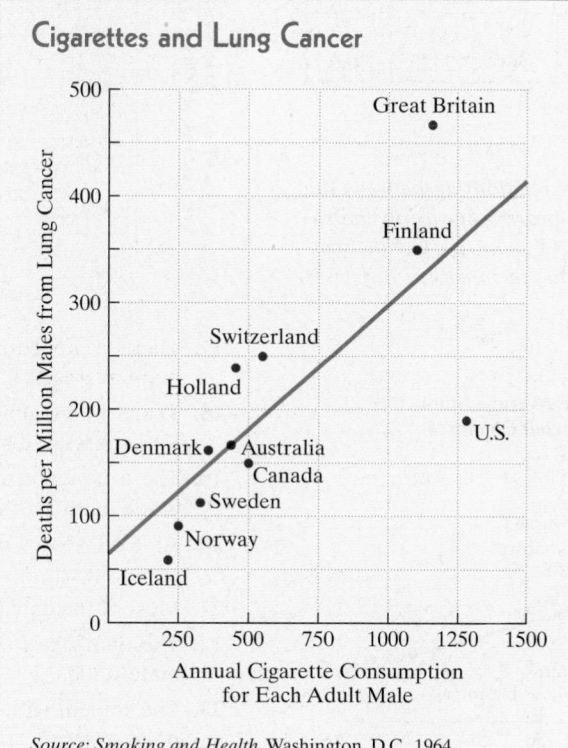

Cigarettes and Lung Cancer

This scatter plot shows a relationship between cigarette consumption among males and deaths due to lung cancer per million males. The data are from 11 countries and date back to a 1964 report by the U.S. Surgeon General. The scatter plot can be modeled by a line whose slope indicates an increasing death rate from lung cancer with increased cigarette consumption. At that time, the tobacco industry argued that in spite of this regression line, tobacco use is not the cause of cancer. Recent data do, indeed, show a causal effect between tobacco use and numerous diseases.

Source: Smoking and Health, Washington, D.C., 1964

Exercise Set 12.5

Practice and Application Exercises

In Exercises 1–8, make a scatter plot for the given data. Use the scatter plot to describe whether or not the variables appear to be related.

1.

x	1	6	4	3	7	2
y	2	5	3	3	4	1

2.

x	2	1	6	3	4
y	4	5	10	8	9

3.

x	8	6	1	5	4	10	3
y	2	4	10	5	6	2	9

4.

x	4	5	2	1
y	1	3	5	4

5.

Respondent	A	B	C	D	E	F	G
Years of education of parent (x)	13	9	7	12	12	10	11
Years of education of child (y)	13	11	7	16	17	8	17

6.

Respondent	A	B	C	D	E
IQ (x)	110	115	120	125	135
Annual income (y) (in thousands of dollars)	30	32	36	40	44

7. The data show the number of registered automatic weapons, in thousands, and the murder rate, in murders per 100,000, for eight randomly selected states.

Automatic weapons, x	11.6	8.3	6.9	3.6	2.6	2.5	2.4	0.6
Murder rate, y	13.1	10.6	11.5	10.1	5.3	6.6	3.6	4.4

Source: FBI and Bureau of Alcohol, Tobacco, and Firearms

8. The data show the number of employed and unemployed male workers, 20 years and older, in thousands, for six selected years in the United States.

Year	1995	1996	1997	1998	1999	2000
Employed, x	64,085	64,897	66,524	67,134	67,761	68,580
Unemployed, y	3239	3147	2826	2580	2433	2350

Source: Bureau of Labor Statistics

The scatter plot in the figure shows the relationship between the percentage of married women of child-bearing age using contraceptives and births per woman in selected countries. Use the scatter plot to determine whether each of the statements in Exercises 9–16 is true or false.

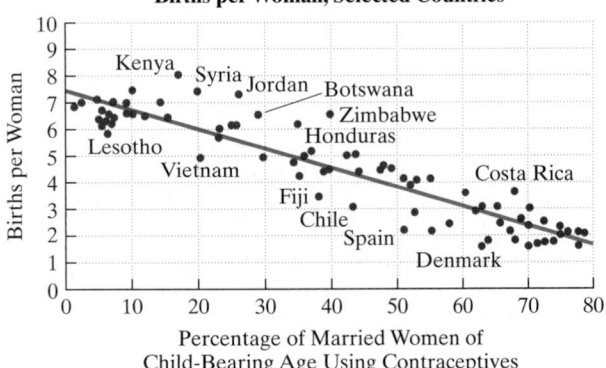

Contraceptive Prevalence and Average Number of Births per Woman, Selected Countries

Source: Population Reference Bureau

9. There is a strong positive correlation between contraceptive use and births per woman.

10. There is no correlation between contraceptive use and births per woman.

11. There is a strong negative correlation between contraceptive use and births per woman.

12. There is a causal relationship between contraceptive use and births per woman.

13. With approximately 43% of women of child-bearing age using contraceptives, there are 3 births per woman in Chile.

14. With 20% of women of child-bearing age using contraceptives, there are 6 births per woman in Vietnam.

15. No two countries have a different number of births per woman with the same percentage of married women using contraceptives.

16. The country with the greatest number of births per woman also has the smallest percentage of women using contraceptives.

The scatter plot in the figure shows the relationship between SAT scores and grade point average for the first year in college for a group of randomly selected college students. Use the scatter plot to determine whether each of the statements in Exercises 17–24 is true or false.

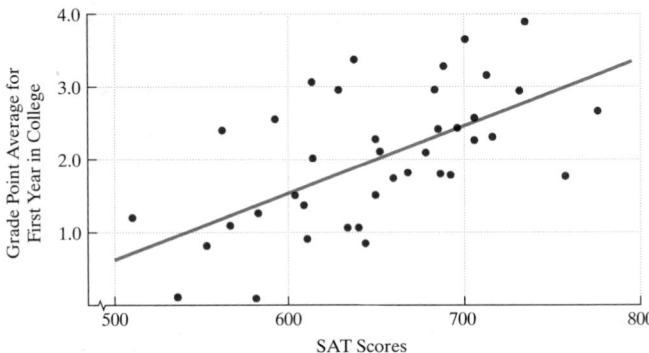

17. There is no correlation between SAT scores and grade point averages.

18. There is an almost-perfect positive correlation between SAT scores and grade point averages.

19. There is a positive correlation between SAT scores and grade point averages.

20. As SAT scores decrease, grade point averages also tend to decrease.

21. Most of the data points lie on the regression line.

22. The number of college students in this sample is approximately 200.

23. The student with the lowest SAT score has the worst grade point average.

24. The student with the highest SAT score has the best grade point average.

The scatter plot shows the relationship between the Olympic year and the men's and women's winning times, in seconds, in the Olympic 100-meter freestyle swimming race. Also shown are the regression lines. Use the information given in the figure to determine whether each statement in Exercises 25–28 is true or false.

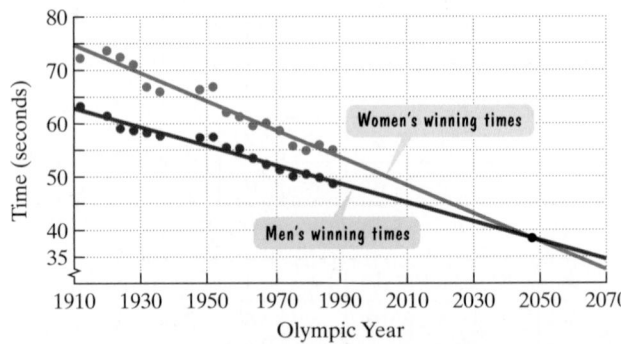

Winning Times in the Olympic 100-Meter Freestyle Swimming Race

25. There is a strong positive correlation between the Olympic year and winning time in the race for both women and men.

26. Even if present trends continue, women will not swim faster than men in this Olympic event.

27. Many of the data points do not fall exactly on the regression lines for the two sets of data.

28. A reasonable estimate of the correlation coefficient for the data for the women is −0.2.

Use the scatter plots shown, labeled (a)–(f), to solve Exercises 29–32.

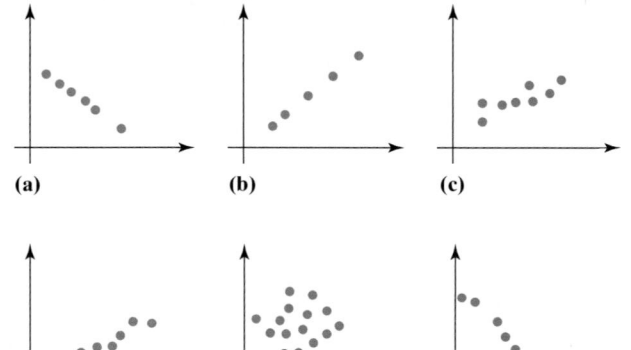

(a)   (b)   (c)

(d)   (e)   (f)

29. Which scatter plot indicates a perfect negative correlation?

30. Which scatter plot indicates a perfect positive correlation?

31. In which scatter plot is $r = 0.9$?

32. In which scatter plot is $r = 0.01$?

Compute r, the correlation coefficient, rounded to the nearest thousandth, for the data in

33. Exercise 1.   **34.** Exercise 2.

35. Exercise 3.   **36.** Exercise 4.

37. Use the data in Exercise 5 to solve this exercise.

    a. Determine the correlation coefficient between years of education of parent and child.

    b. Find the equation of the regression line for years of education of parent and child.

    c. Approximately how many years of education can we predict for a child with a parent who has 16 years of education?

38. Use the data in Exercise 6 to solve this exercise.

    a. Determine the correlation coefficient between IQ and income.

    b. Find the equation of the regression line for IQ and income.

    c. Approximately what annual income can be anticipated by a person whose IQ is 123?

39. Use the data in Exercise 7 to solve this exercise.

    a. Determine the correlation coefficient between the number of automatic weapons and the murder rate.

    b. Find the equation of the regression line.

    c. Approximately what murder rate can we anticipate in a state that has 14 thousand registered weapons?

40. Use the data in Exercise 8 to solve this exercise.

    a. Determine the correlation coefficient between the number of employed males and the number of unemployed males.

    b. Find the equation of the regression line.

    c. Approximately how many unemployed males can we anticipate for a year in which there are 70,000 thousand employed males?

In Exercises 41–47, the correlation coefficient, r, is given for a sample of n data points. Use the $\alpha = 0.05$ column in Table 12.12 on page 690 to determine whether or not we may conclude that a correlation does exist in the population. (Using the $\alpha = 0.05$ column there is a probability of 0.05 that the variables are not really correlated in the population and our results could be attributed to chance. Ignore this possibility when concluding whether or not there is a correlation in the population.)

41. $n = 20, r = 0.5$ **42.** $n = 27, r = 0.4$

43. $n = 12, r = 0.5$ **44.** $n = 22, r = 0.04$

45. $n = 72, r = -0.351$ **46.** $n = 37, r = -0.37$

47. $n = 20, r = -0.37$

48. In the 1964 study on cigarette consumption and deaths due to lung cancer (see the Blitzer Bonus on page 691), $n = 11$ and $r = 0.73$. What can you conclude using the $\alpha = 0.05$ column in Table 12.12?

Writing in Mathematics

49. What is a scatter plot?

50. How does a scatter plot indicate that two variables are correlated?

51. Give an example of two variables with a strong positive correlation and explain why this is so.

52. Give an example of two variables with a strong negative correlation and explain why this is so.

53. What is meant by a regression line?

54. When all points in a scatter plot fall on the regression line, what is the value of the correlation coefficient? Describe what this means.

For the pairs of quantities in Exercises 55–58, describe whether a scatter plot will show a positive correlation, a negative correlation, or no correlation. If there is a correlation, is it strong, moderate, or weak? Explain your answers.

55. Height and weight

56. Number of days absent and grade in a course

57. Height and grade in a course

58. Hours of television watched and grade in a course

59. Explain how to use the correlation coefficient for a sample to determine if there is a correlation in the population.

Critical Thinking Exercises

60. Which one of the following is true?

    a. A scatter plot need not define y as a function of x.

    b. The correlation coefficient and the slope of the regression line for the same set of data can have opposite signs.

    c. When all points in a scatter plot fall on the regression line, the value of the correlation coefficient is 0.

    d. If the same number is subtracted from each x-item, but the y-item stays the same, the correlation coefficient for these new data points decreases.

61. Give an example of two variables with a strong correlation, where each variable is not the cause of the other.

Technology Exercises

62. Enter the data items from Table 12.11 on page 683 into an Excel spreadsheet.

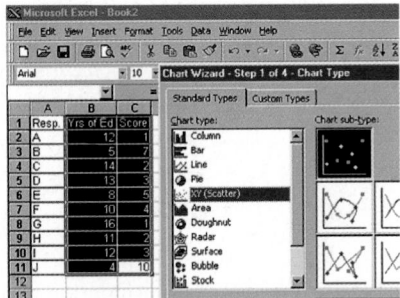

After you have entered the data, use the computer's mouse to highlight the cells containing the data corresponding to the years of education and test scores. Then, activate the Chart Option of the spreadsheet by clicking on the Insert pull-down menu and selecting Chart. Select the XY (Scatter) option (see picture). Then, click the "Next" button to work your way through the options, including the labels and cosmetic effects. Once you have decided on all of the options (don't worry, you can go back and change them), click the "Finish" button. Excel should generate a scatter plot that looks like Figure 12.28 on page 684.

63. Use MS Excel (or another spreadsheet program) to create a scatter plot for the data in Exercise 6.

64. Use the linear regression feature of a graphing calculator to verify your work in any two exercises from Exercises 37–40, parts (a) and (b).

Group Exercises

65. The group should select two variables that it believes have a strong positive or negative correlation. Once these variables have been determined,

    a. Collect at least 30 ordered pairs of data (x, y) from a sample of people on your campus.

    b. Draw a scatter plot for the data collected.

    c. Does the scatter plot indicate a positive correlation, a negative correlation, or no relationship between the variables?

    d. Calculate r. Does the value of r reinforce the impression conveyed by the scatter plot?

    e. Find the equation of the regression line.

    f. Use the regression line's equation to make a prediction about a y-value given an x-value.

    g. Are the results of this project consistent with the group's original belief about the correlation between the variables, or are there some surprises in the data collected?

66. Group members should consult an almanac, a magazine, or a newspaper in which data about two variables are presented. (See, for example, Exercises 7 and 8 in this exercise set.)

    a. Draw a scatter plot for the data selected by the group.

    b. Does the scatter plot indicate a positive correlation, a negative correlation, or no correlation between the variables?

    c. Calculate r. Does the value of r reinforce the impression conveyed by the scatter plot?

    d. Find the equation of the regression line and use the equation to make a prediction.

Chapter Summary, Review, and Test

SUMMARY

DEFINITIONS AND CONCEPTS

EXAMPLES

12.1 Sampling, Frequency Distributions, and Graphs

    a. A population is the set containing all objects whose properties are to be described and analyzed. A sample is a subset of the population.

    Ex. 1, p. 632

    b. Random samples are obtained in such a way that each member of the population has an equal chance of being selected.

    Ex. 2, p. 633

    c. Data can be organized and presented in frequency distributions, grouped frequency distributions, histograms, frequency polygons, and stem-and-leaf plots.

    Ex. 3, p. 634;
    Ex. 4, p. 635;
    Figures 12.2 and 12.3,
    p. 637;
    Ex. 5, p. 637

12.2 Measures of Central Tendency

    a. The mean is the sum of the data items divided by the number of items. Mean $= \dfrac{\Sigma x}{n}$.

    Ex. 1, p. 645

b. The mean of a frequency distribution is computed using

$$\text{Mean} = \frac{\Sigma xf}{n}$$

where x is each data value, f is its frequency, and n is the total frequency of the distribution.

Ex. 2, p. 646

c. The median of ranked data is the item in the middle or the mean of the two middlemost items. The median is the value in the $\frac{n+1}{2}$ position in the list of ranked data.

Ex. 3, p. 648;
Ex. 4, p. 649;
Ex. 5, p. 649;
Ex. 6, p. 650

d. The mode of a data set is the value that occurs most often. If there is no such value, there is no mode. If more than one data value has the highest frequency, then each of these data values is a mode.

Ex. 8, p. 653

e. The midrange is computed using

$$\frac{\text{lowest data value } + \text{ highest data value}}{2}.$$

Ex. 9, p. 653

12.3 Measures of Dispersion

a. Range = highest data value − lowest data value

Ex. 1, p. 658

b. Standard deviation $= \sqrt{\dfrac{\Sigma(\text{data item} - \text{mean})^2}{n-1}}$

Ex. 2, p. 659;
Ex. 3, p. 660;
Ex. 4, p. 661

c. As the spread of data items increases, the standard deviation gets larger.

Ex. 5, p. 663

12.4 The Normal Distribution

a. The normal distribution is a theoretical distribution for the entire population. The distribution is bell shaped and symmetric about a vertical line through its center, where the mean, median, and mode are located.

b. The 68–95–99.7 Rule

Ex. 2, p. 668

Approximately 68% of the data items fall within 1 standard deviation of the mean,

95% within 2 standard deviations of the mean, and

99.7% within 3 standard deviations of the mean.

c. A z-score describes how many standard deviations a data item in a normal distribution lies above or below the mean.

Ex. 3, p. 669;
Ex. 4, p. 670;
Ex. 5, p. 671

$$z\text{-score} = \frac{\text{data item} - \text{mean}}{\text{standard deviation}}$$

d. If $n\%$ of the items in a distribution are less than a particular data item, that data item is in the nth percentile of the distribution. A table showing z-scores and their percentiles can be used to find the percentage of data items less than or greater than a given data item in a normal distribution, as well as the percentage of data items between two given items. See the boxed summary on computing percentage of data items on page 676.

Ex. 6, p. 671;
Ex. 7, p. 672;
Ex. 8, p. 674;
Ex. 9, p. 675

e. If a statistic is obtained from a random sample of size n, there is a 95% probability that it lies within $\dfrac{1}{\sqrt{n}}$ of the true population statistic. $\pm\dfrac{1}{\sqrt{n}}$ is called the margin of error.

Ex. 10, p. 677

12.5 Scatter Plots, Correlation, and Regression Lines

a. A plot of data points is called a scatter plot. If the points lie approximately along a line, the line that best fits the data is called a regression line.

b. A correlation coefficient, r, measures the strength and direction of a possible relationship between variables. If $r = 1$, there is a perfect positive correlation, and if $r = -1$, there is a perfect negative correlation. If $r = 0$, there is no relationship between the variables. Table 12.12 on page 690 indicates whether r denotes a correlation in the population.

Ex. 1, p. 686;
Ex. 4, p. 690

c. The formula for computing the correlation coefficient, r, is given in the box on page 687. The equation of the regression line is given in the box on page 689.

Ex. 2, p. 687;
Ex. 3, p. 689

REVIEW EXERCISES

12.1

A telephone poll of 181 randomly selected American mothers with children under 18 in the household was taken by Time/CNN May 21–22, 2003. The data below are from the poll. Use this information to solve Exercises 1–2.

Do you think the war with Iraq has made terrorist attacks in the U.S. more likely or less likely, or hasn't it made any difference?

More likely	Less likely	No difference
47%	19%	31%

1. Describe the population and the sample of this poll.
2. For each person polled, what variable is measured?
3. The government of a large city wants to know if its citizens will support a three-year tax increase to provide additional support to the city's community college system. The government decides to conduct a survey of the city's residents before placing a tax increase initiative on the ballot. Which one of the following is most appropriate for obtaining a sample of the city's residents?

   a. Survey a random sample of persons within each geographic region of the city.
   b. Survey a random sample of community college professors living in the city.
   c. Survey every tenth person who walks into the city's government center on two randomly selected days of the week.
   d. Survey a random sample of persons within each geographic region of the state in which the city is located.

A random sample of ten college students is selected and each student is asked how much time he or she spent on homework during the previous weekend. The following times, in hours, are obtained:

$$8, 10, 9, 7, 9, 8, 7, 6, 8, 7.$$

Use these data items to solve Exercises 4–6.

4. Construct a frequency distribution for the data.
5. Construct a histogram for the data.
6. Construct a frequency polygon for the data.

The 50 grades on a physiology test are shown. Use the data to solve Exercises 7–8.

44	24	54	81	18
34	39	63	67	60
72	36	91	47	75
57	74	87	49	86
59	14	26	41	90
13	29	13	31	68
63	35	29	70	22
95	17	50	42	27
73	11	42	31	69
56	40	31	45	51

7. Construct a grouped frequency distribution for the data. Use 0–39 for the first class, 40–49 for the second class, and make each subsequent class width the same as the second class.
8. Construct a stem-and-leaf plot for the data.
9. The data show the U.S. prison population for five selected years. The graph shown, which appeared in a recent textbook, is based on the data. Describe what is misleading in terms of how each bar represents the prison population.

Year	U.S. Prison Population
1980	329,821
1985	502,507
1990	773,919
1995	1,125,874
2001	1,406,031

Source: Bureau of Justice Statistics

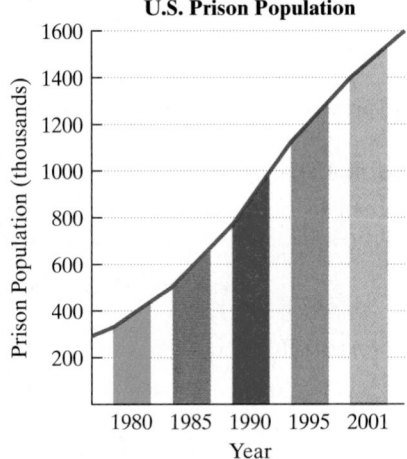

U.S. Prison Population

12.2

In Exercises 10–11, find the mean for each group of data items.

10. 84, 90, 95, 89, 98
11. 33, 27, 9, 10, 6, 7, 11, 23, 27
12. Find the mean for the data items in the given frequency distribution.

Score x	Frequency f
1	2
2	4
3	3
4	1

In Exercises 13–14, find the median for each group of data items.

13. 33, 27, 9, 10, 6, 7, 11, 23, 27
14. 28, 16, 22, 28, 34
15. Find the median for the data items in the frequency distribution in Exercise 12.

In Exercises 16–17, find the mode for each group of data items. If there is no mode, so state.

16. 33, 27, 9, 10, 6, 7, 11, 23, 27

17. 582, 585, 583, 585, 587, 587, 589

18. Find the mode for the data items in the frequency distribution in Exercise 12.

In Exercises 19–20, find the midrange for each group of data items.

19. 84, 90, 95, 88, 98

20. 33, 27, 9, 10, 6, 7, 11, 23, 27

21. Find the midrange for the data items in the frequency distribution in Exercise 12.

22. In 2001, the mean salary for major league baseball players was $2,264,403. Explain how this figure was obtained.

23. A student took seven tests in a course, scoring between 90% and 95% on three of the tests, between 80% and 89% on three of the tests, and below 40% on one of the tests. In this distribution, is the mean or the median more representative of the student's overall performance in the course? Explain your answer.

24. The data items below are the ages of U.S. presidents at the time of their first inauguration.

57 61 57 57 58 57 61 54 68 51 49 64 50 48

65 52 56 46 54 49 51 47 55 55 54 42 51 56

55 51 54 51 60 62 43 55 56 61 52 69 64 46 54

    a. Organize the data in a frequency distribution.

    b. Use the frequency distribution to find the mean age, median age, modal age, and midrange age of the presidents when they were inaugurated.

12.3

In Exercises 25–26, find the range for each group of data items.

25. 28, 34, 16, 22, 28

26. 312, 783, 219, 312, 426, 219

27. The mean for the data items 29, 9, 8, 22, 46, 51, 48, 42, 53, 42 is 35. Find **a.** the deviation from the mean for each data item and **b.** the sum of the deviations in part (a).

28. Use the data items 36, 26, 24, 90, and 74 to find **a.** the mean, **b.** the deviation from the mean for each data item, and **c.** the sum of the deviations in part (b).

In Exercises 29–30, find the standard deviation for each group of data items.

29. 3, 3, 5, 8, 10, 13

30. 20, 27, 23, 26, 28, 32, 33, 35

31. A test measuring anxiety levels is administered to a sample of ten college students with the following results. (High scores indicate high anxiety.)

$$10, 30, 37, 40, 43, 44, 45, 69, 86, 86$$

Find the mean, range, and standard deviation for the data.

32. Compute the mean and the standard deviation for each of the following data sets. Then, write a brief description of similarities and differences between the two sets based on each of your computations.

    Set A: 80, 80, 80, 80   Set B: 70, 70, 90, 90

33. Describe how you would determine

    a. which of the two groups, men or women, at your college has a higher mean grade point average.

    b. which of the groups is more consistently close to its mean grade point average.

12.4

The scores on a test are normally distributed with a mean of 70 and a standard deviation of 8. In Exercises 34–36, find the score that is

34. 2 standard deviations above the mean.

35. $3\frac{1}{2}$ standard deviations above the mean.

36. $1\frac{1}{4}$ standard deviations below the mean.

The ages of people living in a retirement community are normally distributed with a mean age of 68 years and a standard deviation of 4 years. In Exercises 37–43, use the 68–95–99.7 Rule to find the percentage of people in the community whose ages

37. are between 64 and 72.

38. are between 60 and 76.

39. are between 68 and 72.

40. are between 56 and 80.

41. exceed 72.

42. are less than 72.

43. exceed 76.

A set of data items is normally distributed with a mean of 50 and a standard deviation of 5. In Exercises 44–48, convert each data item to a z-score.

44. 50   **45.** 60   **46.** 58

47. 35   **48.** 44

49. A student scores 60 on a vocabulary test and 80 on a grammar test. The data items for both tests are normally distributed. The vocabulary test has a mean of 50 and a standard deviation of 5. The grammar test has a mean of 72 and a standard deviation of 6. On which test did the student have the better score? Explain why this is so.

The number of miles that a particular brand of car tires lasts is normally distributed with a mean of 32,000 miles and a standard deviation of 4000 miles. In Exercises 50–52, find the data item in this distribution that corresponds to the given z-score.

50. $z = 1.5$ **51.** $z = 2.25$ **52.** $z = -2.5$

The mean cholesterol level for all men in the United States is 200 and the standard deviation is 15. In Exercises 53–56, use Table 12.10 on page 672 to find the percentage of U.S. men whose cholesterol level

53. is less than 221.   **54.** is greater than 173.

55. is between 173 and 221.   **56.** is between 164 and 182.

Use the percentiles for the weights of adult men over 40 to solve Exercises 57–59.

Weight	Percentile
235	86
227	75
180	50
173	25

Find the percentage of men over 40 who weigh

57. less than 227 pounds.

58. more than 235 pounds.

59. between 227 and 235 pounds.

60. "The Man Poll" of 1014 randomly selected American men ages 18 and older was taken by *Newsweek* June 2–4, 2003. The poll indicated that 33% of the men responded "yes" to the question

Have you ever suffered from depression or anxiety disorder?

a. Find the margin of error for this percent.

b. Write a statement about the percentage of American men who have ever suffered from depression or anxiety disorder.

12.5

In Exercises 61–62, make a scatter plot for the given data. Use the scatter plot to describe whether or not the variables appear to be related.

61.

x	1	3	4	6	8	9
y	1	2	3	3	5	5

62.

The scatter plot in the figure shows the relationship between the percentage of adult females in a country who are literate and the mortality of children under five. Also shown is the regression line. Use this information to determine whether each of the statements in Exercises 63–69 is true or false.

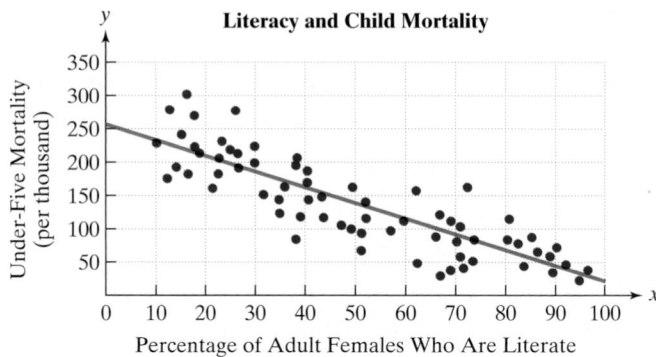

Literacy and Child Mortality

Source: United Nations

63. There is a perfect negative correlation between the percentage of adult females who are literate and under-five mortality.

64. As the percentage of adult females who are literate increases, under-five mortality tends to decrease.

65. The country with the least percentage of adult females who are literate has the greatest under-five mortality.

66. No two countries have the same percentage of adult females who are literate but different under-five mortalities.

67. There are more than 20 countries in this sample.

68. There is no correlation between the percentage of adult females who are literate and under-five mortality.

69. The country with the greatest percentage of adult females who are literate has an under-five mortality rate that is less than 50 children per thousand.

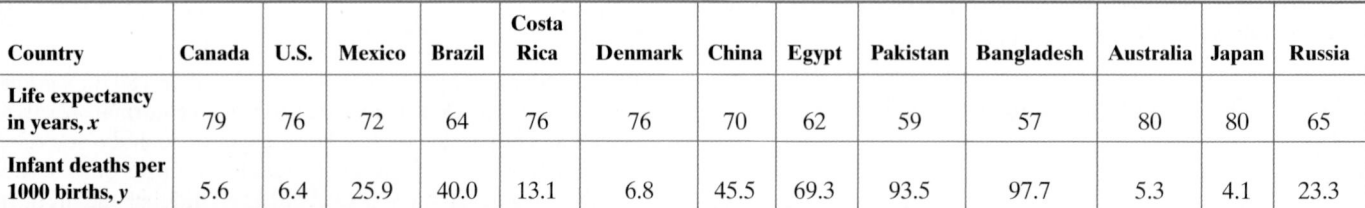

Country	Canada	U.S.	Mexico	Brazil	Costa Rica	Denmark	China	Egypt	Pakistan	Bangladesh	Australia	Japan	Russia
Life expectancy in years, x	79	76	72	64	76	76	70	62	59	57	80	80	65
Infant deaths per 1000 births, y	5.6	6.4	25.9	40.0	13.1	6.8	45.5	69.3	93.5	97.7	5.3	4.1	23.3

Source: U.S. Bureau of the Census International Database

70. Which one of the following scatter plots indicates a correlation coefficient of approximately -0.9?

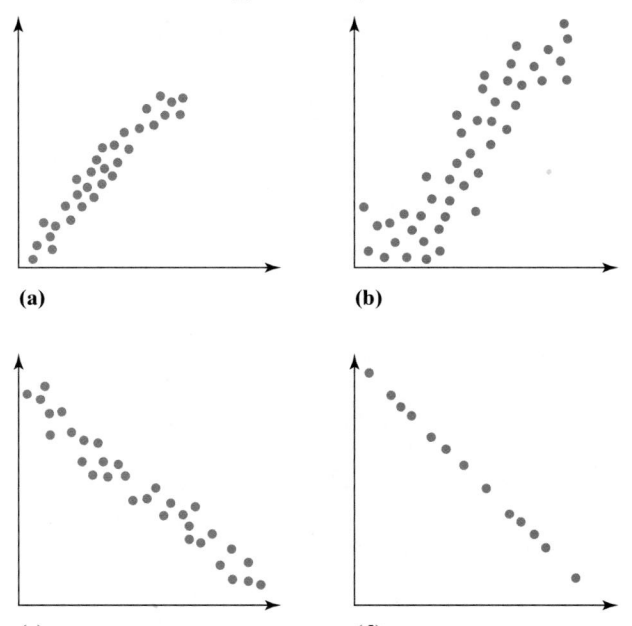

(a)   **(b)**

(c)   **(d)**

Number of Movie Screens in the U.S.

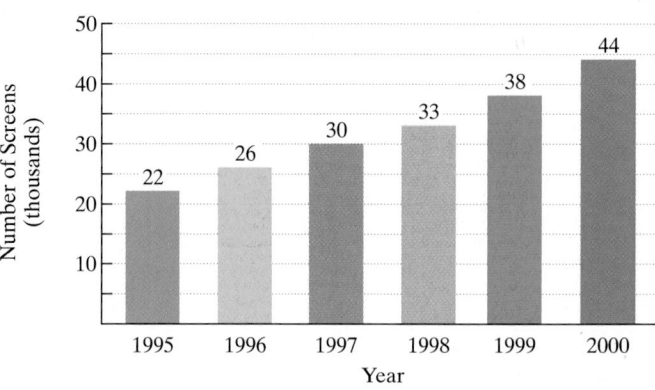

Source: Motion Picture Association of America

b. Let x represent the number of years after 1995 (so $x = 0$ corresponds to 1995, $x = 1$ to 1996, and so on) and y represent the thousands of screens. Calculate the correlation coefficient.

c. Using Table 12.12 on page 690 and the $\alpha = 0.01$ column, determine whether there is a correlation between time and the number of screens.

71. Use the data in Exercise 61 to solve the exercise.

a. Compute r, the correlation coefficient, rounded to the nearest thousandth.

b. Find the equation of the regression line.

72. Films may not be getting any better, but in this era of moviegoing, the number of screens available for new films and the classics has exploded. The graph in the next column shows the number of screens in the United States from 1995 through 2000.

a. Construct a scatter plot with the year on the horizontal axis.

Group Exercise

73. What is the opinion of students on your campus about … ? Group members should begin by deciding on some aspect of college life around which student opinion can be polled. The poll should consist of the question, "What is your opinion of … ?" Be sure to provide options such as excellent, good, average, poor, horrible, or a 1 to 10 scale, or possibly grades of A, B, C, D, F. Use a random sample of students on your campus and conduct the opinion survey. After collecting the data, present and interpret it using as many of the skills and techniques learned in this chapter as possible

CHAPTER 12 TEST

1. Politicians in the Florida Keys need to know if the residents of Key Largo think the amount of money charged for water is reasonable. The politicians decide to conduct a survey of a sample of Key Largo's residents. Which procedure would be most appropriate for a sample of Key Largo's residents?

a. Survey all water customers who pay their water bills at Key Largo City Hall on the third day of the month.

b. Survey a random sample of executives who work for the water company in Key Largo.

c. Survey 5000 individuals who are randomly selected from a list of all people living in Georgia and Florida.

d. Survey a random sample of persons within each neighborhood of Key Largo.

Use these scores on a ten-point quiz to solve Exercises 2–4.

$$8, 5, 3, 6, 5, 10, 6, 9, 4, 5, 7, 9, 7, 4, 8, 8$$

2. Construct a frequency distribution for the data.

3. Construct a histogram for the data.

4. Construct a frequency polygon for the data.

Use the 30 test scores listed below to solve Exercises 5–6.

79	51	67	50	78
62	89	83	73	80
88	48	60	71	79
89	63	55	93	71
41	81	46	50	61
59	50	90	75	61

5. Construct a grouped frequency distribution for the data. Use 40–49 for the first class and use the same width for each subsequent class.

6. Construct a stem-and-leaf display for the data.

Use the six data items listed below to solve Exercises 7–10.

$$3, 6, 2, 1, 7, 3$$

7. Find the mean.   **8.** Find the median.

9. Find the midrange.   **10.** Find the standard deviation.

Use the frequency distribution shown to solve Exercises 11–13.

Score x	Frequency f
1	3
2	5
3	2
4	2

11. Find the mean.

12. Find the median.

13. Find the mode.

14. The annual salaries of four salespeople and the owner of a bookstore are

$$\$17,500, \$19,000, \$22,000, \$27,500, \$98,500.$$

Is the mean or the median more representative of the five annual salaries? Briefly explain your answer.

According to the American Freshman, *the number of hours that college freshmen spend studying each week is normally distributed with a mean of 7 hours and a standard deviation of 5.3 hours. In Exercises 15–16, use the 68–95–99.7 Rule to find the percentage of college freshmen who study*

15. between 7 and 12.3 hours each week.

16. more than 17.6 hours each week.

17. IQ scores are normally distributed in the population. Who has a higher IQ: a student with a 120 IQ on a scale where 100 is the mean and 10 is the standard deviation, or a professor with a 128 IQ on a scale where 100 is the mean and 15 is the standard deviation? Briefly explain your answer.

18. Use the *z*-scores and the corresponding percentiles shown below to solve this exercise. Test scores are normally distributed with a mean of 74 and a standard deviation of 10. What percentage of the scores are above 88?

z-score	Percentile
1.1	86.43
1.2	88.49
1.3	90.32
1.4	91.92
1.5	93.32

19. Use the percentiles in the table shown below to find the percentage of scores between 630 and 690.

Score	Percentile
780	99
750	87
720	72
690	49
660	26
630	8
600	1

20. Using a random sample of 100 students from a campus of approximately 12,000 students, 60% of the students in the sample said they were very satisfied with their professors.

a. Find the margin of error in this percent.

b. Write a statement about the percentage of the entire population of students from this campus who are very satisfied with their professors.

21. Make a scatter plot for the given data. Use the scatter plot to describe whether or not the variables appear to be related.

x	1	4	3	5	2
y	5	2	2	1	4

The scatter plot shows the number of minutes each of 16 people exercise per week and the number of headaches per month each person experiences. Use the scatter plot to determine whether each of the statements in Exercises 22–24 is true or false.

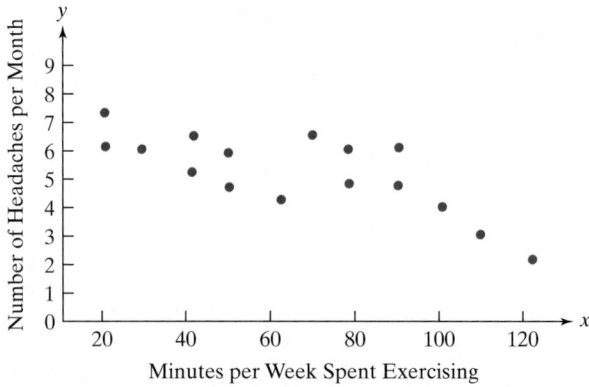

22. An increase in the number of minutes devoted to exercise causes a decrease in headaches.

23. There is a perfect negative correlation between time spent exercising and number of headaches.

24. The person who exercised most per week had the least number of headaches per month.

25. Is the relationship between the price of gas and the number of people visiting our national parks a positive correlation, a negative correlation, or is there no correlation? Explain your answer.

13 Mathematical Systems

s beauty in the eye of the beholder? Or are there certain objects (or people) that are so well balanced and proportioned that they are universally pleasing to the eye? In this chapter you will learn to use mathematics to describe the characteristics of forms such as the passionflower that many people find alluring.

It took a while to get your passionflower vine going, but now it's finally taken off. Flowers have appeared and you are overwhelmed by their beauty. What is it about the flower that makes it so visually appealing?

This problem appears as Exercise 50 in Exercise Set 13.2 and Exercise 35 in the review exercises.

SECTION 13.1   MATHEMATICAL SYSTEMS

OBJECTIVES

1. Understand what is meant by a mathematical system.

2. Understand properties of certain mathematical systems.

FIGURE 13.1 A three-way switch on the table lamp

 Understand what is meant by a mathematical system.

TABLE 13.1

Number of clockwise turns				
⊕	0	1	2	3
0	0	1	2	3
1	1	2	3	0
2	2	3	0	1
3	3	0	1	2

Starting positions

⊕	0	1	2	3
0	0	1	2	3
1	1	2	3	0
2	2	3	0	①
3	3	0	1	2

FIGURE 13.2 Using Table 13.1 to find 2 ⊕ 3. The intersecting lines show that 2 ⊕ 3 = 1.

Thanks for the gift. What, exactly, is it? Oh, a lamp. It's—uh—very nice. (What am I ever going to do with this hideous monument to bad taste?) What's that? You want to tell me about some of its special features? I'm all ears. We're going to start with the three-way switch. Looks pretty high-tech: Does it come with a manual? Only kidding. Do go on. So, if I turn the switch once to the right, I get low-intensity light. Another turn to the right gives medium-intensity light. Still another turn results in high-intensity light. Finally, one more right turn puts the switch in the off position, resulting in no light. You've even made me a diagram of the switch? (See Figure 13.1.) Nothing like capping off this fascinating tutorial with a visual aid. Life is good.

Despite the sarcasm of the lamp's recipient, there is something mathematical happening. We begin with the obvious: The amount of light that you get depends on the position of the three-way switch when you begin and the number of turns to the right you move the switch. Using ⊕ to represent turning the switch to the right (clockwise), here are some examples:

$$0 \quad ⊕ \quad 2 \quad = \quad 2$$

The switch is in the off position.	Turn the switch clockwise **2** turns.	The final position of the switch is **2**, giving medium light.

$$1 \quad ⊕ \quad 2 \quad = \quad 3$$

The switch is in the low position.	Turn the switch clockwise **2** turns.	The final position of the switch is **3**, giving high-intensity light.

$$3 \quad ⊕ \quad 1 \quad = \quad 0.$$

The switch is in the high position.	Turn the switch clockwise **1** turn.	The final position of the switch is **0**, the off position with no light.

We can use a table to show all possible starting and ending positions for the three-way switch. In Table 13.1, the starting positions are shown in the left column. The number of clockwise turns are shown across the top. The 16 entries in the table represent all possible final positions of the switch. For example, to find $2 ⊕ 3$, find 2 in the left column and 3 across the top. Assuming a horizontal line through 2 and a vertical line through 3, as in Figure 13.2, the point of intersection of these lines is 1. Thus, $2 ⊕ 3 = 1$. This means that if the switch starts in the 2, medium, position and is turned 3 positions clockwise, its final position is 1, the setting for low-intensity light.

Table 13.1 consists of a set of four elements, namely $\{0, 1, 2, 3\}$. The table provides a rule, designated by ⊕, that can be performed on two elements of the set at a time. Such a rule is called a *binary operation*. **A binary operation** is a rule that can be used to combine any two elements of a set, resulting in a single element. A **mathematical system** consists of a set of elements and at least one binary operation. Many mathematical systems are presented in the form of a table, such as Table 13.1.

EXAMPLE 1 | AN EXAMPLE OF A MATHEMATICAL SYSTEM

If two even or two odd numbers are added, the sum is even. For example, $4 + 4 = 8$ (even + even = even). Likewise, $5 + 5 = 10$ (odd + odd = even). By contrast, if one number is even and the other odd, the sum is odd. For example, $4 + 3 = 7$ (even + odd = odd). Likewise, $7 + 6 = 13$ (odd + even = odd). If we let E represent any even number and O any odd number, these observations are summarized in Table 13.2.

a. What is the set of elements of this mathematical system?

b. What is the binary operation of this mathematical system?

c. Use Table 13.2 to find $E + O$. Explain what this means.

Solution

a. The set of elements of this mathematical system is $\{E, O\}$.

b. The binary operation used to combine any two elements of $\{E, O\}$ is $+$.

c. We find $E + O$ by drawing a horizontal line through E in the left column and a vertical line through O in the top row. These lines intersect at O, shown in Figure 13.3. Thus, $E + O = O$. This means that the sum of an even number and an odd number is an odd number.

TABLE 13.2

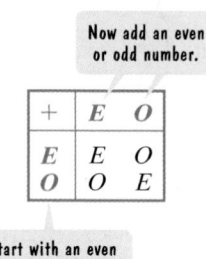

Now add an even or odd number.

+	E	O
E	E	O
O	O	E

Start with an even or odd number.

+	E	O
E	E	O
O	O	E

FIGURE 13.3 Using Table 13.2 to find $E + O$

 Check 1 Point Use Table 13.2 to find $O + O$. Explain what this means.

② Understand properties of certain mathematical systems.

Properties of Some Mathematical Systems

We now look at five properties that apply to certain mathematical systems. You may recognize some of these properties from our work in Chapter 5, where we studied the set of real numbers with two binary operations: addition and multiplication.

The Closure Property

A mathematical system is **closed** under a binary operation if the answer to any possible combination of two elements of the set is an element that is in the set. For example, take a second look at Table 13.1, which shows all possible starting and ending positions for the three-way switch. The answers in the body of the table are all elements of the set $\{0, 1, 2, 3\}$. Thus, the system is closed under $\oplus$. If an element other than 0, 1, 2, or 3 had appeared in the body of the table, the system would not have been closed under the binary operation.

CLOSED SETS; SETS HAVING CLOSURE

Suppose a binary operation is performed on any two elements of a set. If the result is an element of the set, then that set is **closed** (or has **closure**) under the binary operation.

EXAMPLE 2 UNDERSTANDING THE CLOSURE PROPERTY

Is the set $\{0, 1, 2, 3\}$ closed under the operation of addition?

Solution Shown on the right is an addition table for $\{0, 1, 2, 3\}$. Is every entry in the table an element of $\{0, 1, 2, 3\}$? The answer is no. The numbers 4, 5, and 6 are not elements of $\{0, 1, 2, 3\}$. The set $\{0, 1, 2, 3\}$ is not closed under addition.

+	0	1	2	3
0	0	1	2	3
1	1	2	3	4
2	2	3	4	5
3	3	4	5	6

✔ Check Point **2** Is the set $\{0, 1, 2\}$ closed under the operation of addition?

EXAMPLE 3 UNDERSTANDING THE CLOSURE PROPERTY

Consider the set of integers

$$\{\ldots, -3, -2, -1, 0, 1, 2, 3, \ldots\}.$$

Is this set closed under the operation of multiplication?

Solution If any two integers are multiplied, the product will be an integer. Thus, the set of integers is closed under multiplication.

✔ Check Point **3** Is the set of natural numbers, $\{1, 2, 3, 4, \ldots\}$, closed under the operation of multiplication?

EXAMPLE 4 UNDERSTANDING THE CLOSURE PROPERTY

Is the set of integers closed under the operation of division?

Solution If any two integers are divided, the quotient might not be an integer. For example,

$$\boxed{\text{−3 and 9 are integers.}} \quad \frac{-3}{9} = -\frac{1}{3}. \quad \boxed{-\tfrac{1}{3} \text{ is not an integer.}}$$

Thus, the set of integers is not closed under division.

✔ Check Point **4** Is the set of natural numbers, $\{1, 2, 3, 4, \ldots\}$, closed under the operation of division?

The Commutative Property

Suppose that ∘ represents a binary operation for the elements of a set. The set is **commutative** under the operation ∘ if for any two elements of the set a and b, $a \circ b = b \circ a$. The order of the two elements being combined in a commutative binary operation can be switched without changing the answer.

In Chapter 5, we saw that the integers are commutative under the operations of addition and multiplication. For example,

$$3 + 7 = 7 + 3 \quad \text{and} \quad 3 \cdot 7 = 7 \cdot 3.$$

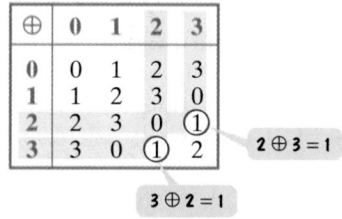

FIGURE 13.4 Showing that $2 \oplus 3 = 3 \oplus 2$.

$\oplus$	0	1	2	3
0	0	1	2	3
1	1	2	3	0
2	2	3	0	1
3	3	0	1	2

FIGURE 13.5 Showing that $\{0, 1, 2, 3\}$ is commutative under $\oplus$

By contrast, the integers are not commutative under the operations of subtraction and division. For example,

$$8 - 4 \neq 4 - 8 \quad \text{and} \quad 8 \div 4 \neq 4 \div 8.$$

In Figure 13.4, we use the table for the three-way lamp switch to show that $2 \oplus 3 = 3 \oplus 2$. Observe that $2 \oplus 3$ is found by first locating 2 on the left and 3 across the top. The green intersecting lines show that $2 \oplus 3 = 1$. We find $3 \oplus 2$ by locating 3 on the left and 2 across the top, again using intersecting lines. These yellow lines show that $3 \oplus 2 = 1$. Thus, $2 \oplus 3 = 3 \oplus 2$ because both binary operations give 1.

To verify that $\{0, 1, 2, 3\}$ is commutative under the operation $\oplus$, we need to verify all possibilities, not just $2 \oplus 3 = 3 \oplus 2$. This might take a while. Fortunately, there is a way to determine if a mathematical system given by a table is commutative under the operation shown in the table. Draw a diagonal line from the upper left corner to the lower right corner, as in Figure 13.5. Observe that for each number in the table, the same number is repeated the same number of times in the same configuration on each side of the green diagonal. For any mathematical system presented in the form of a table, if the part of the table above the diagonal is a mirror image of the part below the diagonal, the system is commutative under the operation given in the table.

The Associative Property

A set of elements is **associative** under a given operation $\circ$ if for any three elements a, b, and c of the set,

$$(a \circ b) \circ c = a \circ (b \circ c).$$

If a binary operation is associative, the answer does not change if we group the first two elements together or the last two elements together.

In Chapter 5, we saw that the integers are associative under the operations of addition and multiplication. For example,

$$(3 + 4) + 2 = 3 + (4 + 2) \qquad\qquad (3 \cdot 4) \cdot 2 = 3 \cdot (4 \cdot 2)$$
$$7 + 2 = 3 + 6 \qquad\qquad 12 \cdot 2 = 3 \cdot 8$$
$$9 = 9 \qquad\qquad 24 = 24.$$

By contrast, the integers are not associative under the operations of subtraction and division. For example,

$$(6 - 3) - 2 \neq 6 - (3 - 2) \qquad\qquad (6 \div 3) \div 2 \neq 6 \div (3 \div 2)$$
$$3 - 2 \neq 6 - 1 \qquad\qquad 2 \div 2 \neq 6 \div 1.5$$
$$1 \neq 5 \qquad\qquad 1 \neq 4.$$

EXAMPLE 5 VERIFYING ONE CASE FOR THE ASSOCIATIVE PROPERTY

TABLE 13.3

$\oplus$	0	1	2	3
0	0	1	2	3
1	1	2	3	0
2	2	3	0	1
3	3	0	1	2

Use Table 13.3, the table for the three-way lamp switch, to verify the associative property for $a = 3$, $b = 2$, and $c = 1$.

Solution We must show that $(3 \oplus 2) \oplus 1 = 3 \oplus (2 \oplus 1)$.

Locate 3 on the left and 2 on the top of Table 13.3. Intersecting lines show $3 \oplus 2 = 1$.

$$(3 \oplus 2) \oplus 1 = 3 \oplus (2 \oplus 1)$$
$$1 \oplus 1 = 3 \oplus 3$$
$$2 = 2$$

Locate 2 on the left and 1 across the top of Table 13.3. Intersecting lines show $2 \oplus 1 = 3$.

Changing the location of the parentheses does not change the answer, 2.

Check Point 5 Use Table 13.3 to verify the associative property for $a = 1$, $b = 3$, and $c = 2$.

If every choice of three elements from the set $\{0, 1, 2, 3\}$ were checked, it would show that the mathematical system in Table 13.3 is associative under the operation $\oplus$.

The Identity Property

I see your hand on my three-way lamp switch. Please do not turn it because I like the light intensity just the way that it is. If it's off, leave it off: $0 \oplus 0 = 0$. If it's in the low position, leave it there: $1 \oplus 0 = 1$. If it's in the medium position, don't change anything: $2 \oplus 0 = 2$. Finally, if the light intensity is high, on high it will remain: $3 \oplus 0 = 3$.

Does not changing anything remind you of the role that the number 0 plays in addition and the number 1 plays in multiplication? If a represents any real number,

$$a + 0 = a \quad \text{and} \quad 0 + a = a$$

$$a \cdot 1 = a \quad \text{and} \quad 1 \cdot a = a.$$

In each equation, the number a is unchanged when we add with 0 or multiply with 1. We call 0 the **identity element of addition** and 1 the **identity element of multiplication**.

In a mathematical system, the **identity element** (if there is one) is an element from the set such that when a binary operation is performed on any element in the set and the identity element, the result is the given element. If a represents any element in the set and the binary operation is represented by $\circ$:

$$a \circ \text{identity element} = a \quad \text{and} \quad \text{identity element} \circ a = a.$$

An element does not change when we perform an operation with an identity element.

EXAMPLE 6 | IDENTIFYING THE IDENTITY ELEMENT

Consider the set $\{a, b, c\}$. A binary operation on this set is given in Table 13.4. What is the identity element for this mathematical system?

TABLE 13.4

$\circ$	a	b	c
a	c	a	b
b	a	b	c
c	b	c	a

Solution Look for the element in $\{a, b, c\}$ that doesn't change anything when used in the binary operation $\circ$. Notice that the column under b is the same as the column with the blue letters on the left.

$$a \circ b = a \quad b \circ b = b \quad c \circ b = c$$

Notice, too, that the row to the right of b is the same as the top row with the blue letters.

$$b \circ a = a \quad b \circ b = b \quad b \circ c = c$$

When any element is combined with b under the operation $\circ$, the result is that element. The element b does not change anything. Thus, b is the identity element for the mathematical system.

Check
6
Point

Consider the set $\{e, f, g\}$. A binary operation for this set is shown in the table on the right. What is the identity element for this mathematical system?

$\circ$	e	f	g
e	f	g	e
f	g	e	f
g	e	f	g

The Inverse Property

It's time to get this lamp with the three-way switch out of my life. Hope I don't get charged with design pollution while driving it to the dump. I'll start by turning it off. (See Figure 13.6.) If it's already off, I'll leave it there: $0 \oplus 0 = 0$. If it's on low, 1, I'll move it to the off position by 3 clockwise turns: $1 \oplus 3 = 0$. If it's on medium, 2, two turns to the right will quiet the beast: $2 \oplus 2 = 0$. Finally, if it's on high, 3, one right turn will slay the dragon: $3 \oplus 1 = 0$.

$0 \oplus 0 = 0$

$1 \oplus 3 = 0$

$2 \oplus 2 = 0$

$3 \oplus 1 = 0$

FIGURE 13.6 Slaying the dragon

Buried in this lamp rant is a property that applies to many, but not all, mathematical systems. Do you see that the answer to each of the four binary operations in Figure 13.6 is the identity element 0? For example,

identity element

$$1 \oplus 3 = 0.$$

In a situation like this, we say that 3 is the *inverse* of 1. Similarly, since $3 \oplus 1 = 0$, 1 is the *inverse* of 3. In general, suppose that a binary operation is performed on two elements in a set and the result is the identity element for the binary operation. Under these conditions, each element is called the **inverse** of the other. Inverses restore the identity.

0 is its own inverse.
2 is its own inverse.

Element	$\oplus$	Inverse	=	Identity
0	$\oplus$	0	=	0
1	$\oplus$	3	=	0
2	$\oplus$	2	=	0
3	$\oplus$	1	=	0

1 and 3 are inverses of each other.

EXAMPLE 7 | IDENTIFYING AN INVERSE

Consider the set of integers

$$\{\ldots, -3, -2, -1, 0, 1, 2, 3, \ldots\}.$$

What is the inverse of 17 under the operation of addition?

Solution When an element operates on its inverse, the result is the identity element. Because the operation is addition, the identity element is 0. To find the inverse of 17, we answer the question: Which integer must be added to 17 to obtain a sum of 0?

$$17 + ? = 0$$

The required integer is -17: $17 + (-17) = 0$. Thus, the inverse of 17 under the operation of addition is -17.

Check 7 Point

Consider the set of integers. What is the inverse of 12 under the operation of addition?

We can generalize from Example 7. If a is an integer, the inverse for the operation of addition is $-a$.

$$a + (-a) = 0$$

$-a$ is often called the **additive inverse** of a.

In Chapter 5, we defined the set of *rational numbers* as the set of all numbers in the form $\frac{a}{b}$, where a and b are integers and $b \neq 0$. Our next example involves a mathematical system with rational numbers.

EXAMPLE 8 IDENTIFYING AN INVERSE

Consider the set of rational numbers. What is the inverse of 17 under the operation of multiplication?

Solution When an element operates on its inverse, the result is the identity element. Because the operation is multiplication, the identity element is 1. To find the inverse of 17, we answer the question: 17 times which rational number gives 1?

$$17 \cdot ? = 1$$

The required rational number is $\frac{1}{17}$: $17 \cdot \frac{1}{17} = 1$. Thus, the inverse of 17 under the operation of multiplication is $\frac{1}{17}$.

 **Check Point 8** Consider the set of rational numbers. What is the inverse of 12 under the operation of multiplication?

We can generalize from Example 8. If a is any rational number *other than zero*, the inverse for the operation of multiplication is $\frac{1}{a}$.

$$a \cdot \frac{1}{a} = 1$$

$\frac{1}{a}$ is often called the **multiplicative inverse** of a. Zero has no multiplicative inverse because no rational number can be multiplied by 0 to give 1.

EXAMPLE 9 IDENTITY ELEMENT AND INVERSE ELEMENTS

For the mathematical system defined by Table 13.5,

a. What is the identity element?

b. Find the inverse for each element in $\{S, L, A, R\}$.

Solution

a. The identity element is the element in $\{S, L, A, R\}$ that does not change any element with which it is combined in the binary operation $\circ$. Figure 13.7 shows that the column under S is the same as the column with the blue letters on the left and the row next to S is the same as the row with the blue letters on the top. The element S does not change anything. Thus, S is the identity element.

b. When an element operates on its inverse, the result is the identity element:

$$\text{element} \circ \text{inverse} = \text{identity element.}$$

Inverses restore the identity.

Because the identity element is S, we can find the inverse of each element in $\{S, L, A, R\}$ by replacing the question mark in

$$\text{element} \circ ? = S.$$

TABLE 13.5

$\circ$	S	L	A	R
S	S	L	A	R
L	L	A	R	S
A	A	R	S	L
R	R	S	L	A

$\circ$	S	L	A	R
S	S	L	A	R
L	L	A	R	S
A	A	R	S	L
R	R	S	L	A

FIGURE 13.7 S does not change any element with which it is combined.

∘	S	L	A	R
S	S	L	A	R
L	L	A	R	S
A	A	R	S	L
R	R	S	L	A

FIGURE 13.8 Locating inverses

How do we replace this question mark? Look at the entries on the left, find S in the body of the table in each row, and then locate the entry across the top that gives S. Figure 13.8 shows that

$$S \circ S = S, \text{ so } S \text{ is its own inverse.}$$
$$L \circ R = S, \text{ so } R \text{ is the inverse of } L.$$
$$A \circ A = S, \text{ so the inverse of } A \text{ is } A.$$
$$R \circ L = S, \text{ so } L \text{ is the inverse of } R.$$

 **Check Point 9**

For the mathematical system defined by the table on the right,

a. What is the identity element?

b. Find the inverse for each element in $\{j, k, l, m\}$.

∘	j	k	l	m
j	m	j	k	l
k	j	k	l	m
l	k	l	m	j
m	l	m	j	k

Exercise Set 13.1

Practice Exercises

Use the mathematical system given by the table shown to solve Exercises 1–14.

∘	e	a	b	c
e	e	a	b	c
a	a	e	c	b
b	b	c	e	a
c	c	b	a	e

1. What is the set of elements of this mathematical system?

2. What is the binary operation of this mathematical system?

In Exercises 3–13, find each result.

3. $a \circ b$ **4.** $c \circ a$ **5.** $b \circ c$

6. $c \circ b$ **7.** $e \circ e$ **8.** $e \circ a$

9. $b \circ e$ **10.** $c \circ e$ **11.** $a \circ a$

12. $(b \circ b) \circ c$ **13.** $(c \circ c) \circ a$

14. Is the mathematical system closed under the binary operation? Explain your answer.

15. Is the set $\{0, 1, 2, 3, 4\}$ closed under the operation of addition? Explain your answer.

16. Is the set $\{0, 1, 2, 3, 4, 5\}$ closed under the operation of addition? Explain your answer.

17. Is the set of natural numbers $\{1, 2, 3, 4, 5, \ldots\}$ closed under the operation of addition? Explain your answer.

18. Is the set of natural numbers $\{1, 2, 3, 4, 5, \ldots\}$ closed under the operation of multiplication? Explain your answer.

19. Is the set of natural numbers closed under the operation of subtraction? Explain your answer.

20. Is the set of natural numbers closed under the operation of division? Explain your answer.

In Exercises 21–22, use the table shown to explain whether or not the mathematical system is closed under the binary operation.

21.

*	a	b
a	a	b
b	c	a

22.

*	d	e	f
d	d	e	g
e	e	d	f
f	f	e	d

Use the mathematical system given by the table shown to solve Exercises 23–32.

⊕	0	1	2	3	4
0	0	1	2	3	4
1	1	2	3	4	0
2	2	3	4	0	1
3	3	4	0	1	2
4	4	0	1	2	3

In Exercises 23–36, show that

23. $2 \oplus 4 = 4 \oplus 2$.   **24.** $3 \oplus 4 = 4 \oplus 3$.

25. $4 \oplus 1 = 1 \oplus 4$.   **26.** $2 \oplus 3 = 3 \oplus 2$.

27. What property is illustrated in Exercises 23–26? Use the table to explain how to verify this property for all possible cases without having to consider one case at a time.

In Exercises 28–31, use the table at the bottom of page 709 to show that

28. $(3 \oplus 2) \oplus 4 = 3 \oplus (2 \oplus 4)$.

29. $(4 \oplus 3) \oplus 2 = 4 \oplus (3 \oplus 2)$.

30. $(2 \oplus 2) \oplus 3 = 2 \oplus (2 \oplus 3)$.

31. $(4 \oplus 4) \oplus 2 = 4 \oplus (4 \oplus 2)$.

32. What property is illustrated in Exercises 28–31?

33. Use the table shown below to find each result.

   a. $(b \circ c) \circ b$ **b.** $b \circ (c \circ b)$

   c. Is the mathematical system associative under the operation $\circ$? Explain your answer.

$\circ$	a	b	c
a	a	b	c
b	b	a	c
c	c	b	a

34. Use the table shown below to find each result.

   a. $(b \circ b) \circ c$ **b.** $b \circ (b \circ c)$

   c. Is the mathematical system associative under the operation $\circ$? Explain your answer.

$\circ$	a	b	c
a	a	b	c
b	b	b	a
c	c	a	c

35. Provide an original example to show that the integers are not associative under subtraction.

36. Provide an original example to show that the integers are not associative under division.

In Exercises 37–40, consider the set of rational numbers.

37. What is the inverse of 29 under the operation of addition?

38. What is the inverse of 43 under the operation of addition?

39. What is the inverse of 29 under the operation of multiplication?

40. What is the inverse of 43 under the operation of multiplication?

Use the mathematical system given by the table shown to solve Exercises 41–56.

$\circ$	a	b	c	d	e
a	a	b	c	d	e
b	b	c	d	e	a
c	c	d	e	a	b
d	d	e	a	b	c
e	e	a	b	c	d

Find each result:

41. $a \circ a$ **42.** $a \circ b$ **43.** $a \circ c$

44. $a \circ d$ **45.** $a \circ e$ **46.** $b \circ a$

47. $c \circ a$ **48.** $d \circ a$ **49.** $e \circ a$

50. What can you conclude about the element a in this mathematical system?

Find the element represented by a question mark.

51. $a \circ ? = a$ **52.** $b \circ ? = a$

53. $c \circ ? = a$ **54.** $d \circ ? = a$

55. $e \circ ? = a$

56. What is true about the relationship between the elements on the left sides of the equations in Exercises 51–55?

Use the mathematical system given by the table shown to solve Exercises 57–62.

$\oplus$	0	1	2	3	4
0	0	1	2	3	4
1	1	2	3	4	0
2	2	3	4	0	1
3	3	4	0	1	2
4	4	0	1	2	3

57. What is the identity element?

Find the inverse for each element.

58. 0   **59.** 1   **60.** 2   **61.** 3   **62.** 4

63. a. Use the table shown below to find the identity element for this mathematical system.

   b. Find the inverse for each element of the set that has one. Which elements do not have inverses?

$\circ$	a	b	c	d
a	a	a	a	a
b	a	b	b	b
c	a	b	c	c
d	a	b	c	d

64. a. Use the table shown below to find the identity element for this mathematical system.

   b. Find the inverse for each element of the set that has one. Which elements do not have inverses?

$\circ$	a	b	c	d	e
a	d	e	a	b	b
b	e	a	b	a	d
c	a	b	c	d	e
d	b	a	d	e	c
e	b	d	e	c	a

Application Exercises

65. Suppose that a lamp has a four-way switch. The positions of the switch are 0 : off, 1 : dim, 2 : low, 3 : medium, and 4 : high. Describe the relationship between this switch and the mathematical system given by the table shown for Exercises 57–62.

66. An army drill instructor gives you four different commands. They are S : stand still, L : left face (turn 90° to your left), A : about face (turn 180° to your right), and R : right face (turn 90° to your right). The operation ∘ represents two commands in succession. Describe the relationship between this situation and the mathematical system given by Table 13.5 on page 708.

Let E represent any even number, let O represent any odd number, and let × represent the operation of multiplication. In order to solve Exercises 67–72, begin by filling in the four missing entries in the table shown.

×	E	O
E	__	__
O	__	__

67. Is the set $\{E, O\}$ closed under the operation of multiplication? Explain your answer.

68. Verify one case of the commutative property.

69. Verify one case of the associative property.

70. What is the identity element?

71. Find the inverse of E.   **72.** Find the inverse of O.

Writing in Mathematics

73. Describe what is meant by a mathematical system.

74. What is the closure property? Give an example.

75. What is the commutative property? Give an example.

76. What is the associative property? Give an example.

77. What is the identity property? Give an example.

78. What is the inverse property? Give an example.

79. Describe the difference between a nonzero number's additive inverse and its multiplicative inverse. Provide an example.

Critical Thinking Exercises

80. Give an example of a mathematical system that is different than any seen in this section.

81. If $\begin{bmatrix} a & b \\ c & d \end{bmatrix} \times \begin{bmatrix} e & f \\ g & h \end{bmatrix} = \begin{bmatrix} ae + bg & af + bh \\ ce + dg & cf + dh \end{bmatrix}$, find

a. $\begin{bmatrix} 2 & 3 \\ 4 & 7 \end{bmatrix} \times \begin{bmatrix} 0 & 1 \\ 5 & 6 \end{bmatrix}$.

b. $\begin{bmatrix} 0 & 1 \\ 5 & 6 \end{bmatrix} \times \begin{bmatrix} 2 & 3 \\ 4 & 7 \end{bmatrix}$.

c. Draw a conclusion about one of the properties discussed in this section in terms of these arrays of numbers under multiplication.

SECTION 13.2   ROTATIONAL SYMMETRY, GROUPS, AND CLOCK ARITHMETIC

OBJECTIVES

1. Recognize rotational symmetry.

2. Determine if a mathematical system is a group.

3. Understand clock systems as groups.

4. Understand congruence in a modulo m system.

5. Perform additions in a modulo m system.

FIGURE 13.9

Mathematics is about the patterns that arise in the world about us. Mathematicians look at a snowflake in terms of its underlying structure. Notice that if the snowflake in Figure 13.9 is rotated by any multiple of 60° ($\frac{1}{6}$ of a rotation), it will always look about the same. A **symmetry** of an object is a motion that moves the object back onto itself. In a symmetry, you cannot tell, at the end of the motion, that the object has been moved.

1 Recognize rotational symmetry.

A two-blade propeller with twofold rotational symmetry

The logo from a Korean bank with fivefold rotational symmetry

The seedpot of the autograph tree (*Clusia Rosea*) with eightfold rotational symmetry

The snowflake in Figure 13.9 has **sixfold rotational symmetry**. After six 60° turns, the snowflake is back to its original position. If it takes *m* equal turns to restore an object to its original position and each of these turns is a figure that is identical to the original figure, the object has ***m*-fold rotational symmetry**. Consider, for example, a pinwheel. Figure 13.10 shows that as the pinwheel rotates, at each quarter turn it is in a new position, yet it looks exactly the same as it did at the start. If all its petals are the same color, we cannot distinguish the original position from the new position after one quarter turn, two quarter turns, or three quarter turns. After four quarters (that is, after a full rotation), we are back to the starting position. The pinwheel has **fourfold rotational symmetry**. If the blades are colored or numbered, it is possible to keep track of the pinwheel's rotation.

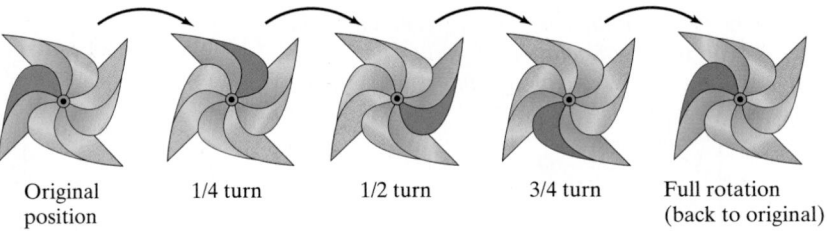

| Original position | 1/4 turn | 1/2 turn | 3/4 turn | Full rotation (back to original) |

FIGURE 13.10

Does the pinwheel remind you of anything? Its symmetry is exactly the same as that of the infamous light switch with its off, low, medium, and high positions. Special kinds of mathematical systems, called *groups*, are used to study the symmetries and hidden structure of our world. Objects that appear to be totally different, such as a pinwheel and a lamp switch, can be unified by the definition of a group. We begin with this definition.

Groups

I went to the movies with a *group* of friends. I bought an interesting *group* of paintings for my living room wall. We use the word *group* in everyday speech when referring to any collection of persons or things. However, in mathematics *group* means something entirely different. A group is a mathematical system meeting four requirements.

DEFINITION OF A GROUP

A mathematical system that meets the following four requirements is called a **group**:

1. The set of elements in the mathematical system is *closed* under the given binary operation, represented in this box by ∘ .
2. The set of elements is *associative* under the given binary operation. If *a, b*, and *c* are any three elements of the set,

$$(a \circ b) \circ c = a \circ (b \circ c).$$

3. The set of elements contains an *identity* element. When the binary operation ∘ is performed on the identity element and each element in the set, the result is the given element.
4. Each element of the set has an *inverse* that lies within the set. When an element operates on its inverse, the result is the identity element.

Do you notice anything missing from these four requirements? The commutative property does not need to hold for a mathematical system to be a group. If a group does satisfy the commutative property (that is, if $a \circ b = b \circ a$ for any two elements of the set), it is called a **commutative group**.

2 Determine if a mathematical system is a group.

TABLE 13.6

·	−1	1
−1	1	−1
1	−1	1

Fivefold rotational symmetry

Sixfold rotational symmetry
Flowers with rotational symmetries

EXAMPLE 1   SHOWING THAT A MATHEMATICAL SYSTEM IS A GROUP

Table 13.6 is a mathematical system for the set $\{-1, 1\}$ under the binary operation of multiplication.

a. Show that the mathematical system is a group.

b. Show that the group is commutative.

Solution

a. We begin by checking the four requirements—closure, the associative property, the identity property, and the inverse property—for a mathematical system to qualify as group.

  1. The Closure Property. The set $\{-1, 1\}$ is closed under the binary operation of multiplication because the entries in the body of Table 13.6 are all elements of the set.

  2. The Associative Property. Is $(a \cdot b) \cdot c = a \cdot (b \cdot c)$ for all elements a, b, and c of the set? Let us select some values for a, b, and c. Let $a = 1$, $b = -1$ and $c = -1$. Then

  $$[1 \cdot (-1)] \cdot (-1) = -1 \cdot (-1) = 1$$

  and

  $$1 \cdot [(-1) \cdot (-1)] = 1 \cdot 1 = 1.$$

  The associative property holds for $a = 1$, $b = -1$, and $c = -1$. This particular case does not prove that the mathematical system is associative. However, we could check every possible choice of three elements from the set. Such a check would verify that the system does satisfy the associative property.

  3. The Identity Property. Look for the element in $\{-1, 1\}$ that does not change anything when used in multiplication. The element 1 does not change anything:

  $$-1 \cdot 1 = -1 \quad 1 \cdot (-1) = -1 \quad 1 \cdot 1 = 1$$

  An element does not change when we multiply with 1.

  Thus, 1 is the identity element. The identity property is satisfied because 1 is contained in the set $\{-1, 1\}$.

  4. The Inverse Property. When an element operates on its inverse, the result is the identity element. Because the identity element is 1, we can find the inverse of each element in $\{-1, 1\}$ by answering the question: What must each element be multiplied by to obtain 1?

  $$\text{element} \cdot ? = 1$$

  Figure 13.11 illustrates how we answer the question. If each element in $\{-1, 1\}$ has an inverse that restores the identity element 1, then a 1 will appear in every row and column of the table. This is, indeed, the case. Using 1 in each row, we see that

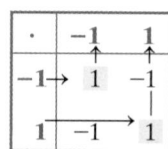

FIGURE 13.11

  • $-1 \cdot -1 = 1$, so the inverse of -1 is -1.
  • $1 \cdot 1 = 1$, so the inverse of 1 is 1.

  Because each element in $\{-1, 1\}$ has an inverse within the set, the inverse property is satisfied.

  In summary, the set $\{-1, 1\}$ satisfies the closure property, the associative property, the identity property, and the inverse property under the operation of multiplication. Thus, the mathematical system is a group.

b. To show that the group is commutative, we must show that the commutative property is satisfied. Is $a \cdot b = b \cdot a$ for all elements a and b of $\{-1, 1\}$? We need to examine only one case, the case where a and b have different values.

$$1 \cdot (-1) = -1 \quad \text{and} \quad -1 \cdot 1 = -1$$

The order of the factors in the multiplication does not matter. Thus, $\{-1, 1\}$ is a commutative group under multiplication.

 Check 1 Point

The table on the right is a mathematical system for $\{E, O\}$ under the binary operation $\circ$.

$\circ$	E	O
E	E	O
O	O	E

a. Show that the mathematical system is a group. Just verify two examples for the associative property.

b. Show that the group is commutative.

If a mathematical system fails to meet even one of the four requirements for a group, that system is not a group. For example, consider the set of integers $\{\ldots, -3, -2, -1, 0, 1, 2, 3, \ldots\}$ under the operation of multiplication. Although the closure property, the associative property, and the identity property are satisfied, the inverses of the integers are not contained within the set. For example, the multiplicative inverse of 3 is $\frac{1}{3}\left(3 \cdot \frac{1}{3} = 1\right)$. However, $\frac{1}{3}$ is not an integer. Because there is at least one element, 3, in the set of integers that fails to have an inverse that lies within the set, the integers do not satisfy the inverse property under multiplication. The failure to meet one of a group's four requirements means that the integers are not a group under multiplication.

Clock Arithmetic and Groups

3 Understand clock systems as groups.

Figure 13.12 looks like the ordinary face of a clock, with two differences. The minute hand is left off, and 12 is replaced with 0. If we replace each number with a small tick-mark, we can observe 12-fold rotational symmetry. If the clock is rotated $\frac{1}{12}$ of a rotation, the tick-marks will fall in the same positions and you cannot tell that the clock's face has been moved. Because Figure 13.12 has numbers instead of tick-marks, it takes 12 turns to restore 0 to its original position.

Consider the mathematical system formed by the set of numbers on the 12-hour clock:

$$\{0, 1, 2, 3, 4, 5, 6, 7, 8, 9, 10, 11\}$$

FIGURE **13.12** A 12-hour clock

and a binary operation called *clock addition*. **Clock addition** is defined as follows: Add by moving the hour hand in a clockwise direction. The symbol $+$ is used to designate clock addition. Using the definition of clock addition, Figure 13.13 illustrates that $5 + 2 = 7$, $8 + 9 = 5$, and $11 + 4 = 3$.

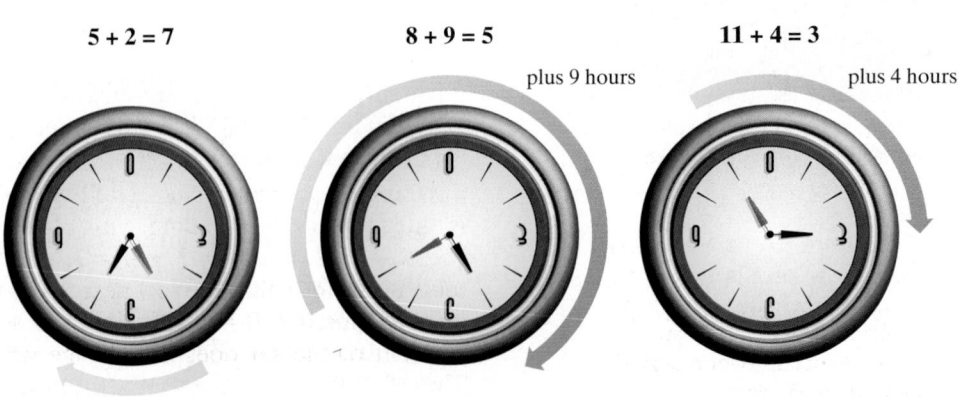

FIGURE **13.13** Addition in a 12-Hour Clock System

Friendship
with Morocco
1787-1987

USA **22**

This U.S. stamp celebrating our friendship with Morocco shows a geometric form with 12-fold rotational symmetry. The form, and all objects with such symmetry, can be analyzed by the properties and patterns of the 12-hour clock system.

Table 13.7 is the addition table for clock addition in a 12-hour clock system.

TABLE 13.7 TABLE FOR 12-HOUR CLOCK ADDITION

+	0	1	2	3	4	5	6	7	8	9	10	11
0	0	1	2	3	4	5	6	7	8	9	10	11
1	1	2	3	4	5	6	7	8	9	10	11	0
2	2	3	4	5	6	7	8	9	10	11	0	1
3	3	4	5	6	7	8	9	10	11	0	1	2
4	4	5	6	7	8	9	10	11	0	1	2	3
5	5	6	7	8	9	10	11	0	1	2	3	4
6	6	7	8	9	10	11	0	1	2	3	4	5
7	7	8	9	10	11	0	1	2	3	4	5	6
8	8	9	10	11	0	1	2	3	4	5	6	7
9	9	10	11	0	1	2	3	4	5	6	7	8
10	10	11	0	1	2	3	4	5	6	7	8	9
11	11	0	1	2	3	4	5	6	7	8	9	10

EXAMPLE 2   THE 12-HOUR CLOCK SYSTEM AS A GROUP

Show that the mathematical system given by Table 13.7 is a commutative group.

Solution   We must check the five requirements—the closure property, the associative property, the identity property, the inverse property, and the commutative property—for a commutative group.

1. **The Closure Property.** The set $\{0, 1, 2, 3, 4, 5, 6, 7, 8, 9, 10, 11\}$ is closed under the operation of clock addition because the entries in the body of Table 13.7 are all elements of the set.

2. **The Associative Property.** Is $(a + b) + c = a + (b + c)$ for all elements a, b, and c of the set? Let us select some values for a, b, and c. Let $a = 4$, $b = 7$, and $c = 9$.

Locate 4 on the left and 7 on the top of Table 13.7. Intersecting lines show $4 + 7 = 11$.

$$(4 + 7) + 9 = 4 + (7 + 9)$$
$$11 + 9 = 4 + 4$$
$$8 = 8$$

Locate 7 on the left and 9 across the top of Table 13.7. Intersecting lines show $7 + 9 = 4$.

The associative property holds for $a = 4$, $b = 7$, and $c = 9$. As always, this particular case does not prove that the mathematical system is associative under clock addition. However, we could check every possible choice of three elements from the set. Such a check would verify that the system does satisfy the associative property.

3. **The Identity Property.** Look for the element in Table 13.7 that does not change anything when used in clock addition. Table 13.7 shows that the column under 0 is the same as the column with blue numbers on the left. Thus, $0 + 0 = 0$, $1 + 0 = 1$, $2 + 0 = 2$, $3 + 0 = 3$, and so on, up through $11 + 0 = 11$. The table also shows that the row next to 0 is the same as the row with blue numbers on top. Thus, $0 + 0 = 0$, $0 + 1 = 1$, $0 + 2 = 2$, up through $0 + 11 = 11$. Each element of the set does not change when we perform clock addition with 0. Thus, 0 is the identity element. The identity property is satisfied because 0 is contained in the given set.

BLITZER BONUS

The Longest Proof in Mathematics

The *Guinness Book of World Records* mentions the "enormous theorem" that deals with classifying groups into "families." Sounds simple? Not so. Completed in the 1980s, the proof is spread over more than 14,000 pages in nearly 500 articles in mathematical journals, contributed by more than 100 mathematicians over a period of more than 35 years.

4. **The Inverse Property.** When an element operates on its inverse, the result is the identity element. Because the identity element is 0, we can find the inverse of each element in $\{0, 1, 2, 3, 4, 5, 6, 7, 8, 9, 10, 11\}$ by answering the question: What must be added to each element to obtain 0?

$$\text{element} + ? = 0$$

Figure 13.14 illustrates how we answer the question. If each element in the set has an inverse that restores the identity element 0, then a 0 will appear in every row and column of the table. This is, indeed, the case. Use the 0 in each row. Because each element in $\{0, 1, 2, 3, 4, 5, 6, 7, 8, 9, 10, 11\}$ has an inverse within the set, the inverse property is satisfied.

+	0	1	2	3	4	5	6	7	8	9	10	11
0	0	1	2	3	4	5	6	7	8	9	10	11
1	1	2	3	4	5	6	7	8	9	10	11	0
2	2	3	4	5	6	7	8	9	10	11	0	1
3	3	4	5	6	7	8	9	10	11	0	1	2
4	4	5	6	7	8	9	10	11	0	1	2	3
5	5	6	7	8	9	10	11	0	1	2	3	4
6	6	7	8	9	10	11	0	1	2	3	4	5
7	7	8	9	10	11	0	1	2	3	4	5	6
8	8	9	10	11	0	1	2	3	4	5	6	7
9	9	10	11	0	1	2	3	4	5	6	7	8
10	10	11	0	1	2	3	4	5	6	7	8	9
11	11	0	1	2	3	4	5	6	7	8	9	10

0 + 0 = 0: The inverse of 0 is 0.

1 + 11 = 0: The inverse of 1 is 11.

2 + 10 = 0: The inverse of 2 is 10.

3 + 9 = 0: The inverse of 3 is 9.

4 + 8 = 0: The inverse of 4 is 8.

5 + 7 = 0: The inverse of 5 is 7.

6 + 6 = 0: The inverse of 6 is 6.

7 + 5 = 0: The inverse of 7 is 5.

8 + 4 = 0: The inverse of 8 is 4.

9 + 3 = 0: The inverse of 9 is 3.

10 + 2 = 0: The inverse of 10 is 2.

11 + 1 = 0: The inverse of 11 is 1.

FIGURE 13.14 Locating inverses. The arrows illustrate that $1 + 11 = 0$ and $8 + 4 = 0$.

5. **The Commutative Property.** Is $a + b = b + a$ for all elements a and b of $\{0, 1, 2, 3, 4, 5, 6, 7, 8, 9, 10, 11\}$? Let us check a few cases. $6 + 9$ and $9 + 6$ both yield 3. Also, $5 + 8$ and $8 + 5$ both yield 1. Order of clock addition does not matter for these cases.

We can use the method discussed in the previous section to avoid checking the commutative property one case at a time. Draw a diagonal line from upper left to lower right, as in Figure 13.15. The part of the table above the diagonal is a mirror image of the part below the diagonal. For each number in the table, the same number is repeated the same number of times on each side of the diagonal in an identical configuration. Thus, the system satisfies the commutative property.

+	0	1	2	3	4	5	6	7	8	9	10	11
0	0	1	2	3	4	5	6	7	8	9	10	11
1	1	2	3	4	5	6	7	8	9	10	11	0
2	2	3	4	5	6	7	8	9	10	11	0	1
3	3	4	5	6	7	8	9	10	11	0	1	2
4	4	5	6	7	8	9	10	11	0	1	2	3
5	5	6	7	8	9	10	11	0	1	2	3	4
6	6	7	8	9	10	11	0	1	2	3	4	5
7	7	8	9	10	11	0	1	2	3	4	5	6
8	8	9	10	11	0	1	2	3	4	5	6	7
9	9	10	11	0	1	2	3	4	5	6	7	8
10	10	11	0	1	2	3	4	5	6	7	8	9
11	11	0	1	2	3	4	5	6	7	8	9	10

FIGURE 13.15 Showing that a system is commutative by observing symmetry about the diagonal

FIGURE 13.16 Symmetry in Native American design

FIGURE 13.17 Mod 7 {0, 1, 2, 3, 4, 5, 6}

FIGURE 13.18 Mod 5 {0, 1, 2, 3, 4}

 4 Understand congruence in a modulo m system.

Because the five properties listed on pages 715-716 are satisfied, the 12-hour clock system is a commutative group.

 **Check 2 Point**

a. Use Table 13.7 on page 715 to verify that $(a + b) + c = a + (b + c)$ for $a = 8, b = 5,$ and $c = 11$.

b. Verify that $9 + 4 = 4 + 9$.

Modular Systems

The 12-hour clock system and the operation of clock addition form what is called a **modulo 12** system, or a **mod 12** system. Although this system is ideal for illuminating the structure of 12-fold symmetry, it would not be a system used to describe the structure of the design in Figure 13.16. Native American pottery has decorations with a wide variety of different kinds of rotational symmetry. This design, similar to those found on Pueblo pottery, has sevenfold rotational symmetry. Its pattern can be studied with a modulo 7 system consisting of seven elements, 0 through 6, and the binary operation of clock addition. The seven-hour clock for the mod 7 system is shown in Figure 13.17.

We perform clock addition in the mod 7 system exactly as we did in the 12-hour clock system. For example, the figure on the right shows how we can find $6 + 3$. If we move the hour hand forward 6 hours from 0, and then forward another 3 hours, this puts the hour hand at 2. The traditional sum, $6 + 3 = 9$, suggests that 9 and 2 are, in a sense, equivalent. We express this by writing

Plus 3 hours

$$9 \equiv 2 \ (\text{mod } 7).$$

$6 + 3 = 2$ in mod 7

The symbol $\equiv$ is read "is congruent to." Thus, 9 is congruent to 2 in a modulo 7 system. Notice that if 9 is divided by 7, the remainder is 2. We can also say that the same remainder, 2, is obtained when 9 and 2 are divided by 7.

Generalizing from this situation, we describe m-fold rotational symmetry with a **modulo m** system. Such a system consists of m integers, starting with 0 and ending with $m - 1$. For example, a modulo 5 system (Figure 13.18) consists of five integers, starting with 0 and ending with $5 - 1$, or 4.

The concept of congruence can be applied to any modulo m system.

> ### CONGRUENCE MODULO m
>
> **a is congruent to b modulo m**, written
> $$a \equiv b \ (\text{mod } m)$$
> means that if a is divided by m, the remainder is b. Equivalently,
> $$a \equiv b \ (\text{mod } m)$$
> if and only if a and b have the same remainder when divided by m.

EXAMPLE 3 | UNDERSTANDING CONGRUENCE

Label each statement as true or false:

a. $22 \equiv 4 \ (\text{mod } 6)$ **b.** $12 \equiv 4 \ (\text{mod } 3)$ **c.** $49 \equiv 5 \ (\text{mod } 11)$.

Solution

a. Is $22 \equiv 4 \ (\text{mod } 6)$? If 22 is divided by 6, is the remainder 4? $22 \div 6 = 3$, remainder 4. Because the remainder is 4, the statement is true.

b. Is $12 \equiv 4 \pmod 3$? If 12 is divided by 3, is the remainder 4? $12 \div 3 = 4$, remainder 0. Because the remainder is not 4, the statement is false. A true statement is $12 \equiv 0 \pmod 3$.

c. Is $49 \equiv 5 \pmod{11}$? If 49 is divided by 11, is the remainder 5? $49 \div 11 = 4$, remainder 5. Because the remainder is 5, the statement is true.

 Check **3** Point

Label each statement as true or false:

a. $61 \equiv 5 \pmod 7$

b. $36 \equiv 0 \pmod 6$

c. $57 \equiv 3 \pmod{11}$.

 5 Perform additions in a modulo m system.

Understanding congruence gives us a procedure for performing addition in any modulo system without having to draw a clock.

> **HOW TO ADD IN A MODULO m SYSTEM**
>
> **1.** Add the numbers using ordinary arithmetic.
> **2.** If the sum is less than m, the answer is the sum obtained.
> **3.** If the sum is greater than or equal to m, the answer is the remainder obtained upon dividing the sum in step 1 by m.

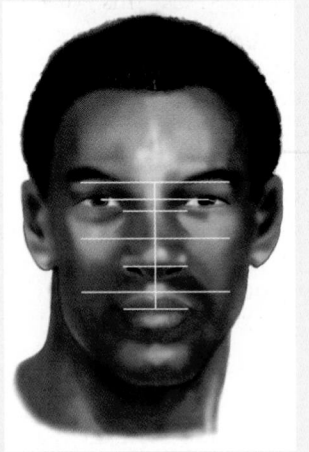
EXAMPLE 4   MODULAR ADDITIONS

Find each of the following sums:

a. $(2 + 4)\pmod 7$ **b.** $(4 + 4)\pmod 5$ **c.** $(9 + 10)\pmod{12}$.

Solution

a. To find $(2 + 4)\pmod 7$, first add $2 + 4$ to get 6. Because this sum is less than 7,

$$2 + 4 \equiv 6 \pmod 7.$$

b. To find $(4 + 4)\pmod 5$, first add $4 + 4$ to get 8. Because 8 is greater than 5, divide 8 by 5.

$$8 \div 5 = 1, \text{remainder } 3$$

The remainder, 3, is the desired sum.

$$4 + 4 \equiv 3 \pmod 5$$

c. To find $(9 + 10)\pmod{12}$, first add $9 + 10$ to get 19. Because this sum is greater than 12, divide 19 by 12.

$$19 \div 12 = 1, \text{remainder } 7$$

The remainder, 7, is the desired sum.

$$9 + 10 \equiv 7 \pmod{12}$$

 Check **4** Point

Find each of the following sums:

a. $(1 + 3)\pmod 5$ **b.** $(5 + 4)\pmod 7$ **c.** $(8 + 10)\pmod{13}$.

Being able to add in a modular system without having to draw a clock means that we can construct an addition table for the system fairly quickly. All such tables represent mathematical systems that are commutative groups. We can show that any modular system is a commutative group under addition by verifying each of the five properties (closure, associative, identity, inverse, commutative) as we did for the modulo 12 system in Example 2.

Modulo systems are also used to solve problems involving cyclical changes, such as calendar problems. Because there are 7 days in a week, we can consider this as a modulo 7 system, in which Sunday is 0, Monday is 1, Tuesday 2, Wednesday 3, Thursday 4, Friday 5, and Saturday 6.

EXAMPLE 5 | SOLVING A CALENDAR PROBLEM

Suppose today is Thursday, July 2. Which day of the week will it be 47 days from now?

Solution   Dates that differ by a multiple of 7 fall on the same day of the week. This means that we can divide 47 by 7 and keep the remainder: $47 \div 7 = 6$, remainder 5. Equivalently,

$$47 \equiv 5 \ (\text{mod } 7).$$

Thus, the desired day of the week is 5 days past Thursday, or Tuesday.

✓ **Check Point 5**   Suppose today is Wednesday, May 4. Which day of the week will it be 97 days from now?

In the exercise set that follows, there is a problem in a modulo 27 system that illustrates how modular arithmetic is used in coding. Modular arithmetic is central to the security of the Internet, which utilizes calculations in systems whose clock faces contain more numbers than there are atoms in the observable universe.

Exercise Set 13.2

Practice Exercises

Name the kind of rotational symmetry shown in Exercises 1–4.

1.

2.

3.

4.

In Exercises 5–19, use the mathematical system given by the table shown.

∘	e	p	q	r	s	t
e	e	p	q	r	s	t
p	p	q	e	s	t	r
q	q	e	p	t	r	s
r	r	t	s	e	q	p
s	s	r	t	p	e	q
t	t	s	r	q	p	e

5. Explain why the closure property is satisfied for the given binary operation.

6. Show that $(p \circ s) \circ t = p \circ (s \circ t)$.

7. Show that $(r \circ t) \circ q = r \circ (t \circ q)$.

8. As in Exercises 6–7, if every choice of three elements is checked, changing the location of the parentheses will not change the answer. What property is satisfied by the mathematical system?

9. What is the identity element?

Find the inverse of each element.

10. e 11. p 12. q 13. r 14. s 15. t

16. Based on Exercises 5–15, what can you conclude about this mathematical system?

17. Find $r \circ p$.

18. Find $p \circ r$.

19. Based on Exercises 17–18, what can you conclude about this mathematical system?

In Exercises 20–22, give a reason to support each statement.

20. The natural numbers $\{1, 2, 3, 4, 5, \ldots\}$ are not a group under addition.

21. The natural numbers $\{1, 2, 3, 4, 5, \ldots\}$ are not a group under multiplication.

22. The integers $\{\ldots, -3, -2, -1, 0, 1, 2, 3, \ldots\}$ are not a group under subtraction.

23. Consider the 6-hour, modulo 6, clock system under the operation of clock addition.

   a. Complete the clock addition table.

+	0	1	2	3	4	5
0						
1						
2						
3						
4						
5						

   b. Show that this mathematical system is a commutative group. Just verify two examples for the associative property.

24. Consider the 7-hour, modulo 7, clock system under the operation of clock addition.

   a. Complete the clock addition table.

+	0	1	2	3	4	5	6
0							
1							
2							
3							
4							
5							
6							

   b. Show that this mathematical system is a commutative group. Just verify two examples for the associative property.

In Exercises 25–34, determine if each statement is true or false. If the statement is false, change the first number on the right side of the congruence to make it true.

25. $7 \equiv 2 \pmod 5$

26. $8 \equiv 3 \pmod 5$

27. $41 \equiv 6 \pmod 7$

28. $77 \equiv 5 \pmod{12}$

29. $84 \equiv 1 \pmod 7$

30. $21 \equiv 5 \pmod 7$

31. $23 \equiv 2 \pmod 4$

32. $29 \equiv 3 \pmod 4$

33. $55 \equiv 0 \pmod{11}$

34. $75 \equiv 0 \pmod{25}$

In Exercise 35–42, find each sum.

35. $(3 + 2)\pmod 6$

36. $(3 + 4)\pmod 8$

37. $(4 + 5)\pmod 6$

38. $(5 + 6)\pmod 8$

39. $(6 + 5)\pmod 7$

40. $(8 + 7)\pmod 9$

41. $(49 + 49)\pmod{50}$

42. $(75 + 75)\pmod{100}$

Application Exercises

To avoid A.M. and P.M. designations, the military uses a 24-hour clock. For example, 11 A.M. is designated 1100 hours, and 11 P.M. (12 noon + 11 hours) is designated 2300 hours. In this system, the first two digits represent hours and the last two digits represent minutes. In Exercises 43–46, find each sum in the 24-hour clock system.

43. $1200 + 0600$

44. $1300 + 1900$

45. $0830 + 1550$

46. $1315 + 0945$

47. Suppose today is Wednesday, July 8. Which day of the week will it be 67 days from now?

48. Suppose today is Tuesday, October 12. Which day of the week will it be 147 days from now?

49. The figure shows a modulo 27 system used for coding. Each number on the clock represents a letter of the alphabet and 0 represents a space between words. Add 20, mod 27, called the *code key*, to each of the following numbers and use the figure to decipher this code:

$$9 \quad 12 \quad 8 \quad 20 \quad 7 \quad 20 \quad 12 \quad 7 \quad 1 \quad 23.$$

Writing in Mathematics

50. Describe the three kinds of rotational symmetry exhibited by different parts of the passionflower shown in the introduction to this chapter (page 701).

51. Look around and find an object that exhibits rotational symmetry. Describe the object and the kind of rotational symmetry shown.

52. Under what conditions is a mathematical system a group?

53. Describe two ways of finding $8 + 7$ in a modulo 12, 12-hour clock system.

54. Describe two ways of finding $4 + 4$ in a modulo 5 system.

55. Describe the meaning of $a \equiv b \pmod{m}$ and give an example.

56. The days of the week can be considered a modulo 7 system. Describe another phenomenon that occurs in a cycle. What modulo system is appropriate for this phenomenon?

57. A fascinating collection of readings in mathematics can be found in the four-volume set *The World of Mathematics*, published by Simon and Schuster. Read pages 1552–1557 in volume 3, beginning with the header "A Question for Psychologists." In this part of the selection, the author presents a model in which the "mind" might be viewed within the framework of the group definition. After reading the article, describe an example of nonclosure within your own learning experiences.

Critical Thinking Exercises

58. Consider the following operations on a conditional statement:

E:	leaves the conditional unchanged
C:	changes the conditional into its converse
I:	changes the conditional into its inverse
CP:	changes the conditional into its contrapositive

The operation ∘ means "followed by."

a. Show that $C \circ CP = I$ and $C \circ I = CP$.

b. Fill in the following table.

∘	E	C	I	CP
E				
C				
I				
CP				

c. Show that this mathematical system is a commutative group. Just verify two examples for the associative property.

59. A mathematical system consists of the set of integers and the operation ∘ defined by $a \circ b = a - b + ab$. Use an example involving the associative property and explain why the mathematical system is not a group.

60. Give an example of a mathematical system with just one element which is a group.

61. If it is now 5:00 P.M., what time will it be 99,999,999 hours from now?

62. Clock subtraction is defined by moving a clock's hour hand in a *counterclockwise direction*. Use this definition to find each difference:

a. $(3 - 6) \pmod 7$

b. $(2 - 4) \pmod 5$

In Exercises 63–64, find the first six whole numbers for which each congruence is true.

63. $x + 3 \equiv 5 \pmod 7$ **64.** $3x \equiv 3 \pmod 9$

Group Exercise

65. Group members should read the Blitzer Bonus on page 718. Then, members should find two pictures of human faces, one which they find attractive and one which they find unattractive. Do an analysis of the facial symmetry, much like that shown in the essay, for each face. Summarize the observations. Are there other characteristics of an attractive face that have nothing to do with symmetry? What are some of these characteristics?

Chapter Summary, Review, and Test

SUMMARY

DEFINITIONS AND CONCEPTS	EXAMPLES

13.1 Mathematical Systems

a. A mathematical system consists of a set of elements and at least one binary operation.

Ex. 1, p. 703

b. Certain mathematical systems satisfy the closure, associative, identity, inverse, and commutative properties. These properties are described in the box on page 712. The commutative property is described directly under the box.

Exs. 2–4, p. 704;
Ex. 5, p. 705;
Ex. 6, p. 706;
Exs. 7–9, pp. 707–708

13.2 Rotational Symmetry, Groups, and Clock Arithmetic

a. If it takes m equal turns to restore an object to its original position and each of these turns is a figure that is identical to the original figure, the object has m-fold rotational symmetry. The object will look the same after each turn.

b. A group is a mathematical system meeting the requirements for the closure, associative, identity, and inverse properties. See the box on page 712.

c. A group that meets the requirements for the commutative property is called a commutative group.

Ex. 1, p. 713

d. Clock addition is defined by moving a clock's hour hand in a clockwise direction.

e. Our usual 12-hour clock system is a modulo 12 system. It is a commutative group under clock addition.

Ex. 2, p. 715

f. A modulo m system consists of m integers, starting with 0 and ending with $m - 1$. A modulo m system describes the properties of objects having m-fold rotational symmetry. All such systems are commutative groups under clock addition.

g. The integer a is congruent to b modulo m, denoted $a \equiv b \pmod{m}$, means that if a is divided by m, the remainder is b.

Ex. 3, p. 717

h. Addition in a modulo m system can be performed using a clock or the rule given in the box on page 718.

Ex. 4, p. 718

REVIEW EXERCISES

13.1

Use the mathematical system given by the table shown to solve Exercises 1–18.

∘	e	c	f	r
e	e	c	f	r
c	c	f	r	e
f	f	r	e	c
r	r	e	c	f

1. What is the set of elements of this mathematical system?

2. Is the mathematical system closed under the binary operation? Explain your answer.

Find each result.

3. $c \circ f$ **4.** $r \circ r$ **5.** $e \circ c$

6. $c \circ r$ and $r \circ c$ **7.** $f \circ r$ and $r \circ f$ **8.** $f \circ e$ and $e \circ f$

9. What property is illustrated in Exercises 6–8?

10. Use the table to explain how the property you named in Exercise 9 can be verified for all possible cases without having to consider one case at a time.

Find each result.

11. $(c \circ r) \circ f$ and $c \circ (r \circ f)$

12. $(r \circ e) \circ c$ and $r \circ (e \circ c)$

13. What property is illustrated in Exercises 11–12?

14. What is the identity element for the mathematical system?

Find the inverse for each element.

15. e **16.** c **17.** f **18.** r

19. Is the set $\{0, 1\}$ closed under the operation of ordinary addition? Explain your answer.

20. Is the set $\{0, 1\}$ closed under the operation of multiplication? Explain your answer.

21. Is the set $\{1, 2, 4\}$ closed under division? Explain your answer.

In Exercises 22–23, consider the set of rational numbers.

22. What is the additive inverse of 123?

23. What is the multiplicative inverse of 123?

24. Suppose that $a \circ b$ means to select the larger of the two numbers a and b, or if the numbers are the same, select either number.

a. Complete the following table.

$\circ$	0	1	2	3	4
0					
1					
2					
3					
4					

b. What is the identity element?

c. Does 2 have an inverse? Explain your answer.

13.2

Name the kind of rotational symmetry shown in Exercises 25–26.

25.

Mercedes Benz symbol

26.

Based on the form of a sea urchin

27. Consider the 5-hour, modulo 5, clock system under the operation of clock addition.

a. Complete the clock addition table.

+	0	1	2	3	4
0					
1					
2					
3					
4					

b. Show that this mathematical system is a commutative group. Just verify one example for the associative property.

In Exercises 28–30, determine if each statement is true or false. If the statement is false, change the first number on the right side of the congruence to make it true.

28. $17 \equiv 2 \pmod 8$ **29.** $37 \equiv 3 \pmod 5$

30. $60 \equiv 0 \pmod{10}$

In Exercises 31–34, find each sum.

31. $(4 + 3)\pmod 6$ **32.** $(7 + 7)\pmod 8$

33. $(4 + 3)\pmod 9$ **34.** $(3 + 18)\pmod{20}$

Group Exercise

35. One of the reasons that a passionflower is so visually appealing is that it has three parts with different kinds of rotational symmetry: three-, five-, and ten-fold. The flower has three stigmas, five stamens, and ten petals. For this activity, each member of the group should find an object or picture of an object that exhibits rotational symmetry. Once each member has shared the object with the group, the group should

a. Construct a clock addition table for the object's symmetry.

b. Describe how each of the properties for a commutative group applies to or describes something about the object.

CHAPTER 13 TEST

Use the mathematical system given by the table shown to solve Exercises 1–5.

$\circ$	x	y	z
x	x	y	z
y	y	z	x
z	z	x	y

1. Is the mathematical system closed under the binary operation? Explain your answer.

2. Find $z \circ y$ and $y \circ z$. What property is illustrated?

3. Find $(x \circ z) \circ z$ and $x \circ (z \circ z)$. What property is illustrated?

4. What is the identity element for the mathematical system?

5. Give the inverse for each element in the mathematical system.

6. Is the set $\{-1, 0, 1\}$ closed under the operation of ordinary addition? Explain your answer.

7. In the set of rational numbers, what is the multiplicative inverse of 5? Explain your answer.

8. Name the kind of rotational symmetry shown in the design on the right. Describe why the design has this particular kind of rotational symmetry.

In Exercises 9–10, consider the 4-hour, modulo 4, clock system under the operation of clock addition.

9. Complete the clock addition table.

+	0	1	2	3
0				
1				
2				
3				

10. Show that this mathematical system is a commutative group. Present your work clearly, naming and explaining one property at a time. Just verify one example for the associative property.

In Exercises 11–12, determine if each statement is true or false. If the statement is false, change the first number on the right side of the congruence to make it true.

11. $39 \equiv 3 \pmod 6$

12. $14 \equiv 2 \pmod 7$

In Exercises 13–14, find each sum.

13. $(9 + 1)(\bmod 11)$

14. $(9 + 6)(\bmod 10)$

14 Voting and Apportionment

In the 2000 presidential election, there were 204 statewide ballot measures that presented voters with politically diverse, emotional, and controversial issues. Voters were asked to make choices on issues not typically dealt with in the legislative process, including animal protection, drug policy reform, medical marijuana, gun control, physician-assisted suicide, same-sex marriage, and bilingual education. Because we live in a democracy and do not think alike on these issues, we have elections.

As citizens of a free society, we have both a right and a duty to vote. The political, social, financial, and environmental choices we make in elections can affect every aspect of our lives. However, voting is only part of the story. The second part—methods for counting our votes—lies at the heart of the democratic process.

In this chapter, you will learn the role that mathematics plays in finding our collective voice when faced with more than two options in an election. You will see that determining the winning option may depend on the method used for counting votes, as well as on the issues and the preferences of the voters. Is there a method for counting votes that is always fair? In exploring voting theory, as well as the ideal of "one person, one vote," you will begin to understand some of the mathematical paradoxes of our democracy. This will increase your ability to function as a more fully aware citizen.

Mumble.
Grumble.
Complain.
Wallow.
Hope.
Despair.
Worry.

Vote.

You were somewhat confused over the 2000 presidential election, the closest in U.S. history. How are votes counted so that a candidate loses the popular vote yet gets elected president? Has this occurred at other times in our history? As you watched the election returns, you heard a political scientist, an expert on the mathematics of voting, speak of how a candidate can win an election by requesting, "Please vote against me." Now you're really confused.

This problem appears as Example 3 in Section 14.2.

SECTION 14.1 VOTING METHODS

OBJECTIVES

1. Understand and use preference tables.
2. Use the plurality method to determine an election's winner.
3. Use the Borda count method to determine an election's winner.
4. Use the plurality-with-elimination method to determine an election's winner.
5. Use the pairwise comparison method to determine an election's winner.

Orson Welles (center), star and director of the highly innovative *Citizen Kane*

In the dramatic arts, ours is the era of the movies. As individuals and as a nation, we've grown up with them. Our images of love, war, family, country—even of things that terrify us—owe much to what we've seen on screen.

To celebrate one hundred years of movie making (1896–1996), more than 1500 ballots were sent to leaders in the industry to rank the top American films. The winner? *Citizen Kane* (1941). Controversial outcome? You bet. In this section, you will see how an outcome can be greatly influenced by the voting method used. We begin our discussion of voting theory with a group of students who love movies and are attempting to elect a president of their club. We will use this example throughout the section. Its importance is not in the example itself, but in its implications for conducting fair elections in a democracy.

Preference Tables

1 Understand and use preference tables.

Four candidates are running for president of the Student Film Institute: Paul (P), Rita (R), Sarah (S), and Tim (T). Each of the club's 37 members submits a secret ballot indicating his or her first, second, third, and fourth choice for president. The 37 ballots are shown in Figure 14.1.

	Ballot 1	Ballot 2	Ballot 3	Ballot 4	Ballot 5	Ballot 6	Ballot 7	Ballot 8	Ballot 9	Ballot 10	Ballot 11
1st	P	S	R	P	R	S	R	S	P	S	P
2nd	R	R	T	R	T	R	T	R	R	R	R
3rd	S	T	S	S	S	T	S	T	S	T	S
4th	T	P	P	T	P	P	P	P	T	P	T

	Ballot 12	Ballot 13	Ballot 14	Ballot 15	Ballot 16	Ballot 17	Ballot 18	Ballot 19	Ballot 20	Ballot 21	Ballot 22	Ballot 23	Ballot 24
1st	T	P	T	S	P	T	R	P	S	P	P	T	P
2nd	S	R	S	R	R	S	T	R	T	R	R	S	R
3rd	R	S	R	T	S	R	S	S	R	S	S	R	S
4th	P	T	P	P	T	P	P	T	P	T	T	P	T

	Ballot 25	Ballot 26	Ballot 27	Ballot 28	Ballot 29	Ballot 30	Ballot 31	Ballot 32	Ballot 33	Ballot 34	Ballot 35	Ballot 36	Ballot 37
1st	T	P	P	S	P	S	T	S	P	T	T	S	S
2nd	S	R	R	R	R	R	S	R	R	S	S	R	R
3rd	R	S	S	T	S	T	R	T	S	R	R	T	T
4th	P	T	T	P	T	P	P	P	T	P	P	P	P

FIGURE 14.1 The 37 ballots for the Student Film Institute election

Do you see that several club members had identical ballots? For example, ballots 3, 5, 7, and 18 ranked the candidates exactly the same way, namely RTSP. We can group together these identical ballots. Figure 14.2 shows the 37 ballots placed in identical stacks.

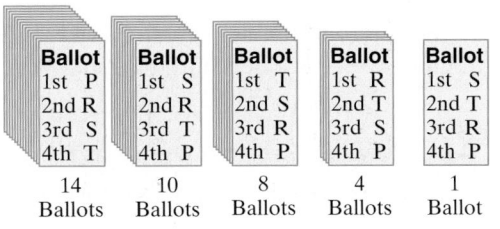

Ballot	Ballot	Ballot	Ballot	Ballot
1st P	1st S	1st T	1st R	1st S
2nd R	2nd R	2nd S	2nd T	2nd T
3rd S	3rd T	3rd R	3rd S	3rd R
4th T	4th P	4th P	4th P	4th P
14 Ballots	10 Ballots	8 Ballots	4 Ballots	1 Ballot

FIGURE 14.2 The 37 election ballots placed into identical stacks

Preference ballots are ballots in which a voter is asked to rank all the candidates in order of preference. Rather than placing the preference ballots in identical stacks, we can use a *preference table* to summarize the results of the election. **A preference table** shows how often each particular outcome occurred. Table 14.1 shows a preference table for the 37 ballots cast for president of the Student Film Institute.

TABLE 14.1 PREFERENCE TABLE FOR THE STUDENT FILM INSTITUTE ELECTION

There were 14 + 10 + 8 + 4 + 1, or 37 voters in the election.

Number of Votes	14	10	8	4	1
First Choice	P	S	T	R	S
Second Choice	R	R	S	T	T
Third Choice	S	T	R	S	R
Fourth Choice	T	P	P	P	P

| 14 ballots rank the candidates P, R, S, T. | 10 ballots rank the candidates S, R, T, P. | 8 ballots rank the candidates T, S, R, P. | 4 ballots rank the candidates R, T, S, P. | 1 ballot ranks the candidates S, T, R, P. |

EXAMPLE 1 UNDERSTANDING A PREFERENCE TABLE

Four candidates are running for mayor of Smallville: Antonio (A), Bob (B), Carmen (C), and Donna (D). The voters were asked to rank all the candidates in order of preference. Table 14.2 shows the preference table for this election.

TABLE 14.2 PREFERENCE TABLE FOR THE SMALLVILLE MAYORAL ELECTION

Number of Votes	130	120	100	150
First Choice	A	D	D	C
Second Choice	B	B	B	B
Third Choice	C	C	A	A
Fourth Choice	D	A	C	D

a. How many people voted in the election?

b. How many people selected the candidates in this order: D, B, A, C?

c. How many people selected Donna (D) as their first choice for mayor?

Solution

a. We find the number of people who voted in the election by adding the numbers in the row labeled *Number of Votes*:

$$130 + 120 + 100 + 150 = 500.$$

Thus, 500 people voted in the election.

TABLE 14.2, repeated

Number of Votes	130	120	100	150
First Choice	A	D	D	C
Second Choice	B	B	B	B
Third Choice	C	C	A	A
Fourth Choice	D	A	C	D

b. We find the number of people selecting the candidates in the order D, B, A, C by referring to the third column of letters in the preference table. Above this column is the number 100. Thus, 100 people voted for the candidates in the order D, B, A, C.

c. We find the number of people who voted for D as their first choice by reading across the row that says *First Choice*. When you see a D in this row, write the number above it. Then find the sum of the numbers:

$$120 + 100 = 220.$$

Thus, 220 voters selected Donna as their first choice for mayor.

 **Check 1 Point**

Four candidates are running for student body president: Alan (A), Bonnie (B), Carlos (C), and Samir (S). The students were asked to rank all the candidates in order of preference. Table 14.3 shows the preference table for this election.

TABLE 14.3 PREFERENCE TABLE FOR THE ELECTION OF STUDENT BODY PRESIDENT

Number of Votes	2100	1305	765	40
First Choice	S	A	S	B
Second Choice	A	S	A	S
Third Choice	B	B	C	A
Fourth Choice	C	C	B	C

a. How many students voted in the election?

b. How many students selected the candidates in this order: B, S, A, C?

c. How many students selected Samir (S) as their first choice for student body president?

Now that you know how to summarize the results of an election using a preference table, it's time to decide the outcome of elections in general and the Student Film Institute election in particular. We will discuss four of the most popular voting methods:

1. the plurality method

2. the Borda count method

3. the plurality-with-elimination method

4. the pairwise comparison method.

 2 Use the plurality method to determine an election's winner.

The Plurality Method

If there are three or more candidates in an election, it often happens that no single candidate receives a majority of first-place votes—that is, more than 50% of the votes. Using the **plurality method**, the candidate with the most first-place votes wins. The only information that we use from the preference ballots are the votes for first place.

THE PLURALITY METHOD

The candidate (or candidates, if there is more than one) with the most first-place votes is the winner.

EXAMPLE 2   USING THE PLURALITY METHOD

Table 14.1, repeated below, shows the preference table for the four candidates running for president of the Student Film Institute: Paul (P), Rita (R), Sarah (S), and Tim (T). Who is declared the winner using the plurality method?

Solution   The candidate with the most first-place votes is the winner. When using Table 14.1, we only need to look at the row indicating the number of first-choice votes.

TABLE 14.1 PREFERENCE TABLE FOR THE STUDENT FILM INSTITUTE ELECTION

P received 14 first-place votes.
S received 10 + 1 = 11
   first-place votes.
T received 8 first-place votes.
R received 4 first-place votes.

Number of Votes	14	10	8	4	1
First Choice	P	S	T	R	S
Second Choice	R	R	S	T	T
Third Choice	S	T	R	S	R
Fourth Choice	T	P	P	P	P

The voice balloon to the left of the preference table indicates that P (Paul) is the winner and the new president of the Student Film Institute. Although Paul received the most first-place votes, he received only $\frac{14}{37}$, or approximately 38%, of the first-place votes, which is less than a majority.

 Check 2 Point   Table 14.2 on page 727 shows the preference table for the four candidates running for mayor of Smallville: Antonio (A), Bob (B), Carmen (C), and Donna (D). Who is declared the winner using the plurality method?

3 Use the Borda count method to determine an election's winner.

The Borda Count Method

The **Borda count method**, developed by the French mathematician and naval captain Jean-Charles de Borda (1733–1799), assigns points to each candidate based on how they are ranked by the voters.

> ### THE BORDA COUNT METHOD
>
> Each voter ranks the candidates from the most favorable to the least favorable. Each last-place vote is given 1 point, each next-to-last-place vote is given 2 points, each third-from-last-place vote is given 3 points, and so on. The points are totaled for each candidate separately. The candidate with the most points is the winner.

When using the plurality method, all the information in the preference table not related to first place is ignored. Example 3 illustrates that this is not the case when using the Borda count method. The Borda count method takes into account all the information provided by the voters' preferences.

EXAMPLE 3   USING THE BORDA COUNT METHOD

Table 14.1, repeated below, shows the preference table for the four candidates running for president of the Student Film Institute: Paul (P), Rita (R), Sarah (S), and Tim (T). Who is declared the winner using the Borda count method?

TABLE 14.1 PREFERENCE TABLE FOR THE STUDENT FILM INSTITUTE ELECTION

Number of Votes	14	10	8	4	1
First Choice	P	S	T	R	S
Second Choice	R	R	S	T	T
Third Choice	S	T	R	S	R
Fourth Choice	T	P	P	P	P

Solution   Because there are four candidates, a first-place vote is worth 4 points, a second-place vote is worth 3 points, a third-place vote is worth 2 points, and a fourth-place vote is worth 1 point. Table 14.4 shows the points produced by the votes in the preference table.

TABLE 14.4 POINTS FOR THE STUDENT FILM INSTITUTE ELECTION

Number of Votes	14	10	8	4	1
First Choice: 4 points	P:$14 \times 4 = 56$ pts	S:$10 \times 4 = 40$ pts	T:$8 \times 4 = 32$ pts	R:$4 \times 4 = 16$ pts	S:$1 \times 4 = 4$ pts
Second Choice: 3 points	R:$14 \times 3 = 42$ pts	R:$10 \times 3 = 30$ pts	S:$8 \times 3 = 24$ pts	T:$4 \times 3 = 12$ pts	T:$1 \times 3 = 3$ pts
Third Choice: 2 points	S:$14 \times 2 = 28$ pts	T:$10 \times 2 = 20$ pts	R:$8 \times 2 = 16$ pts	S:$4 \times 2 = 8$ pts	R:$1 \times 2 = 2$ pts
Fourth Choice: 1 point	T:$14 \times 1 = 14$ pts	P:$10 \times 1 = 10$ pts	P:$8 \times 1 = 8$ pts	P:$4 \times 1 = 4$ pts	P:$1 \times 1 = 1$ pt

Now we read down each column and total the points for each candidate separately.

$$P \text{ gets }\quad 56 + 10 + 8 + 4 + 1 = 79 \quad \text{points.}$$

$$R \text{ gets }\ 42 + 30 + 16 + 16 + 2 = 106 \quad \text{points.}$$

$$S \text{ gets }\quad 28 + 40 + 24 + 8 + 4 = 104 \quad \text{points.}$$

$$T \text{ gets }\ 14 + 20 + 32 + 12 + 3 = 81 \quad \text{points.}$$

Because R (Rita) has received the most points, she is the winner and the new president of the Student Film Institute.

In Examples 2 and 3, the plurality method and the Borda count method produced different election winners. This illustrates the importance of choosing a voting system prior to an election.

 **Check Point 3**   Table 14.2 on page 727 shows the preference table for the four candidates running for mayor of Smallville: Antonio (A), Bob (B), Carmen (C), and Donna (D). Who is declared the winner using the Borda count method?

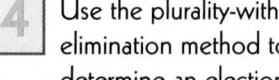 **4**   Use the plurality-with-elimination method to determine an election's winner.

The Plurality-with-Elimination Method

The **plurality-with-elimination method** is based on "survival of the fittest" and may involve a series of elections. If a candidate receives a majority of votes, that candidate is the "fittest" and is declared the winner. If no candidate receives a majority of

votes, the candidate with the least number of votes is eliminated and a second election is held. The process is repeated until a candidate receives a majority.

Multiple elections are not necessary when using preference ballots. We remove the eliminated candidate from the preference table and assume that the relative preferences of a voter are not affected by the elimination of this candidate.

THE PLURALITY-WITH-ELIMINATION METHOD USING PREFERENCE BALLOTS

The candidate with the majority of first-place votes is the winner. If no candidate receives a majority of first-place votes, eliminate the candidate (or candidates if there is a tie) with the fewest first-place votes from the preference table. Move the candidates in each column below the eliminated candidate up one place. The candidate with the majority of first-place votes in the new preference table is the winner. If no candidate receives a majority of first-place votes, repeat this process until a candidate receives a majority.

EXAMPLE 4   USING THE PLURALITY-WITH-ELIMINATION METHOD

Table 14.1, repeated below, shows the preference table for the four candidates running for president of the Student Film Institute: Paul (P), Rita (R), Sarah (S), and Tim (T). Who is declared the winner using the plurality-with-elimination method?

TABLE 14.1 PREFERENCE TABLE FOR THE STUDENT FILM INSTITUTE ELECTION

Number of Votes	14	10	8	4	1
First Choice	P	S	T	R	S
Second Choice	R	R	S	T	T
Third Choice	S	T	R	S	R
Fourth Choice	T	P	P	P	P

Solution   Recall that there are $14 + 10 + 8 + 4 + 1$, or 37, people voting. In order to receive a majority, a candidate must receive more than 50% of the first-place votes, meaning 19 or more votes. The number of first-place votes for each candidate is

$$P(\text{Paul}) = 14 \qquad S(\text{Sarah}) = 10 + 1 = 11 \qquad T(\text{Tim}) = 8 \qquad R(\text{Rita}) = 4.$$

We see that no candidate receives a majority of first-place votes. Because Rita received the fewest first-place votes, she is eliminated in the first round.

The new preference table with R eliminated is shown in Table 14.5. Compare this table with Table 14.1. For each column, each candidate below R moves up one place while the positions of candidates above R remain unchanged. Thus, once Rita is eliminated, the order of preference is preserved.

TABLE 14.5 PREFERENCE TABLE FOR THE STUDENT FILM INSTITUTE ELECTION WITH R ELIMINATED

Number of Votes	14	10	8	4	1
First Choice	P	S	T	T	S
Second Choice	S	T	S	S	T
Third Choice	T	P	P	P	P

TABLE 14.5, repeated

Number of Votes	14	10	8	4	1
First	P	S	T	T	S
Second	S	T	S	S	T
Third	T	P	P	P	P

Recall that to receive a majority, a candidate must get 19 or more first-place votes. The number of first-place votes for each candidate is now

$$P(\text{Paul}) = 14 \qquad S(\text{Sarah}) = 10 + 1 = 11 \qquad T(\text{Tim}) = 8 + 4 = 12.$$

Once again, no candidate receives a majority of first-place votes. Because Sarah received the fewest first-place votes, she is eliminated in the second round.

The new preference table with S eliminated is shown in Table 14.6. Compare this table with Table 14.5. For each column, each candidate below S moves up one place while the positions of candidates above S remain unchanged.

TABLE 14.6 PREFERENCE TABLE FOR THE STUDENT FILM INSTITUTE ELECTION WITH R AND S ELIMINATED

Number of Votes	14	10	8	4	1
First Choice	P	T	T	T	T
Second Choice	T	P	P	P	P

Keep in mind that a candidate must get 19 or more first-place votes to receive a majority. The number of first-place votes for each candidate is now

$$P(\text{Paul}) = 14 \qquad T(\text{Tim}) = 10 + 8 + 4 + 1 = 23.$$

Because T (Tim) has received the majority of first-place votes, he is the winner and the new president of the Student Film Institute.

We have now used three different voting methods in the same election. Each method resulted in a different winner. Can you now see how important it is to choose a voting system before an election takes place?

 Table 14.2 on page 727 shows the preference table for the four candidates running for mayor of Smallville: Antonio (A), Bob (B), Carmen (C), and Donna (D). Who is declared the winner using the plurality-with-elimination method?

BLITZER BONUS

Choosing a New Voting System in San Francisco

A little-noticed proposition approved in 2003 by San Francisco voters offers a glimpse of how we might vote in the future. Instead of casting their ballots for just one candidate, San Franciscans will rank the candidates in most local races according to their first, second, and third choices. The election's winner will be declared using the plurality-with-elimination method, also known as instant-runoff voting. San Francisco approved this new voting system after a runoff election for city attorney drew an abysmally low 16.6% of registered voters.

5 Use the pairwise comparison method to determine an election's winner.

The Pairwise Comparison Method

Using the **pairwise comparison method**, every candidate is compared one-on-one with every other candidate. For example, the candidates running for president of the Student Film Institute are Paul, Rita, Sarah, and Tim. Thus, we make the following comparisons:

Paul vs. Rita	Paul vs. Sarah	Paul vs. Tim
Rita vs. Sarah	Rita vs. Tim	Sarah vs. Tim.

We must make 6 comparisons.

Is there a formula for how many comparisons are needed when there are n candidates? The answer is yes. We are basically counting the number of combinations possible if 2 candidates are taken from n candidates. In Section 11.3, we saw that the number of possible combinations if r items are taken from n items is

$$_nC_r = \frac{n!}{(n-r)!r!}.$$

The number of comparisons needed if 2 candidates are taken from n candidates is

$$_nC_2 = \frac{n!}{(n-2)!2!} = \frac{n(n-1)(n-2)!}{(n-2)!2 \cdot 1} = \frac{n(n-1)}{2}.$$

> ### THE NUMBER OF COMPARISONS USING THE PAIRWISE COMPARISON METHOD
>
> In an election with n candidates, the number of comparisons, C, that must be made is
>
> $$C = \frac{n(n-1)}{2}.$$

How do we make each of these comparisons? We use the preference table and award points, as described in the following box:

> ### THE PAIRWISE COMPARISON METHOD
>
> Voters rank all the candidates and the results are summarized in a preference table. The table is used to make a series of comparisons in which each candidate is compared to each of the other candidates. For each pair of candidates, X and Y, use the table to determine how many voters prefer X to Y and vice versa. If a majority prefer X to Y, then X receives 1 point. If a majority prefer Y to X, then Y receives 1 point. If the candidates tie, then each receives $\frac{1}{2}$ point. After all comparisons have been made, the candidate receiving the most points is the winner.

EXAMPLE 5   USING THE PAIRWISE COMPARISON METHOD

Table 14.1, repeated below, shows the preference table for the four candidates running for president of the Student Film Institute: Paul (P), Rita (R), Sarah (S), and Tim (T). Who is declared the winner using the pairwise comparison method?

TABLE 14.1 PREFERENCE TABLE FOR THE STUDENT FILM INSTITUTE ELECTION

Number of Votes	14	10	8	4	1
First Choice	P	S	T	R	S
Second Choice	R	R	S	T	T
Third Choice	S	T	R	S	R
Fourth Choice	T	P	P	P	P

STUDY TIP

You can list the comparisons by first listing all the candidates. Then pair each candidate with every other person whose name follows in the list.

Solution Because there are 4 candidates, $n = 4$ and the number of comparisons we must make is

$$C = \frac{n(n-1)}{2} = \frac{4(4-1)}{2} = \frac{4 \cdot 3}{2} = \frac{12}{2} = 6.$$

Although you can make each of the 6 comparisons using Table 14.1, we will show the table 6 times to make certain you can follow the details behind each comparison. With P, R, S, and T, the comparisons are P vs. R, P vs. S, P vs. T, R vs. S, R vs. T, and S vs. T.

P vs. R

14	10	8	4	1
P̄	S	T	R̄	S
Ⓡ	R̄	S	T	T
S	T	R̄	S	Ⓡ
T	Ⓟ	Ⓟ	Ⓟ	Ⓟ

14 voters prefer P to R. **10 + 8 + 4 + 1, or 23, voters prefer R to P.**

Conclusion: R wins this comparison and gets 1 point.

P vs. S

14	10	8	4	1
P̄	S̄	T	R	S̄
R	R	S̄	T	T
Ⓢ	T	R	S̄	R
T	Ⓟ	Ⓟ	Ⓟ	Ⓟ

14 voters prefer P to S. **10 + 8 + 4 + 1, or 23, voters prefer S to P.**

Conclusion: S wins this comparison and gets 1 point.

P vs. T

14	10	8	4	1
P̄	S	T̄	R	S
R	R	S	T̄	T̄
S	T̄	R	S	R
Ⓣ	Ⓟ	Ⓟ	Ⓟ	Ⓟ

14 voters prefer P to T. **10 + 8 + 4 + 1, or 23, voters prefer T to P.**

Conclusion: T wins this comparison and gets 1 point.

R vs. S

14	10	8	4	1
P	S̄	T	R̄	S̄
R̄	Ⓡ	S̄	T	T
Ⓢ	T	Ⓡ	Ⓢ	Ⓡ
T	P	P	P	P

14 + 4, or 18, voters prefer R to S. **10 + 8 + 1, or 19, voters prefer S to R.**

Conclusion: S wins this comparison and gets 1 point.

R vs. T

14	10	8	4	1
P	S	T̄	R̄	S
R̄	R̄	S	Ⓣ	T̄
S	Ⓣ	Ⓡ	S	Ⓡ
Ⓣ	P	P	P	P

14 + 10 + 4, or 28, voters prefer R to T. **8 + 1, or 9, voters prefer T to R.**

Conclusion: R wins this comparison and gets 1 point.

S vs. T

14	10	8	4	1
P	S̄	T̄	R	S̄
R	R	Ⓢ	T̄	Ⓣ
S̄	Ⓣ	R	Ⓢ	R
Ⓣ	P	P	P	P

14 + 10 + 1, or 25, voters prefer S to T. **8 + 4, or 12, voters prefer T to S.**

Conclusion: S wins this comparison and gets 1 point.

BLITZER BONUS

Using Democracy to Ensure a Candidate's Election

Donald Saari, author of *Chaotic Elections! A Mathematician Looks at Voting*, often begins his lectures by joking with the audience, while simultaneously telling them that a voting method can affect an election's outcome:

For a price, I will serve as a consultant for your group for your next election. Tell me who you want to win and I will design a "democratic procedure" which ensures the election of your candidate.

Now let's use each of the six conclusions and add points for the six comparisons.

R gets $1 + 1 = 2$ points.

S gets $1 + 1 + 1 = 3$ points.

T gets 1 point.

After all comparisons have been made, the candidate receiving the most points is S (Sarah). Sarah is the winner and the new president of the Student Film Institute.

Check 5 Point Table 14.2 on page 727 shows the preference table for the four candidates running for mayor of Smallville: Antonio (A), Bob (B), Carmen (C), and Donna (D). Make 6 comparisons (A vs. B, A vs. C, A vs. D, B vs. C, B vs. D, and C vs. D) and determine the winner using the pairwise comparison method.

If an election results in a tie, a tie-breaking method should be determined before the election takes place. For example, prior to an election it can be announced that the plurality method will be used to determine the winner and the pairwise comparison method will decide the winner in the event that the plurality method leads to a tie.

In this section we have seen that the chosen voting method can affect the election results. Table 14.7 summarizes the four voting methods we discussed. Which of these methods is the best? Unfortunately, each method has serious flaws, as you will see in the next section.

TABLE 14.7 SUMMARY OF VOTING METHODS

Voting Method	How the Winning Candidate Is Determined
Plurality Method	The candidate with the most first-place votes is the winner.
Borda Count Method	Voters rank all candidates from the most favorable to the least favorable. Each last-place vote receives 1 point, each next-to-last-place vote 2 points, and so on. The candidate with the most points is the winner.
Plurality-with-Elimination Method	The candidate with the majority (over 50%) of first-place votes is the winner. If no candidate receives a majority, eliminate the candidate with the fewest first-place votes. Either hold another election or adjust the preference table. Continue this process until a candidate receives a majority of first-place votes. That candidate is the winner.
Pairwise Comparison Method	Voters rank all the candidates. A series of comparisons is made in which each candidate is compared to each of the other candidates. The preferred candidate in each comparison receives 1 point; in case of a tie, each receives $\frac{1}{2}$ point. The candidate with the most points is the winner.

STUDY TIP

All candidates must be ranked by voters when using the Borda count and pairwise comparison methods. Ranking is not necessary with the plurality method, in which voters need only select one candidate. Ranking is optional in the plurality-with-elimination method. If voters do not rank all the candidates and simply vote for one candidate, multiple elections can result. If they do rank the candidates, multiple elections can be avoided.

Exercise Set 14.1

Practice and Application Exercises

In Exercises 1–2, the preference ballots for three candidates (A, B, and C) are shown. Fill in the number of votes in the first row of the given preference table.

1.

ABC	BCA	BCA	CBA
CBA	ABC	ABC	BCA
BCA	CBA	ABC	ABC
BCA	ABC	ABC	CBA

Number of Votes			
First Choice	A	B	C
Second Choice	B	C	B
Third Choice	C	A	A

2.

ABC	CBA	BCA	BCA
ABC	BCA	BCA	ABC
ABC	ABC	BCA	CBA
BCA	ABC	ABC	ABC

Number of Votes			
First Choice	A	B	C
Second Choice	B	C	B
Third Choice	C	A	A

In Exercises 3–4, four students are running for president of the Let's-Talk-Politics Club: Ann (A), Barbara (B), Charlie (C), and Devon (D). The preference ballots for the four candidates are shown. Construct a preference table to illustrate the results of the election.

3. ABCD   BDCA   CBDA   ABCD   CBDA   CBDA
    ABCD   ABCD   CBAD   CBAD   ABCD   CBDA

4. ABCD   CBAD   CBAD   ABCD   BDCA   CBAD
    CBAD   ABCD   CBDA   CBAD   ABCD   CBDA

5. Your class is given the option of choosing a day for the final exam. The students in the class are asked to rank the three available days, Monday (M), Wednesday (W), and Friday (F). The results of the election are shown in the following preference table.

Number of Votes	14	8	3	1
First Choice	F	F	W	M
Second Choice	W	M	F	W
Third Choice	M	W	M	F

a. How many students voted in the election?

b. How many students selected the days in this order: F, M, W?

c. How many students selected Friday as their first choice for the final?

d. How many students selected Wednesday as their first choice for the final?

6. Students at your college are given the option of choosing a topic for which a speaker will be selected. Students are asked to rank three topics: Technology (T), Environmental Issues (E), and Terrorism in the Name of Religion (R). The results of the election are shown in the following preference table.

Number of Votes	70	30	10	5
First Choice	R	T	T	E
Second Choice	E	R	E	T
Third Choice	T	E	R	R

   a. How many students voted?

   b. How many students selected the topics in this order: T, E, R?

   c. How many students selected technology as their first choice for a speaker's topic?

   d. How many students selected environmental issues as their second choice for a speaker's topic?

7. The theater society members are voting for the kind of play they will perform next semester: a comedy (C), a drama (D), or a musical (M). Their votes are summarized in the following preference table.

Number of Votes	10	6	6	4	2	2
First Choice	M	C	D	C	D	M
Second Choice	C	M	C	D	M	D
Third Choice	D	D	M	M	C	C

   Which type of play is selected using the plurality method?

8. The travel club members are voting for the American city they will visit next semester: New York (N), San Francisco (S), or Chicago (C). Their votes are summarized in the following preference table.

Number of Votes	16	8	6	4
First Choice	S	N	N	C
Second Choice	N	S	C	N
Third Choice	C	C	S	S

   Which city is selected using the plurality method?

9. Four professors are running for chair of the Natural Science Division: Professors Darwin (D), Einstein (E), Freud (F), and Hawking (H). The votes of the professors in the natural science division are summarized in the following preference table.

Number of Votes	30	22	20	12	2
First Choice	D	E	F	H	H
Second Choice	H	F	E	E	F
Third Choice	F	H	H	F	D
Fourth Choice	E	D	D	D	E

   Who is declared the new division chair using the plurality method?

10. Four professors are running for president of the League of Innovation: Professors Disney (D), Ford (F), Gates (G), and Sarnoff (S). The votes of the members in the League of Innovation are summarized in the following preference table.

Number of Votes	30	22	18	10	2
First Choice	F	G	S	D	G
Second Choice	D	D	G	S	S
Third Choice	G	S	D	G	D
Fourth Choice	S	F	F	F	F

   Who is declared the new president using the plurality method?

11. Use the preference table shown in Exercise 7. Which type of play is selected using the Borda count method?

12. Use the preference table shown in Exercise 8. Which city is selected using the Borda count method?

13. Use the preference table shown in Exercise 9. Who is declared the new division chair using the Borda count method?

14. Use the preference table shown in Exercise 10. Who is declared the new president using the Borda count method?

15. Use the preference table shown in Exercise 7. Which type of play is selected using the plurality-with-elimination method?

16. Use the preference table shown in Exercise 8. Which city is selected using the plurality-with-elimination method?

17. Use the preference table shown in Exercise 9. Who is declared the new division chair using the plurality-with-elimination method?

18. Use the preference table shown in Exercise 10. Who is declared the new president using the plurality-with-elimination method?

In Exercises 19–22, suppose that the pairwise comparison method is used to determine the winner in an election.

19. If there are 5 candidates, how many comparisons must be made?

20. If there are 6 candidates, how many comparisons must be made?

21. If there are 8 candidates, how many comparisons must be made?

22. If there are 9 candidates, how many comparisons must be made?

23. Use the preference table shown in Exercise 7. Which type of play is selected using the pairwise comparison method?

24. Use the preference table shown in Exercise 8. Which city is selected using the pairwise comparison method?

25. Use the preference table shown in Exercise 9. Who is declared the new division chair using the pairwise comparison method?

26. Use the preference table shown in Exercise 10. Who is declared the new president using the pairwise comparison method?

In Exercises 27–30, 72 voters are asked to rank four brands of soup: A, B, C, and D. The votes are summarized in the following preference table.

Number of Votes	34	30	6	2
First Choice	A	B	C	D
Second Choice	B	C	D	B
Third Choice	C	D	B	C
Fourth Choice	D	A	A	A

27. Determine the winner using the plurality method.

28. Determine the winner using the Borda count method.

29. Determine the winner using the plurality-with-elimination method.

30. Determine the winner using the pairwise comparison method.

The programmers at the Theater Channel need to select a live musical to introduce their new network. The five choices are Cabaret (C), The Producers (P), Rent (R), Sweeney Todd (S), or West Side Story (W). The 22 programmers rank their choices, summarized in the following preference table. Use the table to solve Exercises 31–34.

Number of Votes	5	5	4	3	3	2
First Choice	C	S	C	W	W	P
Second Choice	R	R	P	P	R	S
Third Choice	P	W	R	R	S	C
Fourth Choice	W	P	S	S	C	R
Fifth Choice	S	C	W	C	P	W

31. Determine which musical is selected using the Borda count method.

32. Determine which musical is selected using the plurality method.

33. Determine which musical is selected using the pairwise comparison method.

34. Determine which musical is selected using the plurality-with-elimination method.

35. Five candidates, A, B, C, D, and E, are running for chair of the Psychology Department. The votes of the department members are summarized in the following preference table.

Number of Votes	5	5	3	3	3	2
First Choice	A	C	D	A	B	D
Second Choice	B	E	C	D	E	C
Third Choice	C	D	B	B	A	B
Fourth Choice	D	A	E	C	C	A
Fifth Choice	E	B	A	E	D	E

   a. Who is declared the new department chair using the Borda count method?

   b. Suppose that Professor E withdraws from the running before the votes are counted. Write the new preference table. Now who is declared the new department chair using the Borda count method?

36. Rework parts (a) and (b) of Exercise 35 using the pairwise comparison method.

37. Three candidates, A, B, and C, are running for mayor. Election rules stipulate that the plurality method will determine the winner. In the event that the plurality method leads to a tie, the Borda count method will decide the winner. The election results are summarized in the following preference table. Under these rules, which candidate becomes the new mayor?

Number of Votes	12,000	7500	4500
First Choice	C	A	A
Second Choice	B	B	C
Third Choice	A	C	B

38. Three candidates, A, B, and C, are running for mayor. Election rules stipulate that the pairwise comparison method will determine the winner. In the event that the pairwise comparison method leads to a tie, the Borda count method will decide the winner. The election results are summarized in the following preference table. Under these rules, which candidate becomes the new mayor?

Number of Votes	60,000	40,000	40,000	20,000	20,000
First Choice	A	C	B	A	C
Second Choice	B	A	C	C	B
Third Choice	C	B	A	B	A

Writing in Mathematics

39. What is a preference ballot?

40. Describe what is contained in a preference table. What does the table show?

41. Describe the plurality method. Why is ranking not necessary when using this method?

42. Describe the Borda count method. Is it possible to use this method without ranking the candidates? Explain.

43. What is the plurality-with-elimination method? Why is it advantageous to rank the candidates when using this method?

44. What is the pairwise comparison method? Is it possible to use this method without ranking the candidates? Explain.

45. Describe the process used to determine how many comparisons must be made with the pairwise comparison method.

46. Why is it important to choose a voting system before an election takes place?

47. Playwright Tom Stoppard wrote, "It's not the voting that's democracy; it's the counting." Explain what he meant by this.

48. Based on the answers to Exercises 27–30, if you were at the market, which brand of soup, A, B, C, or D, would you select? Explain your choice.

Critical Thinking Exercises

49. Select the false statement or statements from the following list. If all statements are true, so state.

   a. A candidate with the majority always wins under the plurality method.

   b. A candidate with the majority always wins under the Borda count method.

   c. A candidate with the majority always wins under the plurality-with-elimination method.

   d. A candidate with the majority always wins under the pairwise comparison method.

50. An election with 20 candidates is to be determined using the pairwise comparison method. If it takes two minutes to calculate a pairwise comparison, how long will it take to calculate the election results?

In Exercises 51–54, construct a preference table for an election with candidates A, B, and C satisfying the given condition.

51. A wins using the Borda count method.

52. B wins using the plurality-with-elimination method.

53. C wins using the pairwise comparison method.

54. B wins using all four methods discussed in this section.

Group Exercises

55. Research and present a group report on how voting is conducted for the Academy Awards. Describe the single transferable voting method, a variation of the plurality-with-elimination method, in the nomination stage for best picture. (Members of the Irish Senate are also elected by this method.) Be sure to describe some of the more bizarre occurrences at the Oscar ceremonies.

56. Research and present a group report on how voting is conducted for one or more of the following awards: the Heisman Trophy, the Nobel Prize, the Grammy, the Tony, the Emmy, the Pulitzer Prize, or any event or award that the group finds intriguing. Be sure to discuss how the nominees are selected, who participates in the voting, the voting system or systems used, and who counts the results.

SECTION 14.2   FLAWS OF VOTING METHODS

OBJECTIVES

1. Use the majority criterion to determine a voting system's fairness.

2. Use the head-to-head criterion to determine a voting system's fairness.

3. Use the monotonicity criterion to determine a voting system's fairness.

4. Use the irrelevant alternatives criterion to determine a voting system's fairness.

5. Understand Arrow's Impossibility Theorem.

It's not fair! The Olympic Games can generate $10 billion in spending for the host city. With all that money at stake, no wonder the selection process has a questionable past. In 1998, evidence surfaced to indicate what many Olympic insiders had been whispering for years: that bid cities like Salt Lake City, host of the 2002 winter games, had showered members of the International Olympic Committee with gifts—fully paid shopping trips for spouses, college scholarships for children, and even cash in envelopes for members.

   The voting method used to select the winning Olympic host city is a variation of the plurality-with-elimination method. If the International Olympic Committee can eliminate the abuse and bribery that has plagued its selection process, can a winning city be fairly elected? Not necessarily. In the early 1950s, the mathematical economist Kenneth Arrow (1921–) proved that **a method for determining election results that is democratic and always fair is a mathematical impossibility**.

   What exactly is meant by "democratic" and "fair"? In this section, we will look at four criteria that mathematicians and political scientists have agreed a fair voting system should meet. These four **fairness criteria** include the *majority criterion*, the *head-to-head criterion*, the *monotonicity criterion*, and the *irrelevant alternatives criterion*.

Use the majority criterion to determine a voting system's fairness.

STUDY TIP

Remember that a *majority* refers to receiving more than half the first-place votes. A *plurality* refers to receiving the most first-place votes.

The Majority Criterion

Most people would agree that if a single candidate receives more than half of the first-place votes in an election, then that candidate should be declared the winner. This requirement for a fair and democratic election is called the **majority criterion**.

THE MAJORITY CRITERION

If a candidate receives a majority of first-place votes in an election, then that candidate should win the election.

Example 1 shows that the Borda count method may not satisfy the majority criterion.

EXAMPLE 1 | THE BORDA COUNT METHOD VIOLATES THE MAJORITY CRITERION

TABLE 14.8 PREFERENCE TABLE FOR SELECTING A NEW COLLEGE PRESIDENT

Number of Votes	6	3	2
First Choice	E	G	F
Second Choice	F	H	G
Third Choice	G	F	H
Fourth Choice	H	E	E

The 11 members of the Board of Trustees of your college must hire a new college president. The four finalists for the job, E, F, G, and H, are ranked by the 11 members. The preference table is shown in Table 14.8. The board members agree to use the Borda count method to determine the winner.

a. Which candidate has a majority of first-place votes?

b. Which candidate is declared the new college president using the Borda count method?

Solution

a. There are 11 first-place votes. Because a majority involves more than half of these votes, any candidate with 6 or more first-place votes receives a majority. The first-choice row in Table 14.8 shows that candidate E received 6 first-place votes. Thus, candidate E has a majority of first-place votes. By the majority criterion, candidate E should be the new college president.

b. Using the Borda count method with four candidates, a first-place vote is worth 4 points, a second-place vote 3 points, a third-place vote 2 points, and a fourth-place vote 1 point. Table 14.9 shows the points produced by the votes in the preference table.

TABLE 14.9 POINTS FOR THE COLLEGE PRESIDENT ELECTION

Number of Votes	6	3	2
First Choice: 4 points	E: $6 \times 4 = 24$ pts	G: $3 \times 4 = 12$ pts	F: $2 \times 4 = 8$ pts
Second Choice: 3 points	F: $6 \times 3 = 18$ pts	H: $3 \times 3 = 9$ pts	G: $2 \times 3 = 6$ pts
Third Choice: 2 points	G: $6 \times 2 = 12$ pts	F: $3 \times 2 = 6$ pts	H: $2 \times 2 = 4$ pts
Fourth Choice: 1 point	H: $6 \times 1 = 6$ pts	E: $3 \times 1 = 3$ pts	E: $2 \times 1 = 2$ pts

Now we read down the columns and total the points for each candidate.

E gets $24 + 3 + 2 = 29$ points.   F gets $18 + 6 + 8 = 32$ points.

G gets $12 + 12 + 6 = 30$ points.   H gets $6 + 9 + 4 = 19$ points.

Because candidate F has received the most points, candidate F is declared the new college president using the Borda count method.

In Example 1, although a majority of members of the Board of Trustees preferred candidate E, the Borda count method yielded a different choice, candidate F. In a situation like this, the Borda count method is said to *violate* the majority criterion.

Check 1 Point

The 14 members of the school board must hire a new principal. The four finalists for the job, A, B, C, and D, are ranked by the 14 members. The preference table is shown in Table 14.10. The board members agree to use the Borda count method to determine the winner.

TABLE 14.10 PREFERENCE TABLE FOR SELECTING A NEW PRINCIPAL

Number of Votes	6	4	2	2
First Choice	A	B	B	A
Second Choice	B	C	D	B
Third Choice	C	D	C	D
Fourth Choice	D	A	A	C

a. Which candidate has a majority of first-place votes?

b. Which candidate is declared the new principal using the Borda count method?

We have seen that an advantage of the Borda count method is that it takes into account all the information about the voters' preferences in the preference table. However, a candidate who has a majority of first-place votes can lose an election with the Borda count method. The other three voting methods that we have discussed—the plurality method, the plurality-with-elimination method, and the pairwise comparison method—never violate the majority criterion. Unfortunately, there are other fairness criteria that are violated by each of these three systems.

The Head-to-Head Criterion

Most people would agree that if there is one candidate who is favored by the voters when compared, in turn, to each of the other candidates, then that candidate should win the election. This requirement of a fair and democratic election is called the **head-to-head criterion**.

2 Use the head-to-head criterion to determine a voting system's fairness.

> ### THE HEAD-TO-HEAD CRITERION
> If a candidate is favored when compared separately—that is, head-to-head—with every other candidate, then that candidate should win the election.

Example 2 shows that the plurality method may violate the head-to-head criterion.

EXAMPLE 2 THE PLURALITY METHOD VIOLATES THE HEAD-TO-HEAD CRITERION

Twenty-two people are asked to taste-test and rank three different brands, A, B, and C, of tuna fish. The results are summarized in the preference table in Table 14.11.

a. Which brand is favored over all others using a head-to-head comparison?

b. Which brand wins the taste test using the plurality method?

TABLE 14.11 PREFERENCE TABLE FOR THREE BRANDS OF TUNA FISH

Number of Votes	8	6	4	4
First Choice	A	C	C	B
Second Choice	B	B	A	A
Third Choice	C	A	B	C

Solution

a. We begin by comparing brands A and B using Table 14.11. A is favored over B in columns 1 and 3, giving A 8 + 4, or 12, votes. B is favored over A in columns 2 and 4, giving B 6 + 4, or 10, votes. Because A has 12 votes and B has 10 votes, A is favored when compared to B.

Now let's use Table 14.11 to compare brands A and C. A is favored over C in columns 1 and 4, giving A 8 + 4, or 12, votes. C is favored over A in columns 2 and 3, giving C 6 + 4, or 10, votes. Because A has 12 votes and C has 10 votes, A is favored when compared to C.

We see that A is favored over the other two brands using a head-to-head comparison.

b. Using the plurality method, the brand with the most first-place votes is the winner. Look at the row indicating the first choice. A received 8 votes, B received 4 votes, and C received 6 + 4, or 10, votes. Brand C wins the taste test using the plurality method.

In Example 2, although brand A was favored over the other two brands using a head-to-head comparison, the plurality method yielded a different winner, brand C. Thus, in this situation, the plurality method violates the head-to-head criterion.

Is the plurality method the only voting method with the potential to violate the head-to-head criterion? The answer is no. The Borda count method and the plurality-with-elimination method both have the potential to violate the head-to-head criterion. Can this occur with the pairwise comparison method? No. A candidate who defeats each of the other candidates in a head-to-head vote wins every pairwise comparison and thus gets the greatest number of points under this method.

 **Check 2 Point** Seven people are asked to listen to and rate three different pairs of stereo speakers, A, B, and C. The results are summarized in Table 14.12.

a. Which brand is favored over all others using a head-to-head comparison?

b. Which brand wins the listening test using the plurality method?

TABLE 14.12 PREFERENCE TABLE FOR THREE PAIRS OF STEREO SPEAKERS

Number of Votes	3	2	2
First Choice	A	B	C
Second Choice	B	A	B
Third Choice	C	C	A

3 Use the monotonicity criterion to determine a voting system's fairness.

The Monotonicity Criterion

Elections are generally preceded by a campaign and discussions of the candidates' strengths and weaknesses. In some situations, a preliminary election in which votes do not count, called a **straw vote**, is taken as a measure of voters' intentions. Most people would agree that if a candidate wins the straw poll, which we can consider the first election, and then gains additional support without losing any of the original support, then that candidate should win the second election. This requirement for a fair and democratic election is called the **monotonicity criterion**.

> ### THE MONOTONICITY CRITERION
>
> If a candidate wins an election and, in a reelection, the only changes are changes that favor the candidate, then that candidate should win the reelection.

Example 3 shows that the plurality-with-elimination method may not satisfy the monotonicity criterion.

BLITZER BONUS

Choosing the Olympic Host City

Like cardinals sequestered to vote for a new pope, the members of the International Olympic Committee lock themselves into a conference room and do not leave until they can proclaim a winner from among the candidate cities. Think of it as a sort of papal consistory on steroids. The voting method used to select the winner is the plurality-with-elimination method with a slight difference: Instead of indicating their preferences all at once, the members let their preferences be known one round at a time.

(*continued on page 743*)

EXAMPLE 3 | THE PLURALITY-WITH-ELIMINATION METHOD VIOLATES THE MONOTONICITY CRITERION

The 58 members of the Student Activity Council are meeting to elect a keynote speaker to launch student involvement week. The choices are Bill Gates (G), Jesse Ventura (V), or Oprah Winfrey (W). A straw vote is taken and the results are given in Table 14.13. After a lengthy discussion, all but eight students vote in exactly the same way. The eight voters, shown in the last column of Table 14.13, all changed their ballots to make Oprah Winfrey (W) their first choice. The results of the second election are given in Table 14.14. Notice that the first column has increased by 8 votes.

TABLE 14.13 PREFERENCE TABLE FOR THE STRAW VOTE

Number of Votes	20	16	14	8
First Choice	W	V	G	G
Second Choice	G	W	V	W
Third Choice	V	G	W	V

TABLE 14.14 PREFERENCE TABLE FOR THE SECOND ELECTION

Number of Votes	28	16	14
First Choice	W	V	G
Second Choice	G	W	V
Third Choice	V	G	W

These 8 voters changed their ballots to make Oprah Winfrey (W) their first choice.

a. Using the plurality-with-elimination method, which speaker wins the first election?

b. Using the plurality-with-elimination method, which speaker wins the second election?

c. Does this violate the monotonicity criterion?

Solution

a. There are 58 people voting. No speaker receives a majority of votes (30 or more votes) in Table 14.13, the first election (the straw vote). Using the plurality-with-elimination method, because V (Jesse Ventura) receives the fewest first-place votes, he is eliminated in the first round. The new preference table with V eliminated is shown on the next page in Table 14.15. Because W (Oprah Winfrey) has received a majority of first-place votes, she is the winner of the straw poll.

(continued)

Here is the voting that took place in the selection of the site for the 2000 Summer Olympics. In each round, the delegates voted for just one city, and the city with the fewest votes was eliminated.

First Round	
Beijing	32
Sydney	30
Manchester	11
Berlin	9
Istanbul	7

Istanbul eliminated.

Second Round	
Beijing	37
Sydney	30
Manchester	13
Berlin	9

Berlin eliminated.

Third Round	
Beijing	40
Sydney	37
Manchester	11
Abstentions	1

Manchester eliminated.

Fourth Round	
Sydney	45
Beijing	43
Abstentions	1

Beijing eliminated.

Sydney became the host city for the 2000 Summer Olympics.

While the voting takes only a day, cities campaign for years to be selected. Atlanta had the games in 1996, Nagano in 1998, Sydney in 2000, Salt Lake City in 2002, and they will be in Athens in 2004.

b. Now we focus on Table 14.14, the preference table for the second election. No speaker receives a majority of votes (30 or more votes) in this election. Using the plurality-with-elimination method, because G (Bill Gates) receives the fewest first-place votes, he is eliminated in the first round. The new preference table with G eliminated is shown in Table 14.16. Because V (Jesse Ventura) has received a majority of first-place votes, he is the winner of the second election.

TABLE 14.15 PREFERENCE TABLE FOR THE STRAW VOTE WITH V ELIMINATED

Number of Votes	20	16	14	8
First Choice	W	W	G	G
Second Choice	G	G	W	W

TABLE 14.16 PREFERENCE TABLE FOR THE SECOND ELECTION WITH G ELIMINATED

Number of Votes	28	16	14
First Choice	W	V	V
Second Choice	V	W	W

c. Oprah Winfrey (W) won the first election. She then gained additional support with the 8 voters who changed their ballots to make her their first choice. However, she lost the second election. This violates the monotonicity criterion.

Example 3 illustrates the bizarre possibility that you can actually do worse by doing better! Of the four voting methods we have studied, the only one that can violate the monotonicity criterion is the plurality-with-elimination method.

Check 3 Point An election with 120 voters and three candidates, A, B, and C, is to be decided using the plurality-with-elimination method. Table 14.17 shows the results of a straw poll. After the straw poll, 12 voters all changed their ballots to make candidate A their first choice. The results of the second election are given in Table 14.18.

TABLE 14.17 PREFERENCE TABLE FOR THE STRAW VOTE

Number of Votes	42	34	28	16
First Choice	A	C	B	B
Second Choice	B	A	C	A
Third Choice	C	B	A	C

TABLE 14.18 PREFERENCE TABLE FOR THE SECOND ELECTION

Number of Votes	54	34	28	4
First Choice	A	C	B	B
Second Choice	B	A	C	A
Third Choice	C	B	A	C

a. Using the plurality-with-elimination method, which candidate wins the first election?

b. Using the plurality-with-elimination method, which candidate wins the second election?

c. Does this violate the monotonicity criterion? Explain your answer.

 Use the irrelevant alternatives criterion to determine a voting system's fairness.

The Irrelevant Alternatives Criterion

A candidate wins an election. Then one or more of the other candidates are removed from the ballot and a recount is done. Most people would agree that the previous winner should still be declared the winner. This requirement for a fair and democratic election is called the **irrelevant alternatives criterion**.

> **THE IRRELEVANT ALTERNATIVES CRITERION**
>
> If a candidate wins an election and, in a recount, the only changes are that one or more of the other candidates are removed from the ballot, then that candidate should still win the election.

Each of the four voting methods we have discussed may violate the irrelevant alternatives criterion. Example 4 shows that the pairwise comparison method may not satisfy the criterion.

EXAMPLE 4 | THE PAIRWISE COMPARISON METHOD VIOLATES THE IRRELEVANT ALTERNATIVES CRITERION

Four candidates, E, F, G, and H, are running for mayor of Bolinas. The election results are shown in Table 14.19.

TABLE 14.19 PREFERENCE TABLE FOR THE MAYOR OF BOLINAS

Number of Votes	160	100	80	20
First Choice	E	G	H	H
Second Choice	F	F	E	E
Third Choice	G	H	G	F
Fourth Choice	H	E	F	G

a. Using the pairwise comparison method, who wins this election?

b. Prior to the announcement of the election results, candidates F and G both withdraw from the running. Using the pairwise comparison method, which candidate is declared mayor of Bolinas with F and G eliminated from the preference table?

c. Does this violate the irrelevant alternatives criterion?

Solution

a. Because there are 4 candidates, $n = 4$ and the number of comparisons we must make is

$$C = \frac{n(n-1)}{2} = \frac{4(4-1)}{2} = \frac{4 \cdot 3}{2} = \frac{12}{2} = 6.$$

The following table shows the results of these 6 comparisons. Use Table 14.19 to verify each of these results.

Comparison	Vote Results	Conclusion
E vs. F	260 voters prefer E to F. 100 voters prefer F to E.	E wins and gets 1 point.
E vs. G	260 voters prefer E to G. 100 voters prefer G to E.	E wins and gets 1 point.
E vs. H	160 voters prefer E to H. 200 voters prefer H to E.	H wins and gets 1 point.
F vs. G	180 voters prefer F to G. 180 voters prefer G to F.	It's a tie. F gets $\frac{1}{2}$ point. G gets $\frac{1}{2}$ point.
F vs. H	260 voters prefer F to H. 100 voters prefer H to F.	F wins and gets 1 point.
G vs. H	260 voters prefer G to H. 100 voters prefer H to G.	G wins and gets 1 point.

TABLE 14.20 PREFERENCE TABLE FOR THE MAYOR OF BOLINAS WITH F AND G REMOVED

Number of votes	160	100	80	20
First choice	E	H	H	H
Second choice	H	E	E	E

160 voters prefer E to H.   200 voters prefer H to E.

Thus, E gets 2 points, F and G each get $1\frac{1}{2}$ points, and H gets 1 point. Therefore, E is the winner when candidates F and G are included.

b. Once F and G both withdraw from the running, only two candidates, E and H, remain. Keep in mind that E was the winning candidate prior to this withdrawl. The new preference table is shown in Table 14.20. Using the pairwise comparison test, there are 2 candidates and the number of comparisons we must make is

$$C = \frac{n(n-1)}{2} = \frac{2(2-1)}{2} = \frac{2 \cdot 1}{2} = 1.$$

The one comparison we must make is E to H. The voice balloons below Table 14.20 show that H wins, defeating E by 200 votes to 160 votes. H gets 1 point, E gets 0 points, and candidate H is the new mayor of Bolinas.

c. The first election count produced E as the winner. Then, even though F and G were not the winning candidates, their removal from the ballots produced H, not E, as the winner. This violates the irrelevant alternatives criterion.

TABLE 14.21 PREFERENCE TABLE FOR MAYOR

Number of Votes	150	90	90	30
First Choice	A	C	D	D
Second Choice	B	B	A	A
Third Choice	C	D	C	B
Fourth Choice	D	A	B	C

Check Point 4

Four candidates, A, B, C, and D, are running for mayor. The election results are shown in Table 14.21.

a. Using the pairwise comparison method, who wins this election?

b. Prior to the announcement of the election results, candidates B and C both withdraw from the running. Using the pairwise comparison method, which candidate is declared mayor with B and C eliminated from the preference table?

c. Does this violate the irrelevant alternatives criterion? Explain your answer.

Table 14.22 summarizes the four fairness criteria we have discussed. Table 14.23 shows which voting methods satisfy the criteria.

TABLE 14.22 SUMMARY OF FAIRNESS CRITERIA

Criterion	Description
Majority Criterion	If a candidate receives a majority of first-place votes in an election, then that candidate should win the election.
Head-to-Head Criterion	If a candidate is favored when compared head-to-head with every other candidate, then that candidate should win the election.
Monotonicity Criterion	If a candidate wins an election and, in a reelection, the only changes are changes that favor the candidate, then that candidate should win the reelection.
Irrelevant Alternatives Criterion	If a candidate wins an election and, in a recount, the only changes are that one or more of the other candidates are removed from the ballot, then that candidate should still win the election.

TABLE 14.23 VOTING METHODS AND WHETHER THEY SATISFY THE FAIRNESS CRITERIA

Fairness Criteria	Voting Method			
	Plurality Method	Borda Count Method	Plurality-with-Elimination Method	Pairwise Comparison Method
Majority Criterion	Always satisfies	May not satisfy	Always satisfies	Always satisfies
Head-to-Head Criterion	May not satisfy	May not satisfy	May not satisfy	Always satisfies
Monotonicity Criterion	Always satisfies	Always satisfies	May not satisfy	Always satisfies
Irrelevant Alternatives Criterion	May not satisfy	May not satisfy	May not satisfy	May not satisfy

5 Understand Arrow's Impossibility Theorem.

Kenneth Arrow (1921–). Awarded 1972 Nobel Prize in Economics

The Search for a Fair Voting System

We have seen that none of the voting methods discussed in this chapter always satisfies each of the fairness criteria. Is there another democratic voting method that does satisfy all four criteria—a perfectly fair voting system? For elections involving more than two candidates, the answer is no. No perfectly fair voting method exists. How do we know? In 1951, economist Kenneth Arrow proved the now famous **Arrow's Impossibility Theorem**: There does not exist, and will never exist, any democratic voting system that satisfies all of the fairness criteria.

ARROW'S IMPOSSIBILITY THEOREM

It is mathematically impossible for any democratic voting system to satisfy each of the four fairness criteria.

In 1972, Arrow was awarded the Nobel Prize in Economics for his pioneering work. Arrow's discipline is now known as *social-choice theory*, a field that combines mathematics, economics, and political science.

BLITZER BONUS

The Electoral College

The framers of the Constitution believed that the opinion of the majority sometimes had to be tempered by the wisdom of elected representatives. Consequently, the President and the Vice President are the only elected U.S. officials who are not chosen by a direct vote of the people, but rather by the Electoral College.

Here is an outline of how the Electoral College works.

- There are a total of 538 electoral votes.

- The votes are divided by the states and among the District of Columbia. The number of electoral votes for each state is equal to its total number of Congressional senators and representatives.

- The candidate who wins a majority of popular votes in a given state earns all the votes of that state's Electoral College members.

- The candidate who wins the majority, not merely a plurality, of the electoral votes is declared the winner. A presidential candidate needs at least 270 electoral votes to win. If no candidate wins, then the election is turned over to the House of Representatives, who vote for one of the top three contenders.

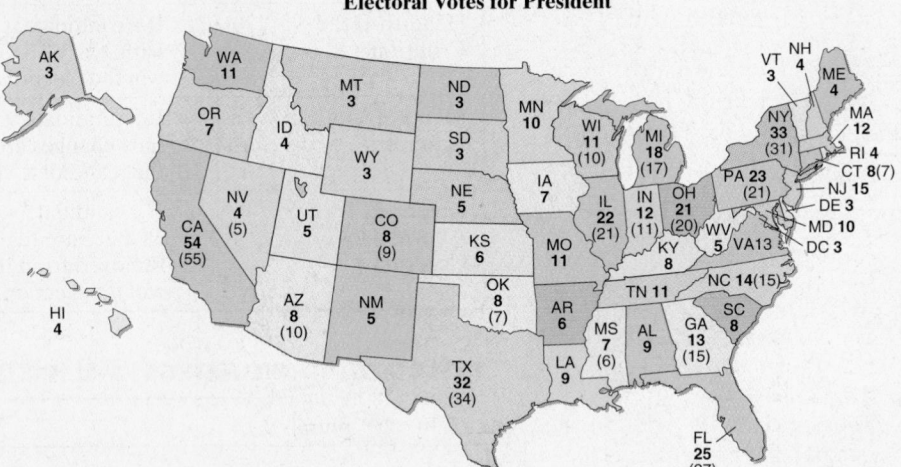

Electoral Votes for President

Figures in boldface, based on the 1990 census, were used in the 1992, 1996, and 2000 elections. Some of the electoral votes for the 2004 election, based on the 2000 census, are different. These are given in parentheses.

Source: Federal Election Commission

As we know from the 2000 election, it is possible for a presidential candidate who has not won the popular vote to win the election. If a candidate wins the popular vote in large states (ones with lots of electoral votes) by only a slim margin, and loses the popular vote in smaller states by a wide margin, it is possible that the popular vote winner will actually lose the election. Is George W. Bush the only U.S. president who lost the popular vote but won the election? The answer is no. In 1888, Benjamin Harrison won the popular vote over Grover Cleveland, but Cleveland won the electoral vote and became president. Also, in the section opener on page 763, we describe how the "flawed" electoral college system led to an election debacle in 1876 reminiscent of the bitter divisiveness that tainted the 2000 election.

Exercise Set 14.2

Practice and Application Exercises

1. Voters in a small town are considering four proposals, A, B, C, and D, for the design of affordable housing. The winning design is to be determined by the Borda count method. The preference table for the election is shown.

   a. Which design has a majority of first-place votes?

   b. Using the Borda count method, which design will be used for the affordable housing?

   c. Is the majority criterion satisfied? Explain your answer.

Number of Votes	300	120	90	60
First Choice	D	C	C	A
Second Choice	A	A	A	D
Third Choice	B	B	D	B
Fourth Choice	C	D	B	C

2. Fifty-three people are asked to taste-test and rank three different brands of yogurt, A, B, and C. The preference table shows the rankings of the 53 voters.

   a. Which brand has a majority of first-place votes?

   b. Suppose that the Borda count method is used to determine the winner. Which brand wins the taste test?

   c. Is the majority criterion satisfied? Explain your answer.

Number of Votes	27	24	2
First Choice	A	B	C
Second Choice	C	C	B
Third Choice	B	A	A

3. MTV's *Real World* is considering three cities for their new season: Amsterdam (A), Rio de Janeiro (R), or Vancouver (V). Programming executives and the show's production team vote to decide where the new season will be taped. The winning city is to be determined by the plurality method. The preference table for the election is shown.

   a. Which city is favored over all others using a head-to-head comparison?

   b. Which city wins the vote using the plurality method?

   c. Is the head-to-head criterion satisfied? Explain your answer.

Number of Votes	12	9	4	4
First Choice	A	V	V	R
Second Choice	R	R	A	A
Third Choice	V	A	R	V

4. A computer company is considering opening a new branch in Atlanta (A), Boston (B), or Chicago (C). Senior managers vote to decide where the new branch will be located. The winning city is to be determined by the plurality method. The preference table for the election is shown.

   a. Which city is favored over all others using a head-to-head comparison?

   b. Which city wins the vote using the plurality method?

   c. Is the head-to-head criterion satisfied? Explain your answer.

Number of Votes	20	19	5
First Choice	A	B	C
Second Choice	B	C	B
Third Choice	C	A	A

5. A town is voting on an ordinance dealing with smoking in public spaces. The options are (A) permit unrestricted smoking; (B) permit smoking in designated areas only; and (C) ban all smoking in public places. The winner is to be determined by the Borda count method. The preference table for the election is shown.

   a. Which option is favored over all others using a head-to-head comparison?

   b. Which option wins the vote using the Borda count method?

   c. Is the head-to-head criterion satisfied? Explain your answer.

Number of Votes	120	60	30	30	30
First Choice	A	C	B	C	B
Second Choice	C	B	A	A	C
Third Choice	B	A	C	B	A

6. A town is voting on an ordinance dealing with nudity at its public beaches. The options are (A) make clothing optional at all beaches; (B) permit nudity at designated beaches only; and (C) permit no nudity at public beaches. The winner is to be determined by the Borda count method. The preference table for the election is shown.

   a. Which option is favored over all others using a head-to-head comparison?

   b. Which option wins the vote using the Borda count method?

   c. Is the head-to-head criterion satisfied? Explain your answer.

Number of Votes	200	80	80
First Choice	C	B	A
Second Choice	A	A	B
Third Choice	B	C	C

7. The preference table gives the results of a straw vote among three candidates, A, B, and C.

  a. Using the plurality-with-elimination method, which candidate wins the straw vote?

  b. In the actual election, the four voters in the last column who voted A, C, B, in that order, change their votes to C, A, B. Using the plurality-with-elimination method, which candidate wins the actual election?

  c. Is the monotonicity criterion satisfied? Explain your answer.

Number of Votes	10	8	7	4
First Choice	C	B	A	A
Second Choice	A	C	B	C
Third Choice	B	A	C	B

8. The preference table gives the results of a straw vote among three candidates, A, B, and C.

  a. Using the plurality-with-elimination method, which candidate wins the straw vote?

  b. In the actual election, the six voters in the last column who voted A, C, B, in that order, change their votes to C, A, B. Using the plurality-with-elimination method, which candidate wins the actual election?

  c. Is the monotonicity criterion satisfied? Explain your answer.

Number of Votes	14	12	10	6
First Choice	C	B	A	A
Second Choice	A	C	B	C
Third Choice	B	A	C	B

9. Members of the Student Activity Committee at a college are considering three film directors to speak at a campus arts festival: Ron Howard (H), Spike Lee (L), and Steven Spielberg (S). Committee members vote for their preferred speaker. The winner is to be selected by the pairwise comparison method. The preference table for the election is shown.

  a. Using the pairwise comparison method, who is selected as the speaker?

  b. Prior to the announcement of the speaker, Ron Howard informs the committee that he will not be able to participate due to other commitments. Construct a new preference table for the election with H eliminated. Using the new table and the pairwise comparison method, who is selected as the speaker?

  c. Is the irrelevant alternatives criterion satisfied? Explain your answer.

Number of Votes	10	8	5
First Choice	H	L	S
Second Choice	S	S	L
Third Choice	L	H	H

10. Members of the student activity committee at a college are considering three actors to speak at a campus festival on women in the arts: Whoopi Goldberg (G), Bette Midler (M), and Julia Roberts (R). Committee members vote for their preferred speaker. The winner is to be selected by the pairwise comparison method. The preference table for the election is shown.

  a. Using the pairwise comparison method, who is selected as the speaker?

  b. Prior to the announcement of the speaker, Bette Midler informs the committee that she will not be able to participate due to other commitments. Construct a new preference table for the election with M eliminated. Using the new table and the pairwise comparison method, who is selected as the speaker?

  c. Is the irrelevant alternatives criterion satisfied? Explain your answer.

Number of Votes	12	8	6
First Choice	M	R	G
Second Choice	G	G	R
Third Choice	R	M	M

In Exercises 11–18, the preference table for an election is given. Use the table to answer the questions that appear below it.

11.

Number of Votes	20	16	10	4
First Choice	D	C	C	A
Second Choice	A	A	B	B
Third Choice	B	B	D	D
Fourth Choice	C	D	A	C

  a. Using the Borda count method, who is the winner?

  b. Is the majority criterion satisfied? Explain your answer.

12.

Number of Votes	20	15	3	1
First Choice	A	B	C	D
Second Choice	B	C	D	B
Third Choice	C	D	B	C
Fourth Choice	D	A	A	A

  a. Using the Borda count method, who is the winner?

  b. Is the majority criterion satisfied? Explain your answer.

13.

Number of Votes	24	18	10	8	8	2
First Choice	D	C	A	A	C	B
Second Choice	A	A	D	B	B	C
Third Choice	B	B	B	C	D	A
Fourth Choice	C	D	C	D	A	D

  a. Using the plurality-with-elimination method, who is the winner?

  b. Is the head-to-head criterion satisfied? Explain your answer.

14.

Number of Votes	18	10	9	7	4	3
First Choice	A	C	C	B	D	D
Second Choice	D	B	D	D	B	C
Third Choice	C	D	B	C	A	A
Fourth Choice	B	A	A	A	C	B

a. Using the plurality-with-elimination method, who is the winner?

b. Is the head-to-head criterion satisfied? Explain your answer.

15.

Number of Votes	14	8	4
First Choice	A	B	D
Second Choice	B	D	A
Third Choice	C	C	C
Fourth Choice	D	A	B

a. Using the Borda count method, who is the winner?

b. Is the majority criterion satisfied? Explain your answer.

c. Is the head-to-head criterion satisfied? Explain your answer.

d. Suppose that candidate C drops out of the race. Using the Borda count method, who among the remaining candidates wins the election? Is the irrelevant alternatives criterion satisfied? Explain your answer.

16.

Number of Votes	14	12	10	6
First Choice	A	B	C	D
Second Choice	B	A	B	C
Third Choice	C	C	A	B
Fourth Choice	D	D	D	A

a. Using the plurality-with-elimination method, who is the winner?

b. The six voters on the right all move candidate A from last place on their preference lists to first place on their preference lists. Construct a new preference table for the election. Using this table and the plurality-with-elimination method, who is the winner? Is the monotonicity criterion satisfied? Explain your answer.

17.

Number of Votes	16	14	12	4	2
First Choice	A	D	D	C	E
Second Choice	B	B	B	A	A
Third Choice	C	A	E	B	D
Fourth Choice	D	C	C	D	B
Fifth Choice	E	E	A	E	C

a. Using the Borda count method, who is the winner?

b. Is the majority criterion satisfied? Explain your answer.

c. Is the head-to-head criterion satisfied? Explain your answer.

18.

Number of Votes	6	6	6	6	2	2	2	2
First Choice	A	B	A	B	C	C	E	B
Second Choice	D	A	C	A	E	B	D	A
Third Choice	E	D	D	C	B	A	A	E
Fourth Choice	C	E	E	D	D	D	C	C
Fifth Choice	B	C	B	E	A	E	B	D

a. Using the pairwise comparison method, who is the winner?

b. Suppose that candidate C drops out, but the winner is still chosen using the pairwise comparison method. Is the irrelevant alternatives criterion satisfied? Explain your answer.

19. The preference table shows the results of an election among three candidates, A, B, and C.

Number of Votes	7	3	2
First Choice	A	B	C
Second Choice	B	C	B
Third Choice	C	A	A

a. Using the plurality method, who is the winner?

b. Is the majority criterion satisfied? Explain your answer.

c. Is the head-to-head criterion satisfied? Explain your answer.

d. The two voters on the right both move candidate A from last place on their preference lists to first place on their preference lists. Construct a new preference table for the election. Using the table and the plurality method, who is the winner?

e. Suppose that candidate C drops out, but the winner is still chosen by the plurality method. Is the irrelevant alternatives criterion satisfied? Explain your answer.

f. Do your results from parts (b) through (e) contradict Arrow's Impossibility Theorem? Explain your answer.

Writing in Mathematics

20. Describe the majority criterion.

21. Describe the head-to-head criterion.

22. Describe the monotonicity criterion.

23. Describe the irrelevant alternatives criterion.

24. In your own words, state Arrow's Impossibility Theorem.

25. Write a realistic voting problem similar to Exercise 5 or 6 that would be relevant to your college or the area of the country in which you live. Describe the voting method that you will use to determine the winning option. What problems might occur due to flaws in this method?

26. Is it possible to have election results using a particular voting method that satisfy all four fairness criteria? If so, does this contradict Arrow's Impossibility Theorem?

27. William Arrow wrote, "Well, okay, since we can't find perfection, let's at least try to find a method that works well most of the time." Explain what he meant by this.

Critical Thinking Exercises

28. Why must the pairwise comparison method always satisfy the majority criterion?

In Exercises 29–32, construct a preference table for an election among three candidates, A, B, and C, with the given characteristics. Do not use any of the tables from this section.

29. The Borda count winner violates the majority criterion.

30. The winner by the plurality method violates the head-to-head criterion.

31. The winner by the plurality method violates the irrelevant alternatives criterion.

32. The winner by the plurality-with-elimination method violates the monotonicity criterion.

Group Exercises

33. Present a group report on voting for the host city by the International Olympic Committee. What steps have Committee members taken to attempt to eliminate the abuse and bribery that tainted its selection process? What current voting methods are used? Have there been changes in the makeup of the people doing the voting? Describe these changes. Discuss the particular fairness criteria that may not be satisfied by the current voting methods.

34. Present a group report on a voting method and a fairness criterion that have not been discussed in this chapter.

SECTION 14.3   APPORTIONMENT METHODS

OBJECTIVES

1. Find standard divisors and standard quotas.
2. Understand the apportionment problem.
3. Use Hamilton's method.
4. Understand the quota rule.
5. Use Jefferson's method.
6. Use Adams's method.
7. Use Webster's method.

The signing of the Constitution

The Senate of the United States shall be composed of two Senators from each state ...

ARTICLE I. SECTION 3. CONSTITUTION OF THE UNITED STATES

Representatives ... shall be apportioned among the several states ... according to their respective numbers ...

ARTICLE I. SECTION 2. CONSTITUTION OF THE UNITED STATES

It is the summer of 1787. Delegates from the 13 states are meeting in Philadelphia to draft a constitution for a new nation. The most heated debate concerns the makeup of the new legislature.

The smaller states, led by New Jersey, insist that all states have the same number of representatives. The larger states, led by Virginia, want some form of proportional representation based on population. The delegates compromise on this issue by creating a Senate, in which each state has two senators, and a House of Representatives, in which each state has a number of representatives based on its population.

According to the Constitution of the United States, "Representatives ... shall be apportioned among the several states ... according to their respective numbers ... ." But how are these calculations to be done, and what sorts of problems can arise? In this section, you will learn to divide, on an equitable basis, those things that are not individually divisible. By doing this, you will get to look at a unique role that mathematics played in United States history.

Standard Divisors and Standard Quotas

Our discussion in the first part of this section deals with the following simple, but important example: The Republic of Margaritaville is composed of four states, A, B, C, and D. According to the country's constitution, the congress will have 30 seats, divided among the four states according to their respective populations. Table 14.24 shows each state's population.

TABLE 14.24 POPULATION OF MARGARITAVILLE BY STATE

State	A	B	C	D	Total
Population (in thousands)	275	383	465	767	1890

Before determining a method to allocate the 30 seats in a fair manner, we introduce two important definitions. The first quantity we define is called the *standard divisor*.

THE STANDARD DIVISOR

The **standard divisor** is found by dividing the total population under consideration by the number of items to be allocated.

$$\text{Standard divisor} = \frac{\text{total population}}{\text{number of allocated items}}$$

For the population of Margaritaville in Table 14.24, the standard divisor is found as follows:

$$\text{Standard divisor} = \frac{\text{total population}}{\text{number of allocated items}} = \frac{1890}{30} = 63.$$

> The congress will have 30 seats.

In situations dealing with allocating congressional seats to states based on their populations, the standard divisor gives the number of people per seat in congress on a national basis. Thus, in Maragaritaville, there are 63 thousand people for each seat in congress.

The second quantity we define is called the *standard quota*.

THE STANDARD QUOTA

The **standard quota** for a particular group is found by dividing that group's population by the standard divisor.

$$\text{Standard quota} = \frac{\text{population of a particular group}}{\text{standard divisor}}$$

In computing standard divisors and standard quotas, we will round to the nearest hundredth—that is, to two decimal places.

1 Find standard divisors and standard quotas.

EXAMPLE 1   FINDING STANDARD QUOTAS

Find the standard quotas for states A, B, C, and D in Margaritaville and complete Table 14.25.

TABLE 14.25 POPULATION OF MARGARITAVILLE BY STATE

State	A	B	C	D	Total
Population	275	383	465	767	1890
Standard Quota					

Solution   The standard quotas are obtained by dividing each state's population by the standard divisor. We previously computed the standard divisor and found it to be 63. Thus, we use 63 in the denominator for each of the four computations for the standard quota.

$$\text{Standard quota for state A} = \frac{\text{population of state A}}{\text{standard divisor}} = \frac{275}{63} \approx 4.37$$

$$\text{Standard quota for state B} = \frac{\text{population of state B}}{\text{standard divisor}} = \frac{383}{63} \approx 6.08$$

$$\text{Standard quota for state C} = \frac{\text{population of state C}}{\text{standard divisor}} = \frac{465}{63} \approx 7.38$$

$$\text{Standard quota for state D} = \frac{\text{population of state D}}{\text{standard divisor}} = \frac{767}{63} \approx 12.17$$

Table 14.26 shows the standard quotas for each state in the Republic of Margaritaville.

STUDY TIP

Due to rounding, the sum of the standard quotas can be slightly above or slightly below the total number of allocated items.

TABLE 14.26 STANDARD QUOTAS FOR EACH STATE IN MARGARITAVILLE

State	A	B	C	D	Total
Population	275	383	465	767	1890
Standard Quota	4.37	6.08	7.38	12.17	30

Notice that the sum of all the standard quotas is 30, the total number of seats in the congress.

The Republic of Amador is composed of five states, A, B, C, D, and E. According to the country's constitution, the congress will have 200 seats, divided among the five states according to their respective populations. Table 14.27 shows each state's population.

TABLE 14.27 POPULATION OF AMADOR BY STATE

State	A	B	C	D	E	Total
Population (in thousands)	1112	1118	1320	1515	4935	10,000
Standard Quota						

a. Find the standard divisor.

b. Find the standard quota for each state and complete Table 14.27.

The Apportionment Problem

The standard quotas in Table 14.26 represent each state's exact fair share of the 30 seats for the congress of Margaritaville. However, what do we mean by saying that state A has 4.37 seats in congress? Seats have to be given out in whole numbers. The

2   Understand the apportionment problem.

STUDY TIP

In the unusual case that a standard quota is a whole number, the lower and upper quotas are both equal to the standard quota.

apportionment problem is to determine a method for rounding standard quotas into whole numbers so that the sum of the numbers is the total number of allocated items.

Can we possibly round standard quotas down to the nearest whole number or up to the nearest whole number and solve the apportionment problem? The **lower quota** is the standard quota rounded down to the nearest whole number. The **upper quota** is the standard quota rounded up to the nearest whole number. Table 14.28 shows the lower and upper quotas.

TABLE 14.28 STANDARD QUOTAS, LOWER QUOTAS, AND UPPER QUOTAS FOR EACH STATE IN MARGARITAVILLE

State	A	B	C	D	Total
Population	275	383	465	767	1890
Standard quota	4.37	6.08	7.38	12.17	30
Lower quota	4	6	7	12	29
Upper quota	5	7	8	13	33

These totals should be 30 because Margaritaville has 30 seats in congress.

STUDY TIP

According to one dictionary, to *apportion* means "to divide and assign in due and proper proportion or according to some plan."

The voice balloon shows that by rounding to lower or upper quotas, we have not solved the apportionment problem. We are giving out either 29 or 33 seats in congress. With 29 seats, where will the extra seat go? With 33 seats, where are those three extra seats going to come from?

We now discuss four different **apportionment methods**—that is, methods for solving the apportionment problem. The methods are called *Hamilton's method, Jefferson's method, Adams's method,* and *Webster's method.*

Hamilton's Method

Hamilton's method of apportionment proceeds in three steps.

 Use Hamilton's method.

> HAMILTON'S METHOD
>
> 1. Calculate each group's standard quota.
> 2. Round each standard quota down to the nearest whole number, thereby finding the lower quota. Initially, give to each group its lower quota.
> 3. Give the surplus items, one at a time, to the groups with the largest decimal parts until there are no more surplus items.

BLITZER BONUS

Alexander Hamilton (1757–1804)

Alexander Hamilton was the nation's first Secretary of the Treasury in George Washington's administration. Hamilton and Thomas Jefferson were bitter political adversaries. In 1804, Hamilton was killed in a duel with Aaron Burr, who served as vice president under Jefferson.

For example, consider the lower quotas in the Margaritaville example. We initially give each state its lower quota, shown in Table 14.29. Because there are 30 seats in congress, can you see that there is one surplus seat?

TABLE 14.29 STANDARD QUOTAS, AND LOWER QUOTAS, FOR EACH STATE IN MARGARITAVILLE

State	A	B	C	D	Total
Population	275	383	465	767	1890
Standard quota	4.37	6.08	7.38	12.17	30
Lower quota	4	6	7	12	29

There are 30 seats in congress. Who will get the extra seat?

The surplus seat goes to the state with the greatest decimal part. The greatest decimal part is 0.38, corresponding to state C. Thus, state C receives the additional seat, and is assigned 8 seats in congress.

Table 14.30 summarizes the congressional seat assignments for the states.

TABLE 14.30 SOLVING MARGARITAVILLE'S APPORTIONMENT PROBLEM: HAMILTON'S METHOD

State	A	B	C	D	Total
Lower Quota	4	6	7	12	29
Hamilton's Apportionment	4	6	8	12	30

EXAMPLE 2 | USING HAMILTON'S METHOD

A rapid transit service operates 130 buses along six routes, A, B, C, D, E, and F. The number of buses assigned to each route is based on the average number of daily passengers per route, given in Table 14.31. Use Hamilton's method to apportion the buses.

TABLE 14.31

Route	A	B	C	D	E	F	Total
Average Number of Passengers	4360	5130	7080	10,245	15,535	22,650	65,000

Before applying Hamilton's method, we must compute the standard divisor.

$$\text{Standard divisor} = \frac{\text{total population}}{\text{number of allocated items}}$$

$$= \frac{\text{total number of passengers}}{\text{number of buses}} = \frac{65,000}{130} = 500$$

The standard divisor, 500, represents the average number of passengers per bus per day. Using this value, Table 14.32 shows the three steps in Hamilton's method.

TABLE 14.32 APPORTIONMENT OF BUSES USING HAMILTON'S METHOD

Route	Passengers	Standard Quota	Lower Quota		Decimal Part	Surplus Buses	Final Apportionment
A	4360	8.72	8	largest	0.72	1	9
B	5130	10.26	10		0.26		10
C	7080	14.16	14		0.16		14
D	10,245	20.49	20	next largest	0.49	1	21
E	15,535	31.07	31		0.07		31
F	22,650	45.30	45		0.30		45
Total	65,000	130	128				130

Step 1 Calculate each standard quota:
standard quota
$= \dfrac{\text{population of a group}}{\text{standard divisor}}$
$= \dfrac{\text{number of passengers}}{500}$

Step 2 Round down and find each lower quota. The sum, 128, indicates we must assign two extra buses.

Step 3 Give the two surplus buses, one at a time, to the routes with the largest decimal parts.

The apportionment of buses, shown in the final column of Table 14.32, is computed by giving each route its lower quota of buses plus any surplus buses.

 Refer to Check Point 1 on page 752. Use Hamilton's method to apportion the 200 congressional seats.

4 Understand the quota rule.

Have you noticed that every apportionment in Hamilton's method is either the lower quota or the upper quota? This is an important criterion for fairness, called the **quota rule**.

> ### THE QUOTA RULE
> A group's apportionment should be either its upper quota or its lower quota. An apportionment method that guarantees that this will always occur is said to **satisfy the quota rule**.

5 Use Jefferson's method.

Jefferson's Method

Congress approved Hamilton's method as the apportionment method to be used following the 1790 census. However, the method was vetoed by President Washington. The veto was sustained, and a method proposed by Thomas Jefferson was adopted. (In the 1850s, Congress resurrected Hamilton's method; it was used until 1900.)

What could possibly concern President Washington about Hamilton's method? Perhaps it was the fact that in allocating congressional seats, some states would be chosen over others for preferential treatment. These would be the states with the greatest fractional parts in their standard quotas. Ideally, we should be able to replace the standard divisor by some other divisor, called the *modified divisor*. After dividing every state's population by this divisor and then rounding the modified quotas down, we should be left with no surplus items. Thus, no group (or state) could receive preferential treatment.

> ### JEFFERSON'S METHOD
> 1. Find a **modified divisor**, d, such that when each group's **modified quota**, (group's population divided by d) is rounded down to the nearest whole number, the sum of the whole numbers for all the groups is the number of items to be apportioned. The modified quotients that are rounded down are called **modified lower quotas**.
> 2. Apportion to each group its modified lower quota.

How do we find the modified divisor, d, that makes Jefferson's method work? Before turning to this issue, let's illustrate the method using calculations based on a given value of the modified divisor.

EXAMPLE 3 | USING JEFFERSON'S METHOD

As in Example 2, a rapid transit service operates 130 buses along six routes, A, B, C, D, E, and F. The number of buses assigned to each route is based on the average number of daily passengers per route, repeated in Table 14.31. Use Jefferson's method with $d = 486$ to apportion the buses.

TABLE 14.31, repeated

Route	A	B	C	D	E	F	Total
Average Number of Passengers	4360	5130	7080	10,245	15,535	22,650	65,000

Solution Using $d = 486$, Table 14.33 illustrates Jefferson's method.

TABLE 14.33 APPORTIONMENT OF BUSES USING JEFFERSON'S METHOD WITH $d = 486$

Route	Passengers	Modified Quota	Modified Lower Quota	Final Apportionment
A	4360	8.97	8	8
B	5130	10.56	10	10
C	7080	14.57	14	14
D	10,245	21.08	21	21
E	15,535	31.97	31	31
F	22,650	46.60	46	46
Total	65,000		130	130

Modified quota $= \dfrac{\text{number of passengers}}{486}$

Round each modified quota down to the nearest whole number. The sum, 130, indicates there are no surplus buses to assign.

The apportionment of buses, shown in the final column of Table 14.33, is determined by giving each route a number of buses equal to its modified lower quota.

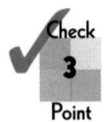

Check Point 3 Refer to Check Point 1 on page 752. Use Jefferson's method with $d = 49.3$ to apportion the 200 congressional seats.

Unfortunately, there is no formula for the modified divisor, d. Furthermore, there is usually more than one modified divisor that will make Jefferson's method work. You can find one of these values using trial and error. Begin with the fact that **in Jefferson's method, a modified divisor is slightly less than the standard divisor**. Pick a number d that you think might work. Carry out the calculations required by Jefferson's method: Divide each group's population by d, round down to the nearest whole number, and find the sum of the whole numbers. With luck, the sum is the number of items to be apportioned. Otherwise, change the value of d (make it greater if the sum is too high and less if the sum is too low) and try again. In most cases, with two or three guesses, you will find a modified divisor, d, that works.

Jefferson's method was adopted in 1791. In the apportionment of 1832, New York, with a standard quota of 38.59, received 40 seats in the House of Representatives. Because 40 is neither the upper quota nor the lower quota of 38.59, it was discovered that Jefferson's method violated the quota rule. The apportionment of 1832 was the last time the House of Representatives was apportioned using Jefferson's method.

Adams's Method

6 Use Adams's method.

At the same time that Jefferson's method was found to be problematic because of its violation of the quota rule, John Quincy Adams proposed a method that is the mirror image of it. Rather than using modified lower quotas as in Jefferson's method, Adams suggested using modified upper quotas.

In Example 4, we will use trial and error to find a modified divisor, d, that satisfies Adams's method. **When using Adams's method, begin with a modified divisor that is slightly greater than the standard divisor.**

ADAMS'S METHOD

1. Find a **modified divisor**, d, such that when each group's **modified quota**, (group's population divided by d) is rounded up to the nearest whole number, the sum of the whole numbers for all the groups is the number of items to be apportioned. The modified quotas that are rounded up are called **modified upper quotas**.

2. Apportion to each group its modified upper quota.

EXAMPLE 4   USING ADAMS'S METHOD

Consider, again, the rapid transit service that operates 130 buses along six routes, A, B, C, D, E, and F. The number of buses assigned to each route is based on the average number of daily passengers per route, repeated in Table 14.31. Use Adams's method to apportion the buses.

TABLE 14.31, repeated

Route	A	B	C	D	E	F	Total
Average Number of Passengers	4360	5130	7080	10,245	15,535	22,650	65,000

Solution   In Example 2, we found that the standard divisor is 500. We begin by guessing at a possible modified divisor, d, that we hope will work. This value of d must be greater than 500 so that the modified quotas will be less than the standard quotas. When rounded up, the sum of these whole numbers should be 130, the number of buses to be apportioned. Perhaps a good guess might be $d = 512$. Table 14.34 shows the calculations using Adams's method with $d = 512$.

TABLE 14.34 CALCULATIONS USING ADAMS'S METHOD WITH $d = 512$

Route	Passengers	Modified Quota	Modified Upper Quota
A	4360	8.52	9
B	5130	10.02	11
C	7080	13.83	14
D	10,245	20.01	21
E	15,535	30.34	31
F	22,650	44.24	45
Total	65,000		131

Modified quota
$$= \frac{\text{number of passengers}}{512}$$

The sum should be 130, not 131.

Because the sum of modified upper quotas is too high, we need to lower the modified quotas a bit. We try a higher divisor, $d = 513$, and obtain 129 for the sum of modified upper quotas. Because this sum is too low, we increase the modified divisor a bit and try $d = 512.7$. Table 14.35 shows the calculations required by Adams's method using this value.

TABLE 14.35 CALCULATIONS USING ADAMS'S METHOD WITH $d = 512.7$

Route	Passengers	Modified Quota	Modified Upper Quota	Final Apportionment
A	4360	8.50	9	9
B	5130	10.01	11	11
C	7080	13.81	14	14
D	10,245	19.98	20	20
E	15,535	30.30	31	31
F	22,650	44.18	45	45
Total	65,000		130	130

Modified quota
$$= \frac{\text{number of passengers}}{512.7}$$

This sum is precisely the number of buses to be apportioned.

The apportionment of buses, shown in the final column of Table 14.35, is determined by giving each route a number of buses equal to its modified upper quota.

 **Check 4 Point** Refer to Check Point 1 on page 752. Use Adams's method to apportion the 200 congressional seats. In guessing at a value for d, begin with $d = 50.5$. If necessary, modify this value as we did in Example 4.

Recall that by Jefferson's method, a delighted New York, with a standard quota of 38.59, received 40 seats. This exceeds 39, the upper quota, and is called an **upper-quota violation**. Adams thought he could avoid this violation, which he did. Unfortunately, his method can result in an assignment of seats in the House of Representatives that is below a state's lower quota. This flaw in Adams's method is called a **lower-quota violation**. With Adams's method, all violations of the quota rule are of this kind.

 Use Webster's method.

Webster's Method

In 1832, Daniel Webster suggested an apportionment method that sounds like a compromise between Jefferson's method, where modified quotas are rounded down, and Adams's method, where modified quotas are rounded up. Let's round the modified quotas in the usual way that we round decimals: If the fractional part is less than 0.5, round down to the nearest whole number. If the fractional part is greater than or equal to 0.5, round up to the nearest whole number. Webster believed that this was the only fair way to round numbers.

WEBSTER'S METHOD

1. Find a **modified divisor**, d, such that when each group's **modified quota** (group's population divided by d) is rounded to the nearest whole number, the sum of the whole numbers for all the groups is the number of items to be apportioned. The modified quotas that are rounded are called **modified rounded quotas**.
2. Apportion to each group its modified rounded quota.

In many ways, Webster's method works in a manner that is similar to Jefferson's and Adams's methods. However, **when using Webster's method, the modified divisor, d, can be less than, greater than, or equal to the standard divisor.** Thus, it may take a bit longer to find a modified divisor that satisfies Webster's method.

Daniel Webster (1782–1852)

Daniel Webster served in the U.S. House of Representatives (1813–1817; 1823–1827), where he made his name as one of the nation's best-known orators. In the Senate (1827–1841; 1845–1850) and as Secretary of State (1841–1843; 1850–1852), he was influential in maintaining the Union. His great disappointment was never becoming president.

EXAMPLE 5 | USING WEBSTER'S METHOD

We return, for the last time, to the rapid transit service that operates 130 buses along six routes, A, B, C, D, E, and F. The number of buses assigned to each route is based on the average number of daily passengers per route, repeated in Table 14.31. Use Webster's method to apportion the buses.

TABLE 14.31, repeated

Route	A	B	C	D	E	F	Total
Average Number of Passengers	4360	5130	7080	10,245	15,535	22,650	65,000

Solution From Example 2, we know that the standard divisor is 500. We begin by trying $d = 502$, using a modified divisor greater than the standard divisor. After rounding the modified quotients to the nearest integer and taking the resulting sum, we obtain 129, rather than the desired 130. Because 129 is too low, this suggests that the modified divisor is too large, so we should guess at a modified divisor less than 500, the standard divisor. We try $d = 498$. Table 14.36 shows the calculations required by Webster's method using this value.

TABLE 14.36 CALCULATIONS USING WEBSTER'S METHOD WITH $d = 498$

Route	Passengers	Modified Quota	Modified Rounded Quotas	Final Apportionment
A	4360	8.76	9	9
B	5130	10.30	10	10
C	7080	14.22	14	14
D	10,245	20.57	21	21
E	15,535	31.19	31	31
F	22,650	45.48	45	45
Total	65,000		130	130

Modified quota
$= \dfrac{\text{number of passengers}}{498}$

This sum is precisely the number of buses to be apportioned.

The apportionment of buses, shown in the final column of Table 14.36, is determined by giving each route a number of buses equal to its modified rounded quota.

Montana vs. the U.S. Department of Commerce

In 1992, Montana challenged the constitutionality of the apportionment method being used. Based on the 1990 census, Montana was concerned that it would have to give up one of its two seats in the House of Representatives to the state of Washington. The Supreme Court ruled against Montana and upheld the Huntington-Hill method as constitutional.

Check 5 Point Refer to Check Point 1 on page 752. Use Webster's method with $d = 49.8$ to apportion the 200 congressional seats.

In Examples 2–5, we apportioned the buses using four different methods. Although the final apportionment using Hamilton's method is the same as the one obtained using Webster's method, this is not always the case. Like voting methods, different apportionment methods applied to the same situation can produce different results.

Although Webster's method can violate the quota rule, such violations are rare. Many experts consider Webster's method the best overall apportionment method available. The method currently used to apportion the U.S. House of Representatives is known as the *Huntington-Hill method*. (See Exercise 46.) However, some experts believe that in our lifetimes Webster's method will make a comeback and replace Huntington-Hill.

Table 14.37 summarizes the four apportionment methods discussed in this section.

TABLE 14.37 SUMMARY OF APPORTIONMENT METHODS

Method	Divisor	Apportionment
Hamilton's	Standard divisor = $\dfrac{\text{total population}}{\text{number of allocated items}}$	Round each standard quota down to the nearest whole number. Initially give each group its lower quota. Give surplus items, one at a time, to the groups with the largest decimal parts.
Jefferson's	The modified divisor is less than the standard divisor.	Round each group's modified quota down to the nearest whole number. Apportion to each group its modified lower quota.
Adams's	The modified divisor is greater than the standard divisor.	Round each group's modified quota up to the nearest whole number. Apportion to each group its modified upper quota.
Webster's	The modified divisor is less than, greater than, or equal to, the standard divisor.	Round each group's modified quota to the nearest whole number. Apportion to each group its modified rounded quota.

Exercise Set 14.3

Practice and Application Exercises

Throughout this exercise set, in computing standard divisors, standard quotas, and modified quotas, round to the nearest hundredth when necessary.

A small country is comprised of four states, A, B, C, and D. The population of each state, in thousands, is given in the following table. Use this information to solve Exercises 1–2.

State	A	B	C	D	Total
Population (in thousands)	138	266	534	662	1600

1. According to the country's constitution, the congress will have 80 seats, divided among the four states according to their respective populations.
   a. Find the standard divisor, in thousands. How many people are there for each seat in congress?
   b. Find each state's standard quota.
   c. Find each state's lower quota and upper quota.

2. According to the country's constitution, the congress will have 200 seats, divided among the four states according to their respective populations.
   a. Find the standard divisor, in thousands. How many people are there for each seat in congress?
   b. Find each state's standard quota.
   c. Find each state's lower quota and upper quota.

3. Use Hamilton's method to find each state's apportionment of congressional seats in Exercise 1.

4. Use Hamilton's method to find each state's apportionment of congressional seats in Exercise 2.

5. A university is composed of five schools. The enrollment in each school is given in the following table.

School	Humanities	Social Science	Engineering	Business	Education
Enrollment	1050	1410	1830	2540	3580

There are 300 new computers to be apportioned among the five schools according to their respective enrollments. Use Hamilton's method to find each school's apportionment of computers.

6. A university is composed of five schools. The enrollment in each school is given in the following table.

School	Liberal Arts	Education	Business	Engineering	Sciences
Enrollment	1180	1290	2140	2930	3320

There are 300 new computers to be apportioned among the five schools according to their respective enrollments. Use Hamilton's method to find each school's apportionment of computers.

7. A small country is composed of five states, A, B, C, D, and E. The population of each state is given in the following table. Congress will have 57 seats, divided among the five states according to their respective populations. Use Jefferson's method with $d = 32,920$ to apportion the 57 congressional seats.

State	A	B	C	D	E
Population	126,316	196,492	425,264	526,664	725,264

8. A small country is comprised of four states, A, B, C, and D. The population of each state, in thousands, is given in the following table. Congress will have 400 seats, divided among the four states according to their respective populations. Use Jefferson's method with $d = 7.82$ to apportion the 400 congressional seats.

State	A	B	C	D
Population (in thousands)	424	664	892	1162

9. An HMO has 150 doctors to be apportioned among six clinics. The HMO decides to apportion the doctors based on the average weekly patient load for each clinic, given in the following table. Use Jefferson's method to apportion the 150 doctors. (*Hint:* Find the standard divisor. A modified divisor that is less than this standard divisor will work.)

Clinic	A	B	C	D
Average Weekly Patient Load	1714	5460	2440	5386

10. An HMO has 70 doctors to be apportioned among six clinics. The HMO decides to apportion the doctors based on the average weekly patient load for each clinic, given in the following table. Use Jefferson's method to apportion the 70 doctors. (*Hint:* A modified divisor between 39 and 40 will work.)

Clinic	A	B	C	D	E	F
Average Weekly Patient Load	316	598	396	692	426	486

11. The police department in a large city has 180 new officers to be apportioned among six high-crime precincts. Crimes by precinct are shown in the following table. Use Adams's method with $d = 16$ to apportion the new officers among the precincts.

Precinct	A	B	C	D	E	F
Crimes	446	526	835	227	338	456

12. Four people pool their money to buy 60 shares of stock. The amount that each person contributes is shown in the following table. Use Adams's method with $d = 108$ to apportion the shares of stock.

Person	A	B	C	D
Contribution	$2013	$187	$290	$3862

13. Three people pool their money to buy 30 shares of stock. The amount that each person contributes is shown in the table at the top of the next column. Use Adams's method to apportion the shares of stock. (*Hint:* Find the standard divisor. A modified divisor that is greater than this standard divisor will work.)

Person	A	B	C
Amount	$795	$705	$525

14. Refer to Exercise 9. Use Adams's method to apportion the 150 doctors. (*Hint:* Find the standard divisor. A modified divisor that is greater than this standard divisor will work.)

15. Twenty sections of bilingual math courses, taught in both English and Spanish, are to be offered in introductory algebra, intermediate algebra, and liberal arts math. The preregistration figures for the number of students planning to enroll in these bilingual sections are given in the following table. Use Webster's method with $d = 29.6$ to determine how many bilingual sections of each course should be offered.

Course	Introductory Algebra	Intermediate Algebra	Liberal Arts Math
Enrollment	130	282	188

16. Refer to the populations of the four states in Exercise 8. Unlike Exercise 8, congress now has 314 seats to divide among the states according to their respective populations. Use Webster's method with $d = 9.98$ to apportion the 314 congressional seats.

17. A rapid transit service operates 200 buses along five routes, A, B, C, D, and E. The number of buses assigned to each route is based on the average number of daily passengers per route, given in the following table. Use Webster's method to apportion the buses. (*Hint:* A modified divisor between 55 and 56 will work.)

Route	A	B	C	D	E
Average Number of Passengers	1087	1323	1592	1596	5462

18. Refer to Exercise 11. Use Webster's method to apportion the 180 new officers among the precincts. (*Hint:* A modified divisor between 15 and 16 will work.)

A hospital has a nursing staff of 250 nurses working in four shifts: A (7 : 00 A.M. to 1 : 00 P.M.), B (1 : 00 P.M. to 7 : 00 P.M.), C (7 : 00 P.M. to 1 : 00 A.M.), and D (1 : 00 A.M. to 7 : 00 A.M.). The number of nurses apportioned to each shift is based on the average number of patients per shift, given in the following table. Use this information to solve Exercises 19–22.

Shift	A	B	C	D
Average Number of Patients	453	650	547	350

19. Use Hamilton's method to apportion the 250 nurses among the shifts at the hospital.

20. Use Jefferson's method to apportion the 250 nurses among the shifts in the hospital. (*Hint:* A modified divisor that works should be less than the standard divisor in Exercise 19.)

21. Use Adams's method to apportion the 250 nurses among the shifts in the hospital. (*Hint:* A modified divisor that works should be greater than the standard divisor in Exercise 19.)

22. Use Webster's method to apportion the 250 nurses among the shifts in the hospital. (*Hint:* A modified divisor that works is equal to the standard divisor that you found in Exercise 19.)

The table shows the 1790 United States census. In 1793, at the direction of President George Washington, 105 seats in the House of Representatives were to be divided among the 15 states according to their 1790 populations. Use this information to solve Exercises 23–26.

1790 United States Census			
Connecticut	236,841	New York	331,589
Delaware	55,540	North Carolina	353,523
Georgia	70,835	Pennsylvania	432,879
Kentucky	68,705	Rhode Island	68,446
Maryland	278,514	South Carolina	206,236
Massachusetts	475,327	Vermont	85,533
New Hampshire	141,822	Virginia	630,560
New Jersey	179,570		

23. Use Hamilton's method to find each state's apportionment of congressional seats.

24. Use Jefferson's method with $d = 33,000$ to find each state's apportionment of congressional seats.

25. Use Adams's method with $d = 36,100$ to find each state's apportionment of congressional seats.

26. Use Webster's method with $d = 34,500$ to find each state's apportionment of congressional seats.

Writing in Mathematics

27. Describe how to find a standard divisor.

28. Describe how to determine a standard quota for a particular group.

29. How is the lower quota found from a standard quota?

30. How is the upper quota found from a standard quota?

31. Describe the apportionment problem.

32. In your own words, describe Hamilton's method of apportionment.

33. What is the quota rule?

34. Explain why Hamilton's method satisfies the quota rule.

35. Describe the difference between how modified quotas are rounded using Jefferson's method and Adams's method.

36. Suppose that you guess at a modified divisor, d, using Jefferson's method. How do you determine if your guess satisfies the method?

37. Describe the difference between the modified divisor, d, in terms of the standard divisor using Jefferson's method and Adams's method.

38. In allocating congressional seats, how does Hamilton's method choose some states over others for preferential treatment? Explain how this is avoided in Jefferson's and Adams's methods.

39. How are modified quotas rounded using Webster's method?

40. Why might it take longer to guess at a modified divisor that works using Webster's method than using Jefferson's method or Adams's method?

41. In this exercise set, we have used apportionment methods to divide congressional seats, assign computers to schools, assign doctors to clinics, divide police officers among precincts, divide shares of stock, assign sections of bilingual math, assign buses to city routes, and assign nurses to hospital shifts. Describe another situation that requires the use of apportionment methods.

Critical Thinking Exercises

42. In 1880, there were 300 congressional seats in the U.S. House of Representatives. The population of Alabama was 1,262,505, and its standard quota was 7.671. Find, to the nearest whole number, the U.S. population in 1880.

A small country is composed of three states, A, B, and C. The country's constitution specifies that congressional seats will be divided among the three states according to their respective populations. In Exercises 43–45, write an apportionment problem satisfying the given criterion.

43. Hamilton's method and Jefferson's method result in the same apportionment.

44. Hamilton's method and Adams's method result in the same apportionment.

45. Hamilton's method and Webster's method result in the same apportionment.

Group Exercises

46. The method currently used to apportion the U.S. House of Representatives is known as the **Huntington-Hill method**, and more commonly as the **method of equal proportions**. Research and present a group report on this method. Include the history of how the method came into use and describe how the method works.

47. Research and present a group report on a brief history of apportionment in the United States.

SECTION 14.4   FLAWS OF APPORTIONMENT METHODS

OBJECTIVES

1. Understand and illustrate the Alabama paradox.

2. Understand and illustrate the population paradox.

3. Understand and illustrate the new-states paradox.

4. Understand Balinski and Young's Impossibility Theorem.

Posters from the controversial election of 1876. The candidates were Rutherford B. Hayes (Republican) and Samuel J. Tilden (Democrat).

The very mention of Florida outraged the Democrats. Florida's contested electoral votes helped elect a Republican president who had lost the popular vote.

No, we're not referring to the controversial election of 2000, but rather the 1876 election of Republican Rutherford B. Hayes over Democrat Samuel J. Tilden. Tilden won the popular vote with 4,300,590 votes, whereas Hayes received 4,036,298 votes. In 1872, a power grab among the states resulted in an illegal apportionment of the House of Representatives that was not based on Hamilton's method. Two seats that would have gone to New York and Illinois under Hamilton's method instead went to Florida and New Hampshire. Four years later, Rutherford B. Hayes became President of the United States based on this unconstitutional apportionment, winning the electoral vote by a margin of one vote, 185 to 184. If Hamilton's method had been used in the congressional apportionment, New York would have had one more vote, taking the Hayes votes away from Florida or New Hampshire. In short, if the 1872 apportionment of the House of Representatives had been carried out according to the legal Hamilton's method, Tilden would have won the electoral vote and, therefore, the presidency.

Hamilton's method may appear to be a fair and reasonable apportionment method. It is the only method that satisfies the quota rule—that is, a group's apportionment is always its upper quota or its lower quota. Unfortunately, it is also the only method that can produce some serious problems, called the *Alabama paradox*, the *population paradox*, and the *new-states paradox*. In this section, we discuss each of these flaws. We conclude by seeing if there is an ideal apportionment method that can guarantee the states their fair share of seats in the House of Representatives. At stake is equal representation, avoiding election debacles such as the one that occurred in 1872, and the reality that many government policies are based on the number of representatives for each state.

1 Understand and illustrate the Alabama paradox.

The Alabama Paradox

After the 1880 census, two possible sizes for the House of Representatives were under consideration: 299 members or 300 members. The chief clerk of the U.S. Census Office used Hamilton's method to compute apportionments for both House sizes. He found that adding one more seat to the House in order to have 300 seats would actually decrease the number of seats for Alabama, from 8 seats to 7 seats. This was the first time this paradoxical behavior was observed in Congressional apportionment; as a result, it is called the **Alabama paradox**.

"No invasions of the Constitution are fundamentally so dangerous as the tricks played on their own numbers."

THOMAS JEFFERSON

> ## THE ALABAMA PARADOX
>
> An increase in the total number of items to be apportioned results in the loss of an item for a group.

EXAMPLE 1   ILLUSTRATING THE ALABAMA PARADOX

A small country with a population of 10,000 is composed of three states. According to the country's constitution, the congress will have 200 seats, divided among the three states according to their respective populations. Table 14.38 shows each state's population. Use Hamilton's method and show that the Alabama paradox occurs if the number of seats is increased to 201.

TABLE 14.38

State	A	B	C	Total
Population	5015	4515	470	10,000

Solution We begin with 200 seats in the congress. First we compute the standard divisor.

$$\text{Standard divisor} = \frac{\text{total population}}{\text{number of allocated items}} = \frac{10{,}000}{200} = 50$$

Using this value, Table 14.39 shows the apportionment for each state using Hamilton's method.

TABLE 14.39 APPORTIONMENT OF 200 CONGRESSIONAL SEATS USING HAMILTON'S METHOD

State	Population	Standard Quota	Lower Quota	Decimal Part	Surplus Seats	Final Apportionment
A	5015	100.3	100	0.3		100
B	4515	90.3	90	0.3		90
C	470	9.4	9	0.4	1	10
Total	10,000	200	199	largest		200

Standard quota = $\dfrac{\text{population}}{50}$

This sum indicates we must assign one extra seat.

Now let's see what happens with 201 seats in congress. First we compute the standard divisor.

$$\text{Standard divisor} = \frac{\text{total population}}{\text{number of allocated items}} = \frac{10{,}000}{201} \approx 49.75$$

Using this value, Table 14.40 shows the apportionment for each state using Hamilton's method.

TABLE 14.40 APPORTIONMENT OF 201 CONGRESSIONAL SEATS USING HAMILTON'S METHOD

State	Population	Standard Quota	Lower Quota	Decimal Part	Surplus Seats	Final Apportionment
A	5015	100.80	100	largest 0.80	1	101
B	4515	90.75	90	next largest 0.75	1	91
C	470	9.45	9	0.45		9
Total		201	199			201

Standard quota
$= \dfrac{\text{population}}{49.75}$

This sum indicates we must assign two extra seats.

The final apportionments are summarized in Table 14.41. When the number of seats increased from 200 to 201, state C's apportionment actually decreased, from 10 to 9. This is an example of the Alabama paradox. In this situation, the larger states, A and B, benefit at the expense of the smaller state, C.

TABLE 14.41 ILLUSTRATING THE ALABAMA PARADOX

State	Apportionment with 200 Seats	Apportionment with 201 Seats
A	100	101
B	90	91
C	10	9

State C's apportionment decreased from 10 to 9.

Check 1 Point

Table 14.42 shows the populations of the four states in a country with a population of 20,000. Use Hamilton's method and show that the Alabama paradox occurs if the number of seats in congress is increased from 99 to 100.

TABLE 14.42

State	A	B	C	D	Total
Population	2060	2080	7730	8130	20,000

2 Understand and illustrate the population paradox.

The Population Paradox

The issue of power and representation among the states has been a serious concern since the drafting of the Constitution. In the early 1900s, Virginia was growing much faster than Maine—about 60% faster—yet Virginia lost a seat to Maine in the House of Representatives. This paradox, called the **population paradox**, illustrates another serious flaw of Hamilton's method.

Hamilton's Method in Other Countries

Parliament in session in Brussels, Belgium

The discovery of the Alabama paradox in 1880 resulted in the termination of Hamilton's method in the U.S. House of Representatives. However, Hamilton's method, also known as the method of largest remainders, is still used to apportion at least part of the legislatures in Austria, Belgium, El Salvador, Indonesia, Liechtenstein, and Madagascar.

THE POPULATION PARADOX

Group A loses items to group B, even though the population of group A grew at a faster rate than that of group B.

EXAMPLE 2 ILLUSTRATING THE POPULATION PARADOX

A small country with a population of 10,000 is composed of three states. There are 11 seats in the congress, divided among the three states according to their respective populations. Using Hamilton's method, Table 14.43 shows the apportionment of congressional seats for each state.

TABLE 14.43 APPORTIONMENT OF 11 CONGRESSIONAL SEATS USING HAMILTON'S METHOD

State	Population	Standard Quota	Lower Quota	Hamilton's Apportionment
A	540	0.59	0	0
B	2430	2.67	2	3
C	7030	7.73	7	8
Total	10,000	10.99	9	11

The population of the country increases, shown in Table 14.44.

TABLE 14.44

State	A	B	C	Total
Original Population	540	2430	7030	10,000
New Population	560	2550	7890	11,000

a. Find the percent increase in the population of each state.

b. Use Hamilton's method and show that the population paradox occurs.

Solution

a. Recall that the fraction for percent increase is the amount of increase divided by the original amount. The percent increase in the population of each state is determined as follows:

$$\text{State A:}\quad \frac{560 - 540}{540} = \frac{20}{540} \approx 0.037 = 3.7\%$$

increase in population   original population

$$\text{State B:}\quad \frac{2550 - 2430}{2430} = \frac{120}{2430} \approx 0.049 = 4.9\%$$

$$\text{State C:}\quad \frac{7890 - 7030}{7030} = \frac{860}{7030} \approx 0.122 = 12.2\%$$

All three states had an increase in their populations. State C is increasing at a faster rate than state B, while state B is increasing at a faster rate than state A.

b. We need to use Hamilton's method to find the apportionment for each state with its new population. First we compute the standard divisor.

$$\text{Standard divisor} = \frac{\text{total population}}{\text{number of allocated items}} = \frac{11{,}000}{11} = 1000$$

Using this value, Table 14.45 shows the apportionment for each state using Hamilton's method.

TABLE 14.45 APPORTIONMENT OF 11 CONGRESSIONAL SEATS FOR THE NEW POPULATIONS USING HAMILTON'S METHOD

State	Population	Standard Quota	Lower Quota	Decimal Part	Surplus Seats	Final Apportionment
A	560	0.56	0	0.56	1	1
B	2550	2.55	2	0.55		2
C	7890	7.89	7	0.89	1	8
Total	11,000	11	9			11

> Standard quota $= \dfrac{\text{population}}{1000}$

> This sum indicates we must assign two extra seats.

The final apportionments are summarized in Table 14.46. The good news is that state A now has a congressional seat. The bad news is that state B has lost a seat to state A, even though state B's population grew at a faster rate than that of state A. This is an example of the population paradox.

TABLE 14.46 ILLUSTRATING THE POPULATION PARADOX

State	Growth Rate	Original Apportionment	New Apportionment
A	3.7%	0	1
B	4.9%	3	2
C	12.2%	8	8

> State B lost a seat to state A.

Check 2 Point A small country has 100 seats in the congress, divided among the three states according to their respective populations. Table 14.47 shows each state's population before and after the country's population increase.

TABLE 14.47

State	A	B	C	Total
Original Population	19,110	39,090	141,800	200,000
New Population	19,302	39,480	141,800	200,582

a. Use Hamilton's method to apportion the 100 congressional seats using the original population.

b. Find the percent increase in the population of states A and B. (State C did not have any change in population.)

c. Use Hamilton's method to apportion the 100 congressional seats using the new population. Show that the population paradox occurs.

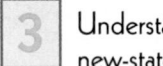

The New-States Paradox

Another flaw in Hamilton's method was discovered in 1907 when Oklahoma became a state. Previously, the House of Representatives had had 386 seats. Based on its population, it was determined that Oklahoma should have 5 seats. Thus, the House size was changed from 386 to 391. The intent was to leave the number of seats unchanged for the other states. However, when the apportionment was recalculated using Hamilton's method, Maine gained a seat (4 instead of 3) and New York lost a seat (from 38 to 37). The addition of Oklahoma, with its fair share of seats, forced New York to give a seat to Maine. The **new-states paradox** occurs when the addition of a new state affects the apportionments of other states.

> **THE NEW-STATES PARADOX**
>
> The addition of a new group changes the apportionments of other groups.

EXAMPLE 3 ILLUSTRATING THE NEW-STATES PARADOX

A school district has two high schools: East High, with an enrollment of 1688 students, and West High, with an enrollment of 7912 students. The school district has a counseling staff of 48 counselors, who are apportioned to the schools using Hamilton's method. This apportionment results in 8 counselors assigned to East High and 40 counselors assigned to West High, shown in Table 14.48. The standard divisor is

$$\frac{\text{total population}}{\text{number of allotted items}} = \frac{9600}{48} = 200.$$

> There are 200 students per counselor.

TABLE 14.48 APPORTIONMENT OF 48 COUNSELORS USING HAMILTON'S METHOD

School	Population	Standard Quota	Lower Quota	Hamilton's Apportionment
East High	1688	8.44	8	8
West High	7912	39.56	39	40
Total	9600	48	47	48

Suppose that a new high school, North High, with a population of 1448 students, is added to the district. Using the standard divisor of 200 students per counselor, the school district decides to hire 7 new counselors for North High. Show that the new-states paradox occurs when the counselors are reapportioned.

Solution The new total population is the previous student population, 9600, plus the student population at North High, 1448: 9600 + 1448 = 11,048. The new number of counselors is the previous number, 48, plus the new counselors hired at North High, 7: 48 + 7 = 55. Thus, when North High is added, the new standard divisor is

$$\frac{\text{total population}}{\text{number of allocated items}} = \frac{11{,}048}{55} \approx 200.87.$$

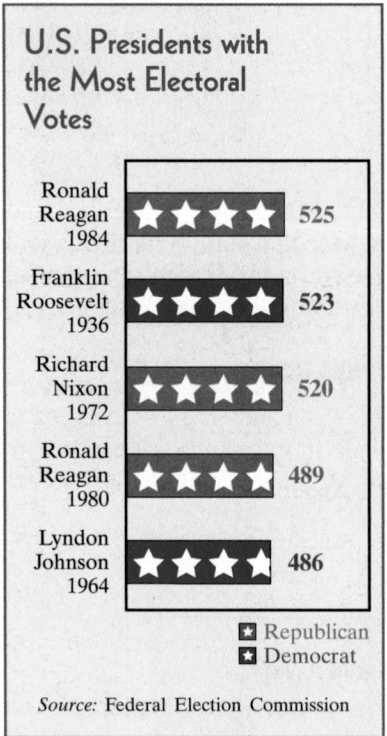

The new standard quotas and Hamilton's apportionment are shown in Table 14.49.

TABLE 14.49 APPORTIONMENT OF 55 COUNSELORS USING HAMILTON'S METHOD

School	Population	Standard Quota	Lower Quota	Decimal Part	Surplus Counselors	Final Apportionment
East High	1688	8.40	8 *largest*	0.40	1	9
West High	7912	39.39	39	0.39		39
North High	1448	7.21	7	0.21		7
Total	11,048	55	54			55

standard quota
$= \dfrac{\text{school population}}{200.87}$

This sum indicates we must assign one extra counselor.

Before North High was added to the district, East High was assigned 8 counselors and West High was assigned 40 counselors. After the addition of a new school and an increase in the total number of counselors to be apportioned, East High was assigned 9 counselors and West High was assigned 39 counselors. Thus, West High ended up losing a counselor to East High. The addition of a new high school changed the apportionments of the other two high schools. This is an example of the new-states paradox.

a. A school district has two high schools, East High, with an enrollment of 2574 students, and West High, with an enrollment of 9426 students. The school district has a counseling staff of 100 counselors. Use Hamilton's method to apportion the counselors to the two schools.

b. Suppose that a new high school, North High, with a population of 750 students, is added to the district. The school district decides to hire 6 new counselors for North High. Use Hamilton's method to show that the new-states paradox occurs when the counselors are reapportioned.

Understand Balinski and Young's Impossibility Theorem.

The Search for an Ideal Apportionment Method

We have seen that although Hamilton's method satisfies the quota rule, it can produce paradoxes. By contrast, Jefferson's, Adams's, and Webster's methods can all violate the quota rule, but do not produce paradoxes. Furthermore, Hamilton's and Jefferson's methods can favor large states, while Adams's and Webster's methods can favor small states. For many years, scholars inside and outside Congress hoped mathematicians would eventually devise an ideal apportionment method—one that satisfies the quota rule, does not produce any paradoxes, and treats large and small states without favoritism.

Is there an ideal apportionment method? The answer, unfortunately, is no. In 1980, mathematicians Michel L. Balinski and H. Peyton Young proved that there is no apportionment method that avoids all paradoxes and at the same time satisfies the quota rule. Their theorem is known as **Balinski and Young's Impossibility Theorem**.

> **BALINSKI AND YOUNG'S IMPOSSIBILITY THEOREM**
>
> There is no perfect apportionment method. Any apportionment method that does not violate the quota rule must produce paradoxes, and any apportionment method that does not produce paradoxes must violate the quota rule.

Because any apportionment method must be flawed, the politics of representation can play as large a role as mathematics when Congress discusses an apportionment method. The United States House of Representatives has been apportioned approximately 20 times, and the choice of an apportionment method can ultimately be a political decision.

Table 14.50 compares the four apportionment methods we discussed and serves as an example of Balinski and Young's Impossibility Theorem.

TABLE 14.50 FLAWS OF APPORTIONMENT METHODS

Flaw	Apportionment Method: Is It Flawed?			
	Hamilton	**Jefferson**	**Adams**	**Webster**
May not satisfy the quota rule (apportionment should always be the upper or lower quota)	No	Yes	Yes	Yes
May produce the Alabama paradox (an increase in total apportioned items results in the loss of an item for a group)	Yes	No	No	No
May produce the population paradox (group A loses items to group B, even though group A's population grew at a faster rate than group B's)	Yes	No	No	No
May produce the new-states paradox (the addition of a new group changes the apportionments of other groups)	Yes	No	No	No
In the House of Representatives, apportionment method favors large states	Yes	Yes	No	No
In the House of Representatives, apportionment method favors small states	No	No	Yes	Yes

BLITZER BONUS

Florida and the Presidential Election of 2000

Results of the 2000 Presidential Election				
Party	Candidate	Popular Vote	Percent	Electoral Vote
Republican	George W. Bush and Dick Cheney	50,455,156	47.87	271
Democrat	Al Gore and Joe Lieberman	50,992,335	48.38	266[1]
Green	Ralph Nader and Winona LaDuke	2,882,897	2.74	—
Reform	Pat Buchanan and Ezola Foster	448,892	0.42	—
Libertarian	Harry Browne	384,429	0.36	—
Other		232,932	0.23	—
	Total Vote	105,396,641		

[1]One elector from the District of Columbia abstained.

Source: Federal Election Commission

On November 7, 2000, Americans expected to end the day knowing who their new president would be. By early in the evening, most states were decided. Because of how the numbers were adding up, the winner of Florida's 25 electoral votes would win the majority of electoral votes and the election. By 8 P.M. in the East, most television networks announced that Al Gore had won Florida. Later that evening, Gore's lead dwindled and George W. Bush was announced as Florida's winner. Still later, Bush's lead shrank. People went to bed not knowing who the next president would be.

The next 36 days brought an unprecedented struggle over Florida's ballots. Bush led the state's popular vote by fewer than one thousand votes out of six million cast. Allegations of confusing ballots, illegal barring of voters, and defective voting machines threw the process of recounting votes into chaos. The race ended on December 12, when the U.S. Supreme Court, largely along party lines, voted 5 to 4 to suspend recounts in Florida. This gave Bush a majority of electoral votes and the White House. Although Gore won the popular vote by more than 500,000 votes, he conceded the election the next day.

Exercise Set 14.4

Practice and Application Exercises

1. The mathematics department has 30 teaching assistants to be divided among three courses, according to their respective enrollments. The table shows the courses and the number of students enrolled in each course.

Course	College Algebra	Statistics	Liberal Arts Math	Total
Enrollment	978	500	322	1800

   a. Apportion the teaching assistants using Hamilton's method.

   b. Use Hamilton's method to determine if the Alabama paradox occurs if the number of teaching assistants is increased from 30 to 31. Explain your answer.

2. A school district has 57 new laptop computers to be divided among four schools, according to their respective enrollments. The table shows the number of students enrolled in each school.

School	A	B	C	D	Total
Enrollment	5040	4560	4040	610	14,250

   a. Apportion the laptop computers using Hamilton's method.

   b. Use Hamilton's method to determine if the Alabama paradox occurs if the number of laptop computers is increased from 57 to 58. Explain your answer.

3. The table shows the populations of three states in a country with a population of 20,000. Use Hamilton's method to show that the Alabama paradox occurs if the number of seats in congress is increased from 40 to 41.

State	A	B	C	Total
Population	680	9150	10,170	20,000

4. The table shows the populations, in thousands, of the three states in a country with a population of 3760 thousand. Use Hamilton's method to show that the Alabama paradox occurs if the number of seats in congress is increased from 24 to 25.

State	A	B	C	Total
Population (in thousands)	530	990	2240	3760

5. A small country has 24 seats in the congress, divided among the three states according to their respective populations. The table shows each state's population, in thousands, before and after the country's population increase.

State	A	B	C	Total
Original Population (in thousands)	530	990	2240	3760
New Population (in thousands)	680	1250	2570	4500

   a. Use Hamilton's method to apportion the 24 congressional seats using the original population.

   b. Find the percent increase, to the nearest tenth of a percent, in the population of each state.

   c. Use Hamilton's method to apportion the 24 congressional seats using the new population. Does the population paradox occur? Explain your answer.

6. A country has 200 seats in the congress, divided among the five states according to their respective populations. The table shows each state's population, in thousands, before and after the country's population increase.

State	A	B	C	D	E	Total
Original Population (in thousands)	2224	2236	2640	3030	9870	20,000
New Population (in thousands)	2424	2436	2740	3130	10,070	20,800

   a. Use Hamilton's method to apportion the 200 congressional seats using the original population.

   b. Find the percent increase, to the nearest tenth of a percent, in the population of each state.

   c. Use Hamilton's method to apportion the 200 congressional seats using the new population. What do you observe about the percent increases for states A and B and their respective changes in apportioned seats? Is this the population paradox?

7. A town has 40 mail trucks and four districts in which mail is distributed. The trucks are to be apportioned according to each district's population. The table shows these populations before and after the town's population increase. Use Hamilton's method to show that the population paradox occurs.

District	A	B	C	D	Total
Original Population	1188	1424	2538	3730	8880
New Population	1188	1420	2544	3848	9000

8. A town has five districts in which mail is distributed and 50 mail trucks. The trucks are to be apportioned according to each district's population. The table shows these populations before and after the town's population increase. Use Hamilton's method to show that the population paradox occurs.

District	A	B	C	D	E	Total
Original Population	780	1500	1730	2040	2950	9000
New Population	780	1500	1810	2040	2960	9090

9. A corporation has two branches, A and B. Each year the company awards 100 promotions within its branches. The table shows the number of employees in each branch.

Branch	A	B	Total
Employees	1045	8955	10,000

a. Use Hamilton's method to apportion the promotions.

b. Suppose that a third branch, C, with the number of employees shown in the table, is added to the corporation. The company adds 5 new yearly promotions for branch C. Use Hamilton's method to determine if the new-states paradox occurs when the promotions are reapportioned.

Branch	A	B	C	Total
Employees	1045	8955	525	10,525

10. A corporation has three branches, A, B, and C. Each year the company awards 60 promotions within its branches. The table shows the number of employees in each branch.

Branch	A	B	C	Total
Employees	209	769	2022	3000

a. Use Hamilton's method to apportion the promotions.

b. Suppose that a fourth branch, D, with the number of employees shown in the table, is added to the corporation. The company adds 5 new yearly promotions for branch D. Use Hamilton's method to determine if the new-states paradox occurs when the promotions are reapportioned.

Branch	A	B	C	D	Total
Employees	209	769	2022	260	3260

11. a. A country has two states, state A, with a population of 9450, and state B, with a population of 90,550. The congress has 100 seats, divided between the two states according to their respective populations. Use Hamilton's method to apportion the congressional seats to the states.

b. Suppose that a third state, state C, with a population of 10,400, is added to the country. The country adds 10 new congressional seats for state C. Use Hamilton's method to show that the new-states paradox occurs when the congressional seats are reapportioned.

12. a. A country has three states, state A, with a population of 99,000, state B, with a population of 214,000, and state C, with a population of 487,000. The congress has 50 seats, divided among the three states according to their respective populations. Use Hamilton's method to apportion the congressional seats to the states.

b. Suppose that a fourth state, state D, with a population of 116,000, is added to the country. The country adds 7 new congressional seats for state D. Use Hamilton's method to show that the new-states paradox occurs when the congressional seats are reapportioned.

13. In Exercise 12, use Jefferson's method with $d = 15,500$ to solve parts (a) and (b). Does the new-states paradox occur? Explain your answer.

Writing in Mathematics

14. What is the Alabama paradox?

15. What is the population paradox?

16. What is the new-states paradox?

17. According to Balinski and Young's Impossibility Theorem, can the democratic ideal of "one person, one vote" ever be perfectly achieved? Explain your answer.

18. Do you think that voters should be angry over either the election of Rutherford Hayes in 1876, described in the section opener, or the election of George W. Bush in 2000? If so, which election should annoy them more? Explain your answers to these very subjective questions.

Critical Thinking Exercises

19. Which one of the following is true?

a. If a country has two states and congressional seats are apportioned by population using Hamilton's method, the Alabama paradox cannot occur if the number of seats increases.

b. Balinski and Young's Theorem shows that "one person, one vote" is mathematically possible.

c. In an apportionment for a country with two states, both Jefferson's and Adams's methods can violate the quota rule.

d. None of the above is true.

20. Give an example of a country with three states in which the Alabama paradox occurs using Hamilton's method when the number of seats in congress is increased from 3 to 4.

Chapter Summary, Review, and Test

SUMMARY

| DEFINITIONS AND CONCEPTS | EXAMPLES |

14.1 Voting Methods

Preference Ballots and Preference Tables
Preference ballots are ballots in which a voter is asked to rank all the candidates in order of preference. Preference tables show how often each particular outcome occurred.

Ex. 1, p. 727

Voting Methods
The four voting methods—the plurality method, the Borda count method, the plurality-with-elimination method, and the pairwise comparison method—are summarized in Table 14.7 on page 735.

Ex. 2, p. 729;
Ex. 3, p. 730;
Ex. 4, p. 731;
Ex. 5, p. 733

14.2 Flaws of Voting Methods

Fairness Criteria
The four fairness criteria for voting methods—the majority criterion, the head-to-head criterion, the monotonicity criterion, and the irrelevant alternatives criterion—are summarized in Table 14.22 on page 745. Table 14.23 on page 745 shows which voting methods satisfy which criteria.

Ex. 1, p. 739;
Ex. 2, p. 741;
Ex. 3, p. 742;
Ex. 4, p. 744

Arrow's Impossibility Theorem
It is mathematically impossible for any democratic voting system to satisfy each of the four fairness criteria.

14.3 Apportionment Methods

Standard Divisors and Standard Quotas

a. Standard divisor $= \dfrac{\text{total population}}{\text{number of allocated items}}$

Ex. 1, p. 752

b. Standard quota $= \dfrac{\text{population of a particular group}}{\text{standard divisor}}$

c. The lower quota is the standard quota rounded down to the nearest whole number. The upper quota is the standard quota rounded up to the nearest whole number.

The Apportionment Problem
The apportionment problem is to determine a method for rounding standard or modified quotas into whole numbers so that the sum of the numbers is the total number of allocated items.

Apportionment Methods
The four apportionment methods discussed in the text—Hamilton's method, Jefferson's method, Adams's method, and Webster's method—are summarized in Table 14.37 on page 760.

Ex. 2, p. 754;
Ex. 3, p. 755;
Ex. 4, p. 757;
Ex. 5, p. 759

The Quota Rule
A group's apportionment should be either its upper quota or its lower quota.

14.4 Flaws of Apportionment Methods

Flaws of Hamilton's Method
Although Hamilton's method satisfies the quota rule, it may produce paradoxes, including the Alabama paradox, the population paradox, and the new-states paradox. These paradoxes are summarized in Table 14.50 on page 770.

Ex. 1, p. 764;
Ex. 2, p. 766;
Ex. 3, p. 768

Flaws of Jefferson's, Adams's, and Webster's Methods
Although these methods do not produce paradoxes, they may not satisfy the quota rule.

Balinski and Young's Impossibility Theorem
There is no perfect apportionment method. Any method that does not violate the quota rule must produce paradoxes, and any method that does not produce paradoxes must violate the quota rule.

REVIEW EXERCISES

14.1

1. The 12 preference ballots for four candidates (A, B, C, and D) are shown. Construct a preference table to illustrate the results of the election.

   ABCD   BDCA   CBDA   ABCD   CBDA   ABCD

   BDCA   BDCA   CBAD   CBAD   ABCD   CBDA

In Exercises 2–5, your class is given the option of choosing a day for the final exam. Students are asked to rank the four available days, Monday (M), Tuesday (T), Wednesday (W), and Friday (F). The results of the election are shown in the preference table.

Number of Votes	9	5	4	2	2	1
First Choice	T	T	M	T	W	W
Second Choice	W	W	T	M	F	M
Third Choice	F	M	F	W	T	F
Fourth Choice	M	F	W	F	M	T

2. How many students voted in the election?
3. How many students selected the days in this order: M, T, F, W?
4. How many students selected Tuesday as their first choice for the final?
5. How many students selected Wednesday as their second choice for the final?

In Exercises 6–9, the Theater Society members are voting for the kind of play they will perform next semester: a comedy (C), a drama (D), or a musical (M). Their votes are summarized in the following preference table.

Number of Votes	10	8	4	2
First Choice	C	M	M	D
Second Choice	D	C	D	M
Third Choice	M	D	C	C

6. Which type of play is selected using the plurality method?
7. Which type of play is selected using the Borda count method?
8. Which type of play is selected using the plurality-with-elimination method?
9. Which type of play is selected using the pairwise comparison method?

In Exercises 10–13, four candidates, A, B, C, and D, are running for chair of the Natural Science Division. The votes of the division members are summarized in the following preference table.

Number of Votes	40	30	6	2
First Choice	A	B	C	D
Second Choice	B	C	D	B
Third Choice	C	D	B	C
Fourth Choice	D	A	A	A

10. Determine the winner using the plurality method.
11. Determine the winner using the Borda count method.
12. Determine the winner using the plurality-with-elimination method.
13. Determine the winner using the pairwise comparison method.

14.2

In Exercises 14–16, voters in a small town are considering four proposals, A, B, C, and D, for the design of affordable housing. Their votes are summarized in the following preference table.

Number of Votes	1500	600	300
First Choice	A	B	C
Second Choice	B	D	B
Third Choice	C	C	D
Fourth Choice	D	A	A

14. Using the Borda count method, which design will be used for the affordable housing?
15. Which design has a majority of first-place votes? Based on your answer to Exercise 14, is the majority criterion satisfied?
16. Which design is favored over all others using a head-to-head comparison? Based on your answer to Exercise 14, is the head-to-head criterion satisfied?

Use the following preference table to solve Exercises 17–18.

Number of Votes	1500	700	300
First Choice	B	A	C
Second Choice	C	B	A
Third Choice	A	C	B

17. Using the plurality-with-elimination method, which candidate wins the election?
18. Which candidate is favored over all others using a head-to-head comparison? Based on your answer to Exercise 17, is the head-to-head criterion satisfied?

Use the following preference table to solve Exercises 19–23.

Number of Votes	180	100	40	30
First Choice	A	B	D	C
Second Choice	B	D	B	B
Third Choice	C	A	C	A
Fourth Choice	D	C	A	D

19. Using the plurality method, which candidate wins the election?
20. Using the Borda count method, which candidate wins the election?
21. Using the plurality-with-elimination method, which candidate wins the election?

22. Using the pairwise comparison method, which candidate wins the election?

23. Which candidate has a majority of first-place votes? Based on your answers to Exercises 19–22 which voting system(s) violate the majority criterion in this election?

Use the following preference table, which shows the results of a straw vote, to solve Exercises 24–25.

Number of Votes	500	400	350	200
First Choice	B	A	C	C
Second Choice	C	B	A	B
Third Choice	A	C	B	A

24. Using the plurality-with-elimination method, which candidate wins the straw vote?

25. In the actual election, the 200 voters in the last column who voted C, B, A, in that order, change their votes to B, C, A. Using the plurality-with-elimination method, which candidate wins the actual election? Based on your answer to Exercise 24, is the monotonicity criterion satisfied?

Use the following preference table to solve Exercises 26–29.

Number of Votes	400	250	200
First Choice	A	C	B
Second Choice	B	B	C
Third Choice	C	A	A

26. Using the plurality method, which candidate wins the election?

27. Suppose that candidate B drops out. Construct a new preference table for the election with B eliminated. Using the new table and the plurality method, which candidate wins the election? Based on your answer to Exercise 26, is the irrelevant alternatives criterion satisfied by the plurality method?

28. Using the Borda count method, which candidate wins the election?

29. If candidate C drops out, which candidate wins the election using the Borda count method? Based on your answer to Exercise 28 is the irrelevant alternatives criterion satisfied by the Borda count method?

14.3

In Exercises 30–36, an HMO has 40 doctors to be apportioned among four clinics. The HMO decides to apportion the doctors based on the average weekly patient load for each clinic, given in the following table.

Clinic	A	B	C	D
Average Weekly Patient Load	275	392	611	724

30. Find the standard divisor.

31. Find each clinic's standard quota.

32. Find each clinic's lower quota and upper quota.

33. Use Hamilton's method to apportion the doctors to the clinics.

34. Use Jefferson's method with $d = 48$ to apportion the doctors to the clinics.

35. Use Adams's method with $d = 52$ to apportion the doctors to the clinics.

36. Use Webster's method with $d = 49.95$ to apportion the doctors to the clinics.

In Exercises 37–40, a country is composed of four states, A, B, C, and D. The population of each state is given in the following table. Congress will have 200 seats, divided among the four states according to their respective populations.

State	A	B	C	D
Population	3320	10,060	15,020	19,600

37. Use Hamilton's method to apportion the congressional seats.

38. Use Jefferson's method to apportion the congressional seats. (*Hint*: A modified divisor that works should be less than the standard divisor in Exercise 37.)

39. Use Adams's method to apportion the congressional seats. (*Hint*: A modified divisor that works should be greater than the standard divisor in Exercise 37.)

40. Use Webster's method to apportion the congressional seats. (*Hint:* A modified divisor that works should lie between 240 and 241.)

14.4

41. A school district has 150 new laptop computers to be divided among three schools, according to their respective enrollments. The table shows the number of students enrolled in each school.

School	A	B	C	Total
Enrollment	370	3365	3765	7500

a. Apportion the laptop computers using Hamilton's method.

b. Use Hamilton's method to determine if the Alabama paradox occurs if the number of laptop computers is increased from 150 to 151.

42. A country has 100 seats in the congress, divided among the three states according to their respective populations. The table shows each state's population before and after the country's population increase.

State	A	B	C	Total
Original Population	143,796	41,090	15,114	200,000
New Population	143,796	41,420	15,304	200,520

a. Use Hamilton's method to apportion the 100 congressional seats using the original population.

b. Find the percent increase, to the nearest tenth of a percent, in the population of state B and state C.

c. Use Hamilton's method to apportion the 100 congressional seats using the new population. Does the population paradox occur?

43. A corporation has two branches, A and B. Each year the company awards 33 promotions within its branches. The table shows the number of employees in each branch.

Branch	A	B	Total
Employees	372	1278	1650

a. Use Hamilton's method to apportion the promotions.

b. Suppose that a third branch, C, with the number of employees shown in the table, is added to the corporation. The company adds 7 new yearly promotions for branch C. Use Hamilton's method to determine if the new-states paradox occurs when the promotions are reapportioned.

Branch	A	B	C	Total
Employees	372	1278	355	2005

44. Is the following statement true or false?

There are perfect voting methods, as well as perfect apportionment methods.

Explain your answer.

Group Exercise

45. Citizen-initiated ballot measures often present voters with controversial issues over which they do not think alike. Here's one your author would like to initiate:

Please rank each of the following options regarding permitting dogs on national park trails.

A. Unleashed dogs accompanied by their caregivers should be permitted on designated national park trails.

B. Leashed dogs accompanied by their caregivers should be permitted on designated national park trails.

C. No dogs should be permitted on any national park trails.

Your author was not happy with the fact that he could not take his dog running with him on the park trails at Point Reyes National Seashore. Of course, that is his issue. For this project, group members should write a ballot measure, perhaps controversial, like the sample above, but dealing with an issue of relevance to your campus and community. Rather than holding an election, use a random sample of students on your campus, administer the ballot, and have them rank their choices.

a. Use each of the four voting methods to determine the winning option for your ballot measure.

b. Check to see if any of the four fairness criteria are violated.

CHAPTER 14 TEST

In Exercises 1–8, three candidates, A, B, and C, are running for mayor of a small town. The results of the election are shown in the following preference table.

Number of Votes	1200	900	900	600
First Choice	A	C	B	B
Second Choice	B	A	C	A
Third Choice	C	B	A	C

1. How many people voted in the election?

2. How many people selected the candidates in this order: B, A, C?

3. How many people selected candidate B as their first choice?

4. How many people selected candidate A as their second choice?

5. Determine the winner using the plurality method.

6. Determine the winner using the Borda count method.

7. Determine the winner using the plurality-with-elimination method.

8. Determine the winner using the pairwise comparison method.

Use the following preference table to solve Exercises 9–10.

Number of Votes	240	160	60
First Choice	A	C	D
Second Choice	B	B	A
Third Choice	C	D	C
Fourth Choice	D	A	B

9. Determine the winner using the Borda count method.

10. Which candidate has a majority of first-place votes? Based on your answer to Exercise 9, is the majority criterion satisfied?

Use the following preference table to solve Exercises 11–12.

Number of Votes	1500	1000	1000
First Choice	A	B	C
Second Choice	B	A	B
Third Choice	C	C	A

11. Determine the winner using the plurality method.

12. Which candidate is favored over all others using a head-to-head comparison? Based on your answer to Exercise 11, is the head-to-head criterion satisfied?

Use the following preference table, which shows the results of a straw vote, to solve Exercises 13–14.

Number of Votes	70	60	50	30
First Choice	C	B	A	A
Second Choice	A	C	B	C
Third Choice	B	A	C	B

13. Determine the winner using the plurality-with-elimination method.

14. In the actual election, the 30 voters in the last column who voted A, C, B, in that order, change their votes to C, A, B. Using the plurality-with-elimination method, which candidate wins the election? Based on your answer to Exercise 13, is the monotonicity criterion satisfied?

15. Suppose that the plurality method is used on the following preference table. If candidate C drops out, is the irrelevant alternatives criterion satisfied? Explain your answer.

Number of Votes	90	75	45
First Choice	B	C	A
Second Choice	C	A	C
Third Choice	A	B	B

In Exercises 16–24, an HMO has 10 doctors to be apportioned among three clinics. The HMO decides to apportion the doctors based on the average weekly patient load for each clinic, given in the following table.

Clinic	A	B	C
Average Weekly Patient Load	119	165	216

16. Find the standard divisor.

17. Find each clinic's standard quota.

18. Find each clinic's lower quota and upper quota.

19. Use Hamilton's method to apportion the doctors to the clinics.

20. Use Jefferson's method to apportion the doctors to the clinics.

21. Use Adams's method to apportion the doctors to the clinics.

22. Use Webster's method with $d = 47.7$ to apportion the doctors to the clinics.

23. Suppose that the HMO hires one new doctor. Using Hamilton's method, does the Alabama paradox occur when the number of doctors is increased from 10 to 11?

24. As given in the original description, assume that the HMO has 10 doctors. Suppose that a third clinic, clinic D, with the average weekly patient load shown in the table, is added to the HMO. The administration hires 2 new doctors for clinic D. Use Hamilton's method to determine if the new-states paradox occurs when the doctors are reapportioned.

Clinic	A	B	C	D
Average Weekly Patient Load	119	165	216	110

25. Write one sentence for a person not familiar with the mathematics of voting and apportionment that summarizes the two impossibility theorems discussed in this chapter.

Graph Theory

- A more user-friendly map for a city's subway system is needed.
- Public works administrators must find the most efficient routes for snowplows, garbage trucks, and street sweepers.
- A sales director traveling to branch offices needs to minimize the cost of the trip.
- You need to run a series of errands and return home using the shortest route.
- After a heavy snow, campus services must clear a minimum number of sidewalks and still ensure that students walking from building to building will be able to do so along cleared paths.
- You need to represent the parent-child relationships for five generations of your family.

These unrelated problems can be solved using special diagrams, called *graphs*. Graph theory was launched by the Swiss mathematician Leonhard Euler (1707–1783). Euler analyzed a puzzle involving a city with seven bridges. Would it be possible to walk through the city and cross each of its bridges exactly once? Although graph theory developed from this puzzle, you will see how graphs are used to solve a diverse series of realistic problems faced by businesses and individuals. Graph theory has saved businesses billions of dollars over the past 30 years. Perhaps it might even help you to find a more efficient way of dealing with those chore days from hell.

You are not looking forward to this day, devoted entirely to errands around town. You need to go to the post office, deposit a check at the bank, drop off dry cleaning, visit a friend at the hospital, and get a flu shot. How can mathematics be used to find the shortest and most efficient route for carrying out these errands?

This problem appears as Exercises 37–40 in Exercise Set 15.3.

SECTION 15.1   GRAPHS, PATHS, AND CIRCUITS

OBJECTIVES

1. Understand relationships in a graph.
2. Model relationships using graphs.
3. Understand and use the vocabulary of graph theory.

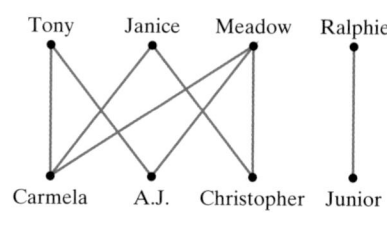

FIGURE 15.1

The Sopranos: Only on cable could you get away with this violent and satiric pro-file of an angst-ridden mafia family man and his dysfunctional domestic clan. In Figure 15.1, suppose that eight members of the Soprano family are at a party. Any line drawn between two people indicates an unpleasant verbal encounter at the party. We can learn quite a bit from this picture, such as the fact that Ralphie and Junior only had an unpleasant encounter with each other. The picture provides a convenient way to describe the numerous eruptions in the family's shrink-wrapped relationships.

The diagram in Figure 15.1 is an example of a *graph*. A graph provides a struc-ture for describing relationships. Relationships between people (friendship, love, unpleasant party encounters, etc.) can be described by means of a graph. These graphs are different from the line, bar, circle, and rectangular coordinate graphs we have seen throughout the book. Instead they tell us how a group of objects are related to each other. Let's begin with the definition of a graph.

STUDY TIP

Graphs were briefly introduced in Section 10.7. However, it is not necessary to read that section before covering this chapter.

1 Understand relationships in a graph.

DEFINITION OF A GRAPH

A **graph** consists of a finite set of points, called **vertices** (singular is *vertex*) and line segments or curves, called **edges**, that start and end at vertices. An edge that starts and ends at the same vertex is called a **loop**.

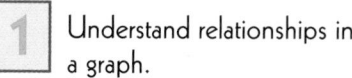

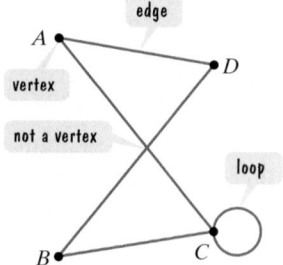

FIGURE 15.2 An example of a graph

Figure 15.2 shows an example of a graph. This graph has four vertices, labeled A, B, C, and D. (We will often use uppercase letters to refer to vertices.) If there is not more than one edge between two vertices, we can use the two vertices to refer to that edge. For example, the edge that connects vertex A to vertex D can be called edge AD or edge DA. The loop on the lower right can be referred to as edge CC or loop CC.

We will always indicate vertices by black dots. In Figure 15.2, the point where edges AC and DB cross has no black dot. Thus, this point is not a vertex. **Not every point where two edges cross is a vertex.** You might find it useful to think of one of the edges as passing above the other.

In a graph, the important information is which vertices are connected by edges. Two graphs are **equivalent** if they have the same number of vertices connected to each other in the same way. The placement of the vertices and the shapes of the edges are unimportant.

EXAMPLE 1   UNDERSTANDING RELATIONSHIPS IN GRAPHS

Explain why Figures 15.3(a) and (b) show equivalent graphs.

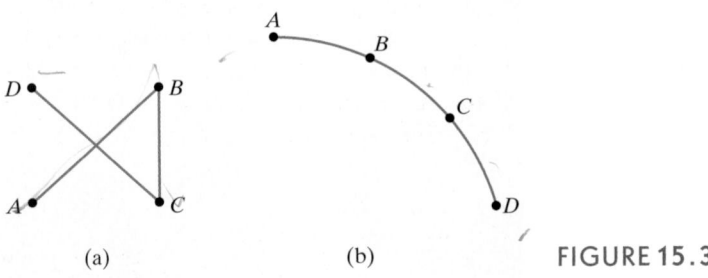

(a)   (b)   FIGURE 15.3

Solution   In both figures, the vertices are A, B, C, and D. Both graphs have an edge that connects vertex A to vertex B (edge AB or BA), an edge that connects vertex B to vertex C (edge BC or CB), and an edge that connects vertex C to vertex D (edge CD or DC). Because the two graphs have the same number of vertices connected to each other in the same way, they are equivalent.

Explain why Figures 15.4(a) and (b) show equivalent graphs.

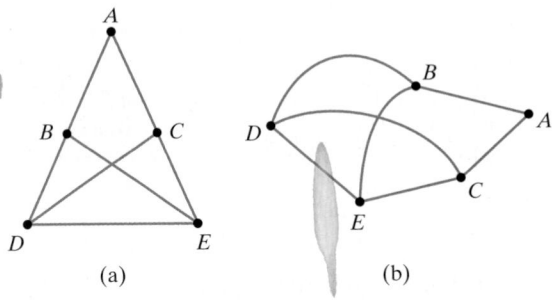

(a)   (b)   FIGURE 15.4

Modeling Relationships Using Graphs

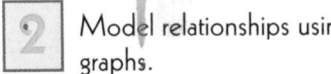

Model relationships using graphs.

Examples 2 through 5 illustrate a number of different situations that we can represent, or model, using graphs. We begin with a situation that was briefly discussed in Chapter 10. Our graph will tell how a group of land masses are related to each other.

EXAMPLE 2   MODELING KÖNIGSBERG WITH A GRAPH

In the early 1700s, the city of Königsberg, Germany, was located on both banks and two islands of the Pregel River. Figure 15.5 shows that the town's sections were connected by seven bridges. Draw a graph that models the layout of Königsberg. Use vertices to represent the land masses and edges to represent the bridges.

Solution   In drawing the graph, the only thing that matters is the relationship between land masses and bridges: which land masses are connected to each other and by how many bridges. We label each of the four land masses with an uppercase letter, shown in Figure 15.6(a).

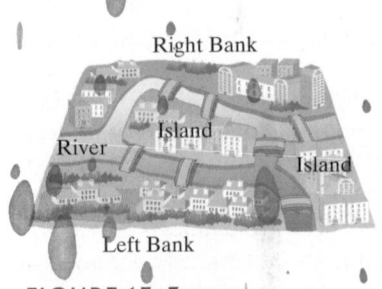

FIGURE 15.5 The layout of Königsberg

Next, we use points to represent the land masses. Figure 15.6(b) shows vertices $A, B, L,$ and $R.$ The precise placement of these vertices is not important.

Now we are ready to draw the edges that represent the bridges. There are two bridges that connect island A to right bank $R.$ Therefore, we draw two edges connecting vertex A to vertex $R,$ shown in the graph in Figure 15.6(c). Can you see that there is only one bridge that connects island A to island B? Thus, we draw one edge connecting vertex A to vertex B in our graph. There are also two bridges connecting island A to left bank $L,$ so we draw two edges connecting vertex A to vertex $L.$ Continuing in this manner, we obtain the graph shown in Figure 15.6(c).

STUDY TIP

When modeling relationships using graphs, keep in mind that it is not the shape of the graph that matters. The important part is how the vertices are connected to each other. Remember that a graph can be drawn in many equivalent ways.

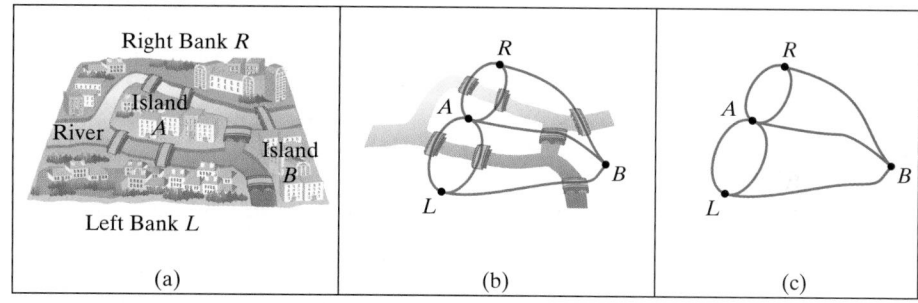

(a)   (b)   (c)

FIGURE 15.6 Creating a graph that models the layout of Königsberg

 **Check Point 2** The city of Metroville is located on both banks and three islands of the Metro River. Figure 15.7 shows that the town's sections are connected by five bridges. Draw a graph that models the layout of Metroville.

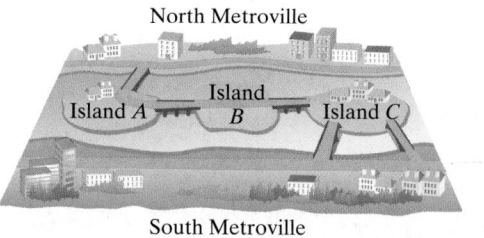

North Metroville

Island A   Island B   Island C

South Metroville

FIGURE 15.7

In the next example, our graph will tell how a group of states are related to each other.

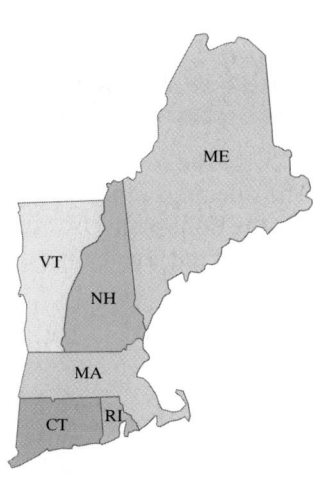

FIGURE 15.8 The New England states

EXAMPLE 3   MODELING BORDERING RELATIONSHIPS FOR THE NEW ENGLAND STATES

The map in Figure 15.8 shows the New England states. Draw a graph that models which New England states share a common border. Use vertices to represent the states and edges to represent common borders.

Solution   Because the states are labeled with their abbreviations, we can use the abbreviations in Figure 15.9(a) to label each vertex. We use points to represent these vertices, shown in Figure 15.9(b). The precise placement of these vertices is not important.

Now we are ready to draw the edges that represent common borders. Whenever two states share a common border, we connect the respective vertices with an edge. For example, Rhode Island shares a common border with Connecticut and with Massachusetts. Therefore, we draw an edge that connects vertex RI to vertex CT and one that connects vertex RI to vertex MA, shown in Figure 15.9(c). Can you see that Massachusetts shares a common border with four states, resulting in edges connecting MA to CT, MA to RI, MA to VT, and MA to NH? Continuing in this manner, we obtain the graph shown in Figure 15.9(c).

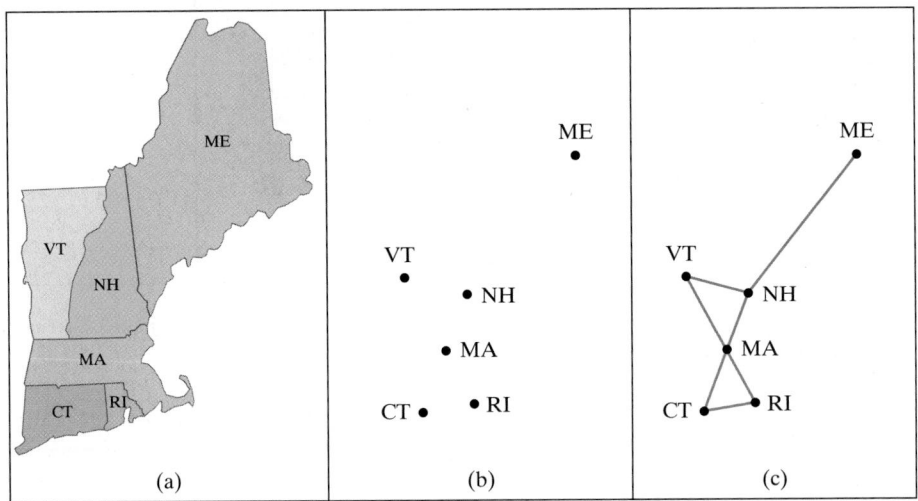

FIGURE 15.9 Creating a graph that models bordering relationships between New England states

 **Check 3 Point**   Create a graph that models the bordering relationships among the five states shown in Figure 15.10.

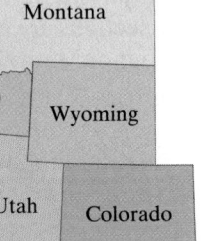

FIGURE 15.10

In the next example, our graph will tell how rooms in a floor plan are related to each other.

EXAMPLE 4   MODELING CONNECTING RELATIONSHIPS IN A FLOOR PLAN

The floor plan of a four-room house is shown in Figure 15.11. The rooms are labeled $A, B, C,$ and D. The outside of the house is labeled E. The openings represent doors. Draw a graph that models the connecting relationships in the floor plan. Use vertices to represent the rooms and the outside, and edges to represent the connecting doors.

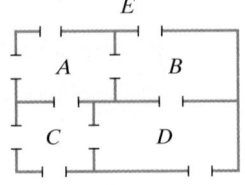

FIGURE 15.11

Solution Because the rooms and outside are labeled, we can use the letters in Figure 15.12(a) to label each vertex. We use points to represent these vertices, shown in Figure 15.12(b).

Now we are ready to draw the edges that represent the doors. Two doors connect the outside, E, with room A, so we draw two edges that connect vertex E to vertex A, shown in Figure 15.12(c). Only one door connects the outside, E, with room B, so we draw one edge from vertex E to vertex B. Two doors connect the outside, E, to room C; we draw two edges from E to C. Counting doors from the outside to each room and doors between the rooms, the graph is completed as shown in Figure 15.12(c).

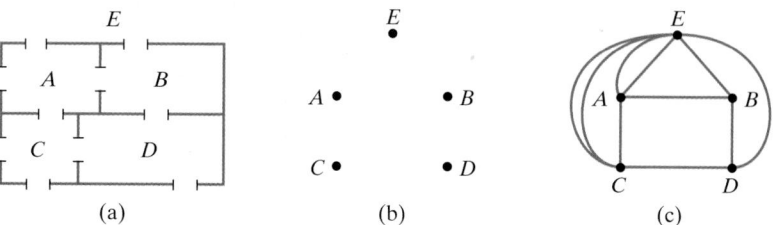

FIGURE 15.12 Creating a graph that models connecting relationships between rooms

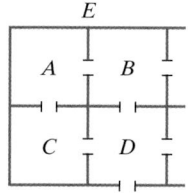

FIGURE 15.13

 The floor plan of a four-room house is shown in Figure 15.13. The rooms are labeled $A, B, C,$ and D. The outside of the house is labeled E. Draw a graph that models the connecting relationships in the floor plan.

In setting up routes for mail carriers, the post office is interested in delivering mail along the route that involves the least amount of walking or driving. Graphs play an important role in problems concerned with finding efficient ways to route services such as mail delivery, garbage collection, and police protection. The first step in the planning and design of delivery routes is creating graphs that model neighborhoods.

EXAMPLE 5 MODELING WALKING RELATIONSHIPS FOR A NEIGHBORHOOD'S STREETS

A mail carrier delivers mail to the four-block neighborhood shown in Figure 15.14. She parks her truck at the intersection shown in the figure and then walks to deliver mail to each of the houses. The streets on the outside of the neighborhood have houses on one side only. By contrast, the interior streets have houses on both sides of the street. On these streets, the mail carrier must walk down the street twice, covering each side separately. Draw a graph that models the streets of the

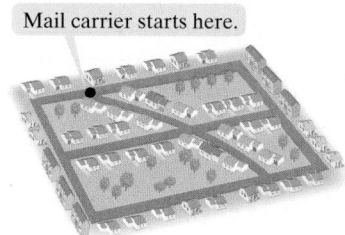

FIGURE 15.14

neighborhood walked by the mail carrier. Use vertices to represent the street intersections and corners. Use one edge if streets must be covered only once and two edges for streets that must be covered twice.

Solution We begin by labeling each of the corners and street intersections with an uppercase letter, shown in Figure 15.15(a) on the next page.

Next, we use points to represent the corners and street intersections. Figure 15.15(b) shows vertices *A* through *I*.

Now we are ready to draw the edges that represent walking relationships between corners and street intersections. The street from *B* to *A* must be covered only once, so we draw an edge that connects vertex *A* to vertex *B*, shown in Figure 15.15(c). The same is true for the street from *A* to *D*. By contrast, the street from *D* to *E* has houses on both sides and must be walked twice. Therefore, we draw two edges that connect vertex *D* to vertex *E*. Continuing in this manner, the graph is completed as shown in Figure 15.15(c).

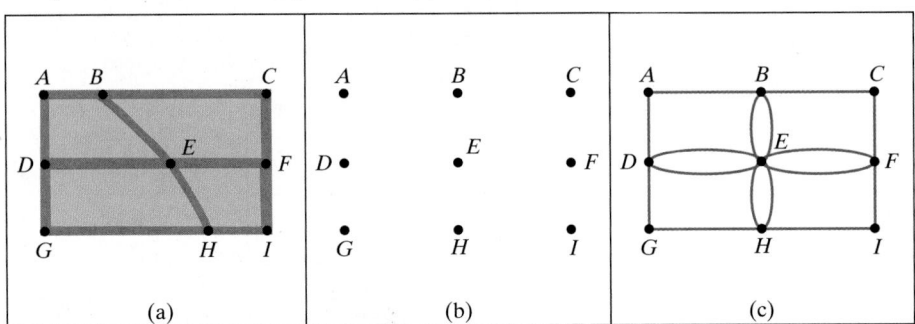

FIGURE 15.15

Check 5 Point A security guard needs to walk the streets of the neighborhood shown in Figure 15.14 on page 783. Unlike the postal worker, the guard would like to walk down each street once, whether or not the street has houses on both sides. Draw a graph that models the streets of the neighborhood walked by the security guard. ■

3 Understand and use the vocabulary of graph theory.

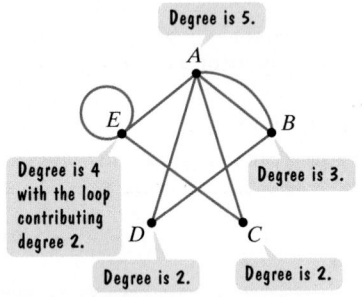

FIGURE 15.16 The degrees of a graph's vertices

STUDY TIP

Vertices that are near each other are not necessarily adjacent. In Figure 15.16, vertices *C* and *D* are not adjacent because there is no edge connecting them.

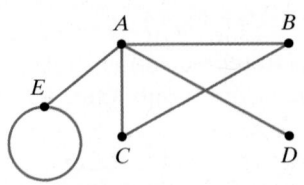

FIGURE 15.17

The Vocabulary of Graph Theory

Every branch of mathematics has its own unique vocabulary. We conclude this section by introducing some of the terms used in graph theory.

The **degree of a vertex** is the number of edges at that vertex. If a loop connects a vertex to itself, that loop contributes 2 to the degree of the vertex. On a graph, the degree of each vertex is found by counting the number of line segments or curves attached to the vertex. Figure 15.16 illustrates a graph and the degree of each vertex.

A vertex with an even number of edges attached to it is an **even vertex**. For example, in Figure 15.16, vertices *E*, *D*, and *C* are even. A vertex with an odd number of edges attached to it is an **odd vertex**. In Figure 15.16, vertices *A* and *B* are odd.

Two vertices in a graph are said to be **adjacent vertices** if there is at least one edge connecting them. It is helpful to think of adjacent vertices as *connected* vertices.

EXAMPLE 6   IDENTIFYING ADJACENT VERTICES

List the pairs of adjacent vertices for the graph in Figure 15.16.

Solution   A systematic way to approach this problem is to first list all the pairs of adjacent vertices involving vertex *A*, then all those involving *B* but not *A*, then all those involving *C* but neither *A* nor *B*, then those involving *D* but not *A*, *B*, or *C*. Finally, check to see if *E* is adjacent to itself. (It is.) Thus, the adjacent vertices are *A* and *B*, *A* and *C*, *A* and *D*, *A* and *E*, *B* and *D*, *C* and *E*, and *E* and *E*.

Check 6 Point List the pairs of adjacent vertices for the graph in Figure 15.17. ■

We can use adjacent vertices to describe movement along a graph. A **path** in a graph is a sequence of adjacent vertices and the edges connecting them. Although a vertex can appear on the path more than once, **an edge can be part of a path only once**. For example, a path along a graph, shown in red, is illustrated in Figure 15.18.

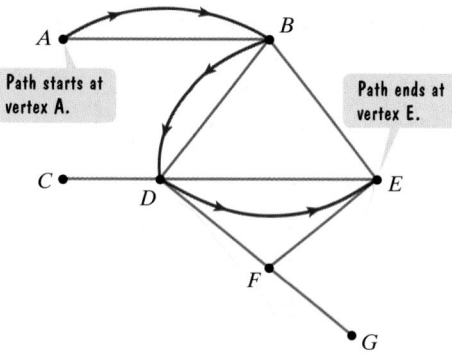

FIGURE 15.18 A path along a graph

You can think of this path as movement from vertex A to vertex B to vertex D to vertex E. We can refer to this path using a sequence of vertices separated by commas. Thus, the path shown in Figure 15.18 is described by A, B, D, E.

The graph in Figure 15.18 has many paths. Can you see that a path does not have to include every vertex and every edge of a graph?

A **circuit** is a path that begins and ends at the same vertex. In Figure 15.19, the path given by B, D, F, E, B, is a circuit. Observe that every circuit is a path. However, because not every path ends at the same vertex where it starts, not every path is a circuit.

The words "connected" and "disconnected" are used to describe graphs. A graph is **connected** if for any two of its vertices there is at least one path connecting them. Thus, a graph is connected if it consists of one piece. If a graph is not connected, it is said to be **disconnected**. A disconnected graph is made up of pieces that are by themselves connected. Such pieces are called the **components** of the graph. Figure 15.20 illustrates connected and disconnected graphs.

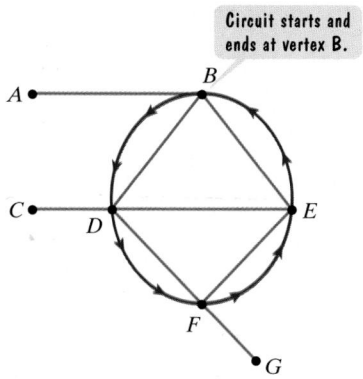

FIGURE 15.19 A circuit along a graph

Connected Graphs

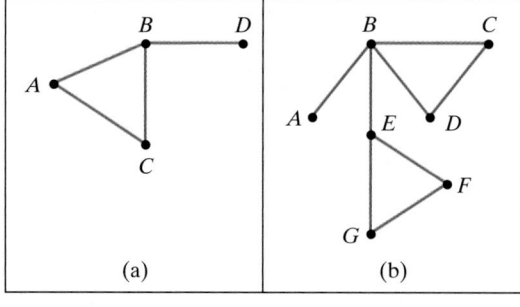

(a)   (b)

Disconnected Graphs

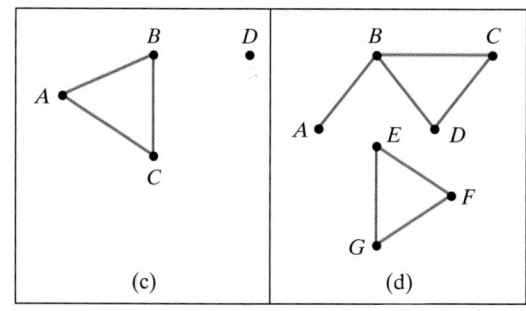

(c)   (d)

FIGURE 15.20 Examples of two connected graphs and two disconnected graphs

STUDY TIP

Do not confuse the definition of *bridge* in graph theory with the *bridges* connecting sections of a city such as Königsberg. In our graph for the layout of Königsberg [Figure 15.6(c) on page 781], none of the edges that model the city's bridges are bridges as defined in graph theory.

A cliché says that if you burn a bridge behind you, you'll never get back to where you were. This cliché tells us something about the word *bridge* in graph theory. A **bridge** is an edge that if removed from a connected graph would leave behind a disconnected graph. For example, compare Figures 15.20(a) and (c). If edge BD were removed from Figure 15.20(a), vertex D would be isolated from the rest of the graph, leaving behind the disconnected graph in Figure 15.20(c). Thus, BD is a bridge for the graph in Figure 15.20(a).

Next, compare Figures 15.20(b) and (d). If edge BE were removed from Figure 15.20(b), the graph would become disconnected, shown by the two components in Figure 15.20(d). Thus, BE is a bridge for the graph in Figure 15.20(b).

BLITZER BONUS

Subway Simplification

London's subway system is as sprawling as the city it serves. Drawing the system on a map made it difficult for users to follow. Planning a trip with it was about as easy as finding one's way into and out of a maze—and more difficult if the trip involved changing trains.

   In 1931, draftsman Henry C. Beck found a solution, which led to the London underground map so widely used today. Beck's proposal for a user-friendly subway map was to abandon the idea that it should be a literal representation of how the lines ran underground. Instead, it should show which vertices (subway stations) are connected by edges (underground routes). Beck knew that two graphs are equivalent if they have the same number of vertices connected to each other in the same way. The important information was the relationship between the stations. The shapes of the edges were unimportant. For simplicity, Beck used horizontal, vertical, and diagonal lines to represent edges. He also enlarged the central, most complex, part of the system in relation to the simple parts at the outskirts. In 1933, a few experimental copies of the simplified graph that modeled the underground system were printed. They were an immediate success.

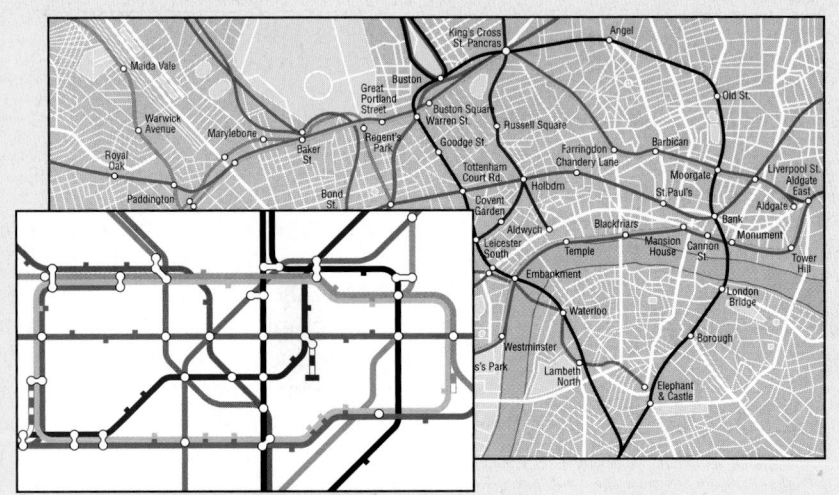

Tangled Lines London's subway system is difficult to use on a conventional map. However, a simple graph that models the system (inset) helps travelers find their way.

Exercise Set 15.1

Practice and Application Exercises

The graph models the baseball schedule for a week. The vertices represent the teams. Each game played during that week is represented as an edge between two teams. Use the information in the graph to solve Exercises 1–4.

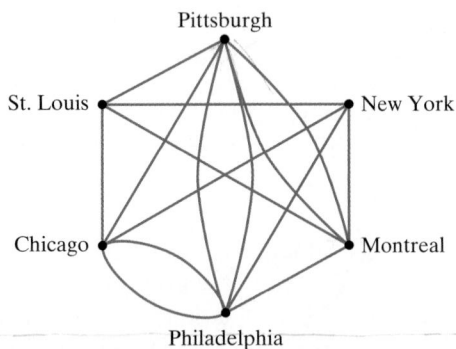

1. How many games are scheduled for Pittsburgh during the week? List the teams that they are playing. How many times are they playing each of these teams?

2. How many games are scheduled for Montreal during the week? List the teams that they are playing. How many times are they playing each of these teams?

3. Do the positions of New York and Montreal correspond to their geographic locations on a map? If not, is the graph drawn incorrectly? Explain your answer.

4. Do the positions of New York and Chicago correspond to their geographic locations on a map? If not, is the graph drawn incorrectly? Explain your answer.

In Exercises 5–6, draw two equivalent graphs for each description.

5. The vertices are $A, B, C,$ and D. The edges are $AB, BC, BD, CD,$ and CC.

6. The vertices are $A, B, C,$ and D. The edges are $AD, BC, DC, BB,$ and DB.

In Exercises 7–8, explain why the two figures show equivalent graphs. Then draw a third equivalent graph.

7.

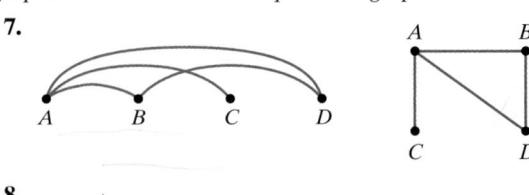

8.

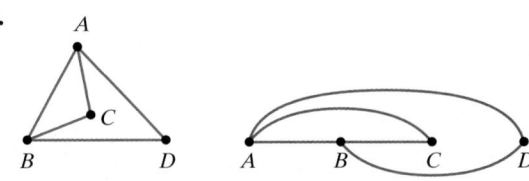

9. Eight students form a math homework group. The students in the group are Zeb, Stryder, Amy, Jed, Evito, Moray, Carrie, and Oryan. Prior to forming the group, Stryder was friends with everyone but Moray. Moray was friends with Zeb, Amy, Carrie, and Evito. Jed was friends with Stryder, Evito, Oryan, and Zeb. Draw a graph that models pairs of friendships among the eight students prior to forming the math homework group.

10. An environmental action group has six members, A, B, C, D, E, and F. The group has three committees: The Preserving Open Space Committee (B, D, and F), the Fund Raising Committee (B, C, and D), and the Wetlands Protection Committee (A, C, D, and E). Draw a graph that models the common members among committees. Use vertices to represent committees and edges to represent common members.

In Exercises 11–12, draw a graph that models the layout of the city shown in each map. Use vertices to represent the land masses and edges to represent the bridges.

11. The City of Gothamville

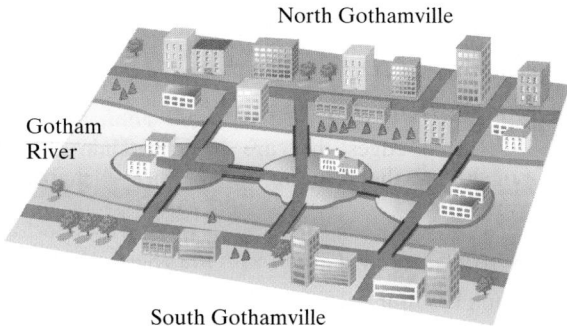

12. The City of Wisdomville

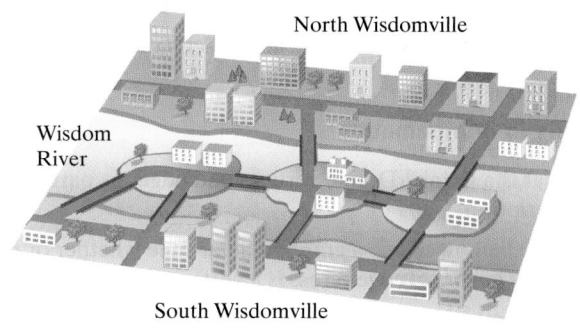

In Exercises 13–14, create a graph that models the bordering relationships among the states shown in each map. Use vertices to represent the states and edges to represent common borders.

13.

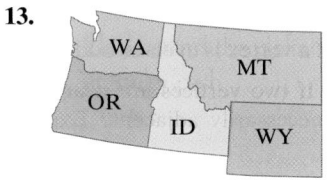

14.

In Exercises 15–18, draw a graph that models the connecting relationships in each floor plan. Use vertices to represent the rooms and the outside, and edges to represent the connecting doors.

15.

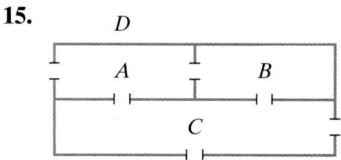

16.

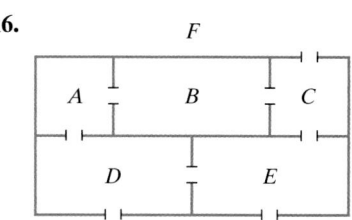

17.

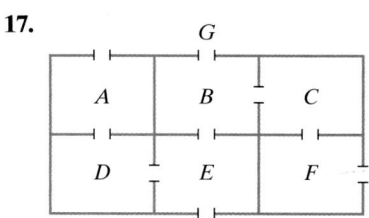

18.

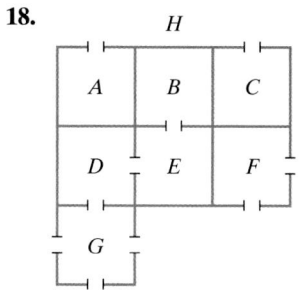

In Exercises 19–20, a security guard needs to walk the streets of the neighborhood shown in each figure. The guard is to walk down each street once, whether or not the street has houses on both sides. Draw a graph that models the neighborhood. Use vertices to represent the street intersections and corners. Use edges to represent the streets the security guard needs to walk.

19.

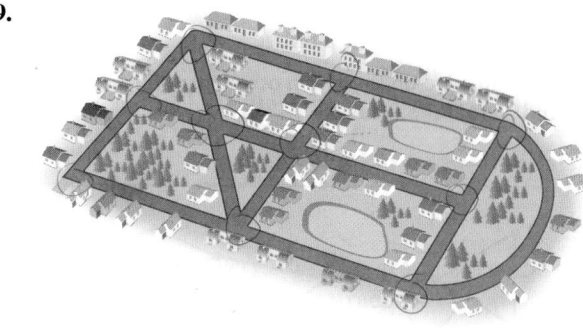

We know that vertices separated by commas can be used to designate paths. As we discuss Euler paths, you can use a pencil to trace these paths. We can also indicate a starting vertex and numbered edges to illustrate Euler paths. For example, examine the graph in Figure 15.21.

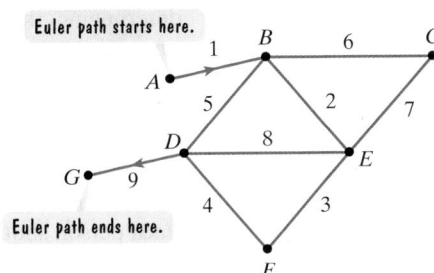

FIGURE 15.21 An Euler path: $A, B, E, F, D, B, C, E, D, G$

The path $A, B, E, F, D, B, C, E, D, G$ is an Euler path because each edge is traveled once. Trace this path with your pencil. Now try using the numbers along the edges. The voice balloon indicates a starting vertex, A, and the arrow shows which way to trace first. When you arrive at the next vertex, B, take the next numbered edge, 2. When you arrive at the next vertex, E, take the next numbered edge, 3. Continue in this manner until numbered edge 9 ends the path at vertex G.

We will use vertices as well as numbered edges to designate Euler paths. Do you see why the Euler path in Figure 15.21 is not a circuit? A circuit must begin and end at the same vertex. The Euler path in Figure 15.21 begins at vertex A and ends at vertex G.

If an Euler path begins and ends at the same vertex, it is called an *Euler circuit*.

2 Understand the definition of an Euler circuit.

DEFINITION OF AN EULER CIRCUIT

An **Euler circuit** is a circuit that travels through every edge of a graph once and only once. Like all circuits, an Euler circuit must begin and end at the same vertex.

The graph in Figure 15.22 shows an Euler circuit.

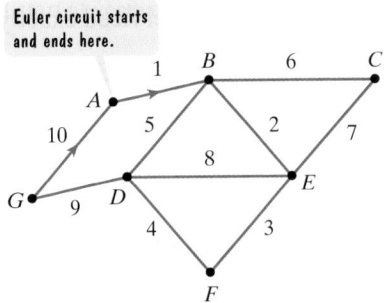

FIGURE 15.22 An Euler circuit: $A, B, E, F, D, B, C, E, D, G, A$

The path $A, B, E, F, D, B, C, E, D, G, A$, shown with numbered edges 1 through 10, is an Euler circuit. Do you see why? Each edge is traced only once. Furthermore, the path begins and ends at the same vertex, A. Notice that **every Euler circuit is an Euler path**. However, **not every Euler path is an Euler circuit**.

Some graphs have no Euler paths. Other graphs have several Euler paths. Furthermore, some graphs with Euler paths have no Euler circuits. **Euler's Theorem** is used to determine if a graph contains Euler paths or Euler circuits.

3 Use Euler's Theorem.

STUDY TIP

In all three parts of Euler's Theorem, it is the *number of odd vertices* that determines if a graph has Euler paths and Euler circuits.

EULER'S THEOREM

The following statements are true for connected graphs:

1. If a graph has exactly two odd vertices, then it has at least one Euler path, but no Euler circuit. Each Euler path must start at one of the odd vertices and end at the other one.

2. If a graph has no odd vertices (all even vertices), it has at least one Euler circuit (which, by definition, is also an Euler path). An Euler circuit can start and end at any vertex.

3. If a graph has more than two odd vertices, then it has no Euler paths and no Euler circuits.

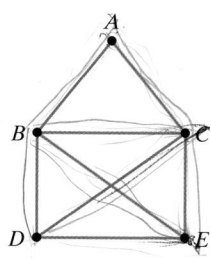

FIGURE 15.23

EXAMPLE 1   USING EULER'S THEOREM

a. Explain why the graph in Figure 15.23 has at least one Euler path.

b. Use trial and error to find one such path.

Solution

a. In Figure 15.24, we count the number of edges at each vertex to determine if the vertex is odd or even. We see that there are exactly two odd vertices, namely D and E. By the first statement in Euler's Theorem, the graph has at least one Euler path, but no Euler circuit.

b. Euler's Theorem tells us that a possible Euler path must start at one of the odd vertices and end at the other one. We will use trial and error to determine an Euler path, starting at vertex D and ending at vertex E. Figure 15.25 shows an Euler path: $D, C, B, E, C, A, B, D, E$. Trace this path and verify the numbers along the edges.

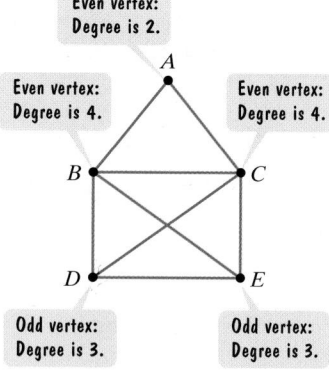

FIGURE 15.24 The graph has two odd vertices.

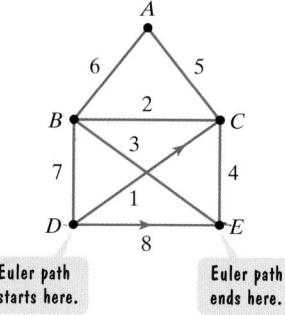

FIGURE 15.25 An Euler path: $D, C, B, E, C, A, B, D, E$

Check Point 1

Refer to the graph in Figure 15.23. Use trial and error to find an Euler path that starts at E and ends at D. Number the edges of the graph to indicate the path. Then use vertex letters separated by commas to name the path.

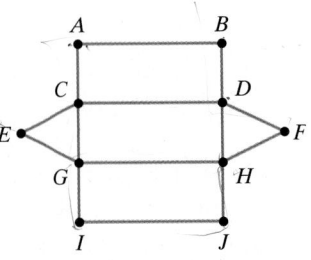

FIGURE 15.26

EXAMPLE 2 | USING EULER'S THEOREM

a. Explain why the graph in Figure 15.26 has at least one Euler circuit.

b. Use trial and error to find one such circuit.

Solution

a. In Figure 15.27, we count the number of edges at each vertex to determine if the vertex is odd or even. We see that the graph has no odd vertices. By the second statement in Euler's Theorem, the graph has at least one Euler circuit.

b. An Euler circuit can start and end at any vertex. We will use trial and error to determine an Euler circuit, starting and ending at vertex H. Remember that you must trace every edge exactly once and start and end at H. Figure 15.28 shows an Euler circuit. Trace this circuit using the vertices in the figure's caption and verify the numbers along the edges.

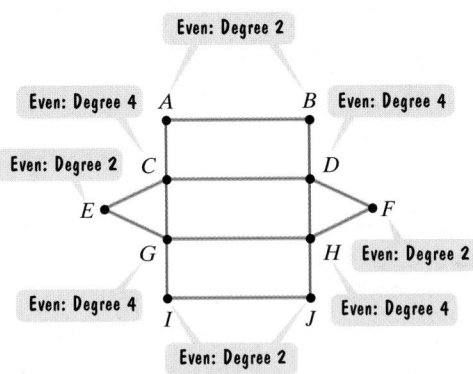

FIGURE 15.27 The graph has no odd vertices.

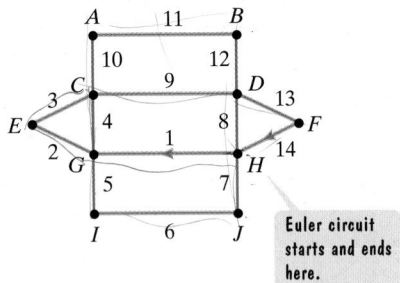

FIGURE 15.28 An Euler circuit: $H, G, E, C, G, I, J, H, D, C, A, B, D, F, H$

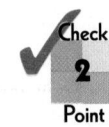 **Check 2 Point**

Refer to the graph in Figure 15.26. Use trial and error to find an Euler circuit that starts and ends at G. Number the edges of the graph to indicate the circuit. Then use vertex letters separated by commas to name the circuit.

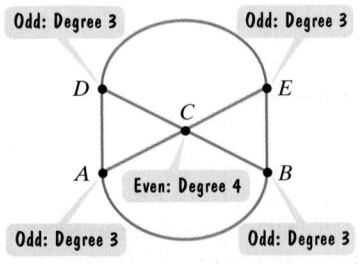

FIGURE 15.29

Figure 15.29 shows a graph with one even vertex and four odd vertices. Because the graph has more than two odd vertices, by the third statement in Euler's Theorem, it has no Euler paths and no Euler circuits.

Applications of Euler's Theorem

We can use Euler's Theorem to solve problems involving situations modeled by graphs. For example, we return to the graph that models the layout of Königsberg, shown in Figure 15.30(c). As we mentioned in Chapter 10, people in the city were interested in finding if it were possible to walk through the city so as to cross each bridge exactly once. Translated into the language of graph theory, we are interested in whether the graph in Figure 15.30(c) has an Euler path. Count the number of odd vertices. The graph has four odd vertices. Because it has more than two odd vertices, it has no Euler paths. Thus, there is no possible way anyone in Königsberg can walk across all of the bridges without having to recross some of the them.

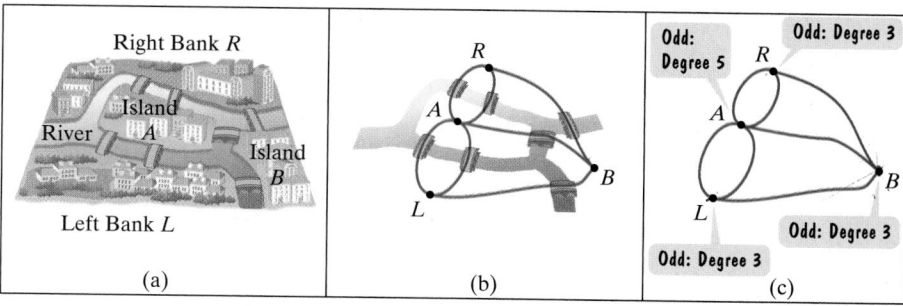

FIGURE 15.30 Revisiting the graph that models the layout of Königsberg

EXAMPLE 3 | APPLYING EULER'S THEOREM

Figure 15.31(c) shows the graph we developed to model the connecting relationships in a floor plan. Recall from our work in Section 15.1 that $A, B, C,$ and D represent the rooms and E represents the outside of the house. The edges represent the connecting doors.

a. Is it possible to find a path that uses each door exactly once?

b. If it is possible, use trial and error to show such a path on the graph in Figure 15.31(c) and on the floor plan in Figure 15.31(a).

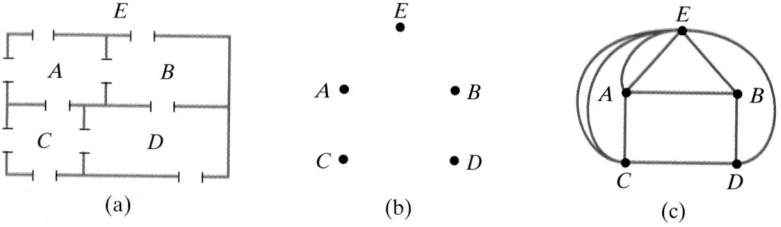

FIGURE 15.31 Revisiting the graph that models the layout of a floor plan

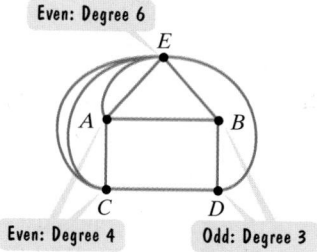

FIGURE 15.32 The graph has two odd vertices.

Solution A path that uses each door (or edge) exactly once means that we are looking for an Euler path or an Euler circuit on the graph in Figure 15.31(c). Figure 15.32 indicates that there are exactly two odd vertices, namely B and D. By Euler's

Theorem, the graph has at least one Euler path, but no Euler circuit. It is possible to find a path that uses each door exactly once. It is not possible to begin and end the walk in the same place.

Euler's Theorem tells us that a possible Euler path must start at one of the odd vertices and end at the other one. We will use trial and error to determine an Euler path, starting at vertex B (room B in the floor plan) and ending at vertex D (room D in the floor plan). Figure 15.33(a) shows an Euler path on the graph. Figure 15.33(b) translates the path into a walk through the rooms.

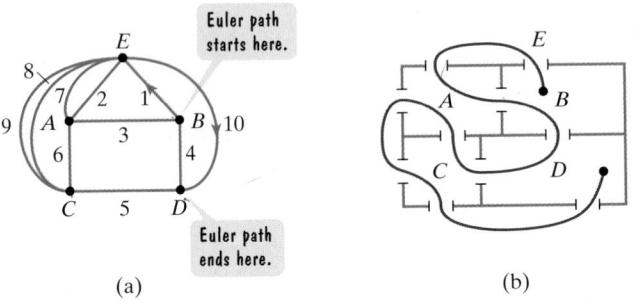

(a)   (b)

FIGURE 15.33 An Euler path:
$B, E, A, B, D, C, A, E, C, E, D$ shown on a graph and on a floor plan

Check 3 Point   The floor plan of a four-room house and a graph that models the connecting relationships in the floor plan are shown in Figure 15.34.

   a. Is it possible to find a path that uses each door exactly once?

   b. If it is possible, use trial and error to show such a path on the graph in Figure 15.34(b) and the floor plan in Figure 15.34(a).

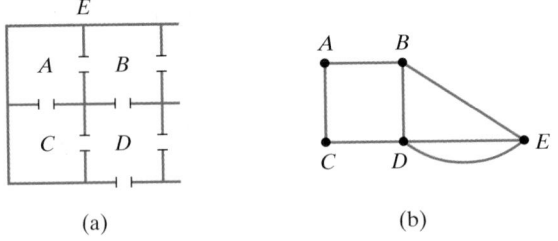

(a)   (b)

FIGURE 15.34 A floor plan and a graph that models its connecting relationships

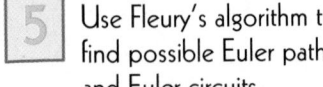

5 Use Fleury's algorithm to find possible Euler paths and Euler circuits.

Finding Possible Euler Paths and Euler Circuits without Using Trial and Error

Euler's Theorem enables us to count a graph's odd vertices and determine if it has an Euler path or an Euler circuit. Unfortunately, the theorem provides almost no assistance in finding an actual Euler path or circuit. There is, however, an algorithm, or procedure, for finding such paths and circuits, called **Fleury's Algorithm**.

FLEURY'S ALGORITHM

If Euler's Theorem indicates the existence of an Euler path or Euler circuit, one can be found using the following procedure:

1. If the graph has exactly two odd vertices (and therefore an Euler path), choose one of the two odd vertices as the starting point. If the graph has no odd vertices (and therefore an Euler circuit), choose any vertex as the starting point.

2. Number edges as you trace through the graph according to the following rules:

   After you have traveled over an edge, erase it. (This is because you must travel each edge exactly once.) Show the erased edges as dashed lines.

   When faced with a choice of edges to trace, choose an edge that is not a bridge. Travel over an edge that is a bridge only if there is no alternative.

STUDY TIP

Don't cross a bridge unless you have to!

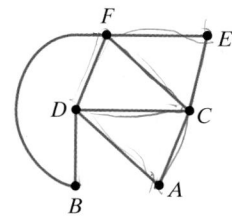

FIGURE 15.35

EXAMPLE 4 | USING FLEURY'S ALGORITHM

The graph in Figure 15.35 has at least one Euler circuit. Find one by Fleury's Algorithm.

Solution   The graph has no odd vertices, so we can begin at any vertex. We choose vertex A as the starting point and proceed as follows.

Step 1

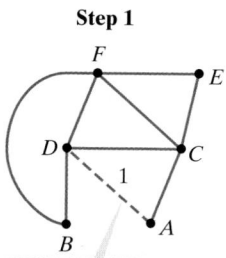

Travel from A to D and erase edge AD.

We can also travel from A to C.

Step 2

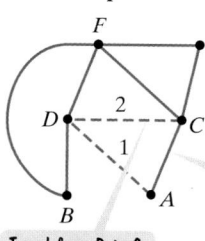

CA is a bridge. If it were removed, vertex A would be isolated from the rest of the graph.

Travel from D to C and erase edge DC.

We can also travel from D to F or B.

Step 3

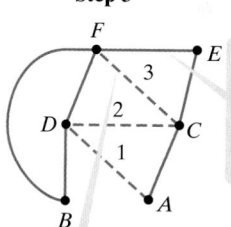

FE is a bridge. If it were removed, the graph would have two disconnected components.

Travel from C to F and erase edge CF.

We can also travel from C to E, but not from C to A. Don't cross the bridge.

Step 4

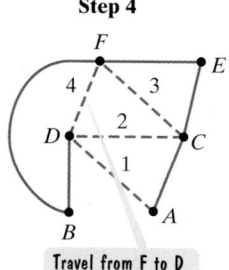

Travel from F to D and erase edge FD.

We can also travel from F to B, but not from F to E. Don't cross the bridge.

Step 5

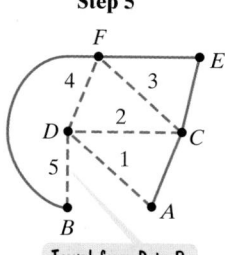

Travel from D to B and erase edge DB.

There are no other choices.

Steps 6, 7, 8, 9

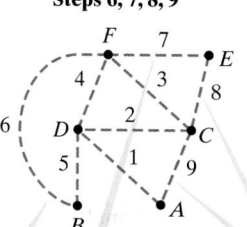

Travel from B to F, F to E, E to C, and C to A, and erase the respective edges.

There are no choices at each step.

The completed Euler circuit is shown in Figure 15.36.

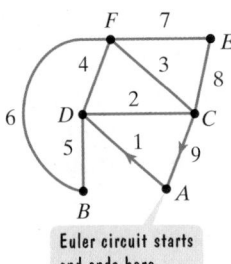

Euler circuit starts and ends here.

FIGURE 15.36 An Euler circuit:
A, D, C, F, D, B, F, E, C, A

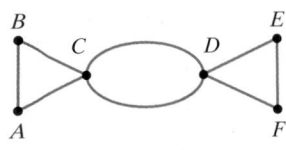

FIGURE 15.37

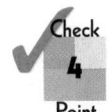 **Check 4 Point**   The graph in Figure 15.37 has at least one Euler circuit. Find one by Fleury's Algorithm.

In the exercise set, we will revisit many of the graph models from the previous section. In solving problems with these models, you can use trial and error or Fleury's Algorithm to find possible Euler paths and Euler circuits. The procedure of Fleury's Algorithm is useful as graphs become more complicated.

Here is a summary, in chart form, of Euler's Theorem that you might find helpful in the exercise set that follows:

SUMMARY OF EULER'S THEOREM

Number of Odd Vertices	Euler Paths	Euler Circuits
None (all even)	at least one	at least one
exactly two	at least one	none
more than two	none	none

Exercise Set 15.2

Practice Exercises

In Exercises 1–6, use the graph shown. In each exercise, a path along the graph is described. Trace this path with a pencil and number the edges. Then determine if the path is an Euler path, an Euler circuit, or neither. Explain your answer.

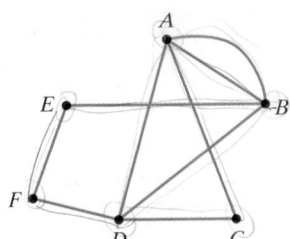

1. *F, E, B, A, C, D, A, B, D*
2. *F, E, B, A, B, D, C, A, D*
3. *F, E, B, A, C, D, A, B, D, F*
4. *F, E, B, A, B, D, C, A, D, F*

5. *A, B, A, C, D, B, E, F, D*
6. *A, B, A, C, D, F, E, B, D*

In Exercises 7–8, a graph is given.

    a. *Explain why the graph has at least one Euler path.*

    b. *Use trial and error or Fleury's Algorithm to find one such path.*

7. 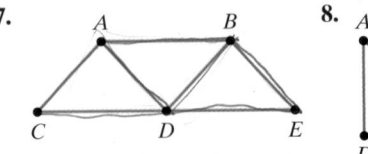   8.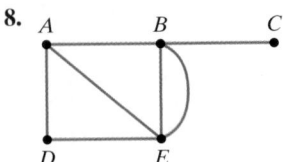

In Exercises 9–10, a graph is given.

    a. *Explain why the graph has at least one Euler circuit.*

    b. *Use trial and error or Fleury's Algorithm to find one such circuit.*

9.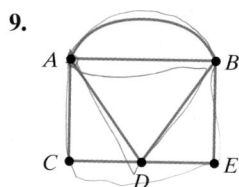

10.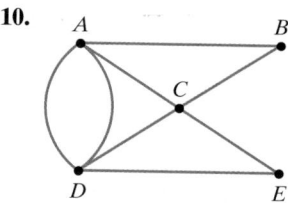

In Exercises 11–12, a graph is given. Explain why each graph has no Euler paths and no Euler circuits.

11.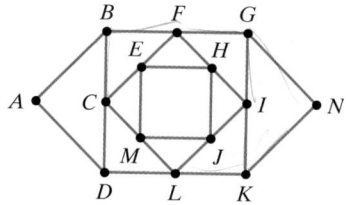

12.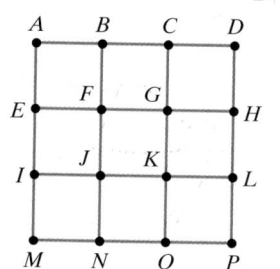

In Exercises 13–18, a connected graph is described. Determine whether the graph has an Euler path (but not an Euler circuit), an Euler circuit, or neither an Euler path nor an Euler circuit. Explain your answer.

13. The graph has 60 even vertices and no odd vertices.

14. The graph has 80 even vertices and no odd vertices.

15. The graph has 58 even vertices and two odd vertices.

16. The graph has 78 even vertices and two odd vertices.

17. The graph has 57 even vertices and four odd vertices.

18. The graph has 77 even vertices and four odd vertices.

In Exercises 19–32, a graph is given.

   a. Determine whether the graph has an Euler path, an Euler circuit, or neither.

   b. If the graph has an Euler path or circuit, use trial and error or Fleury's Algorithm to find one.

19.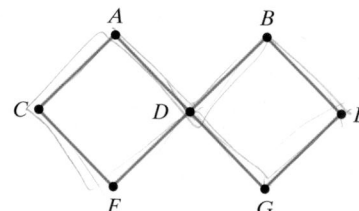

20.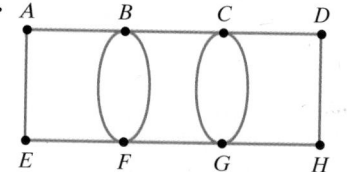

21.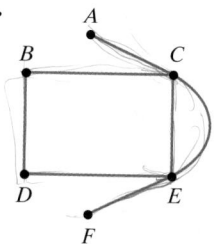

22.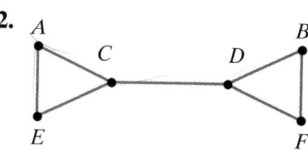

23.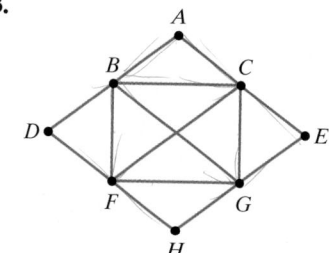

24.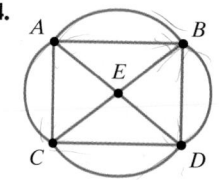

25.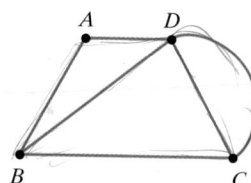

26.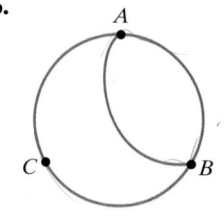

27.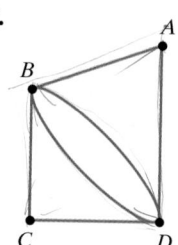

28.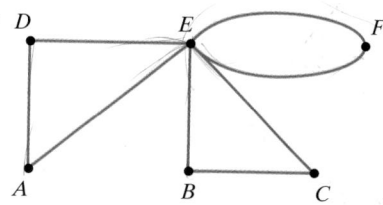

29.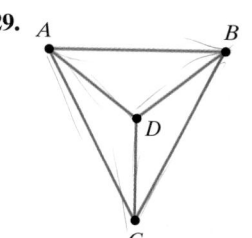

30.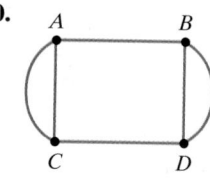

31.

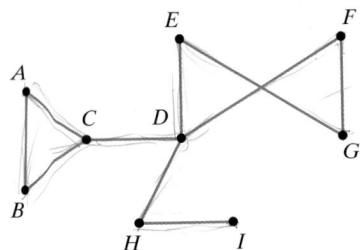

32.

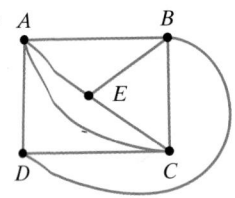

In Exercises 33–36, use Fleury's Algorithm to find an Euler path.

33.

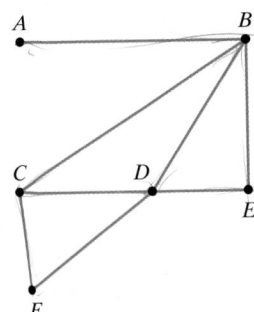

34.

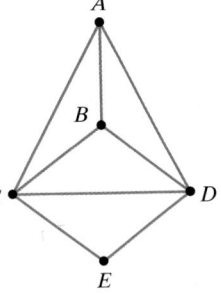

35.

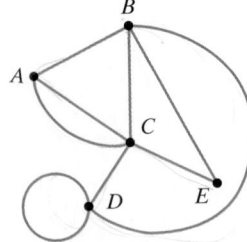

36.

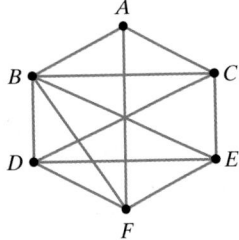

In Exercises 37–40, use Fleury's Algorithm to find an Euler circuit.

37.

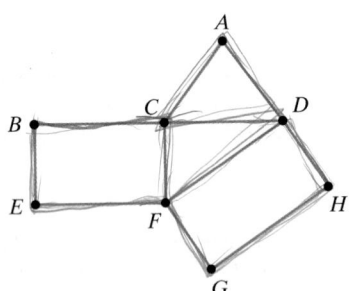

38.

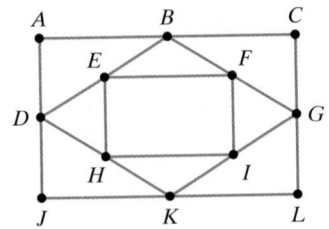

39.

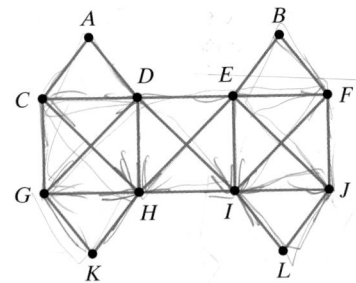

40.

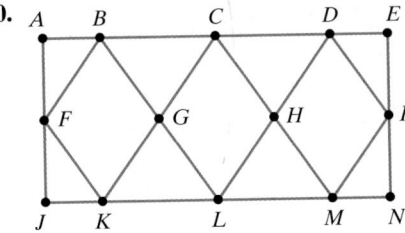

Application Exercises

In Exercises 41–44, we revisit the four-block neighborhood discussed in the previous section. Recall that a mail carrier parks her truck at the intersection shown in the figure and then walks to deliver mail to each of the houses. The streets on the outside of the neighborhood have houses on one side only. The interior streets have houses on both sides of the street. On these streets, the mail carrier must walk down the street twice, covering each side of the street separately. A graph that models the streets of the neighborhood walked by the mail carrier is shown.

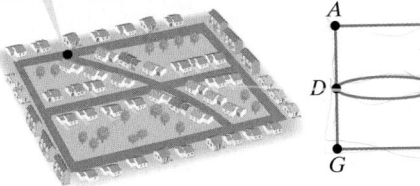

Mail carrier starts here.

41. Use Euler's Theorem to explain why it is possible for the mail carrier to park at *B*, deliver mail to each house without retracing her route, and then return to *B*.

42. Use trial and error or Fleury's Algorithm to find an Euler circuit that starts and ends at vertex *B* on the graph that models the neighborhood.

43. Use your Euler circuit in Exercise 42 to show the mail carrier the route she should follow on the map of the neighborhood. Use the method of your choice, such as red lines with arrows, but be sure that the route is clearly designated.

44. A security guard needs to walk the streets of the neighborhood. Unlike the postal worker, the guard is to walk down each street once, whether or not the street has houses on both sides. Draw a graph that models the streets of the neighborhood walked by the security guard. Then determine whether the residents in the neighborhood will be able to establish a route for the security guard so that each street is walked exactly once. If this is possible, use your map to show where the guard should begin the walk.

45. A security guard needs to walk the streets of the neighborhood shown, walking each street once.

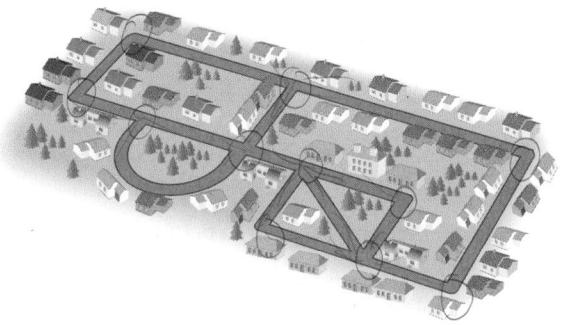

a. Draw a graph that models the streets of the neighborhood walked by the security guard.

b. Determine whether the residents in the neighborhood will be able to establish a route for the security guard so that each street is walked exactly once. If this is possible, use your map to show where the guard should begin the walk.

46. A mail carrier needs to deliver mail to each house in the three-block neighborhood shown. He plans to park at one of the street intersections and walk to deliver mail. All streets have houses on both sides. This means that the mail carrier must walk down every street twice, delivering mail to each side separately.

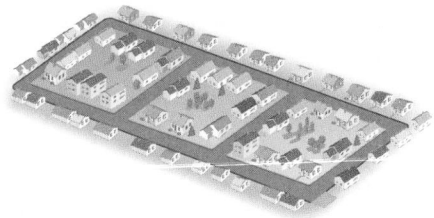

a. Draw a graph that models the streets of the neighborhood walked by the mail carrier.

b. Use the graph to determine if the carrier can park at an intersection, deliver mail to each house without retracing the side of any street, and return to the parked truck.

c. If such a walk is possible, show the path on your graph in part (a). Then trace this route on the neighborhood map in a manner that is clear to the mail carrier.

In Exercises 47–48, the layout of a city with land masses and bridges is shown.

a. *Draw a graph that models the layout of the city. Use vertices to represent the land masses and edges to represent the bridges.*

b. *Use the graph to determine if the city residents would be able to walk across all of the bridges without crossing the same bridge twice.*

c. *If such a walk is possible, show the path on your graph in part (a). Then trace this route on the city map in a manner that is clear to the city's residents.*

47.

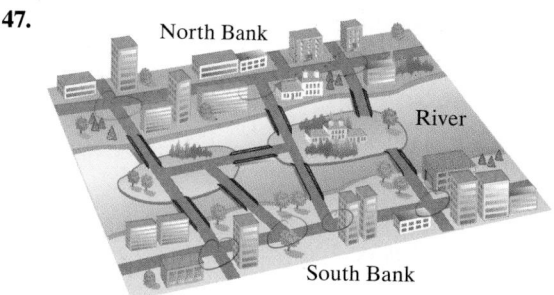

48.

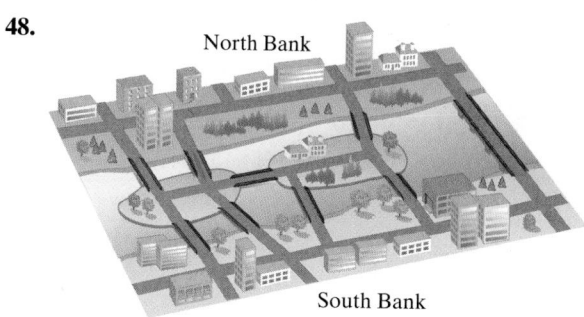

49. The figure shows a map of a portion of New York City with the bridge and tunnel connections. Use a graph to determine if it is possible to visit Manhattan, Long Island, Staten Island, and New Jersey using each bridge or tunnel only once.

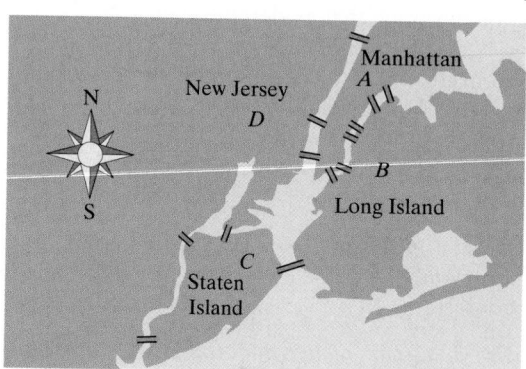

In Exercises 50–51, a floor plan is shown.

a. *Draw a graph that models the connecting relationships in each floor plan. Use vertices to represent the rooms and the outside, and edges to represent the connecting doors.*

b. *Use your graph to determine if it is possible to find a path that uses each door only once.*

c. *If such a path is possible, show it on your graph in part (a). Then trace this route on the floor plan in a manner that is clear to a person strolling through the house.*

50.   **51.**

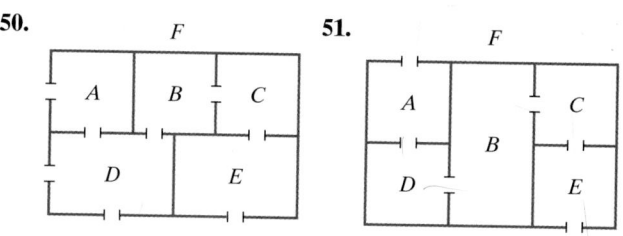

In Exercises 52–54, we revisit the map of the New England states discussed in the previous section. Shown in the figure are the map and the graph that models the bordering relationships among the six states.

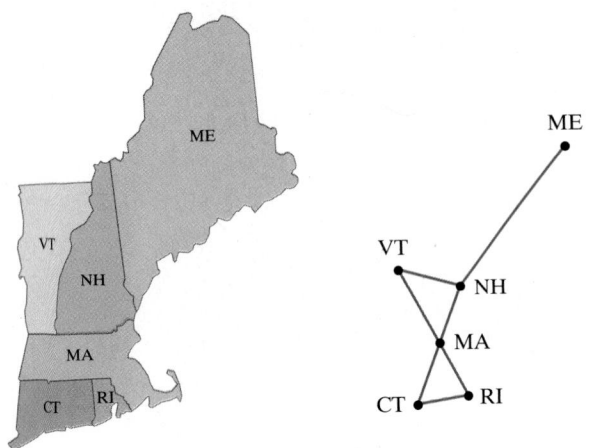

52. Use Euler's Theorem to explain why it is possible to find a path that crosses each common state border exactly once.

53. Use trial and error or Fleury's Algorithm to find an Euler path on the graph that models the map.

54. Use your Euler path in Exercise 53 to show the route on the map of the New England states.

In Exercises 55–56, a map is shown.

   a. *Draw a graph that models each map. Use vertices to represent the states and edges to represent common borders.*

   b. *Use your graph to determine if it is possible to find a path that crosses each common state border exactly once.*

   c. *If such a trip is possible, show the path on your graph in part (a). Then trace this route on the map in a manner that is clear to people driving through these states.*

   d. *Determine if it is possible to find a path that crosses each common state border exactly once, starting and ending the trip in the same state.*

55.

56.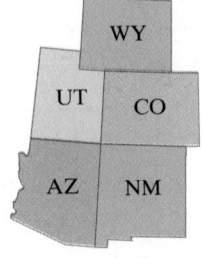

Writing in Mathematics

57. What is an Euler path?

58. What is an Euler circuit?

59. How do you determine if a graph has at least one Euler path, but no Euler circuit?

60. How do you determine if a graph has at least one Euler circuit?

61. How do you determine if a graph has no Euler paths and no Euler circuits?

62. What is the purpose of Fleury's Algorithm?

63. If a graph has at least one Euler path, but no Euler circuit, which vertex should be chosen as the starting point for a path?

64. Explain why it is important that the director of municipal services (police patrols, garbage collection, curb sweeping, snow removal) of a large city have a knowledge of graph theory.

Critical Thinking Exercises

65. Refer to the layout of Königsberg in Figure 15.30 on page 793. Suppose that the citizens of the city decide to build a new bridge. Where should it be located to make it possible to walk through the city and cross each bridge exactly once? If such a location is not possible, so state and explain why.

66. A police car would like to patrol each street in the neighborhood shown on the map exactly once. Use a numbering system from 1 through 31, one number per street, to show the police how this can be done.

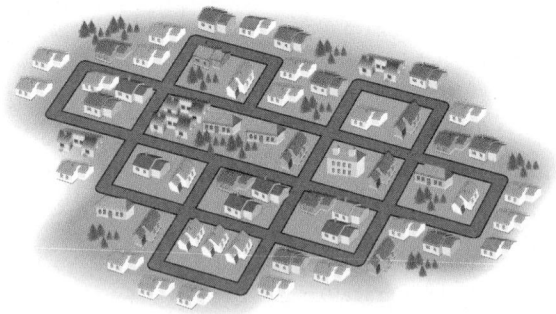

67. Use Fleury's algorithm to find an Euler circuit.

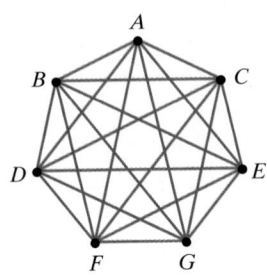

SECTION 15.3   HAMILTON PATHS ~~AND~~ HAMILTON CIRCUITS

OBJECTIVES

1. **Understand the definitions of Hamilton paths and Hamilton circuits.**
2. **Find the number of Hamilton circuits in a complete graph.**
3. **Understand and use weighted graphs.**
4. **Use the Brute Force Method to solve traveling salesperson problems.**
5. **Use the Nearest Neighbor Method to approximate solutions to traveling salesperson problems.**

In the last section, we studied paths and circuits that covered every edge of a graph. Using every edge of a graph exactly once is helpful for garbage collection, curb sweeping, or snow removal. But what about a United Parcel Service (UPS) driver trying to find the best way to deliver packages around town? We can model the delivery locations as the vertices on a graph and the roads between the locations as the edges. The driver is only concerned with finding a path that passes through each vertex of the graph once. It is not important to use every road, or edge, between delivery locations. In this section, we focus on paths and circuits that contain each vertex of a graph exactly once.

Hamilton Paths and Hamilton Circuits

 Understand the definitions of Hamilton paths and Hamilton circuits.

Trying to find the best way to deliver packages around town or planning the best route by which to run a series of errands can be modeled and solved using paths and circuits that pass through each vertex of a graph exactly once. These kinds of paths and circuits are named after the Irish mathematician William Rowan Hamilton (1805–1865).

= vertex

> ### HAMILTON PATHS AND CIRCUITS
>
> A path that passes through each vertex of a graph exactly once is called a **Hamilton path**. If a Hamilton path begins and ends at the same vertex and passes through all other vertices exactly once, it is called a **Hamilton circuit**.

EXAMPLE 1   EXAMPLES OF HAMILTON PATHS AND HAMILTON CIRCUITS

a. Find a Hamilton path for the graph in Figure 15.38.

b. Find a Hamilton circuit for the graph in Figure 15.38.

Solution

a. A Hamilton path must pass through each vertex exactly once. The graph has many Hamilton paths. An example of such a path is

$$A, B, C, D, E.$$

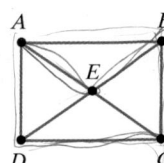

FIGURE 15.38

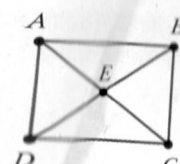

FIGURE 15.38, repeated

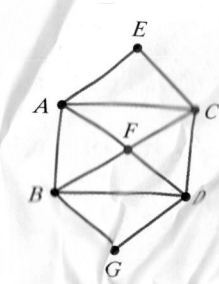

FIGURE 15.39

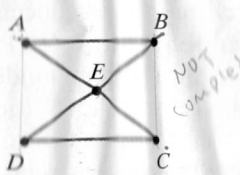

NOT
Complete

FIGURE 15.40

2 Find the number of Hamilton circuits in a complete graph.

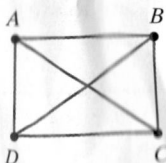

FIGURE 15.41

...uit must pass through every vertex exactly once and begin and

b. A Hamilte vertex. The graph has many Hamilton circuits. An example of end at it is
such a

$$A, B, C, D, E, A.$$

...raph in Figure 15.38 has many Hamilton paths and circuits. However, it has four vertices of odd degree, it has no Euler paths and no Euler becc... When it comes to Hamilton circuits and Euler circuits, a graph can have one circ... other, or both, or neither.

Check
1
Point

a. Find a Hamilton path that begins at vertex E for the graph in Figure 15.39.

b. Find a Hamilton circuit that begins at vertex E for the graph in Figure 15.39.

Consider the graph in Figure 15.40. It has many Hamilton paths, such as

$$A, B, E, C, D \quad \text{or} \quad C, D, E, A, B$$

However, it has no Hamilton circuits. Regardless of the starting point, it is necessary to pass through vertex E more than once to return to the starting point.

A **complete graph** is a graph that has an edge between each pair of its vertices. Can you see that the graph in Figure 15.40 is not complete? There is no edge between vertex A and vertex D. Nor is there an edge between vertex B and vertex C. Complete graphs are significant because **every complete graph with three or more vertices has a Hamilton circuit.** The graph in Figure 15.40 is not complete and has no Hamilton circuits.

The Number of Hamilton Circuits in a Complete Graph

The graph in Figure 15.41 has an edge between each pair of its four vertices. Thus, the graph is complete and has a Hamilton circuit. Actually, it has a complete repertoire of Hamilton circuits. For example, one Hamilton circuit is

$$A, B, C, D, A.$$

Any two circuits that pass through the same vertices in the same order will be considered to be the same. For example, here are four different sequences of letters that produce the same Hamilton circuit on the graph in Figure 15.41:

$$A, B, C, D, A \text{ and } B, C, D, A, B \text{ and } C, D, A, B, C \text{ and } D, A, B, C, D$$

> The Hamilton circuit passing through A, B, C, and D clockwise along the four outside edges can be written in four ways.

In order to avoid this duplication, in forming a Hamilton circuit, we can always assume that it begins at A.

There are six different, or unique, Hamilton circuits for the graph with four vertices in Figure 15.41. They are given as follows:

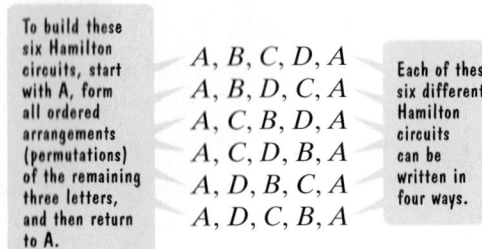

To build these six Hamilton circuits, start with A, form all ordered arrangements (permutations) of the remaining three letters, and then return to A.

A, B, C, D, A
A, B, D, C, A
A, C, B, D, A
A, C, D, B, A
A, D, B, C, A
A, D, C, B, A

Each of these six different Hamilton circuits can be written in four ways.

How many different Hamilton circuits are there in a complete graph? For a complete graph with four vertices, the preceding list indicates that the number of circuits depends upon the number of permutations of the three letters B, C, and D. If a graph has n vertices, once we list vertex A, there are $n - 1$ remaining letters. The number of Hamilton circuits depends upon the number of permutations of the $n - 1$ letters.

Recall from our work in Chapter 11 that factorial notation is useful in determining the number of ordered arrangements, or permutations. If n is a positive integer, then $n!$ (n factorial) is the product of all positive integers from n down to 1. For example, $4! = 4 \cdot 3 \cdot 2 \cdot 1 = 24$ and $6! = 6 \cdot 5 \cdot 4 \cdot 3 \cdot 2 \cdot 1 = 720$.

We can find the number of ordered arrangements of $n - 1$ letters using the Fundamental Counting Principle, discussed in Section 11.1. With $n - 1$ letters to arrange, there are $n - 1$ choices for first position. Once you've chosen the first letter, you'll have $n - 2$ choices for second position. Continuing in this manner, there are $n - 3$ choices for third position, $n - 4$ choices for fourth position, down to only one choice (the one remaining letter) for last position. We multiply the number of choices to find the total number of ordered arrangements:

$$(n - 1)(n - 2)(n - 3) \cdot \ldots \cdot 1.$$

This product, by definition, is $(n - 1)!$. This expression therefore describes the number of Hamilton circuits in a complete graph with n vertices.

THE NUMBER OF HAMILTON CIRCUITS IN A COMPLETE GRAPH

The number of Hamilton circuits in a complete graph with n vertices is

$$(n - 1)!.$$

EXAMPLE 2 DETERMINING THE NUMBER OF HAMILTON CIRCUITS

Determine the number of Hamilton circuits in a complete graph with

a. four vertices.   **b.** five vertices.   **c.** eight vertices.

Solution In each case, we use the expression $(n - 1)!$. For four vertices, substitute 4 for n in the expression. For five and eight vertices, substitute 5 and 8, respectively, for n.

a. A complete graph with four vertices has $(4 - 1)! = 3! = 3 \cdot 2 \cdot 1 = 6$ Hamilton circuits. These are the six circuits that we listed on page 802.

b. A complete graph with five vertices has

$$(5 - 1)! = 4! = 4 \cdot 3 \cdot 2 \cdot 1 = 24 \text{ Hamilton circuits.}$$

c. A complete graph with eight vertices has

$$(8 - 1)! = 7! = 7 \cdot 6 \cdot 5 \cdot 4 \cdot 3 \cdot 2 \cdot 1 = 5040 \text{ Hamilton circuits.}$$

As the number of vertices in a complete graph increases, notice how quickly the number of Hamilton circuits increases.

Check
2
Point

Determine the number of Hamilton circuits in a complete graph with

a. three vertices.   **b.** six vertices.   **c.** ten vertices.

3 Understand and use weighted graphs.

Weighted Graphs and the Traveling Salesperson Problem

Sales directors for large companies are often required to visit regional offices in a number of different cities. How can these visits be scheduled in the cheapest possible way?

For example, a sales director who lives in city A is required to fly to regional offices in cities B, C, and D. Other than starting and ending the trip in city A, there are no restrictions as to the order in which the other three cities are visited. The one-way fares between each of the four cities are given in Table 15.1. A graph that models this information is shown in Figure 15.42. The vertices represent the cities. The airfare between each pair of cities is shown as a number on the respective edge.

TABLE 15.1 ONE-WAY AIRFARES

	A	B	C	D
A	*	$190	$124	$157
B	$190	*	$126	$155
C	$124	$126	*	$179
D	$157	$155	$179	*

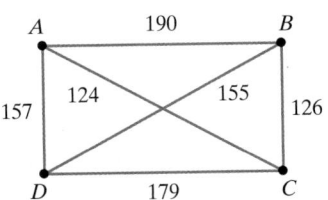

FIGURE 15.42 Modeling Table 15.1 with a graph

A complete graph whose edges have numbers attached to them is called a **weighted graph**. The numbers shown along the edges of a weighted graph are called the **weights** of the edges. Figure 15.42 is an example of a weighted graph. The weight of edge AB is 190, modeling a $190 airfare from city A to city B. The sales director needs to find the least expensive way to visit cities B, C, and D once, and return home to A. Our goal is to find the Hamilton circuit with the lowest associated cost.

EXAMPLE 3 UNDERSTANDING THE INFORMATION IN A WEIGHTED GRAPH

Use the weighted graph in Figure 15.42 to find the cost of the trip for the Hamilton circuit A, B, D, C, A.

Solution The trip described by the Hamilton circuit A, B, D, C, A involves the sum of four costs:

$$\$190 \ + \ \$155 \ + \ \$179 \ + \ \$124 \ = \ \$648.$$

This is the cost from A to B in A, B, D, C, A.

This is the cost from B to D in A, B, D, C, A.

This is the cost from D to C in A, B, D, C, A.

This is the cost from C to A in A, B, D, C, A.

The cost of the trip is $648.

Check 3 Point

Use the weighted graph in Figure 15.42 to find the cost of the trip for the Hamilton circuit A, C, B, D, A.

Use the Brute Force
Method to solve traveling
salesperson problems.

The traveling salesperson problem can be stated as follows:

THE TRAVELING SALESPERSON PROBLEM

The **traveling salesperson problem** is the problem of finding a Hamilton circuit
in a complete weighted graph for which the sum of the weights of the edges is a
minimum. Such a Hamilton circuit is called the **optimal Hamilton circuit** or the
optimal solution.

One method for finding an optimal Hamilton circuit is called the **Brute
Force Method**.

**THE BRUTE FORCE METHOD OF SOLVING TRAVELING SALESPERSON
PROBLEMS**

The optimal solution is found using the following steps:

1. Model the problem with a complete, weighted graph.

2. Make a list of all possible Hamilton circuits.

3. Determine the sum of the weights of the edges for each of these Hamilton
   circuits.

4. The Hamilton circuit with the minimum sum of weights is the optimal
   solution.

EXAMPLE 4   USING THE BRUTE FORCE METHOD

Use the weighted graph in Figure 15.42, repeated in the margin, to find the optimal
solution. Describe what this means for the sales director who starts at A, flies once
to each of B, C, and D, and returns home to A.

Solution The graph has four vertices. Thus, using $(n - 1)!$, there are
$(4 - 1)! = 3! = 6$ possible Hamilton circuits. The six possible Hamilton circuits
and their total costs are shown in Table 15.2.

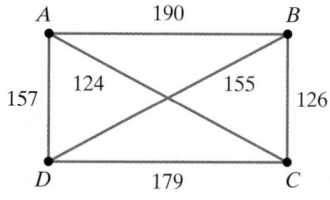

FIGURE 15.42 , repeated

TABLE 15.2 POSSIBLE HAMILTON CIRCUITS AND THEIR TOTAL COSTS

Hamilton Circuit	Sum of the Weights of the Edges	=	Total Cost
A, B, C, D, A	$190 + 126 + 179 + 157$	=	\$652
A, B, D, C, A	$190 + 155 + 179 + 124$	=	\$648
A, C, B, D, A	$124 + 126 + 155 + 157$	=	\$562
A, C, D, B, A	$124 + 179 + 155 + 190$	=	\$648
A, D, B, C, A	$157 + 155 + 126 + 124$	=	\$562
A, D, C, B, A	$157 + 179 + 126 + 190$	=	\$652

Minimum
sums

The voice balloon in the last column indicates that two Hamilton circuits have the
minimum cost of \$562. The optimal solution is either

$$A, C, B, D, A \quad \text{or} \quad A, D, B, C, A.$$

For the sales director, this means that either route shown in Figure 15.43 is the least expensive way to visit the regional offices in cities B, C, and D. Notice that the route in Figure 15.43(b) involves visiting the cities in the reverse order of the route in Figure 15.43(a). Although these are different Hamilton circuits, because the one-way airfares are the same in either direction, the cost is the same regardless of the direction flown.

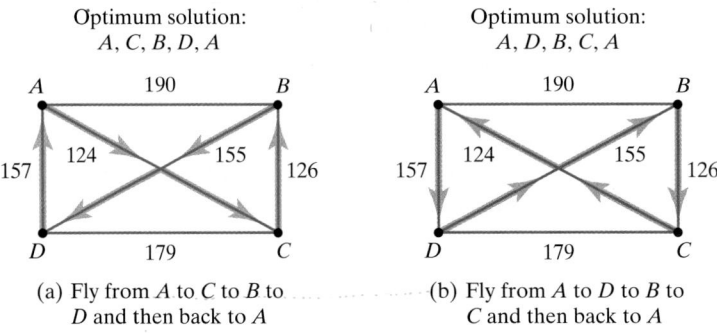

(a) Fly from A to C to B to D and then back to A

(b) Fly from A to D to B to C and then back to A

FIGURE 15.43 Options for an optimal solution

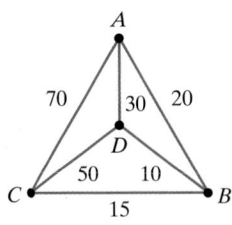

FIGURE 15.44

Check 4 Point

Use the Brute Force Method to find the optimal solution for the complete, weighted graph in Figure 15.44. List Hamilton circuits as in Table 15.2 on page 805.

As the number of vertices increases, the rapidly increasing number of possible Hamilton circuits makes the Brute Force Method impractical. Unfortunately, mathematicians have not been able to establish another method for solving the traveling salesperson problem, and they do not know whether it is even possible to find such a method. There are, however, a number of methods for finding approximate solutions.

BLITZER BONUS

The Brute Force Method and Supercomputers

Suppose that a supercomputer can find the sum of the weights of one billion, or 10^9, Hamilton circuits per second. Because there are 31,536,000 seconds in a year, the computer can calculate the sums for approximately 3.2×10^{16} Hamilton circuits in one year. The table shows that as the number of vertices increases, the Brute Force Method is useless even with a powerful computer.

COMPUTER TIME NEEDED TO SOLVE THE TRAVELING SALESPERSON PROBLEM

Number of Vertices	Number of Hamilton Circuits	Time Needed by a Supercomputer to Find Sums of All Hamilton Circuits
18	$17! \approx 3.6 \times 10^{14}$	≈ 0.01 year ≈ 3.7 days
19	$18! \approx 6.4 \times 10^{15}$	≈ 0.2 year ≈ 73 days
20	$19! \approx 1.2 \times 10^{17}$	≈ 3.8 years
21	$20! \approx 2.4 \times 10^{18}$	≈ 76 years
22	$21! \approx 5.1 \times 10^{19}$	≈ 1597 years
23	$22! \approx 1.1 \times 10^{21}$	$\approx 35,125$ years

5 Use the Nearest Neighbor Method to approximate solutions to traveling salesperson problems.

Suppose a sales director who lives in city A is required to fly to regional offices in ten other cities and then return home to city A. With $(11 - 1)!$, or 3,628,800, possible Hamilton circuits, a list is out of the question. What do you think of this option? Start at city A. From there, fly to the city to which the airfare is cheapest. Then from there fly to the next city to which the airfare is cheapest, and so on. From the last of the ten cities, fly home to city A.

By continually taking an edge with the smallest weight, we can find approximate solutions to traveling salesperson problems. This method is called the **Nearest Neighbor Method**.

THE NEAREST NEIGHBOR METHOD OF FINDING APPROXIMATE SOLUTIONS TO TRAVELING SALESPERSON PROBLEMS

The optimal solution can be approximated using the following steps:

1. Model the problem with a complete, weighted graph.
2. Identify the vertex that serves as the starting point.
3. From the starting point, choose the edge with the smallest weight. Move along this edge to the second vertex. (If there is more than one edge with the smallest weight, choose either one.)
4. From the second vertex, choose the edge with the smallest weight that does not lead to a vertex already visited. Move along this edge to the third vertex.
5. Continue building the circuit, one vertex at a time, by moving along the edge with the smallest weight until all vertices are visited.
6. From the last vertex, return to the starting point.

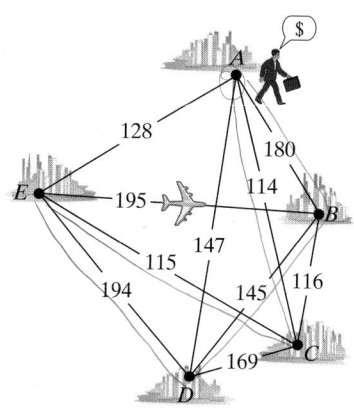

FIGURE 15.45

EXAMPLE 5 USING THE NEAREST NEIGHBOR METHOD

A sales director who lives in city A is required to fly to regional offices in cities B, C, D, and E. The weighted graph showing the one-way airfares is given in Figure 15.45. Use the Nearest Neighbor Method to find an approximate solution. What is the total cost?

Solution The Nearest Neighbor Method is carried out as follows:

- Start at A.
- Choose the edge with the smallest weight: 114. Move along this edge to C. (cost: $114)
- From C choose the edge with the smallest weight that does not lead to A: 115. Move along this edge to E. (cost: $115)
- From E, choose the edge with the smallest weight that does not lead to a city already visited: 194. Move along this edge to D. (cost: $194)
- From D, there is little choice but to fly to B, the only city not yet visited. (cost: $145)
- From B, close the circuit and return home to A. (cost: $180)

An approximate solution is the Hamilton circuit

$$A, C, E, D, B, A.$$

The circuit is shown in Figure 15.46. The total cost is

$$\$114 + \$115 + \$194 + \$145 + \$180 = \$748.$$

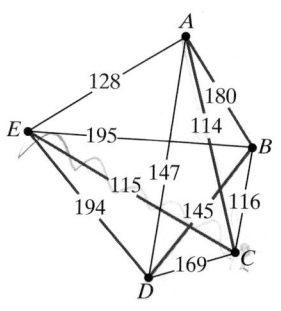

FIGURE 15.46 An approximate solution: A, C, E, D, B, A

The Brute Force Method can be applied to the 24 possible Hamilton circuits in Example 5. How does the actual solution compare with the $748 total cost obtained by the Nearest Neighbor Method? The actual solution is *A, D, B, C, E, A* or the reverse order *A, E, C, B, D, A*, and the cost is $651. This shows that the Nearest Neighbor Method does not always give the optimal solution. Of the 24 Hamilton circuits, eight result in a total cost that is less than the $748 we obtained in Example 5.

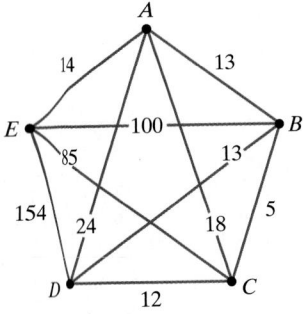

FIGURE 15.47

Check 5 Point Use the Nearest Neighbor Method to approximate the optimal solution for the complete, weighted graph in Figure 15.47. Begin the circuit at vertex *A*. What is the total weight of the resulting Hamilton circuit?

BLITZER BONUS

A Hamilton Circuit for Visiting State Capitals

What is the shortest Hamilton circuit that connects all of the state capitals in the 48 contiguous states? The optimal Hamilton circuit, shown in red, is approximately 12,000 miles long.

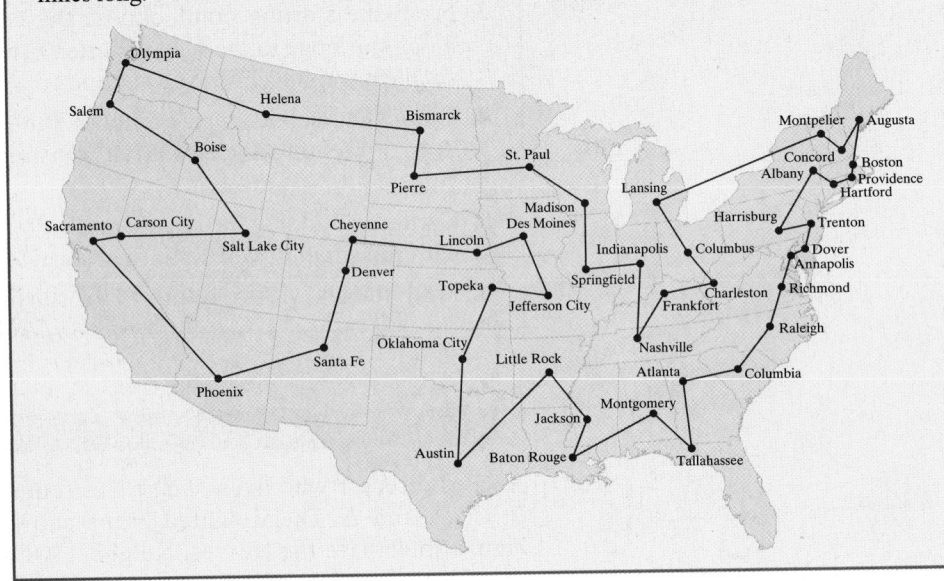

Exercise Set 15.3

Practice Exercises

In Exercises 1–4, use the graph shown.

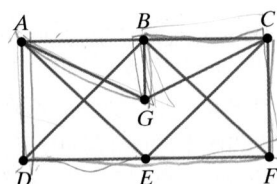

1. Find a Hamilton path that begins at *A* and ends at *B*.

2. Find a Hamilton path that begins at *G* and ends at *E*.

3. Find a Hamilton circuit that begins as *A, B,....*

4. Find a Hamilton circuit that begins as *A, G,....*

In Exercises 5–8, use the graph shown.

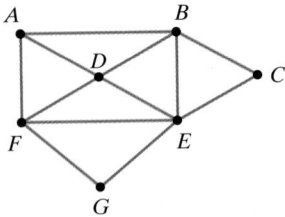

5. Find a Hamilton path that begins at *A* and ends at *D*.

6. Find a Hamilton path that begins at *A* and ends at *G*.

7. Find a Hamilton circuit that begins at *A* and ends with the pair of vertices *D, A*.

8. Find a Hamilton circuit that begins at *F* and ends with the pair of vertices *D, F*.

For each graph in Exercises 9–14,

  a. *Determine if the graph must have Hamilton circuits. Explain your answer.*

  b. *If the graph must have Hamilton circuits, determine the number of such circuits.*

9.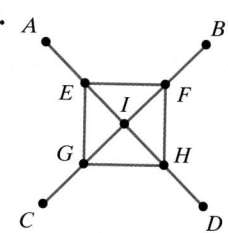

10.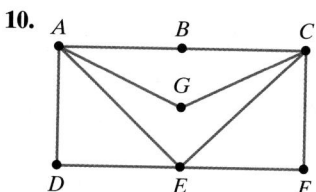

11.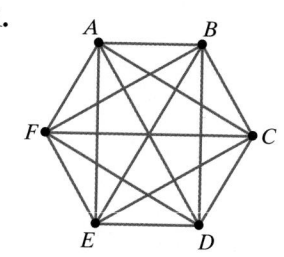

12.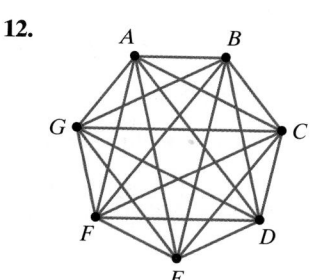

13.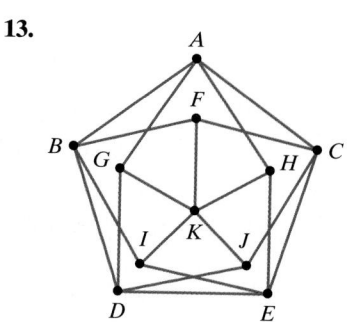

14.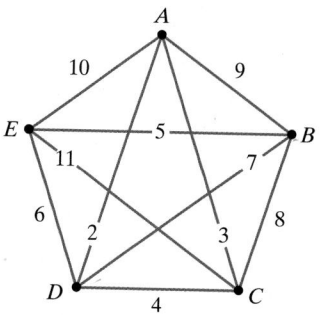

In Exercises 15–18, determine the number of Hamilton circuits in a complete graph with the given number of vertices.

15. 3 $(3-1)!$ **16.** 4   **17.** 12 $(12-1)!$ **18.** 13

$2! = 2 \cdot 1 = 2$ $11!$

In Exercises 19–24, use the complete, weighted graph shown.

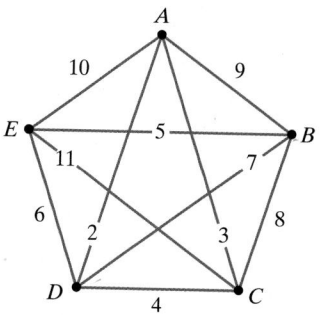

19. Find the weight of edge *CE*.   11

20. Find the weight of edge *BD*.   7

21. Find the total weight of the Hamilton circuit

     A, B, C, E, D, A. = 36

22. Find the total weight of the Hamilton circuit

     A, B, D, C, E, A. = 41

23. Find the total weight of the Hamilton circuit

     A, B, D, E, C, A. = 36

24. Find the total weight of the Hamilton circuit

     A, B, E, C, D, A.

In Exercises 25–34, use the complete, weighted graph shown.

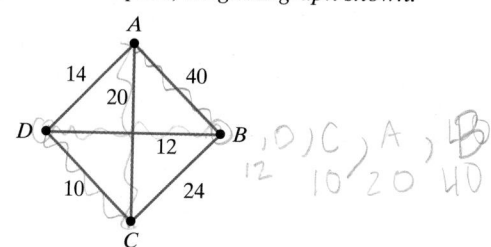

25. Find the total weight of the Hamilton circuit *A, B, C, D, A.* 88

26. Find the total weight of the Hamilton circuit *A, B, D, C, A.*

27. Find the total weight of the Hamilton circuit *A, C, B, D, A.* 70

28. Find the total weight of the Hamilton circuit *A, C, D, B, A.*

29. Find the total weight of the Hamilton circuit *A, D, B, C, A.* 70

30. Find the total weight of the Hamilton circuit *A, D, C, B, A.*

31. Use your answers from Exercises 25–30 and the Brute Force Method to find the optimal solution.

32. Use the Nearest Neighbor Method, with starting vertex A, to find an approximate solution. What is the total weight of the Hamilton circuit?

33. Use the Nearest Neighbor Method, with starting vertex B, to find an approximate solution. What is the total weight of the Hamilton circuit?

34. Use the Nearest Neighbor Method, with starting vertex C, to find an approximate solution. What is the total weight of the Hamilton circuit?

Application Exercises

In Exercises 35–36, a sales director who lives in city A is required to fly to regional offices in cities B, C, D, and E. The weighted graph shows the one-way airfares between any two cities.

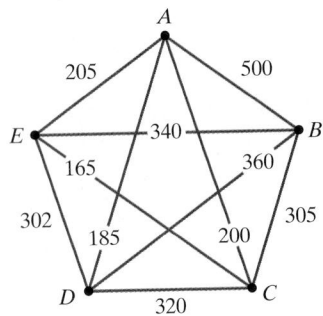

35. Use the Brute Force Method to find the optimal solution. Describe what this means for the sales director. (*Hint:* Because the airfares are the same in either direction, you need only compute the total cost for 12 Hamilton circuits:

A, B, C, D, E, A; 1632 A, B, C, E, D, A; 1457
A, B, D, C, E, A; 1550 A, B, D, E, C, A; 1527
A, B, E, C, D, A; 1510 A, B, E, D, C, A; 1662
A, C, B, D, E, A; 1372 A, C, B, E, D, A; 1332
A, C, D, B, E, A; 1425 A, C, E, B, D, A; 1250
A, D, B, C, E, A; 1220 A, D, C, B, E, A.) 1355

36. Use the Nearest Neighbor Method, with starting vertex A, to find an approximate solution. What is the total cost for this Hamilton circuit?

You have five errands to run around town, in no particular order. You plan to start and end at home. You must go to the post office, deposit a check at the bank, drop off dry cleaning, visit a friend at the hospital, and get a flu shot. The map shows your home and the locations of your five errands. Each block represents one mile. Also shown is a weighted graph with distances on the appropriate edges. In Exercises 37–40, your goal is to run the errands and return home using the shortest route.

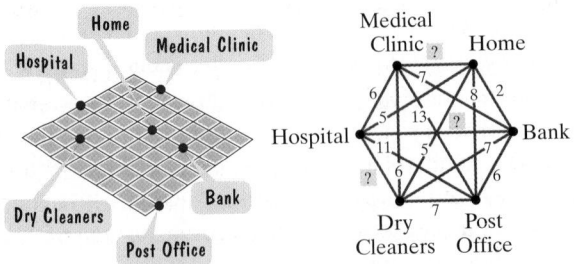

37. Use the map to fill in the three missing weights in the graph.

38. If each Hamilton circuit represents a route to run your errands, how many different routes are possible?

39. Using the Brute Force Method, the optimal solution is Home, Bank, Post Office, Dry Cleaners, Hospital, Medical Clinic, Home. What is the total length of the shortest route?

40. Use the Nearest Neighbor Method to find an approximate solution. What is the total length of the shortest route using this solution? How does this compare with your answer to Exercise 39?

In Exercises 41–43, you have three errands to run around town, although in no particular order. You plan to start and end at home. You must go to the bank, the post office, and the market. Distances, in miles, between any two of these locations are given in the table.

DISTANCES (IN MILES) BETWEEN LOCATIONS

	Home	Bank	Post Office	Market
Home	*	3	5.5	3.5
Bank	3	*	4	5
Post Office	5.5	4	*	4.5
Market	3.5	5	4.5	*

41. Create a complete, weighted graph that models the information in the table.

42. Use the Brute Force Method to find the shortest route to run your errands and return home. What is the minimum distance you can travel?

43. Use the Nearest Neighbor Method to approximate the shortest route to run your errands and return home. What is the minimum distance given by the Hamilton circuit? How does this compare with your answer to Exercise 42?

Writing in Mathematics

44. What is a Hamilton path? How does this differ from an Euler path?

45. What is a Hamilton circuit? How does this differ from an Euler circuit?

46. What is a complete graph?

47. How can you look at a graph and determine if it has a Hamilton circuit?

48. Describe how to determine the number of Hamilton circuits in a complete graph.

49. What is a weighted graph and what are the weights?

50. What is the traveling salesperson problem? What is the optimal solution?

51. Describe the Brute Force Method of solving traveling salesperson problems.

52. Why is the Brute Force Method impractical for large numbers of vertices?

53. Describe the Nearest Neighbor Method for approximating the optimal solution to traveling salesperson problems.

54. Describe a practical example of a traveling salesperson problem other than those involving one-way airfares or distances between errand locations.

55. An efficient solution for solving traveling salesperson problems has eluded mathematicians for more than 50 years. What explanations can you offer for this?

Critical Thinking Exercises

56. A complete graph has 120 distinct Hamilton circuits. How many vertices does the graph have?

57. The answer section in a math textbook states that the number of Hamilton circuits in a particular complete graph is 25. How do you know that there is an error in the answer section?

58. Ambassadors from countries $A, B, C, D, E,$ and F are to be seated around a circular conference table. Friendly relations among the various countries are described as follows: A has friendly relations with B and F. B has friendly relations with $A, C,$ and E. C has friendly relations with $B, D, E,$ and F. E has friendly relations with $B, C, D,$ and F. All friendly relations are mutual. Using vertices to represent countries and edges to represent friendly relations, draw a graph that models the information given. Then use a Hamil-

ton circuit to devise a seating arrangement around the table so that the ambassadors from B and E are seated next to each other, and each ambassador represents a country that has friendly relations with the countries represented by the ambassadors next to him or her.

Group Exercise

59. In this group exercise, you will create and solve a traveling salesperson problem similar to Exercises 35 and 36. Consult the graph given for these exercises as you work on this activity.

a. Group members should agree on four cities to be visited.

b. As shown in the graph for Exercises 35 and 36, assume that you are located at A. Let $B, C, D,$ and E represent each of the four cities you have agreed upon. Consult the Internet and use one-way airfares between cities to create a weighted graph.

c. As you did in Exercise 35, use the Brute Force Method to find the optimal solution to visiting each of your chosen cities and returning home.

d. As you did in Exercise 36, use the Nearest Neighbor Method to approximate the optimal solution. How much money does the group save using the optimal solution instead of the approximation?

SECTION 15.4   TREES

OBJECTIVES

1. Understand the definition and properties of a tree.

2. Find a spanning tree for a connected graph.

3. Find the minimum spanning tree for a weighted graph.

'I need to call the Weather Channel to let them know I just shoveled 3 feet of "partly cloudy" off my driveway. This late spring storm caught all of us by surprise. I was amazed that classes were not cancelled. Somehow the campus groundskeeper managed to shovel the minimum total length of walkways while ensuring that students could walk from building to building.

The general theme of this section is finding an efficient network linking a set of points. Think about the cleared campus sidewalks. Because of the unexpected storm, there is not enough time to shovel every sidewalk. Instead, campus services must clear a select number of sidewalks so that by moving from building to building, students can reach any location without having to trudge through mountains of snow. Finding efficient networks is accomplished using special kinds of graphs, called *trees*.

Trees

The campus groundskeeper is interested in a path that passes though each vertex (campus building) of a graph exactly once with the smallest possible number of edges (cleared sidewalks). A graph with the smallest number of edges that allows all vertices to be reached from all other vertices is called a *tree*.

Figure 15.48 shows examples of graphs that are trees.

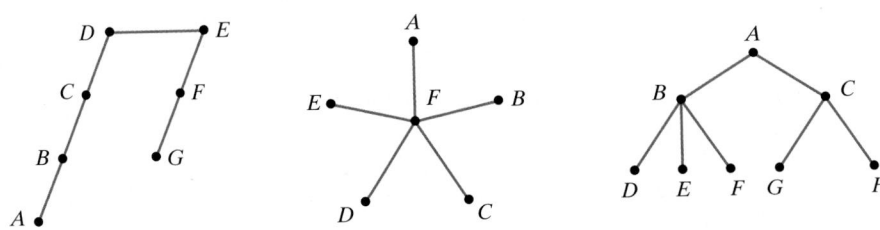

FIGURE 15.48 Examples of graphs that are trees

Notice that each graph is connected. This is a requirement because we must be able to reach any vertex from any other vertex. Furthermore, no graph contains any circuits. This is because vertices must be reached with the smallest number of edges. Thus, the graphs in Figure 15.49 are not trees. The circuits create redundant connections—take away each redundant connection and all vertices can still be reached from all other vertices.

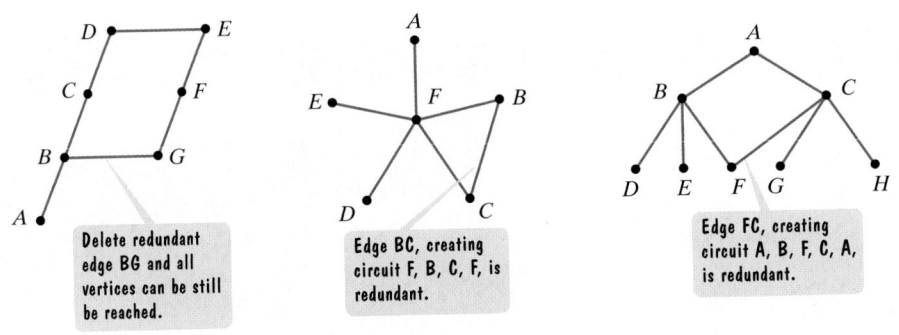

FIGURE 15.49 Examples of graphs that are not trees

Using these ideas, we are now ready to define a tree and list some of its properties.

DEFINITION AND PROPERTIES OF A TREE

A **tree** is a graph that is connected and has no circuits. All trees have the following properties:

1. There is one and only one path joining any two vertices.
2. Every edge is a bridge.
3. A tree with n vertices must have $n - 1$ edges.

Property 3 is a numerical property of trees that relates the number of vertices and the number of edges. The total number of edges is always one less than the number of vertices. For example, a tree with 5 vertices must have $5 - 1$, or 4, edges.

EXAMPLE 1 | IDENTIFYING TREES

Which graph in Figure 15.50 is a tree? Explain why the other two graphs shown are not trees.

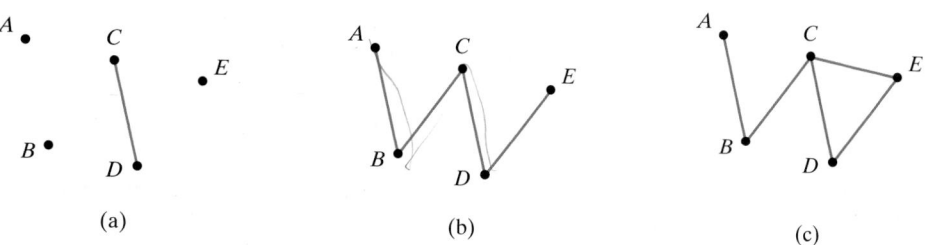

FIGURE 15.50 Identifying a tree

Solution   The graph in Figure 15.50(b) is a tree. It is connected and has no circuits. There is only one path joining any two vertices. Every edge is a bridge; if removed, each edge would create a disconnected graph. Finally, the graph has 5 vertices and $5 - 1$, or 4, edges.

The graph in Figure 15.50(a) is not a tree because it is disconnected. There are five vertices and only one edge; a tree with five vertices must have four edges.

The graph in Figure 15.50(c) is not a tree because it has a circuit, namely C, D, E, C. There are five vertices and five edges; a tree with five vertices must have four edges.

 Check 1 Point

Which graph in Figure 15.51 is a tree? Explain why the other two graphs shown are not trees.

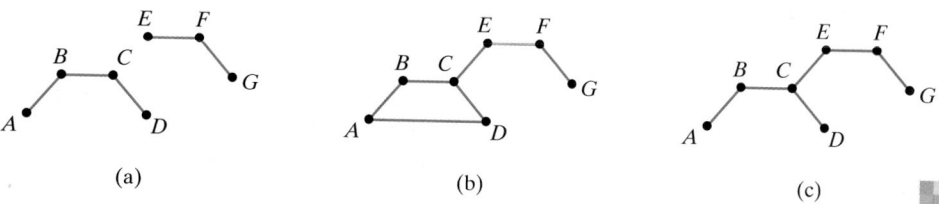

FIGURE 15.51

(a)   (b)   (c)

2 | Find a spanning tree for a connected graph.

Spanning Trees

One way to increase the efficiency of a network is to remove redundant connections. We are interested in a **subgraph**, meaning a set of vertices and edges chosen from among those of the original graph. Figure 15.52 illustrates a graph and two possible subgraphs.

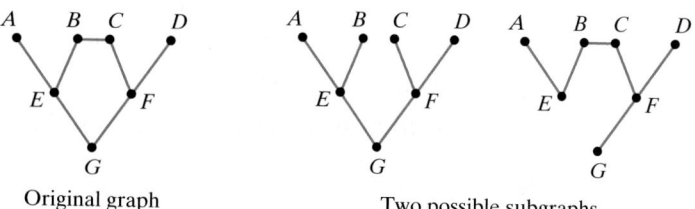

Original graph   Two possible subgraphs

FIGURE 15.52 A graph and two possible subgraphs

The original graph is a connected graph with seven vertices and seven edges. Each subgraph has all seven of the vertices of the connected graph and six of its edges. Although the original graph is not a tree, each subgraph is a tree—that is, each is connected and contains no circuits.

A subgraph that contains all of a connected graph's vertices, is connected, and contains no circuits is called a **spanning tree**. The two subgraphs in Figure 15.52 are spanning trees for the original graph. By removing redundant connections, the spanning trees increase the efficiency of the network modeled by the original graph.

It is possible to start with a connected graph, retain all of its vertices, and remove edges until a spanning tree remains. Being a tree, the spanning tree must have one less edge than it has vertices.

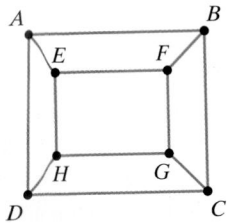

FIGURE 15.53

EXAMPLE 2   FINDING A SPANNING TREE

Find a spanning tree for the graph in Figure 15.53.

Solution   A possible spanning tree must contain all eight vertices shown in the connected graph in Figure 15.53. The spanning tree must have one less edge than it has vertices, so it must have seven edges. The graph in Figure 15.53 has 12 edges, so we need to remove five edges. We break the inner rectangular circuit by removing edge *FG*. We break the outer rectangular circuit by removing all four of its edges while retaining the edges leading to vertices *A*, *B*, *C*, and *D*. This leaves us the spanning tree in Figure 15.54. Notice that each edge is a bridge and no circuits are present.

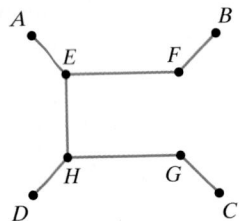

FIGURE 15.54 A spanning tree for the graph in Figure 15.53

Check
2
Point

Find a spanning tree for the graph in Figure 15.55.

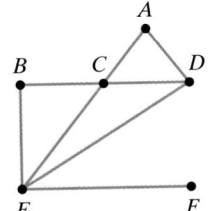

FIGURE 15.55

STUDY TIP

Most connected graphs have many possible spanning trees. You can think of each spanning tree as a bare skeleton holding together the connected graph. There are many such skeletons.

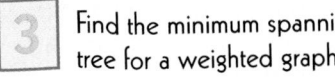

Find the minimum spanning tree for a weighted graph.

Minimum Spanning Trees and Kruskal's Algorithm

Many applied problems involve creating the most efficient network for a weighted graph. The weights often model distances, costs, or time, which we want to minimize. We do this by finding a *minimum spanning tree*.

MINIMUM SPANNING TREES

The **minimum spanning tree** for a weighted graph is a spanning tree with the smallest possible total weight.

Figures 15.56(b) and (c) show two spanning trees for the weighted graph in Figure 15.56(a). The total weight for the spanning tree in Figure 15.56(c), 107, is less than that in Figure 15.56(b), 119. Is this the minimum spanning tree, or should we continue to explore other possible spanning trees whose total weight might be less than 107?

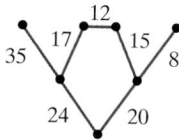
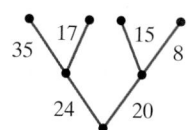
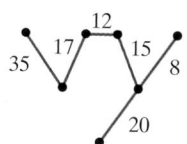

(a) Original
weighted graph

(b) A spanning tree
with weight
35 + 24 + 20 + 8 + 17 + 15
= 119

(c) A spanning tree
with weight
35 + 17 + 12 + 15 + 20 + 8
= 107

FIGURE 15.56 A weighted graph and two spanning trees with their weights

A very simple graph can have many spanning trees. Finding the minimum spanning tree by finding all possible spanning trees and comparing their weights would be too time-consuming. In 1956, the American mathematician Joseph Kruskal discovered a procedure that will always yield a minimum spanning tree for a weighted graph. The basic idea in **Kruskal's Algorithm** is to always pick the smallest available edge but avoid creating any circuits.

KRUSKAL'S ALGORITHM

Here is a procedure for finding the minimum spanning tree from a weighted graph:

1. Find the edge with the smallest weight in the graph. If there is more than one, pick one at random. Mark it in red (or using any other designation).

2. Find the next-smallest edge in the graph. If there is more than one, pick one at random. Mark it in red.

3. Find the next-smallest unmarked edge in the graph that does not create a red circuit. If there is more than one, pick one at random. Mark it in red.

4. Repeat step 3 until all vertices have been included. The red edges are the desired minimum spanning tree.

EXAMPLE 3 | USING KRUSKAL'S ALGORITHM

Seven buildings on a college campus are connected by the sidewalks shown in Figure 15.57. The weighted graph in Figure 15.58 represents buildings as vertices, sidewalks as edges, and sidewalk lengths as weights. A heavy snow has fallen and the

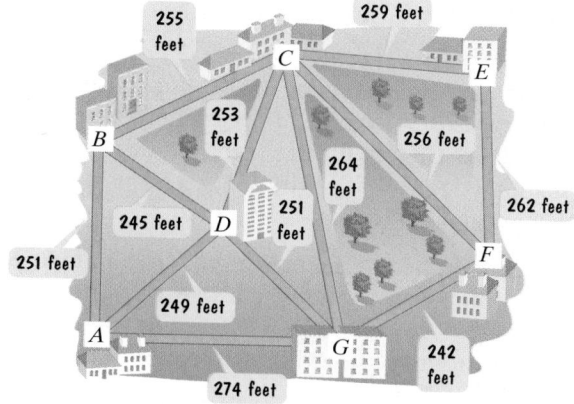

FIGURE 15.57 A campus with seven buildings and connecting sidewalks

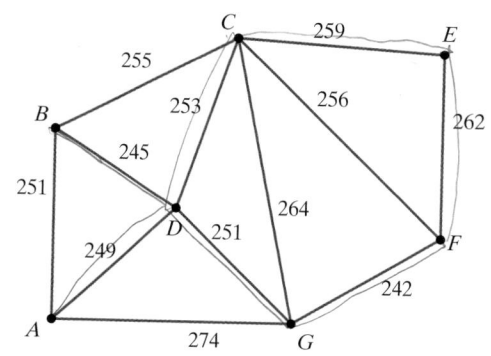

FIGURE 15.58 A weighted graph modeling the campus in Figure 15.57

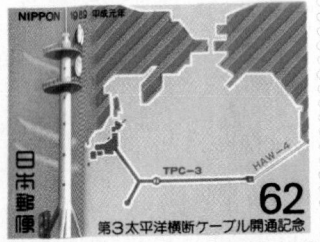
sidewalks need to be cleared quickly. Campus services decides to clear as little as possible and still ensure that students walking from building to building will be able to do so along cleared paths. Determine the shortest series of sidewalks to clear. What is the total length of the sidewalks that need to be cleared?

Solution   Campus services wants to keep the total length of cleared sidewalks to a minimum and still have a cleared path connecting any two buildings. Thus, they are seeking a minimum spanning tree for the weighted graph in Figure 15.58. We find this minimum spanning tree using Kruskal's Algorithm. Refer to Figure 15.59 as you read the steps in the algorithm.

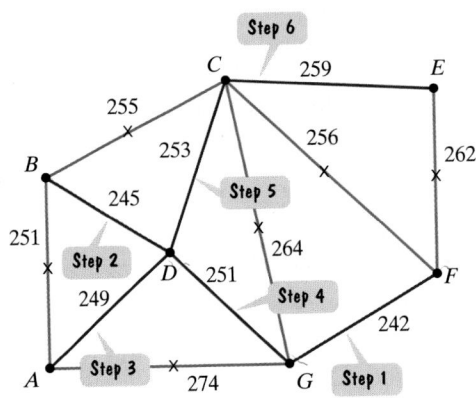

FIGURE 15.59 Finding a minimum spanning tree

STEP 1. Find the edge with the smallest weight. Select edge GF (length: 242 feet) by marking it in red.

STEP 2. Find the next-smallest edge in the graph. Select edge BD (length: 245 feet) by marking it in red.

STEP 3. Find the next-smallest edge in the graph. Select edge AD (length: 249 feet) by marking it in red.

STEP 4. Find the next-smallest edge in the graph that does not create a circuit. The next-smallest edges are AB and DG (length of each: 251 feet). Do not select AB—it creates a circuit. Select edge DG by marking it in red.

STEP 5. Find the next-smallest edge in the graph that does not create a circuit. Select edge CD (length: 253 feet) by marking it in red. Notice that this does not create a circuit.

STEP 6. Find the next-smallest edge in the graph that does not create a circuit. The next-smallest edge is BC (length: 255 feet), but this creates a circuit. Discard BC. The next-smallest edge is CF (length: 256 feet), but this also creates a circuit. Discard CF. The next-smallest edge is CE (length: 259 feet). This does not create a circuit, so select edge CE by marking it in red.

Can you see that the minimum spanning tree in Figure 15.59 is completed? The red subgraph contains all of the graph's seven vertices, is connected, contains no circuits, and has $7 - 1$, or 6, edges. Therefore, the red subgraph in Figure 15.59 shows the shortest series of sidewalks to clear. From the figure, we see that there are

$$242 + 245 + 249 + 251 + 253 + 259,$$

or 1499 feet of sidewalks that need to be cleared.

Use Kruskal's Algorithm to find the minimum spanning tree for the graph in Figure 15.60. Give the total weight of the minimum spanning tree.

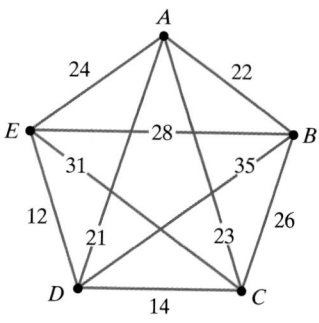

FIGURE 15.60

BLITZER BONUS

A Family Tree: The Sopranos

Genealogists use trees to show the relationship between members of different generations of a family. Figure 15.61 shows three generations of HBO's Soprano family. Figure 15.62 models this information with a tree. The vertices represent some of the pictured people. The edges represent parent-child relationships.

FIGURE 15.61 Three generations of the Soprano family

FIGURE 15.62 A tree modeling Soprano parent-child relationships

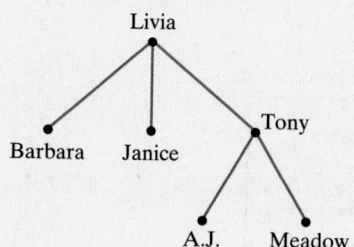

Exercise Set 15.4

Practice Exercises

In Exercises 1–10, determine whether each graph is a tree. If the graph is not a tree, give a reason why.

1.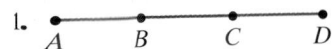

2.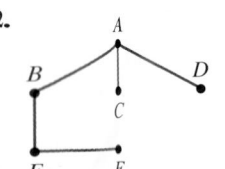

3.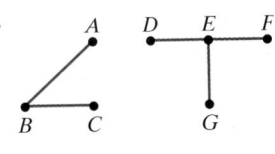

4.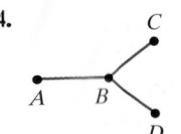

5.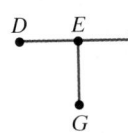

6.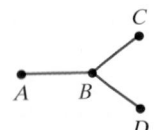

7.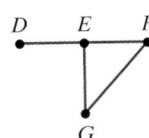

8.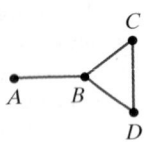

9.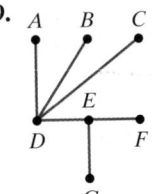

10.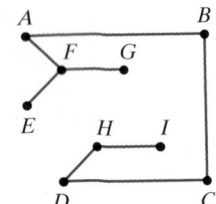

In Exercises 11–18, a graph with no loops or multiple edges is described. Which one of the following applies to the description?

i. The described graph is a tree.

ii. The described graph is not a tree.

iii. The described graph may or may not be a tree.

11. The graph has five vertices, and there is exactly one path from any vertex to any other vertex.

12. The graph has five vertices, is connected, and every edge is a bridge.

13. The graph has eight vertices and five edges.

14. The graph has nine vertices and six edges.

15. The graph has eight vertices, seven edges, and a Hamilton circuit.

16. The graph has nine vertices, eight edges, and a Hamilton circuit.

17. The graph has five vertices and four edges.

18. The graph has four vertices and three edges.

In Exercises 19–24, find a spanning tree for each connected graph. Because many spanning trees are possible, answers will vary. Do not be concerned if your spanning tree is not the same as the sample given in the answer section. Be sure that your spanning tree is connected, contains no circuits, and that each edge is a bridge, with one less edge than vertices.

19.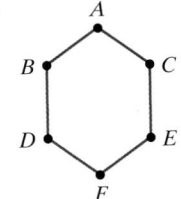

20.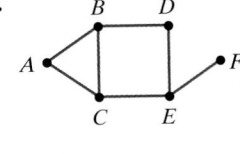

21.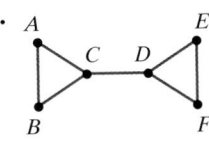

22.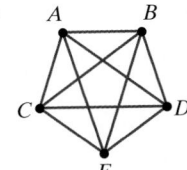

23.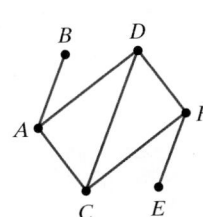

24.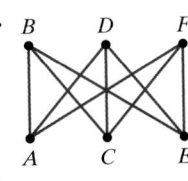

In Exercises 25–32, use Kruskal's Algorithm to find the minimum spanning tree for each weighted graph. Give the total weight of the minimum spanning tree.

25.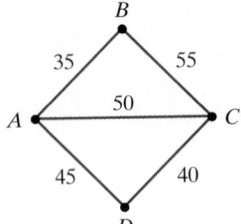

26.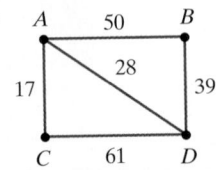

27.

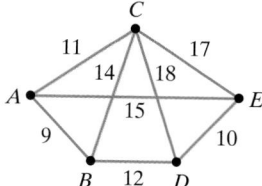

28.

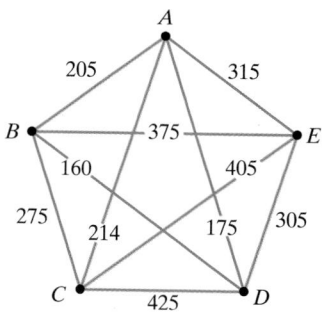

29.

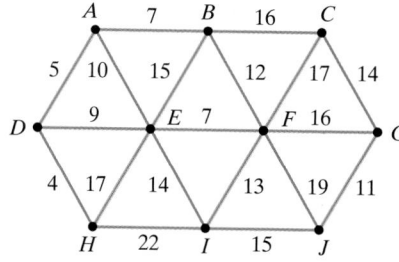

30.

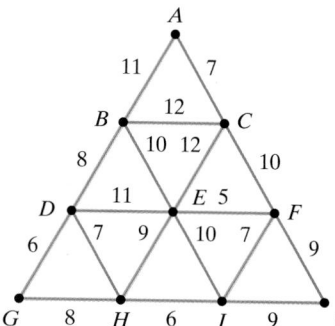

31.

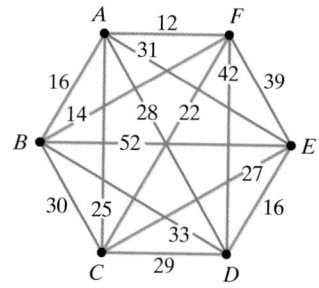

32.

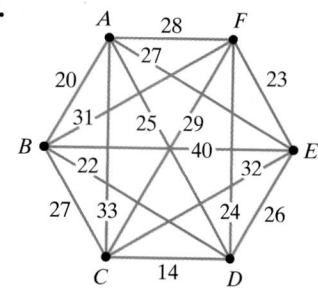

Application Exercises

33. Use a tree to model the parent-child relationships in the following family:

> Peter has three children: Zoila, Keanu, and Sandra. Zoila has two children: Sean and Helen. Keanu has no children. Sandra has one child: Martin.

Use vertices to model the people and edges to represent the parent-child relationships.

34. Use a tree to model the employee relationships among the chief administrators of a large community college system:

> Three campus vice presidents report directly to the college president. On two campuses, the academic dean, the dean for administration, and the dean of student services report directly to the vice president. On the third campus, only the academic dean and the dean for administration report directly to the vice president.

35. A college campus plans to provide awnings above its sidewalks to shelter students from the rain as they walk from the parking lot and between buildings. To save money, awnings will not be placed over all of the sidewalks shown in the figure. Just enough awnings will be placed over a select number of sidewalks to ensure that students walking from building to building will be able to do so without getting wet.

a. Use a weighted graph to model the given map. Represent buildings as vertices, sidewalks as edges, and sidewalk lengths as weights.

b. Use Kruskal's Algorithm to find a minimum spanning tree that allows students to move between the parking lot and any buildings shown without getting wet. What is the total length of the sidewalks that need to be sheltered by awnings?

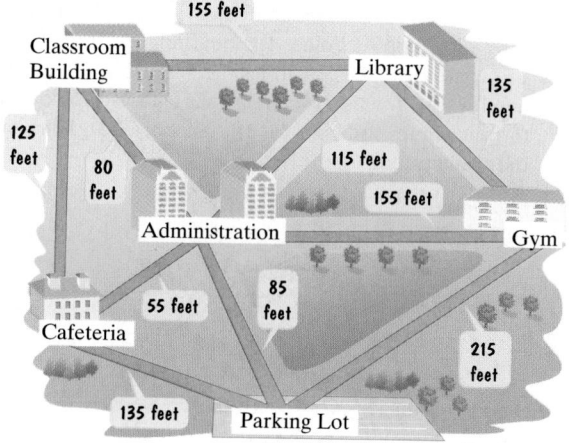

36. A fiber-optic cable system is to be installed along highways connecting six cities. The cities include Boston, Cincinnati, Cleveland, Detroit, New York, and Philadelphia. The highway distances, in miles, between the cities are given in the table.

a. Model this information with a weighted, complete graph.

b. Use Kruskal's Algorithm to find a minimum spanning tree that would connect each city using the smallest amount of cable. Determine the total length of cable needed.

	Boston	Cincinnati	Cleveland	Detroit	New York	Philadelphia
Boston	*	840	628	734	222	296
Cincinnati	840	*	244	269	647	567
Cleveland	628	244	*	170	473	413
Detroit	734	269	170	*	641	573
New York	222	647	473	641	*	101
Philadelphia	296	567	413	573	101	*

37. The graph shows a proposed layout of an irrigation system. The main water source is located at vertex *C*. The other vertices represent the proposed sprinkler head locations. The edges indicate all possible choices for installing underground pipes. The weights show the distances, in feet, between proposed sprinkler head locations. The sprinkler system will work if each sprinkler head is connected to the main water source through one or more underground pipes. Use Kruskal's Algorithm to determine the layout of the irrigation system and the smallest number of feet of underground pipes.

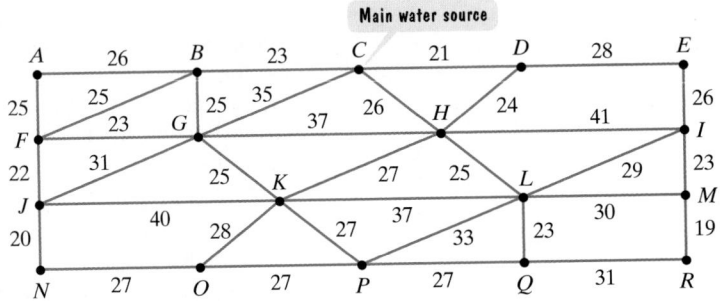

38. The graph shows a proposed layout of a bike trail system to be installed between the towns shown. The vertices represent the ten towns, designated *A* through *J*. The edges indicate all possible choices for building the trail. The weights show the distances, in miles, between bike trailheads connecting the towns. Use Kruskal's Algorithm to determine the minimum mileage for the bike trail and the layout of the trail system.

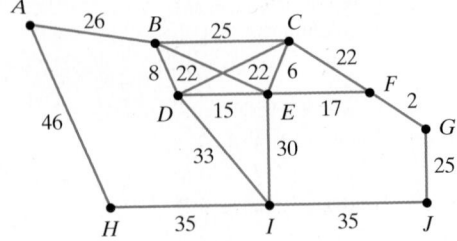

Writing in Mathematics

39. What is a tree?

40. If a graph is given, how do you determine whether or not it is a tree?

41. Describe the relationship between the number of vertices and the number of edges in a tree.

42. What is a subgraph?

43. What is a spanning tree?

44. Describe how to obtain a spanning tree for a connected graph.

45. What is the minimum spanning tree for a weighted graph?

46. In your own words, briefly describe how to find the minimum spanning tree using Kruskal's Algorithm.

47. Describe a practical problem that can be solved using Kruskal's Algorithm.

Critical Thinking Exercises

48. Modify Kruskal's algorithm to find a *maximum* spanning tree for the given graph.

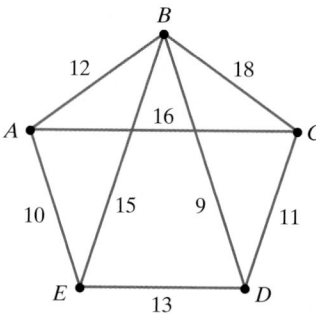

49. Find all the possible spanning trees for the given graph.

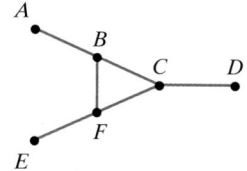

50. Give an example of a tree with 6 vertices whose degrees are 1, 1, 2, 2, 2, and 2.

Group Exercise

51. Group members should determine a project whose installation would enhance the quality of life on campus or in your community. For example, the project might involve installing awnings over campus sidewalks, building a community bike path, creating a community hiking trail, or installing a metrorail system providing easy access to your community's most desirable locations. The project that you determine should be one that can be carried out most efficiently using a minimum spanning tree. Begin by defining the project and its locations (vertices). Group members should then research the distances between the various locations in the project. Once these distances have been determined, the group should reassemble and create a graph that models the project. Then find a minimum spanning tree that serves as the most efficient way to carry out your project.

Chapter Summary, Review, and Test

SUMMARY

DEFINITIONS AND CONCEPTS	EXAMPLES

15.1 Graphs, Paths, and Circuits

Definition of a Graph

A graph consists of a finite set of points, called vertices, and line segments or curves, called edges, that start and end at vertices. A loop is an edge that starts and ends at the same vertex. Equivalent graphs have the same number of vertices connected to each other in the same way.

Ex. 1, p. 780;
Ex. 2, p. 780;
Ex. 3, p. 781;
Ex. 4, p. 782;
Ex. 5, p. 783

The Vocabulary of Graph Theory

a. The degree of a vertex is the number of edges at that vertex. A loop connecting a vertex to itself contributes 2 to the degree of that vertex.

b. A vertex with an even number of edges attached is an even vertex; one with an odd number of edges attached is an odd vertex.

c. Two vertices are adjacent if there is at least one edge connecting them.

Ex. 6, p. 784

d. A path in a graph is a sequence of adjacent vertices and the edges connecting them. An edge can be part of a path only once.

Figure 15.18, p. 785

e. A circuit is a path that begins and ends at the same vertex.

Figure 15.19, p. 785

f. A graph is connected if for any two of its vertices there is at least one path connecting them. A connected graph consists of one piece. A graph that is not connected, made up of components, is called disconnected.

Figure 15.20, p. 785

g. A bridge is an edge that if removed from a connected graph would leave behind a disconnected graph.

15.2 Euler Paths and Euler Circuits

Euler Paths and Euler Circuits

   a. An Euler path is a path that travels through every edge of a graph once and only once.

   b. An Euler circuit is a circuit that travels through every edge of a graph once and only once.

Figure 15.21, p. 790
Figure 15.22, p. 790

Euler's Theorem

   a. A graph with two odd vertices has at least one Euler path, but no Euler circuit. Each Euler path starts at an odd vertex and ends at the other odd vertex.

   b. A graph with no odd vertices has at least one Euler circuit that can start and end at any vertex.

   c. A graph with more than two odd vertices has no Euler paths and no Euler circuits.

Ex. 1, p. 791;
Ex. 2, p. 792;
Ex. 3, p. 793

Fleury's Algorithm

If Euler's Theorem indicates the existence of Euler paths or Euler circuits, trial and error or Fleury's Algorithm can be used to find such paths or circuits. Fleury's Algorithm is explained in the box on page 795.

Ex. 4, p. 795

15.3 Hamilton Paths and Hamilton Circuits

Hamilton Paths and Hamilton Circuits

A path that passes through each vertex of a graph exactly once is called a Hamilton path. If a Hamilton path begins and ends at the same vertex and passes through all other vertices exactly once, it is called a Hamilton circuit.

Ex. 1, p. 801

Complete Graphs and Hamilton Circuits

   a. A complete graph is a graph that has an edge between each pair of its vertices.

   b. A complete graph with n vertices has $(n - 1)!$ Hamilton circuits.

Ex. 2, p. 803

Weighted Graphs and the Traveling Salesperson Problem

   a. A complete graph whose edges have numbers, called weights, attached to them is called a weighted graph.

   b. The traveling salesperson problem is the problem of finding a Hamilton circuit in a complete, weighted graph for which the sum of the weights of the edges is a minimum. Such a Hamilton circuit is called the optimal Hamilton circuit or the optimal solution.

   c. The Brute Force Method can be used to solve the traveling salesperson problem. The method, given in the box on page 805, involves determining the sum of the weights of the edges for all possible Hamilton circuits.

   d. The Nearest Neighbor Method can be used to approximate the optimal solution to traveling salesperson problems. The method, given in the box on page 807, involves continually taking an edge with the smallest weight.

Ex. 3, p. 804

Ex. 4, p. 805

Ex. 5, p. 807

15.4 Trees

Trees

   a. A tree is a graph that is connected and has no circuits.

   b. Properties of Trees

   1. There is one and only one path joining any two vertices.

   2. Every edge is a bridge.

   3. A tree with n vertices has $n - 1$ edges.

Ex. 1, p. 813

Spanning Trees

   a. A subgraph is a set of vertices and edges chosen from among those of the original graph.

   b. A spanning tree is a subgraph that contains all of a connected graph's vertices, is connected, and contains no circuits.

Ex. 2, p. 814

Minimum Spanning Trees and Kruskal's Algorithm

   a. The minimum spanning tree for a weighted graph is a spanning tree with the smallest possible total weight.

   b. Kruskal's Algorithm, described in the box on page 815, gives a procedure for finding the minimum spanning tree from a weighted graph. The basic idea is to always pick the smallest available edge, but avoid creating any circuits.

Ex. 3, p. 815

REVIEW EXERCISES

15.1

1. Explain why the two figures show equivalent graphs. Then draw a third equivalent graph.

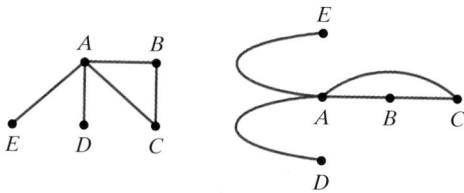

In Exercises 2–8, use the following graph.

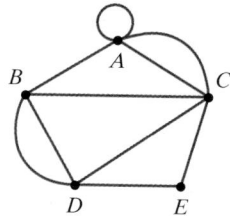

2. Find the degree of each vertex in the graph.

3. Identify the even vertices and identify the odd vertices.

4. Which vertices are adjacent to vertex D?

5. Use vertices to describe two paths that start at vertex E and end at vertex A.

6. Use vertices to describe a circuit that begins and ends at vertex E.

7. Is the graph connected? Explain your answer.

8. Is any edge in the graph a bridge? Explain your answer.

9. List all edges that are bridges in the graph shown.

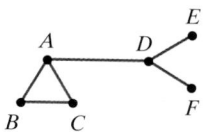

10. Draw a graph that models the layout of the city shown in the map. Use vertices to represent the land masses and edges to represent the bridges.

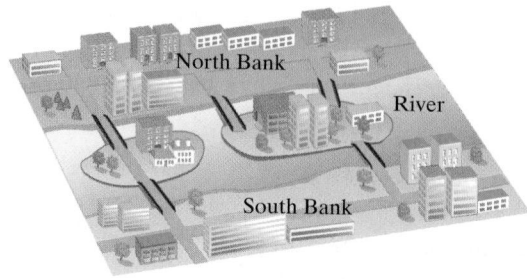

11. Draw a graph that models the bordering relationships among the states shown in the map. Use vertices to represent the states and edges to represent common borders.

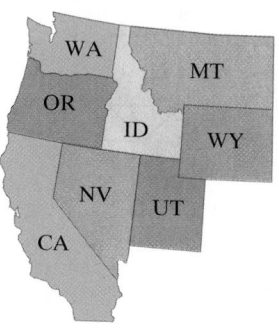

12. Draw a graph that models the connecting relationships in the floor plan. Use vertices to represent the rooms and the outside, and edges to represent the connecting doors.

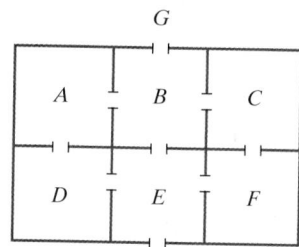

15.2

In Exercises 13–15, a graph is given.

   a. *Determine whether the graph has an Euler path, an Euler circuit, or neither.*

   b. *If the graph has an Euler path or circuit, use trial and error or Fleury's Algorithm to find one.*

13.

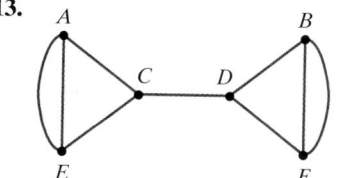

14.   **15.**

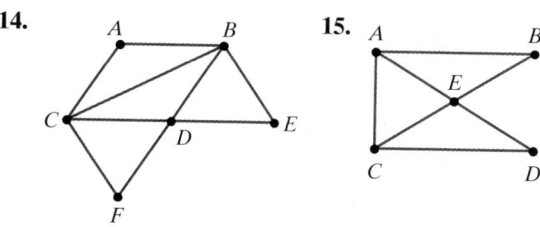

16. Use Fleury's Algorithm to find an Euler path.

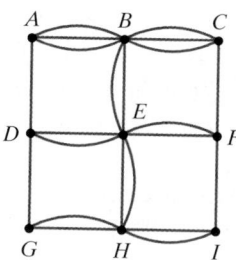

17. Use Fleury's Algorithm to find an Euler circuit.

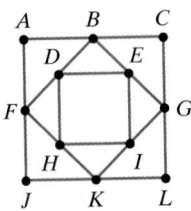

18. Refer to Exercise 10.

   a. Use your graph to determine if the city residents would be able to walk across all of the bridges without crossing the same bridge twice.

   b. If such a walk is possible, show the path on your graph. Then trace this route on the city map in a manner that is clear to the city's residents.

   c. Use your graph to determine if there is a path that crosses each bridge exactly once and begins and ends on the same island. Explain your answer.

19. Refer to Exercise 11. Use your graph to determine if it is possible to find a path that crosses each common state border exactly once. Explain your answer.

20. Refer to Exercise 12.

   a. Use your graph to determine if it is possible to find a path that uses each door exactly once.

   b. If such a path is possible, show it on your graph. Then trace this route on the floor plan in a manner that is clear to a person strolling through the house.

21. A security guard needs to walk the streets of the neighborhood in the figure shown. The guard is to walk down each street once, whether or not the street has houses on both sides.

   a. Draw a graph that models the streets of the neighborhood walked by the security guard.

   b. Use your graph to determine if the security guard can walk each street exactly once. Explain your answer.

   c. If so, use the map to show where the guard should begin and where the guard should end the walk.

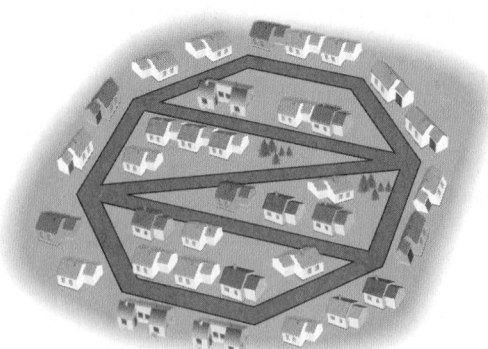

15.3

In Exercises 22–23, use the graph shown.

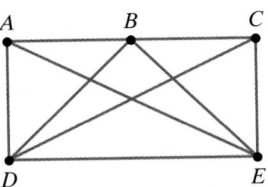

22. Find a Hamilton circuit that begins as $A, E, \ldots$.

23. Find a Hamilton circuit that begins as $D, B, \ldots$.

For each graph in Exercises 24–27,

   a. *Determine if the graph must have Hamilton circuits. Explain your answer.*

   b. *If the graph must have Hamilton circuits, determine the number of such circuits.*

24.

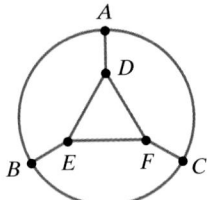

25.

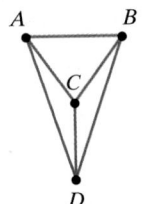

26.

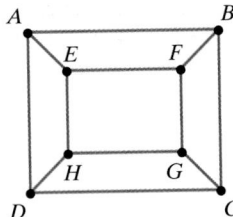

27.

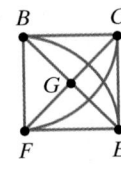

In Exercises 28–30, use the complete, weighted graph shown.

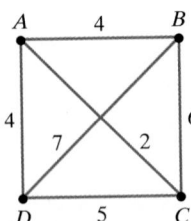

28. Find the total weight of each of the six possible Hamilton circuits:

$$A, B, C, D, A \qquad A, B, D, C, A \qquad A, C, B, D, A$$

$$A, C, D, B, A \qquad A, D, B, C, A \qquad A, D, C, B, A.$$

29. Use your answers from Exercise 28 and the Brute Force Method to find the optimal solution.

30. Use the Nearest Neighbor Method, with starting vertex A, to find an approximate solution. What is the total weight of the Hamilton circuit?

31. Use the Nearest Neighbor Method to find a Hamilton circuit that begins at vertex A in the given graph. What is the total weight of the Hamilton circuit?

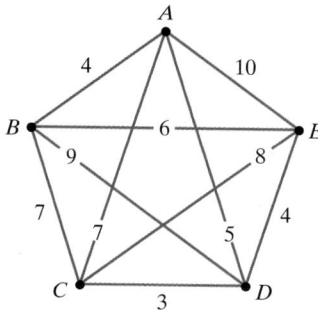

The table shows the one-way airfares between cities. Use this information to solve Exercises 32–33.

	A	**B**	**C**	**D**	**E**
A	*	$220	$240	$290	$430
B	$220	*	$260	$320	$360
C	$240	$260	*	$180	$300
D	$290	$320	$180	*	$250
E	$430	$360	$300	$250	*

32. Draw a graph that models the information in the table. Use vertices to represent the cities and weights on appropriate edges to show airfares.

33. A sales director who lives in city A is required to fly to regional offices in cities $B, C, D,$ and E, and then return to city A. Use the Nearest Neighbor Method to approximate the optimal route. What is the total cost for this Hamilton circuit?

15.4

In Exercises 34–36, determine whether each graph is a tree. If the graph is not a tree, give a reason why.

34.

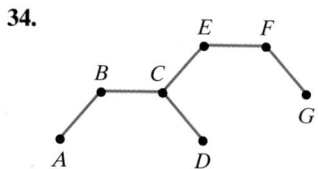

35.

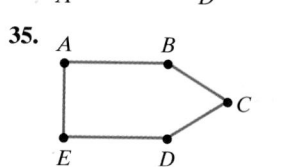

36.

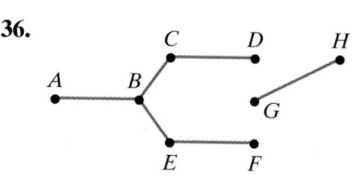

In Exercises 37–38, find a spanning tree for each connected graph.

37.

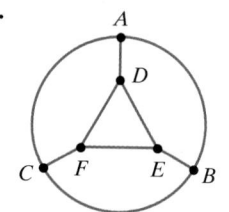

38.

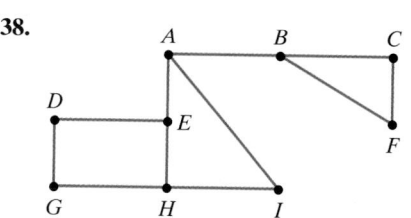

In Exercises 39–40, use Kruskal's Algorithm to find the minimum spanning tree for each weighted graph. Give the total weight of the minimum spanning tree.

39.

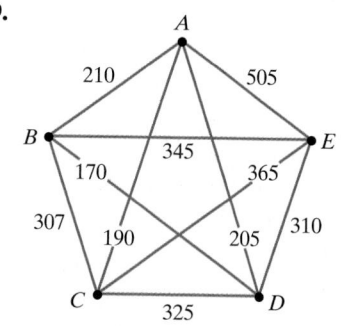

40.

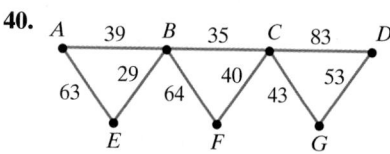

41. A fiber-optic cable system is to be installed along highways connecting nine cities. The weighted graph shows the highway distances, in miles, between the cities, designated A through I. Use Kruskal's Algorithm to determine the layout of the cable system and the smallest length of cable needed.

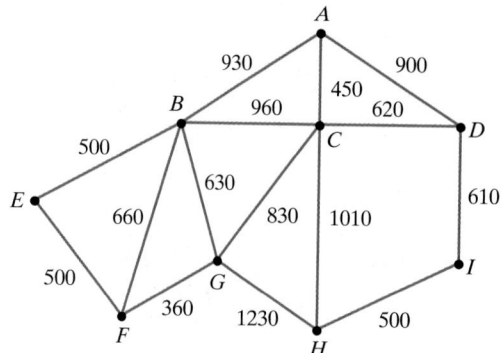

Group Exercise

42. Each group member should list two errands he or she has been putting off that need to be done. The group should select any six of these errands. Suppose that you will leave from campus in a large van, run each of the six errands for the appropriate person, and return to campus.

   a. Create a table similar to the one on page 810 showing the distances, in miles, between the locations you cover by van.

   b. Create a complete, weighted graph that models the information in the table.

   c. Use the Nearest Neighbor Method to approximate the shortest route for the group to run the six errands and return to campus. What is the minimum distance given by the Hamilton circuit?

CHAPTER 15 TEST

In Exercises 1–4, use the following graph.

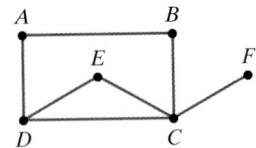

1. Find the degree of each vertex.

2. Use vertices to describe two paths that start at vertex *A* and end at vertex *E*.

3. Use vertices to describe a circuit that starts at vertex *B* and passes through vertex *E*.

4. List all edges that are bridges.

5. Draw a graph that models the bordering relationships among the countries shown in the map. Use vertices to represent the countries and edges to represent common borders.

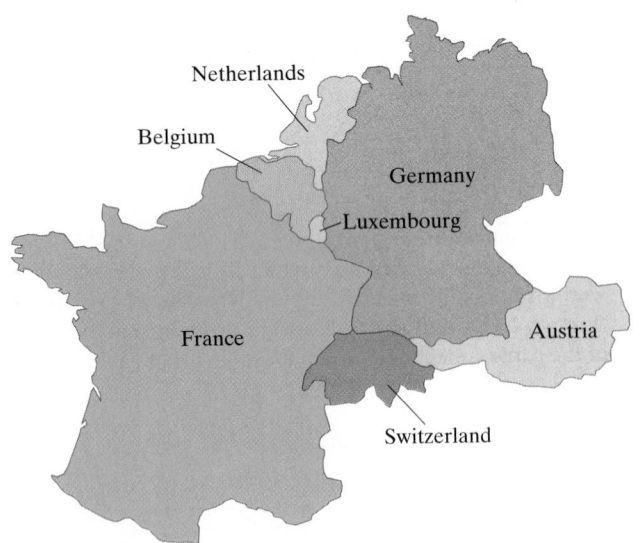

In Exercises 6–8, a graph is given.

   a. *Determine whether the graph has an Euler path, an Euler circuit, or neither.*

   b. *If the graph has an Euler path or circuit, use trial and error or Fleury's Algorithm to find one.*

6.

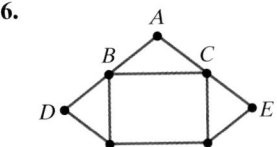

7.

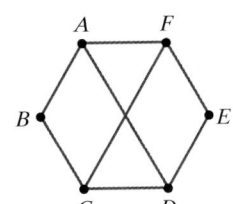

8.

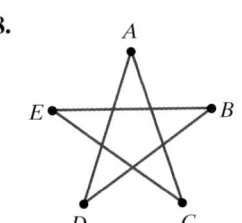

9. Use Fleury's Algorithm to find an Euler circuit.

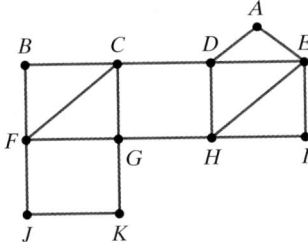

10. a. Draw a graph that models the layout of the city shown in the map.

   b. Use your graph to determine if the city residents would be able to walk across all of the bridges without crossing the same bridge twice. Explain your answer.

   c. If such a walk is possible, where should it begin?

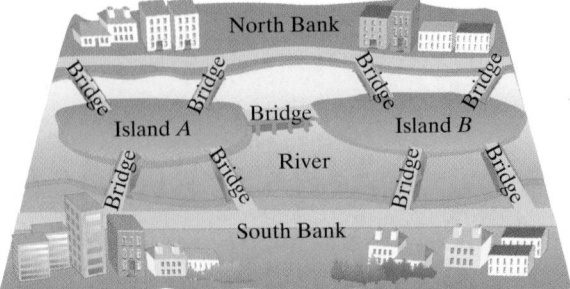

11. a. Draw a graph that models the connecting relationships in the floor plan.

b. Use your graph to determine if it is possible to find a path that uses each door exactly once. Explain your answer.

c. If such a path is possible, where should it begin?

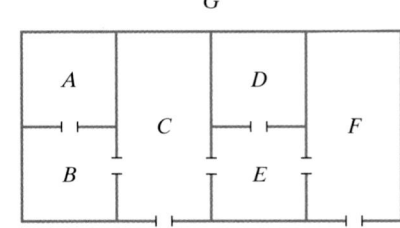

12. a. Draw a graph that models the streets of the neighborhood for a police officer who is to walk each street once.

b. Use your map to determine if the officer can walk each street exactly once. Explain your answer.

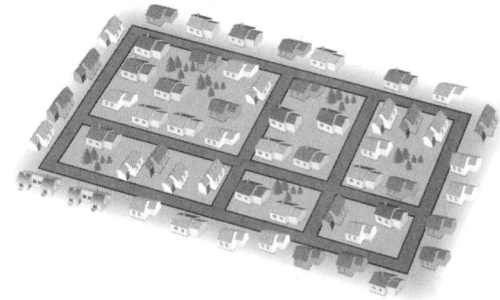

13. Find two Hamilton circuits in the graph shown. One circuit should begin as $A, B, \ldots$. The second circuit should begin as $A, F, \ldots$.

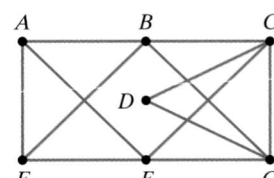

14. How many Hamilton circuits are there in a complete graph with five vertices?

The table shows the one-way airfares between cities. Use this information to solve Exercises 15–16.

	A	B	C	D
A	*	$460	$200	$210
B	$460	*	$720	$680
C	$200	$720	*	$105
D	$210	$680	$105	*

15. Draw a graph that models the information in the table. Use vertices to represent the cities and weights on appropriate edges to show airfares.

16. A sales director who lives in city A is required to fly to regional offices in cities B, C, and D, and then return to city A. Use the Brute Force Method to approximate the optimal route. What is the total cost for this Hamilton circuit? (*Hint:* The six possible Hamilton circuits are

$A, B, C, D, A;$ $A, B, D, C, A;$
$A, C, B, D, A;$ $A, C, D, B, A;$
$A, D, B, C, A;$ $A, D, C, B, A.$)

17. Use the Nearest Neighbor Method to find a Hamilton circuit that begins at vertex A in the given graph. What is the total weight of the Hamilton circuit?

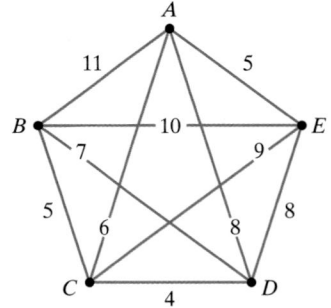

18. Is the graph shown a tree? Explain your answer.

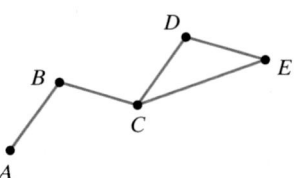

19. Find a spanning tree for the graph shown.

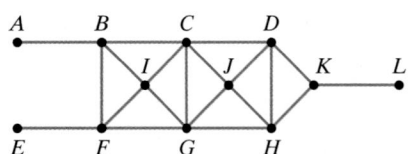

20. Use Kruskal's Algorithm to find the minimum spanning tree for the weighted graph shown. Give the total weight of the minimum spanning tree.

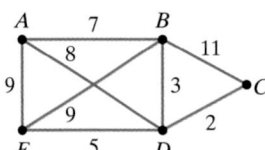

Answers to Selected Exercises

CHAPTER 1

Section 1.1
Check Point Exercises
1. inductive reasoning; Answers will vary. **2.** Answers will vary. **3. a.** Each number in the list is obtained by adding 6 to the previous number.; 33 **b.** Each number in the list is obtained by multiplying the previous number by 5.; 1250
4. The figures alternate between rectangles and triangles, and the number of appendages follows the pattern: one, two, three, one, two, three, etc.;
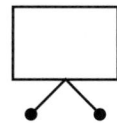
5. a. The result of the process is two times the original number selected.
  b. Representing the original number as n, we have

Select a number: n
Multiply the number by 4: $4n$
Add 6 to the product: $4n + 6$
Divide this sum by 2: $\dfrac{4n + 6}{2} = 2n + 3$
Subtract 3 from the quotient: $2n + 3 - 3 = 2n.$

Exercise Set 1.1
1. inductive reasoning; Answers will vary. **3.** deductive reasoning; Answers will vary.
5. Answers will vary; an example is: Bill Clinton was younger than 65 at the time of his inauguration.
7. Answers will vary; an example is: When the number 5 is multiplied by itself, the result, 25, is not even.
9. Answers will vary; an example is: When 1 is added to the top and bottom of $\frac{1}{2}$, the new fraction $\frac{2}{3}$ is not equivalent to the original fraction $\frac{1}{2}$. **11.** Answers will vary; an example is $0.0 + 0$ is not greater than 0.
13. Each number in the list is obtained by adding 4 to the previous number.; 28
15. Each number in the list is obtained by subtracting 5 from the previous number.; 12
17. Each number in the list is obtained by multiplying the previous number by 3.; 729
19. Each number in the list is obtained by multiplying the previous number by 2.; 32
21. The numbers in the list alternate between 1 and numbers obtained by multiplying the number prior to the previous number by 2.; 32
23. Each number in the list is obtained by subtracting 2 from the previous number.; -6
25. Each number in the list is obtained by adding 4 to the denominator of the previous fraction.; $\frac{1}{22}$
27. Each number in the list is obtained by multiplying the previous number by $\frac{1}{3}$.; $\frac{1}{81}$
29. Each number in the list is obtained by multiplying the previous number by $-\frac{1}{4}$.; $\frac{1}{4}$
31. For each pair in the list, the second number is obtained by subtracting 4 from the first number.; -1
33. The pattern is: square, triangle, circle, square, triangle, circle, etc.;
35. Each figure contains the letter of the alphabet following the letter in the previous figure with one more occurrence than in the previous figure.;

d	d	d
d	d	

37. a. The result of the process is two times the original number selected.
  b. Representing the original number as n, we have
Select a number: n
Multiply the number by 4: $4n$
Add 8 to the product: $4n + 8$
Divide this sum by 2: $\dfrac{4n + 8}{2} = 2n + 4$
Subtract 4 from the quotient: $2n + 4 - 4 = 2n.$

39. a. The result of the process is 3.
  b. Representing the original number as n, we have
Select a number: n
Add 5: $n + 5$
Double the result: $2(n + 5) = 2n + 10$
Subtract 4: $2n + 10 - 4 = 2n + 6$
Divide by 2: $\dfrac{2n + 6}{2} = n + 3$
Subtract n: $n + 3 - n = 3$

41. $1 + 2 + 3 + 4 + 5 + 6 = \dfrac{6 \times 7}{2}; 21 = 21$ **43.** $1 + 3 + 5 + 7 + 9 + 11 = 6 \times 6; 36 = 36$

45. a. The successive differences are consecutive counting numbers beginning with 2. **b.** 28, 36, 45, 55, 66
47. a. 14, 17, 20 **b.** 17 in. **49–55.** Answers will vary. **57.** Beginning with the third number, each number in the list is the sum of the previous two numbers.; 55 **59.** $165 \times 3367 = 555{,}555$; correct **61.** Answers will vary; an example is: 1, 5, 1, 10, 1, … .

63. a. 10,101; 20,202; 30,303; 40,404   **b.** In the multiplications, the first factor is always 3367, and the second factors are consecutive muliples of 3, beginning with $3 \times 1 = 3$.; The second and fourth digits of the products are always 0; the first, third, and last digits are the same within each product; this digit is 1 in the first product and increases by 1 in each subsequent product.
c. $3367 \times 15 = 50,505$; $3367 \times 18 = 60,606$ **d.** inductive reasoning; Answers will vary.

Section 1.2
Check Point Exercises
1. 60   **2. a.** $2, $1, $5, $4, $1, $2, and $3; $\approx$ $18   **b.** no   **3.** 24   **4. a.** $\approx$ $2000 per wk   **b.** $\approx$ $100,000 per yr   **5.** $\approx$ 3 hr
6. a. 13.6% of the U.S. population in 2050 will be African American.   **b.** $\approx$ 32,000,000 people of Asian descent   **7. a.** $\approx$ 22%
b. 1,200,000 cars   **8.** $\approx$ 84,000

Exercise Set 1.2
1. $\approx$ $2.40   **3.** $\approx$ 350 mi   **5.** $\approx$ $43   **7.** $\approx$ $100   **9.** $\approx$ $40,000 per yr   **11.** $\approx$ 68   **13.** $\approx$ 10,000 hr
15. $\approx$ 800,000 hr   **17.** $\approx$ $30 per hr   **19.** $\approx$ $1000   **21.** $\approx$ $24,000   **23.** $\approx$ $3.00   **25. a.** $\approx$ 240 mi   **b.** $\approx$ 6 hr

27. $\approx$ 180,000,000 Protestants   **29.** $\approx$ 6000 hate crimes   **31.** $\approx$ 33%   **33.** historical places and museums, outdoor recreation, and shopping   **35.** $\approx$ 48 yr   **37.** $\approx$ 8 yr   **39.** $\approx$ $470   **41.** $\approx$ $20   **43.** 1960; 22%   **45.** 36,000,000   **47.** 1832; $\approx$ 20%
49. $\approx$ 52%   **51–59.** Answers will vary.   **61.** The graphs project that the Israeli and Palestinian populations will be the same in 2030.;
9 million   **63.** $\approx \dfrac{5}{6}$ mi

65. a. Answers will vary.;

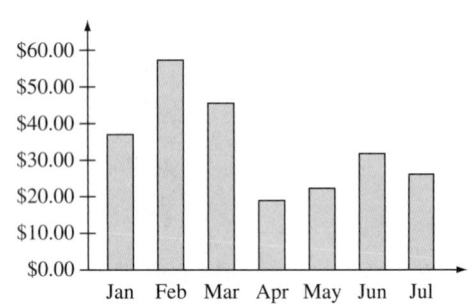

Utility Bill

b. Answers will vary.;

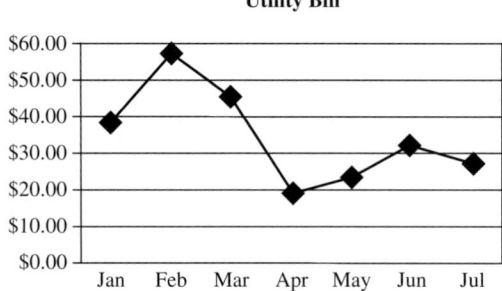

Utility Bill

c. Answers will vary.;

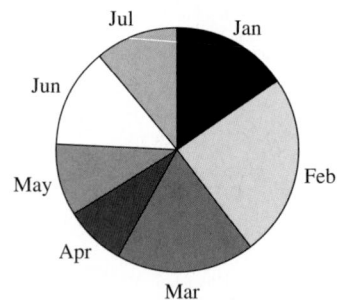

Utility Bill

Section 1.3
Check Point Exercises
1. the amount of money given to the cashier   **2.** The 128-ounce bottle at approximately 4¢ an ounce is the better value.
3. 14 months   **4.** 5 combinations   **5.** 6 outfits   **6.** A route that will cost less than $1460 is A, D, E, C, B, A.   **7.** Answers will vary.

Trick Questions
1. 12   **2.** 12   **3.** sister and brother   **4.** match

Exercise Set 1.3
1. the price of the computer   **3.** the number of words on the page   **5.** unnecessary information: weekly salary of $350; extra pay: $180
7. unnecessary information: $20 given to the parking attendant; charge: $4.50   **9. a.** 24-ounce box for $4.59 **b.** 22¢ per ounce for the 15.3-ounce box and $3.06 per pound for the 24-ounce box **c.** no; Answers will vary.   **11.** $3000   **13.** $50   **15.** $90   **17.** $4525
19. $565   **21.** 4 mi   **23.** $104   **25.** 5 ways   **27.** 4 ways   **29.** Andy, Darnell, Caleb, Beth, Ella   **31.** Home, Bank, Post Office, Dry Cleaners, Home   **33.** Sample answer: CO, WY, UT, AZ, NM, CO, UT   **35.** Bob's major is psychology.   **37.**

$$\begin{array}{r} 156 \\ 28\overline{)4368} \\ 28 \\ \hline 156 \\ 140 \\ \hline 168 \\ 168 \\ \hline 0 \end{array}$$

39.

10	35	30
45	25	5
20	15	40

41. 14   **43.**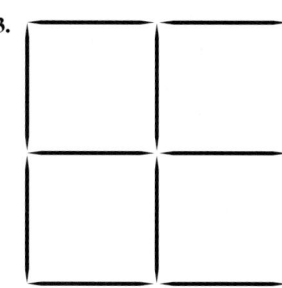

45. ESCAPE TONIGHT   **47–53.** Answers will vary.   **55.** the dentist with poor dental work   **57.** He takes the goat across first. Then he returns and gets either the wolf or the cabbage. On his second return trip, he takes the goat back across the river, and leaves it, and picks up the cabbage or the wolf, whichever remains. He returns a third time to get the goat.

59. Sample answer:

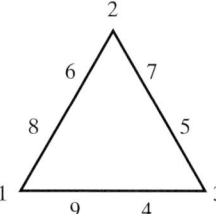

Review Exercises

1. deductive reasoning: Answers will vary.   **2.** inductive reasoning; Answers will vary.

3. Answers will vary; an example is: Vermont is a U.S. state for which the first letter of its name does not come before u in the alphabet.

4. Answers will vary; an example is: When 5 is multiplied by $5 - 2 = 3$, the answer, 15, is not an even number.

5. $-3 > -4$ but $(-3)^2 < (-4)^2$ **6.** Each number in the list is obtained by adding 5 to the previous number.; 24

7. Each number in the list is obtained by multiplying the previous number by 2.; 112

8. The successive differences are consecutive counting numbers beginning with 2.; 21

9. Each number in the list is obtained by writing a fraction with a denominator that is one more than the denominator of the previous fraction before it is reduced to lowest terms.; $\frac{3}{8}$.

10. Each number in the list is obtained by multiplying the previous number by $-\frac{1}{2}$.; $\frac{5}{2}$

11. Each number in the list is obtained by subtracting 60 from the previous number; -200.

12. The figures alternate between squares and circles, and in each figure the tick mark has been rotated 90° clockwise from its position in the previous figure.;

13. a. The result is the original number.

   b. Representing the original number as n, we have

Select a number:	n
Double the number:	$2n$
Add 4 to the product:	$2n + 4$
Divide the sum by 2:	$\dfrac{2n + 4}{2} = n + 2$

   Subtract 2 from the quotient: $n + 2 - 2 = n$.

14. $\approx \$18.00$ **15.** $\approx \$560$ **16.** $\approx 80{,}000$ sec   **17.** ≈ 4 **18.** $\approx \$60$ **19.** 2700 gallons   **20.** $\approx \$1$ billion

21. $\approx \$8$ billion   **22.** $\approx \$8$ billion   **23.** dogs: 1991; cats: 1996 and 2000   **24.** ≈ 7 murders   **25.** 1980; ≈ 10 murders

26. Answers will vary.; An example is ≈ 9 murders.   **27.** the weight of the child   **28.** unnecessary information: $20 given to driver; cost of trip: $8.00   **29.** 8 lb   **30.** $885   **31.** Healthy Bodies; $40   **32.** 5.75 hr or 5 hr 45 min   **33.** $15,500   **34.** 6 combinations

Chapter 1 Test

1. Answers will vary; an example is: When 25, a two-digit number, is multiplied by 4, a one-digit number, the answer, 100, is not a two-digit number.   **2.** Each number in the list is obtained by adding 5 to the previous number.; 20

3. Each number in the list is obtained by multiplying the previous number by $\dfrac{1}{2}$; $\dfrac{1}{96}$

4. The outer figure is always a square; the inner figure follows the pattern: triangle, circle, square, triangle, circle, square, etc.; the number of appendages on the outer square alternates between 1 and 2.;

5. a. The original number is doubled.

   b. Representing the original number as n, we have

Select a number:	n
Multiply the number by 4:	$4n$
Add 8 to the product:	$4n + 8$
Divide the sum by 2:	$\dfrac{4n + 8}{2} = 2n + 4$

   Subtract 4 from the quotient: $2n + 4 - 4 = 2n$.

6. ≈ $90   **7.** ≈ $25,000 per person   **8.** ≈ 60   **9.** Answers will vary.; An example is 2.4 billion.   **10.** 250
11. Carter, Reagan, and Bush   **12.** 50   **13.** 1980; ≈ 54 million   **14.** Estes Rental; $12   **15.** $14,080   **16.** 26 weeks

CHAPTER 2

Section 2.1
Check Point Exercises
1. L is the set of the first six letters of the alphabet.   **2.** $D = \{3, 7, 4, 1, 8\}$ **3.** $L = \{m, a, r, c, h\}$ **4. a.** true   **b.** true   **c.** false
5. a. $A = \{1, 2, 3\}$ **b.** $B = \{15, 16, 17, 18, ...\}$ **c.** $O = \{1, 3, 5, 7, ...\}$ **d.** $\{12, 13\}$ **6. a.** $n(A) = 5$ **b.** $n(B) = 1$
c. $n(C) = 8$ **d.** $n(D) = 0$ **7. a.** true   **b.** false   **8. a.** no; A and B do not contain exactly the same elements.
b. yes; A and B contain the same number of elements.

Exercise Set 2.1
1. the set of planets in our solar system   **3.** the set of months that begin with J   **5.** the set of odd natural numbers less than 100
7. {winter, spring, summer, fall}   **9.** {September, October, November, December}   **11.** $\{1, 2, 3\}$ **13.** $\{1, 3, 5, 7, 9, 11\}$
15. $\{101, 102, 103, 104, ...\}$ **17.** $\{11, 12, 13, 14, 15\}$ **19.** $\{10, 11, 12, 13, 14, 15, 16\}$ **21.** $\{2\}$
23. $\{x \mid x$ is a day of the week that begins with T$\}$ **25.** $\{x \mid x \in \mathbf{N}$ and x is less than 6$\}$ **27.** $\{x \mid x \in \mathbf{N}$ and x lies between 3 and 5, inclusive$\}$
29. $\{x \mid x \in \mathbf{N}$ and x is greater than 2$\}$ **31.** true   **33.** true   **35.** false   **37.** true   **39.** false   **41.** false   **43.** false   **45.** false
47. ∉   **49.** ∈   **51.** ∉   **53.** ∉   **55.** 5   **57.** 25   **59.** 0   **61.** 1   **63.** true   **65.** false   **67.** true   **69. a.** not equal;
Answers will vary.   **b.** not equivalent; Answers will vary.   **71. a.** not equal; Answers will vary.   **b.** equivalent; Answers will vary.
73. a. equal; Answers will vary.   **b.** equivalent; Answers will vary.   **75.** {amusement parks, gardening, movies, exercise}
77. {home improvement, amusement parks, gardening}   **79.** {Chicago, Newark, LaGuardia}   **81.** {Atlanta, Philadelphia, Boston}
83. {75}   **85–93.** Answers will vary.   **95.** a

Section 2.2
Check Point Exercises
1. the set of TV game shows   **2.** {b, c, e}   **3.**   **4. a.** ⊄   **b.** ⊆   **c.** ⊆   **5. a.** ⊆   **b.** ⊆, ⊂

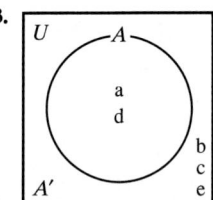

6. yes   **7. a.** 8; the number of different combinations of the books that you can take on vacation   **b.** Bring no books: ∅; Bring one book:
{*The Color Purple*}, {*Hannibal*}, {*The Royals*}; Bring two books: {*The Color Purple, Hannibal*}, {*The Color Purple, The Royals*},
{*Hannibal, The Royals*}; Bring three books: {*The Color Purple, Hannibal, The Royals*}   **c.** 7

Exercise Set 2.2
1. the set of all composers   **3.** the set of all brands of soft drinks   **5.** {c, d, e}   **7.** {b, c, d, e, f}   **9.** $\{6, 7, 8, 9, ..., 20\}$
11. $\{2, 4, 6, 8, ..., 20\}$ **13.** $\{21, 22, 23, 24, ...\}$ **15.** $\{1, 3, 5, 7, ...\}$ **17.** $\{3, 4, 6, 7\}$;   **19.** {c, d, e};

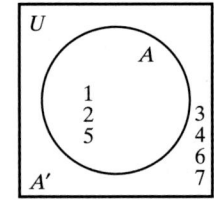 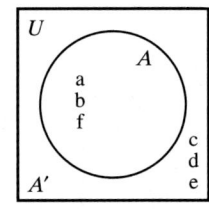

21. $\{0, 2, 4, 6, 8, 10, 11, 12\}$ **23.** ⊆   **25.** ⊄   **27.** ⊄   **29.** ⊄   **31.** ⊆   **33.** ⊄   **35.** both   **37.** ⊆   **39.** neither
41. true   **43.** false; Answers will vary.   **45.** true   **47.** false; Answers will vary.   **49.** true   **51.** false; Answers will vary.
53. true   **55.** ∅ {border collie}, {poodle}, {border collie, poodle},   **57.** ∅, {t}, {a}, {b}, {t, a}, {t, b}, {a, b}, {t, a, b}   **59.** ∅ and {0}
61. 16; 15   **63.** 64; 63   **65.** 128; 127   **67.** 8; 7   **69.** 32   **71.** 64   **73.** 256   **75–85.** Answers will vary.
87. $0.00, $0.05, $0.10, $0.15, $0.25, $0.30, $0.35, and $0.40

Section 2.3
Check Point Exercises
1. a. $\{7, 10\}$ **b.** ∅   **c.** ∅   **2. a.** $\{1, 3, 5, 6, 7, 10, 11\}$ **b.** $\{1, 2, 3, 4, 5, 6, 7\}$ **c.** $\{1, 2, 3\}$ **3. a.** {a, d}   **b.** {a, d}

Exercise Set 2.3
1. $\{1, 3\}$ **3.** $\{1, 2, 3, 5, 7\}$ **5.** $\{2, 4, 6\}$ **7.** $\{4, 6\}$ **9.** $\{1, 3, 5, 7\}$ or A **11.** $\{1, 2, 4, 6, 7\}$ **13.** $\{1, 2, 4, 6, 7\}$ **15.** $\{4, 6\}$
17. $\{1, 3, 5, 7\}$ or A **19.** ∅   **21.** $\{1, 2, 3, 4, 5, 6, 7\}$ or U **23.** $\{1, 3, 5, 7\}$ or A **25.** {g, h}   **27.** {a, b, g, h}   **29.** {b, c, d, e, f} or C
31. {c, d, e, f}   **33.** {a, g, h} or A **35.** {a, b, c, d, e, f, g, h} or U **37.** {a, b, c, d, e, f, g, h} or U **39.** {c, d, e, f}   **41.** {a, g, h} or A
43. ∅   **45.** {a, b, c, d, e, f, g, h} or U **47.** {a, g, h} or A **49.** $\{1, 2, 3, 4, 5, 6, 7, 8\}$ or U **51.** $\{1, 3, 5, 7\}$ or A **53.** $\{1, 7\}$
55. $\{1, 3, 5, 6, 7, 8\}$ **57.** $\{1, 3, 4, 7\}$ **59.** $\{1, 2, 3, 4, 5, 6, 7, 8, 9\}$ **61.** $\{3, 7\}$ **63.** $\{1, 4, 8, 9\}$ **65.** $\{8, 9\}$ **67.** {spacial-temporal
toys, sports equipment, toy cars and trucks}   **69.** {dollhouses, spacial-temporal toys, sports equipment, toy cars and trucks}   **71.** ∅

73–77. Answers will vary. **79.** b **81.** Answers will vary; an example is: If $U = \{1,2\}, A = \{1\}$, and $B = \{2\}$, then $(A \cup B)' = \varnothing$ but $A' \cup B' = \{1,2\}$.

Section 2.4
Check Point Exercises
1. a. $\{a,b,c,d,f\}$ **b.** $\{a,b,c,d,f\}$ **c.** $\{a,b,d\}$ **2. a.** IV, V, VI, and VII **b.** II, III, IV, V, VI, and VII **c.** IV and V
d. I, II, III, and VIII **e.** I, II, III, IV, V, VI, and VII **3. a.** IV **b.** IV **c.** $(A \cup B)' = A' \cap B'$ **4. a.** II, IV, and V
b. II, IV, and V **c.** $A \cap (B \cup C) = (A \cap B) \cup (A \cap C)$

Exercise Set 2.4
1. $\{1,2,3,5,7\}$ **3.** $\{1,2,3,5,7\}$ **5.** $\{2\}$ **7.** $\{2\}$ **9.** $\varnothing$ **11.** $\{4,6\}$ **13.** $\{a,b,g,h\}$ **15.** $\{a,b,g,h\}$ **17.** $\{b\}$
19. $\{b\}$ **21.** $\varnothing$ **23.** $\{c,d,e,f\}$ **25.** II, III, V, and VI **27.** I, II, IV, V, VI, and VII **29.** II and V **31.** I, IV, VII, and VIII
33. $\{1,2,3,4,5,6,7,8\}$ **35.** $\{1,2,3,4,5,6,7,8,9,10,11\}$ **37.** $\{12,13\}$ **39.** $\{4,5,6\}$ **41.** $\{6\}$ **43.** $\{1,2,3,4,5,7,8,9,10,11,12,13\}$
45. a. II **b.** II **c.** $A \cap B = B \cap A$ **47. a.** I, III, and IV **b.** IV **c.** no; Answers will vary. **49.** not equal **51.** not equal
53. equal **55. a.** II, IV, V, VI, and VII **b.** II, IV, V, VI, and VII **c.** $(A \cap B) \cup C = (A \cup C) \cap (B \cup C)$ **57. a.** II, IV, and V
b. I, II, IV, V, and VI **c.** no; Answers will vary. **59.** not equal **61.** equal **63.** equal **65.** I **67.** II **69.** V **71.** V
73. IV **75.** Answers will vary. **77.** AB^+ **79.** no

Section 2.5
Check Point Exercises
1. a. 75 **b.** 90 **c.** 20 **d.** 145 **e.** 55 **f.** 70 **g.** 30 **h.** 175 **2.** 55 **3. a.** 500 **b.** 250 **4. a.** 34 **b.** 10

Exercise Set 2.5
1. 26 **3.** 17 **5.** 37 **7.** 7 **9.** 31 **11.** 27 **13.** 18 **15.** 9 **17.** 9 **19. a–d.**

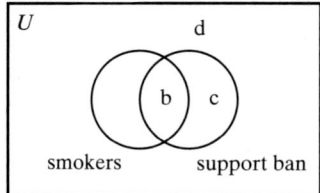

21. a–d.

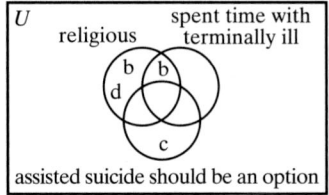

e. Answers will vary.

23.

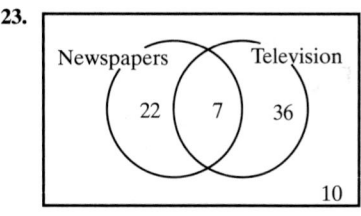

a. 22 **b.** 36 **c.** 65 **d.** 10

25.

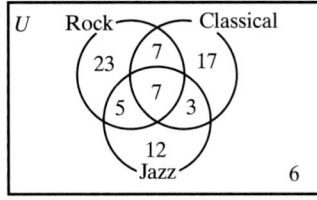

a. 23 **b.** 3 **c.** 32 **d.** 52 **e.** 15 **f.** 6

27. Answers will vary. **29.** b **31.** Under the conditions given concerning enrollment in math, chemistry, and psychology courses, the total number of students is 100, not 90.

Review Exercises
1. Answers will vary; an example is: the set of natural numbers less than 11. **2.** $\{\text{Tuesday, Thursday}\}$
3. Answers will vary. An example is: $\{x|x \in \mathbf{N} \text{ and } x \text{ is between } 4 \text{ and } 8\}$ **4.** $\in$ **5.** $\notin$ **6.** 12
7. 15 **8.** $\neq$ **9.** $\neq$ **10.** equivalent **11.** both
12. $A' = \{1,2,8,9\}$;

13. $\{\text{January, February, March, May, July, August, October, December}\}$
14. $\subseteq$ **15.** $\not\subseteq$ **16.** $\subseteq$ **17.** $\subseteq$ **18.** both **19.** false; Answers will vary.
20. false; Answers will vary. **21.** true **22.** false; Answers will vary.
23. true **24.** false; it has $2^1 = 2$ subsets. **25.** true
26. $\varnothing, \{1\}, \{5\}, \{1,5\}; \{1,5\}$ **27.** 32; 31 **28.** 8; 7 **29.** $\{1,2,4\}$
30. $\{1,2,3,4,6,7,8\}$ **31.** $\{5\}$ **32.** $\{6,7,8\}$ **33.** $\{6,7,8\}$ **34.** $\{4,5,6\}$
35. $\{2,3,6,7\}$ **36.** $\{1,4,5,6,8,9\}$ **37.** $\{4,5\}$ **38.** $\{1,2,3,6,7,8,9\}$
39. $\{2,3,7\}$ **40.** $\{6\}$ **41.** $\{1,2,3,4,5,6,7,8,9\}$ **42.** $\{1,2,3,4,5\}$
43. $\{1,2,3,4,5,6,7,8\}$ or U **44.** $\{c,d,e,f,k,p,r\}$ **45.** $\{f,p\}$ **46.** $\{c,d,f,k,p,r\}$
47. $\{c,d,e\}$ **48.** $\{a,b,c,d,e,g,h,p,r\}$ **49.** $\{f\}$

50. Use Figure 2.9 on page 70.

Set	Regions in the Venn Diagram
A	I, II
B	II, III
$A \cup B$	I, II, III
$(A \cup B)'$	IV

Set	Regions in the Venn Diagram
A'	III, IV
B'	I, IV
$A' \cap B'$	IV

Since $(A \cup B)'$ and $A' \cap B'$ represent the same region, $(A \cup B)' = A' \cap B'$

51. Use Figure 2.10 on page 71.

Set	Regions in the Venn Diagram
A	I, II, IV, V
B	II, III, V, VI
C	IV, V, VI, VII
$B \cup C$	II, III, IV, V, VI, VII
$A \cap (B \cup C)$	II, IV, V

Set	Regions in the Venn Diagram
A	I, II, IV, V
B	II, III, V, VI
C	IV, V, VI, VII
$B \cap C$	V, VI
$A \cup (B \cap C)$	I, II, IV, V, VI

Since $A \cap (B \cup C)$ and $A \cup (B \cap C)$ do not represent the same regions, $A \cap (B \cup C) = A \cup (B \cap C)$ is false.

52. 33   **53. a–c.**   **54. a.** 250   **b.** 800   **c.** 200   **55. a.** 50   **b.** 26   **c.** 130   **d.** 46   **e.** 0

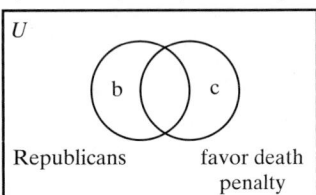
U
b   c
Republicans   favor death penalty

Chapter 2 Test

1. $\{18, 19, 20, 21, 22, 23, 24, 25\}$ **2.** 12   **3.** false; Answers will vary.   **4.** true   **5.** true   **6.** false; Answers will vary.
7. true   **8.** false; Answers will vary.   **9.** false; Answers will vary.   **10.** false; Answers will vary.
11. $\varnothing, \{6\}, \{9\}, \{6, 9\}; \{6, 9\}$ **12.** $\{a, b, c, d, e, f\}$ **13.** $\{a, b, c, d, f, g\}$ **14.** $\{b, c, d\}$ **15.** $\{a, e\}$ **16.** I, II, IV, V, and VI
17. a.   **b.** 30   **c.** 57   **d.** 36   **e.** 120

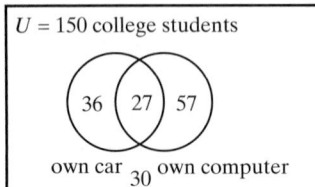
$U = 150$ college students
36   27   57
own car   30   own computer

CHAPTER 3

Section 3.1

Check Point Exercises

1. a. Paris is not the capital of Spain.   **b.** July is a month.   **2. a.** $\sim p$ **b.** $\sim q$ **3.** Tom Hanks is not a jazz singer.
4. Some new tax dollars will not be used to improve education.; At least one new tax dollar will not be used to improve education.

Exercise Set 3.1

1. statement   **3.** statement   **5.** not a statement   **7.** not a statement   **9.** statement   **11.** statement   **13.** It is not raining.
15. *Macbeth* is a comedy show on television.   **17.** $\sim p$ **19.** $\sim r$ **21.** "I Love Lucy" is not a Broadway musical.   **23.** Benjamin Franklin invented the rocking chair.   **25.** The United States is not the country with the most Internet users.   **27.** Michael is the most common boys' name.

29. a. There are no whales that are not mammals. **b.** Some whales are not mammals.. **31. a.** At least one student is a business major. **b.** No students are business majors. **33. a.** At least one thief is not a criminal. **b.** All thieves are criminals. **35. a.** All Democratic presidents have not been impeached. **b.** Some Democratic presidents have been impeached. **37.** b **39–47.** Answers will vary.

Section 3.2
Check Point Exercises
1. a. $p \wedge q$ **b.** $\sim p \wedge q$ **2. a.** $p \vee \sim q$ **b.** $q \vee p$ **3. a.** $p \rightarrow q$ **b.** $\sim p \rightarrow \sim q$ **c.** $\sim q \rightarrow \sim p$
4. a. $q \leftrightarrow p$ **b.** $\sim p \leftrightarrow \sim q$ **5. a.** It is not true that Steven Spielberg is an actor and a director.
b. Steven Spielberg is not an actor and he is a director. **c.** Steven Spielberg is neither an actor nor a director.
6. a. If the plant is fertilized and watered, then it will not wilt. **b.** The plant is fertilized, and if it is watered then it will not wilt.

Exercise Set 3.2
1. $p \wedge q$ **3.** $p \wedge \sim q$ **5.** $p \vee q$ **7.** $p \vee \sim q$ **9.** $p \rightarrow q$ **11.** $\sim p \rightarrow \sim q$ **13.** $p \leftrightarrow q$ **15.** $\sim q \leftrightarrow \sim p$
17. The heater is not working and the house is cold. **19.** The heater is working or the house is not cold. **21.** If the heater is working, then the house is not cold. **23.** The heater is working if and only if the house is not cold. **25.** It is July 4th and we are not having a barbecue.
27. It is not July 4th or we are having a barbecue. **29.** If we are having a barbecue, then it is not July 4th. **31.** It is not July 4th if and only if we are having a barbecue. **33.** It is not true that Romeo loves Juliet and Juliet loves Romeo. **35.** Romeo does not love Juliet and Juliet loves Romeo. **37.** Neither Juliet loves Romeo nor Romeo loves Juliet. **39.** Juliet does not love Romeo or Romeo loves Juliet.
41. Romeo does not love Juliet and Juliet does not love Romeo. **43.** $(p \wedge q) \vee r$ **45.** $(p \vee \sim q) \rightarrow r$ **47.** $r \leftrightarrow (p \wedge \sim q)$
49. If the temperature is above 85° and we have finished studying, then we will go to the beach.
51. The temperature is above 85°, and if we have finished studying then we will go to the beach.
53. If we do not go to the beach, then the temperature is not above 85° or we have not finished studying.
55. If we do not go to the beach then we have not finished studying, or the temperature is above 85°.
57. We will go to the beach if and only if the temperature is above 85° and we have finished studying.
59. The temperature is above 85° if and only if we have finished studying, and we go to the beach.
61. If we do not go to the beach, then it is not true that both the temperature is above 85° and we have finished studying.
63. $p \rightarrow q$ **65.** $(p \rightarrow q) \wedge \sim q$ **67.** $((p \rightarrow q) \wedge \sim p) \rightarrow \sim q$ **69–75.** Answers will vary. **77.** Answers will vary.

Section 3.3
Check Point Exercises
1. true **2.** false **3.**

p	q	$p \vee q$	$\sim(p \vee q)$
T	T	T	F
T	F	T	F
F	T	T	F
F	F	F	T

$\sim(p \vee q)$ is true when both p and q are false; otherwise, $\sim(p \vee q)$ is false.

4.

p	q	$\sim p$	$\sim q$	$\sim p \wedge \sim q$
T	T	F	F	F
T	F	F	T	F
F	T	T	F	F
F	F	T	T	T

$\sim p \wedge \sim q$ is true when both p and q are false; otherwise, $\sim p \wedge \sim q$ is false.

5. Will Smith is not an actor or Bob Dylan is not an actor. **6.**

p	q	$\sim q$	$p \wedge \sim q$	$\sim p$	$(p \wedge \sim q) \vee \sim p$
T	T	F	F	F	F
T	F	T	T	F	T
F	T	F	F	T	T
F	F	T	F	T	T

$(p \wedge \sim q) \vee \sim p$ is false when both p and q are true; otherwise, $(p \wedge \sim q) \vee \sim p$ is true.

Exercise Set 3.3
1. true **3.** false **5.** false **7.** true **9.** true **11.** false

13.

p	$\sim p$	$p \vee \sim p$
T	F	T
F	T	T

15.

p	q	$\sim p$	$\sim p \wedge q$
T	T	F	F
T	F	F	F
F	T	T	T
F	F	T	F

17.

p	q	$p \vee q$	$\sim(p \vee q)$
T	T	T	F
T	F	T	F
F	T	T	F
F	F	F	T

19.

p	q	$\sim p$	$\sim q$	$\sim p \wedge \sim q$
T	T	F	F	F
T	F	F	T	F
F	T	T	F	F
F	F	T	T	T

21.

p	q	$\sim q$	$p \vee \sim q$
T	T	F	T
T	F	T	T
F	T	F	F
F	F	T	T

23.

p	q	$\sim p$	$\sim p \vee q$	$\sim(\sim p \vee q)$
T	T	F	T	F
T	F	F	F	T
F	T	T	T	F
F	F	T	T	F

25.

p	q	$\sim p$	$p \vee q$	$(p \vee q) \wedge \sim p$
T	T	F	T	F
T	F	F	T	F
F	T	T	T	T
F	F	T	F	F

27.

p	q	$\sim p$	$\sim q$	$p \wedge \sim q$	$\sim p \vee (p \wedge \sim q)$
T	T	F	F	F	F
T	F	F	T	T	T
F	T	T	F	F	T
F	F	T	T	F	T

29.

p	q	$\sim p$	$\sim q$	$p \vee q$	$\sim p \vee \sim q$	$(p \vee q) \wedge (\sim p \vee \sim q)$
T	T	F	F	T	F	F
T	F	F	T	T	T	T
F	T	T	F	T	T	T
F	F	T	T	F	T	F

31.

p	q	$\sim q$	$p \wedge \sim q$	$p \wedge q$	$(p \wedge \sim q) \vee (p \wedge q)$
T	T	F	F	T	T
T	F	T	T	F	T
F	T	F	F	F	F
F	F	T	F	F	F

33. Texas is not a state or Paris is not a state.   **35.** *Romeo and Juliet* was not written by Shakespeare or *Jurassic Park* was not written by Shakespeare.   **37.** true   **39.** true   **41.** The chance of divorce peaks during the fourth year of marriage and the chance of divorce does not decrease after four years of marriage.; false   **43.** The chance of divorce does not peak during the fourth year of marriage and more than 2% of all divorces occur during the 25th year of marriage.; false   **45.** The chance of divorce peaks during the fourth year of marriage or the chance of divorce does not decrease after four years of marriage.; true   **47.** The chance of divorce does not peak during the fourth year of marriage or more than 2% of all divorces occur during the 25th year of marriage.; false   **49.** The chance of divorce peaks during the fourth year of marriage and the chance of divorce decreases after four years of marriage, or more than 2% of all divorces occur during the 25th year of marriage.; true   **51–59.** Answers will vary.   **61.** p is false and q is false.

Section 3.4

Check Point Exercises

1.

p	q	$\sim p$	$\sim q$	$\sim p \rightarrow \sim q$
T	T	F	F	T
T	F	F	T	T
F	T	T	F	F
F	F	T	T	T

$\sim p \rightarrow \sim q$ is false when p is false and q is true; otherwise, $\sim p \rightarrow \sim q$ is true.

2. If Shakespeare is not the author, then the play is not *Macbeth*.

3. The table shows that $[(p \rightarrow q) \wedge \sim q] \rightarrow \sim p$ is always true; therefore, it is a tautology.

p	q	$\sim p$	$\sim q$	$p \rightarrow q$	$(p \rightarrow q) \wedge \sim q$	$[(p \rightarrow q) \wedge \sim q] \rightarrow \sim p$
T	T	F	F	T	F	T
T	F	F	T	F	F	T
F	T	T	F	T	F	T
F	F	T	T	T	T	T

4.

p	q	$p \rightarrow q$	$\sim p$	$\sim q$	$\sim p \rightarrow \sim q$	$(p \rightarrow q) \wedge (\sim p \rightarrow \sim q)$
T	T	T	F	F	T	T
T	F	F	F	T	T	F
F	T	T	T	F	F	F
F	F	T	T	T	T	T

$(p \rightarrow q) \wedge (\sim p \rightarrow \sim q)$ is true when p and q are both true or both false; otherwise, $(p \rightarrow q) \wedge (\sim p \rightarrow \sim q)$ is false.

5.

p	q	$p \vee q$	$\sim p$	$\sim p \rightarrow q$	$(p \vee q) \leftrightarrow (\sim p \rightarrow q)$
T	T	T	F	T	T
T	F	T	F	T	T
F	T	T	T	T	T
F	F	F	T	F	T

Because all cases are true, the statement is a tautology.

6.

p	q	r	$p \rightarrow q$	$(p \rightarrow q) \wedge r$
T	T	T	T	T
T	T	F	T	F
T	F	T	F	F
T	F	F	F	F
F	T	T	T	T
F	T	F	T	F
F	F	T	T	T
F	F	F	T	F

$(p \rightarrow q) \wedge r$ is false when r is false or when p is true and q is false; otherwise, $(p \rightarrow q) \wedge r$ is true. It is not a tautology.

7. $(T \vee F) \leftrightarrow \sim T$
    $T \leftrightarrow F$
      F
The statement is false.

Exercise Set 3.4

1.

p	q	$\sim q$	$p \rightarrow \sim q$
T	T	F	F
T	F	T	T
F	T	F	T
F	F	T	T

3.

p	q	$q \rightarrow p$	$\sim(q \rightarrow p)$
T	T	T	F
T	F	T	F
F	T	F	T
F	F	T	F

5.

p	q	$p \wedge q$	$p \vee q$	$(p \wedge q) \rightarrow (p \vee q)$
T	T	T	T	T
T	F	F	T	T
F	T	F	T	T
F	F	F	F	T

7.

p	q	$p \rightarrow q$	$\sim q$	$(p \rightarrow q) \wedge \sim q$
T	T	T	F	F
T	F	F	T	F
F	T	T	F	F
F	F	T	T	T

9. If I don't live in Washington, then I don't live in Seattle.

11. If a plant doesn't wilt, then it is being watered.

13. not a tautology

p	q	$p \rightarrow q$	$(p \rightarrow q) \wedge q$	$[(p \rightarrow q) \wedge q] \rightarrow p$
T	T	T	T	T
T	F	F	F	T
F	T	T	T	F
F	F	T	F	T

15. tautology

p	q	$\sim p$	$\sim q$	$p \rightarrow q$	$(p \rightarrow q) \wedge \sim q$	$[(p \rightarrow q) \wedge \sim q] \rightarrow \sim p$
T	T	F	F	T	F	T
T	F	F	T	F	F	T
F	T	T	F	T	F	T
F	F	T	T	T	T	T

17. not a tautology

p	q	$\sim q$	$p \vee q$	$(p \vee q) \wedge p$	$[(p \vee q) \wedge p] \rightarrow \sim q$
T	T	F	T	T	F
T	F	T	T	T	T
F	T	F	T	F	T
F	F	T	F	F	T

19. tautology

p	q	$\sim p$	$p \rightarrow q$	$\sim p \vee q$	$(p \rightarrow q) \rightarrow (\sim p \vee q)$
T	T	F	T	T	T
T	F	F	F	F	T
F	T	T	T	T	T
F	F	T	T	T	T

21. not a tautology

p	q	$\sim p$	$\sim q$	$p \wedge q$	$\sim p \vee \sim q$	$(p \wedge q) \wedge (\sim p \vee \sim q)$
T	T	F	F	T	F	F
T	F	F	T	F	T	F
F	T	T	F	F	T	F
F	F	T	T	F	T	F

23.

p	q	$\sim q$	$p \leftrightarrow \sim q$
T	T	F	F
T	F	T	T
F	T	F	T
F	F	T	F

25.

p	q	$p \leftrightarrow q$	$\sim(p \leftrightarrow q)$
T	T	T	F
T	F	F	T
F	T	F	T
F	F	T	F

27.

p	q	$p \leftrightarrow q$	$(p \leftrightarrow q) \rightarrow p$
T	T	T	T
T	F	F	T
F	T	F	F
F	F	T	F

29.

p	q	$\sim p$	$\sim p \leftrightarrow q$	$\sim p \rightarrow q$	$(\sim p \leftrightarrow q) \rightarrow (\sim p \rightarrow q)$
T	T	F	F	T	T
T	F	F	T	T	T
F	T	T	T	T	T
F	F	T	F	F	T

31.

p	q	$p \wedge q$	$q \rightarrow p$	$(p \wedge q) \wedge (q \rightarrow p)$	$[(p \wedge q) \wedge (q \rightarrow p)] \leftrightarrow (p \wedge q)$
T	T	T	T	T	T
T	F	F	T	F	T
F	T	F	F	F	T
F	F	F	T	F	T

33. not a tautology

p	q	$\sim p$	$\sim q$	$p \wedge q$	$\sim(p \wedge q)$	$\sim p \wedge \sim q$	$\sim(p \wedge q) \leftrightarrow (\sim p \wedge \sim q)$
T	T	F	F	T	F	F	T
T	F	F	T	F	T	F	F
F	T	T	F	F	T	F	F
F	F	T	T	F	T	T	T

35. not a tautology

p	q	$p \rightarrow q$	$q \rightarrow p$	$(p \rightarrow q) \leftrightarrow (q \rightarrow p)$
T	T	T	T	T
T	F	F	T	F
F	T	T	F	F
F	F	T	T	T

37. tautology

p	q	$\sim p$	$p \rightarrow q$	$\sim p \vee q$	$(p \rightarrow q) \leftrightarrow (\sim p \vee q)$
T	T	F	T	T	T
T	F	F	F	F	T
F	T	T	T	T	T
F	F	T	T	T	T

39. tautology

p	q	$p \leftrightarrow q$	$p \rightarrow q$	$q \rightarrow p$	$(q \rightarrow p) \wedge (p \rightarrow q)$	$(p \leftrightarrow q) \leftrightarrow [(q \rightarrow p) \wedge (p \rightarrow q)]$
T	T	T	T	T	T	T
T	F	F	F	T	F	T
F	T	F	T	F	F	T
F	F	T	T	T	T	T

41.

p	q	r	$\sim q$	$p \wedge \sim q$	$(p \wedge \sim q) \vee r$
T	T	T	F	F	T
T	T	F	F	F	F
T	F	T	T	T	T
T	F	F	T	T	T
F	T	T	F	F	T
F	T	F	F	F	F
F	F	T	T	F	T
F	F	F	T	F	F

43. not a tautology

p	q	r	$p \vee q$	$(p \vee q) \rightarrow r$
T	T	T	T	T
T	T	F	T	F
T	F	T	T	T
T	F	F	T	F
F	T	T	T	T
F	T	F	T	F
F	F	T	F	T
F	F	F	F	T

45. tautology

p	q	r	$p \wedge q$	$p \vee r$	$(p \wedge q) \rightarrow (p \vee r)$
T	T	T	T	T	T
T	T	F	T	T	T
T	F	T	F	T	T
T	F	F	F	T	T
F	T	T	F	T	T
F	T	F	F	F	T
F	F	T	F	T	T
F	F	F	F	F	T

47. not a tautology

p	q	r	$p \wedge q$	$r \rightarrow (p \wedge q)$
T	T	T	T	T
T	T	F	T	T
T	F	T	F	F
T	F	F	F	T
F	T	T	F	F
F	T	F	F	T
F	F	T	F	F
F	F	F	F	T

49. tautology

p	q	r	$p \rightarrow q$	$q \rightarrow r$	$(p \rightarrow q) \wedge (q \rightarrow r)$	$p \rightarrow r$	$[(p \rightarrow q) \wedge (q \rightarrow r)] \rightarrow (p \rightarrow r)$
T	T	T	T	T	T	T	T
T	T	F	T	F	F	F	T
T	F	T	F	T	F	T	T
T	F	F	F	T	F	F	T
F	T	T	T	T	T	T	T
F	T	F	T	F	F	T	T
F	F	T	T	T	T	T	T
F	F	F	T	T	T	T	T

51. false **53.** true **55.** false **57.** true **59.** true **61.** false **63.** true **65.** true **67.** true **69.** false
71–75. Answers will vary. **77.** d **79.** Answers will vary; examples are: $p \rightarrow q, \sim p, (p \rightarrow q) \vee \sim p$, and $\sim p \rightarrow [(p \rightarrow q) \vee \sim p]$. **81.** 2^n

Section 3.5
Check Point Exercises

1. a.

p	q	$\sim q$	$p \vee q$	$\sim q \rightarrow p$
T	T	F	T	T
T	F	T	T	T
F	T	F	T	T
F	F	T	F	F

The statements are equivalent since their truth values are the same.

2.

p	$\sim p$	$\sim(\sim p)$	$\sim[\sim(\sim p)]$
T	F	T	F
F	T	F	T

Since the truth values are the same, $\sim[\sim(\sim p)] \equiv \sim p$.

b. If I don't lose my scholarship, then I attend classes.
3. c   **4. a.** If you do not pay a fine, then the book is not overdue.   **b.** If we use the pool, then it is not cold.
c. If supervision is needed, then some students do not take exams honestly.   **d.** $q \rightarrow \sim p$
5. Converse: If it can fly, then it's a bird.; Inverse: If it's not a bird, then it can't fly.   **6.** The triangle is isosceles and it does not have two equal sides.
7. Kelsey Grammer is not an actor or Katie Couric is not an actor.   **8.** Oprah Winfrey is not a jazz musician and Oprah Winfrey is not a presidential candidate.   **9.** You do not leave by 5 P.M. and you will arrive home on time.   **10.** If we cannot swim or we can sail, it is windy.

Exercise Set 3.5

1. a.

p	q	$\sim p$	$p \rightarrow q$	$\sim p \vee q$
T	T	F	T	T
T	F	F	F	F
F	T	T	T	T
F	F	T	T	T

b. A number is not even or it is divisible by 2.

3. a   **5.** c   **7.** If I am not in Illinois, then I am not in Chicago.   **9.** If I can hear you, then the stereo is not playing.
11. If you don't die, you laugh.   **13.** If some troops were not withdrawn, then the president was not telling the truth.
15. If no people suffer, then some institutions do not place profit above human need.   **17.** $r \rightarrow q$
19. Converse: If you get a skin rash, then you have touched poison oak.; Inverse: If you do not touch poison oak, then you will not get a skin rash.
21. Converse: If Shakespeare is the author, then the play is *Macbeth*.; Inverse: If the play is not *Macbeth*, then Shakespeare is not the author.
23. Converse: If you are not sleeping, then you are driving the car.; Inverse: If you are not driving the car, then you are sleeping.
25. Converse: If I am not in the West, then I am in Charleston.; Inverse: If I am not in Charleston, then I am in the West.
27. Converse: If some people wear green, then it's St. Patrick's Day.; Inverse: If it's not St. Patrick's Day, then no people wear green.
29. Converse: $\sim r \rightarrow \sim q$; Inverse: $q \rightarrow r$.   **31.** I am in Los Angeles and not in California.   **33.** It is purple and it is a carrot.
35. He doesn't and I won't.   **37.** There is a blizzard and some schools are not closed.   **39.** $\sim q \wedge r$
41. Australia is not an island or China is not an island.   **43.** My high school did not encourage creativity or it did not encourage diversity.
45. Babe Ruth was neither a writer nor a lawyer.   **47.** The United States has eradicated neither poverty nor racism.   **49.** $p \vee \sim q$
51. If you do not succeed, you did not attend lecture or did not study.   **53.** If his wife does not and his child does not cook, then he does.
55. $(\sim q \wedge r) \rightarrow \sim p$ **57.** I'm going to neither Seattle nor San Francisco.   **59.** I do not study and I pass.   **61.** I am going or he is not going.
63. A bill does not become law or it receives majority approval.   **65.** $\sim p \wedge q$ **67.** $\sim p \vee (\sim q \wedge \sim r)$ **69.** none   **71.** none
73. none   **75.** a and b   **77. a.** true   **b.** If it is not the case that 96% of the patients are women, then the procedure is not a chemical peel.; true   **c.** If 96% of the patients are women, then the procedure is a chemical peel.; not necessarily true   **d.** If the procedure is not a chemical peel, then it is not the case that 96% of the patients are women.; not necessarily true   **e.** The procedure is a chemical peel and it is not the case that 96% of the patients are women.; false   **79. a.** true   **b.** There are not 32,283 face lifts each year and it is not the case that 8% of the face-lift patients are men.   **c.** false   **81–87.** Answers will vary.   **89.** b   **91.** We will replace or repair the roof, or we will not sell the house.

Section 3.6
Check Point Exercises
1. valid   **2.** invalid   **3.** invalid   **4.** It is not midnight.

Exercise Set 3.6
1. invalid   **3.** valid   **5.** valid   **7.** invalid

9.
$$\begin{array}{c} p \rightarrow \sim q \\ q \\ \hline \therefore \sim p \end{array}$$
valid

11.
$$\begin{array}{c} p \vee q \\ q \\ \hline \therefore \sim p \end{array}$$
invalid

13.
$$\begin{array}{c} p \rightarrow q \\ \sim p \\ \hline \therefore \sim q \end{array}$$
invalid

15. invalid   **17.** valid   **19.** valid

21.
$$\begin{array}{c} (p \wedge q) \rightarrow r \\ p \wedge \sim r \\ \hline \therefore \sim q \end{array}$$
valid

23.
$$\begin{array}{c} (p \vee q) \rightarrow r \\ \sim r \\ \hline \therefore \sim p \wedge \sim q \end{array}$$
valid

25.
$$\begin{array}{c} (p \vee q) \rightarrow r \\ r \\ \hline \therefore p \vee q \end{array}$$
invalid

27. $(p \wedge q) \to r$
$\underline{\sim p \vee \sim q}$
$\therefore \sim r$
invalid

29. $p \to q$
$\underline{q}$
$\therefore p$
invalid

31. $p \vee q$
$\underline{\sim q}$
$\therefore p$
valid

33. $p \to q$
$\underline{q \to r}$
$\therefore p \to r$
valid

35. $p \to q$
$q \to r$
$\overline{\therefore \sim p \to \sim r}$
invalid

37. My best friend is not a chemist.
39. They were dropped from prime time.
41. Some electricity is not off.

43. If I vacation in Paris, I gain weight. **45.** If the power comes on, Mr. Scott is still with us.

47. $p \to q$
$\underline{q}$
$\therefore p$
invalid

49–55. Answers will vary.

57. $p \to q$
$\sim p \to \sim r$
$\underline{r}$
$\therefore q$
valid

Section 3.7
Check Point Exercises
1. valid **2.** invalid **3.** valid **4.** invalid **5.** invalid **6.** invalid

Exercise Set 3.7
1. valid **3.** invalid **5.** valid **7.** invalid **9.** invalid **11.** valid **13.** valid **15.** invalid **17.** invalid **19.** valid
21. valid **23.** valid **25.** valid **27–29.** Answers will vary. **31.** All opera singers take voice lessons.

Review Exercises
1. If the temperature is below 32° and we have finished studying, we will go to the movies.
2. If we do not go to the movies, then the temperature is not below 32° or we have not finished studying.
3. The temperature is below 32°, and if we finish studying, we will go to the movies.
4. We will go to the movies if and only if the temperature is below 32° and we have finished studying.
5. It is not true that both the temperature is below 32° and we have finished studying.
6. We will not go to the movies if and only if the temperature is not below 32° or we have not finished studying.
7. $(p \wedge q) \vee r$ **8.** $(p \vee \sim q) \to r$ **9.** $q \to (p \leftrightarrow r)$ **10.** $r \leftrightarrow (p \wedge \sim q)$ **11.** Some houses are not made with wood.
12. Some students major in business. **13.** No crimes are motivated by passion. **14.** All Democrats are registered voters.
15. Some new taxes will not be used for education. **16.** false **17.** true **18.** true

19.

p	q	$\sim p$	$\sim p \wedge q$	$p \vee (\sim p \wedge q)$
T	T	F	F	T
T	F	F	F	T
F	T	T	T	T
F	F	T	F	F

20.

p	q	$\sim p$	$\sim q$	$\sim p \vee \sim q$
T	T	F	F	F
T	F	F	T	T
F	T	T	F	T
F	F	T	T	T

21.

p	q	$\sim p$	$\sim p \vee q$	$p \to (\sim p \vee q)$
T	T	F	T	T
T	F	F	F	F
F	T	T	T	T
F	F	T	T	T

22.

p	q	$\sim q$	$p \leftrightarrow \sim q$
T	T	F	F
T	F	T	T
F	T	F	T
F	F	T	F

23.

p	q	$\sim p$	$\sim q$	$p \vee q$	$\sim(p \vee q)$	$\sim p \wedge \sim q$	$\sim(p \vee q) \to (\sim p \wedge \sim q)$
T	T	F	F	T	F	F	T
T	F	F	T	T	F	F	T
F	T	T	F	T	F	F	T
F	F	T	T	F	T	T	T

24.

p	q	r	$\sim r$	$p \vee q$	$(p \vee q) \to \sim r$
T	T	T	F	T	F
T	T	F	T	T	T
T	F	T	F	T	F
T	F	F	T	T	T
F	T	T	F	T	F
F	T	F	T	T	T
F	F	T	F	F	T
F	F	F	T	F	T

25.

p	q	r	$p \wedge q$	$p \wedge r$	$(p \wedge q) \leftrightarrow (p \wedge r)$
T	T	T	T	T	T
T	T	F	T	F	F
T	F	T	F	T	F
T	F	F	F	F	T
F	T	T	F	F	T
F	T	F	F	F	T
F	F	T	F	F	T
F	F	F	F	F	T

26. true **27.** false **28.** false **29.** true **30.** true

31. a.

p	q	$\sim p$	$\sim p \vee q$	$p \rightarrow q$
T	T	F	T	T
T	F	F	F	F
F	T	T	T	T
F	F	T	T	T

32. c **33.** If I am not in the South, then I am not in Atlanta. **34.** If today is a holiday, then I am not in class. **35.** If I do not pass some course, then I did not work hard. **36.** Converse: If classes are cancelled, then there is a storm.; Inverse: If there is not a storm, then classes are not cancelled. **37.** Converse: If we do not talk, then the television is on.; Inverse: If the television is not on, then we talk. **38.** I am in Bogota and I am not in Columbia. **39.** I do not work hard and I succeed.

40. Chicago is not a city or Maine is not a city. **41.** Ernest Hemingway was neither a musician nor an actor. **42.** If a number is not 0, the number is positive or negative. **43.** I do not work hard and I succeed. **44.** She is using her car or she is not taking a bus. **45.** $p \wedge \sim q$
46. a and c **47.** a and b **48.** a and c **49.** invalid **50.** valid

51. $p \rightarrow q$
$\sim q$
$\therefore \sim p$
valid

52. $p \rightarrow q$
q
$\therefore p$
invalid

53. $p \vee q$
q
$\therefore \sim p$
invalid

54. $p \vee q$
$\sim p$
$\therefore q$
valid

55. $p \rightarrow q$
$q \rightarrow r$
$\therefore r \rightarrow p$
invalid

56. invalid **57.** valid **58.** valid **59.** invalid **60.** invalid **61.** valid

Chapter 3 Test

1. If I'm registered and I'm a citizen, then I vote. **2.** I don't vote if and only if I'm not registered or I'm not a citizen.
3. I'm neither registered nor a citizen. **4.** $(p \wedge q) \vee \sim r$ **5.** $(\sim p \vee \sim q) \rightarrow \sim r$ **6.** Some numbers are not divisible by 5.
7. No people wear glasses.

8.

p	q	$\sim p$	$\sim p \vee q$	$p \wedge (\sim p \vee q)$
T	T	F	T	T
T	F	F	F	F
F	T	T	T	F
F	F	T	T	F

9.

p	q	$\sim p$	$\sim q$	$p \wedge q$	$\sim(p \wedge q)$	$(\sim p \vee \sim q)$	$\sim(p \wedge q) \leftrightarrow (\sim p \vee \sim q)$
T	T	F	F	T	F	F	T
T	F	F	T	F	T	T	T
F	T	T	F	F	T	T	T
F	F	T	T	F	T	T	T

10.

p	q	r	$q \vee r$	$p \leftrightarrow (q \vee r)$
T	T	T	T	T
T	T	F	T	T
T	F	T	T	T
T	F	F	F	F
F	T	T	T	F
F	T	F	T	F
F	F	T	T	F
F	F	F	F	T

11. true **12.** true
13. b **14.** If it snows, then it is not August.
15. Converse: If I cannot concentrate, then the radio is playing.; Inverse: If the radio is not playing, then I can concentrate.
16. It is cold and we use the pool.
17. The test is not today and the party is not tonight.
18. The banana is not green or it is ready to eat.
19. a and b **20.** a and c
21. invalid **22.** valid **23.** invalid
24. invalid **25.** valid **26.** invalid

CHAPTER 4

Section 4.1

Check Point Exercises
1. a. 100,000 **b.** 1,000,000 **2. a.** 64 **b.** 216 **c.** 32 **d.** 18
3. a. $(4 \times 10^3) + (0 \times 10^2) + (2 \times 10^1) + (6 \times 1)$ or $(4 \times 1000) + (0 \times 100) + (2 \times 10) + (6 \times 1)$
b. $(2 \times 10^4) + (4 \times 10^3) + (2 \times 10^2) + (3 \times 10^1) + (2 \times 1)$ or $(2 \times 10,000) + (4 \times 1000) + (2 \times 100) + (3 \times 10) + (2 \times 1)$
4. a. 6073 **b.** 80,900 **5.** 12,031 **6.** 80,293

Exercise Set 4.1
1. 25 **3.** 8 **5.** 81 **7.** 100,000 **9.** $(3 \times 10^1) + (6 \times 1)$ **11.** $(2 \times 10^2) + (4 \times 10^1) + (9 \times 1)$
13. $(7 \times 10^2) + (0 \times 10^1) + (3 \times 1)$ **15.** $(4 \times 10^3) + (8 \times 10^2) + (5 \times 10^1) + (6 \times 1)$
17. $(3 \times 10^3) + (0 \times 10^2) + (7 \times 10^1) + (0 \times 1)$ **19.** $(3 \times 10^4) + (4 \times 10^3) + (5 \times 10^2) + (6 \times 10^1) + (9 \times 1)$
21. $(2 \times 10^8) + (3 \times 10^7) + (0 \times 10^6) + (0 \times 10^5) + (0 \times 10^4) + (7 \times 10^3) + (0 \times 10^2) + (0 \times 10^1) + (4 \times 1)$
23. 73 **25.** 385 **27.** 528,743 **29.** 7002 **31.** 600,002,007 **33.** 23 **35.** 1262 **37.** 11,523 **39.** 2,416,271
41. 14 **43.** 6846 **45.** 3048 **47.** 14,411 **49.** 75,610 **51.** 1776; Declaration of Independence **53.** Answers will vary.
55. base: 10; digits: 0, 1, 2, 3, 4, 5, 6, 7, 8, and 9 **57–59.** Answers will vary. **61.**  **63.** $<\vee \ <<<<<\vee\vee\vee\vee\vee\vee\vee ; <\vee\vee \lessgtr$
65. 625 **67.** 592,704

Section 4.2

Check Point Exercises
1. 487 **2.** 51 **3.** 2772 **4.** 11_{five} **5.** 1031_{seven} **6.** 42023_{five}

Exercise Set 4.2

1. 23 **3.** 42 **5.** 30 **7.** 11 **9.** 455 **11.** 28,909 **13.** 8342 **15.** 53 **17.** 44,261 **19.** 12_{five} **21.** 14_{seven}
23. 10_{two} **25.** 31_{four} **27.** 101_{six} **29.** 322_{five} **31.** 1230_{four} **33.** 10011_{two} **35.** 12010_{three} **37.** 1442_{six} **39.** 4443_{seven}
41. 21 weeks and 6 days **43.** 35 quarters, 0 nickels, and 4 pennies **45.** Answers will vary. **47.** 887_{nine}; 1000_{nine}
49. 11111011_{two}, 673_{eight}, $3A6_{\text{twelve}}$ **51.** 37 **53.** 1000110111_{two}

Section 4.3

Check Point Exercises

1. 131_{five} **2.** 1110_{two} **3.** 13_{five} **4.** 1605_{seven} **5.** 201_{seven} **6.** 23_{four}

Exercise Set 4.3

1. 102_{four} **3.** 110_{two} **5.** 1310_{five} **7.** 1302_{seven} **9.** 15760_{nine} **11.** 23536_{seven} **13.** 3_{four} **15.** 4_{five} **17.** 206_{eight}
19. 366_{seven} **21.** 10_{two} **23.** 111_{three} **25.** 152_{six} **27.** 11_{two} **29.** 4011_{seven} **31.** 3114_{eight} **33.** 312_{four} **35.** 20_{four}
37. 41_{five} remainder of 1_{five} **39–41.** Answers will vary. **43.** 1367_{eight} **45.** $12F_{\text{sixteen}}$

Section 4.4

Check Point Exercises

1. 300,222 **2.** [Egyptian hieroglyphic numeral] **3.** 1361 **4.** 1447 **5.** CCCXCIX **6.** [Chinese numeral] **7.** 885

Exercise Set 4.4

1. 322 **3.** 300,423 **5.** 132 **7.** [Egyptian hieroglyphic numeral] **9.** [Egyptian hieroglyphic numeral]
11. [Egyptian hieroglyphic numeral] **13.** 11 **15.** 16 **17.** 40 **19.** 59 **21.** 146 **23.** 1621 **25.** 2677
27. 9466 **29.** XLIII **31.** CXXIX **33.** MDCCCXCVI **35.** $\overline{\text{V}}$DCCCXCII **37.** 88 **39.** 527 **41.** 2776
43. [Chinese numeral] **45.** [Chinese numeral] **47.** [Chinese numeral] **49.** 12 **51.** 234 **53.** $\mu\gamma$
55. $\upsilon\pi\gamma$ **57.** 1776; Declaration of Independence
59–63. Answers will vary.
65. Preceding: [Egyptian hieroglyphic numeral]
Following: [Egyptian hieroglyphic numeral]

Review Exercises

1. 121 **2.** 343 **3.** $(4 \times 10^2) + (7 \times 10^1) + (2 \times 1)$ **4.** $(8 \times 10^3) + (0 \times 10^2) + (7 \times 10^1) + (6 \times 1)$
5. $(7 \times 10^4) + (0 \times 10^3) + (3 \times 10^2) + (2 \times 10^1) + (9 \times 1)$ **6.** 706,953 **7.** 740,000,306 **8.** 673 **9.** 8430 **10.** 2331
11. 65,536 **12.** Each position represents a particular value. The symbol in each position tells how many of that value are represented.
13. 19 **14.** 6 **15.** 325 **16.** 805 **17.** 4051 **18.** 560 **19.** 324_{five} **20.** 10101_{two} **21.** 122112_{three} **22.** 26452_{seven}
23. 111140_{six} **24.** $3A2_{\text{twelve}}$ **25.** 132_{seven} **26.** 1401_{eight} **27.** 110000_{two} **28.** $AD0_{\text{sixteen}}$ **29.** 5_{six} **30.** 345_{seven} **31.** 11_{two}
32. 2304_{five} **33.** 222_{four} **34.** 354_{seven} **35.** 1102_{five} **36.** 133_{four} **37.** 12_{five} **38.** 1246 **39.** 12,432
40. [Egyptian hieroglyphic numeral] **41.** [Egyptian hieroglyphic numeral] **42.** 2314
43. DDDDDCCCCBBBBBBBBBBAA **44.** Answers will vary. **45.** 163 **46.** 1034 **47.** 1990 **48.** XLIX
49. MMCMLXV **50.** If symbols increase in value from left to right, subtract the value of the symbol on the left from the symbol on the right.
Answers will vary. **51.** 554 **52.** 8253 **53.** [Chinese numeral] **54.** [Chinese numeral] **55.** 365 **56.** 4520 **57.** G **58.** F **59.** Answers will vary.
Y Z
I H
X Y
C E
X
D
60. 653 **61.** 678 **62.** $\upsilon\nu\gamma$ **63.** $\pi\beta$ **64.** 357 **65.** 37,894 **66.** 80,618 **67.** WRG **68.** IfVQC **69.** Answers will vary.

Chapter 4 Test

1. 729 **2.** $(5 \times 10^2) + (6 \times 10^1) + (7 \times 1)$ **3.** $(6 \times 10^4) + (3 \times 10^3) + (0 \times 10^2) + (2 \times 10^1) + (8 \times 1)$ **4.** 7493 **5.** 400,206
6. A number represents "How many?" whereas a numeral is a symbol used to write a number. **7.** A symbol for zero is needed for a place holder when there are no values for a position. **8.** 72,731 **9.** 1560 **10.** 113 **11.** 223 **12.** 53 **13.** 2212_{three} **14.** 111000_{two}
15. 24334_{five} **16.** 1212_{five} **17.** 414_{seven} **18.** 250_{six} **19.** 221_{five} **20.** 20,303 **21.**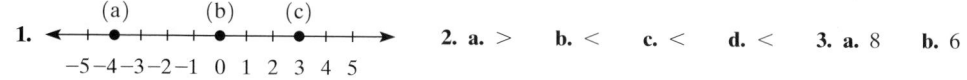
22. 1994 **23.** CDLIX **24.** Answers will vary.

CHAPTER 5

Section 5.1
Check Point Exercises
1. b **2.** $2^3 \cdot 3 \cdot 5$ **3.** 75 **4.** 96 **5.** 90 **6.** 5:00 P.M.

Exercise Set 5.1
1. a. yes **b.** no **c.** yes **d.** no **e.** no **f.** yes **g.** no **h.** no **i.** no **3. a.** yes **b.** yes **c.** yes **d.** no
e. yes **f.** yes **g.** no **h.** no **i.** yes **5. a.** yes **b.** no **c.** yes **d.** no **e.** no **f.** no **g.** no **h.** no **i.** no
7. a. yes **b.** yes **c.** yes **d.** no **e.** yes **f.** yes **g.** yes **h.** no **i.** yes **9. a.** yes **b.** yes **c.** yes **d.** yes
e. yes **f.** yes **g.** yes **h.** yes **i.** yes **11.** true; $5 + 9 + 5 + 8 = 27$ which is divisible by 3. **13.** true; the last two digits of 10,612 form the number 12 and 12 is divisible by 4. **15.** false **17.** true; 104,538 is an even number so it is divisible by 2. $1 + 0 + 4 + 5 + 3 + 8 = 21$ which is divisible by 3 so it is divisible by 3. Any number divisible by 2 and 3 is divisible by 6. **19.** true; the last three digits of 20,104 form the number 104 and 104 is divisible by 8. **21.** false **23.** true; $5 + 1 + 7 + 8 + 7 + 2 = 30$ which is divisible by 3 so it is divisible by 3. The last two digits of 517,872 form the number 72 and 72 is divisible by 4 so it is divisible by 4. Any number divisible by 3 and 4 is divisible by 12.
25. $3 \cdot 5^2$ **27.** $2^3 \cdot 7$ **29.** $3 \cdot 5 \cdot 7$ **31.** $2^2 \cdot 5^3$ **33.** $3 \cdot 13 \cdot 17$ **35.** $3 \cdot 5 \cdot 59$ **37.** $2^5 \cdot 3^2 \cdot 5$ **39.** $2^2 \cdot 499$ **41.** $3 \cdot 5^2 \cdot 7^2$
43. $2^3 \cdot 3 \cdot 5^2 \cdot 11 \cdot 13$ **45.** 14 **47.** 2 **49.** 12 **51.** 24 **53.** 38 **55.** 15 **57.** 168 **59.** 336 **61.** 540 **63.** 360
65. 3420 **67.** 4560 **69.** The numbers are the prime numbers between 1 and 97. **71.** 12 **73.** 10 **75.** July 1 **77.** 90 min or $1\frac{1}{2}$ hr
79–89. Answers will vary. **91.** GCD: $2^{14} \cdot 3^{25} \cdot 5^{30}$; LCM: $2^{17} \cdot 3^{37} \cdot 5^{31}$ **93.** 2:20 A.M. on the third day **95.** no

Section 5.2
Check Point Exercises
1.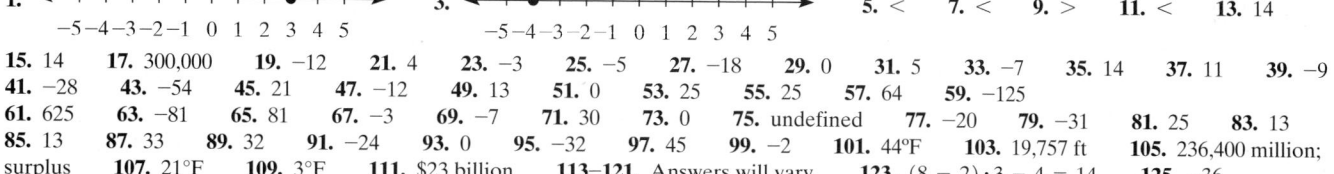

(a) (b) (c)

$-5\,-4\,-3\,-2\,-1\;\;0\;\;1\;\;2\;\;3\;\;4\;\;5$ **2. a.** > **b.** < **c.** < **d.** < **3. a.** 8 **b.** 6
4. a. 37 **b.** −4 **c.** −24 **5.** 19,763 m **6. a.** 64 **b.** 64 **c.** 343 **d.** −343 **e.** 81 **7.** 36 **8.** 82

Exercise Set 5.2
1. $-5\,-4\,-3\,-2\,-1\;\;0\;\;1\;\;2\;\;3\;\;4\;\;5$ **3.** $-5\,-4\,-3\,-2\,-1\;\;0\;\;1\;\;2\;\;3\;\;4\;\;5$ **5.** < **7.** < **9.** > **11.** < **13.** 14
15. 14 **17.** 300,000 **19.** −12 **21.** 4 **23.** −3 **25.** −5 **27.** −18 **29.** 0 **31.** 5 **33.** −7 **35.** 14 **37.** 11 **39.** −9
41. −28 **43.** −54 **45.** 21 **47.** −12 **49.** 13 **51.** 0 **53.** 25 **55.** 25 **57.** 64 **59.** −125
61. 625 **63.** −81 **65.** 81 **67.** −3 **69.** −7 **71.** 30 **73.** 0 **75.** undefined **77.** −20 **79.** −31 **81.** 25 **83.** 13
85. 13 **87.** 33 **89.** 32 **91.** −24 **93.** 0 **95.** −32 **97.** 45 **99.** −2 **101.** 44°F **103.** 19,757 ft **105.** 236,400 million;
surplus **107.** 21°F **109.** 3°F **111.** $23 billion **113–121.** Answers will vary. **123.** $(8 - 2) \cdot 3 - 4 = 14$ **125.** −36

Section 5.3
Check Point Exercises
1. $\frac{4}{5}$ **2.** $\frac{21}{8}$ **3.** $1\frac{2}{3}$ **4. a.** 0.375 **b.** $0.\overline{45}$ **5. a.** $\frac{9}{10}$ **b.** $\frac{43}{50}$ **c.** $\frac{53}{1000}$ **6.** $\frac{2}{9}$ **7.** $\frac{79}{99}$ **8. a.** $\frac{8}{33}$ **b.** $\frac{3}{2}$ or $1\frac{1}{2}$
c. $\frac{51}{10}$ or $5\frac{1}{10}$ **9. a.** $\frac{36}{55}$ **b.** $-\frac{4}{3}$ or $-1\frac{1}{3}$ **c.** $\frac{3}{2}$ or $1\frac{1}{2}$ **10. a.** $\frac{2}{3}$ **b.** $\frac{3}{2}$ or $1\frac{1}{2}$ **c.** $-\frac{9}{4}$ or $-2\frac{1}{4}$ **11.** $\frac{19}{20}$ **12.** $-\frac{17}{60}$
13. $\frac{5}{12}$ **14.** $2\frac{4}{5}$ eggs; 3 eggs

Exercise Set 5.3
1. $\frac{2}{3}$ **3.** $\frac{5}{6}$ **5.** $\frac{4}{7}$ **7.** $\frac{5}{9}$ **9.** $\frac{9}{10}$ **11.** $\frac{14}{19}$ **13.** $\frac{19}{8}$ **15.** $-\frac{38}{5}$ **17.** $\frac{199}{16}$ **19.** $4\frac{3}{5}$ **21.** $-8\frac{4}{9}$ **23.** $35\frac{11}{20}$ **25.** 0.75
27. 0.35 **29.** 0.875 **31.** $0.\overline{81}$ **33.** $3.\overline{142857}$ **35.** $0.\overline{285714}$ **37.** $\frac{3}{10}$ **39.** $\frac{2}{5}$ **41.** $\frac{39}{100}$ **43.** $\frac{41}{50}$ **45.** $\frac{29}{40}$ **47.** $\frac{5399}{10,000}$
49. $\frac{7}{9}$ **51.** 1 **53.** $\frac{4}{11}$ **55.** $\frac{257}{999}$ **57.** $\frac{21}{88}$ **59.** $-\frac{7}{120}$ **61.** $\frac{3}{2}$ **63.** 6 **65.** $\frac{10}{3}$ **67.** $-\frac{14}{15}$ **69.** 6 **71.** $\frac{5}{11}$
73. $\frac{2}{3}$ **75.** $\frac{2}{3}$ **77.** $\frac{7}{10}$ **79.** $\frac{9}{10}$ **81.** $\frac{53}{120}$ **83.** $\frac{1}{2}$ **85.** $\frac{7}{12}$ **87.** $-\frac{71}{150}$ **89.** $1\frac{5}{12}$ **91.** $\frac{4}{15}$ **93.** $\frac{7}{24}$ **95.** $\frac{7}{12}$
97. $-\frac{3}{4}$ **99.** Both are equal to $\frac{169}{36}$. **101.** $\frac{86}{192}; \frac{43}{96}$ **103.** $\frac{4}{5}$ yr **105.** $-\frac{9}{10}$ yr **107.** $\frac{3}{8}$ cup **109.** $3\frac{2}{3}$ c **111.** $\frac{1}{3}$ of the business

113. $1\frac{3}{20}$ mi; $\frac{7}{20}$ mi **115.** $\frac{1}{10}$ **117–127.** Answers will vary.

129.

say does that Star-span-gled Ban-ner yet wave O'er the

131. a. 0.24625 **b.** 1.45248 **c.** 0.00112

Section 5.4
Check Point Exercises
1. a. $2\sqrt{3}$ **b.** $2\sqrt{15}$ **c.** cannot be simplified **2. a.** $\sqrt{30}$ **b.** 10 **c.** $2\sqrt{3}$ **3. a.** 4 **b.** $2\sqrt{2}$
4. a. $18\sqrt{3}$ **b.** $-5\sqrt{13}$ **c.** $8\sqrt{10}$ **5. a.** $3\sqrt{3}$ **b.** $-13\sqrt{2}$ **6. a.** $\frac{5\sqrt{10}}{2}$ **b.** $\frac{\sqrt{14}}{7}$ **c.** $\frac{5\sqrt{2}}{6}$

Exercise Set 5.4
1. 3 **3.** 5 **5.** 8 **7.** 11 **9.** 13 **11. a.** 13.2 **b.** 13.15 **c.** 13.153 **13. a.** 133.3 **b.** 133.27 **c.** 133.270
15. a. 1.8 **b.** 1.77 **c.** 1.772 **17.** $2\sqrt{5}$ **19.** $4\sqrt{5}$ **21.** $5\sqrt{10}$ **23.** $14\sqrt{7}$ **25.** $\sqrt{42}$ **27.** 6 **29.** $3\sqrt{2}$
31. $2\sqrt{13}$ **33.** 3 **35.** $3\sqrt{5}$ **37.** $-4\sqrt{3}$ **39.** $13\sqrt{3}$ **41.** $-2\sqrt{13}$ **43.** $2\sqrt{5}$ **45.** $7\sqrt{2}$ **47.** $3\sqrt{5}$
49. $2\sqrt{2}$ **51.** $34\sqrt{2}$ **53.** $-\frac{3}{2}\sqrt{3}$ **55.** $11\sqrt{3}$ **57.** $\frac{5\sqrt{3}}{3}$ **59.** $3\sqrt{7}$ **61.** $\frac{2\sqrt{30}}{5}$ **63.** $\frac{5\sqrt{3}}{2}$ **65.** $\frac{\sqrt{10}}{5}$
67. $\frac{7\sqrt{2 \cdot 2 \cdot 3}}{6} = \frac{7 \cdot \sqrt{4} \cdot \sqrt{3}}{6} = \frac{7 \cdot 2 \cdot \sqrt{3}}{6} = \frac{14 \cdot \sqrt{3}}{6} = \frac{7}{3}\sqrt{3}$ **69.** $8\sqrt{3}$ mi/hr; 13.9 mi/hr **71.** $2\sqrt{10}$ sec; 6.3 sec
73. 42.6 in.; The formula models the actual median height well. **75.** 32.7%; The formula models the actual data well. **77.** 44%
79–83. Answers will vary. **85.** c **87.** > **89.** The square root is multiplied by $\sqrt{2}$. **91.** $\frac{3\sqrt{2}}{2}$

Section 5.5
Check Point Exercises
1. a. $\sqrt{9}$ **b.** $0, \sqrt{9}$ **c.** $-9, 0, \sqrt{9}$ **d.** $-9, -1.3, 0, 0.\overline{3}, \sqrt{9}$ **e.** $\frac{\pi}{2}, \sqrt{10}$ **f.** $-9, -1.3, 0, 0.\overline{3}, \frac{\pi}{2}, \sqrt{9}, \sqrt{10}$
2. a. associative property of multiplication **b.** commutative property of addition **c.** distributive property of multiplication over addition
d. commutative property of multiplication **3. a.** yes **b.** no

Exercise Set 5.5
1. a. $\sqrt{100}$ **b.** $0, \sqrt{100}$ **c.** $-9, 0, \sqrt{100}$ **d.** $-9, -\frac{4}{5}, 0, 0.25, 9.2, \sqrt{100}$ **e.** $\sqrt{3}$ **f.** $-9, -\frac{4}{5}, 0, 0.25, \sqrt{3}, 9.2, \sqrt{100}$
3. a. $\sqrt{64}$ **b.** $0, \sqrt{64}$ **c.** $-11, 0, \sqrt{64}$ **d.** $-11, -\frac{5}{6}, 0, 0.75, \sqrt{64}$ **e.** $\sqrt{5}, \pi$ **f.** $-11, -\frac{5}{6}, 0, 0.75, \sqrt{5}, \pi, \sqrt{64}$
5. 0 **7.** Answers will vary; an example is: $\frac{1}{2}$. **9.** Answers will vary; an example is: 1. **11.** Answers will vary; an example is: $\sqrt{2}$.
13. 4 **15.** 6 **17.** 4; 3 **19.** 7 **21.** $30 + 5\sqrt{2}$ **23.** $3\sqrt{7} + \sqrt{14}$ **25.** $5\sqrt{3} + 3$ **27.** $2\sqrt{3} + 6$
29. commutative property of addition **31.** associative property of addition **33.** commutative property of addition **35.** distributive property of multiplication over addition **37.** associative property of multiplication **39.** Answers will vary; an example is: $1 - 2 = -1$.
41. Answers will vary; an example is: $4 \div 8 = \frac{1}{2}$. **43.** Answers will vary; an example is: $\sqrt{2} \cdot \sqrt{2} = 2$. **45.** no
47–55. Answers will vary. **57.** c

Section 5.6
Check Point Exercises
1. a. 1 **b.** 1 **c.** 1 **d.** -1 **2. a.** $\frac{1}{81}$ **b.** $\frac{1}{216}$ **c.** $\frac{1}{12}$ **3. a.** 7,400,000,000 **b.** 0.000003017 **4. a.** 7.41×10^9
b. 9.2×10^{-8} **5.** 1.062×10^{11} **6.** 520,000 **7.** 0.0000023 **8. a.** 1.872×10^4 **b.** 18,720 **9.** $\approx \$12.86$

Exercise Set 5.6
1. $2^5 = 32$ **3.** $4^3 = 64$ **5.** $2^6 = 64$ **7.** $1^{20} = 1$ **9.** $4^2 = 16$ **11.** $2^4 = 16$ **13.** 1 **15.** 1 **17.** -1 **19.** $\frac{1}{4}$
21. $\frac{1}{64}$ **23.** $\frac{1}{32}$ **25.** $3^2 = 9$ **27.** $3^{-2} = \frac{1}{9}$ **29.** $2^{-4} = \frac{1}{16}$ **31.** 270 **33.** 912,000 **35.** 80,000,000
37. 100,000 **39.** 0.79 **41.** 0.0215 **43.** 0.000786 **45.** 0.00000318 **47.** 3.7×10^2 **49.** 3.6×10^3 **51.** 3.2×10^4
53. 2.2×10^8 **55.** 2.7×10^{-2} **57.** 3.7×10^{-3} **59.** 2.93×10^{-6} **61.** 8.2×10^7 **63.** 4.1×10^5 **65.** 2.1×10^{-6} **67.** 600,000
69. 60,000 **71.** 0.123 **73.** 30,000 **75.** 3,000,000 **77.** 0.03 **79.** 0.0021 **81.** $(8.2 \times 10^7)(3.0 \times 10^9) = 2.46 \times 10^{17}$
83. $(5.0 \times 10^{-4})(6.0 \times 10^6) = 3 \times 10^3$ **85.** $\frac{9.5 \times 10^6}{5 \times 10^2} = 1.9 \times 10^4$ **87.** $\frac{8 \times 10^{-5}}{2 \times 10^2} = 4 \times 10^{-7}$ **89.** $\frac{4.8 \times 10^{11}}{1.2 \times 10^{-4}} = 4 \times 10^{15}$
91. 5.33×10^7 people **93.** 4.17×10^7 people **95.** $\$4 \times 10^{11}$ **97.** \$6800 **99.** $\$1.12 \times 10^{12}$ **101.** 1.06×10^{-18} g
103–111. Answers will vary. **113.** b **115.** Answers will vary.

Section 5.7
Check Point Exercises
1. 100, 120, 140, 160, 180, and 200　**2.** 8, 5, 2, −1, −4, and −7　**3.** −34　**4. a.** $a_n = 9700n + 149{,}300$　**b.** \$304,500

5. $12, -6, 3, -\dfrac{3}{2}, \dfrac{3}{4},$ and $-\dfrac{3}{8}$　**6.** 3645　**7.** $a_n = 3(2)^{n-1}; 384$

Exercise Set 5.7
1. 8, 10, 12, 14, 16, and 18　**3.** 200, 220, 240, 260, 280, and 300　**5.** −7, −3, 1, 5, 9, and 13　**7.** −400, −100, 200, 500, 800, and 1100

9. 7, 4, 1, −2, −5, and −8　**11.** 200, 140, 80, 20, −40, and −100　**13.** $\dfrac{5}{2}, 3, \dfrac{7}{2}, 4, \dfrac{9}{2},$ and 5　**15.** $\dfrac{3}{2}, \dfrac{7}{4}, 2, \dfrac{9}{4}, \dfrac{5}{2},$ and $\dfrac{11}{4}$

17. 4.25, 4.55, 4.85, 5.15, 5.45, and 5.75　**19.** 4.5, 3.75, 3, 2.25, 1.5, and 0.75　**21.** 33　**23.** 252　**25.** 67　**27.** 955

29. 82　**31.** −142　**33.** −43　**35.** −248　**37.** $\dfrac{23}{2}$　**39.** 1.75　**41.** $a_n = 1 + (n-1)4; 77$　**43.** $a_n = 7 + (n-1)(-4); -69$

45. $a_n = 9 + (n-1)2; 47$　**47.** $a_n = -20 + (n-1)(-4); -96$　**49.** 4, 8, 16, 32, 64, and 128　**51.** 1000, 1000, 1000, 1000, 1000, and 1000
53. 3, −6, 12, −24, 48, and −96　**55.** 10, −40, 160, −640, 2560, and −10,240　**57.** 2000, −2000, 2000, −2000, 2000, and −2000

59. −2, 6, −18, 54, −162, and 486　**61.** −6, 30, −150, 750, −3750, and 18,750　**63.** $\dfrac{1}{4}, \dfrac{1}{2}, 1, 2, 4,$ and 8　**65.** $\dfrac{1}{4}, \dfrac{1}{8}, \dfrac{1}{16}, \dfrac{1}{32}, \dfrac{1}{64},$ and $\dfrac{1}{128}$

67. $-\dfrac{1}{16}, \dfrac{1}{4}, -1, 4, -16,$ and 64　**69.** 2, 0.2, 0.02, 0.002, 0.0002, and 0.00002　**71.** 256　**73.** $2{,}324{,}522{,}934 \approx 2.32 \times 10^9$　**75.** 50

77. 320　**79.** −2　**81.** 486　**83.** $\dfrac{3}{64}$　**85.** $-\dfrac{2}{27}$　**87.** $\approx -1.82 \times 10^{-9}$　**89.** 0.1　**91.** $a_n = 3(4)^{n-1}; 12{,}288$

93. $a_n = 18\left(\dfrac{1}{3}\right)^{n-1}; \dfrac{2}{81}$　**95.** $a_n = 1.5(-2)^{n-1}; 96$　**97.** $a_n = 0.0004(-10)^{n-1}; 400$　**99.** arithmetic; 18 and 22

101. geometric; 405 and 1215　**103.** arithmetic; 13 and 18　**105.** geometric; $\dfrac{3}{16}$ and $\dfrac{3}{32}$　**107.** arithmetic; $\dfrac{5}{2}$ and 3

109. geometric; 7 and −7　**111.** arithmetic; −49 and −63　**113.** geometric; $25\sqrt{5}$ and 125　**115. a.** $a_n = 1.7n + 148.3$　**b.** 211.2 lbs
117. Company A; \$1400　**119.** \$16,384　**121.** \$3,795,957　**123. a.** 1.013, 1.013, 1.013, 1.013, 1.013, 1.013, 1.013; the population is increasing
geometrically with $r \approx 1.013$.　**b.** $a_n = 29.76(1.013)^{n-1}$;　**c.** ≈ 33.86 very well　**125–131.** Answers will vary.　**133.** d
135. −287,309　**137.** $\approx 3.595 \times 10^{25}$

Review Exercises
1. 2: yes; 3: yes; 4: yes; 5: no; 6: yes; 8: yes; 9: no; 10: no; 12: yes　**2.** 2: yes; 3: yes; 4: no; 5: yes; 6: yes; 8: no; 9: yes; 10: yes; 12: no　**3.** $3 \cdot 5 \cdot 47$
4. $2^6 \cdot 3 \cdot 5$　**5.** $3 \cdot 5^2 \cdot 7 \cdot 13$　**6.** GCD: 6; LCM: 240　**7.** GCD: 6; LCM: 900　**8.** GCD: 2; LCM: 27,432　**9.** 12　**10.** 11:48 A.M.
11. <　**12.** >　**13.** 860　**14.** 53　**15.** 0　**16.** −3　**17.** −11　**18.** −15　**19.** 1　**20.** 99　**21.** −15　**22.** 9　**23.** −4

24. −16　**25.** −16　**26.** 10　**27.** −2　**28.** 17　**29.** $\dfrac{8}{15}$　**30.** $\dfrac{6}{25}$　**31.** $\dfrac{11}{12}$　**32.** $\dfrac{64}{11}$　**33.** $-\dfrac{23}{7}$　**34.** $5\dfrac{2}{5}$　**35.** $-1\dfrac{8}{9}$

36. 0.8　**37.** $0.\overline{428571}$　**38.** 0.625　**39.** 0.5625　**40.** $\dfrac{3}{5}$　**41.** $\dfrac{17}{25}$　**42.** $\dfrac{147}{250}$　**43.** $\dfrac{21}{2500}$　**44.** $\dfrac{5}{9}$　**45.** $\dfrac{34}{99}$　**46.** $\dfrac{113}{999}$

47. $\dfrac{21}{50}$　**48.** $\dfrac{35}{6}$　**49.** $\dfrac{8}{3}$　**50.** $-\dfrac{1}{4}$　**51.** $\dfrac{2}{3}$　**52.** $\dfrac{43}{36}$　**53.** $\dfrac{37}{60}$　**54.** $\dfrac{11}{15}$　**55.** $\dfrac{5}{16}$　**56.** $\dfrac{15}{112}$　**57.** $\dfrac{27}{40}$

58. $11\dfrac{1}{4}$ or about 11 pounds　**59.** $\dfrac{5}{12}$ of the tank　**60.** $2\sqrt{7}$　**61.** $6\sqrt{2}$　**62.** $5\sqrt{6}$　**63.** $10\sqrt{3}$　**64.** $4\sqrt{3}$　**65.** $5\sqrt{2}$

66. $2\sqrt{3}$　**67.** 3　**68.** $5\sqrt{5}$　**69.** $-6\sqrt{11}$　**70.** $7\sqrt{2}$　**71.** $-17\sqrt{3}$　**72.** $12\sqrt{2}$　**73.** $6\sqrt{5}$　**74.** $\dfrac{\sqrt{6}}{3}$

75. $20\sqrt{2}$ mi/hr; 28 mi/hr　**76. a.** $\sqrt{81}$　**b.** $0, \sqrt{81}$　**c.** $-17, 0, \sqrt{81}$　**d.** $-17, -\dfrac{9}{13}, 0, 0.75, \sqrt{81}$　**e.** $\sqrt{2}, \pi$

f. $-17, -\dfrac{9}{13}, 0, 0.75, \sqrt{2}, \pi, \sqrt{81}$　**77.** Answers will vary; an example is: 0.　**78.** Answers will vary; an example is: $\dfrac{1}{2}$.

79. Answers will vary; an example is: $\sqrt{2}$.　**80.** commutative property of addition　**81.** associative property of multiplication

82. distributive property of multiplication over addition　**83.** commutative property of multiplication　**84.** commutative property of multiplication

85. commutative property of addition　**86.** Answers will vary; an example is: $2 \div 6 = \dfrac{1}{3}$.

87. Answers will vary; an example is: $0 - 2 = -2$.　**88.** 216　**89.** 64　**90.** 16　**91.** 729　**92.** 25　**93.** 1　**94.** 1　**95.** $\dfrac{1}{216}$

96. $\dfrac{1}{16}$　**97.** $\dfrac{1}{49}$　**98.** 27　**99.** 460　**100.** 37,400　**101.** 0.00255　**102.** 0.0000745　**103.** 7.52×10^3　**104.** 3.59×10^6
105. 7.25×10^{-3}　**106.** 4.09×10^{-7}　**107.** 4.2×10^{13}　**108.** 9.7×10^{-5}　**109.** 390　**110.** 1,150,000　**111.** 0.023　**112.** 40

113. $(6.0 \times 10^4)(5.4 \times 10^5) = 3.24 \times 10^{10}$　**114.** $(9.1 \times 10^4)(4 \times 10^{-4}) = 3.64 \times 10^1$　**115.** $\dfrac{8.4 \times 10^6}{4 \times 10^3} = 2.1 \times 10^3$

116. $\dfrac{3 \times 10^{-6}}{6 \times 10^{-8}} = 5 \times 10^1$　**117.** 1000 yr　**118.** $\$4.20 \times 10^{10}$　**119.** 1.22×10^{10}　**120.** 7, 11, 15, 19, 23, and 27

121. −4, −9, −14, −19, −24, and −29　**122.** $\dfrac{3}{2}, 1, \dfrac{1}{2}, 0, -\dfrac{1}{2},$ and −1　**123.** 20　**124.** −30　**125.** −38

126. $a_n = -7 + (n-1)4; 69$ **127.** $a_n = 200 + (n-1)(-20); -180$ **128.** $3, 6, 12, 24, 48,$ and 96

129. $\dfrac{1}{2}, \dfrac{1}{4}, \dfrac{1}{8}, \dfrac{1}{16}, \dfrac{1}{32},$ and $\dfrac{1}{64}$ **130.** $16, -8, 4, -2, 1,$ and $-\dfrac{1}{2}$ **131.** 54 **132.** $\dfrac{1}{2}$ **133.** -48 **134.** $a_n = 1(2)^{n-1}; 128$

135. $a_n = 100\left(\dfrac{1}{10}\right)^{n-1}; \dfrac{1}{100,000}$ **136.** arithmetic; 24 and 29 **137.** geometric; 162 and 486 **138.** geometric; $\dfrac{1}{256}$ and $\dfrac{1}{1024}$

139. arithmetic; -28 and -35 **140. a.** $a_n = 0.20 + (n-1)(0.0052)$ **b.** 77.2% **141. a.** $\dfrac{20.72}{19.96} \approx 1.038, \dfrac{21.51}{20.72} \approx 1.038; \dfrac{22.33}{21.51} \approx 1.038;$

the population is increasing geometrically with $r \approx 1.038$.   **b.** $a_n = 19.96(1.038)^{n-1}$ **c.** ≈ 28.98 million people

Chapter 5 Test

1. $2, 3, 4, 6, 8, 9,$ and 12 **2.** $2^2 \cdot 3^2 \cdot 7$ **3.** GCD: 24; LCM: 144 **4.** 1 **5.** -4 **6.** -32 **7.** $0.58\overline{3}$ **8.** $\dfrac{64}{99}$ **9.** $\dfrac{1}{5}$

10. $\dfrac{37}{60}$ **11.** $-\dfrac{19}{2}$ **12.** $\dfrac{7}{12}$ **13.** $5\sqrt{2}$ **14.** $9\sqrt{2}$ **15.** $3\sqrt{2}$ **16.** $-7, -\dfrac{4}{5}, 0, 0.25, \sqrt{4},$ and $\dfrac{22}{7}$

17. commutative property of addition   **18.** distributive property of multiplication over addition   **19.** 243 **20.** 64 **21.** $\dfrac{1}{64}$ **22.** 7500

23. $\dfrac{4.9 \times 10^4}{7 \times 10^{-3}} = 7 \times 10^6$ **24.** 4 times   **25.** $1, -4, -9, -14, -19,$ and -24 **26.** 22 **27.** $16, 8, 4, 2, 1,$ and $\dfrac{1}{2}$ **28.** 320

CHAPTER 6

Section 6.1

Check Point Exercises

1. 81 **2.** -2 **3.** 9 **4.** 108 beats/min   **5. a.** $19y$ **b.** $-13x$ **c.** $100y$ **d.** $-10x^2$ **6. a.** $18x + 10$
b. $6x - 14y - 1$ **7.** $3x - 21$ **8.** $-8x - 2y + 1$ **9.** 2311 gray wolves; The formula models the data quite well.

Exercise Set 6.1

1. 27 **3.** 23 **5.** 29 **7.** -2 **9.** 140 **11.** 217 **13.** 176 **15.** 30 **17.** 27 **19.** -12 **21.** 12 **23.** -2 **25.** 69
27. 3 **29.** -142 **31.** $3x + 15$ **33.** $16x + 24$ **35.** $4 + 2r$ **37.** $5x + 5y$ **39.** $3x - 6$ **41.** $8x - 10$
43. $-8x + 12$ **45.** $6x - 12$ **47.** $17x$ **49.** $8x$ **51.** $10x$ **53.** $-2y$ **55.** $-4y$ **57.** $7x^2$ **59.** $-3x^2$ **61.** $11x - 3$
63. $8y + 10$ **65.** $10x + 8y - 2$ **67.** $15x + 16$ **69.** $27x - 10$ **71.** $29y - 29$ **73.** $-3x + 12y + 2$ **75.** $8y - 12$
77. 13 lb   **79.** $78.2; 80$; The formula models that data quite well.   **81.** $65,000$; The formula models the data quite well.
83–89. Answer will vary.   **91.** d   **93. a.** $\$50.50, \$5.50,$ and $\$1.00$ per clock, respectively   **b.** no; Answers will vary.
95.

Years from 1985	Model 1	Difference from graph	Model 2	Difference from graph	Model 3	Difference from graph
0	505	2	485	−18	503	0
5	803	29	817	43	772	−2
10	1100	−26	1119	−7	1126	0
12	1219	−23	1230	−12	1251	9
13	1279	−22	1284	−17	1305	4
14	1338	−29	1337	−30	1350	−17
15	1398	16	1389	7	1386	4
16	1457	51	1439	33	1411	5

The third model consistently gives the closest values to the graph.

Section 6.2

Check Point Exercises

1. $\{17\}$ **2.** $\{-15\}$ **3.** $\{36\}$ **4.** $\{-21\}$ **5.** $\{6\}$ **6.** $\{-3\}$ **7.** $\{2\}$ **8.** $\{5\}$ **9.** $\{-2\}$ **10.** 31 yr after 1980 or in 2011
11. $y = -\dfrac{4}{9}x + 3$ **12.** $m = \dfrac{T - D}{p}$ **13.** $\varnothing$ **14.** all real numbers

Exercise Set 6.2

1. $\{10\}$ **3.** $\{-17\}$ **5.** $\{12\}$ **7.** $\{9\}$ **9.** $\{-3\}$ **11.** $\left\{-\dfrac{1}{4}\right\}$ **13.** $\{3\}$ **15.** $\{11\}$ **17.** $\{-17\}$ **19.** $\{11\}$ **21.** $\left\{\dfrac{7}{5}\right\}$
23. $\left\{\dfrac{25}{3}\right\}$ **25.** $\{8\}$ **27.** $\{-3\}$ **29.** $\{2\}$ **31.** $\{-4\}$ **33.** $\left\{-\dfrac{1}{5}\right\}$ **35.** $\{-4\}$ **37.** $\{5\}$ **39.** $\{6\}$ **41.** $\left\{-\dfrac{7}{2}\right\}$
43. $\{1\}$ **45.** $\{24\}$ **47.** $\{20\}$ **49.** $\{5\}$ **51.** $y = -3x + 6$ **53.** $y = -\dfrac{1}{2}x + 3$ **55.** $y = -\dfrac{1}{3}x + 2$ **57.** $y = \dfrac{1}{2}x$
59. $y = \dfrac{2}{3}x - \dfrac{5}{3}$ **61.** $L = \dfrac{A}{W}$ **63.** $b = \dfrac{2A}{h}$ **65.** $P = \dfrac{I}{rt}$ **67.** $m = \dfrac{E}{c^2}$ **69.** $m = \dfrac{y - b}{x}$ **71.** $a = 2A - b$ **73.** $r = \dfrac{S - P}{Pt}$

75. all real numbers   **77.** no solution   **79.** all real numbers   **81.** 85 mi/hr   **83.** 120 chirps/min   **85.** 409.2 ft   **87.** 40 yr old; The
solution is the age that is directly below the point on the line representing females which is directly to the right of 117 on the axis representing

heart rate.   **89.** 3.7; The solution is the intensity that is directly below the point on the line representing low-humor group which is directly to the right of 10 on the axis representing level of depression.   **91.** 2002   **93–101.** Answers will vary.   **103.** c   **105.** Answers will vary.

Section 6.3
Check Point Exercises
1. 12   **2.** 1050 mi   **3.** *Jagged Little Pill:* 16 million copies; *Saturday Night Fever:* 11 million copies   **4.** 26 yr; $33,600

Exercise Set 6.3
1. $x + 9$ **3.** $20 - x$ **5.** $8 - 5x$ **7.** $\dfrac{15}{x}$ **9.** $2x + 20$ **11.** $7x - 30$ **13.** $4(x + 12)$ **15.** $x + 40 = 450; 410$

17. $x - 13 = 123; 136$ **19.** $7x = 91; 13$ **21.** $\dfrac{x}{18} = 6; 108$ **23.** $4 + 2x = 36; 16$ **25.** $5x - 7 = 123; 26$ **27.** $x + 5 = 2x; 5$

29. $2(4 + x) = 36; 14$ **31.** $9x = 3x + 30; 5$ **33.** $\dfrac{3x}{5} + 4 = 34; 50$ **35.** 800 mi   **37.** 2005   **39.** births per day: 375,000;

deaths per day: 146,000   **41.** Hugo: $7 billion; Northridge earthquake: $12.5 billion; Andrew: $20 billion   **43.** 11 hr   **45.** 300 min
47. 28 times   **49.** length: 12 ft; height: 4 ft   **51–53.** Answers will vary.   **55.** 17 min   **57.** uncle: 60 yr old; woman: 20 yr old

Section 6.4
Check Point Exercises
1. $\dfrac{8}{9}$, or 8:9   **2. a.** {15}   **b.** {140}   **3.** $1500   **4.** 720 deer   **5.** 137.5 lb per in^2 **6.** 556 ft   **7.** ≈ 4.36 lb per in^2

Exercise Set 6.4
1. $\dfrac{1}{2}$ **3.** $\dfrac{12}{5}$ **5.** $\dfrac{3}{4}$ **7.** $\dfrac{2}{1}$; 2:1   **9.** $\dfrac{1}{3}$; 1:3   **11.** {14}   **13.** {27}   **15.** $\left\{ -\dfrac{9}{4} \right\}$ **17.** {−15}   **19.** $\left\{ \dfrac{7}{2} \right\}$ **21.** {10}

23. $\dfrac{1}{2}$; 1:2   **25.** $1115.38   **27.** $\approx 20,489$ fur seal pups   **29.** $950   **31.** 154.1 in. or ≈ 12.8 ft   **33.** 2442 mi/hr   **35.** ≈ 607 lb
37. 0.5 hr   **39.** 6.4 lb   **41–45.** Answers will vary.   **47.** The front sprocket can be replaced by a 100-tooth sprocket, or the rear sprocket can be replaced with a 12-tooth sprocket.   **49.** The destructive power is 4 times as strong.

Section 6.5
Check Point Exercises
1. a.

c.   **2.** $\{x \mid x \le 4\}$;

3. a. $\{x \mid x < 8\}$;   **b.** $\{x \mid x > -3\}$;

4. $\{x \mid x < -6\}$;   **5.** $\{x \mid x \ge 1\}$;   **6.** at least 83%

Exercise Set 6.5
1. 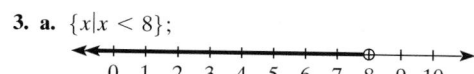   **3.** 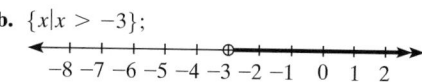   **5.**

7. 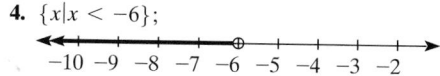   **9.** 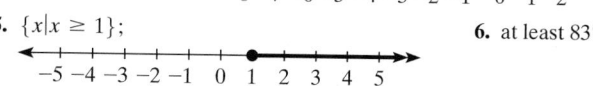   **11.**

13. $\{x \mid x > 5\}$;   **15.** $\{x \mid x \le 5\}$;   **17.** $\{x \mid x < 3\}$;

19. $\{x \mid x < 5\}$;   **21.** $\{x \mid x \ge -5\}$;   **23.** $\{x \mid x > 5\}$;

25. $\{x \mid x < 5\}$;   **27.** $\{x \mid x < 8\}$;   **29.** $\{x \mid x > -6\}$;

31. $\{x \mid x > -5\}$;   **33.** $\{x \mid x \le 5\}$;   **35.** $\{x \mid x \le 3\}$;

37. $\{x \mid x < 16\}$;   **39.** $\{x \mid x > -3\}$;   **41.** $\{x \mid x \ge -2\}$;

43. $\{x|x > -4\}$;

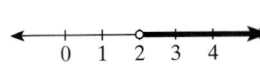

45. $\{x|x \geq 4\}$;

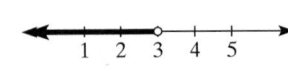

47. $\left\{x|x > \dfrac{11}{3}\right\}$;

49. $\{x|x > 2\}$;

51. $\{x|x < 3\}$;

53. $\left\{x|x > \dfrac{5}{3}\right\}$;

55. Denmark, Netherlands, Norway   **57.** Japan, Mexico   **59.** United States, Canada, Norway, Netherlands   **61. a.** 88; no   **b.** at least 68
63. after 20 yr; all years after 2008   **65.** at most 1280 mi   **67.** at most 29 bags   **69–73.** Answers will vary.   **75.** more than 1200 packages

Section 6.6

Check Point Exercises

1. $x^2 + 11x + 30$ **2.** $28x^2 - x - 15$ **3.** $(x + 2)(x + 3)$ **4.** $(x + 5)(x - 2)$ **5.** $(5x - 4)(x - 2)$ **6.** $(3y - 1)(2y + 7)$

7. $\{-6, 3\}$ **8.** $\{-2, 8\}$ **9.** $\left\{-4, \dfrac{1}{2}\right\}$ **10.** $\left\{-\dfrac{1}{2}, \dfrac{1}{4}\right\}$ **11.** $\left\{\dfrac{3 + \sqrt{7}}{2}, \dfrac{3 - \sqrt{7}}{2}\right\}$

Exercise Set 6.6

1. $x^2 + 8x + 15$ **3.** $x^2 - 2x - 15$ **5.** $2x^2 + 3x - 2$ **7.** $12x^2 - 43x + 35$ **9.** $(x + 2)(x + 3)$ **11.** $(x - 5)(x + 3)$
13. $(x - 5)(x - 3)$ **15.** $(x - 12)(x + 3)$ **17.** prime   **19.** $(x + 1)(x + 16)$ **21.** $(2x + 1)(x + 3)$ **23.** $(2x - 5)(x - 6)$
25. $(3x + 2)(x - 1)$ **27.** $(3x - 28)(x + 1)$ **29.** $(3x - 4)(2x - 1)$ **31.** $(2x + 3)(2x + 5)$ **33.** $\{-3, 8\}$
35. $\left\{-\dfrac{5}{4}, 2\right\}$ **37.** $\{-5, -3\}$ **39.** $\{-3, 5\}$ **41.** $\{-3, 7\}$ **43.** $\{-8, -1\}$ **45.** $\{6\}$ **47.** $\left\{-\dfrac{1}{2}, 4\right\}$ **49.** $\left\{-2, \dfrac{9}{5}\right\}$
51. $\left\{-\dfrac{7}{2}, -\dfrac{1}{3}\right\}$ **53.** $\{-5, -3\}$ **55.** $\left\{\dfrac{-5 + \sqrt{13}}{2}, \dfrac{-5 - \sqrt{13}}{2}\right\}$ **57.** $\{-2 + \sqrt{10}, -2 - \sqrt{10}\}$ **59.** $\{-2 + \sqrt{11}, -2 - \sqrt{11}\}$
61. $\{-3, 6\}$ **63.** $\left\{-\dfrac{2}{3}, \dfrac{3}{2}\right\}$ **65.** $\{1 + \sqrt{11}, 1 - \sqrt{11}\}$ **67.** $\left\{\dfrac{1 + \sqrt{57}}{2}, \dfrac{1 - \sqrt{57}}{2}\right\}$ **69.** $\left\{\dfrac{-3 + \sqrt{3}}{6}, \dfrac{-3 - \sqrt{3}}{6}\right\}$ **71.** $\left\{\dfrac{3}{2}\right\}$
73. 9 teams   **75.** 2048   **77.** 33-yr-olds and 58-yr-olds; The formula models the trend in the data quite well.
79–83. Answers will vary.   **85.** 8 and 16   **87.** $(x^n + 9)(x^n + 11)$

Review Exercises

1. 33   **2.** 87   **3.** 15   **4.** $6x - 3$ **5.** $30x - 2$ **6.** $39x - 4$ **7.** 135; If you can stay in the sun 15 minutes without burning
with no lotion, then you can stay in the sun 135 minutes or 2.25 hours without burning with a number 9 lotion.
8. 1800; If the regular price of the computer is $2400, the sale price is $1800.   **9.** $453.4 billion; The formula describes the cost very well.

10. $\{6\}$ **11.** $\{2\}$ **12.** $\{12\}$ **13.** $\left\{-\dfrac{13}{3}\right\}$ **14.** $\{2\}$ **15.** $y = -3x + 9$ **16.** $y = -2x + 8$ **17.** $T = \dfrac{D}{R}$ **18.** $w = \dfrac{P - 2l}{2}$

19. $h = \dfrac{2A}{b}$ **20.** $B = 2A - C$ **21.** 8 yr after 1989; 1997   **22.** $17 - 9x$ **23.** $3x - 7$ **24.** $5x + 8$ **25.** $\dfrac{6}{x} + 3x$
26. $17 - 4x = 5; 3$ **27.** $5x + 2 = x + 22; 5$ **28.** $7x - 1 = 5x + 9; 5$ **29.** $8(x - 5) = 56; 12$ **30.** 9 yr after 2000; 2009

31. preschool: $28,080; fitness: $50,950   **32.** Los Angeles: 213 days; New York City: 55 days   **33.** 500 min   **34. a.** $\dfrac{15}{11}$; 15:11

b. $\dfrac{13}{9}$; 13:9   **35.** $\{5\}$ **36.** $\{-65\}$ **37.** $\left\{\dfrac{2}{5}\right\}$ **38.** $\{3\}$ **39.** 324 teachers   **40.** 287 trout   **41.** $154   **42.** 1600 ft   **43.** 5 hr
44. $\{x|x < 4\}$;

45. $\{x|x > -8\}$;

46. $\{x|x \geq -3\}$;

47. $\{x|x > 6\}$;

48. $\{x|x \geq 4\}$;

49. $\{x|x \leq 2\}$;

50. at least 64   **51.** $x^2 + 4x - 45$ **52.** $12x^2 - 13x - 14$
53. $(x - 4)(x + 3)$ **54.** $(x - 5)(x - 3)$ **55.** prime   **56.** $(3x - 2)(x - 5)$ **57.** $(2x - 5)(3x + 2)$ **58.** prime
59. $\{-7, 2\}$ **60.** $\{-4, 8\}$ **61.** $\left\{-8, \dfrac{1}{2}\right\}$ **62.** $\{-5, -2\}$ **63.** $\{1, 3\}$ **64.** $\left\{\dfrac{5 + \sqrt{41}}{2}, \dfrac{5 - \sqrt{41}}{2}\right\}$ **65.** $\left\{-3, \dfrac{1}{2}\right\}$

66. $\left\{ \dfrac{3 + 2\sqrt{6}}{3}, \dfrac{3 - 2\sqrt{6}}{3} \right\}$ **67.** 4835 alligators   **68.** 10 yr

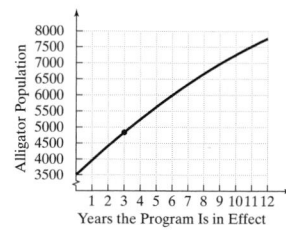

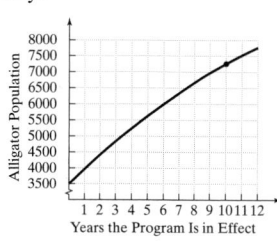

Chapter 6 Test

1. 64   **2.** $22x - 10$ **3.** \$1074 thousand or \$1,074,000   **4.** $\{8\}$ **5.** $\{2\}$ **6.** 24 yr after 1980; 2004   **7.** $y = -\dfrac{1}{2}x + 2$

8. $P = 2L + 2W$ **9.** 63.8   **10.** Reagan: 69 yr old; Buchanan: 65 yr old   **11.** 600 min   **12.** $\dfrac{2}{5}$; 2:5   **13.** $\left\{ \dfrac{15}{2} \right\}$

14. $\{3\}$ **15.** 6000 tule elk   **16.** 137.5 lb per in^2 **17.** 52.5 amps   **18.** $\{x \mid x \le -3\}$;   **19.** $\{x \mid x > 7\}$;

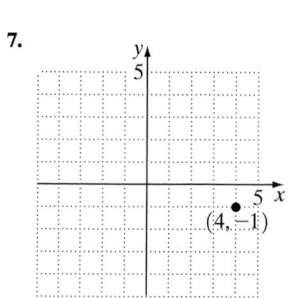

20. at least 92   **21.** $6x^2 - 7x - 20$ **22.** $(2x - 5)(x - 2)$ **23.** $\{-9, 4\}$ **24.** $\left\{ \dfrac{-2 + \sqrt{2}}{2}, \dfrac{-2 - \sqrt{2}}{2} \right\}$

CHAPTER 7

Section 7.1

Check Point Exercises

1.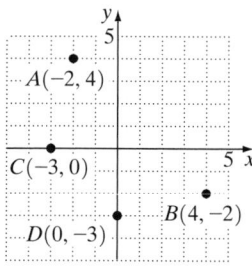

2.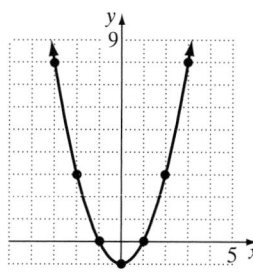

3. a. $y = 2x$

x	(x, y)
0	$(0, 0)$
2	$(2, 4)$
4	$(4, 8)$
6	$(6, 12)$
8	$(8, 16)$
10	$(10, 20)$
12	$(12, 24)$

$y = 10 + x$

x	(x, y)
0	$(0, 10)$
2	$(2, 12)$
4	$(4, 14)$
6	$(6, 16)$
8	$(8, 18)$
10	$(10, 20)$
12	$(12, 22)$

b.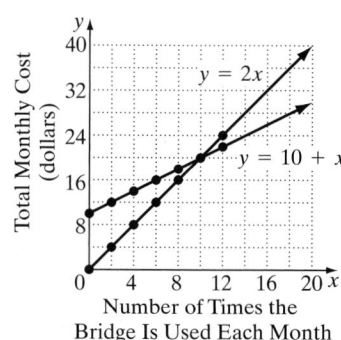

c. $(10, 20)$; When the bridge is used 10 times during a month, the cost is \$20 with or without the coupon book.   **4. a.** $f(x) = -0.05x^2 + 4.2x - 26$ **b.** 38; 38% of 20-year-old coffee drinkers become irritable if they do not have coffee at their regular time.

5. 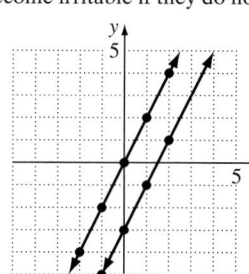 ; The graph of g is the graph of f shifted down 3 units.

6. a. function   **b.** function   **c.** not a function

7. $f(10) \approx 16$

Exercise Set 7.1

1.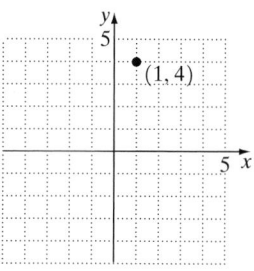

3.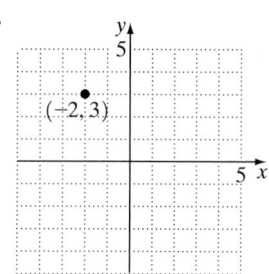

5.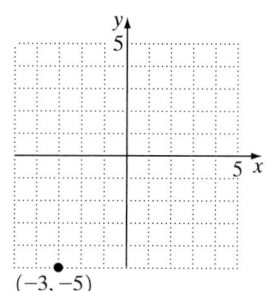

7.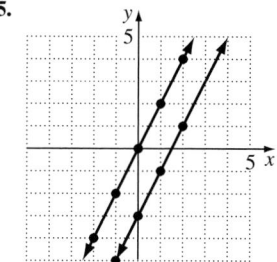

9.

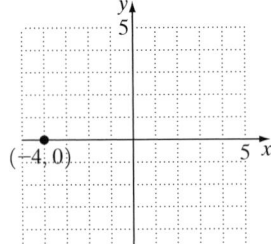

11.

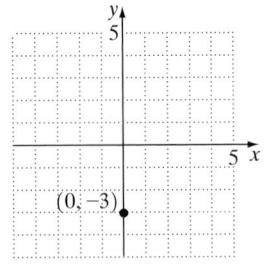

13.

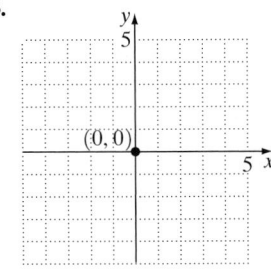

15.

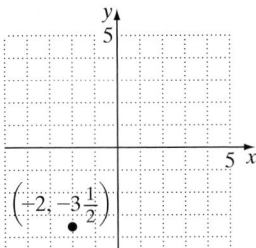

17.

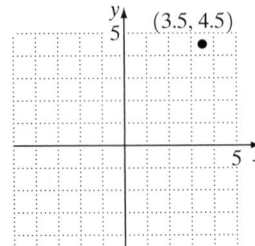

19.

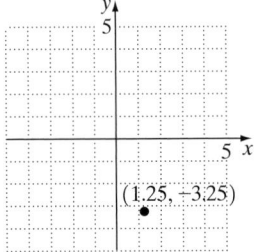

21.

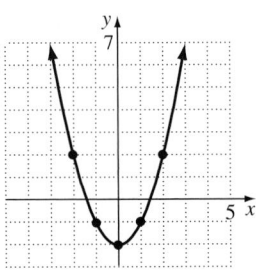

23.

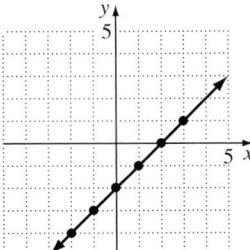

25.

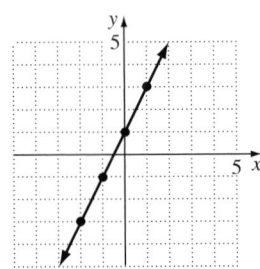

27.

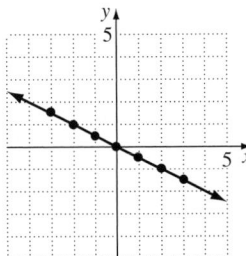

29.

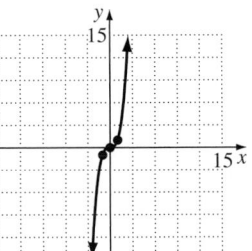

31. a. 4   **b.** −3   **33. a.** 19   **b.** −2   **35. a.** 5   **b.** 5   **37. a.** 53   **b.** 8   **39. a.** 26   **b.** 19

41.

x	$f(x) = x^2 - 1$
−2	3
−1	0
0	−1
1	0
2	3

43.

x	$f(x) = x - 1$
−2	−3
−1	−2
0	−1
1	0
2	1

45.

x	$f(x) = (x - 2)^2$
0	4
1	1
2	0
3	1
4	4

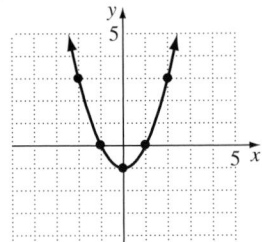

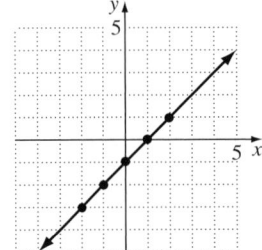

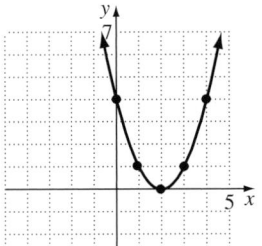

47.

x	$f(x) = x^3 + 1$
-3	-26
-2	-7
-1	0
0	1
1	2

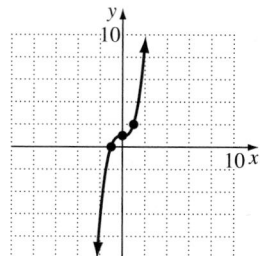

49. function   **51.** function   **53.** not a function   **55.** function   **57.** $(2, 7)$; The ball is 7 feet above the ground when it is 2 yards from the quarterback.   **59.** $(6, 9.25)$; The ball is 9.25 feet above the ground when it is 6 yards from the quarterback.   **61.** maximum height: 12 ft; distance from quarterback: 15 yd   **63.** 186.6; The cholesterol level of a 20-year-old American man is 186.6.   **65.** 27; In 1994, 27 million people received food stamps.   **67.** $(4, 27)$; This is the highest point on the graph.   **69.** 4 divorces; In 1970, the divorce rate for the United States was 4 divorces per 1000 persons ages 15 and older.   **71.** 1945 and 1970; 4 divorces   **73.** No vertical line intersects the graph in more than one point.

75. a.

x	$f(x) = 0.1x^2 - 0.4x + 0.6$
0	0.6
1	0.3
2	0.2
3	0.3
4	0.6
5	1.1

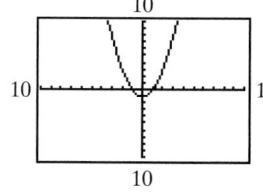

b. between 10 A.M. and 12 P.M.
77–81. Answers will vary.
83. a. $f(x) = -\dfrac{x}{2}$ **b.** $f(x) = \dfrac{x+1}{2}$ **c.** 3

85.

Exercise 41   Exercise 43   Exercise 45   Exercise 47

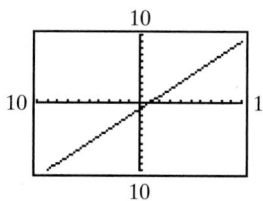
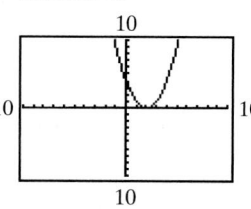
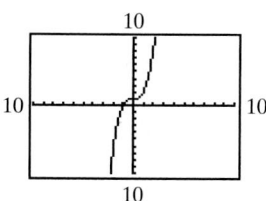

Section 7.2
Check Point Exercises

1.

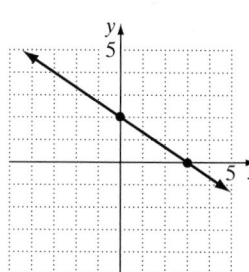

2. a. $m = 6$ **b.** $m = -\dfrac{7}{5}$ **3.**

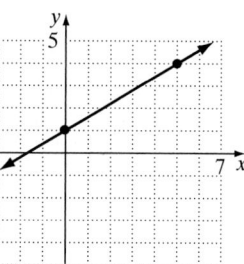

4.

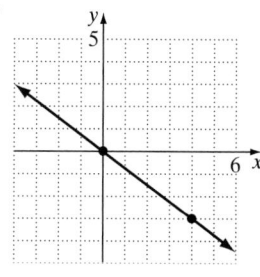

5.

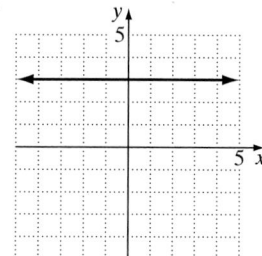

6.

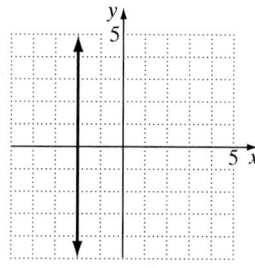

7. $\dfrac{2}{15} \approx 0.13$; The slope indicates that the number of U.S. men living alone is projected to increase by 0.13 million each year.

Exercise Set 7.2

1.
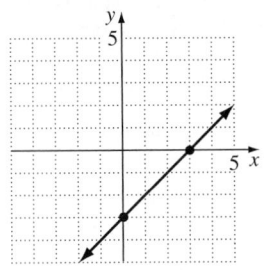

3.
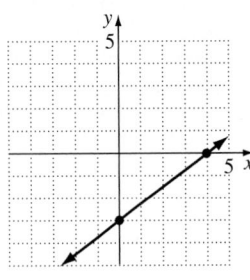

5.
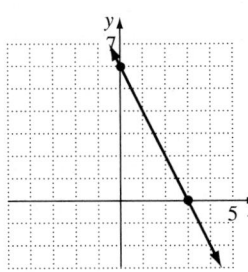

7.
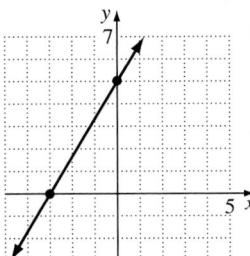

9. -1; falls **11.** $\frac{1}{4}$; rises **13.** -5; falls **15.** undefined; vertical **17.** -4; falls **19.** 0; horizontal

21.
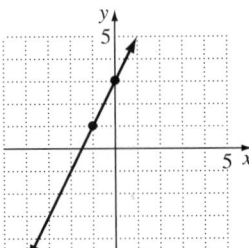

23.
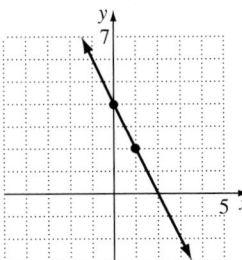

25.
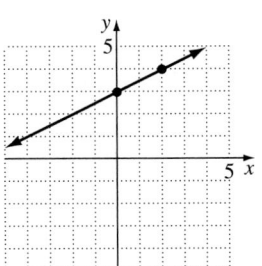

27.
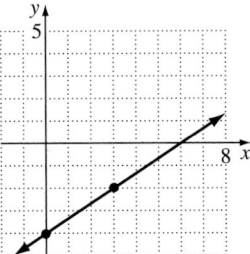

29.
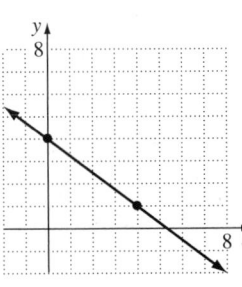

31.
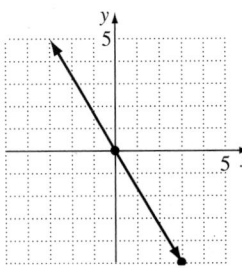

33. a. $y = -3x$ or $y = -3x + 0$
b. slope $= -3$; y-intercept $= 0$
c.
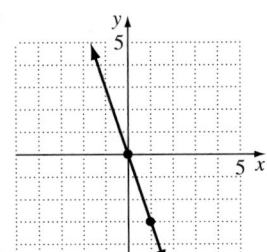

35. a. $y = \frac{4}{3}x$ or $y = \frac{4}{3}x + 0$

b. slope $= \frac{4}{3}$; y-intercept $= 0$

c.
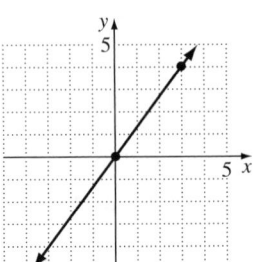

37. a. $y = -2x + 3$

b. slope $= -2$; y-intercept $= 3$

c.
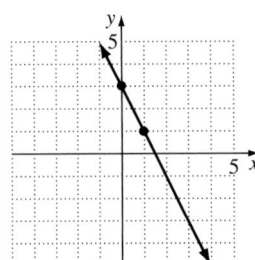

39. a. $y = -\frac{7}{2}x + 7$

b. slope $= -\frac{7}{2}$; y-intercept $= 7$

c.
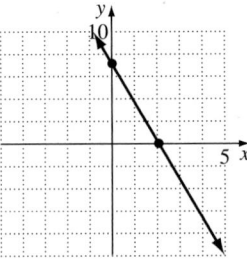

41.
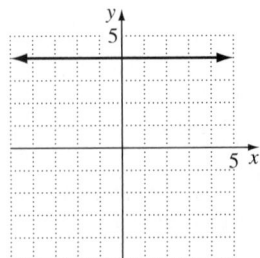

43.
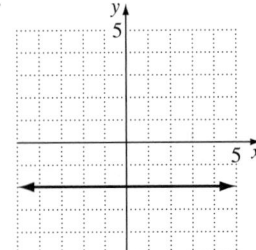

45.
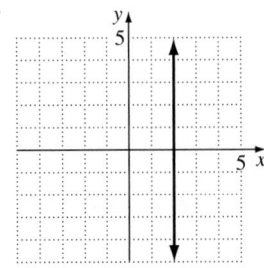

47.
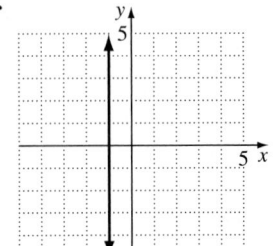

49. $m = 250$; The amount spent online per U.S. online household increased $250 per year from 1999 to 2001.

51. $m = -\dfrac{112}{3} \approx -37.33$; The federal budget surplus is expected to decrease \$37.33 billion per year from 2001 through 2010.

53. a. 24; In 1991, the average cost of a retail prescription was \$24. **b.** $m = 2$; The average cost increased \$2 per year from 1991 to 1998.

c. $y = 2x + 24$ **d.** \$52 **55–63.** Answers will vary. **65.** coefficient of x: -6; coefficient of y: 3 **67.** $m = -3$ **69.** $m = \dfrac{3}{4}$

Section 7.3
Check Point Exercises
1. 280; 30-year-old drivers have 280 accidents per 50 million miles driven **2. a.** upward **b.** 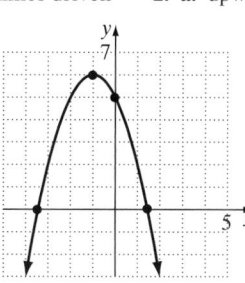 **3.** 2 and 4
4. 8 **5.** $(3, -1)$ **6.** 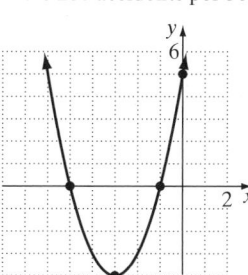 **7.**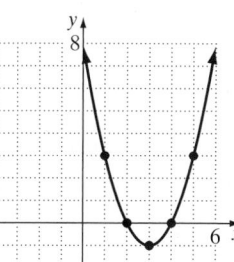

8. $(4, 27)$; The number of people receiving food stamps reached a maximum of 27 million in 1994, 4 years after 1990.

Exercise Set 7.3
1. upward **3.** downward **5.** 1 and 3 **7.** 2 and 6 **9.** -3.2 and 1.2 **11.** 3 **13.** -12 **15.** -4 **17.** 0
19. $(2, -1)$; minimum **21.** $(-1, -8)$; minimum **23.** $(-3, -9)$; minimum

25. 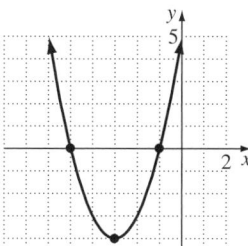 **27.** 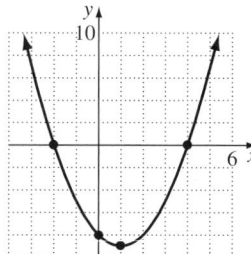 **29.** 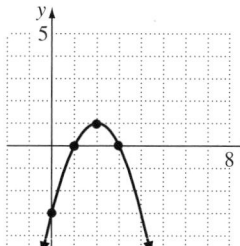 **31.**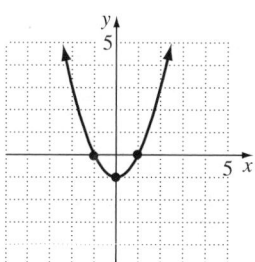

33. 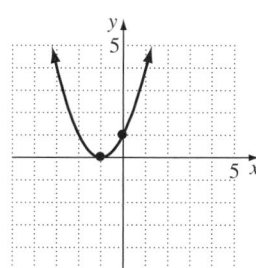 **35.**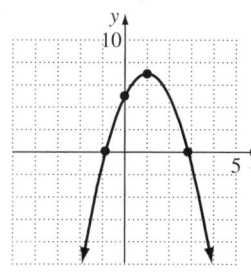

37. 8.14; There are 8.14 deaths per year per thousand people among 50 year olds. **39.** $(50, 8.14)$ **41.** 57 years old **43.** 6.25 sec; 629 ft **45.** 40 trees/acre; 16 bushels per tree **47–53.** Answers will vary. **55.** 2 **57.** 0

Section 7.4
Check Point Exercises
1. 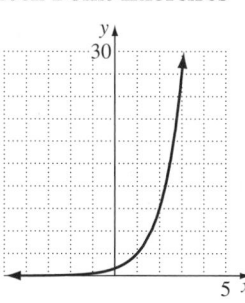 **2.** 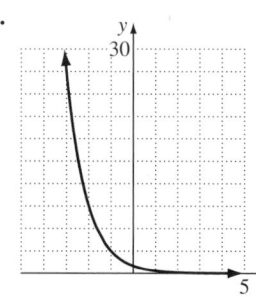 **3.** 1 **4.** Since $f(90) = 125 > 100$, people will not be able to live there 90 years after the explosion. **5.** ≈ 11.49 billion

Exercise Set 7.4

1.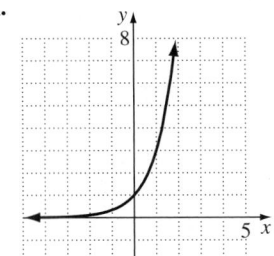

3.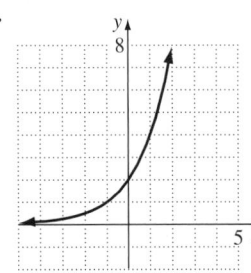

5.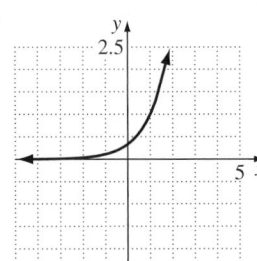

7.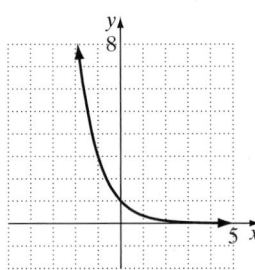

9.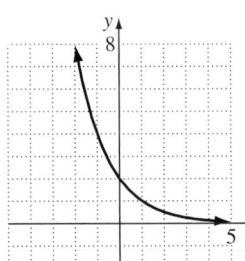

11. a. 574 million   **b.** 1148 million   **c.** 2295 million   **d.** 4590 million   **e.** It appears to double.
13. $116,405.10   **15. a.** $5.5 million; It models the data quite well.   **b.** 63%
17. 11.3; About 11.3% of 30-year-olds have some coronary heart disease.
19–23. Answers will vary.   **25.** Answers will vary.

Section 7.5
Check Point Exercises

1. not a solution   **2.**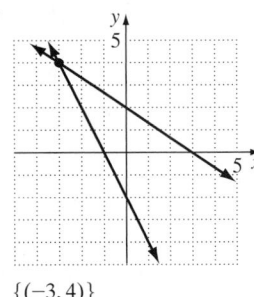

$\{(-3, 4)\}$

3. $\{(6, 11)\}$ **4.** $\{(1, -2)\}$ **5.** $\{(3, -4)\}$ **6.** $\{(2, -1)\}$
7. $\{(2, -1)\}$ **8.** $\varnothing$ **9.** $\{(x,y) \mid y = 4x - 4\}$ or $\{(x,y) \mid 8x - 2y = 8\}$
10. Quarter Pounder: 420 calories; Whopper: 589 calories

Exercise Set 7.5

1. solution   **3.** not a solution

5.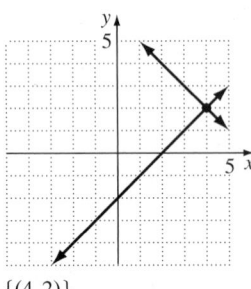

$\{(4, 2)\}$

7.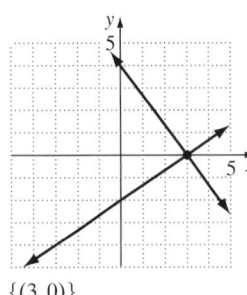

$\{(3, 0)\}$

9.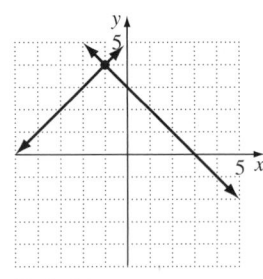

$\{(-1, 4)\}$

11.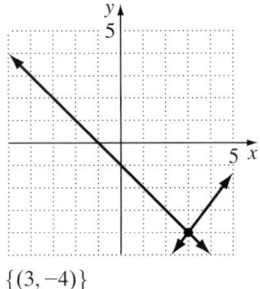

$\{(3, -4)\}$

13. $\{(1, 3)\}$ **15.** $\{(5, 1)\}$ **17.** $\{(2, 1)\}$ **19.** $\{(-1, 3)\}$ **21.** $\{(4, 5)\}$ **23.** $\left\{\left(-4, \dfrac{5}{4}\right)\right\}$ **25.** $\{(2, -1)\}$ **27.** $\{(3, 0)\}$

29. $\{(-4, 3)\}$ **31.** $\{(3, 1)\}$ **33.** $\{(1, -2)\}$ **35.** $\{(-5, -2)\}$ **37.** $\varnothing$ **39.** $\{(x,y) \mid y = 3x - 5\}$ or $\{(x,y) \mid 21x - 35 = 7y\}$
41. $\{(1, 4)\}$ **43.** $\{(x,y) \mid x + 3y = 2\}$ or $\{(x,y) \mid 3x + 9y = 6\}$ **45.** 3 and 4   **47.** 2 and 5   **49.** pan pizza: 1120 calories; beef burrito:
430 calories   **51.** 2004; $\approx$ 12.24 deaths per 100,000 persons; The lines intersect at approximately the point (2004, 12.24).
53. 26 yr after 1985; 2011; college grads: $1158; high school grads: $579   **55–65.** Answers will vary.   **67.** downstairs: 3 people; upstairs: 5 people

Section 7.6
Check Point Exercises

1.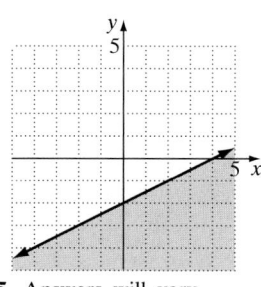

2.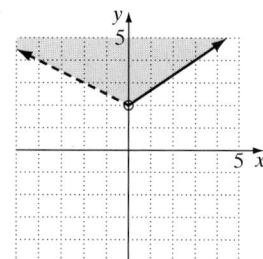

3.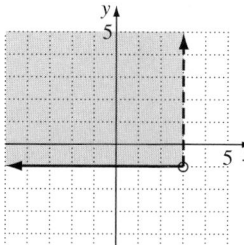

4.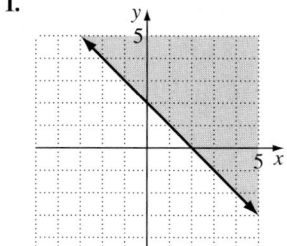

5. Answers will vary.

Execise Set 7.6

1.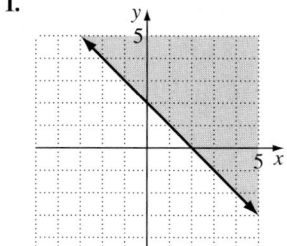

3.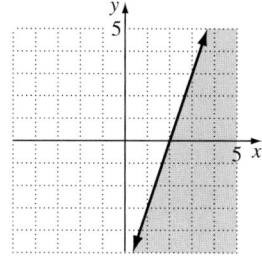

5.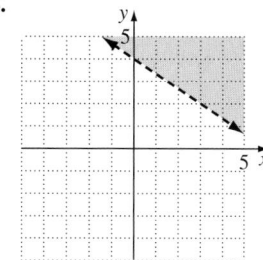

7.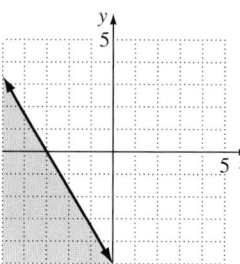

9.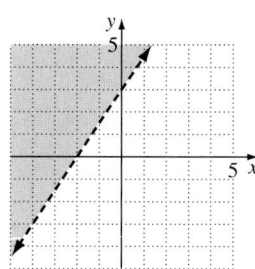

11.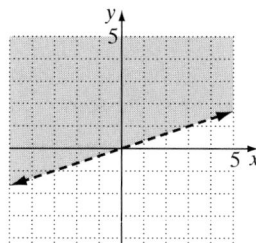

13.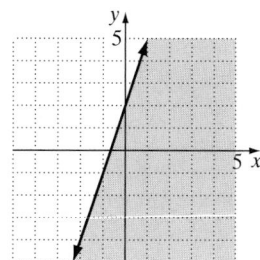

15.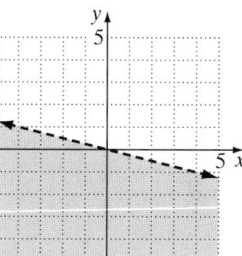

17.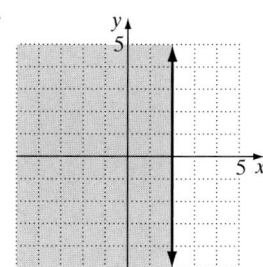

19.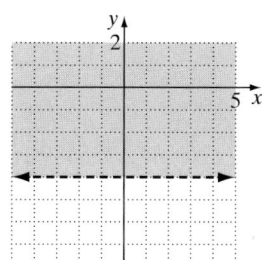

21.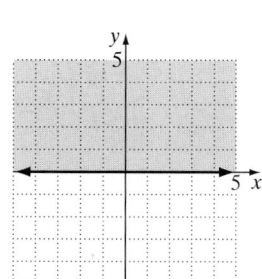

23.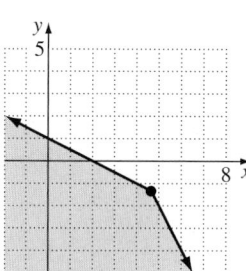

25.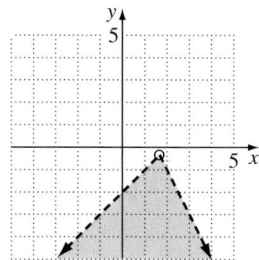

27.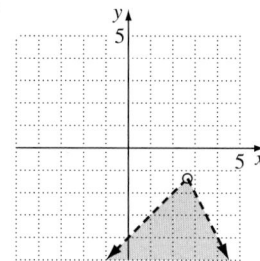

29.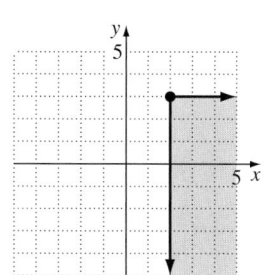

31.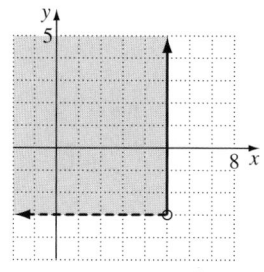

33. 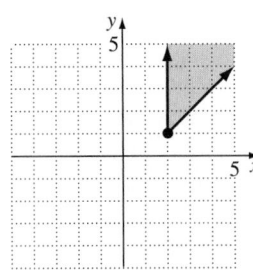 **35.** 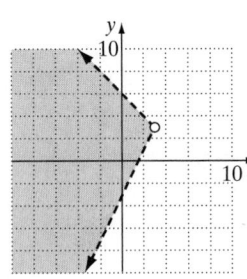 **37.**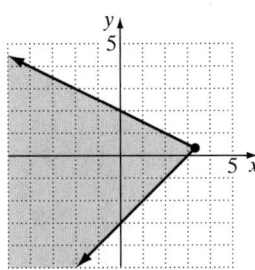

39. $(50, 30)$; Using $T = 50$ and $P = 30$, both of the inequalities for forests are true: $50 \geq 35$, true; $5(50) - 7(30) < 70$, true.
41. a. $50x + 150y > 2000$ **43. a.** 27.1 **b.** borderline overweight **45–49.** Answers will vary.
b.
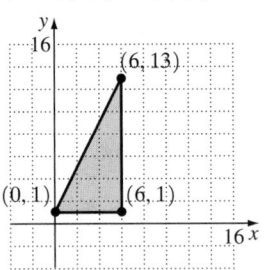 **51.** 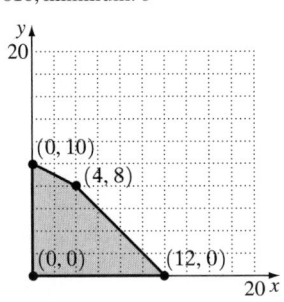 **53.**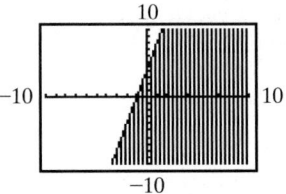

c. Answers will vary.

Section 7.7

Check Point Exercises
1. $z = 25x + 55y$ **2.** $x + y \leq 80$ **3.** $30 \leq x \leq 80, 10 \leq y \leq 30$; $z = 25x + 55y$, $x + y \leq 80$, $30 \leq x \leq 80, 10 \leq y \leq 30$
4. 50 bookshelves and 30 desks; $2900

Exercise Set 7.7
1. $(2, 10)$: 70; $(7, 5)$: 65; $(8, 3)$: 58; $(1, 2)$: 17; maximum: 70; minimum: 17
3. $(0, 8)$: 400; $(4, 9)$: 610; $(8, 0)$: 320; $(0, 0)$: 0; maximum: 610; minimum: 0

5. a. 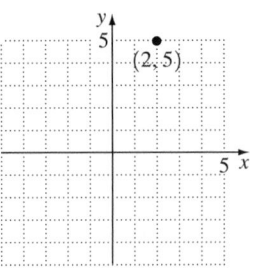 **7. a.** 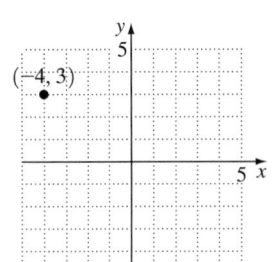 **9. a.** $z = 10x + 7y$
b. $x + y \leq 20; x \geq 3; x \leq 8$
c.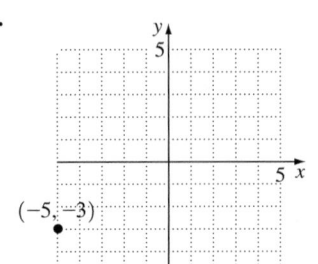

b. $(0, 1)$: 1; $(6, 13)$: 19; $(6, 1)$: 7
c. maximum of 19 at $x = 6$ and $y = 13$; minimum of 1 at $x = 0$ and $y = 1$

b. $(0, 10)$: 100; $(4, 8)$: 104; $(12, 0)$: 72; $(0, 0)$: 0
c. maximum of 104 at $x = 4$ and $y = 8$; minimum of 0 at $x = 0$ and $y = 0$

d. $(3, 0)$: 30; $(8, 0)$: 80; $(3, 17)$: 149; $(8, 12)$: 164
e. 8; 12; 164

11. 300 boxes of food and 200 boxes of clothing **13.** 100 parents and 50 students **15–19.** Answers will vary.

Review Exercises
1. 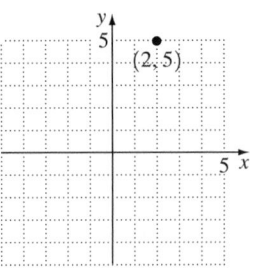 **2.** 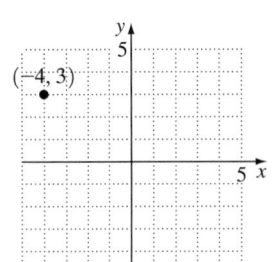 **3.**   **4.**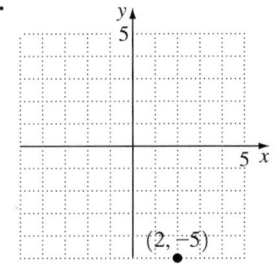

5. 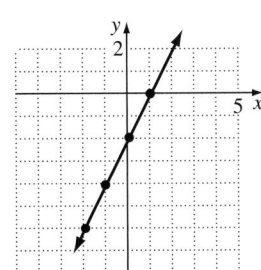 **6.** 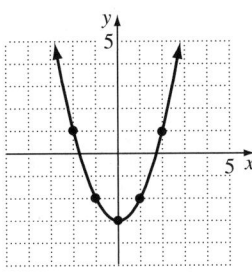 **7.**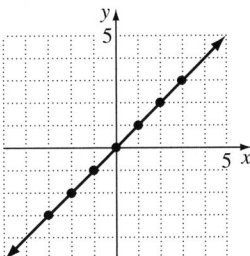

8. 3   **9.** 26   **10.** 30   **11.** −64

12.

x	$f(x) = \dfrac{1}{2}x$
−6	−3
−4	−2
−2	−1
0	0
2	1
4	2
6	3

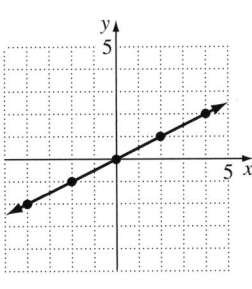

13.

x	$f(x) = x^2 - 2$
−2	2
−1	−1
0	−2
1	−1
2	2

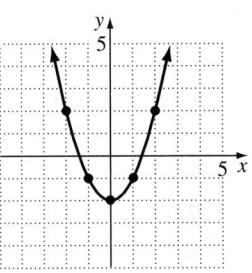

14. not a function   **15.** function   **16.** not a function   **17. a.** 3.1; In 1960, Jewish Americans comprised about 3.1% of the U.S. population.   **b.** 1940; 3.7%   **c.** 1900; 1.4%   **d.** No vertical line intersects the graph in more than one point.
e. The percentage of Jewish Americans in the U.S. population was increasing until 1940 and has been decreasing ever since.

18. 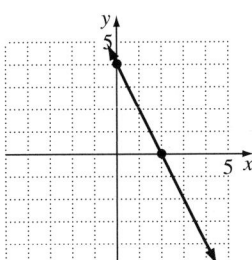 **19.** 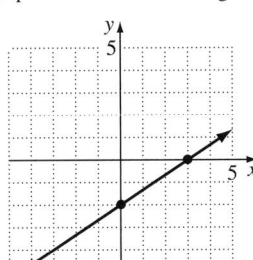 **20.**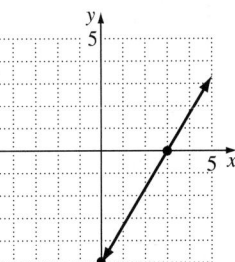

21. $-\dfrac{1}{2}$; falls   **22.** 3; rises   **23.** 0; horizontal   **24.** undefined; vertical

25. 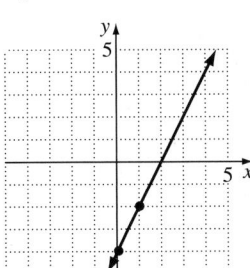 **26.** 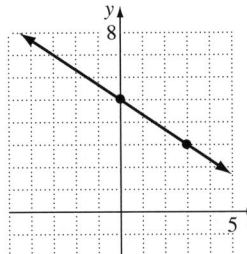 **27.** 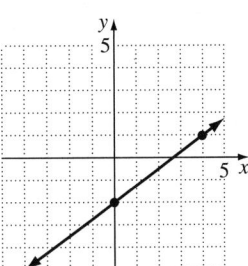 **28.**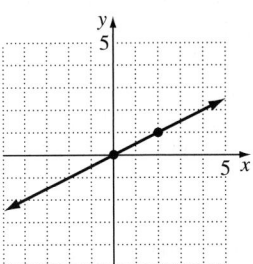

29. a. $y = -2x$

b. slope: −2; y-intercept: 0

30. a. $y = \dfrac{5}{3}x$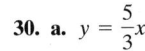

b. slope: $\dfrac{5}{3}$; y-intercept: 0

31. a. $y = -\dfrac{3}{2}x + 2$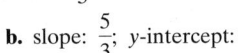

b. slope: $-\dfrac{3}{2}$; y-intecept: 2

c. 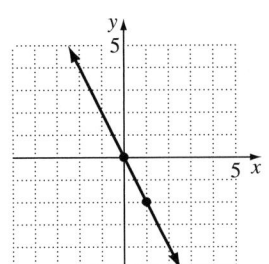   **c.** 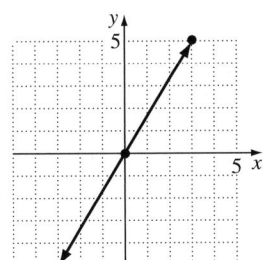   **c.**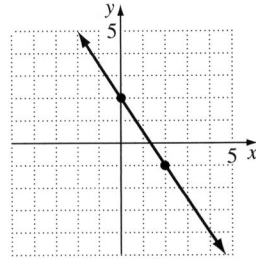

32.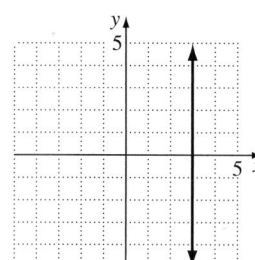

33.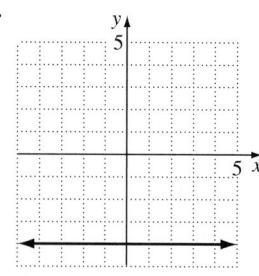

34.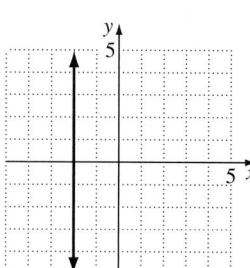

35. a. 4

b. $\dfrac{4}{11}$

c. $y = \dfrac{4}{11}x + 4$

d. 12 million

e. $y = 8$

36. a. upward
b. −1 and 7
c. −7
d. $(3, -16)$
e.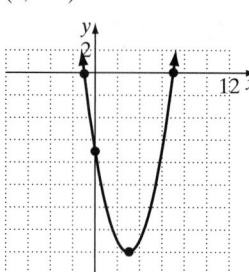

37. a. downward
b. −3 and 1
c. 3
d. $(-1, 4)$
e.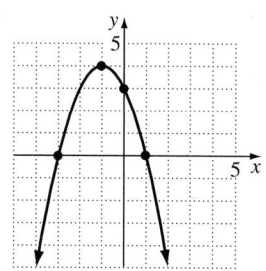

38. a. downward
b. ≈ −0.2 and ≈ 2.2
c. 1
d. $(1, 4)$
e.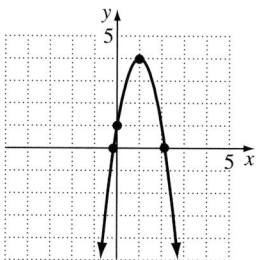

39. $(42, 62.2)$; The maximum percentage of irritable coffee drinkers is 62.2 for 42-year-olds.

40.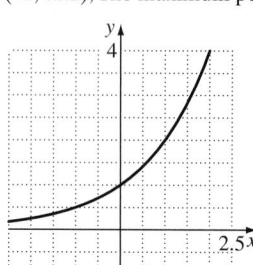

41.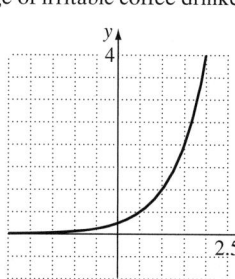

42.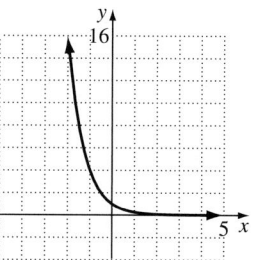

43. 559 ppm; nearly double   **44.** 55.1 million   **45.** $\{(2, 3)\}$ **46.** $\{(-2, -3)\}$ **47.** $\{(3, 2)\}$ **48.** $\{(4, -2)\}$

49. $\{(1, 5)\}$ **50.** $\{(2, 3)\}$ **51.** $\{(-9, 3)\}$ **52.** $\{(3, 4)\}$ **53.** $\{(2, -3)\}$ **54.** $\varnothing$ **55.** $\{(3, -1)\}$

56. $\{(x,y) \mid 3x - 2y = 6\}$ or $\{(x,y) \mid 6x - 4y = 12\}$ **57.** −3 and 1   **58.** shrimp: 42 mg; scallops: 15 mg

59.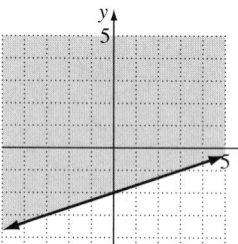

60.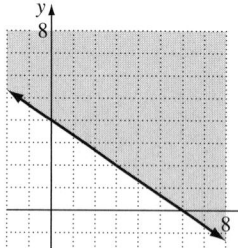

61.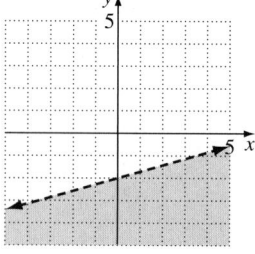

62.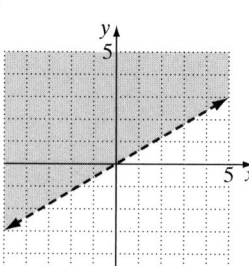

63.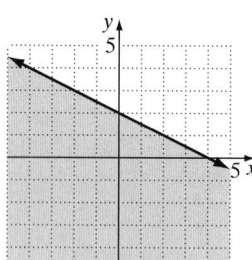

64.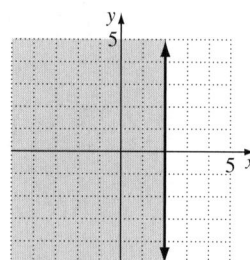

65.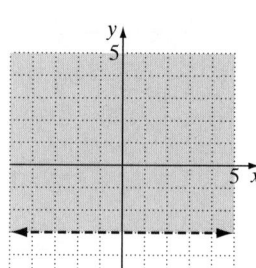

66.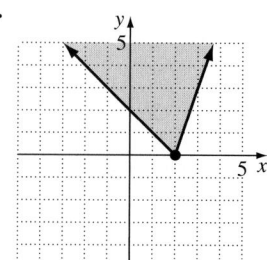

67.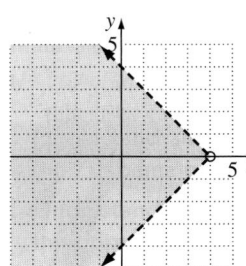

68.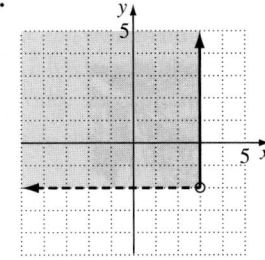

69.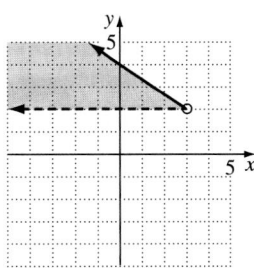

70.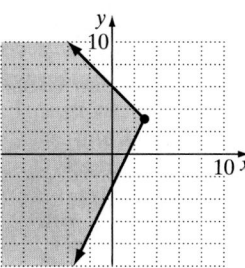

71.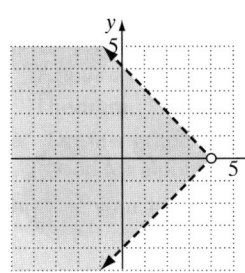

72. $(0, 1): 2; \left(\dfrac{1}{2}, \dfrac{1}{2}\right): \dfrac{5}{2};$
$(2, 2): 10; (4, 0): 8;$
maximum: 10; minimum: 2

73. a.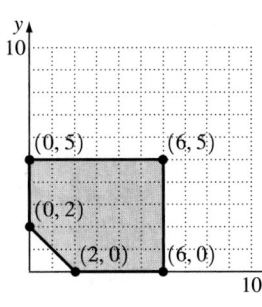

b. $(0, 2): 6; (0, 5): 15; (6, 5): 27;$
$(6, 0): 12; (2, 0): 4$

c. maximum of 27 at $x = 6$ and
$y = 5$; minimum of 4 at
$x = 2$ and $y = 0$

74. a. $z = 500x + 350y$ **b.** $x + y \leq 200; x \geq 10; y \geq 80$ **c.**

d. $(10, 80): 33,000; (10, 190): 71,500; (120, 80): 88,000$

e. $120; 80; 88,000$

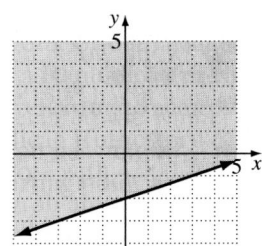

Chapter 7 Test

1.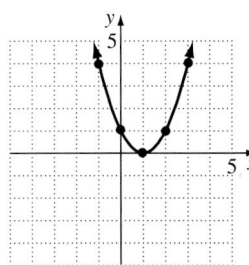

2. 21   **3.**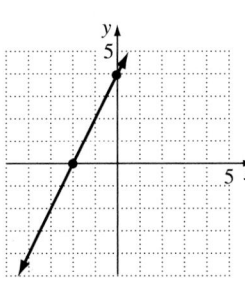

4. $m = 3$ **5.** $m = 42$; One-way airline fares are
increasing $42 per year.

6.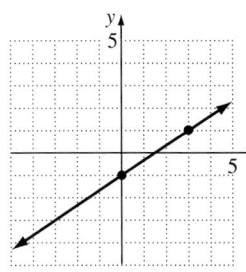

7.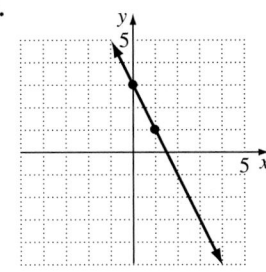

8. a. upward   **b.** $(1, -9)$ **c.** -2 and 4   **d.** -8
e.

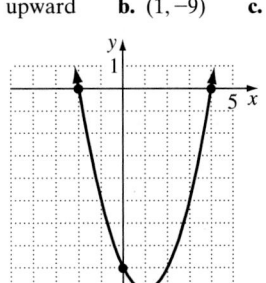

9. after 3 sec; 147 ft   **10.**

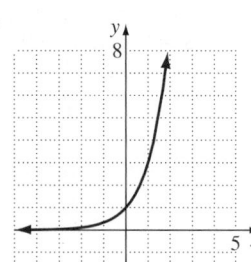

11. 17%   **12.** $\{(2,4)\}$ **13.** $\{(1,-3)\}$
14. $\{(6,-5)\}$ **15.** hotel: 80; car: 60

16. 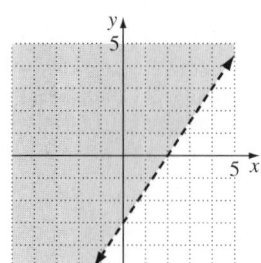   **17.** 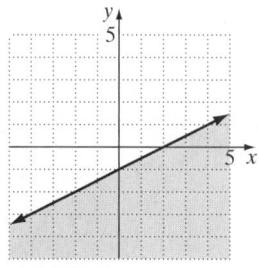   **18.**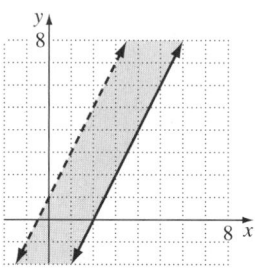

19. $(2,0)$: 6; $(2,6)$: 18; $(6,3)$: 24; $(8,0)$: 24; maximum: 24; minimum: 6   **20.** 26   **21.** 50 regular skis, 100 deluxe skis; $35,000

CHAPTER 8

Section 8.1
Check Point Exercises
1. 12.5%   **2.** 2.3%   **3. a.** 0.67   **b.** 2.5   **4.** 1040   **5.** 15   **6.** 5%   **7. a.** $75.60   **b.** $1335.60   **8. a.** $133
b. $247   **9. a.** $66\frac{2}{3}$%   **b.** 40%   **10.** 35%   **11.** 20%   **12. a.** $1152   **b.** 4% decrease

Exercise Set 8.1
1. 40%   **3.** 25%   **5.** 37.5%   **7.** 2.5%   **9.** 11.25%   **11.** 59%   **13.** 38.44%   **15.** 287%   **17.** 1487%   **19.** 10,000%
21. 0.72   **23.** 0.436   **25.** 1.3   **27.** 0.02   **29.** 0.005   **31.** 0.00625   **33.** 0.625   **35.** 6   **37.** 7.2   **39.** 5   **41.** 170
43. 20%   **45.** 12%   **47.** 408   **49.** 31,050,000   **51.** 12.5%   **53.** 31%   **55.** $9   **57. a.** $1008   **b.** $17,808   **59. a.** $103.20
b. $756.80   **61.** 173.6%   **63.** 16.6%   **65.** 15%   **67.** no; 2% decrease   **69.** no; Answers will vary; an example is: You need to know
the total number of dollars spent on health expenditures before you can calculate dollars.   **71–75.** Answers will vary.   **77.** $2588.93

Section 8.2
Check Point Exercises
1. $150   **2.** $336   **3.** $2091   **4.** 18%   **5.** $3846.16   **6. a.** $1200   **b.** $3800   **c.** 15.8%

Exercise Set 8.2
1. $240   **3.** $10.80   **5.** $318.75   **7.** $426.25   **9.** $3420   **11.** $38,350   **13.** $9390   **15.** 7.5%   **17.** 9%   **19.** 31.3%
21. $5172.42   **23.** $8917.20   **25.** $4509.59   **27. a.** $93.33   **b.** $1906.67   **c.** 7.3%   **29. a.** $1560   **b.** $10,440   **c.** 7.5%
31. a. $247.50   **b.** $4247.50   **33.** 21.4%   **35.** $2654.87   **37–39.** Answers will vary.   **41. a.** $A = 275t + 5000$
b. Answers will vary; an example is: The rate of change of the future value per year is $275.

Section 8.3
Check Point Exercises
1. a. $1216.65   **b.** $216.65   **2. a.** $6253.23   **b.** $2053.23   **3.** $6189.55   **4. a.** $6628.28   **b.** 10.5%   **5.** $\approx 8.24\%$

Exercise Set 8.3
1. a. $10,816   **b.** $816   **3. a.** $3655.21   **b.** $655.21   **5. a.** $12,795.12   **b.** $3295.12   **7. a.** $5149.12   **b.** $649.12
9. a. $1855.10   **b.** $355.10   **11. a.** $49,189.33   **b.** $29,189.33   **13.** $8374.84   **15.** $7528.59   **17. a.** $10,457.65   **b.** 4.6%
19. 6.1%   **21.** 6.2%   **23.** 6.2%   **25.** 8.3%; 8.25%; 8% compounded monthly   **27.** 5.6%; 5.5%; 5.5% compounded semiannually
29. a. $\approx$ $3,052,400,000   **b.** $\approx$ $3,169,200,000   **31.** $65,728.51   **33.** $21,679.39   **35.** $94,460.79   **37–41.** Answers will vary.
43. $8544.49

Section 8.4
Check Point Exercises
1. a. $13,720   **b.** $19,180   **c.** $5180   **2.** $\approx$ 13.5%   **3. a.** $\approx$ $2057.85   **b.** $9597.15   **4. a.** $\approx$ $1885.18   **b.** $9769.82
5. a. $59.52   **b.** $4849.52   **c.** $135   **6.** unpaid balance method: $113.40; previous balance method: $122.40; average daily balance
method: $115.43

Exercise Set 8.4
1. a. $22,000   **b.** $29,600   **c.** $2600   **3. a.** $1000   **b.** $1420   **c.** $320   **5.** 12.5%   **7.** 14.5%   **9.** 13.0%
11. a. $17,000   **b.** $22,133   **c.** $4633   **d.** 10.0%   **13. a.** $1805.73   **b.** $11,534.62   **c.** $1686.11   **d.** $11,654.24

15. a. $11.05   **b.** $1013.05   **c.** ≈ $28   **17. a.** $4.96   **b.** $618.61   **c.** $61.86   **19. a.** $7.50   **b.** $45   **c.** $13.75
21–29. Answers will vary.   **31.** b   **33.** Answers will vary.   **35. a.** 31 days; $376.26   **b.** $4.89   **c.** $471.04

Section 8.5
Check Point Exercises
1. a. $24,000   **b.** $216,000   **c.** $6480   **d.** $1365.12   **e.** $275,443.20   **2.** $1881.59; $152,757
3.

Payment Number	Interest Payment	Principal Payment	Balance of Loan
1	$1166.67	$383.33	$199,616.67
2	$1164.43	$385.57	$199,231.10

Exercise Set 8.5
1. a. $25,000   **b.** $100,000   **c.** $3000   **d.** $665   **e.** $139,400   **3.** 20-year option: $38,243.20; 30-year option: $62,411.20; $24,168
5. 20-year mortgage at 7.5%   **7.** $68,784,914   **9. a.** $1238.50   **b.** $167,433.30   **c.** $109,563.15   **11. a.** $758.44   **b.** $97,826.12
c. $102,173.88   **13–19.** Answers will vary.

21. $$PV = 1238.50 \, \frac{1 - \left(1 + \dfrac{0.064}{12}\right)^{x-360}}{\dfrac{0.064}{12}}$$

Section 8.6
Check Point Exercises
1. a. high: $63.38; low: $42.37   **b.** $2160   **c.** 1.5%; This is much lower than a bank account paying 3.5%.   **d.** 7,203,200 shares
e. high: $49.94; low: $48.33   **f.** $49.50   **g.** The closing price is up $0.03 from the previous day's closing price.   **h.** ≈ $1.34
2. a. $6140   **b.** $239.50

Exercise Set 8.6
1. a. high: $73.25; low: $45.44   **b.** $840   **c.** 2.2%; Answers will vary.   **d.** 591,500 shares   **e.** high: $56.38; low: $54.38   **f.** $55.50
g. $1.25 increase   **h.** ≈ $3.26   **3. a.** $6485   **b.** $406.25   **5. a.** $15,000   **b.** $300   **7. a.** $4260   **b.** $111.50
9–19. Answers will vary.

Review Exercises
1. 80%   **2.** 12.5%   **3.** 75%   **4.** 72%   **5.** 0.35%   **6.** 475.6%   **7.** 0.65   **8.** 0.997   **9.** 1.50   **10.** 0.03   **11.** 0.0065
12. 0.0025   **13.** 9.6   **14.** 200   **15.** 48%   **16.** 2000   **17.** 400   **18.** 70%   **19. a.** $1.44   **b.** $25.44
20. a. $297.50   **b.** $552.50   **21.** 12.5% increase   **22.** 35% decrease   **23.** no; $9900; 1% decrease   **24.** $180   **25.** $2520
26. $1200   **27.** $900   **28. a.** $122.50   **b.** $3622.50   **29.** $12,738   **30.** 7.5%   **31.** $13,389.12   **32.** $9287.93   **33.** 40%
34. a. $94.50   **b.** $1705.50   **c.** 7.4%   **35. a.** $8114.92   **b.** $1114.92   **36. a.** $38,490.80   **b.** $8490.80   **37. a.** $5556.46
b. $3056.46   **38.** $16,653.64   **39.** $13,175.19   **40. a.** $2122.73   **b.** 6.1%   **41.** 5.6%; Answers will vary; an example is: The same
amount of money would earn 5.6% in a simple interest account for a year.   **42.** 6.25% compounded monthly   **43. a.** $16,000
b. $21,500   **c.** $5000   **d.** 11.5%   **44. a.** $250.40   **b.** $4229.60   **45. a.** $213.11   **b.** $4336.89   **46.** Answers will vary.
47. a. $16.50   **b.** $1756.50   **c.** $49   **48. a.** $28.80   **b.** $64.80   **c.** $34.61   **49. a.** $29,000   **b.** $116,000   **c.** $2320
d. $771.40   **e.** $161,704   **50.** fixed-rate 20-year mortgage at 8%; Answers will vary.
51.

Payment Number	Interest Payment	Principal Payment	Balance of Loan
1	$500	$59.20	$79,940.80
2	$499.63	$59.57	$79,881.23

52. option A: $769; option B: $699; Answers will vary.   **53.** high: $64.06; low: $26.13   **54.** $144   **55.** 0.3%
56. 545,800 shares   **57.** high: $61.25; low: $59.25   **58.** $61.00   **59.** $1.75 increase   **60.** $1.49
61. a. $9750   **b.** $656.25   **62–63.** Answers will vary.

Chapter 8 Test
1. 19.24   **2.** 160   **3.** 22.5%   **4. a.** $18   **b.** $102   **5.** 20%   **6.** $2697.70; $697.70   **7.** $72; $2472   **8.** $14,204.57
9. 25%   **10.** $6698.57   **11.** 4.58%; Answers will vary; an example is: The same amount of money would earn 4.58% in a simple interest account
for a year.   **12.** $13,000   **13.** $21,000   **14.** $5000   **15.** 13.5%   **16.** $819.67   **17.** $6680.33   **18.** $15.60
19. $1335.60   **20.** $37   **21.** $16.80   **22.** $12,000   **23.** $108,000   **24.** $2160   **25.** $830.52   **26.** $190,987.20
27. 30-year mortgage: $699; 15-year mortgage: $899; Answers will vary.   **28.** high: $25.75; low: $25.50   **29.** $2030   **30.** $386.25

CHAPTER 9

Section 9.1
Check Point Exercises

1. a. 6.5 ft   **b.** 3.25 mi   **c.** $\frac{1}{12}$ yd   **2. a.** 8 km   **b.** 53,000 mm   **c.** 0.0604 hm   **d.** 6720 cm

3. a. 243.84 cm   **b.** ≈ 22.22 yd   **c.** ≈ 1181.1 in.   **4.** 37.5 mi/hr

Exercise Set 9.1

1. 2.5 ft   **3.** 360 in.   **5.** $\frac{1}{6}$ yd ≈ 0.17 yd   **7.** 216 in.   **9.** 18 ft   **11.** 2 yd   **13.** 4.5 mi   **15.** 3960 ft   **17.** 500 cm   **19.** 1630 m
21. 0.03178 hm   **23.** 0.000023 m   **25.** 21.96 dm   **27.** 35.56 cm   **29.** ≈ 5.51 in.   **31.** 424 km   **33.** 165.625 mi   **35.** ≈ 13.33 yd
37. ≈ 55.12 in.   **39.** 0.4064 dam   **41.** 1.524 m   **43.** ≈ 16.40 ft   **45.** 60 mi/hr   **47.** 72 km/hr   **49.** meter   **51.** millimeter
53. meter   **55.** millimeter   **57.** millimeter   **59.** b   **61.** a   **63.** c   **65.** a   **67.** 0.216 km
69. 148.8 million km   **71–77.** Answers will vary.   **79.** 9 hm   **81.** 11 m or 1.1 dam   **83.** 457.2 cm

Section 9.2
Check Point Exercises

1. 8 square units   **2.** ≈ 9809.5 people per square mile   **3. a.** 0.72 ha   **b.** $\approx \$576,389$ per hectare   **4.** 9 cubic units
5. 74,800 gal   **6.** 220 L

Exercise Set 9.2

1. 16 square units   **3.** 8 square units   **5.** ≈ 2.15 in.^{2} **7.** 37.5 yd^2 **9.** 25.5 acres   **11.** 91 cm^2 **13.** 24 cubic units
15. 74,800 gal   **17.** 1600 gal   **19.** 9 gal   **21.** 13.5 yd^3 **23.** 45 L   **25.** 17 mL   **27.** 1500 cm^3 **29.** 150 cm^3 **31.** 12,000 dm^3
33. 6.2 people per square mile   **35. a.** 20 acres   **b.** \$12,500 per acre   **37.** square centimeters   **39.** square kilometers   **41.** b
43. b   **45.** 336,600 gal   **47.** 4 L   **49–53.** Answers will vary.   **55.** 12,139.4 people per square mile   **57.** Answers will vary.
59. 6.5 liters; Answers will vary.

Section 9.3
Check Point Exercises

1. a. 420 mg   **b.** 6.2 g   **2.** 145 kg   **3. a.** 83.7 kg   **b.** 100.44 mg   **4.** 122°F   **5.** 15°C

Exercise Set 9.3

1. 740 mg   **3.** 0.87 g   **5.** 800 cg   **7.** 18,600 g   **9.** 0.000018 g   **11.** 50 kg   **13.** 4200 cm^3 **15.** 1100 t   **17.** 40,000 g
19. 2.25 lb   **21.** 1008 g   **23.** 243 kg   **25.** 36,000 g   **27.** 1200 lb   **29.** ≈ 222.22 T   **31.** 50°F   **33.** 95°F   **35.** 134.6°F
37. 23°F   **39.** 20°C   **41.** 5°C   **43.** 22.2°C   **45.** −5°C   **47.** 176.7°C   **49.** −30°C   **51.** milligram   **53.** gram   **55.** kilogram
57. kilogram   **59.** b   **61.** a   **63.** c   **65.** 13.28 kg   **67.** \$1.06   **69.** Purchasing the regular size is more economical.
71. 90 mg   **73. a.** 43 mg   **b.** 516 mg   **75.** a   **77.** c   **79.** 19.4°F   **81–85.** Answers will vary.   **87.** b

Review Exercises

1. 5.75 ft   **2.** 0.25 yd   **3.** 7 yd   **4.** 2.5 mi   **5.** 2280 cm   **6.** 70 m   **7.** 1920 m   **8.** 0.0144 hm   **9.** 0.0005 m   **10.** 180 mm
11. 58.42 cm   **12.** ≈ 7.48 in.   **13.** 528 km   **14.** 375 mi   **15.** ≈ 15.56 yd   **16.** ≈ 39.37 ft   **17.** ≈ 28.13 mi/hr
18. 96 km/hr   **19.** 0.024 km, 24,000 cm, 2400 m   **20.** 4.8 km   **21.** 24 square units   **22.** 206.9 people per square mile; In California,
there is an average of 206.9 people for each square mile.   **23.** 18 acres   **24.** ≈ 333.33 ft^2 **25.** 31.2 km^2 **26.** a   **27.** 24 cubic units
28. 251,328 gal   **29.** 76 L   **30.** c   **31–32.** Answers will vary.   **33.** 1240 mg   **34.** 1200 cg   **35.** 0.000012 g   **36.** 0.00045 kg
37. 50,000 cm^3 **38.** 4000 dm^3, 4,000,000 g   **39.** 94.5 kg   **40.** 14 oz   **41.** kilograms; Answers will vary.   **42.** 2.25 lb   **43.** a
44. c   **45.** 59°F   **46.** 212°F   **47.** 41°F   **48.** 32°F   **49.** −13°F   **50.** 15°C   **51.** 5°C   **52.** 100°C   **53.** 37°C
54. ≈ -17.8°C   **55.** −10°C   **56.** more; Answers will vary.

Chapter 9 Test

1. 0.00807 hm   **2.** 250 in.   **3.** 4.8 km   **4.** mm   **5.** cm   **6.** km   **7.** 128 km/hr   **8.** 9 times   **9.** 200.8 people per square
mile; In Spain, there is an average of 200.8 people for each square mile.   **10.** 45 acres   **11.** b   **12.** Answers will vary.; 1000 times
13. 74,800 gal   **14.** b   **15.** 0.137 kg   **16.** 40.5 kg   **17.** kg   **18.** mg   **19.** 86°F   **20.** 80°C   **21.** d

CHAPTER 10

Section 10.1
Check Point Exercises

1. 30°   **2.** 71°   **3.** 46°, 134°   **4.** $m\angle 1 = 57°$, $m\angle 2 = 123°$, and $m\angle 3 = 123°$
5. $m\angle 1 = 29°$, $m\angle 2 = 29°$, $m\angle 3 = 151°$, $m\angle 4 = 151°$, $m\angle 5 = 29°$, $m\angle 6 = 151°$, and $m\angle 7 = 151°$

Exercise Set 10.1

1. 150°   **3.** 90°   **5.** 20°; acute   **7.** 160°; obtuse   **9.** 180°; straight   **11.** 65°   **13.** 146°   **15.** 42°; 132°   **17.** 1°; 91°
19. 52.6°; 142.6°   **21.** $x + x + 12° = 90°$; 39°, 51°   **23.** $x + 3x = 180°$; 45°, 135°   **25.** $m\angle 1 = 108°$; $m\angle 2 = 72°$; $m\angle 3 = 108°$
27. $m\angle 1 = 50°$; $m\angle 2 = 90°$; $m\angle 3 = 50°$ **29.** $m\angle 1 = 68°$; $m\angle 2 = 68°$; $m\angle 3 = 112°$; $m\angle 4 = 112°$; $m\angle 5 = 68°$; $m\angle 6 = 68°$; $m\angle 7 = 112°$
31. $m\angle 1 = 38°$; $m\angle 2 = 52°$; $m\angle 3 = 142°$ **33–41.** Answers will vary.   **43.** yes; Answers will vary.   **45.** 90°

Section 10.2
Check Point Exercises

1. 49°   **2.** $m\angle 1 = 90°$; $m\angle 2 = 54°$; $m\angle 3 = 54°$; $m\angle 4 = 68°$; $m\angle 5 = 112°$ **3.** 32 in.   **4.** 32 yd   **5.** 25 ft   **6.** 12 in.

Exercise Set 10.2

1. 67°   **3.** 32°   **5.** $m\angle 1 = 50°$; $m\angle 2 = 130°$; $m\angle 3 = 50°$; $m\angle 4 = 130°$; $m\angle 5 = 50°$
7. $m\angle 1 = 50°$; $m\angle 2 = 50°$; $m\angle 3 = 80°$; $m\angle 4 = 130°$; $m\angle 5 = 130°$ **9.** $m\angle 1 = 55°$; $m\angle 2 = 65°$; $m\angle 3 = 60°$; $m\angle 4 = 65°$; $m\angle 5 = 60°$;
$m\angle 6 = 120°$; $m\angle 7 = 60°$; $m\angle 8 = 60°$; $m\angle 9 = 55°$; $m\angle 10 = 55°$ **11.** 5 in.   **13.** 6 m   **15.** 16 in.   **17.** 5   **19.** 9   **21.** 17 m

23. 39 m **25** 12 cm **27.** no; 0.3° **29.** 71.7 ft **31.** 6 km **33.** $5\sqrt{7}$ ft $\approx$ 13.2 ft **35.** $750,000 **37–43.** Answers will vary.
45. 70° **47.** no

Section 10.3
Check Point Exercises
1. $3120 **2. a.** 1800° **b.** 150°

Exercise Set 10.3
1. quadrilateral **3.** pentagon **5.** a: square; b: rhombus; d: rectangle; e: parallelogram **7.** a: square; d: rectangle **9.** c: trapezoid
11. 30 cm **13.** 46 cm **15.** 1000 in. **17.** 27 ft **19.** 18 yd **21.** 84 yd **23.** 32 ft **25.** 540° **27.** 360° **29.** 108°; 72°
31. a. 540° **b.** $m \angle A = 140°; m \angle B = 40°$ **33.** no entry **35.** stop, yield, and deer crossing **37.** equilateral triangle **39.** yield
41. $5600 **43.** 48 **45–51.** Answers will vary. **53.** $6a$

Section 10.4
Check Point Exercises
1. 66 ft^2; yes **2.** $672 **3.** 60 in.^{2} **4.** 30 ft^2 **5.** 105 ft^2 **6.** 10π in.; 31.4 in. **7.** 49.7 ft **8.** large pizza

Exercise Set 10.4
1. 18 m^2 **3.** 16 in.^{2} **5.** 2100 cm^2 **7.** 56 in.^{2} **9.** 20.58 yd^2 **11.** 30 in.^{2} **13.** 91 m^2 **15.** C: 8π cm, $\approx$ 25.1 cm;
A: 16π cm^2, $\approx$ 50.3 cm^2 **17.** C: 12π yd, $\approx$ 37.7 yd; A: 36π yd^2, $\approx$ 113.1 yd^2 **19.** 72 m^2 **21.** 300 m^2 **23.** $100 + 50\pi$; 257.1 cm^2
25. $556.50 **27.** 148 ft^2 **29. a.** 23 bags **b.** $575 **31. a.** 3680 ft^2 **b.** 15 cans of paint **c.** $404.25 **33.** $933.40
35. 40π m; 125.7 m **37.** 377 plants **39.** large pizza **41–45.** Answers will vary. **47.** by a factor of $\dfrac{9}{4}$ **49.** $13,032.73

Section 10.5
Check Point Exercises
1. 105 ft^3 **2.** 8 yd^3 **3.** 48 ft^3 **4.** 302 in.^{3} **5.** 101 in.^{3} **6.** no **7.** 632 yd^2

Exercise Set 10.5
1. 36 in.^{3} **3.** 64 cm^3 **5.** 175 yd^3 **7.** 56 in.^{3} **9.** 150π cm^3, 471 cm^3 **11.** 3024π in.^{3}, 9500 in.^{3} **13.** 48π m^3, 151 m^3
15. 15π yd^3, 47 yd^3 **17.** 288π m^3, 905 m^3 **19.** 972π cm^3, 3054 cm^3 **21.** 62 m^2 **23.** 96 ft^2 **25.** 324π cm^3, 1018 cm^3 **27.** $40
29. no **31. a.** 3,386,880 yd^3 **b.** 2,257,920 blocks **33.** yes **35.** $27 **37.** Answers will vary. **39.** The volume is multiplied by 8.
41. 168 cm^3 **43.** 84 cm^2

Section 10.6
Check Point Exercises
1. $\sin A = \dfrac{3}{5}$; $\cos A = \dfrac{4}{5}$; $\tan A = \dfrac{3}{4}$ **2.** 263 cm **3.** 298 cm **4.** 994 ft **5.** 54°

Exercise Set 10.6
1. $\sin A = \dfrac{3}{5}$; $\cos A = \dfrac{4}{5}$; $\tan A = \dfrac{3}{4}$ **3.** $\sin A = \dfrac{20}{29}$; $\cos A = \dfrac{21}{29}$; $\tan A = \dfrac{20}{21}$ **5.** $\sin A = \dfrac{5}{13}$; $\cos A = \dfrac{12}{13}$; $\tan A = \dfrac{5}{12}$
7. $\sin A = \dfrac{4}{5}$; $\cos A = \dfrac{3}{5}$; $\tan A = \dfrac{4}{3}$ **9.** 188 cm **11.** 182 in. **13.** 7 m **15.** 22 yd **17.** 40 m **19.** $m \angle B = 50°, a = 18$ yd,
$c = 29$ yd **21.** $m \angle B = 38°, a = 43$ cm, $b = 33$ cm **23.** 37° **25.** 28° **27.** 529 yd **29.** 2879 ft **31.** 2059 ft **33.** 695 ft
35. 36° **37.** 1376 ft **39.** 15.1° **41–49.** Answers will vary. **51. a.** 357 ft **b.** 394 ft

Section 10.7
Check Point Exercises
1. Answers will vary.

Exercise Set 10.7
1. a. traversable **b.** sample path: D, A, B, D, C, B **3. a.** traversable **b.** sample path: A, D, C, B, D, E, A, B **5.** not traversable
7. 3 **9.** 2 **11.** pitcher and wrench **13.** The sum of the angles in such a triangle is greater than 180°.
15. yes **17.** no **19.** Each carbon atom is even; each hydrogen atom is odd. **21.** 2 **23.** yes; There are two odd vertices, B and D.
25–41. Answers will vary.

Review Exercises
1. $\angle 3$ **2.** $\angle 5$ **3.** $\angle 4$ and $\angle 6$ **4.** $\angle 1$ and $\angle 6$ **5.** $\angle 4$ and $\angle 1$ **6.** $\angle 2$ **7.** $\angle 5$ **8.** 65° **9.** 49° **10.** 17° **11.** 134°
12. $m \angle 1 = 110°; m \angle 2 = 70°; m \angle 3 = 110°$ **13.** $m \angle 1 = 138°; m \angle 2 = 42°; m \angle 3 = 138°; m \angle 4 = 138°; m \angle 5 = 42°; m \angle 6 = 42°; m \angle 7 = 138°$
14. 72° **15.** 51° **16.** $m \angle 1 = 90°; m \angle 2 = 90°; m \angle 3 = 140°; m \angle 4 = 40°; m \angle 5 = 140°$ **17.** $m \angle 1 = 80°; m \angle 2 = 65°; m \angle 3 = 115°$;
$m \angle 4 = 80°; m \angle 5 = 100°; m \angle 6 = 80°$ **18.** 5 ft **19.** 3.75 ft **20.** 10 ft **21.** 7.2 in. **22.** 10.2 cm
23. 12.5 ft **24.** 15 ft **25.** 12 yd **26.** rectangle, square **27.** rhombus, square **28.** parallelogram, rhombus, trapezoid
29. 30 cm **30.** 4480 yd **31.** 44 m **32.** 1800° **33.** 1080° **34.** $m \angle 1 = 135°; m \angle 2 = 45°$ **35.** $132 **36.** 32.5 ft^2 **37.** 20 m^2
38. 50 cm^2 **39.** 135 yd^2 **40.** $C = 20\pi$ m $\approx$ 62.8 m; $A = 100\pi$ m^2 $\approx$ 314.2 m^2 **41.** 192 in.^{2} **42.** 28 m^2 **43.** $787.50
44. $650 **45.** 31 yd **46.** 60 cm^3 **47.** 960 m^3 **48.** 128π yd^3, 402 yd^3 **49.** $\dfrac{44,800\pi}{3}$ in.^{3}, 46,914 in.^{3} **50.** 288π m^3, 905 m^3

51. 126 m² **52.** 4800 m³ **53.** 651,775 m³ **54.** $80 **55.** $\sin A = \dfrac{3}{5}$; $\cos A = \dfrac{4}{5}$; $\tan A = \dfrac{3}{4}$ **56.** 42 mm **57.** 23 cm

58. 37 in. **59.** 58° **60.** 772 ft **61.** 31 m **62.** 56° **63.** not traversable
64. traversable; sample path: $A, B, C, D, A, B, C, D, A$ **65.** 0 **66.** 2 **67.** 1 **68.** 2 **69–70.** Answers will vary.

Chapter 10 Test
1. 36°; 126° **2.** 133° **3.** 70° **4.** 35° **5.** 3.2 in. **6.** 10 ft **7.** 1440° **8.** 40 cm **9.** d **10.** 517 m² **11.** 525 in.²
12. a. 12 cm **b.** 30 cm **c.** 30 cm² **13.** $C: 40\pi$ m, 125.7 m; $A: 400\pi$ m², 1256.6 m² **14.** 108 tiles **15.** 18 ft³ **16.** 16 m³
17. 175π cm³, 550 cm³ **18.** 85 cm **19.** 70 ft **20.** traversable; sample path: B, C, A, E, C, D, E **21.** Answers will vary.

CHAPTER 11

Section 11.1
Check Point Exercises
1. 150 **2.** 40 **3.** 30 **4.** 160 **5.** 729 **6.** 90,000

Exercise Set 11.1
1. 80 **3.** 12 **5.** 6 **7.** 40 **9.** 144; Answers will vary. **11.** 8 **13.** 96 **15.** 243 **17.** 144 **19.** 676,000 **21.** 2187
23–25. Answers will vary. **27.** 720 hr

Section 11.2
Check Point Exercises
1. 2 **2.** 120 **3. a.** 504 **b.** 524,160 **c.** 100 **4.** 840 **5.** 15,120 **6.** 420

Exercise Set 11.2
1. 720 **3.** 120 **5.** 120 **7.** 362,880 **9.** 6 **11.** 4 **13.** 504 **15.** 570,024 **17.** 3,047,466,240 **19.** 600 **21.** 10,712
23. 5034 **25.** 24 **27.** 6 **29.** 42 **31.** 1716 **33.** 3024 **35.** 6720 **37.** 720 **39.** 1 **41.** 720 **43.** 8,648,640
45. 120 **47.** 15,120 **49.** 180 **51.** 831,600 **53.** 105 **55.** 280 **57–63.** Answers will vary. **65.** 360 **67.** 14,400
69. $\dfrac{n(n-1)\cdots 3 \cdot 2 \cdot 1}{2} = n(n-1)\cdots 3$

Section 11.3
Check Point Exercises
1. a. combinations **b.** permutations **2.** 35 **3.** 1820 **4.** 23,049,600

Exercise Set 11.3
1. permutations **3.** combinations **5.** combinations **7.** permutations **9.** permutations **11.** 6 **13.** 126 **15.** 330
17. 8 **19.** 1 **21.** 4060 **23.** 1 **25.** 7 **27.** 20 **29.** 495 **31.** 24,310 **33.** 4495 **35.** 22,957,480 **37.** 735
39. 4,839,570,450 **41–43.** Answers will vary. **45.** The 5/36 lottery is easier to win.; Answers will vary.
47. 570 sec or 9.5 min; 2340 sec or 39 min

Section 11.4
Check Point Exercises
1. a. $\dfrac{1}{6}$ **b.** $\dfrac{1}{2}$ **c.** 0 **d.** 1 **2. a.** $\dfrac{1}{13}$ **b.** $\dfrac{1}{2}$ **c.** $\dfrac{1}{26}$ **3.** $\dfrac{1}{2}$ **4.** 0.23

Exercise Set 11.4
1. $\dfrac{1}{6}$ **3.** $\dfrac{1}{2}$ **5.** $\dfrac{1}{3}$ **7.** 1 **9.** 0 **11.** $\dfrac{1}{13}$ **13.** $\dfrac{1}{4}$ **15.** $\dfrac{3}{13}$ **17.** $\dfrac{1}{52}$ **19.** 0 **21.** $\dfrac{1}{4}$ **23.** $\dfrac{1}{2}$ **25.** $\dfrac{1}{2}$ **27.** $\dfrac{3}{8}$

29. $\dfrac{3}{8}$ **31.** $\dfrac{7}{8}$ **33.** 0 **35.** $\dfrac{1}{4}$ **37.** $\dfrac{1}{9}$ **39.** 0 **41.** $\dfrac{3}{10}$ **43.** $\dfrac{1}{5}$ **45.** $\dfrac{1}{2}$ **47.** 0 **49.** $\dfrac{1}{4}$ **51.** $\dfrac{1}{4}$ **53.** $\dfrac{1}{2}$

55. $\dfrac{4}{25} = 0.16$ **57.** $\dfrac{3}{25} = 0.12$ **59.** $\dfrac{23}{89} \approx 0.258$ **61.** $\dfrac{36}{89} \approx 0.404$ **63.** $\dfrac{6}{89} \approx 0.067$ **65–71.** Answers will vary. **73.** $\dfrac{3}{8} = 0.375$

Section 11.5
Check Point Exercises
1. $\dfrac{1}{60}$ **2.** $\dfrac{1}{13,983,816} \approx 0.0000000715$ **3. a.** $\dfrac{1}{6}$ **b.** $\dfrac{1}{2}$

Exercise Set 11.5
1. a. 120 **b.** 6 **c.** $\dfrac{1}{20}$ **3. a.** $\dfrac{1}{6}$ **b.** $\dfrac{1}{30}$ **c.** $\dfrac{1}{720}$ **d.** $\dfrac{1}{3}$ **5. a.** 84 **b.** 10 **c.** $\dfrac{5}{42}$

7. $\dfrac{1}{18,009,460} \approx 0.0000000555$; $\dfrac{100}{18,009,460} \approx 0.00000555$ **9. a.** $\dfrac{1}{177,100} \approx 0.00000565$ **b.** $\dfrac{27,132}{177,100} \approx 0.153$ **11.** $\dfrac{3}{10} = 0.3$

13. a. 2,598,960 **b.** 1287 **c.** $\dfrac{1287}{2,598,960} \approx 0.000495$ **15.** $\dfrac{11}{1105} \approx 0.00995$ **17.** $\dfrac{36}{270,725} \approx 0.000133$ **19–21.** Answers will vary.

23. $9,004,730 **25.** $\dfrac{1}{10}$

Section 11.6

Check Point Exercises

1. $\frac{3}{4}$ **2.** $\frac{120}{137}$ **3.** $\frac{1}{3}$ **4.** $\frac{27}{50}$ **5.** $\frac{3}{4}$ **6.** $\frac{122}{137}$ **7. a.** 2:50 or 1:25 **b.** 50:2 or 25:1 **8.** 199:1 **9.** 1:15; $\frac{1}{16}$

Exercise Set 11.6

1. $\frac{12}{13}$ **3.** $\frac{3}{4}$ **5.** $\frac{10}{13}$ **7.** $\frac{2,598,924}{2,598,960} \approx 0.999986$ **9.** $\frac{2,595,216}{2,598,960} \approx 0.998559$ **11. a.** 0.10 **b.** 0.90 **13.** 0.533 **15.** $\frac{2}{13}$

17. $\frac{1}{13}$ **19.** $\frac{1}{26}$ **21.** $\frac{9}{22}$ **23.** $\frac{5}{6}$ **25.** $\frac{7}{13}$ **27.** $\frac{11}{26}$ **29.** $\frac{3}{4}$ **31.** $\frac{5}{8}$ **33.** $\frac{33}{40}$ **35.** $\frac{4}{5}$ **37.** $\frac{7}{10}$ **39.** $\frac{43}{58}$ **41.** $\frac{50}{87}$

43. $\frac{113}{174}$ **45.** 15:43; 43:15 **47.** 2:1 **49.** 1:2 **51. a.** 9:91 **b.** 91:9 **53.** 1:3 **55.** 1:1 **57.** 12:1 **59.** 25:1 **61.** 47:5

63. 49:1 **65.** 1:19 **67.** $\frac{3}{7}$ **69.** $\frac{4}{25}$; 84 **71–77.** Answers will vary. **79.** 0.06

Section 11.7

Check Point Exercises

1. $\frac{1}{361} \approx 0.00277$ **2.** $\frac{1}{16}$ **3. a.** $\frac{625}{130,321} \approx 0.005$ **b.** $\frac{38,416}{130,321} \approx 0.295$ **c.** $\frac{91,905}{130,321} \approx 0.705$ **4.** $\frac{1}{221} \approx 0.00452$

5. $\frac{11}{850} \approx 0.0129$ **6.** $\frac{1}{2}$ **7. a.** $\frac{39}{100} = 0.39$ **b.** $\frac{5}{9} \approx 0.556$

Exercise Set 11.7

1. $\frac{1}{6}$ **3.** $\frac{1}{36}$ **5.** $\frac{1}{4}$ **7.** $\frac{1}{36}$ **9.** $\frac{1}{8}$ **11.** $\frac{1}{36}$ **13.** $\frac{1}{3}$ **15.** $\frac{3}{52}$ **17.** $\frac{1}{169}$ **19.** $\frac{1}{4}$ **21.** $\frac{1}{64}$ **23.** $\frac{1}{6}$

25. a. $\frac{1}{256} \approx 0.00391$ **b.** $\frac{1}{4096} \approx 0.000244$ **c.** ≈ 0.524 **d.** ≈ 0.476 **27.** ≈ 0.00234 **29.** $\frac{7}{29}$ **31.** $\frac{5}{87}$ **33.** $\frac{2}{21}$ **35.** $\frac{4}{35}$

37. $\frac{11}{21}$ **39.** $\frac{1}{57}$ **41.** $\frac{8}{855}$ **43.** $\frac{11}{57}$ **45.** $\frac{1}{5}$ **47.** $\frac{2}{3}$ **49.** $\frac{3}{4}$ **51.** $\frac{3}{4}$ **53.** $\frac{1}{125} = 0.008$ **55.** $\frac{9}{10} = 0.9$ **57.** $\frac{45}{479} \approx 0.094$

59. $\frac{83}{100}$ **61.** $\frac{67}{100}$ **63.** $\frac{29}{50}$ **65.** $\frac{13}{55}$ **67.** $\frac{13}{33}$ **69.** $\frac{68}{2475}$ **71.** $\frac{1}{6}$ **73–79.** Answers will vary. **81.** 0.59049

83. a. Answers will vary. **b.** $\frac{365}{365} \cdot \frac{364}{365} \cdot \frac{363}{365} \approx 0.992$ **c.** ≈ 0.008 **d.** 0.411 **e.** 23 people

Section 11.8

Check Point Exercises

1. 2.5 **2.** 2 **3. a.** $8000; In the long run, the average cost of a claim is $8000. **b.** $8000 **4.** 0; no; Answers will vary.
5. table entries: $998, $48, and −$2; expected value: −$0.90; In the long run, a person can expect to lose an average of $0.90 for each ticket purchased.; Answers will vary. **6.** −$0.20; In the long run, a person can expect to lose an average of $0.20 for each card purchased.

Exercise Set 11.8

1. 1.75 **3. a.** $29,000; In the long run, the average cost of a claim is $29,000. **b.** $29,000 **c.** $29,050 **5.** $0; Answers will vary.
7. $0.73 **9.** $\frac{1}{16} = 0.0625$; yes **11.** the second mall **13. a.** $140,000 **b.** no **15.** $\approx -$0.17$; In the long run, a person can expect to lose an average of about $0.17 for each game played. **17.** $\approx -$0.05$; In the long run, a person can expect to lose an average of about $0.05 for each game played. **19.** −$0.50; In the long run, a person can expect to lose an average of $0.50 for each game played.
21–25. Answers will vary. **27.** $160

Chapter Review

1. 800 **2.** 20 **3.** 9900 **4.** 125 **5.** 243 **6.** 60 **7.** 720 **8.** 120 **9.** 120 **10.** 240 **11.** 800 **12.** 114
13. 990 **14.** 151,200 **15.** 9900 **16.** 32,760 **17.** 1,860,480 **18.** 420 **19.** 60 **20.** combinations **21.** permutations

22. combinations **23.** 330 **24.** 2002 **25.** 210 **26.** 1287 **27.** 1140 **28.** 55,440 **29.** $\frac{1}{6}$ **30.** $\frac{2}{3}$ **31.** 1 **32.** 0

33. $\frac{1}{13}$ **34.** $\frac{3}{13}$ **35.** $\frac{3}{13}$ **36.** $\frac{1}{52}$ **37.** $\frac{1}{26}$ **38.** $\frac{1}{2}$ **39.** $\frac{1}{3}$ **40.** $\frac{1}{6}$ **41. a.** $\frac{1}{2}$ **b.** 0 **42.** $\frac{6}{175} \approx 0.03$

43. $\frac{103}{175} \approx 0.59$ **44.** $\frac{1}{24}$ **45.** $\frac{1}{6}$ **46.** $\frac{1}{30}$ **47.** $\frac{1}{720}$ **48.** $\frac{1}{3}$ **49. a.** $\frac{1}{15,504} \approx 0.0000645$ **b.** $\frac{100}{15,504} \approx 0.00645$

50. a. $\frac{1}{14}$ **b.** $\frac{3}{7}$ **51.** $\frac{3}{26}$ **52.** $\frac{5}{6}$ **53.** $\frac{1}{2}$ **54.** $\frac{1}{3}$ **55.** $\frac{2}{3}$ **56.** 1 **57.** $\frac{10}{13}$ **58.** $\frac{3}{4}$ **59.** $\frac{2}{13}$ **60.** $\frac{1}{13}$ **61.** $\frac{7}{13}$

62. $\frac{11}{26}$ **63.** $\frac{5}{6}$ **64.** $\frac{5}{6}$ **65.** $\frac{1}{2}$ **66.** $\frac{2}{3}$ **67.** 1 **68.** $\frac{5}{6}$ **69.** $\frac{4}{5}$ **70.** $\frac{3}{4}$ **71.** $\frac{18}{25}$ **72.** $\frac{6}{7}$ **73.** $\frac{3}{5}$ **74.** $\frac{12}{35}$

75. in favor: 1:12; against: 12:1 **76.** 99:1 **77.** $\frac{3}{4}$ **78.** $\frac{2}{9}$ **79.** $\frac{1}{36}$ **80.** $\frac{1}{9}$ **81.** $\frac{1}{36}$ **82.** $\frac{8}{27}$ **83.** $\frac{1}{32}$ **84. a.** 0.04

b. 0.008 **c.** 0.4096 **d.** 0.5904 **85.** $\frac{1}{9}$ **86.** $\frac{1}{12}$ **87.** $\frac{1}{196}$ **88.** $\frac{1}{3}$ **89.** $\frac{3}{10}$ **90. a.** $\frac{1}{2}$ **b.** $\frac{2}{7}$ **91.** $\frac{27}{29}$ **92.** $\frac{4}{29}$

93. $\dfrac{144}{145}$ **94.** $\dfrac{11}{20}$ **95.** $\dfrac{11}{135}$ **96.** $\dfrac{1}{125}$ **97.** $\dfrac{1}{232}$ **98.** $\dfrac{19}{1044}$ **99.** 3.125 **100. a.** $0.50; In the long run, the average cost of a claim is $0.50. **b.** $10.00 **101.** $4500; Answers will vary. **102.** −$0.25; In the long run, a person can expect to lose an average of $0.25 for each game played.

Chapter 11 Test

1. 240 **2.** 24 **3.** 720 **4.** 990 **5.** 210 **6.** 420 **7.** $\dfrac{6}{25}$ **8.** $\dfrac{17}{25}$ **9.** $\dfrac{11}{25}$ **10.** $\dfrac{5}{13}$ **11.** $\dfrac{1}{210}$ **12.** $\dfrac{10}{1001} \approx 0.00999$

13. $\dfrac{1}{2}$ **14.** $\dfrac{1}{16}$ **15.** $\dfrac{1}{8000} = 0.000125$ **16.** $\dfrac{8}{13}$ **17.** $\dfrac{3}{5}$ **18.** $\dfrac{1}{19}$ **19.** $\dfrac{1}{256}$ **20.** 3:4 **21. a.** 4:1 **b.** $\dfrac{4}{5}$

22. $\dfrac{3}{5}$ **23.** $\dfrac{39}{50}$ **24.** $\dfrac{3}{5}$ **25.** $\dfrac{9}{19}$ **26.** $\dfrac{7}{150}$ **27.** $1000; Answers will vary.

28. −$12.75; In the long run, a person can expect to lose an average of $12.75 for each game played.

CHAPTER 12

Section 12.1

Check Point Exercises

1. a. the set containing all the city's homeless **b.** no: People already in the shelters are probably less likely to be against mandatory residence in the shelters. **2.** By selecting people from a shelter, homeless people who do not go to the shelters have no chance of being selected. An appropriate method would be to randomly select neighborhoods of the city and then randomly survey homeless people within the selected neighborhood.

3.

Grade	Frequency
A	3
B	5
C	9
D	2
F	1
	20

4.

Class	Frequency
40–49	1
50–59	5
60–69	4
70–79	15
80–89	5
90–99	7
	37

5.

Stem	Leaves
4	1
5	8 2 8 0 7
6	8 2 9 9
7	3 5 9 9 7 5 5 3 3 6 7 1 7 1 5
8	7 3 9 9 1
9	4 6 9 7 5 8 0

Exercise Set 12.1

1. a. population: all American men aged 18 or older; sample: those men actually polled **b.** each man's impression of his own health; qualitative
3. c **5.** 7; 31 **7.** 151

9.

Time Spent on Homework (in hours)	Number of Students
15	4
16	5
17	6
18	5
19	4
20	2
21	2
22	0
23	0
24	2
	30

11. 0, 5, 10, ..., 40, 45 **13.** 5
15. 13 **17.** the 5–9 class
19.

Age at Inauguration	Number of Presidents
41–45	2
46–50	8
51–55	15
56–60	9
61–65	7
66–70	2
	43

21. a.

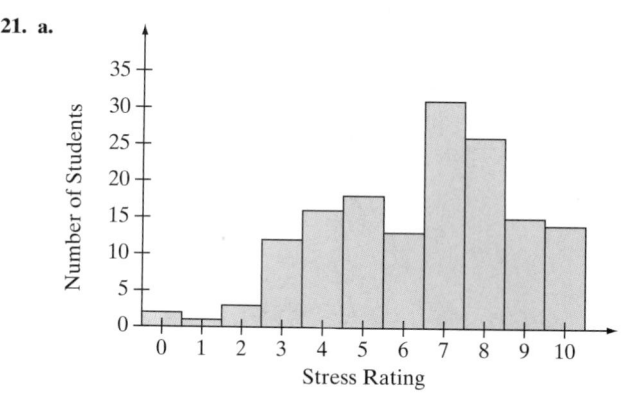

b.

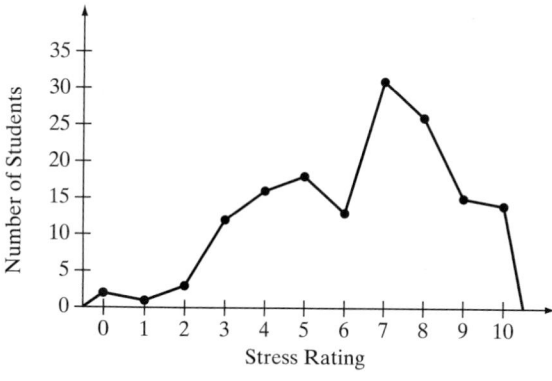

23.

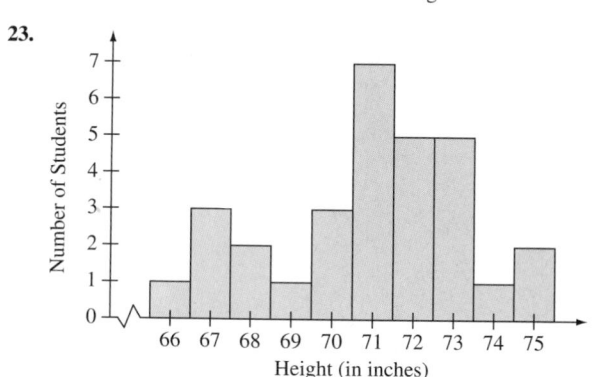

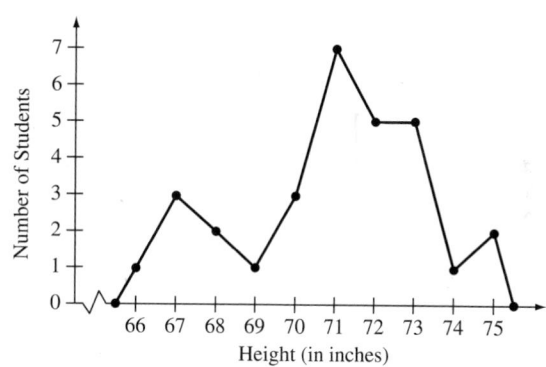

25. b   **27.**

Stem	Leaves
2	8 8 9 5
3	8 7 0 1 2 7 6 4 0 5
4	8 2 2 1 4 5 4 6 2 0 8 2 7 9
5	9 4 1 9 1 0
6	3 2 3 6 6 3

The greatest number of college professors are in their 40s.

29. Answers will vary.   **31.** Unequal intervals on horizontal scale make the increase appear less dramatic.   **33–43.** Answers will vary.

Section 12.2

Check Point Exercises

1. a. 30   **b.** 30   **2.** 36   **3. a.** 35   **b.** 82   **4.** 5   **5.** 24.5   **6.** 54.5   **7.** ≈ $4.1 million; The mean is much greater than the median.; The considerably higher salaries of O'Neal and Bryant pulls the mean to a higher value than the median.   **8.** 8   **9.** $105,750
10. mean: 158.6 cal; median: 153 cal; mode: 138 cal; midrange: 151 cal

Exercise Set 12.2

1. 4.125   **3.** 95   **5.** 62   **7.** ≈ 3.45   **9.** ≈ 4.71   **11.** ≈ 6.26   **13.** 3.5   **15.** 95   **17.** 60   **19.** 3.6   **21.** 5   **23.** 6
25. 3   **27.** 95   **29.** 40   **31.** 2.5, 4.2 (bimodal)   **33.** 5   **35.** 6   **37.** 4.5   **39.** 95   **41.** 70   **43.** 3.3   **45.** 4.5   **47.** 5.5
49. a. ≈ 69.2 yr   **b.** 67 yr   **c.** 67 yr   **d.** 69 yr   **51. a.** $18.24 million   **b.** $17.55 million   **c.** 17, 18 (bimodal)
d. $20.35 million   **53.** 176.875 lb   **55.** 175 lb   **57.** ≈ $2.4 million; This is close to the median.   **59.** ≈ 2.76
61–71. Answers will vary.   **73.** All 30 students had the same grade.

Section 12.3

Check Point Exercises

1. 9   **2.** mean: 6;   **3.** ≈ 3.92   **4.** sample A: 3.74; sample B: 28.06   **5. a.** stocks
b. stocks; Answers will vary.

Data item	Deviation
2	−4
4	−2
7	1
11	5

Exercise Set 12.3

1. 4 **3.** 8 **5.** 2

7. a.

Data item	Deviation
3	−9
5	−7
7	−5
12	0
18	6
27	15

b. 0

9. a.

Data item	Deviation
29	−20
38	−11
48	−1
49	0
53	4
77	28

b. 0

11. a. 91

b.

Data item	Deviation
85	−6
95	4
90	−1
85	−6
100	9

c. 0

13. a. 155

b.

Data item	Deviation
146	−9
153	−2
155	0
160	5
161	6

c. 0

15. a. 2.70

b.

Data item	Deviation
2.25	−0.45
3.50	0.80
2.75	0.05
3.10	0.40
1.90	−0.80

c. 0

17. ≈ 1.58 **19.** ≈ 3.46 **21.** ≈ 0.89 **23.** 3 **25.** ≈ 2.14
27. *Sample A*: mean: 12; range: 12; standard deviation: ≈ 4.32; *Sample B*: mean: 12; range: 12; standard deviation: ≈ 5.07;
Sample C: mean: 12; range: 12; standard deviation: 6; The samples have the same mean and range, but different standard deviations.
29. mean: 196; standard deviation: ≈ 7.29 **31.** mean: ≈ 5.09; standard deviation: ≈ 4.76 **33–39.** Answers will vary.
41. Answers will vary. **43.** a

Section 12.4

Check Point Exercises

1. a. 77.5 in. **b.** 65 in. **2. a.** 95% **b.** 47.5% **c.** 16% **3. a.** 2 **b.** 0 **c.** −1 **4.** ACT **5.** 137.5
6. The student did better than about 62% of all those who took the exam. **7.** 88.49% **8.** 8.08% **9.** 83.01% **10. a.** ± 3.0%
b. We can be 95% confident that between 61.0% and 67.0% of Americans find the price increase to be acceptable.

Exercise Set 12.4

1. 120 **3.** 160 **5.** 150 **7.** 60 **9.** 90 **11.** 68% **13.** 34% **15.** 47.5% **17.** 49.85% **19.** 16% **21.** 2.5%
23. 95% **25.** 47.5% **27.** 16% **29.** 2.5% **31.** 0.15% **33.** 1 **35.** 3 **37.** 0.5 **39.** 1.75 **41.** 0 **43.** −1
45. −1.5 **47.** −3.25 **49.** 1.5 **51.** 2.25 **53.** −0.5 **55.** −1.5 **57.** math **59.** 500 **61.** 475
63. 250 **65.** 275 **67. a.** 72.57% **b.** 27.43% **69. a.** 88.49% **b.** 11.51% **71. a.** 24.2% **b.** 75.8% **73. a.** 11.51%
b. 88.49% **75.** 33.99% **77.** 15.74% **79.** 86.64% **81.** 28.57% **83.** $z = 1.5$; 93.32% **85.** $z = 0.6$; 27.43%
87. $z = -1.2$; 88.49% **89.** $z = 1.4, z = 2.1$; 6.29% **91.** $z = -0.6, z = 0.6$; 45.14% **93.** 6.68% **95.** 24.17% **97. a.** ±5.0%
b. We can be 95% confident that between 21% and 31% of all parents feel that crime is a bad thing about being a kid. **99. a.** ±1.6%
b. We can be 95% confident that between 58.6% and 61.8% of all TV households watched the final episode of "M*A*S*H".
101. 0.2% **103.** no; Answers will vary. **105–117.** Answers will vary. **119.** 62 in.

Section 12.5

Check Point Exercises

1. This indicates a moderate relationship. **2.** −0.84 **3.** $y = -22.97x + 260.56$; 30.86 deaths per 100,000 people **4.** yes

Exercise Set 12.5

1.
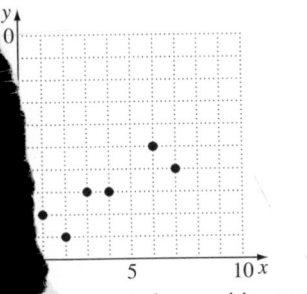
e appears to be a positive correlation.

3.
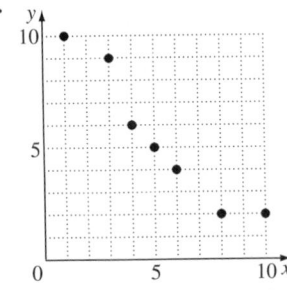
There appears to be a negative correlation.

5.
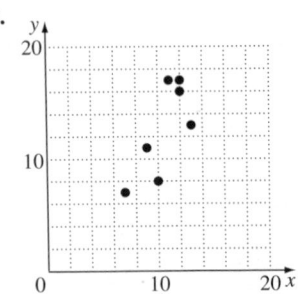
There appears to be a positive correlation.

7.

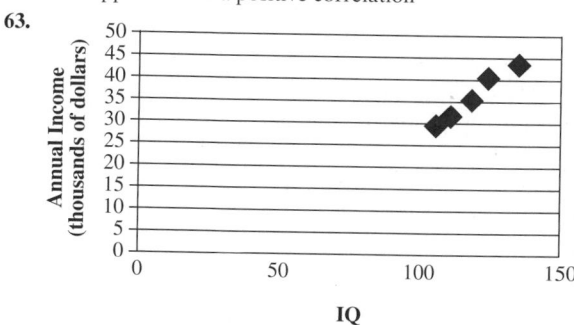

There appears to be a positive correlation

9. false   **11.** true   **13.** true   **15.** false   **17.** false   **19.** true   **21.** false   **23.** false
25. false   **27.** true   **29.** a   **31.** d   **33.** 0.855   **35.** −0.954   **37. a.** 0.75
b. $y = 1.52x - 3.38$ **c.** 21 years   **39. a.** 0.885   **b.** $y = 0.85x + 4.05$
c. 16 murders per 100,000 people   **41.** A correlation does exist.   **43.** A correlation does not exist.
45. A correlation does exist.   **47.** A correlation does not exist.   **49–59.** Answers will vary.
61. Answers will vary.

63.

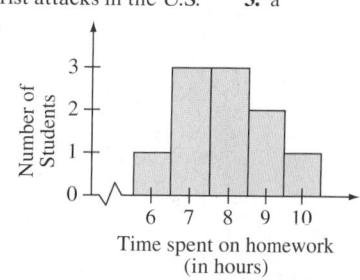

Review Exercises

1. population: American mothers with children under 18; sample: those mothers actually polled.
2. belief about likelihood of future terrorist attacks in the U.S.   **3.** a

4.

Time Spent on Homework (in hours)	Number of Students
6	1
7	3
8	3
9	2
10	1
	10

5.

6.

7.

Grades	Number of Students
0–39	19
40–49	8
50–59	6
60–69	6
70–79	5
80–89	3
90–100	3
	50

8.

Stem	Leaves
1	8 4 3 3 7 1
2	4 6 9 9 2 7
3	4 9 6 1 5 1 1
4	4 7 9 1 2 2 0 5
5	4 7 9 0 6 1
6	3 7 0 8 3 9
7	2 5 4 0 3
8	1 7 6
9	1 0 5

9. Left edge of each bar lines up with appropriate value, but each right edge is well above correct value.   **10.** 91.2   **11.** 17   **12.** 2.3
13. 11   **14.** 28   **15.** 2   **16.** 27   **17.** 585, 587 (bimodal)   **18.** 2   **19.** 91   **20.** 19.5   **21.** 2.5   **22–23.** Answers will vary.

24. a.

Age at First Inauguration	Number of Presidents
42	1
43	1
44	0
45	0
46	2
47	1
48	1
49	2
50	1
51	5
52	2
53	0
54	5
55	4
56	3
57	4
58	1
59	0
60	1
61	3
62	1
63	0
64	2
65	1
66	0
67	0
68	1
69	1
	43

b. mean: ≈ 54.84 yr; median: 55 yr; mode: 51 yr, 54 yr (bimodal); midrange: 55.5 yr   **25.** 18   **26.** 564

27. a.

Data item	Deviation
29	–6
9	–26
8	–27
22	–13
46	11
51	16
48	13
42	7
53	18
42	7

b. 0

28. a. 50

b.

Data item	Deviation
36	–14
26	–24
24	–26
90	40
74	24

c. 0

29. ≈ 4.05 **30.** ≈ 5.13 **31.** mean: 49; range: 76; standard deviation: ≈ 24.32
32. Set A: mean: 80; standard deviation: 0; Set B: mean: 80; standard deviation: ≈ 11.55; Answers will vary.   **33.** Answers will vary.
34. 86   **35.** 98   **36.** 60   **37.** 68%   **38.** 95%   **39.** 34%   **40.** 99.7%   **41.** 16%   **42.** 84%   **43.** 2.5%   **44.** 0
45. 2   **46.** 1.6   **47.** –3   **48.** –1.2   **49.** vocabulary test   **50.** 38,000 miles   **51.** 41,000 miles   **52.** 22,000 miles
53. 91.92%   **54.** 96.41%   **55.** 88.33%   **56.** 10.69%   **57.** 75%   **58.** 14%   **59.** 11%   **60. a.** $\pm 3.1\%$
b. We can be 95% confident that between 29.9% and 36.1% of American men have suffered from depression or anxiety disorder

61.

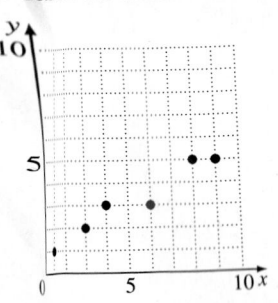

There appears to be a positive correlation.

62.

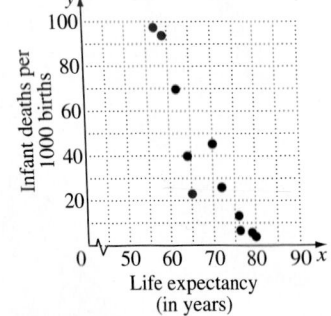

There appears to be a negative correlation.

63. false   **64.** true   **65.** false   **66.** false   **67.** true   **68.** false   **69.** true   **70.** c

71. a. 0.972   **b.** $y = 0.509x + 0.537$ **72. a.**

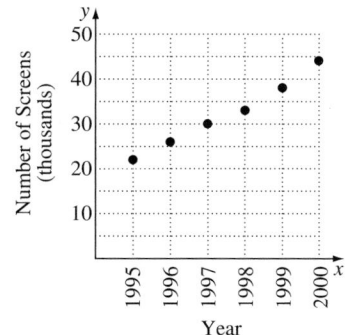

b. 0.994   **c.** yes

Chapter 12 Test

1. d   **2.**

Score	Frequency
3	1
4	2
5	3
6	2
7	2
8	3
9	2
10	1
	16

3.

4.

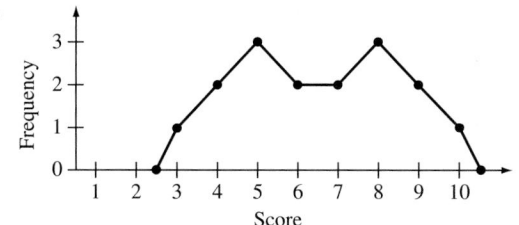

5.

Class	Frequency
40–49	3
50–59	6
60–69	6
70–79	7
80–89	6
90–99	2
	30

6.

Stem	Leaves
4	8 1 6
5	1 0 5 0 9 0
6	7 2 0 3 1 1
7	9 8 3 1 9 1 5
8	9 3 0 8 9 1
9	3 0

7. ≈ 3.67   **8.** 3   **9.** 4   **10.** ≈ 2.34   **11.** 2.25   **12.** 2   **13.** 2   **14.** Answers will vary.
15. 34%   **16.** 2.5%   **17.** student   **18.** 8.08%   **19.** 41%
20. a. ±10%   **b.** We can be 95% confident that between 50% and 70% of all students are very satisfied with their professors.
21.

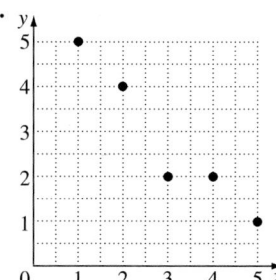

; There appears to be a strong negative correlation.
22. false   **23.** false   **24.** true   **25.** Answers will vary.

CHAPTER 13

Section 13.1

Check Point Exercises

1. E; The sum of two odd numbers is an even number.   **2.** not closed   **3.** closed   **4.** not closed

5. $(1 \oplus 3) \oplus 2 = 1 \oplus (3 \oplus 2)$ **6.** g **7.** -12 **8.** $\frac{1}{12}$ **9. a.** k **b.** l is the inverse of j, k is its own inverse; j is the inverse of l;
$$0 \oplus 2 = 1 \oplus 1$$
$$2 = 2$$
$$2 = 2$$
    m is its own inverse.

Exercise Set 13.1

1. $\{e, a, b, c\}$ **3.** c **5.** a **7.** e **9.** b **11.** e **13.** a **15.** not closed; Answers will vary.   **17.** closed; The sum of any two
natural numbers is a natural number.   **19.** not closed; Answers will vary.   **21.** not closed; $b * a = c$ is not an element of the set $\{a, b\}$.

23. Both sides of the equality equal 1.   **25.** Both sides of the equality equal 0.   **27.** commutative property; The entries in the table are mirror images about the diagonal line that extends from the upper left corner to the lower right corner.   **29.** Both sides of the equality equal 4.
31. Both sides of the equality equal 0.   **33. a.** b **b.** a **c.** not associative; When the grouping changes, the answer changes.
35. Answers will vary. Sample answer: $(5 - 2) - 1 = 3 - 1 = 2$ but $5 - (2 - 1) = 5 - 1 = 4$.   **37.** -29 **39.** $\dfrac{1}{29}$ **41.** a **43.** c
45. e **47.** c **49.** e **51.** a **53.** d **55.** b **57.** 0   **59.** 4   **61.** 2   **63. a.** d **b.** d is its own inverse.; $a, b,$ and c do not have inverses.   **65.** The mathematical system shows all possible starting and ending position for the four-way switch. The starting positions are shown in the left column. The number of clockwise turns are shown across the top. The 25 entries in the table represent the final position of the switch for each starting position and number of clockwise turns.

67–71.

$\times$	E	O
E	E	E
O	E	O

67. closed; All entries in the body of the table are elements of the set $\{E, O\}$.
69. Answers will vary. Sample answer: $(O \times E) \times O = E \times O = E$ and $O \times (E \times O) = O \times E = E$.
71. E does not have an inverse.   **73–79.** Answers will vary.
81. a. $\begin{bmatrix} 15 & 20 \\ 35 & 46 \end{bmatrix}$ **b.** $\begin{bmatrix} 4 & 7 \\ 34 & 57 \end{bmatrix}$ **c.** This multiplication is not commutative.

Section 13.2
Check Point Exercises
1. a. 1. All entries in the table are elements of the set $\{E, O\}$; thus, the system is closed.
 2. Two examples of the associative property follow. Answers will vary.
  $(O \circ E) \circ O = O \circ O = E$ and $O \circ (E \circ O) = O \circ O = E$
  $(E \circ O) \circ E = O \circ E = O$ and $E \circ (O \circ E) = E \circ O = O$
 3. E is the identity element.
 4. E is its own inverse, and O is its own inverse. Each element has an inverse. Since the system meets the four requirements, the system is a group.
 b. Since $E \circ O = O$ and $O \circ E = O$, the order does not matter. Thus, the group is commutative.
2. a. $(8 + 5) + 11 = 1 + 11 = 0$ and $8 + (5 + 11) = 8 + 4 = 0$ **b.** Both sides of the equality equal 1.
3. a. true   **b.** true   **c.** false   **4. a.** 4 (mod 5)   **b.** 2 (mod 7)   **c.** 5 (mod 13)   **5.** Tuesday

Exercise Set 13.2
1. 8-fold rotational symmetry   **3.** 18-fold rotational symmetry   **5.** All entries in the body of the table are elements of the set $\{e, p, q, r, s, t\}$.
7. $(r \circ t) \circ q = p \circ q = e$ and $r \circ (t \circ q) = r \circ r = e$ **9.** e **11.** q **13.** r **15.** t **17.** t **19.** The operation is not commutative.
21. Answers will vary. Sample answer: The element 2 does not have an inverse.

23. a.

+	0	1	2	3	4	5
0	0	1	2	3	4	5
1	1	2	3	4	5	0
2	2	3	4	5	0	1
3	3	4	5	0	1	2
4	4	5	0	1	2	3
5	5	0	1	2	3	4

b. 1. Since all entries in the body of the table are elements of the set, the closure property is satisfied.
2. Since the result of using clock addition to add 3 elements is the same regardless of whether the first two elements are grouped or the last two elements are grouped, the associative property is satisfied.
Example 1:
$(1 + 2) + 3 = 3 + 3 = 0$
$1 + (2 + 3) = 1 + 5 = 0$
Example 2:
$(2 + 4) + 5 = 0 + 5 = 5$
$2 + (4 + 5) = 2 + 3 = 5$
3. Since 0 is an element of the set and $0 + e = e$ and $e + 0 = e$ for all elements e in the set, the identity property is satisfied.   **4.** Since each element has an inverse within the set, the inverse property is satisfied.   **5.** Since the entries in the table are symmetric about the diagonal from the upper left corner to the lower right corner, we know that the order of clock addition does not matter. The commutative property is satisfied.
Since all 5 properties are satisfied, this mathematical system is a commutative group.

25. true   **27.** true   **29.** false; $84 \equiv 0 \pmod 7$ **31.** false; $23 \equiv 3 \pmod 4$ **33.** true   **35.** 5 (mod 6)   **37.** 3 (mod 6)
39. 4 (mod 7)   **41.** 48 (mod 50)   **43.** 1800   **45.** 0020   **47.** Sunday   **49.** *beam me up*   **51–57.** Answers will vary.
59. Answers will vary. In general, $(a \cdot b) \cdot c = (a - b + ab) \cdot c$
$= a - b + ab - c + (a - b + ab)c$
$= a - b + ab - c + ac - bc + abc$
$= a - b - c + ab + ac - bc + abc$
$a \cdot (b \cdot c) = a \cdot (b - c + bc)$
$= a - (b - c + bc) + a(b - c + bc)$
$= a - b + c - bc + ab - ac + abc$
Since $(a \cdot b) \cdot c \neq a \cdot (b \cdot c)$, the associate property is not satisfied. Thus the mathematical system is not a group.
61. 8:00 A.M.   **63.** $2, 9, 16, 23, 30, 37$

Review Exercises
1. $\{e, c, f, r\}$ **2.** closed; All entries in the body of the table are elements of the set $\{e, c, f, r\}$.   **3.** r **4.** f **5.** c
6. Both expressions equal e.   **7.** Both expressions equal c.   **8.** Both expressions equal f.   **9.** commutative property
10. The entries in the table are mirror images about the diagonal from the upper left corner to the lower right corner.
11. Both expressions equal f.   **12.** Both expressions equal e.   **13.** associative property   **14.** e **15.** e **16.** r
17. f **18.** c **19.** not closed; $1 + 1 = 2$ which is not in the set.   **20.** closed; All products result in either 0 or 1.
21. not closed; $1 \div 2 = \dfrac{1}{2}$ which is not in the set.   **22.** -123 **23.** $\dfrac{1}{123}$

24. a.

∘	0	1	2	3	4
0	0	1	2	3	4
1	1	1	2	3	4
2	2	2	2	3	4
3	3	3	3	3	4
4	4	4	4	4	4

b. 0   **c.** no inverse; No element combined with 2 results in 0.

25. 3-fold rotational symmetry   **26.** 18-fold rotational symmetry

27. a.

+	0	1	2	3	4
0	0	1	2	3	4
1	1	2	3	4	0
2	2	3	4	0	1
3	3	4	0	1	2
4	4	0	1	2	3

b. 1. Since all entries in the body of the table are elements of the set, the closure property is satisfied. **2.** Since the result of using clock addition to add 3 elements is the same regardless of whether the first two elements are grouped or the last two elements are grouped, the associative property is satisfied.
Example 1:
$(1 + 2) + 3 = 3 + 3 = 1$
$1 + (2 + 3) = 1 + 0 = 1$
3. Since 0 is an element of the set and $0 + e = e$ and $e + 0 = e$ for all elements e in the set, the identity property is satisfied. **4.** Since each element has an inverse within the set, the inverse property is satisfied. **5.** Since the entries in the table are symmetric about the diagonal from the upper left corner to the lower right corner, we know that the order of clock addition does not matter. The commutative property is satisfied.
Since all 5 properties are satisfied, this mathematical system is a commutative group.

28. false; $17 \equiv 1 \pmod 8$ **29.** false; $37 \equiv 2 \pmod 5$ **30.** true   **31.** 1 (mod 6)   **32.** 6 (mod 8)   **33.** 7 (mod 9)   **34.** 1 (mod 20)

Chapter 13 Test

1. closed; All entries in the body of the table are elements of the set $\{x, y, z\}$.   **2.** Both expressions equal x.; commutative property
3. Both expressions equal y.; associative property   **4.** x **5.** x is its own inverse. z is the inverse of y. y is the inverse of z.
6. not closed; $1 + 1 = 2$ which is not in the set.   **7.** $\frac{1}{5}$; $5 \times \frac{1}{5} = 1$ **8.** 6-fold rotational symmetry; Since it takes 6 equal turns to restore the design to its original position and each of these turns is a design that is identical to the original, the design has 6-fold rotational symmetry.

9.

+	0	1	2	3
0	0	1	2	3
1	1	2	3	0
2	2	3	0	1
3	3	0	1	2

10. Check that all properties are satisfied:
1. Since all entries in the body of the table are elements of the set, the closure property is satisfied.
2. Since the result of using clock addition to add 3 elements is the same regardless of whether the first two elements are grouped or the last two elements are grouped, the associative property is satisfied.
Example 1:
$(1 + 2) + 3 = 3 + 3 = 2$
$1 + (2 + 3) = 1 + 1 = 2$
3. Since 0 is an element of the set and $0 + e = e$ and $e + 0 = e$ for all elements e in the set, the identity property is satisfied. **4.** Since each element has an inverse within the set, the inverse property is satisfied. **5.** Since the entries in the table are symmetric about the diagonal from the upper left corner to the lower right corner, we know that the order of clock addition does not matter. The commutative property is satisfied.
Since all 5 properties are satisfied, this mathematical system is a commutative group.

11. true   **12.** false; $14 \equiv 0 \pmod 7$ **13.** 10 (mod 11)   **14.** 5 (mod 10)

CHAPTER 14

Section 14.1

Check Point Exercises
1. a. 4210   **b.** 40   **c.** 2865   **2.** Donna   **3.** Bob   **4.** Carmen   **5.** Bob

Exercise Set 14.1
1. 7; 5; 4   **3.**

Number of Votes	5	1	4	2
First Choice	A	B	C	C
Second Choice	B	D	B	B
Third Choice	C	C	D	A
Fourth Choice	D	A	A	D

5. a. 26   **b.** 8   **c.** 22   **d.** 3   **7.** musical   **9.** Darwin
11. Comedy   **13.** Freud   **15.** Comedy   **17.** Einstein   **19.** 10
21. 28   **23.** Comedy   **25.** Hawking   **27.** A   **29.** B   **31.** Rent
33. Rent   **35. a.** C   **b.** A   **37.** C   **39–47.** Answers will vary.
49. b   **51–53.** Answers will vary.

Section 14.2
Check Point Exercises
1. a. A **b.** B **2. a.** B **b.** A **3. a.** A **b.** C **c.** yes **4. a.** A **b.** D **c.** yes

Exercise Set 14.2
1. a. D **b.** A **c.** no **3. a.** A **b.** V **c.** no **5. a.** A **b.** C **c.** no **7. a.** C **b.** B **c.** no
9. a. S **b.**

Number of Votes	15	8
First Choice	S	L
Second Choice	L	S

; S **c.** yes

11. a. A **b.** no **13. a.** C **b.** no **15. a.** B **b.** no **c.** no **d.** A; no **17. a.** B **b.** no **c.** no **19. a.** A **b.** yes
c. yes **d.**

Number of Votes	7	3	2
First Choice	A	B	A
Second Choice	B	C	C
Third Choice	C	A	B

; A **e.** yes **f.** no **21–31.** Answers will vary.

Section 14.3
Check Point Exercises
1. a. 50 **b.** 22.24; 22.36; 26.4; 30.3; 98.7 **2.** 22; 22; 27; 30; 99 **3.** 22; 22; 26; 30; 100 **4.** 22; 23; 27; 30; 98 **5.** 22; 22; 27; 30; 99

Exercise Set 14.3
1. a. 20; 20,000 **b.** 6.9; 13.3; 26.7; 33.1 **c.**

State	Lower Quota	Upper Quota
A	6	7
B	13	14
C	26	27
D	33	34

3. 7; 13; 27; 33 **5.** 30; 41; 53; 73; 103

7. 3; 5; 12; 15; 22 **9.** 17; 55; 24; 54 **11.** 28; 33; 53; 15; 22; 29 **13.** 12; 10; 8 **15.** 4; 10; 6 **17.** 20; 24; 29; 29; 98 **19.** 57; 81; 68; 44
21. 57; 81; 68; 44 **23.** 7; 2; 2; 2; 8; 14; 4; 5; 10; 10; 13; 2; 6; 2; 18 **25.** 7; 2; 2; 2; 8; 14; 4; 5; 10; 10; 12; 2; 6; 3; 18 **27–45.** Answers will vary.

Section 14.4
Check Point Exercises
1. B's apportionment decreases from 11 to 10. **2. a.** 10; 19; 71 **b.** State A: 1.005%; State B: 0.998%
c. State A loses a seat to State B, even though the population of State A is increasing at a faster rate.
3. a. 21; 79 **b.** With North High added to the district, West High loses a counselor to East High.

Exercise Set 14.4
1. a. 16; 8; 6 **b.** 17; 9; 5; Liberal Arts Math loses a teaching assistant when the total number of teaching assistants is raised from 30 to 31.
This is an example of the Alabama paradox. **3.** State A loses a seat when the total number of seats increases from 40 to 41.
5. a. 4; 6; 14 **b.** 28.3%; 26.3%; 14.7% **c.** 3; 7; 14; A loses a seat while B gains, even though A has a faster increasing population.
The population paradox does occur. **7.** C loses a truck to B even though C increased in population faster than B. **9. a.** 10; 90 **b.** Branch
B loses a promotion when branch C is added. This means the new-states paradox has occurred. **11. a.** 9; 91 **b.** State B loses a seat when
state C is added. **13. a.** 6; 13; 31 **b.** The new-states paradox does not occur. As long as the modified divisor, d, remains the same, adding a
new state cannot change the number of seats held by existing states. **15–17.** Answers will vary. **19.** d

Review Exercises
1.

Number of Votes	4	3	3	2
First Choice	A	B	C	C
Second Choice	B	D	B	B
Third Choice	C	C	D	A
Fourth Choice	D	A	A	D

2. 23 **3.** 4 **4.** 16 **5.** 14 **6.** Musical **7.** Comedy **8.** Musical
9. Musical **10.** A **11.** B **12.** A **13.** A **14.** B **15.** A; no
16. A; no **17.** B **18.** B; yes **19.** A **20.** B **21.** A **22.** A
23. A; Borda count method **24.** B **25.** A; no **26.** A

27.

Number of Votes	400	450
First Choice	A	C
Second Choice	C	A

; C; no **28.** B **29.** B still wins. The irrelevant alternatives criterion is satisfied.
30. 50.05 **31.** 5.49; 7.83; 12.21; 14.47 **32.** A: 5, 6; B: 7, 8; C: 12, 13; D: 14, 15
33. 6; 8; 12; 14 **34.** 5; 8; 12; 15 **35.** 6; 8; 12; 14 **36.** 6; 8; 12; 14
37. 14; 42; 62; 82 **38.** 13; 42; 63; 82 **39.** 14; 42; 63; 81 **40.** 14; 42; 62; 82
41. a. 8; 67; 75 **b.** The Alabama paradox occurs. **42. a.** 72; 20; 8 **b.** 0.8%; 1.3% **c.** The population paradox occurs.
43. a. 7; 26 **b.** The new-states paradox does not occur. **44.** False. Answers will vary.

Chapter 14 Test
1. 3600 **2.** 600 **3.** 1500 **4.** 1500 **5.** B **6.** B **7.** A **8.** A **9.** A **10.** A; yes **11.** A **12.** B; no **13.** C
14. B; no **15.** no **16.** 50 **17.** 2.38; 3.3; 4.32 **18.** A: 2, 3; B: 3, 4; C: 4, 5 **19.** 3; 3; 4 **20.** 2; 3; 5 **21.** 3; 3; 4 **22.** 2; 3; 5
23. yes **24.** The new-states paradox does not occur. **25.** Answers will vary.

CHAPTER 15

Section 15.1
Check Point Exercises

1. The two graphs have the same number of vertices connected to each other in the same way.

2.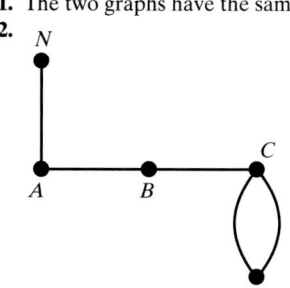

3.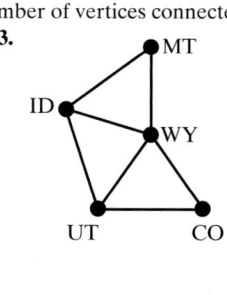

4.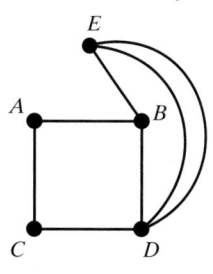

5.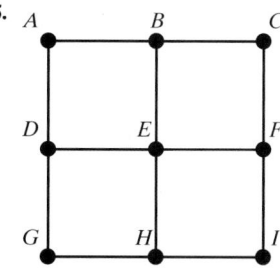

6. *A* and *B*, *A* and *C*, *A* and *D*, *A* and *E*, *B* and *C*, and *E* and *E*

Exercise Set 15.1

1. 6; St. Louis, Chicago, Philadelphia, Montreal; 1, 1, 2, 2   **3.** No; no; Answers will vary.

5. Possible answers:

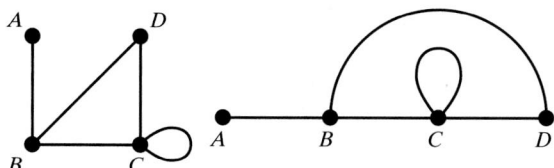

7. The two graphs have the same number of vertices connected in the same way, so they are the same. Possible answer:

9.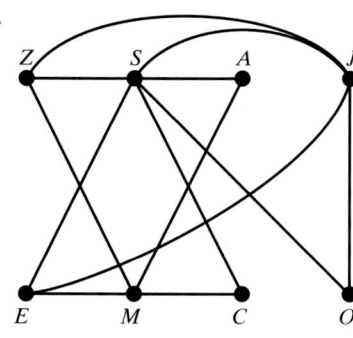

11.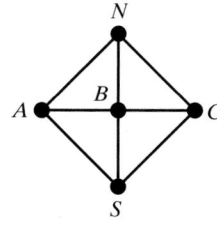

13.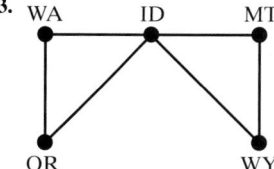

15.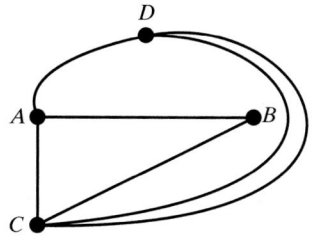

17.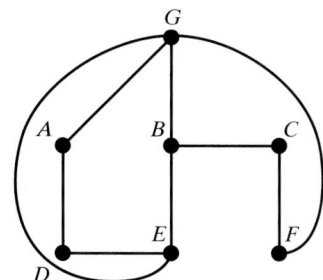

19.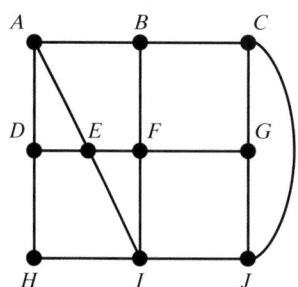

21.
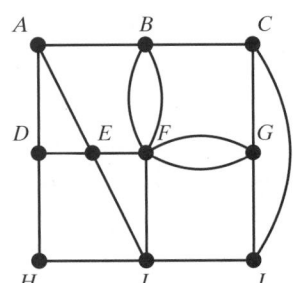

23. *A*: 2; *B*: 2; *C*: 3; *D*: 3; *E*: 3; *F*: 1
25. *B* and *C*   **27.** *A*, *C*, *D*; *A*, *B*, *C*, *D*
29. *AC* and *DF*   **31.** While edge *CD* is included, the graph is connected. If we remove *CD*, the graph will be disconnected.   **33.** *DF*   **35.** Even: *A*, *B*, *G*, *H*, *I*; odd: *C*, *D*, *E*, *F*
37. *D*, *G*, and *I*   **39.** Possible answers: *B*, *C*, *D*, *F*; *B*, *A*, *C*, *D*, *F*   **41.** *G*, *F*, *I*, *H*, *G*
43. *A*, *B*, *C*, *C*, *D*, *F*, *G*, *H*, *I*
45. *G*, *F*, *D*, *E*, *D* requires that edge *DE* be traversed twice. This is not allowed within a path.
47. No edge connects *F* and *E*.

49. Possible answer:

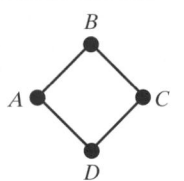

51. Possible answer:

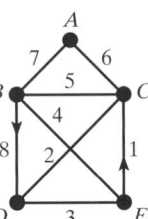

53–65. Answers will vary.   **67.**

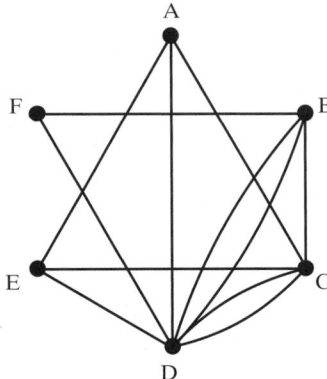

Section 15.2
Check Point Exercises

1.
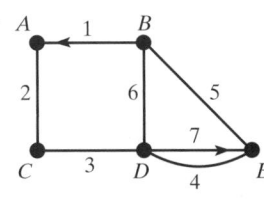
; *E*, *C*, *D*, *E*, *B*, *C*, *A*, *B*, *D*

2.
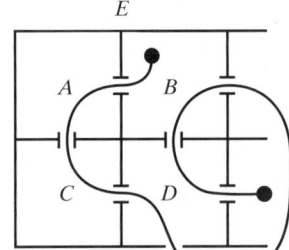
; *G*, *E*, *C*, *A*, *B*, *D*, *C*, *G*, *H*, *D*, *F*, *H*, *J*, *I*, *G*

3. a. yes   **b.**
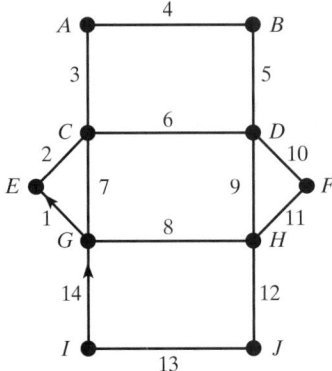

4.
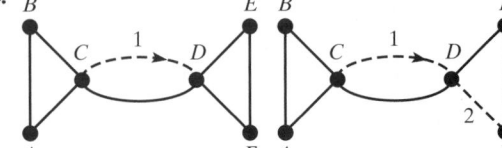 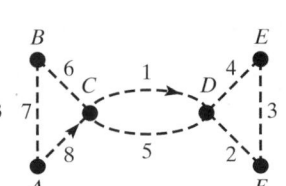

Exercise Set 15.2

1.

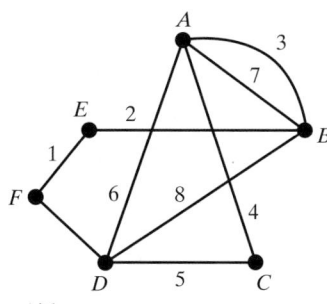

neither

3.

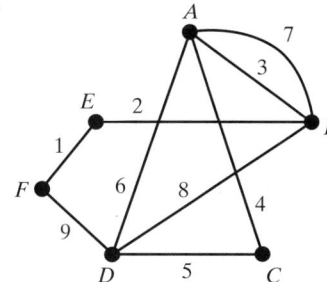
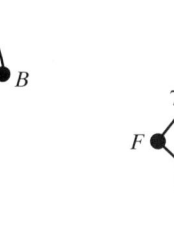

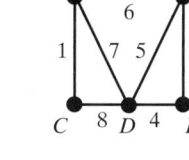

Euler circuit

5.

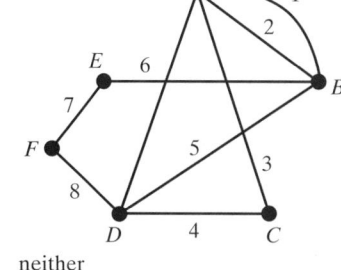

neither

7. a. There are two odd vertices.
b. Possible answer:

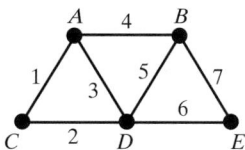

9. a. There are no odd vertices.
b. Possible answer:

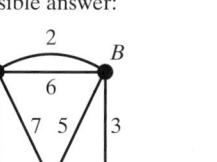

11. There are more than two odd vertices.
13. Euler circuit   **15.** Euler path, but no Euler circuit
17. neither an Euler path nor an Euler circuit

19. a. Euler circuit
b. Possible answer:

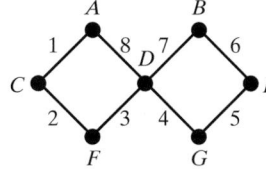

21. a. Euler path
b. Possible answer:

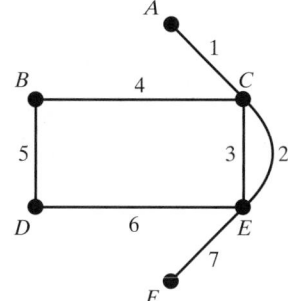

23. a. neither

25. a. Euler path
b. Possible answer:

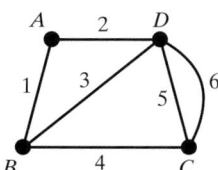

27. a. Euler circuit
b. Possible answer:

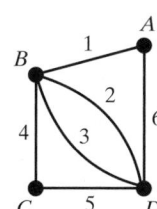

29. a. neither

31. a. Euler path
b. Possible answer:

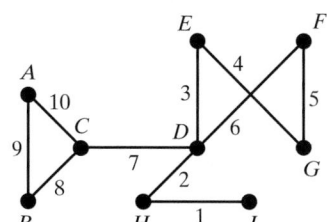

33. Possible answer:

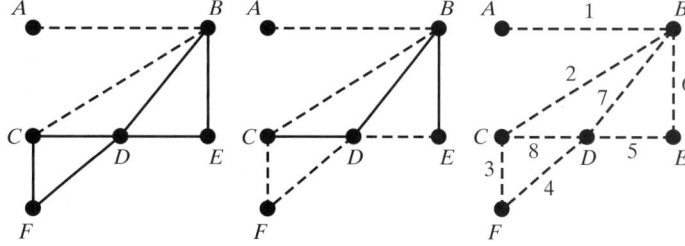

35. Possible answer:

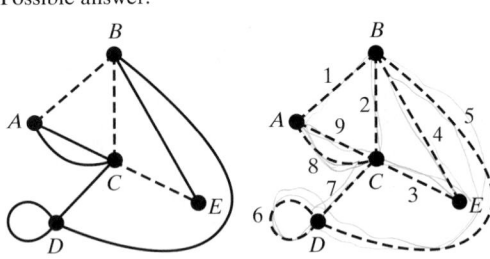

37. Possible answer:

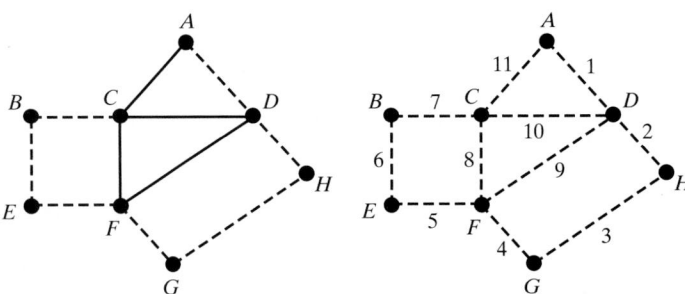

39. Possible answer:

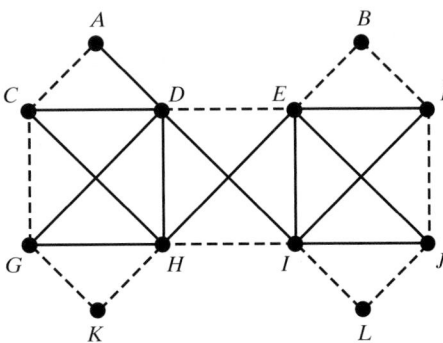

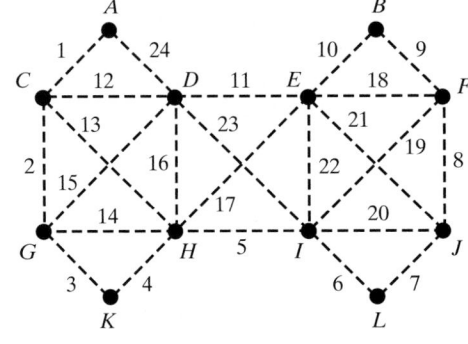

41. The graph has an Euler circuit.

43.

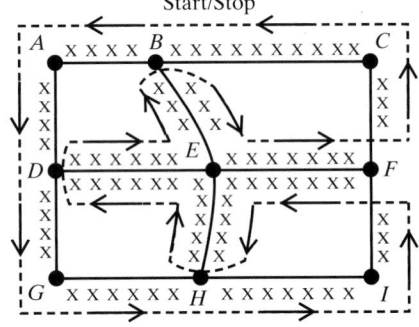

45. a.

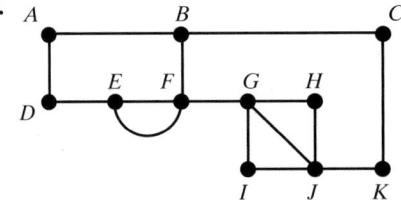

b. Yes; begin at *B* or *E*.

47. a.

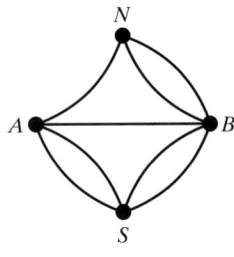

b. Yes, they can.

c. Possible answer:

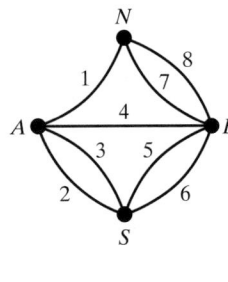

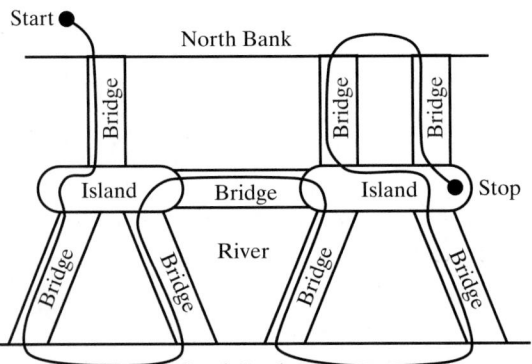

49.

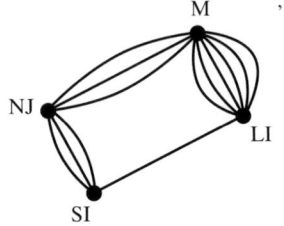

; It is possible.

51. a.

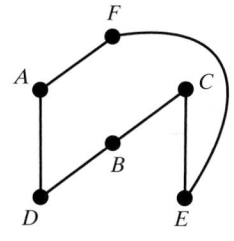

b. It is possible.
c. Possible answer:

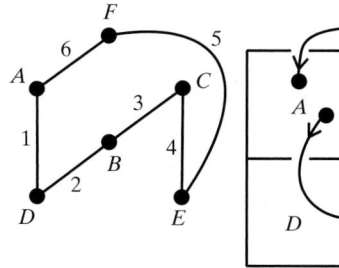 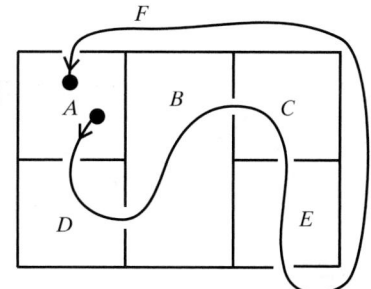

53. Possible answer:

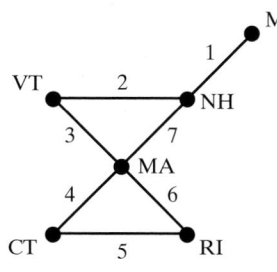

55. a.

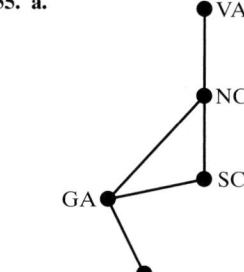

b. not possible
d. not possible

57–67. Answers will vary.

Section 15.3
Check Point Exercises

1. a. Possible answer: E, C, D, G, B, A, F **b.** Possible answer: E, C, D, G, B, F, A, E **2. a.** 2   **b.** 120   **c.** 362,880   **3.** $562

4.

Hamilton Circuit	Sum of the Weights of the Edges	=	Total Cost
A, B, C, D, A	$20 + 15 + 50 + 30$	=	$115
A, B, D, C, A	$20 + 10 + 50 + 70$	=	$150
A, C, B, D, A	$70 + 15 + 10 + 30$	=	$125
A, C, D, B, A	$70 + 50 + 10 + 20$	=	$150
A, D, B, C, A	$30 + 10 + 15 + 70$	=	$125
A, D, C, B, A	$30 + 50 + 15 + 20$	=	$115

$A, B, C, D, A; A, D, C, B, A$

5. A, B, C, D, E, A; 198

Exercise Set 15.3

1. Possible answer: A, G, C, F, E, D, B **3.** Possible answer: A, B, G, C, F, E, D, A **5.** Possible answer: A, F, G, E, C, B, D **7.** Possible answer: A, B, C, E, G, F, D, A **9. a.** no   **11. a.** yes   **b.** 120   **13. a.** no   **15.** 2   **17.** 39,916,800   **19.** 11   **21.** 36   **23.** 36   **25.** 88   **27.** 70   **29.** 70   **31.** A, C, B, D, A and A, D, B, C, A **33.** B, D, C, A, B; 82   **35.** A, D, B, C, E, A

37.

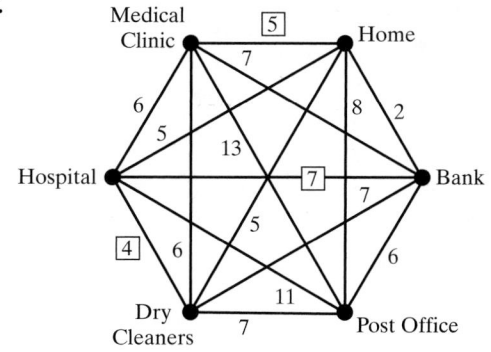

39. 30
41.

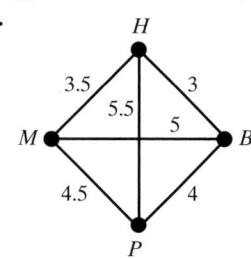

43. H, B, P, M, H; 15 mi; same
45–55. Answers will vary.
57. Answers will vary.

Section 15.4
Check Point Exercises
1. Figure 15.51(c)   **2.** Possible answer:

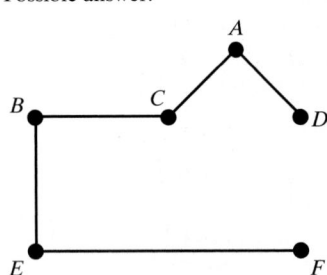

3.

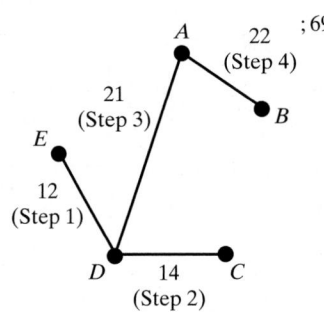

; 69

Exercise Set 15.4
1. yes   **3.** no   **5.** yes   **7.** no   **9.** yes   **11.** i.   **13.** ii.   **15.** ii.   **17.** iii.

19.

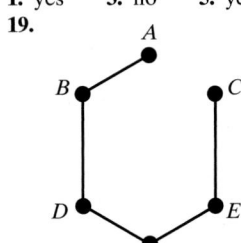

21.

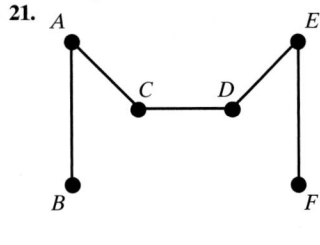

23.

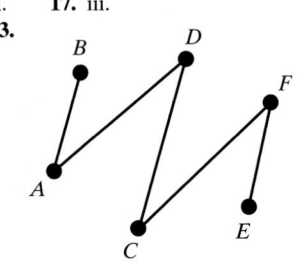

25.

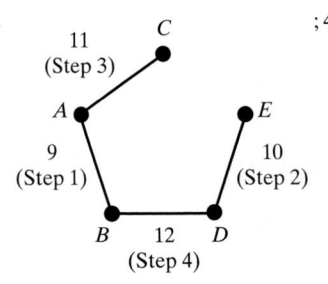

; 120

27.

; 42

29.

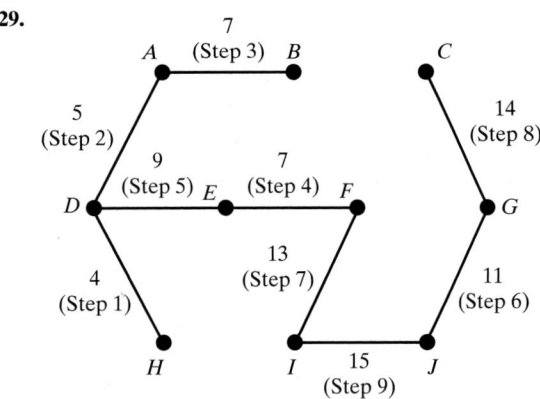

; 85

31.

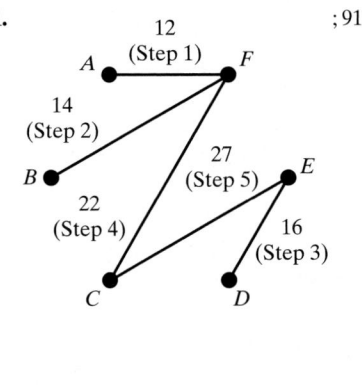

; 91

33.

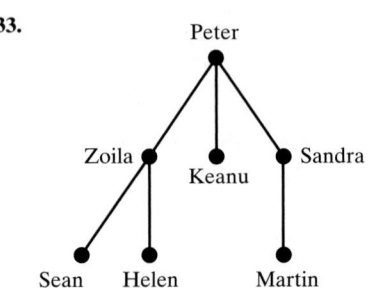

35. a.

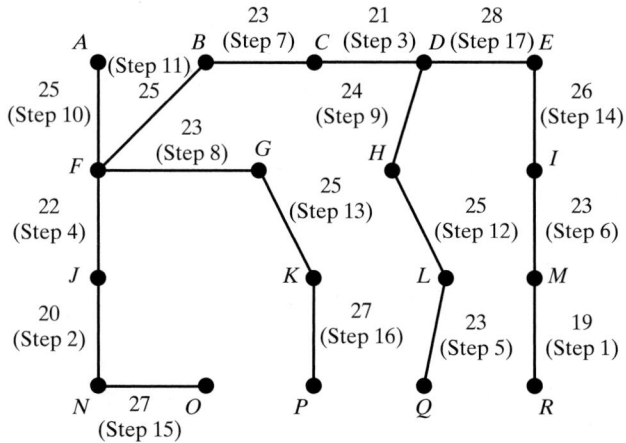

Classroom Building —155— Library
80, 115, 135, 125, 155 Administration, Gym, 55, 85, 215
Cafeteria —135— Parking Lot

b. ; 470 ft

Classroom Building
80 (Step 2)
Library
135 (Step 5)
115 (Step 4)
55 (Step 1) Administration Gym
85 (Step 3)
Cafeteria Parking Lot

37. ; 406 mi

A (Step 11) B —23 (Step 7)— C —21 (Step 3)— D (Step 17) E
25 (Step 10), 25, 23 (Step 8), 24 (Step 9), 28, 26 (Step 14)
F, G, H, I
22 (Step 4), 25 (Step 13), 25 (Step 12), 23 (Step 6)
J, K, L, M
20 (Step 2), 27 (Step 16), 23 (Step 5), 19 (Step 1)
N —27 O (Step 15)— P, Q, R

39–47. Answers will vary.   **49.** Remove BC; remove BF; remove CF

Review Exercises

1. Both graphs have the same number of vertices, and these vertices are connected in the same ways. Possible answer:

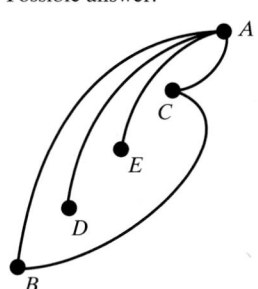

2. A: 5; B: 4; C: 5; D: 4; E: 2   **3.** Even: B, D, E; odd: A, C
4. B, C, and E **5.** Possible answer: E, D, B, A and E, C, A
6. Possible answer: E, D, C, E **7.** yes   **8.** no   **9.** AD, DE, and DF

10.

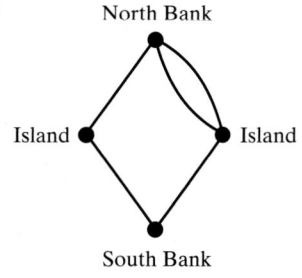

North Bank
Island   Island
South Bank

11.

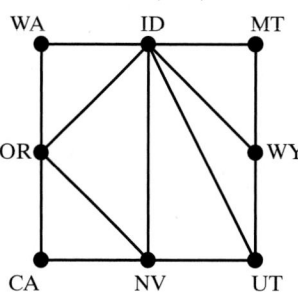

WA   ID   MT
OR   WY
CA   NV   UT

12.

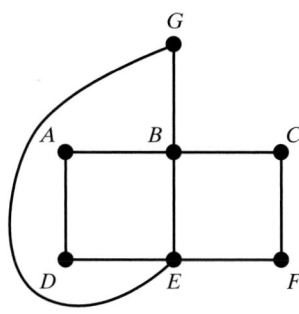

G
A   B   C
D   E   F

13. a. neither

14. a. Euler circuit
b. Possible answer:

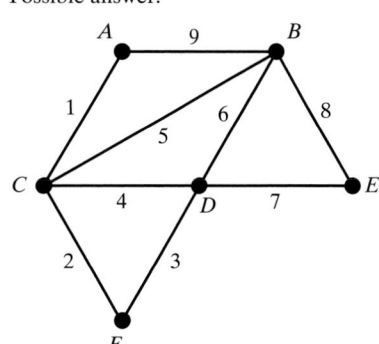

A —9— B
1, 5, 6, 8
C —4— D —7— E
2, 3
F

15. a. Euler path
b. Possible answer:

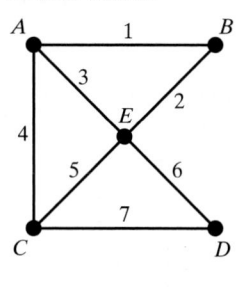

A —1— B
3, 2
4 E
5, 6
C —7— D

16.

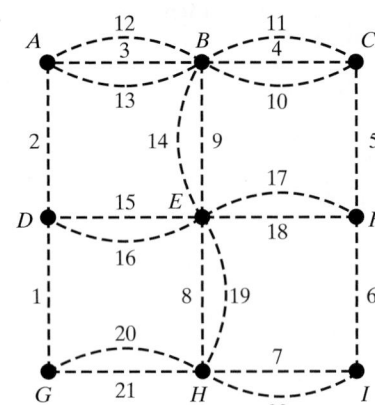

17.

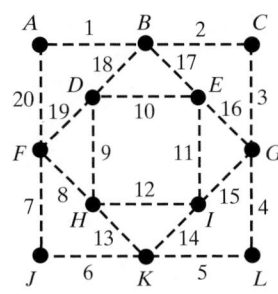

18. a. yes **b.**

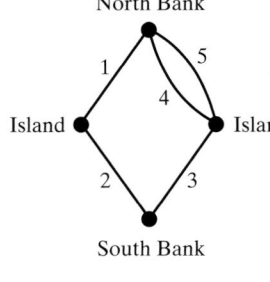

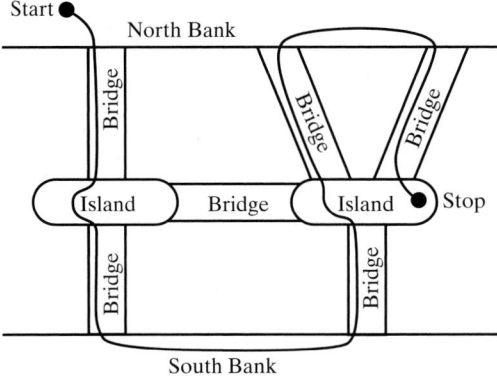

c. no

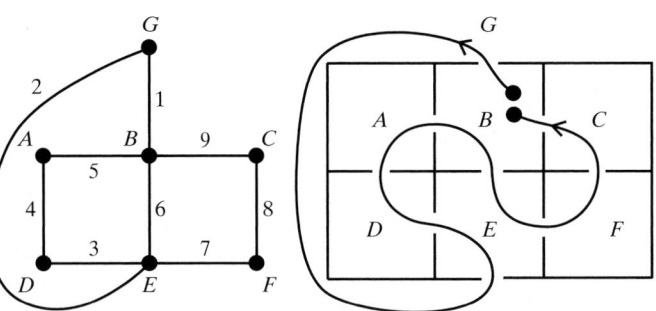

19. It is possible.   **20. a.** yes   **b.** Possible answer:

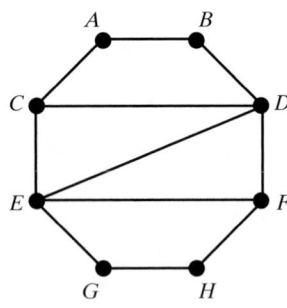

21. a.

b. yes
c. The guard should begin at C and end at F, or vice versa.

22. Possible answer: A, E, C, B, D, A
23. Possible answer: D, B, A, E, C, D
24. a. no
25. a. yes   **b.** 6
26. a. no
27. a. yes   **b.** 24

28. A, B, C, D, A: 19; A, B, D, C, A: 18; A, C, B, D, A: 19; A, C, D, B, A: 18; A, D, B, C, A: 19; A, D, C, B, A: 19
29. A, B, D, C, A and A, C, D, B, A **30.** A, C, D, B, A; 18   **31.** A, B, E, D, C, A; 24

32.

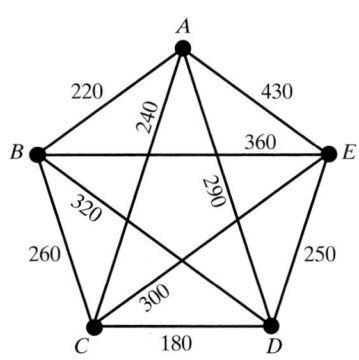

33. A, B, C, D, E; \$1340   **34.** yes   **35.** No. It has a circuit.   **36.** No. It is disconnected.
37. Possible answer: **38.** Possible answer:

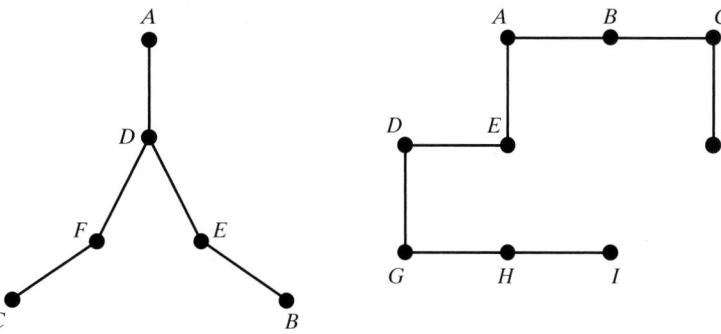

39.   ; 875   **40.**   ; 239

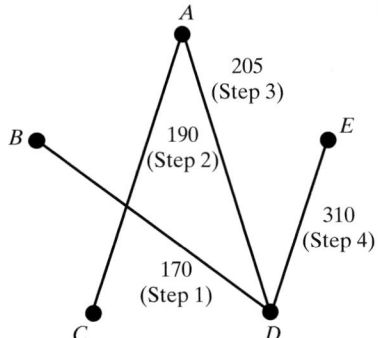

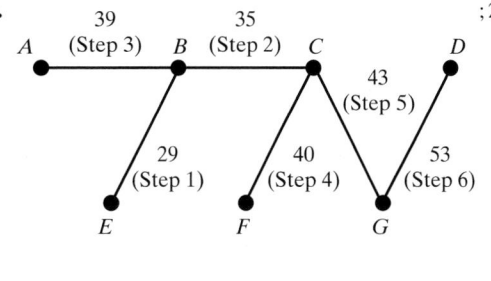

41.   ; 4370 mi

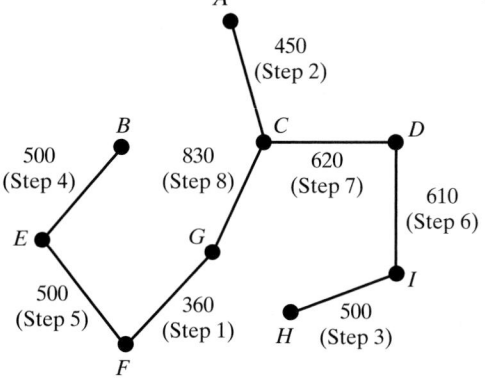

Chapter 15 Test
1. A: 2; B: 2; C: 4; D: 3; E: 2; F: 1   **2.** Possible answer: A, D, E and A, B, C, E **3.** B, A, D, E, C, B **4.** CF
5.   **6. a.** Euler path   **b.** Possible answer: **7. a.** neither

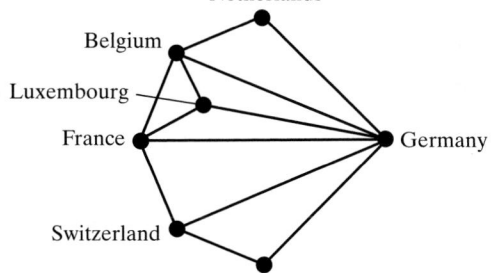

8. a. Euler circuit   **b.** Possible answer:

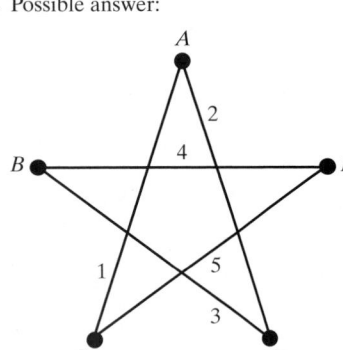

9.

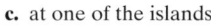

10. a.

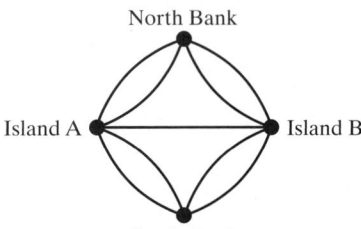

    b. yes   **c.** at one of the islands

11. a.

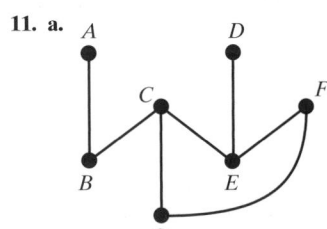

   b. no   **12. a.**

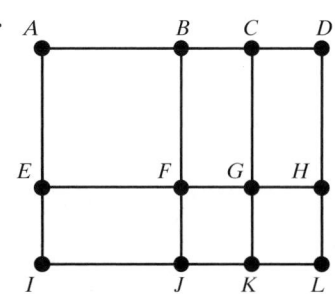

   b. no

13. A, B, C, D, G, F, E, A and A, F, G, D, C, B, E, A **14.** 24

15.

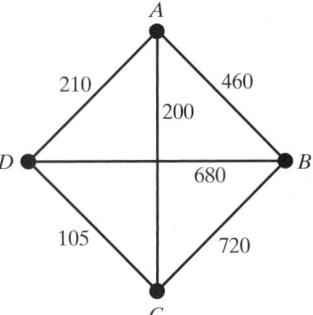

16. A, B, D, C, A or A, C, D, B, A; $1445

17. A, E, D, C, B, A; 33

18. no

19. Possible answer:

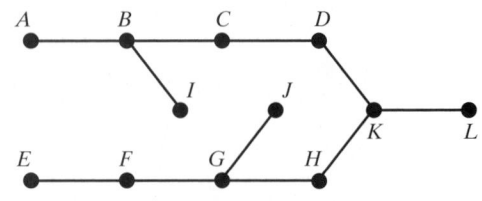

20.

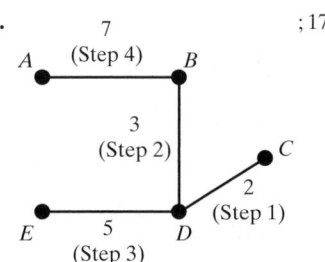

; 17

Subject Index

PHOTO CREDITS

Index of Applications

(continued from inside front cover)